Excel with

Objective Questions in ORGANIC CHEMISTRY

For Competitive Examinations

Excel with

Objective Questions in ORGANIC CHEMISTRY

For Competitive Examinations

By

Prof. S.K. KHANNA
(Formerly, Department of Chemistry)
D.A.V. College
Chandigarh

Dr. N.K. VERMA
(Department of Chemistry)
D.A.V. College
Chandigarh

Dr. B. KAPILA
(Department of Chemistry)
D.A.V. College
Chandigarh

GOLDEN BELLS

(An Imprint of Laxmi Publications Pvt. Ltd.)

BANGALORE • CHENNAI • COCHIN • GUWAHATI • HYDERABAD
JALANDHAR • KOLKATA • LUCKNOW • MUMBAI • PATNA
RANCHI • NEW DELHI

Published by :
GOLDEN BELLS
(An Imprint of Laxmi Publications Pvt. Ltd.)
113, Golden House, Daryaganj,
New Delhi-110002

Phone : 011-43 53 25 00
Fax : 011-43 53 25 28

www.laxmipublications.com
info@laxmipublications.com

Price : ₹ **450.00** *Only.*

New Edition

OFFICES

© **Bangalore**	080-26 75 69 30	© **Chennai**	044-24 34 47 26
© **Cochin**	0484-237 70 04, 405 13 03	© **Guwahati**	0361-251 36 69, 251 38 81
© **Hyderabad**	040-24 65 23 33	© **Jalandhar**	0181-222 12 72
© **Kolkata**	033-22 27 43 84	© **Lucknow**	0522-220 99 16
© **Mumbai**	022-24 91 54 15, 24 92 78 69	© **Patna**	0612-230 00 97
© **Ranchi**	0651-221 47 64		

GBC-5018-450-EXCEL WITH ORGANIC CHEM

C-6483/011/04

Typeset at : Monu Printographics, Delhi.

Printed at : Ajit Printers, Delhi.

SALIENT FEATURES

- Each chapter of the book contains a rich collection of objective questions/problems.
- Various underlying principles and concepts have been summarized under the heading, **Concepts in Brief** is given at the start of each chapter.

  Important Facts and Terms are given to make the students aware of certain important topics. These are not easily available in other books.
- 15-20 MCQs have been solved in most of the chapters under the heading **Solved Problems,** so as to illustrate the problem solving techniques even if a question is on the tougher side.
- In each chapter the **MCQs** have been divided into *five different levels.* This is being done to ensure that a student systematically grows in confidence to develop a logic required for solving MCQs/Problems. *This is a feature which no book carries upto date.*
  - **Question Bank :** Contains topic wise questions of the chapter.
  - **Examiner's Viewpoint :** Contains collection of questions from previous years' (1990 onwards) competitive exams.
  - **MCQs for IIT JEE :** Contains **tricky questions** to prepare the students for IIT JEE. The questions have been prepared especially for IIT JEE. Previous year questions of IIT exams have also been given to give a complete know how of IIT examinations.
  - **Assertion—Reason Questions :** It has been devised to cater to the needs of students preparing for IIT JEE, AIIMS and Rohtak PMT examinations. This section contains *Assertion—Reason type questions.*
  - **Self Evaluation Test :** A **self evaluation test** is given in each chapter so that a student can properly assess himself. The test contain questions from important topics which are required to be learnt by a student.

  Hints/Solutions are given for almost all the questions at the end of the each sub level.
- Various **Model Papers** according to style of different competitive examinations, including CBSE-PMT, AIEEE Orissa CET, PU CET/PMT, Pb. CET/PMT and many more examinations are given with each section.

Excel with

Objective Questions

in

ORGANIC CHEMISTRY

For Competitive Examinations

This Book is essential for all the students preparing for various Engineering and Medical Entrance Examinations given below and also for students preparing for other State Service Exams.

- IIT—JEE
- AIIMS
- DPMT
- BVP, Pune
- AFMC, Pune
- BVP, Pune
- JEE, Orissa
- CEE, Assam
- CEE, Kerala
- JEE, Tripura
- CMC, Vellore
- BCECE, Bihar
- CBSE—PMT
- CBSE—AIEEE
- UPSEAT, U.P.
- BHU, Varanasi
- PET, Rajasthan
- PET/PMT, M.P.
- CMC, Ludhiana
- CEET, Haryana
- CET, Karnataka
- JET, Pondicherry
- JEE, West Bengal
- TNPCEE, Tamil Nadu
- MGIMS, Wardha
- Aligarh Entrance Exam.
- GGSIP University, Delhi
- Manipal Entrance Exam.
- EAMCET, Andhra Pradesh
- CEEE, (DCE, NSIT) Delhi
- Indian School of Mines, Dhanbad
- MLNR, Engineering Entrance Exam.
- CET/PMT, Punjab & UT Chandigarh
- Birla Institute of Technology Mesra, Ranchi
- **Other State Level Competitive Examinations**

PREFACE

We have the pleasure and a deep sense of satisfaction to present the **revised** edition of the book **Objective Questions in Organic Chemistry** for Competitive Examinations. The present edition of the book has been written with the aim of meeting the needs and interests of students appearing in various competitive exams such as AIEEE, CET, CBSE-PMT, AIIMS, AFMC, CPMT, EAMCET and IIT JEE etc.

In this book an emphasis has been made on theoretical/conceptual approach and the application of underlying basic concepts and principles. It also empowers the students with the latest trends in the competitive examinations in a simple manner. Thus, the spectrum of its users ranges from class XI to class XII and beyond ! The book has been tailored with a technical treatment of the subject. Besides quality, it is also the quantity of contents that matters at such specialized academic levels. The readers will appreciate the fact that this book delivers the contents in all possible and relevent, details of the subject matter.

Each unit of the book has been comprehensively written to accommodate the latest syllabi and specific changes therein. A systematic and graded approach is followed in the presentation of the constituent chapters.

We firmly believe that this edition of the book will also enable the students to grasp the subject systematically, easily and more clearly. We hope that the book will also help in various qualifying examinations. The students and teachers will find the present book easier, more logical and more appropriate.

Every care has been taken to ensure that the book is free of errors. But nobody can claim to be infallible. Hence, suggestions for further improvement of the book will be gratefully acknowledged. We will be grateful for any constructive criticism of this book. We hope this book serves well the purpose of those for whom it is intended.

—Authors

Contents

1 NOMENCLATURE, STEREOCHEMISTRY AND GENERAL CONCEPTS

1.1. CONCEPTS IN BRIEF

INTRODUCTION

- **Organic Chemistry** is defined as the chemistry of compounds of carbon and hydrogen *i.e.*, study of hydrocarbons and their derivatives.
- **Vital Force Theory.** According to this theory, which was proposed by Berzelius, only the living bodies could produce organic compounds under the influence of some mysterious force (vital force) present in them.

  Vital force theory was abandoned by Wohler. He obtained an important organic compound urea by heating two inorganic salts-*potassium cyanate* (KCNO) and *ammonium chloride* (NH_4Cl) in the laboratory

$$\underset{\text{Ammonium Chloride}}{NH_4Cl} + \underset{\text{Potassium Cyanate}}{KCNO} \longrightarrow \underset{\text{Ammonium Cyanate}}{NH_4CNO} + KCl$$

$$NH_4CNO \xrightarrow[\text{Isomerization}]{\text{Heat}} \underset{\text{Urea}}{NH_2\text{—}\overset{\overset{\displaystyle O}{\|}}{C}\text{—}NH_2}$$

Urea was the first organic compound synthesised in laboratory. Later on with the development of synthetic methods a large number of organic compounds like *methanol*, *methanoic acid*, acetic acid etc. were prepared.

- Carbon exhibits **catenation** to maximum extent due strong C—C bond and tetravalency. Due to this unique property of carbon, the number of organic compounds is very large.

HYBRID STATE OF CARBON IN ORGANIC COMPOUNDS

- Carbon atom can undergo *sp*, sp^2 and sp^3 hybridisation.
- Carbon atom is *sp* hybridised if it is directly attached to two atoms, sp^2 hybridised if it is attached to three atoms and sp^3 hybridised if it is directly attached to four atoms.

Illustration

The structures of 6-Methylhepta-1, 5-dien-3-yne and 3, 4-Dimethylpenta-1, 2-diene are given below to illustrate the above point.

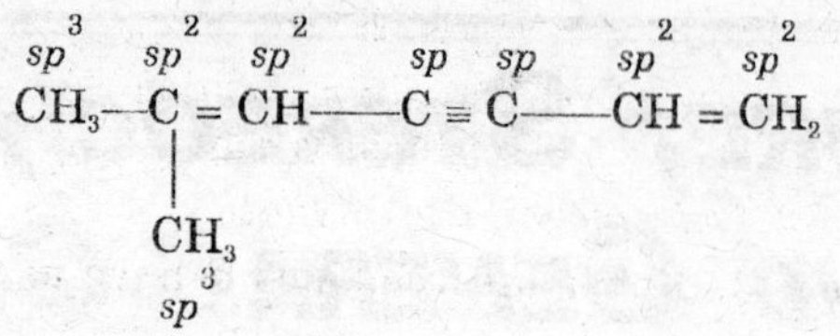

6-Methyl hepta-1, 5-diene-3-yne

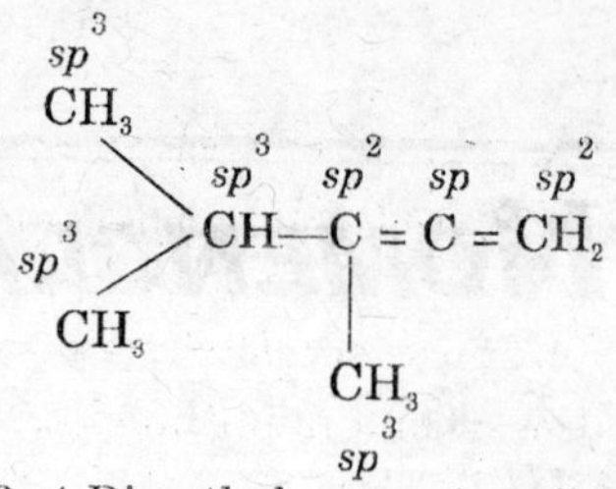

3, 4-Dimethyl penta-1, 2-diene

CLASSIFICATION OF HYDROCARBONS

HYDROCARBONS

Open-chain or Aliphatic or Acyclic compounds
In these compounds, *the constituent atoms are linked together in open-chains*

- **Alkanes** — $CH_3—CH_3$ Ethane
- **Alkenes** — $CH_2═CH_2$ Ethylene (Ethene)
- **Alkynes** — $CH \equiv CH$ Acetylene (Ethyne)

Carbocyclic Compounds
They contains ring(s) made up of carbon atoms

- **Alicyclic Compounds** — *They are cyclic compounds which resemble aliphatic compounds in many ways* (the name *alicyclic* is derived by combining *ali* of aliphatic with *cyclic*). For example, cyclopentane
- **Aromatic Compounds** — *They usually contain at least one benzene ring.* For example, benzene, toluene

- **Alkanes.** General formula of alkanes is C_nH_{2n+2}. All the carbon atoms in alkanes are sp^3 hybridised.
- **Alkenes.** Alkenes are unsaturated hydrocarbons. Their general formula is C_nH_{2n}. In alkenes, the carbon atoms involved in C = C are in sp^2 hybrid state.
- **Alkynes.** Alkynes are unsaturated hydrocarbons having carbon-carbon triple bond. Their general formula is C_nH_{2n-2}. The carbon atoms involved in triple bond are in *sp*-hybrid state.
- **Cycloalkanes.** Cycloalkanes (monocyclic) have general formula C_nH_{2n} which is same as that of alkenes.
- **Aromatic hydrocarbons.** These are unsaturated cyclic compounds which behave as saturated hydrocarbons and undergo substitution reations in preference to addition reactions.

HUCKEL RULE OR (4n + 2) RULE

- According to this rule **cyclic planar systems** containing a total of $(4n + 2)$ delocalized π electrons possess a high degree of stability and behave as aromatic compounds. Thus, benzene molecule having six π electrons (*aromatic sextent*) behaves as an aromatic compound. Some examples of aromatic compounds (species) illustrating Huckel rule are given below :

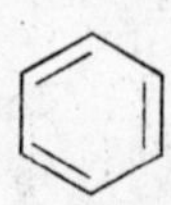

Benzene

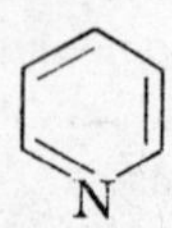

Pyridine

Pyrrole

Furan

Thiophene

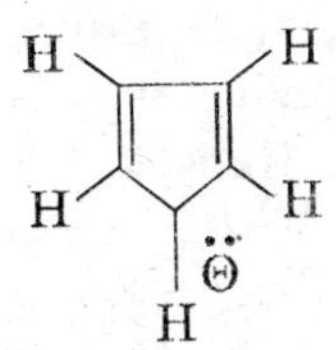

Cyclopentadienyl anion

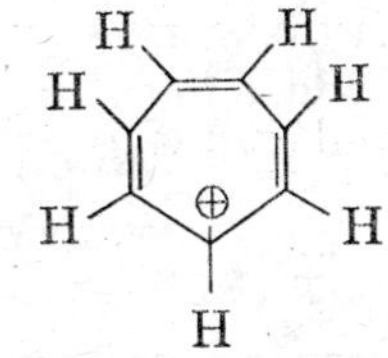

Cycloheptatrienyl cation
(Tropylium cation)

All these species have a system of 6π electrons and are aromatic.

The polycyclic hydrocarbons naphthalene, anthracene and phenanthrene also behave as aromatic compounds, as they contain $(4n + 2)$ π electrons.

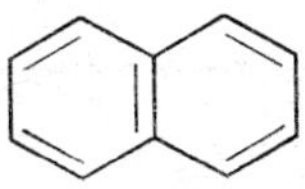

Naphthalene
(10π electrons)
Aromatic

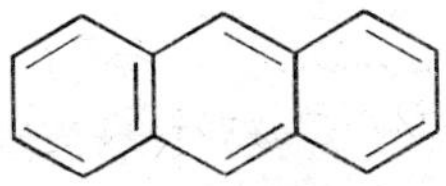

Anthracene
(14π electrons)
Aromatic

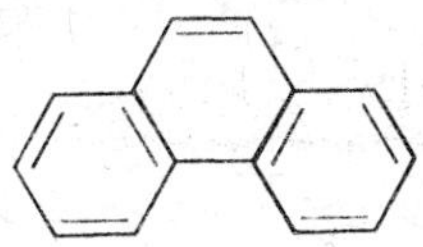

Phenanthrene
(14π electrons)
Aromatic

ANNULENES

- *Annulenes are the conjugated monocyclic polyenes containing an even number of carbon atoms in their molecules.* They may or may not be aromatic. For example,

Cyclobutadiene
[4]-Annulene

Benzene
[6]-Annulene

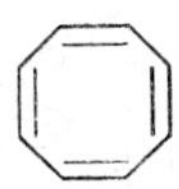

Cyclooctatetraene
[8]-Annulene

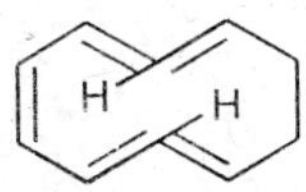

[10]-Annulene

[6]–, [14]–, [18]– and [22]–annulenes are aromatic.

[4]–, [8]–, [10]–, [12]– and [16]– annulenes are non-aromatic.

[10]– annulene contains $(4n + 2)$ π electrons but is non-aromatic due to non-planar structure.

NOMENCLATURE OF ORGANIC COMPOUNDS

- The term nomenclature means the system of naming. The systematic nomenclature of organic compounds is known as IUPAC (International Union of Pure and Applied Chemistry) system, which is being discussed briefly below :

  According to IUPAC system, the name of organic compound, in general, consists of the following parts :

  (i) Word root *(ii) Primary suffix* *(iii) Secondary suffix* *(iv) Prefix(es).*

(*i*) **Word root.** The word root represents the number of carbon atoms in the *parent chain*. The general word root for different aliphatic compounds is **ALK**. The word root for different lengths of carbon chains are given below in Table 1.1.

Table 1.1. Word Roots for different lengths of Carbon Chains

Chain Length	*Word root*	*Chain Length*	*Word root*
C_1	Meth-	C_9	Non-
C_2	Eth-	C_{10}	Dec-
C_3	Prop-	C_{11}	Undec-
C_4	But-	C_{12}	Dodec-
C_5	Pent-	C_{13}	Tridec-
C_6	Hex-	C_{18}	Octadec-
C_7	Hept-	C_{20}	Icos-
C_8	Oct-	C_{30}	Tricont-

(*ii*) **Primary Suffix.** Primary suffix is used to represent saturation or unsaturation in the carbon chain. While writin the name, primary suffix is added to the word root. Some of the primary suffixes are given below in Table 1.2.

Table 1.2. Some Primary Suffixes

Nature of Carbon Chain	*Primary Suffix*
Saturated Carbon Chain	*ane*
Unsaturated Carbon Chains :	
one C = C bond	*ene*
two C = C bonds	*adiene*
three C = C bonds	*atriene*
one C ≡ C bond	*yne*
two C ≡ C bonds	*adiyne*

(*iii*) **Secondary Suffix.** Secondary suffix is used to indicate the functional group in the organic compounds. *The terminc* **'e'** *of the primary suffix is dropped if it is followed by a suffix beginning with* **'a', 'i', 'o', 'u'** *or* **'y'.** Secondary suffixes fc various functional groups are given in Table 1.3.

Table 1.3. Some Organic Families and Secondary Suffixes

Class of organic compound	*General formula*	*Functional Group*	*Suffix*	*IUPAC name of the family* (*Word root* + *p. suffix* + *sec. suffix*)
Alcohols	R—**OH**	**—OH**	-ol	ALKAN**OL**
Thioalcohols	R—**SH**	**—SH**	-thiol	ALKANE**THIOL**
Amines	R—**NH_2**	**—NH_2**	-amine	ALKAN**AMINE**
Aldehydes	R—**CHO**	**—CHO**	-al	ALKAN**AL**
Ketones	R—**COR**	**>CO**	-one	ALKAN**ONE**
Carboxylic acids	R—**COOH**	**—COOH**	-oic acid	ALKAN**OIC ACID**
Amines	R—**$CONH_2$**	**—$CONH_2$**	-amide	ALKAN**AMIDE**
Acid chlorides	R—**COCl**	**—COCl**	-oyl chloride	ALKAN**OYL CHLORIDE**
Esters	R—**COOR**	**—COOR**	-oate	ALKYL ALKAN**OATE**
Nitriles	R—**C ≡ N**	**—C ≡ N**	-nitrile	ALKANE **NITRILE**

(*iv*) **Prefix**. Prefix is a part of the name which appears before the word root. Prefixes are used to represent the names of alkyl groups (branched chains) or some functional groups which are regarded as substituents. It is described below

(*a*) **Alkyl groups** are formed by the removal of H atom from the alkanes. These are represented by the generə formula C_nH_{2n+1} or R-. Some alkyl groups along with their prefixes are given in Table 1.4.

Table 1.4. Some Alkyl Groups along with their Prefixes

Alkane	*Alkyl Group*	*Abbreviation*	*Prefix*
CH_4	CH_3—	Me-	Methyl
C_2H_6	CH_3CH_2—	Et-	Ethyl
C_3H_8	$CH_3CH_2CH_2$—	*n*-Pr-	*n*-Propyl
C_3H_8	CH_3—CH— (with CH_3 on CH)	Iso-Pr-	Isopropyl

Some **unsaturated groups** are given below :

$CH_2 = CH$—	$CH_2 = CH—CH_2$—	$HC \equiv C—CH_2$—	$CH_3—CH = CH—CH_2$—
Vinyl (Ethenyl)	Allyl (2-Propenyl)	Propargyl (2-Propynyl)	Crotyl (2-Butenyl)

(*b*) **Some functional groups** are always indicated by the prefixes instead of secondary suffixes. These functional groups along with their prefixes are listed in Table 1.5.

Table 1.5. Functional groups which are always represented by prefixes

Functional Group	*Prefix*	*Family*	*IUPAC Name*
—NO_2	Nitro	R—NO_2	**NITRO**ALKANE
—OR′	alkoxy	R—OR′	**ALKOXY**ALKANE
—Cl	Chloro	R—Cl	**CHLORO**ALKANE
—Br	Bromo	R—Br	**BROMO**ALKANE
—I	Iodo	R—I	**IODO**ALKANE
—F	Fluoro	R—F	**FLUORO**ALKANE
—NO	Nitroso	R—NO	**NITROSO**ALKANE

(*c*) **In polyfunctional compounds,** *i.e.*, compounds with more than one functional group, one of the functional groups is treated as *principal functional group* and is indicated by the *secondary suffix* whereas *other functional groups* are treated as *substituents* and are indicated by the prefixes. The prefixes for various functional groups are given in Table 1.6.

Table 1.6. Prefixes for functional groups in polyfunctional compounds

Functional Group	*Prefix*	*Functional Group*	*Prefix*
—OH	Hydroxy	—COOH	Carboxy
—CN	Cyano	—COOR	Carbalkoxy
—NC	Isocyano	—COCl	Chloroformyl
—CHO	Formyl or Aldo	—$CONH_2$	Carbamoyl
—SH	Mercapto	—NH_2	Amino
—SR	Alkylthio	>CO	Keto or Oxo

Arrangement of Prefixes, Word root and Suffixes. The prefixes, word root and suffixes are arranged as follows while writing the name

Prefix(es) + Word root + p. suffix + sec. suffix

RULES FOR WRITING IUPAC NAMES

A. NOMENCLATURE OF HYDROCARBONS

- **Rule 1. Longest Chain Rule**

Select the longest continuous chain of carbon atoms as the parent chain. If some carbon-carbon multiple bond is present, the parent chain must contain the carbon atoms involved in it. The number of carbon atoms in the parent chain determines the *word root.* The carbon atoms which are not included in the parent chain are considered as alkyl substituents and are named as *prefixes.*

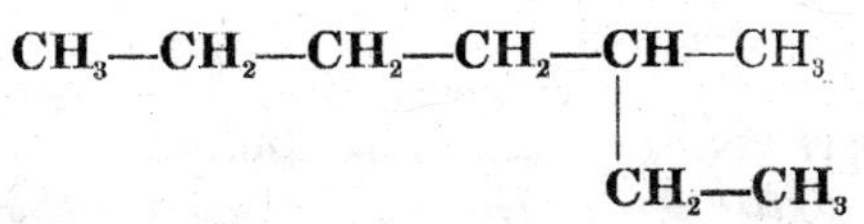

Prefix : *methyl*
Word root : *hept*
p. suffix : *ane*

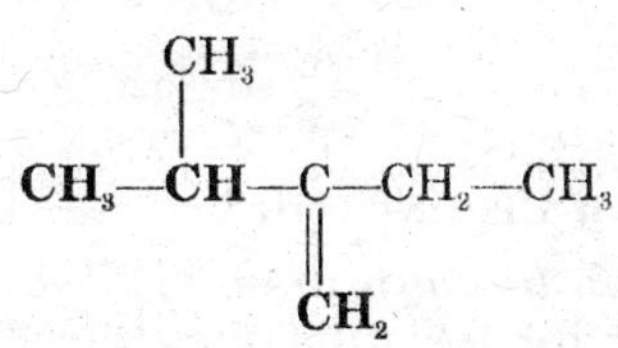

Prefix : *ethyl, methyl*
Word root : *but*
p. suffix : *ene*

If two equally long chains are possible, the chain with maximum number of side chains is selected as parent chain. F example,

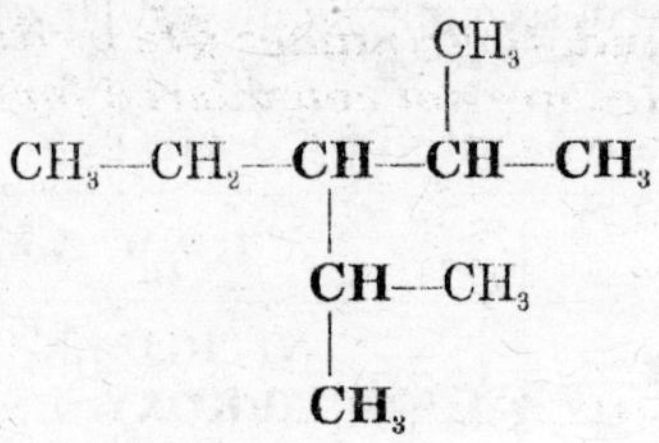

(*Correct*)
(Three side chains)

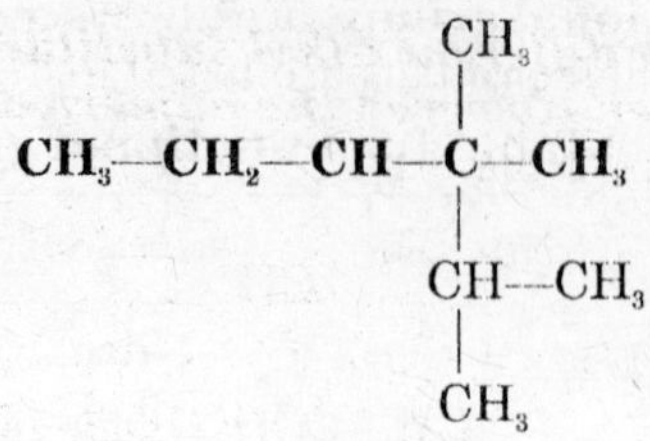

(*Wrong*)
(Two side chains)

- **Rule 2. Lowest Number or Lowest Sum Rule**

The selected parent chain is numbered using Arabic numerals and the positions of the alkyl groups are indicated by the number of the carbon atom to which the alkyl group is attached. ***The numbering is done in such a way that the substituted carbon atoms have the lowest possible numbers. When series of the locants are present then the numbering of chain is done in different ways and the position of locants is compared term by term. The series which contains the lowest number on the occasion of first difference represents the selected chain.*** Some examples are given below to illustrate the above idea :

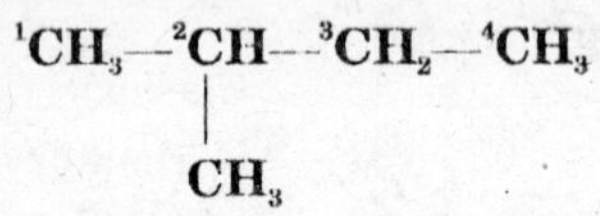

2-Methyl butane and not 3-Methyl butane

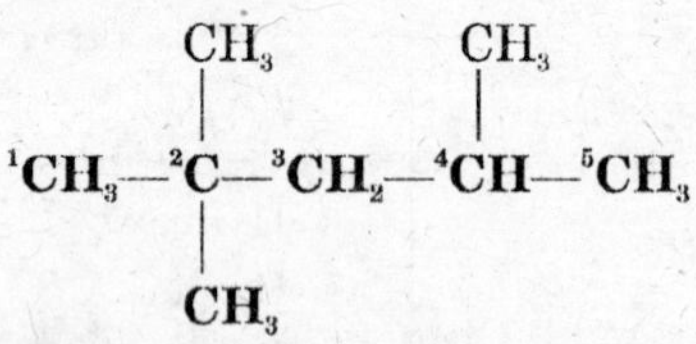

2, 2, 4-Trimethyl pentane and not 2, 4, 4-Trimethyl pentane

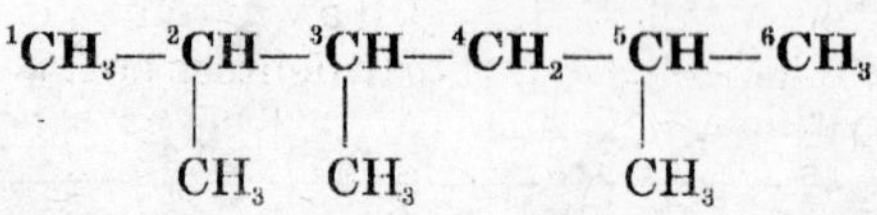

2, 3, 5-Trimethylhexane and not 2, 4, 5-Trimethylhexane

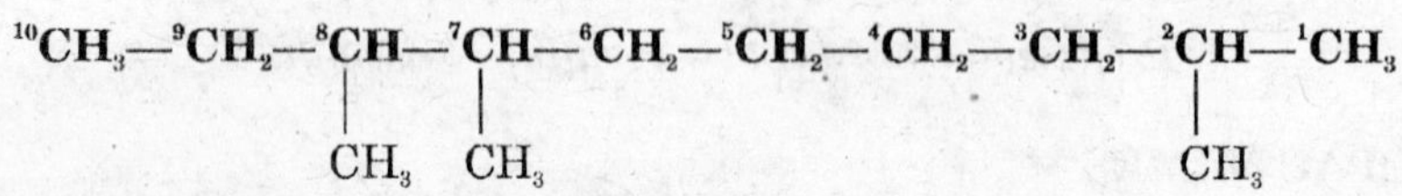

2, 7, 8-Trimethyldecane and not 3, 4, 9-Trimethyldecane

In case of unsaturated hydrocarbons, the carbon atoms involved in the multiple bond should get the lowest possible number. For example,

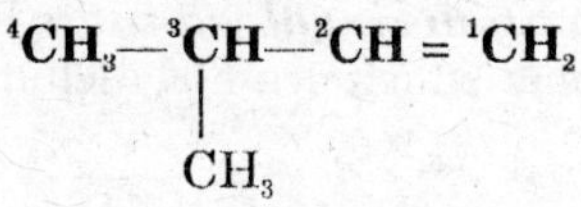

3-Methylbut-1-ene
(*Correct*)

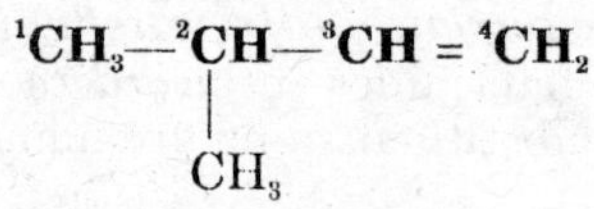

2-Methylbut-3-ene
(*Wrong*)

- **Rule 3. Use of Prefixes Di, Tri, etc.**

If the compound contains more than one similar alkyl groups, their positions are indicated separately and an appropriate numerical prefix : di, tri, etc., is attached to the name of the substituents. The positions of the substitutents are separated by commas. For example,

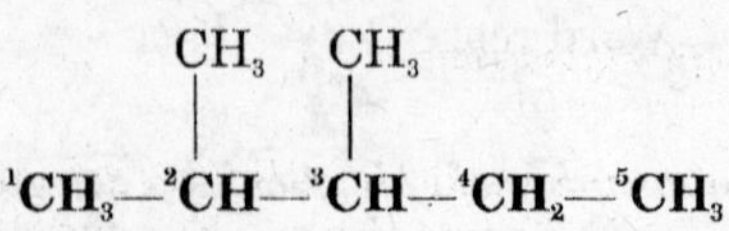

2, 3-Dimethylpentane

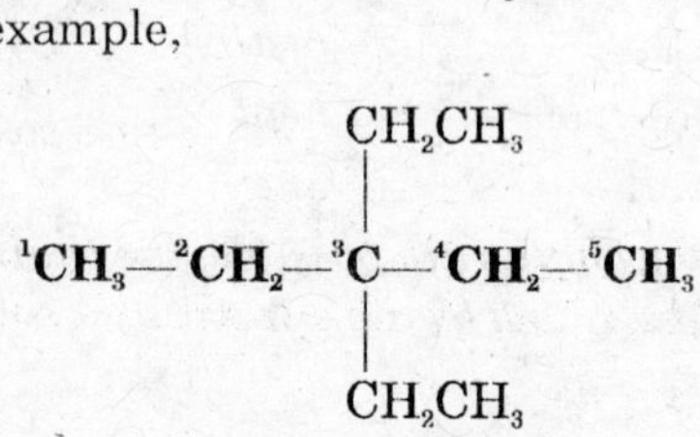

3, 3-Diethylpentane

- **Rule 4. Alphabetical Arrangement of Prefixes**

If there are different alkyl substituents present in the compound, their names are written in the alphabetical order. However, the numerical prefixes such as di, tri, etc., are not considered for the alphabetical order. For example,

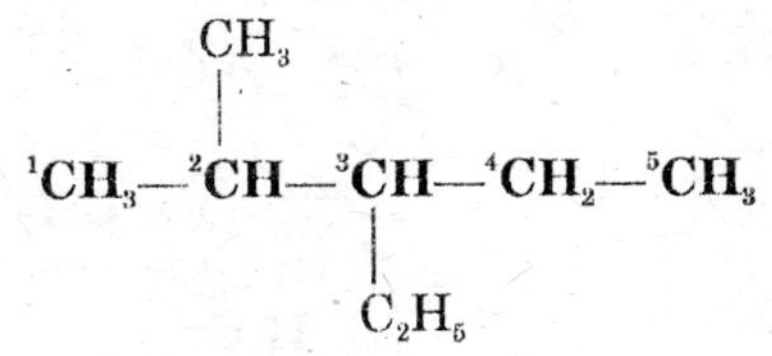

3-Ethyl-2-methyl pentane

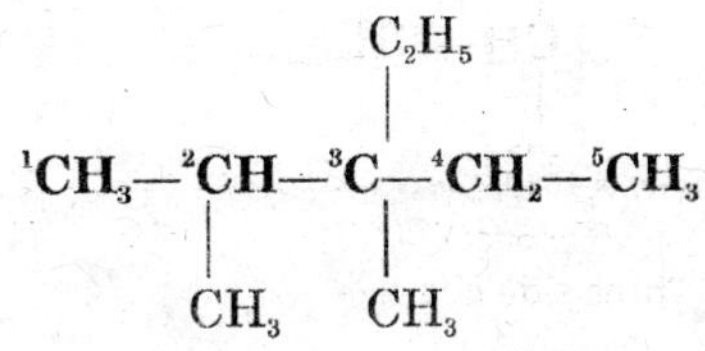

3-Ethyl-2, 3-dimethylpentane

Naming different alkyl substituents at the equivalent positions. If two different alkyl groups are located at the equivalent positions, then numering of the chain is done in such a way that the alkyl group which comes first in alphabetic order gets the lower position. For example, if ethyl and methyl groups are present at equivalent positions, then carbon bearing ethyl group should get the lower number as illustrated in the following example.

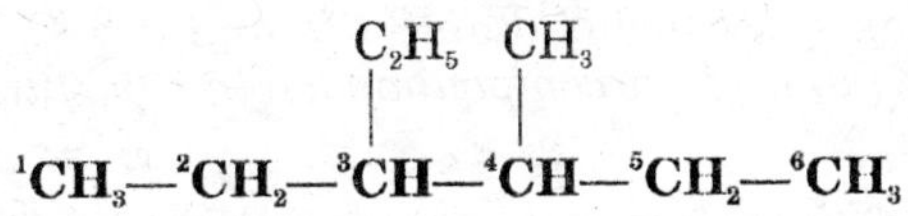

3-Ethyl-4-methylhexane

- **Rule 5. Naming the Complex Alkyl Substituents**

If the alkyl substituent is further branched it is named as substituted alkyl group. For this purpose, the carbon atoms of the alkyl group are separately numbered in such a way, that the carbon atom directly attached to the parent chain is given number 1. The prefix/name of such a substituents is enclosed in brackets as illustrated in the following examples :

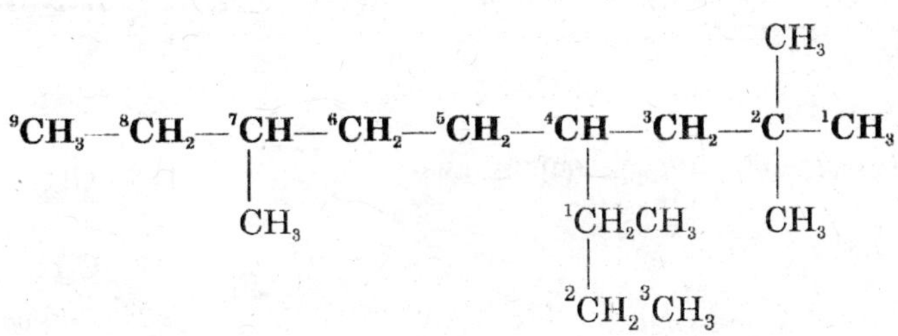

2, 2, 7-Trimethyl-4-(1-methyl propyl) nonane

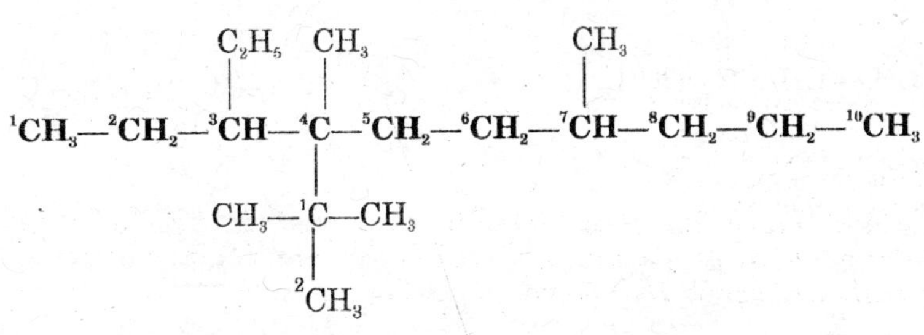

4-(1, 1-Dimethylethyl)-3-ethyl-4, 7-dimethyldecane

The numerical prefixes 'bis-', tris-', 'tetrakis-', 'pentakis-', etc. are used to indicate a multiplicity of substituted substituents. The name of the substituted substituent is enclosed in parentheses. For example,

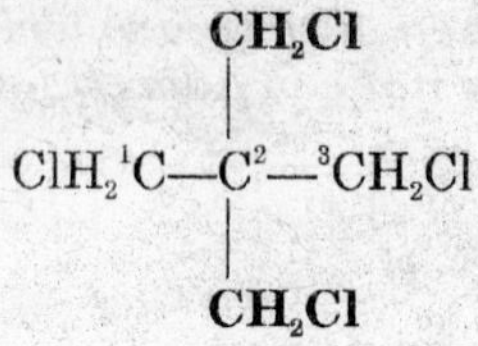

1, 3-Dichloro-2, 2-bis (chloromethyl) propane

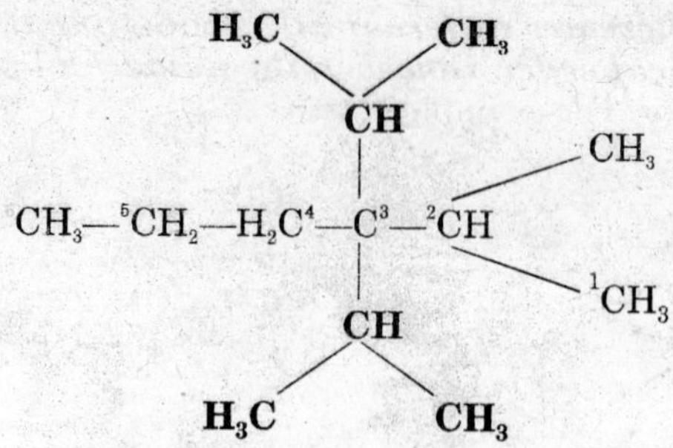

2-Methyl-3, 3-bis (1-methylethyl) hexane

B. NOMENCLATURE OF COMPOUNDS HAVING ONE FUNCTIONAL GROUP

It has been pointed out in the beginning that the functional groups (other than C = C or C ≡ C) present in the molecule are indicated by secondary suffixes. Such functional groups have been listed in Table 1.3. However, some of the functional groups are always regarded as substituents (Table 1.5) and are indicated by the prefixes. The various rules which are followed while writing the name of the compounds having one functional groups are described below :

- **Rule 1. Longest chain rule**

  Select the longest continuous chain of the carbon atoms as **parent chain.** *The selected chain must include the carbon atoms involved in the functional groups like —COOH, —CHO, —CN, etc., or those which carry the functional groups like —OH, —NH_2, —Cl, —NO_2, etc. The number of carbon atoms in parent chain decides the word root.*

- **Rule 2. Lowest number rule**

  The carbon atoms of the parent chain are numbered in such a way so that the carbon atom of the functional group gets the lowest possible number. In case the functional group does not have the carbon atom, then the carbon atom of the parent chain attached to the functional group should get the lowest possible number.

- **Rule 3, Rule 4 and Rule 5** are used in the similar way as described in case of hydrocarbons earlier.

Illustration

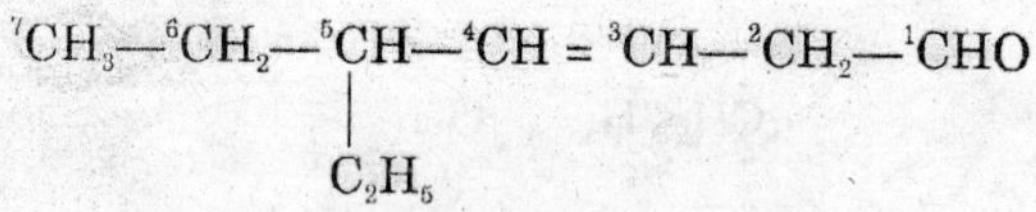

5-Ethylhept-3-enal

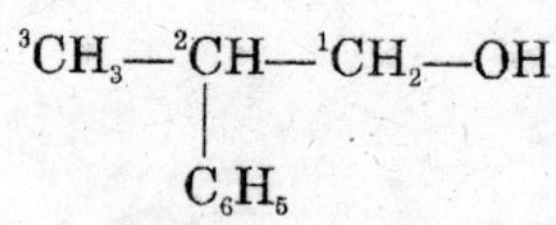

2-Phenylpropan-1-ol

Some more examples of the compounds containing one functional group are being given below :

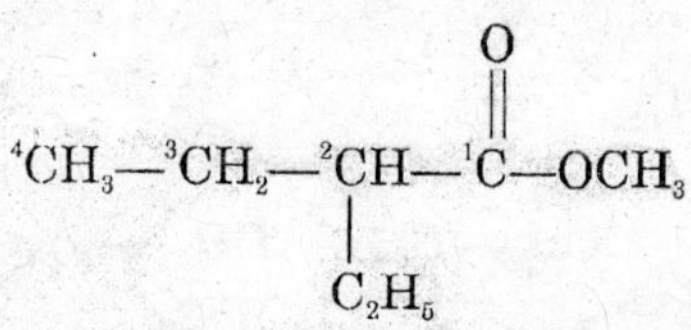

Methyl 2-ethylbutanoate

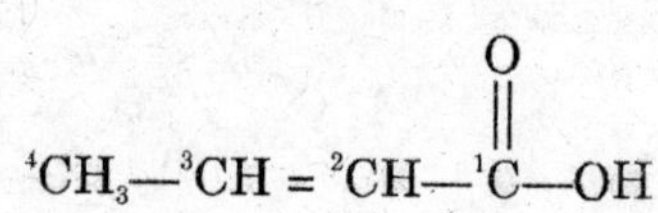

But-2-en-1-oic acid

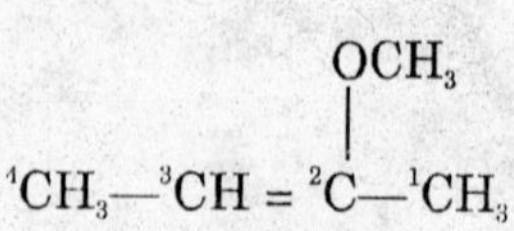

2-Methoxybut-2-ene

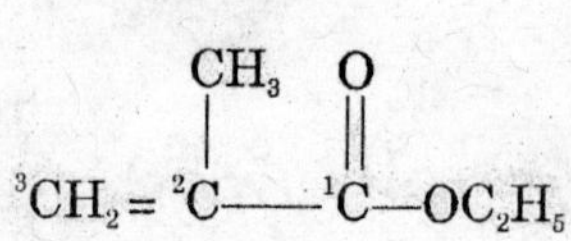

Ethyl 2-methylprop-2-enoate

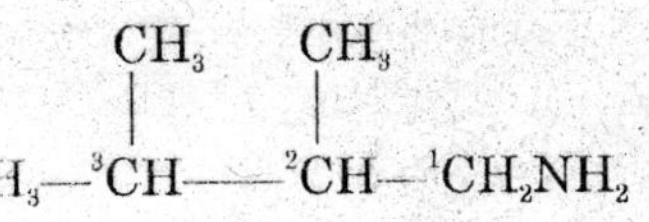

2, 3-Dimethylbutan-1-amine

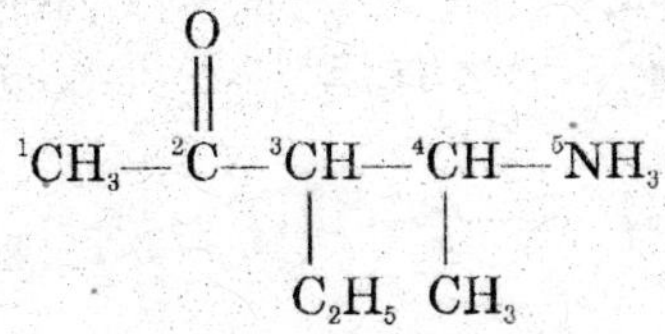

3-Ethyl-4-methylpentan-2-one

C. NAMING THE COMPOUNDS WITH MORE THAN ONE SIMILAR FUNCTIONAL GROUPS

If the organic molecule contains more than one similar functional groups, then in addition to various rules, the numerical prefixes *di* (for 2), *tri* (for 3), etc. are added before the secondary suffix which indicates the functional group. For example

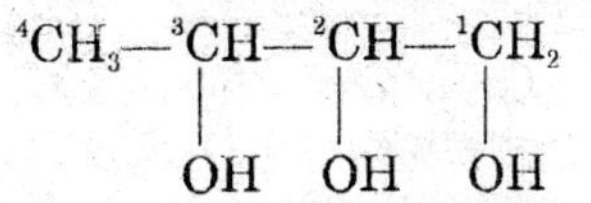

Butane-1, 2, 3-triol

Some more examples of the compounds containing more than one similar functional groups are given below :

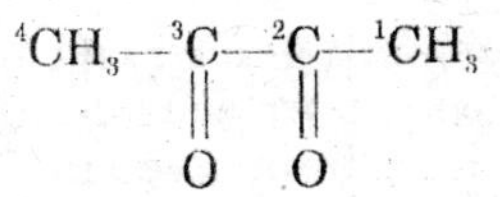

Butane-2, 3-dione

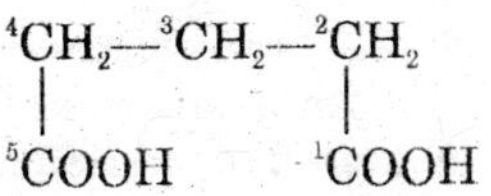

Pentane-1, 5-dioic acid

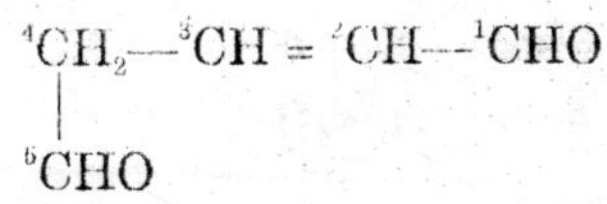

Pent-2-ene-1, 5-dial

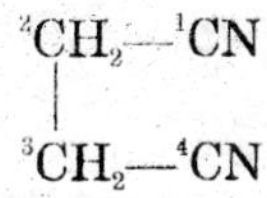

Butane-1, 4-dinitrile

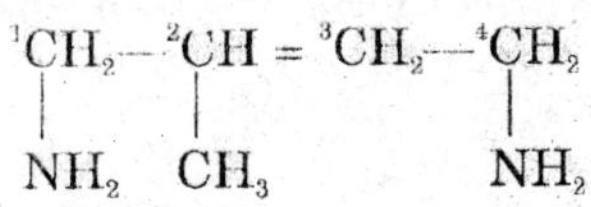

2-Methylbut-2-ene-1, 4-diamine

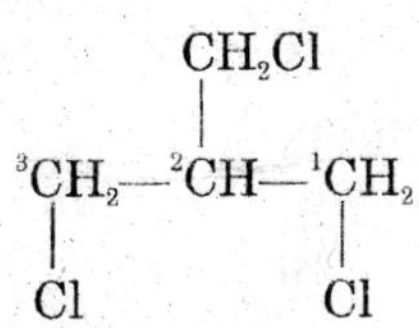

1, 3-Dichloro-2-(chloromethyl) propane

Note. *If an unbranched chain is linked directly to more than two carboxyl groups, these groups are named by substitutive use of suffix "tricarboxylic acid", etc. The principal chain selected should be linked directly to the maximum possible number of carboxyl groups.*

Illustration

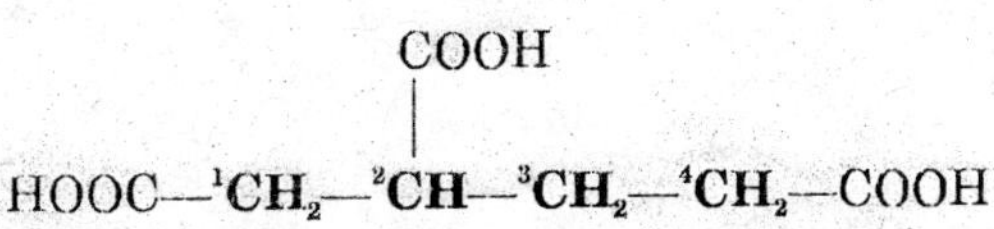

Butane-1, 2, 4-tricarboxylic acid

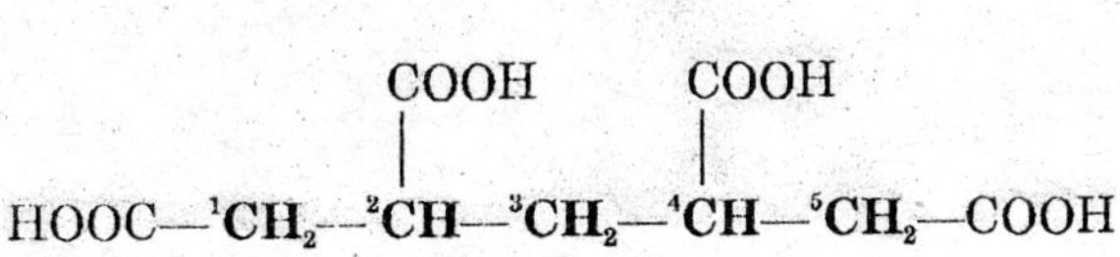

Pentane-1, 2, 4, 5-tetracarboxylic acid

The carboxyl groups which are not directly linked to the principal chain are expressed by *carboxyalkyl* prefixes.

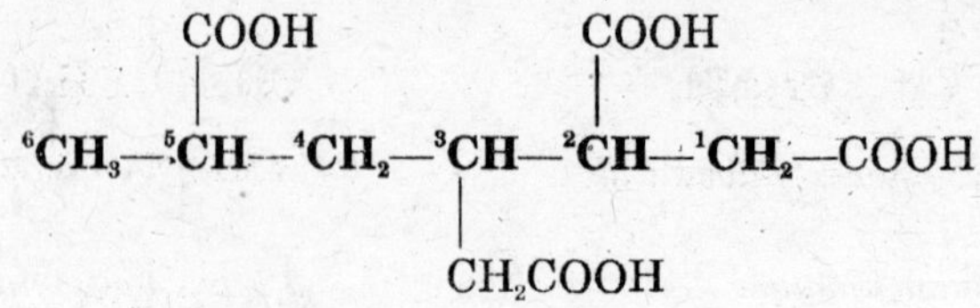

3-(Carboxymethyl) hexane-1, 2, 5-tricarboxylic acid

Similarly, the substitutive prefixes for aldehydes, cyanides, acylchlorides and amides are *carbaldehyde*, *carbonitrile*, *chlorocarbonyl* and *carboxamide* respectively.

D. NAMING THE COMPOUNDS WITH TWO OR MORE DIFFERENT FUNCTIONAL GROUPS

If the molecule contains more than one dissimilar functional groups, the parent chain must contain maximum possible number of functional groups. The numbering of the parent chain is done in such a way so that the functional group of higher priority gets the lower number. **The order of priority** of various groups for the sake of numbering is :

sulphonic acids > carboxylic acids > acid anhydrides > esters > acid chlorides > amides > nitriles, isocyanides > aldehydes > ketones > alcohols > amines > alkenes, alkynes > halo, nitro, alkoxy > alkyl.

The functional group which gets priority is treated as **principal functional group** and is indicated by the secondary suffix. On the other hand, the other functional groups in the compound are considered as substituents and are indicated by suitable prefixes. The prefixes for such functional groups have been given in *Table 1.6*. Let us write the IUPAC name of the following poly-functional compound.

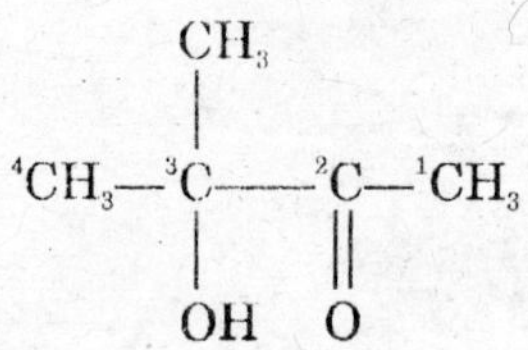

w. root = **but**

primary suffix = **ane**

functional groups = **CO, OH**

sec. suffix = **one**

prefixes = **methyl, hydroxy**

Thus, the name is : *3-Hydroxy-3-methylbutan-2-one*

It may be noted that if two groups of same preference appear at identical positions from the either end of the chain, then the group whose prefix comes first in alphabetic order gets the lower number. For example,

$$Br—{}^1CH_2—{}^2CH_2—{}^3CH_2—{}^4CH_2—{}^5CH_2—I$$

1-Bromo-5-iodopentane

Some more examples of the poly functional compounds are given below :

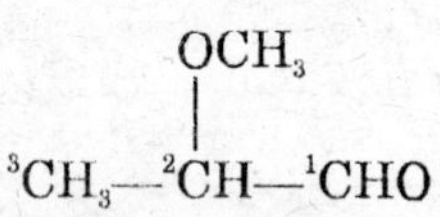

2-Methoxy propanal

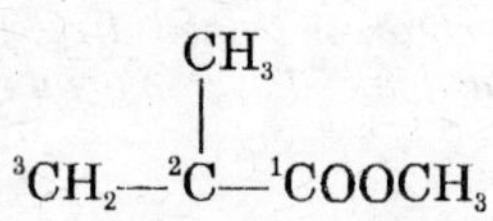

Methyl 2-methyl prop-2-enoate

$${}^4CH_3—{}^3CH = {}^2CH—{}^1COOH$$

But-2-en-1-oic acid

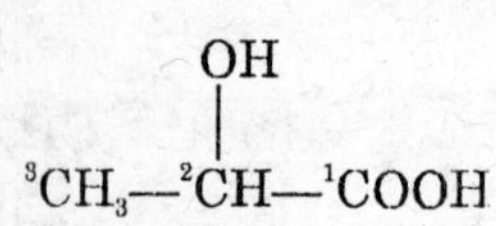

2-Hydroxy propanoic acid

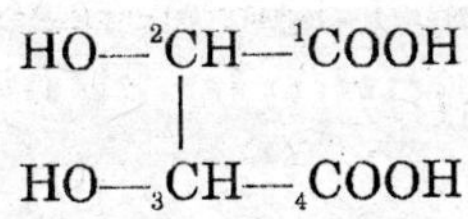

2, 3-Dihydroxybutane-1, 4-dioic acid

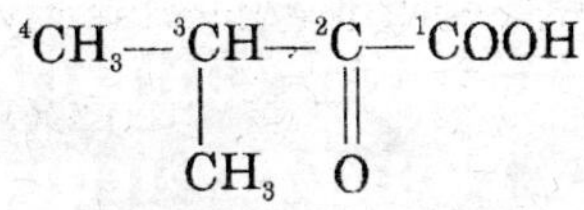

3-Methyl-2-oxobutanoic acid

$NC-{}^{3}CH_2-{}^{2}CH_2-{}^{1}COOH$

3-Cyanopropanoic acid

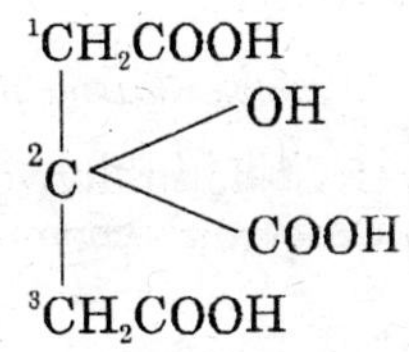

2-Hydroxypropane-1, 2, 3-tricarboxylic acid

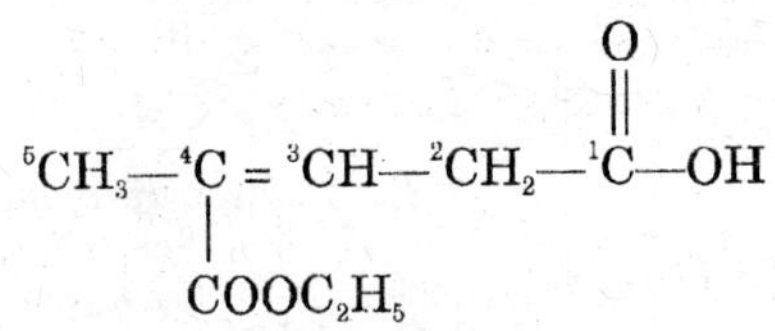

4-Carbethoxypent-3-enoic acid

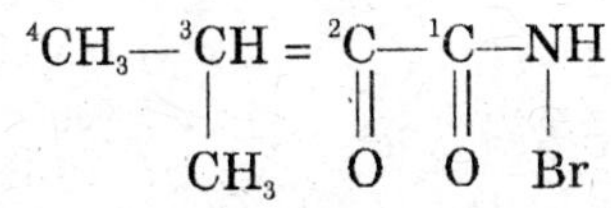

2-keto-3-methyl-(N-bromo) butanamide

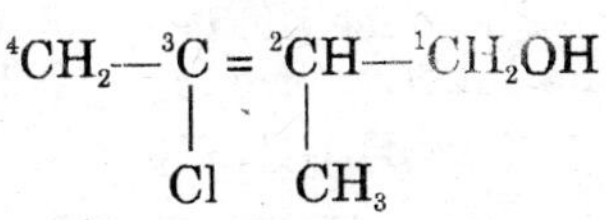

4-Chloro-3-methyl-but-2-en-1-ol

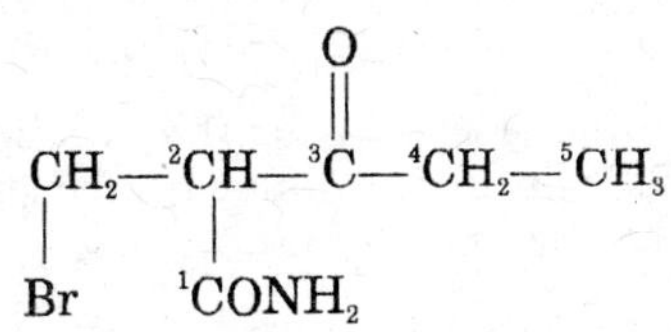

2-(Bromomethyl)-3-oxopentanamide

Some Important 1993 Recommendations of IUPAC Nomenclature of Organic Compounds

1. *Locants (numerals and/or letters) are placed immediately before the part of the name to which they relate.* For example,

$CH_3CH_2CH = CH_2$ should be named as but-1-ene

$CH_3CH_2CH_2OH$ should be named as propan-1-ol

Similarly, a few more examples are given as following :

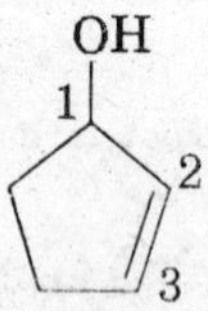

Cyclopent-2-en-1-ol

$^4CH_3—^3CH = ^2C(CH_3)—^1CH_2OH$

2-Methylbut-2-en-1-ol

$^3CH_3—^2C(CH_3)_2—^1CH_2OH$

2, 2-Dimethylpropan-1-ol

2. *The locant 1 is often omitted when there is no ambiguity. For example,*

$CH_3CH_2CH_2COOH$
Butanoic acid

CH_3CH_2CHO
Propanal

$CH_3CH_2CH_2CN$
Butane nitrile

$Cl - CH_2CH_2 - OH$
2-Chloroethanol

In all the above examples locant 1 for the functional group is omitted because the position of the functional group is unambiguous. However, in the following cases the position of the function group must be mentioned.

$CH_3CH_2CH_2OH$
Propan-1-ol

$CH_3CH_2CH_2NH_2$
Propan-1-amine

Here, we cannot write simply propanol (or propanamine) because there are two propanols ; propan-1-ol and propan-2-ol.

3. Arrangement of Prefixes

(*i*) *Simple prefixes such as methyl, ethyl, chloro, nitro, hydroxy, etc. are arranged alphabetically. The prefixes di, tri, etc. are however not considered for comparison.*

For example, $\overset{1}{C}H_3\overset{2}{C}H_2\overset{3}{C}H(CH_3)\overset{4}{C}H_2\overset{5}{C}H(C_2H_5)\overset{6}{C}H_2\overset{7}{C}H_2\overset{8}{C}H_3$

5-Ethyl-3-methyl octane

$\overset{2}{C}H_2(Cl)\overset{1}{C}H_2(Br)$

1-Bromo-3-chloroethane

(*ii*) *The name of a prefix for a substituted substituent is considered to begin with the first letter of its complete name.*

For example, $\overset{8}{C}H_3—\overset{7}{C}H_2—\overset{6}{C}H_2—\overset{5}{C}H_2—\overset{4}{C}H(\overset{1}{C}H(Cl)—\overset{2}{C}H_2—\overset{3}{C}H_3)—\overset{3}{C}H(CH_3)—\overset{2}{C}H_2—\overset{1}{C}H_3$

4-(1-Chloropropyl)-3-methyloctane

For the substituted substituent 1-chloropropyl, 'C' is taken as the first letter.

(*iii*) *When two or more prefixes consist of identical letters priority for citation is given to the group which contains the lowest locant at the first point of difference.*

For example,

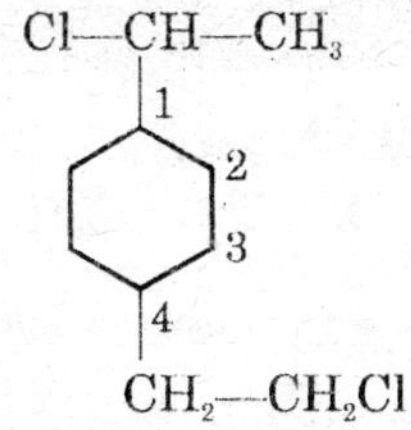

1-(1-Chloroethyl)-4-(2-chloroethyl) cyclohexane

Here, 1-chloroethyl gets priority over 2-chloroethyl.

BOND LINE NOTATIONS OF ORGANIC MOLECULES

It is a simple, brief and convenient method of representing organic molecules. In these notations, the bonds between the carbon atoms are represented by lines. A single line (–) represents a single bond, two parallel lines (=) represent a double bond and three parallel lines (≡) represent a triple bond. The intersection of lines represents carbon atoms carrying appropriate number of H atoms. For example, buta-1, 3-diene (CH_2 = CH—CH = CH_2) can be represented as follows :

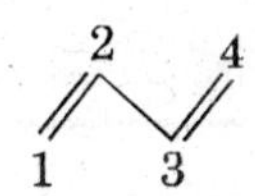

Some bond line structures along with their IUPAC names are given below :

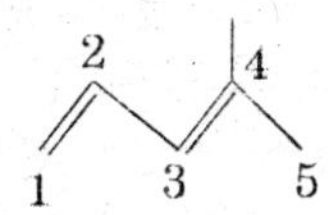

4-Methyl penta-1, 3-diene

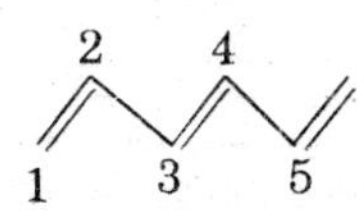

Hexa-1, 3, 5-triene

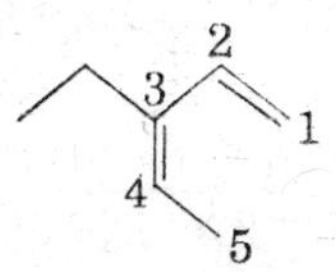

3-Ethyl penta-1, 3-diene

NAMING OF ALICYCLIC COMPOUNDS

Names of alicyclic compounds are derived by putting another prefix '*cyclo*' before the word root, which depends upon the number of carbon atoms in the ring. The suffixes *ane, ene* or *yne* are written depending upon saturation or unsaturation in the ring.

Cyclobutane

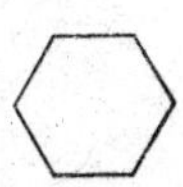

Cyclohexane

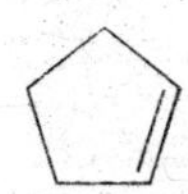

Cyclopentene

If some substituent or functional group is present, it is indicated by some appropriate prefix or suffix and its position is indicated by numbering the carbon atoms of the ring. *The numbering is done in such a way so as to assign least possible number to the functional group or substituent in accordance with the rules already discussed.* Some examples are :

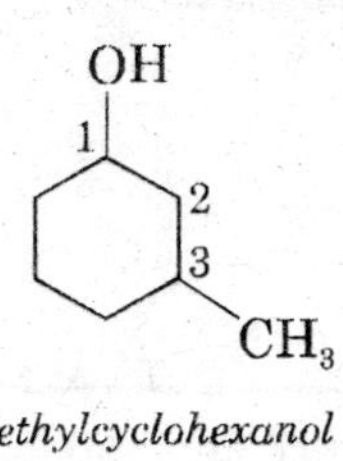

3-Methylcyclohexanol

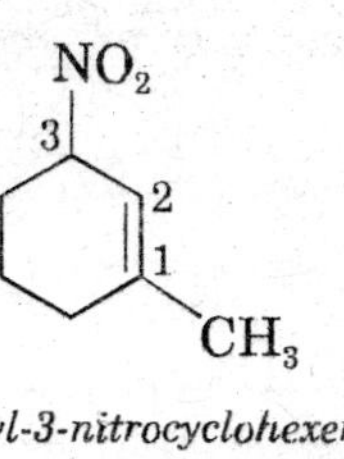

1-Methyl-3-nitrocyclohexene

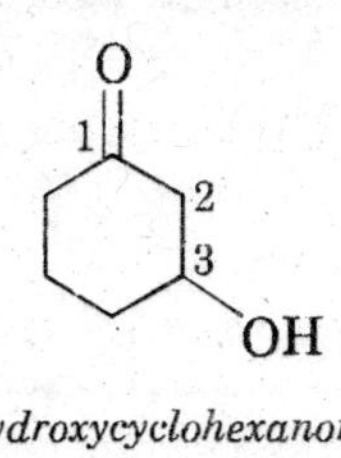

3-Hydroxycyclohexanone

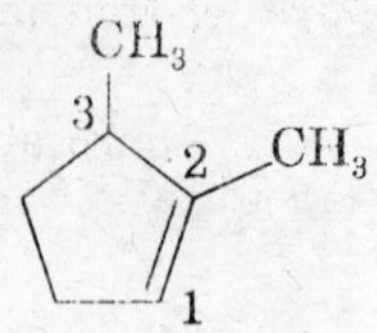

2, 3-Dimethylcyclopentene

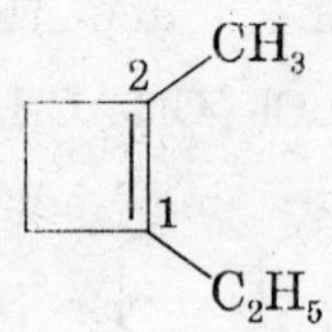

1-Ethyl-2-methylcyclobutene

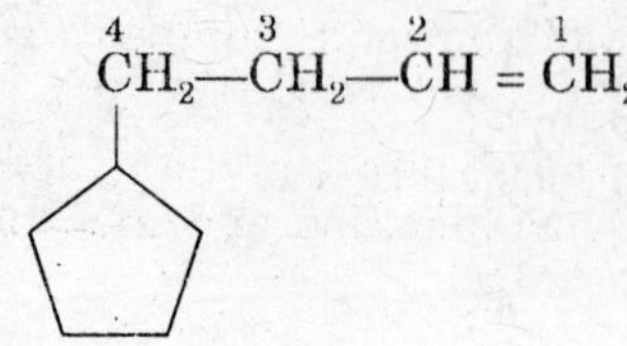

3-Cyclopenyl-1-butene

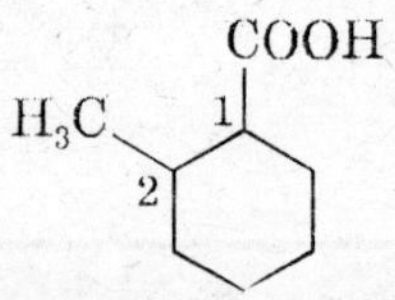

2-Methylcyclohexanecarboxylic acid

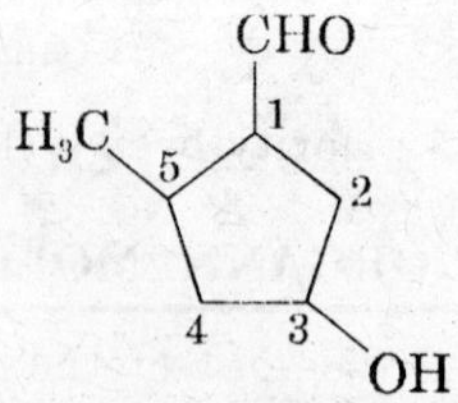

3-Hydroxy-5-methylcyclopentane-carbaldehyde

NAMES OF BRIDGED BICYCLIC HYDROCARBONS

Saturated bicyclic systems having two or more atoms in common are named by prefixing '*bicyclo*' to the name of the cyclic parent hydrocarbon system containing the same total number of carbon atoms in the skeleton. The number of carbon atoms in each of the three bridges, connecting the two tertiary carbon atoms is indicated in parentheses, in descending order and arabic numerals are used to indicate the number of carbon atoms and the numbers are separated by full stops.

The bicyclic system is numbered starting with one of the tertiary bridging carbon and proceeding through longest bridge to the second bridging carbon continuing back to the first bridging carbon through the second longest chain. Numbering is completed by numbering the shortest bridge beginning with the atom next to the first bridging carbon. Some illustrations are given below :

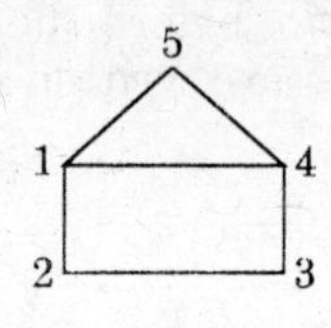

Bicyclo [2.1.0] pentane

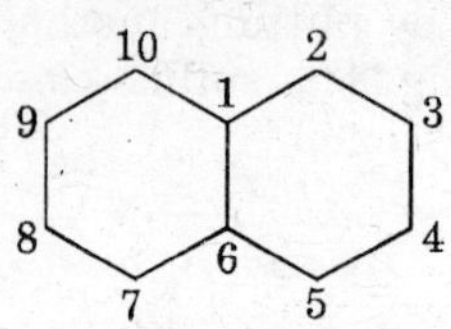

Bicyclo [4.4.0] decane

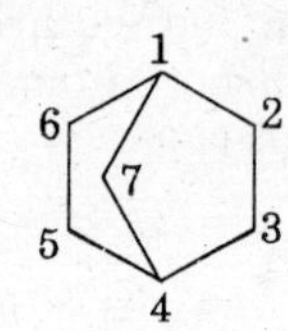

Bicyclo [2.2.1] heptane

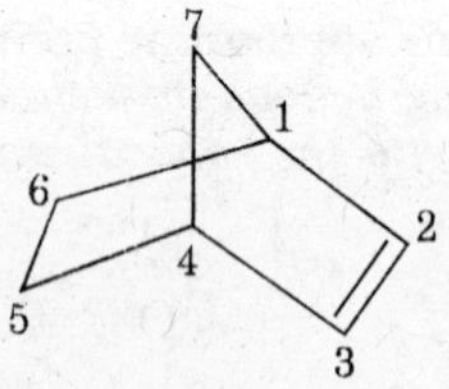

Bicyclo [2.2.1] hept-2-ene

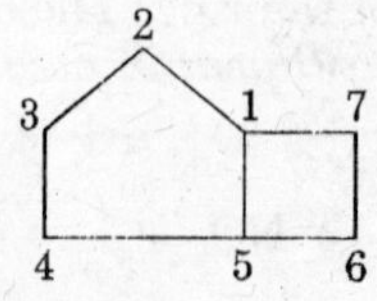

Bicyclo [3.2.0] heptane

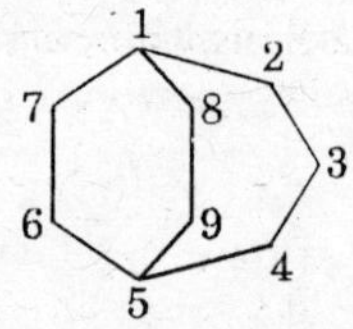

Bicyclo [3.2.2] Nonane

NAMES OF SPIRO BICYCLIC HYDROCARBONS

Spiro bicyclic hydrocarbons contain two rings consisting of carbon atoms only and the two rings are linked by a common carbon. These compounds are named by placing prefix '*spiro*' before the name of the acyclic parent hydrocarbon with same number of skeletal carbon atoms. The numbers of skeletal atoms linked to the spiro atom are indicated by arabic numbers, separated by a full stop. The numbers are written in ascending orde1 and enclosed in square brackets.

Numbering of a spiro bicyclic hydrocarbon starts with a ring carbon next to the spiro atom and proceeds first through the smaller ring and then through the spiro atom and around the second ring. Some examples are given below :

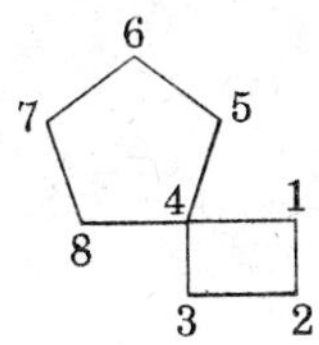

Spiro [3.4] octane

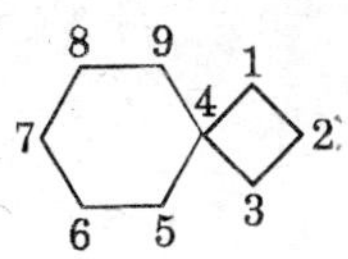

Spiro [3.5] nonane

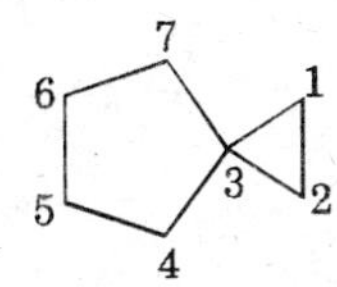

Spiro [2.4] heptane

NAMES OF SOME HETEROCYCLIC COMPOUNDS

Heterocyclic compounds are the cyclic compounds which contain atoms of some other element than carbon in their rings in addition to the atoms of carbon. Some important heterocyclic compounds along with their names are given below. The hetero-atom in a heterocyclic compound is given the lowest possible number consistent with an orderly progression around the ring system.

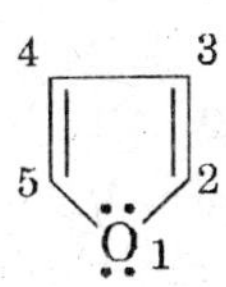

Furan

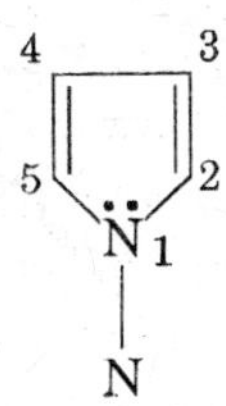

Pyrrole

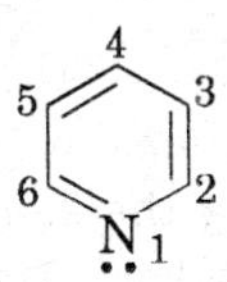

Pyridine

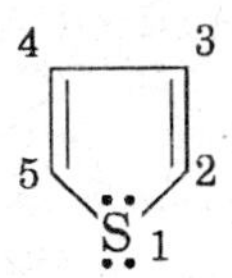

Thiophene

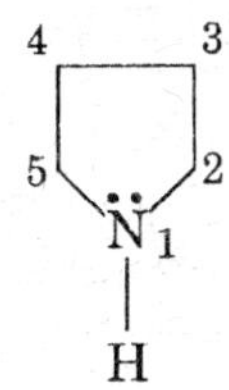

Pyrrolidine

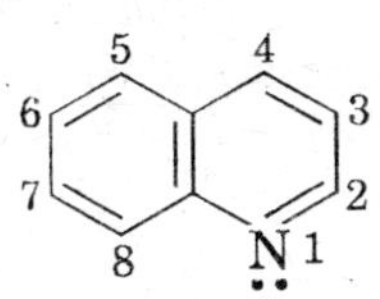

Quinoline

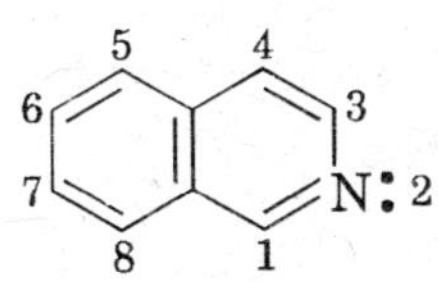

Isoquinoline

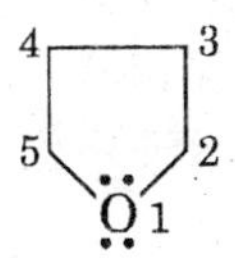

Tetrahydrofuran

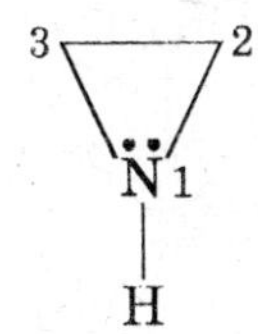

Aziridine

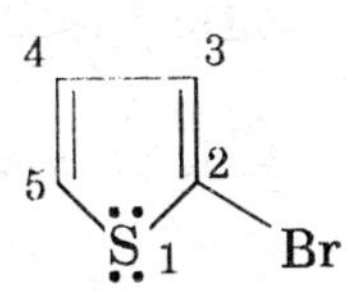

2-Bromothiophene

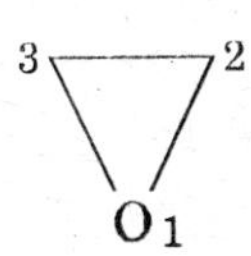

Oxirane

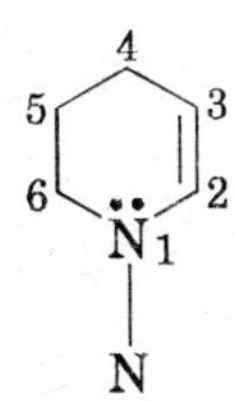

Piperidine

FOR PRACTICE

Give IUPAC names of the following compounds :

1. 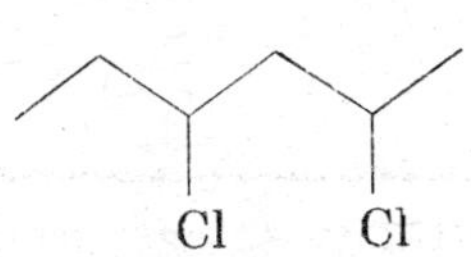

2. 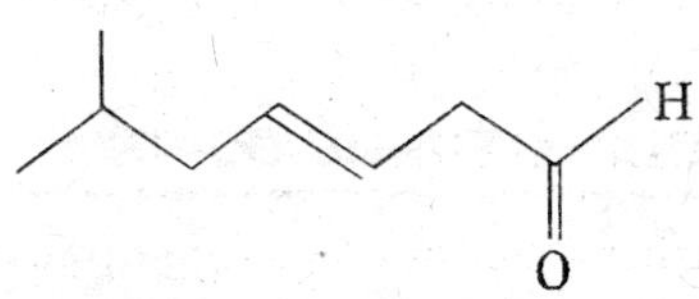

3. 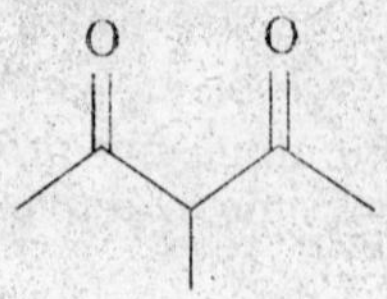

4. 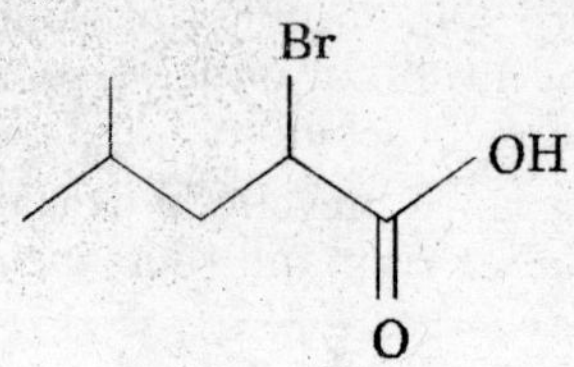

5. 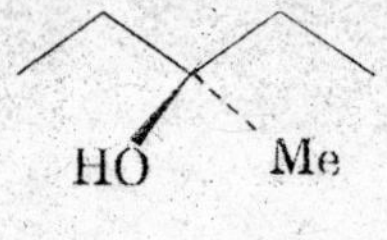

6. 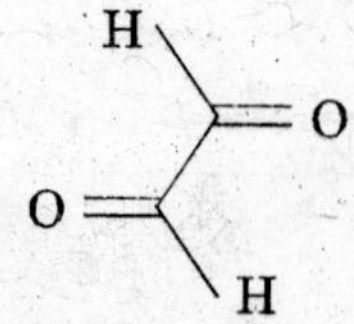

7. 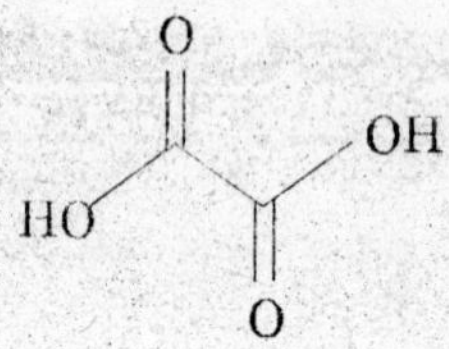

8. 

9. 

10. 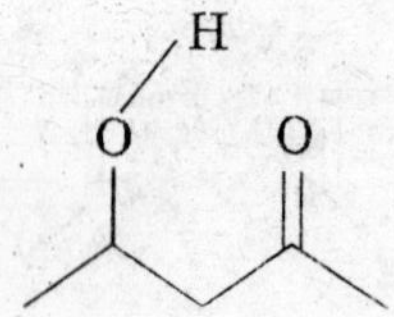

11. 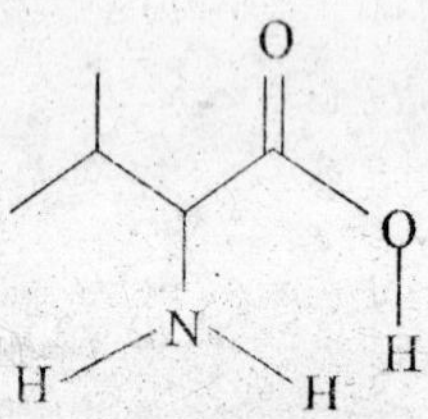

12. 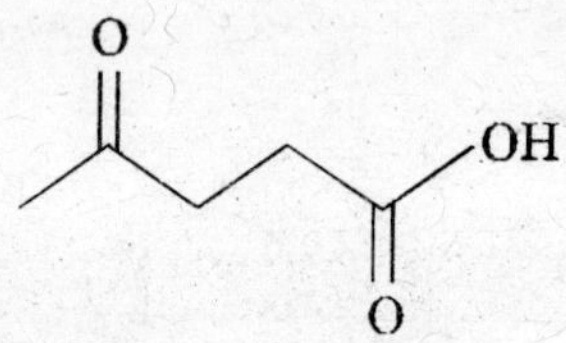

13. 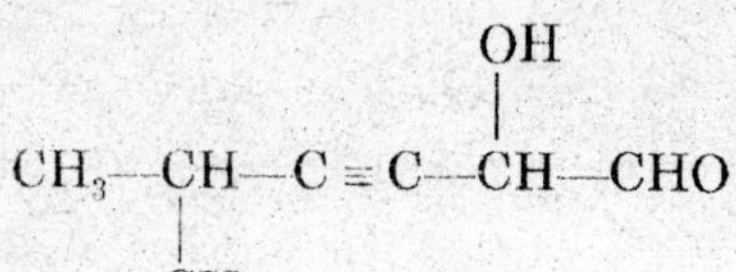

14. 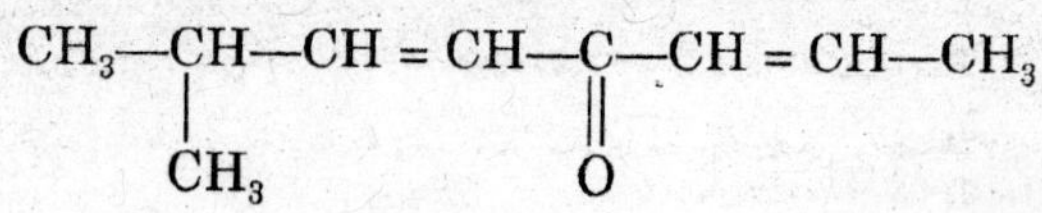

15. 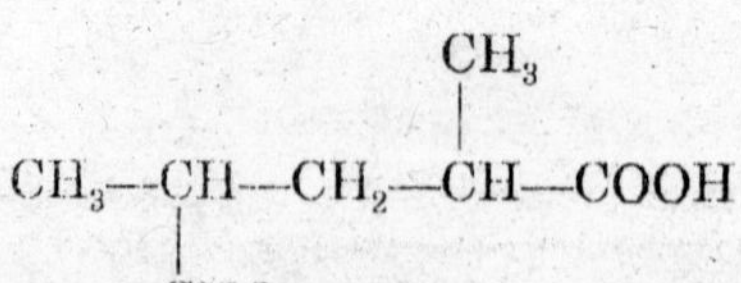

16. 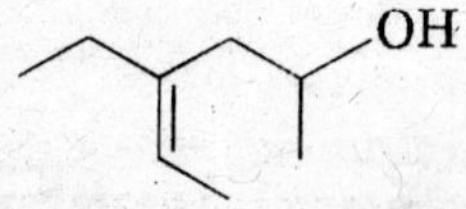

17. 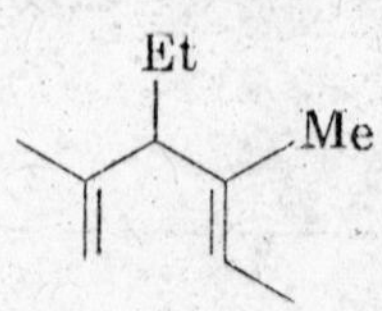

18. 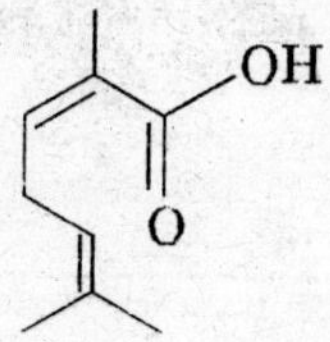

19. 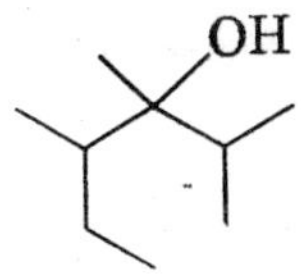

20. 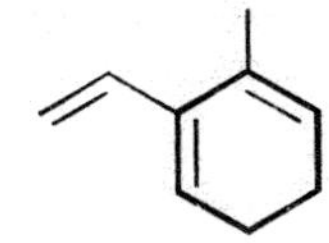

21. 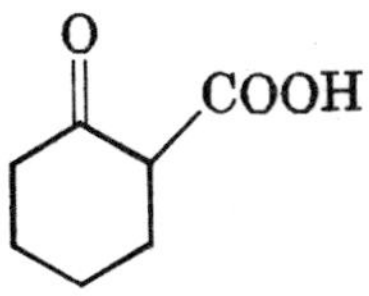

22. 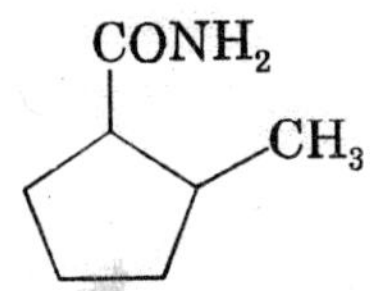

23. 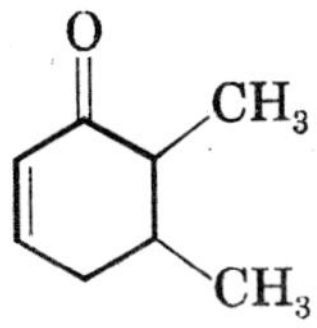

24. 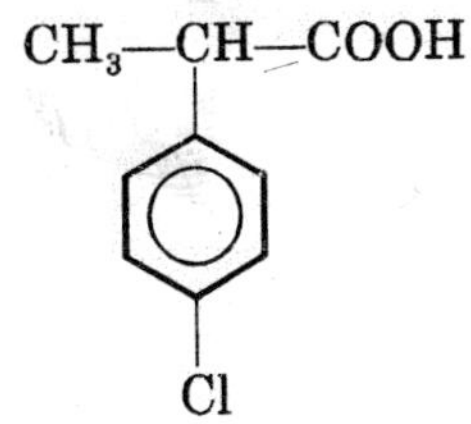

25. 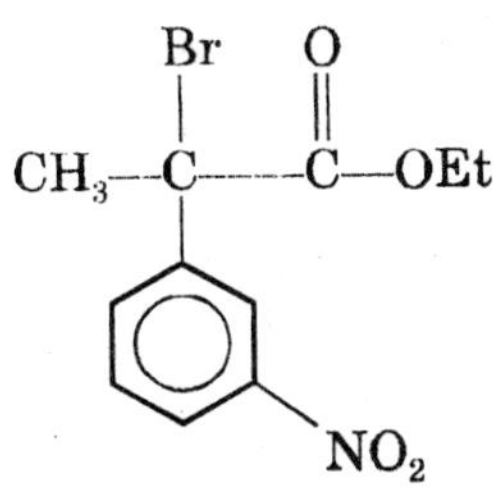

26. 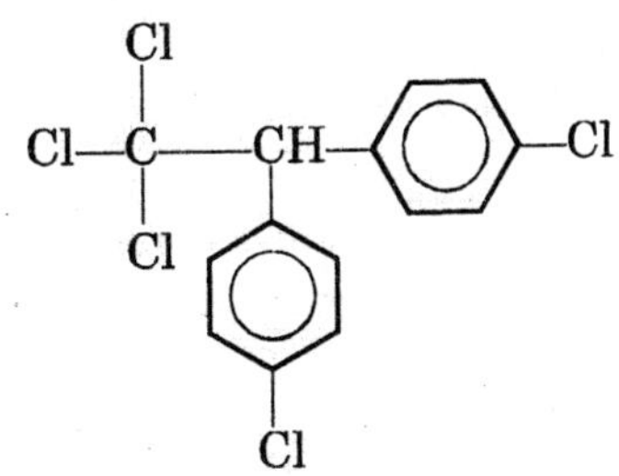

27. 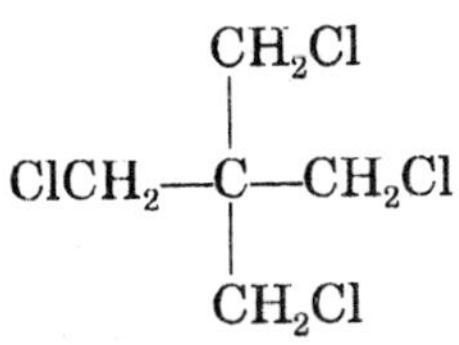

ANSWERS

1. *2, 4-Dichlorohexane*
2. *6-Methylhept-3-enal*
3. *3-Methylpentane-2, 4-dione*
4. *2-Bromo-5-methylhexanoic acid*
5. *3-Methylpentan-3-ol*
6. *Ethanedial*
7. *Ethanedioic acid*
8. *Pent-1-en-4-yne*
9. *Pent-3-en-1-yne*
10. *4-Hydroxypentan-2-one*
11. *2-Amino-3-methylbutanoic acid*
12. *4-Ketopentanoic acid*
13. *2-Hydroxy-5-methylhex-3-ynal*
14. *7-Methylocta-2, 5-dien-4-one*
15. *4-Formyl-2-methylpentanoic acid*
16. *4-Ethylhex-4-en-2-ol*
17. *3-Ethyl--methylhex-4-en-2-one*
18. *2, 6-Dimethylhepta-2, 5-dienoic acid*
19. *2, 3, 4-Trimethylhexan-3-ol*
20. *3-Ethenyl-2-methylcyclohexa-1, 3-diene*
21. *2-Ketocyclohexanecarboxylic acid*
22. *2-Methylcyclopentanecarboxamide*
23. *5, 6-Dimethylcyclohex-2-en-1 one*
24. *2-(4-Chlorophenyl) propanoic acid*
25. *Ethyl 2-bromo-2-(3-nitrophenyl) propanoate*
26. *1, 1, 1-Trichloro-2, 2-bis (4-chlorophenyl) ethane*
27. *1, 3-Dichloro-2, 2-bis (chloromethyl) propane.*

SOME IMPORTANT COMMON NAMES

Molecular Formula	*Common Name*	*IUPAC Name*
$(CH_3)_2CHCH_2CH_3$	*iso*-Pentane	2-Methylbutane
$CH_3-C(CH_3)_2-CH_3$	*neo*-Pentane	2, 2-Dimethylpropane
$CH_3C \equiv CH$	Methylacetylene (Allylene)	Propyne
$CH_2 = C = CH_2$	Allene	Propadiene
$CH_2 = CH - OH$	Vinyl alcohol	Ethenol
$CH_2 = CH - CH_2 - OH$	Allyl alcohol	Prop-2-en-1-ol
$CH \equiv C - CH_2 - OH$	Propargyl alcohol	Prop-2-yn-1-ol
$CH_2 = CH - COOH$	Acrylic acid	2-Propenoic acid
$COOH-COOH$	Oxalic acid	Ethanedioic acid
$H_2C(COOH)_2$	Malonic acid	Propanedioic acid
$H_2C - COOH$ / $H_2C - COOH$	Succinic acid	Butanedioic acid
$HOOC-(CH_2)_3 - COOH$	Glutaric acid	Pentanedioic acid
$HOOC-(CH_2)_4 - COOH$	Adipic acid	Hexanedioic acid
$CH_2 = CH - CHO$	Acrolein	Propenal
$CH_3CH = CH - CHO$	Crotonaldehyde	2-Butenal
CH_3CN	Acetonitrile	Ethanenitrile

ISOMERISM AND STEREOCHEMISTRY

- ***The branch of chemistry which deals with the study of molecular structures in three dimensional space is called stereochemistry.***
- ***If two compounds have same molecular formula but different structural formulae or physical or chemical properties then they are called isomers and the phenomenon is called isomerism.*** Isomerism is mainly of two types

  i.e., structural isomerism and *stereoisomerism.*

  The **flowsheet** of various categories of isomerism is given below.

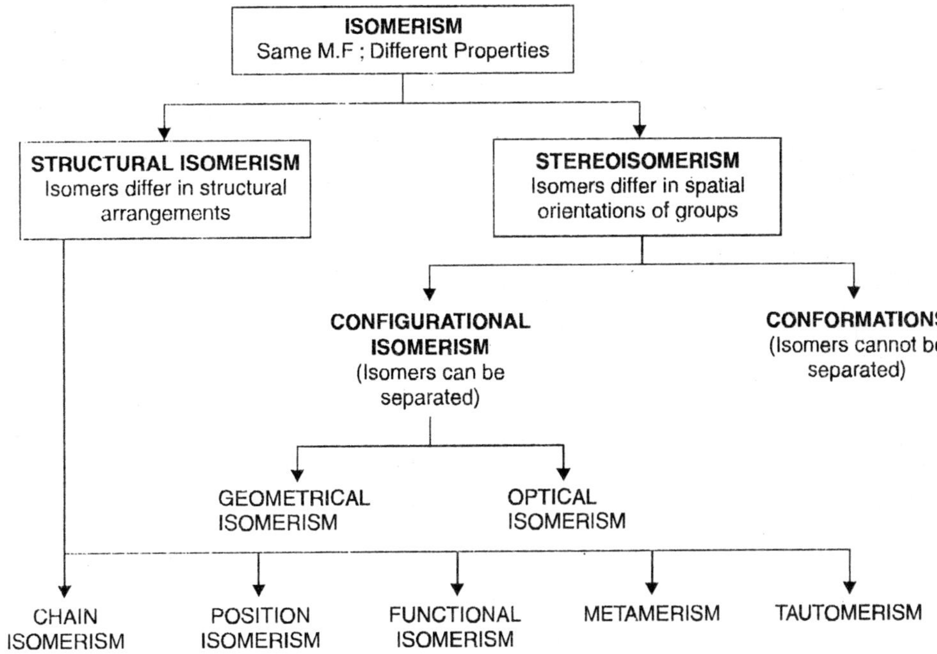

(Flowsheet of different categories of isomerism)

STRUCTURAL ISOMERISM AND ITS VARIOUS TYPES

In this type of isomerism the two isomers have different arrangement of atoms within the molecule. This type of isomers are called structural isomers. Structural isomerism is of following types :

- **Chain isomerism.** *In this type of isomerism, the isomers have different skeletons of carbon atoms.* For example, Pentane (C_5H_{12}) has three chain isomers :

(*i*) $CH_3—CH_2—CH_2—CH_2—CH_3$

n-Pentane
(Pentane)

(*ii*) 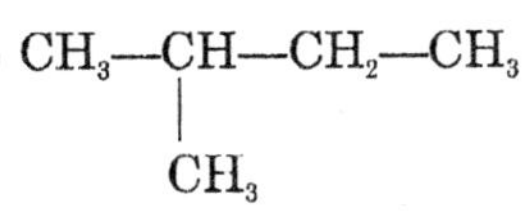

2-Methylbutane
(Isopentane)

(*iii*) 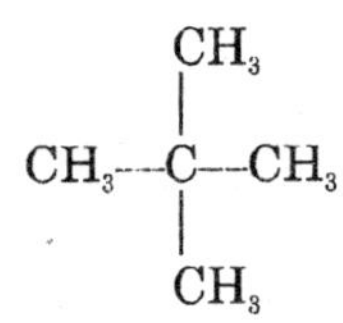

2, 2-Dimethylpropane
(Neopentane)

Molecular formula C_6H_{14} (hexane) represents five isomers.

(*i*) $CH_3—CH_2—CH_2—CH_2—CH_2—CH_3$

n-Hexane
(Hexane)

(*ii*) 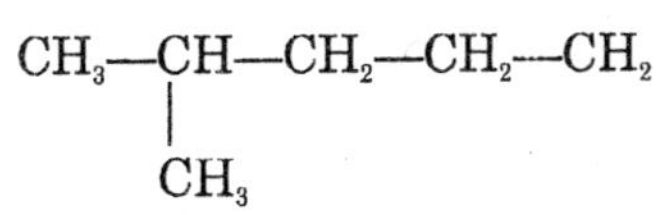

2-Methylpentane

(*iii*) 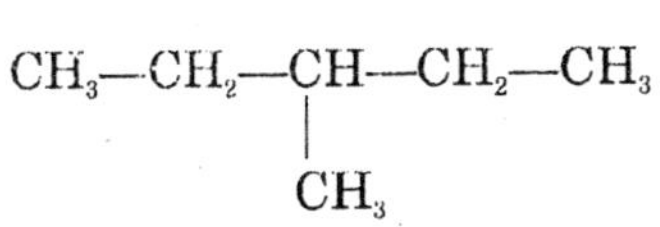

2-Methylpentane

(*iv*) 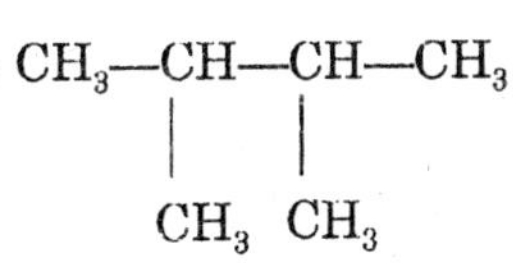

2, 2-Dimethylbutane

(*v*) 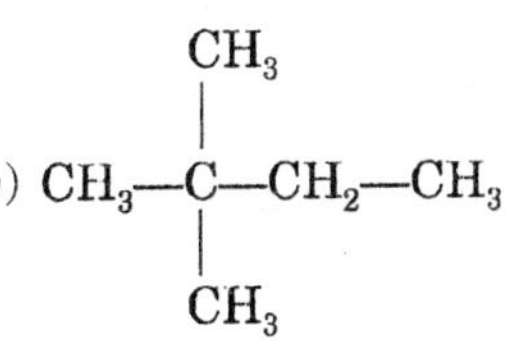

2, 2-Dimethylbutane

- **Position isomerism.** *In this type of isomerism, the isomers have different positions of functional group or multiple bond.* For example,

(*i*) $CH_3CH = CHCH_3$ (But-2-ene) and $CH_3CH_2CH = CH_2$ (But-1-ene)

(*ii*) $CH_3CH_2CH_2CH_2OH$ (Butan-1-ol) and $CH_3CH(OH)CH_2CH_3$ (Butan-2-ol)

- **Functional isomerism.** *In this type of isomerism, the isomers have different functional groups and thus, belong to different families.* For example,

(*i*) $CH_2 = CH - CH = CH_2$ (Buta-1, 3-diene) and $CH \equiv C—CH_2—CH_3$ (But-1-yne)

(*ii*) $CH_3—CO—CH_2—CH_3$ (Butan-2-one) and $CH_3—CH_2—CH_2—CHO$ (Butanal)

(*iii*) $CH_3—CH_2—COOH$ (Propanoic acid) and $CH_3—CO—OCH_3$ (Methylethanoate)

- **Metamerism.** *In this type of isomerism structural formulae of the isomers differ in structure due to difference in distribution of carbon atoms around the functional groups.* For example,

(*i*) $C_2H_5 - O - C_2H_5$ (Diethyl ether) and $CH_3 - O - CH_2 - CH_2 - CH_3$ (Methyl *n*-propyl ether)

(*ii*) $C_2H_5 - NH - C_2H_5$ (Diethyl amine) and $CH_3 - NH - CH_2 - CH_2 - CH_3$ (Methyl *n*-propyl amine)

- **Tautomerism**

This is a special type of functional isomerism where isomers exist simultaneously in dynamic equilibrium with each other. Two forms of a compound which are interconvertible into each other by migration of an atom from one part of the molecule to other part are known as **tautomers** and the phenomenon is called **tautomerism.** It is represented by symbol '$\rightleftharpoons$'

e.g.,

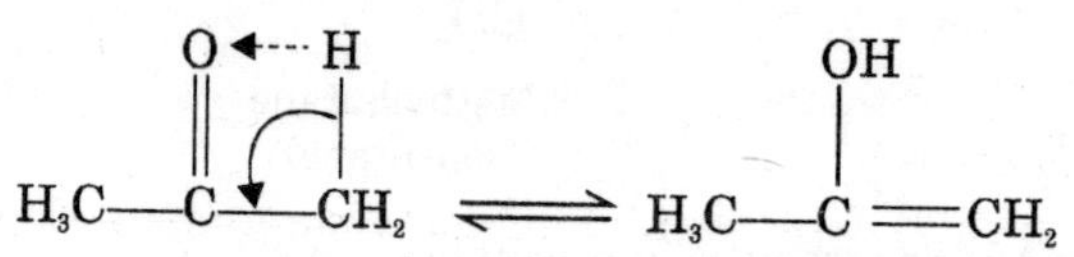

Keto form $\rightleftharpoons$ *Enolic form*

Tautomers of Acetone

Since this involves keto and enol form, it is also known as *keto-enol* tautomerism. The *keto* form is more stable (by about 15 kcal mol^{-1}) than the *enolic* form. Some other examples of tautomerism are as follows :

Example 1. $CH_3—CO—CH_2—CO—OC_2H_5 \rightleftharpoons CH_3—C(OH) = CH—CO—OC_2H_5$

Acetoacetic ester — *Enolic* form

Example 2. $CH_3—CO—CH_2—CO—CH_3 \rightleftharpoons CH_3—C(OH) = CH—CO—OC_2H_5$

Keto form — *Enol* form

In the above examples the enol form is more stable. The stability of enol form in these cases is due to the reason that double bonds are in conjugation and also hydrogen bonds are possible between —OH and C = O groups of the molecule as shown below :

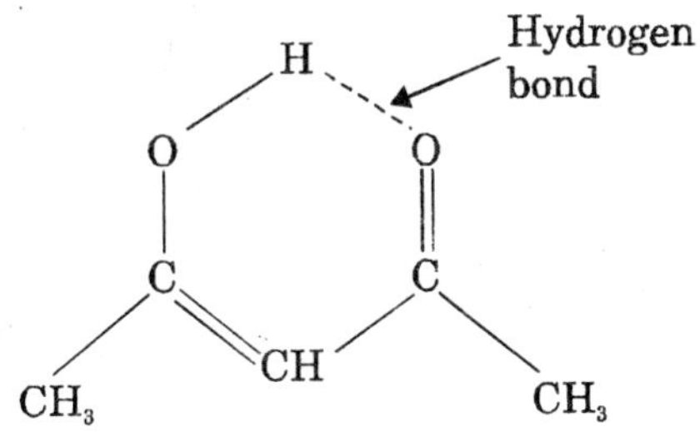

Essential condition for keto-enol tautomerism

- The Carbonyl group of aldehydes, ketone or ester should have α-H atom(s)
- The percentage of enol content in the mixture depends on many factors :

% Enol content ∝ Conjugation

$$\propto \frac{1}{\text{Temperature}} \propto \frac{1}{\text{Hydrogen bonding}} \propto \frac{1}{\text{Base strength}}$$

In simple words if the carbonyl compound is represented as 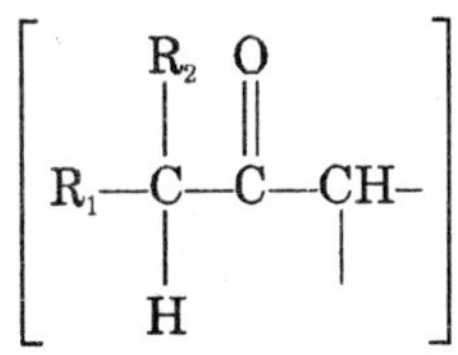 then greater the number of π bonds in R_1 and R_2, greater is the enol content.

i.e., % enol content in the compounds like acetaldehyde, acetone and their acetyl derivatives is as follows :

$$CH_3coch_2CHO > CH_3COCH_2COCH_3 > CH_3CHO > CH_3COCH_3$$

The % enol content of some compounds is shown below :

Compound	*% Enol content*
Acetone	6×10^{-7}
Acetaldehyde	1×10^{-3}
$(CH_3)_2CHCHO$	1.3×10^{-2}

The tautomerism has been **classified** in two categories :

(*a*) **Cationotropy**, if a cation migrates, and

(*b*) **Anionotrophy,** if a anion migrates.

We cannot separate tautomers easily. Relative abundance of tautomers depends upon their stability. The more stable tautomer exists in higher proportion.

Some more **examples of tautomers** are :

$CH_2 = CH - OH \rightleftharpoons CH_3 - CHO$

Vinyl alcohol — Acetaldehyde

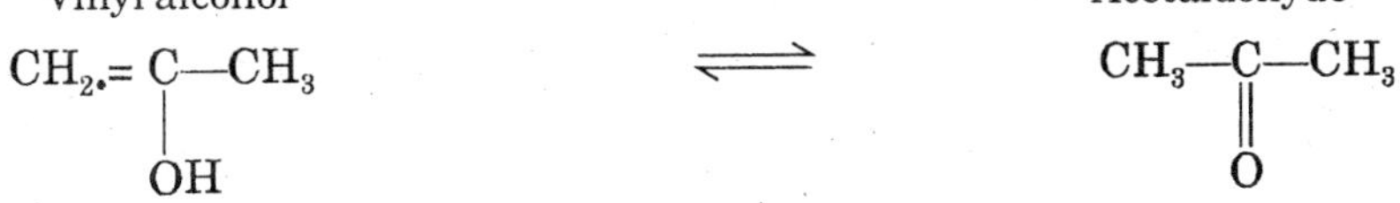

Prop-1-en-2-ol — Propanone

- **Nitro-Aci tautomerism**

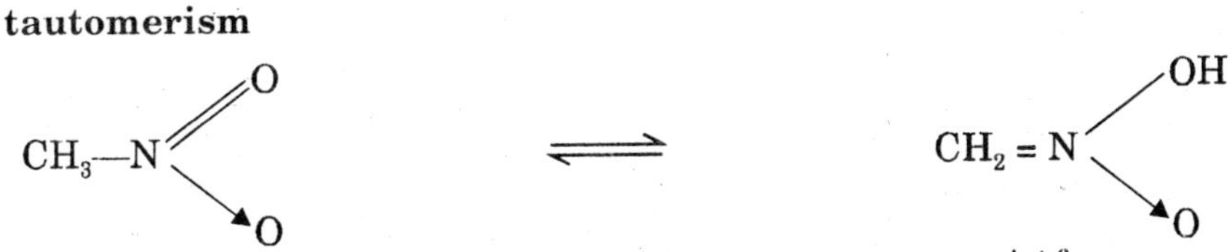

Nitro form — *Aci*-form

- **Nitro-oximino tautomerism**

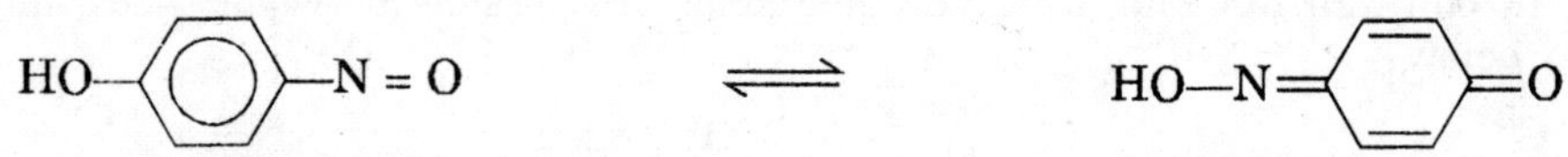

p-nitrosophenol *p*-Benzoquinone monoxime

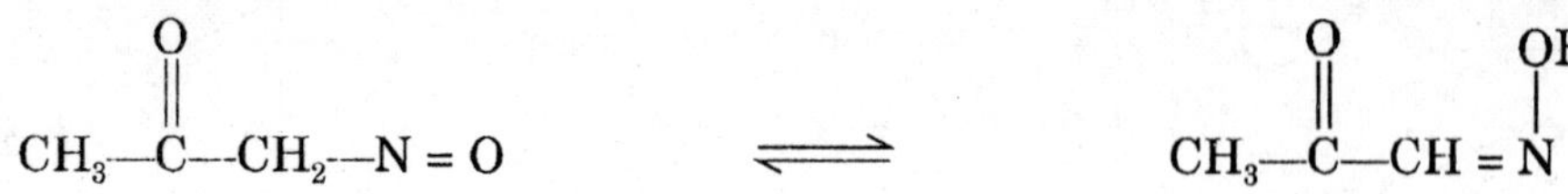

Nitrosoacetone Oximinoderivative

- **Amido-imidol tautomerism**

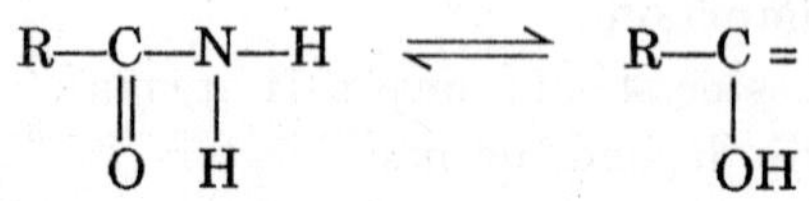

- **Azo-Hydrazone tautomerism**

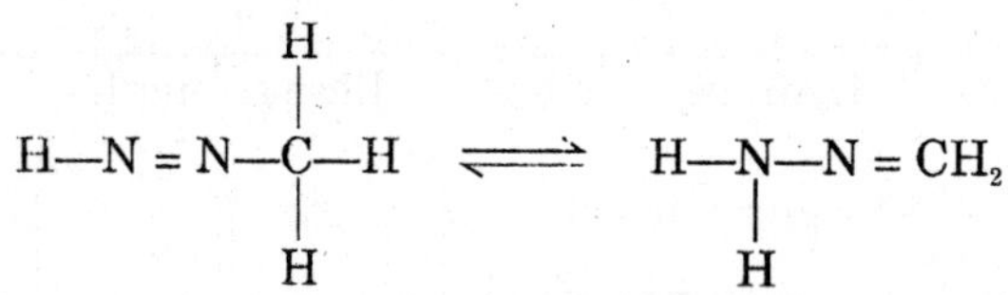

STEREOISOMERISM AND ITS VARIOUS TYPES

In this type of isomerism, isomers have same arrangement of atoms but their orientation in space is different. *The isomers which have same arrangement of atoms but differ only due to different orientation of atoms in space within the molecule are called* **stereoisomers** and *the phenomenon is called* **stereoisomerism.** For example,

(*a*) *Enantiomerism or optical isomerism* (*b*) *Geometric isomerism* (*c*) *Conformations.*

- **Elements of Symmetry**

Let us become familiar with some symmetry elements which will be used in understanding different concepts in the following sections.

1. **Centre of Symmetry.** A centre of symmetry is an imaginary point in a molecule such that a line drawn from any atom/group of the molecule to this point when extended to an equal distance beyond this point meets the mirror image of this group.
2. **Axis of Symmetry.** A molecule is said to possess *n-fold* axis of symmetry when same arrangement is repeated before your eyes *n times* when the molecule is rotated about this axis through 360°. For example, NH_3 has three fold axis of symmetry because rotation of molecule about this axis causes same arrangement three times.

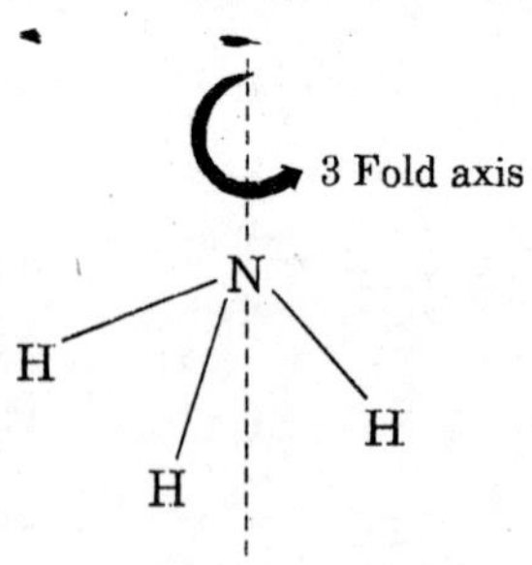

3. **Plane of Symmetry or Mirror Plane.** It is a real or imaginary plane which divides the molecule (or object) into two identical halves. It is also called a mirror plane because it cuts the molecule into two halves which are mirror images of each other.

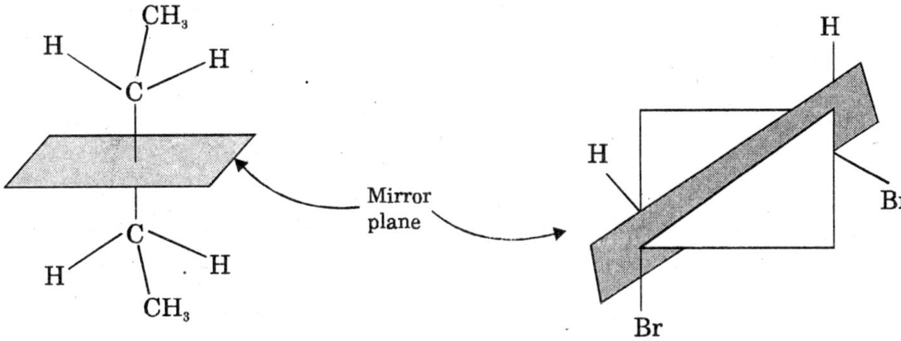

CONFORMATIONS

- The different spatial arrangements of atoms which arise due to rotation around carbon-carbon single bond are called **conformations.**
- Out of infinite number of possible conformations of ethane, the two extreme conformations are **staggered** conformation and **eclipsed** conformation. The infinite intermediate conformations are known as **skew** conformations.
- The ***Sawhorse projection formulae*** or ***Andiron formulae*** of the two extreme conformations of ethane are shown in Fig. 1.1.

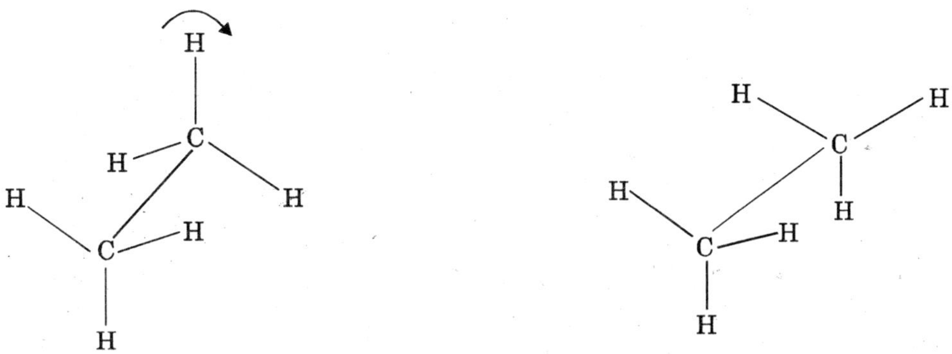

(*a*) *Staggered form* (*b*) *Eclipsed form*

Fig. 1.1. Staggered and Eclipsed conformations of ethane.

- The ***Newman's projection formulae*** for staggered and eclipsed conformations of ethane are shown in Fig. 1.2.

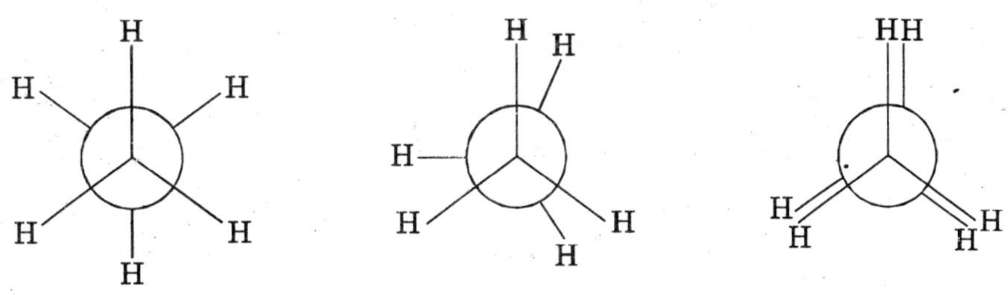

I Staggered *Skew* *II Eclipsed*

Fig. 1.2. Newman's projection formulae of conformation of ethane.

Relative Stabilities of the Conformations of Ethane

The conformations of ethane do not have same stability. The staggered conformation is relatively more stable than the other conformations. It is because *the repulsive interactions between the H atom attached to the two carbon atoms are minimum* due to the maximum distance between them. On the other hand, the **eclipsed conformation** is associated with *maximum energy because the repulsive interactions between H atoms on adjacent carbon atoms are maximum* due to minimum distance between them. The difference in the energy content of staggered and eclipsed conformation is 12.5 kJ mol^{-1}.

The variation of energy with rotation about the C—C bond in ethane has been shown in figure below :

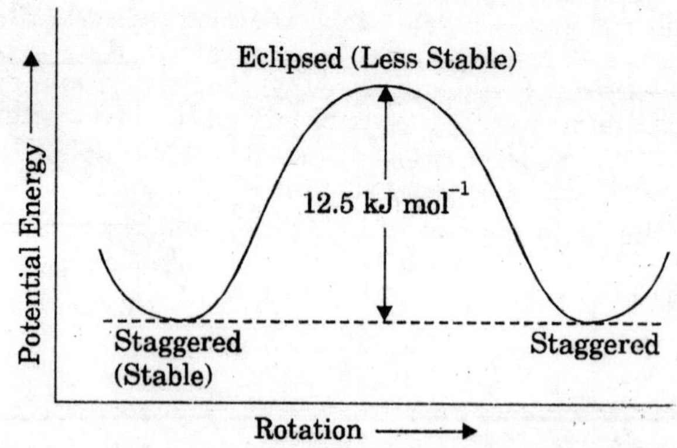

Fig. 1.3. Changes in energy during rotation about C—C bond in ethane.

The difference in the energy of various conformers constitutes an energy barrier to rotation. The energy required to rotate the ethane molecule about carbon-carbon single bond is called **torsional energy.** But this energy barrier is not large enough to prevent the rotation. Even at ordinary temperature the molecules possess sufficient thermal and kinetic energy to overcome the energy barrier through molecular collisions. Thus, conformations keep on changing from one form to another very rapidly and cannot be isolated as separate conformers.

- **Conformations of Butane**

Butane molecule can be represented as derivative of ethane as given below :

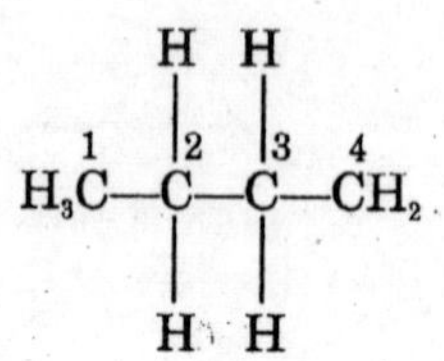

Considering the rotation around single bond between C-2 and C-3 we get many conformations. Some of them have been given as Newman Projections in Fig. 1.4.

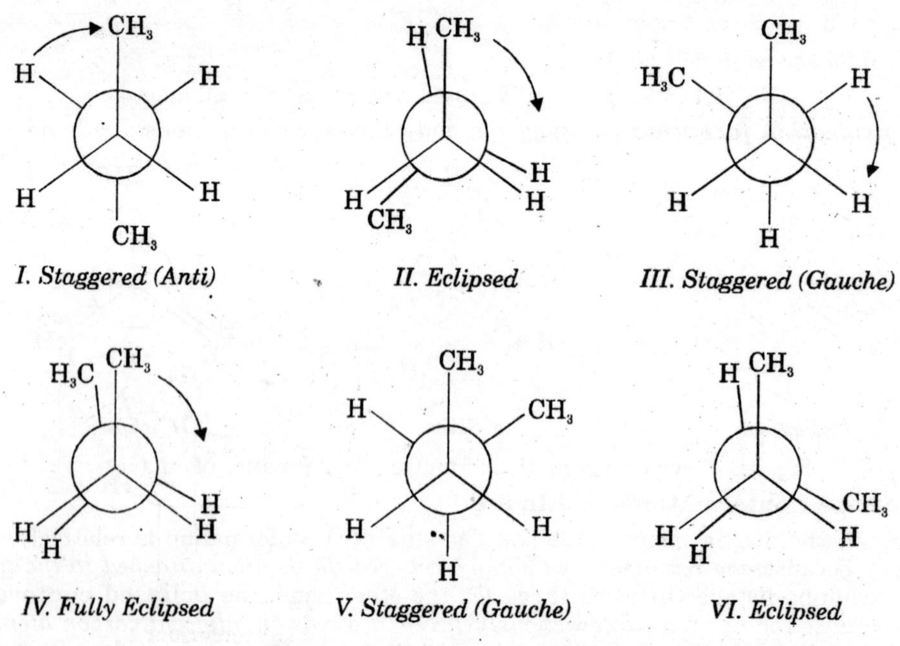

Fig. 1.4. Conformations of Butane

Staggered conformation (I), in which methyl groups are as far apart as possible, is most stable due to minimum repulsion between methyl groups. This conformation is also called ***anti*** conformation. This on rotation through 60° gives eclipsed conformation (II), in which methyl group on one carbon is overlapped by the hydrogen atom on the other carbon. Further rotation through 60° gives another staggered conformation (III) in which methyl groups on two carbon are 60° apart. This conformation is also called ***gauche conformation***. Gauche conformation on further rotation through 60° gives ***fully eclipsed conformation*** (IV) in which methyl groups on two carbons are just opposite each other. In this conformation steric strain is maximum hence this conformation is most unstable. Further rotation through 60° gives again gauche conformation (V) which is mirror image of gauche conformation (III). Conformation (V), on rotation through 60° gives conformation (VI) which is again eclipsed conformation.

The energy profile diagram for the conformations of butane is given below along with the difference of energy between various conformations of butane.

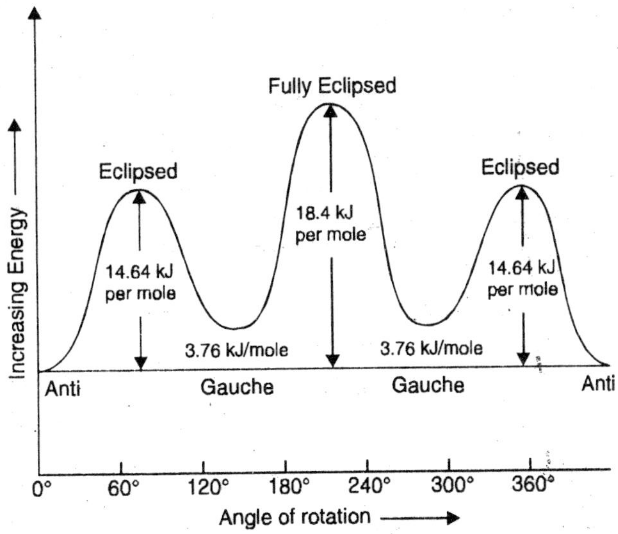

Fig. 1.5. Energy difference between various conformations of *n*-butane.

- **Conformations of Cyclohexane**

If cyclohexane molecule were to be planar, the c-c-c angle would have been 120°. Therefore, ring would be highly strained. Cyclohexane avoids this strain by assuming conformations in which all bond angles between carbon atoms are close to tetrahedral angle, 109° 38′. The two most important conformations of cyclohexane are the **chair form** and the **boat form** as shown in Fig. 1.6. Of the two, the chair form is more stable than the boat form, the energy difference being about 30 kJ mol^{-1}. However, the energy barrier between the two conformations is of about 44 kJ mol^{-1}. This is because in the boat form, may hydrogens on adjacent carbons correspond to the unfavourable eclipsed conformation of ethane. Also the two hydrogens marked H_a also called **flagpole hydrogen** in the boat form are quite close and repel one another.

On the other hand, the chair form does not have these unfavourable interactions and all the hydrogens corresponds to the more stable staggered conformation of ethane.

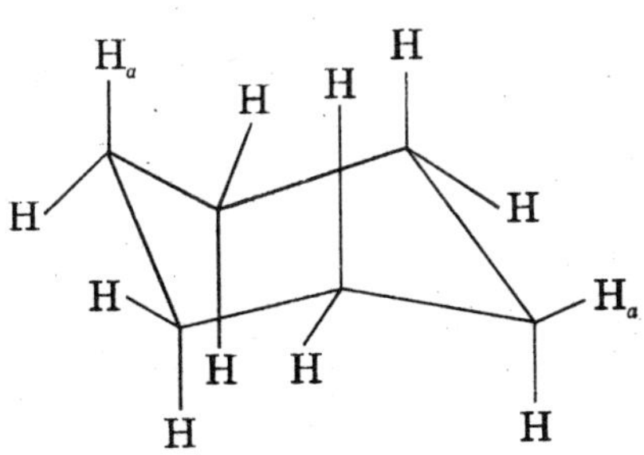

Chair conformer

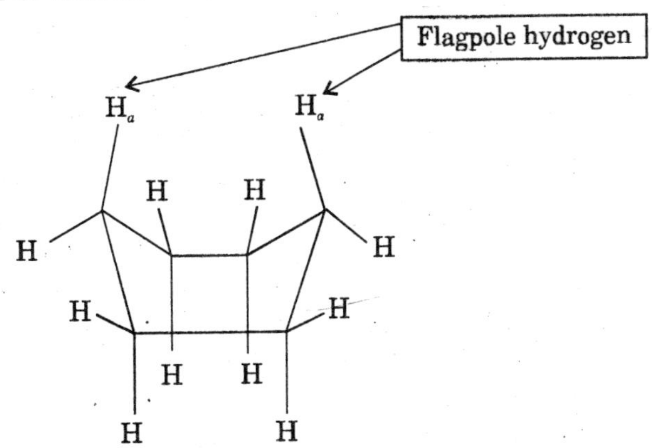

Boat conformer

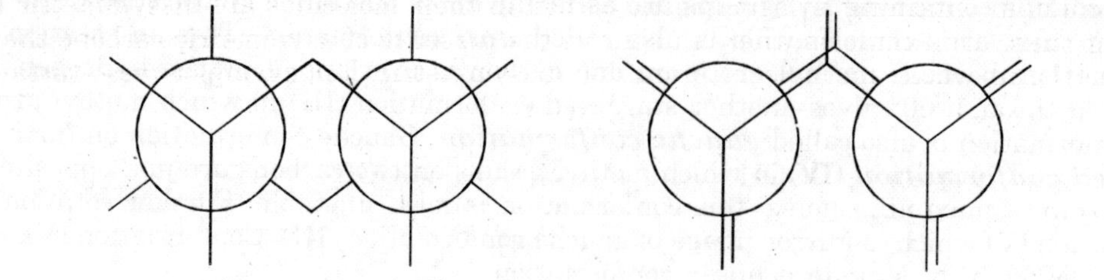

Chair conformation Boat conformation

Fig. 1.6. Chair and boat comformations of cyclohexane.

ENANTIOMERISM OR OPTICAL ISOMERISM

- ***The stereoisomerism which arises due to non-superimposability of mirror image structures is called enantiomerism or optical isomerism.*** The mirror image isomers are called enantiomers.
- ***Dissymmetry or chirality is the necessary and sufficient conditions for the enantiomerism.*** A carbon atom bonded to four different atoms/groups in a molecule is called stereocentre or a chiral carbon (*). In other words, a compound whose molecules are dissymmetric can exist as enantiomers. ***A molecule is said to be dissymmetric or chiral if it is not superimposable on its mirror image.*** On the other hand, a molecule which is superimposable on its mirror image is called ***non-dissymmetric* or *achiral.*** Non dissymmetric molecules do not show enantiomerism.

  The following examples will illustrate the above idea :

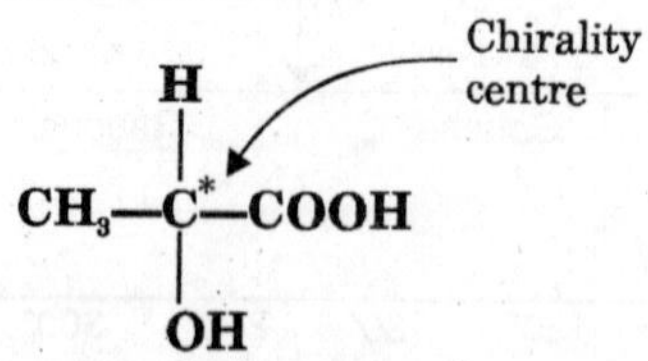

Some more examples of molecules having **one** chirality centre are

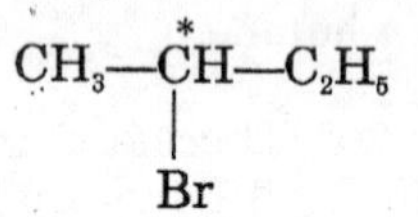

2-Bromobutane

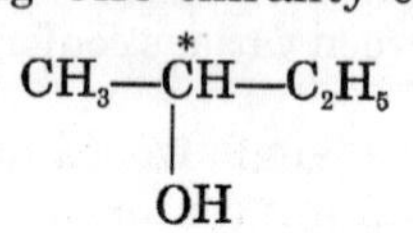

2-Butanol

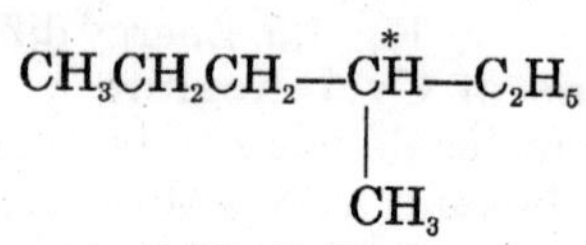

3-Methylhexane

Some examples of molecules having **two** chirality centres as follows :

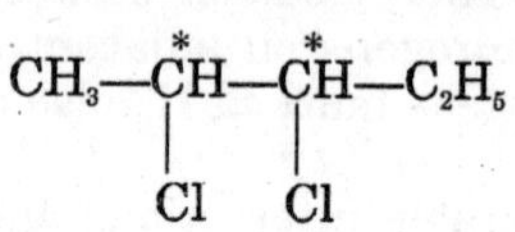

2, 3-Dichloropentane

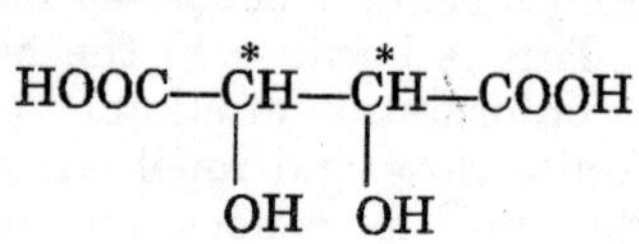

Tartaric acid

Glucose molecule ($C_6H_{12}O_6$) has four chirality centres, whereas fructose has **three** chirality centres.

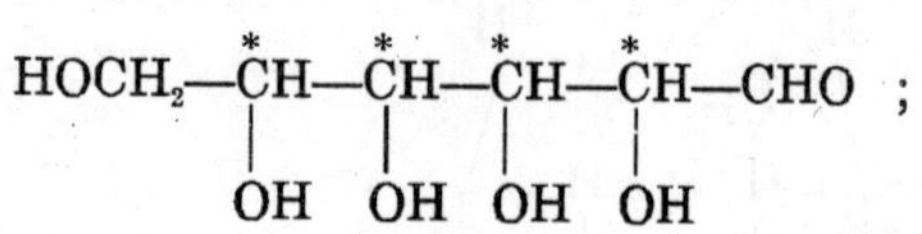

Glucose

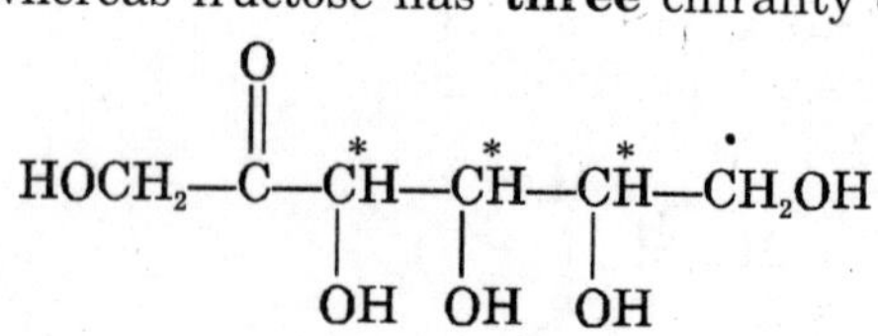

Fructose

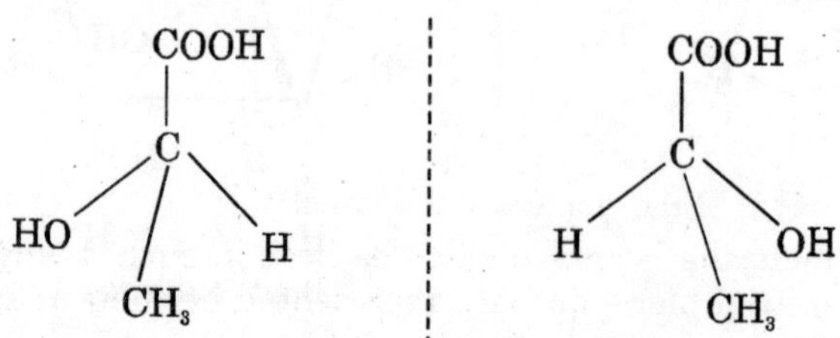

Non-superimposable : exist as enantiomers

- Generally, the compounds containing an asymmetric carbon in their molecules are dissymmetric and hence show enantiomerism. But there are examples where in spite of the presence of asymmetric carbons the compound does not show enantiomerism because its molecules are non-dissymmetric. For example, *meso*-tartaric acid.

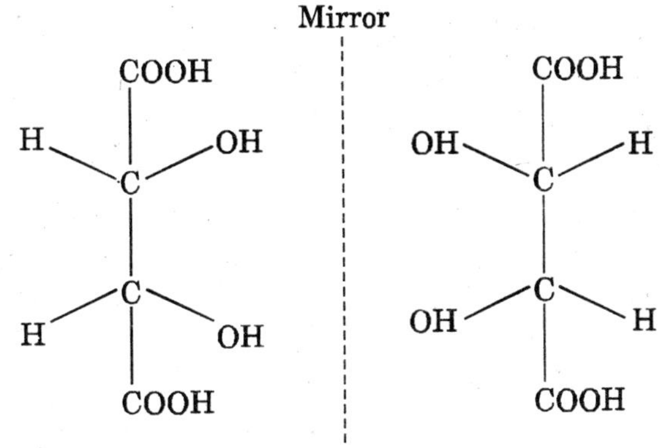

meso-Tartaric acid

Superimposable : does not show enantiomerism

- There are compounds which do not contain any asymmetric carbon atom but still show enantiomerism and optical activity because their molecules are dissymmetric. For example, substituted allenes and substituted bi-phenyls.

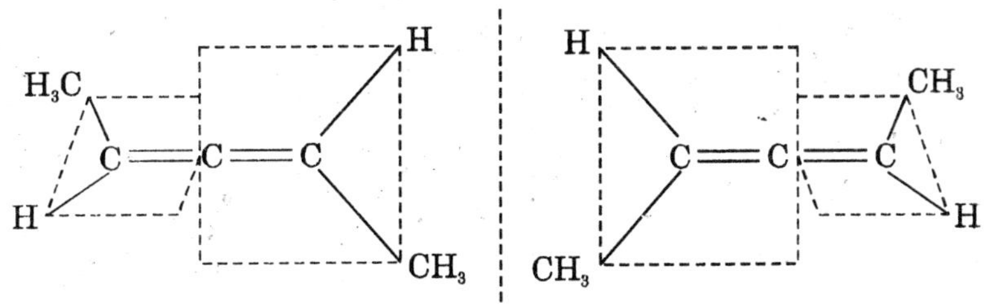

Non-superimposable : Shows enantiomerism

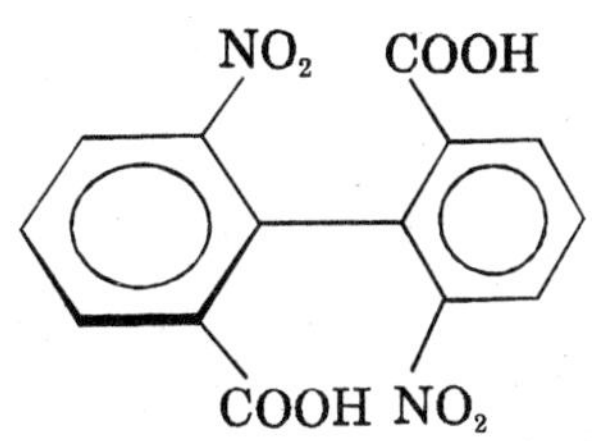

DISSYMMETRIC MOLECULE

(Shows Enantiomerism)

- **Representation of Enantiomers**

1. **Perspective formula.** In this method a normal line means bonds lying in the plane of paper and broken line means bonds going behind the plane of paper. A solid wedge represents bond projected out towards viewer.

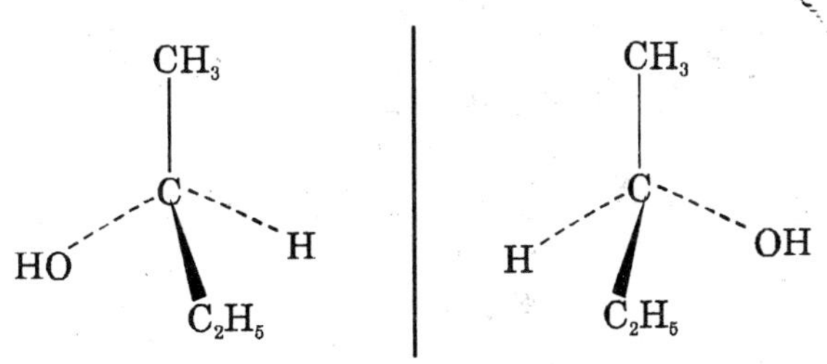

(Enantiomers of 2-butanol)

2. **Projection formulae.** Here the molecule is imagined to be held in such a way that two of the bonds coming towards the viewer are in the horizontal plane and are represented by solid wedges. The other two bonds going

behind the plane of paper are in vertical plane which are represented by broken lines. The projection formulae of enantiomers of 2-butanol are

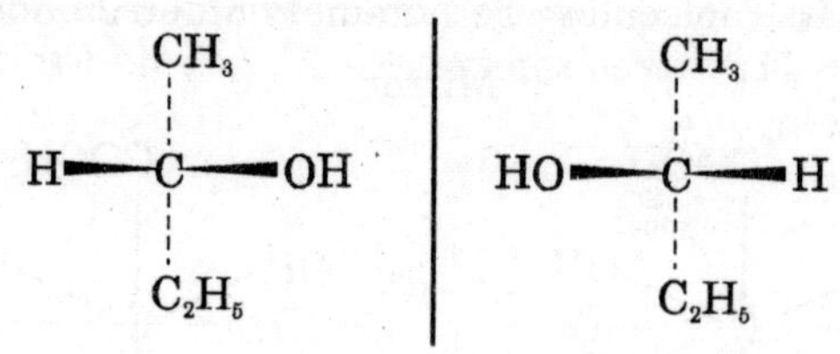

3. **Fischer Projection Formulae.** Envil Fischer devised a most simple and convenient method to represent the three dimensional arrangement of groups bonded to chirality centre. He used the point of intersection of two perpendicular lines to represent the chirality centre. Horizontal lines represent the bonds projected out of the plane of the paper towards viewer. Vertical lines on the other hand, represent the bonds projected back from the plane of the paper away from the viewer. The Fischer projection formulae of enantiomers of 2-butanol and lactic acid are as under :

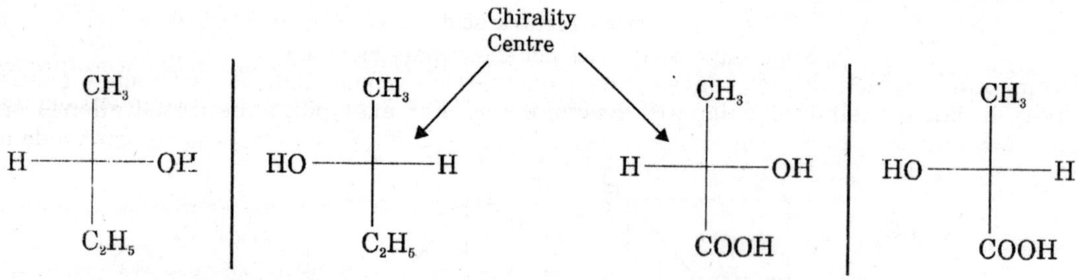

Enantiomers of 2-butanol Enantiomers of lactic acid

- **Characteristics of Enantiomers**

Some of the important characteristics of enantiomers are as given below :

1. Enantiomers have chiral structures.
2. Enantiomers have **identical physical properties** such as melting point, boiling point, density, refractive index etc.
3. Enantiomers are optically active substances. They rotate the plane of polarized light in opposite directions but to the equal extent. This is clear by the physical constants of enantiomeric forms of 2-butanol given as under :

	(+)-2-Methyl-1-butanol	***(–)-2-Methyl-1-butanol***
Specific rotation :	+ 5.90°	– 5.90°
Relative density :	0.8193	0.8193
Boiling point :	128.9°C	128.9°C
Refractive index :	1.4107	1.4107

4. Enantiomers have **identical chemical properties.** This means that they form same products as a result of chemical combination. However, their reactivity, *i.e.*, rates of reaction with other optically active substances are different.
5. Enantiomers have different **biological properties.** For example (+)-sugar plays significant role in animal metabolism. On the other (–)-sugar does not play any role in metabolism.

STEREOISOMERS WITH MORE THAN ONE CHIRALITY CENTRES

We have already studied that organic compounds having one chirality centre always possess chiral structures and they exhibit enantiomerism. However, organic molecules having more than one chirality centres may have chiral as well as achiral structures. The number of possible stereoisomers also increases with the increase in the number of chirality centres, in the molecule. In general, *a compound having* **n** *chiral carbon* atoms *can have* 2^n *stereoisomers.* In this section we shall study some aspects of organic compounds having *two* chirality centres.

(I) DIASTEREOMERS

The stereoisomers which are non superimposable and do not bear mirror-image relationship are called **diastereomers.** Let us understand it by taking the example of 3-chloro-2-butanol. The molecule has 2 chirality centres. Thus, the number of possible stereoisomers are 2^2, *i.e.*, 4. The four possible stereoisomers are shown below by Fischer projection formulae.

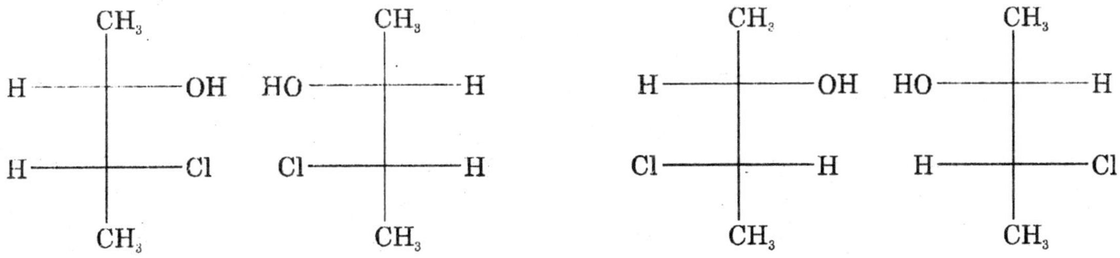

I II III IV

Erythro enantiomers Threo enantiomers

- Stereoisomers I and II are non-superimposable mirror images. Hence, they are called *enantiomers.*
- Stereoisomers III and IV are also non-superimposable mirror images. They are also enantiomers.
- I and II also represent enantiomers in which similar groups (H atoms) are present on same side of carbon chain such enantiomers are called **erythro enantiomers.**
- In III and IV similar groups (H atoms) are present on opposite sides of carbon chain. Such enantiomers are called threo enantiomers.
- Stereoisomers I and III are neither identical nor mirror images. Such stereoisomers are called **diastereomers.**
- In the similar way I and IV ; II and III ; II and IV also represent **diastereomeric pairs.**

Thus, diastereomers are configurational isomers which are not enantiomers (*It may be noted that cis-trans geometrical isomers which are configurational isomers but not enantiomers can also be* called *diastereomers*).

- **Characteristics of Diastereomers**

(*i*) The diastereomers have different physical properties such as *melting points, boiling points, densities, solubilities*, and *values of specific rotation.*

(*ii*) They can be separated from one another by physical means like *fractional distillation, fractional crystallisation, chromatography,* etc.

(*iii*) They are generally optically active. However, geometrical isomers are exceptions.

(*iv*) They exhibit similar but not identical chemical behaviour.

(II) MESO COMPOUNDS

According to 2^n rule each compound having two chirality centres should have (2^2 = 4) four stereoisomers. However, there are some compounds with two chiral carbons which have only three stereomers. For example, tartaric acid molecule has two chirality centres but it has only three stereoisomers. For example, tartaric acid molecule has two chirality centres but it has only three stereoisomers as shown below :

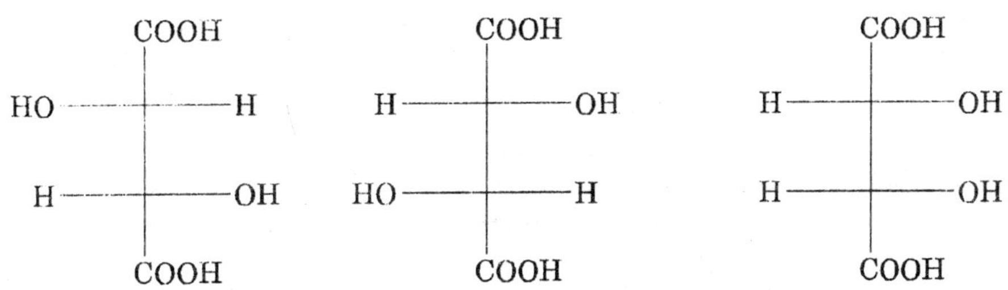

(2S, 3S) Tartaric acid I

(2R, 3R) Tartaric acid II

(2S, 3R) Tartaric acid III

Pair of enantiomers

Optically inactive

- Structure I and II are non-identical mirror images and thus represent pair of enantiomers.
- Structure III is a **chiral** inspite of the presence of two chirality centres. This is because of the presence of plane of symmetry in the molecule. If we draw the mirror image of structure III we shall find that it is superimposable on its mirror image (IV).

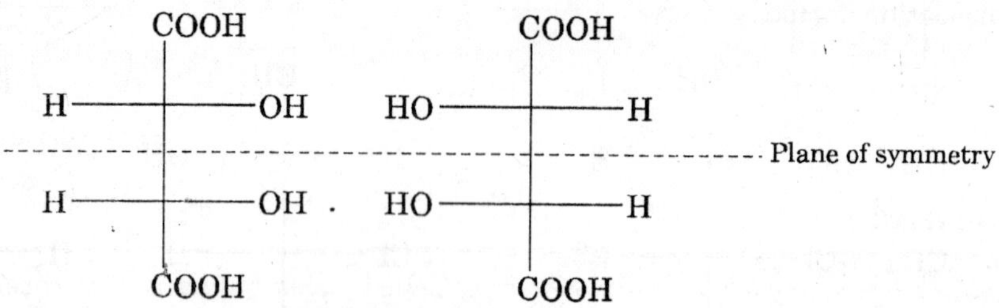

(Identical Mirror Images)

- Structure III or IV which are identical are called **meso compounds** (From Greek : *meso* means *middle*). Thus, *the compounds containing two or more chirality centres but possessing achiral molecular structure* because of having plane of symmetry, are called **meso compound.**

  Some examples of meso compounds are given below :

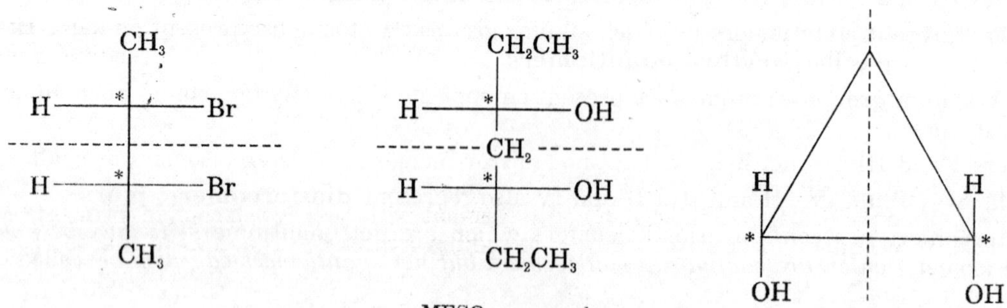

MESO compounds
(Plane of symmetry is represented by dotted line --------)

- **Meso Compounds are optically inactive**

  Meso compounds do not rotate the plane of polarized light in any direction, *i.e.*, they are **optically inactive.** This is because of achiral nature of their molecules. Because of the present of plane of symmetry the optical rotation caused by half of the molecule is compensated by the rotation caused by the other half. This cancellation of rotation within the molecule is referred to as **internal compensation.** In short, the ***meso compounds are optically inactive due to internal compensation.***

  It may be *noted that if a compound with two chirality centres has the same four groups bonded to each of the chiral carbon, then one of its stereoisomer will be a meso compound.*

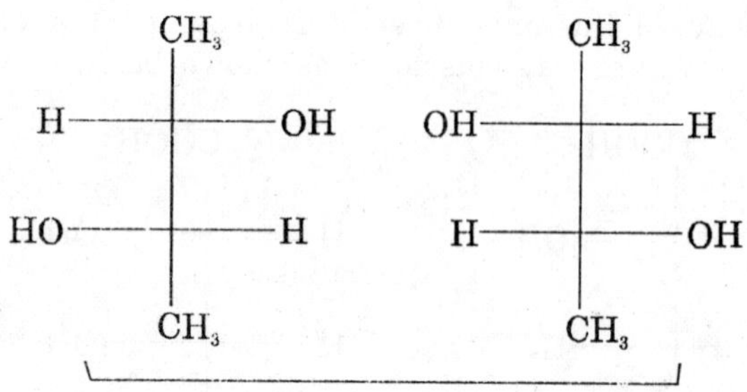

Pair of enantiomers

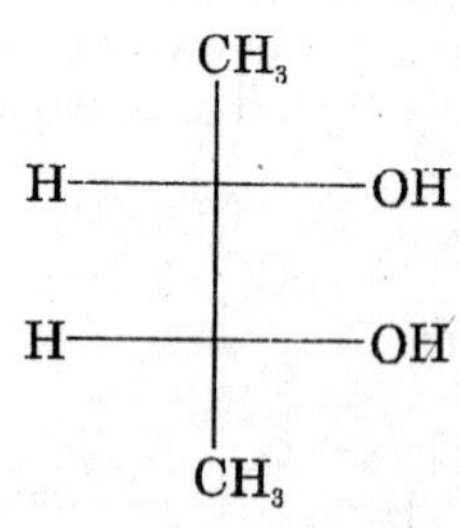

Meso compound

In case of cyclic compounds, the *cis isomer* will be meso compound and the *trans* isomer will exist as a pair of enantiomers.

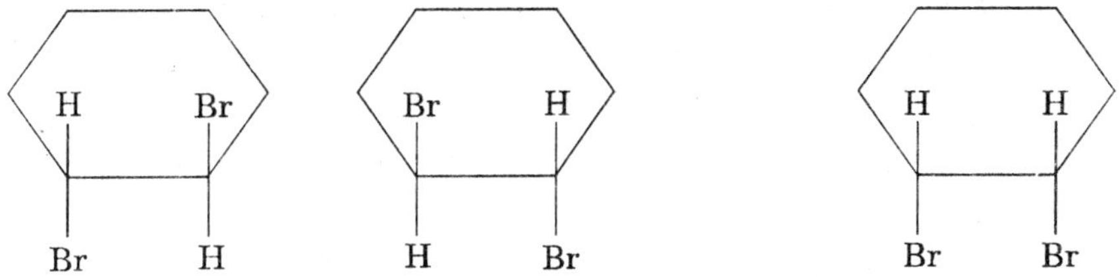

Pair of enantiomers — Meso compound

RACEMIC MODIFICATION

An equimolecular mixture of a pair of enantiomers is called *racemic mixture* or *racemic modification.* A racemic mixture is **optically inactive.** This is because of the fact that in equimolecular mixture of enantiomeric pairs, the rotation caused by the molecules of one enantiomer is cancelled by the rotation caused the molecules of other enantiomer. This type of compensation of optical rotation in a racemic mixture is referred to as **external compensation.** Thus, ***racemic mixture becomes optically inactive because of external compensation.***

Representation of a racemic mixture. The racemic mixture of a particular sample is indicated by using the prefix (dl) or (±). For example, racemic mixture of lactic acid is represented as ***(±) lactic acid.***

- **Racemisation**

It is a process of conversion of an optically active compound into the racemic modification. Both (+) and (–) forms of the compound are capable of racemisation under the influence of heat, light or chemical reagents as discussed below :

(*i*) **Action of heat.** Increase in temperature generally favours racemisation. Amyl alcohol, lactic acid, tartaric acid, etc., gives racemic mixtures on heating.

(*ii*) **Use of chemical reagents.** Many substances undergo racemisation on treatment with chemical reagents. For example, mandelic ($C_6H_5CH(OH)COOH$) on treatment with hydrogen bromide form (±) bromo acid. Similarly, (–) lactic acid is racemised in aqueous solution of sodium hydroxide. The change can be represented as

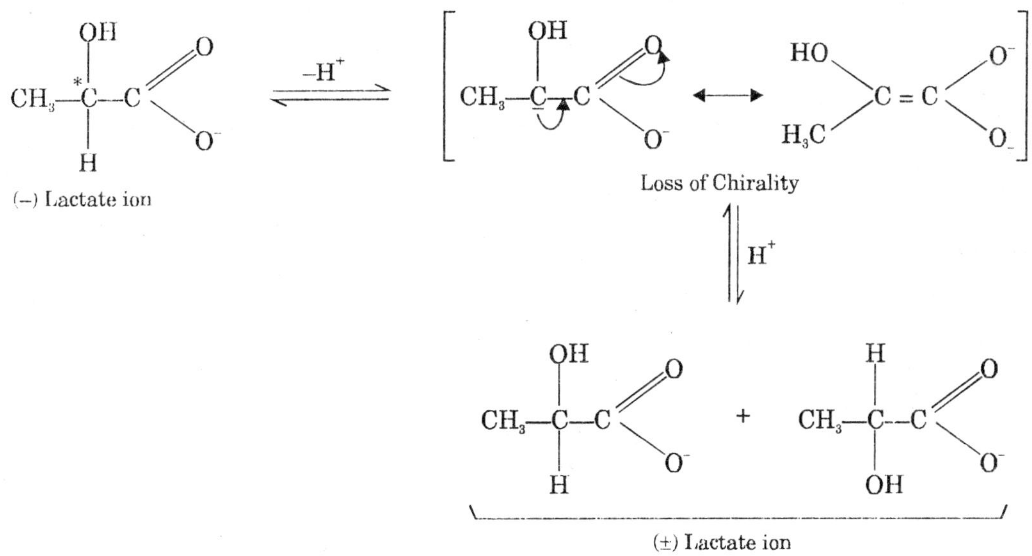

ASYMMETRIC SYNTHESIS

Synthesis of optically active compound in the laboratory generally results in the formation of racemic mixture which has to be resolved further to get optically active constituents. However, by using special means we can prepare optically active isomer in the laboratory without the necessity of resolution. **Asymmetric Synthesis** involves the *production of an optically active isomer from an optically inactive compound by using another opti-*

cally active substance which gets recovered at the end of the reaction. The role of optically active reagent in asymmetric synthesis is just like that of a catalyst. Some examples of asymmetric synthesis are as follows :

1. Synthesis of (–) lactic acid from pyruvic acid

Direct reduction of pyruvic acid produces racemic mixture. Preparation of (–) lactic acid involves the treatment of pyruvic acid with optically active alcohol (menthyl alcohol) to form optically active ester, the ester is further reduced and then hydrolysed to get (–) lactic acid.

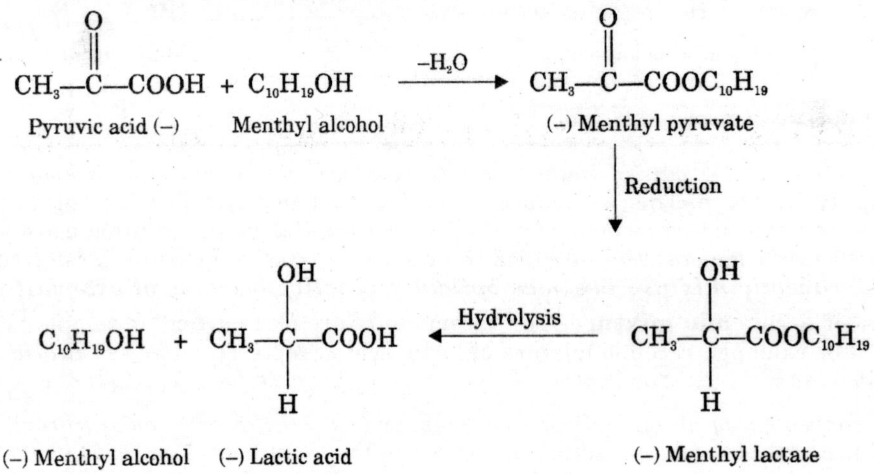

CALCULATION OF d-, l- FORMS AND MESO FORMS

	For molecules having no plane of symmetry	*For molecules having plane of symmetry*	
No. of a symmetric carbon atoms = n	n	When n is even	When n is odd
No. of *d*- and *l*-forms	2^n	2^{n-1}	$2^{n-1} - 2^{\left(\frac{n-1}{2}\right)}$
No. of meso forms	0	$2^{(n/2)-1}$	$2^{\left(\frac{n-1}{2}\right)}$
Total no. of stereoisomers	2^n	$2^{n-1} + 2^{(n/2)-1}$	2^{n-1}

CONFIGURATION OF STEREOISOMERS AND ITS SPECIFICATION

- **Configuration**

The arrangement of atoms or groups in space around the chirality centre that characterises a particular stereoisomer is referred to as it **configuration.** We know that lactic acid has two enantiomers which rotate the plane of polarized light in opposite directions.

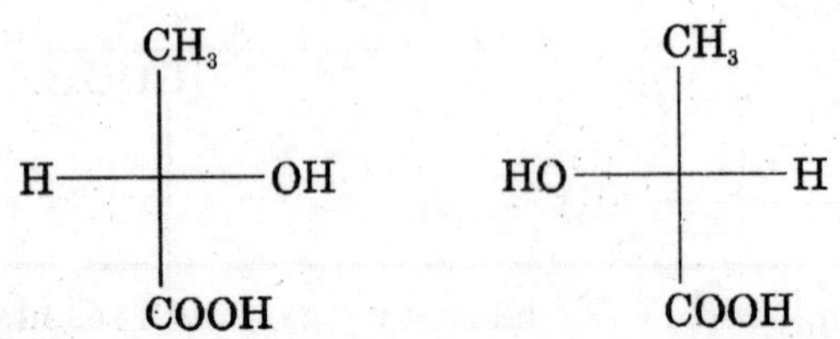

It is not possible to know from the structure as to which arrangement refers to (+) lactic acid and which one belongs to (–) lactic acid. The problem of specification of configuration to a stereoisomer could not be solved till 1951. However, in 1951, Bijvolt suggested a method of assigning the configurations to the stereoisomers by comparing their molecular structures with the structure whose configuration were already known.

- **Stereochemical Notations**

Two common stereochemical notations are generally used for the specification of configurations.

- **D and L Notations**

In this system the standard chosen for assigning relative configurations was *glyceraldehyde.* The absolute configurations of (+) and (–) glyceraldehyde are as

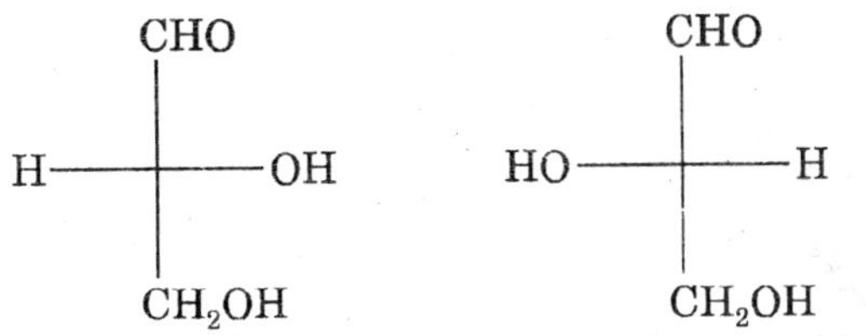

D (+)-Glyceraldehyde L (–)-Glyceraldehyde

The molecular structure with —OH group on the *left side* is denoted as L-glyceraldehyde while, the structure with —OH group on the *right side* is denoted as D-glyceraldehyde.

The other stereoisomers can be assigned D or L notation by comparison with the structural arrangements of D- and L-glyceraldehyde. For the purpose of comparison the following steps may be followed :

(*i*) The most oxidised carbon attached to the chiral center is placed on the top of vertical line

(*ii*) The group with carbon atom forming a part of the chain is kept at the bottom of vertical line.

(*iii*) The remaining groups are assumed to be projected towards the viewer and are placed along horizontal line.

Now comparison can be made with the configurations of glyceraldehyde. For example, the two enantiomeric forms of *lactic acid* can be compared as follows :

Glyceraldehyde

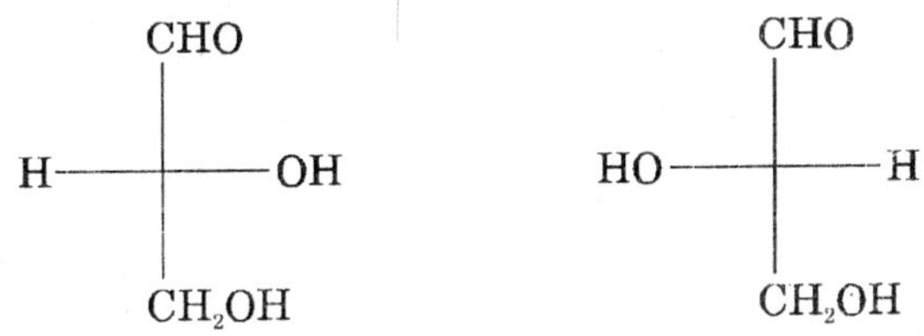

D (+)-Glyceraldehyde L (–)-Glyceraldehyde

Lactic acid

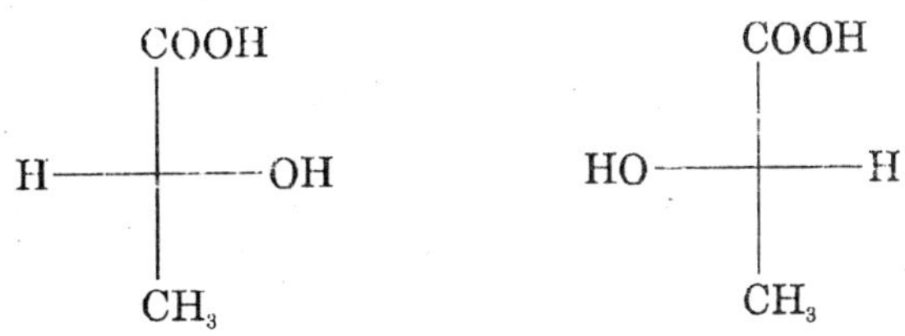

D (+)-Lactic acid L (–)-Lactic acid

It may be noted that the capital letters D and L do not pertains to dextro or laevo rotatory. Specific rotation of a stereoisomer is an experimentally determined property whereas D, L notations are based on comparison. For example, L-alanine is dextro rotatory and L-phenylalanine is laevorotatory. Their structural arrangements are as follows :

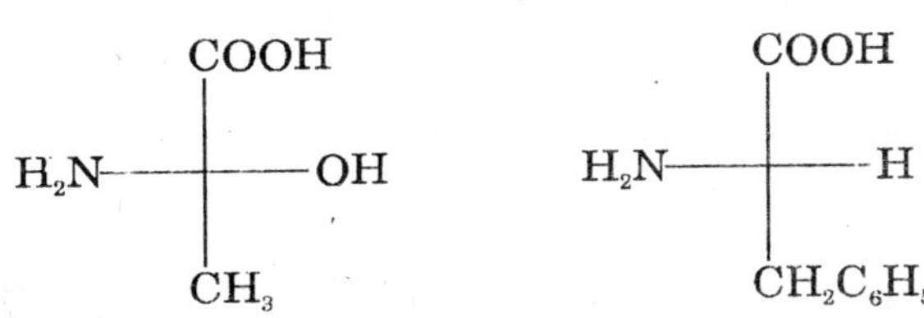

L(+)-Alanine L(+)-Phenylalanine

- **R and S Notations of STEREOISOMERS**

D.L. system of specifying the configuration was not universally applicable because it was limited to those stereoisomers which could be obtained from D and L glyceraldehyde. RS system of nomenclature of stereoisomers was devised by **Cahn, Ingold** and **Prelog**. In this system, the configuration of a stereoisomer is designated by using prefix **R** derived from Greek word **rectus** (means-**right**) and **S** derived from Greek word **sinister** (means-**left**). The procedure involves **two main steps** as described below :

- **Step 1**

Rank the groups (or atoms) which are bonded to the chirality centre in order of priority. The criterion of priority is based on certain set of rules known as **sequence rules** given as under :

Sequence rules

Rule I. *If all the four atoms attached to chirality centre are different, then sequence of priority is decided by atomic number. An atom with higher atomic number has higher priority.* The highest to lowest priority in this book will be represented as 1, 2, 3, 4.

Examples. The sequence of priority in bromochloro iodomethane (CHClBrI) and in chloroiodo methane sulphonic acid is an under

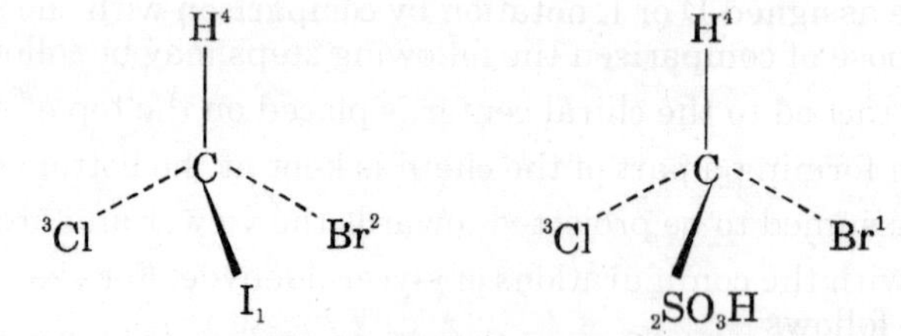

Rule II. *If two of the atoms attached to chirality centre are isotopes of the same element, then atom of higher mass number gets higher priority.*

Example. In α-deutro ethyl bromide ($CH_3CH(D)Br$) the sequence of priority is

Br(1), C(2), D(3), H(4)

Rule III. *If two atoms bonded to chirality centre are same, rule I is unable to decide the priority sequence. In such cases, the relative priorities can be decided by comparing the second (or even third) atom of the respective group.*

Illustration

In 2-chlorobutane the chiral carbon is bonded to Cl—, H—, CH_3—, C_2H_5— groups. The priorities of CH_3— and C_2H_5— groups cannot be decided by rule I. Here, the priorities of CH_3— and C_2H_5— group are decided by application of rule III.

In CH_3— group, the second atoms are H, H, H but in CH_3—CH_2— group, the second atoms are C, H, H. Now, by comparison, C has higher atomic number than H, therefore C_2H_5— group gets priority over methyl group. Hence sequence of priority is Cl(1), C_2H_5(2), CH_3(3) and H(4). The priorities in **2-Chlorobutane** and also in **3-chloro-2-methylpentane** are indicated as follows :

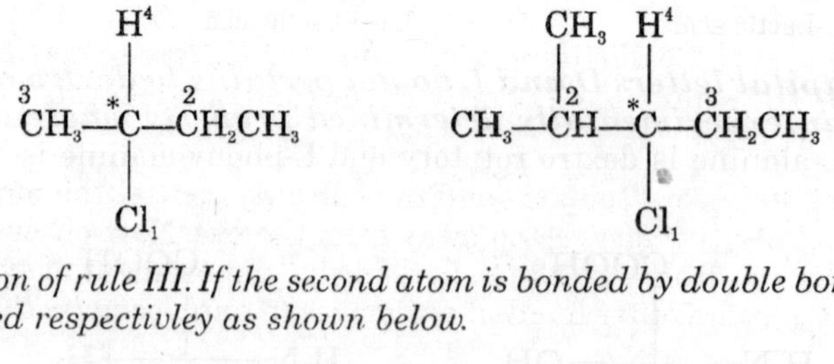

Rule IV. *This rule is extension of rule III. If the second atom is bonded by double bond or by triple bond, it is considered to be duplicated or triplicated respectivley as shown below.*

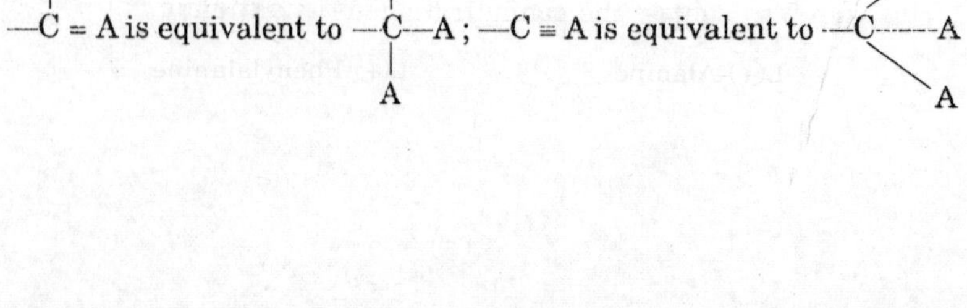

For example, in glyceraldehyde, 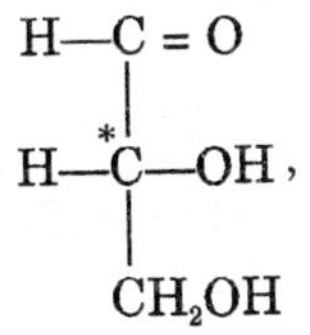, 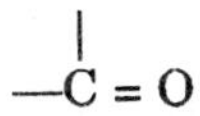 is equivalent to 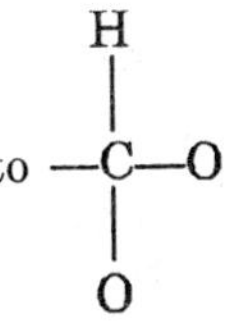

Thus, O, O, H of —CHO gets priority over O, H, H of CH_2OH. The sequence of priority is therefore, —OH(1), —CHO(2), —CH_2OH(3), H(4).

It may be noted that the actual system 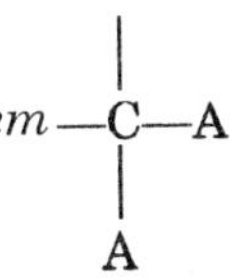 gets priority over 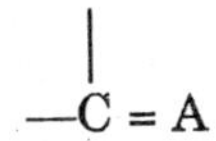.

For example, in the molecule, 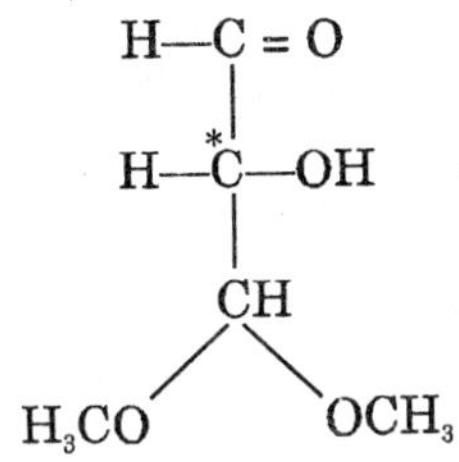

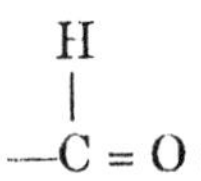 equals 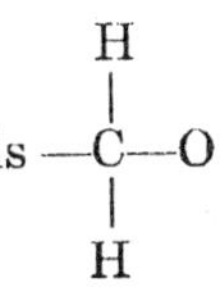 but —$CH(OCH_3)_2$ gets priority over —CHO because in —$CH(OCH_3)_2$C is actually bonded to two O atoms. Hence the sequence of priority is —OH(1), —$CH(OCH_3)_2$(2), —CHO(3), H(4).

The phenyl group, C_6H_5—is handled as follows

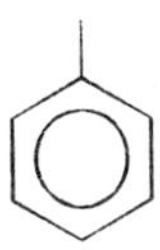 is equivalent to 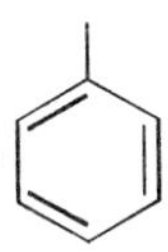 which is further equivalent to 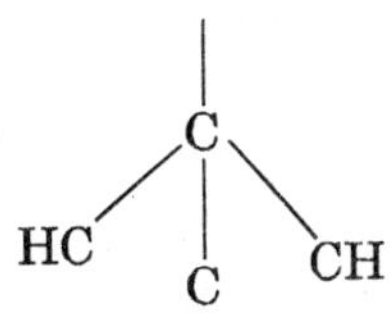

For example, *in 2-methyl-1-phenylpropanamine*

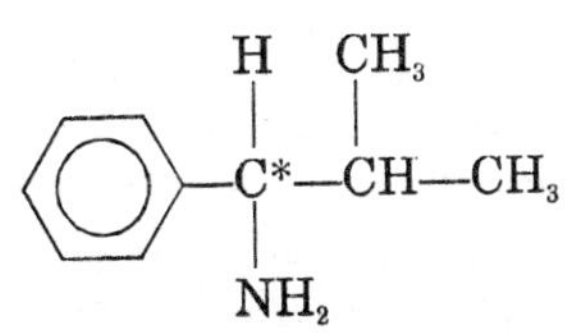

C, C, C of phenyl gets priority over C, C, H of isoprophyl but N of NH_2^- group gets priority over phenyl group because of higher atomic number of N. The sequence of priority in the compound is NH_2(1), C_6H_5(2), $CH(CH_3)_2$(3), H(4).

- **Step 2**

  After deciding the priorities of atoms or group around the chirality centre, the molecule is visualised in such a way that atom or group with lowest priority is directed away from the eye. Now, observe the remaining atoms or groups is decreasing order of priorities, *i.e.*, from 1 to 2 to 3. In doing so, if movement of eye occurs in *clockwise direction* the configuration is specified as **R.** On the other hand, if eye moves in *anticlockwise direction,* the configuration is specified on S.

Illustration

Let us now, use step 1 and step 2 to specify the configurations of enantiomers of (*i*) *2-butanol* and (*ii*) *lactic acid.*

(*i*) ***2-Butanol*** ($CH_3—\overset{\bullet}{C}H(OH)—C_2H_5$). The sequence of priority of groups around the chirality centre is OH(1), C_2H_5(2), CH_3(3), H(4). Visualising the enantiomeric forms with H directed away from the eye

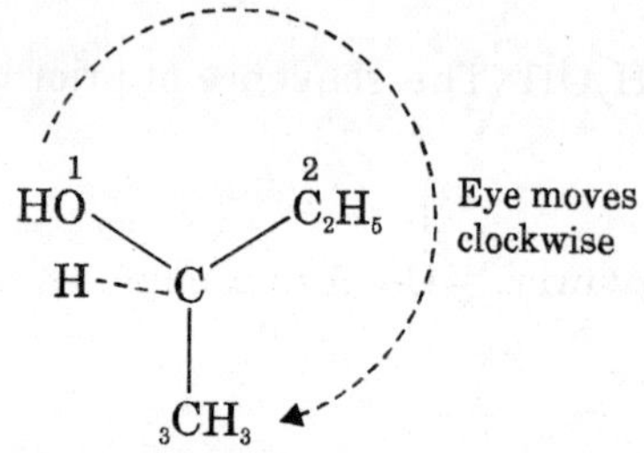

R-2-Butanol

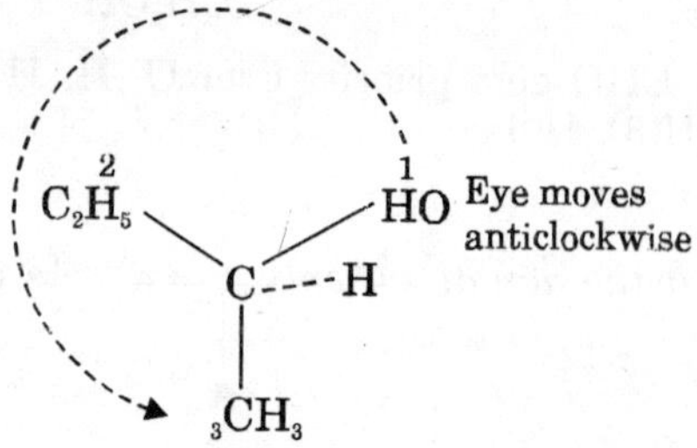

S-2-Butanol

(*ii*) ***Lactic acid*** ($CH_3—\overset{*}{C}H(OH)—COOH$). The sequence of priority of various groups around the chirality centre is OH(1), COOH(2), CH_3(3), H(4). Visualising the formulae with H directed away from eye.

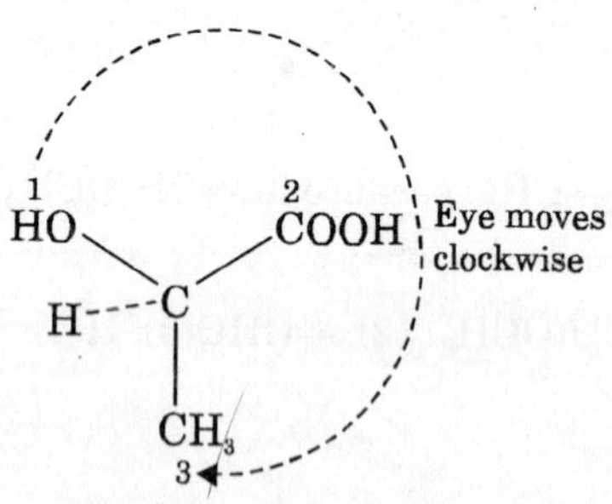

R-Lactic acid

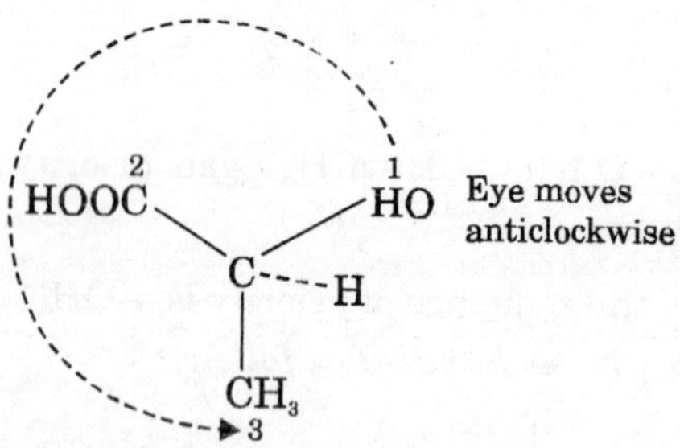

S-Lactic acid

It may be noted that the knowledge of R and S configuration does not tell us the direction in which the compound rotates the plane of polarised light because some compounds with **R**-configuration rotate the plane towards right whereas some compounds with R-configuration rotates the plane towards left. This implies that movement of eye towards left or right has no relationship with rotation of plane of polarised light towards left or right. The dextro or laevo rotatory nature of the compound can be judged by putting the compound in polarimeter and carrying out the experimental studies. For example, S-lactic acid and S-sodium lactate have same configuration but S-lactic acid is *dextrorotatory* whereas S-sodium lactate is *laevo-rotatory.*

Example 1. ***Aginomoto, a Chinese salt used in fast food preparations is mono sodium glucomate (MSG) and is dextrotatory with specific rotation + 24°. MSG has a structure given below. Find out its configuration. Draw the structure of its enantiomeric form and decide its configuration.***

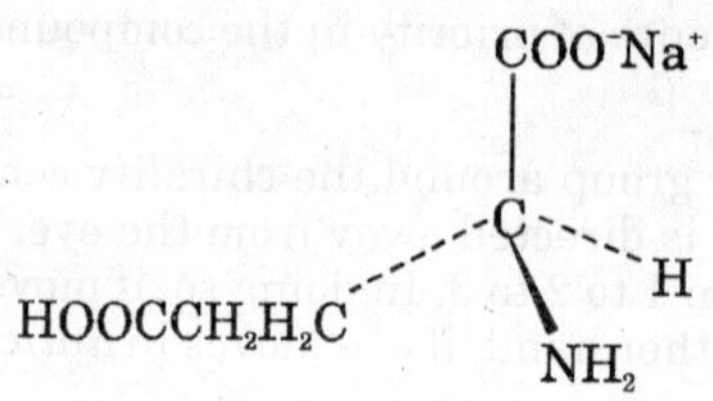

Sol.

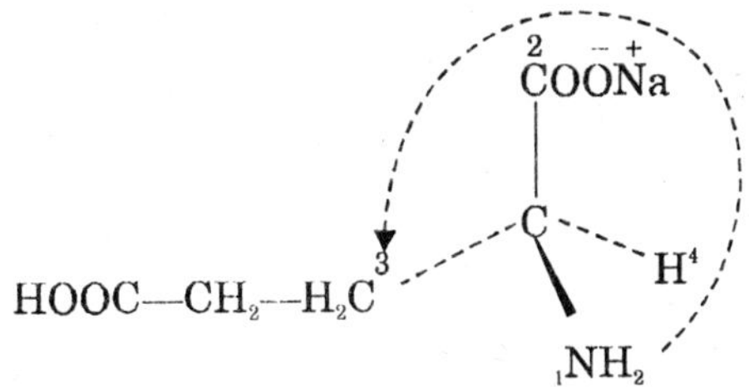

Eye moves anticlockwise. The configuration in (S). Thus, it S (+) monosodium glucomate. The structure of enantiomeric form of S(+) MSG is

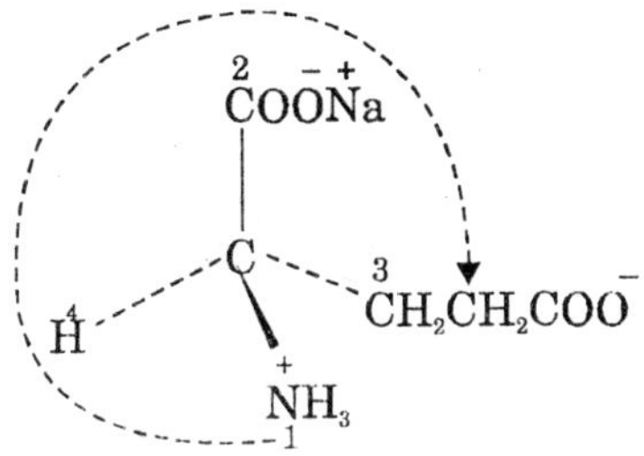

Enantiomer of S (+) MSG

Eye moves clockwise and the configuration is (R).

Example 2. ***It has been found that S-lactic acid is dextrorotatory and S-sodium lactate is laevorotatory. Draw the structures of R-lactic acid and R-sodium lactate. In which direction will they rotate the plane of polarised light ?***

Sol. The structures of R-lactic acid and R-sodium lactate are

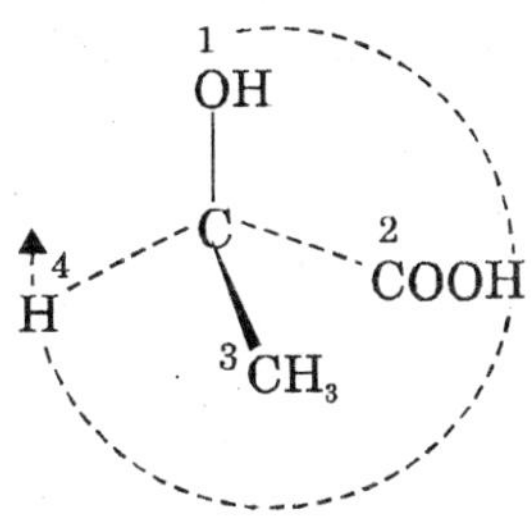

R-lactic acid
(Laevorotatory)

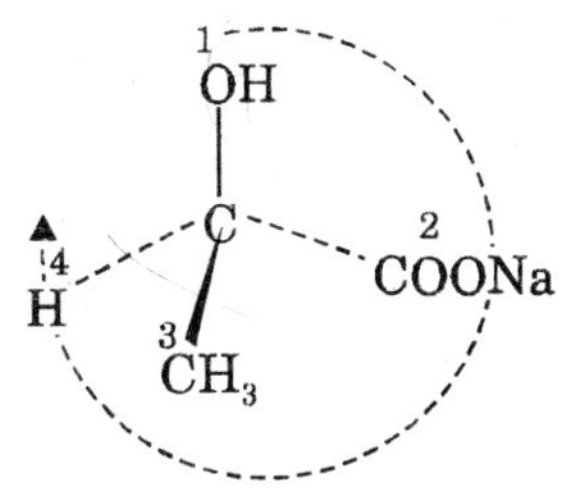

R-sodium lactate
(dextrorotatory)

- **Specification of R and S configuration from Fischer Projection Formulae**

Step 1. Decide the sequence of priority of various atoms or groups around the chirality centre following sequence rule already discussed.

Step 2. This step is carried out differently in four different situations as discussed below :

I. *When atom or group of lowest priority is lying at the bottom of vertical line.* Move the eye in order of decreasing priority and determine the configuration.

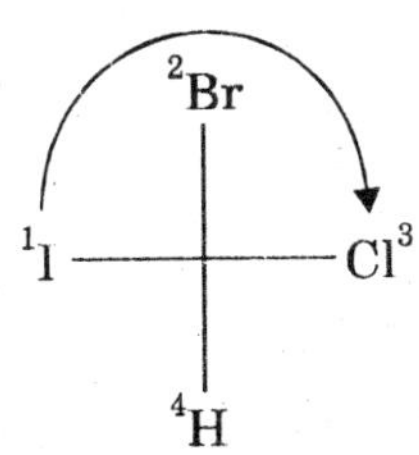

Eye moves in clockwise direction.
Hence, configuration is (R)

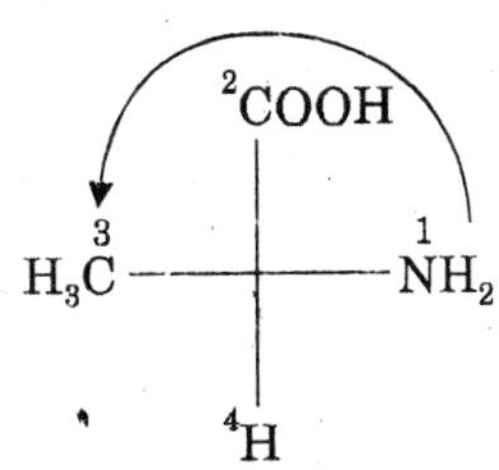

Eye moves in anticlockwise
direction configuration is (S)

II. *When atom or group of lowest priority is at the top of vertical line.* In such a case rotate the entire molecule through 180° so that atom or group of lowest priority comes at the bottom. Now, move the eye in order of decreasing priority to determine the configuration.

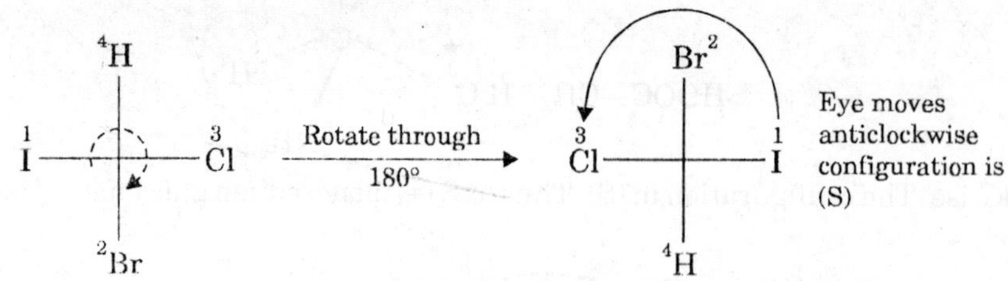

III. *When the atom or group of lowest priority appears at right hand side on the horizontal line.* Then without changing the position at the top of vertical line, change the positions of other atoms or groups till the group having lowest priority comes at the bottom of vertical line. Now, note the movement of eye in order of decreasing priority to determine the configuration.

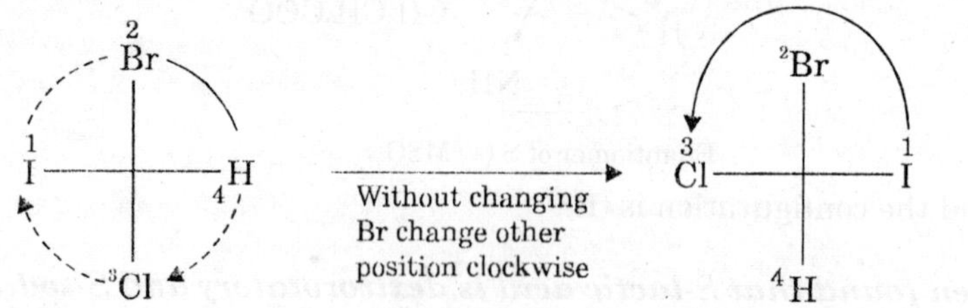

The configuration is (S)

IV. *When the atom or group of lowest priority appears at the left side on the horizontal line.* Then without changing the at the top of vertical line, change the positions of other atoms or groups in anticlockwise direction till the group of lowest priority comes at the bottom of vertical line. Then determine the configuration in a usual manner.

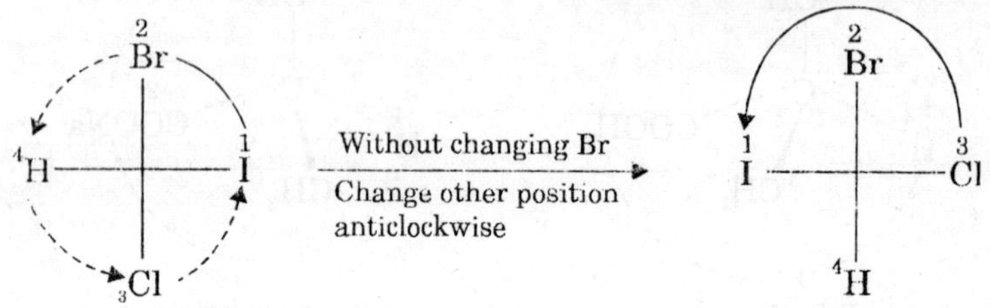

The configuration is (R)

GEOMETRICAL ISOMERISM

In this type of isomerism, the isomers differ in the spatial arrangements of atoms or groups around the double bonded carbon atoms. For example, the arrangements of 2-butene are given below :

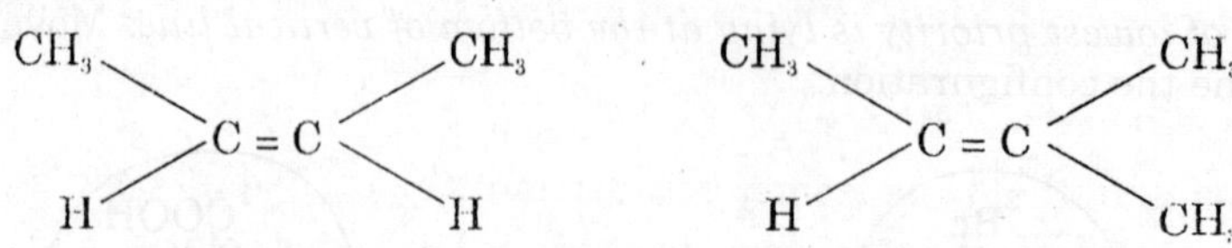

cis-2-butene I | *trans*-2-butene II

Structures I and II are not interconvertible because rotation about carbon-carbon double bond is hindered. So they represent two different isomers of 2-butene. Such isomers are called **geometrical isomers.** The geometrical isomers which have similar groups on same side are called *cis* (structure I) whereas the geometric isomers which have similar groups on opposite sides are called *trans* (structure II).

Hindered rotation about carbon-carbon double bond is responsible for the existence of geometric isomers. As we know C = C double bond consists of a σ bond and a π bond. π bond is formed by sideways overlap of *p*-orbitals of two carbon atoms. If we carry out rotation about C = C double bond then the molecule passes through a stage in which *p*-orbitals are perpendicular to each other (Fig. 1.7). In this stage, sideways overlap of *p*-orbitals is not possible. In other words, π bond would break. But this requires about 284 kJ/mole of energy which is not available at room temperature. So under oridinary conditions, the rotation around carbon-carbon double bond is hindered.

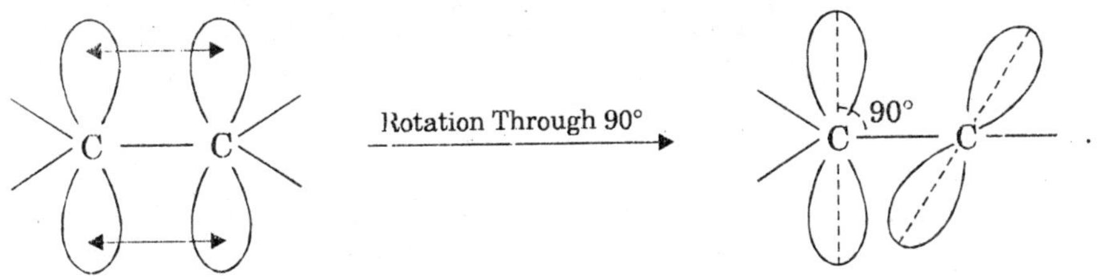

Fig. 1.7. Hindered rotation about C = C bond.

cis-2-Butene *trans*-2-Butene

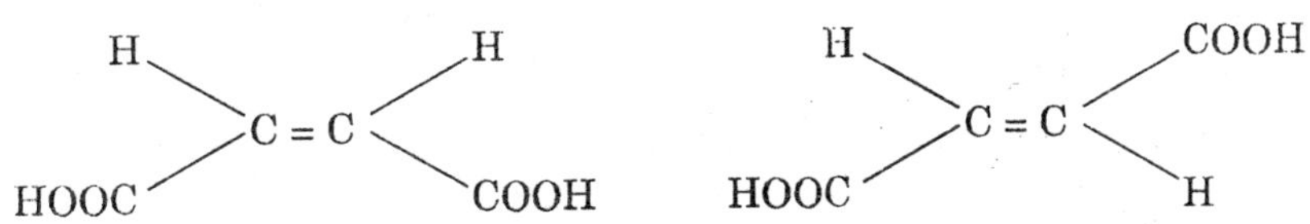

Maleic acid (*cis*) Fumaric acid (*trans*)

- **Conditions for geometrical isomerism**

1. *There should be a double bond in the molecule.*
2. *The two atoms or groups attached to each doubly bonded carbon atom should be different. If one of the two doubly bonded carbon atoms carries two identical groups then the molecule does not exhibit geometrical isomerism.*

Propane 2-Methyl-but-2-ene

No geometric isomerism

Geometrical isomers possess different physical properties. Their chemical properties are similar but not identical.

It may be mentioned here that :

(*i*) *cis*-isomer is more polar than the *trans*-isomer because in it the individual bond dipoles do not cancel each other.

(*ii*) *cis*-isomer has larger boiling point than the *trans*-isomer due to greater polarity and consequent stronger interparticle forces.

(*iii*) *trans*-isomer has higher melting point than the *cis*-isomer because its molecules are more closely packed in solid state due to symmetry.

(*iv*) *trans*-isomer is more stable than the *cis*-isomer.

- **(E)-(Z)—NOTATION OF GEOMETRICAL ISOMERS**

The stereochemistry of highly substituted alkenes cannot be defined as cis or trans. For this purpose a new notation known as E-Z notation is used. Following a set of rules (Cahn-Ingold-Prelog rules) the substituents on a double bond are assigned priorities. The double bond is assigned the configuration E (From *entgegen,* the german word for opposite) if the two groups of **higher priority** are on the **opposite sides** of the double bond. On the other hand, the double bond is assigned the configuration Z (From *zusammen*, the German word for together) if the two groups of **higher priority** are on the same side of the double bond.

Cahn-Ingold-Prelog Rules

1. The higher the atomic number of the atom bonded directly to the stereocentre, the higher is the priority of the substituent. Among isotopes, the one with higher mass number gets the higher priority.
2. If two identical atoms are attached to the stereocentre, the relative priority of the two groups is decided by a similar comparison of the next atoms in the group moving away from the stereocentre.
3. A double bond is counted as two single bonds for both the atoms involved. Thus

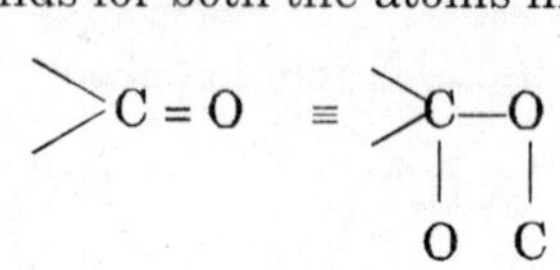

Some examples are being given below :

1.

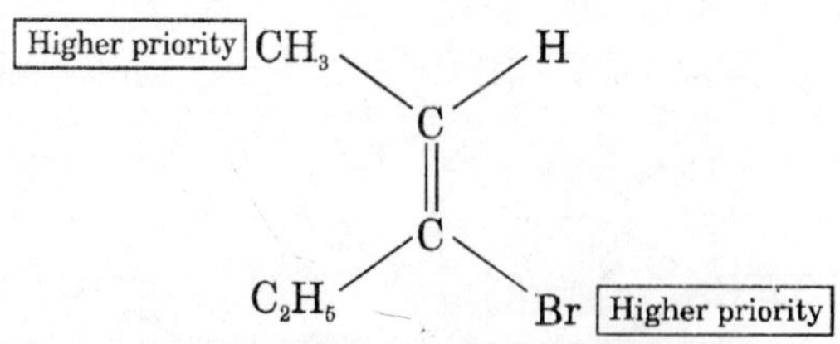

The groups of higher priorities are on opposite sides.
Hence it is (E) – 3-Bromopent-2-ene

2.

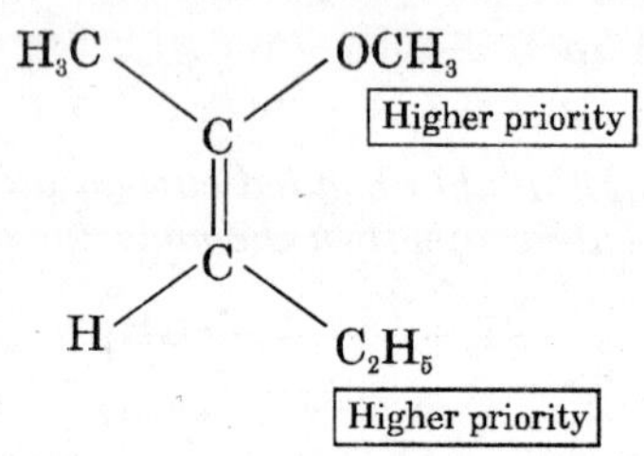

The groups of higher priorities are on same side.
Hence it is (Z) – 2-Methoxypent-2-ene

3.

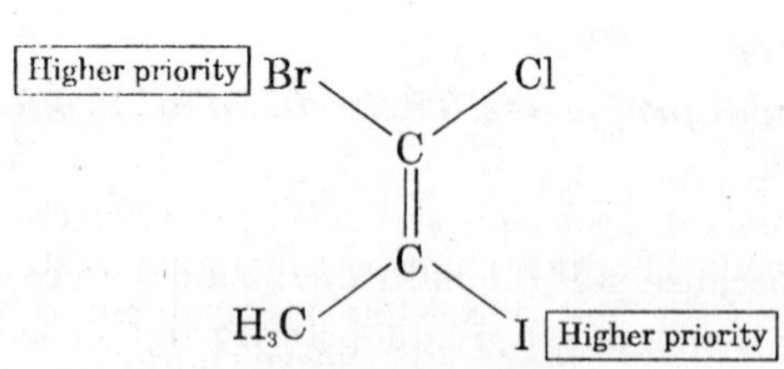

Groups of higher priority
are on opposite sides
(E) 1-Bromo-1-chloro-2-Iodo propene

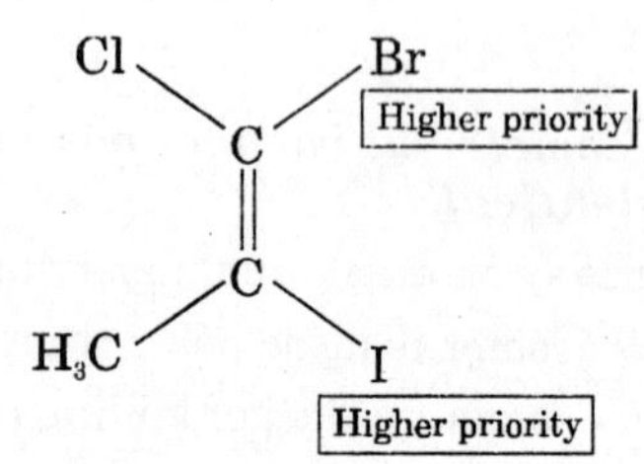

Groups of higher priority
are on same sides
(Z) 1-Bromo-1-chloro-2-Iodo propene

4.

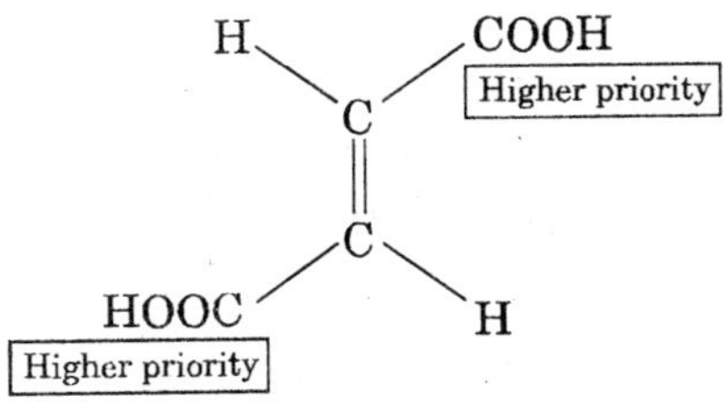

(E)-But-2-ene-1, 4-dioic acid
(*Fumeric acid*)

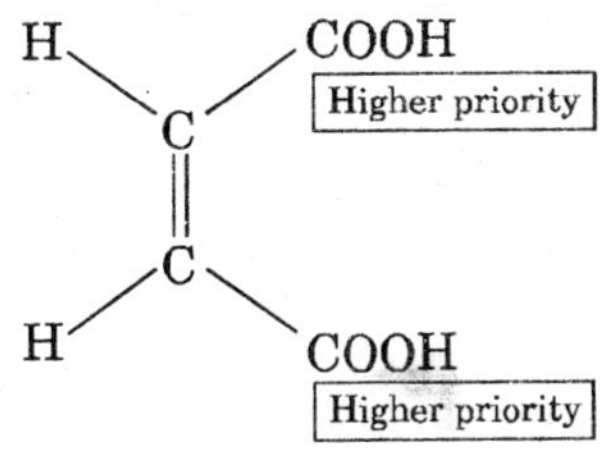

(Z)-But-2-ene-1, 4-dioic acid
(*Maleic acid*)

5.

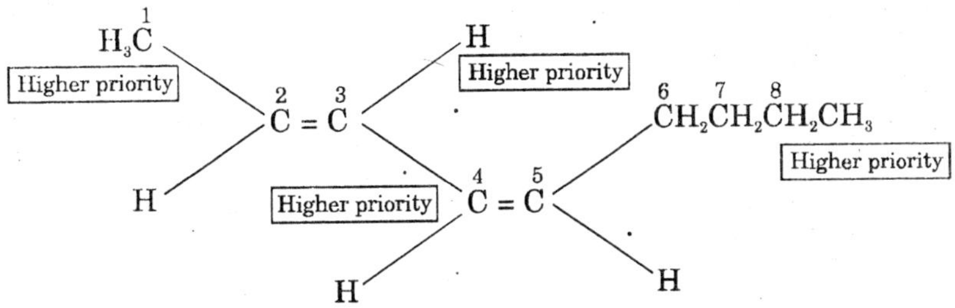

(E, Z)-Nona-2, 4-diene

- **Geometrical isomerism in cyclic compounds**

  Cyclic compounds also exhibit geometrical isomerism because cyclic systems also prevent free rotation about the carbon-carbon single bonds. The substituents on the same side are *cis*-isomers while the substituents on opposite sides represent *trans*-isomers. Some examples are :

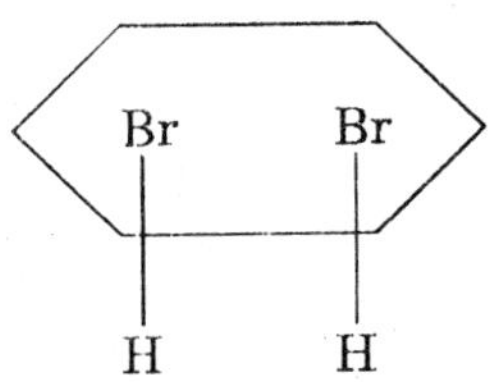

cis-1, 2-Dibromocyclohexane

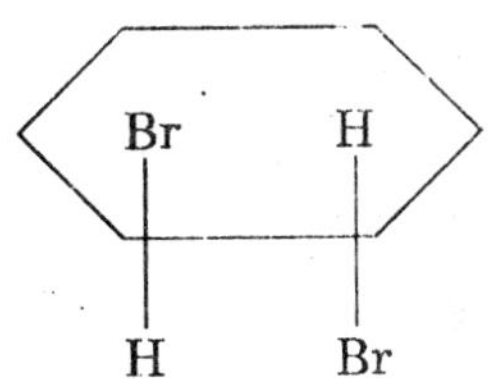

trans-1, 2-Dibromocyclohexane

GENERAL CONCEPTS

- **Inductive Effect**

  Permanent displacement of electrons along a carbon chain when some atom or group of atoms with different electronegativity than carbon is attached to carbon chain is called inductive effect. For illustration consider a carbon chain which has a chlorine atom attached to one end.

$$-C_4 \rightarrow \overset{\delta''+}{C_3} \longrightarrow \overset{\delta'+}{C_2} \longrightarrow \overset{\delta+}{C_1} \rightarrow \overset{\delta-}{Cl} \qquad (-\text{I})\text{-effect}$$

Since chlorine atom has higher electronegativity than carbon so the shared pair of electrons between C_1 and chlorine atom lies more closer to chlorine atom, as a result of which chlorine atom acquires a small negative charge and carbon atom C_1 acquires small + ve charge. As C_1 is now somewhat positively charged it attracts towards itself shared pair of electrons between C_1 and C_2. This results in a small positive charge on C_2 also but this charge is less than that on C_1. In this way electrons are displaced towards chlorine in the carbon chain. This type of displacement of electrons along a carbon chain is known as **Electron Withdrawing Inductive Effect** or **(– I) effect.**

If the electronegativity of the atom or group of atoms attached to the carbon chain is less than the electronegativity of the carbon then the displacement of electrons takes place away from group along the C-chain and this effect is called **Electron Releasing Inductive Effect** or **(+ I)-effect.**

Inductive effect is a permanent effect and decreases rapidly as the distance from the source increases. Some of the groups which show inductive effect are given below :

– I effect : $-NO_2 > -F > -Cl > -Br > -I > -OCH_3 > -C_6H_5$

+ I effect : $-C(CH_3)_3 > -CH(CH_3)_2 > -C_2H_5 > -CH_3$.

- **Electrometric Effect**

The complete transfer of the shared pair of π electrons of a multiple bond to one of the atoms in the presence of the attacking agent is called electrometric effect. For example, in the presence of some nucleophile the π electron pair of carbonyl group in aldehydes and ketones is transferred to oxygen. This effect is a temporary effect and takes place only in the presence of the attacking agent. If the attacking agent moves away without reacting with the molecule then the molecule returns to its original state.

If the transferrence of electrons takes place towards the attacking agent it is called (+ E) effect and if it takes place away from the attacking agent then it is called (– E) effect. Electrons are always transferred towards the more electronegative atom of the two atoms between which the π bond exists.

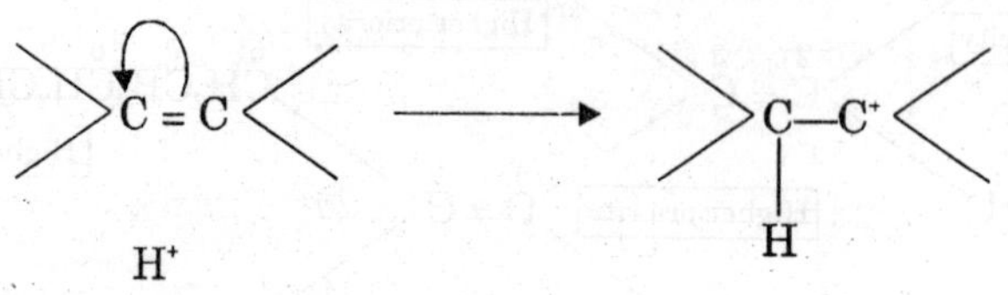

(+ E)-Effect

(– E)-Effect

- **Resonance**

Sometimes it is not possible to assign a single electronic structure to a molecule that may account for all its properties. In such a case the molecule is represented by two or more electronic structures but none of these represents the actual structure of the molecule. The actual structure of the molecule lies somewhere in between these structures but cannot be expressed on paper. Such a molecule is said to exhibit resonance. The various structures assigned to the molecule are called ***contributing structures*** or ***canonical structures*** whereas the intermediate structure is called ***resonance hybrid.*** For example, benzene (C_6H_6) molecule may be assigned following two structures :

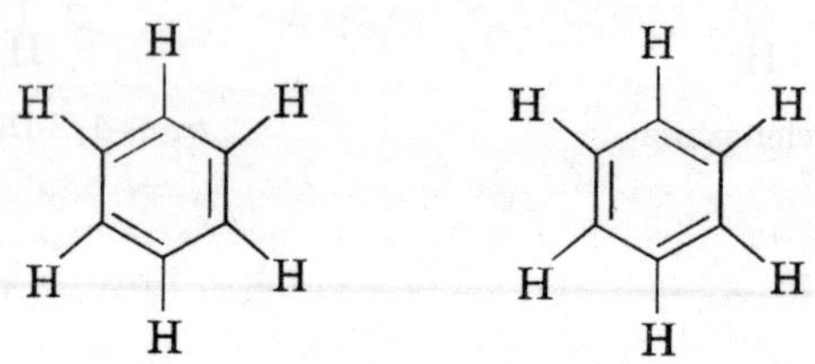

(I) (II)

Any of these structures alone cannot explain all the properties of benzene. According to these structures there should be three single bonds (1.54 Å) and three double bonds (1.34 Å) between carbon atoms. But actually it has been found that all the six carbon-carbon bonds in benzene are equivalent and have bond length 1.39 Å. The actual structure of benzene is resonance hybrid of the above two structures and may be represented as under :

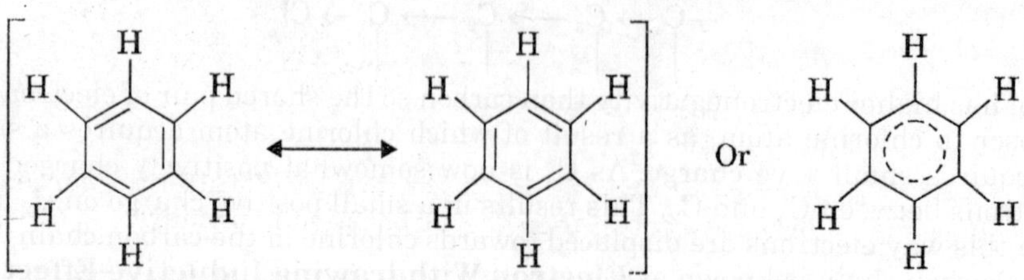

The resonance hybrid is more stable than any of the contributing structures and the difference between the energy of the most stable contributing structure and resonance hybrid is called ***resonance energy.***

Some important features of resonance are :

1. The various contributing structures may differ only in electronic arrangement but should have same arrangement of atoms.

2. The number of unpaired electrons should be same in all the contributing structures.
3. All contributing stuctures should have almost same energy.
4. The more stable contributing structure makes more contribution.
5. The bond distances of hybrid structure are intermediate of those of resonating forms.

Resonance Effect

It may be noted that in **conjugated systems** (*having alternate single and double bonds*) resonance causes displacement of electrons from one part of the system to another part creating centres of high and low electron density. This is called resonance effect.

- **Cleavage of Bonds**

1. Homolytic Cleavage or Homolytic Fission

In this type of cleavage each fragment formed as the result of cleavage of bond gets one electron from the shared pair of electrons. This type of cleavage results in the formation of species with odd or unpaired electrons. Such species are called free radicals.

Homolytic cleavage is favoured by conditions such as *non-polar nature of bond, high temperature or presence of high energy radiations.*

$$Cl \bullet \vdots \bullet Cl \xrightarrow[\text{or } 525\text{—}725\text{ K}]{h\nu} Cl^{\bullet} + Cl^{\bullet}$$

2. Heterolytic Cleavage

In this type of cleavage both the shared electrons of the covalent bond are taken away by one of the fragments (while the other fragment does not get any of the shared electrons).

$$A : :B \longrightarrow A^{+} + :B^{-}$$

This type of cleavage results in the formation of charged species. Heterolytic cleavage is favoured by the conditions such as *polar nature of bond* and *presence of polar solvent.*

REACTION INTERMEDIATES

Reaction intermediates are the species which are produced during the reaction but are lost before the reaction is complete. They are very reactive species and hence exist only for a short time (10^{-6} seconds to a few second). Therefore, these species cannot be isolated. Examples of reaction intermediates are : *carbocations, carbanions, free radicals* and *carbenes.*

CARBOCATIONS

Carbocation may be defined as a group of atoms that contains a carbon atom bearing positive charge and hence containing only six electrons in its valence shell. These ions are formed by heterolytic cleavage in which the leaving group takes along with it the shared pair of electrons.

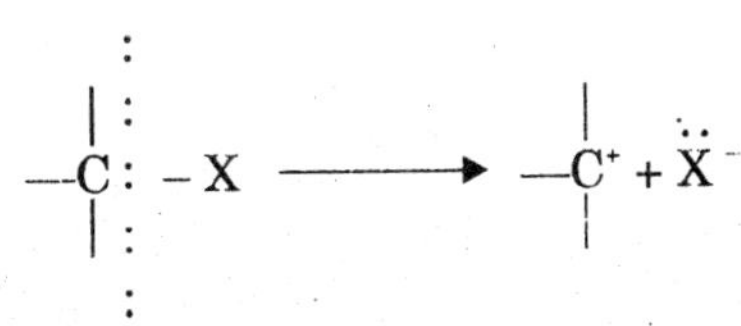

- Carbocations are classified as primary (1°), secondary (2°) or tertiary (3°) after the carbon bearing the positive charge *e.g.* if the positive charge is on primary carbon then the carbocation is primary.

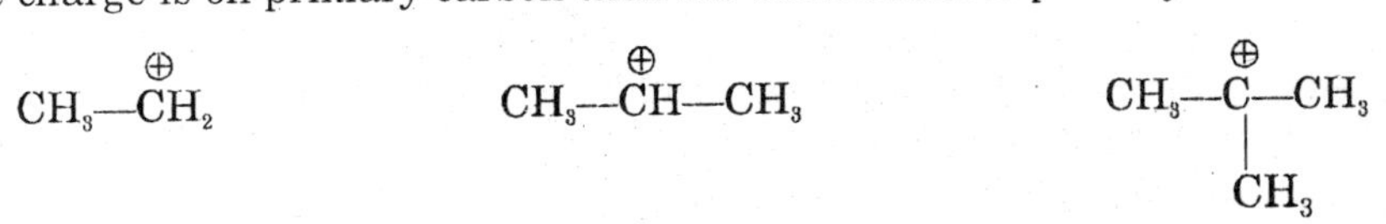

Ethyl carbocation (1°) *Iso* propyl carbocation (2°) *Tert.* butyl carbocation (3°)

- **Decreasing order of stability of carbocations is 3° > 2° > 1°.** This order of stability can be explained on the basis dispersal of positive charge by electron releasing inductive effect of alkyl groups. Greater the number of alkyl groups on the carbon carrying + ve charge greater would be dispersal of charge and hence more stable the carbocation would be 3° carbocations with three alkyl groups are most stable due to maximum dispersal of charge followed by 2° carbocations (2 alkyl groups) and 1° carbocations (1 alkyl group) as illustrated below :

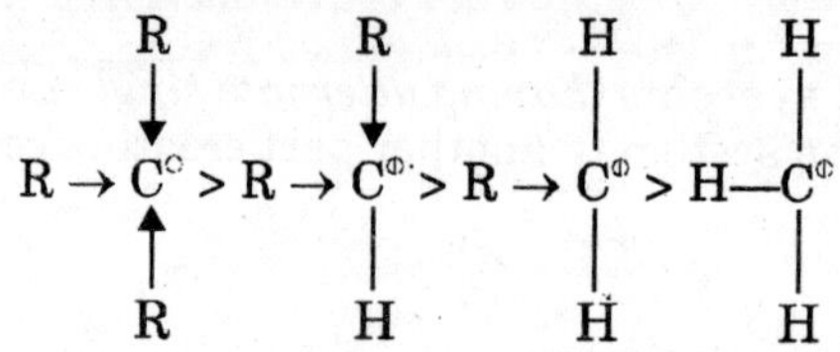

- **Orbital structure of carbocations.** Electron deficient carbon in carbocations is bonded to three other atoms directly and for this bonding it uses sp^2 hybridised orbitals. This part of the carbocation is therefore planar. In addition to this there is a vacant *p*-orbital perpendicular to this plane with one lobe above the plane and the other below the plane as illustrated.

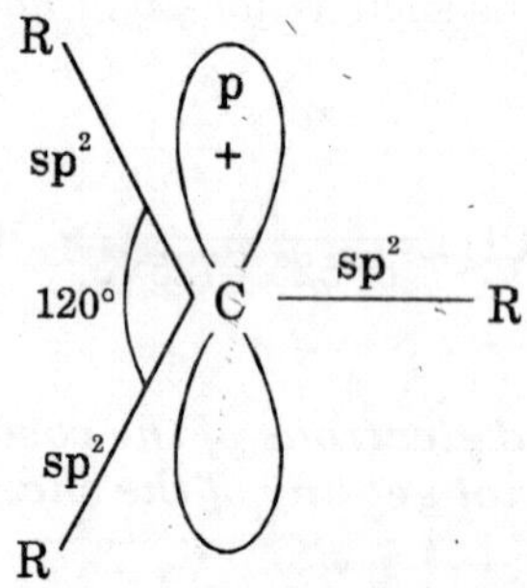

- **CARBANIONS**

  Carbanion may be defined as a group of atoms that contains a carbon atom bearing negative charge. Carbanions are formed by heterolytic cleavage in which shared pair remains with carbon atom.

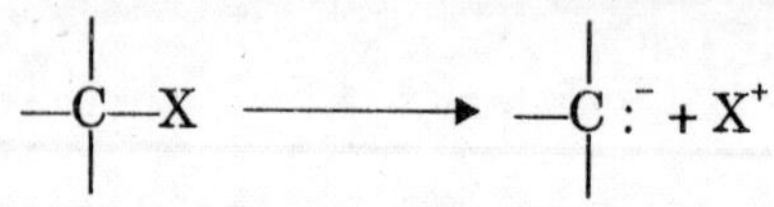

- Carbocations are classified as primary (1°), secondary (2°) or tertiary (3°) after the carbon bearing the negative charge. *e.g.*, if the negative charge is on primary carbon then the carbanion is primary.
- ***Decreasing order of stability of carbanions is 1° > 2° > 3°***

  This order of stability can be explained on the basis of electron releasing inductive effective effect of alkyl groups. Greater the number of alkyl groups on the carbon bearing negative charge, greater would be the concentration of negative charge on carbon resulting in instability. 1° carbanion with only one alkyl group is maximum stable followed by 2° (with two alkyl groups) and 3° (with three alkyl groups).

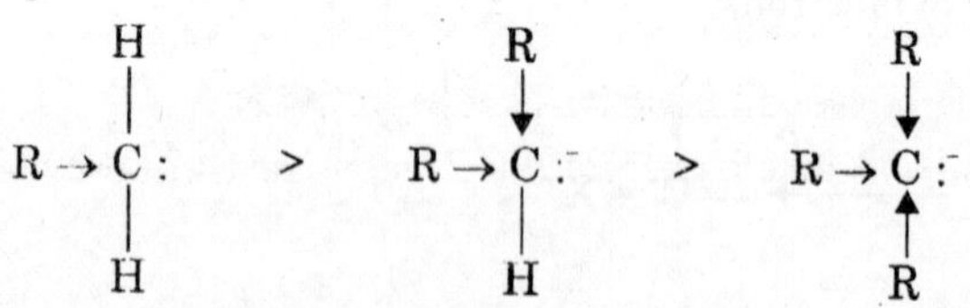

(1°) (2°) (3°)

- **Orbital structure of carbanions**

  In carbanions the carbon carrying negative charge is sp^3 hybridised. Three sp^3 hybrid orbitals are used to form three σ bonds with three different atoms whereas the fourth sp^3 hybrid orbital accommodates lone pair.

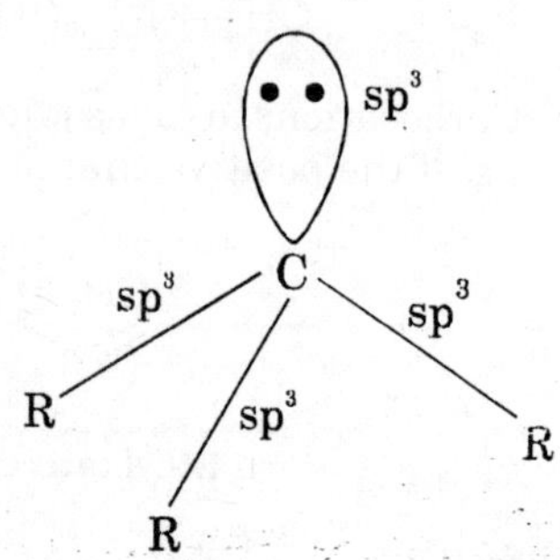

- **FREE RADICALS**

A free radicals may be defined as an atom or group of atoms which contains an unpaired electron. Free radicals are formed by homolytic cleavage of a covalent bond. For example, in the presence of U.V. light, chlorine gives chlorine free radicals.

$$Cl:Cl \xrightarrow{U.V} 2Cl^{\cdot}$$

- If the unpaired electron is on carbon atom then they are called alkyl free radicals. Alkyl free radicals are classified as primary (1°), secondary (2°) or tertiary (3°) after the carbon carrying unpaired electrons.

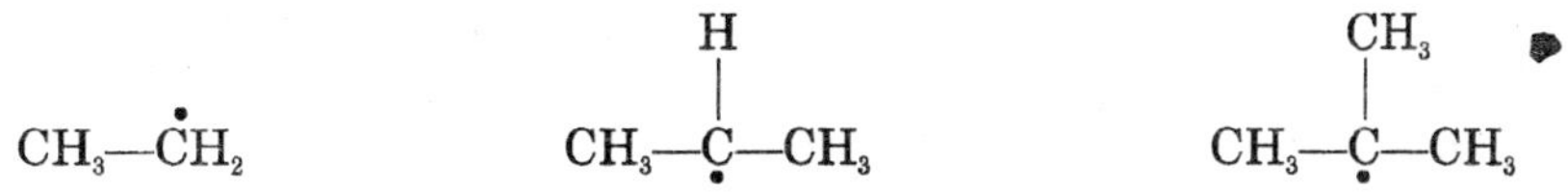

Ethyl free radical (1°) *iso*-Propyl free radical (2°) *tert*-Butyl free radical (3°)

- **Decreasing order of stability of alkyl free radicals relative to alkane from which they are formed is *3° > 2° > 1°*.**

This order of stability can be explained in terms of delocalisation of unpaired electron through hyperconjugation. In hyperconjugation delocalisation takes place for through overlap of *p*-orbital containing unpaired electron with σ orbitals of alkyl groups.

In 3° alkyl group radical there are three alkyl groups on the carbon carrying unpaired electron. Therefore, delocalisation of the unpaired electron through hyperconjugation is maximum and hence the radical is most stable followed by 2° radical (with two alkyl groups) and 1° radical (with one alkyl group).

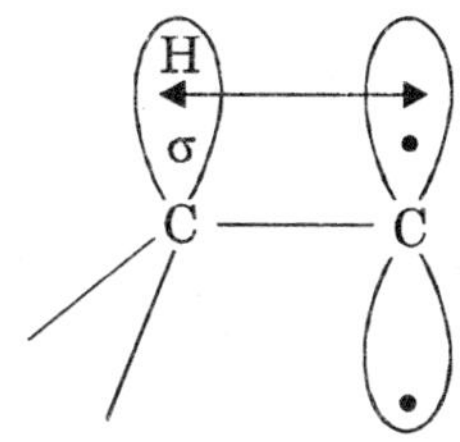

Hyper conjugation

- **Orbital structure**

Structure of alkyl free radicals is not known with certainty but it is proposed that carbon carrying unpaired electron is sp^2 hybridised. These three sp^2 hybrid orbitals are used in forming three σ bonds with other atoms. The unpaired electron is present in the unhybridised *p*-orbital which is perpendicular to the plane of σ bonds.

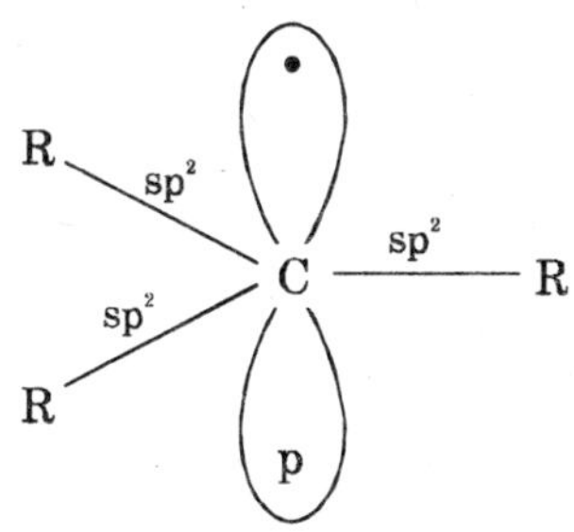

Allylic free radical

- **EXCEPTIONAL STABILITY OF ALLYLIC AND BENZYLIC FREE RADICALS**

Allylic and benzylic free radicals are exceptionally stable intermediates because the odd electron in them is delocalized through resonance.

$$CH_2—CH—CH_2 \longleftrightarrow CH_2—C—CH_2$$

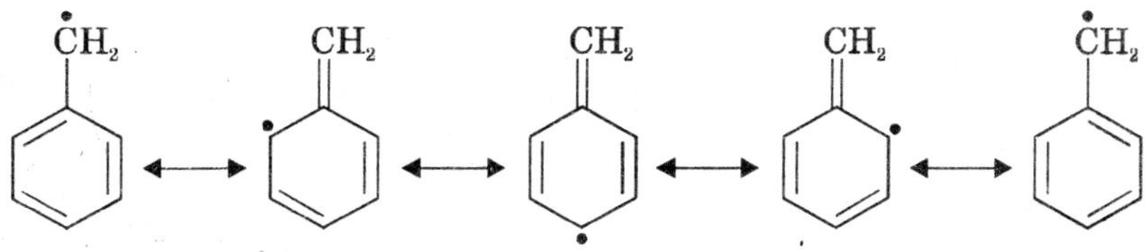

Benzyl free radical

Just like benzyl free radical diphenylmethyl and triphenylmethyl free radicals are also stabilized by resonance. The ***order of stability*** of these free radicals is

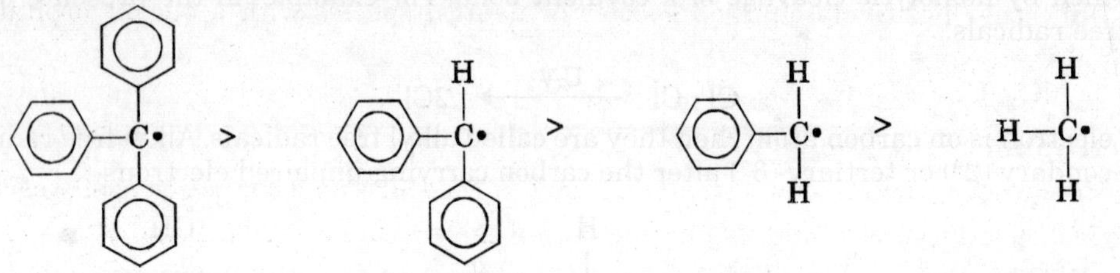

Triphenylmethyl free radical > Diphenylmethyl free radical > Benzyl free radical > Methyl free radical

Because of their extra stability and benzylic free radicals are formed relatively more easily. Therefore, reactions involving allylic and benzylic free radicals take place at faster rate.

- **CARBENES**

A carbene may be defined as a group of atoms which contains a carbon atom with only six electrons in its valence shell, out of which two electrons are unshared. These are neutral species and are formed by removal of two monovalent groups or a divalent group along with one of the two bonding pairs of electrons. Carbenes are very unstable and highly reactive species because they contain a carbon atom which is deficient in electrons and has tendency to complete it octet. **Carbenes are,** therefore, **electrophiles.**

- Due to their high reactivity, carbenes are generally produced in *situ.* Carbenes can be generated by photolysis of diazomethane or ketene.

$$\underset{\text{Diazomethane}}{CH_2 = \overset{+}{N} = \overset{-}{N}} \xrightarrow{\text{U.V.}} :CH_2 + N_2 \qquad \underset{\text{Ketene}}{CH_2 = C = O} \xrightarrow{\text{U.V.}} :CH_2 + CO$$

- Carbenes exist in two different states, singlet and triplet. In singlet state unshared electrons are paired whereas in triplet state they are not paired. ***Carbene in triplet state is more stable than carbene in singlet state.*** In singlet carbene, carbon atom carrying six valence electrons is sp^2 hybridised whereas in triplet it is sp hybridised.

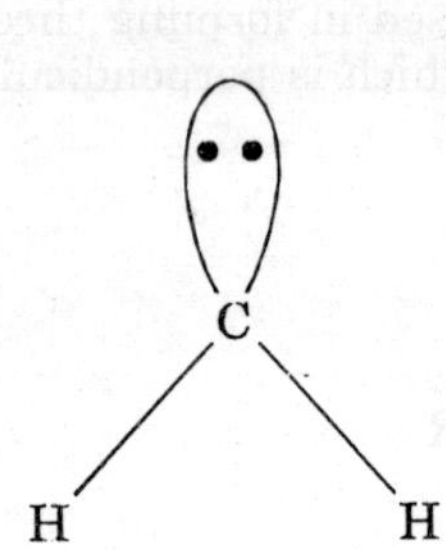

Singlet

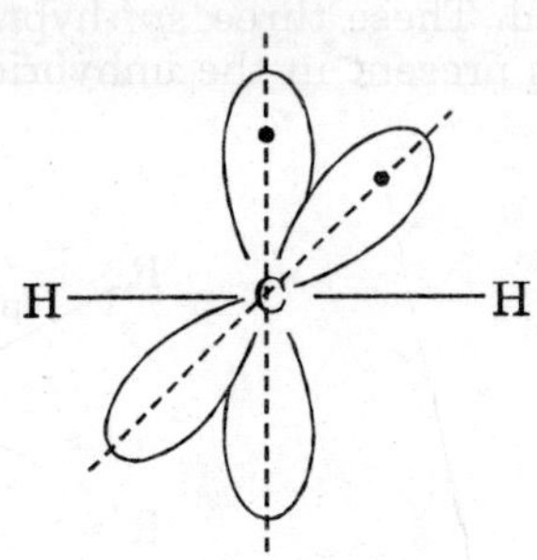

Triplet

TYPES OF ATTACKING SPECIES

- **Nucleophiles**

A nucleophile may be defined as an electron rich species which contains an atom which has an unshared pair of electrons. They are either negatively charged species or neutral species with unshared pair of electrons. For example,

$$:\ddot{\underset{\cdot\cdot}{O}}H^-,\ C\ddot{N}^-,\ \ddot{O}R^-,\ H_2\ddot{O}:,\ \ddot{N}H_3,\ R-\ddot{O}H,\ R-\ddot{O}-R,\ R-\ddot{N}H_2,\ R_2\ddot{N}H$$

Since nucleophiles contain an electron rich atom they have great tendency to react with electron deficient species (electrophiles). Reactions involving nucleophilic reagents are called nucleophilic reactions.

- **Electrophiles**

  An electrophile may be defined as an electron deficient species which contains an atom which is deficient in electrons. They are either positively charged species or neutral species with a vacant orbital. For example,

$$\overset{+}{N}O_2,\ \overset{+}{C}l,\ \overset{+}{C}H_3,\ \overset{+}{H},\ SO_3,\ BF_3,\ AlCl_3,\ SO_3,\ :CH_2$$

- **Hard/Soft Electrophiles and Nucleophiles**

  Hard electrophiles are characterised by small size and valence electrons cannot be easily excited *e.g.* $H^{\oplus}$.

  Soft electrophiles have large size and valence electrons can be easily excited, *e.g.*, I_2, Fe^{+2}, Cu^{+2}, etc.

  Hard nucleophiles are those which contain donor atom with high electronegativity.

  Soft nucleophiles are those in which the electronegativity of donor atom is low.

- ***HSAB Principle***

  According to this principle a soft electrophile reacts with soft nucleophile and a hard electrophile reacts with hard nucleophiles.

1.2. IMPORTANT FACTS AND TERMS

Add to Knowledge

- Name of an amine is always written as one word. For example,

  CH_3NH_2 is written as Methylamine and not Methyl amine.

- When two like carbon containing functional groups are present at the two ends of the carbon chain, the carbon atoms of both the functional groups are counted towards the parent chain. For example,

  $\overset{3}{HOOC}—\overset{2}{CH_2}—\overset{1}{COOH}$ Propane-1, 3-dioic acid

  $\overset{4}{NC}—\overset{3}{CH}=\overset{2}{CH}—\overset{1}{CN}$ But-2-en-1, 4-dinitrile.

- (*a*) If two different carbon containing functional groups are present at the two ends of the carbon chain, only the carbon atom of the principal functional group is to be counted towards the parent chain while the other functional group is treated as the substituents group. For example,

  $NC—CH_2—CH_2—CH_2—COOH$

  4-Cyanobutanoic acid

  (*b*) If, however, the substituent group is named as an oxygen derivative of CH_3 or CH_2 group, the carbon atom of the substituent group is also counted towards the parent chain. For example,

  $O=CH—CH_2—COOH$ 3-Oxopropanoic acid

  $O=CH—COOH$ 2-Oxoethanoic acid

- (*a*) The aryl group obtained by removing one hydrogen atom from benzene is named as phenyl and not benzyl.

  (*b*) The aryl group obtained by removing a hydrogen atom from side chain in case of toluene is named as benzyl.

  (*c*) The aryl group obtained by the removal of one hydrogen atom from ring of toluene is named as tolyl.

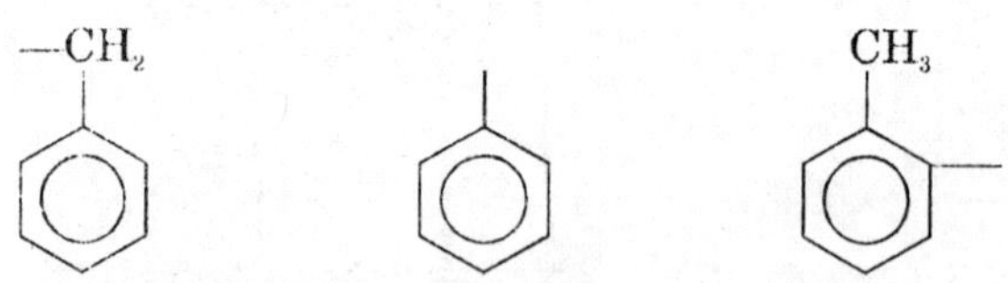

Benzyl Phenyl o-Tolyl

- When two or more prefixes consist of identical words, the priority for citation is given to that group which contains the lowest locant at the first point of difference. For example,

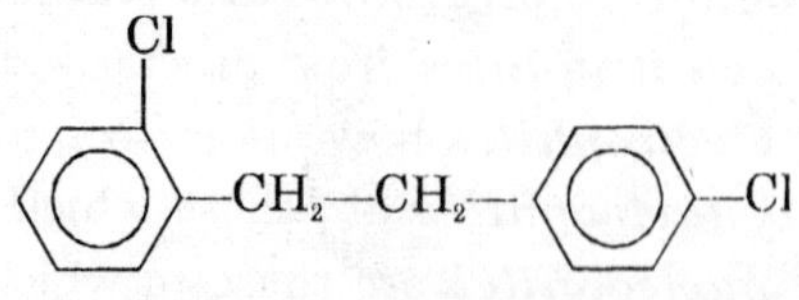

1-(2-Chlorophenyl)-2-(4-chlorophenyl) ethane

If a compound contains a benzene ring coupled to an alicyclic ring, it is named as a derivative of benzene, *i.e.*, compound having lowest state of hydrogenation. For example,

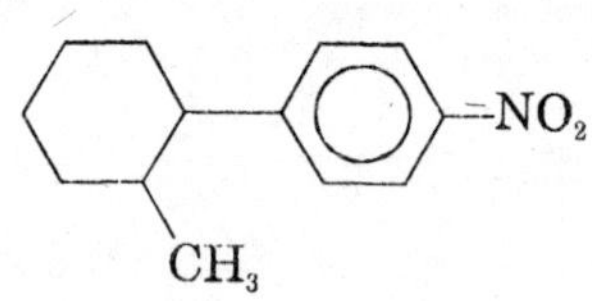

1-(2-Methylcyclohexyl)-4-nitrobenzene

- Bond line notation for alkynes

  Bond line notation for propyne is  and not 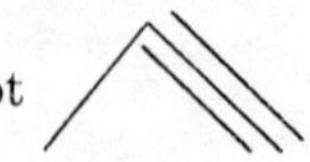

  Bond line notation for but-1-yne is 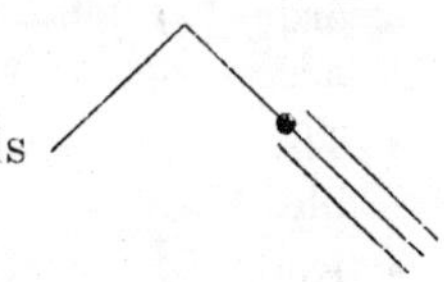

  and not 

  This is because the bond angle involved in *sp*-hybridised carbon atom is 180° and not 120°.

- (*a*) If a hydrocarbon contains both double and triple bonds, it is always named as a derivative of an alkyne rather than an alkene, *i.e.*, it is named as an alkyne. While deciding the locants for double and triple bonds, lowest sum rule is always followed. For example,

  $$\overset{5}{CH_3}—\overset{4}{CH}=\overset{3}{CH}—\overset{2}{C}\equiv\overset{1}{CH}$$

  Pent-3-en-1-yne

  $$\overset{5}{CH_3}—\overset{4}{C}\equiv\overset{3}{C}—\overset{2}{CH}=\overset{1}{CH_2}$$

  Pent-1-en-3-yne

(b) If, however, there is a choice in numbering, the double bond is always given preference over the triple bond. For example,

$$\overset{5}{HC}\equiv\overset{4}{C}-\overset{3}{CH_2}-\overset{2}{CH}=\overset{1}{CH_2}$$

Pent-1-en-4-yne

$$\overset{6}{HC}\equiv\overset{5}{C}-\overset{4}{CH}=\overset{3}{CH}-\overset{2}{CH}=\overset{1}{CH_2}$$

Hexa-1, 3-dien-5-yne

- If two atoms/groups of same preference occupy identical positions from either end of the parent chain, number must be given to the atom/group whose prefix comes first in the alphabetical order.

$$Cl-CH_2-CH_2-CH_2-CH_2-Br$$

1-Bromo-4-chlorobutane

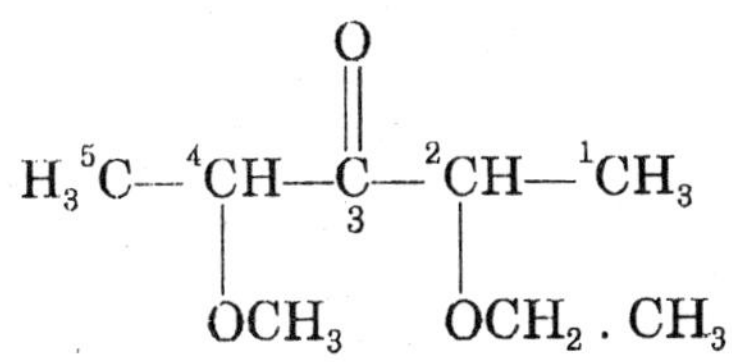

2-Ethoxy-4-methoxypentan-3-one

- Metamers can also be chain isomers or position isomers.
- TAUTOMERS always exist in equilibrium.
- The term tautomerism was introduced by Laar in order to explain the chemical reactivity of a substance according to two possible structures.
- The tautomers differ in their stability. The less stable form is also sometime referred to as Labile form.
- Keto-enol tautomerism is exhibited only by such aldehydes and ketones which contain at least one α-hydrogen. For example

(i) Tautomerism is possible in propionaldehyde (CH_3CH_2CHO) and acetyl acetone $CH_3COCH_2COCH_3$ as these molecules contain α-hydrogen.

(ii) Tautomerism is not possible in benzaldehye (C_6H_5CHO), benzophenone ($C_6H_5COC_6H_5$), trimethyl acetaldehyde, $(CH_3)_3C$-CHO and chloral CCl_3-CHO as they do not carry α-hydrogen.

- In general, polar protic solvents, such as water, methanol, acetic acid etc., which tend to stabilize the keto form relative to the enol form, will reduce the enol content. On the other hand, non-polar solvents such as hexane, benzene etc., which do not form H-bonds tend to increase the enol content. For example the amount of enol content of acetyl acetone in ethyl alcohol is only 76% whereas in hexane, it is 92%.
- Presence of keto form of acetoacetic ester is evidenced by the usual reaction of carbonyl group with hydroxyl amine, phenyl hydrazine etc., whereas the presence of enolic form is evidenced by the usual reaction of —OH group with PCl_5, NH_3 etc.
- 3° Nitro alkanes such as $H_3C-C(CH_3)_2-NO_2$ do not show nitro-acinitro tautomerism due to absence of α-H atom.
- ***Double Bond Equivalents (D.B.E.) or index number.*** It is a numerical number on the basis of which we can have an idea about the different type of equivalent structures we can have for a given formula. Knowing the D.B.E. or index number (I) the problems on structural isomerism can be more easily solved. Mathematically,

$$\text{D.B.E.} = \frac{2(c)+2-H}{2} \quad \text{or} \quad I$$

Here, c = number of carbon atoms and H is number of hydrogen atoms

For example D.B.E. of C_4H_6 is

$$I \quad \text{or} \quad \text{D.B.E.} = \frac{2\times 4+2-6}{2} = 2$$

The index number is 2. This implies that the molecule may contain,

(i) two double bonds or one triple bond

(ii) one ring and one double bond or

(iii) two rings

Thus, by knowing the D.B.E. total number of structural isomers can be formulated.

- Certain amides such as urea, thiourea and nitrosophenol also show tautomerism.

$$H_2N-\overset{O}{\overset{\|}{C}}-NH_2 \rightleftharpoons HN=\overset{OH}{\overset{|}{C}}-NH_2\ ;$$

Urea (*keto* form) — *Enolic* form

$$H_2N-\overset{S}{\overset{\|}{C}}-NH_2 \rightleftharpoons HN=\overset{SH}{\overset{|}{C}}-NH_2$$

Thio urea (*keto* form) — (*Enolic* form)

(*Enolic* form) (*keto* form)

- Fully eclipsed form of Butane has 15.06 kJ mol^{-1} higher energy than that of anti form.

- Terminal alkenes like propene, isobutylene, 2, 3-Dimethyl-2-butene do not show geometrical isomerism.
- Geometrical isomers are also called configurational isomers as these have fixed configurations.
- Ring structures also exhibit geometrical isomerism. This is due to fixed positions of group in the ring. For example,

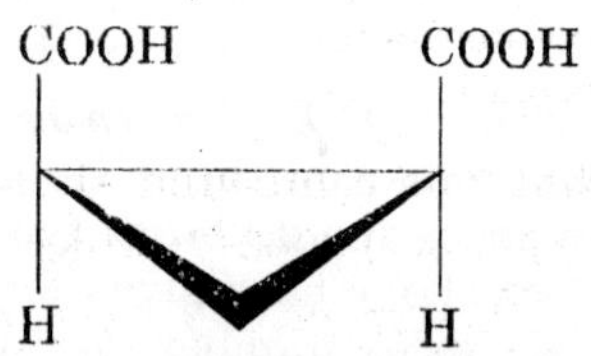

cis-isomer

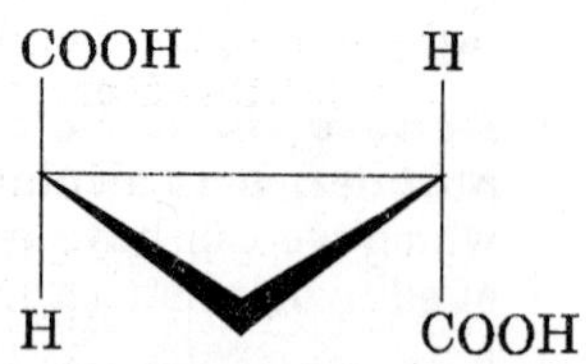

trans-isomer

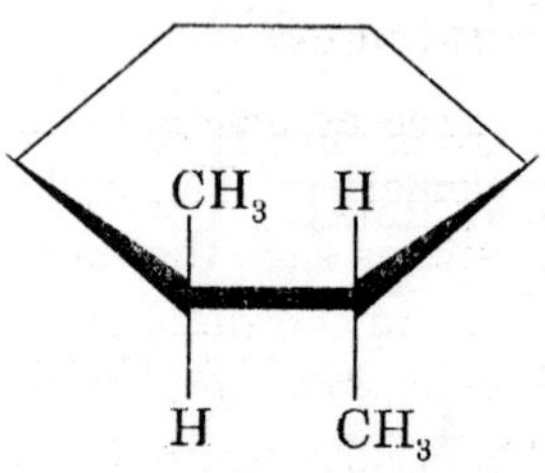

cis-1, 2-Dimethyl Cyclohexane

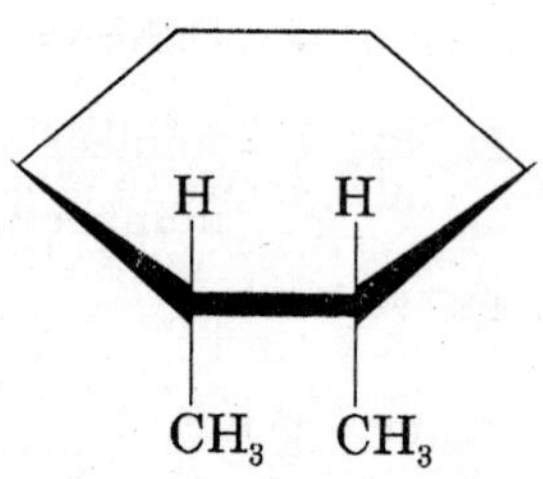

trans-1, 2-Dimethyl Cyclohexane

- ***Auwers-skita rule.*** This is an empirical rule according to which out of the pair of geometrical isomers (cis and trans), the cis isomers has higher boiling point, higher density and higher refractive index.
- ***Dihedral angle.*** The angle between C—H and C—C bonds on adjacent carbon atoms in any conformation is called dihedral angle. For example, the dihedral angle between C—H bonds on adjacent carbons in staggered and eclipsed conformation of ethane are 180° and 0° respectively whereas between the two methyl groups in the gauche and anti conformation of *n*-butane are 60° and 180° respectively.
- ***Asymmetric synthesis.*** The synthesis of an optically active compound from an optically inactive molecule in the presence of a suitable chiral reagent is called Asymmetric synthesis *e.g.*

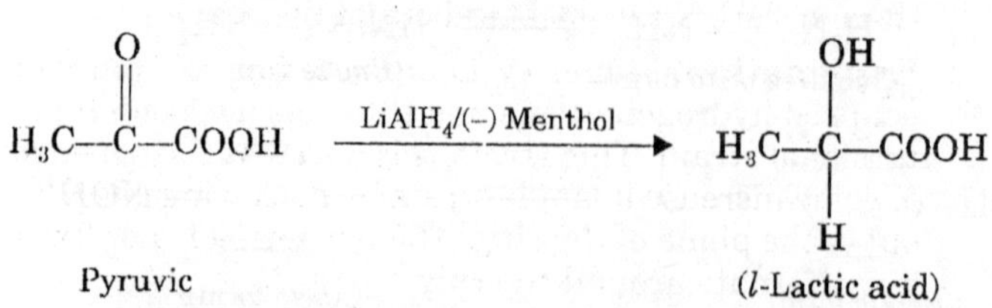

Pyruvic (*l*-Lactic acid)

- ***Walden Inversion.*** The conversion of *d*-form of an optically active compound into *l*-form of the same or of different compound or vice versa is known as Walden inversion.

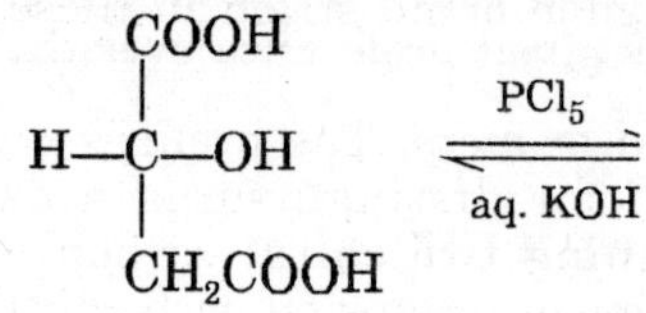

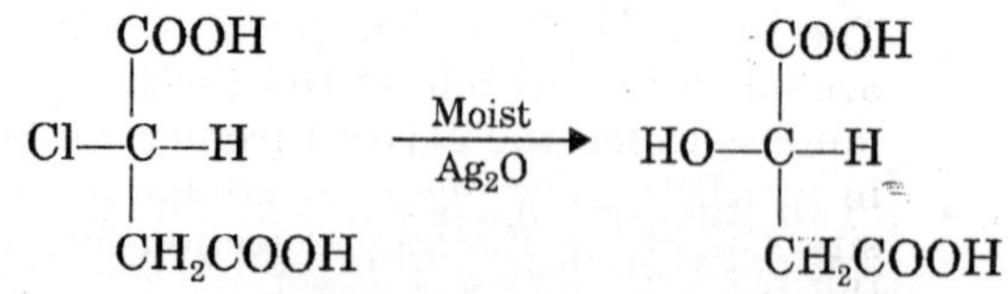

l-malic acid

- ***Resolution.*** The separation of *d* and *l* isomers from the *dl* mixture is referred to as resolution. This can be done by (*a*) Mechanical separation (*b*) Biochemical separation and (*c*) Chemical method (best method).
- Quartz was the first substance to exhibit the phenomenon of optical activity. Other crystalline substances showing optical activity are Benzil, Sodium chlorate, Barium formate.
- Molecular dissymmetry is the essential condition for optical activity.
- ***Baeyer strain theory***

  The relative stability of the different cyclo alkanes is explained on the basis of this theory. The main points of the theory are :

  (*i*) The angle of 109°28¢ is considered to be a strain less angle. Any deviation in bond angle from this ideal value introduces a strain in the molecule.

  (*ii*) The molecules of cyclo alkanes are planar.

  (*iii*) Due to planar geometry, the C—C bond angle can never be 109°28¢ but it may be either less or more than this.

  (*iv*) Greater is the deviation from this bond angle more is the strain and hence lesser is the stability. The angle strain is calculated mathematically as,

Angle strain = 1/2 [109°28′ – observed bond angle]. For example,

(*a*) Angle strain in cyclopropane

$$= 1/2\ [109°28' - 60°] = +\ 24°44'$$

(*b*) Angle strain in cyclohexane

$$= 1/2[109°28' - 120°] = -\ 5°16'$$

The positive value of angle strain suggests that bond angles are compressed while the negative value suggests the expansion of bond angle. In simple words, the magnitude of angle strain and not its sign determines the stability of compound.

Limitations of Baeyer* Strain *theory

This theory wrongly assumes that rings of all sizes are planar. In fact in six numbered and larger rings each carbon atom retains its tetrahedral angle and hence there is negligible or no strain in these molecules..

- **Erythro** and **Threo** *isomers*. The prefixes *erythro* and *threo* are used for distinguishing a pair of enantiomers containing two chiral carbon atoms when two of the atoms or groups on each chiral carbon atom are the same while the third is different. The stereo isomer in which the two identical groups are on the same side of the Fischer projection formula is called the erythro isomer while the isomer in which the two identical groups are on the opposite side of the Fischer projection formula is called the threo isomer. For example,

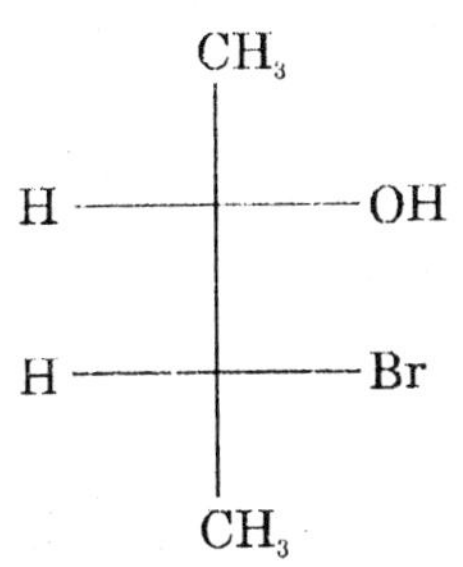

Erythro-3-bromo-2-butanol

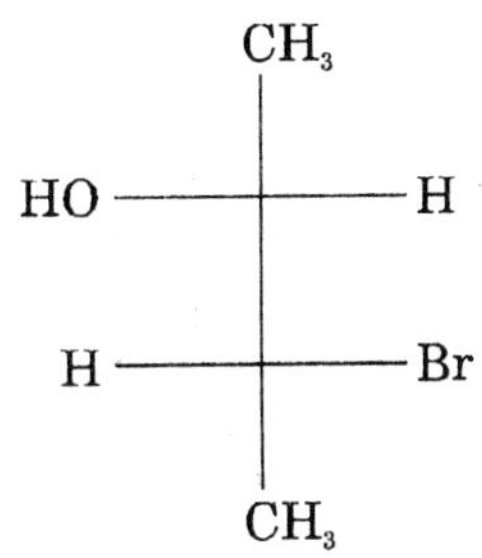

Threo-3-bromo-2-butanol

- **Calculated entropy for ethane.** Is much greater than experimentally determined value. (It is this discrepancy which led to the concept of conformations with different energies).

  Explanation. The calculated value of entropy for ethane, incorrectly assumes unrestricted free rotation about the carbon-carbon single bond. As such all the conformations are equally probable. Since most molecules of ethane have the staggered conformation, the structural randomness is less than calculated and as such actual observed entropy is less than the calculated entropy is less than the calculated entropy. In fact this discrepancy led to the concept of conformations with different energies.

- **Chirality and optical activity in alkanes**

  Lowest molecular mass alkane which is chiral is 3-methylhexane. There are two isomeric alkanes with molecular formula C_7H_{16} which are chiral.

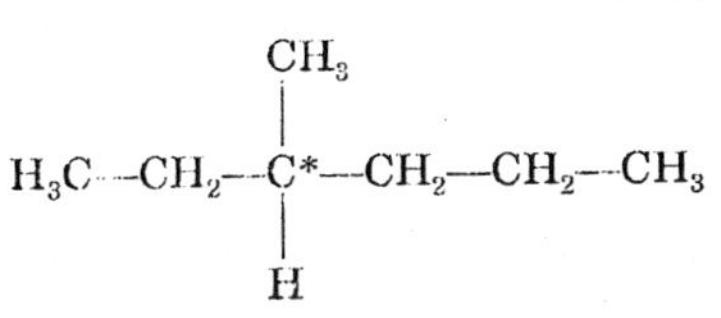

3-Methylhexane

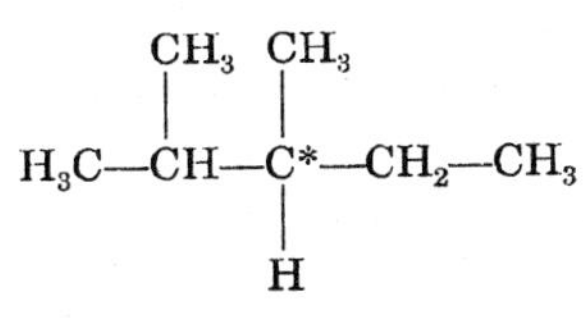

2, 3-Dimethylpentane

However none of these chiral compounds are optically active. This is clear from the following explanation.

Explanation. Even if a molecule is chiral, the magnitude, and hence the detect ability of the optical rotation depends on the structure of the particular molecule concerned. For example, in the following compound, the four groups attached to the chiral centre differ only in chain length.

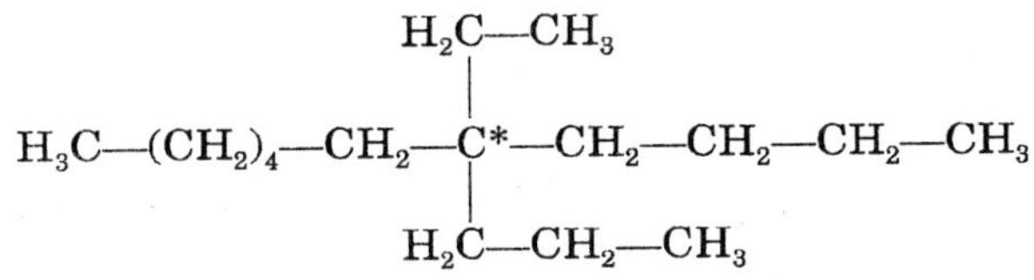

Ethyl-*n*-propyl-*n*-butyl-*n*-hexylmethane (5-Ethyl-5-propylundecane)

It has been calculated that this compound should have a small specific rotation of 0.00001°, which is far below the limits of detection of any existing polarimeter. In 1965, enantiomerically pure samples of both enantiomers of this compound were prepared and each was found to be optically inactive.

- **Isomerism shown by cumulatrienes**

  Cumulatrienes (trienes with three adjacent double bonds) shows only geometric isomerism. This is because their molecule is planar, and as such the terminal —CH_3 groups and H-atoms lie in the same plane. Therefore, in this case their planar structure can exist in two diastereoisomeric forms, *cis*- and *trans*- but no enantiomeric forms are possible.

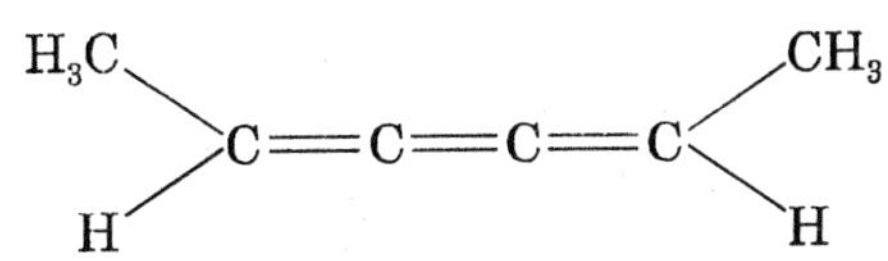

cis-Hexa-2, 3, 4-triene

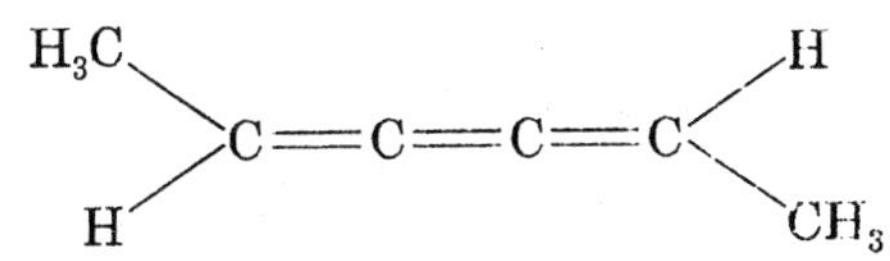

trans-Hexa-2, 3, 4-triene

- Even though planar cyclo pentane has a stable internal bond angle of 109°, yet at any given instant one of the five carbons is puckered out of the plane.

  Explanation. Planar cyclo pentane has five pairs of eclipsed hydrogen atoms and thus has a considerable torsional strain. This strain is reduced at the expense of some increase in angle strain by puckering one CH_2 out of the plane of the ring. The puckering is not fixed but alternate around the ring.

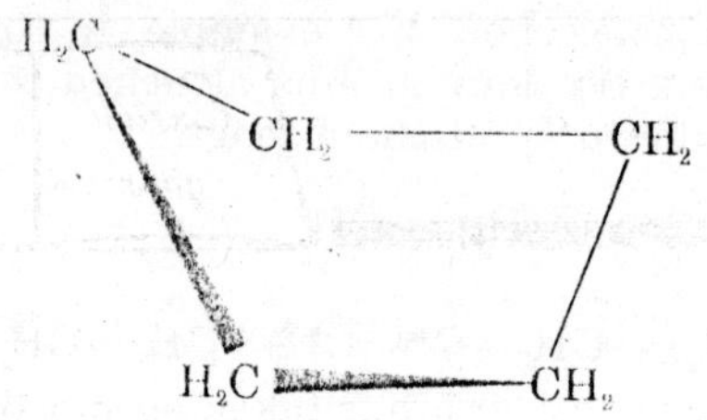

Envelope form of cyclo pentane

Though it is not possible to separate or isolate a particular conformational isomer, yet in it is possible. One such example is badly crowded 1, 1, 2, 2-tetrabromoethane. At low temperature where collision energies are small, two isomeric forms of badly crowded 1, 1, 2, 2-tetrabromoethane have been isolated fractional crystallization. Further it has been found that none of these two materials so isolated by fractional crystallization are optically active.

Explanation. The forms isolated at low temperatures must be the most stable forms of 1, 1, 2, 2-tetrabromoethane *i.e.*, the forms with lowest energy. Newman projection formulae of these forms are given below :

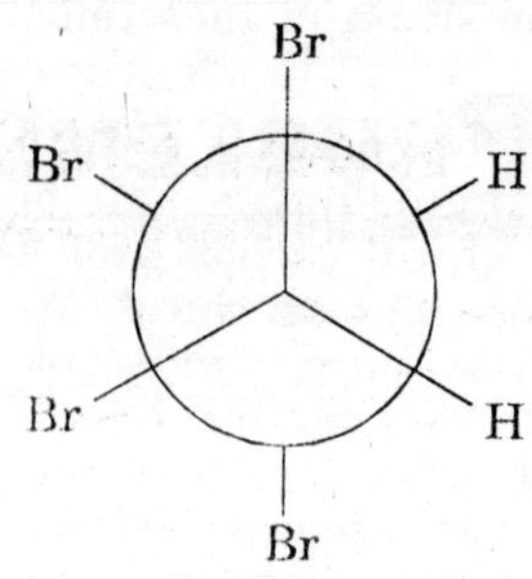

I

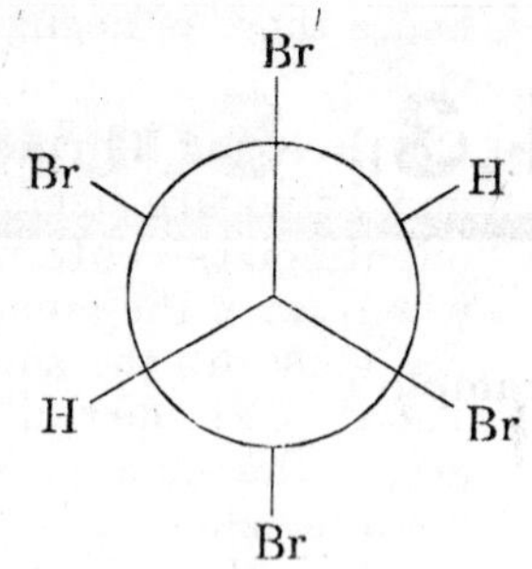

II

Here, form II has relatively lesser energy and hence more stability than form I. None of the two materials isolated by fraction crystallisation will be optically active. Out of the two forms I and II, form II is achiral. As such this form would be optically inactive. However, the form I is chiral and as such can show optical activity. The absence of optical activity in this case is due to the fact that the material obtained by crystallisation (containing form I) will be a racemic mixture of the two enantiomeric forms of of I, as enantiomers cannot be separated by fractional crystallisation.

1.3. Solved Problems

Enrich your Skill

Example 1. *Give the IUPAC name of the following compound*

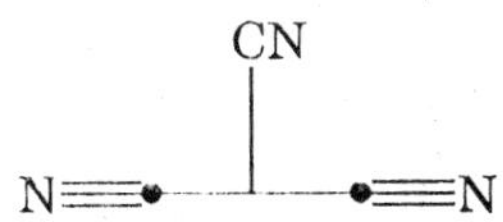

(*a*) *1, 2, 3-tricyanopropane*

(*b*) *2-cyano-1, 5-propanedinitrile*

(*c*) *Propane-1, 2, 3-tricarbonitrile*

(*d*) *none of these.*

Solution. Similar functional groups are treated alike.

Hence the answer is **(*c*).**

Example 2. *The IUPAC name of the following compound*

 *is*

(*a*) *Dicyclobutane*

(*b*) *Bicyclo [2.2.0] hexane*

(*c*) *Bicyclo [2.2.1] hexane*

(*d*) *none of these.*

Solution. It is a bicyclo compound. The carbon bridges have 2, 2 and 0 carbon atoms.

Hence the answer is **(*b*).**

Example 3. *Give the IUPAC name of the following compound*

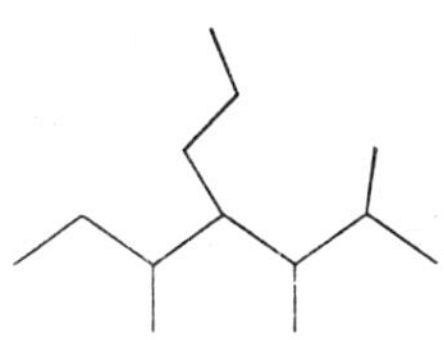

(*a*) *4-secbutyl-2, 3 dimethyl heptane*

(*b*) *2, 3, 5-trimethyl-4-propylheptanez*

(*c*) *6-isopropyl-3-methyl-4-propyl hexane*

(*d*) *none of these.*

Solution. Longest chain of seven carbon atoms to be selected.

Hence the answer is **(*b*).**

Example 4. *The correct IUPAC name of the following compound, $CH_3COCH = CHCH_2C_6H_5$ is*

(*a*) *4-Oxo-pent-2-enyl benzene*

(*b*) *Me(CO benzyl) vinyl ketone*

(*c*) *1-Phenyl-pent 2-ene-3-one*

(*d*) *5-Phenyl pent-3-en-2-one.*

Solution. The phenyl group is taken as substituent on the chain.

Hence the answer is **(*d*).**

Example 5. *The correct IUPAC name of the following given compound is*

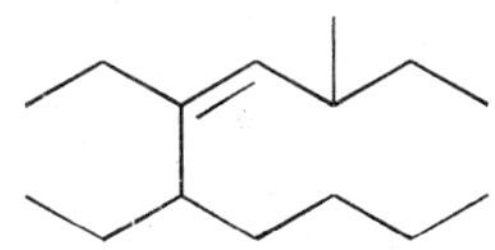

(*a*) *5, 6-Diethyl-3-methyl dec-4-ene*

(*b*) *7-Methyl, 2, 4, 6-trieneoctanal*

(*c*) *6-Methyl heptene*

(*d*) *3, 3-Diethyl-5-ethyl-4-decene.*

Solution. Select the longest chain and least number should be given to double bond.

Hence the answer is **(*a*).**

Example 6. *Give the IUPAC name of the following compound*

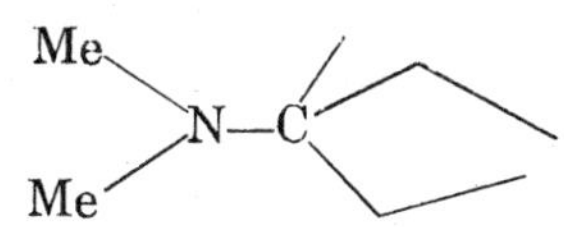

(*a*) *3-Methyl-3-N, N-dimethyl amino pentane*

(*b*) *3-ethyl-3-N, N-dimethyl pentane*

(*c*) *3-Methyl-3-N, N-dimethyl butane*

(*d*) *C-Ethyl-c-methyl propane nitrile-N, N-dimethyl.*

Solution. The name of substituent 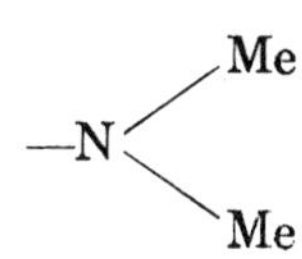

is N, N-dimethyl amino.

Hence the answer is **(*a*).**

Example 7. *Give the IUPAC name of the following compound*

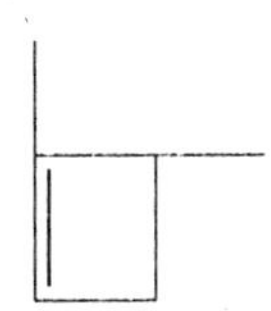

(*a*) *2, 3-dimethyl cyclobutene-1*

(*b*) *1, 2-dimethyl cyclobutene-1*

(*c*) *1, 4-dimethyl cyclobutene-1*

(*d*) *1, 2-dimethyl cyclobutene-2.*

Solution. The double bond should be given lowest position.

Hence the answer is **(*a*).**

Example 8. *Assign the IUPAC name for the following compound*

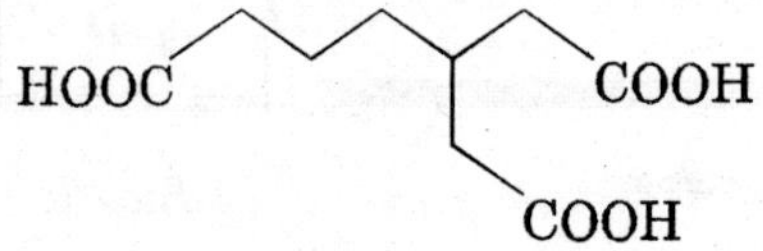

(*a*) *5, 5-bis (carboxymethyl) pentanedioic acid*

(*b*) *Pentane-1, 2, 5-tricarboxylic acid*

(*c*) *3-(Carboxymethyl) heptane dioic acid*

(*d*) *5, 5-bis (acetic acid) pentanoic acid.*

Solution. —CH_2COOH group is named as carboxyl methyl (as a substituent). The longest chain is selected. The carbon atoms of the carboxylic group are counted in the chain.

Hence the answer is **(*c*)**.

Example 9. *Assign the IUPAC name for the following compound*

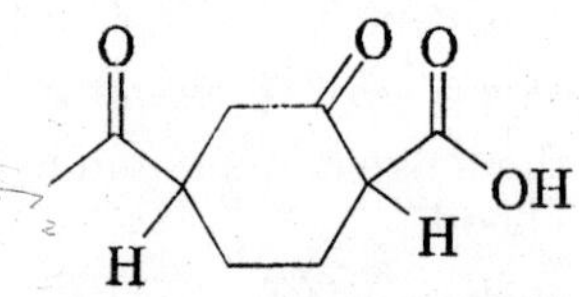

(*a*) *2, 4 dioxocyclohexane-1-carboxylic acid*

(*b*) *4-formyl-2-oxocylohexane-1-carboxylic acid*

(*c*) *2, 4-dioxocyclohexanoic acid*

(*d*) *2-oxo cyclohexane-1, 4-dicarboxylic acid.*

Solution. Amongst the functional groups COOH is given priority. Thus numbering starts from the carbon of the ring holding this group.

Hence the answer is **(*b*)**.

Example 10. *Give the IUPAC name of the following compound*

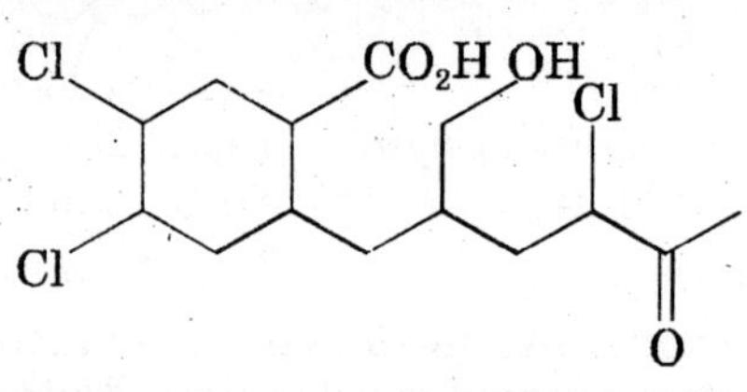

(*a*) *3, 4-Dichloro-6-[4-chloro-2-(hydroxymethyl)-5-oxohexyl] cyclohexane-1-carboxylic acid*

(*b*) *4, 5-Dichloro-2-[4-acetyl-4-chloro-2-(hydroxymethyl) butyl] cyclohexane-1-carboxylic acid*

(*c*) *4, 5-Dichloro-2-[4-chloro-2-(hydroxymethyl)-5-oxohexyl] cyclohexane-1-carboxylic acid*

(*d*) *None of the above.*

Solution. The carbon of the ring holding COOH group is given no. 1. Then keeping in mind term by term comparison rule the name is formulated.

Hence the answer is **(*c*)**.

Example 11. *Assign IUPAC name to the following compound*

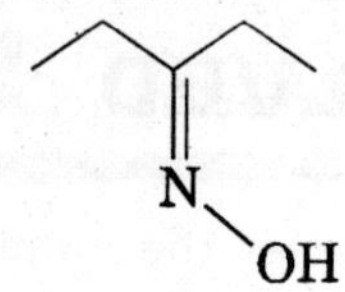

(*a*) *Pentan-3-one oxime*

(*b*) *N-hydroxyamino pentane*

(*c*) *N-hydroxyimino pentane*

(*d*) *none of these.*

Solution. This compound is obtained by reaction of C = O compounds with NH_2OH and therefore substitutive name is used.

Hence the answer is **(*a*)**.

Example 12. *Assign IUPAC name to the following compound*

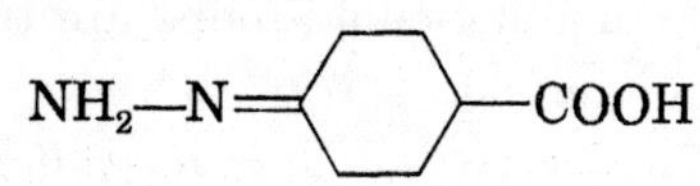

(*a*) *4-Hydrazono cyclohexanoic acid*

(*b*) *4-Hydrazono benzoic acid*

(*c*) *4-Hydrazono cyclohexane-1-carboxylic acid*

(*d*) *None of these.*

Solution. The compound is obtained by the reaction of C = O gr with hydrazine and hence the group is called hydrazono. As carbon of COOH cannot be counted in the ring.

Therefore the answer is **(*c*)**.

Example 13. *Give the IUPAC name of the following compound*

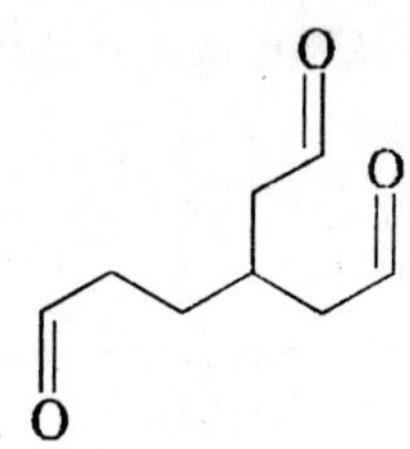

(*a*) *3-methyl hexanetrial*

(*b*) *3-(Formylmethyl) hexanedial*

(*c*) *3-(Oxomethyl) hexanedial*

(*d*) *3-(Methyl formyl) hexanedial.*

Solution. The group CH_2CHO is called formyl methyl if it is not the part of parent chain.

Hence the answer is **(*a*)**.

Example 14. *Give the IUPAC name of the following compound is*

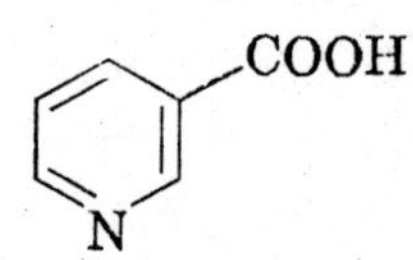

(a) *Pyridine-3-carboxylic acid*

(b) *Carboxyl pyridine*

(c) *Pyridine-1-carboxylic acid*

(d) *None of these.*

Solution. In heterocycles the numbering starts from the hetero atom. Since the compound is obtained from parent pyridine.

Hence the answer is **(*a*).**

Example 15. *Write the IUPAC name of the following compound*

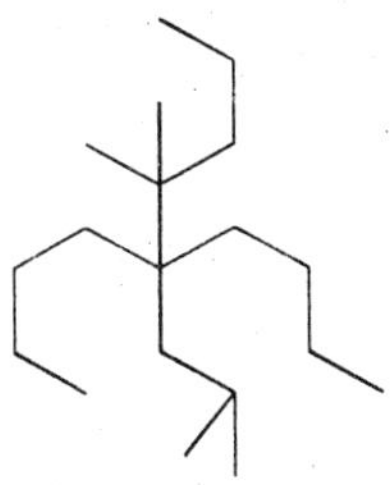

(a) *5-(2-Methylpropyl)-5-(1, 1-dimethylpropyl) nonane*

(b) *5, 5-Dibutyl-3, 3, 6-trimethyl heptane*

(c) *5-(1, 1-Dimethylpropyl)-5-(2 methylpropyl) nonane*

(d) *None of the above.*

Solution. Select the longest chain and name accordingly.

Hence the answer is **(*c*).**

Example 16. *The IUPAC name of* 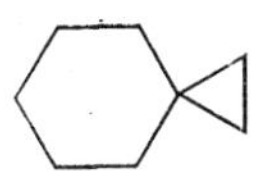 *is*

(a) *Bicyclo [5.5.0] nonane*

(b) *Biphenyl*

(c) *Cyclo propyl Cyclohexane*

(d) *Spiro [2.5] octane.*

Solution. It is a spiro compound.

Hence the answer is **(*d*).**

Example 17. *The IUPAC name of*  *is*

(a) *Bicyclo [2.2.0] hexane* (b) *Spiro [2.2] hexane*

(c) *Spiro [2.2] pentane* (d) *none of these.*

Solution. It is a spiro compound.

Hence the answer is **(*c*).**

Example 18. *The IUPAC name of the compound given below is*

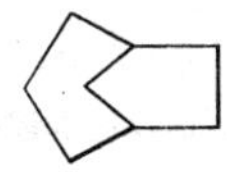

(a) *Bicyclo [3.2.1] octane* (b) *Bicyclo [3.2.2.] octane*

(c) *Spiro [2.2] octane* (d) *none of these.*

Solution. It is a bicyclo compound. See rules for naming these compounds.

Hence the answer is **(*a*).**

Example 19. *The structure of spiro [3.3] heptane is*

(a)  (b) 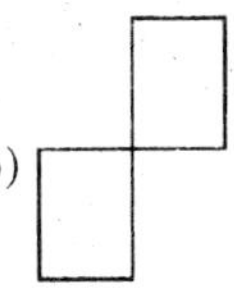

(c) 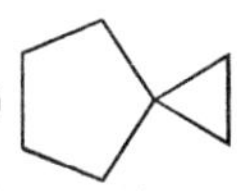 (d) 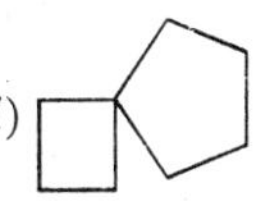

Solution. The spiro atom is fused to two rings with each having 3 carbon atoms (other than spiro).

Hence the answer is **(*b*).**

Example 20. *The structure of bicyclo [1.1.0] butane is*

(a) 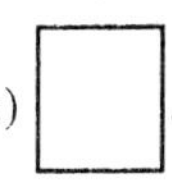 (b) 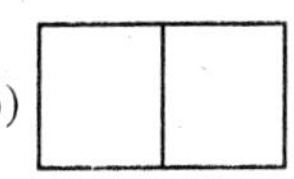

(c)  (d) 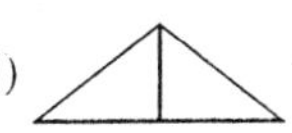

Solution. See rules of naming bicyclic compounds.

Hence the answer is **(*c*).**

Example 21. *The IUPAC name of* 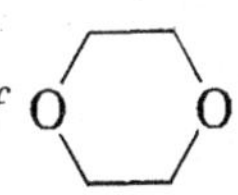 *is*

(a) *1, 4-Dioxane* (b) *Oxirane*

(c) *Oxirene* (d) *1, 2-Dioxane.*

Solution. Oxane is the name assigned to hetero cycle with six atoms and containing oxygen atom. As the above compound has two oxygen atoms.

Therefore the answer is **(*a*).**

1.4. QUESTION BANK

General Questions

1. All the members of a homologous series have same
   (a) Molecular mass
   (b) Molecular formula
   (c) Empirical formula
   (d) General molecular formula.

2. Which of the following compounds has least dipole moment ?
   (a) 1, 2-Dichlorobenzene
   (b) 1, 3-Dichlorobenzene
   (c) 1, 4-Dichlorobenzene
   (d) All have same dipole moment.

3. In vinyl cyanide, the hybrid states of C-1, C-2 and C-3 are respectively
   (a) sp^3, sp^2, sp (b) sp, sp^2, sp^2
   (c) sp, sp^2, sp^3 (d) sp^2, sp^2, sp^3.

4. Paraffins is the common name used for
   (a) Alkanes (b) Alkenes
   (c) Alkynes (d) Arenes.

5. The C—C—C bond angle in propane is expected to be about
   (a) 180° (b) 120°
   (c) 109° (d) 150°.

6. In *iso*-octane (2, 2-4-trimethylpentane), the number of primary hydrogen atoms is
   (a) 12 (b) 15
   (c) 13 (d) 14.

7. Which of the following molecules does not contain any secondary hydrogen atom ?
   (a) *n*-Pentane (b) *neo*-Pentane
   (c) *neo*-Hexane (d) *iso*-Pentane.

8. The first organic compound was synthesized in laboratory by
   (a) Wöhler (b) Kolbe
   (c) Berzelius (d) Neil Bartlett.

9. The vital force theory was proposed by
   (a) Wöhler (b) Berthlot
   (c) Berzelius (d) Kolbe.

10. Organic compounds are generally
    (a) Soluble in water
    (b) Stable
    (c) Soluble in non-polar solvents
    (d) Insoluble in non-polar solvents.

11. Chemically similar compounds differing by a CH_2 are called
    (a) Isomers (b) Isomorphous
    (c) Homologous (d) None of these is correct.

12. The general formula of alkynes is
    (a) C_nH_{2n+2} (b) C_nH_{2n-2}
    (c) C_nH_{2n} (d) C_nH_{2n+1}.

13. Select the molecule which has three pi bond
    (a) $CH \equiv CH$
    (b) $CH_2 = CH—CH = CH—CHO$
    (c) $CH_3CH = CHCOOH$
    (d) $CH_3CH = CH_2$.

14. The shortest C—C bond distance is found in
    (a) Diamond (b) Ethane
    (c) Benzene (d) Acetylene.

15. Number of *sec*-H atoms in *iso*-butane is
    (a) 1 (b) 9
    (c) 4 (d) none.

16. The tetrahedral structure of carbon was proposed by
    (a) Kekule (b) Wöhler
    (c) Vant Hoff and Le-Bel (d) Berthlot.

17. Marsh gas mainly contains
    (a) C_2H_4 (b) C_2H_2
    (c) CH_4 (d) C_2H_6.

18. C—C bond length is minimum in
    (a) C_2H_6 (b) C_2H_2
    (c) C_2H_4 (d) C_6H_6.

19. C—H bond length is minimum in
    (a) C_2H_6 (b) C_2H_4
    (c) C_2H_2 (d) C_6H_6.

20. The bond order of individual C—C bonds in benzene is
    (a) one
    (b) two
    (c) between one and two
    (d) one and two alternatively.

21. C_nH_{2n+2} is the general formula used to represent
    (a) Alkenes (b) Alkanes
    (c) Alkynes (d) Benzenoids.

22. A compound with molecular formula C_2H_2 must contain

(*a*) All sigma bonds
(*b*) One σ bond and two π bonds
(*c*) Two σ bond and two π bonds
(*d*) Three σ bonds and two π bonds.

23. Which of the following is the most stable ?

(*a*) $CH_2 = CH_2$ (*b*) $CH_3CH = CH(CH_3)_2$
(*c*) $CH_3C(CH_3) = CH—CH_2CH_3$ (*d*) $CH_3CH_2CH = CH_2$.

24. An example of conjugated diene is

(*a*) 1, 4-Pentadiene (*b*) 1, 5-Hexadiene
(*c*) 1, 2-Butadiene (*d*) 1, 3-Pentadiene.

25. The hybridization of carbon atoms in C—C single bond of $H_2C = CH—CH = CH_2$ is

(*a*) $sp^3 - sp^3$ (*b*) $sp^2 - sp^3$
(*c*) $sp^2 - sp^2$ (*d*) $sp^3 - sp$.

26. Which of the following has zero dipole moment ?

(*a*) 1-Butene (*b*) *cis*-2-Butene
(*c*) *trans*-2-Butene (*d*) 2-Methyl-1-propene.

27. What is not true about homologous series ?

(*a*) All the members have similar chemical properties
(*b*) They have identical physical properties
(*c*) They can be represented by a common general formula
(*d*) Adjacent members differ in molecular mass by 14.

28. The number of pi bonds in isoprene is

(*a*) 2 (*b*) 3
(*c*) 1 (*d*) 0.

29. The hybrid states of the four carbon atoms in 1, 2-butadiene are ...... respectively.

(*a*) sp^2, sp^2, sp^2, sp^3 (*b*) sp^2, sp, sp^2, sp^3
(*c*) sp^3, sp^2, sp^2, sp^2 (*d*) $sp^3\ sp, sp, sp^2$.

30. The C—C single bond is shortest in

(*a*) Propane
(*b*) Propene
(*c*) Propyne
(*d*) All have same C—C single bond-length.

31. Which of the following compounds contains maximum number of tertiary hydrogen atoms ?

(*a*) *n*-Hexane (*b*) 2-2-Dimethyl butane
(*c*) 2, 3-Dimethyl butane (*d*) 2-Methyl pentane.

32. Which of the following represents the *iso*-butyl radical ?

(*a*) $CH_3—CH_2—CH(—)—CH_3$ (*b*) $CH_3—CH(CH_3)—CH_2—$
(*c*) $CH_3—C(CH_3)(—)—CH_3$ (*d*) $CH_3CH_2CH_2CH_2—$.

33. Which of the following molecules is planar ?

(*a*) Ethane (*b*) Methane
(*c*) Ethyne (*d*) Benzene.

34. C—C—C bond angle in benzene is about

(*a*) 109°. 28′ (*b*) 180°
(*c*) 120° (*d*) 100°.

35. The number of π electrons in anthracene is

(*a*) 10 (*b*) 12
(*c*) 14 (*d*) 8.

Nomenclature

• Open Chain Compounds

36. The correct IUPAC name of $CH_3CH_2CONHCH_3$ is

(*a*) N-Ethylethanamide (*b*) N-Methylethanamide
(*c*) N-Methylpropanamide (*d*) N-Ethylmethanamide.

37. The functional group in acyl chlorides is

(*a*) —Cl (*b*) —C(=O)—Cl
(*c*) —C(=O)—OCl (*d*) $—CH_2—OCl$.

38. IUPAC name of $HCOOCH_3$ is

(*a*) Methoxy methanal (*b*) Ethanoic acid
(*c*) Methyl methanoate (*d*) Methoxy methane.

39. The secondary suffix in the IUPAC names of ketones is

(*a*) -al (*b*) -one
(*c*) -ol (*d*) -oate.

40. The IUPAC name of crotyl chloride is

(*a*) 1-Chloro-1-butene (*b*) 1-Chloro-2-butene
(*c*) 2-Chloro-2-butene (*d*) 3-Chloro-1-butene.

41. Which of the following represents the benzal group ?

(*a*) 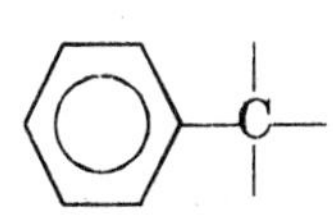

(*b*) 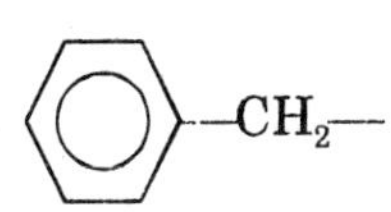

(*c*) 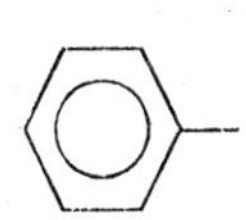
(*d*) 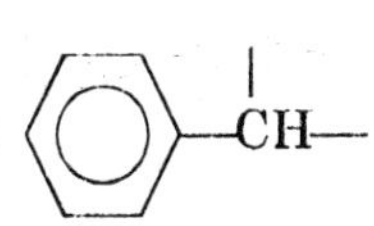

42. The formula of freon—12 is

(*a*) CCl_2F_2 (*b*) CH_2Cl_2
(*c*) CH_2F_2 (*d*) $CClF_3$.

43. IUPAC name of propargyl alcohol is

(*a*) Prop-2-yn-1-ol (*b*) Prop-2-en-1-ol
(*c*) Prop-1-en-2-ol (*d*) Prop-1-yn-2-ol.

44. The correct IUPAC name of the compound is

$$\begin{array}{l} \qquad\qquad\quad CH_3 \\ \qquad\qquad\quad\; | \\ CH_3CH_2—C = CH—CH—CH_2—CH_3 \\ \qquad\qquad\; | \\ CH_3CH_2—CH—CH_2—CH_2—CH_2—CH_3 \end{array}$$

(*a*) 5, 6-Diethyl-8-methyldec-6-ene
(*b*) 5, 6-Diethyl-3-methyldec-4-ene
(*c*) 6-Butyl-5-ethyl-3-methyloct-4-ene
(*d*) 2, 4, 5-Triethyl-3-ene.

45. Which of the following groups is mesyl group (Ms) ?

(*a*) $CH_3—\overset{\overset{O}{\uparrow}}{\underset{\underset{O}{\downarrow}}{S}}—$ (*b*) $CH_3—SO_3^-$

(*c*) $CH_3—O—\overset{\overset{O}{\uparrow}}{\underset{\underset{O}{\downarrow}}{S}}—$ (*d*) $CH_3—S^-$.

46. The IUPAC name of allylene is

(*a*) Propyne (*b*) Propene
(*c*) 2-Butyne (*d*) 1-Butyne.

47. The IUPAC name of isoprene is

(*a*) 1, 3-Butadiene (*b*) 2-Methyl-1, 3-butadiene
(*c*) 2-Methyl-1-butene (*d*) 3-Methyl-1-butene.

48. The IUPAC name of the following compound is

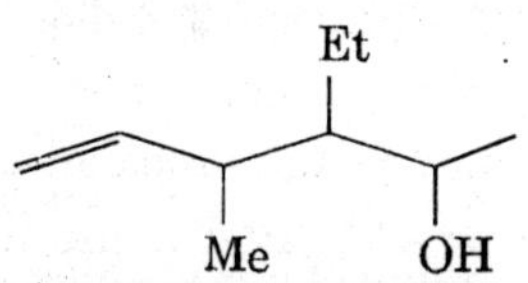

(*a*) 3-Ethyl-4-methyl hex-5-en-2-ol
(*b*) 4-Ethyl-3-methyl hex-1-en-5-ol
(*c*) 3-Methyl-4-ethyl hex-1-en-5-ol
(*d*) 4-Methyl-3-ethyl hex-5-en-2-ol.

49. Which of the following groups is always taken as a substituent in the nomenclature ?

(*a*) $-NH_2$ (*b*) —CN
(*c*) —CHO (*d*) $-NO_2$.

50. The principal functional group of the following compound is

$$HOCH_2\underset{\underset{Cl}{|}}{C}HCH_2-\overset{\overset{O}{||}}{C}-CH_2CHO$$

(*a*) – OH (*b*) – Cl
(*c*) – CHO (*d*) – CO –.

51. The structure of 2-methoxy-2-methyl propane is

(*a*) $H_3C—\overset{\overset{CH_3}{|}}{C}H—O—CH_3$

(*b*) $H_3C—\overset{\overset{CH_3}{|}}{\underset{\underset{CH_3}{|}}{C}}—O—CH_3$

(*c*) $CH_3—O—CH_2—CH_2—CH_3$
(*d*) $CH_3—CH_2—O—CH_2—CH_3$.

52. The IUPAC name of the compound,

$CH \equiv C—CH_2—CH_2—COOH$ is

(*a*) 1-pentyn-4-oic acid (*b*) Pentyn-1-oic acid
(*c*) 5-pentyn-1-oic acid (*d*) Pent-4-yn-1-oic acid.

53. The compound 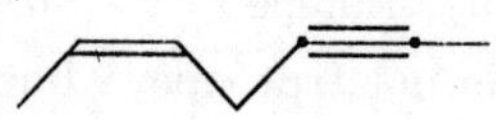 is named in IUPAC as

(*a*) 6-Octene-2-yne
(*b*) 1, 6-Dimethyl hex-1-en-4-yne
(*c*) 7-methyl 2-octene-6-yne
(*d*) oct-2-en-6-yne.

54. The compound 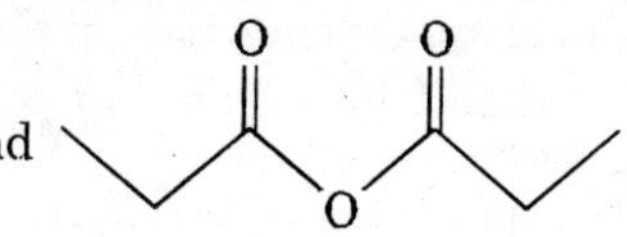 is named in IUPAC as

(*a*) 3, 4-hexane di-one (*b*) 3, 6-hexandione
(*c*) Propoxy propanone (*d*) Propanoic anhydride.

55. The IUPAC name of $CH_2 = CH—CH = CH—C \equiv CH$ is

(*a*) 3, 5-hexadien-5-yne (*b*) 1, 2-hexadien-1-yne
(*c*) 1, 3-hexadien-5-yne (*d*) 3, 5-hexadien-1-yne.

56. The correct IUPAC name of the compound $\begin{array}{c} CHO \\ | \\ CHO \end{array}$ is

(*a*) Glyoxal (*b*) Formyl methanoate
(*c*) Ethane-1, 2-dial (*d*) 1, 2-ethane dione.

57. Urea is named in IUPAC as

(*a*) Amino acetamide (*b*) diamino ketone
(*c*) Amino methanamide (*d*) 1-Amino ethanamide.

58. The IUPAC name of the compound 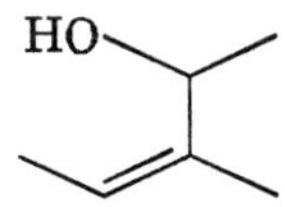 is

(*a*) 3, 4-dimethyl-2-butene-4-ol
(*b*) 1, 2-dimethyl-2-butenol
(*c*) 3-methyl pent-3-en-2-ol
(*d*) 2, 3-dimethyl-3-pentenol.

59. The IUPAC name of the compound

$CH_3—CH = C(Cl)—CH(CH_3)—CH_2—C \equiv CH$ is

(*a*) 5-chloro 4-methyl hept-5-en-1-yne
(*b*) 5-chloro 4-methyl hept-1-en-5-yne
(*c*) 3-chloro 4-methyl hept-2-en-6-yne
(*d*) 3-chloro 4-methyl hept-6-en-2-yne.

60. The IUPAC name of the compound $CBr_3CH_2CH_2CHO$ is

(*a*) 4, 4, 4-tri bromobutanal
(*b*) 3, 3, 3-tri bromobutanal
(*c*) 3, 3, 3-tri bromo butanaldehyde
(*d*) 4, 4, 4-tri bromo butanaldehyde.

61. The IUPAC name of the compound

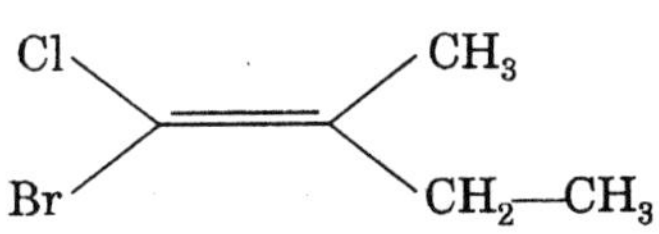

is

(*a*) *cis*-1-bromo-1-chloro-2-methyl but-1-ene.
(*b*) *trans*-1-chloro-1-bromo 2-methyl butene
(*c*) *trans*-4-chloro-4-bromo 2-methyl but-3-ene
(*d*) *cis*-4-chloro-4-bromo 2-methyl but-3-ene.

62. What is the IUPAC name of compound having following structure ?

$CH_2(CN)—CH(CN)—CH_2(CN)$

(*a*) 1, 2, 3-Cyanopropane
(*b*) 1, 2, 3-Tricyanopropane
(*c*) Propane-1, 2, 3-tricarbonitrile
(*d*) 3-Cyano-1, 5 pentane dinitrile.

63. The correct IUPAC name of compound having structural formula $CH_3—CH(OH)—COOH$ is

(*a*) Lactic acid
(*b*) 2-hydroxy propanoic acid
(*c*) α-hydroxy propanoic acid
(*d*) carboxypropanol.

64. The correct IUPAC name of the compound

$HOOC—CH(COOH)—COOH$ is

(*a*) propane 1, 2, 3-tricarboxylic acid
(*b*) propane 1, 2, 3-trioic acid
(*c*) 2-carboxypropane dioic acid
(*d*) methane tricarboxylic acid.

65. The correct IUPAC name of the compound

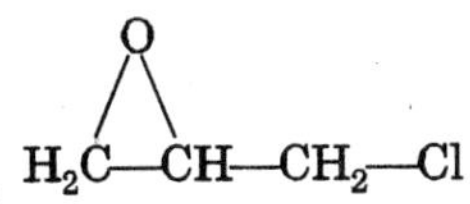

is

(*a*) 2-epoxy-1-chloropropane
(*b*) 2, 3-epoxy-1-chloropropane
(*c*) epoxy 1-chloropropane
(*d*) 3-chloro-1, 2-epoxy propane.

66. What is the IUPAC name of following compound ?

$H_3C—CH_2CH_2CH_2CH[CH(CH_3)CH(CH_3)CH_3]CH_2CH_2CH_2-CH_3$

(*a*) 6, 7-Dimethyl-5-butyl octane
(*b*) 2, 3-Dimethyl-1-4-butyl octane
(*c*) 5-(1, 2-Dimethyl propyl) nonane
(*d*) All these are correct.

67. The compound $OHC—C(=O)—OH$ will be named in IUPAC as

(*a*) Glyoxalic acid (*b*) Formyl methanoic acid
(*c*) 2-ketoethanoic acid (*d*) All these are correct.

68. The correct IUPAC name of the compound $CH_3CH(OH)CH_2C(OH(CH_3)_2$ is

(*a*) 2-methylpentane-2, 4-diol
(*b*) 4-methyl-2, 4-pentane-2, 4-diol
(*c*) 1, 1-dimethyl-1, 3-butanediol
(*d*) 1, 3, 3-trimethyl-1, 3-propanediol.

69. Choose the correct IUPAC name of the following compound

$H_3C—N(Me)—C(Me)(CH_2CH_3)—Et$

(*a*) N-Dimethyl 1-methyl-1-ethyl propane
(*b*) 3-Methyl-3-(N, N-dimethyl amino) pentane
(*c*) N-Dimethyl amino-3-methyl pentane
(*d*) None of the above.

70. The correct IUPAC name of the compound

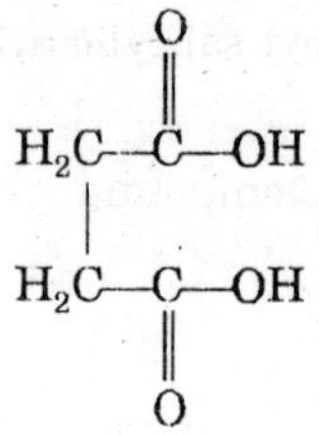

is

(*a*) succinic acid
(*b*) Butane dicarboxylate
(*c*) Butane-1, 4-dioic acid
(*d*) 1, 2-dicarboxy ethane.

71. The IUPAC name of the compound

is

(*a*) 2, 4, 6-heptatriene (*b*) 2, 4, 5-triheptene
(*c*) 2, 4, 6-triheptene (*d*) hepta-1, 3, 5-triene.

72. The IUPAC name of the compound

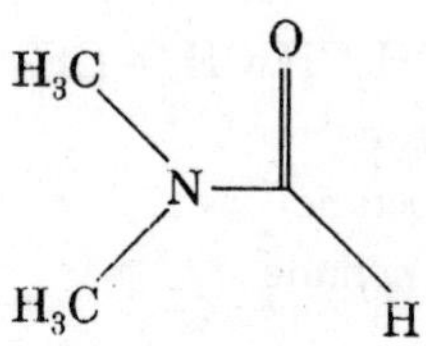

is

(*a*) N, N-dimethyl formamide
(*b*) N, N-dimethyl amino methanol
(*c*) N, N-dimethyl formamine
(*d*) N, N-dimethyl methanamide.

73. The correct IUPAC name of the following compound is

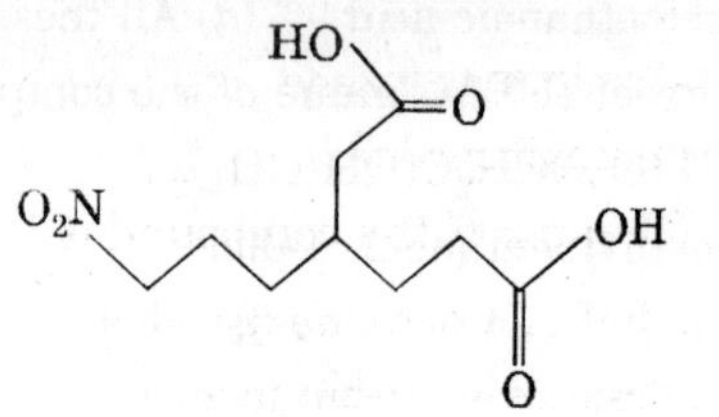

(*a*) 7-nitro-4 (carboxymethyl) heptanoic acid
(*b*) 6-nitro-3-(carboxy ethyl) hexanoic acid
(*c*) 4-(3-nitro propyl) hexane-1, 6-dioic acid
(*d*) 3-(3-nitro propyl) hexane-1, 6-dioic acid.

74. The IUPAC name of the compound

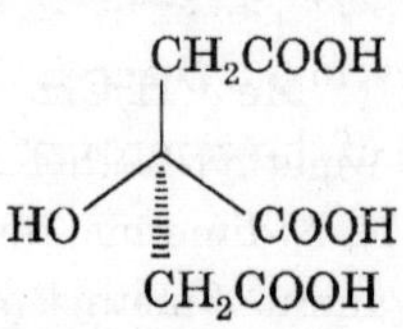

is

(*a*) 1, 2, 3-tricarboxy-2-propanol
(*b*) 3-carboxy-3-hydroxy-1, 5-pentanedioic acid
(*c*) Citric acid
(*d*) 2-hydroxy propane-1, 2, 3-tricarboxylic acid.

75. The IUPAC name of the compound

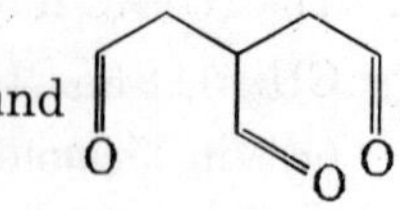

is

(*a*) 1, 2, 3-tri-formyl propane
(*b*) 3-formyl-1, 5-pentane dial
(*c*) Propane-1, 2, 3-trial
(*d*) Propane-1, 2, 3-tricarbaldehyde.

76. The correct IUPAC name of the following compound is

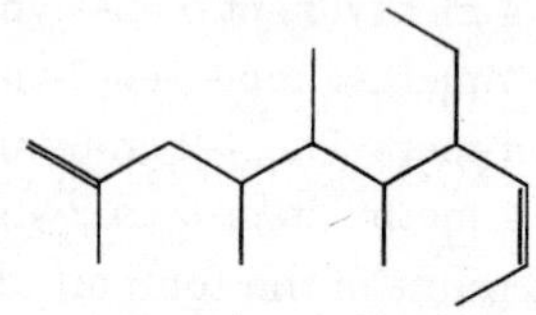

(*a*) 7-ethyl-2, 4, 5, 6-tetra methyl deca-1, 8-diene
(*b*) 4-ethyl-5, 6, 7, 9-tetra methyl deca-2, 9-diene
(*c*) 2, 4, 5, 6-tetramethyl-7-ethyl-deca-1, 7-diene
(*d*) None of these.

77. The IUPAC of the compound

$$H_3C—\overset{O}{\overset{\|}{C}}—\overset{OH}{\overset{|}{CH}}—\overset{NO_2}{\overset{|}{CH}}—CO_2H$$ is

(*a*) 2-nitro-3-hydroxy-4-oxopentanoic acid
(*b*) 2-nitro-3-hydroxy-4-pentanon-1-oic acid
(*c*) 4-amino-3-hydroxy-2-oxopentanoic acid
(*d*) 3-Hydroxy-2-nitro-4-oxopentanoic acid.

78. The correct IUPAC name of

$$H_3C—CH_2—\overset{CH_2}{\overset{\|}{C}}—CH_2—CH_2OH$$ is

(*a*) 2-ethyl prop-1-ene-2-ol
(*b*) 2-ethyl prop-2-ene-1-ol
(*c*) 2-allyl-1-butanoı
(*d*) 3-Ethyl but-3-en-1-ol.

79. The IUPAC name of the compound

$$H_2C = \overset{CH_3}{\overset{|}{C}}—CH_2—\overset{O}{\overset{\|}{C}}—OC_2H_5$$ is

(*a*) Ethyl aceto ethanoate
(*b*) ethyl methyl butenoate
(*c*) Ethyl acrylate
(*d*) Ethyl 3-methylbut-3-en-oate

80. The IUPAC name of the compound

is

(*a*) 1, 2-dicyano ethene (*b*) but-2-ene-1, 4-dinitrile
(*c*) di-but-2-en-1. 4-nitrile (*d*) None of these.

81. The IUPAC name of the compound
$CH_2 = CH—CN$ is
(*a*) Vinyl cyanide (*b*) acrylonitrile
(*c*) prop-2-ene nitrile (*d*) 2-propene-2-nitrile.

82. The correct IUPAC name of the compound $(C_2H_5)_3N$ is
(*a*) Ethyl tri-amine
(*b*) N, N, N-tri-ethyl amine
(*c*) N, N-diethyl ethanamine
(*d*) none of these.

83. The IUPAC name of the
$NH_2—CH—CH_2OH$ compound is
|
COOH
(*a*) 1-amino-2-hydroxy propanoic acid
(*b*) 2-amino-2-carboxy pentanol
(*c*) 2-amino-3-hydroxy propanoic acid
(*d*) 1-hydroxy-2-amino-3 propanoic.

84. The compound 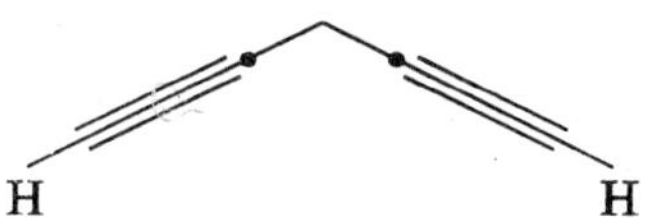 has the IUPAC name
(*a*) 1, 3-penta-di-yne (*b*) penta-1, 4-diyne
(*c*) Propane-1, 3-di-yne (*d*) pentane-1, 3-di-yne.

85. The correct IUPAC name of tartaric acid is
(*a*) 1, 4-dihydroxybutane-2, 3-dioic acid
(*b*) 2, 3-dihydroxybutane-1, 4-dioic
(*c*) 1, 4-dicarboxy-2, 3-dihydroxy ethane
(*d*) α, α′-dihydroxy butane-1, 4-dioic.

Cyclic Compounds

86. IUPAC name of 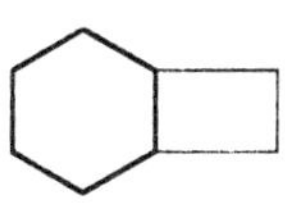 is
(*a*) Bicyclo [4.2.0] octane
(*b*) Bicyclo [4.2.2] octane
(*c*) Bicyclo [6.2.0] octane
(*d*) Bicyclo [4.2.2] decane.

87. IUPAC name of D.D.T. is
(*a*) Dichlorodiphenyl trichloro ethane
(*b*) Dichlorodiphenyl trichloro benzene
(*c*) Dichlorodiphenyl trichloro propane
(*d*) 1, 1, 1-Trichloro-2, 2-bis (*p*-chlorophenyl) ethane.

88. Apsirin is
(*a*) Methyl salicylic acid (*b*) Acetyl salicylic acid
(*c*) Acetyl salicylate (*d*) Phenyl salicylic acid.

89. Picric acid is a
(*a*) Phenol
(*b*) Carboxylic acid
(*c*) Hydroxy aliphatic compound
(*d*) None of the above.

90. The IUPAC name of the following compound is

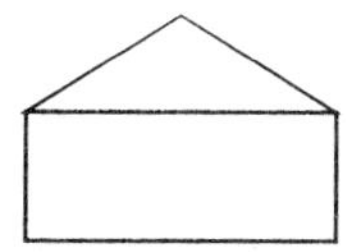

(*a*) Bicyclo [2.1.0] pentane
(*b*) Cyclopropyl cyclobutane
(*c*) Bicyclo [4.2.1] pentane
(*d*) Cyclopentene.

91. The IUPAC name of the compound,

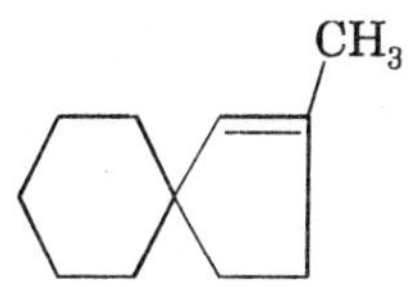

is
(*a*) 2-Methyl spiro [4.5] dec-1-ene
(*b*) 2--Methyl spiro [5.4] dec-1-ene
(*c*) 3-Methyl spiro [4.5] dec-2-ene
(*d*) 3-Methyl spiro [5.4] dec-2-ene.

92. IUPAC name of mesitylene is
(*a*) 1, 2, 3-Trimethylbenzene
(*b*) 1, 3, 5-Trimethylbenzene
(*c*) 1, 2, 4-Trimethylbenzene
(*d*) 1, 4-Dimethylbenzene.

93. The IUPAC name of the compound is

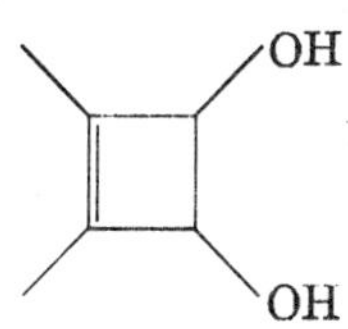

(*a*) 1, 2-dimethyl cyclo 3, 4-butene-di-ol
(*b*) 1, 2-dihydroxy 3, 4-dimethyl cyclo but-3-ene
(*c*) 2, 3-dimethyl cyclo but-2-ene-1, 4-diol
(*d*) 3, 4-dimethyl cyclo but-3-ene-1, 2-diol.

94. The IUPAC name of the compound

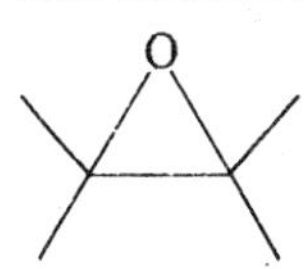

is

(a) Tetra methyl propylene oxide
(b) 1, 1, 2, 2-tetramethyl oxopropane
(c) 1, 1, 2, 2-tetramethyl epoxy propane
(d) 2, 3-Dimethyl-2, 3-epoxybutane.

95. What is correct IUPAC name of the following

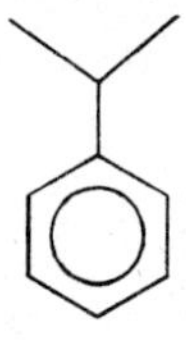

compound ?

(a) Cumene (b) Isopropyl benzene
(c) Phenyl isopropane (d) 2-Phenyl propane.

96. The correct IUPAC name of the compound

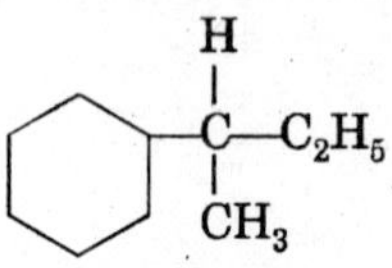

is

(a) 3-Cyclohexyl butane (b) 2-phenyl butane
(c) 3-phenyl butane (d) 2-cyclohexyl butane.

97. The correct IUPAC name of the compound

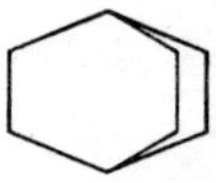

is

(a) bicyclo-(1, 1, 1) octane (b) bicyclo-(1, 2, 1) octane
(c) bicyclo-(2, 2, 1) octane (d) bicyclo-(2, 2, 2) octane.

98. The correct IUPAC name of the compound

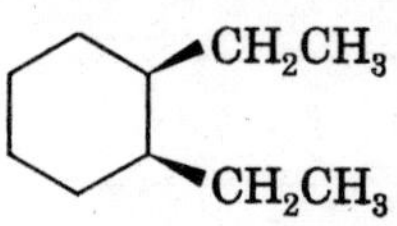

is

(a) *cis*-1, 2-diethyl cyclohexane
(b) *trans*-1, 2-diethyl cyclohexane
(c) *cis*-2, 3-diethyl cyclohexane
(d) *trans*-2, 3-diethyl cyclohexane.

99. The correct IUPAC name of the compound

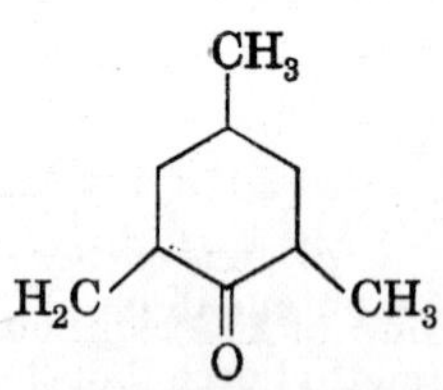

is

(a) 2, 4, 6-trimethyl cyclohexanone
(b) 1, 3, 5-trimethyl benzophenone
(c) Trimethyl cyclobutanone
(d) 1-keto-2, 4, 6-trimethyl cyclohexene.

100. The IUPAC name of the compound having following structure is

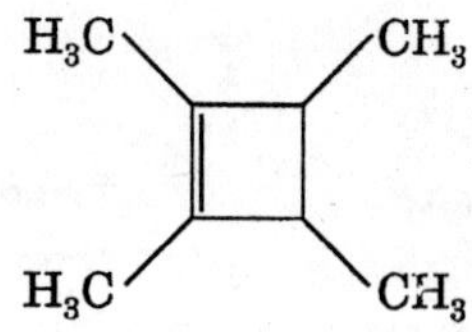

(a) 1, 2, 3, 4-tetramethyl cyclo but-1-ene
(b) 1, 2, 3, 4-tetramethyl-2-cyclobutene
(c) 1, 2, 3, 4-tetramethylene cyclobutane
(d) None of the above.

101. The IUPAC name of given structure is

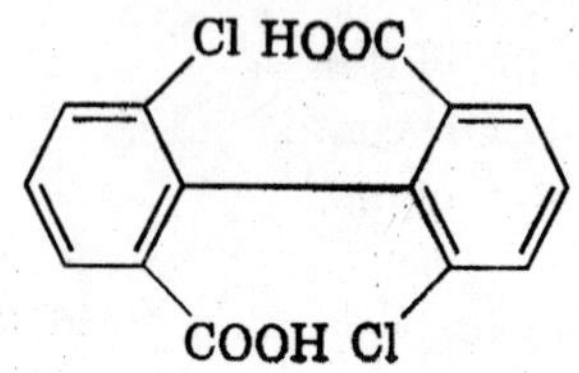

(a) Dichloro dicarboxyl biphenyl
(b) 2, 2′-dicarboxy diphenyl-6, 6′-dichlorine
(c) 6, 6′-dichlorobiphenyl-2, 2′-dicarboxylic acid
(d) None of the above.

102. The correct IUPAC name of the compound

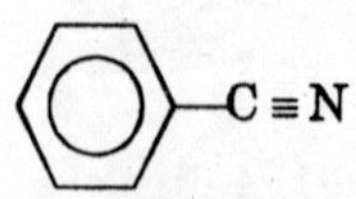

is

(a) Benzene carbonitrile (b) phenyl nitrile
(c) phenyl carbocyanide (d) benzene nitrile.

103. Cyclohexyl alcohol has which of the following structural formula ?

(a) 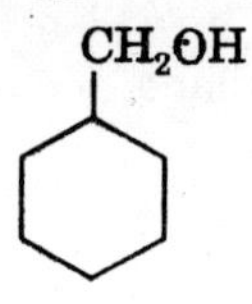

(b) 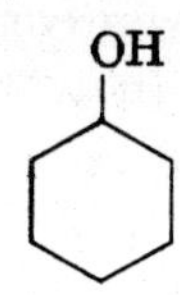

(c) 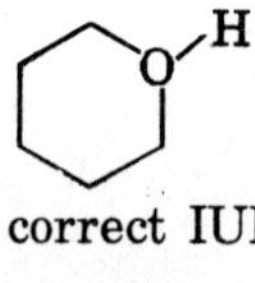

(d) none of these.

104. The correct IUPAC name of the compound

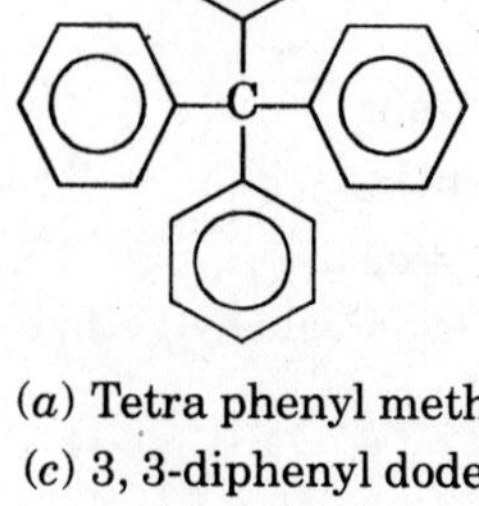

is

(a) Tetra phenyl methane (b) neophenyl
(c) 3, 3-diphenyl dodecane (d) none of these.

105. The IUPAC name of the compound 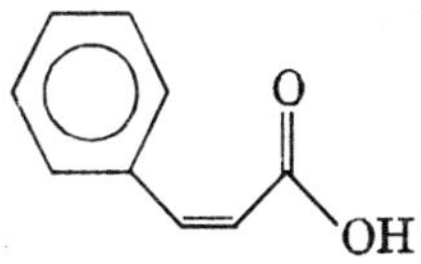 is

(*a*) Cinnamic acid
(*b*) 2-methyl-prop-2-enoic acid
(*c*) 2-phenyl-prop-2-enoic acid
(*d*) 3-phenyl-prop-2-enoic acid.

106. The IUPAC name of the following compound 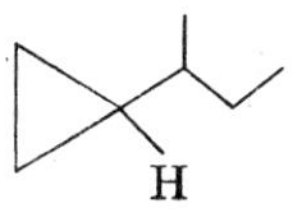 is

(*a*) 2-cyclopropyl butane
(*b*) 1-methyl-1-cyclopropyl propane
(*c*) 2-propyl butane
(*d*) propenyl butane.

107. The compound  has been named in IUPAC as

(*a*) 2-hydroxy cyclopentene
(*b*) cyclopent-2-en-1-ol
(*c*) hydroxy cyclopentene
(*d*) cyclopenten-1-ol.

108. The compound 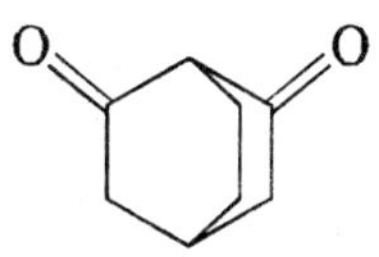 has been named in IUPAC as

(*a*) Bicyclo (1, 2, 1)-octane-2, 6-dione
(*b*) bicyclo (1, 2, 2)-octane-2, 6-dione
(*c*) Bicyclo (2, 2, 2)-octane-2, 6-dione
(*d*) bicyclo (1, 1, 1)-octane-2, 6-dione.

109. The compound 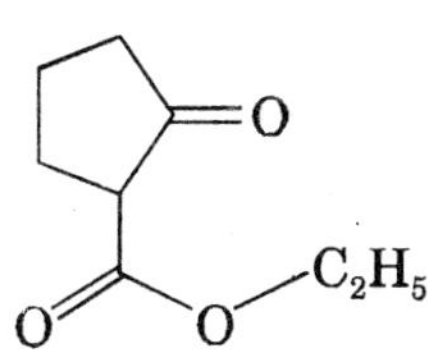 has been named in IUPAC as

(*a*) 2-carbethoxy cyclopentan-1-one
(*b*) 1-oxo-2-carbethoxy cyclopentane
(*c*) Carbethoxy cyclo pentanone
(*d*) none of these.

110. The IUPAC name of 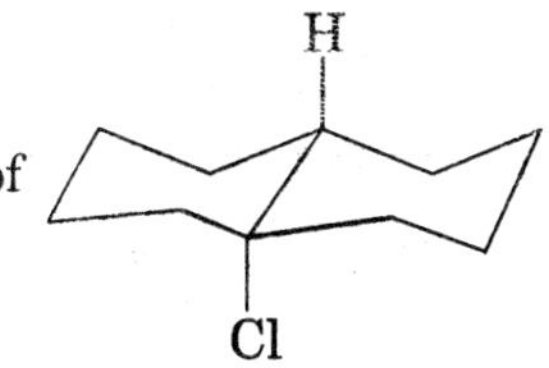 is

(*a*) 2-Chloro heptane
(*b*) Cis-chloro decalin
(*c*) Trans-chloro decalin
(*d*) 2-exa chloro bicyclo [2.2.0] octane.

111. The IUPAC name of 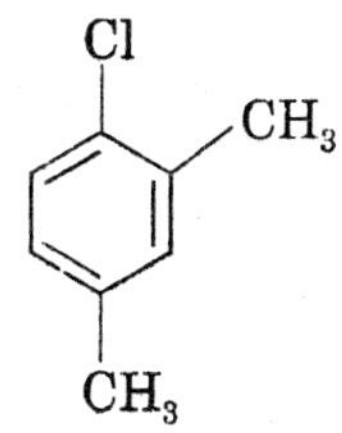 is

(*a*) 4-chloro-meta-xylene
(*b*) 2-Chloro-1, 4-dimethyl benzene
(*c*) 1-chloro-2, 4-dimethyl benzene
(*d*) 4-chloro-3-methyl toluene.

112. Which of the following is Cumene ?

(*a*) 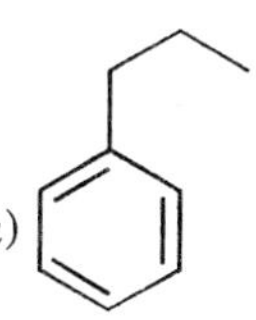 (*b*) 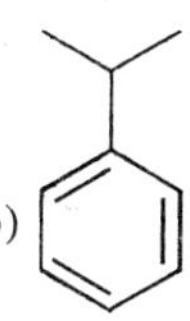

(*c*) 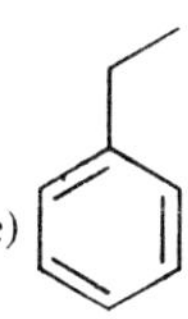 (*d*) 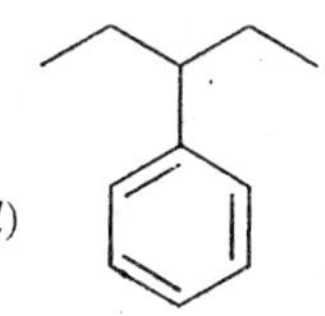

113. The IUPAC name of 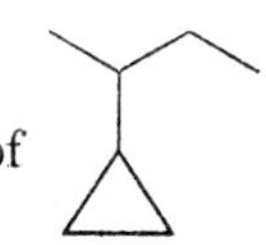 is

(*a*) 2-Δ-butane (*b*) 2-butyl cyclopropane
(*c*) 2-cyclopropyl butane (*d*) none of these.

114. The correct IUPAC name of the compound 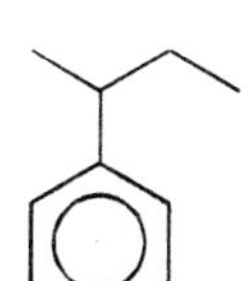 is

(*a*) 2-cyclohexyl butane (*b*) 3-cyclohexylbutane
(*c*) 2-phenylbutane (*d*) 3-phenylbutane.

115. Give the IUPAC name of the following compound

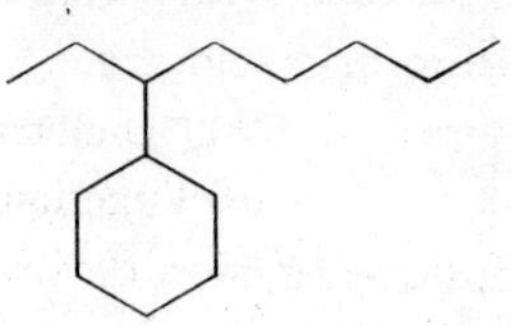

(*a*) Octyl benzene (*b*) Octyl cyclo hexane
(*c*) 3-cyclohexyl octane (*d*) 3-phenyl octane.

Stereochemistry

116. Which of the following compounds can exhibit enantiomerism ?
(*a*) 3-Hydroxypropanoic acid
(*b*) 3-Hydroxybutanoic acid
(*c*) 4-Hydroxybutanoic acid
(*d*) None of these.

117. Which isomer of C_6H_{14} has two isopropyl groups ?
(*a*) 2-Methylpentane (*b*) 3-Methylpentane
(*c*) 2, 3-Dimethylbutane (*d*) 2, 2-Dimethylbutane.

118. How many isomers are possible for methylanthracene ?
(*a*) 2 (*b*) 3
(*c*) 4 (*d*) 5.

119. Ethoxyethane and 1-methoxy propane are
(*a*) Chain isomers (*b*) Position isomers
(*c*) Metamers (*d*) Tautomers.

120. The total number of alkyl radical possible for the formula C_4H_9— is
(*a*) 3 (*b*) 4
(*c*) 5 (*d*) 6.

121. The simplest alkane capable of exhibiting chain isomerism has ...... carbon atoms.
(*a*) three (*b*) five
(*c*) four (*d*) six.

122. The compound which is not isomeric with diethyl ether is
(*a*) *n*-Propyl methyl ether (*b*) 1-Butanol
(*c*) 2-Methyl-2-propanol (*d*) Butanone.

123. *n*-Butane and *iso*-butane are .... isomers
(*a*) Chain (*b*) Position
(*c*) Geometric (*d*) Optical.

124. A straight chain hydrocarbon has the molecular formula C_8H_{10}. The hybridisation for the carbon atoms from one end of the chain to the other are respectively sp^3, sp^2, sp^2, sp^3, sp^2, sp^2, sp and sp. The structural formula of the hydrocarbon would be
(*a*) $CH_3-C\equiv C—CH_2—CH=CH—CH=CH_2$
(*b*) $CH_3—CH_2—CH=CH—CH_2—C\equiv C—CH=CH_2$
(*c*) $CH_3—CH=CH—CH_2—C\equiv C—CH=CH_2$
(*d*) $CH_3—CH=CH—CH_2—CH=CH—C\equiv CH$.

125. How many structural isomers are of $C_4H_{10}O$?
(*a*) 6 (*b*) 7
(*c*) 4 (*d*) 5.

126. Number of dibromo derivatives possible for propane are
(*a*) 2 (*b*) 3
(*c*) 1 (*d*) 4.

127. How many isomers are there of C_5H_{12} ?
(*a*) 4 (*b*) 3
(*c*) 5 (*d*) 2.

128. How many isomeric pentylenes (C_5H_{10}) are possible ?
(*a*) 3 (*b*) 4
(*c*) 5 (*d*) 6.

129. The maximum number of isomeric alkenes with the molecular formula C_4H_8 is
(*a*) 2 (*b*) 3
(*c*) 4 (*d*) 5.

130. Which of the following alkanes can form only two monochloro derivatives ?
(*a*) Hexane (*b*) 2-Methyl pentane
(*c*) 2, 3-Dimethylbutane (*d*) 3-Methyl pentane.

131. The number of monosubstitution derivatives possible for naphthalene is
(*a*) one (*b*) two
(*c*) three (*d*) four.

132. Which of the following is the functional isomer of methyl acetate ?
(*a*) Ethyl acetate (*b*) Propanoic acid
(*c*) Ethyl ethanoate (*d*) Propanone.

133. The functional isomers of ethers are
(*a*) Ketones (*b*) Aldehydes
(*c*) Alcohols (*d*) Esters.

134. The number of dichloro derivatives of naphthalene is
(*a*) Ten (*b*) Fourteen
(*c*) Six (*d*) Five.

135. The functional isomer of ethane nitrile is
(*a*) Ethyl carbylamine (*b*) Methane nitrile
(*c*) Methyl carbylamine (*d*) Propane nitrile.

136. Which of the following pairs of compounds are tautomers ?
(*a*) Propanol and propanone
(*b*) 1-Propanol and 2-propanol
(*c*) Ethanol and vinyl alcohol
(*d*) Vinyl alcohol and ethanal.

137. Anthracene is isomeric with
(*a*) Phenanthrene (*b*) Naphthalene
(*c*) Benzene (*d*) None of these.

138. Which of the following compounds is isomeric with trimethyl amine ?
(*a*) 1-Propanamine (*b*) 2-Propanamine
(*c*) Both (*b*) None.

Conformations

139. Which of the following conformations of *n*-butane is the least stable ?
(*a*) Gauche (*b*) Anti
(*c*) Eclipsed (*d*) Fully eclipsed.

140. The dihedral angle between two methyl groups in partially eclipsed conformation of *n*-butane is
(*a*) 180° (*b*) 120°
(*c*) 90° (*d*) 109° 28′.

141. Total number of conformations of ethane are
(*a*) 2 (*b*) 3
(*c*) 4 (*d*) Infinite.

142. Which of the following conformations of butane is most stable ?
(*a*) Skew (*b*) Anti
(*c*) Gauche (*d*) Eclipsed.

143. Dihedral angle between two methyl groups in gauche conformation of *n*-butane is
(*a*) 109° 28′ (*b*) 120°
(*c*) 180° (*d*) 60°.

144. The least stable conformation of cyclohexane is
(*a*) Boat (*b*) Chair
(*c*) Twist boat (*d*) Half chair.

145. Which of the following compounds is expected to be most reactive ?
(*a*) Cyclohexane (*b*) Cyclopropane
(*c*) Cyclobutane (*d*) Cyclopentene.

146. The energy difference between boat and chair conformations of cyclohexane is about
(*a*) 150 kJ/mole (*b*) 30 kJ/mole
(*c*) 12.5 kJ/mole (*d*) 50 kJ/mole.

147. The energy difference between staggered and eclipsed conformations of ethane is about
(*a*) 25 kJ/mole (*b*) 30 kJ/mole
(*c*) 100 kJ/mole (*d*) 12.5 kJ/mole.

Geometrical isomers

148. *cis*-2-Butene and *trans*-2-butene can be distinguished on the basis of
(*a*) their hydrogenation products
(*b*) their optical rotation
(*c*) their ozonolysis products
(*d*) their addition products with bromine.

149. Maleic acid and fumaric acid are
(*a*) Position isomers (*b*) Geometric isomers
(*c*) Enantiomers (*d*) Functional isomers.

150. Which of the following isomerisms is shown by alkenes but not by alkanes
(*a*) Conformational (*b*) Optical
(*c*) Geometric (*d*) Chain.

151. Geometric isomerism will be exhibited by
(*a*) 1-Pentene (*b*) 3-Methyl-1-butene
(*c*) 2-Pentene (*d*) All of these compounds.

152. Which of the following compounds can show geometric isomerism ?
(*a*) Vinyl chloride (*b*) 1, 1-Dichloroethene
(*c*) 1, 2-Dichloroethene (*d*) Trichloroethene.

153. The simplest alkene which is capable of exhibiting geometric isomerism has
(*a*) 3 carbon atoms (*b*) 5 carbon atoms
(*c*) 6 carbon atoms (*d*) 4 carbon atoms.

154. Which of the following compounds does not exhibit geometric isomerism ?
(*a*) 1, 1-Dichloro-2-butene (*b*) 1, 2-Dichloro-2-butene
(*c*) 1, 1-Dichloro-1-butene (*d*) 2, 3-Dichloro-2-butene.

155. Which of the following compounds can exhibit geometrical isomerism ?
(*a*) 1-Hexene (*b*) 2-Methyl-2-pentene
(*c*) 3-Methyl-1-pentene (*d*) 2-Hexene.

Optical isomers

156. A racemic mixture contains equal ...... of dextrorotatory and laevorotatory isomers
(*a*) No. of molecules (*b*) Masses
(*c*) No. of moles (*d*) All the above are correct.

157. Which of the following can exist as diastereomers ?
(*a*) Lactic acid (*b*) 1-Butene
(*c*) 2-Butene (*d*) Ethane.

158. Which of the following can exist as enantiomers ?
(*a*) CH_3COOH (*b*) $CH_3 . CH(OH) . COOH$
(*c*) $CH_3 . CH_2 . COOH$ (*d*) $HOOC . CH_2 . COOH$.

159. Which of the following compounds can exist as enantiomers ?
(*a*) α-Chloro propionic acid
(*b*) β-Chloro propionic acid
(*c*) Isobutyric acid
(*d*) Propionic acid.

160. A compound whose molecules are superimposable on their mirror images even though they contain asymmetric carbon atoms is called

(a) A threo compound
(b) An erythro compound
(c) A dissymmetric compound
(d) A meso compound.

161. Which of the following statements is false ?
(a) Enantiomers have same m.p. and b.p.
(b) A mixture containing equal amounts of enantiomers is optically inactive
(c) Enantiomers have identical chemical properties
(d) A mixture containing two enantiomers can be separated into fractions containing pure enantiomers.

162. The simplest alkane which can exhibit enantiom-erism has
(a) 5 carbon atoms (b) 6 carbon atoms
(c) 7 carbon atoms (d) 4 carbon atoms.

163. Which of the following compounds can exhibit optical activity ?

(a) 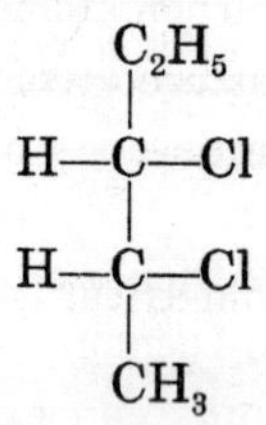

(b) 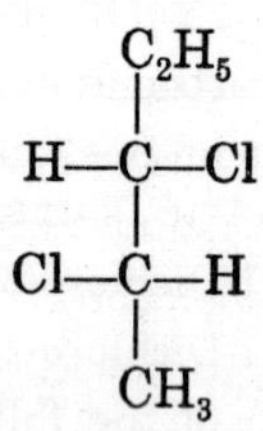

(c) 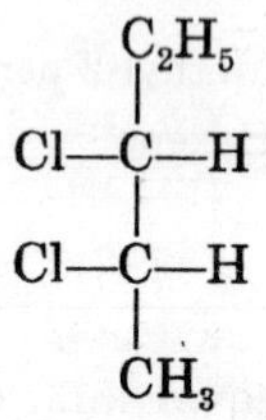

(d) All of these.

164. The number of stereoisomers possible for 2, 3-pentanediol is
(a) two (b) three
(c) four (d) five.

165. Which of the following does not contain any asymmetric carbon but can show enantiomerism ?
(a) Lactic acid (b) 1, 3-Pentadiene
(c) Tartaric acid (d) 2, 3-Pentadiene.

166. The specific rotation of α-D-glucose is + 112° and that of β-D-glucose is + 19°. When either of these two forms of glucose is dissolved in water, the specific rotation changes to + 52°. This phenomenon is called
(a) Inversion (b) Resolution
(c) Mutarotation (d) Mutation.

167. Which of the following is a chiral molecule ?
(a) 1-Chloropropane (b) 2-Chloropropane
(c) 1-Chlorobutane (d) 2-Chlorobutane.

168. Enantiomerism was discovered by
(a) Vant Hoff and Le Bell (b) Louis Pasteur
(c) Berzelius (d) Kolbe.

169. Which of the following represents same enantiomer as

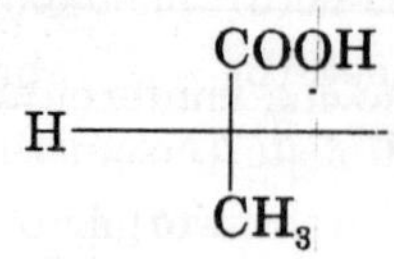

(a) 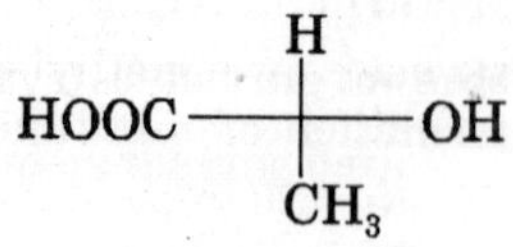

(b) 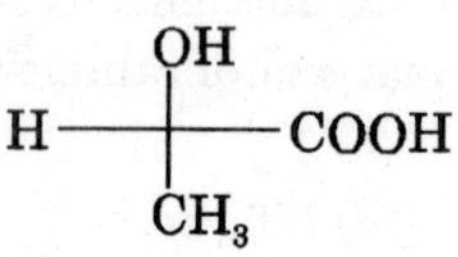

(c) 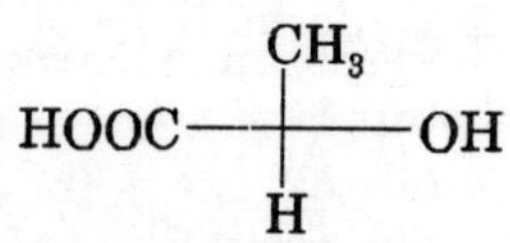

(d) None of these represents the same enantiomer.

170. (+)-Tartaric acid and *meso*-tartaric acid are
(a) Enantiomers (b) Diastereomers
(c) Geometric isomers (d) None of these.

171. Which of the following compounds can exist in *meso*-form ?
(a) 1, 2-Dichlorobutane (b) 2, 3-Dichloropentane
(c) 2, 3-Dichlorobutane (d) 1, 2-Dichloropentane.

172. The simplest alcohol that can exhibit enantiomerism is
(a) 1-Propanol (b) 1-Butanol
(c) 2-Propanol (d) 2-Butanol.

173. Which of the following can exhibit enantiomerism ?
(a) *cis*-1, 2-Dichlorocyclobutane
(b) *trans*-1, 2-Dichlorocyclobutane
(c) *cis*-1, 3-Dichlorocyclobutane
(d) *trans*-1, 3-Dichlorocyclobutane.

174. The angle of rotation of plane polarized light in polarimeter depends upon
(a) Conc. of the substance
(b) Length of the tube
(c) Nature of the substance
(d) All of these.

175. Which of the following compounds exhibit tautomerism ?

(a) Chloroethane (b) Ethanol
(c) Ethoxyethane (d) Nitroethane.

176. The instrument used for measuring specific rotation is
(a) Spectrometer (b) Polarimeter
(c) Lactometer (d) Ammeter.

177. A certain compound contain three chirality centres. The maximum number of stereoisomers it can have is
(a) 8 (b) 6
(c) 9 (d) 3.

178. *d* and *l* forms of tartaric acid are mixed in equal proportions. The mixture is called
(a) Eutectic mixture (b) Azeotropic mixture
(c) Diastereomers (d) Racemic mixture.

179. Which of the following compound will show stereoisomerism ?
(a) 4-Chlorobutan-1-ol (b) Tertiary butyl alcohol
(c) *sec*-butyl amine (d) *n*-butyl alcohol.

180. Which of the following is responsible for the inability of meso compound to show optical activity ?
(a) Absence of chirality centre
(b) Presence of more than one chirality centres
(c) Disymmetric nature of its structure
(d) Internal compensation.

181. The process of separation of (*d*, *l*)-mixture into its contituents is referred to as
(a) Fractional distillation (b) Resolution
(c) Fragmentation (d) Sedimentation.

182. Which of the following compound has chiral structure ?
(a) 2-Methylpentane (b) 2, 3-Dimethylpentane
(c) Isoamyl alcohol (d) Trimethylamine.

183. Among the following structures which one is called Erythro isomer ?

(i)
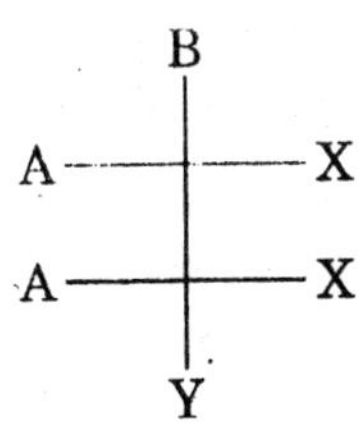

(ii)
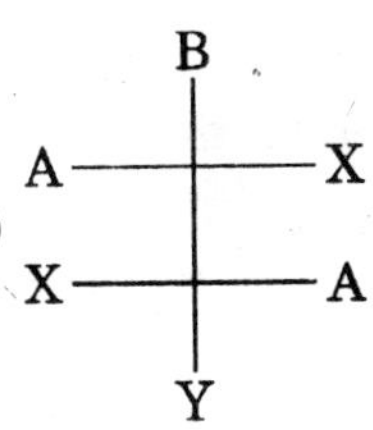

(iii)
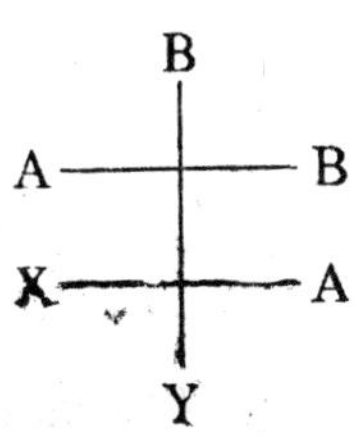

(a) (i) and (ii) (b) (ii)
(c) (i) (d) only (iii).

184. A compound with formula $CH_3(CHOH)_2\,COOH$ can have a maximum of four stereoisomers out of which the meso structures would be
(a) 2 (b) 1
(c) 0 (d) 3.

185. The structures given below are

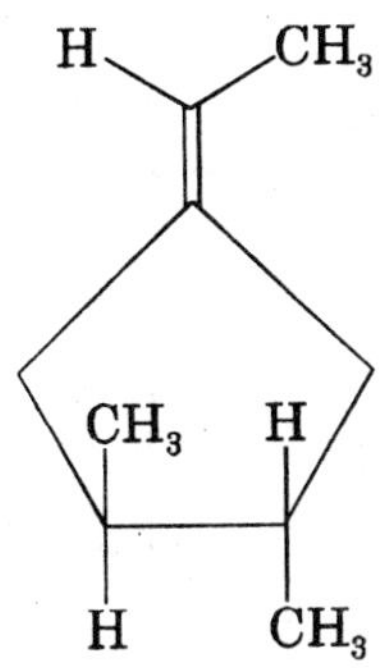

and

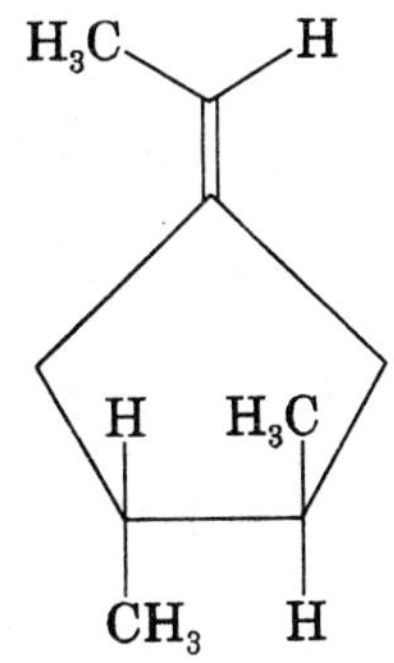

(a) Enantiomers (b) Diastereomers
(c) Geometrical forms (d) Identical.

186. Maleic acid and fumaric acid are
(a) Homologous (b) Geometrical isomers
(c) Enantiomers (d) Diastereomers.

187. How many chirality centres are present in the structure 1-chloro-2-methylcyclohexane ?
(a) 2 (b) 3
(c) 4 (d) 1.

188. The compound
$CH_3C(Cl) = CH(CH_2)_4\,CH(OH)CH_3$ can have
(a) 4-geometrical, 4-optical isomers
(b) 4-geometrical, 2-optical isomers
(c) 2-geometrical, 4-optical isomers
(d) 2-geometrical, 2-optical isomers.

189. Which of the following compound has chiral structure ?
(a) $CH_3CH(OH)CH_3$
(b) $CH_2 = CH — CH(NH_3)CH_3$
(c) $CH_3CH_2\,CH(Br)CH_2CH_3$
(d) $(C_2H_5)_2\,CH—Br$.

190. Which of the following structures are chiral ?

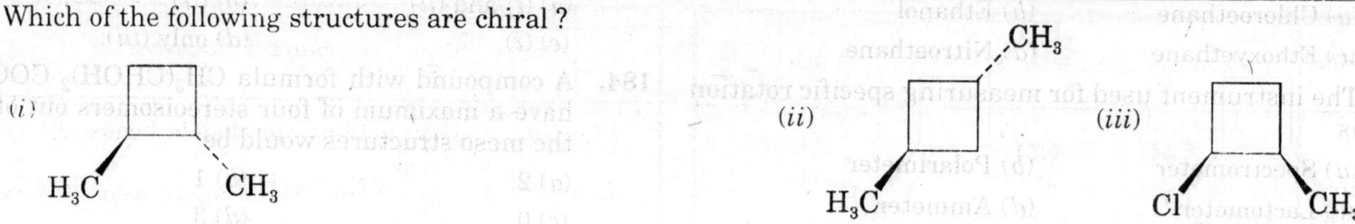

(a) (i) and (iii) (b) all the three
(c) (i) and (ii) (d) (ii) and (iii).

ANSWERS

1. (c)	**2.** (c)	**3.** (b)	**4.** (a)	**5.** (c)	**6.** (b)	**7.** (b)	**8.** (a)	**9.** (c)
10. (c)	**11.** (c)	**12.** (b)	**13.** (b)	**14.** (d)	**15.** (d)	**16.** (c)	**17.** (c)	**18.** (b)
19. (c)	**20.** (c)	**21.** (a)	**22.** (d)	**23.** (c)	**24.** (d)	**25.** (c)	**26.** (c)	**27.** (b)
28. (a)	**29.** (b)	**30.** (c)	**31.** (c)	**32.** (b)	**33.** (d)	**34.** (c)	**35.** (c)	**36.** (c)
37. (b)	**38.** (c)	**39.** (b)	**40.** (b)	**41.** (d)	**42.** (a)	**43.** (a)	**44.** (b)	**45.** (a)
46. (a)	**47.** (b)	**48.** (a)	**49.** (d)	**50.** (c)	**51.** (b)	**52.** (d)	**53.** (d)	**54.** (d)
55. (c)	**56.** (c)	**57.** (c)	**58.** (c)	**59.** (a)	**60.** (a)	**61.** (a)	**62.** (c)	**63.** (b)
64. (d)	**65.** (d)	**66.** (c)	**67.** (b)	**68.** (a)	**69.** (b)	**70.** (c)	**71.** (d)	**72.** (d)
73. (d)	**74.** (d)	**75.** (d)	**76.** (a)	**77.** (d)	**78.** (d)	**79.** (d)	**80.** (b)	**81.** (c)
82. (c)	**83.** (c)	**84.** (b)	**85.** (b)	**86.** (a)	**87.** (a)	**88.** (b)	**89.** (a)	**90.** (a)
91. (c)	**92.** (b)	**93.** (d)	**94.** (d)	**95.** (d)	**96.** (d)	**97.** (d)	**98.** (a)	**99.** (a)
100. (a)	**101.** (c)	**102.** (a)	**103.** (b)	**104.** (a)	**105.** (d)	**106.** (a)	**107.** (c)	**108.** (c)
109. (a)	**110.** (b)	**111.** (c)	**112.** (b)	**113.** (c)	**114.** (c)	**115.** (c)	**116.** (b)	**117.** (c)
118. (b)	**119.** (c)	**120.** (b)	**121.** (c)	**122.** (d)	**123.** (a)	**124.** (d)	**125.** (b)	**126.** (d)
127. (b)	**128.** (d)	**129.** (c)	**130.** (c)	**131.** (b)	**132.** (b)	**133.** (c)	**134.** (a)	**135.** (a)
136. (d)	**137.** (a)	**138.** (c)	**139.** (d)	**140.** (b)	**141.** (d)	**142.** (b)	**143.** (d)	**144.** (d)
145. (b)	**146.** (b)	**147.** (d)	**148.** (d)	**149.** (b)	**150.** (c)	**151.** (c)	**152.** (c)	**153.** (d)
154. (c)	**155.** (d)	**156.** (d)	**157.** (c)	**158.** (b)	**159.** (a)	**160.** (d)	**161.** (c)	**162.** (c)
163. (d)	**164.** (c)	**165.** (d)	**166.** (c)	**167.** (d)	**168.** (b)	**169.** (c)	**170.** (b)	**171.** (c)
172. (d)	**173.** (b)	**174.** (d)	**175.** (d)	**176.** (b)	**177.** (a)	**178.** (d)	**179.** (c)	**180.** (d)
181. (b)	**182.** (b)	**183.** (c)	**184.** (c)	**185.** (a)	**186.** (b)	**187.** (a)	**188.** (c)	**189.** (b)
190. (a).								

HINTS/SOLUTIONS

Q. No.	Ans.	Hints & Solutions
1.	**(c)**	Self explanatary (definition).
2.	**(c)**	 ; In 1, 4-dichlorobenzene, the individual dipoles of two C—Cl bonds cancel each other.
3.	**(b)**	$\overset{sp^2}{CH_2} = \overset{sp^2}{CH} — \overset{sp}{C} \equiv N.$
4.	**(a)**	Paraffins means less reactive.
5.	**(c)**	All the carbon atoms in propane are in sp^3 hybrid state and hence C—C—C bond angle should be about 109°.28′.

Q. No.	Ans.	Hints & Solutions
6.	(b)	$H_3C-C(\mathbf{CH_3})_2-CH_2-CH(\mathbf{CH_3})-\mathbf{CH_3}$ H atoms indicated in bold type are 1° or primary H-atoms.
7.	(b)	neopentane is $CH_3-C(CH_3)_2-CH_3$ and contains only primary carbons and hence primary H atoms.
8.	(a)	It was wohler who accidently prepared urea by heating ammonium cyanate.
9.	(c)	It was proposed by Berzelius in 1815.
10.	(c)	Organic compounds being non-polar are soluble in non-polar solvents.
11.	(c)	Homologues differ by CH_2 group.
12.	(b)	Alkynes contain a triple bond and have general formula C_nH_{2n-2}.
13.	(b)	$CH_2=CH-CH=CH-C(H)=O$
14.	(d)	Acetylene has a triple bond between two carbon atoms.
15.	(d)	$CH_3-C(CH_3)_2-H$, There are 9 primary H-atoms and 1 tertiary H-atom. There is no secondary H-atom.
16.	(c)	It was Vant Hoff and Le-Bel.
17.	(c)	Methane is named Marsh gas.
18.	(b)	C_2H_2, because it has carbon-carbon triple bond.
19.	(c)	C_2H_2, because it involves overlap of *sp* hybrid orbital of C with 1*s* orbital of hydrogen atom.
20.	(c)	Individual C—C bond order in benzene is between one and two because of resonance.
21.	(a)	C_nH_{2n} represents alkene as well as cycloalkane.
22.	(d)	$H-C\equiv C-H$ (π, π) 2π; $H-C-C-H$ (σ σ σ) 3σ
23.	(c)	More alkyl substituted alkenes are more stable due to hyperconjugation.
24.	(d)	Conjugated means alternate double bonds.

Q. No.	Ans.	Hints & Solutions
25.	(c)	Carbon holding the double bonds are sp^2 hybridised.
26.	(c)	In *trans*-2-butene, the bond dipoles of individual bonds cancel each other.
27.	(b)	Physical properties of various homologues show regular gradation with increase in molecular mass.
28.	(a)	Isoprene is $CH_2=C(CH_3)-CH=CH_2$. It is a diene.
29.	(b)	1, 2-butadiene is $CH_2=C=CH-CH_3$.
30.	(c)	In propyne C—C single bond is formed by $sp-sp^3$ overlap as compared with sp^3-sp^3 (in propane) and sp^2-sp^3 (in propene).
31.	(c)	$CH_3-CH(CH_3)-CH(CH_3)-CH_3$ It contains two tertiary hydrogen atoms.
32.	(b)	See concept oriented details.
33.	(d)	In benzene all carbon atoms are sp^2 hybridised and hence it is planar.
34.	(c)	Since each carbon in benzene is sp^2 hybridised.
35.	(c)	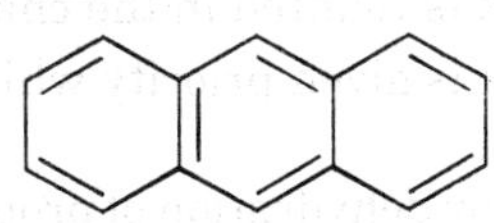 14 π-electrons
36.	(c)	It is a substituted amide with longest chain of three carbons.
37.	(b)	Acid chlorides are called acyl halides.
38.	(c)	$H-C(=O)-O-CH_3$ Methylmethanoate
39.	(b)	It is one.
40.	(b)	$\overset{4}{C}H_3-\overset{3}{C}H=\overset{2}{C}H-\overset{1}{C}H_2Cl$ (Crotyl chloride) 1-Chloro-2-butene
41.	(d)	Benzal group is $C_6H_5-\overset{\mid}{C}H-$ (divalent)
42.	(a)	CCl_2F_2 is Freon-12, CCl_3F is Freon-11 $CClF_3$ is Freon-13 and CF_4 is Freon-14.
43.	(a)	$H-\overset{3}{C}\equiv\overset{2}{C}-\overset{1}{C}H_2OH$ Prop-2-yn-1-ol.
44.	(b)	Longest chain rule is followed along with term by term comparison.
45.	(a)	Factual question.
46.	(a)	Allylene is common name of propyne.

Q. No.	Ans.	Hints & Solutions
47.	(*b*)	$CH_2 = C(CH_3)—CH = CH_2$ (*isoprene*) 2-Methyl-1, 3-butadiene
48.	(*a*)	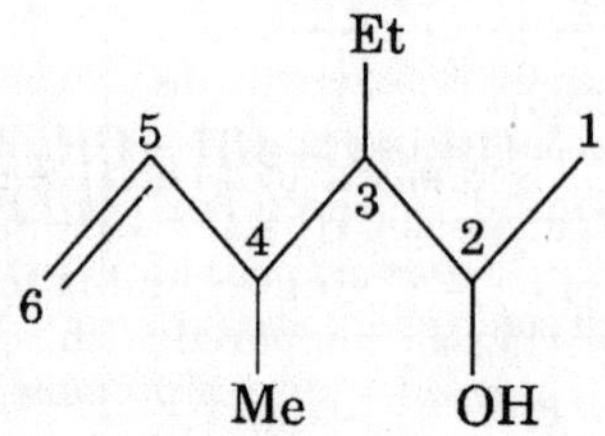
49.	(*d*)	$—NO_2$ group is always taken as substituent and is indicated by prefix nitro.
50.	(*c*)	Among the given groups, —CHO has the highest priority.
51.	(*b*)	Longest chain rule is followed and OCH_3 is taken as substituent on the chain.
52.	(*d*)	COOH group is given priority and the carbon of this group is counted in the chain.
53.	(*d*)	Double bond is given priority while numbering of chain.
54.	(*d*)	It is formed by dehydration of propanoic acid.
55.	(*c*)	Double bond is preferred over triple bond.
56.	(*c*)	Like group like treatment rule.
57.	(*c*)	Urea is $NH_2—\overset{\overset{O}{\parallel}}{C}—NH_2$.
58.	(*c*)	The chain should have double bond as well as OH group.
59.	(*a*)	In this case triple bond is preferred over double bond according to locant rule.
60.	(*a*)	CHO group is given priority.
61.	(*a*)	Carbon having substituted halogen atoms is given preference.
62.	(*c*)	Like group like treatment rule.
63.	(*b*)	Carbon of COOH group is numbered 1.
64.	(*d*)	Like group like treatment rule.
65.	(*d*)	Bridged oxygen is preferred and the name is written in alphabetic order.
66.	(*c*)	Long chain rule.
67.	(*b*)	CHO is taken as substituent.
68.	(*a*)	Locant rule is followed.
69.	(*b*)	Longest chain of five carbon is taken.
70.	(*c*)	Like group like treatment rule.
71.	(*d*)	It is a triene.
72.	(*d*)	The parent functional group is amide.
73.	(*d*)	Smaller groups are treated alike and also substituent 3-nitro propyl should get lowest position.
74.	(*d*)	Smaller groups are treated alike.
75.	(*d*)	Like group like treatment.
76.	(*a*)	Longest chain is taken as with both the double bonds in the chain.
77.	(*d*)	See concepts in brief.
78.	(*d*)	See concepts in brief.
79.	(*d*)	It is an unsaturated ester.
80.	(*b*)	The carbon of both CN groups is counted in chain.
81.	(*c*)	The carbon of CN group is numbered 1 as per IUPAC rules.
82.	(*c*)	Two ethyl group are taken as substituent on the N-atom.
83.	(*c*)	COOH group is parent and longest chain of three carbon atoms is taken.
84.	(*b*)	The given compound is $\overset{1}{HC}≡\overset{2}{C}—\overset{3}{CH_2}—\overset{4}{C}≡CH$
85.	(*b*)	Tartaric acid is $CH(OH)COOH$ | $CH(OH)COOH$
86.	(*a*)	It is bicyclic compound with bridges having 4, 2 and 0 carbon atoms.
87.	(*a*)	DT is 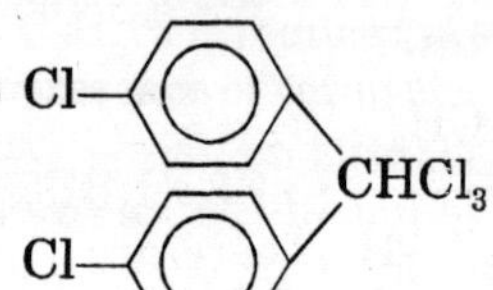
88.	(*b*)	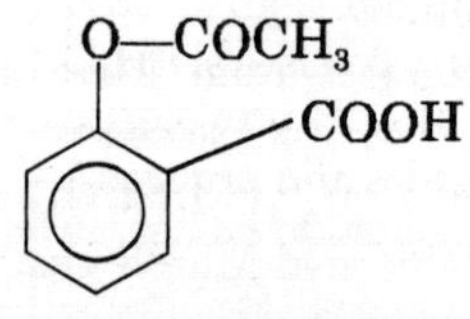 Acetyl salicylic acid [*Aspirin*]
89.	(*a*)	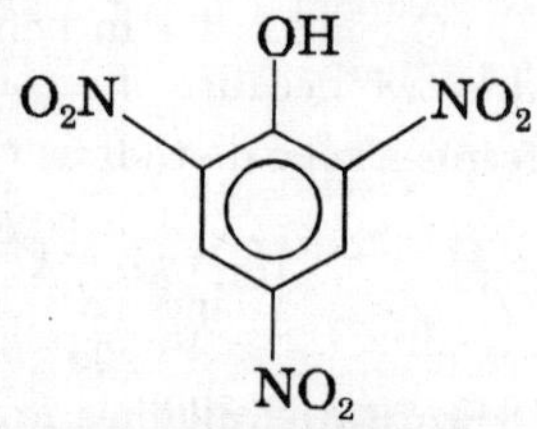 2, 4, 6-Trinitrophenol [*Picric acid*]

Q. No.	Ans.	Hints & Solutions
90.	(*a*)	It is a bicyclic compound, Bicyclo [2.1.0] pentane.
91.	(*c*)	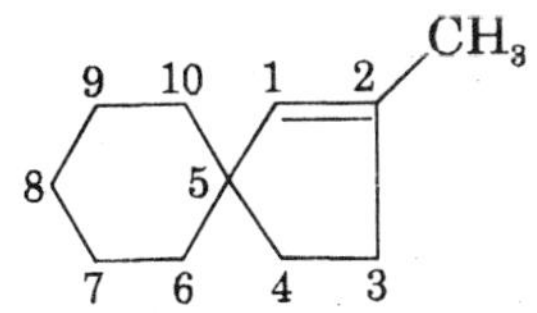
92.	(*b*)	Mesitylene is 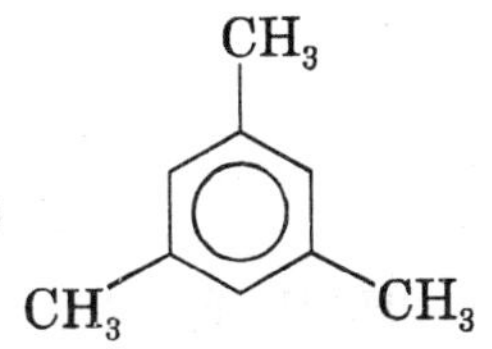
93.	(*d*)	The OH groups are given preference.
94.	(*d*)	Longest chain of 4 carbons is taken.
95.	(*d*)	The benzene ring is taken as substituent on the aliphatic chain.
96.	(*d*)	Cyclohexyl group is substituent on the chain.
97.	(*d*)	It is bicyclic compound.
98.	(*a*)	Wedge means groups point towards you out of plane of paper.
99.	(*a*)	The carbon of C=O group is numbered 1.
100.	(*a*)	The double bond is given preference while numbering.
101.	(*c*)	The carbon where the two rings are connected is numbered 1 and the carbon holding carboxylic group as 2 and so on.
102.	(*a*)	It is benzene carbonitrile.
103.	(*b*)	Cyclohexyl alcohol has a ring with OH group attached to it.
104.	(*a*)	The H atoms of CH_4 are replaced by benzene ring.
105.	(*d*)	Benzene ring is taken as substituent on the aliphatic chain.
106.	(*a*)	As aliphatic chain has more no. of carbon atoms than the ring therefore ring is taken as substituent on the chain.
107.	(*c*)	The carbon of ring carrying OH group is labelled as no. 1.
108.	(*c*)	The ring is numbered as : 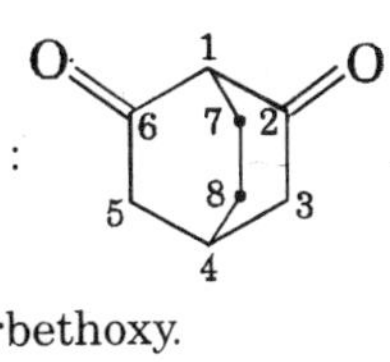
109.	(*a*)	$—COOC_2H_5$ is called carbethoxy.
110.	(*b*)	Two six membered saturated rings fused together are named decalin in IUPAC system.
111.	(*c*)	The carbon holding Cl is numbered 1.
112.	(*b*)	Cumene is isopropyl benzene.
113.	(*c*)	When no. of carbon atoms in the st. chain are more than those in the ring, the ring is taken as a substituent on the chain.

Q. No.	Ans.	Hints & Solutions
114.	(*c*)	Phenyl group is taken as substituent on aliphatic chain.
115.	(*c*)	Description is same as Q. No. 113.
116.	(*b*)	$CH_3—\overset{*}{C}H(OH)—CH_2COOH$; C—3 in 3-hydroxybutanoic acid is chiral centre.
117.	(*c*)	$CH_3—CH(CH_3)—CH(CH_3)—CH_3$
118.	(*b*)	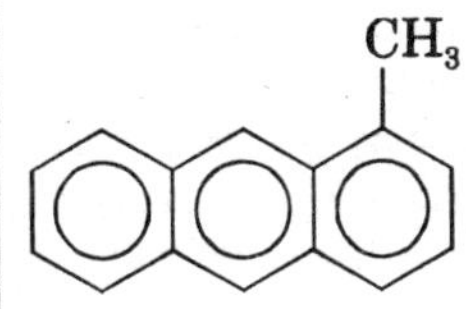 (I) 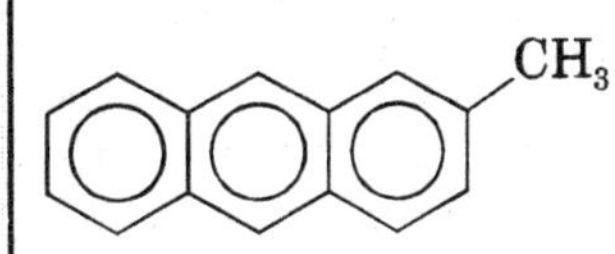 (II)
119.	(*c*)	Metamers differ in the no. of carbon atoms in the chain around the functional atoms.
120.	(*b*)	The four alkyl radicals for $C_4H_9—$ are (*i*) *n*-Butyl (*ii*) *sec*-Butyl (*iii*) *iso*-Butyl (*iv*) *tert*-Butyl.
121.	(*c*)	Only compounds with 4 or more carbons can exhibit chain isomerism.
122.	(*d*)	$CH_3CH_2COCH_3$ Butanone C_4H_8O; $CH_3CH_2-O-CH_2CH_3$ Diethyl ether $C_4H_{10}O$
123.	(*a*)	These are chain isomers.
124.	(*d*)	The carbon of C=C is sp^2 hybridised and —C≡C— is *sp* hybridised.
125.	(*b*)	(*i*) $CH_3CH_2CH_2CH_2OH$ (*ii*) $CH_3CH_2CH(OH)CH_3$ (*iii*) $(CH_3)_2CH—CH_2—OH$

Q. No.	Ans.	Hints & Solutions
		(iv) 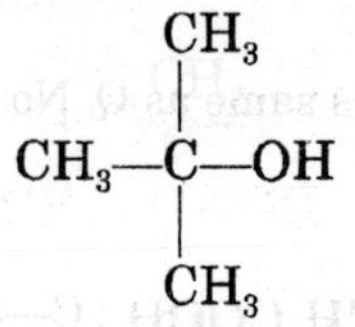
		(v) $CH_3CH_2—O—CH_2CH_3$
		(vi) $CH_3CH_2CH_2—O—CH_3$
		(vii) 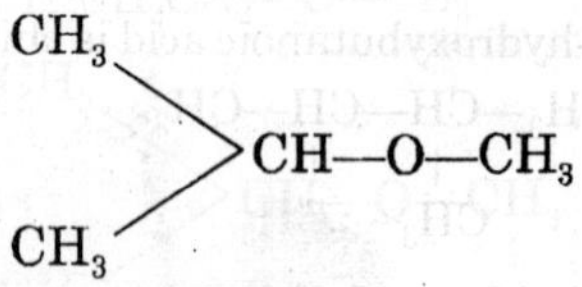
126.	(d)	(i) 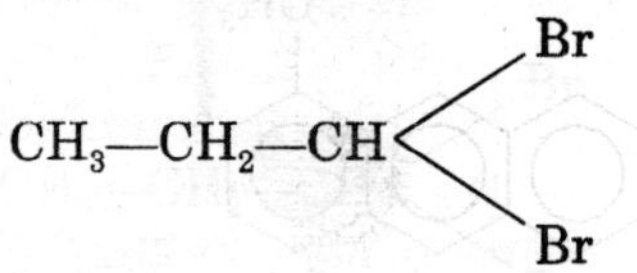
		(ii) 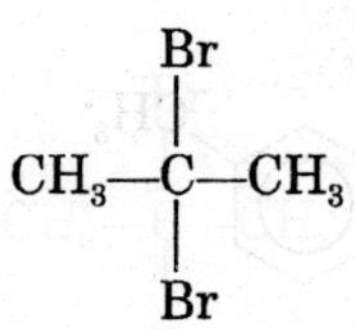
		(iii) 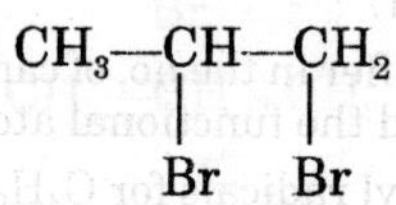
		(iv) 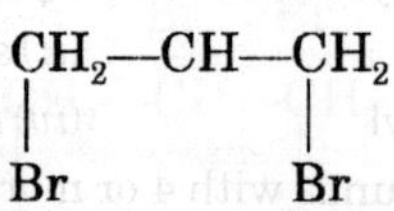
127.	(b)	There are the three isomers of C_5H_{12}
		(i) $CH_3CH_2CH_2CH_2CH_3$
		(ii) 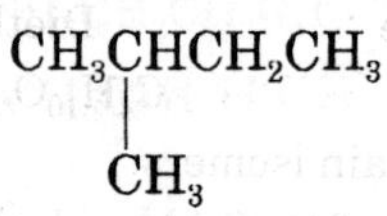
		(iii) 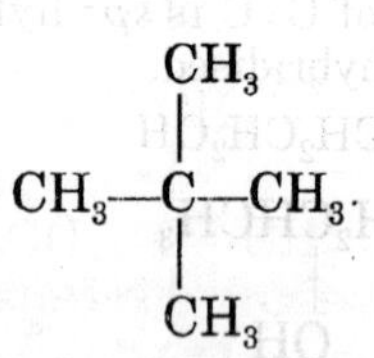
128.	(d)	Six isomeric pentylenes are possible.
		$CH_3CH_2CH_2CH = CH_2$
		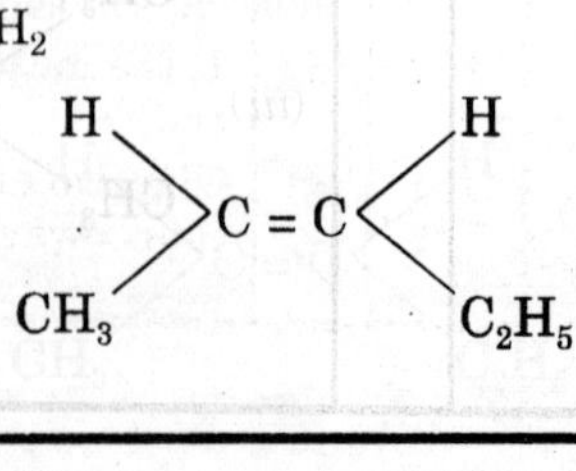
		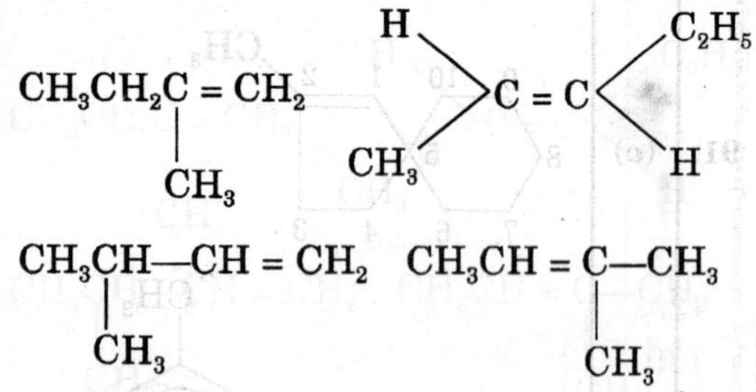
129.	(c)	$CH_3CH_2CH = CH_2$ 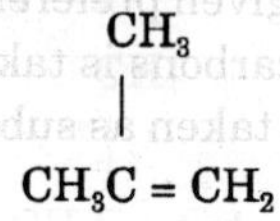 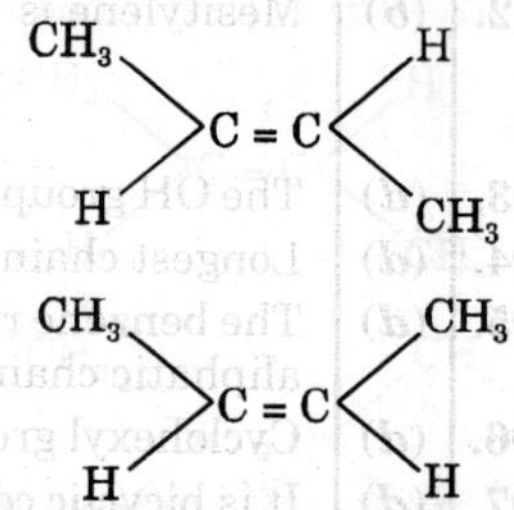
130.	(c)	2, 3-dimethyl butane has only two types of H atoms 1° and 2° and hence conform two types of derivatives.
		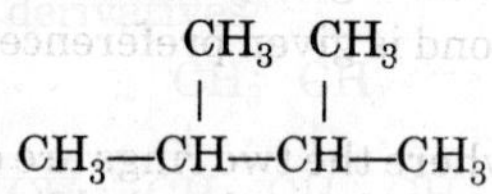
131.	(b)	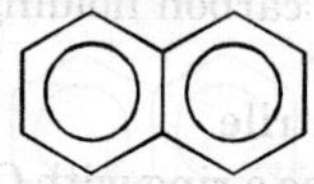 The number of isomers depend upon the environment in the molecule and hence there are three isomers.
132.	(b)	CH_3CH_2COOH Propanoic acid 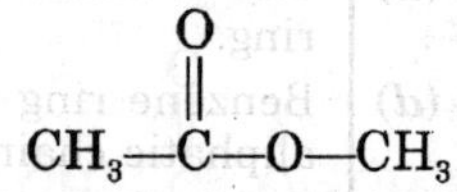 Methyl acetate
		$HCOOCH_2CH_3$ Ethyl formate
133.	(c)	Alcohols and ethers are functional isomers.
134.	(a)	Napthalene is 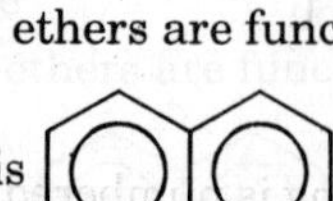
135.	(a)	Cyanides and isocyanides are functional isomers of each other.
136.	(d)	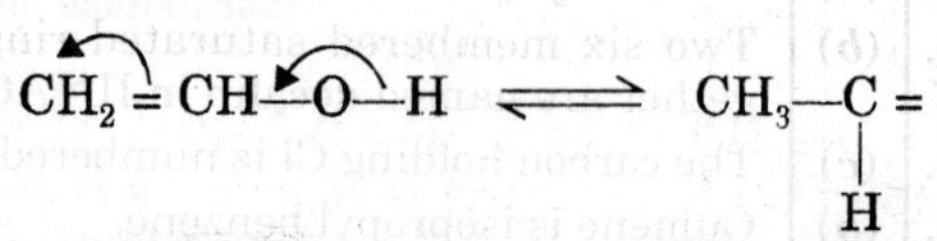
		Vinyl alcohol Ethanal

Q. No.	Ans.	Hints & Solutions
137.	(*a*)	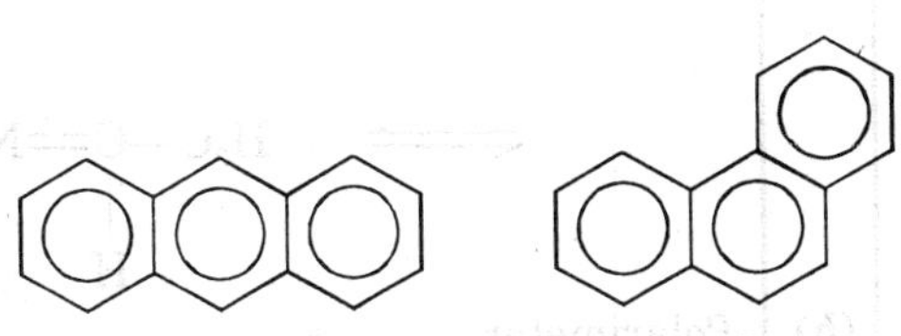 Anthracene Phenanthrene
138.	(*c*)	$(CH_3)_3N$ is isomeric with 1° as well as 2° amine.
139.	(*d*)	Steric repulsions are more.
140.	(*b*)	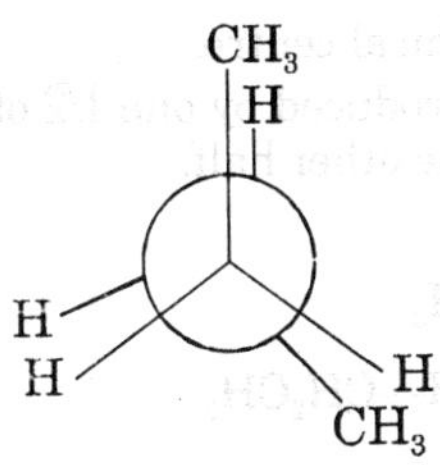 The dihedral bond angle in partially eclipsed conformation is 120°.
141.	(*d*)	It give rise to many conformations.
142.	(*b*)	Due to minimum repulsions
143.	(*d*)	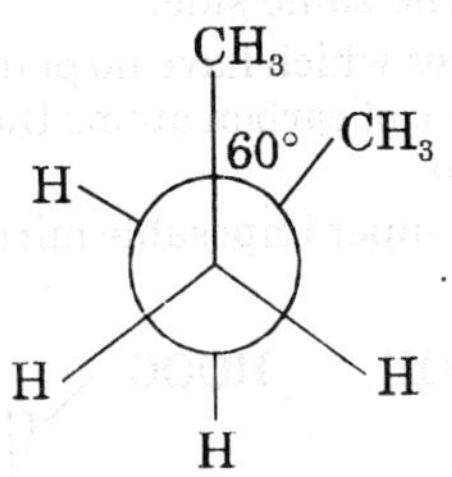
144.	(*d*)	Repulsive interactions are more.
145.	(*b*)	In cyclopropane, the ring is most strained
146.	(*b*)	It is 30 kJ mol^{-1}.
147.	(*d*)	It is 12.5 kJ mol^{-1}.
148.	(*d*)	One produces optically active product but the other gives racemic product.
149.	(*b*)	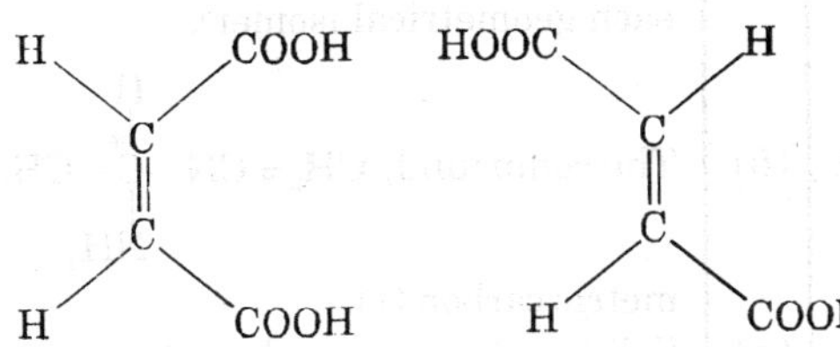 Maleic acid (*cis*-isomer) Fumaric acid (*trans*-isomer)
150.	(*c*)	Geometric isomerism. Because for geometric isomerism one of the condition is that the compound should have C = C.

Q. No.	Ans.	Hints & Solutions
151.	(*c*)	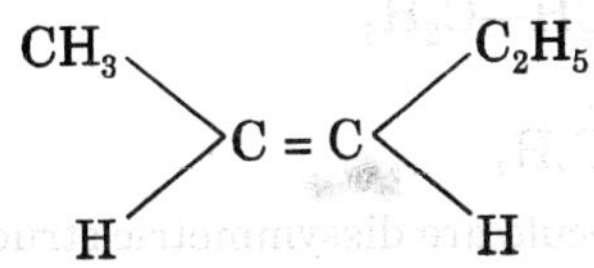 2-Pentene Only 2-pentene will show geometric isomerism because only in this case both the carbons involved in the double bond have two different groups.
152.	(*c*)	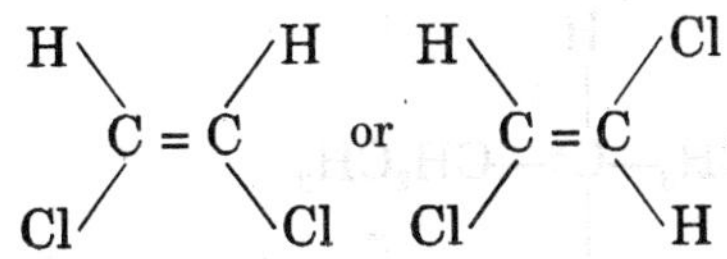
153.	(*d*)	CH_3—CH = CH—CH_3.
154.	(*c*)	Two similar groups are attached to carbon holding the double bond.
155.	(*d*)	It satisfies the conditions required for geometrical isomerism.
156.	(*d*)	Recemic mixture is optically in active solution.
157.	(*c*)	Diastereomers are stereoisomers which are not mirror images of each other. 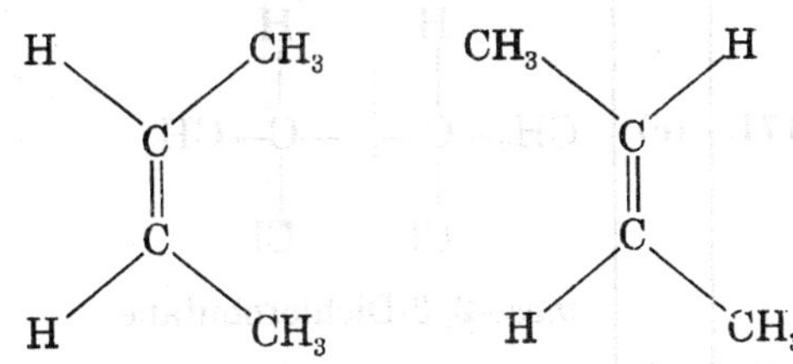 *cis*-2-Butene *trans*-2-Butene Not mirror image of each other.
158.	(*b*)	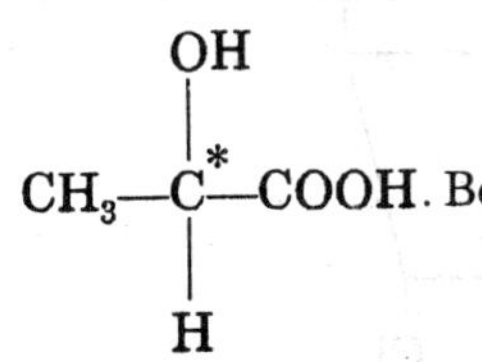 Because it has an asymmetric carbon atom.
159.	(*a*)	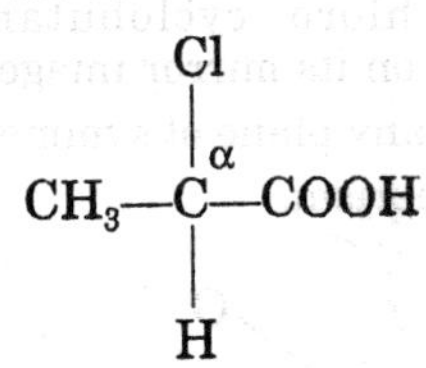 α-carbon is asymmetric.
160.	(*d*)	It is a meso compound as it has plane of symmetry.
161.	(*c*)	The two enantiomers differ in chemical behaviour towards other optically active compounds.

Q. No.	Ans.	Hints & Solutions
162.	(c)	$CH_3-CH(C_3H_7)-C_2H_5$.
163.	(d)	All molecule are dissymmetric structures.
164.	(c)	See concepts in brief 2^n (n = no. of assymmetric carbons).
165.	(d)	2, 3-Pentadiene is a substituted allene. It possesses chirality.
166.	(c)	It is called mutarotation.
167.	(d)	$CH_3-C^*H(Cl)-CH_2CH_3$ 2-Chlorobutane
168.	(b)	The credit goes to Louis Pasteur.
169.	(c)	The given compound has R configuration *i.e.*, same as choice C.
170.	(b)	Diastereomers is a pair of compounds which are neither mirror images nor superimposable.
171.	(c)	$CH_3-CH(Cl)-CH(Cl)-CH_3$ *meso*-2, 3-Dichlorobutane
172.	(d)	It contains chiral carbon.
173.	(b)	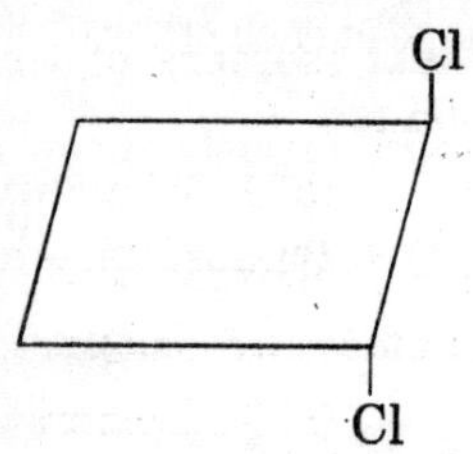 *trans*-1, 2-dichloro cyclobutane is not superimposable on its mirror image. It doesnot have any plane of symmetry.
174.	(d)	See optical isomerism.
175.	(d)	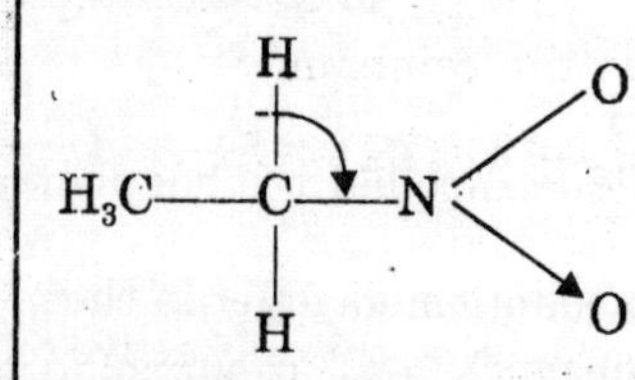 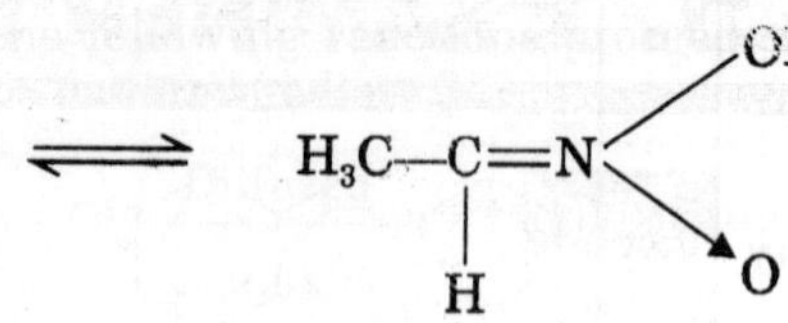
176.	(b)	Polarimeter.
177.	(a)	2^n where n = no. of chiral carbons.
178.	(d)	When equimolar amounts of two optically active isomers of the same compound are mixed, a racemic mixture is formed.
179.	(c)	It contains a chiral centre.
180.	(d)	The rotation produced by one 1/2 of molecule is cancelled by the other half.
181.	(b)	Resolution.
182.	(b)	$CH_3-CH(CH_3)-C^*H(CH_3)-CH_2CH_3$ It has asymmetric carbon (*) and no plane of symmetry.
183.	(c)	Erythro isomers is the one in which the two similar groups or two H atoms on the two carbon atoms are on the same side.
184.	(c)	In the molecules which have no plane of symmetry and even no. of carbon atoms the no. of meso isomers is zero.
185.	(a)	These are non super imposable mirror images of each other.
186.	(b)	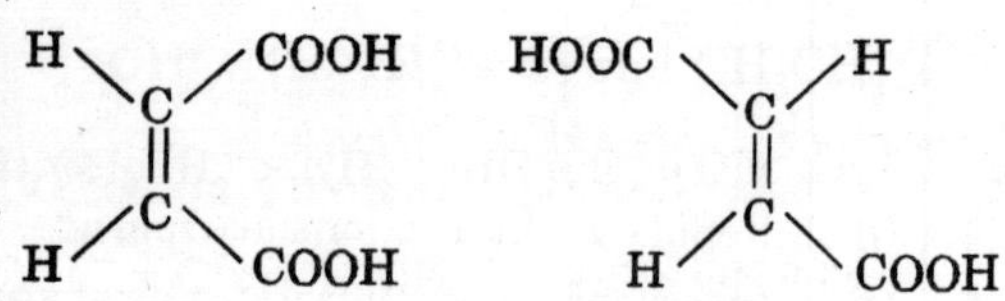 *Maleic acid* *Fumaric acid* These are geometrical isomers.
187.	(a)	Draw the structure and check.
188.	(c)	It will have 2 geometrical isomers *i.e., cis* and *trans* but will have 4 optical isomers (two for each geometrical isomer).
189.	(b)	The compound, $CH_2=CN-C^*H(NH_2)-CH_3$, has asymmetric carbon (*).
190.	(a)	Solid wedge means above the plane and dotted line means below the plane. (*i*) and (*iii*) are dissymmetric structures.

1.5. EXAMINER'S VIEWPOINT

Test Your Knowledge

1996

1. Which of the following has zero dipole moment ?
(*a*) *cis*-2-Butene (*b*) *trans*-2-Butene
(*c*) 1-Butene (*d*) 2-Methyl-1-propene.
(CBSE PMT)

2. Which of the following has sp^2 hybridisation ?
(*a*) C_2H_6 (*b*) C_2H_4
(*c*) $BeCl_2$ (*d*) C_2H_2. ***(CBSE PMT)***

3. IUPAC name of $(CH_3)_2CH-CH_2-CH_2Br$ is
(*a*) 1-Bromopentane
(*b*) 2-Methyl-4-bromopentane
(*c*) 1-Bromo-3-methylbutane
(*d*) 2-Methyl-3-bromo-propane. ***(CBSE PMT)***

4. The order of decreasing stability of the carbonions

$$(CH_3)_3\bar{C}\,(1)\,;\,(CH_3)_2\bar{C}H\,(2)\,;$$

$$CH_3\bar{C}H_2\,(3)\,;\,C_6H_5\bar{C}H_2\,(4)\text{ is}$$

(*a*) 1 > 2 > 3 > 4 (*b*) 4 > 3 > 2 > 1
(*c*) 4 > 1 > 2 > 3 (*d*) 1 > 2 > 4 > 3.
(CEET Haryana)

5. Which of the following will show aromatic behaviour ?

(*a*) 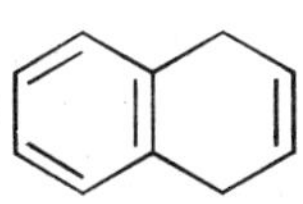 (*b*) 

(*c*)  (*d*) None.

(CEET Haryana)

6. Which one of the following compounds will not exhibit optical isomerism ?
(*a*) $CH_3CH(OH)Br$
(*b*) $CH_3CH(OH)CH_3$
(*c*) $CH_3CH_2-CHBrCH(CH_3)CH_3$
(*d*) $CH_3-CH(OH)-CH(Br)CH_2OH$ ***(Pb. CET)***

7. The most stable carbocation amongst the following is
(*a*) $C_6H_5-\overset{+}{C}H-C_6H_5$ (*b*) $C_6H_5\overset{+}{C}H_2$
(*c*) $CH_3\overset{+}{C}H_2$ (*d*) $C_6H_5CH_2\overset{+}{C}H_2$.
(Pb. CET)

8. IUPAC name of CCl_3CH_2CHO is
(*a*) chloral
(*b*) 3, 3, 3-trichloropropanal
(*c*) 3, 3, 3-trichloropropanol
(*d*) 2, 2, 2-trichloropropanol. ***(Pb. CET)***

1997

9. The IUPAC name of $(CH_3)_3C-CH=CH_2$ is
(*a*) 2, 2-Dimethyl but 3-ene
(*b*) 2, 2-Dimethyl pent-3-ene
(*c*) 3, 3-Dimethyl but-1-ene
(*d*) Hex-1-ene. ***(AIIMS)***

10. Which of the following is not aromatic ?
(*a*) Benzene
(*b*) Cyclooctatetraenyl anion
(*c*) Tropylium cation
(*d*) Cyclopentadienyl cation. ***(AIIMS)***

11. The number of optical isomers of the compound $CH_3CHBrCHBrCOOH$ is
(*a*) 0 (*b*) 1
(*c*) 3 (*d*) 4. ***(AIIMS)***

12. Carboxylic acids and esters are isomers of the type
(*a*) Position isomers (*b*) Tautomers
(*c*) Functional isomers (*d*) None of these.

13. The number of isomers of C_4H_9- is
(*a*) 2 (*b*) 3
(*c*) 4 (*d*) None of these. ***(AFMC)***

14. IUPAC name of the compoun

$$CH_3-\underset{\displaystyle CH_2CH_3}{\underset{|}{CH}}-CH_2-CH(OH)-CH_3\text{ is}$$

(*a*) 4-Methyl-3-hexanol (*b*) Heptanol
(*c*) 4-Methyl-2-hexanol (*d*) None of these. ***(AFMC)***

15. IUPAC name of $(CH_3)_2CHCH_2CH_2Cl$ is

(*a*) 1-chloropentane

(*b*) 1-chloro-3-methyl butane

(*c*) 2-methyl-3-chloro propane

(*d*) None. ***(Pb. CET)***

16. Tautomerism is exhibited by

(*a*) $(CH_3)_3CNO$ (*b*) $(CH_3)_2NH$

(*c*) R_3CNO_2 (*d*) RCH_2NO_2. ***(CBSE PMT)***

17. The cylindrical shape of alkynes is due to

(*a*) Three σ carbon-carbon bonds

(*b*) Three π carbon-carbon bonds

(*c*) Two σ and one π carbon-carbon bonds

(*d*) One σ and two π carbon-carbon bonds. ***(CBSE PMT)***

18. The most stable conformation of *n*-butane is

(*a*) Skew boat (*b*) Eclipsed

(*c*) Gauche (*d*) Staggered-anti. ***(CBSE PMT)***

19. The isomers which can be converted into another form by rotation of molecule around single bond are

(*a*) Geometrical isomers (*b*) Enantiomers

(*c*) Conformers (*d*) Diastereomers. ***(AIIMS)***

20. Which of the following shows geometrical isomerism ?

(*a*) But-1-ene (*b*) But-2-ene

(*c*) Propene (*d*) 1, 1-Dichlorobutane. ***(AFMC)***

21. IUPAC name of $CH_3CH = CH - C \equiv CH$ is

(*a*) Pent-3-en-1-yne (*b*) Pent-2-en-4-yne

(*c*) Pent-3-yn-1-ene (*d*) Pent-2-yn-1-ene. ***(UP CPMT)***

22. In which of the following the bond length between carbon and carbon atoms is equal ?

(*a*) 2-Butene (*b*) 1-Butene

(*c*) 1-Propyne (*d*) Benzene. ***(UP CPMT)***

1998

23. The homologue of ethyne is

(*a*) C_2H_4 (*b*) C_3H_6

(*c*) C_3H_8 (*d*) C_3H_4. ***(EAMCET)***

24. The hybridization involved in the six carbon atoms of benzene is

(*a*) Three sp^3 and three sp^2

(*b*) Three sp^3 and three *sp*

(*c*) All six *sp*

(*d*) All six sp^2. ***(EAMCET)***

25. The IUPAC name of the compound having the formula $(CH_3)_3C - CH = CH_2$ is

(*a*) 3, 3-Dimethyl-1-butene

(*b*) 3, 3, 3-Trimethyl-1-propene

(*c*) 1, 1, 1-Trimethyl-3-propene

(*d*) 1, 1-Dimethyl-3-butene. ***(EAMCET)***

26. The IUPAC name of the compound  is

(*a*) Bicyclo [2.1.0] pentane

(*b*) 1, 2-Cyclopropyl cyclobutane

(*c*) Cyclopentane [4.3] annulene

(*d*) 1, 2-Methylene cyclobutane. ***(Pb. CET)***

27. Which of the following molecules contains asymmetric carbon atom ?

(*a*) $CH_3CHClCOOH$ (*b*) CH_3CH_2COOH

(*c*) $ClCH_2CH_2COOH$ (*d*) $Cl_2CHCOOH$. ***(CEET)***

28. Which of the following has zero dipole moment ?

(*a*) PH_3 (*b*) $CHCl_3$

(*c*) CH_4 (*d*) CH_2Cl_2. ***(AFMC)***

29. The hybridisation of carbon in diamond, graphite and acetylene is in the order

(*a*) sp^3, sp, sp^2 (*b*) sp^3, sp^2, sp

(*c*) sp, sp^2, sp^3 (*d*) sp^2, sp^3, sp. ***(AFMC)***

30. Among the following compounds which has only one type of hybridisation for carbon atom

(*i*) $CH_3CH_2CH_2CH_3$ (*ii*) $CH_3CH = CH - CH_3$

(*iii*) $CH_2 = CH - C \equiv CH$ (*iv*) $CH \equiv CH$.

(*a*) (*ii*) and (*iii*) (*b*) (*ii*)

(*c*) (*iii*) and (*iv*) (*d*) (*iv*). ***(AFMC)***

31. Which one of the following orders is correct regarding the – I effect of the substituents ?

(*a*) $-NR_2 < -OR < -F$ (*b*) $-NR_2 > -OR > -F$

(*c*) $-NR_2 > -OR < -F$ (*d*) $-NR_2 < -OR > -F$. ***(CBSE PMT)***

32. Which of the following molecules is not chiral ?

(*a*) $DCH_2CH_2CH_2Cl$ (*b*) CH_3CHDCH_2Cl

(*c*) $CH_3CHClCH_2D$ (*d*) CH_3CH_2CHDCl. ***(CBSE PMT)***

33. IUPAC name for the compound

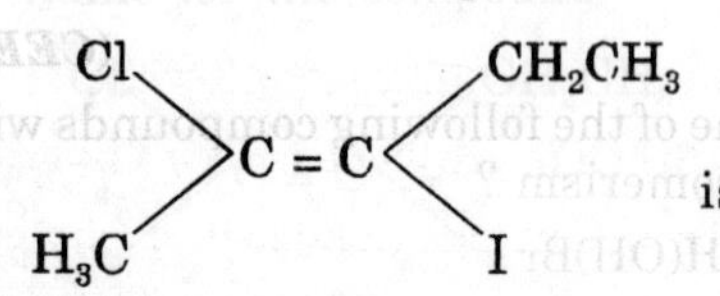

is

(*a*) *trans*-2-Chloro-3-iodopentene-2

(*b*) *cis*-2-Chloro-3-iodo-2-pentene

(*c*) *trans*-3-Iodo-4-chloro-3-pentene

(*d*) *cis*-3-Iodo-4-chloro-3-pentene. ***(CBSE PMT)***

34. The shape of methane molecule is

(*a*) Linear (*b*) Triangular

(*c*) Tetrahedral (*d*) Octahedral. ***(MP CEE)***

35. The number of possible structural isomers of $C_4H_{10}O$ is

(a) 5 (b) 6
(c) 7 (d) 8. **(MP CEE)**

36. Which of the following compounds will exhibit geometrical isomerism ?

(a) 2-Butene (b) 2-Butyne
(c) 2-Butanol (d) Butanal. **(MP CEE)**

37. The compound  is known by which of the following names

(a) Bicyclo-[2.2.2] octane (b) Bicyclo-[2.2.1] octane
(c) Bicyclo-[1.2.1] octane (d) Bicyclo-[1.1.1] octane. **(MP CEE)**

38. The total number of possible isomeric trimethyl benzenes is

(a) 2 (b) 3
(c) 4 (d) 6. **(MP CEE)**

39. The IUPAC name of

$CH_3—C(Cl)=C(CH_3)—CH(C_2H_5)—CH_2—C\equiv CH$ is

(a) 6-Chloro-4-ethyl-5-methyl hept-5-en-1-yne
(b) 6-Chloro-4-ethyl-5-methyl hept-1-yn-5-ene
(c) 2-Chloro-4-ethyl-3-methyl hept-2-ene-6-yne
(d) 2-Chloro-4-ethyl-3-methyl hept-6-yn-2-ene. **(MP CEE)**

40. Point out incorrect statement about resonance :

(a) Resonance structures should have equal energy
(b) In resonance structures, the constituent atoms should be in the same position
(c) In resonance structures, there should not be same number of electron pairs
(d) Resonance structures should differ only in the location of electrons around the constituent atoms. **(MP CEE)**

41. Which one is false in the following statements ?

(a) Each carbon in ethylene is in sp^2 hybridisation
(b) Each carbon in acetylene is in sp^3 hybridisation
(c) Each carbon in benzene is in sp^2 hybridisation
(d) Each carbon in ethane is in sp^3 hybridisation. **(MP CEE)**

42. Which of the following is an electrophile ?

(a) H_2O (b) NH_3
(c) $AlCl_3$ (d) $C_2H_5NH_2$. **(BHU)**

43. IUPAC name of $CH_3—C(H)(C_2H_5)—C(C_4H_9)(CH_3)—CH_3$ is

(a) 2-Butyl-2-methyl-3-ethyl-butane
(b) 2-Ethyl, 3-3-dimethyl heptane
(c) 3, 4, 4-Trimethyl heptane
(d) 3, 4, 4-Trimethyl octane. **(BHU)**

44. *cis* and *trans*-2-butene are

(a) conformational isomers
(b) optical isomers
(c) position isomers
(d) geometrical isomers. **(BHU)**

45. Which is the chiral molecule ?

(a) CH_3Cl (b) CH_2Cl_2
(c) $CHBr_3$ (d) CHClBrI. **(BHU)**

46. The IUPAC name for the formula

$CH_3—C(CH_3)=C(H)—COOH$ is

(a) 2-Methyl-2-butenoic acid
(b) 3-Methyl-3-butenoic acid
(c) 3-Methyl-2-butenoic acid
(d) 2-Methyl-3-butenoic acid. **(Pb. PMT)**

47. Select the molecule which has only one π-bond

(a) $CH\equiv CH$
(b) $CH_2=CH—CHO$
(c) $CH_3—CH=CH_2$
(d) $CH_3—CH=CH—COOH$. **(Pb. PMT)**

48. Heterolytic fission of a covalent bond in organic molecule gives

(a) Free radicals (b) Cations and anions
(c) Only cations (d) Only anions. **(MP PMT)**

1999

49. IUPAC name of the following compound will be

$CH_3—CH=C(CH_2—CH_2—CH_3)—CH_2—CH_3$

(a) 3-Ethyl-2-hexene (b) 3-Propyl-2-hexene
(c) 3-Propyl-3-hexene (d) 4-Ethyl-4-hexene. **(UP CPMT)**

50. Bond lengths of ethane (I), ethene (II), acetylene (III) and benzene (IV) follow the order

(a) I > II > III > IV (b) I > II > IV > III
(c) I > IV > II > III (d) III > IV > II > I. **(UP CPMT)**

51. Which of the following possesses highest melting point ?

(*a*) Chlorobenzene (*b*) *o*-Dichlorobenzene
(*c*) *m*-Dichlorobenzene (*d*) *p*-Dichlorobenzene.

(UP CPMT)

52. Which of the following carbon atoms is likely to possess three-dimensional geometry ?

$$\overset{1}{H_2C} = \overset{2}{CH}—\overset{3}{CH_2}—\overset{4}{COOH}$$

(*a*) 1 (*b*) 2
(*c*) 3 (*d*) 4. ***(UP CPMT)***

53. The IUPAC name of the compound

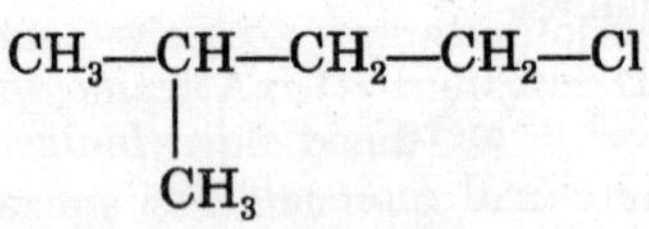 is

(*a*) 1-Chloro-3-methylbutane
(*b*) 2-Methyl-4-chlorobutane
(*c*) 2-Methyl-1-chlorobutane
(*d*) 1-Chloropentane. ***(MP CEE)***

54. Optically active isomers but not mirror images are called

(*a*) Enantiomers (*b*) Mesomers
(*c*) Tautomers (*d*) Diastereoisomers.

(MP CEE)

55. The arrangement of hydrogen atoms around carbon in the methane molecule is

(*a*) Triangular (*b*) Square planar
(*c*) Tetrahedral (*d*) Hexagonal.

(MP CEE)

56. An alkyne has the general formula

(*a*) C_nH_{2n} (*b*) C_nH_{2n+2}
(*c*) C_nH_{2n-2} (*d*) C_nH_{2n+1}. ***(MP CEE)***

57. An organic compound will show optical isomerism if

(*a*) Four groups attached to C atom are different
(*b*) Three groups attached to C atom are different
(*c*) Two groups attached to C atom are different
(*d*) All the groups attached to C atom are same.

(MP CET)

58. The IUPAC name of

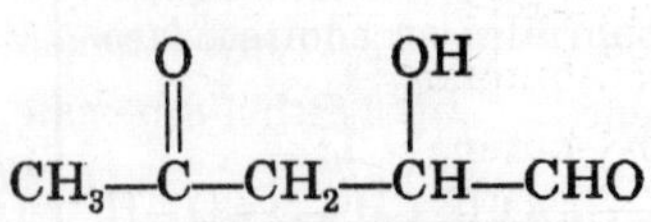 is

(*a*) 5-Oxo-4-hydroxoy-2-pentanone
(*b*) 4-Hydroxy-5-al-2-pentanone
(*c*) 2-Hydroxy-4-oxopentanal
(*d*) 1-Al-4-oxo-2-pentanol. ***(JIPMER)***

2000

59. The IUPAC name of

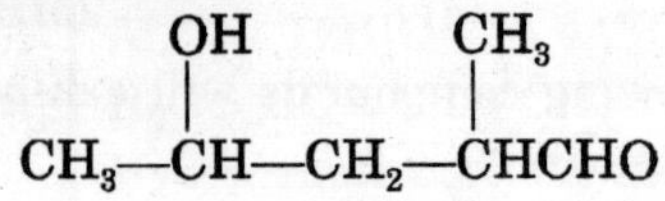 is

(*a*) 4-Hydroxy-2-methylpentanal
(*b*) 2-Hydroxy-4-methylpentanal
(*c*) 2-Methylpent-4-ol-1-al
(*d*) None of these. ***(JIPMER)***

60. The IUPAC name of acraldehyde is

(*a*) Prop-2-en-1-al (*b*) Propenylaldehyde
(*c*) But-2-en-1-al (*d*) Propenal. ***(MP PMT)***

61. $CH_3CH_2CH_2CH(CH = CH_2)CH_2CH_2CH_3$ is

(*a*) 4-Enthenylheptane
(*b*) 3-Propylhex-1-ene
(*c*) 4-Ethenylhexane
(*d*) 3-Ethyenylheptane. ***(Kerala EEE)***

62. The IUPAC name of *tert*-butyl chloride is

(*a*) 4-Chlorobutane
(*b*) 2-Chlorobutane
(*c*) 1-Chloro-3-methylpropane
(*d*) 2-Chloro-2-methylpropane. ***(KCET ; CPMT)***

63. The IUPAC name of

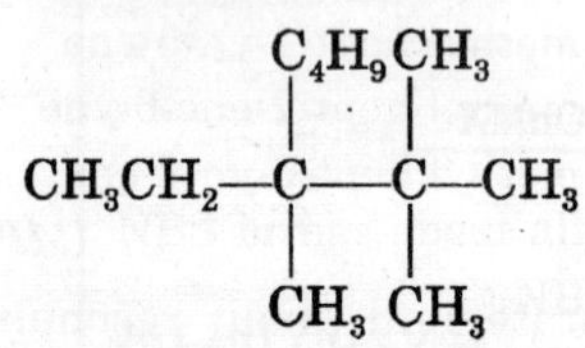 is

(*a*) 3-Ethyl-2, 2, 3-trimethyl heptane
(*b*) 3, 4, 4-Trimethyloctane
(*c*) 2-Butyl-2-methyl-3-ethylbutane
(*d*) 2-Ethyl-3, 3-dimethylheptane. ***(KCET)***

64. The IUPAC name of $CH_3CH = CHCOOC_2H_5$ is

(*a*) Ethyl but-1-enoate
(*b*) Ethyl but-2-enoate
(*c*) Ethyl prop-2-enoate
(*d*) None of these. ***(Haryana CEET)***

65. The IUPAC name of the compound

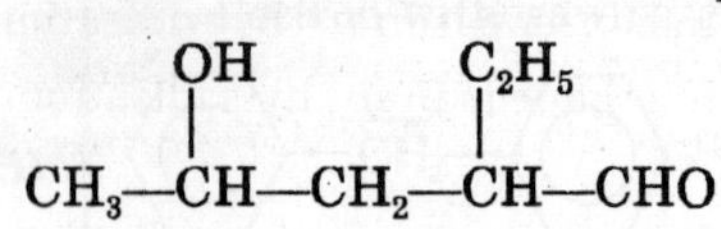 is

(*a*) 4-Hydroxy-1-ethylpentanal
(*b*) 4-Hydroxy-2-ethylpentanal
(*c*) 2-Hydroxy-4-ethylpentanal
(*d*) 2-Hydroxy-2-ethylpentanal. ***(JIPMER)***

66. The IUPAC name of $CH_3OC_2H_5$ is
(*a*) Methyl ethyl ether
(*b*) Ethyl methyl ether
(*c*) Methoxyethane
(*d*) Ethoxymethane. ***(MP CEE)***

67. What is the order of stability for
I. 1-butene ; II. cis-2-butene ;
III. trans-2-butene.
(*a*) I < II < III (*b*) I > II > III
(*c*) III > I > II (*d*) I > II > III. ***(CBSE PMT)***

68. R and S pairs of enantiomers differ from one another in
(*a*) Optical rotation of polarised light
(*b*) Solubility in racemic mixture
(*c*) Reaction with racemic mixture
(*d*) None of these. ***(CBSE PMT)***

69. The dihedral-angle HCH in staggered conformation of C_2H_6 is
(*a*) 120° (*b*) 60°
(*c*) 0° (*d*) 90°. ***(CBSE PMT)***

70. The total number of acyclic isomers including stereoisomers of C_4H_7Cl are
(*a*) 10 (*b*) 9
(*c*) 12 (*d*) 13. ***(BHU)***

71. Which of the following will exhibit geometrical isomerism ?
(*a*) 1-Phenyl-2-butene
(*b*) 3-Phenyl-1-butene
(*c*) 2-Phenyl-1-butene
(*d*) 1, 1-Diphenyl-1-propene. ***(MP PMT)***

72. Which of the following is an isomer of propanal ?
(*a*) Acetone (*b*) Propane
(*c*) Propanol (*d*) Propionic acid.
(Kerala EEE)

2001

73. IUPAC name of the given organic compound $(CH_3)_2C(CH_2CH_3)CH_2CH(Cl)CH_3$ will be
(*a*) 5-Chloro-3, 3-dimethyl hexane
(*b*) 4-Chloro-2-ethyl-2-methylpentane
(*c*) 2-Chloro-4-ethyl-4-methylpentane
(*d*) 2-Chloro-4, 4-dimethylhexane. ***(Bihar CEE)***

74. The correct nomenclature (IUPAC) for the following alcohol

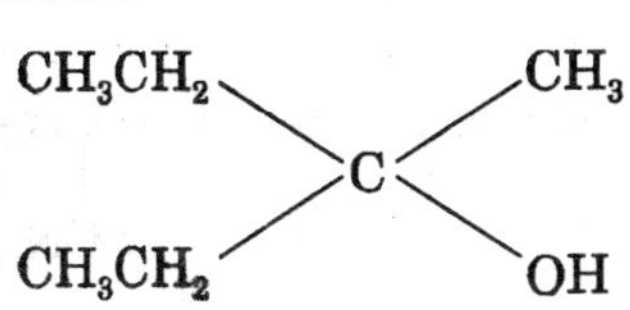

is
(*a*) 2-Ethyl-2-butanol
(*b*) 3-Methyl-3-pentanol
(*c*) 3-Ethyl-3-methyl-2-pentanol
(*d*) 1, 1-Dimethylanol. ***(UP SEAT)***

75. Vinyl carbinol is
(*a*) $HO-CH_2CH=CH_2$
(*b*) $CH_3CH(OH)=CH_2$
(*c*) $CH_3CH=CH-OH$
(*d*) $CH_3-C(CH_2OH)=CH_2$. ***(UP SEAT)***

76. The IUPAC name of

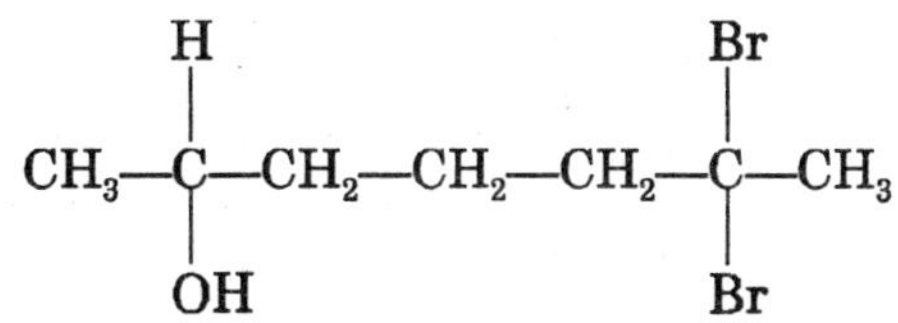

is
(*a*) 6, 6-Dibromoheptan-2-ol
(*b*) 2, 2-Dibromoheptan-2-ol
(*c*) 6, 6-Dibromoheptan-2-al
(*d*) None of these. ***(DCE)***

77. The IUPAC name of 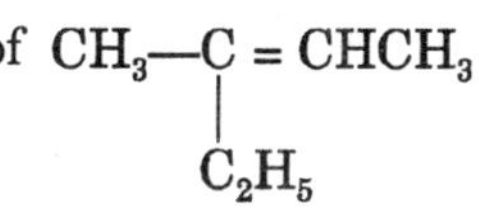 is

(*a*) 2-Ethylbutene
(*b*) 2-Ethylbut-2-ene
(*c*) 3-Methylpent-2-ene
(*d*) 3-Methylpent-3-ene
(*e*) 3-Ethylbut-2-ene. ***(Kerala EEE)***

78. The structural formula of 2-methyl-2-butene is
(*a*) $CH_3-CH(CH_3)-CH=CH_2$
(*b*) $CH_3-CH_2-C(CH_3)=CH_2$
(*c*) $CH_3CH=CH-CH_3$
(*d*) $CH_3CH=C(CH_3)-CH_3$. ***(EAMCET)***

79. IUPAC name of $CH_2=CH-CN$ is
(*a*) Ethenenitrile (*b*) Vinyl cyanide
(*c*) Cyanoethene (*d*) 2-Propenenitrile. ***(NSE)***

80. The IUPAC name of the following $CH_3C(CH_3)_2CH_2CH=CH_2$ is
(*a*) 2, 2-Dimethyl-4-pentene
(*b*) 4, 4-Dimethyl-1-pentene
(*c*) 1, 1, 1-Trimethyl-3-butene
(*d*) 4, 4, 4-Trimethyl-1-butene. ***(CPMT)***

81. IUPAC name of 4-isopropyl-*m*-xylene is
(*a*) 1-Isopropyl-2, 4-dimethylbenzene
(*b*) 4-Isopropyl-*m*-xylene
(*c*) 4-Isopropyl-3, 5-dimethyl benzene
(*d*) 4-Isopropyl-4, 6-dimethyl benzene. ***(DPMT)***

82. The incorrect IUPAC name is

(a) $CH_3—C(=O)—CH(CH_3)—CH_3$
2-Methylbutan-3-one

(b) $CH_3CH(CH_3)—CH(CH_2CH_3)—CH_3$
2, 3-Dimethylpentane

(c) $CH_3 - C \equiv CCH(CH_3)_2$
4-Methylpent-2-yne

(d) $CH_3—CH(Cl)—CH(Br)—CH_3$ **(CBSE PMT)**
2-Bromo-3-chlorobutane

83. The IUPAC name of $H—C(=O)—CH = O$ is
(a) Formylmethanal
(b) 1, 2-Ethanedione
(c) Formly methanoate
(d) Ethane-1, 2-dial. **(MP PET)**

84. The name of $ClCH_2—C(Br) = C(Br)—CH_2Cl$ according to IUPAC nomenclature system is
(a) 2, 3-Dibromo-1, 4-dichlorobutene-2
(b) 1, 4-Dichloro-2, 3-dibromobutene-2
(c) Dichlorodibromobutene
(d) Dichlorodibromobutane. **(MP PMT)**

85. How many carbon atoms in the molecule HOOC – $(CHOH)_2$ – COOH are asymmetric ?
(a) 1 (b) 2
(c) 3 (d) None of these. **(MP PMT)**

86. Which of the following is not isomeric with diethyl ether ?
(a) *n*-propyl methyl ether
(b) 1-butanol
(c) 2-methyl-2-propanol
(d) butanone. **(MP PMT)**

87. Which of the following will show geometrical isomerism ?
(a) 1-Butene
(b) 1, 2-Dibromobut-1-ene
(c) Propene
(d) Isopropylene. **(DPMT)**

88. The number of ether metamers represented by formula $C_4H_{10}O$ is
(a) 4 (b) 3
(c) 2 (d) 1. **(Tamil Nadu CET)**

89. $CH_3CH_2CH_2OH$ and $CH_3CH_2OCH_3$ are
(a) Position isomers (b) Chain isomers
(c) Geometrical isomers (d) Functional isomers
(e) Optical isomers. **(Kerala EEE)**

90. The compound without a chiral atom is
(a) $BrCH_2CH(CH_3)CH_2Br$
(b) $C_2H_5CH_2CH(CH_3)CH_2Br$
(c) $CH_3—CH_2CH(CH_3)CH_2Br$
(d) $HOOC—CH_2—CH(CH_3)—COOH$
(e) OHC – CH(OH)CH_2OH. **(Kerala EEE)**

91. 1-Chloropropane and 2-chloropropane are
(a) Position isomers
(b) Chain isomers
(c) Functional isomers
(d) Optical isomers
(e) Geometrical isomers. **(KCMET)**

92. An organic compound
$\overset{1}{C}H_3—\overset{2}{C}H_2—\overset{3}{C}H_2—\overset{4}{C}H_2—\overset{5}{C}H_2—\overset{6}{C}H_2—\overset{7}{C}H_3$
To make it chiral the attack should be on carbon
(a) 1 (b) 3
(c) 4 (d) 7. **(DCE)**

93. A compound with molecular formula C_7H_{16} shows optical isomerism, the compound will be
(a) 2, 3-Dimethylpentane
(b) 2, 2-Dimethylpentane
(c) 2-Methyl heptane
(d) None of these. **(CBSE PMT)**

2002

94. IUPAC name of $CH_3—CH(CH_3)—CH_2—CH(CN)—CH_3$ is

(a) 2-Cyano, 3-methylhexane
(b) 2-Methyl, 4-cyanopentane
(c) 2, 4-Dimethyl pentane nitrile
(d) 2-dimethyl, 4-cyanopentane. **(AIIMS)**

95. Which of the following is the most stable ?
(a) 1-butene (b) 2-butene
(c) 1-pentene (d) 2-pentene. **(AIIMS)**

96. Which is strongest nucleophile ?
(a) CH_3COO^- (b) CH_3O^-
(c) Cl^- (d) CH_3^-. **(UP SEAT)**

97. Which is stable carbanion ?

(a) 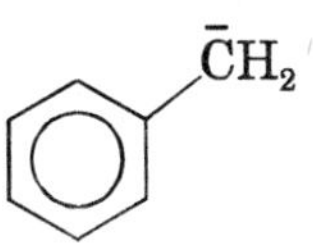
(b) $p\text{-}NO_2C_6H_4CH_2$
(c) $p\text{-}OCH_3\text{–}C_6H_4CH_2$ (d) $p\text{-}CH_3\text{–}C_6H_4CH_2$. **(UP SEAT)**

98. Which of the following doesn't show geometrical isomerism ?
(a) $(CH_3)_3C-C=CH_2$ (b) $CH_3CH=N-OH$
(c) $CH_3CH=CHCH_3$ (d) none of these. **(UP SEAT)**

99. IUPAC name of $CH_3CH=CH-C\equiv CH$ is
(a) Pent-2-ene-4-yne (b) Pent-1-yne-3-ene
(c) Pent-3-ene-1-yne (d) None of these. **(UP SEAT)**

100. Name of the following compound is

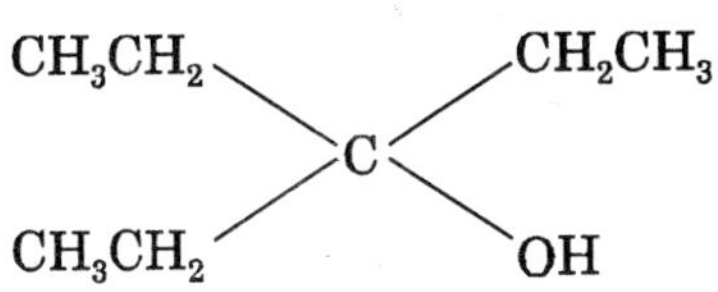

(a) 2-ethyl butanol-2
(b) 1-ethyl, 1-methyl-propanol-1
(c) 3-ethyl-pentanol-3
(d) diethyl ethanol. **(UP SEAT)**

101. The IUPAC name of the following compound $CH_3-C(CH_3)_2-CH=C(CH_3)_2$ is
(a) 1, 1, 3, 3-Tetramethyl-but-1-ene
(b) 1, 3, 3-Trimethyl-pent-2-ene
(c) 2, 2, 4-Trimethylbut-4-ene
(d) 2, 4, 4-Trimethylpent-2-ene. **(AIIMS)**

102. Which of the following compound has wrong IUPAC name ?
(a) $CH_3CH_2-CH_2COO-CH_2CH_3$ (Ethyl butanoate)
(b) 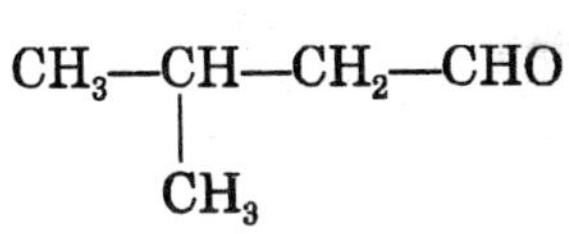 (3-Methylbutanal)
(c) 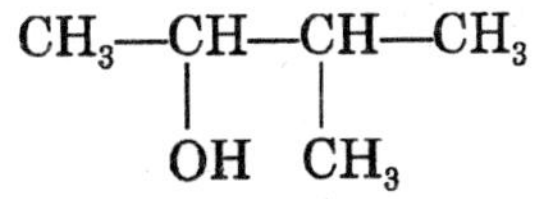 (2-Methyl-3-butanol)
(d) 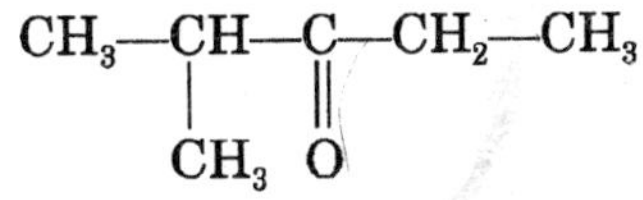 (2-Methyl-3-pentanone) **(AIEEE)**

103. Which of the following does not show geometrical isomerism ?
(a) 1, 2-Dichloro-1-pentene
(b) 1, 3-Dichloro-1-pentene
(c) 1, 1-Dichloro-1-pentene
(d) 1, 4-Dichloro-2-pentene. **(AIEEE)**

104. A similarity between optical and geometrical isomerism is that
(a) each forms equal number of isomers for a given compound
(b) if in a compound, one is present then so is the other
(c) both are included in stereoisomerism
(d) they have no similarity. **(AIEEE)**

105. Isomers of propionic acid are
(a) $HCOOC_2H_5$ and CH_3COOCH_3
(b) $HCOOC_2H_5$ and C_3H_7COOH
(c) CH_3COOCH_3 and C_3H_7COOH
(d) C_3H_7OH and CH_3COOCH_3. **(MP PMT)**

106. Which is a pair of geometrical isomers ?

I. 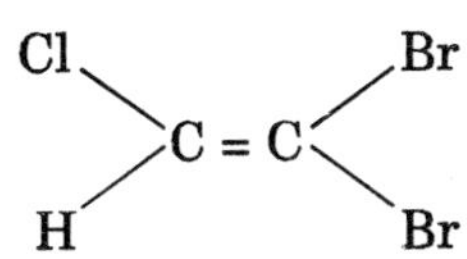

II. 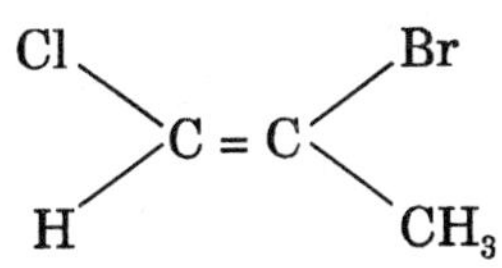

III. 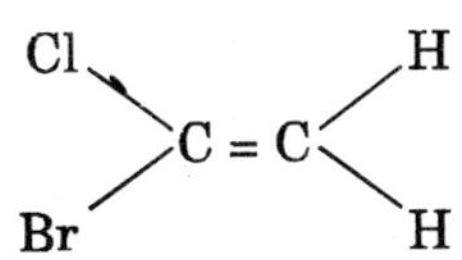

IV. 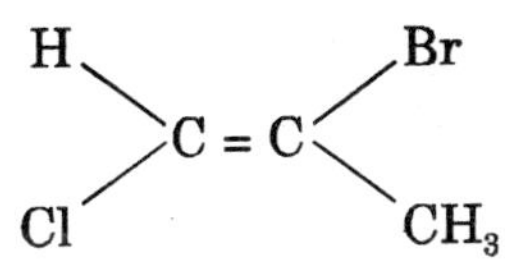

(a) I and II (b) I and III
(c) II and IV (d) III and IV. **(DPMT)**

107. A chiral compound is
(a) 2, 3, 4-Trimethylhexane
(b) n-Hexane

(c) Methane

(d) n-Butane. **(AIIMS)**

108. The compound having highest dipole moment is

(a) 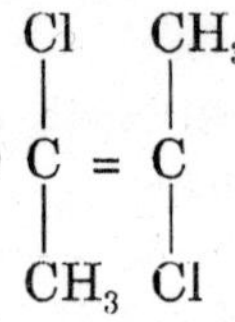 (b) 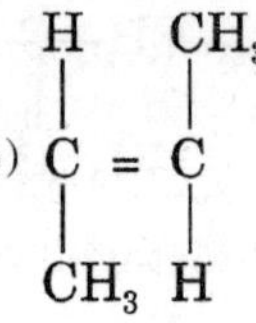

(c) 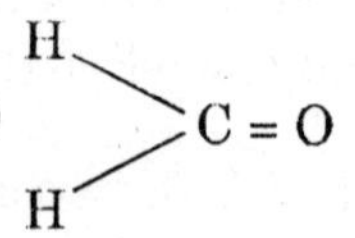 (d) 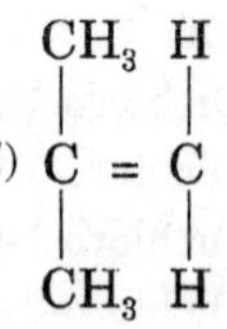

(AIIMS)

109. The prefixes *syn* and *anti* are used to denote

(a) Structural isomers

(b) Conformational isomers

(c) Geometrical isomers

(d) Optical isomers. **(DPMT)**

2003

110. The IUPAC name of 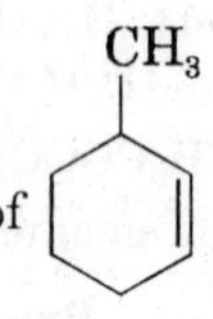 is

(a) 3-methyl cyclohexene (b) 1-methyl cyclohex-2-ene

(c) 6-methyl cyclohexene (d) 1-methyl cyclohex-5-ene. **(AIIMS)**

111. The most reactive nucleophile among the following is

(a) $CH_3\ddot{\underset{..}{O}}:^-$ (b) $C_6H_5\ddot{\underset{..}{O}}:^-$

(c) $(CH_3)_2CH{-}\ddot{\underset{..}{O}}:^-$ (d) $(CH_3)_3{-}C{-}\ddot{\underset{..}{O}}:^-$ **(AIIMS)**

112. The absolute configuration of the following compound is

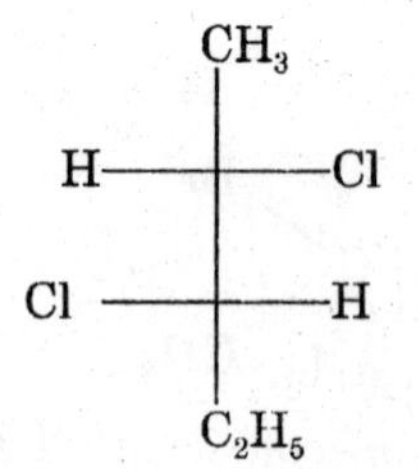

(a) 2S, 3R (b) 2S, 3S

(c) 2R, 3S (d) 2R, 3R. **(AIIMS)**

113. Among the following, the achiral amino acid is

(a) 2-Ethylalanine

(b) 2-Methylglycine

(c) 2-Hydroxymethylserine

(d) Tryptophan. **(AIIMS)**

114. In this reaction,

$$CH_3CHO + HCN \longrightarrow CH_3CH(OH)CN \xrightarrow{H_2O} CH_3CH(OH)COOH.$$

an asymmetric centre is generate. The acid obtained would be

(a) D-isomer

(b) L-isomer

(c) 50% D + 50% L-isomer

(d) 20% D + 80% L-isomer. **(CBSE PMT)**

115. Which of the following pairs of compounds are enantiomers ?

(a) 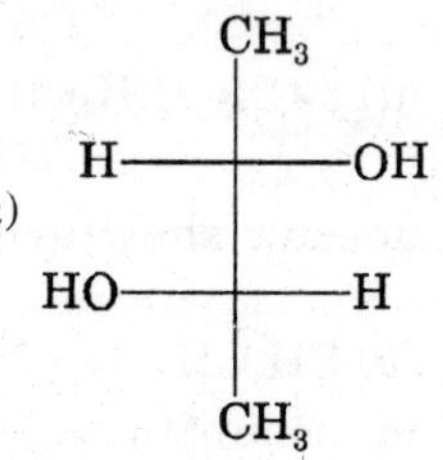 and 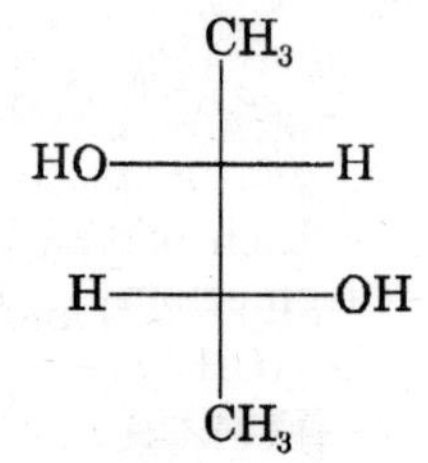

(b) 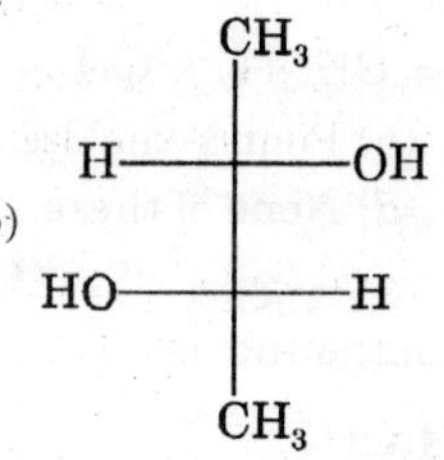 and 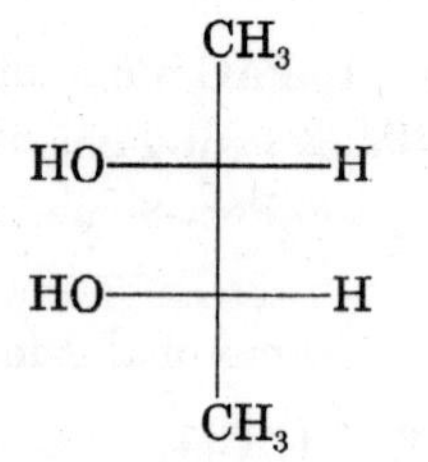

(c) 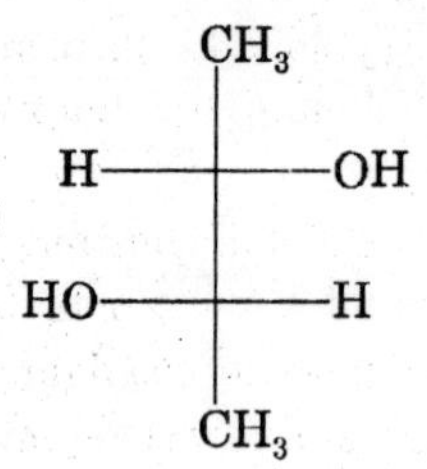 and 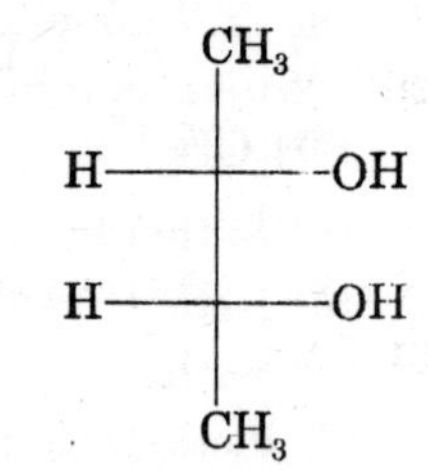

(d) 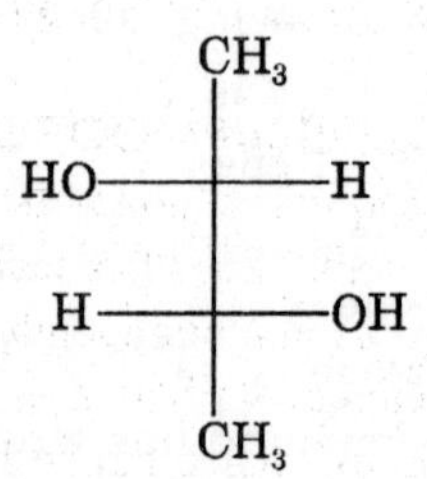 and 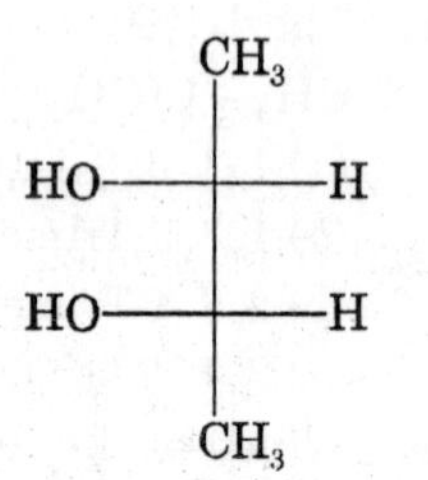

(CBSE PMT)

116. The IUPAC name of $CH_3COCH(CH_3)_2$ is

(a) isopropyl methyl ketone

(b) 2-methyl-3-butanone

(c) 4-methylisopropyl ketone

(d) 3-methyl-2-butanone. **(AIEEE)**

117. The general formula $C_nH_{2n}O_2$ could be for open chain
(*a*) Diketones (*b*) Carboxylic acids
(*c*) Diols (*d*) Dialdehydes. ***(AIEEE)***

118. Among the following structures I to IV,

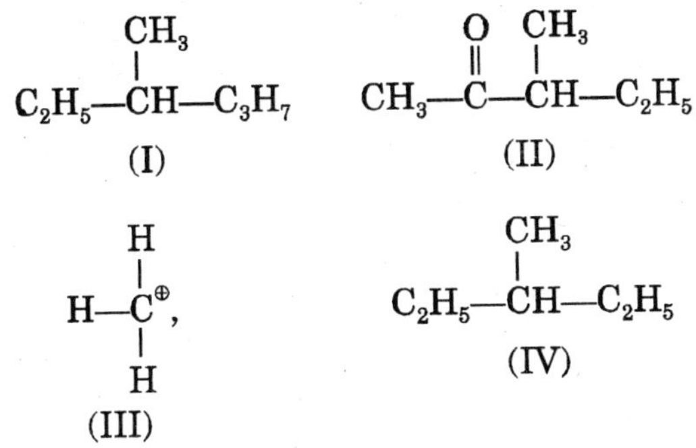

It is true that
(*a*) All four are chiral compounds
(*b*) Only I and II are chiral compounds
(*c*) Only III is a chiral compound
(*d*) Only II and IV are chiral compounds. ***(AIEEE)***

2004

119. The IUPAC name of 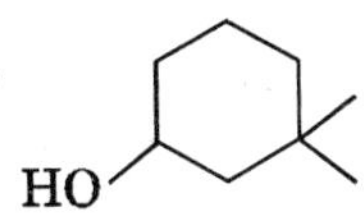 is
(*a*) 3, 3-dimethyl-1-hydroxy cyclohexane
(*b*) 1, 1-dimethyl-3-cyclohexanol
(*c*) 3, 3-dimethyl-1-cyclohexanol
(*d*) 1, 1-dimethyl-3-hydroxy cyclohexane. ***(AIEEE)***

120. Which one of the following does not have sp^2 hybridised carbon ?
(*a*) Acetone (*b*) Acetamide
(*c*) Acetonitrile (*d*) Acetic acid. ***(AIEEE)***

121. Which of the following will have a meso isomer also ?
(*a*) 2-chlorobutane
(*b*) 2-Hydroxypropanoic acid
(*c*) 2, 3-Dichloropentane
(*d*) 2, 3-dichloro butane. ***(AIEEE)***

122. The geometrical isomerism is shown by
(*a*) 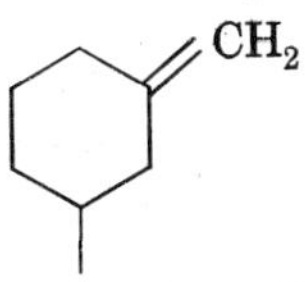
(*b*) 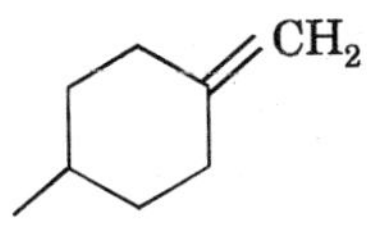
(*c*) 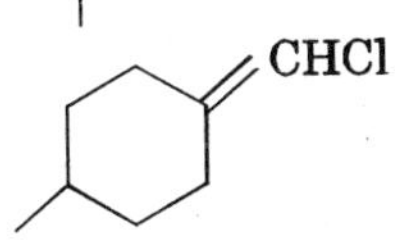
(*d*) 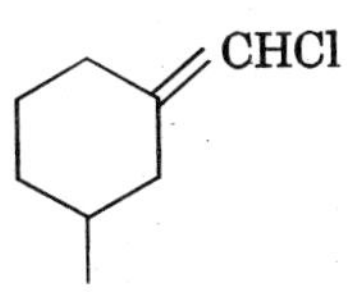
(AIIMS)

123. The compound having only primary hydrogen atoms is :
(*a*) Isobutene (*b*) 2, 3-Dimethyl butene
(*c*) Cyclohexane (*d*) Propyne. ***(AIIMS)***

124. The dipole moment is highest for :
(*a*) *Trans*-2-butene (*b*) 1, 3-Dimethyl benzene
(*c*) Acetophenone (*d*) Ethanol. ***(AIIMS)***

125. The number of chiral carbons in β-D-(+)-glucose is
(*a*) Five (*b*) Six
(*c*) Three (*d*) Four. ***(CBSE PMT)***

126. The molecular formula of diphenyl methane is,
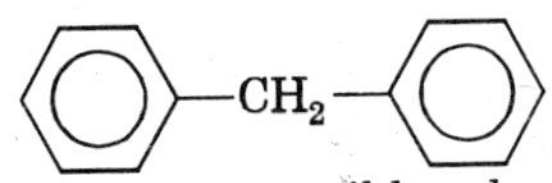. How many structural isomers are possible when one of the hydrogens is replaced by a chlorine atom.
(*a*) 6 (*b*) 4
(*c*) 8 (*d*) 7.

2005

127. The number and type of bonds between two carbon atoms in calcium carbide are
(*a*) Two sigma, two pi (*b*) Two sigma, one pi
(*c*) One sigma, two pi (*d*) One sigma, one pi.
(AIEEE)

128. Which type of isomerism is shown by 2, 3-dichloro butane ?
(*a*) Structural (*b*) Geometric
(*c*) Optical (*d*) Diastereo. ***(AIEEE)***

129. Of the five isomeric hexane, the isomer which can give two monochlorimated compound is
(*a*) 2-methyl pentane (*b*) 2, 2-dimethyl butane
(*c*) 2, 3-dimethyl butane (*d*) *n*-hexane. ***(AIEEE)***

130. The IUPAC name for $CH_3CO—CH_3$ is
(*a*) Dimethyl ketone (*b*) Acetone
(*c*) Propanal (*d*) Propanone.
(J & K CET)

131. Which is the most stable carbocation ?
(*a*) *iso*-propyl cation
(*b*) Triphenylmethyl cation
(*c*) Ethyl cation
(*d*) *n*-propyl cation. ***(J & K CET)***

132. What kind of isomerism is possible for 1-chloro-2-nitroethene ?
(*a*) Functional group isomerism
(*b*) Position isomerism
(*c*) *E*/*Z* isomerism
(*d*) Optical isomerism. ***(J & K CET)***

133. Which of the following will be chiral ?

(*a*) CH_3CHCl_2 (*b*) $CH_3CHBrCl_2$
(*c*) CD_2Cl_2 (*d*) CH_2ClBr. ***(J & K CET)***

134. Which of the following is an electrophile ?

(*a*) H_2O (*b*) SO_3
(*c*) NH_3 (*d*) ROR. ***(J & K CET)***

135. An aromatic compound among other things should have a π – electron cloud containing $(4n + 2)\pi$ electrons where n can't be

(*a*) 1/2 (*b*) 3
(*c*) 3 (*d*) 1. ***(J & K CET)***

136. The compound which contains all the four 1°, 2°, 3° and 4° carbon atoms is

(*a*) 2, 3-dimethyl pentane
(*b*) 3-chloro-2, 3-dimethylpentane
(*c*) 2, 3, 4-trimethylpentane
(*d*) 3, 3-dimethylpentane. ***(J & K CET)***

137. Enol content is highest in

(*a*) Acetone (*b*) Acetophenone
(*c*) Acetic acid (*d*) Acetyl acetone. ***(Orissa JEE)***

138. Number of bonds in benzene

(*a*) 6σ and 3π (*b*) 12σ and 3π
(*c*) 3σ and 12π (*d*) 6σ and 6π. ***(DPMT)***

139. 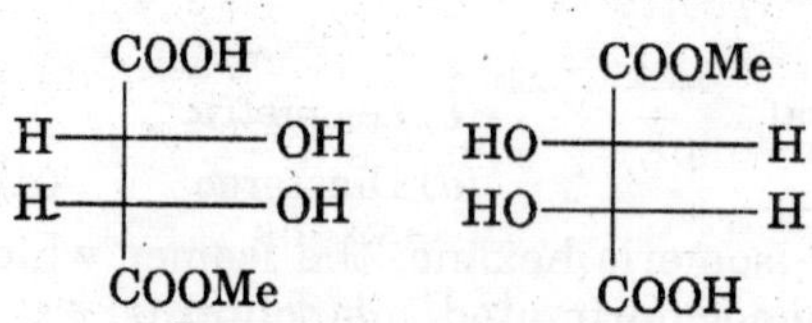

I II

I and II are

(*a*) Enantiomers (*b*) Diastereomers
(*c*) Mesocompound (*d*) Identical. ***(DPMT)***

140. Which will give chiral molecule ?

(*a*) $CH_3COCl \xrightarrow{LiAlH_4}$

(*b*) $C_2H_5CHO \xrightarrow[H^+/H_2O]{CH_3MgBr}$

(*c*) $(CH_3)_2CHC_2H_5 \xrightarrow{Cu}$

(*d*) 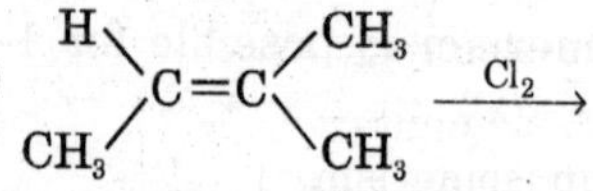 ***(DPMT)***

141. $H_3C—C(Cl) = CH—CH(CH_3)—CH_3$

(*a*) 2-chloro-4-methyl-2-pentene
(*b*) 4-chloro-2-methyl-3-pentene
(*c*) 4-methyl-2-chloro-2-pentene
(*d*) 2-chloro-4, 4-dimethyl-2-butene. ***(Karnataka CET)***

142. The IUPAC name of the compound is

$HOOC—CH_2—CH(COOH)—CH_2—CH_2—COOH$

(*a*) 2(carboxymethyl)-pentane-1, 5-dioic acid
(*b*) 3-carboxyhexane-1, 6-dioic acid
(*c*) butane-1, 2, 4-tricarboxylic acid
(*d*) 4-carboxyhexane-1, 6-dioic acid
(*e*) 1, 2-dicarboxy-pentanoic acid. ***(Kerala PET)***

143. How many sigma and pi bonds are present in toluene ?

(*a*) $3\pi + 8\sigma$ (*b*) $3\pi + 6\sigma$
(*c*) $3\pi + 10\sigma$ (*d*) $6\pi + 3\sigma$
(*e*) $6\pi + 6\sigma$. ***(Kerala PET)***

144. The name of the compound is

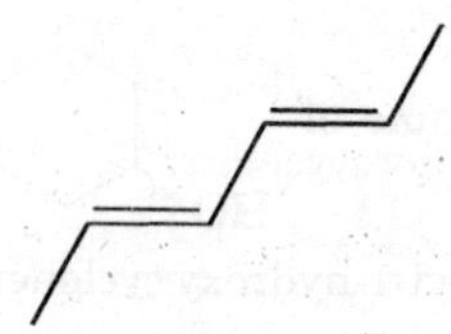

(*a*) (2Z, 4Z)-2, 4-hexadiene
(*b*) (2Z, 4E)-2, 4-hexadiene
(*c*) (2E, 4Z)-2, 4-hexadiene
(*d*) (4E, 4Z)-2, 4-hexadiene
(*e*) (2E, 4E)-2, 4-hexadiene. ***(J & K CET)***

145. Which of the following Fischer's projection formula is identical to D-glyceraldehyde ?

(*a*) 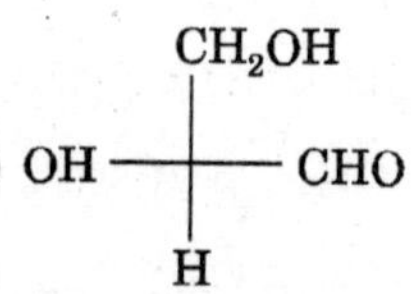 (*b*) 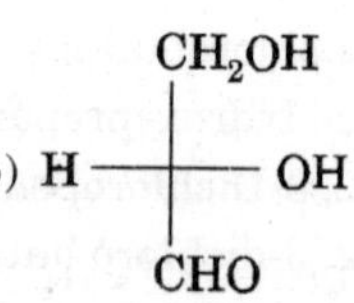

(*c*) 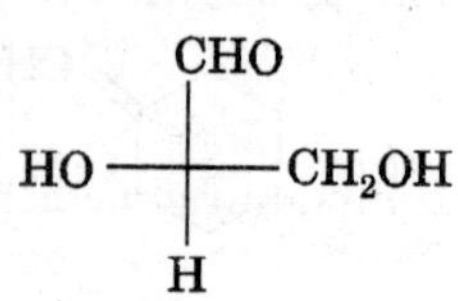 (*d*) 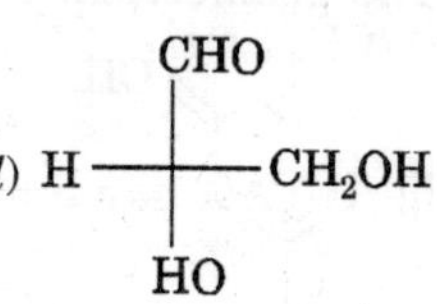

(*e*) 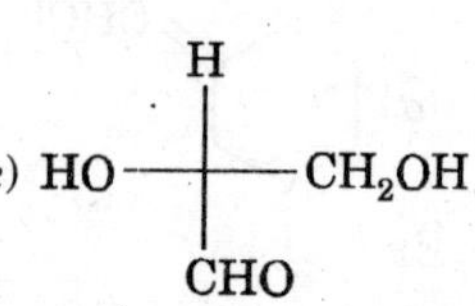 ***(Kerala PET)***

2006

146. Which of the following is ***not*** chiral ?
(*a*) 2-Hydroxypropanoic acid
(*b*) 2-Butanol
(*c*) 2, 3-Dibromopentane
(*d*) 3-Bromopentane. ***(CBSE PMT)***

147. The IUPAC name of 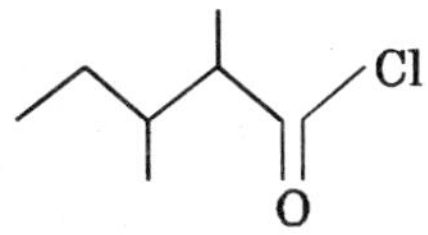 is
(*a*) 2, 3-dimethylpentanoyl chloride
(*b*) 3, 4-dimethylpentanoyl chloride
(*c*) 1-chloro-1-oxo-2, 3-dimethylpentane
(*d*) 2-ethyl-3-methylbutanoyl chloride. ***(CBSE PMT)***

148. The IUPAC name of the compound shown below is

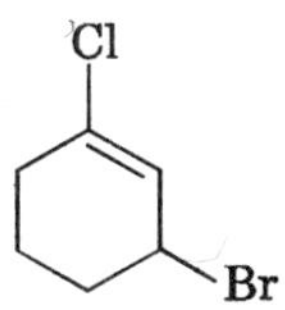

(*a*) 2-bromo-6-chlorocyclohex-1-ene
(*b*) 6-bromo-2-chlorocyclohexene
(*c*) 3-bromo-1-chlorocyclohexene
(*d*) 1-bromo-3-chlorocyclohexene. ***(AIEEE)***

149. Increasing order of stability among the three main conformations (*i.e.,* Eclipse, Anti, Gauche) of 2-fluoroethanol is
(*a*) Eclipse, Gauche, Anti
(*b*) Gauche, Eclipse, Anti
(*c*) Eclipse, Anti, Gauche
(*d*) Anti, Gauche, Eclipse. ***(AIEEE)***

150. The increasing order of stability of the following free radicals is
(*a*) $(CH_3)_2\dot{C}H < (CH_3)_3\dot{C} < (C_6H_5)_2\dot{C}H < (C_6H_5)_3\dot{C}$
(*b*) $(C_6H_5)_3\dot{C} < (C_6H_5)_2\dot{C}H < (CH_3)_3\dot{C} < (CH_3)_2\dot{C}H$
(*c*) $(C_6H_5)_2\dot{C}H < (C_6H_5)_3\dot{C} < (CH_3)_3\dot{C} < (CH_3)_2\dot{C}H$
(*d*) $(CH_3)_2\dot{C}H < (CH_3)_3\dot{C} < (C_6H_5)_3\dot{C} < (C_6H_5)_2\dot{C}H$. ***(AIEEE)***

151. The IUPAC name of the compound shown below is

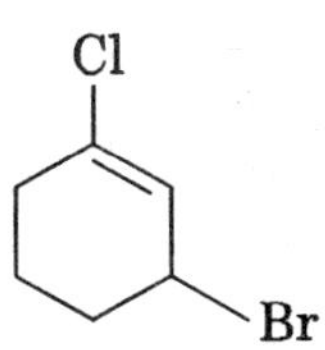

(*a*) 2-bromo-6-chlorocyclohex-1-ene
(*b*) 6-bromo-2-chlorocyclohexane
(*c*) 3-bromo-1-chlorocyclohexane
(*d*) 1-bromo-3-chlorocyclohexane. ***(AIEEE)***

152. Which species are more resonance stabilized in the following pairs ?
$[C_6H_5Cl, C_6H_5CH_3]$, $[CH_2 = CHCl_5]$ $CH_2 = CHCH_2Cl]$, $[C_6H_5Br, C_6H_5CH_2Br]$, $[CH_3COOH, CH_3COO^-]$
(*a*) $C_6H_5Cl, CH_2 = CHCH_2Cl, C_6H_5CH_2Br, CH_3COO^-$
(*b*) $C_6H_5CH_3, CH_2 = CHCH_2Cl, C_6H_5CH_2Br, CH_3COO^-$
(*c*) $C_6H_5CH_3, CH_2 = CHCH_2Cl, C_6H_5CH_2Br, CH_3COOH$
(*d*) $C_6H_5Cl, CH_2 = CHCl, C_6H_5Br, CH_3COOH$
(*e*) $C_6H_5Cl, CH_2 = CHCl, C_6H_5Br, CH_3COO^-$. ***(Kerala MEE)***

153. Consider the following compound

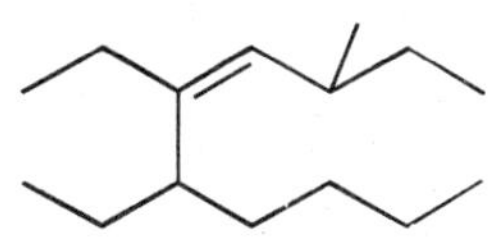

IUPAC name of this compound is
(*a*) 5, 6-diethyl-3-methyldecane
(*b*) 5, 6-diethyl-3-methylec-4-ene
(*c*) 3, 5. 6-triethyldec-6-ene
(*d*) 3, 5, 6-trimethyldec-4-ene
(*e*) 5, 6-diethyl-8-methyldec-6-ene. ***(Kerala MEE)***

154. Consider the following compounds :
A. chloroethene B. benzene
C. 1, 3-butadiene D. 1, 3, 5-hexatriene
All the carbon atoms are sp^2-hybridised in
(*a*) A, C, D only (*b*) A, B only
(*c*) B, C, D only (*d*) C, D only
(*e*) A, B, C and D. ***(Kerala MEE)***

155. An organic compound which produces a bluish green coloured flame on heating in presence of copper is
(*a*) Chlorobenzene (*b*) Benzaldehyde
(*c*) Aniline (*d*) Benzoic acid. ***(KCET)***

156. Inductive effect involves
(*a*) Displacement of σ-electrons
(*b*) Delocalisation of π-electrons
(*c*) Delocalisation of σ-electrons
(*d*) Displacement of π-elements. ***(KCET)***

157. The general molecular formula, which represents the homologous series of alkanols is
(*a*) $C_nH_{2n+2}O$ (*b*) $C_nH_{2n}O_2$
(*c*) $C_nH_{2n}O$ (*d*) $C_nH_{2n+1}O$. ***(KCET)***

158. The IUPAC name of

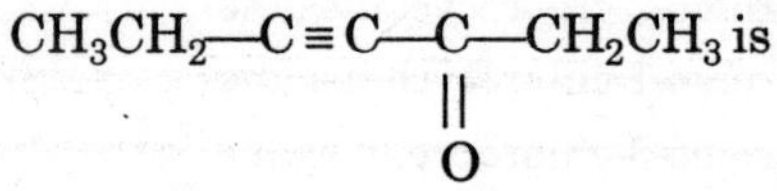

(a) 3-oxo-2-hyptyne (b) hept-3-yn-4-oxone
(c) hept-4-yn-3-one (d) hept-3-yn-4-one.

(Orissa JEE)

2007

159. If there is no rotation of plane polarized light by a compound in a specific solvent, thought to be chiral, it may means that

(a) The compound may be racemic mixture
(b) The compound is certainly a chiral
(c) The compound is certainly meso
(d) There is no compound in the solvent.

(CBSE-AIPMT Screening)

160. $CH_3—CHCl—CH_2—CH_3$ has a chiral centre. Which of the following represents its R configuration ?

(a) 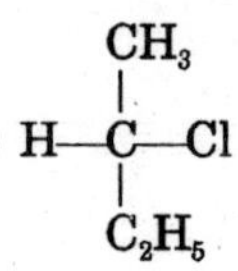 (b) 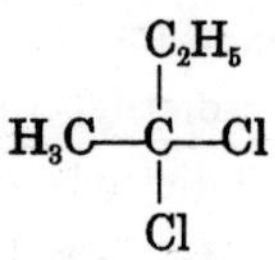

(c) 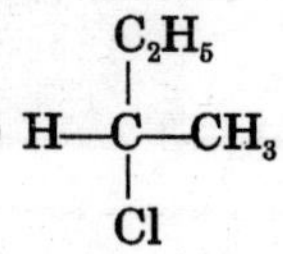 (d) 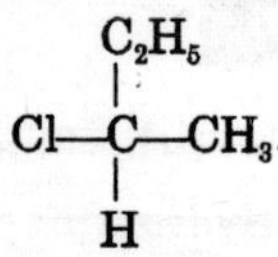

(CBSE-AIPMT Screening)

161. Which of the following molecules is expected to rotate the plane of plane polarized light ?

(a) 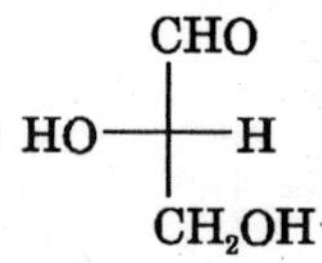 (b) 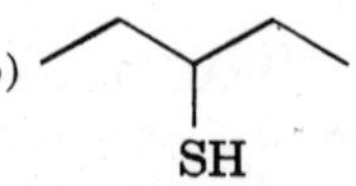

(c) 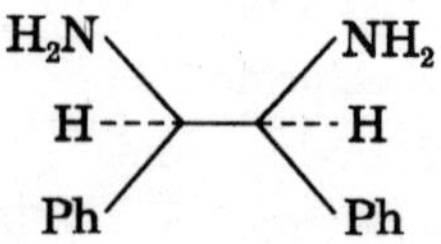 (d) 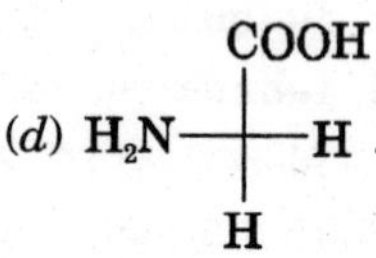 *(AIEEE)*

162. Which one of the following conformations of cyclohexane is chiral ?

(a) Thrust boat (b) Regid
(c) Chair (d) Boat. *(AIEEE)*

163. The IUPAC name of 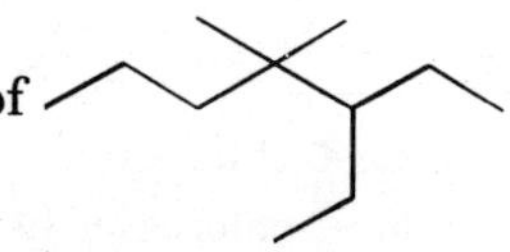 is

(a) 1, 1-diethyl-2, 2-dimethyl pentane
(b) 4, 4-dimethyl-5, 5-diethyl pentane
(c) 5, 5-diethyl-4, 4-dimethyl pentane
(d) 3-ethyl-4, 4-dimethyl heptane. *(AIEEE)*

ANSWERS

1. (b)	2. (b)	3. (c)	4. (b)	5. (b)	6. (b)	7. (a)	8. (b)	9. (c)
10. (d)	11. (d)	12. (c)	13. (c)	14. (c)	15. (a)	16. (d)	17. (d)	18. (d)
19. (c)	20. (b)	21. (a)	22. (d)	23. (d)	24. (d)	25. (a)	26. (a)	27. (a)
28. (c)	29. (b)	30. (d)	31. (a)	32. (a)	33. (a)	34. (c)	35. (c)	36. (a)
37. (a)	38. (b)	39. (a)	40. (a)	41. (b)	42. (c)	43. (d)	44. (d)	45. (d)
46. (c)	47. (c)	48. (b)	49. (a)	50. (c)	51. (d)	52. (c)	53. (a)	54. (d)
55. (c)	56. (c)	57. (a)	58. (c)	59. (a)	60. (a)	61. (b)	62. (d)	63. (a)
64. (b)	65. (b)	66. (c)	67. (a)	68. (a)	69. (b)	70. (a)	71. (a)	72. (a)
73. (d)	74. (b)	75. (a)	76. (a)	77. (c)	78. (d)	79. (d)	80. (b)	81. (a)
82. (a)	83. (d)	84. (a)	85. (b)	86. (d)	87. (b)	88. (b)	89. (d)	90. (a)
91. (a)	92. (b)	93. (a)	94. (c)	95. (b)	96. (d)	97. (b)	98. (a)	99. (c)
100. (c)	101. (d)	102. (c)	103. (c)	104. (c)	105. (a)	106. (c)	107. (a)	108. (c)
109. (c)	110. (a)	111. (d)	112. (b)	113. (b)	114. (c)	115. (a)	116. (d)	117. (b)
118. (b)	119. (c)	120. (c)	121. (d)	122. (d)	123. (b)	124. (d)	125. (d)	126. (b)
127. (c)	128. (c)	129. (c)	130. (d)	131. (b)	132. (c)	133. (b)	134. (b)	135. (a)
136. (b)	137. (d)	138. (b)	139. (d)	140. (b)	141. (a)	142. (c)	143. (c)	144. (e)
145. (c)	146. (d)	147. (a)	148. (c)	149. (a)	150. (b)	151. (c)	152. (e)	153. (b)
154. (e)	155. (a)	156. (a)	157. (a)	158. (c)	159. (c)	160. (d)	161. (a)	162. (a)
163. (d).								

HINTS/SOLUTIONS

Q. No.	Ans.	Hints & Solutions
1.	(*b*)	In *trans*-2-butene the individual bond dipoles cancel each other. 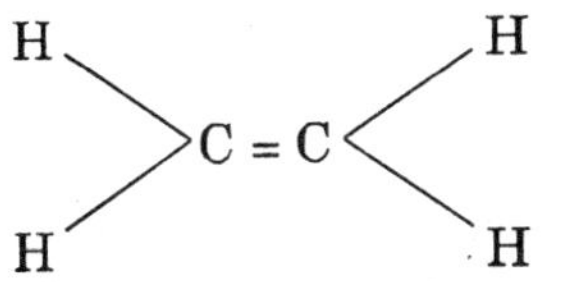
2.	(*b*)	Each carbon in ethene is directly linked to three other atoms and hence is sp^2 hybridized.
3.	(*c*)	$\overset{4}{C}H_3-\overset{3}{C}H(CH_3)-\overset{2}{C}H_2-\overset{1}{C}H_2-Br$ 1-Bromo-3-methylbutane
4.	(*b*)	C_6H_5—(phenyl group) is electron withdrawing group. It withdraws electron density through resonance as well as through inductive effect, disperses the negative charge and hence stabilizes the carbanion. On the other hand CH_3— groups destabilize the carbanion by concentration of negative charge.
5.	(*b*)	 is a cyclic system having 6 π-electrons which obeys $(4n + 2)$ Huckel rule.
6.	(*b*)	$CH_3-CH(OH)-CH_3$ (H and OH on the central C) This molecule does not contain any chiral centre and hence is a chiral and does not exhibit optical isomerism.
7.	(*a*)	In $C_6H_5 - C^+H - C_6H_5$, positive charge is dispersed on both the benzene rings through resonance. Due to maximum dispersal of charge it is the most stable. 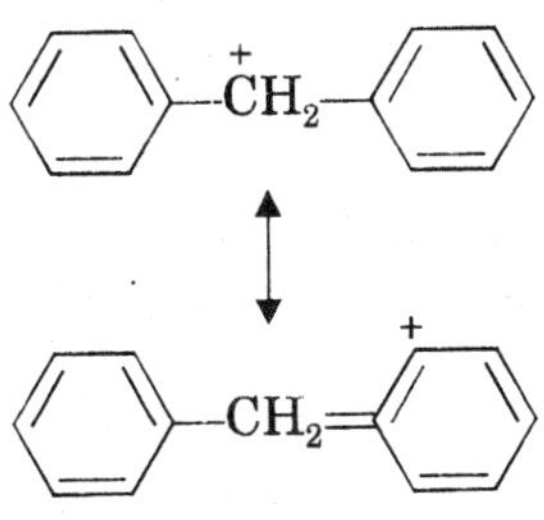 .... and so on
8.	(*b*)	$Cl-CCl_2-\overset{2}{C}H_2-\overset{1}{C}HO$ 3, 3, 3-Trichloropropanal
9.	(*c*)	The carbon of double bond is numbered 1 on the chain.
10.	(*d*)	Cyclopentadienyl cation has only 4 π-electrons, whereas for aromatic character the species should have $(4n + 2)$ π electrons *i.e.*, 2, 6, 10, etc.
11.	(*d*)	There are two asymmetric carbon atoms. Hence, the number of optical isomers is $(2)^n = (2)^2 = 4$.
12.	(*c*)	These are functional isomers.
13.	(*c*)	*n*-Butyl, *iso*-butyl, *sec*-butyl and *tert*-butyl.
14.	(*c*)	$CH_3-\overset{4}{C}H(\overset{5}{C}H_2-\overset{6}{C}H_3)-\overset{3}{C}H_2-\overset{2}{C}H(OH)-\overset{1}{C}H_3$ 4-Methyl-2-hexanol
15.	(*a*)	Follow longest chain rule as well as locant rule.
16.	(*d*)	Primary and secondary nitro compounds exhibit tautomerism.
17.	(*d*)	Presence of triple bond.
18.	(*d*)	In anti or staggered configuration the repulsions are minimum.
19.	(*c*)	Conformers arise due to free rotation of C—C single bond.
20.	(*b*)	Because two different groups are attached to C = C.
21.	(*a*)	$\overset{5}{C}H_3-\overset{4}{C}H=\overset{3}{C}H-\overset{2}{C}\equiv\overset{1}{C}H$ Pent-3-en-1-yne When double and triple bond are at unequivalent positions lowest sum rule is followed.
22.	(*d*)	In benzene molecule all the C—C bonds have same bond length due to resonance.
23.	(*d*)	$C_2H_2 + CH_2 \rightarrow C_3H_4$
24.	(*d*)	All the carbon atoms are sp^2 hybridised.

Q. No.	Ans.	Hints & Solutions
25.	(*a*)	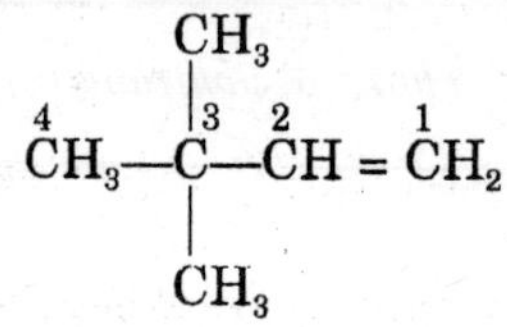 3, 3-Dimethyl-1-butene
26.	(*a*)	It is a bicyclic compound.
27.	(*a*)	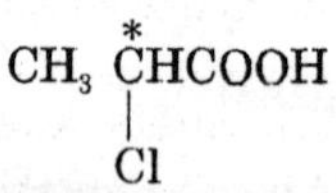
28.	(*c*)	CH_4 has a regular T_d structure with zero dipole moment.
29.	(*b*)	C in diamond (sp^3), in graphite (sp^2) and in acetylene (sp).
30.	(*d*)	It has only *sp* hybridised carbon atoms.
31.	(*a*)	It is $-NR_2 < -OR < -F$.
32.	(*a*)	In $DCH_2CH_2CH_2Cl$ there is no chiral carbon. It has no asymmetric carbon.
33.	(*a*)	Select longest chain and also follow locant rule.
34.	(*c*)	*n*-Butyl alcohol, *iso*-butyl alcohol, *sec*-butyl alcohol, *tert*-butyl alcohol, diethyl ether, ethyl *n*-propyl ether, ethyl *iso*-propyl ether.
35.	(*c*)	It forms 7 isomers.
36.	(*a*)	$CH_3CH = CH—CH_3$ shows geometrical isomerism.
37.	(*a*)	It is bicyclic compound. (See concept oriented details).
38.	(*b*)	The three isomeric trimethylbenzenes are 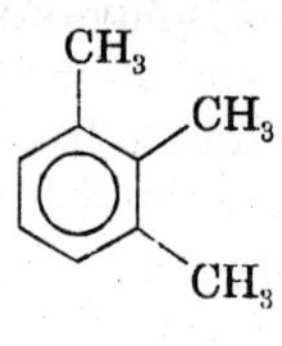 1, 2, 3-Trimethylbenzene 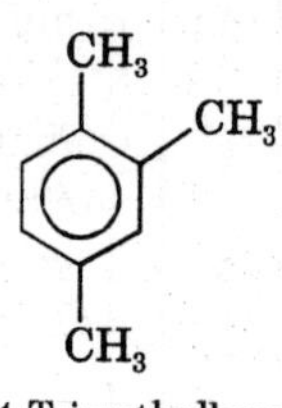 1, 2, 4-Trimethylbenzene 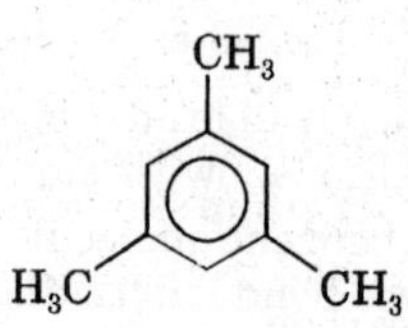 1, 3, 5-Trimethylbenzene
39.	(*a*)	Triple bond at corner of chain is preferred over double bond (see details of nomenclature).
40.	(*a*)	Resonance structures may or may not have equal energies.
41.	(*b*)	In acetylene both the carbon atoms are *sp*-hybridized.
42.	(*c*)	Electron deficient species.
43.	(*d*)	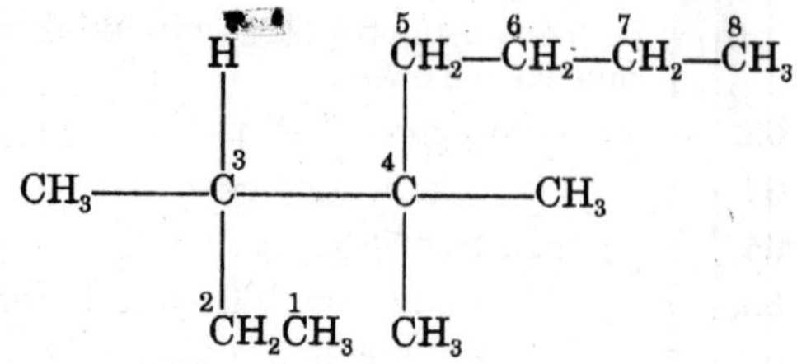 3, 4, 4-Trimethyloctane
44.	(*d*)	Factual question.
45.	(*d*)	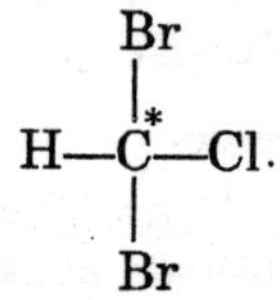 This molecule has a chiral carbon.
46.	(*c*)	It is an acid hence suffix is oic acid.
47.	(*c*)	One double bond means one π bond.
48.	(*b*)	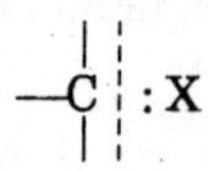 It forms cations and anions.
49.	(*a*)	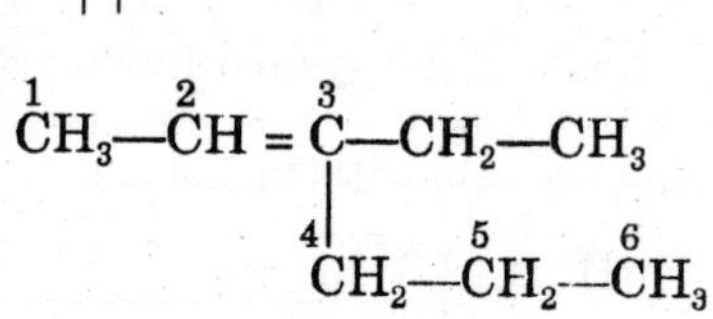 3-Ethyl-2-hexene
50.	(*c*)	See concepts in brief.
51.	(*d*)	The molecules of *p*-dichlorobenzene, due to their symmetry pack well in solid state.
52.	(*c*)	C-3 is sp^3-hybridized whereas other carbon atoms are in sp^2 hybrid state.
53.	(*a*)	Chlorine group is functional group and hence it is given priority.
54.	(*d*)	Such a pair is known as a pair of distereo isomers.
55.	(*c*)	C in CH_4 is sp^3 hybridised.
56.	(*c*)	C_nH_{2n-2}.
57.	(*a*)	Factual. Reason itself is answer.
58.	(*c*)	CHO group is higher in priority.
59.	(*a*)	See details in nomenclature.
60.	(*a*)	$CH_2 = CH - CHO$ is acraldehyde.
61.	(*b*)	Double bond to be taken in chain.

Q. No.	Ans.	Hints & Solutions
62.	(*d*)	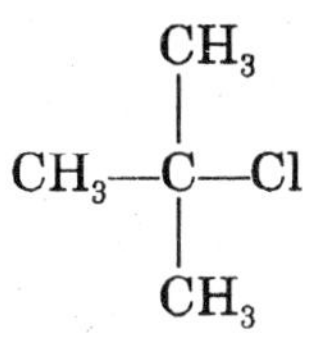 is terbutyl chloride.
63.	(*a*)	Select longest chain. Open C_4H_9 group.
64.	(*b*)	It is an saturated ester
65.	(*b*)	The carbon of CHO is numbered as 1.
66.	(*c*)	It is an ether. Parent chain should have more no. of carbon atoms.
67.	(*a*)	Non terminal alkenes are more stable than terminal alkenes. Out of cis and trans, trans is more stable.
68.	(*a*)	They rotate the plane polarised light in opposite directions.
69.	(*b*)	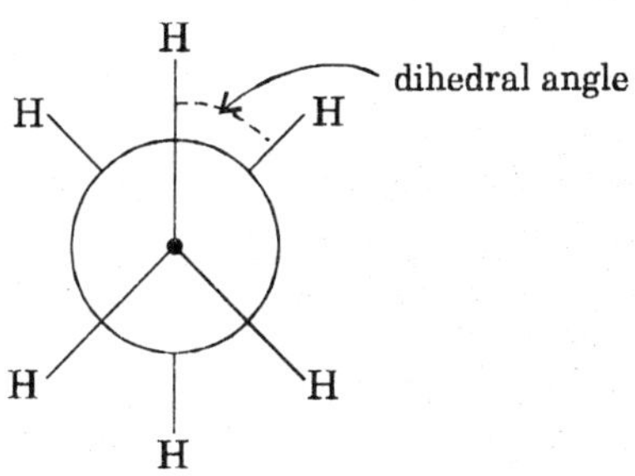
70.	(*a*)	Draw and check.
71.	(*a*)	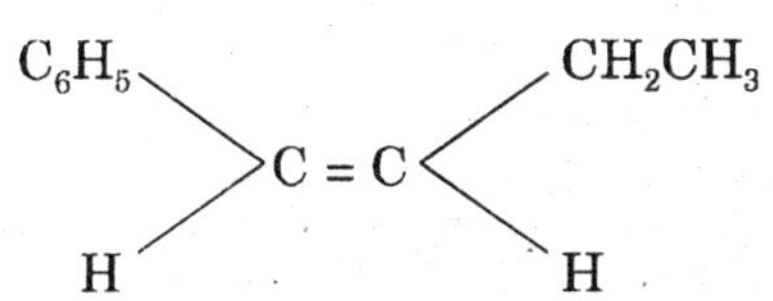 This compound fulfils conditions of geometrical isomerism.
72.	(*a*)	Aldehydes and ketones are functional isomers.
73.	(*d*)	See concepts in brief. Select longest chain of six carbon atoms.
74.	(*b*)	Follow longest chain rule.
75.	(*a*)	$CH_2 = CH - CH_2OH$ is vinyl carbinol. ∵ CH_3OH is called carbinol.
76.	(*a*)	It is an alcohol with seven carbon atoms.
77.	(*c*)	Select longest chain and number it so that double bond gets least number.
78.	(*d*)	It is the correct structure.
79.	(*d*)	Carbon of CN group is numbered as 1.
80.	(*b*)	The 4th carbon has two methyl groups as substituents.
81.	(*a*)	It is 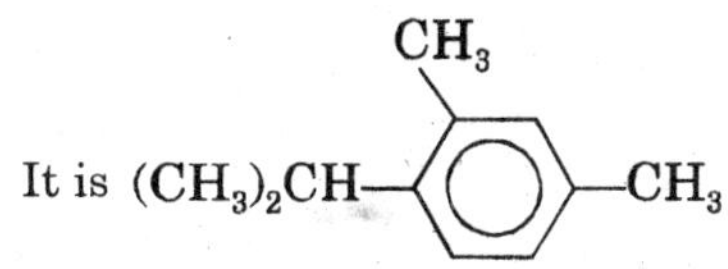 The similar groups are to be given lower priority.
82.	(*a*)	CO group is of more priority than CH_3 – group.
83.	(*d*)	See concepts in brief.
84.	(*a*)	Name of substituents are written in alphabetical order.
85.	(*b*)	It has 2 asymmetric carbon atoms.
86.	(*d*)	Ethers are isomeric with only alcohols and other ethers of same formula.
87.	(*b*)	If fulfils conditions of geometrical isomerism.
88.	(*b*)	$CH_3OCH_2CH_2CH_3$, $CH_3CH_2OCH_2CH_3$, 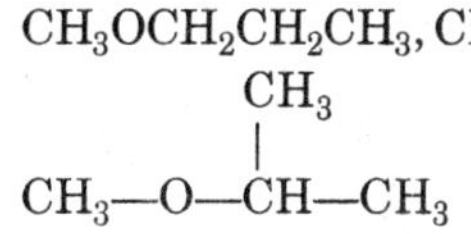
89.	(*d*)	Alcohols and ethers are functional isomers of each other.
90.	(*a*)	It has two similar groups CH_2Br attached to same carbon.
91.	(*a*)	The position of chlorine atom is different.
92.	(*b*)	Attack at C(3) produces an asymmetric carbon.
93.	(*a*)	It contains chiral carbon.
94.	(*c*)	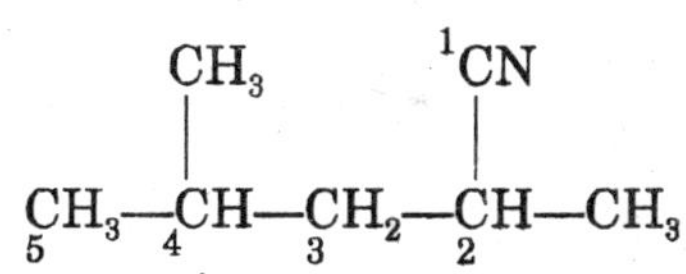
95.	(*b*)	It is due to more number of hypoconjugation structure formed in this case.
96.	(*d*)	A weak acid gives a strongest nucleophile. CH_4 is weak acid.
97.	(*b*)	Presence of electron with drawing NO_2 group stabilizes the anion.
98.	(*a*)	Terminal alkenes donot give geometrical isomers.
99.	(*c*)	Here ≡ bond is preferred over = bond (see concept oriented details).
100.	(*c*)	Select longest chain.
101.	(*d*)	Bond should get least number.
102.	(*c*)	OH group should get higher priority.
103.	(*c*)	It has two similar chlorine atoms on the same carbon.
104.	(*c*)	Both are space isomers.
105.	(*a*)	Acids are isomeric with esters.
106.	(*c*)	(*ii*) and (*iv*) are geometrical isomers.

Q. No.	Ans.	Hints & Solutions
107.	(a)	It has chiral carbon.
108.	(c)	>C = O is highly polar.
109.	(c)	It is used for geometrical isomers.
110.	(a)	Double bond is preferred.
111.	(d)	The nucleophile in which the charge density increases.
112.	(b)	It is S at carbon 2 and carbon 3.
113.	(b)	$NH_2—CH(CH_3)—COOH$ (2-methyl glycine).
114.	(c)	It can exist into two forms D & L hence a racemic mixture is formed.
115.	(a)	Enantiomers are non super imposable mirror images of each other.
116.	(d)	3-methyl butan-2-one.
117.	(b)	It is for carboxylic acids.
118.	(b)	A chiral centre or carbon has four different groups attached to it.
119.	(c)	OH group is given preference.
120.	(c)	Acetonitrite is $CH_3—C \equiv N$.
121.	(d)	2, 3-Dichloro butane will have a plane of symmetry.
122.	(d)	If fulfils the conditions of geometrical isomerism.
123.	(b)	All the H atoms are primary.
124.	(d)	C—O bond is highly polar.
125.	(d)	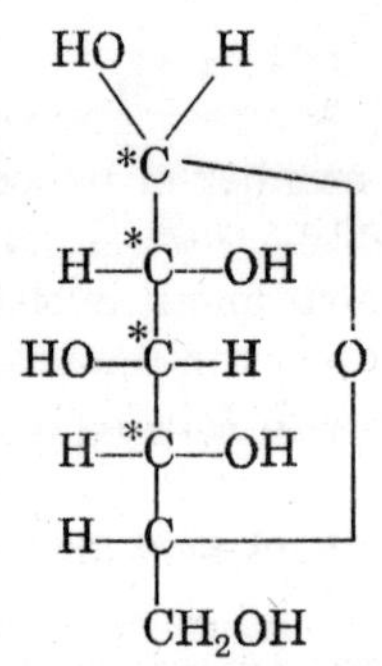 The marked carbon atoms (*) are asymmetric.
126.	(b)	It has 4 different types of H atoms in different environment.
127.	(c)	CaC_2 is $Ca^{2+} (C \equiv C)^{2-}$.
128.	(c)	$CH_3—\overset{*}{C}H(Cl)—\overset{*}{C}H(Cl)—CH_3$. It has two asymmetric carbon atoms. (*)

Q. No.	Ans.	Hints & Solutions
129.	(c)	$CH_3—CH(CH_3)—CH(CH_3)—CH_3$. It has two types of H atoms *i.e.*, 1° and 3° and hence two monochlorinated compounds.
130.	(d)	Propanone. It is a ketone.
131.	(b)	Stability of carbonium ions : tertiary alkyl > secondary alkyl > primary alkyl > methyl. More the number of alkyl groups, the greater the dispersal of positive charge and therefore, more the stability of carbonium ion is observed. 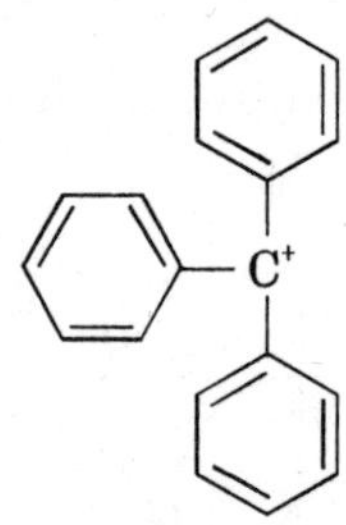 triphenylmethyl cation. Dispersal of positive charge increases with the increase in the number of benzene ring.
132.	(c)	$O_2NHC = CHCl$ 1-chloro-2-nitroethene. For highly substituted alkenes E and Z system of nomenclature is used which is based on a priority system developed by Cahn, Ingold and Prelog.
133.	(b)	$Br—C^*(CH_3)(H)—Cl$ *asymmetric or chiral carbon. All the four valencies of carbon are satisfied with different atoms/substituents.
134.	(b)	$H—\ddot{O}—H$, $:NH_3$, $R—\ddot{O}—R$-nucleophiles. SO_3 has electron deficient centre (a reagent which can accept an electron pair in a reaction, is called an electrophile).
135.	(a)	According to Huckel rule, all aromatic compounds must have $(4n + 2)\pi$ electrons where n is an integer, *i.e.*, n = 0, 1, 2, 3, ... and possesses unusual stability due to the complete delocalisation of π-electrons.

Q. No.	Ans.	Hints & Solutions
136.	(*b*)	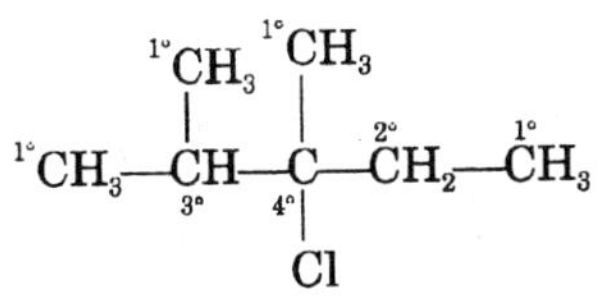 3-Chloro-2, 3-dimethylpentane
137.	(*d*)	The amount of enolic form is highest (76%) in acetyl acetone because keto group is a much better electron-withdrawing group. 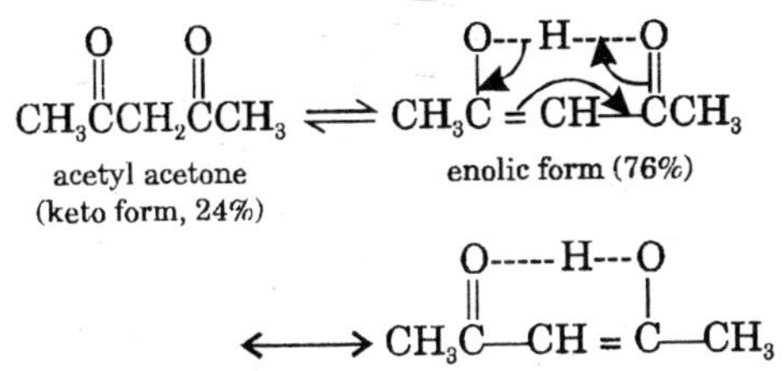
138.	(*b*)	Benzene  has 12σ and 3π bonds.
139.	(*d*)	Both are mirror images of each other I and II are identical.
140.	(*b*)	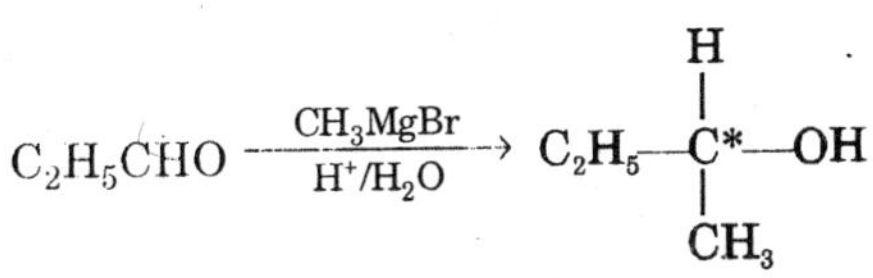 C*-chiral carbon as all the four valencies are attached with different substituents or groups.
141.	(*a*)	2-chloro-4-methyl-2-pentene.
142.	(*c*)	Butane 1, 2, 4-tricarboxylic acid (like group like treatment rule).
143.	(*c*)	10σ and 3π.
144.	(*e*)	If the atoms or groups of higher priority on each carbon atom of the double bond are on the opposite sides of the double bond, the configuration is designated as *E* and if they are on the same side, the configuration is designated as *Z*. 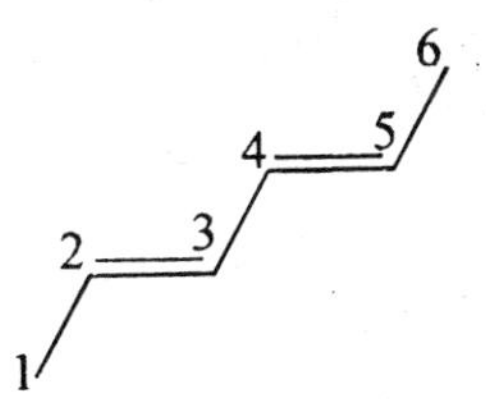 (2*E*, 4*E*)-2, 4-hexadiene
145.	(*c*)	The configuration, in which OH group lies towards right and the H-atom towards left, the CHO lies at the top and CH_2OH lies at the bottom of the Fischer projection, is called *D*-configuration. 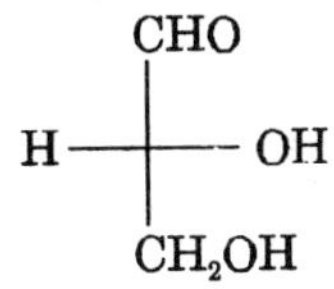 D-glyceraldehyde 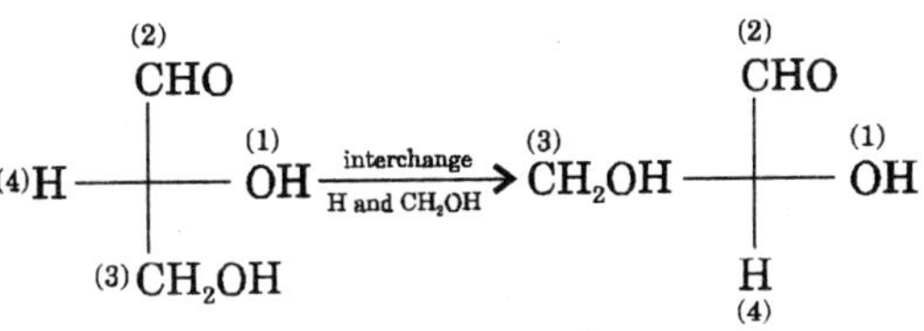 or, 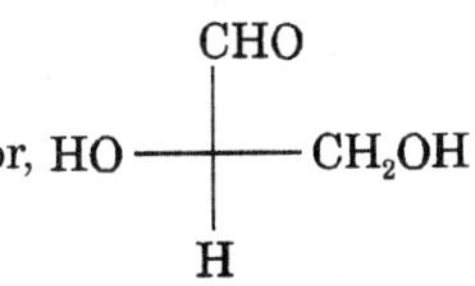
146.	(*d*)	3-Bromo pentane has not chiral carbon.
147.	(*a*)	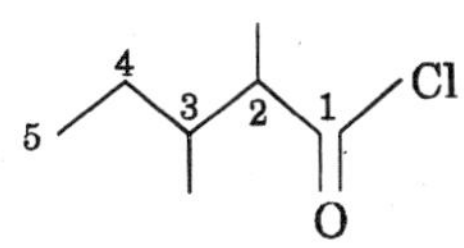 2, 3-dimethyl pentanoyl chloride
148.	(*c*)	Double bond is given preference over halogens which are taken as substituents.
149.	(*a*)	The stability depends upon the magnitude of repulsion of bond pairs. Farther the bond pairs of bulky groups greater is the stability.
150.	(*b*)	Greater the number of hyper conjugation structures more is stability. Also 3° free radicals are more stable than sec. free radicals.
151.	(*c*)	Alphabetical order of preference is followed and priority is given to the C = C of the ring while numbering of ring.
152.	(*e*)	Each of the given compound is resonance stabilized either due to π electrons of ring on due to C = C in conjugation with two pairs or due to rearrangement of electron an in 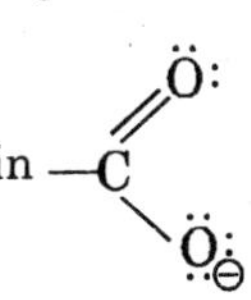 g.
153.	(*b*)	The double bond in the chain should get least number.

Q. No.	Ans.	Hints & Solutions
154.	(e)	Each carbon is sp^2 hybridised as it holds the double bond $CH_2 = CH—Cl$; 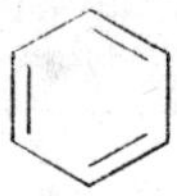 ; $CH_2 = CH—CH = CH_2$; $CH_2 = CH—CH = CH—CH = CH_2$
155.	(a)	Halogen produce a bluish greem coloured flame.
156.	(a)	Inductive effect is and a permanent effect that operates in a molecule and involves displacement of σ electrons.
157.	(a)	Alcohols are represented by the general formula ; $C_nH_{2n+2}O$.
158.	(c)	The $—\overset{O}{\overset{\|\|}{C}}—$ group is preferred as compared to $—C = C—$ and hence it is a ketone.
159.	(c)	A chiral compound is optically inactive due to presence of plane of symmetry. Such a compound is me compound.
160.	(d)	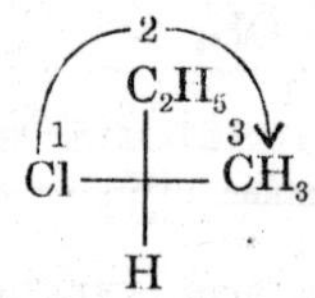 It is a R configuration.
161.	(a)	A chiral compound can rotate the plane of polarized light.
162.	(a)	A twiest boat has an unsymmetrical structure and hence can rotate p.p. light.
163.	(d)	Longest chain with max. no. of substituents is selected. Substituents are written in alphabetical order.

1.6. MCQs FOR IIT JEE

Brain Twister

A. Multiple Choice Questions With One Correct Answer

1. The compound given below is named as (IUPAC)

(a) Vinyl acetylene (b) But-1-en-3-yne
(c) 1-butyne-3-ene (d) Ethyne ethene.

2. The compound given below will be named in IUPAC as

(a) 7, 7-Dimethyl bicyclo [2.2. 1] heptane
(b) 5, 5-Dimethyl bicyclo [2.2. 1] heptane
(c) 7-iso propyl bicyclo [2.2. 1] heptane
(d) 2, 2-Dimethyl bicyclo [2.2. 1] heptane.

3. There is free rotation about (C2—C3) bond in *n*-butane. The same most stable form is repeated after rotation of
(a) 60° (b) 120°
(c) 240° (d) 360°.

4. The IUPAC name of

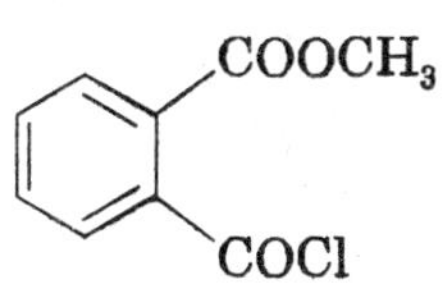

(a) 2-chloro carbonyl ethyl benzoate
(b) 2-carboxy ethyl benzoyl chloride
(c) Methyl-2-(chlorocarbonyl) benzoate
(d) None is correct.

5. The following compound can exhibit

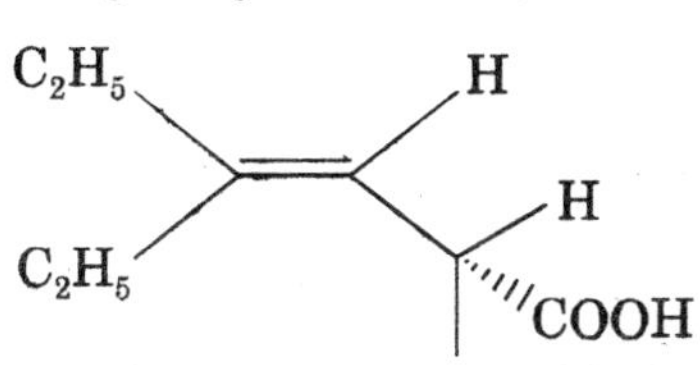

(a) Geometrical isomerism
(b) Geometrical and optical isomerism
(c) Optical isomerism
(d) None is correct.

6. The enol form of acetone, after treatment with D_2O, gives

(a) 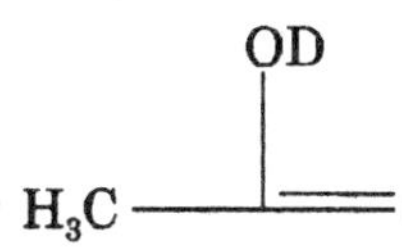 (b) 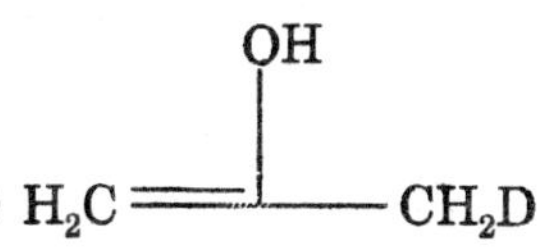

(c) 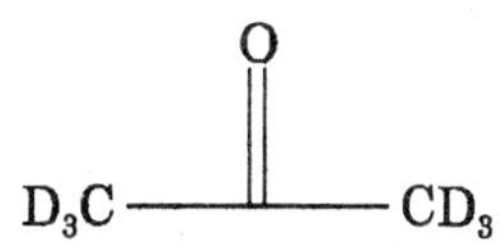 (d) 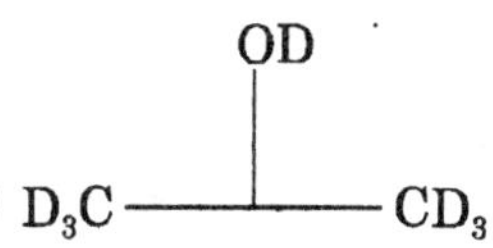

7. Which of the following is chiral ?

(a) 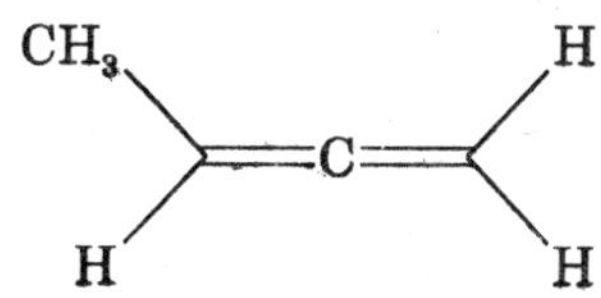

(b) 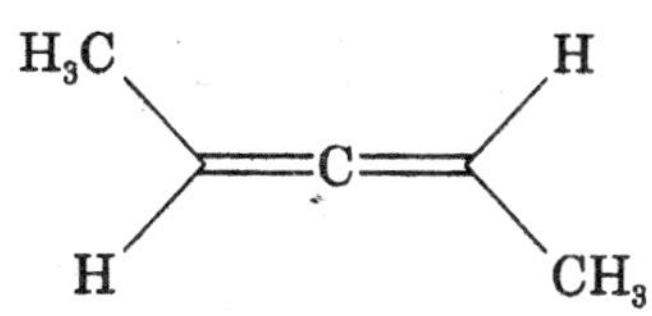

(c) 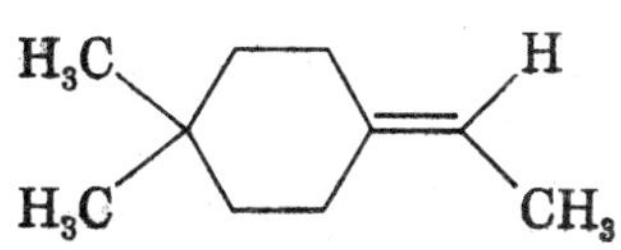

(d) 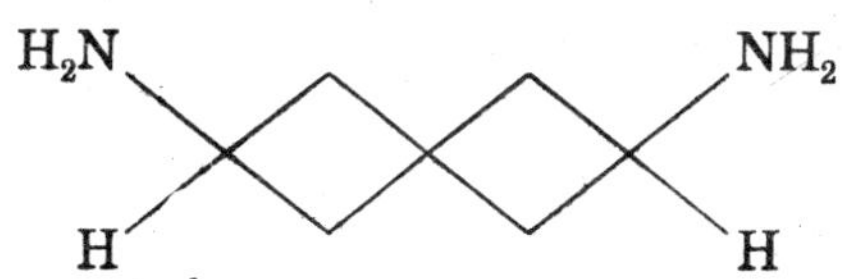

8. Geometrical isomerism is due to
(a) The restricted rotation about a double bond
(b) The presence of keto group
(c) The presence of CH(OH) group
(d) The presence of an asymmetric carbon.

9. Optically active isomers but not mirror images are called
(a) Enantiomers (b) Mesomers
(c) Tautomers (d) Diastereomers.

10. The correct IUPAC name of

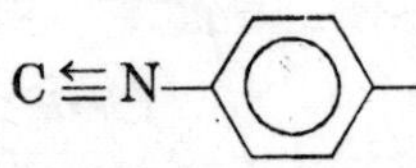 is

(*a*) 4-isocyano benzene carboxylic acid
(*b*) 4-carbylamino benzoic acid
(*c*) 1-carboxylic-4-isocyano benzene
(*d*) None of these.

11. A compound can be divided into equal halves and contains even '*n*' asymmetric carbon atoms. The total number of optical isomer is

(*a*) 2^{n} (*b*) $2^{(n-1)}$
(*c*) $2^{(n/2-1)}$ (*d*) $2^{(n-1)} + 2^{(n/2-1)}$.

12. IUPAC name of

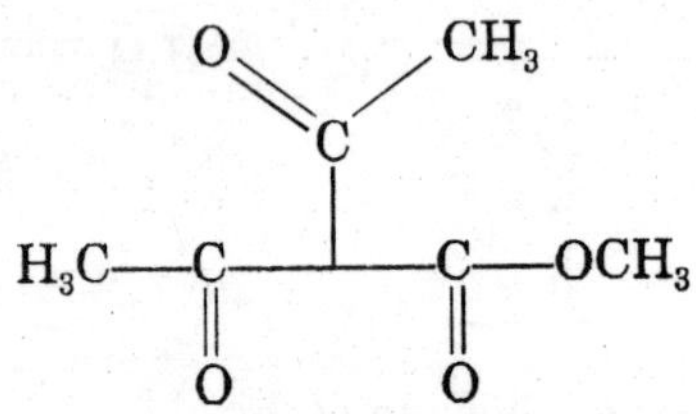

(*a*) methyl-2, 2-acetyl ethanoate
(*b*) 2, 2-acetyl-1-methoxy ethanone
(*c*) methyl-2-acetyl-3-oxobutanoate
(*d*) None is correct.

13. The compound, whose stereo·chemical formula is written below, exhibits *x*-geometrical isomers and *y*-optical isomers the value of *x* and *y* are

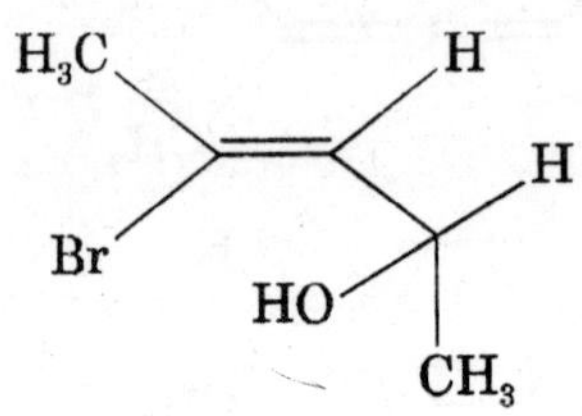

(*a*) 4 and 4 (*b*) 2 and 2
(*c*) 2 and 4 (*d*) 4 and 2.

14. Number of chiral carbon atom in 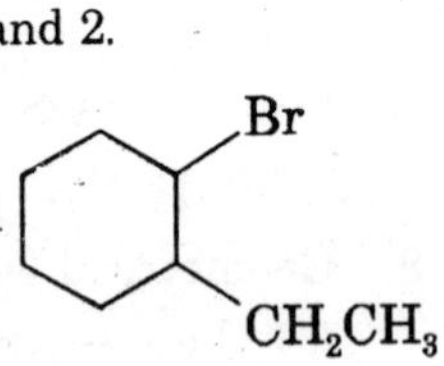

(*a*) 2 (*b*) 4
(*c*) 3 (*d*) 1.

15. The correct IUPAC name for the following compound is

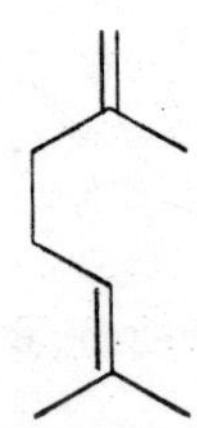

(*a*) 1, 1, 5-trimethyl-1, 6-hexadiene
(*b*) 2, 6-dimethyl-2, 6-heptadiene
(*c*) 2, 6-dimethyl-1, 5-heptadiene
(*d*) 1, 5-dimethyl 1-1, 6-heptadiene.

16. The correct statements about the compounds A, B and C is/are

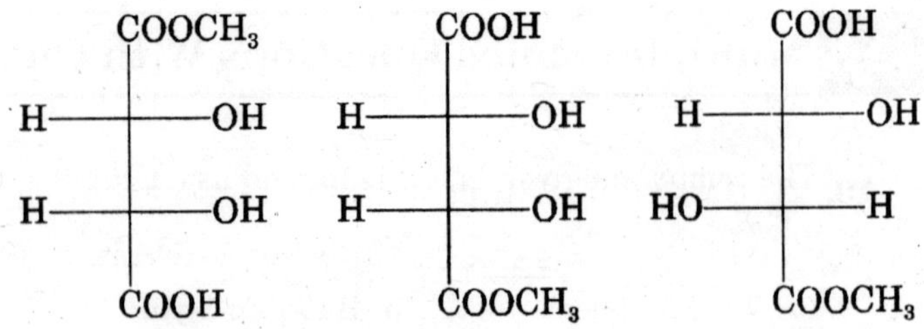

(A) (B) (C)

(*a*) A and B are identical
(*b*) A and B are diastereomers
(*c*) A and C are enantiomers
(*d*) A and B are enantiomers

17. The two isomers given below are

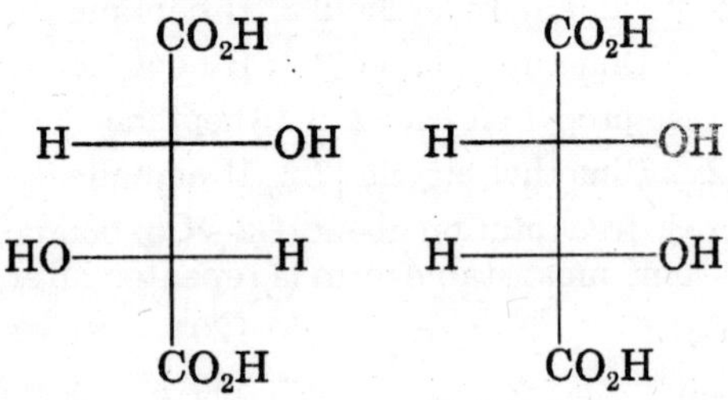

(*a*) Enantiomers (*b*) Diastereomers
(*c*) Regiomers (*d*) Mesomers.

18. Which of the following compounds is not optically active ?

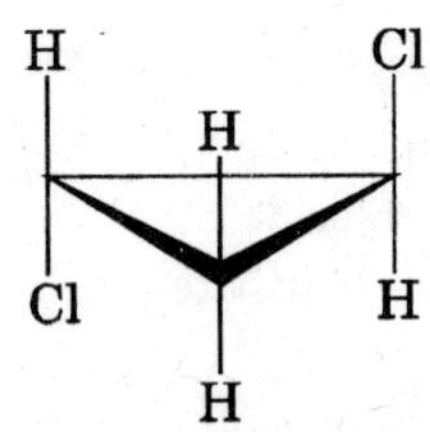

(*a*)

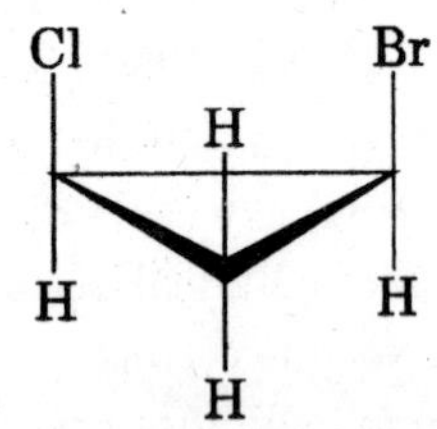

(*b*)

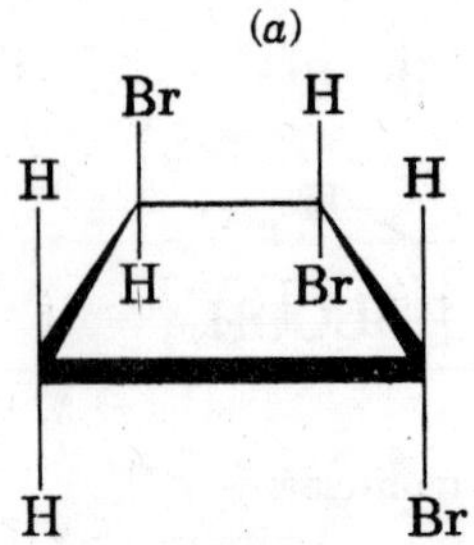

(*c*)

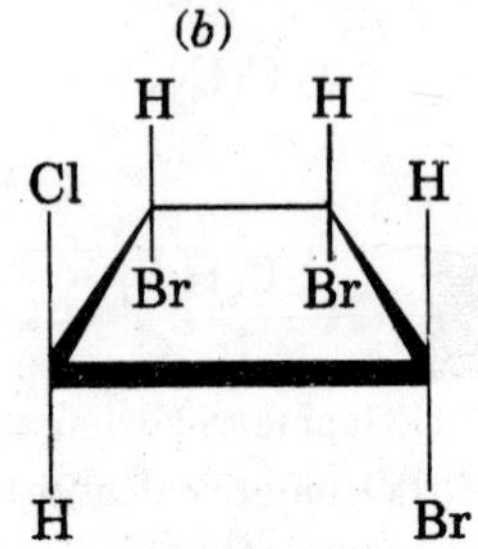

(*d*)

19. The two structures drawn below are

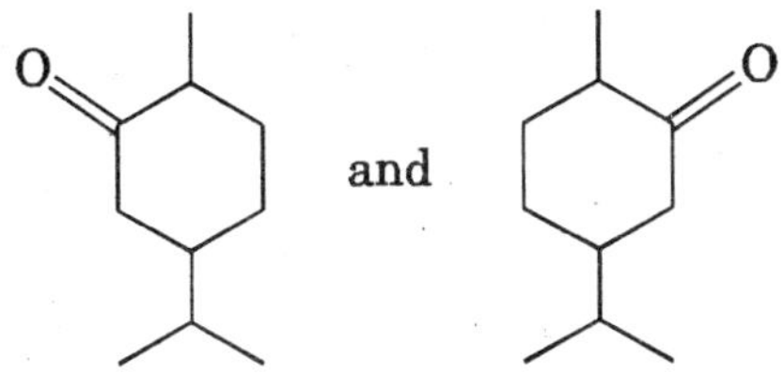

(*a*) Enantiomers (*b*) Mesomers
(*c*) Regiomers (*d*) Epimers.

20. The correct IUPAC name of the compound given below is

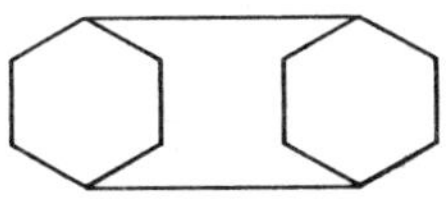

(*a*) Tricyclo [4.2.2.$0^{1,6}$] dodecane
(*b*) Tricyclo [4.2.2.$2^{2,5}$] dodecane
(*c*) Tricyclo [4.2.2.^{1,7}] dodecane
(*d*) Tricyclo [4.2.$0^{2,5}$.$0^{1,6}$] dodecane.

21. Which of the following species has/have electron releasing effect ?
(*a*) – CHO (*b*) – NO_2
(*c*) – CH_3 (*d*) – C_6H_5.

22. Which of the following species is/are electrophilic ?
(*a*) SO_3 (*b*) CCl_4
(*c*) NH_3 (*d*) CH_3OCH_3.

23. Which of the following classes of compounds undergo(es) mainly substitution reactions ?
(*a*) Alkenes (*b*) Arenes
(*c*) Dienes (*d*) Alkynes.

24. C_nH_{2n} is the general formula for
(*a*) Alkanes (*b*) Alkynes
(*c*) Cycloalkenes (*d*) Cycloalkanes.

25. Heterolytic cleavage of C—X bond in an organic compound may result in the formation of a
(*a*) Free radical (*b*) Carbanion
(*c*) Carbocation (*d*) Carbene.

26. Which of the following free radicals is/are resonance stabilized ?
(*a*) Vinyl (*b*) Allyl
(*c*) *iso*-Propyl (*d*) Cyclohexyl.

27. Which of the following species has/have a lone pair of electrons on carbon atom ?
(*a*) Singlet carbene (*b*) Triplet carbene
(*c*) Ethyl free radical (*d*) Carbenium ion.

28. In which of the following molecules all the carbon atoms are sp^2 hybridised ?
(*a*) 1, 4-Pentadiene (*b*) 1, 2-Butadiene
(*c*) 1, 3-Butadiene (*d*) 1, 3-Pentadiene.

29. Which of the following pair of compounds are not isomers ?
(*a*) Propanal and propanone
(*b*) Acetic acid and methyl formate
(*c*) Ethanal and Ethanol
(*d*) Nitroethane and ethyl nitrite.

30. Which of the following compounds can exist in *meso* form ?
(*a*) 2, 3-Pentanediol (*b*) 2, 3-Butanediol
(*c*) 1, 2-Ethanediol (*d*) 1, 2-Propanediol.

31. Which of the following statements is/are not correct ?
(*a*) Only alkenes show gemetric isomerism
(*b*) In alkanes all the carbon atoms are sp^3 hybridised
(*c*) Terminal alkynes give white precipitate with amm. $AgNO_3$ solution
(*d*) Cyclopentadienyl anion has planar structure.

32. Which of the following is not an isomer of acetone ?
(*a*) Propanal (*b*) Methyl vinyl ether
(*c*) 2-Propanol (*d*) Allyl alcohol.

33. Which of the following is/are optically active compound(s) ?
(*a*) 1, 3-Butadiene (*b*) Propadiene
(*c*) 2, 3-Pentadiene (*d*) 1, 3-Pentadiene.

34. An isomer of C_6H_{14} forms three monochloro derivatives. The isomer may be
(*a*) *neo*-pentane (*b*) *n*-Hexane
(*c*) 2, 3-Dimethylbutane (*d*) *iso*-Hexane.

35. Which of the following compounds exhibit(s) geometric isomerism ?
(*a*) 2-Methyl-1-pentene (*b*) 2-Methyl-2-pentene
(*c*) 2-Hexene (*d*) 2, 3-Dimethyl-2-butene.

B. MCQs From Previous IIT Examinations

36. An isomer of ethanol is
(*a*) Methanol (*b*) Diethyl ether
(*c*) Acetone (*d*) Dimethyl ether.
(IIT 1986)

37. Dipole moment is shown by
(*a*) 1, 4-Dichlorobenzene
(*b*) *cis*-1, 2-Dichloroethene
(*c*) *trans*-1, 2-Dichloroethene
(*d*) *trans*-1, 2-Dichloro-1-pentene. ***(IIT 1986)***

38. Only two isomeric monochloro derivatives are possible for

(*a*) *n*-Butane (*b*) 2, 4-Dimethylpentane
(*c*) Benzene (*d*) 2-Methylpropane.
(IIT 1986)

39. The IUPAC name of the compound,
$CH_2 = CH—CH(CH_3)_2$ is

(*a*) 1, 1-Dimethylprop-2-ene
(*b*) 3-Methylbut-1-ene
(*c*) 2-Vinylpropane
(*d*) 1-Isopropylethylene. ***(IIT 1987)***

40. Out of the following compounds, which one would have a zero dipole moment ?

(*a*) 1, 1-Dichloroethylene
(*b*) *cis*-1, 2-Dichloroethylene
(*c*) *trans*-1, 2-Dichloroethylene
(*d*) None of these compounds. ***(IIT 1987)***

41. Which of the following has least hindered rotation about carbon-carbon bond ?

(*a*) Ethane (*b*) Ethylene
(*c*) Acetylene (*d*) Hexachloroethane.
(IIT 1988)

42. Keto enol tautomerism is observed in

(*a*) C_6H_5CHO (*b*) C_6H_5COOH
(*c*) $C_6H_5COC_6H_5$ (*d*) $C_6H_5COCH_2COCH_3$.
(IIT 1988)

43. Which of the following have asymmetric carbon atom ?

(*a*) $CH_2Cl–CH_2Br$ (*b*) CH_3CHCl_2
(*c*) CH_3CHDCl (*d*) $CH_2Br—CHOH—CH_3$.
(IIT 1988)

44. The number of structural and configurational isomers of a bromo compound C_5H_9Br, formed by the addition of HBr to 2-pentyne respectively are

(*a*) 1 and 2 (*b*) 2 and 4
(*c*) 4 and 2 (*d*) 2 and 1. ***(IIT 1988)***

45. The compound which has one isopropyl group is

(*a*) 2, 2, 3, 3-Tetramethylpentane
(*b*) 2, 2-Dimethylpentane
(*c*) 2, 2, 3-Trimethylpentane
(*d*) 2-Methylpentane. ***(IIT 1989)***

46. The IUPAC name of the compound

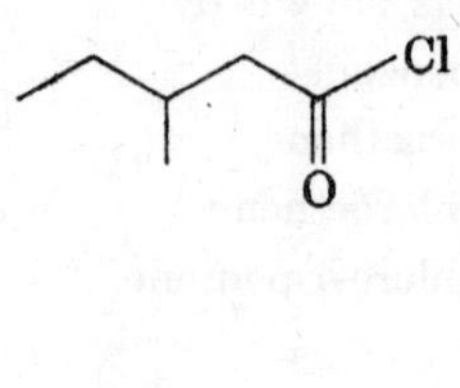

(*a*) 3-Methylpentanoyl chloride
(*b*) 3-Methylbutanoyl chloride
(*c*) 1-Chloro-3-ethylbutanone
(*d*) 1-Chloro-3-methylpentanone. ***(IIT 1990)***

47. The correct IUPAC name of the compound

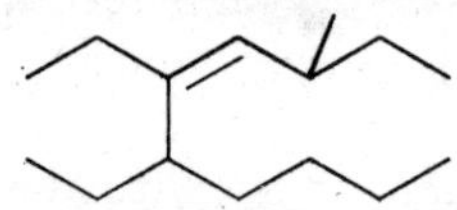

(*a*) 5, 6-Diethyl-8-methyldec-6-ene
(*b*) 6-Butyl-5-ethyl-3-methyloct-4-ene
(*c*) 5, 6-Diethyl-3-methyldec-4-ene
(*d*) 2, 4, 5-Triethylnon-3-ene. ***(IIT 1990)***

48. The enolic form of acetone contains

(*a*) 9 sigma bonds, 1 pi bond and 2 lone pairs
(*b*) 8 sigma bonds, 2 pi bonds and 2 lone pairs
(*c*) 10 sigma bonds, 1 pi bond and 1 lone pair
(*d*) 9 sigma bonds, 2 pi bonds and 1 lone pair.
(IIT 1990)

49. The IUPAC name of

$$OHCCH = CH—\underset{\displaystyle |\atop CH_2CH_2CH_2CH_3}{CH}—CH = CH_2$$ is

(*a*) 5-Vinyloct-3-en-1-al
(*b*) 4-Butylhexa-2, 5-dien-1-al
(*c*) 5-Vinyloct-5-en-8-al
(*d*) 3-Butylhexa-1, 4-dien-8-al. ***(IIT 1991)***

50. The hybridisation of carbon atoms in C—C single bond of $HC \equiv C—CH = CH_2$ is

(*a*) $sp^3–sp^3$ (*b*) $sp^2–sp^3$
(*c*) $sp^2–sp^2$ (*d*) $sp^2–sp$.

51. The C—C bond length of the following molecules is in the order

(*a*) $C_2H_6 > C_2H_4 > C_6H_6 > C_2H_2$
(*b*) $C_2H_2 < C_2H_4 < C_6H_6 < C_2H_6$
(*c*) $C_2H_6 > C_2H_2 > C_6H_6 > C_2H_4$
(*d*) $C_2H_4 > C_2H_6 > C_2H_2 > C_6H_6$. ***(IIT 1991)***

52. Isomers which can be interconverted through rotation around a single bond are

(*a*) conformers (*b*) diastereomers
(*c*) enantiomers (*d*) positional isomers.
(IIT 1992)

53. The molecules that will have dipole moments are

(*a*) 2, 2-Dimethylpropane
(*b*) *trans*-dichloroethene
(*c*) *cis*-3-Hexene
(*d*) 2, 2, 3, 3-Tetramethylbutane. ***(IIT 1992)***

54. An organic molecule necessarily shows optical activity if it
(*a*) contains asymmetric carbon atoms
(*b*) is non-polar
(*c*) is non-superimposable on its mirror image
(*d*) is superimposable on its mirror image. ***(IIT 1993)***

55. A compound with molecular formula C_8H_{14}, that do not contains 12 secondary and two tertiary H atoms is

(*a*)  (*b*) 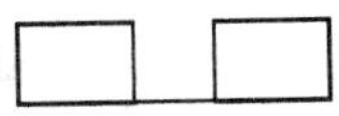

(*c*) 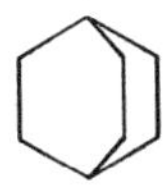 (*d*) 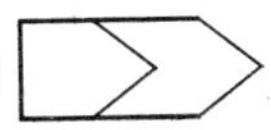 ***(IIT 1993)***

56. Choose the correct IUPAC name for

$$CH_3{-}\underset{\displaystyle|\atop\displaystyle CH_2{-}CH_3}{CH}{-}CHO$$ is

(*a*) Butan-2-aldehyde
(*b*) 2-Methylbutanal
(*c*) 3-Methylisobutyraldehyde
(*d*) 2-Ethylpropanal. ***(IIT 1993)***

57. For the molecule given below, mark the incorrect statement

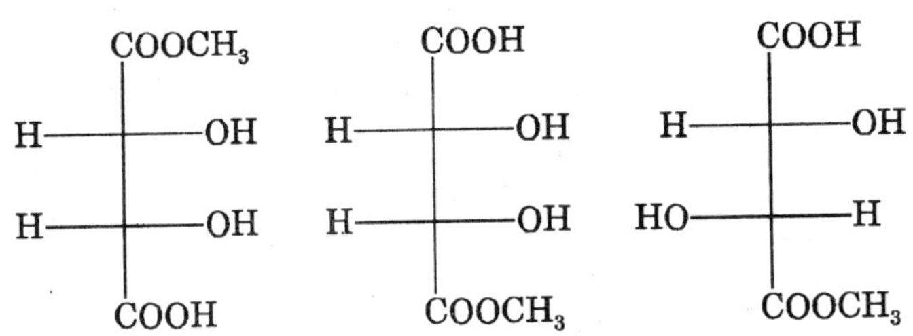

I II III

(*a*) I and II are enantiomers
(*b*) I and III are diastereomers
(*c*) II and III are identical
(*d*) I, is a meso isomer of II and III.

58. An optically active compound is
(*a*) 1-Bromobutane
(*b*) β-Bromobutyric acid
(*c*) 2-Bromo-2-methyl propane
(*d*) 1-Bromo-2-methyl propane.

59. Which is decreasing order of stability ?

$CH_3{-}\overset{+}{C}H{-}CH_3$ (I) $CH_3\overset{+}{C}H{-}O{-}CH_3$ (II) $CH_3{-}\overset{\oplus}{C}H{-}COCH_3$ (III)

(*a*) I < II < III (*b*) I > II > III
(*c*) III > II > I (*d*) II > III > I. ***(IIT 1994)***

60. The correct statement about the compounds I, II and III

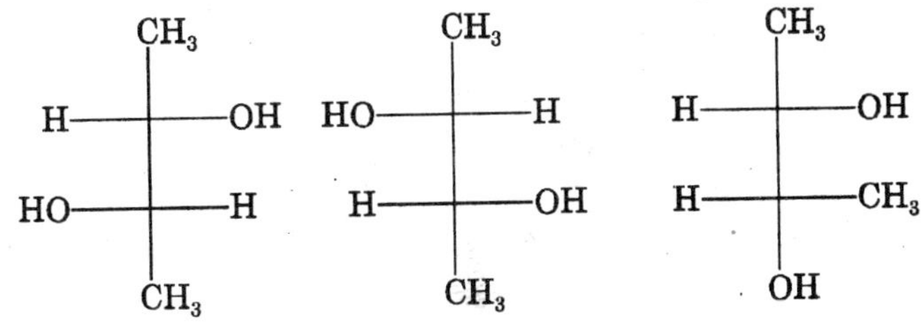

(I) (II) (III)

(*a*) I and II are identical
(*b*) I and II are diastereomers
(*c*) I and III are enantiomers
(*d*) I and II are enantiomers. ***(IIT 1997)***

61. The number of possible enantiomeric pairs that can be produced during monochlorination of 2-methylbutane is
(*a*) 2 (*b*) 3
(*c*) 4 (*d*) 1. ***(IIT 1997)***

62. When cyclohexane is poured on water, it floats because
(*a*) Cyclohexane is in boat form
(*b*) Cyclohexane is in chair form
(*c*) Cyclohexane is in crown form
(*d*) Cyclohexane is less dense than water.

63. Rotation of polarised light can be measured by
(*a*) Monometer (*b*) Galvanometer
(*c*) Polarimeter (*d*) Viscometer. ***(IIT 1998)***

64. Which of the following compounds will show geometrical isomerism ?
(*a*) 2-Butene (*b*) Propene
(*c*) 1-Phenyl propene (*d*) 2-Methyl-2-butene. ***(IIT 1998)***

65. Tautomerism is exhibited by

I. 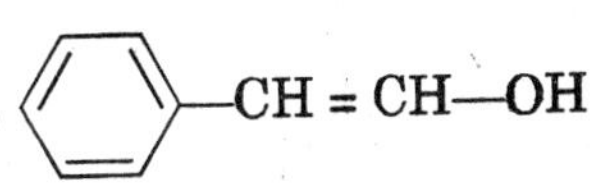

II. 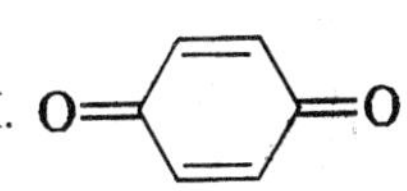

III. 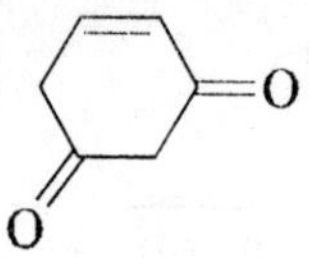

IV. 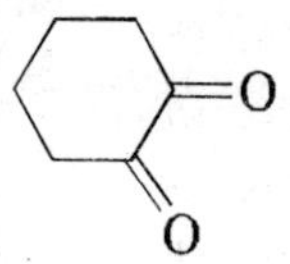

(*a*) All the four (*b*) II, III and IV
(*c*) I, III and IV (*d*) none of these.

(*IIT 1998*)

66. The optically active tartaric acid is named as D-(+)-tartaric acid because it has a positive
(*a*) optical rotation and is derived from D-glucose
(*b*) *p*H in an organic solvent
(*c*) optical rotation and is derived from D-(+)-glyceraldehyde
(*d*) optical rotation only when substituted by deuterium.

(*IIT 1999*)

67. In the compound, $CH_2 = CH—CH_2—CH_2—C \equiv CH$ the $C_2—C_3$ bond is of the type
(*a*) *sp*–sp^2 (*b*) sp^3–sp^3
(*c*) *sp*–sp^3 (*d*) sp^2–sp^3. **(*IIT 1999*)**

68. An aromatic molecule will not
(*a*) have $4n\pi$ electrons
(*b*) have $(4n + 2)\,\pi$ electrons
(*c*) be planar
(*d*) be cyclic. **(*IIT 1999*)**

69. Which of the following compounds will exhibit geometrical is isomerism ?
(*a*) 1-Phenyl-2-butene (*b*) 3-Phenyl-1-butene
(*c*) 2-Phenyl-1-butene (*d*) 1, 1-Diphenyl-1-propene.

(*IIT Screening Test 2000*)

70. The number of isomers for the compound with molecular formula $C_2BrClFI$ is
(*a*) 2 (*b*) 3
(*c*) 5 (*d*) 6.

(*IIT Screening Test 2000*)

71. Which of the following exhibits stereoisomerism
(*a*) 2-Methylbutene-1 (*b*) 3-Methylbutyne-1
(*c*) 3-Methylbutanoic acid (*d*) 2-Methylbutanoic acid.

(*IIT Screening Test 2000*)

72. Hydrogenation of the above compound in the presence of poisoned palladium catalyst gives

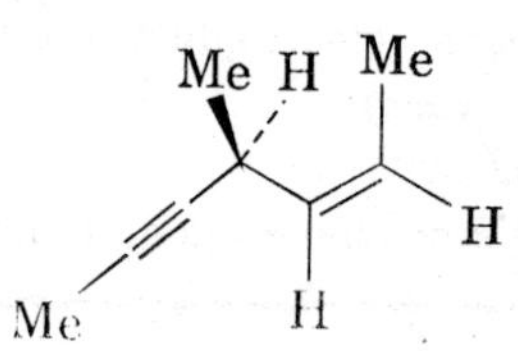

(*a*) An optically active compound
(*b*) An optically inactive compound
(*c*) A racemic mixture
(*d*) A diastereomeric mixture. **(*IIT Screening 2001*)**

73. Which of the following has the lowest dipole moment ?
(*a*) 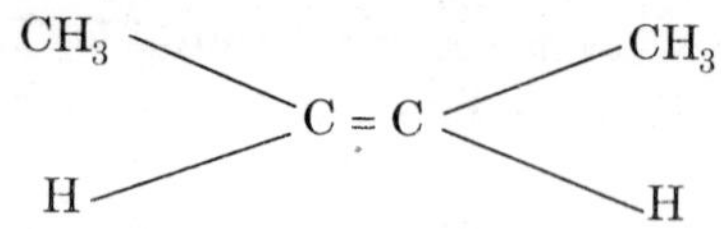

(*b*) $CH_3—C \equiv C—CH_3$
(*c*) $CH_3CH_2C \equiv CH$
(*d*) $CH_2 = CH—C \equiv CH$. **(*IIT Screening Test 2002*)**

74. Which of the following represents the given mode of hybridisation $sp^2 - sp^2 - sp - sp$ from left to right ?
(*a*) $H_2C = CH—C \equiv N$ (*b*) $HC \equiv C—C \equiv CH$
(*c*) $H_2C = C = C = CH_2$ (*d*) 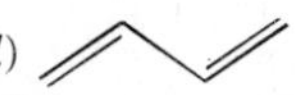

(*IIT Screening Test 2003*)

75. On monochlorination of 2-methylbutane the total number of chiral compounds is
(*a*) 2 (*b*) 4
(*c*) 6 (*d*) 8.

(*IIT Screening Test 2004*)

76. Two forms of D-glucopyranose are called
(*a*) Enantiomers (*b*) Anomers
(*c*) Epimers (*d*) Diastereomers.

(*IIT Screening Test 2005*)

77. The IUPAC name of C_6H_5COCl is
(*a*) Chlorobenzyl ketone
(*b*) Benzene chloroketone
(*c*) Benzenecarbonyl chloride
(*d*) Chlorophenyl ketone. **(*IIT Screening Test 2006*)**

78. Among the following, the least stable resonance structure is

(*a*) 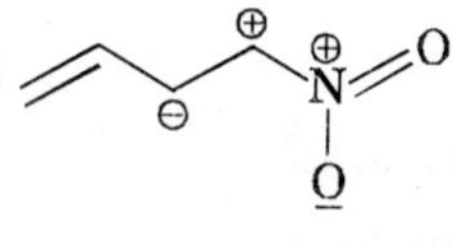 (*b*) 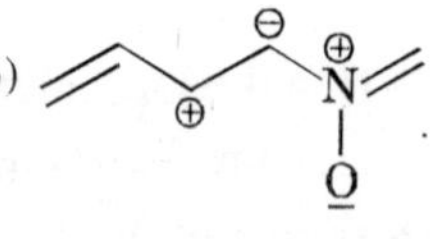

(*c*) 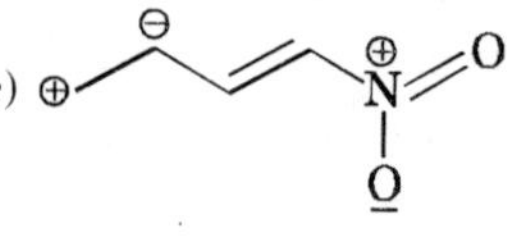 (*d*) 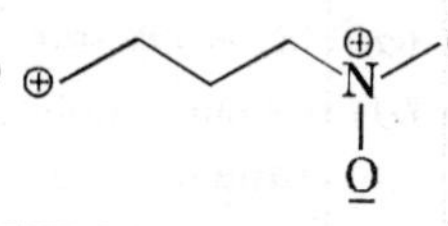

(*IIT Screening Test 2007*)

79. The number of structural isomers for C_6H_{14} is
(*a*) 3 (*b*) 4
(*c*) 5 (*d*) 6.

(*IIT Screening Test 2007*)

ANSWERS

1. (b)	2. (a)	3. (d)	4. (c)	5. (c)	6. (c)	7. (b)	8. (a)	9. (d)
10. (a)	11. (d)	12. (c)	13. (a)	14. (a)	15. (c)	16. (d)	17. (b)	18. (d)
19. (a)	20. (b)	21. (c)	22. (a)	23. (b)	24. (d)	25. (c)	26. (b)	27. (a)
28. (c)	29. (c)	30. (b)	31. (a)	32. (c)	33. (c)	34. (b)	35. (c)	36. (d)
37. (b)	38. (d)	39. (b)	40. (c)	41. (a)	42. (d)	43. (c)	44. (b)	45. (d)
46. (a)	47. (c)	48. (a)	49. (b)	50. (b)	51. (b)	52. (a)	53. (c)	54. (c)
55. (a)	56. (b)	57. (d)	58. (b)	59. (b)	60. (d)	61. (a)	62. (d)	63. (c)
64. (a)	65. (c)	66. (c)	67. (d)	68. (a)	69. (a)	70. (d)	71. (d)	72. (b)
73. (b)	74. (a)	75. (b)	76. (b)	77. (c)	78. (a)	79. (c).		

HINTS/SOLUTIONS

Q. No.	Ans.	Hints & Solutions
1.	(b)	Double bond is preferred over triple bond.
2.	(a)	It is a bicyclic compd. see concepts in brief.
3.	(d)	See conformations of butane.
4.	(c)	Ester is higher in order of preference than acid chloride.
5.	(c)	It has a chiral carbon.
6.	(c)	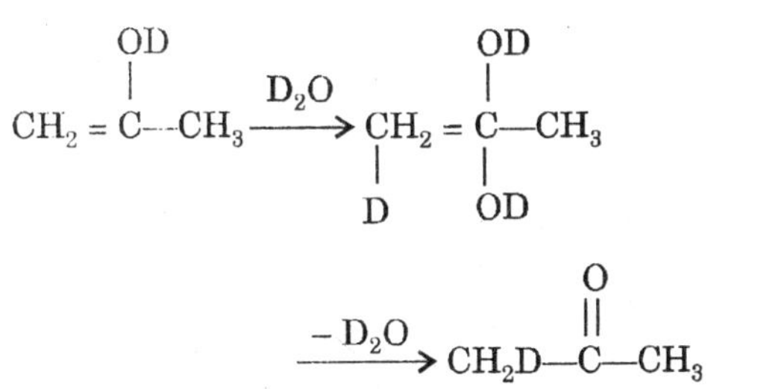 Slowly and gradually all H atoms are replaced by D in the similar manner.
7.	(b)	Substituted allenes are chiral structure.
8.	(a)	It arises due to restriction in rotation.
9.	(d)	Such a pair is known as pair of diastereo isomers.
10.	(a)	See IUPAC nomenclature.
11.	(d)	The formula used to calculate total no. of isomers.
12.	(c)	The parent functional group is esters.
13.	(c)	It will form 2 geometrical and 4 optical isomers (two optical isomers for each geometrical isomer.

Q. No.	Ans.	Hints & Solutions
14.	(a)	It has two chiral carbon atoms.
15.	(c)	See IUPAC nomenclature.
16.	(d)	Enantiomers are non super imposability mirror image of each other.
17.	(b)	A pair which is neither mirror image nor superimposable.
18.	(d)	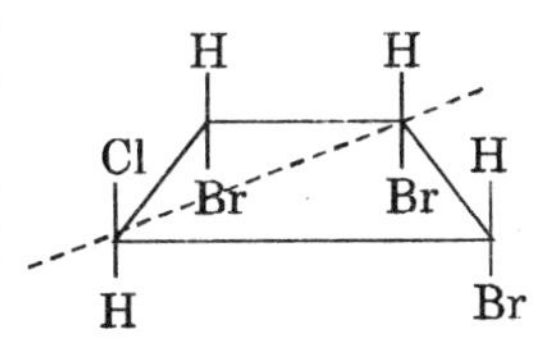 It has a plane of symmetry as shown by dotted line.
19.	(a)	These are enantiomers.
20.	(b)	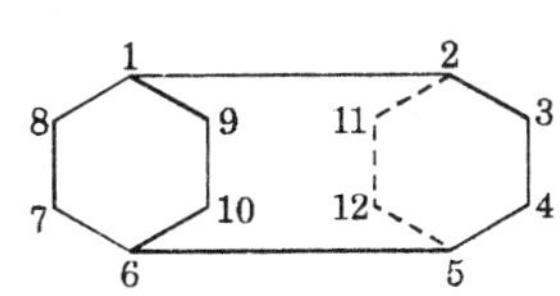 It is a tricyclo compound with two bridges (one with bold and other with dotted line).
21.	(c)	Methyl group is electron releasing.
22.	(a)	SO_3 acts as electrophile due to presence of π bonds.
23.	(b)	Arene is Benzene.
24.	(d)	Cycloalkanes have a general formula C_nH_{2n}.

Q. No.	Ans.	Hints & Solutions
25.	(c)	It forms a carbocation.
26.	(b)	$CH_2 = CH—\dot{C}H_2 \leftrightarrow \dot{C}H_2—CH = CH_2$
27.	(a)	Singlet carbene is $\ddot{C}H_2$.
28.	(c)	$CH_2 = CH—CH = CH_2$. All carbons are sp^2 hybridised.
29.	(c)	Ethanal and ethanol have different molecular formula.
30.	(b)	It has a plane of symmetry.
31.	(a)	All compounds with double bonds *i.e.*, C = C, C = N, C = P etc. can show geometrical isomers. Cyclic compounds can also form geometrical isomers.
32.	(c)	Alcohol is not an isomer of carbonyl compounds.
33.	(c)	It is a substituted allene with no plane of symmetry.
34.	(b)	It has three different types of H atoms $\overset{\alpha}{C}H_3\overset{\beta}{C}H_2\overset{\gamma}{C}H_2\overset{\gamma'}{C}H_2\overset{\beta'}{C}H_2\overset{\alpha'}{C}H_3$.
35.	(c)	$CH_3CH = CH—CH_2\ CH_2\ CH_3$ forms two geometrical isomers.
36.	(d)	Alcohols and ethers are isomers.
37.	(b)	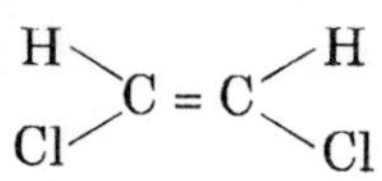 The molecule has net dipole moment.
38.	(d)	$CH_3—\underset{}{CH}(CH_3)—CH_3$ has two types of H atoms *i.e.*, 1° and 3°.
39.	(b)	Carbon of double bond is numbered 1.
40.	(c)	For *trans* isomers the dipole moment cancels and hence net dipole is zero.
41.	(a)	C—C single bond has least hindered rotation.
42.	(d)	It shows keto enol tautomerism 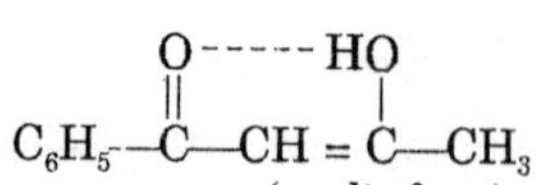 (*enolic form*)
43.	(c)	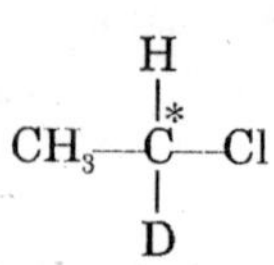
44.	(b)	2 structural isomers and 4 configurational isomers.
45.	(d)	$CH_3—CH(CH_3)CH_2CH_2CH_3$ has one isopropyl group.
46.	(a)	COCl is parent functional group.
47.	(c)	The double bond in the chain should get the least number.
48.	(a)	$CH_2 = C(OH)—CH_3$ is enolic form of Acetone.
49.	(b)	Longest chain rule is followed. The carbon of CHO is numbered 1. Both the double bonds are taken in the longest chain.
50.	(d)	It is sp^2 and sp hybridised because it is numbered as $\overset{4}{H}C \equiv \overset{3}{C}—\overset{2}{C}H = \overset{1}{C}H_2$.
51.	(b)	Triple bond < double bond < bonds is C_6H_6 < bond in alkanes.
52.	(a)	Free rotation results in forming conformers.
53.	(c)	Cis Hex-3-ene has net dipole.
54.	(c)	Compounds having non superimposable mirror images are optically active.
55.	(a)	It has 3 primary, 8 secondary and 3° tertiary H atoms.
56.	(b)	2-methylbutanal.
57.	(d)	A *meso* isomer is the one which has a plane of symmetry.
58.	(b)	It contains chiral carbon.
59.	(b)	More the number of +I groups, greater is the stability.
60.	(d)	I and II are non superimposible mirror images of each other and are enantiomers.
61.	(a)	In the given compound, $\overset{1}{C}H_3—\overset{2}{C}H(CH_3)—\overset{3}{C}H_2—\overset{4}{C}H_3$ the substitution of chlorine at carbon 3 produces enantiomers.
62.	(d)	It has lower density than water.
63.	(c)	Instrument is known as polarimeter.
64.	(a)	$CH_3CH = CHCH_3$.
65.	(c)	Compound I, III and IV show tautomerism.
66.	(c)	(+) means dextrorotatory. The notation D is because it is derived from D glyceraldehyde.

Q. No.	Ans.	Hints & Solutions
		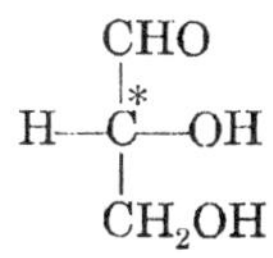 The OH group on C* is on the R.H.S. and hence word D.
67.	(*d*)	$\overset{1}{C}H_2 = \overset{2}{C}H - \overset{3}{C}H_2\overset{4}{C}H_2\overset{5}{C} \equiv \overset{6}{C}H$.
68.	(*a*)	Aromatic compound and should have $(4n + 2)\pi$ electrons (Huckel's Rule).
69.	(*a*)	It fulfills the conditions of geometrical isomerism.
70.	(*b*)	See concepts in brief.
71.	(*a*)	$CH_2 = \underset{\vert}{\overset{CH_3}{C}}—CH_2—CH_3$

Q. No.	Ans.	Hints & Solutions
72.	(*b*)	It is because the hydrogenation of alkynes of Pd/ $BaSO_4$ is a *cis* addition and ∴ It makes the reduced product optically in active.
73.	(*b*)	The molecule is linear as well as symmetrical hence has lowest dipole moment.
74.	(*a*)	$\overset{sp^2}{CH_2} = \overset{sp^2}{CH}—\overset{sp}{C} \equiv \overset{sp}{N}$.
76.	(*b*)	The two form of D-glucosopyranose differ in position of OH group at carbon 1 and are called anomers.
77.	(*c*)	According to latest IUPAC recommendal it is named as benzene carbonyl chloric.
78.	(*a*)	The structure in which the similar charges are present at the nearest position is always less stable.
79.	(*c*)	It gives five structural isomers : *n*-hexane, iso-hexane, 3-methyl pentane, 2-3-dimethyl butane and neo hexane.

1.7. Assertion-Reason Questions

In each of the following questions two statements are given as Assertion A and Reason R. Examine the statements carefully and answer the questions according to the instructions given below :

- *Mark **a** if both **A** and **R** are correct and **R** is the correct reason of **A**.*
- *Mark **b** if both **A** and **R** are correct and **R** is not the correct reason of **A**.*
- *Mark **c** if **A** is correct and **R** is wrong.*
- *Mark **d** if **A** is wrong and **R** is correct.*
- *Mark **e** if both **A** and **R** are wrong.*

1. ***Assertion A.*** Alkyl free radical is more stable than vinyl free radical.

   Reason R. Allyl free radical is stabilized by resonance.

(a)	(b)	(c)	(d)	(e)

2. ***Assertion A.*** Ethanal and vinyl alcohol are tautomers.

   Reason R. Ethanol and vinyl alcohol have same molecular formula.

(a)	(b)	(c)	(d)	(e)

3. ***Assertion A.*** Anti-conformation of *n*-butane is the most stable.

   Reason R. In anti-conformation methyl groups are as close to each other as possible.

(a)	(b)	(c)	(d)	(e)

4. ***Assertion A.*** 1-Butane and 2-butene are geometric isomers.

   Reason R. Geometric isomers are diastereomers.

(a)	(b)	(c)	(d)	(e)

5. ***Assertion A.*** *meso*-Tartaric acid is optically inactive.

   Reason R. *meso*-Tartaric acid contains two asymmetric carbon atoms.

(a)	(b)	(c)	(d)	(e)

6. ***Assertion A.*** *p*-Xylene has zero dipole moment.

   Reason R. In *p*-Xylene all the carbon atoms are in sp^2-hybrid state.

(a)	(b)	(c)	(d)	(e)

7. ***Assertion A.*** Cyclobutane is more reactive than cyclopentane.

   Reason R. The carbocyclic ring in cyclobutane is more strained than in cyclopentane.

(a)	(b)	(c)	(d)	(e)

8. ***Assertion A.*** Isoprene and 1, 3-pentadiene are chain isomers.

   Reason R. Isoprene and 1, 3-pentadiene are conjugated dienes.

(a)	(b)	(c)	(d)	(e)

9. ***Assertion A.*** Anthracene and phenanthrene are isomers.

   Reason R. Anthracene and phenanthrene both have 14π electrons each.

(a)	(b)	(c)	(d)	(e)

10. ***Assertion A.*** In alkanes containing three or more carbon atoms, carbon atoms form *zig-zag* chain.

    Reason R. Carbon atoms in alkanes are in sp^3 hybrid state.

(a)	(b)	(c)	(d)	(e)

11. ***Assertion A.*** Lactic acid can show optical as well as geometrical isomerism.

    Reason R. Lactic acid is prepared in laboratory by treatment of ethanal with HCN followed by hydrolysis of its Racemic modification.

(a)	(b)	(c)	(d)	(e)

12. ***Assertion A.*** All the molecules containing $C = C$ can exhibit geometrical isomerism.

    Reason R. Geometrical isomers have identical physical properties.

(a)	(b)	(c)	(d)	(e)

13. ***Assertion A.*** Racemic mixture is optically inactive.

    Reason R. Racemic mixture is mixture of meso structures.

(a)	(b)	(c)	(d)	(e)

14. ***Assertion A.*** Maleic acid and fumaric acid are constitutional isomers.

    Reason R. Maleic acid and fumaric acid are tautomeric forms.

(a)	(b)	(c)	(d)	(e)

15. ***Assertion A.*** Cyclohexane floats over water.

    Reason R. Cyclohexane generally exists as boat conformers.

(a)	(b)	(c)	(d)	(e)

16. ***Assertion A.*** Both meso-tartaric acid and (±) tartaric acid are optically inactive.

    Reason R. Tartaric acid molecule contains two chirality centres.

(a)	(b)	(c)	(d)	(e)

17. ***Assertion A.*** Chiral objects do not superimpose on their mirror image.

    Reason R. Chiral objects do not have any plane of symmetry.

(a)	(b)	(c)	(d)	(e)

18. ***Assertion A.*** Every optically active substance rotates the flame of polarized light towards right.

    Reason R. Optically active stereomers must possess disymmetric structure.

(a)	(b)	(c)	(d)	(e)

ANSWERS

1. (a)	**2.** (b)	**3.** (c)	**4.** (d)	**5.** (b)	**6.** (c)	**7.** (a)	**8.** (b)	**9.** (b)
10. (a)	**11.** (c)	**12.** (e)	**13.** (d)	**14.** (e)	**15.** (d)	**16.** (b)	**17.** (a)	**18.** (e)

1.8. SELF EVALUATION TEST

Assess Yourself

Choose the correct answer from the four alternatives given in each of the following questions :

1. IUPAC name of cinnammaldehyde is

(*a*) 3-Phenyl prop-2-enal (*b*) 1-Phenyl prop-1-enal
(*c*) 1-Phenyl prop-2-enal (*d*) 3-Phenyl prop-1-enal.

2. Which of the following free radicals is the most stable ?

(*a*) Methyl (*b*) *iso*-Propyl
(*c*) Vinyl (*d*) Allyl.

3. Which of the following species has minimum H—C—H bond angle ?

(*a*) Methane (*b*) Methyl carbocation
(*c*) Methyl carbanion (*d*) Methyl free radical.

4. The number of isomers possible for C_4H_8 is

(*a*) 6 (*b*) 4
(*c*) 8 (*d*) 2.

5. In which of following molecules, all the carbon atoms are not in same hybrid state ?

(*a*) 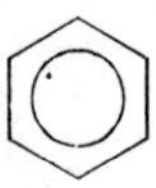 (*b*) 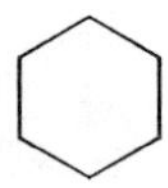

(*c*) $CH_2 = C = CH_2$ (*d*) None of these.

6. In the reaction

$CH_3CHO + HCN \longrightarrow CH_3CH(OH)CN$,

a chiral centre is produced. This product would be

(*a*) Meso compound (*b*) Racemic mixture
(*c*) Laevorotatory (*d*) Dextrorotatory.

7. Rotation of plane polarised light by an optically active compound can be measured by

(*a*) Stalgamometer (*b*) Manometer
(*c*) Polarimeter (*d*) Photometer.

8. A compound has 4 chiral carbon atoms. The number of possible optical isomers it can have

(*a*) 16 (*b*) 2
(*c*) 3 (*d*) 8.

9. Citraconic acid and Mesaconic acid are examples of

(*a*) Tautomers (*b*) Enantiomers
(*c*) Geometric isomers (*d*) Functional isomers.

10. Malic acid exhibits the phenomenon of

(*a*) Geometric isomerism (*b*) Optical isomerism
(*c*) Functional isomerism (*d*) Tautomerism.

11. Which acidity order is correct ?

(*a*) $CH_2FCOOH > CH_2BrCOOH > CH_2ClCOOH$
(*b*) $CH_2ICOOH > CH_2BrCOOH > CH_2ClCOOH$
(*c*) $CH_2ClCOOH > CH_2BrCOOH > CH_2ICOOH$
(*d*) $CH_2ICOOH > CH_2BrCOOH > CH_2ClCOOH$.

12. The shape of methyl carbocation is

(*a*) Planar (*b*) Pyramidal
(*c*) Linear (*d*) V-shaped.

13. The Cl—C—Cl angle in 1, 1, 2, 2-tetrachloroethene and tetrachloromethane respectively are about

(*a*) 120° and 109.5° (*b*) 90° and 109.5°
(*c*) 109.5° and 90° (*d*) 109.5° and 120°.

14. Which of the following statements is not correct ?

(*a*) Methyl carbocation is trigonal planar
(*b*) Methyl carbanion is pyramidal
(*c*) Singlet carbene is linear in shape
(*d*) Triplet carbene is less stable than singlet carbene.

15. The name of $\underset{\displaystyle CHO}{\underset{|}{CH}} = \underset{\displaystyle NH_2}{\underset{|}{CH}}$ is

(*a*) 1-Aminoprop-2-enal
(*b*) 3-Aminoprop-2-enal
(*c*) 1-Amino-2-formylethene
(*d*) 3-Amino-1-oxoprop-2-ene.

16. In which hybridisation, the bond angle is the maximum ?

(*a*) sp^3 (*b*) sp^2
(*c*) sp (*d*) sp^3d.

17. Which of the following species is paramagnetic in nature ?

(*a*) Carbocation (*b*) Free radical
(*c*) Carbanion (*d*) All of them.

18. *s*-character of sp, sp^2 and sp^3 hybrid orbitals follows the decreasing order :

(*a*) $sp^3 > sp^2 > sp$ (*b*) $sp > sp^2 > sp^3$
(*c*) $sp^2 > sp > sp^3$ (*d*) $sp^2 > sp^3 > sp$.

19. Which of the following species has two non-bonding electrons around carbon ?

(*a*) Carbocation (*b*) Alkyl free radical
(*c*) Carbene (*d*) None of these.

20. Which organic compound was synthesised first in laboratory from inorganic compounds ?

(*a*) Urea (*b*) Ethylene
(*c*) Methane (*d*) Acetic acid.

21. The compound which has one isopropyl group is
(*a*) 2, 2, 3, 3-Tetramethyl pentane
(*b*) 2, 2-Dimethyl pentane
(*c*) 2, 2, 3-Trimethyl pentane
(*d*) 2-Methyl pentane.

22. Isomerism exhibited by methyl formate and acetic acid is
(*a*) Tautomerism (*b*) Chain isomerism
(*c*) Functional isomerism (*d*) Position isomerism.

23. The number of geometrical isomers in case of a compound with the structure
$CH_3—CH = CH—CH = CH—C_2H_5$ is
(*a*) 4 (*b*) 3
(*c*) 2 (*d*) 5.

24. The process of conversion of *d* form into *l* form or *vice-versa* is known as
(*a*) Racemisation (*b*) Mutarotation
(*c*) Resolution (*d*) Walden inversion.

25. The number of stereoisomers possible for 2, 3-dichlorobutane is
(*a*) 1 (*b*) 2
(*c*) 3 (*d*) 4.

26. In the keto-enol tautomerism of dicarbonyl compounds, the enol form is preferred in contrast to the keto form ; this is due to
(*a*) Presence of carbonyl group on each side of the – CH_2
(*b*) Resonance stabilisation of the enol form
(*c*) Presence of methylene group
(*d*) Rapid chemical exchange.

27. The hybridisation of carbon atoms in C—C single bond of $HC \equiv C—C \equiv CH$ is
(*a*) $sp^3 - sp^3$ (*b*) $sp^2 - sp^3$
(*c*) $sp - sp$ (*d*) $sp^3 - sp$.

28. The compound in which carbon uses only its sp^3 hybrid orbitals for bond formation is
(*a*) HCOOH (*b*) $(NH_2)_2CO$
(*c*) $(CH_3)_3COH$ (*d*) $(CH_3)_3C.\ CHO$.

29. What hybrid orbitals will form the following compound ?
$H_3C—CH = CH—CH_2—CH_2—CN$
(*a*) sp^2 and sp^3 (*b*) sp, sp^2 and sp^3
(*c*) sp and sp^2 (*d*) only sp^3.

30. In which of the compounds given below is there more than one king of hybridisation (sp, sp^2, sp^3) for carbon ?
(*i*) $CH_3CH_2CH_2CH_3$
(*ii*) $CH_3CH = CHCH_3$
(*iii*) $CH_2 = CH - CH = CH_2$
(*iv*) $CH_3 - C \equiv C - CH_3$
(*a*) (*ii*) and (*iv*) (*b*) (*iii*) and (*iv*)
(*c*) (*i*) and (*iv*) (*d*) (*ii*) and (*iii*).

31. The number of σ and π bonds in but-1-en-3-yne are
(*a*) 5σ and 5π (*b*) 7σ and 3π
(*c*) 8σ and 2π (*d*) 6σ and 4π.

32. IUPAC name for  —C ≡ N is
(*a*) Phenyl cyanide
(*b*) Cyclohexane nitrile
(*c*) Cyclohexane carbonitrile
(*d*) Cyclohexylcyanide.

33. IUPAC name for 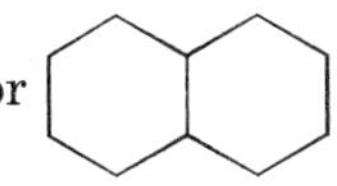 is
(*a*) Naphthalene (*b*) Tetrahydronaphthalene
(*c*) Bicyclo [4.4.2] decane (*d*) Bicyclo [4.4.0] decane.

34. Which of the following compounds can exhibit enantiomerism ?
(*a*) 1, 3-Dichloropropadiene
(*b*) 2, 4-Dimethyl-2, 3-pentadiene
(*c*) 2, 4-Dimethyl-1, 3-pentadiene
(*d*) 2-Methyl-1, 3-pentadiene.

35. IUPAC name of the following compound is

$$\begin{array}{c} \quad\quad\quad CH_3 \;\; NO_2 \\ \quad\quad\quad | \quad\quad | \\ HOOC–CH_2–CH_2–CH—CH–CH_2–CH_2–COOH \end{array}$$

(*a*) 5-Nitro-4-methyloctanedioic acid
(*b*) 4-Methyl-5-nitroctanedioic acid
(*c*) 4-Nitro-5-methyloctanedioic acid
(*d*) 5-Methyl-4-nitroctanedioic acid.

36. Molecular formula of a compound is $C_3H_6Cl_2$. This formula permits the existence of
(*a*) Functional isomers (*b*) geometrical isomers
(*c*) Enantiomers (*d*) Chain isomers.

37. Which of the following show geometric isomersim ?
(*a*) 1-Butane
(*b*) 1, 2-Dimethylcyclo butane
(*c*) Allyl bromide
(*d*) 1, 4-Butanedioic acid.

38. Which of the following can exist as diastereomers ?
(*a*) Lactic acid (*b*) But-1-ene
(*c*) Tartaric acid (*d*) 2-Butyne.

39. Which of the following pairs of compounds are function isomers ?
(*a*) Propanoic acid and ethyl acetate
(*b*) Trimethyl amine and diethyl amine
(*c*) 1, 3-Butadiene and 1-butyne
(*d*) 2-Butene and 2-Butyne.

40. Which of the following compounds do not exhibit tautomerism ?

(*a*) Acetone (*b*) Benzophenone
(*c*) Nitromethane (*d*) Benzaldehyde.

41. Which of the following compounds has atleast one carbon in *sp* hybridised state ?

(*a*) $H_2C = CH—CH = CH_2$ (*b*) $H_2C = CH_2$
(*c*) $H_2C = CHCH_3$ (*d*) $H_2C = C = CH_2$.

42. Which of the following have six electrons in the valence shell of carbon ?

(*a*) Methyl carbanion (*b*) Allylium anion
(*c*) Dichlorocarbene (*d*) Methyl free radical.

43. In which of the compounds given below has more than one kind of hybridisation (sp, sp^2, sp^3) for carbon ?

(*a*) Styrene (*b*) Propene
(*c*) Benzene (*d*) Cyclopentane.

44. Which of the following have sp^2 hybridisation ?

(*a*) Methyl carbanion (*b*) Fluoro boro anion
(*c*) Singlest carbene (*d*) Triplet carbene.

45. Which of the following species is not stabilized by resonance ?

(*a*) $:\overset{-}{C}H_2—\overset{O}{\overset{\|}{C}}—H$ (*b*) 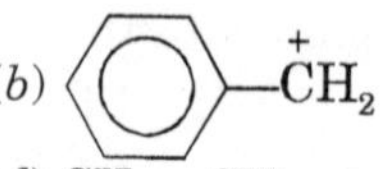

(*c*) $CH_2 = CH - CH_2$ (*d*) $CH_3 - CH_2$.

ANSWERS

1. (*a*)	2. (*d*)	3. (*c*)	4. (*a*)	5. (*c*)	6. (*b*)	7. (*c*)	8. (*a*)	9. (*c*)
10. (*b*)	11. (*c*)	12. (*a*)	13. (*a*)	14. (*d*)	15. (*b*)	16. (*c*)	17. (*b*)	18. (*b*)
19. (*c*)	20. (*a*)	21. (*d*)	22. (*c*)	23. (*a*)	24. (*d*)	25. (*c*)	26. (*b*)	27. (*c*)
28. (*c*)	29. (*b*)	30. (*a*)	31. (*b*)	32. (*c*)	33. (*d*)	34. (*a*)	35. (*b*)	36. (*c*)
37. (*b*)	38. (*c*)	39. (*c*)	40. (*d*)	41. (*d*)	42. (*c*)	43. (*b*)	44. (*d*)	45. (*d*).

HINTS/SOLUTIONS

Q. No.	Ans.	Hints & Solutions
1.	(*a*)	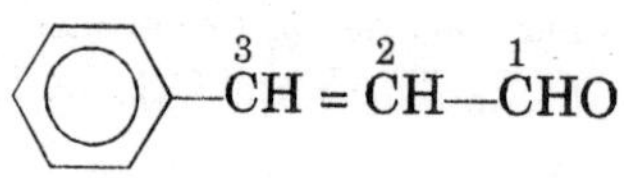 (*Cinnamaldehyde*) 3-Phenyl prop-2-enal
2.	(*d*)	Allyl free radical is resonance stabilized. $CH_2 = CH - CH_2 \leftrightarrow CH_2 - CH = CH_2$.
3.	(*c*)	$\overset{\ominus}{C}H_3$. It has sp^3 hybridised carbon.
4.	(*a*)	There are four alkenes and two cycloalkanes corresponding to the formula C_4H_8.
5.	(*c*)	$\overset{sp^2}{C}H_2 = \overset{sp}{C} = \overset{sp^2}{C}H_2$.
6.	(*b*)	Both *d* and *l* isomers are formed in equal amounts.
7.	(*c*)	It is measured by polarimeter.
8.	(*a*)	Number of optical isomers = $(2)^n = (2)^4 = 16$.
9.	(*c*)	$H_3C - C - COOH$ ‖ $H—C—COOH$ (Citraconic acid); $H_3C - C - COOH$ ‖ $HOOC—C—H$ (Mesaconic acid)
10.	(*b*)	Malic acid is $HOOC—\overset{H}{\underset{OH}{C^*}}—CH_2COOH$
11.	(*a*)	Cl has maximum electron withdrawing inductive effect.
12.	(*a*)	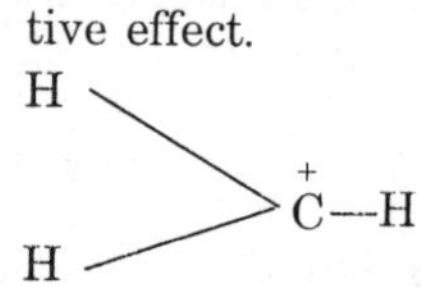 Methyl carbocation (Trigonal planar shape)

Q. No.	Ans.	Hints & Solutions
13.	(*a*)	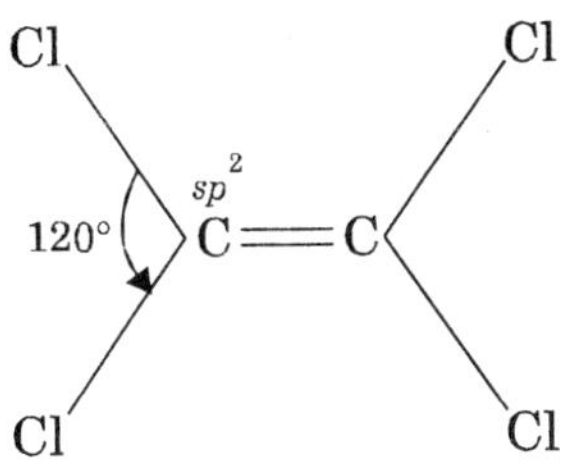 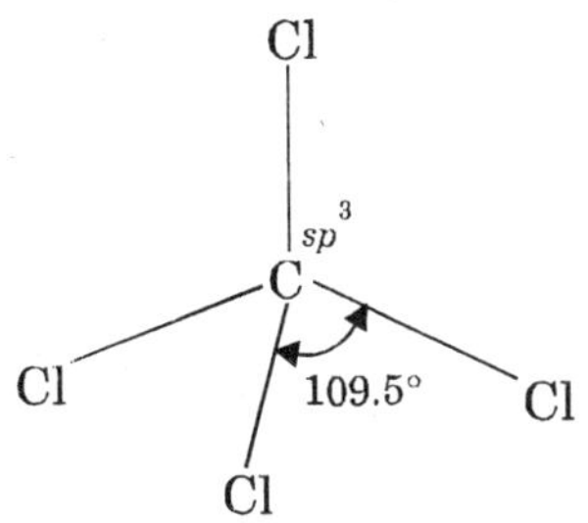
14.	(*d*)	Triplet carbon is more stable than singlet carbene.
15.	(*b*)	$\overset{2}{C}H = \overset{3}{C}H$; $\overset{1}{C}HO$ on C-2, NH_2 on C-3 3-Amino prop-2-enal
16.	(*c*)	In *sp*-hybrid orbitals bond angle is 180°.
17.	(*b*)	Free radical has an odd electron on some atom.
18.	(*b*)	*s* character is 50% (*sp*), 33.33% (sp^2) and 25% in sp^3 hybridised carbon atoms.
19.	(*c*)	$\ddot{C}H_2$ has two non bonding electrons.
20.	(*a*)	Wohler synthesised urea by heating ammonium cyanate.
21.	(*d*)	$CH_3—CH(CH_3)—CH_2—CH_2—CH_3$ 1 2 3 4 5 6 7
22.	(*c*)	These are functional isomers.
23.	(*a*)	$CH_3—CH = CH—CH = CH—C_2H_5$ (*i*) *cis*-(2), *cis*-(4) (*ii*) *cis*-(2), *trans*-(4) (*iii*) *trans*-(2), *cis*-(4) (*iv*) *trans*-(2), *cis*-(4).
24.	(*d*)	It is known as Walden inversion.
25.	(*c*)	$CH_3—\overset{*}{C}H(Cl)—\overset{*}{C}H(Cl)—CH_3$
		The three stereoisomers are (+)–, (–)-and *meso*-2, 3-dichlorobutane.
26.	(*b*)	In dicarbonyl compound resonance stabilisation leads to stability of enol form.
27.	(*c*)	$H—\overset{sp}{C} \equiv C—\overset{sp}{C} \equiv C—H$
28.	(*c*)	$H_3C—C(CH_3)_2—OH$ Each carbon atom in this molecule is linked to four other atoms directly and hence is sp^3 hybridized.
29.	(*b*)	$\overset{sp^3}{H_3C}—\overset{sp^2}{CH} = \overset{sp^2}{CH}—\overset{sp^3}{CH_2}—\overset{sp^3}{CH_2}—\overset{sp}{C} \equiv N$
30.	(*a*)	$\overset{sp^3}{CH_3} - \overset{sp^2}{CH} = \overset{sp^2}{CH} - \overset{sp^3}{CH_3}$ $\overset{sp^3}{}\ \overset{sp}{}\ \overset{sp}{}\ \overset{sp^3}{}$ (*iv*) $CH_3 - C \equiv C - CH_3$
31.	(*b*)	$H - C(H) = C(H) - C \equiv C—H$ But-1-en-3-yne
32.	(*c*)	$—C \equiv N$ group attached to ring is named as carbonitrile.
33.	(*d*)	It is a bicyclic compound.
34.	(*a*)	1, 3-Dichloropropadiene is substituted allene with two different groups on the terminal carbon atoms of allene. It is dissymmetric and hence exhibits enantiomerism.
35.	(*b*)	Alphabetical order of preference is followed.
36.	(*c*)	$CH_3 - CH(Cl) - CH_2Cl$ and $CH_3 - CH_2 - CHCl_2$ are position isomers. $CH_3 - \overset{*}{C}H(Cl) - CH_2Cl$ Contains chiral carbon and hence possesses chirality and exhibits enantiomerism.
37.	(*b*)	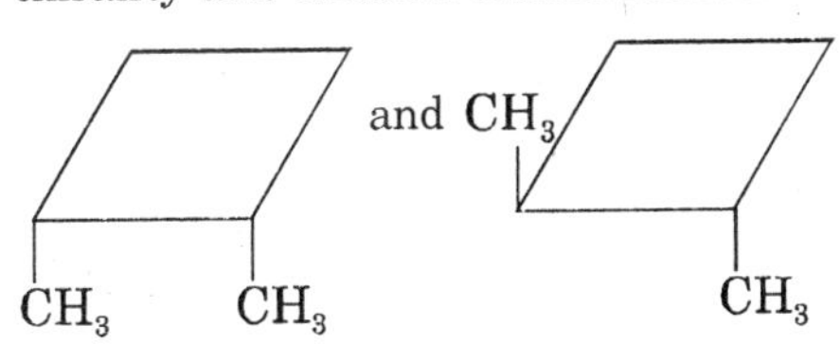 *cis*-isomer *trans*-isomer

Q. No.	Ans.	Hints & Solutions
		(c) 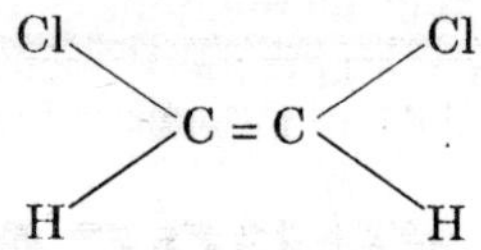 *cis*-isomer and 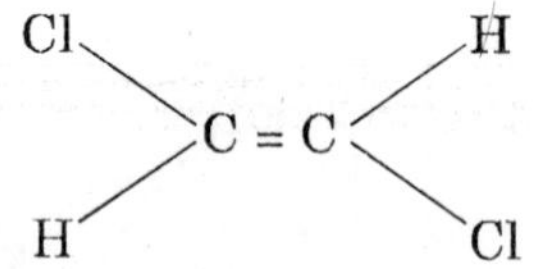 *trans*-isomer
38.	(c)	(b) *cis*-2-Butene and *trans*-2-butene are diastereomers.
		(c) *meso*-Tartaric acid and (+) – tartaric acid are diastereomers.
39.	(c)	$CH_2 = CH—CH = CH_2$ and $HC \equiv C—CH_2\,CH_3$
40.	(d)	Benzaldehyde does not contain α—H atoms for forming tautomer.
41.	(d)	The central carbon is *sp* hybridised.
42.	(c)	Dichlorocarbene *i.e.*, $Cl—\ddot{C}—Cl$.
43.	(b)	sp^2 sp^2 sp^3 $CH_2 = CH—CH_3$ has sp^2 and sp^3 hybridised carbon atoms.
44.	(d)	Triplet carbene has two unpaired electrons and carbon is sp^2 hybridised.
45.	(d)	Ethyl free radical do not stabilise by resonance.

2 HYDROCARBONS

2.1. CONCEPTS IN BRIEF

Develop Logic

SOURCES OF HYDROCARBONS

- The main sources of hydrocarbons are coal, petroleum and natural gas.
- Coal is the major source of aromatic hydrocarbons while petroleum is the major source of aliphatic hydrocarbons.
- Coal on destructive distillation yields coal-tar as one of the products. Further distillation of coal-tar yields following important fractions :

Name of fraction	*Main constituents*
1. **Light oil or crude naphtha**	Benzene, toluene, xylene
2. **Middle oil or carbolic oil**	Naphthalene, phenols, cresols, etc.
3. **Heavy oil or creosote oil**	Naphthol, cresols
4. **Green oil or anthracene oil**	Anthracene, phenathrene
5. **Residue (pitch)**	Carbon

- Petroleum on refining (fractional distillation) yields the following important fractions :

Fraction	Composition	Boiling Range (K)
1. Petroleum gas	C_1-C_4	113—303
2. Petroleum ether (Ligroin)	C_5-C_7	303—363
3. Gasoline	C_7-C_{12}	363—473
4. Kerosene	$C_{12}-C_{15}$	448—548
5. Gas oil, Fuel oil, Diesel oil	$C_{15}-C_{18}$	523—673
6. Lubricating oils	C_{16} onwards	623 and up
7. Paraffin wax	C_{20} onwards	Melts in the range 325-330
8. Petroleum coke		Residue

PROCESSES IN PETROLEUM INDUSTRY

- **Cracking.** The process of decomposing higher hydrocarbons into lower hydrocarbons by heating.

$$CH_3(CH_2)_8CH_3 \xrightarrow{770\ K} C_6H_{14} + C_4H_8$$

- **Reforming or Aromatisation.** This is a process of conversion of aliphatic and alicyclic hydrocarbons into aromatic hydrocarbons in the presence of suitable catalysts such as platinum, palladium or nickel.

$$\underset{n\text{-Hexane}}{C_6H_{14}} \xrightarrow[-3H_2]{Pt,\ 670\ K} \underset{\text{Cyclohexane}}{C_6H_{12}} \xrightarrow[-3H_2]{Pt,\ 670\ K} \underset{\text{Benzene}}{C_6H_6}$$

QUALITY OF GASOLINE—Octane Number

- ***The percentage by volume of iso-octane in the mixture of iso-octane and n-heptane which has the same anti-knock qualities as the fuel under examination is known as octane number.***
- Octane number of n-heptane is zero and that of *iso*-octane (2, 2, 4-Trimethylpentane) is 100.

$$CH_3-CH_2-CH_2-CH_2-CH_2-CH_2-CH_3$$

n-heptane
(Octane number = zero)

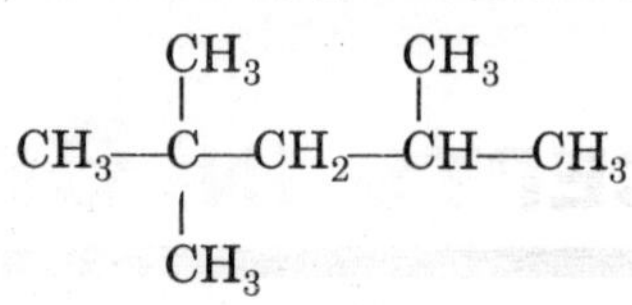

2, 2, 4-Trimethylpentane or *iso*-octane
(Octane number = 100)

- Some compounds have octane number greater than 100 and some have lower than zero. For example, n-nonane has octane number – 45 while 2, 2, 3-Trimethylpentane ***(triptane)*** has octane number 125.
- Branched chain, cyclic, aromatic and unsaturated hydrocarbons have relatively higher octane number as compared with straight chain hydrocarbons.
- Octane number decreases with increase in chain length. For example, octane number of methane is 122 while that of ethane is 101 and that of propane is 96.

QUALITY OF DIESEL —Cetane Number

- ***The percentage by volume of cetane in a mixture of cetane (n-hexadecane) and α-methyl naphthalene which has the same ignition quality as the given sample under examination is known as cetane number.***
- A given fuel is assigned a cetane number of 80 if it has the same ignition quality as a mixture of 80% cetane and 20% α-methyl naphthalene.
- n-Hexadecane has cetane number 100 while α-methyl naphthalene has cetane number zero.

$$CH_3(CH_2)_{14}CH_3$$

n-Hexadecane
(Cetane no. = 100)

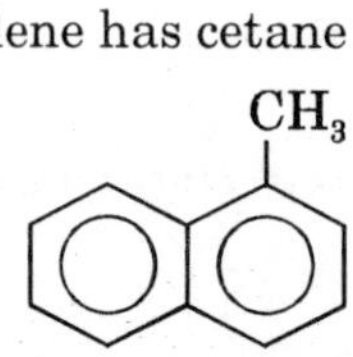

α-Methyl naphthalene
(Cetane no. = 0)

ALKANES

METHODS OF PREPARATION OF ALKANES

1. By Wurtz Reaction

When a solution of an alkyl halide in ether solvent is treated with sodium metal, two alkyl groups combine together. Yields of products are best for primary (~ 60%) halides and poorest for tertiary halides (~ 3%).

$$2CH_3I \xrightarrow[\text{ether}]{2Na} \underset{\text{Ethane}}{CH_3CH_3} + 2NaI$$

$$2\ (CH_3)_2CH{-}Br \xrightarrow[\text{ether}]{2Na} (CH_3)_2CH{-}CH(CH_3)_2 + 2NaBr$$

2, 3-Dimethylbutane

This method is suitable for the preparation of symmetrical alkanes. A mixture of products is obtained if two different alkyl halides are reacted. Methane cannot be prepared by this method.

2. By Reduction of Halides

$$C_2H_5I + H_2 \xrightarrow{Pd} C_2H_6 + HI$$

$$C_2H_5I + 2H \xrightarrow[C_2H_5OH]{Zn/Cu} C_2H_6 + HI$$

$$C_2H_5I + 2H \xrightarrow{Zn,\ HCl} C_2H_6 + HI$$

$$C_2H_5I + HI \xrightarrow{Red\ P} C_2H_6 + I_2$$

3. From Alkyl Halides through Grignard Reagents

$$\underset{\text{Ethyl bromide}}{CH_3CH_2Br + Mg} \xrightarrow{\text{Dry ether}} \underset{\text{Ethyl magnesium bromide}}{CH_3CH_2MgBr} \xrightarrow[\text{Hydrolysis}]{+H_2O} \underset{\text{Ethane}}{C_2H_6} + \underset{\text{Hydroxy magnesium bromide}}{Mg(OH)Br}$$

In these reactions, the alkyl group of alkyl magnesium halide gets converted into alkane.

4. Corey-House Synthesis

$$R{-}X + 2Li \longrightarrow RLi + LiX$$

$$R{-}Li + CuX \longrightarrow R_2CuLi + LiX$$

Lithium dialkyl copper

$$R_2CuLi + R'X \longrightarrow \underset{\text{Alkane}}{R - R'} + RCu + LiX$$

Unlike Wurtz reaction, this reaction can be employed to prepare unsymmetrical alkanes.

5. From unsaturated hydrocarbons (Sabatier and Senderen's reaction)

$$\underset{\text{Ethene}}{CH_2 = CH_2} + H_2 \xrightarrow[523\text{-}573\ K]{Ni} \underset{\text{Ethane}}{CH_3 - CH_3}$$

$$\underset{\text{Ethyne}}{CH \equiv CH} + 2H_2 \xrightarrow[523\text{-}573\ K]{Ni} \underset{\text{Ethane}}{CH_3 - CH_3}$$

6. By Soda-Lime Decarboxylation of Carboxylic Acids

$$\underset{\text{Sodium ethanoate}}{CH_3COONa} + NaOH \xrightarrow[\text{Heat}]{CaO} \underset{\text{Methane}}{CH_4} + Na_2CO_3$$

The alkane formed has one carbon less than the starting material.

7. By Kolbe's Electrolysis

In this method concentrated aqueous solution of sodium or potassium salt of a carboxylic acid is electrolysed using platinum electrodes.

At anode : $2RCOO^- \longrightarrow 2RCOO^\bullet + 2e^-$

$2RC\ddot{O} \longrightarrow 2R + 2CO_2$

$2R^\bullet \longrightarrow R - R$

At cathode : $2H_2O + 2e^- \longrightarrow H_2 + 2OH^-$

If n is the number of carbon atoms in the acid, the alkane formed has 2(n – 1) carbon atoms.

PHYSICAL PROPERTIES OF ALKANES

- **Physical state.** The first four members (C_1 to C_4) are gases ; the next thirteen members, (C_5 to C_{17}) are liquids while the higher members are waxy solids.
- **Boiling points.** Alkanes have generally low boiling points. ***The boiling points of n-alkanes increase regularly with the increase in the number of carbon atoms.*** There is an increase of 20-30 degrees for each carbon atom added to the chain.
- ***Among isomeric alkanes, the branched chain isomers have relatively low boiling points as compared to their corresponding straight chain isomers.***
- **Melting points.** The melting points of alkanes do not show uniform increase with increase in molecular mass as illustrated below :
  The increase in melting point is relatively more in moving from an alkane having odd number of carbon atoms to the higher alkane while it is relatively less in moving from an alkane with even number of carbon atoms to the higher alkane.

Hydrocarbon	C_3H_8	C_4H_{10}	C_5H_{12}	C_6H_{14}	C_7H_{16}	C_8H_{18}
Melting Point (K)	85.9	138	143.3	179	182.5	216.2

- **Solubility.** Hydrocarbons are insoluble in polar solvent like water because they are predominantly non-polar in nature. However, they are soluble in non-polar solvents like ether, carbon tetrachloride, etc.
  The liquid hydrocarbons themselves are good solvents for other non-polar organic molecules.
- **Density.** The densities of alkanes increase with increasing molecular masses but become constant at about 0.8 g cm^{-3}. Thus, all alkanes are lighter than water.

CHEMICAL PROPERTIES OF ALKANES

Alkanes being saturated hydrocarbons do not have any of the reactive sites. Their molecules have all carbon-carbon or carbon-hydrogen single bonds. Thus, alkanes are the least reactive hydrocarbons and are called **paraffins.**

1. Halogenation of Alkanes

$$CH_4 + Cl_2 \xrightarrow[\text{or Heat}]{\text{Diffused sun-light}} \underset{\text{Methyl chloride}}{CH_3Cl} + HCl$$

$$CH_3Cl + Cl_2 \longrightarrow \underset{\text{Methylene chloride}}{CH_2Cl_2} + HCl$$

$$CH_2Cl_2 + Cl_2 \longrightarrow \underset{\text{Chloroform}}{CHCl_3} + HCl$$

$$CHCl_3 + Cl_2 \longrightarrow \underset{\text{Carbon tetrachloride}}{CCl_4} + HCl$$

Bromination, though similar to chlorination, occurs less rapidly.

Iodination is extremely slow and is reversible in nature

$$CH_4 + I_2 \rightleftharpoons \underset{\text{Methyl iodide}}{CH_3I} + HI$$

Iodination is carried out in the presence of some oxidising agent such as HIO_3 or HNO_3 which convert HI to iodine and thus, pushes the reaction in the forward direction

$$5HI + HIO_3 \longrightarrow 3H_2O + 3I_2$$

Fluorination of alkanes takes place violently and hence is seldom done.

The reactivity of halogens towards alkanes varies as :

fluorine > chlorine > bromine > iodine.

Higher alkanes on *halogenation* give different isomers of halogen derivatives.

$$CH_3CH_2CH_3 \xrightarrow[-\,HCl]{Cl_2} \underset{n\text{-Propyl chloride (45\%)}}{CH_3CH_2CH_2Cl} \qquad \underset{iso\text{-Propyl chloride (55\%)}}{CH_3\underset{\displaystyle Cl}{\underset{|}{C}}HCH_3}$$

It may be remembered that in general, ***the order of ease of replacement of hydrogens*** is

tertiary > secondary > primary.

Benzylic hydrogens are replaced even more easily than tertiary hydrogens

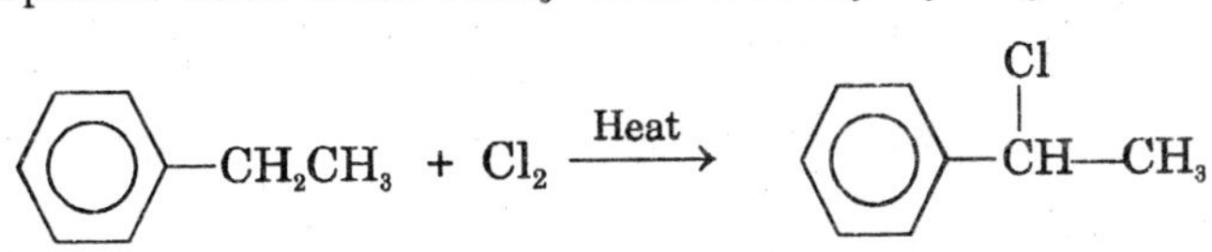

Ethyl benzene 1-Chloro-1-phenylethane

MECHANISM OF HALOGENATION OF ALKANES

Halogenation of alkanes takes place through **free-radical mechanism.** Let us understand it by taking the example of chlorination of methane to give methyl chloride.

$$CH_4 + Cl_2 \xrightarrow{h\nu} CH_3Cl + HCl$$

The substitution occurs through the following steps :

(*a*) **Chain initiation step**

$$Cl : Cl \xrightarrow[\text{Homolysis}]{h\nu} \dot{C}l + \dot{C}l$$

(*b*) **Chain propagation step.** $Cl^{\cdot}$ radicals attack the methane molecules and abstract $H^{\cdot}$ radical to form methyl radicals

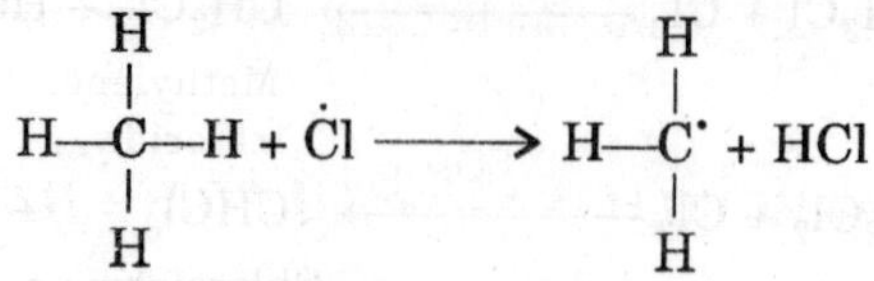

Methyl radical in turn abstract $\dot{C}l$ radical from Cl_2 molecule

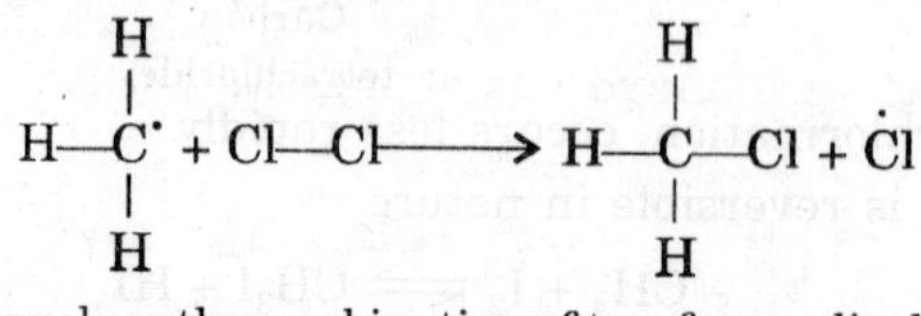

(*c*) **Chain termination step.** This involves the combination of two free radicals

$$\dot{C}l + \dot{C}l \longrightarrow Cl_2$$

$$\dot{C}H_3 + \dot{C}l \longrightarrow CH_3Cl$$

2. **Oxidation**

(*a*) **Combustion**

$$CH_4 + 3O_2 \longrightarrow CO_2 + 2H_2O + \text{Heat}$$

$$C_6H_{14} + \frac{19}{2}O_2 \longrightarrow 6CO_2 + 7H_2O + \text{Heat}$$

(*b*) **Controlled oxidation**

(*i*) On passing *methane and oxygen* in the ratio of 9 : 1 under pressure ($\simeq$ 120 atm.) through copper tube at 475 K, *methane* gets oxidised to *methanol.*

$$\underset{\text{Methane}}{2CH_4} + O_2 \xrightarrow{120 \text{ atm, } 475 \text{ K}} \underset{\text{Methanol}}{2CH_3OH}$$

(*ii*) Methane gets oxidised to *methanal* on passing its mixture with oxygen over heated molybdenum oxide.

$$\underset{\text{Methane}}{CH_4} + O_2 \xrightarrow{Mo_2O_3} \underset{\text{Methanal}}{HCHO} + H_2O$$

(*iii*) Alkanes containing tertiary hydrogen atom get converted to alcohols on reaction with $KMnO_4$.

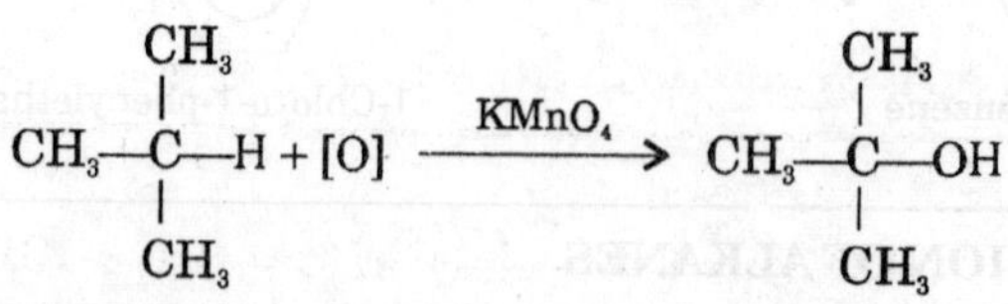

iso-Butane *tert*-Butyl alcohol

3. **Nitration**

In this reaction, a hydrogen atom of an alkane is replaced by nitro (– NO_2) group.

In liquid phase nitration, the alkane is heated for several hours with conc. HNO_3 under pressure. Consider the nitration of *n*-hexane,

$$\underset{\text{Hexane}}{C_6H_{14}} + HONO_2 \xrightarrow{\Delta} \underset{\text{Nitrohexane}}{C_6H_{13}NO_2} + H_2O$$

In vapour phase nitration, the alkane is heated with nitric acid in the vapour phase at 425-750 K.

$$\underset{\text{Propane}}{CH_3CH_2CH_3} + HONO_2 \xrightarrow{675\ K} \underset{\text{1-Nitropropane}}{CH_3CH_2CH_2NO_2} + \underset{\text{2-Nitropropane}}{CH_3\overset{\overset{NO_2}{|}}{C}HCH_3} + \underset{\text{Nitroethane}}{CH_3CH_2NO_2} + \underset{\text{Nitromethane}}{CH_3NO_2}$$

4. Sulphonation

An alkane containing six or more carbon atoms on heating with fuming sulphuric acid for several hours yield the corresponding sulphonic acid,

$$\underset{\text{Hexane}}{C_6H_{14}} + HOSO_3H \xrightarrow{SO_3} \underset{\text{Hexane sulphonic acid}}{C_6H_{13}SO_3H} + H_2O$$

5. Isomerisation

In this process, *n*-alkanes on heating with anhydrous aluminium chloride and hydrogen chloride, get converted into branched chain alkanes.

$$\underset{n\text{-Pentane}}{CH_3CH_2CH_2CH_2CH_3} \xrightarrow{AlCl_3/HCl} \underset{iso\text{-Pentane}}{CH_3\overset{\overset{CH_3}{|}}{C}HCH_2CH_3} + \underset{neo\text{-Pentane}}{CH_3-\overset{\overset{CH_3}{|}}{\underset{\underset{CH_3}{|}}{C}}-CH_3}$$

6. Methylene Insertion

Singlet methylene can insert between C—H bond.

$$CH_2N_2 \xrightarrow{h\nu} H_2C : + N_2 \quad \text{or} \quad H_2C = C = O \xrightarrow{h\nu} H_2C : + CO$$

$$CH_3CH_2CH_2CH_2CH_3 + :CH_2 \longrightarrow \underset{48\%}{CH_3CH_2CH_2CH_2CH_2CH_3} + \underset{35\%}{CH_3\underset{\underset{CH_3}{|}}{C}HCH_2CH_2CH_3} + \underset{17\%}{CH_3CH_2\underset{\underset{CH_3}{|}}{C}HCH_2CH_3}$$

ALKENES

Alkenes are unsaturated hydrocarbons. The characteristic feature of alkenes is the presence of carbon-carbon double bond (C = C) in their molecules. Alkenes are also known as **olefins,** derived from Greek word *olefiant* (means *oil forming*). It is because of the fact that lower members of this series form oily products when treated with chlorine.

METHODS OF PREPARATION OF ALKENES

1. By Dehydrohalogenation of Alkyl Halides

$$\underset{\text{Ethyl bromide}}{CH_3CH_2Br + KOH(alc.)} \longrightarrow \underset{\text{Ethene}}{H_2C = CH_2} + KBr + H_2O$$

In case an alkyl halide can eliminate hydrogen halide in two different ways, the preferred alkene is the one which is maximum alkylated i.e., which carries more number of alkyl groups attached to the doubly bonded carbon atoms. This generalisation is known as **Saytzeff Rule.** For example,

$$\underset{\text{2-Bromobutane}}{CH_3\overset{\beta}{-}CH_2-\underset{Br}{CH}-\overset{\beta}{C}H_3} \xrightarrow{\text{alc. KOH}} \underset{\substack{\text{But-2-ene}\\(80\%)}}{CH_3-CH=CH-CH_3} + \underset{\substack{\text{But-1-ene}\\(20\%)}}{CH_3-CH_2-CH=CH_2}$$

The **reactivity** *of alkyl halides towards elimination reaction follows the order : tertiary > secondary > primary.*

2. By Dehydration of Alcohols

$$\underset{\text{Ethanol}}{CH_3CH_2OH} \underset{440\text{ K}}{\overset{\text{conc. }H_2SO_4}{\rightleftharpoons}} \underset{\text{Ethene}}{CH_2=CH_2} + H_2O$$

$$\underset{\text{Ethanol}}{CH_3CH_2OH} \xrightarrow[623\text{-}633\text{ K}]{Al_2O_3} \underset{\text{Ethene}}{CH_2=CH_2} + H_2O$$

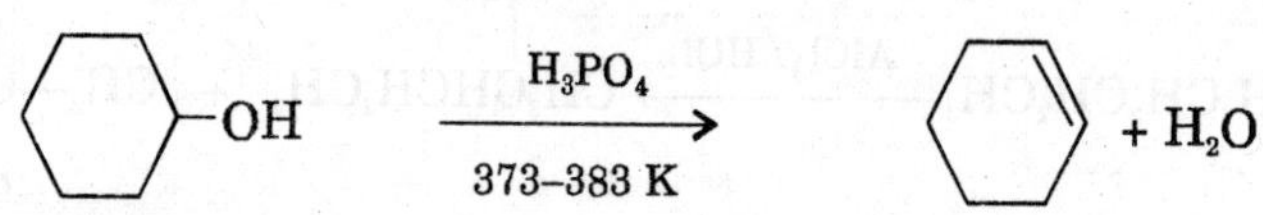

Cyclohexanol Cyclohexane

MECHANISM OF DEHYDRATION OF ALCOHOLS

The mechanism of this reaction involves protonation of the alcohol followed by loss of a water molecule and a proton.

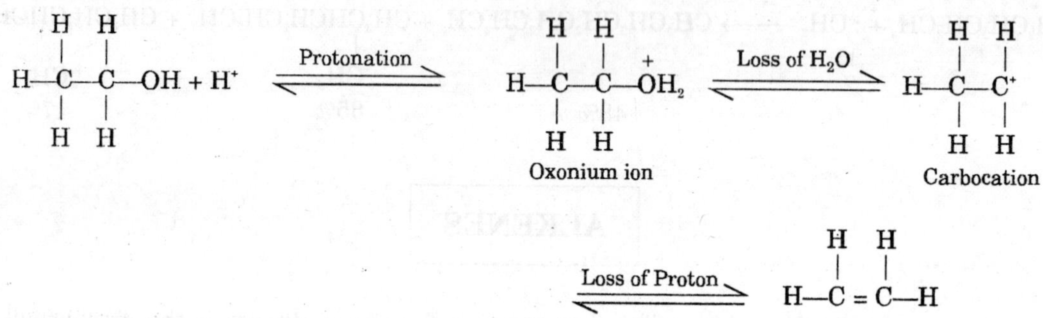

This method is suitable for relatively simple alcohols. In more complex cases rearrangements may take place due to rearrangement of intermediate carbocation.

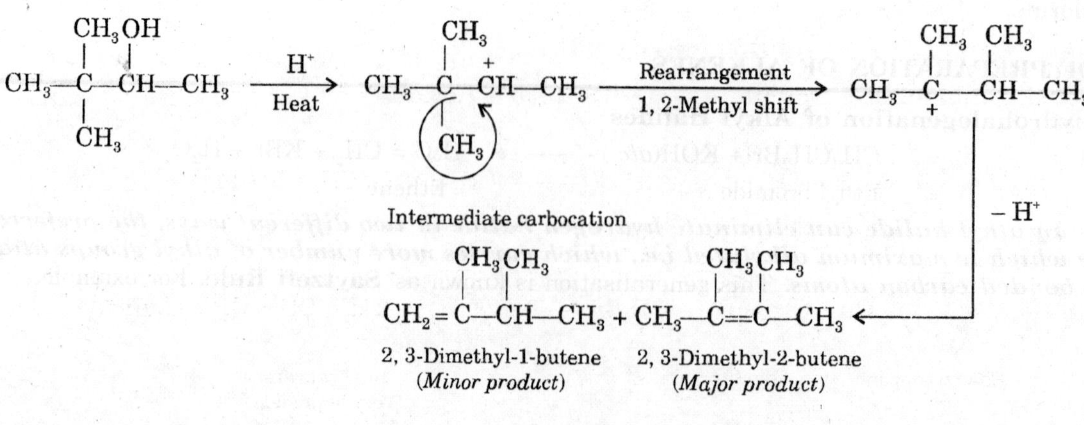

3. By Dehalogenation

Vicinal dihalides on treatment with zinc and alcohol give alkenes.

$$\underset{\text{Ethylene dibromide}}{BrCH_2-CH_2Br} + Zn \xrightarrow{C_2H_5OH} \underset{\text{Ethene}}{CH_2=CH_2} + ZnBr_2$$

4. By Cracking of Alkanes

Alkenes are prepared on large scale by cracking of hydrocarbons.

$$CH_3—CH_3 \xrightarrow{700-900°C} CH_2=CH_2 + H_2$$

$$n\text{-}C_6H_{14} \xrightarrow{700-900°C} CH_4 + CH_2=CH_2 + CH_3CH=CH_2 + \text{others}$$

Ethene and propene are used extensively for the production of polymers.

5. From Alkynes (*By partial hydrogenation of C—C triple bond*)

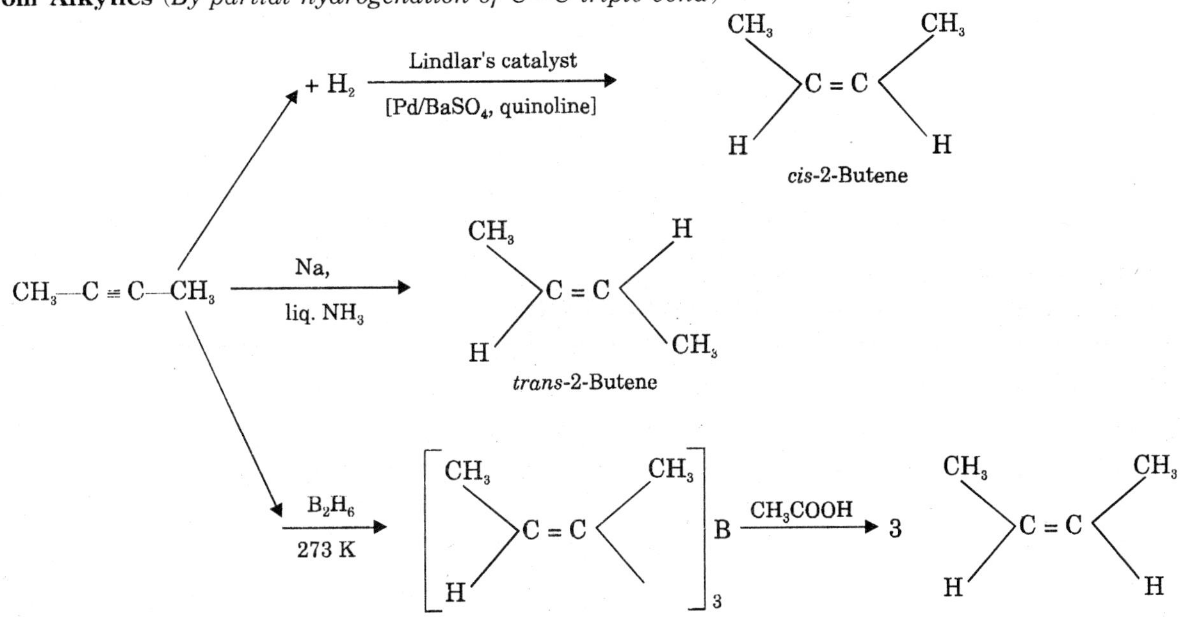

6. By Decomposition of Quaternary Ammonium Hydroxides

$$\left[\begin{array}{c} CH_2CH_3 \\ | \\ CH_3CH_2—N—CH_2CH_3 \\ | \\ CH_2CH_3 \end{array}\right]^+ OH^- \xrightarrow{\Delta} CH_2=CH_2 + (CH_3CH_2)_3N + H_2O$$

7. By use of Organometallic Compounds and Reactive Halides

$$\underset{\substack{\text{Grignard}\\\text{reagent}}}{RMgX} + \underset{\substack{\text{A reactive}\\\text{halide}}}{CH_2=CH-CH_2Cl} \longrightarrow CH_2=CH-CH_2-R + Mg\begin{matrix} X \\ Cl \end{matrix}$$

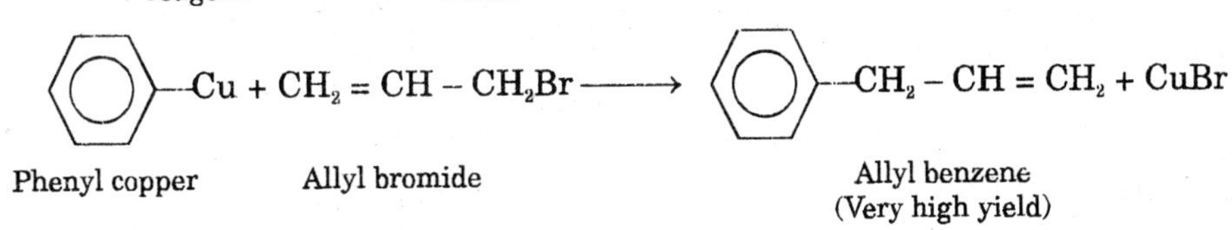

8. By Kolbe Electrolysis of Potassium Succinate

$$\underset{\text{Pot. succinate}}{\begin{matrix} CH_2COOK \\ | \\ CH_2COOK \end{matrix}} + 2H_2O \longrightarrow \underset{\text{Ethene}}{\begin{matrix} CH_2 \\ \| \\ CH_2 \end{matrix}} + 2CO_2 + 2KOH + H_2$$

CHEMICAL PROPERTIES OF ALKENES

1. Addition of Hydrogen

$$CH_2 = CH_2 + H_2 \xrightarrow[\text{Heat}]{\text{Pt, Pd or Ni}} CH_3 - CH_3$$

2. Addition of Bromine

Chlorine and bromine add to alkenes at ordinary temperature to form vicinal dihalides.

$$\underset{\text{Ethylene}}{CH_2 = CH_2} + Br_2 \xrightarrow{CCl_4} \underset{\text{Ethylene dibromide}}{\underset{Br\quad\; Br}{CH_2 - CH_2}}$$

In the reaction of an alkene with bromine, the red colour of bromine is discharged due to the formation of dibromide which is colourless. Therefore, this reaction is used as a **test** for the presence of carbon-carbon double bond in a molecule.

MECHANISM OF ADDITION OF BROMINE TO ALKENES

1.

2.

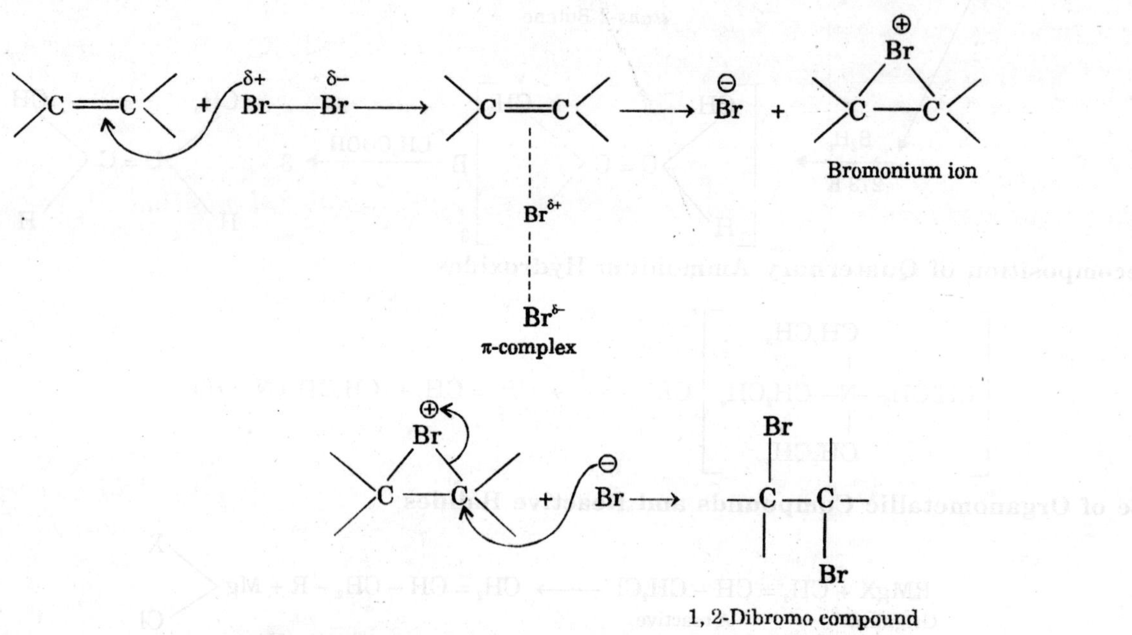

- Addition is *anti* or *trans i.e.*, one Br adds to the face above the π bond while the second Br atom adds to the face below.
- Due to *anti* addition *cis*-2-butene on reaction with Br_2 yields a racemic mixture of 2, 3-dibromobutane while *trans*-2-butene yields *meso*-2, 3-dibromobutane.

3. Addition of Hydrogen Halides

Alkenes readily undergo addition reactions with hydrogen halides (HCl, HBr or HI) to form alkyl halides.

$$\underset{\text{Ethylene}}{CH_2 = CH_2} + HI \longrightarrow \underset{\text{Ethyl iodide}}{CH_3—CH_2—I}$$

MECHANISM OF ADDITION OF HALOGEN ACIDS TO ALKENES

Hydrogen halides easily add to alkenes to form alkyl halides. Addition is believed to proceed through following mechanism.

1. 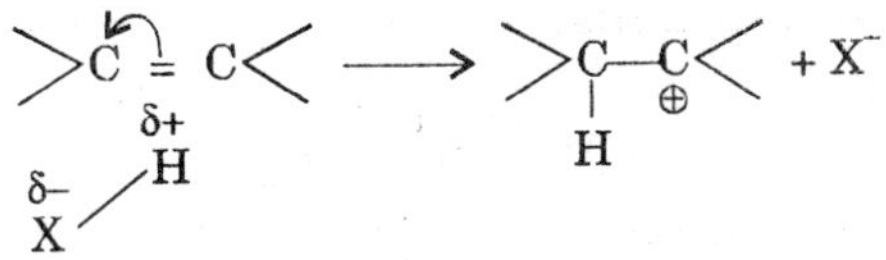 where HX = HCl, HBr, HI

2. 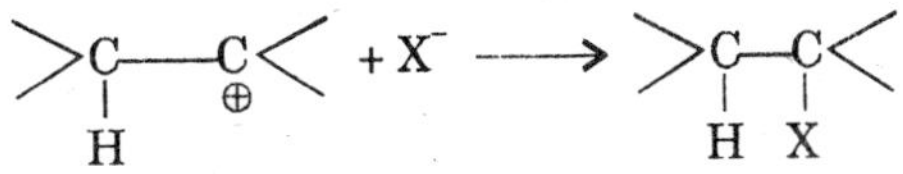

Order of reactivity of different halogen acids in this reaction is :

HI > HBr > HCl

In case of unsymmetrical alkenes such as propene halogen acid can add to give two different products (I) and (II).

$$\underset{\text{(I) Major product}}{CH_3-\underset{X}{\underset{|}{CH}}-CH_3} \xleftarrow{HX} CH_3-CH=CH_2 \xrightarrow{HX} \underset{\text{(II)}}{CH_3-CH_2-CH_2-X}$$

The product actually obtained is the one which obeys **Markownikoffs Rule** which is stated as : ***Electrophillic addition in unsymmetrical alkenes takes place in such a way so that the negative part of the addendum attaches itself to the carbon which carries smaller no. of hydrogen atoms.***

4. **Kharasch Effect or Antimarkownikoff Addition**

Kharasch effect deals with antimarkownikoff addition of HBr across unsymmetrical alkene in the presence of peroxide.

$$CH_3-CH=CH_2+HBr \xrightarrow{\text{Benzoyl peroxide}} \underset{\text{1-Bromopropane}}{CH_3CH_2CH_2Br}$$

In the presence of peroxide the addition of HBr follows free radical mechanism.

$$\underset{\text{Organic peroxide}}{R-O \vdots O-R} \longrightarrow \underset{\text{Free radicals}}{R-\dot{O}+R-\dot{O}}$$

$$R-\dot{O}+H-Br \xrightarrow[\text{of } H^{\cdot}]{\text{Abstraction}} R-OH+Br^{\cdot}$$

$$CH_3-CH=CH_2+\dot{B}r \longrightarrow \underset{\text{1° Free radical (Less stable)}}{CH_3-\underset{Br}{\underset{|}{CH}}-CH_2^{\cdot}}$$

$$CH_3-CH=CH_2+\dot{B}r \longrightarrow \underset{\text{2° Free radical (More stable)}}{CH_3-\dot{C}H-CH_2Br}$$

$$CH_3-\dot{C}H-CH_2Br+HBr \longrightarrow \underset{\text{1-Bromopropane}}{CH_3CH_2CH_2Br}+Br$$

It may be noted that HI and HCl do not follow free radical mechanism even in the presence of peroxide. This is because $\dot{I}$ though readily formed from HI is less reactive and HCl does not readily break to give $\dot{C}l^{\bullet}$. They add by ionic mechanism-forming products in accordance with Markownikoff Rule.

5. Addition of Hypohalous Acids (*Formation of Halohydrins*)

$$\underset{\text{Propene}}{CH_3-CH=CH_2} + HOCl \xrightarrow{Cl_2/H_2O} \underset{\text{1-Chloro-2-propanol}}{CH_3-\underset{\substack{|\\OH}}{CH}-CH_2Cl}$$

6. Addition of Sulphuric Acid

Alkenes add to cold conc. sulphuric acid to form corresponding alkyl hydrogen sulphate.

$$\underset{\text{Propene}}{CH_3-CH=CH_2} + HOSO_2.OH \longrightarrow \underset{\text{Isopropyl hydrogen sulphate}}{CH_3-\underset{\substack{|\\OSO_2OH}}{CH}-CH_3}$$

7. Addition of Water (*Hydration*)

A water molecule gets added to alkenes in the presence of acids to form alochols.

$$\underset{\text{Propene}}{CH_3-CH=CH_2} + H_2O \overset{H^+}{\rightleftharpoons} \underset{\text{2-Propanol}}{CH_3-\underset{\substack{|\\OH}}{CH}-CH_3}$$

Addition to unsymmetrical alkenes follows Markownikoff rule.

During hydration of alkenes many times rearranged products are obtained.

$$CH_3-\overset{\substack{CH_3\\|}}{\underset{\substack{|\\CH_3}}{C}}-CH=CH_2 + H_2O \overset{H^+}{\rightleftharpoons} \underset{\text{(Major product)}}{CH_3-\overset{\substack{CH_3\\|}}{\underset{\substack{|\\OH}}{C}}-\overset{\substack{CH_3\\|}}{CH}-CH_3}$$

8. Oxymercuration-Demercuration

$$CH_3-CH_2-\overset{\substack{CH_3\\|}}{C}=CH_2 \xrightarrow[(ii)\ NaBH_4]{(i)\ Hg(CH_3COO)_2,\ H_2O} CH_3-CH_2-\overset{\substack{CH_3\\|}}{\underset{\substack{|\\OH}}{C}}-CH_3$$

Addition of water here takes place according to Markownikoff's rule but ***no rearrangement takes place during the reaction.***

9. Hydroboration-Oxidation

In this reaction alkene is treated with diborane followed by treatment with alkaline solution of H_2O_2 when alcohol is formed.

$$\underset{\text{Propene}}{3CH_3CH=CH_2} + \underset{\text{Diborane}}{\tfrac{1}{2}B_2H_6} \longrightarrow \underset{\text{Tripropyl borane}}{(CH_3CH_2CH_2)_3B}$$

$$(CH_3CH_2CH_2)_3B \xrightarrow[OH^-,\ H_2O]{H_2O_2} \underset{\text{1-Propanol}}{3CH_3CH_2CH_2OH} + H_3BO_3$$

In this method addition of water occurs in an anti-Markownikoff's way.

10. Addition of Alkanes

Alkenes can also add on alkanes in the presence of suitable acidic **catalysts to** form higher alkanes. The reaction known as **alkylation** is employed commercially for the manufacture of alkanes such as 2, 2, 4-trimethylpentane used as aviation fuel.

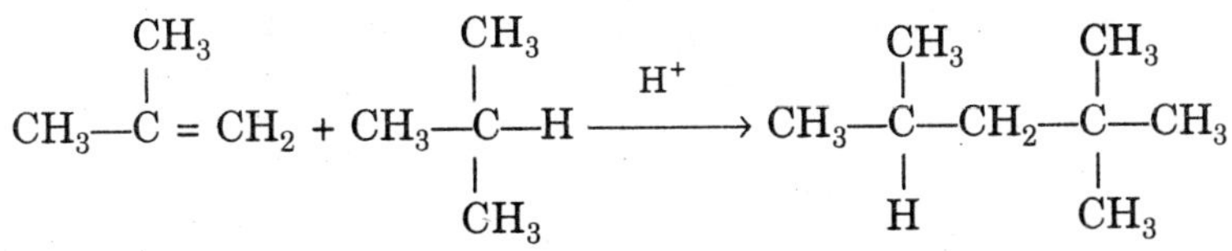

Isobutylene Isobutane 2, 2, 4-Trimethylpentane

11. Dimerisation

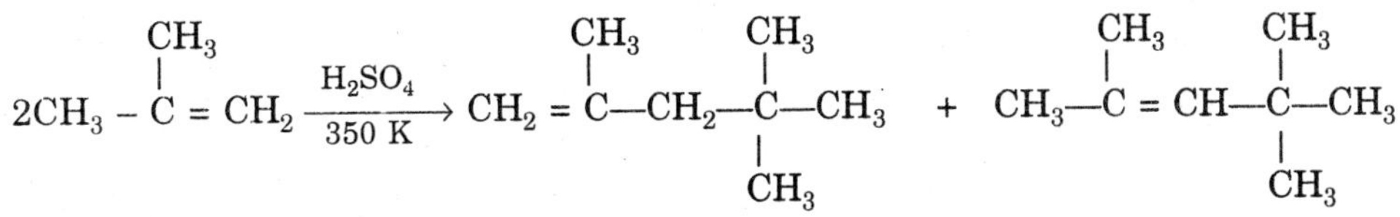

2, 4, 4-Trimethyl-1-pentene 2, 4, 4-Trimethyl-2-pentene

12. Polymerisation

When ethene is heated in presence of traces of oxygen at about 500-675 K under high pressure, n molecules of ethene participate in the reaction as shown below :

$$\underset{\text{Ethylene}}{nCH_2 = CH_2} \longrightarrow \underset{\text{Polythene (Polyethylene)}}{(-CH_2 - CH_2-)_n}$$

13. Addition of Carbenes

Alkenes undergo addition reaction with carbenes (singlet as well as triplet) to form cyclopropanes.

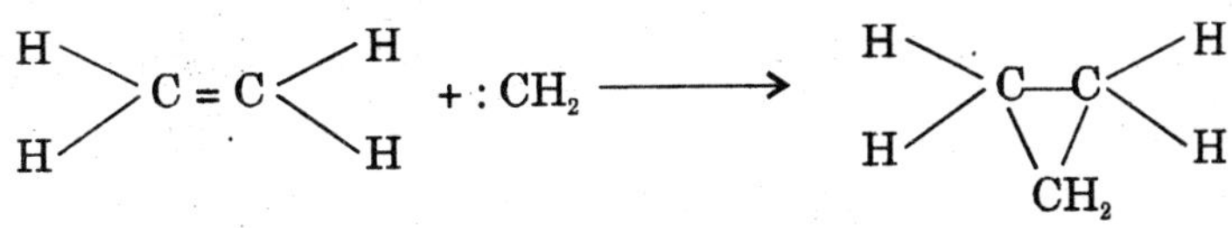

Cyclopropane

14. Addition of Free Radicals

$$R - CH = CH_2 + CCl_4 \xrightarrow{\text{Peroxides}} R-\underset{\underset{Cl}{|}}{CH}-CH_2-CCl_3$$

$$R - CH = CH_2 + CHCl_3 \xrightarrow{\text{Peroxides}} R-\underset{\underset{H}{|}}{CH}-CH_2-CCl_3$$

This reaction follows free radical addition mechanism similar to that in case of Kharasch effect.

15. Oxidation

Oxidation reactions of alkenes are summarised ahead :

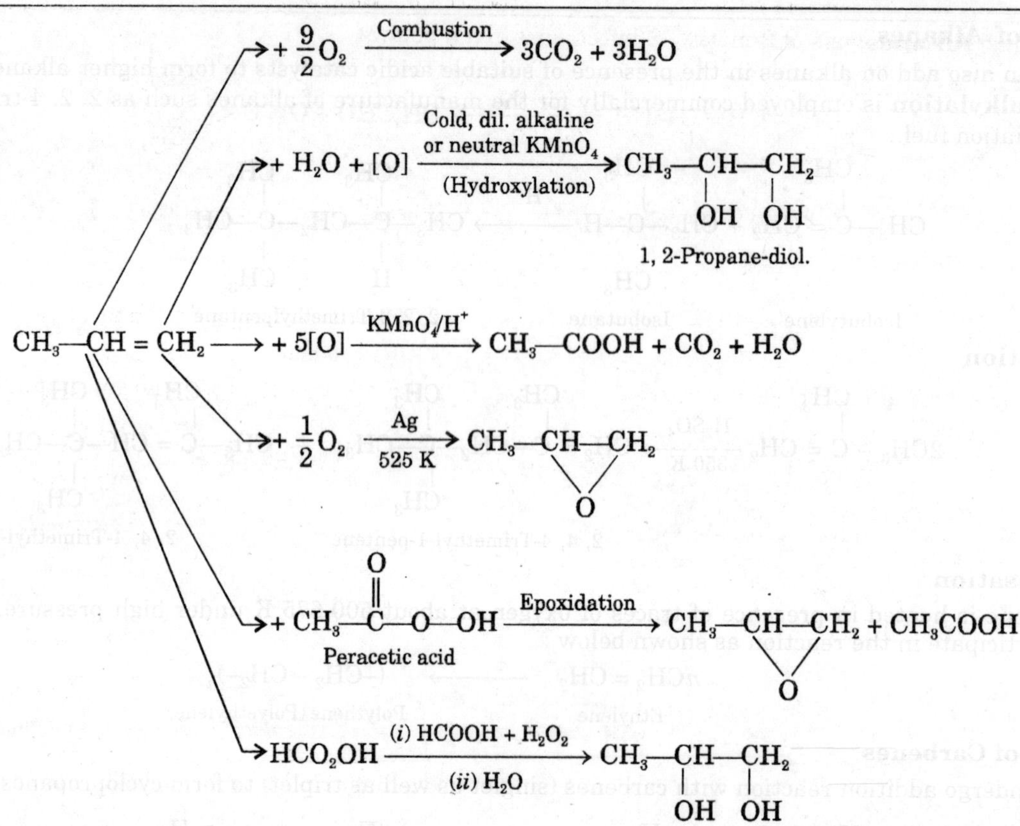

$$2CH_2 = CH_2 + O_2 \xrightarrow[CuCl_2,\ HCl]{PdCl_2} 2CH_3CHO$$ **Wacker Process**

> **Note : 1. Alkaline solution of $KMnO_4$** is known as Baeyer's reagent. During reaction with alkenes (or alkynes) its colour gets discharged. Therefore, this reagent can be used to detect unsaturation in the molecule.
>
> **2.** Tetra-substituted alkenes on reaction with hot, conc. acidified $KMnO_4$ solution give ketones as the products
>
> $$\begin{matrix} R' \\ R \end{matrix}\!\!>C = C<\!\!\begin{matrix} R'' \\ R''' \end{matrix} + 2[O] \xrightarrow{KMnO_4/H^+} \begin{matrix} R' \\ R \end{matrix}\!\!>C = O + O = C<\!\!\begin{matrix} R'' \\ R''' \end{matrix}$$
>
> **3.** Epoxides can be hydrolysed to vicinal diols by boiling water
>
> $$CH_3-\underset{\diagdown O \diagup}{CH-CH_2} + H_2O \xrightarrow{\Delta} CH_3-\underset{OH}{CH}-\underset{OH}{CH_2}$$

16. Ozonolysis

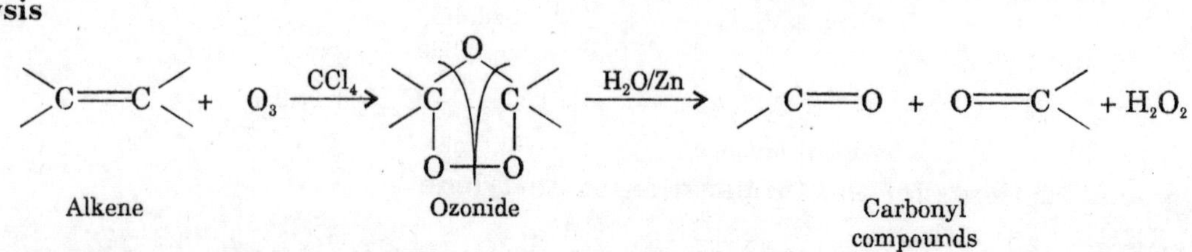

The oxidation of alkenes with ozone followed by the decomposition of the ozonide is called **ozonolysis.** Ozonolysis is the best method for locating the position of the double bond in alkenes.

The products of ozonolysis of 2-methyl-2-butene are given below :

$$\underset{\text{2-Methyl-2-butene}}{CH_3-CH=\underset{\underset{CH_3}{|}}{C}-CH_3} \xrightarrow[(ii)\ H_2O/Zn]{(i)\ O_3} \underset{\text{Ethanal}}{CH_3CHO} + \underset{\text{Propanone}}{O=C(CH_3)_2}$$

17. Allylic Substitution

$$\underset{\text{Propene}}{CH_2=CH-CH_3} + Br_2 \xrightarrow{800\text{ K}} \underset{\text{Allyl bromide}}{CH_2=CH-CH_2-Br} + HBr$$

The allylic substitution can also be carried out by treating the alkene with N-bromo succinimide (NBS)

$$\underset{\text{Propene}}{CH_2=CH-CH_3} + \underset{\text{N-Bromo succinimide}}{(CH_2CO)_2N-Br} \xrightarrow{h\nu} CH_2=CH-CH_2Br + \underset{\text{Succinimide}}{(CH_2CO)_2NH}$$

ALKYNES

Alkynes are unsaturated hydrocarbons. They have carbon-carbon triple bond ($-C\equiv C-$). The carbon atoms involved in triple bond are *sp*-hybridised.

METHODS OF PREPARATION OF ALKYNES

1. By Dehydrohalogenation

$$\underset{\text{Ethylidene dibromide}}{CH_3-CHBr_2} + 2KOH(alc) \longrightarrow \underset{\text{Acetylene}}{H-C\equiv C-H} + 2KBr + 2H_2O$$

$$\underset{\text{Ethylene dibromide}}{BrCH_2-CH_2Br} + 2KOH(alc) \longrightarrow \underset{\text{Acetylene}}{H-C\equiv C-H} + 2KBr + 2H_2O$$

2. By Dehalogenation

$$\underset{\text{1, 1, 2, 2-Tetrabromoethane}}{H-CBr_2-CBr_2-H} + 2\,Zn \xrightarrow{\text{Heat}} \underset{\text{Acetylene}}{H-C\equiv C-H} + 2\,ZnBr_2$$

3. Higher Alkynes from Acetylene

Higher alkynes may be prepared by the action of alkylhalides on sodium acetylide. Sodium acetylide can be obtained from acetylene by the reaction of sodamide.

$$HC\equiv CH + \underset{\text{Sodamide}}{NaNH_2} \xrightarrow{\text{Diethyl ether}} \underset{\text{Sodium acetylide}}{HC\equiv \overset{-\ +}{CNa}} + NH_3$$

$$HC\equiv C^-Na^+ + \underset{\text{Ethyl bromide}}{BrC_2H_5} \longrightarrow \underset{\text{1-Butyne}}{HC\equiv C-C_2H_5} + NaBr$$

4. Some Special Methods for the Preparation of Acetylene

(*i*) $$\underset{\text{Calcium Carbide}}{CaC_2} + H_2O \longrightarrow Ca(OH)_2 + C_2H_2$$

(*ii*) $2C + H_2 \xrightarrow[3000°C]{\text{Electric arc}} H{-}C \equiv C{-}H$

(*iii*) $2CHI_3 + 6Ag \xrightarrow{\Delta} H{-}C \equiv C{-}H + 6AgI$

(*iv*) *By Kolbe electrolysis of potassium fumarate or potassium maleate*

$$\begin{matrix} CHCOOK \\ \| \\ CHCOOK \end{matrix} + 2H_2O \xrightarrow{\text{Electrolysis}} \begin{matrix} CH \\ ||| \\ CH \end{matrix} + 2CO_2 + H_2 + 2KOH$$

(*v*) *By partial oxidation of methane* **(Industrial method)**

$$6CH_4 + O_2 \xrightarrow{1500°C} 2HC \equiv CH + 2CO + 10H_2$$

(*vi*) $2CH_4 \xrightarrow[\text{Electric arc}]{1200°C} HC \equiv CH + 3H_2$

CHEMICAL PROPERTIES OF ALKYNES

1. Addition of Hydrogen

$$H{-}C \equiv C{-}H + H_2 \xrightarrow{\text{Ni, Pt or Pd}} H{-}\overset{H}{\overset{|}{C}} = \overset{H}{\overset{|}{C}}{-}H \xrightarrow{+H_2} CH_3{-}CH_3$$

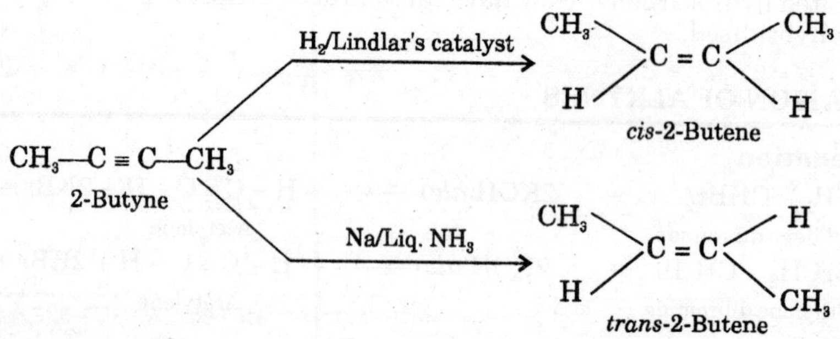

2. Addition of Halogen Acids

$$H{-}C \equiv C{-}H + HCl \longrightarrow \underset{\text{Vinyl chloride}}{CH_2 = CH - Cl} \xrightarrow{+HCl} \underset{\text{Ethylidene dichloride}}{CH_3 - CHCl_2}$$

3. Addition of Halogens

$$H - C \equiv C - H \xrightarrow{+Br_2} \underset{\text{Br Br}}{H{-}C = C{-}H} \xrightarrow{+Br_2} \underset{\text{Br Br}}{\overset{\text{Br Br}}{H{-}C{-}C{-}H}}$$

1, 1, 2, 2-Tetrabromoethane

4. Addition of Hypohalous Acids

$$H{-}C \equiv C{-}H + HOCl \xrightarrow{Cl_2/H_2O} \underset{\text{Cl OH}}{H{-}C = C{-}H} \xrightarrow{Cl_2/H_2O} \left[\underset{\text{Cl OH}}{\overset{\text{Cl OH}}{H{-}C{-}C{-}H}}\right]_{\text{Unstable}} \xrightarrow{-H_2O} \underset{\text{Dichloroacetaldehyde}}{\underset{\text{Cl O}}{\overset{\text{Cl}}{H{-}C{-}C{-}H}}}$$

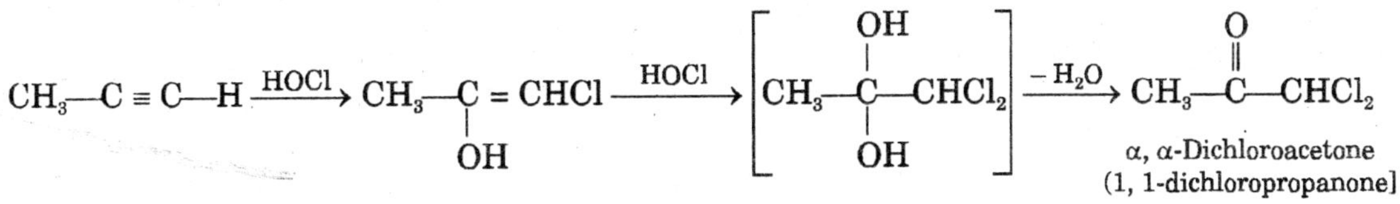

5. Addition of Water

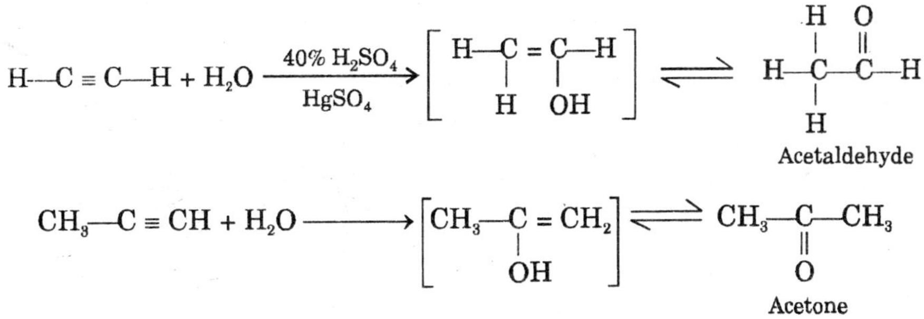

6. Addition of HCN

$$H-C\equiv C-H + HCN \xrightarrow{Ba(CN)_2} H_2C=CH-CN$$

Acrylonitrile

7. Addition of CH_3COOH

$$H-C\equiv C-H + CH_3COOH \xrightarrow{Hg(CH_3COO)_2} H_2C=CH-O-CO-CH_3$$

Vinyl acetate

8. Addition of Alcohols

$$H-C\equiv C-H + CH_3OH \xrightarrow{\overset{+\ -}{KOCH_3}} H_2C=CH-OCH_3$$

Methyl vinyl ether

> *Note : **Reactions 2, 3, 4 are electrophilic addition reactions whereas reactions 5, 6, 7 and 8 are nucleophilic addition reactions. Nucleophilic addition reactions are catalysed by heavy metal ions.***

9. Polymerisation Reactions

(*i*) **Dimerisation.** When passed through a solution of cuprous chloride in ammonium chlroide, acetylene dimerises to give vinyl acetylene.

$$\underset{\text{Acetylene}}{CH\equiv CH + HC\equiv CH} \xrightarrow[NH_4Cl]{CuCl} \underset{\text{Vinyl acetylene}}{CH\equiv C-CH=CH_2} \xrightarrow[NH_4Cl]{CuCl} \underset{\text{Divinyl acetylene}}{CH_2=CH-C\equiv CH=CH_2}$$

(*ii*) **Trimerisation.** When passed through a red hot tube, acetylene is partially converted into benzene.

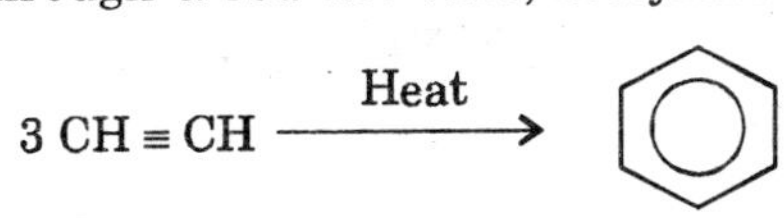

Similarly, propyne trimerises in the presence of sulphuric acid to form *sym*-trimethyl benzene or mesitylene.

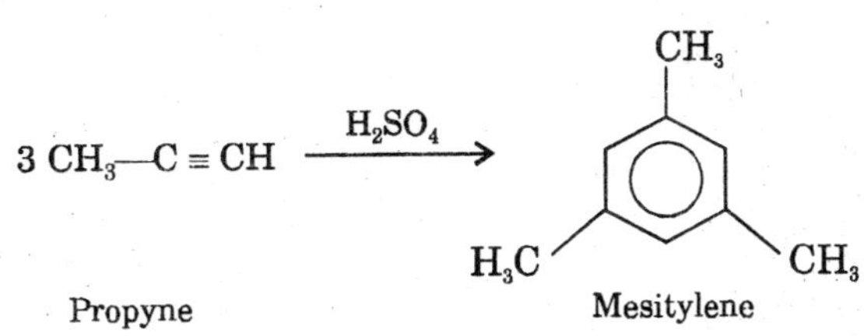

(*iii*) **Tetramerisation.** Acetylene undergoes tetramerisation under high pressure and presence of nickel cyanide catalyst to form cyclo-octatetraene.

$$4CH \equiv CH \xrightarrow{Ni(CN)_2}$$

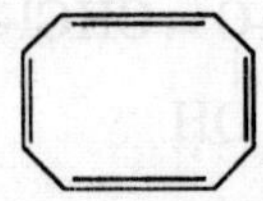

Cyclo-octatetraene

10. Isomerisation

$$\underset{\text{1-Butyne}}{CH_3-CH_2-C\equiv CH} \xrightarrow{\text{alc. KOH}} \underset{\text{2-Butyne}}{CH_3-C\equiv C-CH_3}$$

$$CH_3-C\equiv C-CH_3 \xrightarrow{NaNH_2} CH_3-CH_2-C\equiv CH$$

11. Reaction with Carbonyl Compounds

$$H-C\equiv C-H + HCHO \xrightarrow{\text{High pressure}} \underset{\text{Propargyl alcohol}}{H-C\equiv C-CH_2OH}$$

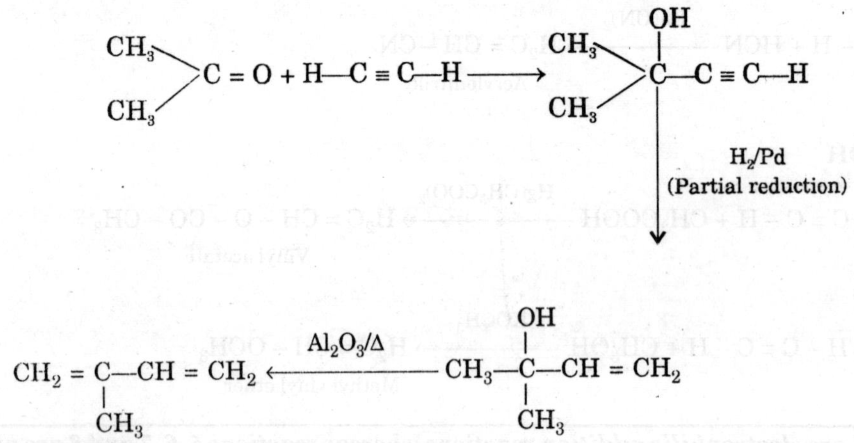

Isoprene

12. Reactions as Acids

The following reactions illustrate the acidic character of terminal alkynes

$$R-C\equiv C-H \xrightarrow{Na} R-C\equiv C^-Na^+ + \frac{1}{2}H_2$$

$$R-C\equiv C-H \xrightarrow{NaH} R-C\equiv C^-Na^+ + H_2$$

$$R-C\equiv C-H \xrightarrow{NaNH_2} R-C\equiv C^-Na^+ + NH_3$$

$$R-C\equiv C-H \xrightarrow{C_2H_5MgBr} R-C\equiv C^-Mg^+Br + C_2H_6$$

Formation of heavy metal acetylides

$$CH\equiv CH + 2AgNO_3 + 2NH_4OH \longrightarrow \underset{\text{White ppt.}}{AgC\equiv CAg} + 2NH_4NO_3 + 2H_2O$$

$$CH\equiv CH + Cu_2Cl_2 + 2NH_4OH \longrightarrow \underset{\text{Red ppt.}}{CuC = CCu} + 2NH_4Cl + 2H_2O$$

The formation of acetylides can be used for distinguishing terminal alkynes from other hydrocarbons and even from disubstituted alkynes (R—C ≡ C—R) which do not undergo this reaction.

These acetylides are quite explosive when in the dry state. Therefore, they should be destroyed, while still wet, by warming with dilute HNO_3.

$$R{-}C \equiv CAg + HNO_3 \longrightarrow R{-}C \equiv C{-}H + AgNO_3$$

13. Oxidation Reactions

Reaction with alkaline $KMnO_4$ (Mild conditions)

$$\underset{\text{Ethyne}}{H{-}C \equiv C{-}H} + 4[O] \xrightarrow{\text{alk. } KMnO_4} \underset{\text{Oxalic acid}}{\begin{array}{c}COOH \\ | \\ COOH\end{array}}$$

$$\underset{\text{Propyne}}{CH_3-C \equiv C-H} + 3[O] \longrightarrow \underset{\text{2-Ketopropanoic acid}}{CH_3-\overset{\overset{\large O}{\|}}{C}-COOH}$$

$$\underset{\text{2-Butyne}}{CH_3-C \equiv C-CH_3} + 2[O] \longrightarrow \underset{\text{2, 3-Butanedione}}{CH_3-\overset{\overset{\large O}{\|}}{C}-\overset{\overset{\large O}{\|}}{C}-CH_3}$$

Reaction with acidified $KMnO_4$ (Drastic conditions)

Molecule of alkyne undergoes cleavage at the carbon-carbon triple bond.

$$CH_3{-}C \equiv C{-}H + 4[O] \xrightarrow{KMnO_4/H^+} CH_3COOH + CO_2$$

Ozonolysis

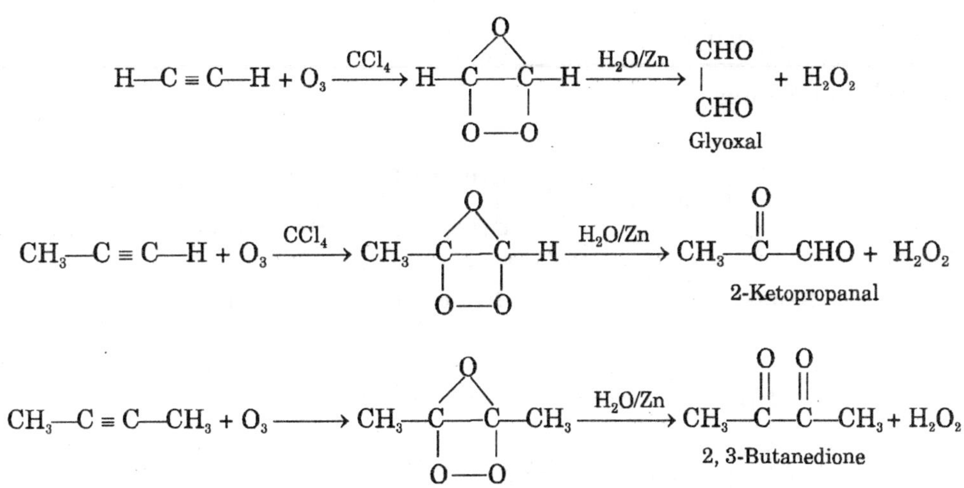

Oxidation with SeO_2

$$H{-}C \equiv C{-}H + 2[O] \xrightarrow{SeO_2} \underset{\text{Glyoxal}}{H-\overset{\overset{\large O}{\|}}{C}-\overset{\overset{\large O}{\|}}{C}-H}$$

$$CH_3{-}C \equiv C{-}CH_3 + 2[O] \xrightarrow{SeO_2} \underset{\text{2, 3-Butanedione}}{CH_3-\overset{\overset{\large O}{\|}}{C}-\overset{\overset{\large O}{\|}}{C}-CH_3}$$

ARENES

METHODS OF PREPARATION OF ARENES

1. Benzene, toluene and xylenes are obtained by fractional distillation of coal tar.
2. **By Friedel-Crafts reaction**

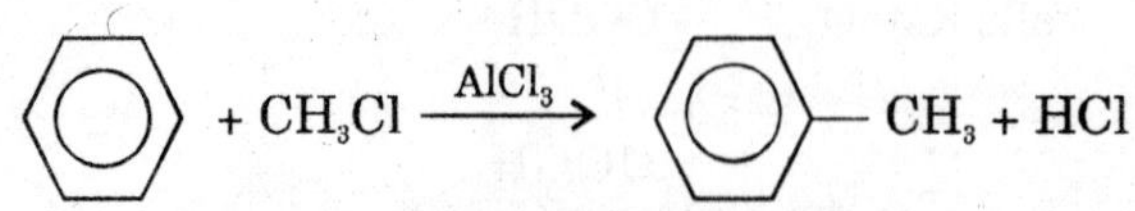

3. **By reaction of alkenes with benzene**

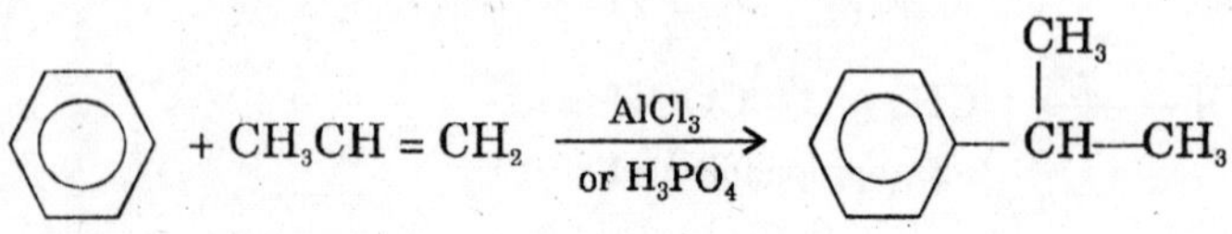

4. **By Wurtz Fittig reaction**

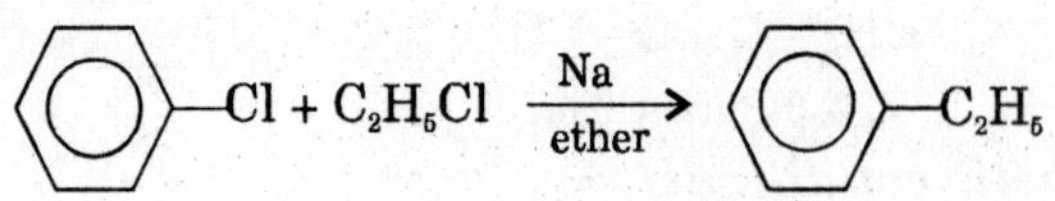

5. **From phenols**

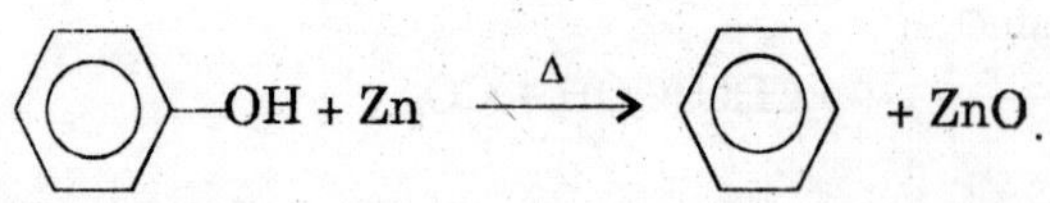

CHEMICAL PROPERTIES OF ARENES

1. **Oxidation Reactions of Arenes**

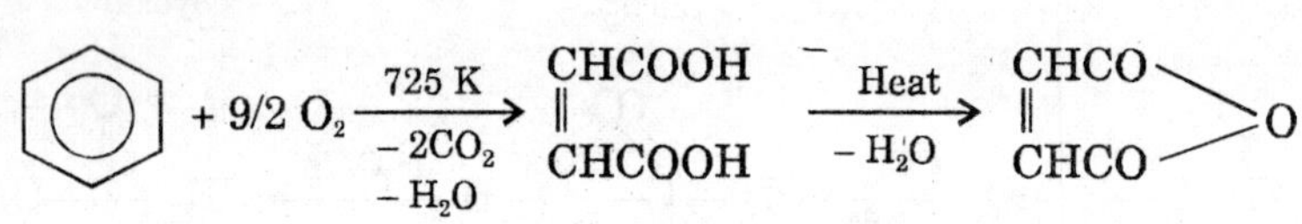

Benzene Maleic acid Maleic anhydride

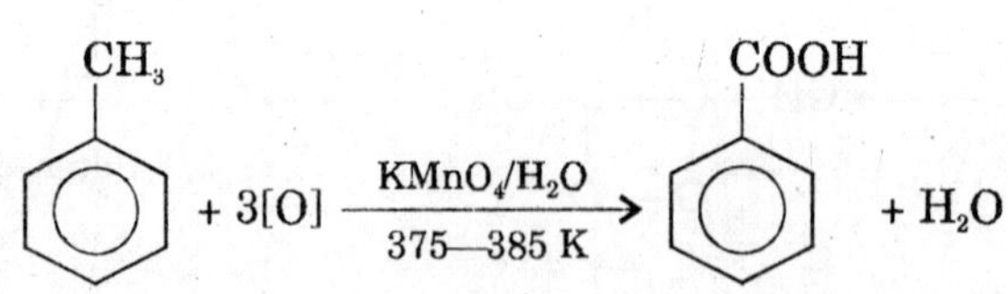

Toluene Benzoic acid

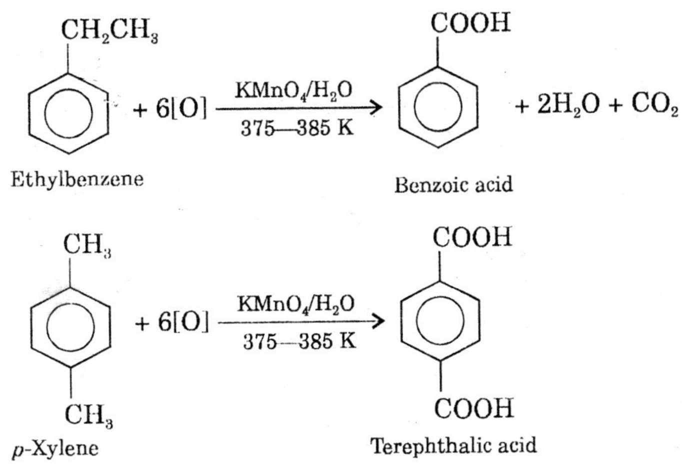

2. Addition Reactions of Benzene

Three moles of H_2, Cl_2 or O_3 add to one mole of benzene under suitable conditions.

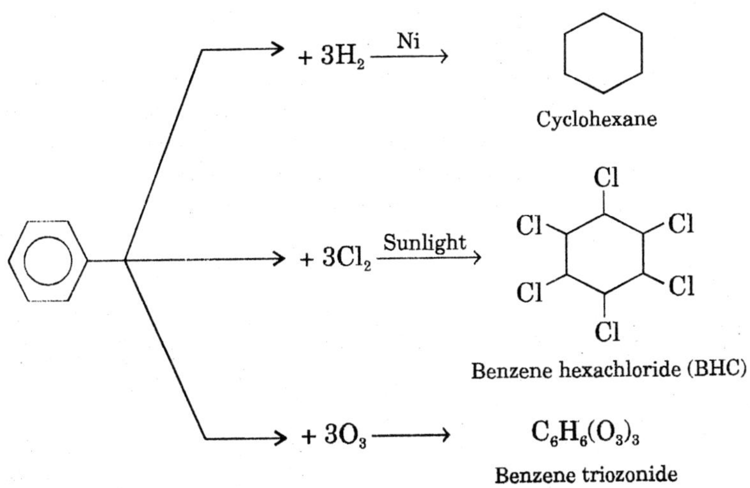

3. Electrophilic Substitution Reactions of Benzene

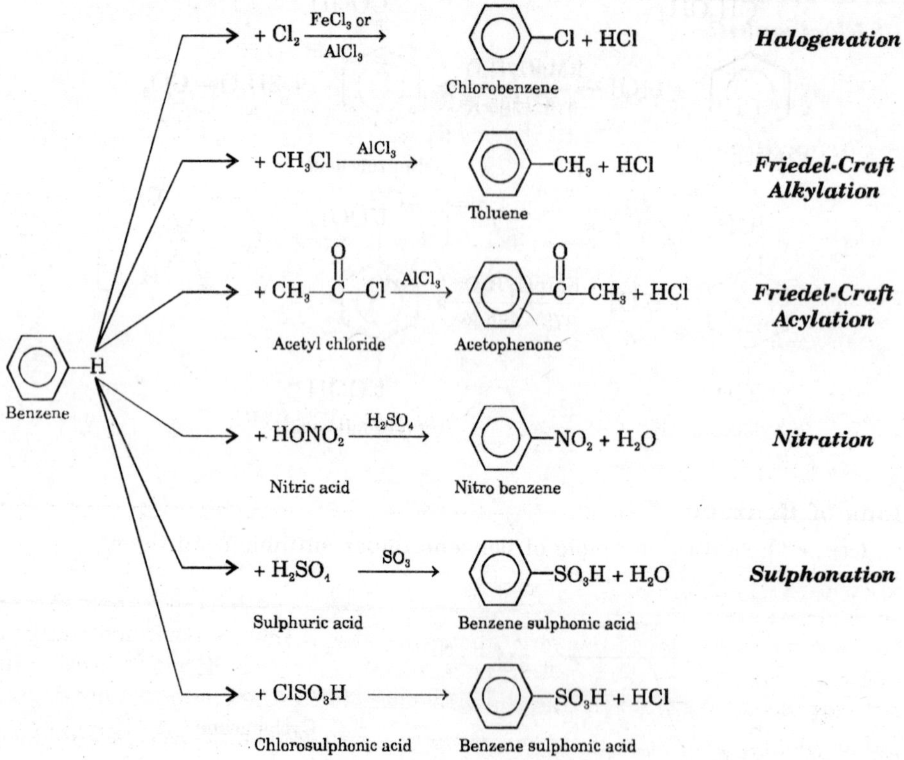

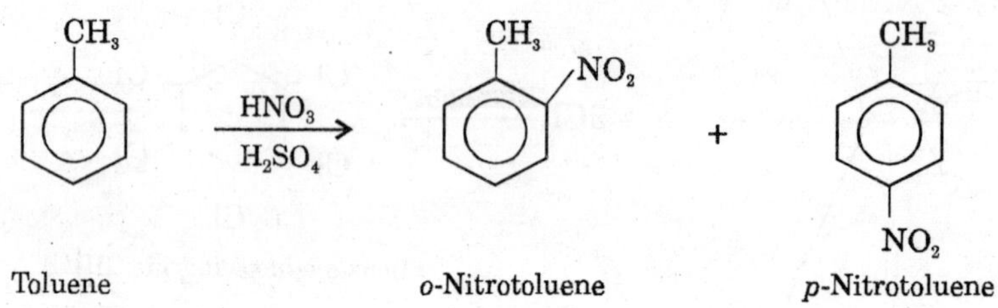

MECHANISM OF HALOGENATION

1. ***Generation of Electrophile***

$$Cl_2 + FeCl_3 \longrightarrow Cl^+ + FeCl_4^-$$

Chloronium ion
(Electrophile)

2. ***Formation of Carbocation***

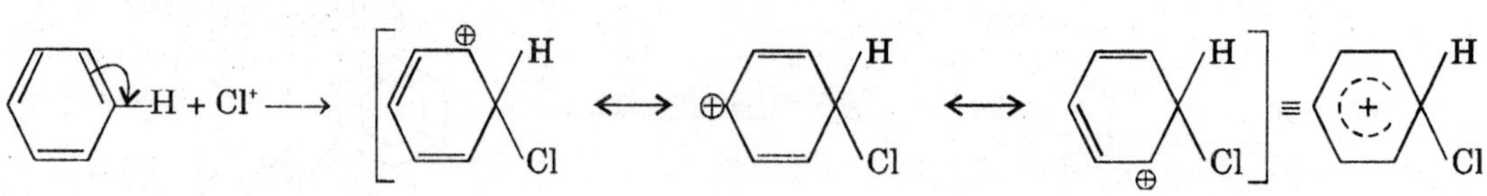

3. ***Abstraction of Proton***

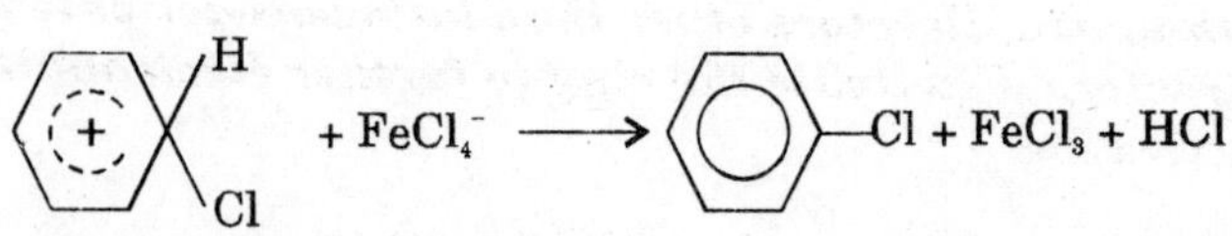

MECHANISM OF FRIEDEL-CRAFT ALKYLATION

1. ***Generation of Electrophile***

$$CH_3Cl + AlCl_3 \longrightarrow \overset{\oplus}{C}H_3 + AlCl_4^-$$

2. ***Formation of Carbocation***

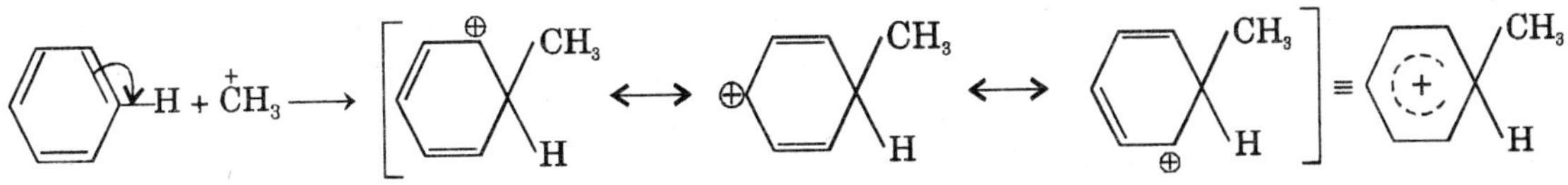

3. ***Abstraction of Proton***

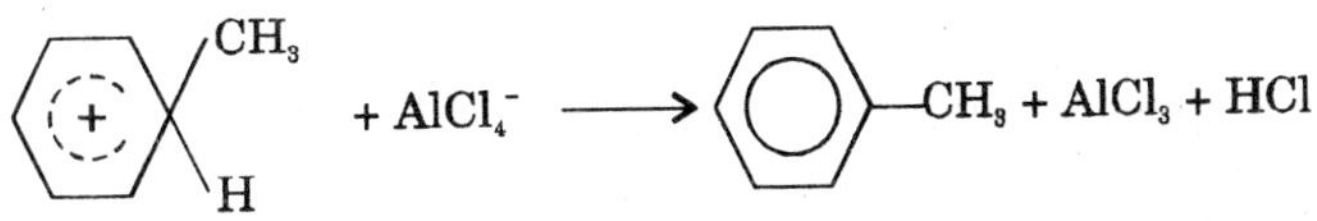

DIRECTIVE INFLUENCE IN ARENES

In benzene all the six hydrogens are equivalent. Therefore, any of the six positions can be occupied during the formation of monosubstituted products. However, when monosubstituted product is to be converted into disubstituted product, the substituent already present in the ring directs the incoming group to a particular position.

1. *The substituents or groups which direct the incoming group to ortho and para positions are called* **ortho and para directing groups.** Some examples are :

$$-CH_3, -C_2H_5, -Cl, -OH, -Br, -NH_2, -OCH_3, \text{etc.}$$

For example, if we carry out nitration of toluene, the mixture of *ortho* and *para* nitro toluene is formed.

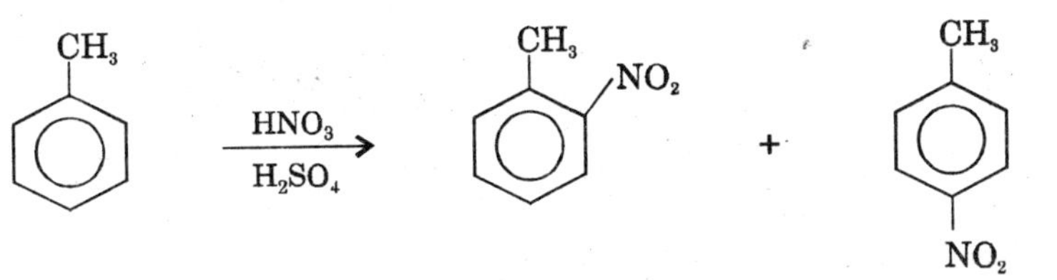

Toluene | *o*-Nitrotoluene | *p*-Nitrotoluene

2. *The substituents or groups which direct the incoming group to meta position* are called **meta directing groups.** Some examples are :

$$-NO_2, -CN, -CHO, -CO-R, -COOH, -COOR, -SO_3H, \text{etc.}$$

For example, nitration of benzoic acid produces *m*-nitro benzene.

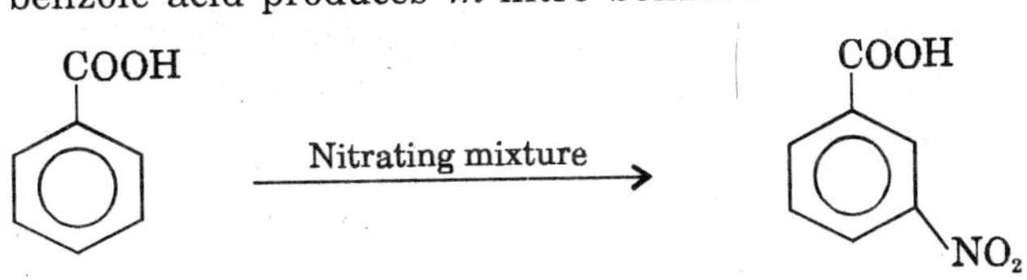

Benzoic acid | *m*-Nitrobenzoic acid

It may be noted that all ortho-para directors other than halogens activate the ring to electrophilic substitution while all meta-directors deactivate the ring to further electrophilic substitution. Halogens are *ortho-para* directing but deactivating.

SOME MISCELLANEOUS IMPORTANT CONCEPTS

- ***During the course of organic reactions, a more stable intermediate is formed in larger proportion.*** For example, reaction of propene ($CH_3CH = CH_2$) and HBr is initiated by the electrophilic attack of H^+ ion. The attack of H^+ ion on propene can give two carbocations primary and secondary, but the secondary carbocation is formed in more proportion. Thus, the final product in major proportion is formed from secondary carbocation.

$$\underset{\substack{1^\circ \text{ carbocation} \\ (\textit{Less stable})}}{CH_3-CH_2-\overset{+}{C}H_2} \xleftarrow{H^+} \underset{\text{Propene}}{CH_3-CH=CH_2} \xrightarrow{H^+} \underset{\substack{2^\circ \text{ carbocation} \\ (\textit{More stable})}}{CH_3-\overset{+}{C}H-CH_3}$$

$$CH_3\overset{+}{C}HCH_3 + Br^- \longrightarrow \underset{\substack{\text{2-Bromopropane} \\ (\textit{Major product})}}{CH_3\overset{\overset{Br}{|}}{C}HCH_3}$$

- ***A less stable intermediate carbocation can rearrange itself to a more stable species to give unexpected product.*** For example, chemical reaction between benzene and $CH_3CH_2CH_2Cl$ in the presence of $AlCl_3$ gives 2-phenylpropane instead of 1-phenylpropane. The reaction occurs as shown below :

$$CH_3CH_2CH_2Cl \xrightarrow{AlCl_3} \underset{1^\circ}{CH_3CH_2\overset{\oplus}{C}H_2} + AlCl_4^-$$

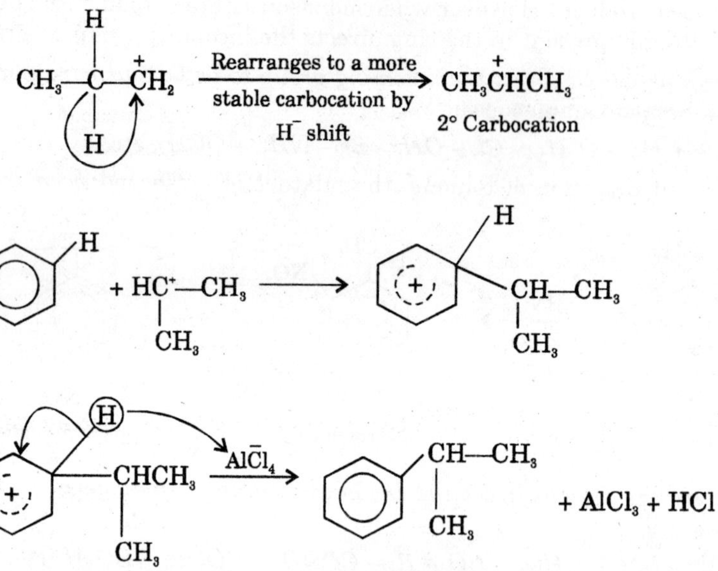

2-Phenylpropane

- ***A compound forming more stable intermediate is relatively more reactive.*** For example, attack of H^+ on ethene produces ethyl carbocation while attack of H^+ ion on ethyne gives vinyl carbocation. Vinyl carbocation is less stable because, here the positive charge is carried by the *sp*-hybridised carbon atom which is more electronegative than the sp^2 hybridised carbon of ethyl carbocation.

$$\underset{\text{Ethene}}{CH_2=CH_2} + H^+ \longrightarrow \underset{\substack{\text{Ethyl carbocation} \\ (\textit{More stable})}}{CH_3-\overset{+}{C}H}$$

$$\underset{\text{Ethyne}}{CH \equiv CH} + H^+ \longrightarrow \underset{\substack{\text{Vinyl carbocation}\\ (\textit{Less stable})}}{CH_2 = \overset{+}{C}H}$$

Since ethene gives more stable intermediate. Hence, ***ethene is more reactive towards electrophilic attack than ethyne.***

In similar way we can compare the reactivity of ethene and propene towards electrophilic addition.

$$\underset{\text{Ethene}}{CH_2 = CH_2} + H^+ \longrightarrow \underset{\text{Ethyl carbocation (1°)}}{CH_3 - \overset{+}{C}H_2}$$

$$\underset{\text{Propene}}{CH_3CH = CH_2} + H^+ \longrightarrow \underset{\text{Isopropyl carbocation (2°)}}{CH_3 - \overset{+}{C}H - CH_3}$$

Since propene gives relatively more stable carbocation (secondary) as compared to that given by ethene, therefore, ***propene is more reactive towards electrophilic attack as compared to ethene.***

AROMATICITY OR AROMATIC CHARACTER

Benzene and many other compounds undergo addition reactions and resist substitution reactions inspite of the high degree of unsaturation in them. This property of benzene and other similar compounds is referred to as aromaticity or aromatic character. The aromaticity was explained by a theory proposed by Eric Huckel. According to this theory cyclic, planar, conjugated systems containing (4n + 2) delocalized π electrons possess aromatic character. Here n is an integer *i.e.,* 1, 2, 3 etc. some examples of **aromatic species** are given below :

Benzene
6π electron
(n = 1)

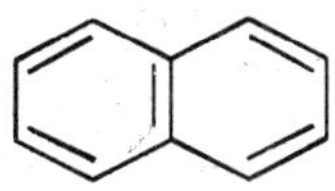

Naphthalene
10π electron
(n = 2)

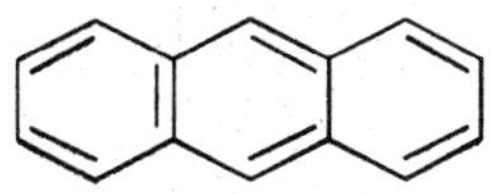

Anthracene
14π electron
(n = 3)

Cyclopentadienyl anion
6π electrons

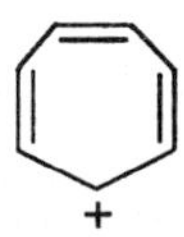

Cycloheptatrienyl cation
(Trophylium ion)
6π electrons

The following examples do not contain ($4n$ + 2) π electrons and hence are **non-aromatic.**

Cyclopentadiene
4π electrons

Cyclobutadiene
4π electrons

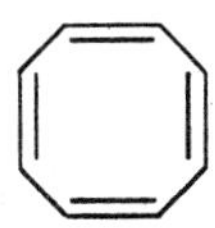

Cyclooctatetraene
8π electron

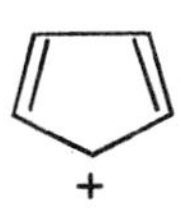

Cyclopentadienyl cation
4π electrons

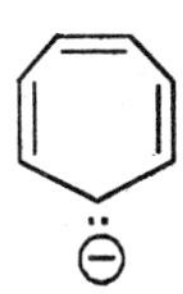

Cycloheptatrienyl anion
8π electrons

2.2. Important Facts and Terms

- **Natural gas** mainly consists of methane (~ 90%) and smaller amounts of ethane ; propane and butane.
- **CNG** is the abbreviation used for **Compressed Natural Gas.** It is usd for running automobiles such as buses, cars, autoricksaws, etc.
- **BTX**. A mixture of benzene, toluene and xylene. It is added to gasoline to improve the octane number of gasoline.
- **TEL.** Tetraethyl lead, $Pb(C_2H_5)_4$. It is an antiknocking agent. Gasoline containing TEL as additive is known as **leaded gasoline** or **ethyl gasoline.**
- **Isooctane.** IUPAC name of iso octane is **2, 2, 4-trimethylpentane.** Its octane number is 100.
- **Triptane.** IUPAC name of triptane is 2, 2, 3-trimethylpentane. Its octane number is greater than 100 (116).
- **Straight run gasoline.** The gasoline obtained by direct distillation. Its octane number is around 74.
- **Sour gasoline.** Sometimes the gasoline obtained by fractional distillation of crude oil has offensive smell due to the presence of sulphur compounds. Such gasoline is known as sour gasoline.
- **Anthracite** form of coal has the highest percentage (~ 90%) of carbon.
- **Mellitic acid or benzene hexacarboxylic acid**, $C_6(COOH)_6$ is formed when graphite is oxidized with fuming nitric acid.
- In Wurtz reaction, if n is the number of carbon atoms in the alkyl halide, the number of carbon atoms in the alkane formed is $2n$.
- Wurtz reaction cannot be used for preparing methane.
- Wurtz reaction is not suitable for preparing alkanes with odd number of carbon atoms.
- Soda-lime decarboxylation yields the alkane with one carbon less than the starting compound.
- **Corey-House reaction** can be employed for obtaining alkanes with odd number of carbon atoms.
- **Kolbe electrolysis** of potassium succinate yields ethene while that of potassium fumarate yields ethyne.
- **Kolbe electrolysis** of an aqueous solution of sodium carboxylate containing n carbon atoms yields alkane having **2(n – 1)** carbon atoms at anode.
- **Zinc-copper couple** is prepared by adding granulated zinc to an aqueous solution of copper sulphate. Zinc displaces copper from copper sulphate. This copper gets deposited on the zinc granules.
- **Raney nickel** is finely divided nickel and is used as a catalyst during hydrogenation. It is prepared by boiling nickel-aluminium alloy with sodium hydroxide solution. Aluminium dissolves leaving behind finely divided nickel.
- **Acetylene has a garlic-like odour** due to the presence of phosphine and hydrogen sulphide as impurities.
- **Allylic bromination** can be carried out by treating alkene with **N-bromosuccinimide** or by heating alkene with **Br_2 at 800 K.**
- **Allylic chlorination** can be carried out by heating alkene with **sulphuryl chloride (SO_2Cl_2)** or by heating alkene with **Cl_2 at 800 K.**
- The **relative acidic character** of ethane, ethene and ethyne is

  $H-C\equiv C-H > H_2C=CH_2 > H_3C-CH_3$
- Hydrocarbons are weaker acids than H_2O.
- **Lindlar's catalyst** is pladium deposited on $BaSO_4$ or $CaCO_3$ and partially poisoned with sulphur or quinoline.
- Hydrogenation of 2-butyne in the presence of Lindlar's catalyst gives *cis*-2-butene.
- Reduction of 2-butyne with sodium and liquid ammonia (Birch reduction) yields *trans*-2-butene.

- **War Gases**

Two important war gases used in World War-I and II are **Mustard gas** and **Lewisite** respectively.

(*a*) **Mustard gas** is obtained by the action of ethylene with sulphur chloride (S_2Cl_2)

$$2\begin{matrix} CH_2 \\ \| \\ CH_2 \end{matrix} + \begin{matrix} Cl \quad Cl \\ | \quad\; | \\ S - S \end{matrix} \longrightarrow S\begin{matrix} \diagup CH_2-CH_2Cl \\ \diagdown CH_2-CH_2Cl \end{matrix} + S$$

Ethylene — Sulphur chloride — Mustard gas (β, β′-*Dichlorodiethyl sulphide*)

(*b*) **Lewisite** is obtained by the action of arsenic trichloride on ethyne in the presence of anhydrous aluminium chloride

$$\begin{matrix} CH \\ ||| \\ CH \end{matrix} + \begin{matrix} Cl \\ | \\ AsCl_2 \end{matrix} \xrightarrow{\text{Anhyd. } AlCl_3} \begin{matrix} CHCl \\ \| \\ CHAsCl_2 \end{matrix}$$

Lewisite

(β-*Chlorovinyl dichlorocosine*)

Lewisite is more poisonous than mustard gas and was used in World War-II. Britain developed antidote to Lewisite in World-War-II. This antidote was named as **British Anti Lewisite (BAL)** and is

$$\begin{matrix} CH_2 & - & CH & - & CH_2 \\ | & & | & & | \\ SH & & SH & & OH \end{matrix}$$

- **More substituted alkenes are more stable.** Correspondingly heat of hydrogenation of unsubstituted alkene (ethene) is maximum. More substituted alkenes have lower value of heat of hydrogenation.
- Benzene on ozonolysis yields glyoxal.

$$C_6H_6 + 3O_3 \longrightarrow C_6H_6(O_3)_3 \xrightarrow{H_2O/Zn} 3 \begin{matrix} CHO \\ | \\ CHO \end{matrix}$$

2.3. Solved Problems

Develop Confidence

Example 1. Which of the following has the lowest boiling point ?

(a)  (b) 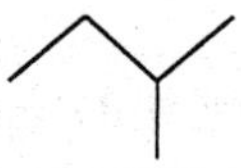

(c) 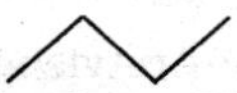 (d) .

Solution. The choice **(c)** is the correct answer because this molecule has least surface area and hence has the lowest boiling point.

Example 2. Pent-1-ene on reaction with HCl gives

(a) 1-Chloropentane (b) 2-Chloropentane

(c) 3-Chloropentane (d) 1, 2-Dichloropentane.

Solution. The addition of HCl to alkenes follows Markovnikov's rule and gives 2-chloropentane as the major product. Hence, **(b)** is the correct answer.

Example 3. Benzene reacts with n-propyl chloride in the presence of anhydrous $AlCl_3$ to give predominantly

(a) *n*-Propylbenzene

(b) *iso*-Propylbenzene

(c) 1-Chloro-3-phenylpropane

(d) 1-chloro-1-phenylpropane.

Solution. The reaction involved is Friedel Craft's alkylation. The intermediate carbocation rearranges through 1, 2-hydride shift to form *iso*-propyl carbocation which on further reaction gives *iso*-propylbenzene as the product. Hence, the answer is **(b)**.

Example 4. Which one of the following will undergo meta-substitution on monochlorination ?

(a) Ethoxybenzene (b) Chlorobenzene

(c) Ethyl benzoate (d) Phenol.

Solution.

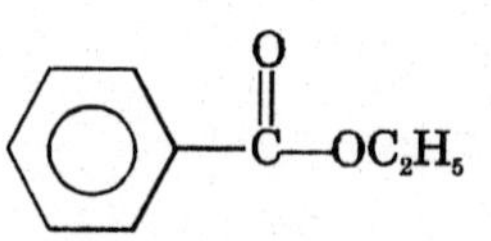

$-\overset{O}{\overset{\|}{C}}-OC_2H_5$ group attached to benzene is deactivating group as well as meta directing due to its electron with drawing resonance effect.

The correct choice is **(c)**.

Example 5. To prepare a pure sample of *n*-hexane using sodium metal as one reactant, the other reactant will be

(a) Ethyl chloride and *n*-butyl chloride

(b) Methyl chloride and *n*-pentyl bromide

(c) Ethyl bromide and *n*-butyl bromide

(d) *n*-Propyl bromide.

Solution. $2CH_3CH_2CH_2Br \xrightarrow[\text{ether}]{\text{Na}} CH_3CH_2CH_2CH_2CH_2CH_3$

n-Propyl bromide

n-Propyl bromide on Wurtz reaction gives *n*-hexane. Hence, the answer is **(c)**.

Example 6. Select the true statement about benzene from amongst the following

(a) Because of unsaturation benzene easily undergoes addition reactions.

(b) There are two types of C—C bonds in benzene molecule.

(c) There is a cyclic delocalisation of π electrons in benzene.

(d) Monosubstitution of benzene group gives three isomeric substances.

Solution. In benzene the six π electrons are uniformly distributed over all the six C—C bonds. Hence, the choice **(c)** is correct.

Example 7. Aqueous sulphuric acids reacts with 2-methyl-1-butene to give predominantly

(a) *iso*-Butyl hydrogen sulphate

(b) 2-Methyl-2-butanol

(c) 2-Methyl-1-butanol

(d) *sec*-Butyl hydrogen sulphate.

Solution. $CH_3-CH_2-\underset{\underset{CH_3}{|}}{C}=CH_2 + H_2O \xrightarrow{H^+}$

$$CH_3-CH_2-\underset{\underset{CH_3}{|}}{\overset{\overset{HO}{|}}{C}}-CH_3$$

2-Methyl-2-butanol

The addition of water molecule takes place in accordance with Markownikoff's rule. Thus, the correct choice is **(b)**.

Example 8. The olefin which on ozonolysis gives CH_3CH_2CHO and CH_3CHO is

(*a*) 1-Butene (*b*) 2-Butene
(*c*) 1-Pentene (*d*) 2-Pentene.

Solution. 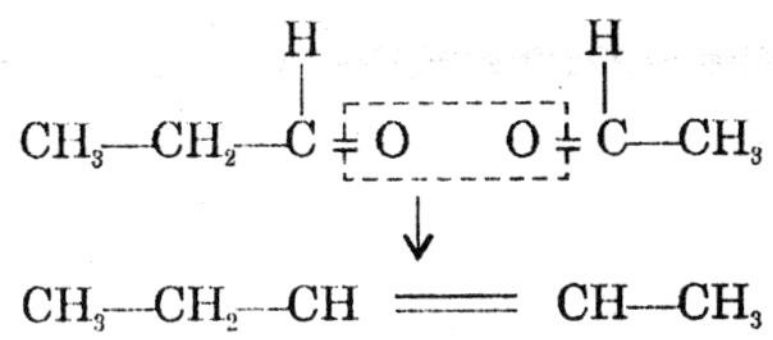

2-Pentene

Thus, the correct choice is (***d***).

Example 9. When HCl is passed through propene in the presence of benzoyl peroxide, it gives

(*a*) *n*-Propyl chloride (*b*) *iso*-Propyl chloride
(*c*) Allyl chloride (*d*) No reaction.

Solution. The addition of HCl to alkenes, even in the presence of peroxides obeys Markownikoff's rule. Hence, the correct answer is (***b***).

$$CH_3CH = CH_2 + HCl \longrightarrow CH_3-\underset{\underset{Cl}{|}}{CH}-CH_3$$

Example 10. Indicate the expected structure of the organic product when ethylmagnesium bromide is treated with heavy water (D_2O).

(*a*) $C_2H_5-C_2H_5$ (*b*) C_2H_5OD
(*c*) C_2H_6 (*d*) C_2H_5D.

Solution.

$$C_2H_5MgBr + D-O-D \longrightarrow \mathbf{C_2H_5D} + MgBr(OD)$$

Thus, the correct answer is (***d***).

Example 11. When CH_3Cl and $AlCl_3$ are used in Friedel-Craft's reaction, the electrophile is

(*a*) Cl^+ (*b*) $AlCl_4^-$
(*c*) CH_3^+ (*d*) $AlCl_2^+$.

Solution. $Cl-Cl + AlCl_3 \longrightarrow Cl^+ + AlCl_4^-$

The electrophile is Cl^+ (chloronium ion). Hence the answer is (***a***).

Example 12. The hydrocarbon which decolourises alkaline $KMnO_4$ solution, but does not give any precipitate with ammonical silver nitrate is

(*a*) Benzene (*b*) Acetylene
(*c*) Propyne (*d*) But-2-yne.

Solution. A non-terminal alkyne decolourises alkaline $KMnO_4$ solution but does not give any precipitate with ammonical $AgNO_3$. Hence, the correct answer is (***d***).

Example 13. The compound which reacts with HBr obeying Markownikoff's rule is

(*a*) $CH_2 = CH_2$ (*b*) 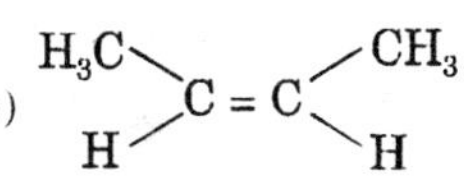

(*c*) 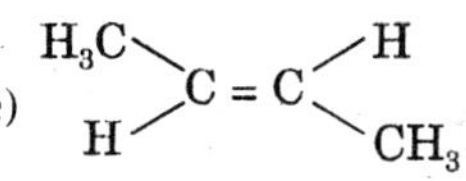 (*d*) 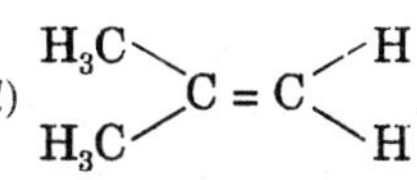

Solution. Addition of HBr to unsymmetrical alkenes obeys Markownikoff's rule. The alkene (*d*) is unsymmetrical. Hence, the answer is (***d***).

2.4. QUESTION BANK

Develop Confidence

Coal and Petroleum

1. During distillation of coal tar, anthracene is mainly present in
(*a*) Light oil (*b*) Heavy oil
(*c*) Middle oil (*d*) Green oil.

2. Gasoline contains hydrocarbons in the range
(*a*) C-7 to C-12 (*b*) C-12 to C-15
(*c*) C-4 to C-6 (*d*) C-5 to C-8.

3. Cetane number of α-methylnaphthalene is
(*a*) 100 (*b*) 0
(*c*) 50 (*d*) 75.

4. Petroleum is mainly a source of
(*a*) Cycloalkanes (*b*) Inorganic compounds
(*c*) Aliphatic compounds (*d*) Aromatic compounds.

5. Coal is mainly a source of
(*a*) Organic compounds (*b*) Inorganic compounds
(*c*) Aliphatic compounds (*d*) Aromatic compounds.

6. Which of the following is not present in crude naphtha ?
(*a*) Petroleum ether (*b*) Paraffin wax
(*c*) Gasoline (*d*) Hexane.

7. Octane number of *iso*-octane is
(*a*) Zero (*b*) 100
(*c*) 80 (*d*) 20.

8. Cetane is
(*a*) *n*-Hexane (*b*) *n*-Octane
(*c*) *n*-Pentadecane (*d*) *n*-Hexadecane.

9. The process of heating coal in the absence of air is called
(*a*) Decarbonisation (*b*) Decarboxylation
(*c*) Cracking (*d*) Carbonisation.

10. Diesel oil having cetane number equal to 80 has the same ignition quality as the mixture containing
(*a*) 80% *iso*-octane and 20% *n*-heptane
(*b*) 80% *n*-Hexadecane and 20% α-methyl napththalene
(*c*) 20% α-Methyl naphthalene 80% *n*-heptane
(*d*) 80% *iso*-octane 20% α-methyl naphthalene.

11. To increase the yield of gasoline from a given sample of crude oil, the chemist carries out
(*a*) Fractional distillation (*b*) Hydrolysis
(*c*) Aromatization (*d*) Cracking.

12. During distillation of coal tar, naphthalene is a constituent of
(*a*) Middle oil (*b*) Green oil
(*c*) Light oil (*d*) Heavy oil.

13. The percentage of carbon is maximum in
(*a*) Anthracite (*b*) Bitumenous
(*c*) Lignite (*d*) Peat.

14. Cetane number of a diesel fuel will increase with the addition of
(*a*) *n*-Decane (*b*) *n*-Hexadecane
(*c*) *n*-Pentane (*d*) *a*-Methyl naphthalene.

15. Coke is obtained by heating coal in the
(*a*) Presence of air
(*b*) Absence of air
(*c*) Restricted supply of air
(*d*) Presence of a catalyst.

16. Which of the following compounds has the highest octane number ?
(*a*) 2, 2, 4-Trimethylpentane
(*b*) 2, 2, 3-Trimethylpentane
(*c*) 2, 4-Dimethylhexane
(*d*) *n*-Octane.

17. Which of the following represents α-methyl naphthalene ?

(*a*) 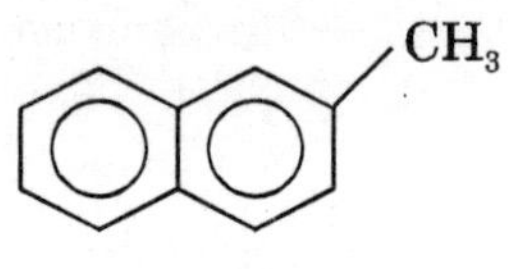

(*b*) 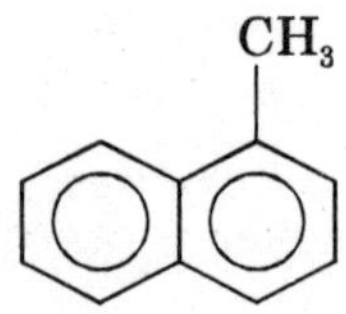

(*c*) 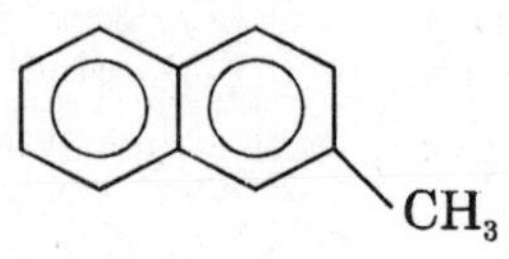

(*d*) 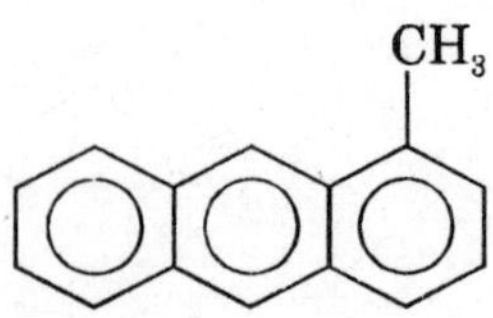

18. Aromatization of n-heptane yields
(*a*) Toluene (*b*) Xylene
(*c*) Benzene (*d*) Methylcyclohexane.

19. Carbonisation of coal is carried out by heating coal at a temperature of about
(*a*) 1000°C (*b*) 1500°C
(*c*) 2000°C (*d*) 500°C.

20. During distillation of coal tar, benzene and toluene are the main components of
(*a*) Middle oil fraction (*b*) Heavy oil fraction
(*c*) Light oil fraction (*d*) Green oil fraction.

Structure, Nomenclature and Isomerism in Hydrocarbons

21. The number of chain isomers possible for an alkane having molecular formula C_5H_{12} is
(*a*) 5 (*b*) 4
(*c*) 3 (*d*) 2.

22. Which of the following compounds can show geometric isomerism ?
(*a*) Vinyl chloride (*b*) 1, 1-Dichloroethene
(*c*) 1, 2-Dichloroethene (*d*) Trichlorethene.

23. Which of the following alkanes can form only two monochloro derivatives ?
(*a*) Hexane (*b*) 2-Methylpentane
(*c*) 2, 3-Dimethylbutane (*d*) 3-Methylpentane.

24. Which of the following alkenes has the maximum value of heat of hydrogenation ?
(*a*) Ethene (*b*) Propene
(*c*) 1-Butene (*d*) *trans*-2-Butene.

25. Which of the following hydrocarbons has the highest value of heat of combustion ?
(*a*) C_4H_{10} (*b*) C_6H_{14}
(*c*) C_5H_{12} (*d*) $C_{10}H_{22}$.

26. An alkyne has the general formula
(*a*) C_nH_{2n} (*b*) C_nH_{2n+2}
(*c*) C_nH_{2n-2} (*d*) C_nH_{2n+1}.

27. Which of the following alkenes can exhibit geometric isomerism ?
(*a*) Pent-1-ene (*b*) Pent-2-ene
(*c*) 2-Methylbut-2-ene (*d*) 3-Methylbut-1-ene.

28. Which of the following is the most stable ?

(*a*)  (*b*) 

(*c*) 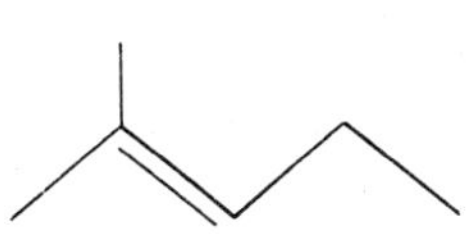 (*d*) 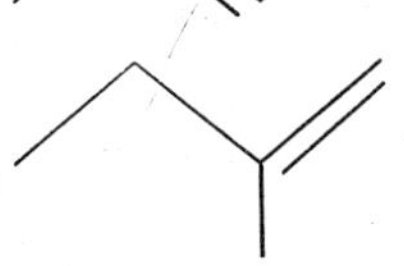

29. Successive alkanes differ by
(*a*) CH_2 (*b*) CH
(*c*) CH_3 (*d*) C_2H_4.

30. Consider the following statements :
A hydrocarbon of molecular formula C_5H_{10} may be a
I. monosubstituted alkene
II. disubstituted alkene
III. trisubstituted alkene
Which of the following statements are correct ?
(*a*) I, II and III (*b*) I and II
(*c*) II and III (*d*) I and III.

Preparation of Aliphatic Hydrocarbons

31. Which of the following reactions can be employed for getting unsymmetrical alkanes in good yield ?
(*a*) Wurtz reaction (*b*) Corey-House reaction
(*c*) Both (*d*) None.

32. *sec*-Butyl chloride on boiling with alc. KOH gives ...... as the main product.
(*a*) 1-Butene (*b*) 2-Butene
(*c*) 1-Butanol (*d*) 2-Butanol.

33. An alkene is formed from a carbocation by
(*a*) Elimination of a H^+ ion
(*b*) Elimination of H^- ion
(*c*) Addition of a H^+ ion
(*d*) Addition of a H^- ion.

34. Alcoholic solution of caustic potash is a specific reagent for
(*a*) Dehydration (*b*) Dehydrohalogenation
(*c*) Dehydrogenation (*d*) Hydration.

35. When two alkenes may be formed by dehydrohalogenation of an alkyl halide, the alkene which is more substituted is the major or preferred product. This generalisation is known as
(*a*) Markownikoff's rule
(*b*) Anti-Markownikoff's rule
(*c*) Saytzeff rule
(*d*) None of these is correct.

36. Which of the following compounds will form a hydrocarbon on reaction with a Grignard reagent ?
(*a*) CH_3CH_2OH (*b*) CH_3CHO
(*c*) CH_3COCH_3 (*d*) $CH_3CO_2CH_3$.

37. Methane is formed when
(*a*) Sodium acetate is heated with soda-lime
(*b*) Iodomethane is reduced

(c) Aluminium carbide reacts with water

(d) All of these.

38. When an alkyl iodide (RI) is treated with Na in dry ether, a symmetrical alkane (R_2) is obtained. The reaction is called

(a) Grignard reaction

(b) Fishcer-Tropsch reaction

(c) Berguis reaction

(d) Wurtz reaction.

39. If n is the number of carbon atoms in the potassium salt of a carboxylic acid, then the alkane formed on electrolysis of aqueous solution of this salt would have carbon atoms equal to

(a) n (b) $n - 1$

(c) $2n - 1$ (d) $2(n - 1)$.

40. Sodium propionate on decarboxylation with soda-lime gives

(a) Propane (b) Ethane

(c) Butane (d) Pentane.

41. Ethyne can be prepared in a single step from

(a) Calcium carbide (b) Ethylidene bromide

(c) Ethylene bromide (d) All of these.

42. 2, 3-Dibromobutane, when heated with zinc dust, yields

(a) 2-Butene (b) 2-Butyne

(c) 1-Butene (d) Butane.

43. Which of the following reactions can be used to prepare an alkane from an alkyl halide ?

(a) Wurtz reaction

(b) Kolbe electrolysis

(c) Sabatier and Senderen reaction

(d) Fittig reaction.

44. For preparing an alkane, a concentrated solution of sodium or potassium salt of a saturated carboxylic acid is subjected to

(a) Hydrolysis (b) Oxidation

(c) Hydrogenation (d) Electrolysis.

45. Which one of the following cannot be prepared by Wurtz reaction ?

(a) CH_4 (b) C_2H_6

(c) C_3H_8 (d) C_4H_{10}.

46. When ethyl alcohol is heated with conc. H_2SO_4 at 443 K, ethylene is formed by

(a) Intramolecular hydration

(b) Intermolecular hydration

(c) Intermolecular dehydration

(d) Intramolecular dehydration.

47. An aqueous solution of potassium salt of fumaric acid is electrolysed. The hydrocarbon produced at anode is

(a) Ethane (b) Ethene

(c) Methane (d) Ethyne.

48. 2-Pentyne can be converted into *trans*-pent-2-ene by reaction with

(a) H_2/Ni (b) H_2/Lindlar's catalyst

(c) Na/Liq : NH_3 (d) Zn/HCl.

49. Which of the following alkanes cannot be produced by Kolbe electrolysis of sodium or potassium salts of carboxylic acids ?

(a) Methane (b) Ethane

(c) Butane (d) Hexane.

50. How many different isomeric compounds having molecular formula $C_5H_{11}Br$ on reaction with Mg, followed by hydrolysis can yield pentane ?

(a) 4 (b) 2

(c) 3 (d) 5.

Physical Properties of Aliphatic Hydrocarbons

51. Which of the following compounds has the highest boiling point ?

(a)  (b) 

(c) 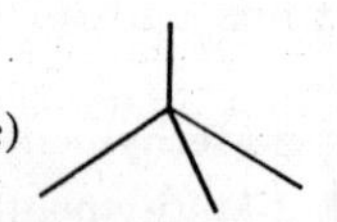 (d) 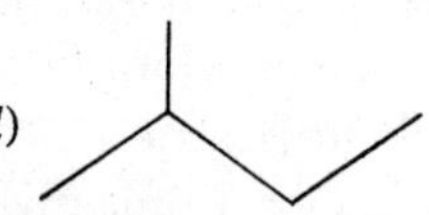

52. Which of the following compounds is liquid under ordinary conditions ?

(a) C_4H_{10} (b) C_3H_8

(c) $C_{15}H_{32}$ (d) $C_{20}H_{42}$.

53. Which of the following compounds has the highest melting point ?

(a) n-Butane (b) n-Pentane

(c) n-Hexane (d) n-Heptane.

54. The highest boiling point is expected for

(a) iso-Octane

(b) n-Octane

(c) 2, 2, 3, 3-Tetramethylbutane

(d) n-Butane.

55. Which of the following compounds has the highest boiling point ?

(a) 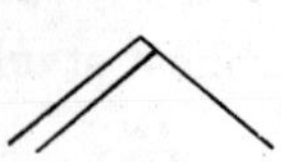 (b) 

(c) 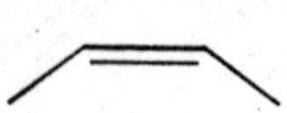 (d) 

Chemical Properties of Alkanes

56. The homolytic fission of hydrocarbon results in the formation of

(*a*) Free radicals (*b*) Carbocations
(*c*) Carbanions (*d*) Carbenes.

57. Abstraction of a *sec*-H atom from propane gives ...... radical.

(*a*) *n*-Propyl (*b*) *iso*-Propyl
(*c*) Ethyl (*d*) None of the above.

58. *n*-Heptane when heated to a temperature of about 800 K under high pressure in the presence of Cr_2O_3/Al_2O_3 catalyst gives

(*a*) 1-Heptene (*b*) 2-Methylhexane
(*c*) Toluene (*d*) Xylene.

59. In the free radical chlorination of methane, the chain initiating step involves the formation of

(*a*) Chlorine atom (*b*) Hydrogen chloride
(*c*) Methyl radical (*d*) Chloromethyl radical.

60. The reaction conditions leading to the best yield of C_2H_5Cl are

(*a*) C_2H_6 (excess) + $Cl_2 \xrightarrow{\text{UV light}}$

(*b*) $C_2H_6 + Cl_2 \xrightarrow[\text{room temperature}]{\text{Dark}}$

(*c*) $C_2H_6 + Cl_2$ (excess) $\xrightarrow{\text{UV light}}$

(*d*) $C_2H_6 + Cl_2 \xrightarrow{\text{UV light}}$

61. Number of isomers which can be theoretically obtained on monochlorination of 2-methylbutane is

(*a*) 1 (*b*) 2
(*c*) 3 (*d*) 4.

62. Chlorination of an alkane involves the attack of

(*a*) An electrophile (*b*) A nucleophile
(*c*) A base (*d*) A free radical.

63. Complete oxidation of ethane yields

(*a*) Ethanol (*b*) Ethanoic acid
(*c*) Ethanal (*d*) CO_2 and H_2O.

64. In *iso*-pentane, the H atom that can be most easily substituted is on

```
  1    2     3     4
CH3—CH—CH2—CH3
       |
      CH3
```

(*a*) C—1 (*b*) C—2
(*c*) C—3 (*d*) C—4.

Chemical Properties of Alkenes

65. A compound on ozonolysis gives glyoxal, acetone and formaldehyde as the products. The compound could be

(*a*) 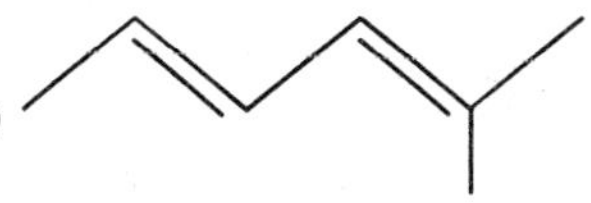

(*b*) 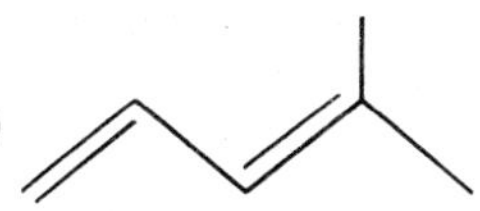

(*c*) 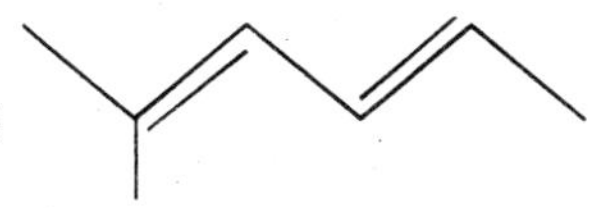

(*d*) 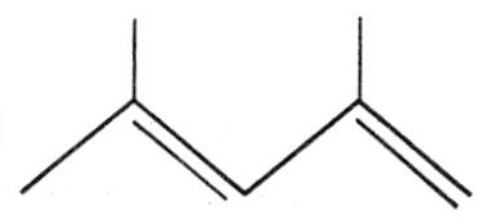

66. 1-Butene and 2-butene can be distinguished by their reaction with

(*a*) Ammonical solution of $AgNO_3$
(*b*) Ammonical solution of Cu_2Cl_2
(*c*) Alkaline solution of $KMnO_4$
(*d*) None of these.

67. On ozonolysis, an alkene gave acetone and acetaldehyde along with H_2O_2. The alkene is

(*a*) 2, 3-Dimethyl-2-butene (*b*) 2-Methyl-2-butene
(*c*) 3-Methyl-2-butene (*d*) 2-Butene.

68. A hydrocarbon of formula C_6H_{12} on ozonolysis gives only one carbonyl product which does not reduce Fehling's solution. The hydrocarbon is

(*a*) 2-Hexene (*b*) 2, 3-Dimethyl-2-butene
(*c*) 2-Methyl-2-pentene (*d*) 2-Hexene.

69. Propene + HBr gives

(*a*) 1, 3-Dibromopropane (*b*) 2-Bromo propane
(*c*) 1-Bromopropane (*d*) 1, 2-Dibromopropane.

70. Addition of Br_2 to 2-butene gives

(*a*) 1, 2-Dibromobutane (*b*) 2, 3-Dibromobutane
(*c*) 1, 3-Dibromobutane (*d*) 1, 4-Dibromobutane.

71. Allylic bromination of an olefin is

(*a*) Nucleophilic substitution
(*b*) Electrophilic substitution
(*c*) Free radical substitution
(*d*) Electrophilic addition.

72. The addition of HBr to styrene gives ...... as the major product

(*a*) 1-Bromo-1-phenylethane
(*b*) 2-Bromo-1-phenylethane

(c) Benzyl chloride
(d) p-Bromophenyl ethylene.

73. When HBr adds to 1-butene in the presence of benzoyl peroxide, the products is
(a) 1-Bromobutane (b) 2-Bromobutane
(c) 1-Bromobutene (d) 2-Bromobutene.

74. A hydrocarbon reacts with hypochlorous acid to give 1-chloro-2-hydroxyethane. The hydrocarbon is
(a) Ethylene (b) Methane
(c) Ethane (d) Acetylene.

75. Reaction of alkenes with halogens is explosive in case of
(a) F_2 (b) Cl_2
(c) Br_2 (d) I_2.

76. Which of the following is the predominant product in the reaction of HOBr with propene ?
(a) 2-Bromo-1-propanol (b) 3-Bromo-1-propanol
(c) 2-Bromo-2-propanol (d) 1-Bromo-2-propanol.

77. To which of the following compounds Br_2 adds most readily.
(a) $CH_2 = CH_2$ (b) $CH_3 - CH = CH_2$
(c) $CH_3 - CH = CH—CH_3$ (d) $CH_3 - \underset{\underset{CH_3}{|}}{C} = CH - CH_3$.

78. An alkene having molecular formula C_7H_{14} was subjected to ozonolysis in the presence of zinc dust. An equimolar amount of the following two compounds was obtained.

$$\begin{matrix} CH_3 \\ CH_3 \end{matrix} \rangle C = O \quad \text{and} \quad \begin{matrix} CH_3 \\ CH_3CH_2 \end{matrix} \rangle C = O$$

The IUPAC name of the alkene is
(a) 3, 4-dimethyl-3-pentene
(b) 3, 4-dimethyl-2-pentene
(c) 2, 3-dimethyl-3-pentene
(d) 2, 3-dimethyl-2-pentene.

79. Which one of the following compounds can decolourise alkaline $KMnO_4$ solution ?
(a) C_3H_8 (b) C_2H_4
(c) CH_4 (d) $CHCl_3$.

80. Ethene is shaken with aqueous solution of Br_2 and NaCl. Which of the following is not the possible product ?

(a) 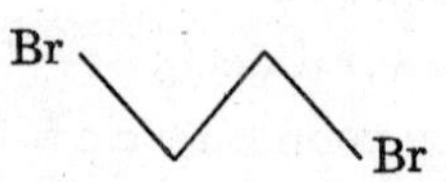

(b) 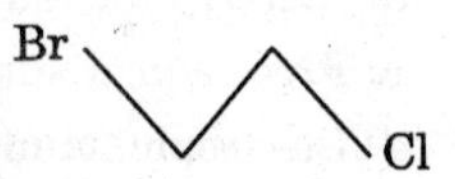

(c) 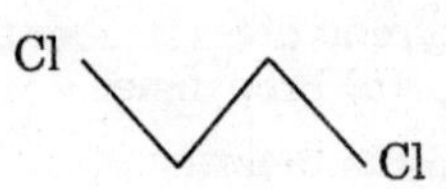

(d) 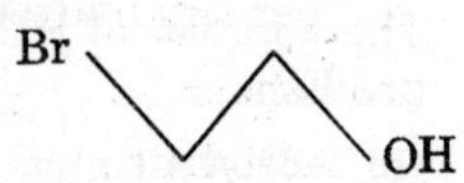

81. Sulphur monochloride reacts with ...... to give mustard gas.
(a) Ethane (b) Ethene
(c) Ethyne (d) Methane.

82. Addition of HBr to *iso*-butylene mainly gives
(a) *iso*-Butyl bromide (b) *n*-Butyl bromide
(c) *tert*-Butyl bromide (d) *sec*-Butyl bromide.

83. Propene is reacted with HBr in presence of peroxides, the product is
(a) 2-Bromopropane (b) 1-Bromopropane
(c) 3-Bromopropane (d) None of the above.

84. Baeyer's reagent is used in the laboratory for
(a) Detection of unsaturation
(b) Reduction process
(c) Oxidation process
(d) Detection of glucose.

85. Ethylene reacts with 1% cold alkaline $KMnO_4$ to form
(a) Oxalic acid (b) Ethylene glycol
(c) Ethyl alcohol (d) HCHO.

86. Markownikoff rule is applicable to
(a) $CH_2 = CH_2$
(b) $CH_3 - CH_2 - CH_3$
(c) $CH_3 - CH = CH - CH_3$
(d) $\begin{matrix} CH_3 \\ CH_3 \end{matrix} \rangle C = C \langle \begin{matrix} H \\ H \end{matrix}$

87. Anti Markownikoff addition of HBr is not observed in
(a) Propene (b) 1-Butene
(c) But-2-ene (d) Pent-2-ene.

88. In methyl alcohol solution bromine reacts with ethylene to yield $BrCH_2CH_2OCH_3$ in addition to 1, 2-dibromoethane because
(a) The ion formed initially may react with Br^- or CH_3OH
(b) The methyl alcohol solvates the bromine
(c) The reaction follows Markownikoff's rule
(d) This is a free-radical mechanism.

89. 2.8 g of pure alkene containing only one double bond per molecule, react completely with 8 g of bromine (in an inert solvent). What is the molecular formula of the alkene ?
(a) C_2H_4 (b) C_4H_8
(c) C_3H_4 (d) C_6H_{12}.

Chemical Properties of Alkynes

90. Which of the following hydrocarbons would give a red precipitate on treatment with ammonical cuprous chloride solution ?
(a) 2-Butene (b) 2-Butyne

91. Reaction of ethyne with HCN in the presence of $Ba(CN)_2$ is

(*a*) Electrophilic addition reaction
(*b*) Nucleophilic addition reaction
(*c*) Free radical addition reaction
(*d*) Electrophilic substitution reaction.

92. $R-CH_2-CCl_2-R \xrightarrow{\text{Reagent}} R-C \equiv C-R$

The reagent is

(*a*) Na (*b*) HCl in H_2O
(*c*) KOH in C_2H_5OH (*d*) Zn in alcohol.

93. Ethyne can be oxidised to oxalic acid by using

(*a*) Chromic acid (*b*) Alkaline $KMnO_4$
(*c*) Hypochlorous acid (*d*) All reagents.

94. Acetylene when treated with dilute HCl at 60°C (333 K) in presence of $HgCl_2$ produces

(*a*) Methyl chloride (*b*) Vinyl chloride
(*c*) Acetaldehyde (*d*) Formaldehyde.

95. Acetylenic hydrogens are acidic because

(*a*) Sigma electron density of C—H bond in acetylene is nearer to carbon, which has 50% *s*-character
(*b*) Acetylene has only one hydrogen on each carbon
(*c*) Acetylene contains least number of hydrogens among the possible hydrocarbons having two carbons
(*d*) Acetylene belongs to the class of alkynes with molecular formula, C_nH_{2n-2}.

96. Lindlar's catalyst is

(*a*) Na in alcohol (*b*) Raney nickel
(*c*) $Pd/CaCO_3$ (*d*) Na/liq. NH_3.

97. When propyne is treated with aqueous H_2SO_4 in the presence of $HgSO_4$, the major product is

(*a*) Acetaldehyde (*b*) Propanal
(*c*) 2-Propanol (*d*) Propanone.

98. Alkaline $KMnO_4$, oxidizes acetylene to

(*a*) Acetic acid (*b*) Glyoxal
(*c*) Oxalic acid (*d*) Ethylene glycol.

99. Which of the following is most acidic ?

(*a*) Methane (*b*) Ethane
(*c*) Ethylene (*d*) Acetylene.

100. Which of the following is most acidic ?

(*a*) Ethyne (*b*) Propyne
(*c*) 1-Butyne (*d*) 2-Butyne.

101. Which of the following is least acidic ?

(*a*) Ethyne (*b*) Propyne
(*c*) 1-Butyne (*d*) 2-Butyne.

102. Which of the following can form metallic derivatives by replacement of H-atoms ?

(*a*) Ethene (*b*) Methane
(*c*) 2-Butyne (*d*) 1-Pentyne.

103. A hydrocarbon of formula C_6H_6 decolourizes bromine water. It also gives precipitate with ammonical $AgNO_3$ solution. The hydrocarbon can possibly be

(*a*) 1, 3, 5-Cyclohexatriene (*b*) 1, 5-Hexadiyne
(*c*) 2, 4-Hexadiyne (*d*) None of the above.

104. When HCl adds to vinyl chloride the product is

(*a*) 1, 2-Dichloroethane (*b*) 1, 2-Dichloroethene
(*c*) 1, 1-Dichloroethene (*d*) 1, 1-Dichloroethane.

105. Ozonolysis of acetylene gives

(*a*) Oxalic acid (*b*) Ethylene glycol
(*c*) Glyoxal (*d*) CH_3CHO.

106. 3-Hexyne reacts with Na/liq. NH_3 to produce

(*a*) *cis*-3-Hexene (*b*) *trans*-3-Hexene
(*c*) 3-Hexylamine (*d*) 2-Hexylamine.

107. The presence of unsaturation in organic compounds can be treated with

(*a*) Schiff's reagent (*b*) Tollen's reagent
(*c*) Fehling's solution (*d*) Baeyer's reagent.

108. Which of the following shows acidic character ?

(*a*) $CH_3-C\equiv CH$ (*b*) $CH_3-C\equiv C-CH_3$
(*c*) CH_3-CH_3 (*d*) $CH_2=CH_2$.

109. Addition of HCl to vinyl chloride gives 1, 1-dichloroethane because of

(*a*) Inductive effect of Cl
(*b*) Mesomeric effect of Cl
(*c*) Restricted rotation around double bond
(*d*) None of these.

110. Acetylene can be converted into methyl vinyl ether on reaction with

(*a*) Vinyl alcohol (*b*) Methyl alcohol
(*c*) Dimethyl ether (*d*) Methanal.

111. Propyne on reaction with aqueous chlorine gives

(*a*) 1, 1, 2, 2-Tetrachloropropane
(*b*) 1, 2-Dichloropropene
(*c*) 1, 1-Dichloropropanone
(*d*) 2, 2-Dichloropropanal.

112. In the reaction,

$$CH_3-C\equiv C-CH_3 \xrightarrow[(ii)\ H_2O/Zn]{(i)\ X} CH_3-\underset{\underset{O}{\|}}{C}-\underset{\underset{O}{\|}}{C}-CH_3$$

X is

(*a*) HNO_3 (*b*) O_2
(*c*) O_3 (*d*) $KMnO_4$.

113. Which of the following compounds reacts with sodium to liberates hydrogen gas ?

(*a*) Ethane (*b*) Propylene
(*c*) Acetylene (*d*) Benzene.

114. Which is the most suitable reagent among the following to distinguish compound (3) from rest of the compounds ?

1. $CH_3-C\equiv C-CH_3$ 2. $CH_3-CH_2-CH_2-CH_3$
3. $CH_3-CH_2-C\equiv CH$ 4. $CH_3-CH=CH_2$.

(*a*) Bromine in carbon tetrachloride
(*b*) Bromine in acetic acid
(*c*) Alk. $KMnO_4$
(*d*) Ammonical silver nitrate.

115. Mesitylene can be obtained by polymerization of
(*a*) Ethyne (*b*) Ethene
(*c*) Propene (*d*) Propyne.

116. The most suitable catalyst for the hydrogenation of 2-hexyne ⟶ *cis*-2-hexene is
(*a*) Pd-$BaSO_4$ (*b*) $(Ph_3P)_3RhCl$
(*c*) 10% Pd—C (*d*) Roney Ni.

117. Which one of the following compounds will react with methylmagnesium iodide ?
(*a*) $CH_3CH_2CH_2CH_2CH_3$
(*b*) $CH_3CH=CH-CH=CH_2$
(*c*) $CH_3-C\equiv C-CH_2CH_3$
(*d*) $CH_3CH_2CH_2C\equiv CH$.

118. Which of the following can distinguish propyne from propene ?
(*a*) Ammonical $AgNO_3$
(*b*) Dil. H_2SO_4
(*c*) Br_2-H_2O
(*d*) Aqueous $KMnO_4$.

119. Excess of CH_3COOH is reacted with $CH\equiv CH$ in presence of Hg^{2+}, the product is
(*a*) $CH_3CH(OOCCH_3)_2$
(*b*) $CH_2=CH(OCOCH_3)$
(*c*) $(CH_3COO)CH_2-CH_2(OOCCH_3)$
(*d*) None of these.

120. A compound is treated with $NaNH_2$ to give sodium salt. Identify the compound
(*a*) C_2H_2 (*b*) C_6H_6
(*c*) C_2H_6 (*d*) C_2H_4.

Arenes or Aromatic Hydrocarbons

121. Benzene does not undergo addition reactions easily because
(*a*) It has a cyclic structure
(*b*) Double bonds in it are very strong
(*c*) Resonance stabilized system is to be preserved
(*d*) It has six hydrogen atoms.

122. The melting points of isomeric xylenes are in the order
(*a*) *p*-xylene > *o*-xylene > *m*-xylene
(*b*) *p*-xylene > *m*-xylene > *o*-xylene
(*c*) *o*-xylene > *m*-xylene > *p*-xylene
(*d*) *p*-xylene > *p*-xylene > *m*-xylene.

123. Hybrid state of all the carbon atoms in benzene is
(*a*) sp^2 (*b*) sp^3
(*c*) sp (*d*) sp and sp^2.

124. The correct order of C—C bond lengths in benzene (I), ethane (II), ethylene (III) and acetylene (IV) is
(*a*) II > III > IV > I
(*b*) IV > III > II > I
(*c*) II > I > III > IV
(*d*) IV > III > I > II.

125. Resonance energy of benzene is approximately
(*a*) 150 kJ mol^{-1} (*b*) 200 kJ mol^{-1}
(*c*) 100 kJ mol^{-1} (*d*) 300 kJ mol^{-1}.

126. Which of the following will not react with benzene in the presence of $AlCl_3$?
(*a*) CCl_4 (*b*) C_2H_5Cl
(*c*) C_6H_5Cl (*d*) $CH_3-CH_2-CH_2-Cl$.

127. Which among the following can be most readily sulphonated ?
(*a*) Benzene (*b*) Toluene
(*c*) Nitrobenzene (*d*) Chlorobenzene.

128. What is the end product obtained by nitration of toluene ?
(*a*) *o*-Nitrotoluene (*b*) *p*-Nitrotoluene
(*c*) 2, 4-Dinitrotoluene (*d*) 2, 4, 6-Trinitrotoluene.

129. Benzene reacts with H_2SO_4 only when the acid is
(*a*) Dilute and cold (*b*) Dilute and hot
(*c*) Conc. and hot (*d*) Mixture with HNO_3.

130. Heating a mixture of sodium benzoate and soda-lime gives
(*a*) Methane (*b*) Phenol
(*c*) Calcium benzoate (*d*) Benzene.

131. Which of the following groups is *ortho* and *para* directing ?
(*a*) $-C(=O)-Cl$ (*b*) —CHO
(*c*) —OH (*d*) $-C(=O)-CH_3$.

132. Which of the following groups is *meta* directing ?
(*a*) $-SO_3H$ (*b*) $-CH_3$
(*c*) –Cl (*d*) –OH.

133. Benzophenone can be prepared from
(*a*) Chlorobenzene and benzaldehyde
(*b*) Benzoyl chloride and chlorobenzene
(*c*) Benzene and benzoyl chloride
(*d*) Benzene and benzoic acid.

134. Ozonolysis of *p*-xylene gives
(*a*) Glyoxal (*b*) Methyl glyoxal
(*c*) Both of these (*d*) None of these.

135. Which of the following is the most reactive towards ring nitration ?
(*a*) Benzene (*b*) Mesitylene
(*c*) Toluene (*d*) *m*-Xylene.

136. During nitration of benzene, the electrophile is
(*a*) NO_3^- (*b*) NO_2^+
(*c*) NO_2^- (*d*) NO_2.

137. During electrophilic substitution in benzene, the intermediate species involves is
(*a*) Carbocation (*b*) Carbanion
(*c*) Free radical (*d*) None of these.

138. In the reaction,

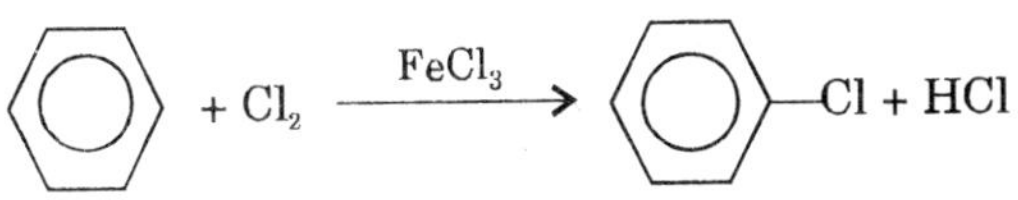

the attacking species is
(*a*) Cl_2 (*b*) Cl^+
(*c*) Cl^- (*d*) $FeCl_4^-$.

139. Benzene on ozonolysis yields
(*a*) Glyoxal (*b*) Acetone
(*c*) Ethanal (*d*) Methanal.

140. In the chlorination of benzene the catalyst used is
(*a*) V_2O_5 (*b*) Al_2O_3
(*c*) $FeCl_3$ (*d*) Cr_2O_3.

141. Benzene reacts with excess of chlorine in sunlight to form
(*a*) Hexachlorobenzene (*b*) *m*-Dichlorobenzene
(*c*) Chlorobenzene (*d*) Benzene hexachloride.

142. Upon oxidation with acidified $KMnO_4$ ethyl benzene gives
(*a*) Benzyl alcohol (*b*) Phenyl acetic acid
(*c*) Benzoic acid (*d*) 1-Phenyl ethanol.

143. On chlorination, nitrobenzene gives ...... as the major product.
(*a*) *o*-Chloro nitrobenzene (*b*) *p*-Chloro nitrobenzene
(*c*) *m*-Chloro nitrobenzene (*d*) All the three.

144. Among the following, the compound that can be most readily sulphonated is
(*a*) Nitrobenzene (*b*) 1, 3-Dinitrobenzene
(*c*) Chlorobenzene (*d*) Toluene.

145. Aromatic hydrocarbons undergo
(*a*) Nucleophilic addition reactions
(*b*) Electrophilic addition reactions
(*c*) Electrophilic substitution reactions
(*d*) None of these.

Miscellaneous Questions

146. Which of the following compounds does not dissolve in conc. H_2SO_4 ?
(*a*) Ethylene (*b*) Aniline
(*c*) Ethyne (*d*) Hexane.

147. Which of the following represents the most oxidised form of hydrocarbons ?
(*a*) R—CH_2OH (*b*) R—COOH
(*c*) CO_2 (*d*) R—CHO.

148. A hydrocarbon with five carbon atoms in the molecule, decolourizes alkaline $KMnO_4$, but does not give a precipitate with ammonical Cu_2Cl_2 solution. The hydrocarbon is possibly
(*a*) 1-Pentyne (*b*) 1, 3-Pentadiyne
(*c*) 2-Pentyne (*d*) 1, 4-Pentadiyne.

149. Isoprene is a valuable substance for making
(*a*) Petrol (*b*) D.D.T.
(*c*) Synthetic rubber (*d*) Ethyl alcohol.

150. An example of conjugated diene is
(*a*) 1, 4-Pentadiene (*b*) 1, 5-Hexadiene
(*c*) 1, 2-Butadiene (*d*) 1, 3-Pentadiene.

151. A mixture of methane, ethylene and ethyne is passed through a solution of ammonical $AgNO_3$. The gas coming out will be
(*a*) Methane
(*b*) Ethyne
(*c*) A mixture of methane and ethylene
(*d*) A mixture of ethylene and ethyne.

152. An organic compound which readily decolourises bromine water and forms an anhydride on heating could be

(*a*) 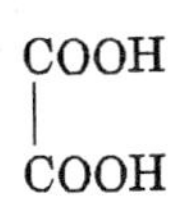

(*b*) 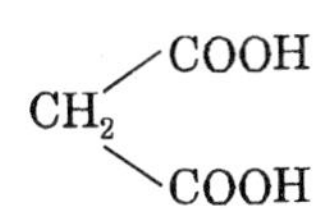

(*c*) 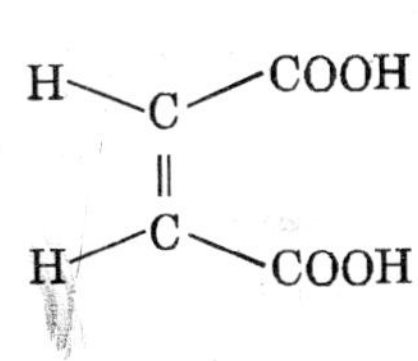

(*d*) 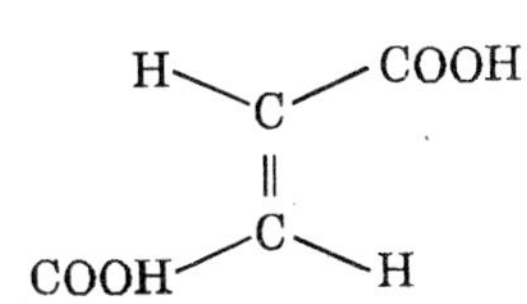

153. Addition of one equivalent of Br_2 to 1, 3-pentadiene produces

(*a*) 4, 5-Dibromo-2-pentene (*b*) 1, 4-Dibromo-2-pentene

(*c*) 3, 4-Dibromopentene (*d*) 3, 4-Dibromo-2-pentene.

154. Heating a mixture of sodium benzoate or benzoic acid and soda-lime gives

(*a*) Benzene (*b*) Methane

(*c*) Sodium benzoate (*d*) Calcium benzoate.

155. Two gases, P and Q decolourise aqueous bromine but only one of them gives a white precipitate with aqueous ammonical silver nitrate solution. P and Q are likely to be

(*a*) Ethane and ethyne

(*b*) But-1-yne and but-2-yne

(*c*) Ethane and but-2-yne

(*d*) Ethyne and propyne.

ANSWERS

1. (*d*)	**2.** (*a*)	**3.** (*b*)	**4.** (*c*)	**5.** (*d*)	**6.** (*b*)	**7.** (*b*)	**8.** (*d*)	**9.** (*d*)
10. (*b*)	**11.** (*d*)	**12.** (*a*)	**13.** (*a*)	**14.** (*b*)	**15.** (*b*)	**16.** (*b*)	**17.** (*b*)	**18.** (*a*)
19. (*a*)	**20.** (*c*)	**21.** (*c*)	**22.** (*c*)	**23.** (*c*)	**24.** (*a*)	**25.** (*d*)	**26.** (*c*)	**27.** (*b*)
28. (*c*)	**29.** (*a*)	**30.** (*a*)	**31.** (*b*)	**32.** (*b*)	**33.** (*a*)	**34.** (*b*)	**35.** (*c*)	**36.** (*a*)
37. (*d*)	**38.** (*d*)	**39.** (*d*)	**40.** (*b*)	**41.** (*d*)	**42.** (*a*)	**43.** (*a*)	**44.** (*d*)	**45.** (*a*)
46. (*d*)	**47.** (*d*)	**48.** (*c*)	**49.** (*a*)	**50.** (*c*)	**51.** (*a*)	**52.** (*c*)	**53.** (*d*)	**54.** (*b*)
55. (*c*)	**56.** (*a*)	**57.** (*b*)	**58.** (*c*)	**59.** (*a*)	**60.** (*a*)	**61.** (*d*)	**62.** (*d*)	**63.** (*d*)
64. (*b*)	**65.** (*b*)	**66.** (*d*)	**67.** (*b*)	**68.** (*b*)	**69.** (*b*)	**70.** (*b*)	**71.** (*c*)	**72.** (*a*)
73. (*a*)	**74.** (*a*)	**75.** (*a*)	**76.** (*d*)	**77.** (*d*)	**78.** (*d*)	**79.** (*b*)	**80.** (*c*)	**81.** (*b*)
82. (*c*)	**83.** (*b*)	**84.** (*a*)	**85.** (*b*)	**86.** (*d*)	**87.** (*c*)	**88.** (*a*)	**89.** (*b*)	**90.** (*d*)
91. (*b*)	**92.** (*c*)	**93.** (*b*)	**94.** (*b*)	**95.** (*a*)	**96.** (*c*)	**97.** (*d*)	**98.** (*c*)	**99.** (*d*)
100. (*a*)	**101.** (*d*)	**102.** (*d*)	**103.** (*b*)	**104.** (*d*)	**105.** (*c*)	**106.** (*b*)	**107.** (*d*)	**108.** (*a*)
109. (*b*)	**110.** (*b*)	**111.** (*c*)	**112.** (*c*)	**113.** (*c*)	**114.** (*d*)	**115.** (*d*)	**116.** (*a*)	**117.** (*d*)
118. (*a*)	**119.** (*a*)	**120.** (*a*)	**121.** (*c*)	**122.** (*a*)	**123.** (*a*)	**124.** (*c*)	**125.** (*a*)	**126.** (*c*)
127. (*b*)	**128.** (*d*)	**129.** (*c*)	**130.** (*d*)	**131.** (*c*)	**132.** (*a*)	**133.** (*c*)	**134.** (*c*)	**135.** (*b*)
136. (*b*)	**137.** (*a*)	**138.** (*b*)	**139.** (*a*)	**140.** (*c*)	**141.** (*d*)	**142.** (*c*)	**143.** (*c*)	**144.** (*d*)
145. (*c*)	**146.** (*d*)	**147.** (*c*)	**148.** (*c*)	**149.** (*c*)	**150.** (*d*).	**151.** (*c*)	**152.** (*c*)	**153.** (*b*)
154. (*a*)	**155.** (*b*).							

HINTS/SOLUTIONS

Q. No.	Ans.	Hints & Solutions
1.	(*d*)	Anthracene is mainly present in green oil fraction.
2.	(*a*)	C—7 to C—12.
3.	(*b*)	Cetane number of α-methyl naphthalene is zero.
4.	(*c*)	Petroleum is mainly source of aliphatic hydrocarbons.
5.	(*d*)	Coal contains large number of aromatic hydrocarbons.
6.	(*b*)	Paraffin wax is high boiling fraction while crude naphtha is low boiling fraction.

Q. No.	Ans.	Hints & Solutions
7.	(*b*)	Octane number of iso-octane is 100.
8.	(*d*)	Cetane is *n*-hexadecane.
9.	(*d*)	The process of heating coal in the absence of air is called carbonisation.
10.	(*b*)	See definition of cetane number.
11.	(*d*)	Cracking converts high boiling fractions into hydrocarbons of gasoline range.
12.	(*a*)	Naphthalene is present in middle oil fraction.
13.	(*a*)	Anthracite form of coal has the highest percentage of carbon.

Q. No.	Ans.	Hints & Solutions
14.	(b)	*n*-Hexadecane has very high cetane number (100).
15.	(b)	Refer to carbonisation of coal.
16.	(b)	2, 2, 3-Trimethylpentane has octane number of greater than 100 (actually 116).
17.	(b)	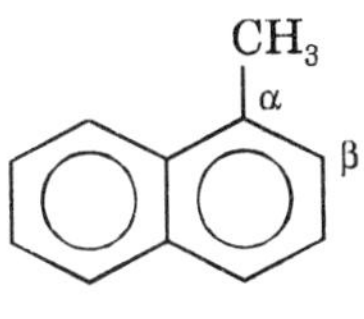
18.	(a)	$CH_3—(CH_2)_5—CH_3 \xrightarrow{-4H_2}$ 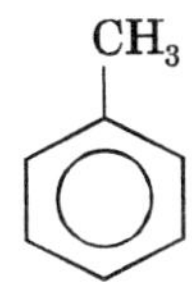
19.	(a)	Carbonisation of coal is carried out by heating coal at temperature of 1000°C in the absence of air.
20.	(c)	Main constituents of light oil fraction are benzene, toluene and xylene.
21.	(c)	The three isomers are *n*-, *iso*- and *neo*-pentane.
22.	(c)	1, 2-Dichloroethene, because in this case on each of the carbon bearing double bond two different groups are attached.
23.	(c)	2, 3-Dimethylbutane has only two types of H-atoms. $CH_3—CH(CH_3)—CH(CH_3)—CH_3$
24.	(a)	The least substituted alkene is least stable and has the highest value of heat of hydrogenation.
25.	(d)	$C_{10}H_{22}$ has maximum number of carbon atoms and hydrogen atoms.
26.	(c)	General formula of alkynes is C_nH_{2n-2}.
27.	(b)	$CH_3-CH_2-C(H)=C(H)-CH_3$. The two groups attached to each doubly bonded carbon atom are different in this case.
28.	(c)	More substituted alkenes are more stable.
29.	(a)	Homologues differ in molecular formula by a $—CH_2—$.
30.	(a)	I. $CH_3—CH_2—CH_2—CH=CH_2$ Monosubstituted II. $CH_3—CH_2—CH=CH—CH_3$ Disubstituted III. $CH_3—C(CH_3)=CH—CH_3$ Trisubstituted
31.	(b)	Wurtz reaction is suitable only for symmetrical alkanes whereas Corey House reaction can be used for obtaining unsymmetrical as well as symmetrical alkanes.
32.	(b)	2-Butene will form. Alcoholic KOH gives elimination product and out of the available choices of alkenes, more substituted alkene is formed (*Saytzeff rule*).
33.	(a)	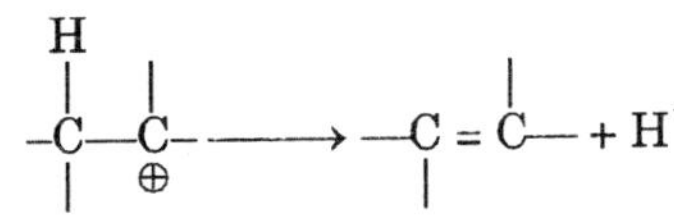 Elimination of $H^{\oplus}$.
34.	(b)	Alcoholic solution of KOH brings about dehydrohalogenation in alkyl halides forming alkenes.
35.	(c)	Self explanatory question.
36.	(a)	The compounds with acidic protons react with Grignard reagent to form a hydrocarbon. $CH_3CH_2\mathbf{OH} + \mathbf{R}MgX \longrightarrow \mathbf{R—H} + MgX(OC_2H_5)$.
37.	(d)	$CH_3COONa + NaOH \xrightarrow{CaO} \mathbf{CH_4} + Na_2CO_3$ $CH_3I + 2[H] \longrightarrow \mathbf{CH_4} + HI$ $Al_4C_3 + 12H_2O \longrightarrow 4Al(OH)_3 + \mathbf{3CH_4}$.
38.	(d)	The reaction of alkyl halides with sodium in the presence of dry ether is known as **Wurtz reaction.**
39.	(d)	$2RCOO^- \longrightarrow R—R + 2CO_2 + 2e^-$ R—R has $2(n-1)$ carbon atoms.
40.	(b)	Decarboxylation with soda lime results in the formation of alkane with one carbon less than the starting compounds.
41.	(d)	$CaC_2 + 2H_2O \longrightarrow Ca(OH)_2 + C_2H_2$ $CH_3CHBr_2 + 2KOH(alc.) \longrightarrow HC\equiv CH + 2KBr + 2H_2O$ $BrCH_2CH_2Br + 2KOH\ (alc.) \longrightarrow HC\equiv CH + 2KBr + 2H_2O$.
42.	(a)	Heating with zinc dust brings about dehalogenation of 2, 3-dibromobutane.
43.	(a)	Wurtz reaction $RX + 2Na \longrightarrow \underset{\text{Alkane}}{R—R} + NaX$.
44.	(d)	Kolbe electrolysis.
45.	(a)	Methane cannot be prepared by Wurtz reaction.

Q. No.	Ans.	Hints & Solutions
46.	(*d*)	Ethene is formed by removal of water molecule within the molecule of ethanol.
47.	(*d*)	$\begin{matrix}\text{CHCOOK}\\ \Vert\\ \text{CHCOOK}\end{matrix} + 2H_2O \xrightarrow{\text{Electrolysis}} \begin{matrix}\text{CH}\\ \vert\vert\vert\\ \text{CH}\end{matrix} + 2CO_2 + H_2 + 2KOH$ Potassium fumarate → Acetylene
48.	(*c*)	Sodium in the presence of liquid ammonia converts alkynes to corresponding *cis*-alkenes.
49.	(*a*)	In Kolbe electrolysis, the alkane is formed by union of two alkyl groups. The alkane formed has, thus, two or more carbon atoms.
50.	(*c*)	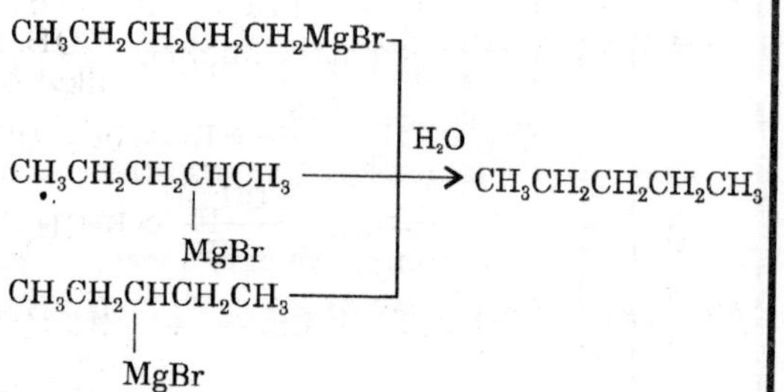
51.	(*a*)	*n*-Hexane () has the longest chain of carbon atoms.
52.	(*c*)	Alkanes having 5 to 17 carbon atoms are liquid at room temperature.
53.	(*d*)	*n*-Heptane has the longest chain of carbon atoms.
54.	(*b*)	*n*-Octane has the longest chain of carbon atoms.
55.	(*c*)	*cis*-Isomer has higher boiling point than *trans* due to its greater polarity.
56.	(*a*)	Homolytic fission results in the formation of free radicals.
57.	(*b*)	$CH_3-CH_2-CH_3 \longrightarrow CH_3-\overset{\bullet}{C}H-CH_3 + H^\bullet$ (2°)
58.	(*c*)	$CH_3-(CH_2)_5-CH_3 \xrightarrow{-4H_2}$ 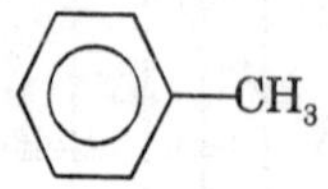 Toluene
59.	(*a*)	$Cl_2 \longrightarrow 2Cl^\bullet$.
60.	(*a*)	C_2H_6 should be used in excess, otherwise polychlorination will take place.
61.	(*d*)	$\overset{1}{C}H_3-\overset{2}{C}H(\overset{1}{C}H_3)-\overset{3}{C}H_2-\overset{4}{C}H_3$ 2-Methylbutane has four types of H-atoms.
62.	(*d*)	Chlorination of alkanes involves attack of free radicals.
63.	(*d*)	$2C_2H_6 + 7O_2 \longrightarrow 4CO_2 + 6H_2O$
64.	(*b*)	Ease of substitution of various types of H atoms is 3° > 2° > 1°.
65.	(*b*)	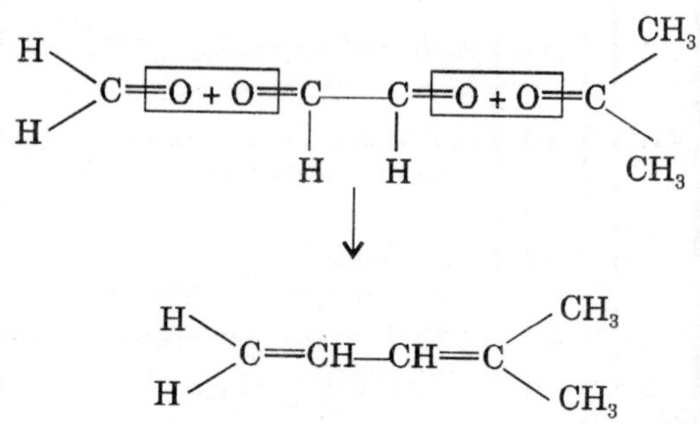
66.	(*d*)	Both react with reagent (*c*) while both do not react with reagents (*a*) and (*b*).
67.	(*b*)	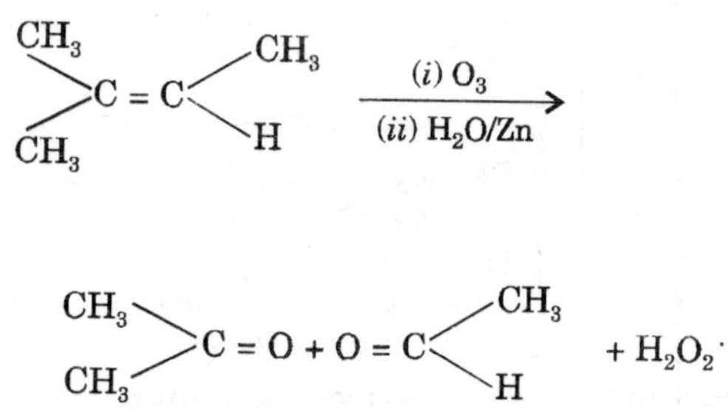
68.	(*b*)	Both the carbon atoms involved in double bond will have alkyl groups attached to them. $CH_3-C(CH_3)=C(CH_3)-CH_3$
69.	(*b*)	2-Bromo propane—Markownikoff's rule.
70.	(*b*)	$CH_3-CH=CH-CH_3 + Br_2 \longrightarrow CH_3-CH(Br)-CH(Br)-CH_3$
71.	(*c*)	Allylic bromination of an alkene is free radical substitution reaction.
72.	(*a*)	1-Bromo-1-phenylethane will form. Addition is according to Markownikoff's rule.

Q. No.	Ans.	Hints & Solutions
73.	(*a*)	1-Bromobutane. Addition of HBr in the presence of peroxides obeys anti-Markownikoff's rule.
74.	(*a*)	$CH_2 = CH_2 + HOCl \longrightarrow ClCH_2 - CH_2OH$ Ethylene
75.	(*a*)	F_2 reacts violently with alkenes.
76.	(*d*)	$CH_3 - CH = CH_2 + HOBr \longrightarrow CH_3 - CH(OH) - CH_2Br$ Markownikoff addition. The electrophile is Br^+ while the negative part is OH^-.
77.	(*d*)	Intermediate carbocation formed in the case of $CH_3 - C(CH_3) = CH - CH_3$ would be most stable, therefore, Br_2 adds most readily to this.
78.	(*d*)	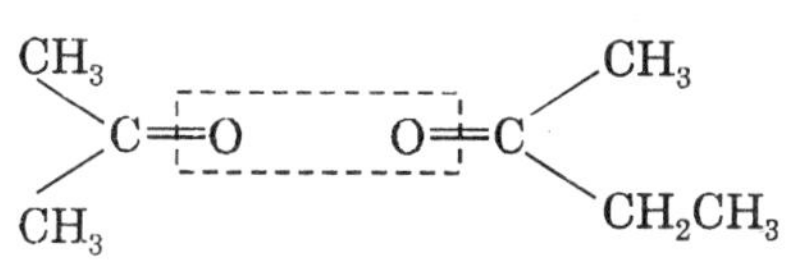 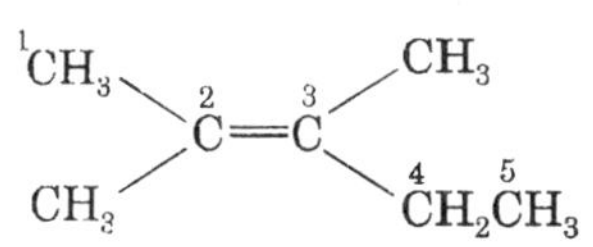 2, 3-Dimethyl-2-pentene
79.	(*b*)	C_2H_4. Because it is unsaturated hydrocarbon.
80.	(*c*)	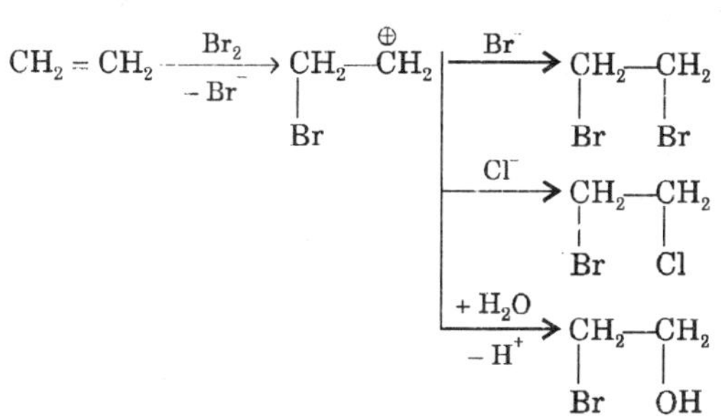 The intermediate carbocation is formed by the initial electrophilic attack of Br_2.
81.	(*b*)	$CH_2 = CH_2 + S_2Cl_2 \longrightarrow ClCH_2CH_2 - S - CH_2CH_2Cl + S.$
82.	(*c*)	$CH_3 - C(CH_3) = CH_2 + HBr \longrightarrow CH_3 - CBr(CH_3) - CH_3$ *tert*-Butyl bromide
83.	(*b*)	anti-Markownikoff addition.
84.	(*a*)	Unsaturated compounds discharge the pink coiour of Baeyer's reagent (alk. $KMnO_4$).
85.	(*b*)	$CH_2 = CH_2 + H_2O + [O] \longrightarrow CH_2(OH) - CH_2(OH)$
86.	(*d*)	Markownikoff's rule is applicable to unsymmetrical alkenes.
87.	(*c*)	But-2-ene is a symmetrical alkene.
88.	(*a*)	Self explanatory answer. $CH_2 = CH_2 \xrightarrow{Br_2} Br{-}CH_2{-}\overset{+}{C}H_2 \xrightarrow[-H^+]{CH_3OH} BrCH_2{-}CH_2{-}OCH_3$
89.	(*b*)	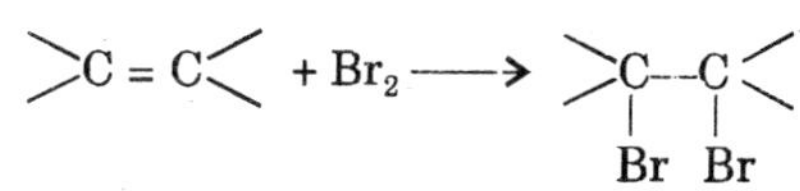 1 mole of alkene reacts with one mole of Br_2. $\therefore$ Molecular mass of alkene $= \frac{2.8}{8} \times 160 = 56$ Hence, alkene is C_4H_8.
90.	(*d*)	Terminal alkynes give red ppt. with ammonical curpous chloride. 1-Butyne is a terminal alkyne.
91.	(*b*)	Nucleophilic addition reaction.
92.	(*c*)	Alcoholic KOH brings about dehyrohalogenation.
93.	(*b*)	Alkaline $KMnO_4$.
94.	(*b*)	$H{-}C \equiv C{-}H + HCl \xrightarrow[60°]{HgCl_2} CH_2 = CH - Cl$ Vinyl chloride
95.	(*a*)	Self explanatory answer.
96.	(*c*)	Lindlar's catalyst is palladium adsorbed on $CaCO_3$ and slightly deactivated.
97.	(*d*)	$CH_3 - C \equiv CH + H_2O \longrightarrow$ $CH_3 - C(OH) = CH_2 \rightleftharpoons CH_3 - C(=O) - CH_3$

Q. No.	Ans.	Hints & Solutions
98.	(c)	$H-C\equiv C-H + 4[O] \longrightarrow \begin{array}{c}COOH\\ \vert\\ COOH\end{array}$
99.	(d)	Alkynes are more acidic than alkenes and alkanes.
100.	(a)	Because ethyne gives most stable anion.
101.	(d)	2-Butyne is least acidic because it is non-terminal alkynes, whereas all the other choices are terminal alkynes.
102.	(d)	1-Pentyne being terminal alkyne is acidic and can form metallic derivatives.
103.	(b)	Terminal alkynes give ppt. with ammonical $AgNO_3$, so it is 1, 5-hexadiyne.
104.	(d)	$CH_2 = CHCl + HCl \longrightarrow CH_3 - CHCl_2$.
105.	(c)	$HC\equiv CH \xrightarrow[(ii)\ H_2O/Zn]{(i)\ O_3} \underset{O\ \ \ O}{\overset{}{H-\underset{\Vert}{C}-\underset{\Vert}{C}-H}}$ Glyoxal
106.	(b)	Na/Liq. NH_3 reduces alkynes to *trans*-alkenes.
107.	(d)	Baeyer's reagent.
108.	(a)	$CH_2 - C\equiv CH$ is a terminal alkyne.
109.	(b)	Intermediate carbocation, $CH_3 - \overset{+}{C}H - Cl$, is stabilized by mesomeric effect. $CH_3 - \overset{+}{C}H - \ddot{C}l: \leftrightarrow CH_3 - CH = \overset{+}{C}l:$
110.	(b)	$H-C\equiv C-H + CH_3OH \xrightarrow{K^+OCH_3^-} \underset{H\ \ \ \ \ OCH_3}{H-\underset{\vert}{C}=\underset{\vert}{C}-H}$
111.	(c)	$CH_3-C\equiv CH + 2HOCl \longrightarrow \left[CH_3-\overset{OH}{\underset{OH}{C}}-CHCl_2\right] \xrightarrow{-H_2O} CH_3-\overset{O}{\overset{\Vert}{C}}-CHCl_2$
112.	(c)	Reaction is ozonolysis.
113.	(c)	Acetylene contains acidic H-atoms.
114.	(d)	Ammonical $AgNO_3$ gives white ppt. with terminal alkynes such as compound 3.
115.	(d)	Propyne on trimerization yields mesitylene $3CH_3C\equiv CH \longrightarrow$ 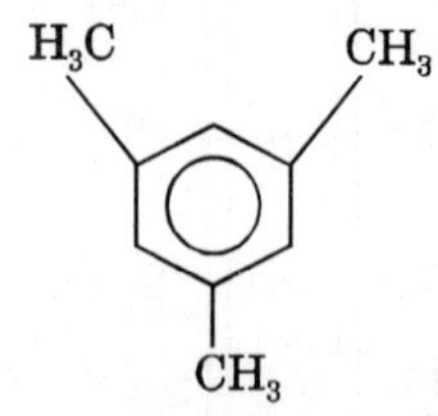 Mesitylene
116.	(a)	Pb-$BaSO_4$ or Lindlar's catayst reduces alkynes to *cis*-alkenes.
117.	(d)	$CH_3CH_2CH_2C\equiv CH$ contains acid hydrogen atom.
118.	(a)	See Hint 114.
119.	(a)	$H-C\equiv C-H + 2CH_3COOH \longrightarrow CH_3 - CH(OCOCH_3)_2$ Both the protons go to same carbon atom.
120.	(a)	Ethyne is acidic in character. $H-C\equiv C-H + NaNH_2 \longrightarrow H-C\equiv C^-Na^+ + \frac{1}{2}H_2$
121.	(c)	Self explanatory question.
122.	(a)	*p*-xylene has highest melting point due to its symmetry. *o*-xylene has higher melting point than *m*-xylene due to greater polarity.
123.	(a)	sp^2 hybridized.
124.	(c)	C-C bond is single in II, double in III, triple in IV and one and half in I.
125.	(a)	Resonance energy of benzene is 147 kJ mol^{-1}.
126.	(c)	Aryl halides do not react with $AlCl_3$ to generate electrophile due to stronger C-Cl bond.
127.	(b)	CH_3-group activates benzene ring by its electron releasing effect.
128.	(d)	All the *ortho* and *para* positions get substituted by nitro group to yield 2, 4, 6-trinitrotoluene.
129.	(c)	Benzene reacts with conc. H_2SO_4 on heating and the product is benzene sulphonic acid.
130.	(d)	Soda-lime decarboxylation.
131.	(c)	Factual question. The groups having lone pair on the atom attached to the benzene ring generally activate the ring and are *ortho-para* directing.
132.	(a)	$-SO_3H$ group is *meta* directing.
133.	(c)	The reaction is Friedel crafts acylation.

Q. No.	Ans.	Hints & Solutions
134.	(c)	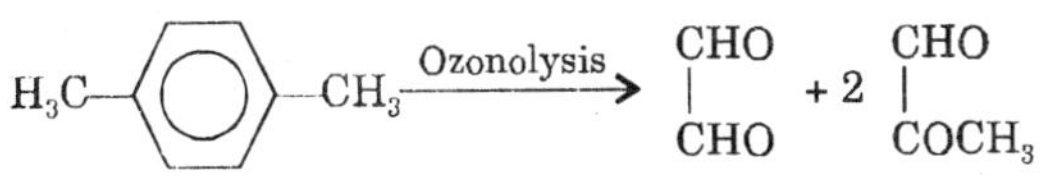 Ozonolysis of *p*-xylene yields two moles of methyl glyoxal and one mole of glyoxal.
135.	(b)	Mesitylene has three methyl groups activating the benzene ring.
136.	(b)	Nitration takes place by electrophilic attack by nitronium (NO_2^+) ion.
137.	(a)	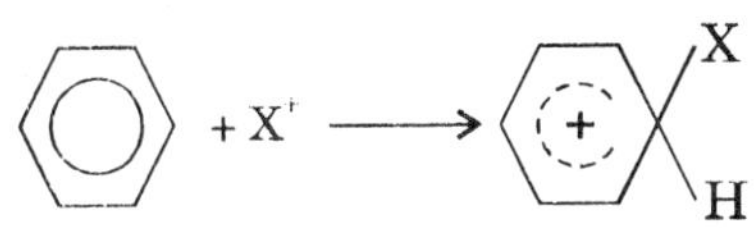
138.	(b)	Chlorination of benzene takes place through electrophilic attack by Cl^+ (Chloronium ion).
139.	(a)	Benzene on ozonolysis yields three moles of glyoxal for each mole of benzene. $C_6H_6 \xrightarrow[(ii)\ H_2O/Zn]{(i)\ O_3} 3\ \begin{array}{c} CHO \\ \vert \\ CHO \end{array}$
140.	(c)	Ferric chloride acts as halogen carrier during chlorination of benzene.
141.	(d)	$C_6H_6 + 3Cl_2 \xrightarrow{h\nu} C_6H_6Cl_6$.
142.	(c)	Side chain attached to benzene ring is oxidized to carboxyl group in the presence of acidified $KMnO_4$ solution.
143.	(c)	Nitro group is meta directing group so, we get *m*-chloronitrobenzene.
144.	(d)	CH_3--group in toluene is activating group.
145.	(c)	Aromatic hydrocarbons undergo electrophilic substitution reactions.
146.	(d)	(*a*), (*b*), (*c*) react with H_2SO_4 and dissolve. Hexane does not react with H_2SO_4.
147.	(c)	Alkane ----→ Alcohol ----→ Aldehyde/ketone ----→ Carboxylic acid ----→ CO_2 ——— **Oxidation**
148.	(c)	Since the compound does not give precipitate with ammonical Cu_2Cl_2 it is not a terminal alkyne. Hence, the choices *a*, *b*, *d* are ruled out.
149.	(c)	Natural rubber is a polymer of isoprene, 2-methyl-1, 3-butadiene.
150.	(d)	$CH_3-CH=CH-CH=CH_2$ is a conjugate diene.
151.	(c)	Ethyne will form a precipitate with ammonical $AgNO_3$. So the gases coming out will be a mixture of methane and ethylene.
152.	(c)	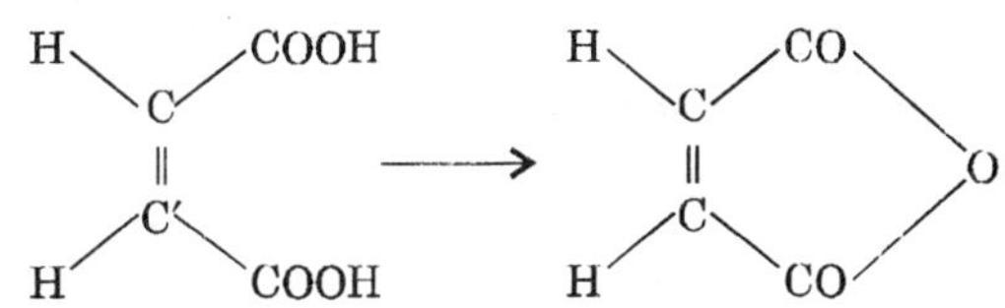 In this case the two carboxyl groups are on same side and on dehydration, the ring formed is five membered which is least strained.
153.	(b)	$CH_3-CH=CH-CH=CH_2 + Br_2$ ↓ 1, 4-Addition $CH_3-\underset{Br}{\underset{\vert}{CH}}-CH=CH-\underset{Br}{\underset{\vert}{CH_2}}$ 1, 4-Dibromo-2-pentene (Major product)
154.	(a)	Benzene is formed through soda-lime decarboxylation.
155.	(b)	Since both gas decolourize bromine water, both of them are unsaturated. Hence choices (*a*) and (*c*) are ruled out. Since only one of them gives white precipitate with ammonical $AgNO_3$, only one of them is a terminal alkyne. Hence, the choice (*b*) is correct.

2.5. EXAMINER'S VIEWPOINT

Test Your Confidence

1996

1. Kerosene oil has compounds having
   (*a*) C_5—C_8 (*b*) C_8—C_{12}
   (*c*) C_{12}—C_{16} (*d*) C_{16}—C_{18}. ***(AFMC)***
2. A war gas Lewisite is formed by reaction of arsenic chloride with
   (*a*) CH_4 (*b*) C_6H_6
   (*c*) C_2H_2 (*d*) C_2H_4. ***(AFMC)***
3. Major constituent of light oil is
   (*a*) Benzene (*b*) Phenol
   (*c*) Aniline (*d*) Anthracene. ***(AFMC)***
4. The reaction
   $CH_2 = CH—CH_3 + HBr \longrightarrow CH_3—CH(Br)—CH_3$ is
   (*a*) Nucleophilic addition
   (*b*) Electrophilic addition
   (*c*) Electrophilic substitution
   (*d*) Free radical addition. ***(CBSE PMT)***
5. Oxidation of toluene to benzaldehyde by chromyl chloride is called
   (*a*) Friedal Craft reaction (*b*) Wurtz reaction
   (*c*) Cannizzaro reaction (*d*) Etard reaction. ***(CBSE PMT)***
6. Which of the following is used as an antiknocking material ?
   (*a*) T.E.L. (*b*) C_2H_5OH
   (*c*) Glyoxal (*d*) Freon. ***(CBSE PMT)***
7. The product formed on passing ethylene through a cold alkaline $KMnO_4$ solution is
   (*a*) Ethyl alcohol (*b*) Ethylene glycol
   (*c*) Ethylene oxide (*d*) Ethane. ***(Pb. CET)***
8. Which of the following compounds has the highest value of Ka ?
   (*a*) Ethyne (*b*) Ethene
   (*c*) Methane (*d*) Ethane.
9. Acetylene reacts with acetic acid in presence of Hg^{++} ions at room temperature to give
   (*a*) Ethyl acetate (*b*) Acetaldehyde
   (*c*) Vinyl acetate (*d*) Methyl acetate. ***(Pb. CET)***
10. The product formed by the action of chlorine on ethene in saturated solution of KBr is/are
   (*a*) $ClCH_2CH_2Cl + ClCH_2CH_2CH_2Br$
   (*b*) $ClCH_2CH_2Cl$
   (*c*) $ClCH_2CH_2Cl + BrCH_2CH_2Cl$
   (*d*) $ClCH_2CH_2Cl + BrCH_2CH_2Br + ClCH_2CH_2Br$. ***(Pb. CET)***
11. The order of reactivity of halogens in aliphatic substitution reactions is
   (*a*) $Br_2 > Cl_2 > F_2$ (*b*) $Cl_2 > Br_2 > F_2$
   (*c*) $F_2 > Cl_2 > Br_2$ (*d*) $F_2 > Br_2 > Cl_2$. ***(Pb. CET)***
12. In which of the following molecules hydrogen is most acidic ?
   (*a*) Acetylene (*b*) Methane
   (*c*) Ethane (*d*) Ethylene. ***(Pb. CET)***
13. When *n*-propyl iodide is heated with alcoholic KOH, one of the products is
   (*a*) Propene (C_3H_6) (*b*) Cyclopropane (C_3H_6)
   (*c*) C_3H_4 (*d*) C_3H_8. ***(DCE)***
14. When propene is treated with HBr in the dark and in absence of peroxide, the main product is
   (*a*) 1-Bromopropane (*b*) 2-Bropropane
   (*c*) 1, 2-Dibromopropane (*d*) 1, 3-Dibromopropane. ***(DCE)***
15. When an alkyne, $RC \equiv CH$, is treated with cuprous ion in an ammoniacal medium, one of the products is
   (*a*) $RC \equiv CCu$ (*b*) $CuC \equiv CH$
   (*c*) $CuC \equiv CCu$ (*d*) $RC \equiv CR$. ***(DCE)***

1997

16. A gas which reacts with aqueous $KMnO_4$ solution but does not give precipitate with ammonical cuprous chloride solution is
   (*a*) Ethylene (*b*) Methane
   (*c*) Ethane (*d*) Acetylene. ***(Pb. CET)***
17. In a substitution reaction of C_6H_5Y, the major product is *m*-isomer, therefore the group Y can be
   (*a*) —COOH (*b*) —Cl
   (*c*) —OH (*d*) $—NH_2$. ***(AIIMS)***
18. Formation of polyethene from calcium carbide takes place as follows
   $CaC_2 + 2H_2O \longrightarrow Ca(OH)_2 + C_2H_2$;
   $C_2H_2 + H_2 \longrightarrow C_2H_4$;
   $nC_2H_4 \longrightarrow (—CH_2—CH_2—)_n$

The amount of polythene obtained from 64 kg of CaC_2 is

(*a*) 7 kg (*b*) 14 kg
(*c*) 21 kg (*d*) 28 kg. **(*AIIMS*)**

19. 1-Butyne reacts with cold alkaline $KMnO_4$ to yield

(*a*) $CH_3CH_2CH_2COOH$
(*b*) CH_3CH_2COOH
(*c*) $CH_3CH_2COOH + CO_2$
(*d*) $CH_3CH_2COOH + HCOOH$. **(*AIIMS*)**

20. Which of the following applies to the reaction,

$$CH_3—CHBr—CH_2CH_3 \xrightarrow{\text{alc. KOH}}$$

(*i*) $CH_3CH = CHCH_3$ (major product)
(*ii*) $CH_2 = CH—CH_2CH_3$ (minor product)

(*a*) Markovnikoff's rule (*b*) Saytzeff's rule
(*c*) Kharasch effect (*d*) Hofmann's rule. **(*AMU*)**

21. Which of the following isomers will have the highest boiling point ?

(*a*) $CH_3—CH_2—CH_2—CH_2—CH_2—CH_3$
(*b*) $CH_3—CH(CH_3)—CH_2—CH_2—CH_3$
(*c*) $CH_3—CH(CH_3)—CH(CH_3)—CH_3$
(*d*) $CH_3—C(CH_3)_2—CH_2—CH_3$. **(*AMU*)**

22. Which of the following will have fastest rate of reaction with $Br_2/FeBr_3$

(*a*) 

(*b*) 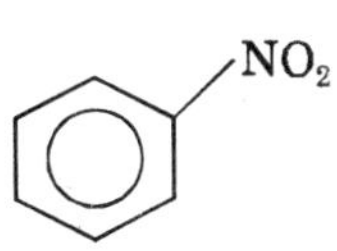

(*c*) 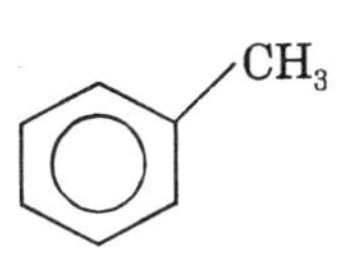

(*d*) 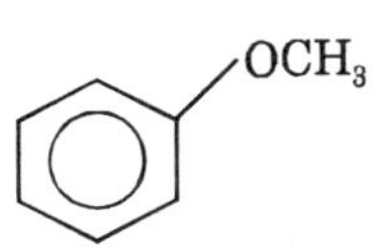

(*AMU*)

23. Which of the following isomers will have the highest octane number ?

(*a*) *n*-octane
(*b*) 2-Methylheptane
(*c*) 2, 3-Dimethylhexane
(*d*) 2, 2, 4-Trimethylpentane. **(*AMU*)**

24. In the commercial gasolines, the type of hydrocarbons which is more desirable is

(*a*) Branched hydrocarbon
(*b*) Straight chain hydrocarbon
(*c*) Linear unsaturated hydrocarbon
(*d*) Toluene and its derivatives. **(*CBSE PMT*)**

25. In the reactions ;

$$CH_2 = CH_2 \xrightarrow[\text{acid}]{\text{Hypochlorus}} M \xrightarrow{R} \begin{array}{c} CH_2OH \\ | \\ CH_2OH \end{array}$$

M and R are respectively

(*a*) CH_3CH_2Cl and NaOH
(*b*) CH_2ClCH_2OH and aq. $NaHCO_3$
(*c*) CH_3CH_2OH and HCl
(*d*) 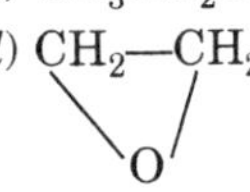 and heat. **(*CBSE PMT*)**

26. Among the following compounds (I—III), the correct order in reaction with electrophile is

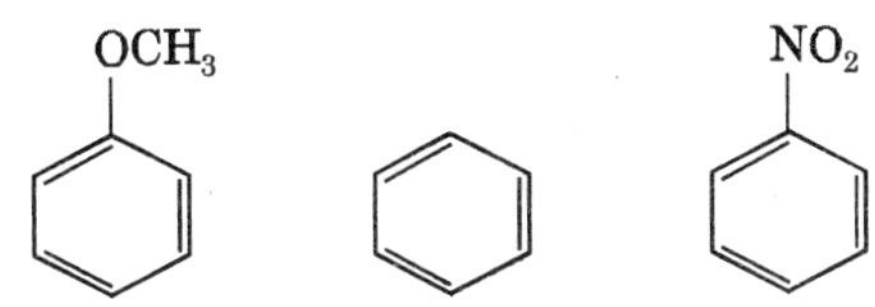

I II III

(*a*) II > III > I (*b*) III < I < II
(*c*) I > II > III (*d*) I = II > III. **(*CBSE PMT*)**

27. Which of the following reactions is expected to readily give a hydrocarbon product in good yield ?

(*a*) $RCOOK \xrightarrow[\text{reduction}]{\text{Electrolytic}}$

(*b*) $RCOOAg \xrightarrow{I_2}$

(*c*) $CH_3—CH_3 \xrightarrow[h\nu]{Cl_2}$

(*d*) $(CH_3)_3CCl \xrightarrow{C_2H_5OH}$ **(*CBSE PMT*)**

28. $C_3H_8 + Cl_2 \xrightarrow{\text{Light}} C_3H_7Cl + HCl$

The above reaction is an example of

(*a*) Elimination (*b*) Substitution
(*c*) Addition (*d*) Rearrangement. **(*AFMC*)**

29. Iodoethane reacts with sodium in the presence of dry ether. The main product is

(*a*) Pentene (*b*) Propyne
(*c*) Butane (*d*) Butene **(*AFMC*)**

30. Marsh gas contains which of the following ?

(*a*) C_2H_2 (*b*) CH_4
(*c*) H_2S (*d*) C_2H_4. **(*AFMC*)**

31. Benzene + Ozone $\longrightarrow$ Y. In this sequence Y is

(*a*) Benzene monoozonide (*b*) Benzene diozonide
(*c*) Benzene triozonide (*d*) Succinic acid. **(*CPMT*)**

32. Toluene on oxidation with chromyl chloride produces

(*a*) Benzoic acid (*b*) Benzaldehyde
(*c*) Chlorobenzene (*d*) None of these. **(*BHU*)**

33. When ethylbromide and *n*-propyl bromide is allowed to react with sodium, in ether, they form

(*a*) Single alkane
(*b*) Mixture of two alkanes
(*c*) Mixture of three alkanes
(*d*) Mixture of four alkanes. **(*BHU*)**

34. $CH_3—CH = CH_2 + HI \longrightarrow$ X, X is

(*a*) $CH_3—CH_2—CH_2—I$ (*b*) $CH_3—CHI—CH_3$
(*c*) $CH_3CH_2CH_3$ (*d*) None. **(*CPMT*)**

1998

35. In Wurtz reaction the reagent used is

(*a*) Na (*b*) Na/Liq. NH_3
(*c*) Na/dry ether (*d*) Na/dry alcohol. **(*EAMCET*)**

36. The purest form of coal is

(*a*) Peat (*b*) Anthracite
(*c*) Lignite (*d*) Bituminous. **(*EAMCET*)**

37. Benzene is obtained by fractional distillation of

(*a*) Light oil (*b*) Middle oil
(*c*) Anthracene oil (*d*) Heavy oil. **(*EAMCET*)**

38. Propene when heated with chlorine at about 500°C forms

(*a*) $ClCH_2—CH = CH_2$ (*b*) $CH_3—CHCl—CH_2Cl$
(*c*) $CH_2Cl—CHCl—CH_2Cl$ (*d*) All of these. **(*MPCEE*)**

39. In the reaction

$$CH_3—C \equiv C—CH_3 \xrightarrow[(ii)\ Zn/H_2O]{(i)\ X} CH_3—\underset{\underset{O}{\|}}{C}—\underset{\underset{O}{\|}}{C}—CH_3,\ \text{X is}$$

(*a*) HNO_3 (*b*) O_2
(*c*) O_3 (*d*) $KMnO_4$. **(*MPCEE*)**

40. In benzene all the C—C bonds have the same length because of

(*a*) Chain isomerism (*b*) Isomerism
(*c*) Hybridization (*d*) Resonance. **(*KCET*)**

41. Conversion of CH_4 to CH_3Cl is an example of ..... reaction

(*a*) Electrophilic substitution
(*b*) Nucleophilic substitution
(*c*) Free radical substitution
(*d*) Free radical addition. **(*KCET*)**

42. When cold aqueous $KMnO_4$ is added to ethylene, it gives

(*a*) Glycerol (*b*) Ethanol
(*c*) Ethylene glycol (*d*) Methanol. **(*AFMC*)**

43. Both methane and ethane can be prepared in one step by the reaction of

(*a*) C_2H_4 (*b*) CH_3OH
(*c*) CH_3Br (*d*) CH_3CH_2OH. **(*AFMC*)**

44. 1-Chlorobutane on reaction with alcoholic potash (KOH) gives

(*a*) 1-Butene (*b*) 1-Butanol
(*c*) 2-Butene (*d*) 2-Butanol. **(*AFMC*)**

45. Reaction of benzene with alkyl halide in the presence of anhydrous $AlCl_3$ is called

(*a*) Friedel-Craft reaction (*b*) Wurtz reaction
(*c*) Williamson's synthesis (*d*) Baeyer's reaction. **(*AFMC*)**

46. Which one of these is not compatible with arenes ?

(*a*) Greater stability
(*b*) Resonance
(*c*) Electrophilic addition
(*d*) Delocalisation of π-electrons. (*AFMC*)

47. Propane is reacted with HBr in presence of peroxide, the product is

(*a*) 2-Bromopropane (*b*) 1-Bromopropane
(*c*) 3-Bromopropane (*d*) None of these. **(*BHU*)**

48. Aromatic compounds undergo

(*a*) Nucleophilic addition reaction
(*b*) Electrophilic addition
(*c*) Electrophilic substitution
(*d*) None of these. **(*BHU*)**

49. Which reaction sequence would be best to prepare 3-choloroaniline from benzene ?

(*a*) Chlorination, nitration, reduction
(*b*) Nitration, chlorination, reduction
(*c*) Nitration, reduction, chlorination
(*d*) Nitration, reduction, acylation, chlorination, hydrolysis. **(*Pb. PMT*)**

50. A group which deactivates the benzene ring towards electrophilic substitution but directs the incoming group towards *o*- and *p*-positions is

(a) —NH_2 (b) —Cl
(c) —NO_2 (d) —C_2H_5. **(Pb. PMT)**

51. Which species represents the electrophile in aromatic nitration ?
(a) NO_2^- (b) NO_2^+
(c) NO_2 (d) NO_3^-. **(Pb. PMT)**

52. The principal organic product formed in the reaction

$CH_2 = CH(CH_2)_8COOH + HBr \xrightarrow{\text{Peroxide}} ...$ is
(a) $CH_3—CHBr(CH_2)_8COOH$
(b) $CH_2—CH(CH_2)_8COBr$
(c) $CH_2BrCH_2(CH_2)_8COOH$
(d) $CH_2 = CH(CH_2)_7CHBrCOOH$. **(Pb. PMT)**

53. The reaction,

Benzene + Methyl halide $\xrightarrow{\text{anhyd. } AlCl_3}$ Toluene, is known as
(a) Kolbe reaction (b) Wurtz reaction
(c) Perkin reaction (d) Friedel-Crafts reaction. **(AIIMS)**

54. The most reactive compound for electrophilic nitration is
(a) Benzene (b) Nitrobenzene
(c) Benzoic acid (d) Toluene. **(AIIMS)**

55. An unknown compound A has a molecular formula C_4H_6. When A is treated with excess of Br_2 a new substance B with formula $C_4H_6Br_4$ is formed. A forms a white ppt. with ammonical silver nitrate solution. A may be
(a) But-1-yne (b) But-2-yne
(c) But-1-ene (d) But-2-ene. **(MP PMT)**

56. Addition of HI on double bond of propane yields isopropyl iodide as major product. It is because the addition proceeds through
(a) More stable carobocation
(b) More stable carbanion
(c) More stable free radical
(d) Homolysis. **(Haryana CEET)**

57. The function of $AlCl_3$ in Friedel Craft's reaction is
(a) To absorb water (b) To absorb HCl
(c) To produce electrophile (d) To produce nucleophile. **(Haryana CEET)**

58. In methanol solution, bromine reacts with ethylene to yield $BrCH_2CH_2OCH_3$ in addition to 1, 2-dibromoethane because
(a) The intermediate carbocation may react with Br^- or CH_3OH
(b) The methyl alcohol solvolates the bromine
(c) The reaction follows Markownikoff's rule
(d) This is a free radical mechanism. **(Haryana CEET)**

59. Which one of the following compounds will be most easily attacked by an electrophile ?

(a) 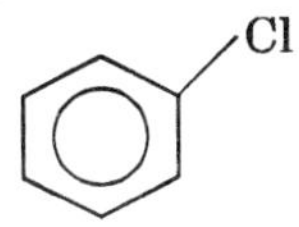 (b) 

(c) 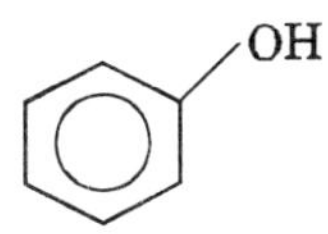 (d) 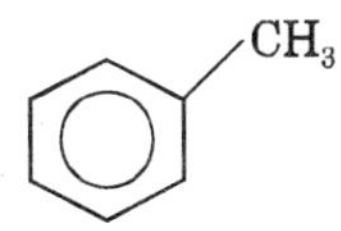

60. Benzene reacts with CH_3COCl to give
(a) Chlorobenzene (b) Toluene
(c) Benzyl chloride (d) Acetophenone. **(Rajasthan PMT)**

61. Acetylene when reacts with two molecules of HBr, then it gives
(a) $Br—C \equiv C—Br$ (b) $Br—CH = CH—Br$
(c) $Br_2C—CBr_2$ (d) $CH_3—CHBr_2$. **(Rajasthan PMT)**

62. Which of the following compound gives similar products obeying Markovnikoff's rule and peroxide effect ?
(a) $CH_3—CH = CH_2$ (b) $CH_3CH = CH—CH_3$
(c) $C_2H_5—CH = CH—CH_3$ (d) $C_2H_5—CH = CH_2$. **(Rajasthan PMT)**

63. Which one of the following compounds would have the highest heat of hydrogenation ?
(a) $CH_2 = CH_2$ (b) $CH_3—CH_2—CH = CH_2$
(c) $CH_3—CH = CHCH_3$ (d) $(CH_3)_2C = C(CH_3)_2$. **(JIPMER)**

1999

64. When acetylene is passed through dil H_2SO_4 in the presence of $HgSO_4$, the compound formed is
(a) Ether (b) Ketone
(c) Acetic acid (d) Acetaldehyde. **(CBSE PMT)**

65. In Friedel Craft's alkylation, besides $AlCl_3$, the other reactants are
(a) $C_6H_6 + NH_3$ (b) $C_6H_6 + CH_4$
(c) $C_6H_6 + CH_3Cl$ (d) $C_6H_6 + CH_3COCl$. **(CBSE PMT)**

66. Propyne on polymerisation gives
(a) Mesitylene (b) Benzene
(c) Ethyl benzene (d) Propyl benzene. **(CPMT)**

67. Which of the following catalysts is used in the polymerisation of $HC \equiv CH$ to C_6H_6 ?

(a) $AlCl_3$ (b) $HgSO_4$
(c) $NbCl_5$ (d) HCl. **(CPMT)**

68. $KMnO_4$, will oxidise acetylene to
(a) Ethylene glycol (b) Ethyl alcohol
(c) Oxalic acid (d) Acetic acid. **(CPMT)**

69. The propene reacts with HBr to form
(a) Propane (b) Ethane
(c) Hexane (d) 2-Bromopropane. **(AIIMS)**

70. Which of the following compounds will be most easily attacked by an electrophile ?

(a) 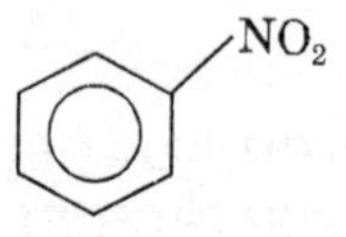 (b) 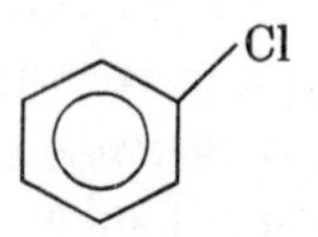

(c) 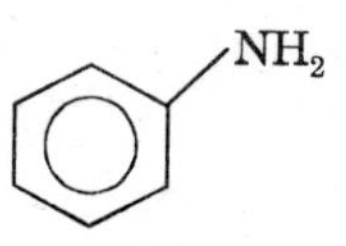 (d) 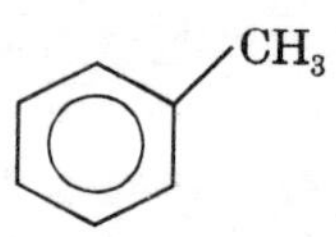

(CBSE PMT)

71. Heating a mixture of sodium benzoate and sodalime gives
(a) Benzene (b) Methane
(c) Sodium benzoate (d) Calcium benzoate. **(MP CEE)**

72. In its reaction with silver nitrate, acetylene shows
(a) Oxidising property (b) Reducing property
(c) Basic property (d) Acidic property. **(MP CEE)**

73. Octane number is zero for
(a) n-Heptane (b) *iso*-Octane
(c) n-Hexane (d) *iso*-Heptane. **(MP CEE)**

74. For preparing an alkane, a concentrated aqueous solution of sodium or potassium salt of a saturated carboxylic acid is subjected to
(a) Hydrolysis (b) Oxidation
(c) Hydrogenation (d) Electrolysis. **(MP CEE)**

75. Ethyl bromide on treatment with alcoholic KOH gives
(a) Ethylene (b) Ethanol
(c) Acetic acid (d) Ethane. **(MP CEE)**

76. Benzene reacts with chlorine to form benzenehexachloride in presence of
(a) Nickel (b) $AlCl_3$
(c) Bright sunlight (d) Zinc. **(MP CEE)**

77. Ozonolysis of 1, 3-butadiene gives
(a) Mixture of an aldehyde and a ketone
(b) Mixture of an aldehyde and an acid
(c) Only aldehydes
(d) Only ketones. **(Haryana CEET)**

78. Which of the following has the highest boiling point ?
(a) n-Pentane (b) Ethane
(c) n-Octane (d) n-Butane. **(AFMC)**

79. The reaction,
$CH_2 = CHCH_3 + HBr \longrightarrow CH_3CHBrCH_3$
is a type of
(a) Electrophilic addition reaction
(b) Nucleophilic addition reaction
(c) Free radical addition reaction
(d) Electrophilic substitution reaction. **(AFMC)**

2000

80. Which of the following methods is most appropriate for the manufacture of methane ?
(a) By reduction of CH_2Cl_2
(b) Wurtz reaction
(c) Liquefaction of natural gas
(d) None of these. **(CPMT)**

81. Which one is used as an antiknock in petrol fuel ?
(a) Basic lead carbonate (b) Lead tetraacetate
(c) Tetraethyl-lead (d) Basic lead sulphate. **(CPMT)**

82. Which of the following has highest octane number
(a) n-Hexane
(b) n-Heptane
(c) n-Pentane
(d) 2, 2, 4-Trimethylpentane. **(MP PMT)**

83. Tetrabromoethane on heating with Zn gives
(a) Ethyl bromide (b) Ethane
(c) Ethene (d) Ethyne. **(BHU)**

84. Arrange the following compounds in increasing order of reactivity towards the addition of HBr
$RCH = CHR$, $CH_2 = CH_2$, $R_2C = CHR$, $R_2C = CR_2$
(a) $CH_2 = CH_2 < RCH = CHR < R_2C = CHR < R_2C = CR_2$
(b) $R_2C = CHR < RCH = CHR < CH_2 = CH_2 < R_2C = CR_2$
(c) $RCH = CHR < R_2C = CR_2 < R_2C = CHR$
(d) $R_2C = CR_2 < R_2C = CHR < RCH = CHR$
(e) $R_2C = CR_2 < CH_2 = CH_2 < RCH = CHR$. **(Kerala CEE)**

85. How many monochlorobutanes will be obtained on chlorination of n-butane ?
(a) 1 (b) 2
(c) 3 (d) 4
(e) 5. **(Kerala CEE)**

86. Which of the following is most reactive towards electrophilic substitution ?
(a) $C_6H_5NO_2$ (b) C_6H_5Cl
(c) $C_6H_5CH_3$ (d) C_6H_6. **(Haryana CEET)**

87. When 2-butyne is treated with dil. $H_2SO_4/HgSO_4$, the product formed is

(*a*) 1-Butanol (*b*) Butanone
(*c*) 2-Butanol (*d*) Butanoic acid.

(Haryana CEET)

88. The order of activity of the various *o*- and *p*-director groups is

(*a*) $-O^- > -OH > -OCOCH_3 > -COCH_3$
(*b*) $OH > -O^- > -OCOCH_3 > -COCH_3$
(*c*) $OH > -O^- > -COCH_3 > -OCOCH_3$
(*d*) $-O^- > -COCH_3 > -OCOCH_3 > -OH$. ***(DCE)***

89. The conversion of 2, 3-dibromobutane to 2-butene with Zn and alcohol is

(*a*) Redox reaction
(*b*) α-Elimination
(*c*) β-Elimination
(*d*) Both α-elimination and redox reaction. ***(DCE)***

90. The conversion of $ClCH = CH—Cl$ to $Cl_2CH—CHCl_2$ can be carried out with

(*a*) Cl_2 (*b*) $Cl_2/h\nu$
(*c*) $Cl_2/AlCl_3$ (*d*) Cl_2/aq. NaOH. ***(DCE)***

91. 1, 2-Dibromopropane on treatment with X moles of $NaNH_2$ followed by treatment with ethyl bromide gave a pentyne. The value of X is

(*a*) One (*b*) Two
(*c*) Three (*d*) Four. ***(DCE)***

92. Among the following compounds, the decreasing order of reactivity towards electrophilic substitution is

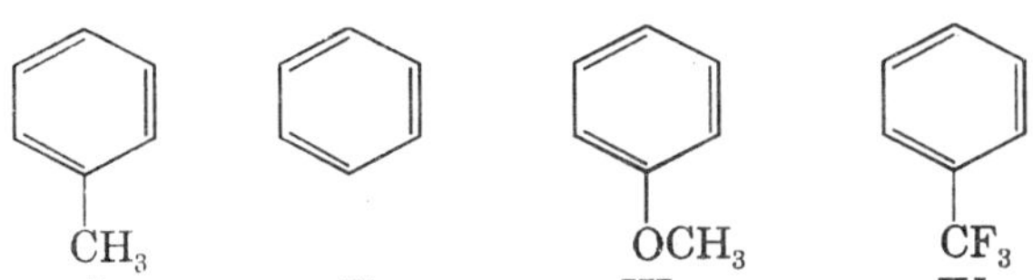

I II III IV

(*a*) III > I > II > IV (*b*) IV > I > II > III
(*c*) I > II > III > IV (*d*) II > I > III > IV.

(C3SE PMT)

93. In the Friedel-Crafts synthesis of toluene, reactants in addition to anhydrous $AlCl_3$ are

(*a*) $C_6H_6 + CH_4$ (*b*) $C_6H_6 + CH_3Cl$
(*c*) $C_6H_5Cl + CH_3Cl$ (*d*) $C_6H_5Cl + CH_4$.

(CBSE PMT)

2001

94. The reaction/method that does not give an alkane is

(*a*) Catalytic hydrogenation of alkenes
(*b*) Wurtz reaction
(*c*) Hydrolysis of alkymagnesium bromide
(*d*) Kolbe's electrolytic method
(*e*) Dehydrohalogenation of an alkyl halide.

(Kerala CEE)

95. A fuel has the same knocking property as a mixture of 70% isooctane (2, 2, 4-trimethylpentane) and 30% *n*-heptane by volume. The octane number of the fuel is

(*a*) 100 (*b*) 70
(*c*) 50 (*d*) 40
(*e*) 30. ***(Kerala PMT)***

96. In the following reaction,

$$C_2H_2 \xrightarrow[HgSO_4/H_2SO_4,]{H_2O,\ 60°C} X \rightleftharpoons CH_3CHO.$$

What is X ?

(*a*) CH_3CH_2OH (*b*) $CH_3—O—CH_3$
(*c*) CH_3CH_2CHO (*d*) $CH_2 = CHOH$.

(EAMCET)

97. Which alkene on ozonolysis gives CH_3CH_2CHO and $CH_3—CO—CH_3$?

(*a*) $CH_3CH_2CH = C(CH_3)_2$
(*b*) $CH_3CH_2CH = CHCH_2CH_3$
(*c*) $CH_3CH_2CH = CHCH_3$
(*d*) $CH_3—C(CH_3) = CHCH_3$ ***(CBSE PMT)***

98. In preparation of alkene from alcohol using Al_2O_3, which is effective factor ?

(*a*) Porosity of Al_2O_3 (*b*) Temperature
(*c*) Concentration (*d*) Surface area of Al_2O_3.

(CBSE PMT)

99. Pure methane can be produced by

(*a*) Wurtz reaction
(*b*) Kolbe's electrolytic method
(*c*) Soda-lime decarboxylation
(*d*) Reduction with H_2. ***(DCE)***

100. An organic compound, on treatment with Br_2 in CCl_4 gives bromoderivative of an alkene. The compound will be

(*a*) $CH_3—CH = CH_2$ (*b*) $CH_3CH = CHCH_3$
(*c*) $HC \equiv CH$ (*d*) $H_2C = CH_2$. ***(DCE)***

101. Position of double bond in alkenes can be identified by

(*a*) Bromine water
(*b*) Ammoniacal silver nitrate solution
(*c*) Ozonolysis
(*d*) None of these. ***(DCE)***

102. Hydrocarbon which is liquid at room temperature is
(*a*) Pentane (b) Butane
(c) Propane (*d*) Ethane.
(*AIIMS*)

103. What is formed when calcium carbide reacts with heavy water ?
(*a*) C_2D_2 (*b*) CaD_2
(*c*) Ca_2D_2O (*d*) CD_2. **(*Manipal*)**

104. Indicate the organic structure for product expected when 2-methylpropene is heated with acetyl chloride in the presence of anhydrous $ZnCl_2$?

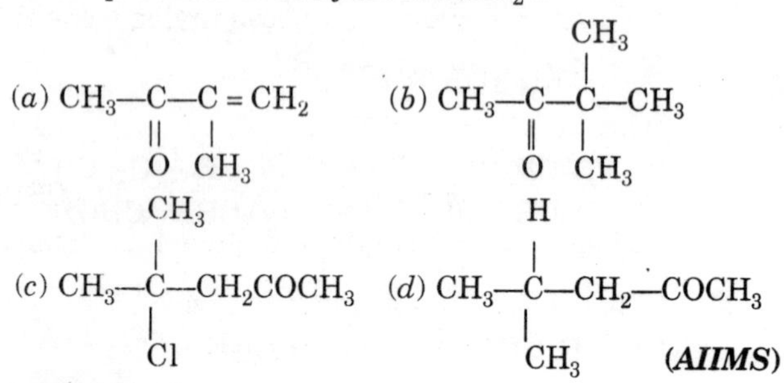

(*AIIMS*)

2002

105. Choose the compound which can react with $[Ag(NH_3)_2]^+$ and on treatment with alk. $KMnO_4$ gives $(CH_3)_3C—COOH$
(*a*) $CH_3CH_2CH_2—C \equiv C—CH_3$
(*b*) $(CH_3)_2CHCH_2—C \equiv CH$
(*c*) $(CH_3)_3C—C \equiv CH$
(*d*) $(CH_3)_3C—C \equiv C—CH_3$. **(*Rajasthan PMT*)**

106. A salt producing hydrocarbon among the compounds is
(*a*) Ethyne (*b*) Ethene
(*c*) Methane (*d*) Ethane. **(*KCET*)**

107. Octane number is zero by
(*a*) iso-heptane (*b*) *n*-Heptane
(*c*) Isoctane (*d*) *n*-Octane.
(*MP PET ; KCET*)

108. When $CH_3CH_2CHCl_2$ is treated with $NaNH_2$, the product formed is
(*a*) $CH_3—CH = CH_2$ (*b*) $CH_3—C \equiv CH$

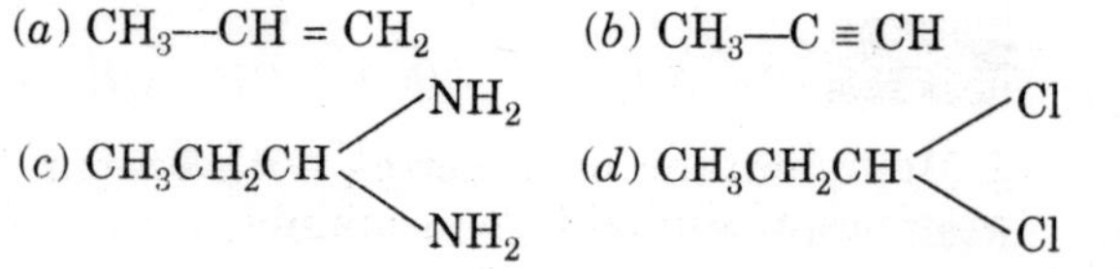

(*CBSE PMT*)

109. LPG contains
(*a*) Methane (*b*) Ethane
(*c*) Butane (d) None of the above.
(*BCEE*)

110. Wurtz reaction involves the interaction of alkyl halides in dry ether with
(*a*) Sodium (*b*) Zinc
(*c*) Copper (*d*) Platinum.
(*MP PET*)

111. Which of the following has lowest octane number ?
(*a*) iso-octane (*b*) *n*-Heptane
(*c*) *n*-Hexane (*d*) *n*-nonane. **(*MP PET*)**

112. The process in which higher hydrocarbons are broken down into lower hydrocarbons by controlled pyrolysis is called
(*a*) Hydrolysis (*b*) Cracking
(*c*) Oxidation (*d*) Redcution. **(*MP PET*)**

113. Ethane is formed by the reaction of methyliodide and sodium metal in dry ether solution. The reaction is known as :
(*a*) Clemmensen reduction (*b*) Kolbe's reaction
(*c*) Wurtz reaction (*d*) Cannizzaro reaction.
(*MP PMT*)

114. Which of these will not react with acetylene ?
(*a*) NaOH (*b*) ammoniacal $AgNO_3$
(*c*) Na (*d*) HCl. **(*AIEEE*)**

115. What is the product formed when acetylene reacts with hypochlorous acid ?
(*a*) CH_3COCl (*b*) $ClCH_2CHO$
(*c*) Cl_2CHCHO (*d*) $ClCH_2COOH$. **(*AIEEE*)**

116. For the reaction,
$CH_3—CH = CH_2 + HOCl \longrightarrow A$,
The product A is
(*a*) $CH_3—CHCl—CH_2OH$
(*b*) $CH_3—CH(OH)—CH_2Cl$
(*c*) $CH_3—CH_2—CH_2—COCl$

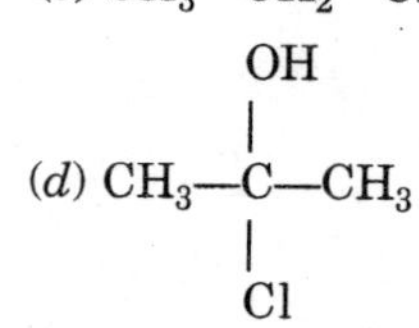

(*Orissa JEE*)

117. PVC is used for manufacture of
(*a*) Tyres (*b*) cosmetics
(*c*) Non-stick pans (*d*) Plastic pipes.
(*Orissa JEE*)

2003

118. Butene-1-may be converted to butane by reaction with
(*a*) Pd/H_2 (*b*) Zn – HCl
(*c*) Sn – HCl (*d*) Zn – Hg. **(*AIEEE*)**

119. On mixing a certain alkane with chlorine and irradiating it with ultraviolet light, it forms only one monochloroalkane. This alkane could be

(*a*) neopentane (*b*) propane
(*c*) pentane (*d*) isopentane. ***(AIEEE)***

120. The correct order of reactivity towards the electrophilic substitution of the compounds, aniline (I), benzene (II) and nitrobenzene (III), is
(*a*) II > III > I (*b*) I < II > III
(*c*) I > II > III (*d*) III > II > I.
(CBSE PMT)

121. Which one of the following is a free radical substitution reaction ?

(*a*) 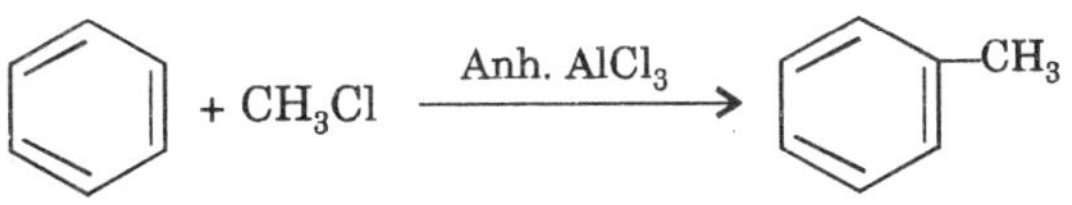

(*b*) 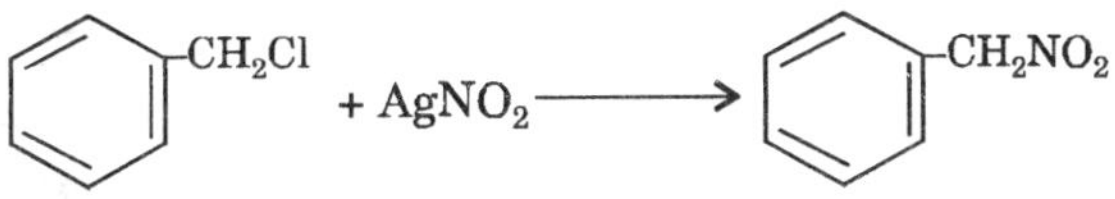

(*c*) $CH_3CHO + HCN \longrightarrow CH_3CH(OH)CN$

(*d*) 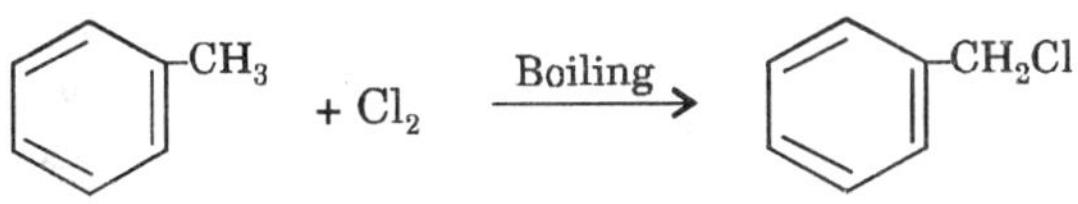

(CBSE PMT)

122. IUPAC name of 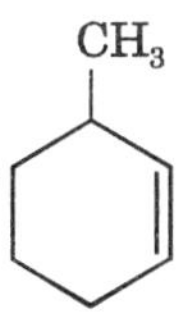 is
(*a*) 3-methyl cyclohexene
(*b*) 1-methyl cyclohex-2-ene
(*c*) 6-methyl cyclohexene
(*d*) 1-methyl cyclohex-5-ene. ***(AIIMS)***

123. The *ortho*/*para* directing group among the following is
(*a*) —COOH (*b*) —CN
(*c*) $—COCH_3$ (*d*) $—NHCOCH_3$. ***(AIIMS)***

124. The treatment of benzene with isobutene in the presence of sulphuric acid gives
(*a*) isobutyl benzene (*b*) *tert*-butyl benzene
(*c*) *n*-Butyl benzene (*d*) No reaction. ***(AIIMS)***

125. Name the alkene with the molecular formula $C_{10}H_{20}$
(*a*) Dodecene (*b*) Undecene
(*c*) Decene (*d*) Heptene
(*e*) Nonene. ***(Kerala MEE)***

126. In electrophilic substitution reactions nitrobenzene is
(*a*) *meta*-directing
(*b*) *ortho*-directing
(*c*) *para*-directing
(*d*) Not reactive and does not undergo any substitution
(*e*) Non-selective. ***(Kerala MEE)***

127. Which one of the following has the minimum boiling point ?
(*a*) *n*-Butane (*b*) isobutane
(*c*) 1-Butene (*d*) 1-Butyne.
(AIEEE)

128. Reaction of HBr with propene in the pressure of peroxide gives
(*a*) 3-bromo propane (*b*) allyl bromide
(*c*) *n*-propyl bromide (*d*) isopropyl bromide
(CBSE PMT)

129. Using anhydrous $AlCl_3$ as catalyst, which one of the following reactions produces ethylbenzene (PhEt) ?
(*a*) $CH_3—CH = CH_2 + C_6H_6$
(*b*) $H_2C = CH_2 + C_6H_6$
(*c*) $H_3C—CH_3 + C_6H_6$
(*d*) $H_3C—CH_2OH + C_6H_6$. ***(CBSE PMT)***

130. The molecular formula of diphenyl methane,

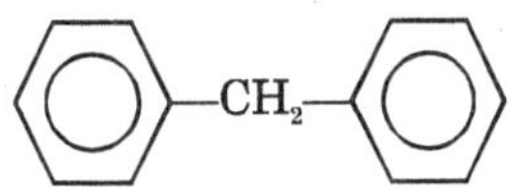 , is $C_{13}H_{12}$.

How many structural isomers are possible when one of the hydrogens is replaced by a chlorine atom ?
(*a*) 4 (*b*) 8
(*c*) 7 (*d*) 6. ***(CBSE PMT)***

131. The compound having only primary hydrogen atoms is :
(*a*) Isobutene (*b*) 2, 3-Dimethylbutene
(*c*) Cyclohexane (*d*) Propyne. ***(AIIMS)***

132. *o*-Toluic acid on reaction with Br_2 + Fe gives :

(*a*) 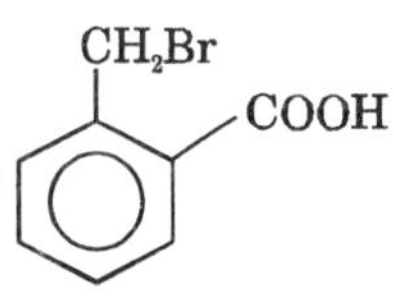 (*b*) 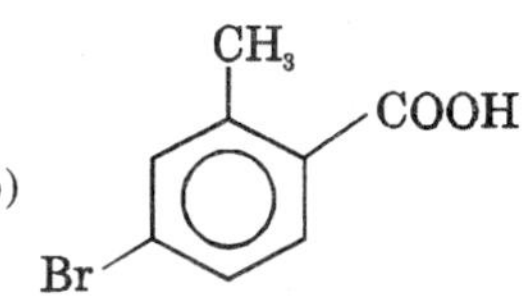

(*c*) 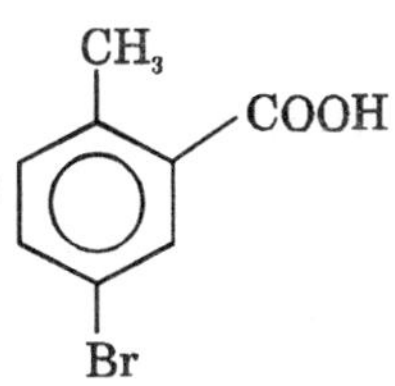 (*d*) 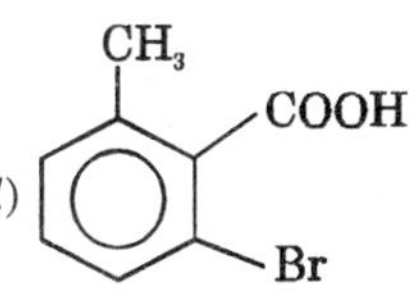

(AIIMS)

2005

133. 2-Methylbutane on reacting with bromine in the presence of sunlight gives mainly
(*a*) 1-bromo-3-methylbutane
(*b*) 2-bromo-3-methylbutane
(*c*) 2-bromo-2-methylbutane
(*d*) 1-bromo-2-methylbutane. ***(AIEEE)***

134. Reaction of one molecule of HBr with one molecule of 1, 3-butadiene at 40°C gives predominantly
(*a*) 1-bromo-2-butene under kinetically controlled conditions
(*b*) 3-bromobutene under thermodynamically controlled conditions
(*c*) 1-bromo-2-butene under thermodynamically controlled conditions
(*d*) 3-bromobutene under kinetically controlled conditions. **(*AIEEE*)**

135. Alkyl halides react with dialkyl copper reagents to give
(*a*) Alkenyl halides (*b*) Alkanes
(*c*) Alkyl copper halides (*d*) Alkenes. **(*AIEEE*)**

136. Of the five isomeric hexanes, the isomer which can give two monochlorinated compounds is
(*a*) 2-methylpentane (*b*) 2, 2-methylbutane
(*c*) 2, 3-dimethylbutane (*d*) n-hexane. **(*AIEEE*)**

137. Which one of the following alkenes will react faster with H_2 catalytic hydrogenation conditions ?

(*a*) 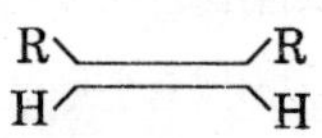 (*b*) 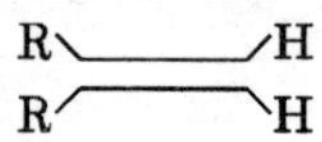

(*c*)  (*d*) 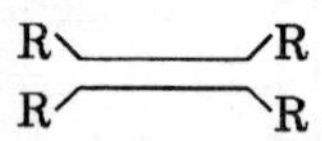

(*CBSE PMT*)

138. Product of the following reaction :

$$CH_3C \equiv C.CH_2CH_3 \xrightarrow[\text{(2) Hydrolysis}]{\text{(1) } O_3} \text{... are :}$$

(*a*) $CH_3COOH + CO_2$
(*b*) $CH_3COOH + HOOC.CH_2CH_3$
(*c*) $CH_3CHO + CH_3CH_2CHO$
(*d*) $CH_3COOH + CH_3COCH_3$. **(*CBSE PMT*)**

139. Which of the following gives propyne on hydrolysis ?
(*a*) Al_4C_3 (*b*) Mg_2C_3
(*c*) B_4C (*d*) La_4C_3. **(*AIIMS*)**

140. 3-Phenylpropene on reaction with HBr gives (as a major product)
(*a*) $C_6H_5CH_2CH(Br)CH_3$ (*b*) $C_6H_5CH(Br)CH_2CH_3$
(*c*) $C_6H_5CH_2CH_2CH_2Br$ (*d*) $C_6H_5CH(Br)CH = CH_2$. **(*AIIMS*)**

141. When toluene is treated with $KMnO_4$, what is produced ?
(*a*) Benzene (*b*) Chlorobenzene
(*c*) Benzaldehyde (*d*) Benzoic acid. **(*AFMC*)**

142. Which gas is liberated when Al_4C_3 is hydrolysed ?
(*a*) CH_4 (*b*) C_2H_2
(*c*) C_2H_6 (*d*) CO_2. **(*AFMC*)**

143. The product obtained on reaction of C_2H_5Cl with hydrogen over palladium carbon is
(*a*) C_3H_8 (*b*) C_4H_{10}
(*c*) C_2H_6 (*d*) C_2H_4. **(*AFMC*)**

144. By which one of the following compounds both CH_4 and CH_3—CH_3 can be prepared in one step ?
(*a*) CH_3I (*b*) CH_3OH
(*c*) CH_3CH_2I (*d*) C_2H_5OH. **(*BHU*)**

145. What will be the product in the following reaction ?

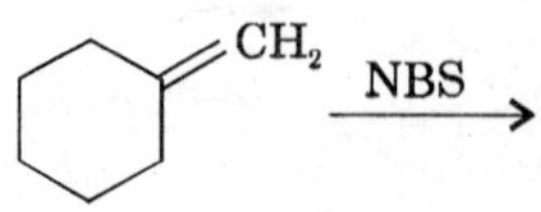

(*a*) 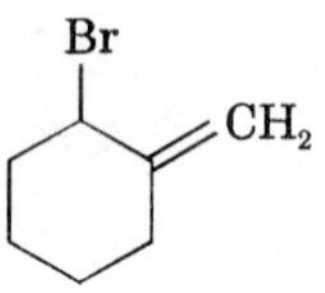 (*b*) 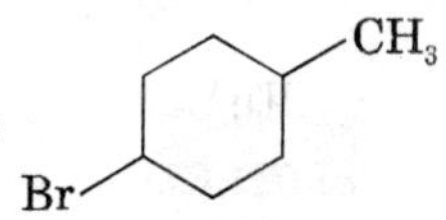

(*c*) 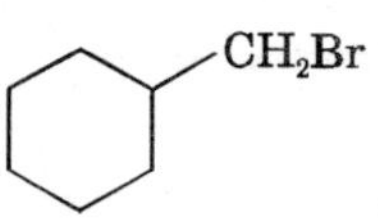 (*d*) 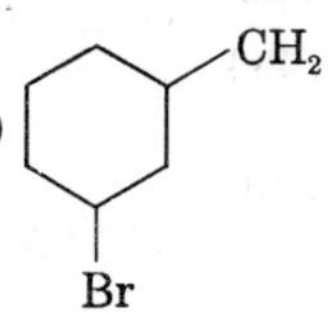 **(*BHU*)**

146. The chemicals and the reaction conditions required for the preparation of ethane are
(*a*) C_2H_5I, Zn-Cu, C_2H_5OH
(*b*) CH_3Cl, Na, H_2O
(*c*) KOOC–CH = CH–COOK, electrolysis
(*d*) CH_3CO_2Na, NaOH, CaO, Δ. **(*EAMCET*)**

147. The compound prepared by a substitution reaction of benzene is
(*a*) Acetophenone (*b*) Glyoxal
(*c*) Cyclohexane (*d*) Hexabromo cyclohexane. **(*EAMCET Engg.*)**

148. 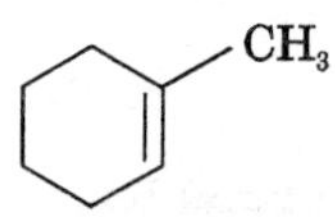 on reductive ozonolysis yields
(*a*) 6-oxoheptanal (*b*) 6-oxoheptanoic acid
(*c*) 6-hydroxyheptanal (*d*) 3-hydroxypentanal. **(*Orissa JEE*)**

149. Which of these does not follow Anti-Markownikoff's rule ?
(*a*) 2-butene (*b*) 1-butene
(*c*) 2-pentene (*d*) 2-hexene. **(*Orissa JEE*)**

150. Which reacts with ammoniacal $AgNO_3$?
(*a*) propyne (*b*) 2-butyne
(*c*) 1, 3-butadiene (*d*) 2-pentene. **(*Orissa JEE*)**

151. Colouration of Br_2/CCl_4 will be discharged by

(a) Cinnamic acid (b) Benzoic acid

(c) o-phthalic acid (d) Acetophenone.

(Orissa JEE)

152. $C_2H_2 + 2HCl \rightarrow C_2H_4Cl_2$ is an example of ...... reaction.

(a) Addition (b) Hydrogenation

(c) Substitution (d) Chlorination.

(EAMCET Medical)

153. 1, 2-Dibromoethane reacts with alcoholic KOH to yield a product X. The hybridisation state of the carbons present in X, respectively are :

(a) sp, sp (b) sp^3, sp^3

(c) sp^2, sp^2 (d) sp^3, sp^2.

(EAMCET Medical)

2006

154. Reaction of *trans* 2-phenyl-1-bromocyclopentane on reaction with alcoholic KOH produces

(a) 4-phenylcyclopentene (b) 2-phenylcyclopentene

(c) 1-phenylcyclopentene (d) 3-phenylcyclopentene.

(AIEEE)

155. Increasing order of stability among the three main conformations (*i.e.*, Eclipse, Anti, Cauche) of 2-fluroethanol is

(a) Eclipse, Gauche, Anti (b) Gauche, Eclipse, Anti

(c) Eclipse, Anti, Gauche (d) Anti, Gauche, Eclipse.

(AIEEE)

156. HBr reacts with $CH_2 = CH—OCH_3$ under anhydrous conditions at room temperature to give

(a) CH_3CHO and CH_3Br

(b) $BrCH_2CHO$ and CH_3OH

(c) $BrCH_2—CH_2—OCH_3$

(d) $H_3C—CHBr—OCH_3$. **(AIEEE)**

157. 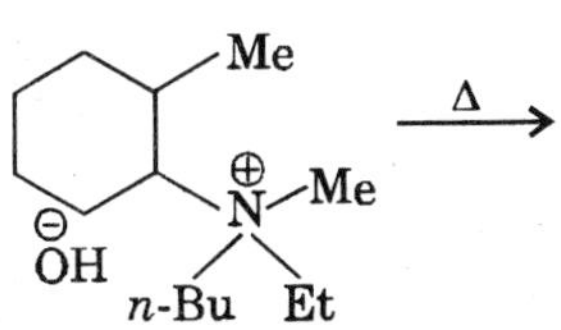

The alkene formed as a major product in the above elimination reaction is

(a) 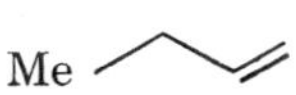 (b) $CH_2 = CH_2$

(c) 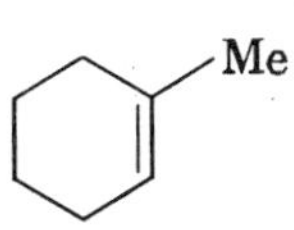 (d) 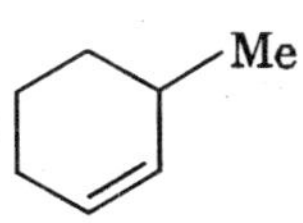 **(AIEEE)**

158. Which cycloalkane has the lowest heat of combustion per CH_2 group ?

(a) Cyclopropane (b) Cyclobutane

(c) Cyclopentane (d) Cyclohexane.

(Karnataka CET)

159. Methyl bromide is converted into ethane by heating it in ether medium with

(a) Al (b) Zn

(c) Na (d) Cu.

(Karnataka CET)

2007

160. The order of decreasing reactivity towards an electrophilic reagent, for the following :

(i) Benzene

(ii) Toluene

(iii) Chlorobenzene and

(iv) Phenol

would be :

(a) (iv) > (ii) > (i) > (iii) (b) (i) > (ii) > (iii) > (iv)

(c) (ii) > (iv) > (i) > (iii) (d) (iv) > (iii) > (ii) > (i).

(CBSE PMT)

161. Predict the product C obtained in the following reaction of butyne-1.

$$CH_3CH_2—C \equiv CH + HCl \longrightarrow B \xrightarrow{HI} C$$

(a) $CH_3CH_2—C(I)(Cl)—CH_3$

(b) $CH_3—CH(Cl)—CH_2CH_2I$

(c) $CH_3—CH_2—CH_2—C(I)(Cl)—H$

(d) $CH_3—CH_2—CH(I)—CH_2Cl$. **(CBSE PMT)**

162. Which of the compounds with molecular formula C_5H_{10} yields acetone on ozonolysis ?

(a) 2-Methyl-1-butene (b) 2-Methyl-2-butene

(c) 3-Methyl-1-butene (d) Cyclopentane.

(CBSE PMT)

163. The IUPAC name of 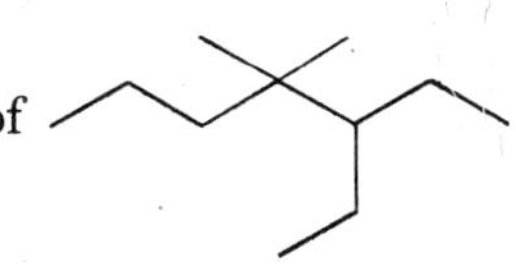 is (*a*) 5, 5-diethyl-4, 4-dimethyl pentane

(*b*) 3-ethyl-4, 4-dimethyl pentane

(*c*) 1, 1-diethyl-2, 2-dimethyl pentane

(*d*) 4, 4-dimethyl-5, 5-diethyl pentane. ***(AIEEE)***

164. The compound formed as a result of oxidation of ethyl benzene by $KMnO_4$ is :

(*a*) Benzoic acid (*b*) Benzyl alcohol

(*c*) Benzophenone (*d*) Acetophenone. ***(AIEEE)***

165. Which of the following reactions will yield 2, 2-dibromopropane ?

(*a*) $CH \equiv CH + 2HBr \longrightarrow$

(*b*) $CH_3—CH = CH_2 + HBr \longrightarrow$

(*c*) $CH_3—C \equiv CH + 2HBr \longrightarrow$

(*d*) $CH_3CH = CHBr + HBr \longrightarrow$ ***(AIEEE)***

166. The reaction of toluence with Cl_2 in presence of $FeCl_3$ gives predominantly :

(*a*) *o*- and *p*-chlorotoluene (*b*) *m*-chlorotoluene

(*c*) Benzoyl chloride (*d*) Benzyl chloride. ***(AIEEE)***

167. Presence of a nitro group in a benzene ring

(*a*) Deactivates the ring towards nucleophilic substitution

(*b*) Deactivates the ring towards electrophilic substitution

(*c*) Activates the ring towards electrophilic substitution

(*d*) Renders the ring basic. ***(AIEEE)***

ANSWERS

1. (*c*)	**2.** (*c*)	**3.** (*a*)	**4.** (*b*)	**5.** (*d*)	**6.** (*a*)	**7.** (*b*)	**8.** (*a*)	**9.** (*c*)
10. (*c*)	**11.** (*c*)	**12.** (*a*)	**13.** (*a*)	**14.** (*b*)	**15.** (*a*)	**16.** (*a*)	**17.** (*a*)	**18.** (*d*)
19. (*c*)	**20.** (*b*)	**21.** (*a*)	**22.** (*d*)	**23.** (*d*)	**24.** (*a*)	**25.** (*b*)	**26.** (*c*)	**27.** (*a*)
28. (*b*)	**29.** (*c*)	**30.** (*b*)	**31.** (*c*)	**32.** (*b*)	**33.** (*c*)	**34.** (*b*)	**35.** (*c*)	**36.** (*b*)
37. (*a*)	**38.** (*a*).	**39.** (*c*)	**40.** (*d*)	**41.** (*c*)	**42.** (*c*)	**43.** (*c*)	**44.** (*a*)	**45.** (*a*)
46. (*c*)	**47.** (*b*)	**48.** (*c*)	**49.** (*b*)	**50.** (*b*)	**51.** (*b*)	**52.** (*c*)	**53.** (*d*)	**54.** (*d*)
55. (*a*)	**56.** (*a*)	**57.** (*c*)	**58.** (*a*)	**59.** (*c*)	**60.** (*d*)	**61.** (*d*)	**62.** (*b*)	**63.** (*a*)
64. (*d*)	**65.** (*c*)	**66.** (*a*)	**67.** (*a*)	**68.** (*c*)	**69.** (*d*)	**70.** (*c*)	**71.** (*c*)	**72.** (*d*)
73. (*a*)	**74.** (*d*)	**75.** (*a*)	**76.** (*c*)	**77.** (*c*)	**78.** (*c*)	**79.** (*a*)	**80.** (*c*)	**81.** (*c*)
82. (*d*)	**83.** (*d*)	**84.** (*a*)	**85.** (*b*)	**86.** (*c*)	**87.** (*b*)	**88.** (*a*)	**89.** (*a*)	**90.** (*a*)
91. (*c*)	**92.** (*a*)	**93.** (*b*)	**94.** (*e*)	**95.** (*b*)	**96.** (*d*)	**97.** (*a*)	**98.** (*d*)	**99.** (*c*)
100. (*c*)	**101.** (*c*)	**102.** (*a*)	**103.** (*a*)	**104.** (*c*)	**105.** (*c*)	**106.** (*a*)	**107.** (*b*)	**108.** (*b*)
109. (*c*)	**110.** (*a*)	**111.** (*d*)	**112.** (*b*)	**113.** (*c*)	**114.** (*a*)	**115.** (*c*)	**116.** (*b*)	**117.** (*d*)
118. (*a*)	**119.** (*a*)	**120.** (*c*)	**121.** (*d*)	**122.** (*a*)	**123.** (*d*)	**124.** (*b*)	**125.** (*c*)	**126.** (*a*)
127. (*b*)	**128.** (*c*)	**129.** (*b*)	**130.** (*a*)	**131.** (*b*)	**132.** (*c*)	**133.** (*c*)	**134.** (*c*)	**135.** (*b*)
136. (*c*)	**137.** (*a*)	**138.** (*b*)	**139.** (*b*)	**140.** (*b*)	**141.** (*d*)	**142.** (*a*)	**143.** (*c*)	**144.** (*a*)
145. (*a*)	**146.** (*a*)	**147.** (*a*)	**148.** (*a*)	**149.** (*a*)	**150.** (*a*)	**151.** (*a*)	**152.** (*a*)	**153.** (*a*)
154. (*d*)	**155.** (*a*)	**156.** (*d*)	**157.** (*b*)	**158.** (*d*)	**159.** (*c*)	**160.** (*a*)	**161.** (*a*)	**162.** (*b*)
163. (*b*)	**164.** (*a*)	**165.** (*c*)	**166.** (*a*)	**167.** (*b*).				

HINTS/SOLUTIONS

Q. No.	*Ans.*	*Hints & Solutions*
1.	**(*c*)**	Factual question. Kerosene contains hydrocarbons having 12 to 16 carbon atoms.
2.	**(*c*)**	$H-C \equiv C-H + AsCl_3 \longrightarrow H-C=C-H$ (with Cl and $AsCl_2$ on the two carbons) Lewisite

Q. No.	*Ans.*	*Hints & Solutions*
3.	**(*a*)**	Light oil mainly contains benzene.
4.	**(*b*)**	The reaction is initiated by attack of electrophile H^+ and involves addition.
5.	**(*d*)**	Etard reaction.
6.	**(*a*)**	T.E.L. is tetraethyl lead.

Q. No.	Ans.	Hints & Solutions
7.	(*b*)	$CH_2=CH_2 + [O] + H_2O \longrightarrow \underset{OH}{CH_2}-\underset{OH}{CH_2}$
8.	(*a*)	Among hydrocarbons terminal alkynes are most acidic and hence have highest value of Ka.
9.	(*c*)	$H{-}C\equiv C{-}H + CH_3COOH \xrightarrow{-Hg^{2+}} \underset{H}{H{-}C}=\underset{OCOCH_3}{C{-}H}$ Vinyl acetate
10.	(*c*)	$CH_2=CH_2 \xrightarrow{Cl_2} \underset{Cl}{CH_2}-\overset{+}{C}H_2 \xrightarrow{Cl^-} \underset{Cl}{CH_2}-\underset{Cl}{CH_2}$; $\xrightarrow{Br^-} \underset{Cl}{CH_2}-\underset{Br}{CH_2}$ The intermediate carbocation may be attacked by Cl^- or Br^-.
11.	(*c*)	$F_2 > Cl_2 > Br_2$.
12.	(*a*)	Terminal alkynes are most acidic. *sp* hybridized carbon is more electronegative than sp^2 (in ethene) and sp^3 hybridized (in methane and ethane) carbon.
13.	(*a*)	$CH_3CH_2CH_2I \xrightarrow[-HI]{alc.\ KOH} CH_3CH=CH_2$
14.	(*b*)	Addition obeys Markownikoff rule.
15.	(*a*)	Cu^+ ion displaces hydrogen attached to the terminal C—C triple bond and gives red ppt. of copper (I) alkynide.
16.	(*a*)	Unsaturated hydrocarbons react with aqueous $KMnO_4$ but only terminal alkynes give precipitate with ammonical cuprous chloride.
17.	(*a*)	– COOH is meta directing group while the remaining are ortho and para directing.
18.	(*d*)	$CaC_2 \equiv C_2H_4$ 64 kg 28 kg
19.	(*c*)	$CH_3CH_2C\equiv CH + 4[O] \xrightarrow{alk.\ KMnO_4} CH_3CH_2COOH + CO_2$
20.	(*b*)	According to Saytzeff rule, the more substituted alkene is the preferred product.
21.	(*a*)	Straight chain isomers have higher boiling point.
22.	(*d*)	$-OCH_3$ is relatively stronger activating group. It releases electrons towards ring through resonance.

Q. No.	Ans.	Hints & Solutions
23.	(*d*)	2, 2, 4-Trimethylpentane is most branched. It has octane number of 100.
24.	(*a*)	Branched hydrocarbons have high octane number.
25.	(*b*)	$CH_2=CH_2 + HOCl \longrightarrow \underset{OH}{CH_2}-\underset{Cl}{CH_2} \xrightarrow{NaHCO_3} \underset{OH}{CH_2}-\underset{OH}{CH_2}$
26.	(*c*)	$-OCH_3$ activates while $-NO_2$ deactivates the ring towards electrophilic reactions.
27.	(*a*)	The given reaction is Kolbe's electrolytic reduction. The major organic product is R—R.
28.	(*b*)	In the reaction one of the H atoms is replaced by Cl atom.
29.	(*c*)	$2C_2H_5I + 2Na \longrightarrow C_2H_5-C_2H_5 + 2NaI$.
30.	(*b*)	Marsh gas is mainly methane.
31.	(*c*)	Factual question.
32.	(*b*)	This is Etard reaction.
33.	(*c*)	Butane, pentane and hexane are the three alkanes.
34.	(*b*)	Markownikoff addition.
35.	(*c*)	Wurtz reaction involves reaction of alkyl halide with sodium in the presence of dry ether.
36.	(*b*)	Anthracite has highest percentage of carbon.
37.	(*a*)	Benzene is present in the light oil fraction of distillation of coal tar.
38.	(*a*)	Under the given conditions allylic substitution takes place.
39.	(*c*)	The given reaction is ozonolysis.
40.	(*d*)	Self explanatory answer.
41.	(*c*)	Self explanatory answer.
42.	(*c*)	$CH_2=CH_2 + H_2O + [O] \longrightarrow HO-CH_2-CH_2-OH$.
43.	(*c*)	$2CH_3Br + 2Na \longrightarrow CH_3-CH_3 + 2NaBr$; $CH_3Br + 2[H] \xrightarrow{Zn/HCl} CH_4 + HBr$.
44.	(*a*)	Dehydrohalogenation of 1-chlorobutane gives 1-butene as the product.
45.	(*a*)	Friedel craft reaction.
46.	(*c*)	Arenes do not undergo addition easily.
47.	(*b*)	Anti-Markownikoff addition.
48.	(*c*)	Aromatic compounds undergo electrophilic substitution reactions.
49.	(*b*)	$-NO_2$ group is meta directing. So nitration is carried out first.
50.	(*b*)	– Cl deactivates the benzene ring due to its electron withdrawing inductive effect.

Q. No.	Ans.	Hints & Solutions
51.	(*b*)	NO_2^+ (Nitronium ion) is the electrophile during nitration.
52.	(*c*)	Anti-Markownikoff addition.
53.	(*d*)	The given reaction is Friedel crafts alkylation.
54.	(*d*)	CH_3-group activates the benzene ring towards electrophilic substitution.
55.	(*a*)	Reaction with ammonical $AgNO_3$ indicates that A is a terminal alkyne. Terminal alkyne with four carbon atoms is 1-butyne.
56.	(*a*)	$CH_3CH=CH_2 \xrightarrow[-I^-]{+HI} CH_3-\overset{+}{C}H-CH_3$ (More stable 2° carbocation) $\xrightarrow{+I^-} CH_3-\underset{\vert}{\underset{I}{C}}H-CH_3$
57.	(*c*)	$R-Cl + AlCl_3 \longrightarrow AlCl_4^- + R^+$.
58.	(*a*)	$CH_2=CH_2 + Br_2 \longrightarrow BrCH_2-\overset{+}{C}H_2 + Br^-$ $BrCH_2-\overset{+}{C}H_2 \xrightarrow{Br^-} BrCH_2CH_2Br$ $BrCH_2-\overset{+}{C}H_2 \xrightarrow{CH_3OH} BrCH_2-CH_2-\overset{+}{\underset{H}{O}}-CH_3 \xrightarrow{-H^+} BrCH_2CH_2OCH_3$
59.	(*c*)	—OH group is highly activating group.
60.	(*d*)	$C_6H_6 + CH_3COCl \xrightarrow{AlCl_3} C_6H_5-COCH_3 + HCl$ Friedel crafts acylation reaction.
61.	(*d*)	$H-C\equiv C-H + 2HBr \longrightarrow H_3C-CHBr_2$. Both the Br atoms go to the same carbon atom.
62.	(*b*)	It is a symmetrical alkene.
63.	(*a*)	$CH_2=CH_2$ being least substituted, is the least stable alkene.
64.	(*d*)	$H-C\equiv C-H \xrightarrow{+H_2O} [H_2C=CH-OH] \rightleftharpoons CH_3CHO$
65.	(*c*)	Benzene reacts with alkyl halides in the presence of $AlCl_3$ (Friedel craft's alkylation).
66.	(*a*)	Trimerization (cyclic) of propyne yields mesitylene (1, 3, 5-trimethyl benzene).
67.	(*a*)	$AlCl_3$.
68.	(*c*)	$HC\equiv CH + 4[O] \longrightarrow \begin{matrix} COOH \\ \vert \\ COOH \end{matrix}$ Oxalic acid
69.	(*d*)	Markownikoff addition.
70.	(*c*)	$—NH_2$ is highly activating group.
71.	(*a*)	$C_6H_5COONa \xrightarrow[(CaO)]{+NaOH} C_6H_6 + Na_2CO_3$ Soda-lime decarboxylation.
72.	(*d*)	In this reaction H^+ is displaced from acetylene.
73.	(*a*)	Octane number of *n*-heptane is zero.
74.	(*d*)	Kolbe's electrolysis.
75.	(*a*)	Dehydrohalogenation reaction.
76.	(*c*)	The addition reaction of benzene with chlorine takes place in the presence of sunlight.
77.	(*c*)	$CH_2=CH-CH=CH_2 \xrightarrow{\text{Ozonolysis}} 2\underset{\vert}{\underset{H}{H-C}}=O + O=\underset{\vert}{\underset{H}{C}}-\underset{\vert}{\underset{H}{C}}=O$
78.	(*c*)	*n*-Octane has the largest number of carbon atoms and hence has the highest boiling point.
79.	(*a*)	The reaction involves initial attack of electrophile and is an addition reaction.
80.	(*c*)	Natural gas is rich in methane and is available in abundance.
81.	(*c*)	Tetraethyl lead is an antiknocking agent.
82.	(*d*)	2, 2, 4-Trimethylpentane is highly branched alkane. The octane number increases with branching.
83.	(*d*)	$Br_2CH-CHBr_2 + 2Zn \longrightarrow CH\equiv CH + 2ZnBr_2$.
84.	(*a*)	More substituted alkene reacts via more stable intermediate carbocation and hence is most reactive.
85.	(*b*)	1-Chlorobutane and 2-chlorobutane. There are only two types of H-atoms in *n*-butane.
86.	(*c*)	CH_3—group is activating group.
87.	(*b*)	$CH_3-C\equiv C-CH_3 \xrightarrow{+H_2O} \left[CH_3-CH=\underset{\vert}{\underset{OH}{C}}-CH_3\right] \rightleftharpoons CH_3-CH_2-\underset{\Vert}{\underset{O}{C}}-CH_3$ Butanone

Q. No.	Ans.	Hints & Solutions
88.	(*a*)	Factual question. —O^- and —OH are activating while —$OCOCH_3$ and —$COCH_3$ are deactivating groups.
89.	(*a*)	$CH_3-CH(Br)-CH(Br)-CH_3 + Zn \longrightarrow CH_3-CH=CH-CH_3 + ZnBr_2$ It is a redox reaction. Zn is oxidized to Zn^{2+}.
90.	(*a*)	$CHCl=CHCl + Cl_2 \longrightarrow CH_2Cl_2\text{–}CHCl_2$.
91.	(*c*)	$CH_3-CH(Br)-CH_2Br + \mathbf{2NaNH_2} \longrightarrow CH_3C\equiv CH_3 + 2NaBr + 2NH_3$ $CH_3C\equiv CH + \mathbf{NaNH_2} \longrightarrow CH_3C\equiv CNa + NH_3$ $CH_3C\equiv CNa + C_2H_5Br \longrightarrow CH_3-C\equiv C-C_2H_5 + NaBr$
92.	(*a*)	III is most reactive because —OCH_3 group is strongly activating due to release of electrons towards ring through resonance. There is only one choice with (III) as the first.
93.	(*b*)	Self explanatory answer.
94.	(*e*)	Dehydrohalogenation of alkyl halides gives alkene as the product.
95.	(*b*)	By definition of octane number.
96.	(*d*)	$H-C\equiv C-H + H_2O \longrightarrow H-C(H)=C(OH)-H$
97.	(*a*)	$CH_3CH_2CH=C(CH_3)_2 \xrightarrow{\text{Ozonolysis}} CH_3CH_2CHO + O=C(CH_3)_2$
98.	(*d*)	Dehydration of alcohols takes place on heating with alumina.
99.	(*c*)	$CH_3COONa + NaOH \xrightarrow{CaO} Na_2CO_3 + CH_4$
100.	(*c*)	$H-C\equiv C-H + Br_2 \longrightarrow H-C(Br)=C(Br)-H$
101.	(*c*)	Ozonolysis helps in detection of position of double bond.
102.	(*a*)	Alkanes with 5 to 17 carbon atoms are liquid at room temperature.
103.	(*a*)	$CaC_2 + D_2O \longrightarrow Ca(OD)_2 + D-C\equiv C-D$.
104.	(*c*)	$CH_3-C(CH_3)=CH_2 + CH_3-C(=O)-Cl \xrightarrow{ZnCl_2} CH_3-C(Cl)(CH_3)-CH_2-C(=O)-CH_3$ This is an example of electrophilic addition. The electrophile that initiates the addition is CH_3-CO^+. Addition obeys Markownikoff's rule.
105.	(*c*)	$(CH_3)_3C-C\equiv CH + 4[O] \longrightarrow (CH_3)_3C-COOH + CO_2$
106.	(*a*)	Ethyne has ionizable H-atoms (acidic H atoms).
107.	(*b*)	Factual question. Octane number of *n*-heptane is zero.
108.	(*b*)	Propyne is formed through removal of two molecules of HBr.
109.	(*c*)	Major component of LPG is butane.
110.	(*a*)	Wurtz reaction takes place by reaction of alkyl halide with sodium in the presence of dry ether.
111.	(*d*)	Octane number decreases with increase in carbon chain length.
112.	(*b*)	Self explanatory answer.
113.	(*c*)	Self explanatory answer.
114.	(*a*)	NaOH cannot remove H^+ from acetylene.
115.	(*c*)	$H-C\equiv C-H + 2HOCl \longrightarrow \left[H-C(Cl)(Cl)-C(OH)(OH)-H\right] \longrightarrow H-C(Cl)(Cl)-C(=O)-H$
116.	(*b*)	Markownikoff's addition. HOCl splits as OH^- and Cl^+.
117.	(*d*)	PVC is used for making plastic pipes.
118.	(*a*)	Hydrogenation of butene-1 in the presence of palladium gives butane.
119.	(*a*)	All the hydrogen atoms in *neo*-pentane are equivalent.
120.	(*c*)	—NH_2 group is activating while —NO_2 group is deactivating.
121.	(*d*)	Benzylic chlorination proceeds via free radical chain mechanism.
122.	(*a*)	3-Methylcyclohexene.

Q. No.	Ans.	Hints & Solutions
123.	(*d*)	$—NHCOCH_3$ is *ortho-para* directing and activating group.
124.	(*b*)	$CH_2{—}\underset{\large CH_3}{\underset{\vert}{C}} = CH_2 + H^+ \longrightarrow CH_3{—}\underset{\large CH_3}{\underset{\vert}{\overset{+}{C}}}{—}CH_3$ Tertiary butyl carbocation attacks on benzene to give tert-butyl benzene.
125.	(*c*)	An alkene with 10 carbon atoms is decene.
126.	(*a*)	$—NO_2$ group is meta directing.
127.	(*b*)	Iso-butane is a branched chain hydrocarbon.
128.	(*c*)	In the presence of peroxides, addition of HBr to alkenes takes place in anti-Markownikoff fashion $CH_3CH = CH_2 + HBr \longrightarrow CH_3CH_2CH_2Br$
129.	(*b*)	$C_6H_6 + CH_2 = CH_2 \xrightarrow{AlCl_3} C_6H_5{—}CH_2CH_3$
130.	(*a*)	Only four structural isomers are possible on monochlorine because the given molecule contains four different types of nonequivalent hydrogen atoms. The four iosmers are : 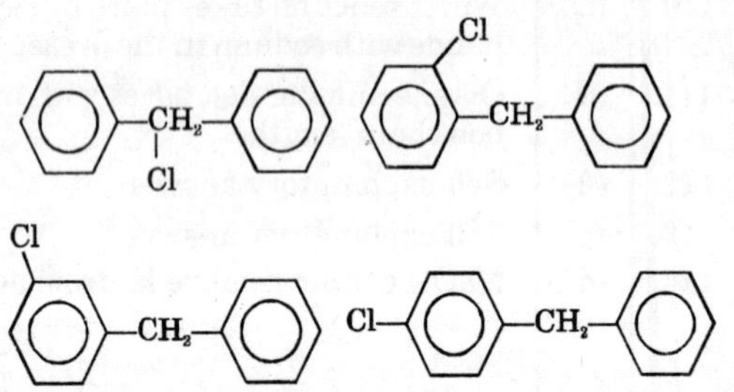
131.	(*b*)	2, 3-Dimethylbutene contains only primary hydrogen atoms $(CH_3)_2C{=}C(CH_3)_2$
132.	(*c*)	Substitution at the ring takes place at a position which is *meta* w.r.t. —COOH group and *para* w.r.t. $—CH_3$ group.
133.	(*c*)	$CH_3{—}\underset{\large CH_3}{\underset{\vert}{C}H}{—}CH_2{—}CH_3 + Br_2 \xrightarrow{\text{Sunlight}}$ $CH_3{—}\underset{\large CH_3}{\underset{\vert}{\overset{\large Br}{\overset{\vert}{C}}}}{—}CH_2{—}CH_3 + HBr$ Ease of substitution of H atoms is 3° > 2° > 1°.
134.	(*c*)	At high temperature the reaction is thermodynamically controlled and 1, 4-addition product is the major product.
		$CH_2{=}CH{—}CH{=}CH_2 \xrightarrow{HBr} CH_3{—}CH{=}CH_2{—}CH_2Br$
135.	(*b*)	$R_2\,CuLi + R'X \longrightarrow R{–}R' + RCu + LiX$ This is Corey-House reaction.
136.	(*c*)	2, 3-Dimethylbutane contains two types of H atoms 1° and 2°. $CH_3{—}\underset{\large H}{\underset{\vert}{\overset{\large CH_3}{\overset{\vert}{C}}}}{—}\underset{\large H}{\underset{\vert}{\overset{\large CH_3}{\overset{\vert}{C}}}}{—}CH_3$
137.	(*a*)	The molecule can approach adsorbed hydrogen more effectively. Moreover, this is less stable alkene.
138.	(*b*)	$C \equiv C$, triple bond undergoes cleavage to form CH_3COOH and CH_3CH_2COOH.
139.	(*b*)	Mg_2C_3 (Magnesium allylide) gives propyne on hydrolysis.
140.	(*b*)	$C_6H_5{—}CH_2{—}CH = CH_2 + HBr$ $\longrightarrow C_6H_5{—}CH_2{—}\underset{\large Br}{\underset{\vert}{C}H}{—}CH_3$ The addition is in accordance with Markownikov's rule.
141.	(*d*)	$KMnO_4$ oxidizes side chains in arenes to —COOH group.
142.	(*a*)	Aluminium carbide on hydrolysis gives methane.
143.	(*c*)	C_2H_5Cl is reduced to ethane (C_2H_6) by H_2 in the presence of catalyst.
144.	(*a*)	$2CH_3I + 2Na \xrightarrow{\text{Wurtz reaction}} CH_3CH_3 + 2NaI$ $CH_3I + H_2 \xrightarrow[\text{Reduction}]{Pd} CH_4 + HI.$
145.	(*a*)	NBS brings about allylic bromination.
146.	(*a*)	C_2H_5I is reduced to ethane on reaction with Zn—Cu couple and ethanol.
147.	(*a*)	Acetophenone is prepared by Friedel Craft acylation of benzene $C_6H_6 + CH_3COCl \xrightarrow{AlCl_3} C_6H_5COCH_3 + HCl.$
148.	(*a*)	$CH_3{—}O{—}CH_2{—}CH_2{—}CH_2{—}CHO$ $\Vert$ C is the ozonolysis product of the given compound.
149.	(*a*)	It is a symmetrical alkene. Both the doubly bonded carbon atoms are equivalent in 2-butene.
150.	(*a*)	Terminal alkynes react with ammoniacal $AgNO_3$.
151.	(*a*)	Cinnamic acid contains a carbon-carbon double bond in the side chain. 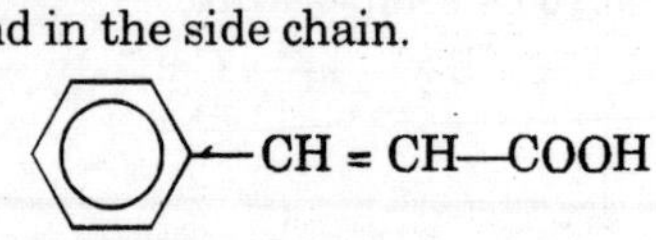

Q. No.	Ans.	Hints & Solutions
152.	(*a*)	It is an addition reaction.
153.	(*a*)	$\underset{\text{Br}}{\underset{\mid}{CH_2}}—\underset{\text{Br}}{\underset{\mid}{CH_2}} \xrightarrow{\text{alc. KOH}} \underset{[X]}{HC \equiv CH}$. Both the carbon atoms is ethyne are *sp* hybridized.
154.	(*d*)	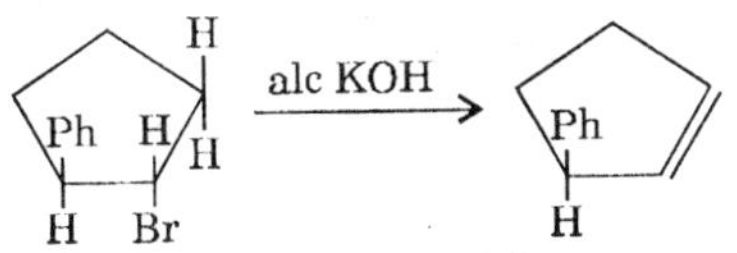 H and Br are removed from opposite sides.
155.	(*a*)	Eclipse < Gauche < Anti.
156.	(*d*)	$H_3C—\underset{\text{Br}}{\underset{\mid}{CH}}—OCH_3$ is formed through more stable intermediate carbocation, $H_3C—\overset{+}{C}H—OCH_3$ which is stabilized by resonance. $H_3C—\overset{+}{C}H—OCH_3 \leftrightarrow H_3C—CH=\overset{+}{O}CH_3$.
157.	(*b*)	This is Hoffmann elimination. The less substituted alkene is the preferred product.

Q. No.	Ans.	Hints & Solutions
158.	(*d*)	Cyclohexane due to its maximum stability.
159.	(*c*)	Wurtz reaction.
160.	(*a*)	Phenol is highly activated ring whereas chlorobenzene is least reactive.
161.	(*a*)	Addition follows Markownikov's rule.
162.	(*b*)	$(CH_3)_2C=CH—CH_3$, 2-Methyl-2-butene.
163.	(*b*)	3-ethyl-4, 4-dimethylheptane.
164.	(*a*)	During oxidation with $KMnO_4$, the side chain in attached to benzene ring is oxidized to —COOH group.
165.	(*c*)	$CH_3—C \equiv CH + 2HBr \longrightarrow CH_3—\underset{\text{Br}}{\overset{\text{Br}}{C}}—CH_3$
166.	(*a*)	CH_3— group in toluene is ortho-para directing. Hence, chlorination of toluene takes place at *ortho-para* positions.
167.	(*b*)	Nitro group deactivates the ring towards electrophilic substitution due to its electron with drawing resonance effect.

2.6. MCQs FOR IIT JEE

Brain Twister

A. Multiple Choice Questions With One Correct Answer

1. Electrolysis of an aqueous solution of potassium fumarate yields
(a) Ethane (b) Ethene
(c) Ethyne (d) Methane.

2. Hydrolysis of Mg_2C_3 yields
(a) Propane (b) Propene
(c) Propyne (d) Ethyne.

3. Addition of oxygen gas to a reaction mixture of chlorine and methane (photochemical chlorination)
(a) Accelerates the reaction
(b) Retards the reaction for sometime
(c) Does not affect the rate of the reaction
(d) Slows down the rate of the reaction permanently.

4. When an aqueous solution containing sodium acetate and sodium propionate is electrolysed, we can get
(a) Ethane (b) Propane
(c) Butane (d) All the above.

5. Which of the following has the highest octane number ?
(a) Triptane (b) Octane
(c) Isooctane (d) Heptane.

6. 1, 2-Dibromoethane when heated with zinc dust yields
(a) Ethane (b) Ethene
(c) Ethyne (d) Methane.

7. $CH_3 - CHCl_2 \xrightarrow{\text{alc. KOH}} X \xrightarrow{NaNH_2} Y$
The compound Y in the above sequence is
(a) Ethyne (b) Vinyl chloride
(c) Ethene (d) Ethyl amine.

8. Amongst the following compounds which can be most easily sulphonated is
(a) Benzene (b) Nitrobenzene
(c) Toluene (d) Chlorobenzene.

9. The compound with highest melting point is
(a) *n*-Hexane (b) *n*-Pentane
(c) 2, 2-Dimethylpropane (d) 2-Methylbutane.

10. A compound $X(C_5H_8)$ reacts with ammonical $AgNO_3$ solution to give a white precipitate, and on oxidation with hot alkaline $KMnO_4$ gives the acid $(CH_3)_2CHCOOH$. Compound X may be
(a) 3-Methyl-1-butyne (b) 3-Methyl-1-butene
(c) 2-Methyl-1, 3-butadiene (d) 1-Pentyne.

11. Which one of the following compounds does not dissolve in conc. H_2SO_4 even on warming ?
(a) Ethylene (b) Benzene
(c) Hexane (d) Aniline.

12. Conjugated double bonds are present in
(a) Isoprene (b) Chloroprene
(c) Both of these (d) None of these.

13. A hydrocarbon on ozonolysis yields 2, 3-butanedione and acetone as the products of ozonolysis. The hydrocarbon may be
(a) 2, 3, 4, 5-Tetramethyl-2, 4-hexadiene
(b) 2, 5-Dimethyl-2, 4-hexadiene
(c) 2, 4-Dimethyl-2, 4-hexadiene
(d) 2, 5-Dimethyl-1, 5-hexadiene.

14. Which the following alkenes is the most reactive towards electrophilic addition reactions ?
(a) $H_2C = CH_2$ (b) $(CH_3)_2C = C(CH_3)_2$
(c) $CH_3CH = CH_2$ (d) $CH_3CH = CHCH_3$.

15. Electrolysis of aqueous solution of potassium succinate yields
(a) Ethane (b) Ethene
(c) Ethyne (d) Butane.

16. Cyclohexene on ozonolysis yields
(a) CH_3CH_2CHO (2 moles)
(b) $CH_3 - \overset{O}{\overset{\|}{C}} - CH_3$ (2 moles)
(c) $CH_3CH_2CHO + CH_2 - \overset{O}{\overset{\|}{C}} - CH_3$
(d) $OHC - (CH_2)_4 - CHO$.

17. In the acid catalysed dehydration of alcohols to alkenes, the intermediate species formed is
(a) Free radical (b) Carbocation
(c) Carbanion (d) Carbene.

18. Acetylene when passed through chlorine water yields
(a) Dicholoroethanal (b) 1, 2-Dichloroethene
(c) Dichloroethanoic acid (d) Ethylidene dichloride.

19. Aluminium carbide on reaction with water yields
(a) Ethene (b) Ethyne
(c) Ethane (d) Methane.

20. Which of the following represents the correct order of acidic character ?

(*a*) $H_2O > HC \equiv CH > C_2H_6 > C_2H_4$

(*b*) $H_2O > HC \equiv CH > C_2H_4 > C_2H_6$

(*c*) $HC \equiv CH > C_2H_4 > C_2H_6 > H_2O$

(*d*) $HC \equiv CH > H_2O > C_2H_4 > C_2H_6$.

21. 2-Butyne on reaction with H_2 in the presence of Lindlar's catalyst yields

(*a*) *n*-Butane (*b*) *cis*-2-Butene

(*c*) *trans*-2-Butene (*d*) Ethane.

22. Which of the following alkenes has the lowest value of heat of hydrogenation ?

(*a*) Ethene (*b*) Propene

(*c*) *cis*-2-Butene (*d*) *trans*-2-Butene.

23. Addition of HBr to propene in the presence of benzoylperoxide takes place via .... as intermediates.

(*a*) Free radicals (*b*) Carbocations

(*c*) Carbanions (*d*) Carbenes.

24. Which of the following represents the correct order of stability of dienes ?

(*a*) Conjugated > Isolated > Cumulated

(*b*) Conjugated > Cumulated > Isolated

(*c*) Isolated > Cumulated > Conjugated

(*d*) Isolated > Conjugated > Cumulated.

25. Which one of the following statements is not true ?

(*a*) Peroxide effect is applicable to HBr and not to other hydrogen halides

(*b*) Chlorination of methane follows free radical chain mechanism

(*c*) All the meta directing groups on the benzene ring are also deactivating groups

(*d*) Propyne is stronger acid than ethyne.

26. Aqueous sulphuric acid reacts with 2-Methyl-1-butene to give predominantly

(*a*) 2-Methyl-2-butanol (*b*) 3-Methyl-2-butanol

(*c*) 3-Methyl-1-butanol (*d*) 3-Methyl-2-butanol.

27. A hydrocarbon X decolourises bromine water. It reacts with one mole of H_2 to give another saturated hydrocarbon. X reacts with acidified $KMnO_4$ to give two moles of propanoic acid. X could be

(*a*) $CH_3CH_2CH_2CH = CHCH_3$

(*b*) $CH_3CH_2CH = CHCH_2CH_3$

(*c*) $CH_3 - \underset{\displaystyle CH_3}{\underset{|}{CH}} - CH = CH - CH_3$

(*d*) $CH_3 - \underset{\displaystyle CH_3}{\underset{|}{C}} = \underset{\displaystyle CH_3}{\underset{|}{C}} - CH_3$

28. $CH_3 - CH_2 - CH = CH_2 \xrightarrow{\text{NBS}} X \xrightarrow{\text{alc. KOH}} Y$
The compound Y is

(*a*) But-3-en-2-ol (*b*) But-3-en-1-ol

(*c*) 1, 3-Butadiene (*d*) 1, 2-Butadiene.

29. Ethylbenzene reacts with Cl_2 in sunlight to give

(*a*) 1-Chloro-1-phenylethane

(*b*) 1-Chloro-2-phenylethane

(*c*) 4-Chloroethylbenzene

(*d*) 3-Chloroethylbenzene.

30. Which of the following is not a meta directing group ?

(*a*) – CN (*b*) – CHO

(*c*) $-NO_2$ (*d*) $-OCH_3$.

31. Which of the following is obtained as the product when excess benzene reacts with CH_2Cl_2 in the presence of anhydrous $AlCl_3$.

(*a*) 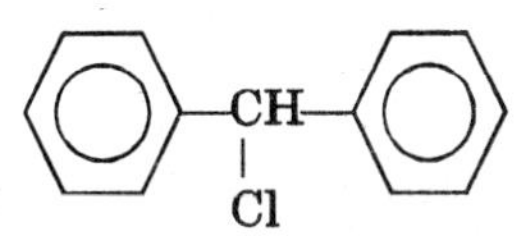

(*b*) 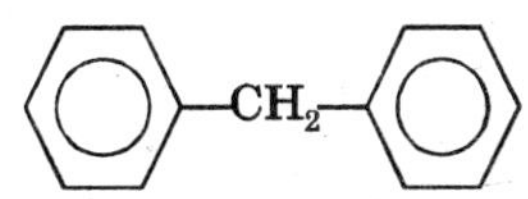

(*c*) 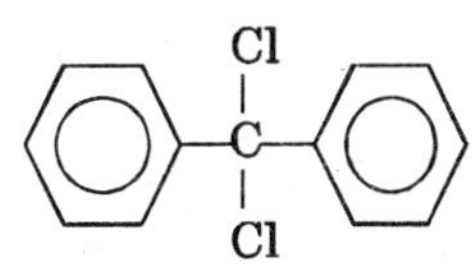

(*d*) 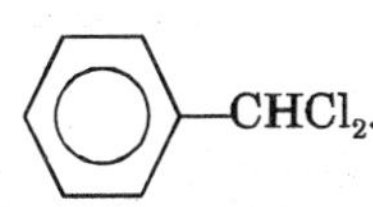

32. 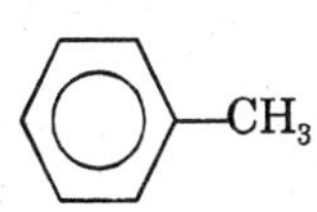 $\xrightarrow{KMnO_4}$ A $\xrightarrow{HNO_3/H_2SO_4}$ B (*Major product*)

The product B is

(*a*) 4-Nitrotoluene (*b*) 3-Nitrotoluene

(*c*) 3-Nitrobenzoic acid (*d*) 4-Nitrobenzoic acid.

33. Benzene reacts with *iso*-butyl chloride in the presence of anhydrous $AlCl_3$ to give

(*a*) *tert*-Butylbenzene (*b*) *iso*-Butylbenzene

(*c*) *n*-Butylbenzene (*d*) *sec*-Butylbenzene.

34. *neo*-Pentylbromide on reaction with alc. KOH yields ...... as the major product

(*a*) 2-Methyl-1-butene

(*b*) 2-Methyl-2-butene

(*c*) 2, 2-Dimethyl-1-propanol

(*d*) 2-Methyl-2-butanol.

35. $CH_3 - CH = CH_2 \xrightarrow[\text{Peroxide}]{\text{HBr}} X \xrightarrow[\text{ether}]{\text{Na}} Y$

The compound Y in the above sequence is

(*a*) Propane (*b*) 2, 3-Dimethylbutane
(*c*) *n*-Hexane (*d*) 2, 4-Hexadiene.

36. *iso*-Butyl chloride on reaction with sodium in the presence of ether yields

(*a*) 3, 4-Dimethylhexane (*b*) 2, 5-Dimethyhexane
(*c*) *n*-Octane (*d*) *iso*-Octane.

37. The reaction of toluene with chlorine in the presence of ferric chloride gives predominantly

(*a*) *m*-chlorotoluene (*b*) *o*- and *p*-chlorotoluene
(*c*) Benzal chloride (*d*) Benzyl chloride.

38. Cyclohexene is treated with hot acidified $KMnO_4$ solution. The product obtained is

(*a*) 1, 2-Cyclohexanediol (*b*) Cyclohexanone
(*c*) Adipic acid (*d*) Benzoic acid.

39. Which isomer of C_5H_{12} gives four monochloro substituted products ?

(*a*) *n*-Pentane (*b*) *iso*-Pentane
(*c*) *neo*-Pentane (*d*) none of these.

40. Which of the following reactions proceeds through carbocations as intermediate ?

(*a*) $CH_3CH_3 + Cl_2 \xrightarrow[\text{Sunlight}]{\text{Diffused}}$

(*b*) $CH_3 - CH = CH_2 + Br_2 \xrightarrow[\text{Peroxides}]{\text{800 K}}$

(*c*) $CH_3CH = CH_2 + HBr \longrightarrow$

(*d*) 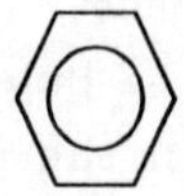 $+ Cl_2 \xrightarrow{FeCl_3}$

B. MCQs From Previous IIT Examinations

41. *n*-Propyl bromide on treatment with ethanolic potassium hydroxide produces

(*a*) Propanol-1 (*b*) Propene
(*c*) Propanol-2 (*d*) Ethyl propyl ether.

(*IIT 1987*)

42. A mixture of ethyl iodide and *n*-propyl iodide is subjected to Wurtz reaction. The hydrocarbon which will not be formed is

(*a*) Butane (*b*) Propane
(*c*) Pentane (*d*) Hexane.

(*IIT 1990*)

43. Towards electrophilic substitution the most reactive will be

(*a*) Nitrobenzene (*b*) Aniline
(*c*) Aniline hydrochloride (*d*) N-Acetylaniline.

(*IIT 1991*)

44. 1-Chlorobutane, on reaction with alcoholic potash (KOH), gives

(*a*) 1-Butene (*b*) 1-Butanol
(*c*) 2-Butene (*d*) 2-Butanol. (*IIT 1991*)

45. Ozonolysis of 2, 3-dimethyl-1-butene followed by reduction with zinc and water gives

(*a*) Methanoic acid and 3-methyl-2-butanone
(*b*) Methanal and 2-methyl-2-butanone
(*c*) Methanal and 3-methyl-2-butanone
(*d*) Methanoic acid and 2-methyl-2-butanone.

(*IIT 1992*)

46. Which one of the following has the smallest heat of hydrogenation per mole ?

(*a*) 1-Butene (*b*) *trans*-2-Butene
(*c*) *cis*-2-Butene (*d*) 1, 3-Butadiene.

(*IIT 1993*)

47. What is the decreasing order of strength of the bases :

$$OH^-, NH_2^-, HC \equiv C^- \text{ and } CH_3CH_2^-$$

(*a*) $CH_3CH_2^- > NH_2^- > HC \equiv C^- > OH^-$

(*b*) $HC \equiv C^- > CH_3CH_2^- > NH_2^- > OH^-$

(*c*) $OH^- > NH_2^- > HC \equiv C^- > CH_3CH_2^-$

(*d*) $NH_2^- > HC \equiv C^- > OH^- > CH_3CH_2^-$. (*IIT 1993*)

48. When cyclohexane is poured on water, it floats, because

(*a*) Cyclohexane is in 'boat' form
(*b*) Cyclohexane is in 'chair' form
(*c*) Cyclohexane is in 'crown' form
(*d*) Cyclohexane is less denser than water. (*IIT 1997*)

49. The intermediate during the addition of HCl to propene in the presence of peroxide is

(*a*) $CH_3\dot{C}HCH_2Cl$ (*b*) $CH_3\overset{+}{C}HCH_3$
(*c*) $CH_3CH_2\dot{C}H_2$ (*d*) $CH_3CH_2\overset{+}{C}H_2$.

(*IIT 1997*)

50. $(CH_3)_3$ CMgCl on reaction with D_2O produces

(*a*) $(CH_3)_3$ CD (*b*) $(CH_3)_3$ COD

(*c*) $(CD)_3$ CD (*d*) $(CD)_3$ COD. ***(IIT 1997)***

51. The reaction of

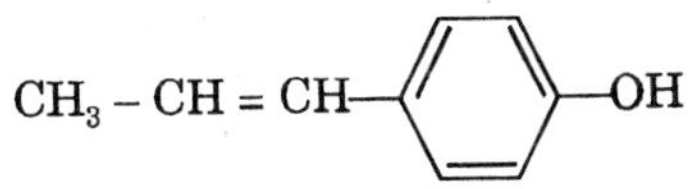

with HBr gives

(*a*) 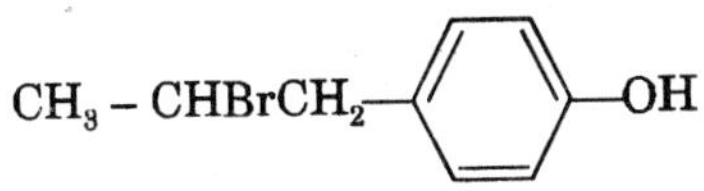

(*b*) 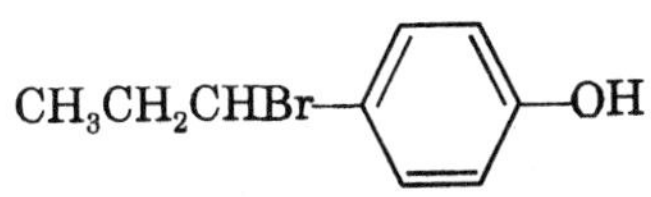

(*c*) 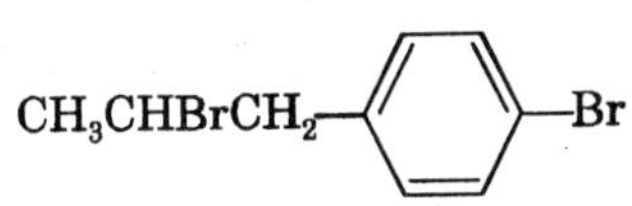

(*d*) 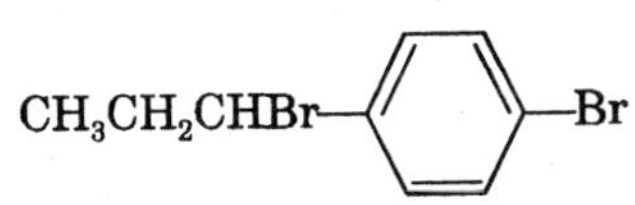

(IIT 1998)

52. Benzyl chloride ($C_6H_5CH_2Cl$) can be prepared from toluene by chlorination with

(*a*) SO_2Cl_2 (*b*) $SOCl_2$

(*c*) Cl_2 (*d*) NaOCl. ***(IIT 1998)***

53. Which of the following is strongest acid ?

(*a*) $CH \equiv CH$ (*b*) C_6H_6

(*c*) C_2H_6 (*d*) CH_3OH. ***(IIT 1998)***

54. The product(s) obtained via oxymercuration ($HgSO_4$ + H_2SO_4) of 1-butyne would be

(*a*) $CH_3—CH_2—\overset{\overset{O}{||}}{C}—CH_3$

(*b*) $CH_3—CH_2—CH_2—CHO$

(*c*) $CH_3—CH_2—CHO—HCHO$

(*d*) CH_3CH_2COOH + HCOOH. ***(IIT 1999)***

55. Toluene, when treated with Br_2/Fe, gives *p*-bromotoluene as the major product because the —CH_3 group

(*a*) is *para* directing

(*b*) is *meta* directing

(*c*) activates the ring by hyperconjugation

(*d*) deactivates the ring. ***(IIT 1999)***

56. Propyne and propene can be distinguished by

(*a*) conc. H_2SO_4 (*b*) Br_2 in CCl_4

(*c*) dil. $KMnO_4$ (*d*) $AgNO_3$ in ammonia.

(IIT Screening Test 2000)

57. The reaction of propene with HOCl proceeds via the addition of

(*a*) H^+ in the first step

(*b*) Cl^+ in the first step

(*c*) OH^- in the first step

(*d*) Cl^+ and OH^- in a single step.

(IIT Screening Test 2001)

58. In presence of peroxide, hydrogen chloride and hydrogen iodide do not give anti-Markovnikov's addition to alkenes because

(*a*) Both are highly ionic

(*b*) One is oxidising and the other is reducing

(*c*) One of the steps is endothermic in both the cases

(*d*) All the steps are exothermic in both the cases.

(IIT Screening Test 2001)

59. Identify the reagent from the following list which can easily distinguish between 1-butyne and 2-butyne

(*a*) Bromine, CCl_4

(*b*) H_2, Lindlar catalyst

(*c*) Dilute H_2SO_4, $HgSO_4$

(*d*) Ammonical Cu_2Cl_2 solution.

(IIT Screening Test 2002)

60. Identify the correct order of reactivity in electrophilic substitution reactions of the following compounds :

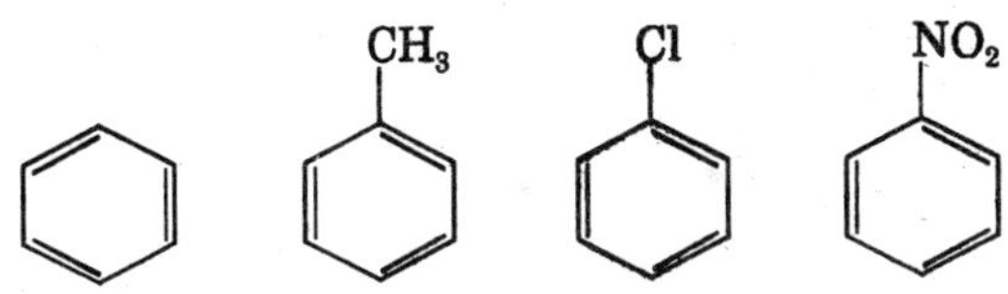

1 2 3 4

(*a*) 1 > 2 > 3 > 4 (*b*) 4 > 3 > 2 > 1

(*c*) 2 > 1 > 3 > 4 (*d*) 2 > 3 > 1 > 4.

(IIT Screening Test 2002)

61. Which of the following has the lowest dipole moment ?

(*a*) 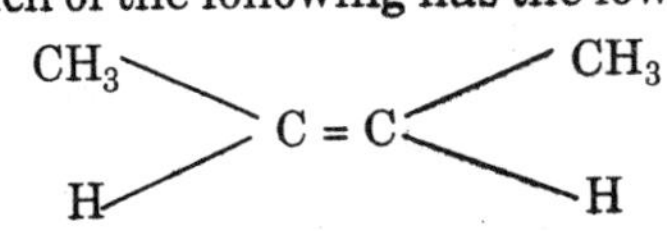

(*b*) $CH_3C \equiv CCH_3$

(*c*) $CH_3CH_2C \equiv CH$

(*d*) $CH_2 = CH—C \equiv CH$. ***(IIT Screening Test 2002)***

62. Consider the following reaction,

$$H_3C—\underset{\underset{D}{|}}{CH}—\underset{\underset{CH_3}{|}}{CH}—CH_3 + \overset{\bullet}{Br} \longrightarrow \text{'X'} + HBr$$

Identify the structure of the major product 'X'

(*a*) $CH_3—\underset{\underset{D}{|}}{CH}—\underset{\underset{CH_3}{|}}{CH}—\overset{\bullet}{C}H_2$

(*b*) $CH_3—CH—\dot{C}—CH_2$ (with D on CH and CH_3 on C)

(*c*) $CH_3—\dot{C}—CH—CH_3$ (with D on C and CH_3 on CH)

(*d*) $CH_3—\dot{C}H—CH—CH_3$ (with CH_3 on CH)

(IIT Screening Test 2002)

63. Which of the following represents the given mode of hybridisation $sp^2 - sp^2 - sp - sp$ from left to right ?

(*a*) $H_2C = CH—C \equiv CH$ (*b*) $HC \equiv C—C \equiv CH$

(*c*) $H_2C = C = C = CH_2$ (*d*) 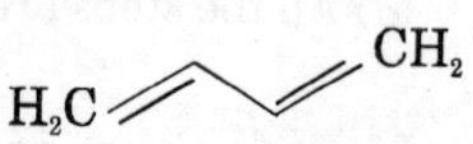

(IIT Screening Test 2003)

64. $Ph—C \equiv C—CH_3 \xrightarrow{Hg^{2+}/H^+} A$. A is :

(*a*) 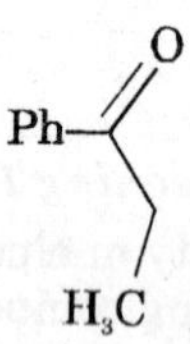 (*b*) 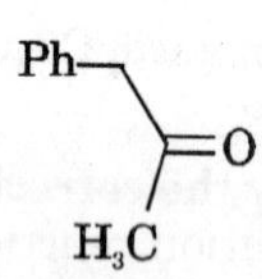

(*c*) 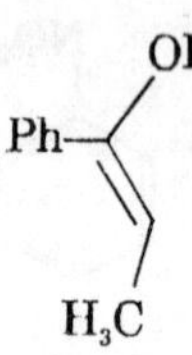 (*d*) 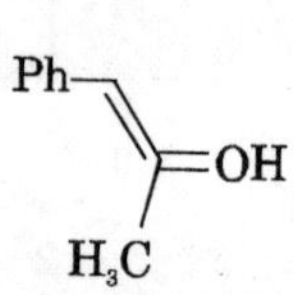

(IIT Screening Test 2003)

65. 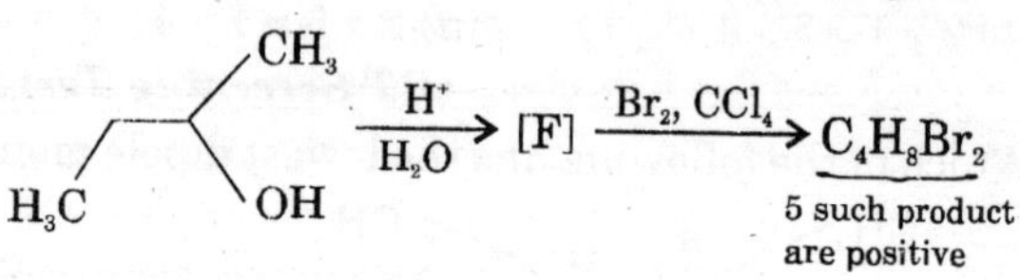

How many structures of F is possible ?

(*a*) 2 (*b*) 5

(*c*) 6 (*d*) 3

(IIT Screening Test 2003)

66. The major product obtained when Br_2/Fe is treated with

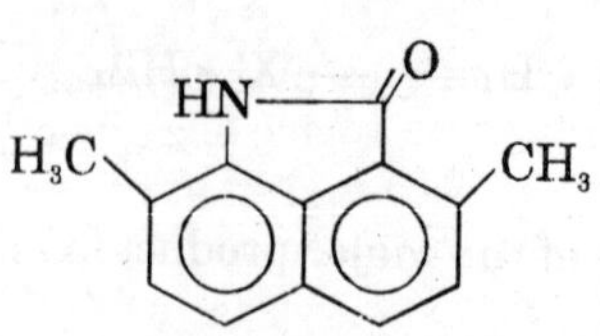 is

(*a*) 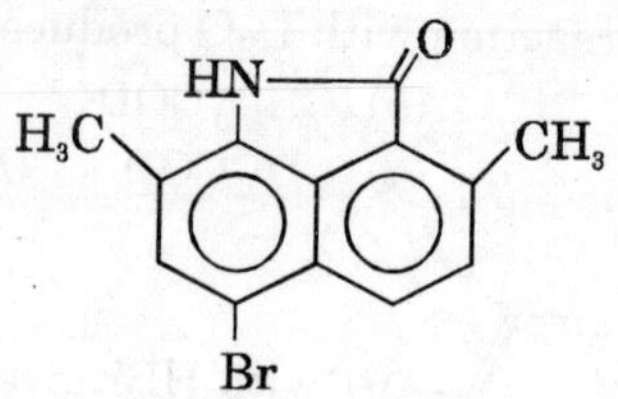

(*b*) 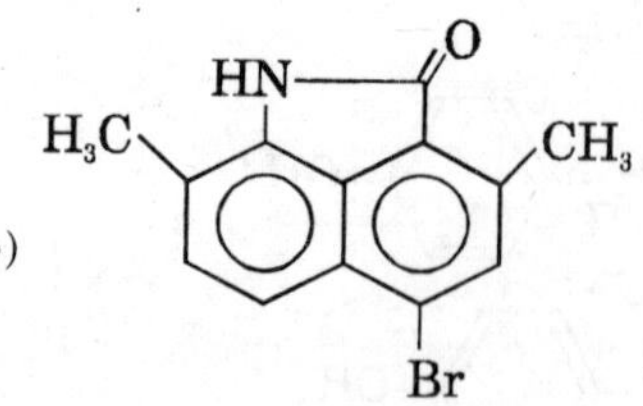

(*c*) 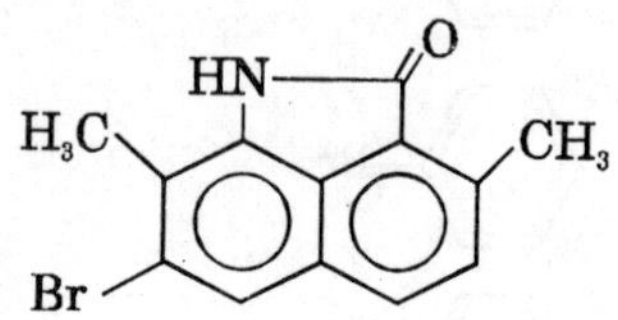

(*d*) 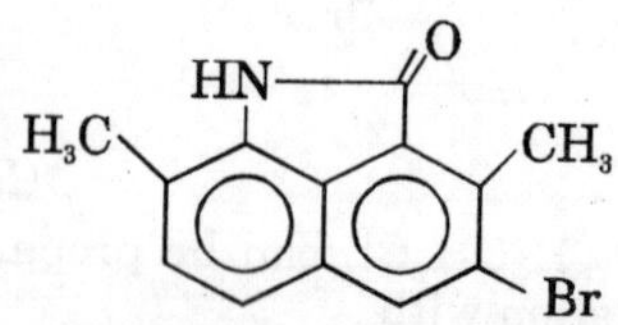

(IIT Screening Test 2003)

67. Which of the following is used for the conversion of 2-hexyne into trans-2-hexene ?

(*a*) $H_2/Pd/BaSO_4$ (*b*) H_2, PtO_2

(*c*) $NaBH_4$ (*d*) $Li-NH_3/C_2H_5OH$.

(IIT Screening Test 2003)

68. Cyclohexane is best prepared from cyclohexanol by which of the following

(*a*) conc. H_3PO_4 (*b*) conc. HCl

(*c*) conc. HBr (*d*) conc. $HCl/ZnCl_2$.

(IIT Screening Test 2005)

69. 1-Bromo-3-chlorocyclobutane when treated with two equivalent of sodium in the presence of dry ether which of the following will be formed ?

(*a*) 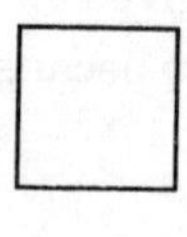 (*b*) 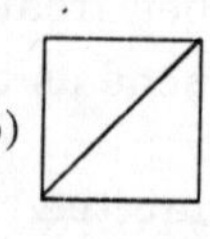

(*c*) 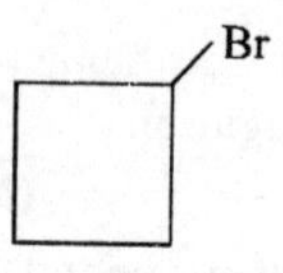 (*d*) 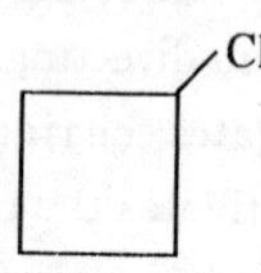

(IIT Screening Test 2005)

70. $CH_3—CH = CH_2 \xrightarrow{NOCl} (A)$

which of the following is the structure of compound (A) ?

(*a*) $CH_3—CH(NO)—CH_2Cl$

(*b*) $ON—CH_2—CH_2—CH_2Cl$

(*c*) $CH_3—CHCl—CH_2NO$

(*d*) $CH_2Cl—CH(NO)—CH_2CH_3$. ***(IIT JEE 2006)***

71. $(CH_3)_2CH—CH_2—CH_3 + Cl_2 \xrightarrow{h\nu}$ 'N' isomers of $C_5H_{11}Cl$ $\xrightarrow{\text{Fractional distillation}}$ 'M' distilled products

The values of 'N' and 'M' are respectively

(*a*) 6, 6 (*b*) 6, 4

(*c*) 4, 4 (*d*) 3, 3. ***(IIT JEE 2006)***

72. The reagent(s) for the following conversion,

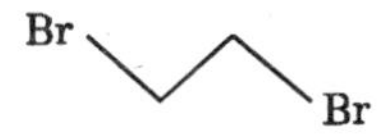

$\xrightarrow{?}$ H—≡—H is/are

(*a*) Alcoholic KOH

(*b*) Alcoholic KOH followed by $NaNH_2$

(*c*) Aqueous KOH followed by $NaNH_2$

(*d*) Zn/CH_3OH. ***(IIT JEE 2007)***

73. The number of stereoisomers obtained by bromination of *trans*-2-butene is

(*a*) 1 (*b*) 2

(*c*) 3 (*d*) 4. ***(IIT JEE 2007)***

74. The number of structural isomers for C_6H_{14} is

(*a*) 3 (*b*) 4

(*c*) 5 (*d*) 6. ***(IIT JEE 2007)***

ANSWERS

1. (*c*)	**2.** (*c*)	**3.** (*b*)	**4.** (*d*)	**5.** (*a*)	**6.** (*b*)	**7.** (*a*)	**8.** (*c*)	**9.** (*c*)
10. (*a*)	**11.** (*c*)	**12.** (*c*)	**13.** (*a*)	**14.** (*b*)	**15.** (*b*)	**16.** (*d*)	**17.** (*b*)	**18.** (*a*)
19. (*d*)	**20.** (*b*)	**21.** (*b*)	**22.** (*d*)	**23.** (*a*)	**24.** (*a*)	**25.** (*d*)	**26.** (*a*)	**27.** (*b*)
28. (*c*)	**29.** (*a*)	**30.** (*d*)	**31.** (*b*)	**32.** (*c*)	**33.** (*a*)	**34.** (*b*)	**35.** (*c*)	**36.** (*b*)
37. (*b*)	**38.** (*c*)	**39.** (*b*)	**40.** (*d*)	**41.** (*b*)	**42.** (*b*)	**43.** (*b*)	**44.** (*a*)	**45.** (*c*)
46. (*d*)	**47.** (*a*)	**48.** (*d*)	**49.** (*b*)	**50.** (*a*)	**51.** (*b*)	**52.** (*a, c*)	**53.** (*d*)	**54.** (*a*)
55. (*a, c*)	**56.** (*d*)	**57.** (*b*)	**58.** (*c*)	**59.** (*d*)	**60.** (*c*)	**61.** (*b*)	**62.** (*b*)	**63.** (*a*)
64. (*a*)	**65.** (*d*)	**66.** (*a*)	**67.** (*d*)	**68.** (*a*)	**69.** (*b*)	**70.** (*c*)	**71.** (*b*)	**72.** (*b*)
73. (*a*)	**74.** (*c*).							

HINTS/SOLUTIONS

Q. No.	*Ans.*	*Hints & Solutions*
1.	**(*c*)**	$\begin{matrix} HC—COOK \\ \| \\ HC—COOK \end{matrix} + 2H_2O \xrightarrow{\text{Electrolysis}} HC \equiv CH + 2CO_2 + 2KOH + H_2$
2.	**(*c*)**	$Mg_2C_3 + 4H_2O \longrightarrow CH_3—C \equiv CH + 2Mg(OH)_2$
3.	**(*b*)**	O_2 reacts with alkyl radical to form a new less reactive radical and hence retards the reaction. Once all the oxygen is consumed, the reaction starts taking place at normal speed. $R + O_2 \longrightarrow R - O_2$.

Q. No.	*Ans.*	*Hints & Solutions*
4.	**(*d*)**	Sodium acetate gives methyl radicals and sodium propionate gives ethyl radicals during electrolysis
5.	**(*a*)**	Triptane (2, 2, 3-Trimethylpentane) has octane no. 116 which is highest among the given hydrocarbons.
6.	**(*b*)**	Dehalogenation of 1, 2-dibromoethane yields ethene $BrCH_2 - CH_2Br + Zn \xrightarrow{\Delta} CH_2 = CH_2 + ZnBr_2$

Q. No.	Ans.	Hints & Solutions
7.	(*a*)	X is vinyl chloride and Y is ethyne. The reaction is dehydrohalogenation.
8.	(*c*)	CH_3 – group activates benzene ring towards electrophilic substitution.
9.	(*c*)	2, 2-Dimethyl propane with spherical molecules has the most close packing in crystalline lattice and hence has the highest melting point.
10.	(*a*)	Formation of $(CH_3)_2CHCOOH$ indicates the presence of $(CH_3)_2CHC\equiv$ group in the compound whereas formation of white ppt. with ammonical $AgNO_3$ indicates the presence of terminal triple bond.
11.	(*c*)	Hexane does not react with conc. H_2SO_4 even on warming.
12.	(*c*)	$CH_2=C(CH_3)-CH=CH_2$ Isoprene; $CH_2=C(Cl)-CH=CH_2$ Chloroprene
13.	(*a*)	$(CH_3)_2C=O \quad O=C(CH_3)-C(CH_3)=O \quad O=C(CH_3)_2$ $(CH_3)_2C=C(CH_3)-C(CH_3)=C(CH_3)_2$ 2, 3, 4, 5-Tetramethyl-2, 4-hexadiene.
14.	(*b*)	The reaction in case of $(CH_3)_2C=C(CH_3)_2$ takes place through most stable intermediate carbocation.
15.	(*b*)	$\begin{matrix}CH_2COOK\\ \vert\\ CH_2COOK\end{matrix} + 2H_2O \xrightarrow{\text{Electrolysis}} \begin{matrix}CH_2\\ \Vert\\ CH_2\end{matrix} + CO_2 + H_2 + 2KOH$
16.	(*d*)	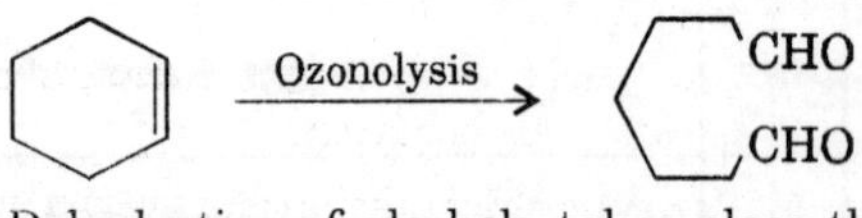
17.	(*b*)	Dehydration of alcohols takes place through carbocations as intermediates.
18.	(*a*)	$H-C\equiv C-H + 2HOCl \longrightarrow H-C(Cl)_2-C(OH)_2-H \xrightarrow{-H_2O} H-C(Cl)_2-C(=O)-H$
19.	(*d*)	$Al_4C_3 + 12H_2O \longrightarrow 4Al(OH)_3 + 3CH_4$.
20.	(*b*)	$H_2O > HC\equiv CH > CH_2=CH_2 > C_2H_6$.
21.	(*b*)	Hydrogenation of alkynes in the presence of Lindlar's catalyst yields *cis*-alkene.
22.	(*d*)	*trans*-2-Butene is most stable among the given alkenes and hence has least value of heat of hydrogenation.
23.	(*a*)	Self explanatory answer.
24.	(*a*)	Conjugated dienes are most stable due to reasonance while cumulated dienes are least stable due to repulsion between π electrons.
25.	(*d*)	Among alkynes, ethyne is the strongest acid. Substituted alkynes are less acidic due to destabilization of carbanion, formed by loss of H^+, through electron releasing inductive effect of alkyl groups.
26.	(*a*)	$CH_3CH_2C(CH_3)=CH_2 + H_2O \xrightarrow{H^+} CH_3CH_2-C(OH)(CH_3)-CH_3$ Addition obeys Markownikoff's rule.
27.	(*b*)	$CH_3CH_2CH \vdots\!= CHCH_2CH_3 + 2[O] \xrightarrow{KMnO_4/H^+} 2CH_3CH_2COOH$
28.	(*c*)	NBS brings about allylic bromination. $CH_3CH_2CH=CH_2 \xrightarrow{NBS} CH_3-CH(Br)-CH=CH_2 \xrightarrow{\text{alc. KOH}} CH_2=CH-CH=CH_2$.
29.	(*a*)	Benzylic hydrogen is more easily subsitituted 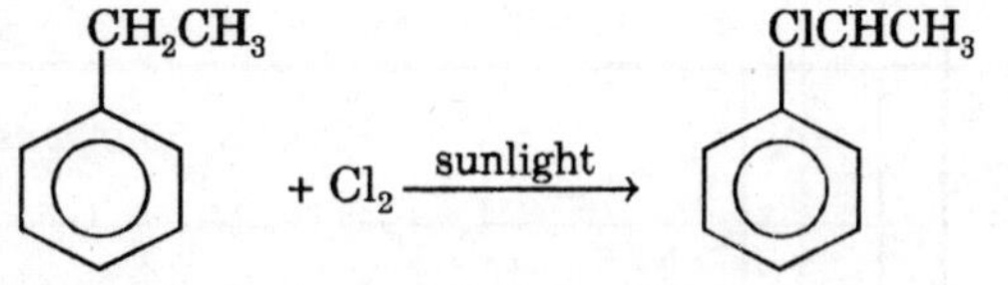
30.	(*d*)	$-OCH_3$ is *o*– and *p*– directing group as it releases electrons through resonance. Others are electron withdrawing groups.
31.	(*b*)	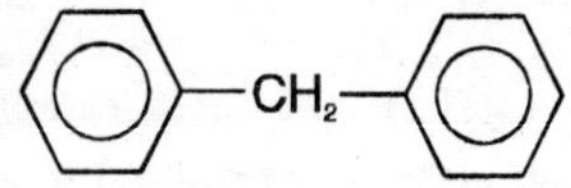
32.	(*c*)	A is  COOH. —COOH directs the electrphilic substitution towards meta position.

Q. No.	Ans.	Hints & Solutions
33.	(*a*)	$CH_3-\underset{\mid\atop CH_3}{CH}-CH_2Cl + AlCl_3 \longrightarrow CH_3-\underset{\mid\atop CH_3}{CH}-\overset{+}{C}H_2 + AlCl_4^-$ 1, 2 Hydride shift ↓ $CH_3-\underset{\mid\atop CH_3}{\overset{+}{C}}-CH_3$ *tert*-Butyl carbocation *tert*-Butyl carbocation on reaction with benzene gives *tert*-butyl benzene.
34.	(*b*)	$CH_3-\underset{CH_3}{\overset{CH_3}{C}}-CH_2Br \longrightarrow CH_3-\underset{CH_3}{\overset{CH_3}{C}}-\overset{+}{C}H_2$ 1, 2-methyl shift ↓ $CH_3-\overset{CH_3}{C}=CH-CH_3 \xleftarrow{-H^+} CH_3-\overset{CH_3}{\underset{+}{C}}-CH_2-CH_3$ 2-Methyl-2-butene
35.	(*c*)	$CH_3CH=CH_2 \xrightarrow[\text{Peroxide}]{HBr} H_3CH_2CH_2Br$ $\xrightarrow[\text{ether}]{Na} CH_3(CH_2)_4CH_3$ *n*-Hexane
36.	(*b*)	$2\,CH_3-\overset{CH_3}{CH}-CH_2Cl \longrightarrow$ *iso*-Butyl chloride $CH_3-\overset{CH_3}{CH}-CH_2-CH_2-\overset{CH_3}{CH}-CH_3$ 2, 5-Dimethyl hexane
37.	(*b*)	$-CH_3$ group attached to benzene ring has ortho-para directing effect in electrophilic substitution.
38.	(*c*)	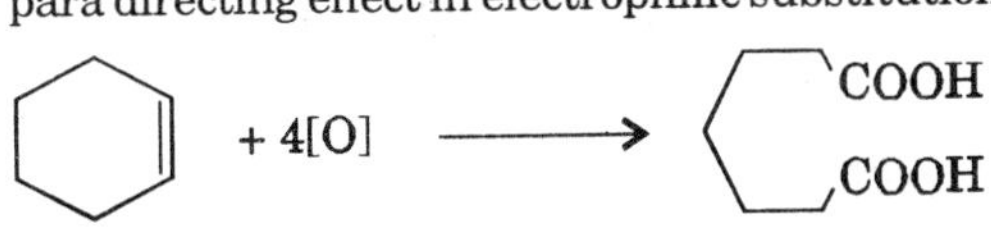 Adipic acid
39.	(*b*)	$\overset{4}{C}H_3-\underset{\mid\atop CH_3}{\overset{3}{C}H}-\overset{2}{C}H_2-\overset{1}{C}H_3$,

Q. No.	Ans.	Hints & Solutions
		iso-Pentane molecule contains four different types of H-atoms.
40.	(*d*)	Electrophilic substitution reactions of benzene take place via resonance stabilized carbocation as intermediate.
41.	(*b*)	$CH_3CH_2CH_2Cl \xrightarrow[-HCl]{\text{alc. KOH}} CH_3CH=CH_2$
42.	(*b*)	In Wurtz reaction, the hydrocarbon is formed by combination of the alkyl groups. Hence, propane cannot be obtained.
43.	(*b*)	$-NH_2$ group is highly activating.
44.	(*a*)	$CH_3CH_2CH_2CH_2Cl \xrightarrow{-HCl} CH_3CH_2CH=CH_2$
45.	(*c*)	$CH_3-\underset{CH_3}{CH}-\underset{CH_3}{C}=CH_2 \xrightarrow[(ii)\ H_2O/Zn]{(i)\ O_3}$ $CH_3-\underset{CH_3}{CH}-\underset{CH_3}{C}=O + O=\underset{H}{C}-H$ 3-Methyl-2-butanone Methanal
46.	(*d*)	Conjugated double bond being most stable has the least heat of hydrogenation.
47.	(*a*)	Acetylene is stronger acid than ethane. Hence, $HC\equiv C^-$ is weaker base than $CH_3CH_2^-$. The choices (*b*), (*c*) and (*d*) are ruled out because in these $HC\equiv C^-$ is shown to be stronger base than $CH_3CH_2^-$.
48.	(*d*)	Cyclohexane is lighter than water.
49.	(*b*)	Peroxide effect is not applicable to addition of HCl. The reaction takes place via more stable intermediate carbocation.
50.	(*a*)	$(CH_3)_3CMgCl + D_2O \longrightarrow (CH_3)_3CD + Mg(OD)Cl$.
51.	(*b*)	The reaction proceeds through most stable benzylic carbocation.
52.	(*a, c*)	Benzylic chlorination can be carried by SO_2Cl_2 or Cl_2 at 800 K.
53.	(*d*)	Alcohols are much stronger acids than hydrocarbons due to the presence of polar —OH group in them.
54.	(*a*)	Oxymercuration of terminal alkynes yields methyl ketones.
55.	(*a, c*)	Self explanatory answer.
56.	(*d*)	Propyne gives white ppt. with ammoniacal $AgNO_3$ while propene does not react.
57.	(*b*)	HOCl ionizes as OH^- and Cl^+. The attacking electrophile is, therefore, Cl^+.

Q. No.	Ans.	Hints & Solutions
58.	(*c*)	In both the cases one of the steps during anti-Markownikoff addition is highly endothermic and henci has high activation energy. The reaction, therefore, follows Markownikoff addition.
59.	(*d*)	Ammoniacal Cu_2Cl_2 gives red ppt with terminal alkynes such as 1-butyne.
60.	(*c*)	CH_3^- group is activating while Cl^- and NO_2^- groups are deactivating. Therefore, 2 should be most reactive, followed by 1.
61.	(*b*)	2-Butyne is symmetrical and linear.
62.	(*b*)	The free radical formed in this case is 3° and hence most stable.
63.	(*a*)	$\overset{sp^2}{H_2C}=\overset{sp^2}{CH}—\overset{sp}{C}\equiv\overset{sp}{CH}$.
64.	(*a*)	$Ph—C\equiv C—CH_3 \longrightarrow Ph—\overset{OH}{\overset{\vert}{C}}=CH—CH_3$ $Ph—\overset{O}{\overset{\Vert}{C}}—CH_2CH_3$ Ketone formed is conjugated with benzene wing.
65.	(*d*)	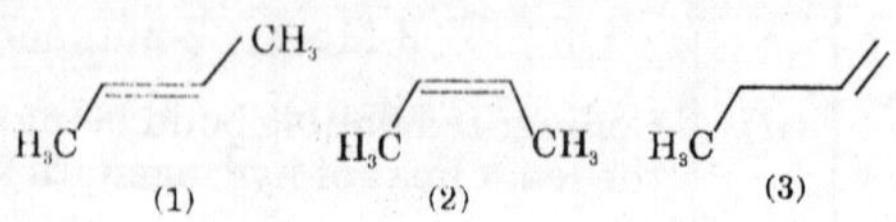
66.	(*a*)	$—\overset{O}{\overset{\Vert}{C}}—$ deactivates while —NH— group activates the benzene ring towards electrophilic substitution. Moreover, —NH— group is *ortho-para* directing. Therefore substitution takes place at para position, relative to —NH—.
67.	(*b*)	Alkynes are converted to trans alkenes by Li-NH_3/C_2H_5OH while $H_2/Pd/BaSO_4$ gives cis alkenes.
68.	(*a*)	Conc. H_3PO_4 brings about dehydration of cyclic alcohols to give alkenes.
69.	(*b*)	It is an example of intramolecular Wurtz reaction.
70.	(*c*)	NOCl adds to propene as NO^+ and Cl^-. The addition follows Markownikov's rule.
71.	(*b*)	The six isomers are 1-chloro-3-methylbutane, (±) 2-chloro-3-methylbutane, 2-chloro-2-methyl butane and (±)-1-chloro-2-methylbutane. Enantiomers cannot beseparated by distillation.
72.	(*b*)	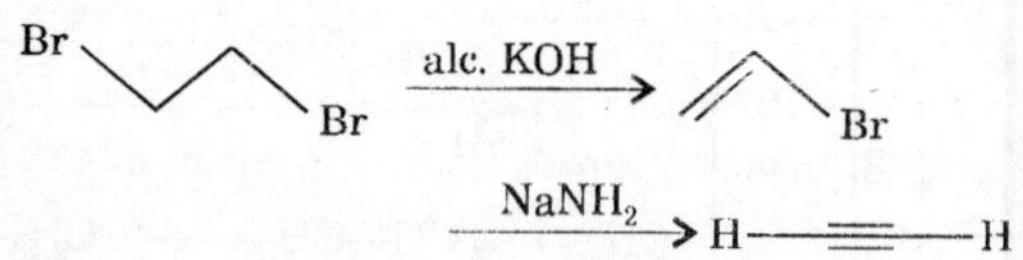
73.	(*a*)	Addition of Br_2 to trans-2-butene yields *meso*-2, 3-dibromobutane by anti addition.
74.	(*c*)	There are five structural isomers for C_6H_{14}.

2.7. ASSERTION-REASON QUESTIONS

Quick Revision

In each of the following questions two statements are given as Assertion A and Reason R. Examine the statements carefully and answer the questions according to the instructions given below :

- *Mark **a** if both **A** and **R** are correct and **R** is the correct reason of **A**.*
- *Mark **b** if both **A** and **R** are correct and **R** is not the correct reason of **A**.*
- *Mark **c** if **A** is correct and **R** is wrong.*
- *Mark **d** if **A** is wrong and **R** is correct.*
- *Mark **e** if both **A** and **R** are wrong.*

1. ***Assertion A.*** Petroleum is a good source of aliphatic hydrocarbons.

   Reason R. Aromatic compounds are obtained by distillation of coal-tar.

(a)	(b)	(c)	(d)	(e)

2. ***Assertion A.*** *neo*-Pentane has higher melting point than *n*-pentane.

   Reason R. In liquid form, interparticle forces are stronger in *n*-pentane than in *neo*-pentane.

(a)	(b)	(c)	(d)	(e)

3. ***Assertion A.*** First four alkanes are gases whereas next thirteen are liquids at room temperature.

   Reason R. Interparticle forces in alkanes increase with increase in molecular mass.

(a)	(b)	(c)	(d)	(e)

4. ***Assertion A.*** Ethyne is stronger acid than ethene.

   Reason R. C—H bond in ethyne is formed by *sp-sp* overlap.

(a)	(b)	(c)	(d)	(e)

5. ***Assertion A.*** *iso*-Butane on monosubstitution with chlorine forms two different products.

   Reason R. In *iso*-butane molecule, there are two different types of H-atoms, primary and tertiary.

(a)	(b)	(c)	(d)	(e)

6. ***Assertion A.*** Aromatic compounds burn with a sootyflame.

   Reason R. Aromatic compounds contain higher percentage of hydrogen.

(a)	(b)	(c)	(d)	(e)

7. ***Assertion A.*** Octane number of a sample of gasoline is more if it contains straight chain hydrocarbons in large proportion.

   Reason R. 2, 2, 3-Trimethylpentane has an octane number of 100.

(a)	(b)	(c)	(d)	(e)

8. ***Assertion A.*** Ethene is more reactive than ethane.

   Reason R. Bond energy of C—C bond in ethene is more than the bond energy of C—C bond in ethane.

(a)	(b)	(c)	(d)	(e)

9. ***Assertion A.*** Hydration of alkenes, in the presence of mineral acids, takes place via carbocations as intermediates.

   Reason R. Hydration of alkenes is initiated by electrophilic attack of H^+ ions on carbon-carbon double bond.

(a)	(b)	(c)	(d)	(e)

10. ***Assertion A.*** Electrophilic substitution in toluene takes place at faster rate than in benzene.

    Reason R. CH_3—group in toluene stabilizes the intermediate carbocation by its electron withdrawing inductive effect.

(a)	(b)	(c)	(d)	(e)

11. ***Assertion A.*** Kerosene oil floats over water.

    Reason R. Hydrocarbons are immiscible with water.

(a)	(b)	(c)	(d)	(e)

12. ***Assertion A.*** Benzene yields only one product on monochlorination.

    Reason R. All the hydrogen atoms in benzene molecule are equivalent.

(a)	(b)	(c)	(d)	(e)

13. ***Assertion A.*** C—C bond length in benzene is shorter than the C—C bond length in ethane.

    Reason R. C—C bond in benzene has partial double bond character.

(a)	(b)	(c)	(d)	(e)

14. ***Assertion A.*** The characteristic reactions of alkenes are electrophilic addition reactions.

    Reason R. Carbon-carbon double bond acts as a source of electrons for the electrophile.

(a)	(b)	(c)	(d)	(e)

15. ***Assertion A.*** 1-Butyne is stronger acid than 2-butyne.

    Reason R. 1-Butyne and 2-butyne are position isomers.

(a)	(b)	(c)	(d)	(e)

16. ***Assertion A.*** Straight chain alkanes have lower cetane number.

    Reason R. Cetane number of *n*-hexadecane is zero.

(a)	(b)	(c)	(d)	(e)

17. ***Assertion A.*** During nitration of benzene attacking electrophile is nitronium ion.

    Reason R. Nitronium ion is formed by dehydration of nitric acid.

(a)	(b)	(c)	(d)	(e)

18. ***Assertion A.*** *neo*-Pentane contains only primary carbon atoms.

    Reason R. *neo*-Pentane forms only one monosubstitution product.

(a)	(b)	(c)	(d)	(e)

ANSWERS

1. (*b*)	**2.** (*b*)	**3.** (*a*)	**4.** (*c*)	**5.** (*a*)	**6.** (*c*)	**7.** (*e*)	**8.** (*b*)	**9.** (*a*)
10. (*c*)	**11.** (*b*)	**12.** (*a*)	**13.** (*a*)	**14.** (*a*)	**15.** (*b*)	**16.** (*e*)	**17.** (*c*)	**18.** (*d*).

HINTS/SOLUTIONS

Q. No.	Ans.	Hints & Solutions
1.	(*b*)	Both these statements are correct.
2.	(*b*)	Statement ***A*** is correct. *neo*-pentane has higher melting point than *n*-pentane because relatively compact molecules of *neo*-pentane pack more closely in solid state and hence have stronger interparticle forces. In liquid state, *n*-pentane has stronger interparticle forces due to greater surface area.
3.	(*a*)	Self explanatory question.
4.	(*c*)	The statement ***R*** is wrong because in ethyne C—H bond is formed by *sp-s* overlap and not *sp-sp* overlap.
5.	(*a*)	The two substitution products are : 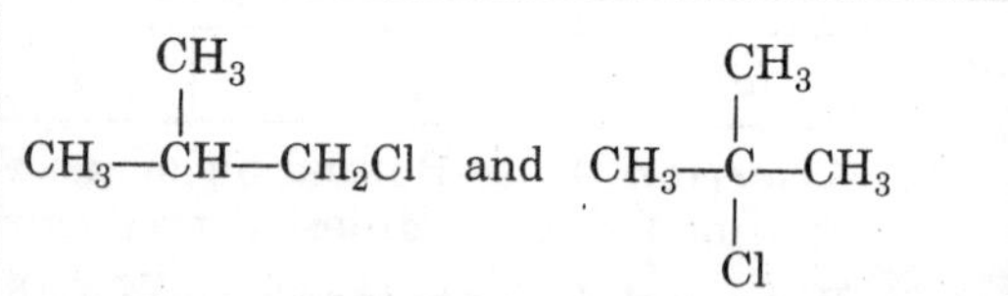
6.	(*c*)	Aromatic compounds burn with sooty flame due to high content of carbon (and not hydrogen). Hence ***R*** is wrong.
7.	(*e*)	Both the statments are wrong. Straight chain hydrocarbons have low octane number. Octane number of 2, 2, 3-trimethyl pentane is 116.
8.	(*b*)	Ethene is more reactive than ethane due weak pi bond. Hence, the statement ***R*** is not the correct explanation of ***A***.
9.	(*a*)	Self explanatory answer.

Q. No.	Ans.	Hints & Solutions
10.	(*c*)	Statement ***R*** is wrong. CH_3 is electron releasing group.
11.	(*b*)	Both the statements are correct but ***R*** is not the explanation of ***A***. Kerosene floats over water because hydrocarbons are lighter than water.
12.	(*a*)	Self explanatory statements.
13.	(*a*)	Self explanatory statements.
14.	(*a*)	Self explanatory statements.
15.	(*b*)	1-Butyne being terminal alkyne is stronger acid than 2-butyne. Statements ***R*** is also correct but is not explanation of the statement ***A***.
16.	(*e*)	Both the statements are wrong. Straight chain alkanes have higher cetane number, cetane number of *n*-hexadecane is 100.
17.	(*c*)	The statement ***R*** is wrong. NO_2^+ ion is not formed by dehydration of HNO_3.
18.	(*d*)	The statement ***A*** is wrong. *neo*-Pentane contains 4 primary and 1 quaternary carbon atom.

2.8. SELF EVALUATION TEST

Assess Yourself

Choose the correct answer from the four alternatives given in each of the following questions :

1. A compound X on treatment with alcoholic KOH gave propane. The compound X may be
(*a*) 1-Propanol (*b*) 1-Bromopropane
(*c*) 2-Propanol (*d*) 3-Bromo-1-propene.

2. Starting from which of the following compounds methane can be prepared in one step from
(*a*) Sodium acetate
(*b*) Aluminium carbide
(*c*) Methyl magnesium bromide
(*d*) All of these.

3. Starting from which of the following compounds methane as well as ethane can be prepared in separate single steps ?
(*a*) Ethyl magnesium bromide
(*b*) Methyl bromide
(*c*) Sodium proanoate
(*d*) Ethyl alcohol.

4. Which of the following compounds has the lowest value of octane number ?
(*a*) *iso*-Octane (*b*) *n*-Heptane
(*c*) *n*-Octane (*d*) *n*-Nonane.

5. When propyne is treated with excess HBr the product obtained is
(*a*) 1, 1-Dibromopropare (*b*) 1, 2-Dibromopropane
(*c*) 1, 3-Dibromopropane (*d*) 2, 2-Dibromopropane.

6. When propene is passed through chlorine water, the product is
(*a*) 1, 2-Dichloropropane (*b*) 1, 1-Dichloropropane
(*c*) 1-Chloro-2-propanol (*d*) 2-Chloro-1-propanol.

7. Which of the following compounds is most acidic ?
(*a*) Propane (*b*) Propene
(*c*) Propyne (*d*) 2-Butyne.

8. Which of the following reagents can be used to distinguish between 1-butyne and 2-butyne ?
(*a*) $KMnO_4/OH^-$
(*b*) Br_2
(*c*) Copper sulphate solution
(*d*) Amm. $AgNO_3$ solution.

9. $C + H_2 \xrightarrow[\text{arc}]{\text{Electric}} X \xrightarrow[HgSO_4]{\text{dil. } H_2SO_4} Y$

The product 'Y' in the above sequence is
(*a*) $(CH_3)_2SO_4$ (*b*) C_2H_5OH
(*c*) $C_2H_5HSO_4$ (*d*) CH_3CHO.

10. Which of the following compounds does not yield ethyne in a single step ?
(*a*) Potassium fumarate (*b*) Ethene
(*c*) Ethylidene dichloride (*d*) Ethylene dichloride.

11. $X \xrightarrow[\text{KOH}]{\text{alc.}} Y \xrightarrow{NaNH_2} H—C \equiv C—H$

The compound X in the above sequence may be
(*a*) Ethyl alcohol (*b*) 1, 2-Dibromoethane
(*c*) 1, 1-Dibromoethene (*d*) Ethyl bromide.

12. Ethyne when treated with excess HBr produces
(*a*) Ethylene bromide (*b*) Ethylidene bromide
(*c*) Vinyl bromide (*d*) Allyl bromide.

13. 1, 1-Dibromoethane when heated with zinc dust produces
(*a*) Ethyl bromide (*b*) Ethene
(*c*) Vinyl bromide (*d*) 2-Butene.

14. When ethyne is treated with chlorine water, the product obtained is
(*a*) 1, 1, 2, 2-Tetrachloroethane
(*b*) Vinyl chloride
(*c*) Ethylidene dichloride
(*d*) Dichloroacetaldehyde.

15. Which of the following alkenes is the most stable ?
(*a*) Ethene (*b*) Propene
(*c*) 1-Butene (*d*) 2-Butene.

16. The correct order of acidic strength is
(*a*) Ethane > Ethene > Ethyne
(*b*) Ethane > Ethyne > Ethene
(*c*) Ethyne > Ethene > Ethane
(*d*) Ethyne > Ethane > Ethene.

17. Propene can be converted into 1-propanol by
(*a*) Hydration
(*b*) Hydroboration-oxidation
(*c*) Reaction with alkaline $KMnO_4$
(*d*) Reaction with dil. NaOH solution.

18. Combustion of which of the following compounds (in the presence of excess of oxygen) does not result in the change in hybrid state of the carbon atom

(a) CH_4 (b) $CH_2 = CH_2$
(c) CH_3-CH_3 (d) $HC \equiv CH$.

19. Which of the following reagents can be used for the conversion of *n*-butane to *iso*-butane ?
(a) $NaBH_4$ (b) Lindlar's catalyst
(c) $AlCl_3$ (d) $LiAlH_4$.

20. When propene is treated with Cl_2 at 800 K the main product formed is
(a) 1, 2-Dichloropropane (b) 1-Chloro-1-propene
(c) 3-Chloro-1-propene (d) 2-Chloro-1-propene.

21. *sec*-Butyl bromide is subjected to Wurtz reaction. The alkane formed is
(a) Octane (b) 3, 4-Dimethylhexane
(c) 3, 4-Dimethyloctane (d) 2, 5-Dimethylhexane.

22. *sec*-Butyl magnesium bromide is treated with ethyl alcohol. The alkane formed is
(a) *n*-Butane (b) *iso*-Butane
(c) Ethane (d) 3-Methylpentane.

23. During electrolysis of an aqueous solution of sodium acetate which of the following gases is not liberated ?
(a) CH_3CH_3 (b) CO_2
(c) H_2 (d) CH_4.

24. When an aqueous solution of sodium propionate is electrolysed the gas/es liberated at anode is/are
(a) Propane (b) Propane and CO_2
(c) Butane and CO_2 (d) Hexane and CO_2.

25. The correct order of melting points of isomeric pentanes is
(a) *n*-Pentane > *iso*-Pentane > *neo*-Pentane
(b) *neo*-Pentane > *iso*-Pentane > *n*-Pentane
(c) *neo*-Pentane > *n*-Pentane > *iso*-Pentane
(d) *n*-Pentane > *neo*-Pentane > *iso*-Pentane.

26. Which of the following alkyl halides on conversion into the Grignard reagent followed by treatment with water would yield 2-methylbutane ?
(a) 2-Chloro-2-methylbutane
(b) 1-Chloro-2-methylbutane
(c) 2-Chloro-3-methylbutane
(d) All of them.

27. How many isomeric bromides having formula $C_6H_{13}Br$ on conversion into the Grignard reagent followed by treatment with water would yield 2, 3-dimethylbutane ?
(a) 1 (b) 2
(c) 3 (d) 4.

28. Consider the reaction

$$CH_3-\underset{\displaystyle CH_3}{\underset{|}{CH}}-CH_3 + Br_2 \xrightarrow[\text{light, 400 K}]{} \underset{\text{(A)}}{CH_3-\underset{\displaystyle CH_3}{\underset{|}{CH}}-CH_2Br} + \underset{\text{(B)}}{CH_3-\overset{\displaystyle CH_3}{\overset{|}{\underset{\displaystyle Br}{\underset{|}{C}}}}-CH_3}$$

Which of the following is true ?
(a) Compounds A and B are formed in nearly equal amounts
(b) Compound A is formed in much larger proportion
(c) Compound B is formed in much larger proportion
(d) Relative amounts of the two cannot be predicted.

29. Which of the following groups directs substitution at meta position in arenes ?
(a) —Br (b) $—OCH_3$
(c) $-COOCH_3$ (d) $-NHCOCH_3$.

30. Which of the following alkenes cannot be prepared in high yield by Kolbe's electrolytic reaction ?
(a) Ethane (b) Propane
(c) Butane (d) All can be prepared.

31. Which of the following alkanes exists in solid state at room temperature ?
(a) *n*-Heptane (b) *n*-Octane
(c) *n*-Decane (d) None of these is solid.

32. Which of the following compounds contains 1°, 2° as well as 3° hydrogen atoms ?
(a) *neo*-Hexane (b) *iso*-Pentane
(c) *n*-Pentane (d) 2, 3-Dimethylbutane.

33. Which of the following is the main component of green oil (a fraction obtained during fractional distillation of coal tar) ?
(a) Xylene (b) Naphthalene
(c) Anthracene (d) Pyridine.

34. Which of the following alkanes cannot be prepared by Wurtz reaction ?
(a) Ethane (b) Propane
(c) Butane (d) Methane.

35. The number of isomers which can be obtained theoretically on monochlorination of *neo*-hexane is
(a) 2 (b) 3
(c) 4 (d) 5.

36. When prepyne is treated with alkaline solution of $KMnO_4$, the product formed is
(a) Acetone (b) Propanoic acid
(c) 2-Keto propanoic acid (d) Propanal.

37. Acetylene on treatment with alkaline $KMnO_4$ yields
(a) Oxalic acid (b) Ethanol
(c) Glyoxal (d) 1, 2-Ethanediol.

38. *m*-Xylene on oxidation with acidified $KMnO_4$ yields
(a) Phthalic acid (b) Benzoic acid
(c) Terephthalic acid (d) Isophthalic acid.

39. Which of the following compounds can be converted to acetylene in a single step ?

(*a*) Chloromethane (*b*) Dichloromethane

(*c*) Trichloromethane (*d*) Tetrachloromethane.

40. Which of the following compounds has the maximum heat of combustion ?

(*a*) CH_4 (*b*) C_2H_6

(*b*) C_3H_8 (*d*) C_4H_{10}.

ANSWERS

1. (*b*)	**2.** (*d*)	**3.** (*b*)	**4.** (*d*)	**5.** (*d*)	**6.** (*c*)	**7.** (*c*)	**8.** (*d*)	**9.** (*d*)
10. (*b*)	**11.** (*b*)	**12.** (*b*)	**13.** (*d*)	**14.** (*d*)	**15.** (*d*)	**16.** (*c*)	**17.** (*b*)	**18.** (*d*)
19. (*c*)	**20.** (*c*)	**21.** (*b*)	**22.** (*a*)	**23.** (*d*)	**24.** (*c*)	**25.** (*c*)	**26.** (*d*)	**27.** (*b*)
28. (*c*)	**29.** (*c*)	**30.** (*b*)	**31.** (*d*)	**32.** (*b*)	**33.** (*b*)	**34.** (*d*)	**35.** (*b*)	**36.** (*c*)
37. (*a*)	**38.** (*d*)	**39.** (*c*)	**40.** (*d*).					

HINTS/SOLUTIONS

Q. No.	Ans.	Hints & Solutions
1.	**(*b*)**	$CH_3CH_2CH_2Br \xrightarrow[KOH]{alc.} CH_3CH=CH_2$ 1-Bromopropane
2.	**(*d*)**	$CH_3COONa + NaOH \xrightarrow{CaO} CH_4 + Na_2CO_3$ $Al_4C_3 + 12H_2O \longrightarrow 4\,Al(OH)_3 + 3CH_4$ $CH_3MgBr + H_2O \longrightarrow CH_4 + Mg(OH)Br$
3.	**(*b*)**	$CH_4 \xleftarrow{Zn/HCl} \underset{\text{Methyl bromide}}{CH_3Br} \xrightarrow{Na/ether} CH_3CH_3$ $CH_4 \xleftarrow[CaO,\ \Delta]{NaOH} \underset{\text{Sod. ethanoate}}{CH_3COONa} \xrightarrow[Electrolysis]{Kolbe} CH_3CH_3$
4.	**(*d*)**	*n*-Nonane has octane number of – 45.
5.	**(*d*)**	$CH_3-C\equiv CH \xrightarrow{2HBr} CH_3-\underset{\underset{Br}{\vert}}{\overset{\overset{CH_3}{\vert}}{C}}-CH_3$
6.	**(*c*)**	$CH_3CH=CH_2 + HOCl \longrightarrow CH_3-\underset{\underset{OH}{\vert}}{CH}-CH_2Cl$
7.	**(*c*)**	Propyne is a terminal alkyne and terminal alkynes are acidic.
8.	**(*d*)**	Ammoniacal $AgNO_3$ gives white ppt. with 1-butyne (a terminal alkyne).
9.	**(*d*)**	$C + H_2 \xrightarrow{\text{Electric arc}} \underset{[X]}{H-C\equiv C-H} \xrightarrow[HgSO_4]{H_2O//H^+}$ $[CH_2=CHOH] \longrightarrow \underset{[Y]}{CH_3CHO}$
10.	**(*b*)**	Ethene cannot be converted into ethyne in a single step.
11.	**(*b*)**	$H_3C-CHBr_2 \xrightarrow[KOH]{alc.} H_2C=CHBr \xrightarrow{NaNH_2} HC\equiv CH$ $BrH_2C-CH_2Br \xrightarrow[KOH]{alc.} H_2C=CHBr \xrightarrow{NaNH_2} HC\equiv CH$
12.	**(*b*)**	$HC\equiv CH \xrightarrow{2HBr} \underset{\text{Ethylidene bromide}}{H_3C-CHBr_2}$
13.	**(*d*)**	$H_3C-CHBr \xrightarrow[\Delta]{Zn} H_3C-CH=CH-CH_3$

Q. No.	Ans.	Hints & Solutions
14.	(*d*)	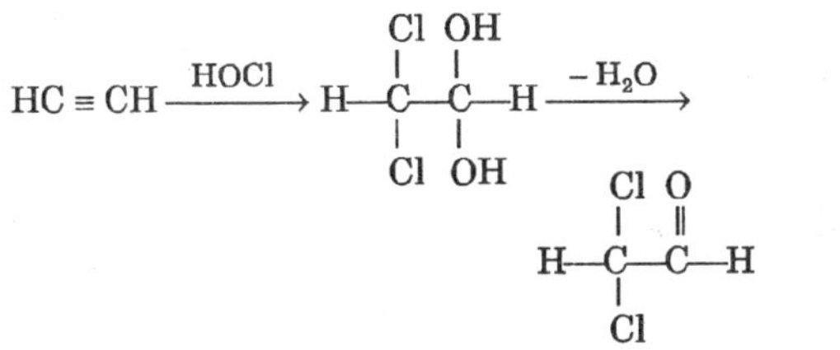
15.	(*d*)	2-Butene is maximum stable because it is maximum substituted.
16.	(*c*)	The electronegativity of carbon in different hybrid states is in the order : $sp > sp^2 > sp^3$.
17.	(*b*)	$CH_3CH = CH_2 \xrightarrow[H_2O_2/OH^-]{(BH_3)_2} CH_3CH_2CH_2OH$ Hydroboration-oxidation results in anti-Markownikoff's addition.
18.	(*d*)	In $HC \equiv CH$ as well as in $O = C = O$, the hybrid state of carbon is *sp*.
19.	(*c*)	Heating with $AlCl_3$ results in isomerisation.
20.	(*c*)	$H_3C - CH = CH_2 \xrightarrow[800\ K]{Cl_2} H_2C(Cl) - CH = CH_2$ [*allylic subsitution*]
21.	(*b*)	$CH_3 - CH(C_2H_5) - Br \xrightarrow[\text{dry ether}]{Na} CH_3 - CH(C_2H_5) - CH(C_2H_5) - CH_3$ 3, 4-Dimethylhexane
22.	(*a*)	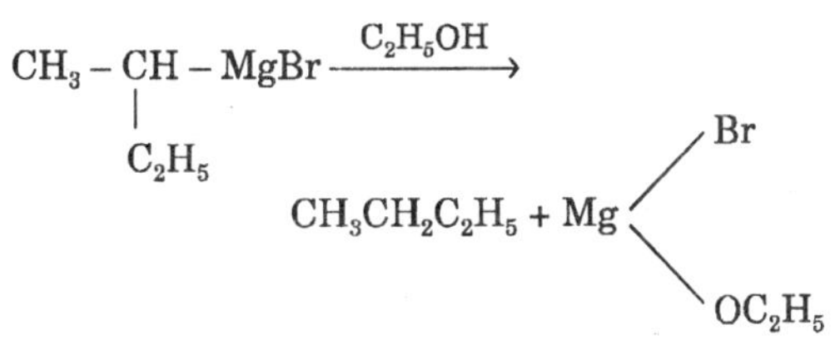
23.	(*d*)	H_2 is liberated at cathode while C_2H_6 and CO_2 are liberated at anode. Methane gas is not liberated.
24.	(*c*)	$2CH_3CH_2COO^- \longrightarrow 2CH_3\dot{C}H_2 + 2CO_2 + 2e^-$ $2CH_3\dot{C}H_2 \longrightarrow CH_3CH_2CH_2CH_3$ At anode gases liberated are **butane** and $\mathbf{CO_2}$.
25.	(*c*)	*neo*-Pentane is spherical and packing of molecules in solid state will be best in this case. Next in symmetry is *n*-pentane, and *iso*-pentane is least symmetrical.
26.	(*d*)	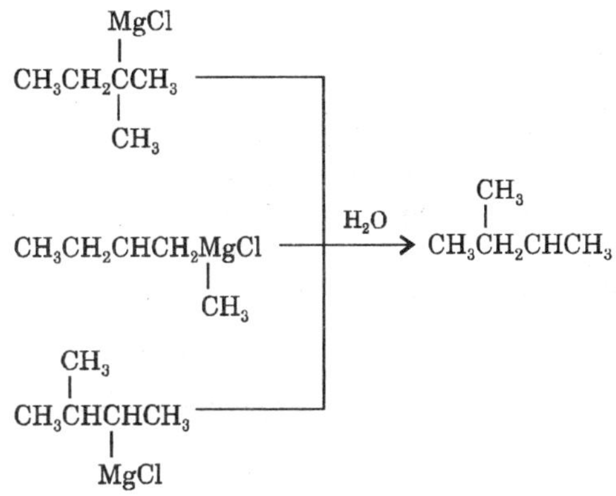
27.	(*b*)	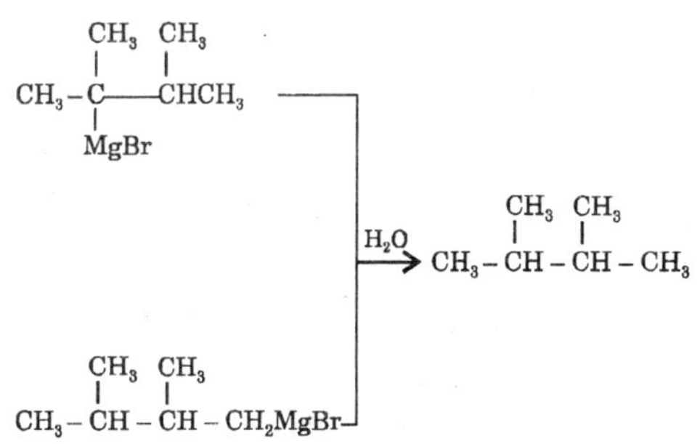
28.	(*c*)	Tertiary hydrogen is easy to replace.
29.	(*c*)	$-\overset{\overset{}{}}{\underset{\underset{O}{\|\|}}{C}}-OCH_3$ is meta directing.
30.	(*b*)	Kolbe electrolytic method gives rise to alkanes with even number of carbon atoms in high yield. While preparing alkanes with odd number of carbon atoms by electrolysis of a mixture of two salts of different carboxylic acids, a number of side products are also formed.
31.	(*d*)	Alkanes having 5—17 carbon atoms are liquids.
32.	(*b*)	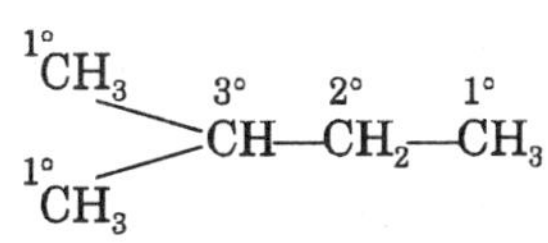
33.	(*c*)	Green oil is mainly anthracene.
34.	(*d*)	In Wurtz reaction alkane is formed by combination of two alkyl groups. Therefore, alkane formed contains at least 2 carbon atoms.

Q. No.	Ans.	Hints & Solutions
35.	(*b*)	*neo*-Hexane is $CH_3—C(CH_3)_2—CH_2CH_3$ There are three different types of hydrogen atoms in this molecule. So, it results in the formation of three monochlorination products.
36.	(*c*)	$CH_3C \equiv CH \xrightarrow{3[O]} CH_3—\overset{\overset{\large O}{\Vert}}{C}—COOH$
37.	(*a*)	$HC \equiv CH \xrightarrow{4[O]} \begin{array}{c} COOH \\ \vert \\ COOH \end{array}$
38.	(*d*)	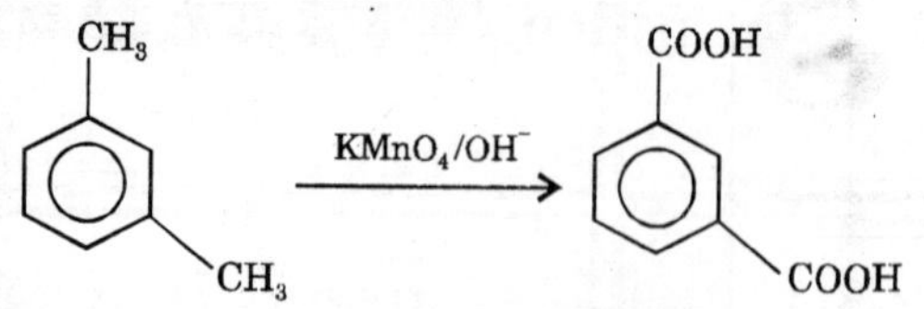 Isophthalic acid
39.	(*c*)	$2CHCl_3 \xrightarrow[\Delta]{Ag} H—C \equiv C—H.$
40.	(*d*)	The hydrocarbon having maximum molecular mass has the maximum value of heat of combustion.

3 HALOALKANES AND HALOARENES

3.1. CONCEPTS IN BRIEF

INTRODUCTION

- Haloalkanes are the compounds in which one hydrogen atom of alkanes is replaced by a halogen atom. They are classified as *primary, secondary* and *tertiary alkyl halides* depending on whether the halogen atom is attached to a *primary, secondary* or *tertiary carbon atom respectively.* For example,

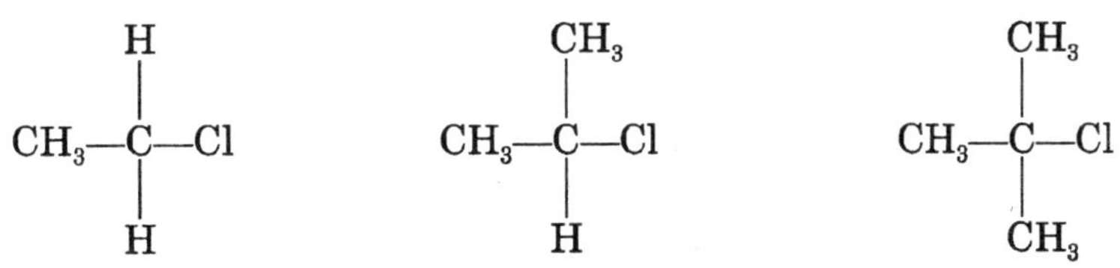

Chloroethane (*Primary* or 1°) | 2-Chloropropane (*Secondary* or 2°) | 2-Chloro-2-methylpropane (*Tertiary* or 3°)

- **Aromatic halogen compounds** are the halogen compounds which contain at least one aromatic ring. Aromatic halogen compounds in which halogen atom is directly attached to benzene ring are called **aryl halides** whereas those in which the halogen atom is attached to the side chain are called **aralkyl halides.**

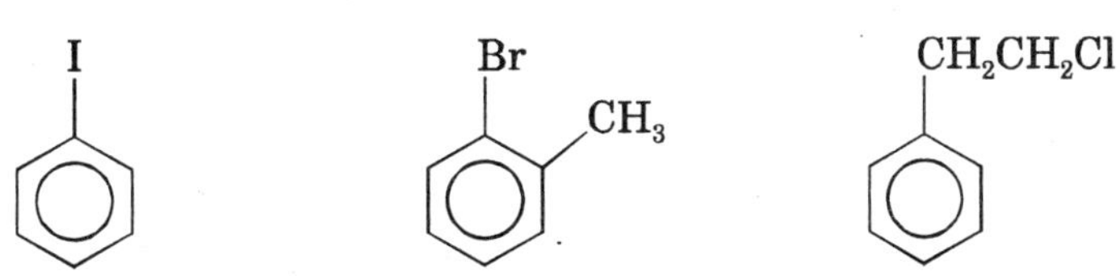

Iodobenzene | 2-Bromotoluene | 1-Chloro-2-phenylethane

Aryl halides (Iodobenzene, 2-Bromotoluene) | Aralkyl halide (1-Chloro-2-phenylethane)

HALOALKANES

METHODS OF PREPARATION OF HALOALKANES

1. From Alcohols

$$ROH \xrightarrow[ZnCl_2]{HCl} R{-}Cl + H_2O$$

$$ROH \xrightarrow[\text{or } (NaBr + H_2SO_4)]{HBr\ (48\%)} R{-}Br + H_2O$$

$$ROH \xrightarrow[\text{or } (KI + H_3PO_4)]{HI\ (57\%)} R{-}I + H_2O$$

$$ROH \xrightarrow{PCl_3} R{-}Cl + H_3PO_3$$

$$ROH \xrightarrow[[P + Br_2]]{PBr_3} R{-}Br + H_3PO_3$$

$$ROH \xrightarrow[[P + I_2]]{PI_3} R{-}I + H_3PO_3$$

$$ROH \xrightarrow{PCl_5} R{-}Cl + POCl_3 + HCl$$

$$ROH \xrightarrow{SOCl_2\ \text{(Thionyl Chloride)}} R{-}Cl + SO_2 + HCl$$

Thionyl chloride is preferred in the preparation of alkyl chlorides because in this case the side products are gaseous and can be easily expelled during distillation.

2. From Alkenes

$$CH_3CH = CH_2 \xrightarrow[X = Cl, Br, I]{HX} CH_3{-}CH(X){-}CH_3$$ **Markownikoff addition**

$$CH_3CH = CH_2 \xrightarrow[\text{Peroxides}]{HBr} CH_3CH_2CH_2Br$$ **Anti-Markownikoff addition**

$$CH_3CH = CH_2 \xrightarrow[X = Cl, Br]{X_2,\ 800K} CH_2(X){-}CH = CH_2 + HX$$ **Allylic substitution**

$$CH_3CH = CH_2 \xrightarrow[X = Cl, Br]{X_2/CCl_4} CH_3{-}CH(X){-}CH_2(X)$$ **Addition of Halogens**

3. From Alkanes

$$CH_4 + Cl_2 \xrightarrow{\text{Light or Heat}} CH_3Cl + HCl$$

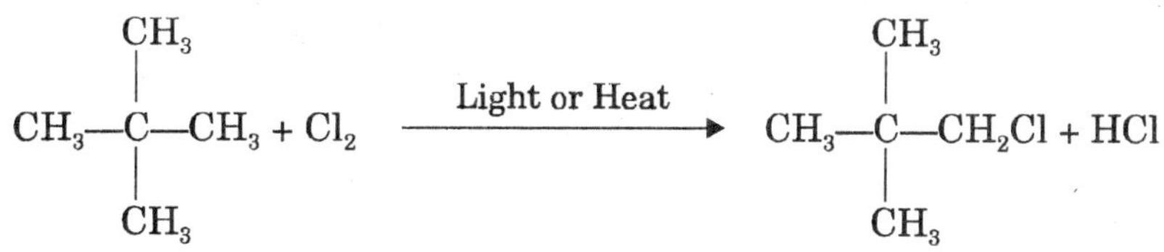

neo-Pentane *neo*-Pentyl chloride

4. By Halide Exchange (Finkelstein Reaction)

$$\underset{\text{Bromoethane}}{C_2H_5Br} + NaI \xrightarrow{\text{Acetone}} \underset{\text{Iodoethane}}{C_2H_5I} + NaBr$$

Finkelstein reaction is suitable for converting alkyl chlorides and bromides into iodides. NaCl and NaBr formed during reaction are almost insoluble in acetone and hence separate out and shift the reaction in forward direction while NaI is soluble in acetone.

5. From Silver Salt of Fatty Acids

$$\underset{\text{Silver propionate}}{CH_3CH_2COOAg} + Br_2 \xrightarrow[\text{Reflux}]{CCl_4} CH_3CH_2Br + CO_2 + AgBr$$

Here, ***the alkyl halide formed has one carbon less than the starting compound.*** The reaction is known as **Hunsdiecker reaction.**

PHYSICAL PROPERTIES

1. Boiling Points

Haloalkanes have higher boiling points as compared to those of corresponding alkanes. This is due to their polarity and strong dipole-dipole attractive interactions between haloalkane molecules.

(*i*) *For a given alkyl group, the boiling points are in the order : RI > RBr > RCl.*

(*ii*) *For a given halogen, the boiling points of alkyl halides increases with increase in the size of alkyl group.*

(*iii*) *For isomeric alkyl halides, the boiling points decrease with branching. This is because branching of the chain decreases the surface area and consequently the inter-molecular forces of attraction.*

2. Solubility

Although polar in nature, alkyl halides are insoluble in water because they are not able to form hydrogen bonds with water molecules. However, they are soluble in organic solvents such as ether, alcohol, etc.

3. Density

Alkyl chlorides are generally lighter than water whereas bromides and iodides are heavier. Their densities are in the order R—I > R—Br > R—Cl. Polychlorides are however, heavier than water.

CHEMICAL PROPERTIES OF HALOALKANES

I. Nucleophillic Substitution Reactions

$$CH_3Br + OH^- \xrightarrow[\text{Moist silver oxide}]{\text{KOH }(aq)\text{ or}} CH_3OH + Br^-$$

$$CH_3Br + CN^- \xrightarrow{\text{KCN (alc.)}} CH_3{-}CN + Br^-$$

$$CH_3Br + :\bar{O}CH_3 \xrightarrow{\overset{+}{Na}\overset{-}{O}CH_3\text{ (alc.)}} CH_3{-}O{-}CH_3 + Br^-$$

$$CH_3Br + :\overset{\ominus}{O}{-}\overset{O}{\overset{||}{C}}{-}CH_3 \xrightarrow{AgOCOCH_3} CH_3{-}O{-}\overset{O}{\overset{||}{C}}{-}CH_3 + Br^-$$

$$CH_3Br + :\overset{\ominus}{C}\equiv C{-}H \xrightarrow{Na^+\bar{C}\equiv C{-}H} CH_3{-}C\equiv C{-}H + Br^-$$

$$CH_3Br + \ddot{N}H_3 \xrightarrow{\text{(alc.) }NH_3} CH_3{-}NH_2 + HBr$$

$$CH_3Br + H_2\ddot{N}{-}CH_3 \xrightarrow{\text{More alkyl halide}} CH_3{-}\underset{..}{N}H{-}CH_3 + HBr$$

$$CH_3Br + CH_3\ddot{N}HCH_3 \longrightarrow CH_3{-}\underset{..}{\overset{CH_3}{\overset{|}{N}}}{-}CH_3 + HBr$$

$$CH_3Br + :N(CH_3)_3 \xrightarrow{\text{Excess alkyl halide}} [H_3C{-}N(CH_3)_2{-}CH_3]^+ Br^-$$

$$CH_3Br + :SH^- \xrightarrow{\text{NaSH (alc.)}} CH_3{-}SH + Br^-$$

$$CH_3Br + :\overset{\ominus}{S}CH_3 \xrightarrow{Na^+\bar{S}CH_3\text{ (alc.)}} CH_3{-}S{-}CH_3 + Br^-$$

$$CH_3Br + AgCN \longrightarrow CH_3{-}NC + AgBr$$

$$CH_3Br + AgNO_2 \longrightarrow \underset{\text{Major product}}{CH_3NO_2} + AgBr$$

$$CH_3Br + NaNO_2 \longrightarrow \underset{\text{Major product}}{CH_3ONO} + NaBr$$

The alkyl cyanides obtained through nucleophilic substitution of alkyl halides can be used for the preparation of a number of other important compounds. The product obtained has one carbon atom more than the starting alkyl halide.

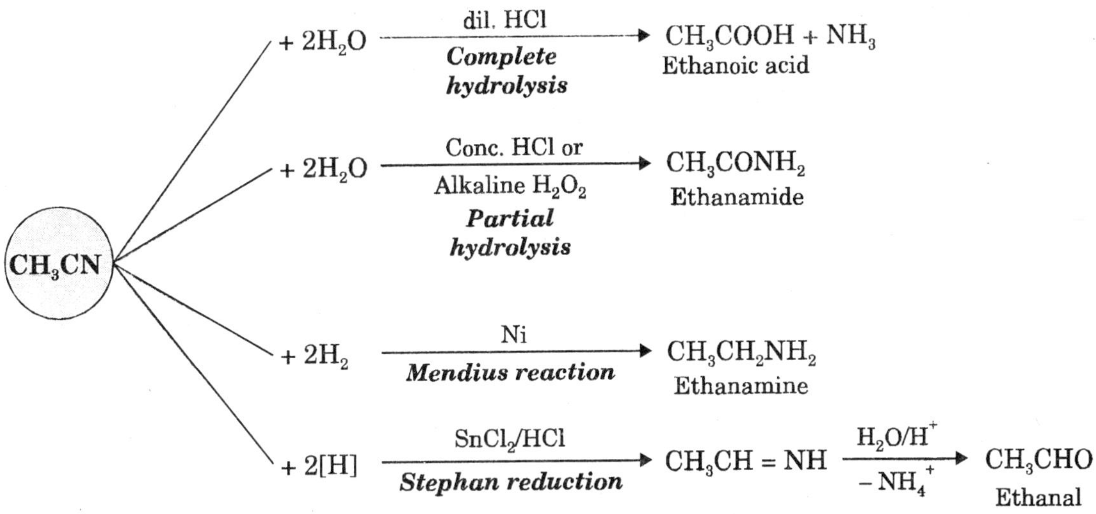

Mechanism of Nucleophilic Substitution Reactions

Nucleophilic substitution may follow either of the two different types of mechanisms. These are S_N^1 and S_N^2.

S_N^1 Mechanism. S_N^1 stands for ***substitution nucleophilic unimolecular***. Consider the hydrolysis of some tertiary alkyl halide such as *tert*-butyl bromide with aq. KOH.

From kinetic studies it is found that the rate of this reaction depends only upon the concentration of the alkyl halide and is independent of conc. of the nucleophile (OH^-)

$$\text{rate} \propto [(CH_3)_3C\,.\,Br]$$

Therefore, the slow rate determining step is unimolecular and involves only *tert*-butyl bromide. The following mechanism is proposed for this reaction :

1 ...*Slow*

2. ...*Fast*

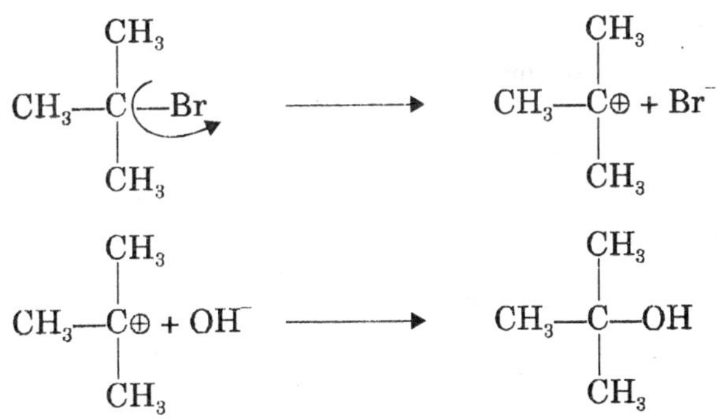

The first step involves the slow ionization of the bromide to form a carbocation. The energy required for breaking C—Br bond becomes available from attractive interactions between solvent molecules and alkyl halide. The second step is a fast step. It involves quick reaction between carbocation and nucleophile (OH^-) to form the final product. The **order of reactivity** of alkyl halides through S_N^1 mechanism is

Tertiary > Secondary > Primary.

Starting with optically active reactant, the reaction through S_N^1 leads to partial racemization.

S_N^2 Mechanism. S_N^2 stands for ***substitution nucleophilic bimolecular***. Consider the hydrolysis of some primary alkyl halide such as methyl bromide with aq. KOH.

From kinetic studies it has been found that rate of this reaction depends both upon the conc. of CH_3Br and nucleophile OH^-

$$\text{rate} \propto [CH_3Br]\,[OH^-]$$

Therefore, the rate determining step involves both CH_3Br and OH^- and is bimolecular. The mechanism proposed for this reaction is S_N^2. The mechanism involves the attack of the nucleophile (OH^-) on the halogen bearing carbon from the opposite side whereby the halogen atom is pushed away.

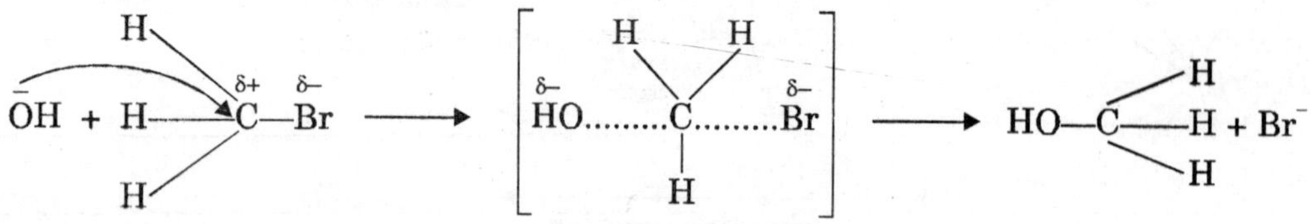

Transition state

In the transition state nucleophile is partially attached to carbon whereas bond between carbon and halogen is partially cleaved.

The ***order of reactivity*** through S_N^2 mechanism is : ***Primary > Secondary > Tertiary***.

Starting with optically active alkyl halide, the reaction through S_N^2 mechanism results in complete inversion of configuration.

FACTORS FAVOURING S_N^1 and S_N^2 MECHANISM

1. **Nature of alkyl halide.** In S_N^1 mechanism, carbocation is formed as intermediate. 3° alkyl halides result in the formation of stable 3° carbocation and hence they react through S_N^1.
   On the other hand, reaction through S_N^2 mechanism is guided by steric factors. In S_N^2 mechanism carbon carrying halogen is bonded to five groups in the transition state. If these groups are larger in size then the transition state becomes unstable and this results in increase in energy of activation. Therefore, 1° alkyl halides with smaller groups generally react through S_N^2.
2. **Nature of nucleophile.** Strong nucleophile favours S_N^2 mechanism whereas weak nucleophile favours S_N^1 mechanism.
3. **Concentration of nucleophile.** Low concentration of nucleophile favours S_N^1 whereas high concentration favours S_N^2.
4. **Polarity of solvent.** Polar solvent favours S_N^1 mechanism because molecules of polar solvent can interact strongly with polar $\overset{\delta+}{C}—\overset{\delta-}{X}$ bond of alkyl halide and this facilitates its ionisation to give carbocation.

- Non-polar solvent favours S_N^2 mechanism.

II. Dehydrohalogenation Reactions or β-Elimination Reactions

When alkyl halides are heated with alcoholic KOH, they undergo dehydrohalogenation to form alkenes. These reactions are called β-elimination because the hydrogen atom present at β position of the alkyl halide is removed.

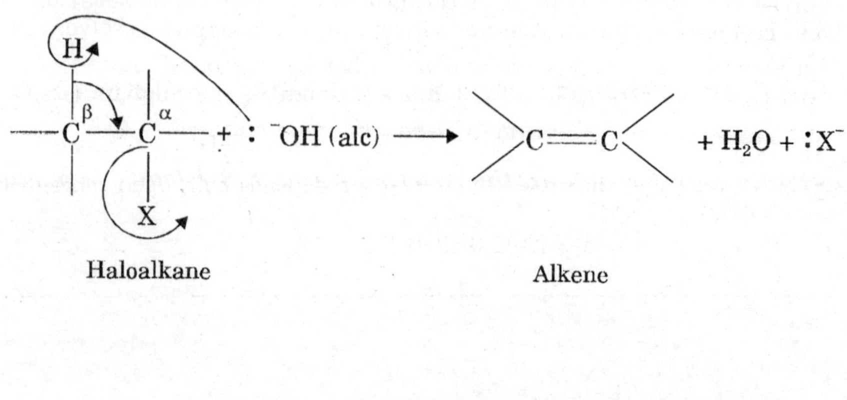

Haloalkane Alkene

$$CH_3-CH_2-Br + KOH\ (alc.) \longrightarrow H_2C = CH_2 + KBr + H_2O$$

Bromoethane   Ethene

In case the alkyl halide can undergo elimination in two different ways, the preferred alkene is the one which carries more number of alkyl groups attached to the doubly bonded carbon atoms. This generalisation is known as **Saytzeff Rule.** For example,

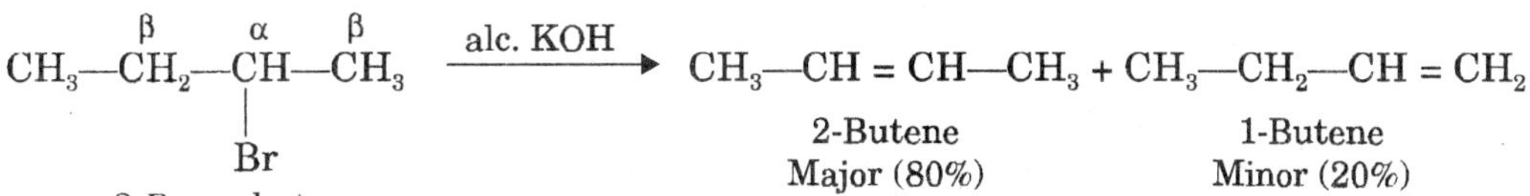

2-Bromobutane

The **reactivity** *of alkyl halides towards elimination reaction follows the order :*

tertiary > secondary > primary.

MECHANISM OF β-ELIMINATION REACTION

β-elimination follows two different mechanisms as discussed under :

E_1-mechanism. E_1 stands for ***elimination unimolecular.*** Let us consider the reaction between some tertiary alkyl halide such as *tert*-butyl bromide and alcoholic KOH solution to form alkene.

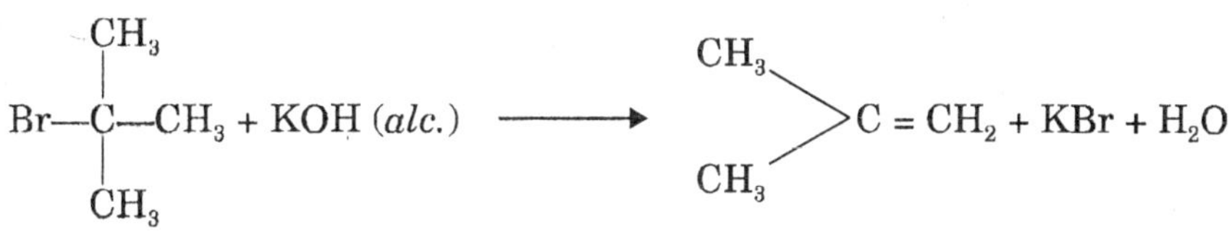

From kinetic studies it is found that rate of the above reaction depends only upon the concentration of the alkyl halide and is independent of the concentration of the nucleophile.

$$\text{rate} \propto [(CH_3)_3\ C\ .\ Br]$$

Therefore, the slow rate determining step is unimolecular and involves only *tert*-butyl bromide. The following mechanism is proposed for the above reaction.

1.

2.

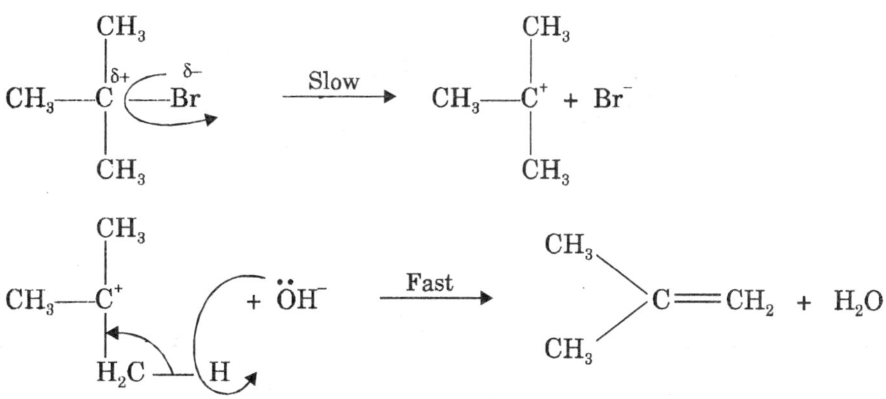

E_2-mechanism. E_2 means ***elimination biomolecular.*** Let us consider the reaction between some primary alkyl halide such as ethyl bromide and alcoholic KOH to form an alkene.

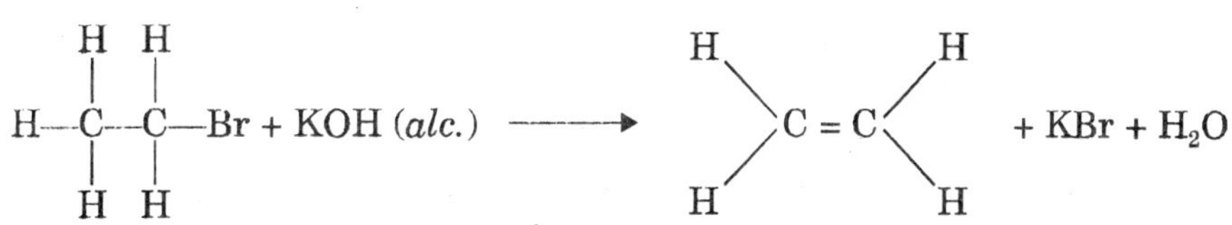

From kinetic studies it has been found that rate of this reaction depends both upon concentration of alkyl halide and concentration of KOH.

$$\text{rate} \propto [C_2H_5Br][OH^-]$$

Therefore, the rate determining step involves both C_2H_5Br and OH^- ions and hence is bimolecular. For this reaction a single step concerted mechanism (known as E_2-mechanism) is proposed.

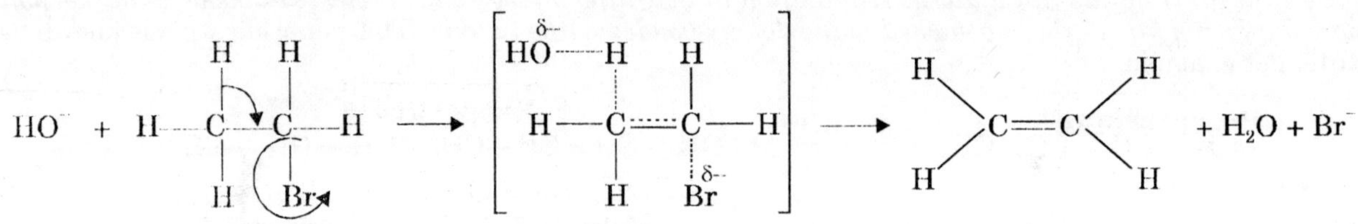

Transition state

III. Reactions with Metals

1. Reaction with sodium (Wurtz reaction)

$$2CH_3I + 2Na \xrightarrow{\text{ether}} \underset{\text{Ethane}}{CH_3 - CH_3} + 2NaI$$

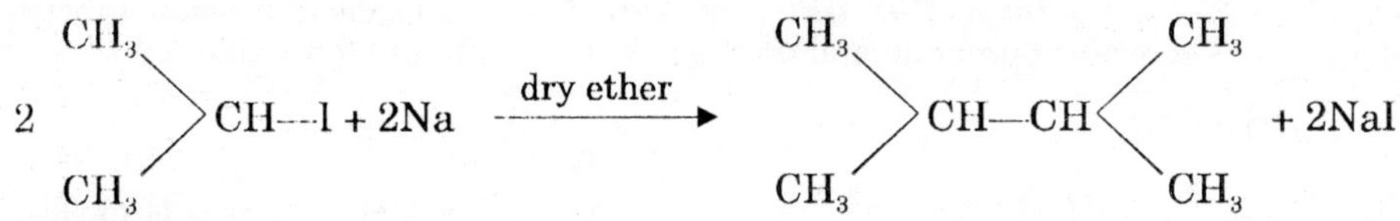

2-Iodopropane 2, 3-Dimethyl butane

The tertiary halides do not undergo this reaction.

2. Reaction with magnesium. Alkyl halides react with magnesium in the presence of dry ether to form alkyl magnesium halides **(Grignard reagents).**

$$CH_3I + Mg \xrightarrow{\text{Dry ether}} \underset{\text{Methyl magnesium iodide}}{CH_3MgI}$$

3. Reaction with Zinc (Frankland reaction)

$$2C_2H_5Br + Zn \longrightarrow C_2H_5 - C_2H_5 + ZnBr_2.$$

IV. Reduction

1. Reaction with H_2/Ni

$$R - X + H_2 \xrightarrow[525\ K]{Ni} R{-}H + HX$$

2. Reaction with zinc-copper couple. In the presence of alcohols (95% ethyl alcohol), zinc-copper couple reduces haloalkanes to alkanes.

$$\underset{(Zn-Cu\ \text{couple})}{Zn} + 2C_2H_5OH \longrightarrow \underset{\text{Zinc ethoxide}}{Zn(OC_2H_5)_2} + 2H$$

$$RX + 2H \longrightarrow \underset{\text{Alkane}}{R{-}H} + HX$$

PREPARATION AND PROPERTIES OF CHLOROFORM

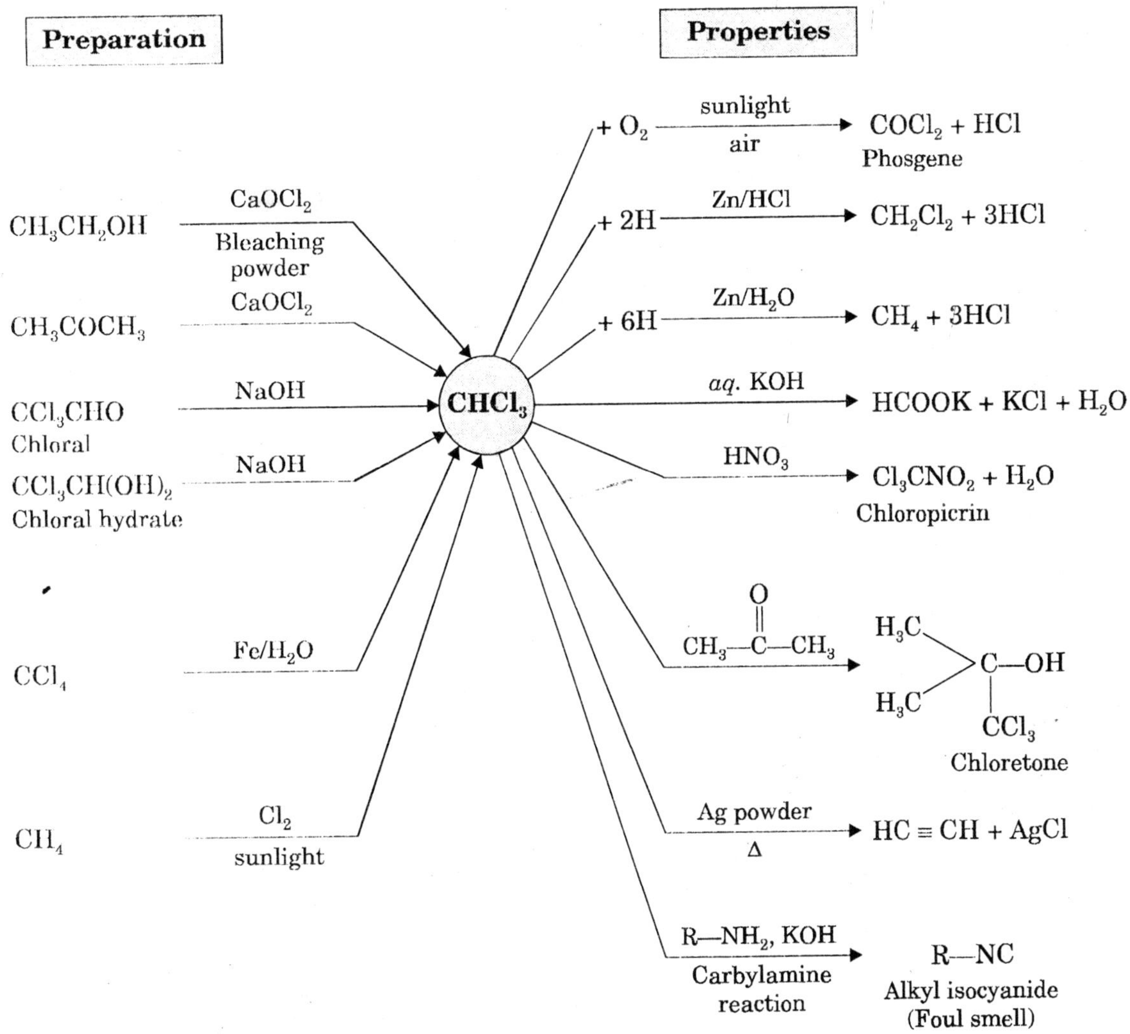

PREPARATION AND PROPERTIES OF CARBON TETRACHLORIDE

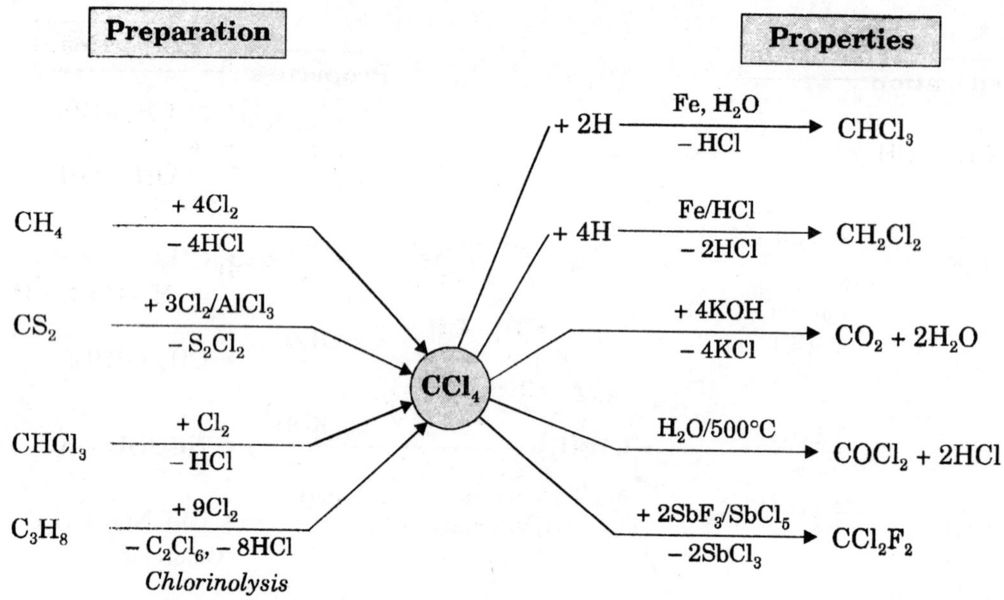

PREPARATION AND PROPERTIES OF ETHYLIDENE DICHLORIDE

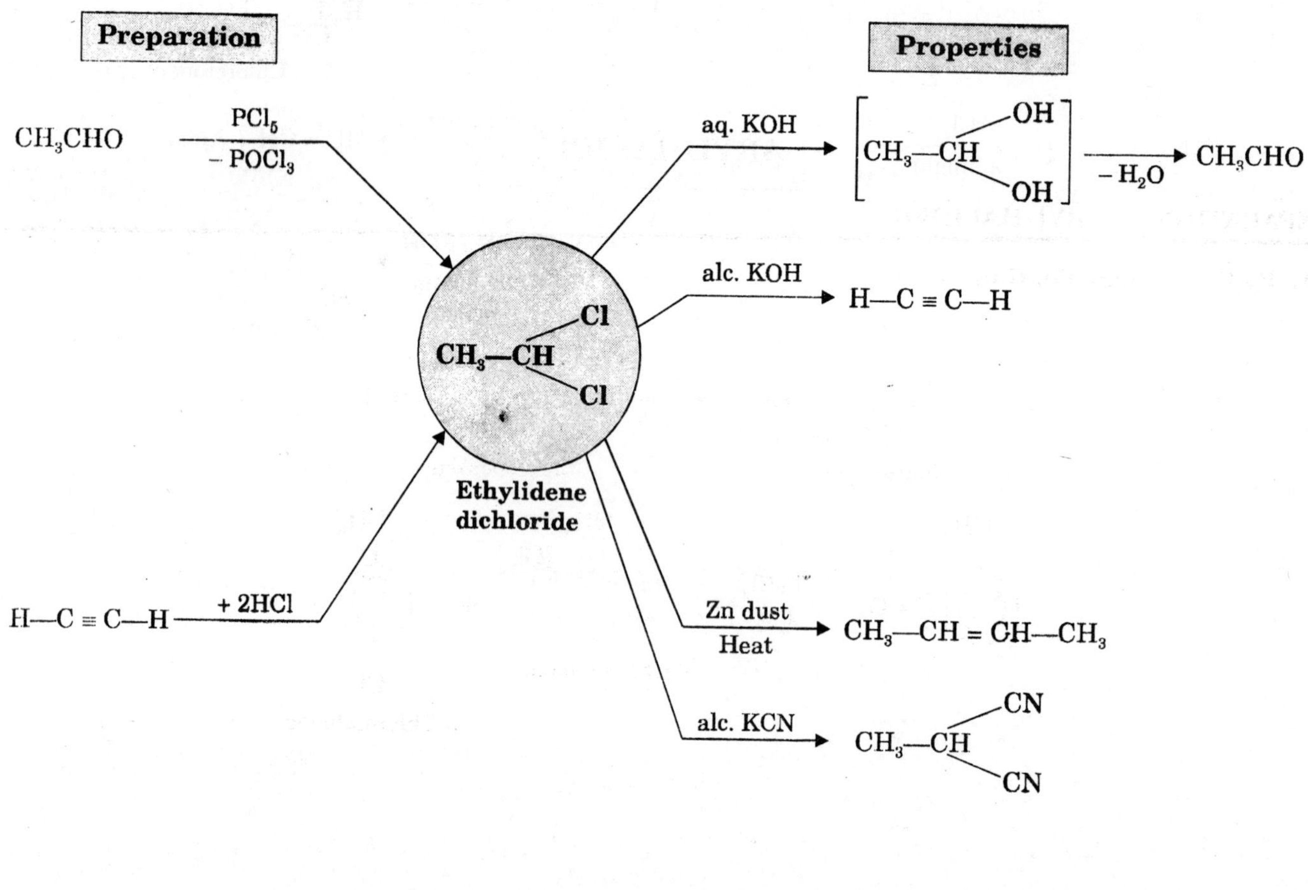

PREPARATION AND PROPERTIES OF ETHYLENE DICHLORIDE

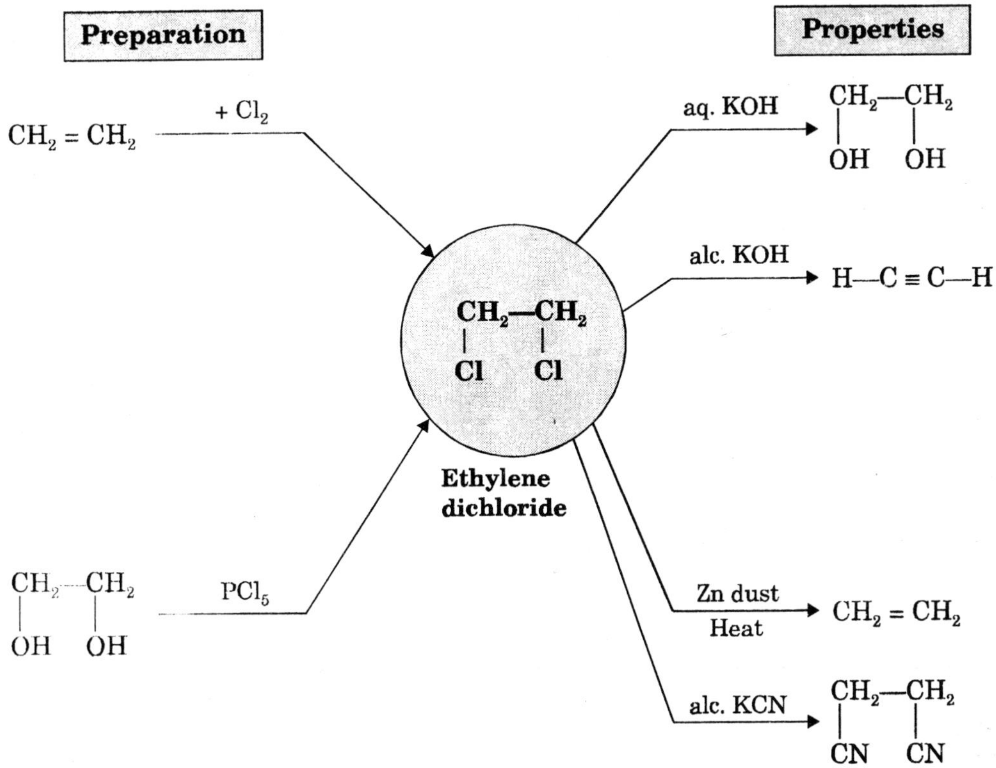

ARYL HALIDES

PREPARATION OF ARYL HALIDES

1. **By Direct Halogenation**

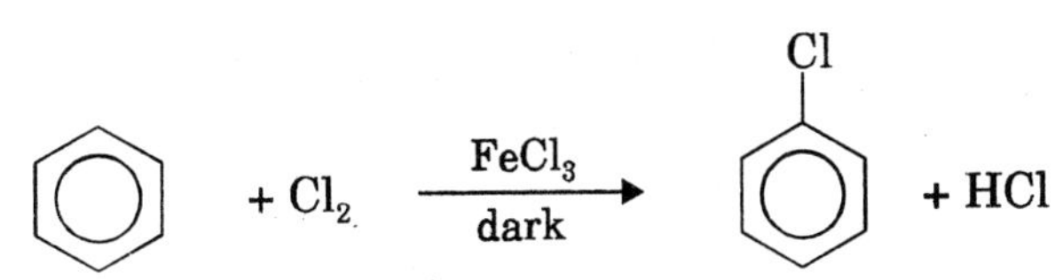

Benzene Chlorobenzene

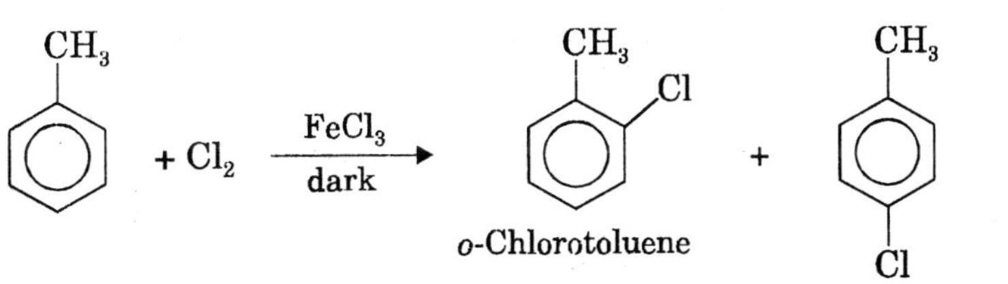

o-Chlorotoluene *p*-Chlorotoluene

2. From Diazonium Compounds

In this method diazonium salt is treated with CuCl and HCl or CuBr and HBr to give the corresponding haloarene.

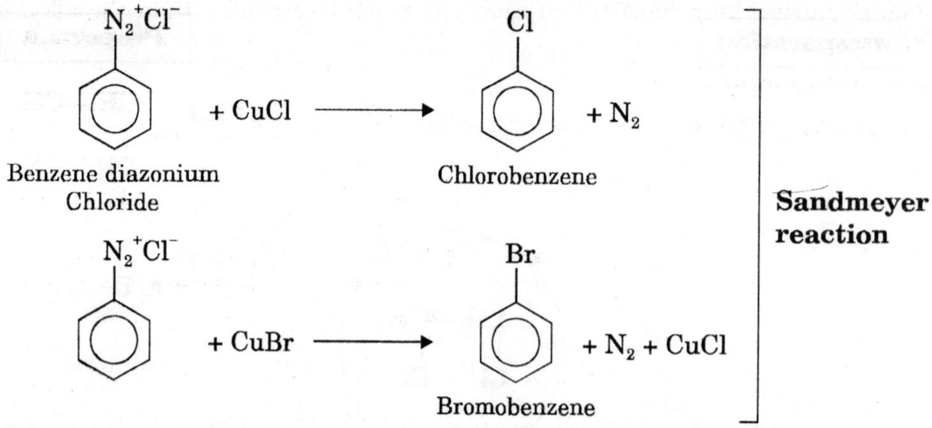

The reaction of diazonium salts with CuCl and CuBr is known as **Sandmeyer reaction.**

If instead of CuCl and CuBr, copper powder and HCl or HBr is used, the reaction is called **Gatterman's reaction.**

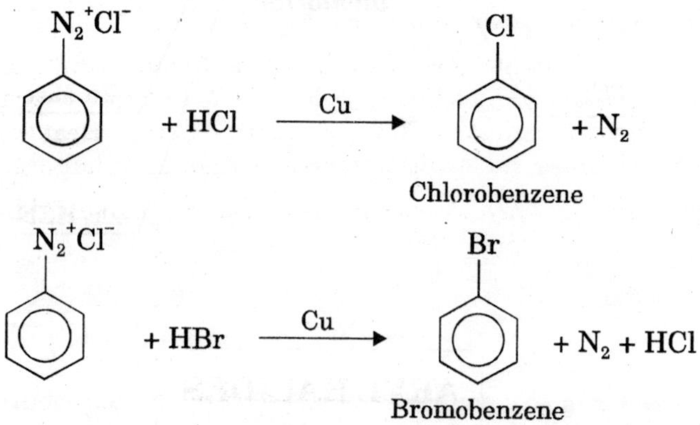

Iodoarenes are obtained by the reaction of benzene diazonium salts with KI.

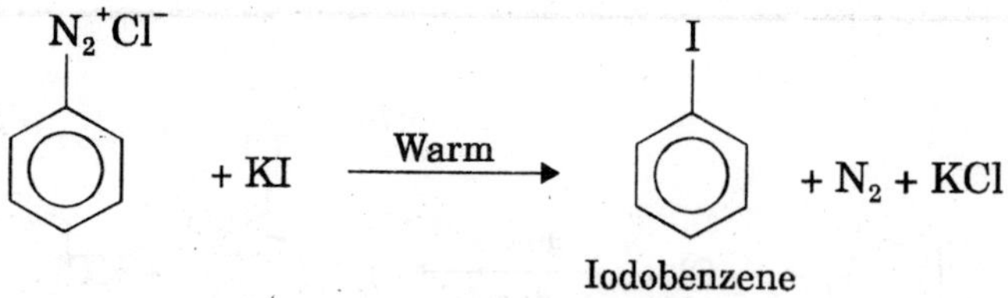

Fluoroarenes are obtained by the reaction of corresponding diazonium salt with fluoboric acid to produce diazonium fluoborate which on heating produces fluorobenzene.

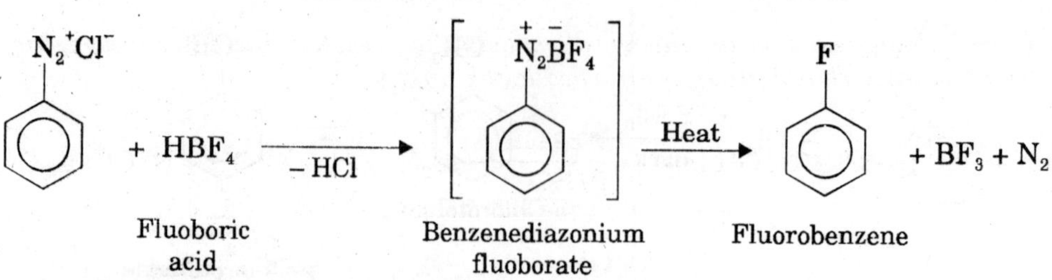

This reaction is called **Balz-Schiemann reaction.**

REACTIONS OF ARYL HALIDES

I. Nucleophilic Substitution Reactions

Aryl halides undergo nucleophilic substitution reactions much less readily than the alkyl halides. This can be explained as follows :

(a) *Stabilization by resonance*

Aryl halides are resonance hybrid of the following structures :

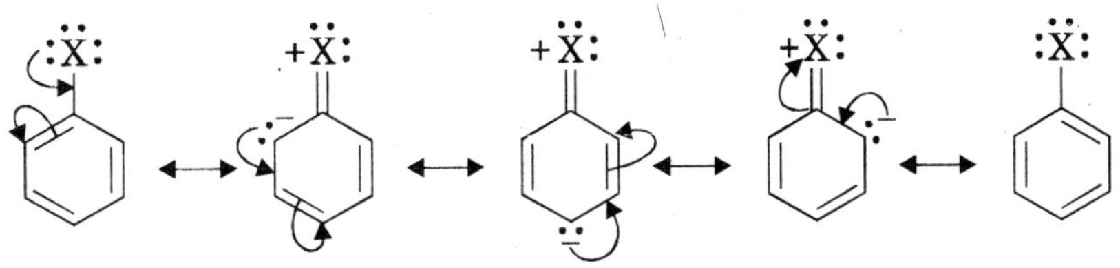

I II III IV V

The contributing structures **II, III** and **IV** indicate that C—X bond has partial double bond character. As a result the C—X bond in halobenzene is shorter and hence, stronger as compared to that in alkyl halides. Thus, cleavage of C—X bond in halobenzene becomes difficult which makes it less reactive towards nucleophilic substitution.

(b) *sp^2 hybrid state of the carbon atom carrying halogen*

In aryl halides carbon atom, carrying halogen is sp^2 hybridized whereas in alkyl halides it is sp^3 hybridized. sp^2 hybrid orbital is of smaller size than sp^3 hybrid orbital because of greater *s*-character and hence the bond formed by its overlap will have smaller bond length and greater strength. Therefore, aryl halides having stronger C—Cl bond are less reactive towards nucleophilic substitution reactions than alkyl halides.

On similar lines we can explain why vinyl halides are less reactive than alkyl halides.

I II

Here also carbon-halogen bond acquires double bond character due to contribution from structure (II).

Because of less reactivity aryl halides undergo nucleophilic substitution reactions under drastic conditions.

1. Replacement by Hydroxy Group

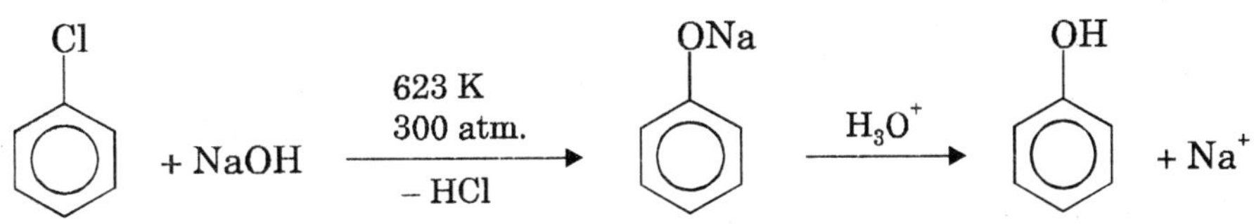

Chlorobenzene Sod. phenoxide Phenol

The reactivity of the haloarenes towards nuleophilic substitution reaction increases if some electron withdrawing group such as nitro, cyano, carboxyl group is attached to the ring.

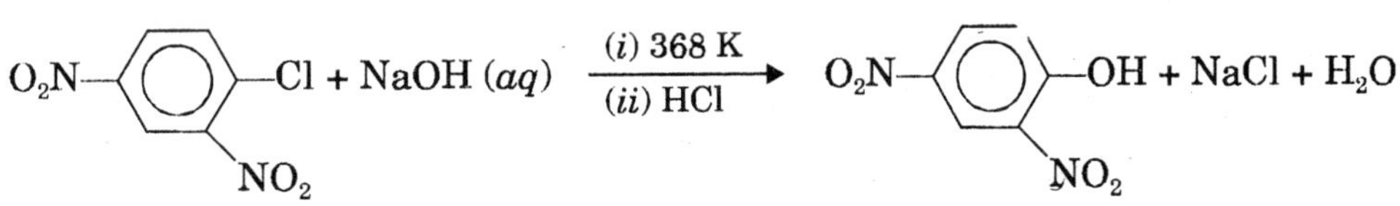

2. Replacement by Cyano Group

Bromobenzene Cyanobenzene

3. Replacement by Amino Group

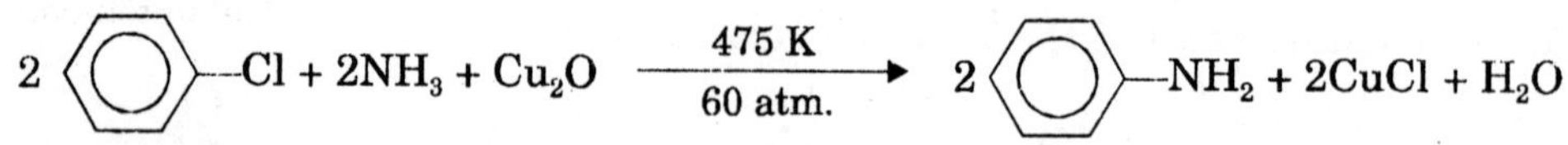

Chlorobenzene Aniline

II. Reactions with Metals

1. Action with Magnesium (*Formation of Grignard Reagents*)

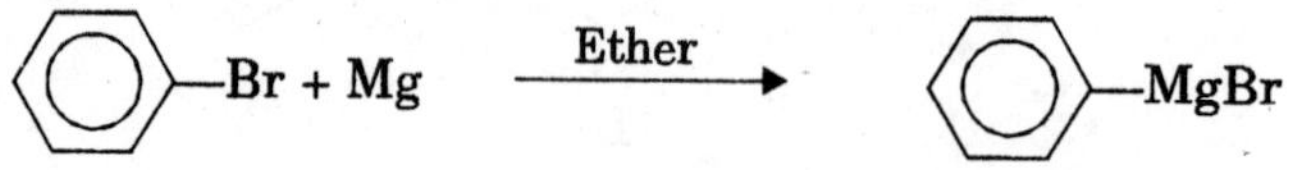

Bromobenzene Phenyl magnesium bromide

Aryl chlorides, however, form Grignard's reagent in the presence of tetrahydrofuran (THF) as solvent.

2. Reaction with Sodium (*Fittig Reaction*)

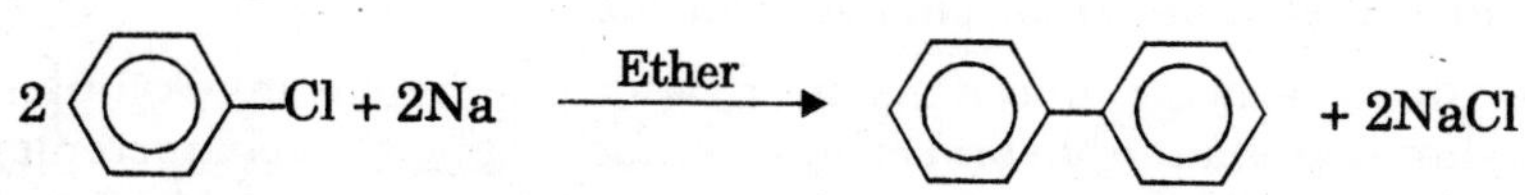

Chlorobenzene Diphenyl

However, aryl halides when treated with haloalkane and sodium in dry ether undergo **Wurtz Fittig reaction**.

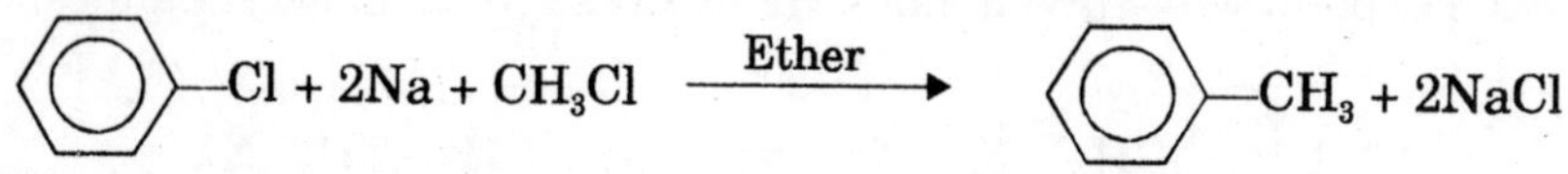

Chlorobenzene Toluene

3. Reaction with Copper Powder (*Ulmann Reaction*)

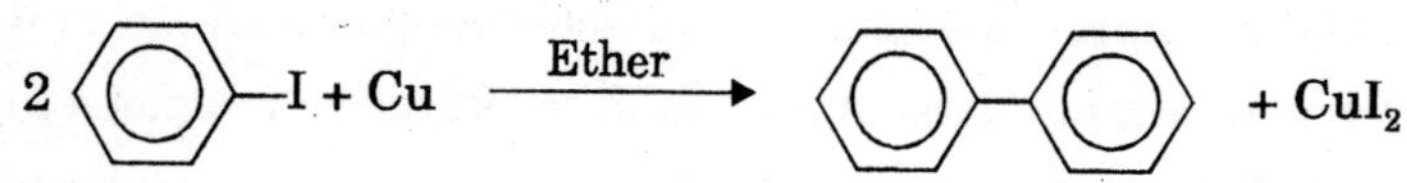

Iodobenzene Diphenyl

III. Reduction

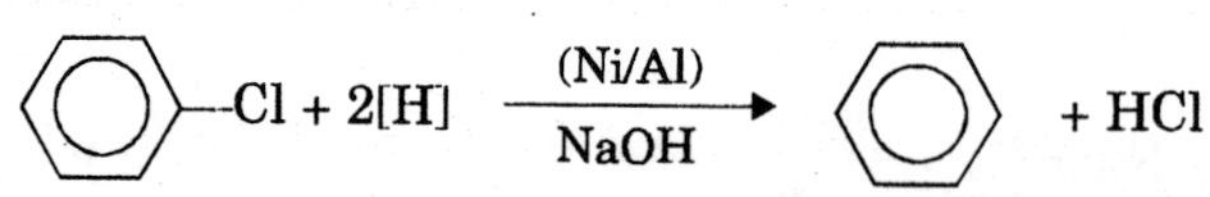

Chlorobenzene Benzene

IV. Electrophilic Substitution Reactions

- ***Aryl halides are less reactive than benzene towards electrophilic substitution reactions.***
- ***In aryl halides electrophilic substitution reactions take place mainly or predominantly at ortho and para positions.***

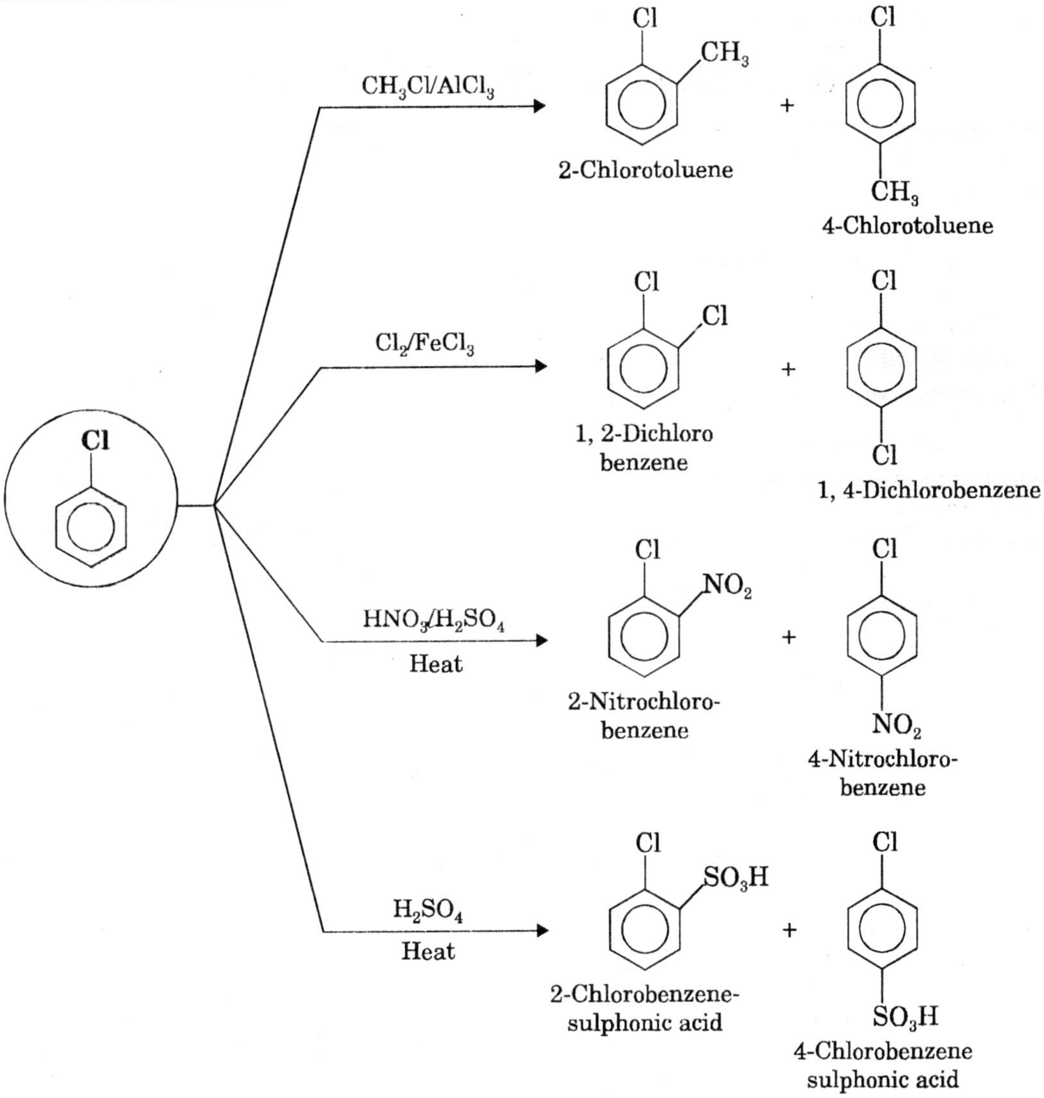

ANALYSIS OF ALKYL AND ARYL HALIDES

1. **Silver nitrate test**

   In this test, the halide is warmed with aqueous or alcoholic KOH. The solution is then acidified with dil. HNO_3 followed by addition of $AgNO_3$ solution.

   Alkyl, benzyl and allyl halides give precipitate.

   Aryl and vinyl halides do not give precipitate.

2. **Lassaigne's Test**

   Lassaigne's extract is boiled with dilute nitric acid to expel all the gases, if evolved. Then silver nitrate solution is added to the resulting solution.

   If any of the halogen is present, a precipitate is formed :

   (*i*) *A white precipitate soluble in ammonium hydroxide* indicates the presence of *chlorine.*

   (*ii*) *A dull yellow precipitate partially soluble in ammonium hydroxide* indicates the presence of *bromine.*

   (*iii*) *A bright yellow precipitate, almost insoluble in ammonium hydroxide*, indicates the presence of *iodine.*

3.2. IMPORTANT FACTS AND TERMS

Add to Knowledge

- For C_4H_9Br, four isomeric bromoalkanes are possible, one of which can exhibit enantiomerism.
- For $C_5H_{11}Br$, eight isomeric bromoalkanes are possible. Among these, three are primary (1°), four are secondary (2°) and one is tertiary (3°).
- Thionyl chloride ($SOCl_2$) is preferred in the preparation of alkyl chlorides from alcohols because in this case side products are gaseous and can be easily expelled during distillation.
- **Finkelstein Reaction** is used to convert alkyl chlorides and bromides into iodides.
- **Swarts reaction** is used to convert alkyl chlorides into fluorides by reacting with inorganic fluorides such as AsF_3, HgF_2, SbF_3, etc.

  $CH_3Cl + HgF_2 \longrightarrow 2CH_3F + HgCl_2$
- **Hunsdiecker Reaction** results in the formation of alkyl halide (chloride or bromide) one carbon less than the starting compound.

$$\underset{\text{3 carbon atoms}}{CH_3CH_2COOAg} + Br_2 \xrightarrow[\text{reflux}]{CCl_4} \underset{\text{2 carbon atoms}}{CH_3CH_2Br} + CO_2 + AgBr$$

- **Sandmeyer's Reaction.** The reaction of benzene diazonium salts with CuCl or CuBr or CuCN in the presence of HCl, HBr or KCN respectively is known as Sandmeyer's reaction.

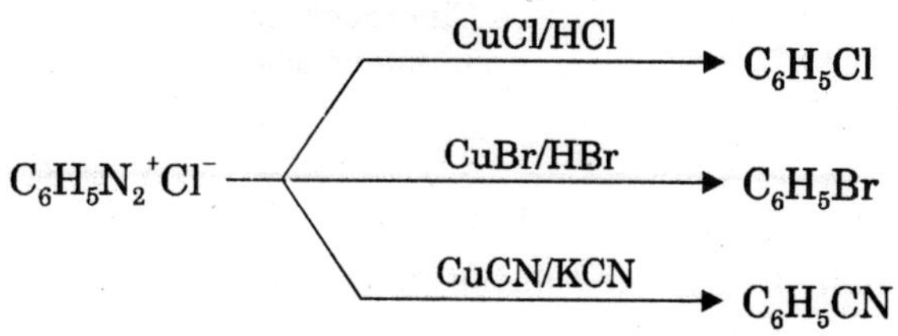

- For same alkyl group, boiling points of alkyl halides are in the order : RI > RBr > RCl > RF.
- For same halogen boiling points of halides increase with increase in size of alkyl group.
- Among isomeric halides boiling point decreases with increase in branching in alkyl group.
- Alkyl halides are insoluble in water due to their inability to form hydrogen bonds with water.
- Density of alkyl halides in the order RI > RBr > RCl. Alkyl chlorides are lighter than water whereas bromides and iodides are heavier than water.
- **Dipole moments of alkyl halides** are in the order :

  R—Cl > R—F > R—Br > R—I

  Dipole moment of R—F is less than that of R—Cl due to smaller size of F atom.
- **Nucleophilic substitution reactions** are the characteristic reactions of alkyl halides.
- These reactions may proceed through S_N^1 and S_N^2 mechanism.

  The order of reactivity of alkyl halides through S_N^1 mechanism is 3° > 2° > 1°. The order of reactivity of alkyl halides through S_N^2 mechanism is 1° > 2° > 3°.
- **Saytzeff Rule.** During dehydrohalogenation the more substituted alkene is the preferred product.
- The order of reactivity of various alkyl halides towards dehydrohalogenation reaction is 3° > 2 ° > 1°.
- Dehydrohalogenation may follow E_1 or E_2 mechanism.
- Silver salts of carboxylic acids on reaction with I_2 yield esters. The reaction is known as **Birn baurn Simonini reaction.**

  $2RCOOAg + I_2 \longrightarrow RCOOR + CO_2 + AgI$
- **Freund Reaction**

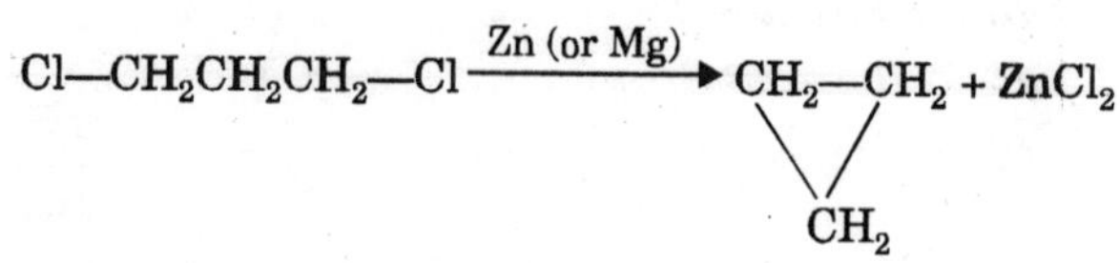

- **Frankland Reaction**

  $2C_2H_5Br + Zn \rightarrow C_2H_5 - C_2H_5 + ZnBr_2$. This reaction is similar to Wurtz reaction.
- **Stereoselective Reaction.** A reaction that yields predominantly one stereoisomer (or one pair of enantiomers) of several possible diastereomers.

Illustration

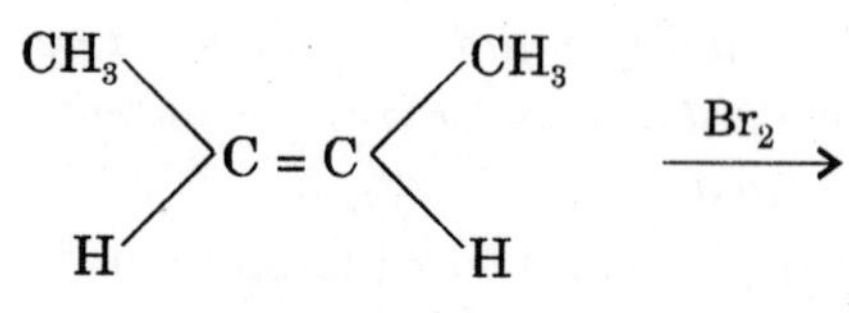

cis-2-Butene

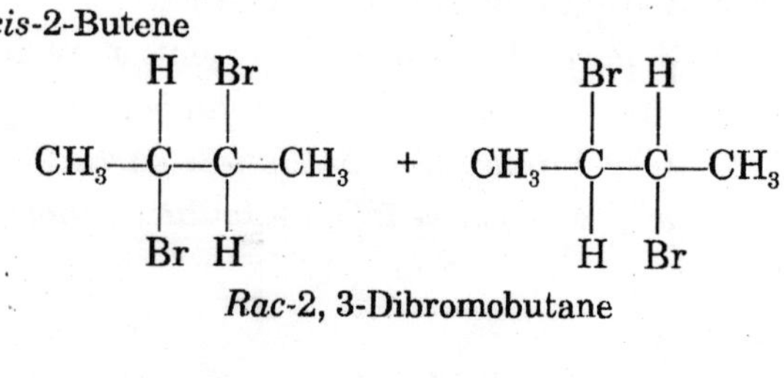

Rac-2, 3-Dibromobutane

- **Stereospecific Reaction.** A reaction in which stereochemically different molecules react differently.

Illustration

cis-2-Butene on reaction with Br_2 gives a *rac*-2, 3-dibromobutane while *trans*-2-butene on reaction with Br_2 give *meso*-2, 3-dibromobutane as the product.

- **Wurtz Reaction.** For preparing symmetrical alkanes.

$$R—X + Na \xrightarrow{\text{ether}} R—R + NaX$$

Methane cannot be prepared by this method.

- C—X bond in aryl halides is shorter, stronger and less polar than in alkyl halides.
- Aryl halides undergo substitution less readily than alkyl halides. Reactivity towards nucleophilic substitution increases if some electron withdrawing group such as $—NO_2$ is present at ortho or para positions with respect to halogen.
- **Ulmann Reaction**

$$2C_6H_5I + Cu \xrightarrow{\Delta} C_6H_5 - C_6H_5 + CuI_2$$

- **Fittig Reaction**

$$2C_6H_5I + 2Na \xrightarrow{\text{ether}} C_6H_5 - C_6H_5 + 2NaI$$

- ***C—X bond in aryl halides is less polar than in alkyl halides.***
- ***C—X bond in aryl halides is shorter than in alkyl halides.***
- Haloarenes are less reactive than benzene towards electrophilic substitution reactions.
- CCl_4 is stable to heat and is used as fire extinguisher under the name **pyrene.**
- 1, 1, 2, 2-Tetrachloroethane (Acetylene tetrachloride) is called **Westron.** It is used as industrial solvent.

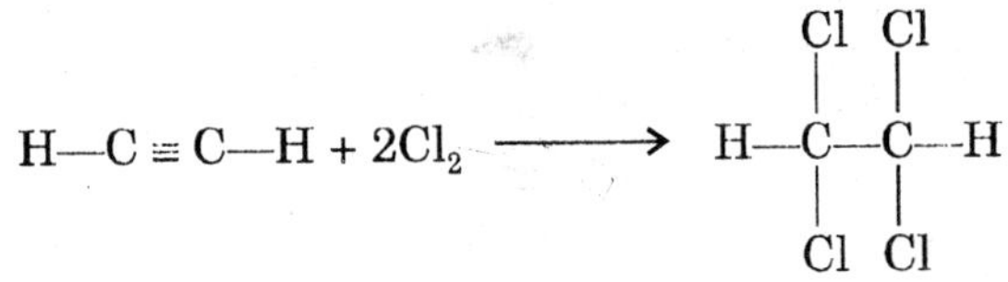

Westron

- Trichloroethene is known as **Westrosol.** It is prepared by dehydrohalogenation of Westron.

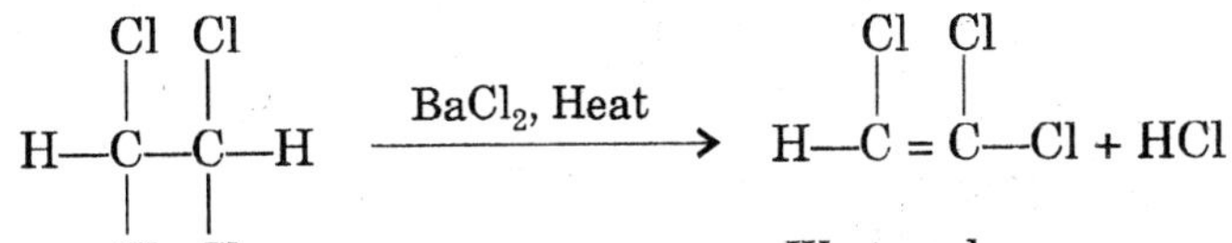

Westrosol

- **D.D.T. :** 1, 1, 1-Trichloro-2, 2-bis (*p*-chlorophenyl) ethane
- $\mathbf{CF_2Cl_2}$ **:** (Freon-12) is used as refrigerant.
- IUPAC name of B.H.C. is 1, 2, 3, 4, 5, 6. hexachlorocyclohexane.
- Aryl and vinyl halides do not give precipitate with $AgNO_3$ solution.

3.3. SOLVED PROBLEMS

Example 1. *When $CH_3CHClCH_2CH_3$ is treated with alcoholic KOH, the major product is*

(a) $CH_3-CH=CH-CH_3$

(b) $CH_2=CH-CH_2-CH_3$

(c) $CH_3-CHOH-CH_2-CH_3$

(d) $CH_3-CH_2-CH_2-CH_3$.

Solution. Alcoholic KOH brings about dehydrohalogenation in accordance with Saytzeff's rule. The more substituted alkene is the major product. Hence, the correct answer is **(*a*)**.

Example 2. *Which of the following is called Sandmeyer reaction ?*

(a) $2HCHO \xrightarrow{NaOH} CH_3OH + HCOONa$

(b) 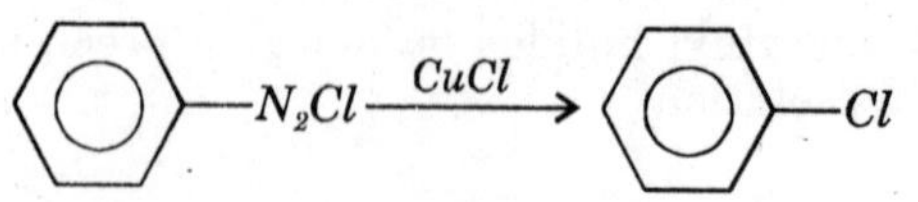

(c) 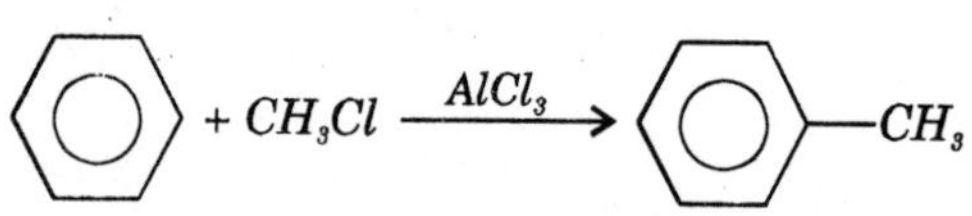

(d) 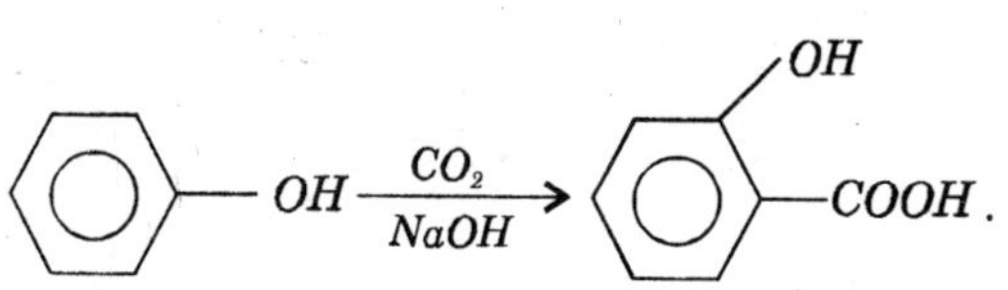

Solution. The correct answer is **(*b*)**. Sandmeyer reaction involves reaction between aromatic diazonium salts and CuCl (or CuBr) to form corresponding aryl halide.

Example 3. *The reaction of toluene with chlorine in the presence of ferric chloride gives mainly*

(a) *m-Chlorotoluene*

(b) *Benzyl chloride*

(c) *o- and p-Chlorotoluene*

(d) *Benzal chloride.*

Solution. Methyl group in toluene is *ortho, para* directing. Hence, the correct answer is **(*c*)**.

Example 4. *The reaction of tert-butylbromide with sodium methoxide produces mainly*

(a) *iso-Butene*

(b) *iso-Butylene*

(c) *tert-Butyl methyl ether*

(d) *Sodium tert-butoxide.*

Solution. Tertiary halides on reaction with strong bases give alkenes as the major product through dehydrohalogenation.

$$CH_3-\underset{CH_3}{\overset{CH_3}{C}}-Br \longrightarrow CH_3-\overset{CH_3}{C}=CH_2$$

Hence, the correct answer is **(*b*)**.

Example 5. *A compound A has a molecular formula C_2Cl_3OH. It reduces Fehling solution and on oxidation gives a monocarboxylic acid B. A is obtained by the action of chlorine on ethyl alcohol. A is*

(a) *Methyl chloride* (b) *Monochloroacetic acid*

(c) *Chloral* (d) *Chloroform.*

Solution. Cl_2 oxidizes C_2H_5OH to chloral

$$C_2H_5OH \xrightarrow{Cl_2} CH_3CHO \xrightarrow{Cl_2} CCl_3.CHO$$

Hence, the correct answer is **(*c*)**.

Example 6. *Chloroform when treated with aniline and alcoholic KOH forms*

(a) *Phenyl cyanide* (b) *Phenyl isocyanide*

(c) *Phenyl cyanate* (d) *Phenyl isocyanate.*

Solution. This is carbylamine reaction. The product is phenyl isocyanide. Hence, the correct answer is **(*b*)**.

Example 7. *C—Cl bond of chlorobenzene in comparison to C—Cl bond of methyl chloride is*

(*a*) *longer and weaker* (*b*) *shorter and weaker*

(*c*) *shorter and stronger* (*d*) *longer and stronger.*

Solution. C—Cl bond in chlorobenzene has partial double bond character due to resonance. Hence, it is shorter and stronger than the C—Cl bond in CH_3Cl.

Hence, the correct answer is **(*c*).**

Example 8. *C—X bond is strongest is*

(*a*) CH_3Cl (*b*) CH_3Br

(*c*) CH_3F (*d*) CH_3I.

Solution. C—X bond is strongest in CH_3F due to smallest size of F atom among halogens.

Hence, the correct answer is **(*c*).**

Example 9. *Which of the following molecules has the highest dipole moment ?*

(*a*) CH_3Br (*b*) CH_3F

(*c*) CH_3Cl (*d*) CH_3I.

Solution. The dipole moment of a bond is determined by the polarity of the bond as well as by the bond length. The maximum dipole moment among the given compounds is for CH_3Cl. Although C—Cl bond is less polar than C—F bond but due to longer C—Cl bond length, the C—Cl bond has greater dipole moment than C—F bond.

Hence, the correct answer is **(*c*).**

Example 10. *Which of the following has the highest boiling point ?*

(*a*) 

(*b*) 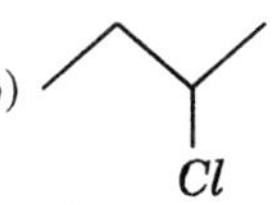

(*c*) 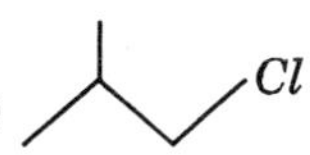

(*d*) 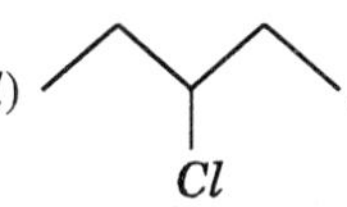

Solution. The correct choice is **(*d*).** The compound in (*d*) has the maximum molecular mass and largest surface area of the molecule. Hence, it would have highest boiling point due to strongest interparticle forces.

3.4. QUESTION BANK

Develop Confidence

Nomenclature ; Isomerism and Classification

1. Which of the following is an example of aralkyl halide ?
(*a*) *p*-Chlorotoluene (*b*) Chlorobenzene
(*c*) Allyl chloride (*d*) Benzyl chloride.

2. In which of the following alkyl halides all the hydrogen atoms are equivalent ?
(*a*) *iso*-Butyl chloride (*b*) *neo*-Pentyl chloride
(*c*) *n*-Butyl chloride (*d*) *tert-Butyl chloride.*

3. How many structural isomers are possible for C_4H_9Br ?
(*a*) 4 (*b*) 3
(*c*) 2 (*d*) 5.

4. How many primary halides are possible for $C_5H_{11}Br$?
(*a*) 1 (*b*) 2
(*c*) 3 (*d*) 4.

5. Which of the following alkyl halides is *iso*-butyl bromide ?
(*a*) $CH_3CH_2\underset{\underset{Br}{|}}{C}HCH_3$ (*b*) $CH_3CH_2CH_2CH_2Br$
(*c*) $CH_3-\overset{\overset{CH_3}{|}}{\underset{\underset{CH_3}{|}}{C}}-Br$ (*d*) $CH_3\underset{\underset{CH_3}{|}}{C}HCH_2Br$

6. How many isomeric halogen derivatives including stereoisomers are possible for $C_2H_2Br_2$?
(*a*) 2 (*b*) 3
(*c*) 4 (*d*) 5.

7.. Which of the following an aralkyl halide ?
(*a*) 4-Chlorotoluene
(*b*) Bromobenzene
(*c*) 1, 4-Dichlorobenzene
(*d*) 1-Chloro-1-phenyl ethane.

8. Among the isomers of $C_5H_{11}Cl$, the one which is chiral is
(*a*) 2, 2-Dimethyl 1-chloropropane
(*b*) 2-Chloropentane
(*c*) 2-Methyl-2-chlorobutane
(*d*) 3-Chloropentane.

9. Which of the following compounds can exhibit optical isomerism ?
(*a*) Chlorocyclopropane
(*b*) *cis*-1, 2-Dichlorocyclopropane
(*c*) *trans*-1, 2-Dichlorocyclopropane
(*d*) *trans*-1, 3-Dichlorocyclobutane.

10. Which of the following is not a primary halide ?
(*a*) Isobutyl bromide (*b*) Neo-Pentyl chloride
(*c*) Isopentyl bromide (*d*) Isopropyl chloride.

11. IUPAC name of crotyl chloride is
(*a*) 1-Chloro-2-butene (*b*) 2-Chloro-2-butene
(*c*) 3-Chloro-1-butene (*d*) 4-Chloro-1-butene.

12. Carbon atom holding halogen in aryl halides is
(*a*) sp^2-Hybridized (*b*) sp^3-Hybridized
(*c*) *sp*-Hybridized (*d*) sp^3d-Hybridised.

13. C—Cl bond length is smallest in
(*a*) Allyl chloride (*b*) Vinyl chloride
(*c*) Ethyl chloride (*d*) Benzyl chloride.

Synthesis of Alkyl Halides

14. When silver propionate is treated with Br_2 in CCl_4 as solvent, the product is
(*a*) Propionyl bromide (*b*) *n*-Propyl bromide
(*c*) *iso*-Propyl bromide (*d*) Ethyl bromide.

15. Hunsdiecker reaction is used for preparation of alkyl chlorides and alkyl bromides starting from
(*a*) Diazonium salts
(*b*) Silver salt of carboxylic acids
(*c*) Sodium salt of carboxylic acids
(*d*) Alcohols.

16. *iso*-Butylene can be converted into *tert*-butyl bromide by its reaction with
(*a*) HBr
(*b*) Br_2
(*c*) HBr in the presence of peroxides
(*d*) HOBr.

17. Methane reacts with limited amount of chlorine in the presence of diffused sunlight to form chloromethane. The reaction takes place through ..... as intermediates.
(*a*) Carbocations (*b*) Free radicals
(*c*) Carbanions (*d*) Carbenes.

18. The best reagent for converting an alcohol into the corresponding chloride is
(*a*) PCl_5 (*b*) PCl_3
(*c*) $HCl/ZnCl_2$ (*d*) $SOCl_2$.

19. Which reagent cannot be used to prepare an alkyl chloride from an alcohol ?
(*a*) HCl + anhyd. $ZnCl_2$ (*b*) NaCl
(*c*) PCl_5 (*d*) $SOCl_2$.

20. Benzyl chloride can be prepared by reacting
(*a*) Toluene with Cl_2 in the presence of $FeCl_3$
(*b*) Benzene with CH_3Cl in the presence of $AlCl_3$
(*c*) Toluene with Cl_2 in the presence of sunlight
(*d*) Benzene with Cl_2 in the presence of $FeCl_3$.

Properties of Alkyl Halides

21. Which of the following has the highest boiling point ?
(*a*) 1-Chloropentane
(*b*) 2-Chloropentane
(*c*) 3-Chloropentane
(*d*) All have equal boiling point.

22. Which of the following alkyl halides has the maximum density ?
(*a*) C_2H_5Cl (*b*) C_2H_5Br
(*c*) C_2H_5I (*d*) C_2H_5F.

23. Dipole moment is shown by
(*a*) 1, 4-Dichlorobenzene
(*b*) *cis*-1, 2-Dichloroethylene
(*c*) *trans*-1, 2-Dichloroethylene
(*d*) Tetrachloromethane.

24. Out of the following compounds which one will have zero dipole moment ?
(*a*) Chloromethane
(*b*) Dichloromethane
(*c*) Trichloromethane
(*d*) None of these compounds.

25. Which of the following is lighter than water ?
(*a*) 1-Chlorobutane (*b*) 1-Bromopropane
(*c*) 2-Bromopropane (*d*) 1-Bromobutane.

26. Which of the following compounds will react with ethanolic KCN ?
(*a*) Chlorobenzene (*b*) Vinyl chloride
(*c*) Allyl chloride (*d*) 4-Chlorotoluene.

27. Treatment of ammonia with excess ethyl chloride will yield
(*a*) Triethyl amine
(*b*) Diethyl amine
(*c*) Ethyl amine
(*d*) Tetraethylammonium chloride.

28. Ethyl bromide reacts with Zn-Cu couple in the presence of ethyl alcohol to form
(*a*) Butane (*b*) Diethyl ether
(*c*) Methane (*d*) Ethane.

29. Identify Z in the following sequence

$$CH_3CH_2I \xrightarrow{KCN} X \xrightarrow{Conc.\ HCl} Y \xrightarrow[\Delta]{dil.\ HCl} Z$$

(*a*) CH_3COCl (*b*) CH_3CONH_2
(*c*) CH_3COOH (*d*) CH_3CH_2COOH.

30. When tert-butyl chloride is made to react with sodium methoxide, the major product is
(*a*) Dimethyl ether
(*b*) Di-tert butyl ether
(*c*) Tert-Butyl methyl ether
(*d*) Isobutylene.

31. $$CH_3 - CH_2 - CH_2 - Cl \xrightarrow[KOH]{alc.} B \xrightarrow{HBr} C \xrightarrow[ether]{Na} D$$

In the above sequence of reaction, the product D is
(*a*) Propane (*b*) 2, 3-Dimethylbutane
(*c*) Hexane (*d*) Allyl bromide.

32. NBS is a specific reagent for
(*a*) Nucleophilic substitution
(*b*) Electrophilic substitution
(*c*) Allylic substitution
(*d*) Electrophilic addition.

33. 2-Chloro-2-methyl butane, on reaction with alc. KOH gives X as the major product. X is
(*a*) 2-Methyl-1-butanol (*b*) 2-Methyl-1-butene
(*c*) 2-Methyl-2-butene (*d*) 2-Methyl-2-butanol.

34. 2-Chloro-2-methyl propane, on reaction with aq. KOH gives X as the major product. X is
(*a*) 2-Butene (*b*) 2-Methyl propene
(*c*) 1-Butene (*d*) 2-Methyl-2-propanol.

35. An alkyl halide may be converted into an alkene by
(*a*) Addition (*b*) Substitution
(*c*) Elimination (*d*) Hydrogenation.

36. Which of the following compounds would be hydrolysed most easily ?
(*a*) C_2H_5Cl (*b*) C_2H_5Br
(*c*) C_2H_5F (*d*) C_2H_5I.

37. Which of the following compounds would be hydrolysed most easily ?
(*a*) C_2H_5Br (*b*) CH_3Br
(*c*) $CH_2 = CH - Br$ (*d*) $CH_2 = CH - CH_2Br$.

38. Ethylene chloride and ethylidene chloride are isomeric compounds. Identify the statement that is not applicable to both of them.
(*a*) Both of them are dihalogen derivatives
(*b*) Both of them give Beilstein test

(c) Both of them react with alc. KOH and give same product

(d) Both of them react with aqueous KOH and give same product.

39. A mixture of ethyl iodide and *n*-propyl iodide is subjected to Wurtz reaction. Which of the following compounds is not formed during the reaction ?

(a) Butane (b) Propane

(c) Pentane (d) Hexane.

40. When aqueous KOH reacts with *n*-propyl chloride, the product is

(a) Propylene (b) 2-Propanol

(c) 1-Propanol (d) 1-Propenol.

41. An alkyl halide on reaction with sodium in the presence of ether gives 2, 2, 5, 5-tetramethyl hexane. The alkyl halide possibly is

(a) 1-Chloropentane

(b) 1-Chloro-2, 2-dimethyl propane

(c) 3-Chloro-2, 2-dimethyl butane

(d) 2-Chloro-2-methyl butane.

42. Which of the following does not form Grignard reagent on reaction with Mg in the presence of ether ?

(a) Chloroethane (b) 1-Chloropropane

(c) Bromobenzene (d) Chlorobenzene.

43. Chlorobenzene on reaction with CH_3Cl in the presence of $AlCl_3$ will give

(a) Toluene

(b) *m*-Chloro toluene

(c) *p*-Chloro toluene

(d) A mixture of *o*- and *p*-chlorotoluene.

44. Most reactive halide towards S_N1 reaction is

(a) *n*-Butyl chloride (b) *sec*-Butyl chloride

(c) *tert*-Butyl chloride (d) Allyl chloride.

45. Mg reacts with RBr best in

(a) $C_2H_5OC_2H_5$

(b) $C_6H_5OCH_3$

(c) $C_6H_5N(CH_3)_2$

(d) Equally in all the three.

46. Photochemical chlorination of an alkane is initiated by a process of

(a) Pyrolysis (b) Peroxidation

(c) Homolysis (d) Rearrangement.

47. Allylic bromination of olefins is generally carried out with

(a) Phenyl magnesium bromide

(b) Dibromo succinic acid

(c) N-Bromo succinimide

(d) Pyridine perbromide.

48. 2, 2-Dichloropropane on alkaline hydrolysis gives

(a) 2, 2-Propanediol (b) Propanol

(c) Propanone (d) Ethanal.

49. Grignard reagent is prepared by the reaction bewteen alkyl halide and

(a) Zinc in the presence of alcohol

(b) Magnesium in the presence of dry ether

(c) Magnesium in the presence of alcohol

(d) Magnesium in the presence of moist ether.

50. Which one of the following is likely to give a precipitate with $AgNO_3$ solution ?

(a) CCl_4 (b) CH_3-CH_2-Cl

(c) $(CH_3)_3CCl$ (d) $CHCl_3$.

51. An alkyl chloride produces a single alkene on reaction with sodium ethoxide and ethanol. The alkene further undergoes hydrogenation to yield-2-methylbutane. Identify the alkyl chloride from amongst the following.

(a) $ClCH_2\ C(CH_3)_2CH_3$

(b) $ClCH_2CH_2CH_2CH_2CH_3$

(c) $ClCH_2CH(CH_3)CH_2CH_3$

(d) $CH_3C(Cl)(CH_3)CH_2CH_3$.

52. Geminal dihalides on hydrolysis give

(a) Ketones (b) Aldehydes

(c) Ketones and aldehydes (d) Alcohols.

53. When ethyl iodide is treated with moist silver oxide, the product formed is

(a) C_2H_5OH (b) $C_2H_5OC_2H_5$

(c) 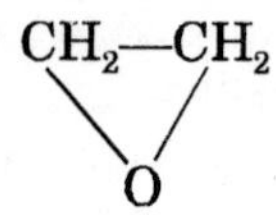

(d) CH_3CHO.

54. The reaction of an alkyl halide with RCOOAg produces

(a) Ester (b) Ether

(c) Aldehyde (d) Ketone.

55. The reaction of 4-bromobenzyl chloride with NaCN in ethanol leads to

(a) 4-Bromobenzyl cyanide

(b) 4-Cyanobenzyl chloride

(c) 4-Cyanobenzyl cyanide

(d) 4-Bromo-2-cyanobenzyl chloride.

56. Grignard reagent is not prepared in aqueous medium but prepared in ether medium, because

(a) The reagent is highly reactive in ether

(b) The reagent is insoluble in water

(c) The reagent becomes inactive in water

(d) The reagent reacts with water.

57. Ethylbromide reacts with sodium-lead alloy to form

(a) Tetraethyl lead (b) Ethyl sodium

(c) Ethane (d) Ethene.

58. Which of the following alkyl halide is *neo*-pentyl chloride ?

(*a*) 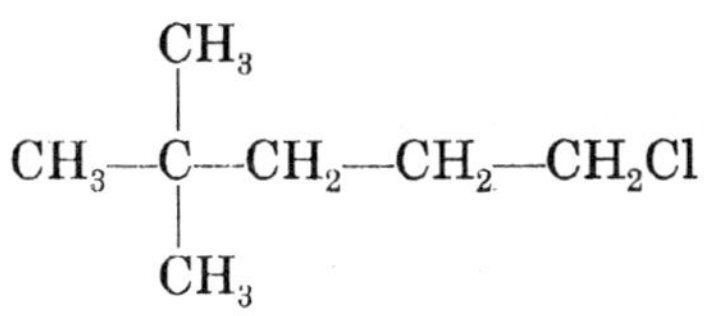

(*b*) 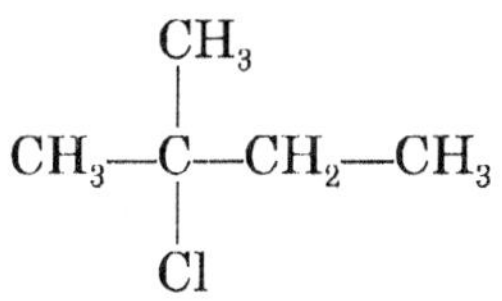

(*c*) 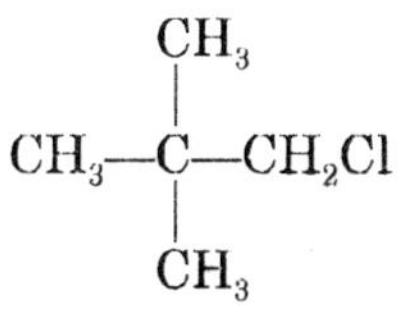

(*d*) 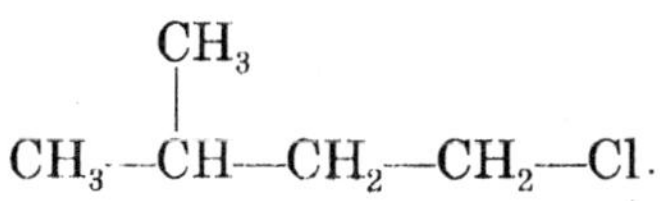

59. An alkyl halide may be converted into an alcohol by

(*a*) Addition (*b*) Elimination

(*c*) Substitution (*d*) Hydrogenation.

60. For the reaction

$C_2H_5OH + HX \longrightarrow C_2H_5X + H_2O$

the order of reactivity is

(*a*) HCl > HBr > HI (*b*) HI > HBr > HCl

(*c*) HBr > HCl > HI (*d*) HI > HCl > HBr.

61. 1-Phenyl-2-chloropropane when treated with alcoholic KOH gives ...... as the major product.

(*a*) 1-Phenylpropene (*b*) 3-Phenylpropene

(*c*) 1-Phenyl-2-propanol (*d*) 3-Phenyl-1-propanol.

62. Which of the following would produce ethyl isocyanide as the major product on reaction with C_2H_5Br ?

(*a*) KCN (*b*) AgCN

(*c*) CH_3CN (*d*) CH_3NC.

63. Which one of the following is most reactive towards nucleophilic substitution reaction ?

(*a*) $CH_2 = CH—Cl$ (*b*) C_6H_5Cl

(*c*) $CH_3CH = CH—Cl$ (*d*) $ClCH_2—CH = CH_2$.

64. Which of the following is not true for S_N1 reactions ?

(*a*) They occur through a single step concerted reaction

(*b*) They are favoured by polar solvents

(*c*) 3° alkyl halides generally react through this mechanism

(*d*) Concentration of nucleophile does not affect the rate of such reactions.

65. The reaction,

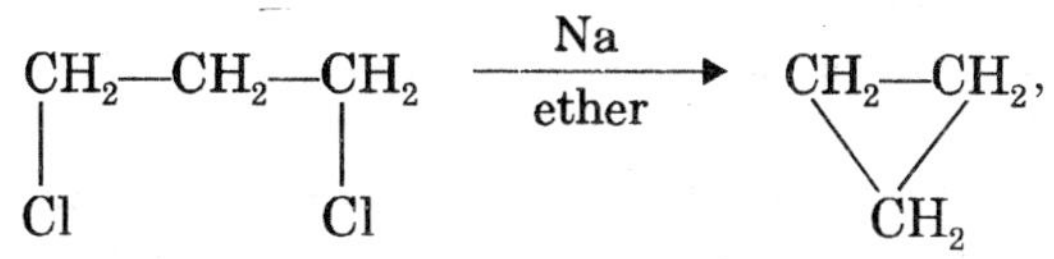

is known as

(*a*) Wurtz reaction (*b*) Fittig reaction

(*c*) Wurtz-Fittig reaction (*d*) Freund reaction.

66. Only two isomeric monochloro derivatives are possible for

(*a*) *n*-Butane (*b*) 2, 4-Dimethylpentane

(*c*) Benzene (*d*) 2, 2-Dimethylpropane.

67. The major product of reaction of alcoholic silver nitrite with ethyl bromide is

(*a*) Ethane (*b*) Nitroethane

(*c*) Ethyl nitrite (*d*) Ethyl alcohol.

68. Isopropylidene on reaction with aqueous KOH yields

(*a*) 1, 2-Propanediol (*b*) Propanone

(*c*) Propanal (*d*) *iso*-Propyl alcohol.

Polyhalogen Compounds

69. IUPAC name of Gammexane is

(*a*) Hexachlorobenzene

(*b*) Benzene hexachloride

(*c*) 1, 2, 3, 4, 5, 6-Hexachlorocyclohexane

(*d*) None of these.

70. Chloral on treatment with aqueous NaOH produces ?

(*a*) Ethanal (*b*) Propanol

(*c*) Chloroform (*d*) Chloroethanal.

71. '*Pyrene*' is the trade name of ...... when used as fire extinguisher

(*a*) CO_2 (*b*) $CHCl_3$

(*c*) CCl_4 (*d*) CH_2Cl_2.

72. '*Westron*' is

(*a*) 1, 1, 2, 2-Tetrachloroethane

(*b*) 1, 1-Dichloropropanone

(*c*) 1, 3-Dichloropropanone

(*d*) 1, 2-Dichloroethene.

73. '*Mustard gas*' is

(*a*) 2, 2′-Dichlorodiethyl sulphide

(*b*) Dichlorodimethyl sulphide

(*c*) Dichlorodiethyl ether

(*d*) None of these.

74. '*Westrosol*' is

(*a*) Trichloroethene (*b*) 1, 1, 1-Trichloroethane

(*c*) 1, 1, 2-Trichloroethane (*d*) Trichloromethanol.

75. Which of the following would not yield iodoform on reaction with NaOH and I_2 ?

(*a*) Ethanol (*b*) Propanal

(*c*) Propanone (*d*) 2-Propanol.

76. On warming with silver powder, chloroform is converted into

(*a*) Acetylene

(*b*) Hexachloroethane

(*c*) 1, 1, 2, 2-tetrachloroethane

(*d*) Ethylene.

77. In the preparation of $CHCl_3$ from ethanol and bleaching powder, the latter provides

(*a*) Cl_2 (*b*) $Ca(OH)_2$

(*c*) Both (*d*) None.

78. The starting substance for the preparation of CHI_3 is any of the following except

(*a*) C_2H_5OH (*b*) CH_3OH

(*c*) CH_3CHO (*d*) $CH_3 - CO - CH_3$.

79. Which of the following with aqueous KOH will give acetaldehyde ?

(*a*) 1, 2-Dichloroethane (*b*) 1, 1-Dichloroethane

(*c*) Chloroacetic acid (*d*) Ethyl chloride.

80. Benzene on reaction with CCl_4 in the presence of $AlCl_3$ gives

(*a*) Benzotrichloride

(*b*) DDT

(*c*) Triphenyl chlorobenzene

(*d*) Triphenyl chloromethane.

81. Which of the following compounds will give a yellow precipitate with iodine and alkali ?

(*a*) Benzyl alcohol (*b*) 1-Phenyl-1-propanol

(*c*) Phenol (*d*) 1-Phenyl-2-propanol.

82. Carbylamine test is performed in alcoholic KOH by heating a mixture of

(*a*) Chloroform and silver powder

(*b*) Chloroform and a primary amine

(*c*) An alkyl halide and a primary amine

(*d*) An alkyl cyanide and a primary amine.

83. Which of the following compounds is used as a refrigerant ?

(*a*) CCl_4 (*b*) CH_2Cl_2

(*c*) CF_4 (*d*) CF_2Cl_2.

84. When ethanol and I_2 are heated in the presence of Na_2CO_3, the yellow crystals obtained are of

(*a*) C_2H_5I (*b*) CH_3I

(*c*) CHI_3 (*d*) CH_2I_2.

85. Iodoform in medicines is used as

(*a*) Antibiotic (*b*) Antiseptic

(*c*) Analgesic (*d*) Antipyretic.

86. DDT is formed from

(*a*) Benzene and chlorobenzene

(*b*) Chloral and chlorobenzene

(*c*) Chloral and benzene

(*d*) Chlorobenzene and chlorine.

87. In the reaction,

$$C_6H_5 - NH_2 \xrightarrow{X} C_6H_5 - NC,$$ the reagent 'X' is

(*a*) Br_2 in chloroform (*b*) KOH in chloroform

(*c*) KCN in alkali (*d*) KCN in chloroform.

88. Reaction of ethylamine with chloroform in alcoholic KOH produces

(*a*) CH_3OH (*b*) CH_3NC

(*c*) C_2H_5NC (*d*) C_2H_5CN.

89. Chloroform on reaction with conc. HNO_3 yields

(*a*) Nitromethane (*b*) Chloropicrin

(*c*) Chloroprene (*d*) CO_2, NO_2 and H_2O.

90. Industrial preparation of chloroform employs acetone and

(*a*) Cl_2 (*b*) $COCl_2$

(*c*) NaCl (*d*) Calcium hypochlorite.

91. Chlorination of CS_2 gives

(*a*) $CHCl_3$ (*b*) CCl_4

(*c*) CS_2Cl_2 (*d*) $COCl_2$.

92. What happens when CCl_4 is treated with $AgNO_3$ solution ?

(*a*) NO_2 will be evolved

(*b*) A white ppt. of AgCl will form

(*c*) CCl_4 will dissolve in $AgNO_3$ solution

(*d*) Nothing will happen.

93. Which of the following compounds gives a positive iodoform test ?

(*a*) Pentanal (*b*) 1-Phenylethanol

(*c*) 2-Phenyethanol (*d*) 3-Pentanol.

94. Pure chloroform is prepared by

(*a*) Distilling chloral hydrate with aqueous sodium hydroxide

(*b*) Heating ethanol with bleaching powder

(*c*) Heating acetone with bleaching powder

(*d*) Reducing carbon tetrachloride.

95. Which of the following reactants will not liberate ethyne ?

(*a*) Ethylene bromide and alc. KOH

(*b*) Ethylidene bromide and alc. KOH

(*c*) Chloroform and silver

(*d*) Aluminium carbide and water.

96. $H{-}C \equiv C{-}H + 2HCl \longrightarrow B \xrightarrow{\text{dil. NaOH}} C$

The product in the above sequence is

(*a*) Ethylidene dichloride (*b*) Ethylene glycol

(*c*) Ethanal (*d*) Ethyl alcohol.

97. Acetylene is treated with excess HBr and the resultant compound is then heated with zinc dust. The final product will be

(*a*) Ethylene (*b*) Acetylene

(*c*) 1-Butene (*d*) 2-Butene.

98. Iodoform is a ...... at room temperature.

(*a*) Gas (*b*) Thick viscous liquid

(*c*) Volatile liquid (*d*) Crystalline solid.

99. If chloroform is left open in the air in the presence of sunlight

(*a*) Explosion takes place

(*b*) Polymerization takes place

(*c*) Poisonous gas phosgene is formed

(*d*) No reaction takes place.

100. When chloroform is exposed to air and sunlight, it gives

(*a*) Mustard gas (*b*) Lewisite

(*c*) Carbonyl chloride (*d*) Carbon tetrachloride.

101. When chloroform reacts with acetone, the product is

(*a*) Chloral (*b*) Dichloroacetone

(*c*) Chloropicrin (*d*) Chloretone.

102. Which of the following is an anaesthetic ?

(*a*) CCl_4 (*b*) CH_3Cl

(*c*) C_2H_5OH (*d*) $CHCl_3$.

103. Which of the following represents a *gem*-dihalide ?

(*a*) Ethylene dichloride (*b*) 2, 2-dichloropropane

(*c*) 1, 3-dichloropropane (*d*) 1, 2-dichloropropane.

104. 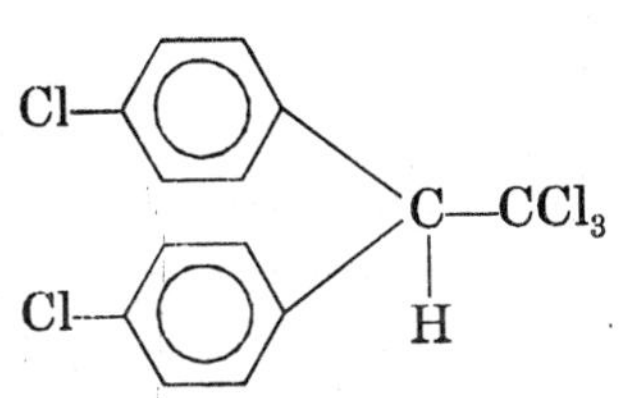

The above structural formula refers to

(*a*) BHC (*b*) DNA

(*c*) DDT (*d*) RNA.

Synthesis of Aryl and Aralkyl Halides

105. Chlorobenzene can be prepared by reacting benzene-diazonium chloride with

(*a*) Hydrochloric acid

(*b*) Cuprous chloride

(*c*) Chloride in presence of anhydrous aluminium chloride

(*d*) Nitrous acid.

106. Chlorination of toluene in the presence of light and heat followed by treatment with aqueous NaOH gives

(*a*) *o*-Cresol (*b*) *p*-Cresol

(*c*) 2, 4-Dihydroxytoluene (*d*) Benzyl alcohol.

107. Chlorobenzene can be prepared by reacting aniline with

(*a*) Hydrochloric acid

(*b*) Cuprous chloride

(*c*) Chlorine in the presence of aluminium chloride

(*d*) Nitrous acid followed by heating with cuprous chloride.

108. Which of the following is called Sandmeyer reaction ?

(*a*) $2CH_3CHO \xrightarrow{NaOH} CH_3{-}\overset{OH}{\overset{|}{C}H}{-}CH_2{-}CHO$

(*b*) 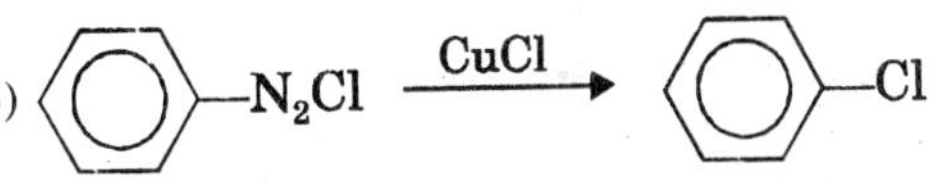

(*c*) 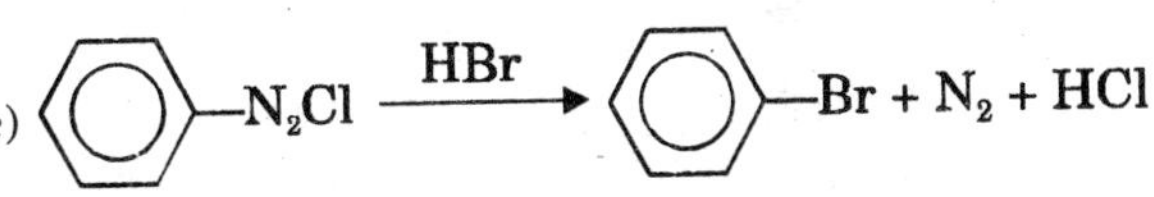

(*d*) 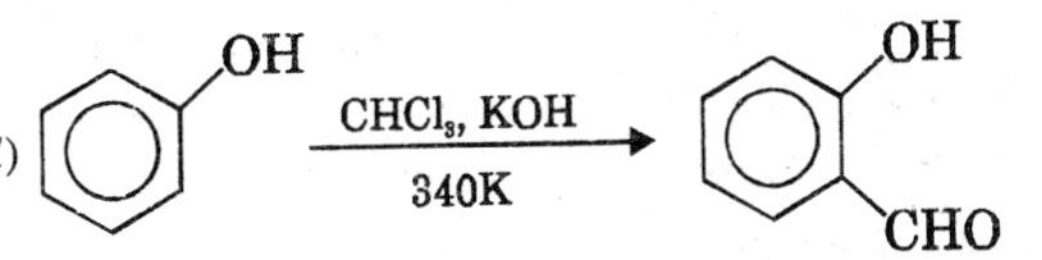

109. Toluene when refluxed with Br_2 in the presence of light mainly gives

(*a*) *o*-Bromotoluene

(*b*) *p*-Bromotoluene

(*c*) Mixture of *o*- and *p*-bromotoluene

(*d*) Benzyl bromide.

110. Benzene diazonium chloride can be converted into fluorobenzene by

(*a*) Sandmeyer's reaction

(*b*) Gattermann's reaction

(c) Balz Shiemann reaction

(d) Fittig reaction.

Properties of Aryl Halides

111. 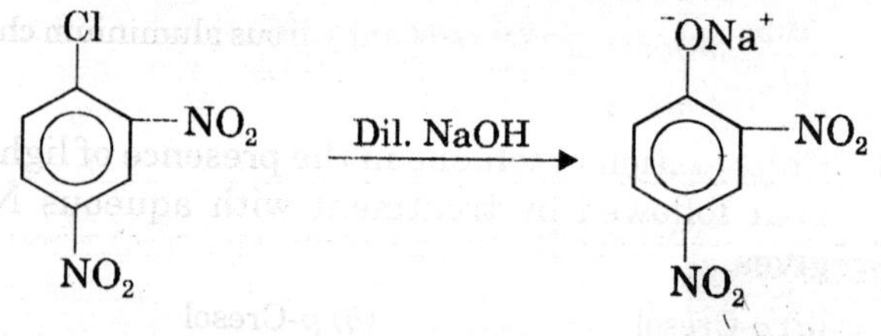

The above transformation proceeds through

(a) Electrophilic-addition

(b) Benzyne intermediate

(c) Activated nucleophilic substitution

(d) Oxirane.

112. Bromination of toluene gives

(a) Only *m*-substituted product

(b) Only *p*-substituted product

(c) Mixture of *ortho*- and *para*-substituted products

(d) Mixture of *ortho* and *m*-substituted products.

113. *o*-Methoxybromobenzene is treated with sodamide and then with ammonia. The product formed is

(a) *o*-Methoxyaniline (b) Aniline

(c) Methoxybenzene (d) *m*-Methoxyaniline.

114. Benzene reacts with Cl_2 in the presence of $FeCl_3$ (or halogen carrier) and in absence of sunlight to form

(a) Benzyl chloride (b) Benzal chloride

(c) Chlorobenzene (d) Benzenehexachloride.

115. Friedel-Craft's reaction of bromobenzene with methyl iodide gives

(a) *o*-Bromotoluene

(b) *p*-Bromotoluene

(c) *o*- and *p*-Bromotoluenes

(d) *m*-Bromotoluene.

116. The decreasing order of reactivity of *m*-nitrobromobenzene (I) ; 2, 4, 6-trinitrobromobenzene (II) ; *p*- nitrobromobenzene (III) ; 2, 4-dinitrobromobenzene (IV), towards OH^- ions is

(a) I > II > III > IV (b) II > IV > III > I

(c) IV > II > III > I (d) II > IV > I > III.

117. Which chloro derivative of benzene among the following would undergo hydrolysis most readily with aqueous sodium hydroxide to furnish the corresponding hydroxy derivative ?

(a) 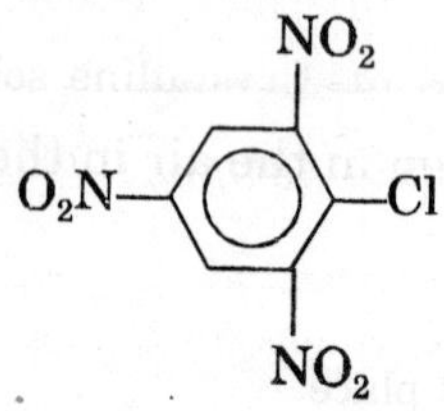 (b) 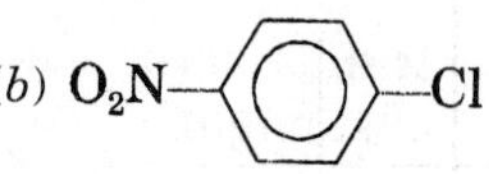

(c) 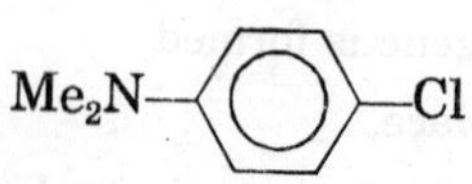 (d) C_6H_5Cl.

118. 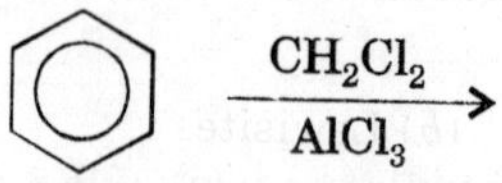 X. The product X is

(a) Benzyl chloride (b) Benzal chloride

(c) Benzo chloride (d) Diphenyl methane.

119. When chloroform reacts with benzene in the presence of aluminium chloride, the product formed is

(a) 1, 1, 1-Trichloroethane

(b) Trichlorophenyl methane

(c) Triphenyl methane

(d) Triphenyl chloromethane.

120. Which of the following groups is an *ortho-para* directing as well as deactivating group ?

(a) —OH (b) $—NH_2$

(c) $—CH_3$ (d) —Cl.

ANSWERS

1. (d)	2. (d)	3. (a)	4. (d)	5. (d)	6. (b)	7. (d)	8. (b)	9. (c)
10. (d)	11. (a)	12. (a)	13. (b)	14. (d)	15. (b)	16. (a)	17. (b)	18. (d)
19. (a)	20. (c)	21. (a)	22. (c)	23. (b)	24. (d)	25. (a)	26. (c)	27. (d)
28. (d)	29. (d)	30. (d)	31. (b)	32. (c)	33. (c)	34. (b)	35. (c)	36. (d)
37. (d)	38. (d)	39. (b)	40. (c)	41. (b)	42. (d)	43. (d)	44. (c)	45. (a)
46. (c)	47. (c)	48. (c)	49. (b)	50. (c)	51. (c)	52. (c)	53. (a)	54. (a)
55. (a)	56. (d)	57. (a)	58. (c)	59. (c)	60. (b)	61. (a)	62. (b)	63. (d)

64. (a)	65. (d)	66. (a)	67. (b)	68. (b)	69. (c)	70. (c)	71. (c)	72. (a)
73. (a)	74. (a)	75. (b)	76. (a)	77. (c)	78. (b)	79. (b)	80. (d)	81. (d)
82. (b)	83. (d)	84. (c)	85. (b)	86. (b)	87. (b)	88. (c)	89. (b)	90. (d)
91. (b)	92. (d)	93. (b)	94. (a)	95. (d)	96. (c)	97. (d)	98. (d)	99. (c)
100. (c)	101. (d)	102. (d)	103. (b)	104. (c)	105. (b)	106. (d)	107. (d)	108. (b)
109. (d)	110. (c)	111. (c)	112. (c)	113. (d)	114. (c)	115. (c)	116. (b)	117. (a)
118. (d)	119. (c)	120. (d).						

HINTS/SOLUTIONS

Q. No.	Ans.	Hints & Solutions
1.	(d)	In benzyl chloride, halogen atom is attached to the side chain.
2.	(d)	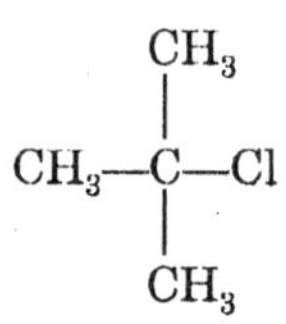 *tert*-Butyl chloride
3.	(a)	*n*-Butyl bromide, *iso*-butyl bromide, *sec*-butyl bromide and *tert*-butyl bromide.
4.	(d)	1. $CH_3CH_2CH_2CH_2CH_2Br$ 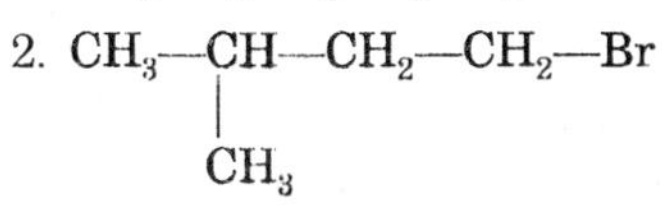 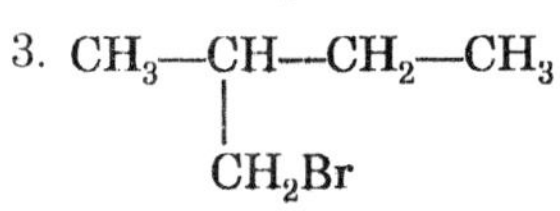 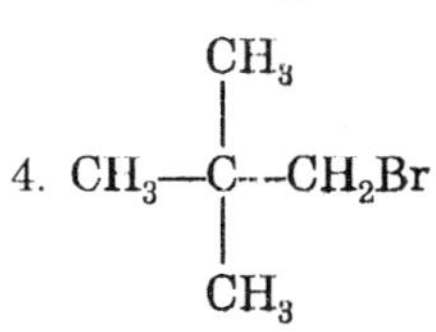
5.	(d)	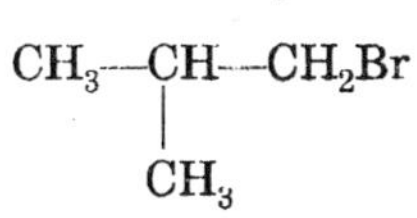 is *iso*-Butyl bromide.
6.	(b)	$CH_2 = CBr_2$ 1, 1-Dibromoethene 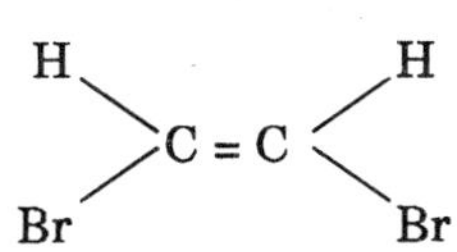 *cis*-1, 2-Dibromoethene and 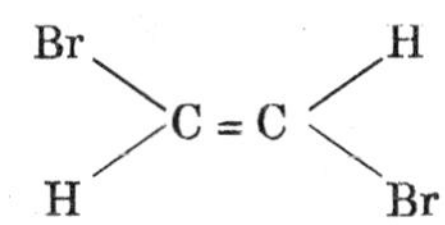 *trans*-1, 2-Dibromoethene
7.	(d)	1-Chloro-1-phenylethane has Cl atom in the side chain.
8.	(b)	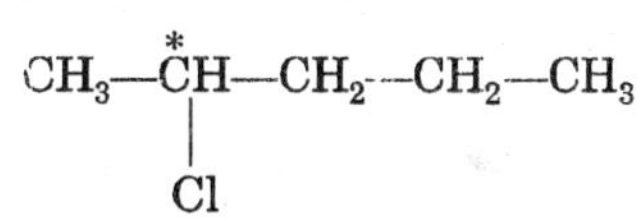 2-Chloropentane contains an asymmetric carbon atom and hence is chiral.
9.	(c)	*trans*-1, 2-Dichlorocyclopropane does not have any plane of symmetry and hence is non superimposable on its mirror image.
10.	(d)	Isopropyl chloride 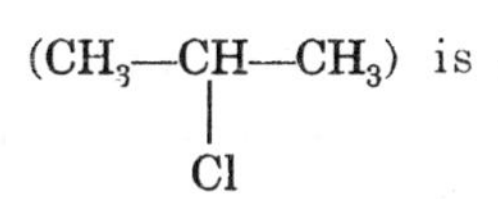 is a secondary halide.
11.	(a)	$\overset{4}{C}H_3\overset{3}{C}H = \overset{2}{C}H—\overset{1}{C}H_2Cl$ Crotyl chloride (1-Chloro-2-butene)
12.	(a)	Carbon atoms in benzene ring are sp^2-hybridized.
13.	(b)	C—Cl bond in vinyl chloride has partial double bond character due to resonance.
14.	(d)	Hunsdiecker reaction, the alkyl halide formed has one carbon less than the starting compound.
15.	(b)	Self explanatory question.

Q. No.	Ans.	Hints & Solutions
16.	(*a*)	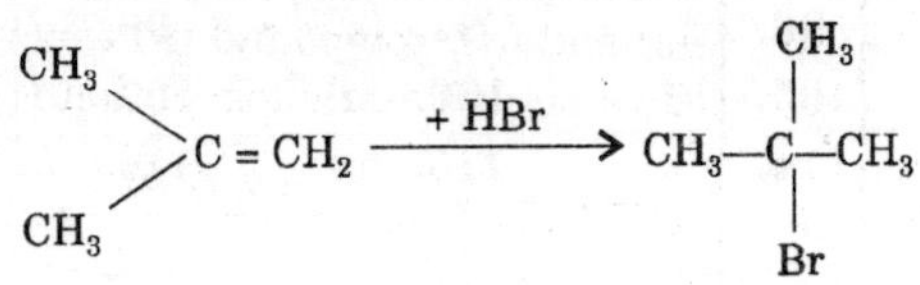 *iso*-Butylene → *tert*-Butyl bromide. The product is in accordance with Markownikoff's rule.
17.	(*b*)	Chlorination of methane proceeds via free radical chain mechanism.
18.	(*d*)	$ROH + SOCl_2 \longrightarrow RCl + SO_2 + HCl.$ Side products in this reaction are volatile and can be easily expelled during distillation.
19.	(*a*)	NaCl cannot replace —OH by —Cl.
20.	(*c*)	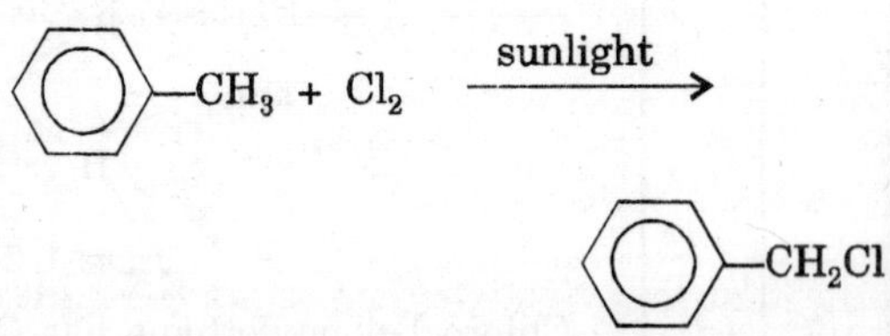 Benzyl chloride
21.	(*a*)	1-Chloropentane molecule has greater surface area and hence has greater interparticle forces.
22.	(*c*)	Density of alkyl halides is in the order R—I > R—Br > R—Cl > R—F.
23.	(*b*)	In *cis*-1,2-dichloroethylene individual bond dipoles of the two C—Cl bonds donot cancel each other.
24.	(*d*)	All these compounds have net dipole moment.
25.	(*a*)	Chloroalkanes are lighter than water while bromo- and iodo-derivatives are heavier than water.
26.	(*c*)	Aryl and vinyl halides donot undergo nucleophilic substitution easily.
27.	(*d*)	$4C_2H_5Cl + NH_3 \longrightarrow [N(C_2H_5)_4]^+ Cl^- + 3HCl.$
28.	(*d*)	$Zn + 2C_2H_5OH \longrightarrow Zn(OC_2H_5)_2 + 2H$ $C_2H_5Br + 2H \longrightarrow \mathbf{C_2H_6} + HBr.$
29.	(*d*)	$X = CH_3CH_2CN,\quad Y = CH_3CH_2CONH_2$ $Z = CH_3CH_2COOH.$
30.	(*d*)	Tertiary halides on treatment with base, such as sodium methoxide, readily undergo elimination resulting in the formation of alkenes.
31.	(*b*)	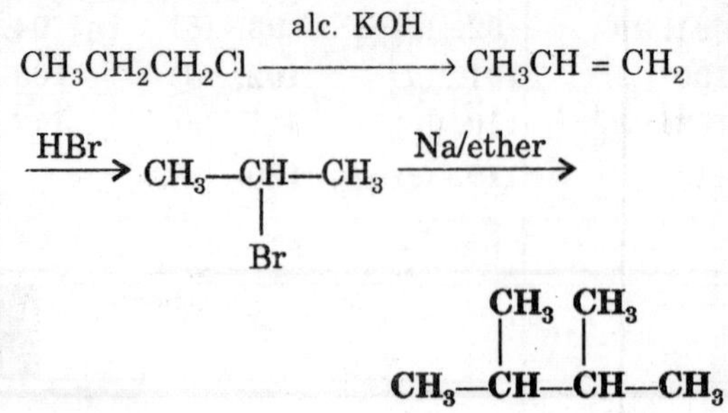
32.	(*c*)	NBS brings about substitution at allylic position.
33.	(*c*)	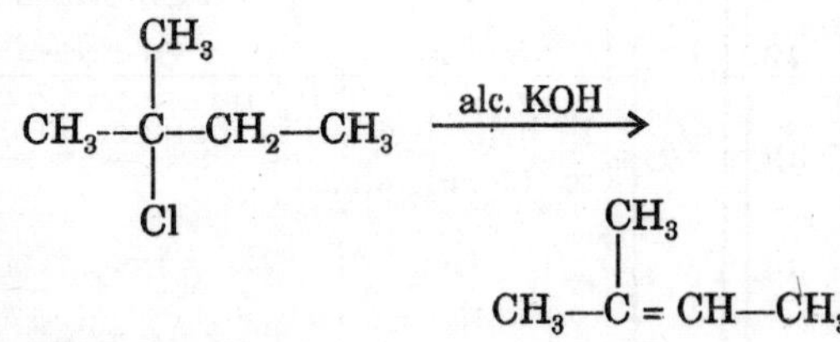 More substituted alkene is the major product (Saytzeff rule).
34.	(*b*)	For tertiary halides, elimination product is the major product on reaction with base.
35.	(*c*)	β-Elimination converts alkyl halide into alkene.
36.	(*d*)	Ease of nucleophilic substitution is R—I > R—Br > R—Cl > R—F.
37.	(*d*)	Allylic halides undergo substitution very easily.
38.	(*d*)	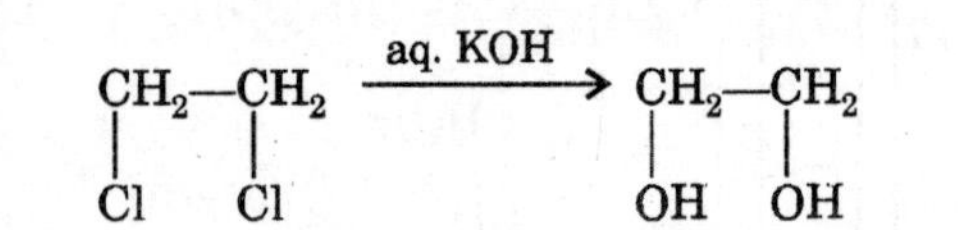 Ethylene chloride 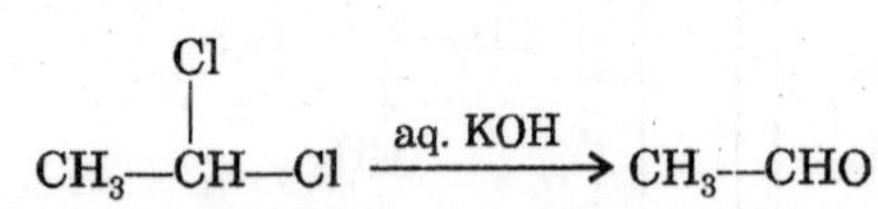 Ethylidene chloride Thus, the statement (*d*) is wrong.
39.	(*b*)	In Wurtz reaction, alkane is formed by combination of two alkyl groups. Combination of ethyl and propyl groups cannot give propane.
40.	(*c*)	Aqueous KOH substitutes —Cl by —OH.

Q. No.	Ans.	Hints & Solutions
41.	(*b*)	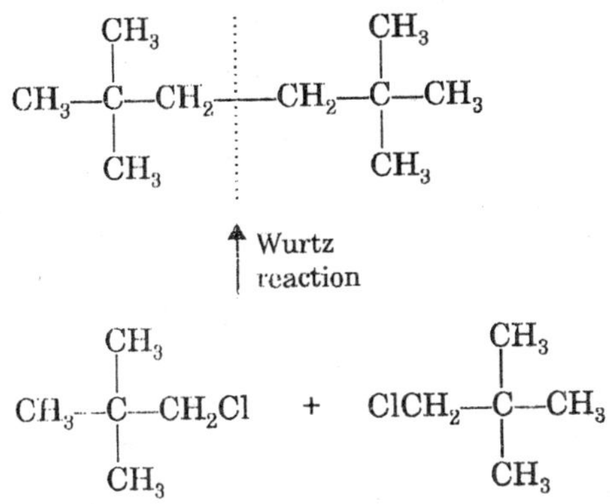
42.	(*d*)	Chlorobenzene forms Grignard reagent in the presence of THF (tetrahydrofuran) as solvent.
43.	(*d*)	—Cl is *ortho* and *para* directing group. The reaction is Friedel Craft alkylation.
44.	(*c*)	The order of reactivity of various alkyl halides towards $S_N{}^1$ reaction is $3° >$ allyl, $2° > 1°$.
45.	(*a*)	Diethyl ether is volatile. Moreover, it can stabilize the Grignard reagent by formation of coordinate bonds with Mg through O atom much more readily.
46.	(*c*)	$Cl\bullet \vdots \bullet Cl \xrightarrow[\text{Homolysis}]{h\nu} 2Cl^{\bullet}$ [*Chain initiation step*]
47.	(*c*)	N-Bromo succinimide.
48.	(*c*)	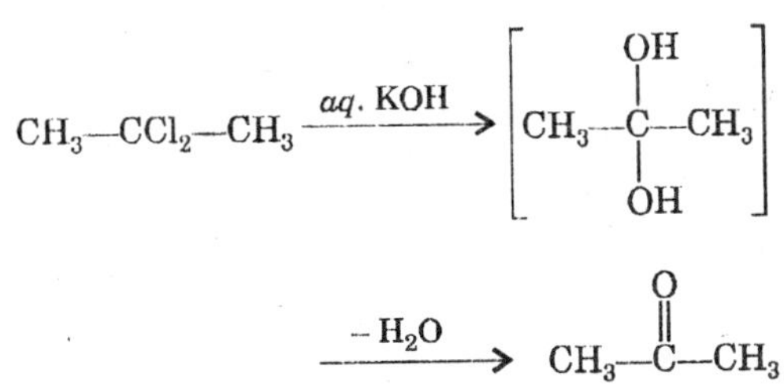
49.	(*b*)	$RX + Mg \xrightarrow{\text{Dry ether}} RMgX$ Grignard reagent
50.	(*c*)	3° halides undergo nucleophilic substitution most readily. $(CH_3)_3CCl$ is a 3° halide.
51.	(*c*)	Formation of 2-methylbutane indicates that the carbon chain skeleton in the chloride is 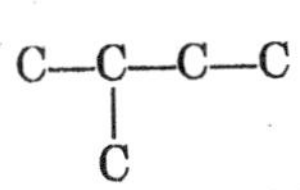 Hence, choices (*a*) and (*b*) are ruled out. Out of (*c*) and (*d*), compound (*d*) can yield two products on dehydrohalogenation.
52.	(*c*)	$CH_3CH_2CHCl_2 \xrightarrow{aq.\ KOH} CH_3CH_2CHO$ 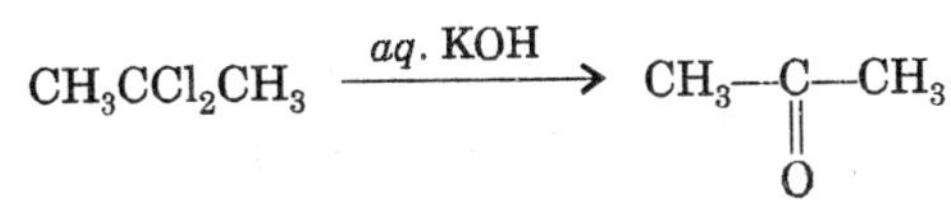
53.	(*a*)	$C_2H_5Br + AgOH \longrightarrow C_2H_5OH + AgBr.$
54.	(*a*)	$R'X + RCOOAg \longrightarrow \underset{\text{Ester}}{RCOOR'} + AgX$
55.	(*a*)	Benzylic halogen is substituted by —CN readily. 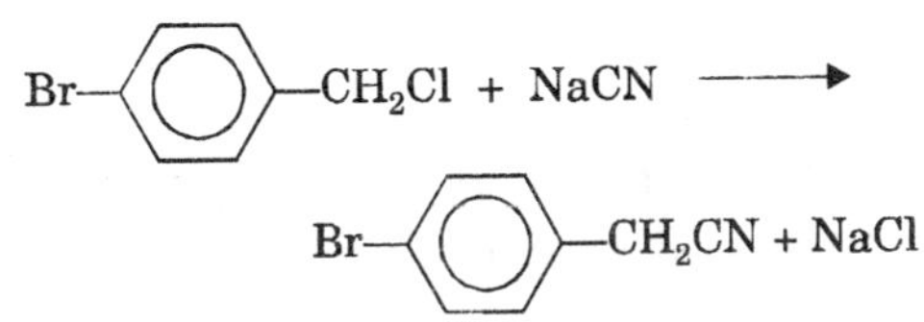
56.	(*d*)	$RMgX + H_2O \longrightarrow RH + Mg(OH)X.$
57.	(*a*)	Tetraethyl lead.
58.	(*c*)	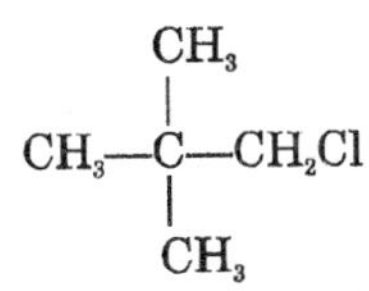
59.	(*c*)	$R - X + OH^- (aq) \longrightarrow R - OH + X^-.$
60.	(*b*)	$HI > HBr > HCl.$
61.	(*a*)	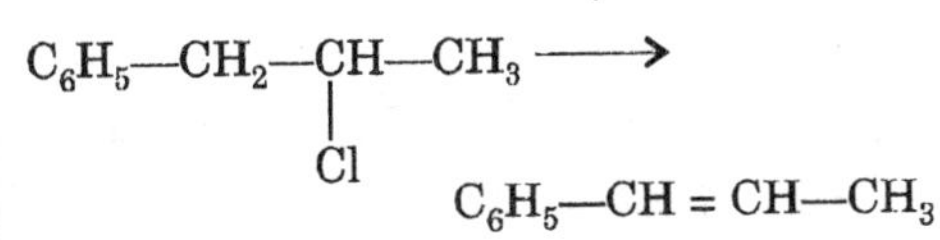 1-Phenyl propene. 1-Phenylpropene is more stable because double bond is in conjugation with benzene ring and is more substituted.
62.	(*b*)	AgCN reacts with alkyl halides to form isocyanides as the major products while KCN gives alkyl cyanide.
63.	(*d*)	Allyl halides are more reactive than aryl and vinyl halides towards nucleophilic substitution reactions.
64.	(*a*)	$S_N{}^1$ reactions occur through a two step mechanism involving intermediate carbocation and not through single step reaction.
65.	(*d*)	Factual question.

Q. No.	Ans.	Hints & Solutions
66.	(*a*)	*n*-Butane gives 1-chlorobutane and 2-chlorobutane on monochlorination.
67.	(*b*)	Ethyl bromide on reaction with alcoholic solution of $AgNO_2$ would yield nitroethane as well as ethyl nitrite. However nitroethane is the major product.
68.	(*b*)	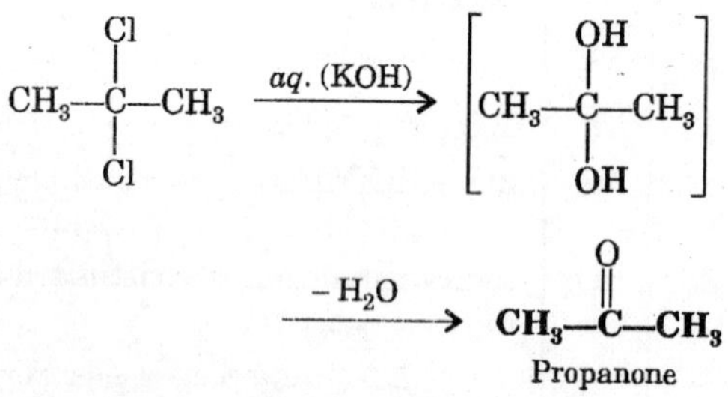
69.	(*c*)	Self explanatory question.
70.	(*c*)	$Cl_3C-CHO + NaOH \longrightarrow CHCl_3 + HCOONa$
71.	(*c*)	Factual question.
72.	(*a*)	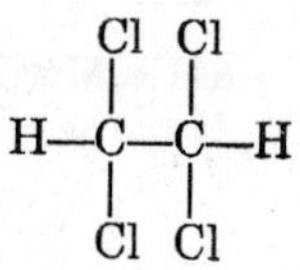 ...*Westron*
73.	(*a*)	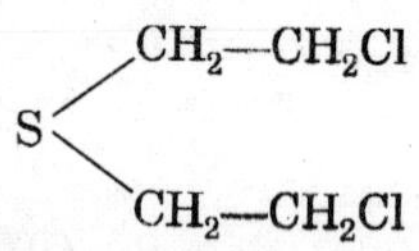 ...*Mustard gas*
74.	(*a*)	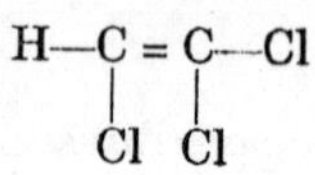 ...*Westrosol*
75.	(*b*)	Propanal (CH_3CH_2CHO) does not have $CH_3-\underset{\underset{O}{\|}}{C}-$ or $CH_3-\underset{\underset{OH}{\vert}}{CH}-$ group.
76.	(*a*)	$2CHCl_3 + 6Ag \xrightarrow{\Delta}$ **HC ≡ CH** $+ 6AgCl$.
77.	(*c*)	$CaOCl_2 + H_2O \longrightarrow Ca(OH)_2 + Cl_2$
78.	(*b*)	Haloform reaction is given by compounds containing $CH_3-\underset{\underset{O}{\|}}{C}-$ or $CH_3-\underset{\underset{OH}{\vert}}{CH}-$ group. CH_3OH does not contain any of these groups.
79.	(*b*)	$CH_3CHCl_2 \xrightarrow{KOH} [CH_3CH(OH)_2] \xrightarrow{-H_2O} CH_3CHO$
80.	(*d*)	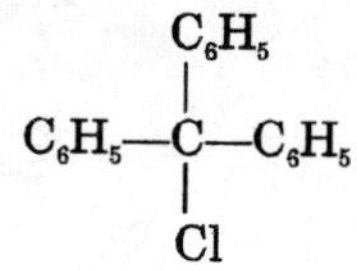
81.	(*d*)	1-Phenyl-2-propanol contains $CH_3-\underset{\underset{OH}{\vert}}{CH}-$ group.
82.	(*b*)	Self explanatory statement.
83.	(*d*)	CF_2Cl_2 (Freon) is used as refrigerant.
84.	(*c*)	Iodoform reaction.
85.	(*b*)	Iodoform releases iodine slowly and has antiseptic action.
86.	(*b*)	D.D.T. is obtained by heating chloral with chlorobenzene in the presence of conc. H_2SO_4.
87.	(*b*)	Carbylamine reaction.
88.	(*c*)	Carbylamine reaction.
89.	(*b*)	$CHCl_3 + HO\,.\,NO_2 \longrightarrow \underset{\text{Chloropicrin}}{CCl_3-NO_2} + H_2O$
90.	(*d*)	Acetone and Bleaching powder.
91.	(*b*)	$CS_2 + 3Cl_2 \xrightarrow{AlCl_3} CCl_4 + S_2Cl_2$.
92.	(*d*)	CCl_4 does not give any precipitate with $AgNO_3$. It does not dissolve in its solution.
93.	(*b*)	1-Phenylethanol contains $CH_3-\underset{\underset{OH}{\vert}}{CH}-$ group.
94.	(*a*)	Pure chloroform is prepared by heating chloral or chloral hydrate with aqueous NaOH.
95.	(*d*)	Aluminium carbide on reaction with water liberates methane.
96.	(*c*)	$H-C\equiv C-H + 2HCl \longrightarrow CH_3-CHCl_2 \xrightarrow{\text{dil. NaOH}} [CH_3CH(OH)_2] \xrightarrow{-H_2O} CH_3CHO$
97.	(*d*)	$H-C\equiv C-H \xrightarrow{2HBr} CH_3-CH(Br)_2 \xrightarrow{\text{Zn dust}} CH_3-CH=CH-CH_3$
98.	(*d*)	Iodoform is a yellow crystalline solid at room temperature.
99.	(*c*)	$2CHCl_3 + O_2 \longrightarrow \underset{\text{Phosgene}}{2COCl_2} + 2HCl$
100.	(*c*)	Phosgene is also known as carbonyl chloride. See hint 99.

Q. No.	Ans.	Hints & Solutions
101.	(*d*)	$CHCl_3 + CH_3-\overset{\overset{O}{\parallel}}{C}-CH_3 \longrightarrow CH_3-\underset{\underset{CCl_3}{\vert}}{\overset{\overset{OH}{\vert}}{C}}-CH_3$ Chloretone
102.	(*d*)	$CHCl_3$ is used as anaesthetic.
103.	(*b*)	In 2, 2-dichloropropane both the Cl atoms are on the same carbon.
104.	(*c*)	The given compound is D.D.T.
105.	(*b*)	Sandmeyer's reaction.
106.	(*d*)	$C_6H_5CH_3 \xrightarrow[h\nu]{Cl_2} C_6H_5-CH_2Cl \xrightarrow{aq.\ NaOH} C_6H_5-CH_2OH$
107.	(*d*)	Sandmeyer's reaction.
108.	(*b*)	Sandmeyer's reaction involves reaction of benzene diazonium salts with CuCl, CuBr or CuCN.
109.	(*d*)	$C_6H_5CH_3 \xrightarrow[h\nu]{Br_2} C_6H_5CH_2Br$
110.	(*c*)	Factual question.
111.	(*c*)	$-NO_2$ groups activate the molecule towards nucleophilic substitution by stabilizing the intermediate carbanion.

Q. No.	Ans.	Hints & Solutions
112.	(*c*)	$-CH_3$ group is *ortho-para* directing group.
113.	(*d*)	Reaction proceeds through benzyne.
114.	(*c*)	$C_6H_6 + Cl_2 \xrightarrow{FeCl_3} C_6H_5Cl + HCl$
115.	(*c*)	Bromo group in bromobenzene is *ortho-para* directing.
116.	(*b*)	Greater the number of $-NO_2$ groups at *ortho* and *para* positions to – Br group, greater is the reactivity of the compound towards OH^- ions due to stabilization of intermediate carbanion. (II) having three $-NO_2$ groups is most reactive while (I) having $-NO_2$ group at *meta* position is least reactive.
117.	(*a*)	Greater the number of electron withdrawing groups at *ortho* and *para* position to $-Cl$, greater is the reactivity with aqueous sodium hydroxide.
118.	(*d*)	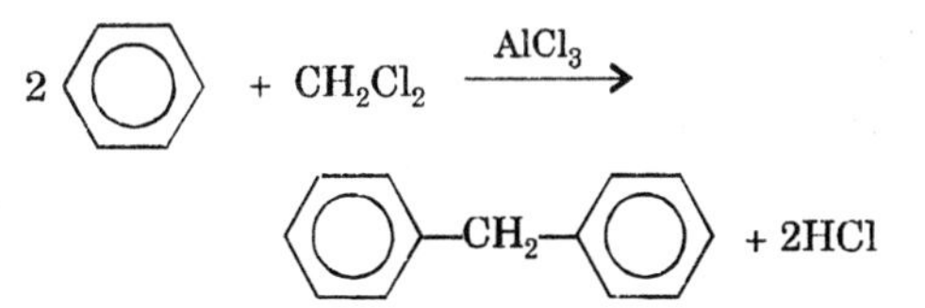
119.	(*c*)	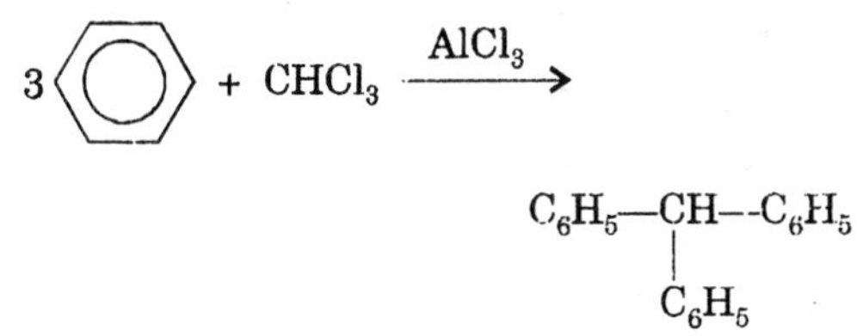
120.	(*d*)	$-Cl$ group is *ortho-para* directing and deactivating.

3.5. EXAMINER'S VIEWPOINT

1996

1. The reaction

$$CH_2 = CH - CH_3 + HBr \longrightarrow CH_3 - \underset{}{\overset{Br}{\overset{|}{C}}}HCH_3$$

(*a*) Nucleophilic addition
(*b*) Electrophilic addition
(*c*) Electrophilic substitution
(*d*) Free radical addition. ***(CBSE PMT)***

2. Which of the following compounds on oxidation gives benzoic acid ?
(*a*) Chlorophenol (*b*) Chlorotoluence
(*c*) Chlorobenzene (*d*) Benzyl chloride. ***(CBSE PMT)***

3. Silver benzoate will react with bromine in acetone to form

(*a*)  (*b*) 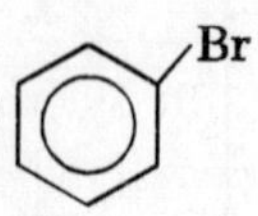

(*c*) 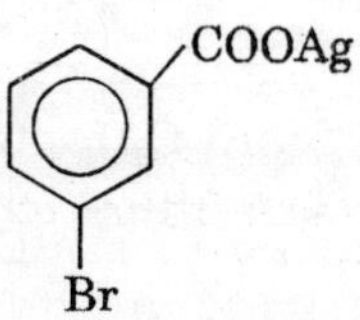 (*d*) 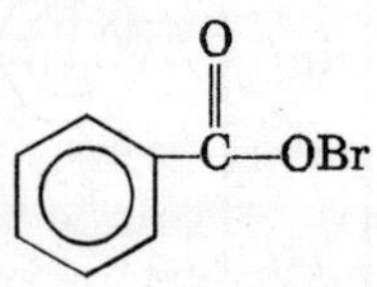

(Haryana CEET)

4. The end product (B) in the following sequence is

$$CH_3Cl \xrightarrow{KCN} A \xrightarrow{H_2O/H^+} B$$

(*a*) CH_3COOH (*b*) $HCOOH$
(*c*) CH_3NH_2 (*d*) CH_3COCH_3. ***(Pb. CET)***

5. Which one gives carbylamine reaction ?
(*a*) CH_3NH_2 (*b*) $C_2H_5NO_2$
(*c*) CH_3CONH_2 (*d*) $(CH_3)_2NH$. ***(Pb. CET)***

6. Chlorobenzene on heating with NH_3 under pressure in the presence of cuprous chloride gives
(*a*) Benzamide
(*b*) Nitrobenzene
(*c*) Aniline
(*d*) *o*- and *p*-chloroaminobenzene. ***(Pb. CET)***

7. The final product formed by distilling ethyl alcohol with excess of Cl_2 and $Ca(OH)_2$ is
(*a*) CH_3CHO (*b*) CCl_3CHO
(*c*) $CHCl_3$ (*d*) $(CH_3)_2O$. ***(MP PET)***

8. Ethane as well as methane can be prepared in single step from
(*a*) CH_3Br (*b*) C_2H_5Br
(*c*) CH_3CHO (*d*) C_2H_5OH. ***(AFMC)***

9. Chloroform on warming with Ag powder gives
(*a*) C_6H_6 (*b*) C_2H_4
(*c*) C_2H_2 (*d*) None. ***(AFMC)***

10. Which of the following compounds is used as a refrigerant ?
(*a*) Acetone (*b*) CCl_4
(*c*) CF_4 (*d*) CCl_2F_2. ***(AFMC)***

11. S_{N^2} mechanism proceeds through intervention of
(*a*) Carbonium ion (*b*) Transition state
(*c*) Free radical (*d*) Carbanion. ***(MLNR)***

12. Substitution of chlorine takes place at higher temperature in
(*a*) $CH_3CH = CH_2$ (*b*) $CH_2 = CH_2$
(*c*) $CH \equiv CH$ (*d*) None of these. ***(MLNR)***

1997

13. The reaction described below is :

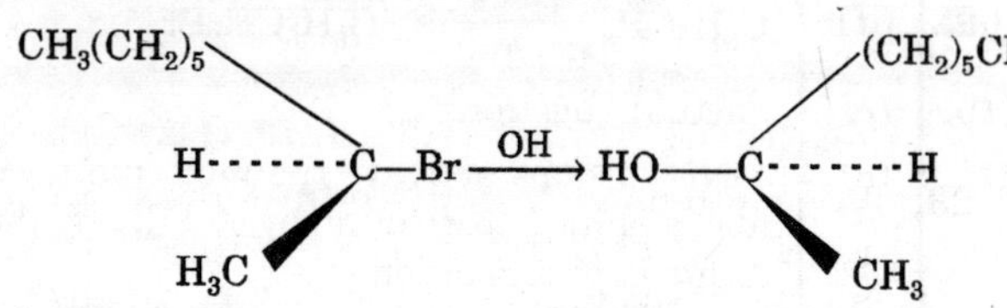

(*a*) E_2 (*b*) $S_N{}^1$
(*c*) $S_N{}^0$ (*d*) $S_N{}^2$. ***(CBSE PMT)***

14. Replacement of Cl of chlorobenzene to give phenol requires drastic conditions but chlorine of 2, 4-dinitrochlorobenzene is readily replaced because
(*a*) – NO_2 makes the ring electron rich at *ortho* and *para* positions
(*b*) – NO_2 withdraws electrons at *meta*-position
(*c*) – NO_2 donate electrons at *meta*-position
(*d*) – NO_2 withdraws electrons at *ortho* and *para* positions. ***(CBSE PMT)***

15. The compound obtained by heating a mixture of a primary amine and chloroform with ethanolic potassium hydroxide (KOH) is
(*a*) an alkyl isocyanide
(*b*) an alkyl halide
(*c*) an amide
(*d*) an amide and nitro compound. ***(CBSE PMT)***

16. Chlorobenzene on heating with aqueous NH_3 under pressure in the presence of cuprous chloride gives
(*a*) Benzamide (*b*) Nitrobenzene
(*c*) Aniline (*d*) Chloroaminobenzene. ***(Pb. CET)***

17. Ethyl alcohol gives ethyl chloride by reaction with
(*a*) $SOCl_2$ (*b*) NaCl
(*c*) Cl_2 (*d*) KCl. ***(BHU)***

18. $R - OH + HX \longrightarrow R - X + H_2O$
In the above reaction, the reactivity of different alcohols is
(*a*) Tertiary > Secondary > Primary
(*b*) Tertiary > Secondary < Primary
(*c*) Tertiary < Secondary > Primary
(*d*) Secondary < Primary < Tertiary. ***(CPMT)***

19. $C_6H_6Cl_6$ can be obtained from :
(*a*) HCl and benzene
(*b*) Cl_2 and benzene and $AlCl_3$
(*c*) Cl_2 and benzene in diffused light
(*d*) NaOCl and benzene. ***(CPMT)***

20. $CH_3 - CH = CH_2 + HI \longrightarrow$ X. Here X is
(*a*) $CH_3CH_2CH_2I$ (*b*) CH_3CHICH_3
(*c*) $CH_3CH_2CH_3$ (*d*) $CH_3CH_3 + CH_4$. ***(CPMT)***

21. 1, 3-Dilbromopropane reacts with metallic zinc to form
(*a*) Propene (*b*) Propane
(*c*) Cyclopropane (*d*) Hexane. ***(AIIMS)***

22. The chloroform reacts with NaOH to give
(*a*) CH_3COONa (*b*) Sodium oxalate
(*c*) CH_3OH (*d*) HCOONa. ***(BHU)***

23. When ethyl iodide and *n*-propyl iodide are allowed to react with sodium metal in ether, the number of alkanes that could be produced is
(*a*) Only one (*b*) Two alkanes
(*c*) Three alkanes (*d*) Four alkanes. ***(BHU)***

24. The alkyl halide is converted into an alcohol by
(*a*) Addition (*b*) Substitution
(*c*) Dehydrohalogenation (*d*) Elimination. ***(Haryana CEET)***

25. $C_3H_8 + Cl_2 \xrightarrow{\text{light}} C_3H_7Cl + HCl$ is an example of
(*a*) Elimination (*b*) Substitution
(*c*) Addition (*d*) None of these. ***(Haryana CEET)***

26. Iodoethane reacts with sodium in ether, the product formed is
(*a*) Pentene (*b*) Propyne
(*c*) Butene (*d*) Butane. ***(AFMC)***

27. Which of the following reacts with chloroform and base to form phenyl isocyanide ?
(*a*) Nitrobenzene (*b*) Phenol
(*c*) Chlorobenzene (*d*) Aniline. ***(AFMC)***

28. Under which of the following conditions would toluene, $C_6H_5—CH_3$ be converted into bromomethylbenzene, $C_6H_5 - CH_2Br$?
(*a*) Reaction with $Br_2/FeBr_3$
(*b*) Reaction with Br_2 in sunlight
(*c*) Reaction with Br_2 in dark
(*d*) Reaction with HBr. ***(AMU)***

1998

29. 2-Bromopentane is heated with potassium ethoxide in ethanol. The major product obtained is
(*a*) 2-Ethoxypentane (*b*) 1-Pentane
(*c*) *cis*-2-Pentane (*d*) *trans*-2-Pentene. ***(CBSE PMT)***

30. Iodoform test is not given by
(*a*) 2-Pentanone (*b*) Ethanol
(*c*) Ethanal (*d*) 3-Pentanone. ***(CBSE PMT)***

31. The reaction of benzyl chloride with sodium cyanide followed by reduction with hydrogen in the presence of nickel gives
(*a*) β-Phenylethylamine (*b*) N-Isobutylaniline
(*c*) Benzylamine (*d*) Aniline. ***(Pb. CET)***

32. Which one of the following does not participate in the carbylamine reaction ?
(*a*) Aniline (*b*) Chloroform
(*c*) Ethanol (*d*) KOH. ***(EAMCET)***

33. Alkyl halide is not obtained from one of the following reactions. Which one is that ?
(*a*) Diethyl ether + HBr → (*b*) Diethyl ether + HI →
(*c*) Diethyl ether + PCl_5 → (*d*) Diethyl ether + Cl_2 → ***(EAMCET)***

34. Freon-12 is commonly used as
(*a*) Insecticide (*b*) Refrigerant
(*c*) A solvent (*d*) A fire extinguisher. ***(Haryana CEET)***

35. Chlorination of toluene in the presence of light and heat followed by treatment with NaOH (*aq*) gives
(*a*) *o*-Cresol (*b*) *p*-Cresol
(*c*) 2, 4-Dihydroxytoluene (*d*) Benzoic acid. ***(Haryana CEET)***

36. Which of the following has highest chlorine content ?
(*a*) Pyrene (*b*) D.D.T
(*c*) Chloral (*d*) Gammexane. ***(BHU)***

37. Both butane and ethane can be prepared in single steps by the use of

(a) C_2H_4 (b) CH_3OH
(c) CH_3CH_2COONa (d) CH_3CHO. **(Pb. PMT)**

38. 2-Chlorobutane on reaction with alcoholic potash gives
(a) 2-Butene (b) 1-Butene
(c) 2-Butanol (d) 1-Butanol. **(Pb. PMT)**

39. Propene when heated with Cl_2 at 500°C forms
(a) $CH_2Cl - CH = CH_2$
(b) $CH_3 - CHCl - CH_2Cl$
(c) $CH_2Cl - CHCl - CH_2Cl$
(d) All of these. **(MP CEE)**

40. Identify Z in the following series

$$C_2H_5I \xrightarrow{\text{Alc. KOH}} X \xrightarrow{Br_2} Y \xrightarrow{KCN} Z$$

(a) CH_3CH_2CN (b) $NCCH_2 - CH_2CN$
(c) $BrCH_2 - CH_2CN$ (d) $BrCH = CHCN$. **(AFMC)**

1999

41. Ethyl chloride is converted into diethyl ether by
(a) Wurtz reaction (b) Grignard reaction
(c) Perkin's reaction (d) Williamson synthesis. **(CBSE PMT)**

42. Which of the following compounds will be most easily attacked by an electrophile ?

(a)  (b) 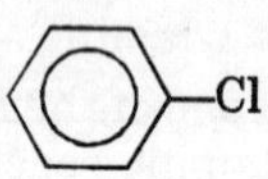

(c) 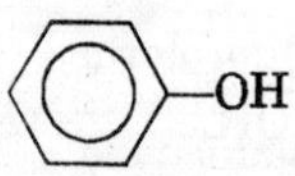 (d) 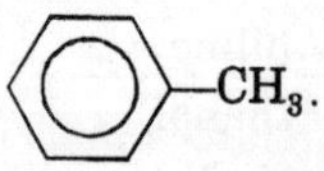

(CBSE PMT)

43. The reaction of $C_6H_5N_2^+C^-$ with CuCl gives
(a) C_6H_5Cl (b) C_6H_6
(c) $C_6H_5 - C_6H_5$ (d) $C_6H_4Cl_2$. **(Pb. CET)**

44. Isobutyl bromide may be obtained from isobutylene and HBr in the presence of
(a) Peroxide (b) Hydroquinone
(c) Diphenylamine (d) All the above. **(Pb. CET)**

45. Butanenitrile is formed by reaction of KCN with
(a) Propyl alcohol (b) Butyl chloride
(c) Butyl alcohol (d) Propyl chloride. **(Pb. CET)**

46. Which of the following is example of S_N2 reaction ?
(a) $CH_3Br + OH^- \longrightarrow CH_3OH + Br^-$
(b) $CH_3 - \underset{\underset{Br}{|}}{CH} - CH_3 + OH^- \longrightarrow CH_3 - \underset{\underset{OH}{|}}{CH} - CH_3 + Br^-$
(c) $CH_3 - CH_2 - OH \xrightarrow{-H_2O} H_2C = CH_2$
(d) $CH_3 - \overset{\overset{CH_3}{|}}{\underset{\underset{Br}{|}}{C}} - CH_3 + OH^- \longrightarrow CH_3 - \overset{\overset{CH_3}{|}}{\underset{\underset{OH}{|}}{C}} - CH_3 + Br^-$ **(CPMT)**

47. Which of the following possesses highest melting point ?
(a) Chlorobenzene (b) o-Dichlorobenzene
(c) m-Dichlorobenzene (d) p-Dichlorobenzene. **(CPMT)**

2000

48. Chloroform when reacts with nitric acid, it gives
(a) $C_2H_5C(NO)_5SH$ (b) CCl_3CHOH
(c) CCl_3NO_2 (d) CCl_3NO_3. **(AFMC)**

49. Which of the following alkyl halides is used as a methylating agent ?
(a) C_2H_5Br (b) C_8H_5Cl
(c) CH_3I (d) C_2H_5Cl. **(KCET)**

50. Which of the following is not used in Friedel-Crafts reaction ?
(a) N-Phenyl acetanilide (b) Bromobenzene
(c) Benzene (d) Chlorobenzene. **(KCET)**

51. On sulphonation of C_6H_5Cl
(a) m-Chlorobenzenesulphonic acid is formed
(b) Benzenesulphonic acid is formed
(c) o-Chlorobenzenesulphonic acid is formed
(d) o- and p-Chlorobenzenesulphonic acid is formed. **(CPMT)**

52. Iodoform on heating with KOH gives
(a) CH_3CHO (b) CH_3COOK
(c) HCOOK (d) HCHO. **(MP PMT)**

53. Which among MeX, RCH_2X, R_2CHX, R_3CX is most reactive towards S_{N^2} reaction ?
(a) MeX (b) RCH_2X
(c) R_2CHX (d) R_3CX. **(JIPMER)**

54. S_{N^1} reaction of alkyl halides leads to
(a) Retention of configuration
(b) Racemisation
(c) Inversion of configuration
(d) None of these. **(JIPMER)**

55. When chlorine is passed through propene at 400°C, which of the following is formed ?

(a) PVC (b) Allyl chloride
(c) Vinyl chloride (d) 1, 2-Dichloroethane.
(Haryana CEET)

56. The compound

$$C_7H_8 \xrightarrow{3Cl_2/\Delta} A \xrightarrow{Br_2/Fe} B \xrightarrow{Zn/HCl} C$$

The compound C is
(a) o-Bromotoluene
(b) m-Bromotoluene
(c) p-Bromotoluene
(d) 3-Bromo-2, 4, 6-trichlorotoluene. **(DCE)**

57. The pesticide DDT slowly changes to
(a) CCl_3-CHO and chlorobenzene
(b) p, p′-Dichlorodiphenylethene
(c) p, p′-Dichlorodiphenyldichloroethane
(d) p, p′-Dichlorodiphenyldichloroethene. **(DCE)**

2001

58. $CH_3—CH_2—CH(Cl)—CH_3$ obtained by chlorination of n-butane, will be
(a) meso-form (b) racemic mixture
(c) d-form (d) l-form. **(CBSE PMT)**

59. An organic compound A(C_4H_9Cl) on reaction with Na/diethyl ether gives a hydrocarbon which on monochlorination gives only one chloro derivative then, A is
(a) t-butyl chloride
(b) secondary butyl chloride
(c) iso-butyl chloride
(d) n-butyl chloride. **(CBSE PMT)**

60. Unpleasant smell of carbylamine is obtained when chloroform and alcoholic KOH are heated with
(a) any aromatic amine (b) any primary amine
(c) any amine (d) any aliphatic amine.
(KCET)

61. In S_N1 (substitution, nucleophilic unimolecular) reaction, the racemization takes place. It is due to
(a) Inversion of configuration
(b) Retention of configuration
(c) Conversion of configuration
(d) Both (a) and (b). **(Manipal)**

62. Alkyl halides react with Mg in dry ether to form
(a) Magnesium halide (b) Grignard's reagent
(c) Alkene (d) Alkyne. **(MP PET)**

63. The reaction between an alkyl halide and sodium metal in dry ether is called
(a) Kolbe's reaction (b) Cannizzaro's reaction
(c) Williamson synthesis (d) Wurtz reaction.
(MP PET)

64. Which of the following will give iodoform test ?
(a) $CH_3CH_2CH_2OH$
(b) $CH_3CH_2COCH_2CH_3$
(c) $CH_3—CH(OH)—CH_2—CH_2—CH_3$
(d) $CH_3—CH(CH_3)—CO—CH_2—CH_3$ **(CPMT)**

65. Which of the following is optically active ?
(a) 1-Chloropropane (b) 2-Chloropropane
(c) 1-Chlorobutane (d) 2-Chlorobutane.
(CPMT)

2002

66. The reactivity order of halides for dehydrohalogenation is
(a) R—F > R—Cl > R—Br > R—I
(b) R—I > R—Br > R—Cl > R—F
(c) R—I > R—Cl > R—Br > R—F
(d) R—F > R—I > R—Br > R—Cl. **(CBSE PMT)**

67. When primary amine reacts with chloroform in ethanolic KOH, then the product is
(a) An isocyanide (b) An aldehyde
(c) A cyanide (d) An alcohol. **(AIEEE)**

68. A set of compounds in which the reactivity of halogen atom in the ascending order is
(a) Chlorobenzene, vinyl chloride, chloroethane
(b) Chloroethane, chlorobenzene, vinyl chloride
(c) Vinyl chloride, chlorobenzene, chloroethane
(d) Vinyl chloride, chloroethane, chlorobenzene.
(KCET)

69. The order of reactivity of alkyl halides towards elimination reaction is
(a) 3° > 2° > 1° (b) 2° > 1° > 3°
(c) 3° > 1° > 2° (d) 1° > 2° > 3°. **(Manipal)**

70. 1-Chlorobutane on reaction with alcoholic potash gives
(a) 1-Butene (b) 1-Butanol
(c) 2-Butene (d) 2-Butanol.
(Rajasthan PMT)

71. Ethyl iodide when heated with sodium is dry ether gives

(a) C_2H_6 (b) C_4H_{10}
(c) C_3H_8 (d) C_2H_5OH. **(Orissa JEE)**

72. Alkyl halide on heating with alc. NH_3 in a sealed tube yields
(a) 1° amine (b) 2° amine
(c) 3° amine (d) All of these.
(Orissa JEE)

2003

73. Which of the following are arranged in the decreasing order of dipole moment ?
(a) CH_3Cl, CH_3Br, CH_3F
(b) CH_3Cl, CH_3F, CH_3Br
(c) CH_3Br, CH_3Cl, CH_3F
(d) CH_3Br, CH_3F, CH_3Cl. **(AIIMS)**

74. Allyl chloride on dehydrochlorination gives
(a) Propadiene (b) Propylene
(c) Allyl alcohol (d) Acetone
(e) Acetyl chloride. **(Kerala MEE)**

75. Alkyl halides react with metallic sodium in dry ether producing
(a) Alkanes with same number of carbon atoms
(b) Alkanes with double the number of carbon atoms
(c) Alkenes with same number of carbon atoms
(d) Alkenes with double the number of carbon atoms.
(e) Alkenes with triple the number of carbon atoms.
(Kerala MEE)

2004

76. The compound formed on heating chlorobenzene with chloral in the presence of concentrated sulphuric acid, is
(a) Gammaxene (b) Hexachloroethane
(c) Freon (d) DDT
(AIEEE)

77. Which of the following is responsible for depletion of the ozone layer in the upper strata of the atmosphere ?
(a) Ferrocene (b) Fullerenes
(c) Freons (d) Polyhalogens
(CBSE PMT)

78. Which of the following is least reactive in a nucleophilic substitution reaction ?
(a) $CH_2 = CHCl$ (b) CH_3CH_2Cl
(c) $CH_2 = CHCH_2Cl$ (d) $(CH_3)_3C—Cl$
(CBSE PMT)

79. Chloropicrin is obtained by the reaction of
(a) Nitric acid on chlorobenzene
(b) Chlorine on picric acid
(c) Nitric acid on chloroform
(d) Steam on carbon tetrachloride. **(CBSE PMT)**

80. Among the following the most reactive towards alcoholic KOH is
(a) $CH_2 = CHBr$ (b) $CH_3COCH_2CH_2Br$
(c) CH_3CH_2Br (d) $CH_3CH_2CH_2Br$
(AIIMS)

81. Aromatic nitriles (ArCN) are **not** prepared by reaction :
(a) ArX + KCN (b) ArN_2^+ + CuCN
(c) $ArCONH_2 + P_2O_5$ (d) $ArCONH_2 + SOCl_2$
(AIIMS 2004)

2005

82. Tertiary alkyl halides are practically inert to substitution by S_N2 mechanism because of
(a) Steric hindrance (b) Inductive effect
(c) Instability (d) Insolubility
(AIEEE)

83. Elimination of bromine from 2-bromobutane results in the formation of
(a) Predominantly 2-butyne
(b) Predominantly 1-butene
(c) Predominantly 2-butene
(d) Equimolar mixture of 1 and 2-butene **(AIEEE)**

84. Which of the following undergoes nucleophilic substitution exclusively by S_N1 mechanism ?
(a) Ethyl chloride (b) Isopropyl chloride
(c) Chlorobenzene (d) Benzyl chloride.
(CBSE PMT)

85. In which of the following reactions, the product is an ether ?
(a) $C_6H_6 + CH_3COCl$/anhydrous $AlCl_3$
(b) C_2H_5Cl + aq. KOH
(c) $C_6H_6 + C_6H_5COCl$/anhydrous $AlCl_3$
(d) $C_2H_5Cl + C_2H_5ONa$. **(EAMCET Engg.)**

86. Identify A and B in the following reactions :

$$A \xrightarrow[\Delta]{\text{aq. NaOH}} C_2H_5OH \xleftarrow{\text{AgOH}} B$$

(a) $A = C_2H_2, B = C_2H_6$ (b) $A = C_2H_5Cl, B = C_2H_4$
(c) $A = C_2H_4, B = C_2H_5Cl$ (d) $A = C_2H_5Cl, B = C_2H_5Cl$.
(EAMCET Engg.)

87. Consider the following reaction :

$$C_2H_5OK + C_2H_5I \xrightarrow{\Delta} X. \text{ (Pleasant smelling liquid)}$$

Which of the following reagents converts ethylbromide to X ?

(a) Sodium (b) Dry silver oxide
(c) Ethyl chloride (d) Dry silver powder.

(EAMCET Medical)

88. What are the reagent and reaction conditions used for converting ethylchloride to ethylnitrite (as the major product) ?

(a) KNO_2, C_2H_5OH, H_2O, Δ
(b) $NaNO_2$, HCl, 0°C
(c) KCN, H_2O, Δ
(d) $AgNO_2$, C_2H_5OH, H_2O, Δ. **(EAMCET Medical)**

89. Which of the following is most reactive towards nucleophilic substitution reaction ?

(a) $CH_2 = CH—Cl$ (b) $CH_2 = CH—CH_2—Cl$
(c) $CH_3—CH = CH—Cl$ (d) $C_6H_5—Cl$. **(DCE)**

90. Which of the following haloalkanes is most reactive ?

(a) 1-chloropropane (b) 1-Bromopropane
(c) 2-chloropropane (d) 2-Bromopropane.

(Karnataka CET)

91. Ethyl chloride on heating with AgCN forms a compound X. The functional isomer of X is

(a) C_2H_5NC (b) $C_2H_5NH_2$
(c) C_2H_5CN (d) None of these.

(Karnataka CET)

92. Which of the following applies is the reaction ;

$$CH_3CHBrCH_2CH_3 \xrightarrow{\text{alc. KOH}} \underset{\text{(major product)}}{CH_3CH = CHCH_3} + \underset{\text{(minor product)}}{CH_2 = CHCH_2CH_3}$$

(a) Markovnikov's rule (b) Saytzeff's rule
(c) Kharasch effect (d) Hoffmann's rule.

(Orissa JEE)

93. Identify X and Y in the following sequence :

$$C_2H_5Br \xrightarrow{X} \text{Product} \xrightarrow{Y} C_3H_7NH_2$$

(a) X = KCN, Y = $LiAlH_4$
(b) X = KCN, Y = H_3O^+
(c) X = CH_3Cl, Y = $AlCl_3/HCl$
(d) X = CH_3NH_2, Y = HNO_2. **(Orissa JEE)**

94. Which of the following cannot undergo nucleopholic substitution under ordinary conditions ?

(a) Chlorobenzene (b) *tert*-Butyl chloride
(c) *iso*-Propyl chloride (d) None of these.

(J & K CET)

95. Observe the following reactions and predict the nature of A and B.

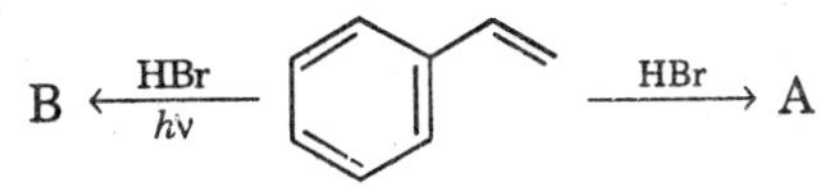

(a) A and B are both 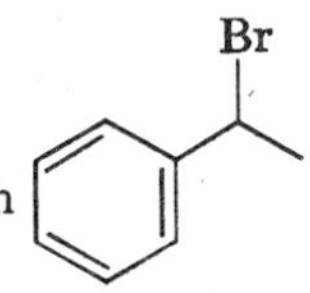

(b) A and B are both 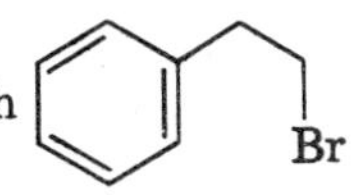

(c) A is 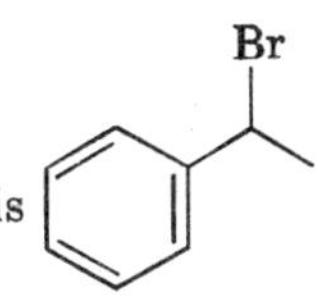 and B is 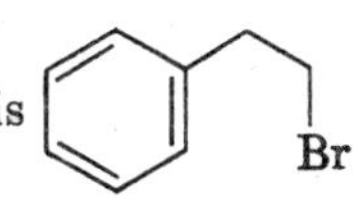

(d) A is 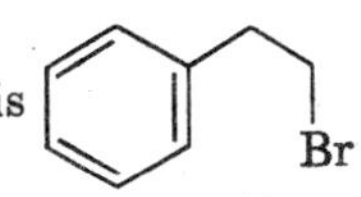 and B is 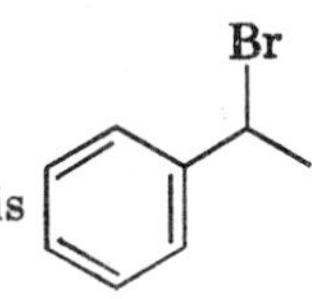

(e) A is 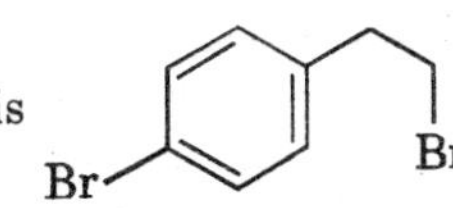 and B is 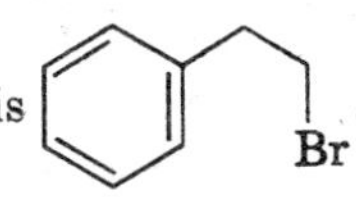

(Kerala PET)

96. 3-Phenylpropene on reaction with HBr gives (as a major product)

(a) $C_6H_5—CH_2CH(Br)CH_3$
(b) $C_6H_5—CH(Br)CH_2CH_3$
(c) $C_6H_5—CH_2CH_2CH_2Br$
(d) $C_6H_5—CH(Br)—CH = CH_2$. **(AIIMS)**

97. In Welliamson synthesis, ethoxyethane is prepared by

(a) Passing ethanol over heated aluminia
(b) Heating sodium ethoxide with ethyl bromide
(c) Heating ethanol with dry silver oxide
(d) Treating ethyl alcohol with excess of H_2SO_4 at 430–440 K. **(BHU)**

2006

98. The structure of the major product formed in the following reaction is

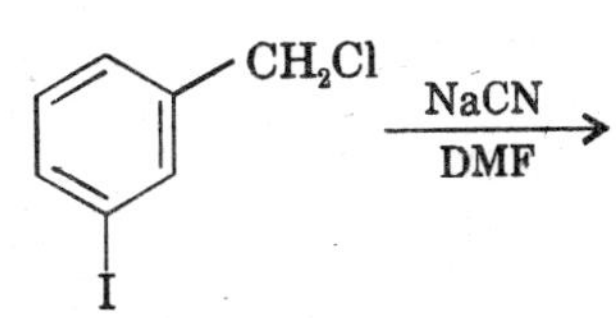

(a) 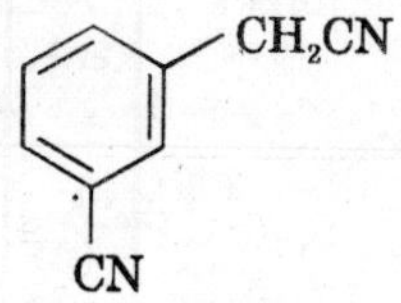

(b) 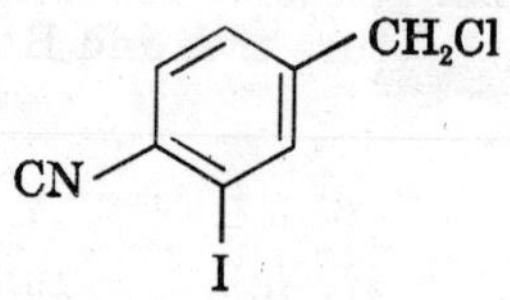

(c) 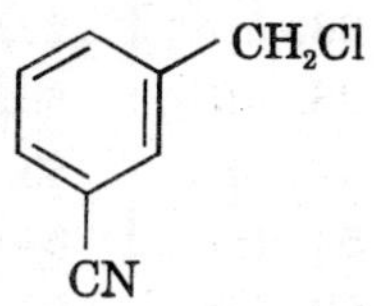

(d) 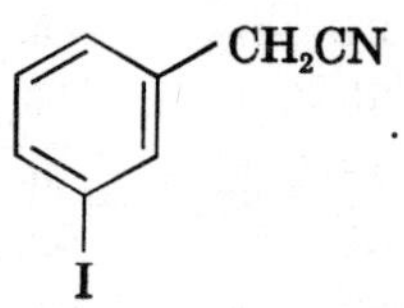

(AIEEE)

99. The IUPAC name of the compound shown below is

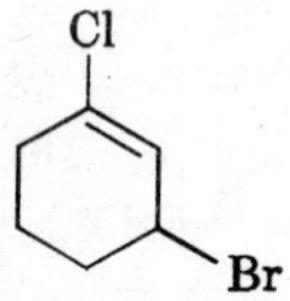

(a) 2-bromo-6-chlorocyclohex-1-ene
(b) 6-bromo-2-chlorocyclohexene
(c) 3-bromo-1-chlorocyclohexene
(d) 1-bromo-3-chlorocyclohexene. **(AIEEE)**

100. Fluorobenzene (C_6H_5F) can be synthesized in the laboratory
(a) By heating phenol with HF and KF
(b) From aniline by diazotisation followed by heating the diazonium salt with HBF_4
(c) By direct fluorination of benzene with F_2 gas
(d) By reacting bromobenzene with NaF solution. **(AIEEE)**

101. Phenyl magnesium bromide reacts with methanol to give
(a) A mixture of anisole and Mg(OH)Br
(b) A mixture of benzene and Mg(OMe)Br
(c) A mixture of toluene and Mg(OH)Br
(d) A mixture of phenol and Mg(Me)Br. **(AIEEE)**

102. $CH_3Br + Nu^- \longrightarrow CH_3—Nu + Br^-$
The decreasing order of the rate of the above reaction with nucleophiles (Nu^-) A to D is
[Nu^- = (A) PhO^-, (B) AcO^-, (C) HO^-, (D) CH_3O^-]
(a) D > C > A > B (b) D > C > B > A
(c) A > B > C > D (d) B > D > C > A. **(AIEEE)**

103. The catalyst used in the preparation of an alkyl chloride by the action of dry HCl on an alcohol is
(a) Anhydrous $AlCl_3$ (b) $FeCl_3$
(c) Anhydrous $ZnCl_2$ (d) Cu. ***(Karnataka CET)***

104. In the reaction,

$$R—X \xrightarrow{\text{alcoholic KCN}} A \xrightarrow{\text{dilute HCl}} B$$

the product B is
(a) Alkyl chloride (b) Aldehyde
(c) Carboxylic acid (d) Ketone.
(Karnataka CET)

105. Which of the following sequences of reactions (reagents) can be used for the conversion of $C_6H_5CH_2CH_3$ into $C_6H_5CH = CH_2$?
(a) $SOCl_2$; H_2O (b) SO_2Cl_2 ; alc. KOH
(c) Cl_2/hv ; H_2O (d) $SOCl_2$; alc. KOH.
(AIIMS)

106. The correct increasing order of the reactivity of halides for S_N^1 reaction is
(a) $CH_3—CH_2—X < (CH_3)_2\,CH—X < CH_2 = CH—CH_2—X < PhCH_2—X$
(b) $(CH_3)_2CH—X < CH_3—CH_2—X < CH_2 = CH—CH_2—X < PhCH_2—X$
(c) $PhCH_2—X < (CH_3)_2CH—X < CH_3—CH_2—X < CH_2 = CH—CH_2—X$
(d) $CH_2 = CH—CH_2—X < Ph—CH_2—X < (CH_3)_2\,CH—X < CH_3—CH_2—X$. **(AIIMS)**

107. Which of the following compounds has the highest boiling point ?
(a) $CH_3CH_2CH_2Cl$ (b) $CH_3CH_2CH_2CH_2Cl$
(c) $CH_3CH(CH_3)CH_2Cl$ (d) $(CH_3)_3CCl$. **(AIIMS)**

2007

108. In the chemical reaction,
$CH_3CH_2NH_2 + CHCl_3 + 3KOH \longrightarrow (A) + (B) + 3H_2O$,
the compounds (A) and (B) are respectively :
(a) C_2H_5NC and K_2CO_3 (b) C_2H_5NC and 3KCl
(c) C_2H_5CN and 3KCl (d) $CH_3CH_2CONH_2$ and 3KCl
(AIEEE)

109. The reaction of toluene with Cl_2 in presence of $FeCl_3$ gives predominantly :
(a) o- and p-chlorotoluene (b) m-chlorotoluene
(c) Benzyl chloride (d) Benzyl chloride.
(AIEEE)

110. Which of the following is the correct order of decreasing S_N^2 reactivity ?
(a) $R_3CX > R_2CHX > RCH_2X$
(b) $R_2CHX > R_3CX > RCH_2X$
(c) $RCH_2X > R_3CX > R_2CHX$
(d) $RCH_2X > R_2CHX > R_3CX$. **(AIEEE)**

111. For the following
(i) I^- (ii) Cl^- (iii) Br^-
the increasing order of nucleophilicity would be :
(a) $Br^- < Cl^- < I^-$ (b) $I^- < Br^- < Cl^-$
(c) $Cl^- < Br^- < I^-$ (d) $I^- < Cl^- < Br^-$.
(CBSE PMT)

ANSWERS

1. (*b*)	**2.** (*d*)	**3.** (*b*)	**4.** (*a*)	**5.** (*a*)	**6.** (*c*)	**7.** (*c*)	**8.** (*a*)	**9.** (*c*)
10. (*d*)	**11.** (*b*)	**12.** (*a*)	**13.** (*d*)	**14.** (*d*)	**15.** (*a*)	**16.** (*c*)	**17.** (*a*)	**18.** (*a*)
19. (*c*)	**20.** (*b*)	**21.** (*c*)	**22.** (*d*)	**23.** (*c*)	**24.** (*b*)	**25.** (*b*)	**26.** (*d*)	**27.** (*d*)
28. (*b*)	**29.** (*d*)	**30.** (*c*)	**31.** (*a*)	**32.** (*c*)	**33.** (*d*)	**34.** (*b*)	**35.** (*d*)	**36.** (*a*)
37. (*c*)	**38.** (*a*)	**39.** (*a*)	**40.** (*b*)	**41.** (*d*)	**42.** (*c*)	**43.** (*a*)	**44.** (*a*)	**45.** (*d*)
46. (*a*)	**47.** (*d*)	**48.** (*c*)	**49.** (*c*)	**50.** (*a*)	**51.** (*d*)	**52.** (*c*)	**53.** (*a*)	**54.** (*b*)
55. (*b*)	**56.** (*b*)	**57.** (*b*)	**58.** (*b*)	**59.** (*a*)	**60.** (*b*)	**61.** (*d*)	**62.** (*b*)	**63.** (*d*)
64. (*c*)	**65.** (*d*)	**66.** (*b*)	**67.** (*a*)	**68.** (*a*)	**69.** (*a*)	**70.** (*a*)	**71.** (*b*)	**72.** (*d*)
73. (*b*)	**74.** (*a*)	**75.** (*b*)	**76.** (*d*)	**77.** (*c*)	**78.** (*a*)	**79.** (*c*)	**80.** (*b*)	**81.** (*a*)
82. (*a*)	**83.** (*c*)	**84.** (*d*)	**85.** (*d*)	**86.** (*d*)	**87.** (*b*)	**88.** (*b*)	**89.** (*b*)	**90.** (*d*)
91. (*c*)	**92.** (*b*)	**93.** (*a*)	**94.** (*a*)	**95.** (*a*)	**96.** (*a*)	**97.** (*b*)	**98.** (*d*)	**99.** (*c*)
100. (*b*)	**101.** (*b*)	**102.** (*a*)	**103.** (*c*)	**104.** (*c*)	**105.** (*b*)	**106.** (*a*)	**107.** (*b*)	**108.** (*b*)
109. (*a*)	**110.** (*d*)	**111.** (*c*).						

HINTS/SOLUTIONS

Q. No.	Ans.	Hints & Solutions
1.	**(*b*)**	The reaction is initiated by attack of electrophile H^+ on carbon-carbon double bond.
2.	**(*d*)**	Side chain ($-CH_2Cl$) in benzyl chloride is oxidized to $-COOH$.
3.	**(*b*)**	Hunsdiecker reaction.
4.	**(*a*)**	A is CH_3CN and B is CH_3COOH.
5.	**(*a*)**	Primary amines give carbylamine reaction. CH_3NH_2 is a primary amine.
6.	**(*c*)**	Aniline is the product.
7.	**(*c*)**	This a method for the preparation of chloroform.
8.	**(*a*)**	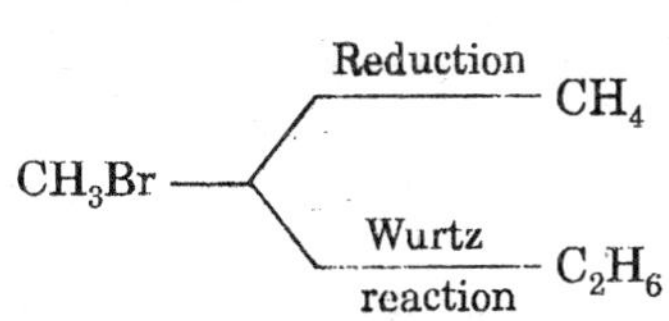
9.	**(*c*)**	$2CHCl_3 + 6Ag \longrightarrow 6AgCl + HC \equiv CH$
10.	**(*d*)**	From (CCl_2F_2) is used as refrigerant.
11.	**(*b*)**	S_N2 mechanism proceeds through a transition state.
12.	**(*a*)**	Allylic substitution takes place at high temperature.
13.	**(*d*)**	Inversion of configuration indicates S_N2 mechanism.
14.	**(*d*)**	$-NO_2$ group withdraws electrons at ortho-para positions and stabilizes the intermediate carbanion.
15.	**(*a*)**	Carbylamine reaction.
16.	**(*c*)**	Aniline is formed under the given conditions.
17.	**(*a*)**	Thionyl chloride ($SOCl_2$) converts C_2H_5OH into ethyl chloride.
18.	**(*a*)**	$3° > 2° > 1°$.
19.	**(*c*)**	$C_6H_6 + 3Cl_2 \xrightarrow{\text{Sunlight}} C_6H_6Cl_6$.
20.	**(*b*)**	Markownikoff addition.
21.	**(*c*)**	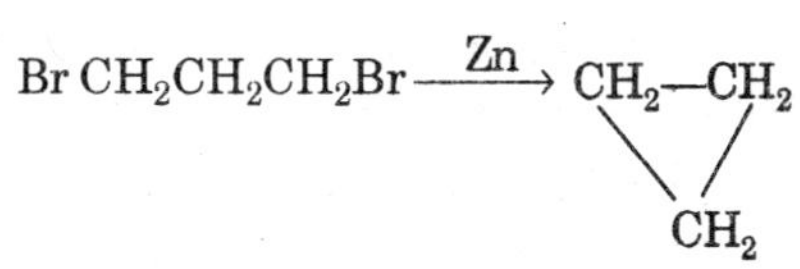
22.	**(*d*)**	$CHCl_3 \xrightarrow{NaOH} [CH(OH)_3] \xrightarrow{-H_2O}$ $HCOOH \xrightarrow{NaOH} HCOONa$
23.	**(*c*)**	Butane, pentane and hexane.
24.	**(*b*)**	$-X$ is replaced by $-OH$.
25.	**(*b*)**	H is replaced by Cl in C_3H_8.
26.	**(*d*)**	Wurtz reaction.

Q. No.	Ans.	Hints & Solutions
27.	(*d*)	$C_6H_5NH_2 \xrightarrow{CHCl_3,\ KOH} C_6H_5NC$ Carbylamine reaction
28.	(*b*)	Br_2, in the presence of sunlight brings about subsititution at benzylic position.
29.	(*d*)	With 2° and 3° alkyl halides elimination takes place readily. *trans*-2-pentene being most stable is formed as the major product.
30.	(*c*)	3-Pentanone does not contain $CH_3{-}\overset{\overset{\large O}{\|\|}}{C}{-}$ or $CH_3{-}\underset{\underset{\large OH}{\|}}{CH}{-}$ group.
31.	(*a*)	$C_6H_5{-}CH_2Cl \xrightarrow{NaCN} C_6H_5{-}CH_2CN \xrightarrow{H_2/Ni} C_6H_5{-}CH_2{-}CH_2NH_2.$
32.	(*c*)	The three reactants in carbylamine reaction are *p*-amine, chloroform and KOH.
33.	(*d*)	$C_2H_5{-}O{-}C_2H_5 + 10Cl_2 \longrightarrow C_2Cl_5{-}O{-}C_2Cl_5 + 10HCl$ Perchlorodiethylether
34.	(*b*)	Freon-12 (CF_2Cl_2) is used as refrigerant.
35.	(*d*)	$C_6H_5-CH_3 \longrightarrow C_6H_5-CCl_3 \xrightarrow{NaOH} [C_6H_5C(OH)_3] \xrightarrow{-H_2O} C_6H_5COOH$ Benzoic acid
36.	(*a*)	Pyrene is CCl_4.
37.	(*c*)	Ethane by soda-lime decarboxylation. Butane by Kolbe electrolysis.
38.	(*a*)	Dehydrohalogenation according to Saytzeff rule.
39.	(*a*)	Allylic subsititution.
40.	(*b*)	$C_2H_5I \xrightarrow{alc.\ KOH} CH_2=CH_2 \xrightarrow{Br_2} \underset{Br}{CH_2}-\underset{Br}{CH_2} \xrightarrow{KCN} \underset{CN}{CH_2}-\underset{CN}{CH_2}$
41.	(*d*)	$C_2H_5Cl + C_2H_5ONa \rightarrow C_2H_5-O-C_2H_5 + NaCl$ Williamson synthesis.
42.	(*c*)	—OH group is highly activating group.
43.	(*a*)	Sandmeyer reaction.
44.	(*a*)	$(CH_3)_2C=CH_2 + HBr \xrightarrow{Peroxide} (CH_3)_2CH-CH_2Br$ Anti-Markownikoff addition.
45.	(*d*)	$CH_3CH_2CH_2Cl \xrightarrow{KCN} CH_3CH_2CH_2CN$ Butane nitrile.
46.	(*a*)	1° halides react through S_N2.
47.	(*d*)	*p*-Dichlorobenzene, due to its symmetry and consequent close, packing of its molecules has highest melting point.
48.	(*c*)	$CHCl_3 + HO.NO_2 \longrightarrow CCl_3{-}NO_2 + H_2O.$
49.	(*c*)	CH_3-I is used as methylating agent.
50.	(*a*)	N-Phenyl acetanilide.
51.	(*d*)	—Cl is ortho and para directing group.
52.	(*c*)	$CHI_3 \xrightarrow{KOH} [CH(OH)_3] \xrightarrow{-H_2O} HCOOH \xrightarrow{KOH} HCOOK$
53.	(*a*)	The order of reactivity of halides towards S_N2 is $CH_3{-}X > 1° > 2° > 3°$.
54.	(*b*)	Factual question. The intermediate carbocation can undergo nucleophilic attack from either side resulting in racemic product.
55.	(*b*)	Allylic subsititution.
56.	(*b*)	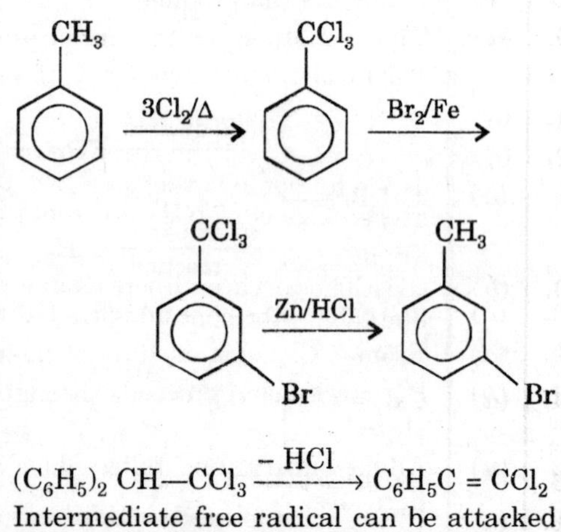
57.	(*b*)	$(C_6H_5)_2\,CH{-}CCl_3 \xrightarrow{-HCl} C_6H_5C=CCl_2$
58.	(*b*)	Intermediate free radical can be attacked by Cl_2 from either side resulting in the racemic mixture.

Q. No.	Ans.	Hints & Solutions
59.	(a)	$2CH_3\text{—}C(CH_3)_2\text{—}Cl \longrightarrow CH_3-C(CH_3)_2-C(CH_3)_2-CH_3$ All H-atoms in the product are equivalent.
60.	(b)	Primary amines undergo carbylamine reaction.
61.	(d)	Self explanatory question.
62.	(b)	$RX + Mg \longrightarrow \underset{\text{Grignard reagent}}{RMgX}$
63.	(d)	Wurtz reaction.
64.	(c)	Compound (c) contains $CH_3\text{—}\underset{\vert}{CH}\text{—}OH$ group.
65.	(d)	$CH_3-CH_2-\overset{*}{C}H(Cl)-CH_3$ 2-Chlorobutane has a chiral centre.
66.	(b)	R—I > R—Br > R—Cl > R—F. This order can be explained in terms of strength of C—X bond. C—I bond is weakest whereas C—F bond is strongest.
67.	(a)	Carbylamine reaction.
68.	(a)	Chlorobenzene and vinyl chloride are stabilized by resonance and hence are less reactive. Chlorobenzene is more stabilized by resonance.
69.	(a)	3° > 2° > 1°. The product formed from 3° halides is more stable.
70.	(a)	$CH_3CH_2CH_2Cl \xrightarrow{-\,HCl} CH_3CH_2CH = CH_2$ Dehydrohalogenation.
71.	(b)	Wurtz reaction.
72.	(d)	A mixture of all the three amines is formed depending on whether one, two or three hydrogen atoms of NH_3 are replaced by alkyl groups.
73.	(b)	CH_3Cl has greater dipole moment than CH_3F due to greater bond length.
74.	(a)	$CH_2 = CH\text{—}CH_2Cl \xrightarrow{-\,HCl} \underset{\text{Propadiene}}{CH_2 = C = CH_2}$
75.	(b)	Wurtz reaction.
76.	(d)	The product formed is DDT.
		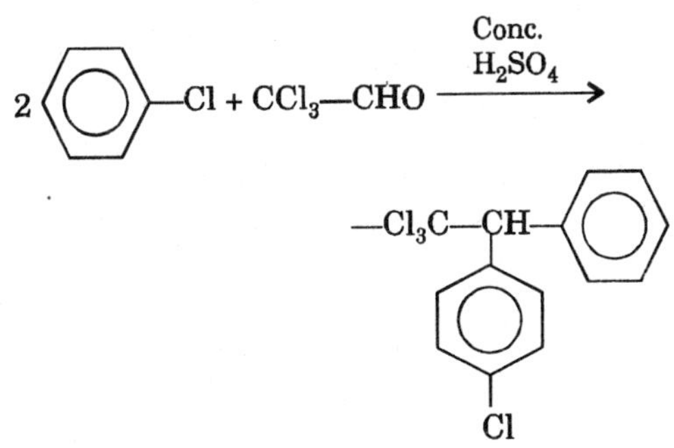 DDT
77.	(c)	Freons are chlorofluorocarbons and are responsible for depletion of ozone layer.
78.	(a)	In vinyl chloride (CH_2 = CH—Cl), the C—Cl bond has partial double bond character due to resonance.
79.	(c)	$CHCl_3 + HONO_2 \longrightarrow \underset{\text{Chloropicrin}}{CCl_3\text{—}NO_2} + H_2O$
80.	(b)	$CH_3\text{—}CO\text{—}CH_2CH_2Br$ on dehydrohalogenation gives an alkene in which C—C double bond is in conjugation with carbonyl group. $CH_3\text{—}\underset{\underset{O}{\Vert}}{C}\text{—}CH_2CH_2Br \xrightarrow{\text{alc. KOH}} CH_3\text{—}\underset{\underset{O}{\Vert}}{C}\text{—}CH = CH_2$
81.	(a)	In Ar—X (Aryl halides) C—X bond has partial double bond character and hence is difficult to break.
82.	(a)	Steric hindrance.
83.	(c)	More substituted alkene is formed predominantly (Saytzeff rule).
84.	(d)	Benzyl carbocation is stabilized by resonance and hence is formed, readily, guiding the reaction through S_N1 mechanism.
85.	(d)	It is Williamson synthesis of ethers.
86.	(d)	Alkyl halides are converted to alcohols on reaction with aqueous NaOH or moist silver oxide (AgOH).
87.	(b)	$2C_2H_5Br + Ag_2O \longrightarrow C_2H_5OC_2H_2 + 2AgBr.$
88.	(a)	KNO_2 on reaction with ethyl chloride gives ethyl nitrite as the major product.
89.	(b)	Allylic halides are much more reactive than vinylic and aryl halides.
90.	(d)	The order of reactivity of alkyl halides is 3° > 2 ° > 1° Iodides > Bromides > Chlorides.

Q. No.	Ans.	Hints & Solutions
91.	(c)	AgCN on reaction with AgCN gives C_2H_5—NC. Its functional isomer is C_2H_5CN.
92.	(b)	It is in accordance with Saytzeff's rule.
93.	(a)	C_2H_5Br on reaction with KCN gives cyanides which is reduced to an amine with $LiAlH_4$.
94.	(a)	In chlorobenzene C—Cl bond has partial double bond character due to resonance and hence it does not undergo nucleophilic substitution under ordinary conditions.
95.	(a)	A is formed in accordance with Markovnikov's rule through formation of carbocation as intermediate. On the other hand, in the presence of light the reaction takes place through free radical as intermediate and the product is formed in accordance with anti-Markovnikov's rule.
96.	(a)	Addition follows Markovnikov's rule.
97.	(b)	Williamson synthesis involves reaction of alkyl halides with sodium alkoxide.
98.	(d)	Alkyl and benzyl halides undergo nucleophilic substitution readily whereas aryl halides are less reactive.
99.	(c)	3-bromo-1-chlorocyclohexene.
100.	(b)	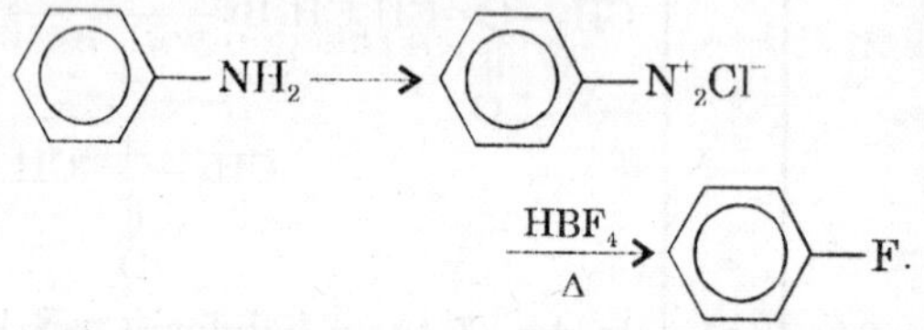
101.	(b)	$C_6H_5MgBr + CH_3OH \longrightarrow C_6H_6 + Mg(OCH_3)Br$
102.	(a)	D > C > A > B D is the strongest nucleophilic while B is the weakest nucleophile.
103.	(c)	Anhydrous $ZnCl_2$.
104.	(c)	$R—X \xrightarrow{\text{alc. KCN}} R—CN \xrightarrow{\text{dil. HCl}} RCOOH$
105.	(b)	$C_6H_5—CH_2CH_3 \xrightarrow{SO_2Cl_2} C_6H_5—CH(Cl)—CH_3 \xrightarrow{\text{alc. KOH}} C_6H_5—CH = CH_2$
106.	(a)	The order of reactivity of halides for S_N1 reaction is : Benzylic > Allylic > 2° > 1°.
107.	(b)	$CH_3—CH_2—CH_2—CH_2—Cl$ has the longest chain of atoms.
108.	(b)	This is carbylamine reaction.
109.	(a)	—CH_3 group has *ortho-para* directing effect.
110.	(d)	For S_N2 mechanism, the order of reactivity of alkyl halides is 1° > 2° > 3° or $RCH_2X > R_2CHX > R_3CX$.
111.	(d)	$Cl^- < Br^- < I^-$ I^- has maximum nucleophilicity due to poor solvation.

3.6. MCQs FOR IIT JEE

Brain Twister

A. Multiple Choice Questions With One Correct Answer

1. The correct order of dipole moment is
   (a) $CH_3Cl > CH_3F > CH_3Br$
   (b) $CH_3F > CH_3Cl > CH_3Br$
   (c) $CH_3Br > CH_3Cl > CH_3F$
   (d) $CH_3Br > CH_3F > CH_3Cl$.
2. The number of dibromoderivatives possible for butane is
   (a) 8 (b) 4
   (c) 6 (d) 5.
3. Which of the following reactions is maximum exothermic ?
   (a) Chlorination of benzene
   (b) Iodination of benzene
   (c) Flurination of benzene
   (d) Bromination of benzene.
4. Which of the following has the highest nucleophilicity ?
   (a) H_2O
   (b) OH^-
   (c) CH_3O^-
   (d) CH_3COO^-.
5. Which of the following represents the correct order of nucleophilicity ?
   (a) $F^- > {}^-OH\ NH_2^- > CH_3^-$
   (b) ${}^-OH > NH_2^- > CH_3^- > F^-$
   (c) ${}^-OH > NH_2^- > F^- > CH_3^-$
   (d) $CH^- > NH^- > {}^-OH > F^-$.
6. When aniline is heated with chloroform and alcoholic caustic potash, a compound with offensive smell is obtained. The compound is a/an
   (a) Primary amine (b) Secondary amine
   (c) Isocyanide (d) Cyanide.
7. Which of the following compounds will react with ethanolic KCN ?
   (a) Chlorobenzene (b) Allyl bromide
   (c) Vinyl bromide (d) p-Chlorotoluene.
8. Benzene diazonium chloride can be converted into chlorobenzene by
   (a) Gattermann reaction (b) Sandmeyer reaction
   (c) Both (d) None.
9. The major product of reaction of ethyl bromide with silver cyanide is
   (a) C_2H_5CN (b) CH_3CN
   (c) CH_3NC (d) C_2H_5NC.
10. Pure chloroform is obtained by treating
    (a) Ethanol with bleaching powder
    (b) Acetone with bleaching powder
    (c) Chloral with NaOH
    (d) CCl_4 with moist iron.
11. The reaction of chlorobenzene with chlorine in the presence of ferric chloride gives mainly
    (a) o-Dichlorobenzene
    (b) p-Dichlorobenzene
    (c) m-Dichlorobenzene
    (d) A mixture of o- and p-Dichlorobenzene.
12. Which of the following statements is not correct ?
    (a) C—Cl bond in vinyl chloride is less polar than in CH_3Cl
    (b) C—Cl bond in vinyl chloride is stronger than in CH_3Cl
    (c) C—Cl bond in vinyl chloride is shorter than in CH_3Cl
    (d) Vinyl chloride undergoes nucleophilic substitution more readily than CH_3Cl.
13. 1-Bromopropane on reaction with $LiAlH_4$ yields
    (a) Propane (b) Hexane
    (c) Propene (d) Propyne.
14. When ethyl iodide is heated with dry silver oxide, the product formed is
    (a) CH_3CH_2OH (b) 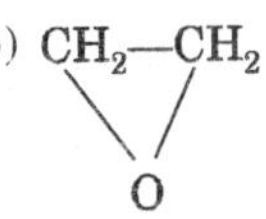
    (c) CH_3CHO (d) $CH_3CH_2OCH_2CH_3$.
15. Acetonitrile may be prepared by heating
    (a) Ethyl alcohol and KCN
    (b) Methyl alcohol and KCN
    (c) Ethyl chloride and KCN
    (d) Methyl chloride and KCN.
16. 2-Chlorobutane on reaction with alcoholic potash gives .... as the major product.
    (a) 1-Butene (b) 1-Butanol
    (c) 2-Butene (d) 2-Butanol.
17. Among the following compounds, the one with least percentage of chlorine is
    (a) Chloroform (b) Pyrene
    (c) Gammexane (d) Vinyl chloride.
18. Chloroform on heating with silver powder gives
    (a) Ethene (b) Ethane
    (c) Ethyne (d) Benzene.

19. Which of the following does not give white precipitate of AgCl when boiled with alcoholic solution of $AgNO_3$?

(*a*) Crotylchloride (*b*) Allyl chloride

(*c*) Benzyl chloride (*d*) Vinyl chloride.

20. In which of the following compounds —Cl is most easily replaceable ?

(*a*) C_6H_5Cl (*b*) $C_6H_5-CH_2-CH_2Cl$

(*c*) $C_6H_5-\underset{\underset{Cl}{|}}{CH}-CH_3$ (*d*) $C_6H_4Cl_2$.

21. A dihalogen derivative 'X' of a hydrocarbon with three carbon atoms in a molecule, reacts with alc. KOH giving another hydrocarbon which gives red precipitate with amm. CuCl. 'X' gives an aldehyde when treated with aqueous KOH. The compound 'X' is

(*a*) 1, 3-Dichloropropane (*b*) 1, 2-Dichloropropane

(*c*) 2, 2-Dichloropropane (*d*) 1, 1-Dichloropropane.

22. Treatment of ethyl bromide with excess NH_3 yields ...... as the major product.

(*a*) Ethyl amine

(*b*) Diethyl amine

(*c*) Triethyl amine

(*d*) Tetraethyl ammonium bromide.

23. Hydrolysis of acetonitrile with conc. HCl gives

(*a*) CH_3NC (*b*) CH_3COOH

(*c*) CH_3CONH_2 (*d*) CH_3CH_2COOH.

24. Monochlorination of ethyl benzene in the presence of light and heat yields ...... as the major product. The missing item is

(*a*) *m*-Chlorotoluene

(*b*) A mixture of *o*- and *p*-chlorotoluene

(*c*) 1-Chloro-2-phenylethane

(*d*) 1-Chloro-1-phenylethane.

25. Iodoform reacts with NaOH to give

(*a*) Sodium acetate (*b*) Acetic acid

(*c*) Sodium oxalate (*d*) Sodium formate.

26. The order of reactivity of various alkyl halides towards S_N1 reaction is

(*a*) 3° > 2° > 1° (*b*) 1° > 2° > 3°

(*c*) 3° = 2° = 1° (*d*) 1° > 3° > 2°.

27. Alkyl halides can be converted to ethers through

(*a*) Frankland reaction (*b*) Williamson synthesis

(*c*) Fittig reaction (*d*) Grignard reaction.

28. Which of the following groups is *ortho-para* directing as well as deactivating, when attached to benzene ring ?

(*a*) $-NO_2$ (*b*) —OH

(*c*) OCH_3 (*d*) —Cl.

29. The order of reactivity of various alkyl halides towards nucleophilic substitution follows the order

(*a*) R—I > R—Br > R—Cl > R—F

(*b*) R—F > R—Cl > R—Br > R—I

(*c*) R—Cl > R—Br > R—I > R—F

(*d*) R—Br > R—I > R—Cl > R—F.

30. $$C_6H_5-\underset{\underset{H}{|}}{\overset{\overset{CH_3}{|}}{C}}-Br + H_2O \longrightarrow HO-\underset{\underset{H}{|}}{\overset{\overset{CH_3}{|}}{C}}-C_6H_5 + HBr$$

The above reaction results in 98% racemization. The reaction mainly follows

(*a*) S_N1 mechanism (*b*) S_N2 mechanism

(*c*) E^1 mechanism (*d*) E^2 mechanism.

31. The correct order of rate of S_N2 reactivity of H_2O, $\bar{O}CH_3$, $\bar{O}H$ and CH_3COO^- is

(*a*) $^-OH > ^-OCH_3 > CH_3COO^- > H_2O$

(*b*) $H_2O > ^-OH > ^-OCH_3 > CH_3COO^-$

(*c*) $^-OCH_3 > ^-OH > CH_3COO^- > H_2O$

(*d*) $H_2O > CH_3COO^- > ^-OH > ^-OCH_3$.

32. A compound that will form an offensive smell when heated with chloroform and alcoholic potash is

(*a*) $C_2H_5NH_2$ (*b*) $(C_2H_5)_2NH$

(*c*) $(CH_3)_3N$ (*d*) CH_3CN.

33. Toluene on reaction with N-bromosuccinimide gives

(*a*) *p*-Bromomethylbenzene

(*b*) *o*-Bromomethylbenzene

(*c*) Phenyl bromomethane

(*d*) *m*-Bromomethylbenzene

34. The correct order of boiling points is

(*a*) $CH_3I > CH_3Br > CH_3Cl > CH_3F$

(*b*) $CH_3F > CH_3Cl > CH_3Br > CH_3I$

(*c*) $CH_3I > CH_3F > CH_3Cl > CH_3Br$

(*d*) $CH_3Br > CH_3Cl > CH_3F > CH_3I$.

35. Dehalogenation of *meso*-2, 3-dibromobutane results in

(*a*) *cis*-2-Butene

(*b*) *trans*-2-Butene

(*c*) *n*-Butane

(*d*) A mixture of *cis*- and *trans*-2-Butene.

36. Which of the following statements is/are not correct ?

(*a*) Alkyl iodides are heavier than water

(*b*) Alkyl bromides are lighter than water

(*c*) Ethyne reacts with excess HCl to form ethylidene dichloride

(*d*) Vinyl chloride does not undergo nucleophilic substitution easily.

37. Which of the following reactions do not result in the formation of a new C—C bond ?
(*a*) Ullmann reaction (*b*) Fittig reaction
(*c*) Williamson synthesis (*d*) Wurtz reaction.

38. Which of the following represents a geminal dihalide ?
(*a*) Ethylene dichloride (*b*) 2, 2-Dichloropropane
(*c*) 1, 3-Dichloropropane (*d*) 1, 2-Dichloropropane.

39. The reactivity of the compounds
(*i*) MeBr (*ii*) $PhCH_2Br$ (*iii*) MeCl (*iv*) *p*-$MeOC_6H_4Br$
decreases as
(*a*) (*i*) > (*ii*) > (*iii*) > (*iv*) (*b*) (*iv*) > (*ii*) > (*i*) > (*iii*)
(*c*) (*iv*) > (*iii*) > (*i*) > (*ii*) (*d*) (*ii*) > (*i*) > (*iii*) > (*iv*).

40. Arrange the following halides in the decreasing order of S_N1 reactivity : $CH_3CH_2CH_2Cl$ (I), $CH_2=CHCH(Cl)CH_3$ (II), $CH_3CH_2CH(Cl)CH_3$ (III)
(*a*) I > II > III (*b*) II > I > III
(*c*) II > III > I (*d*) III > II > I.

41. The reaction of toluene with chlorine in presence of ferric chloride give predominanly
(*a*) Benzoyl chloride
(*b*) *m*-Chlorotoluene
(*c*) Benzyl chloride
(*d*) *o*- and *p*-Cholorotoluene.

42. Carbon tetrachloride does not have net dipole moment because of
(*a*) Its planar structure
(*b*) Its regular tetrahedral structure
(*c*) Similar sizes of carbon and chlorine atoms
(*d*) Similar electron affinities of carbon and chlorine.

43. Carbylamine test is performed in alcoholic KOH by heating a mixture of
(*a*) Chloroform and silver powder
(*b*) Trihalogenated methane and a primary amine
(*c*) An alkyl halide and a primary amine
(*d*) An alkyl cyanide and a primary amine.

B. MCQs From Previous IIT Examinations

44. Chlorobenzene can be prepared by reacting aniline with
(*a*) Hydrochloric acid
(*b*) Cuprous chloride
(*c*) Chlorine in the presence of anhydrous aluminium chloride
(*d*) Nitrous acid followed by heating with cuprous chloride. ***(IIT 1984)***

45. Out of the following compounds, which one will have zero dipole moment ?
(*a*) 1, 1-Dichloroethylene
(*b*) *cis*-1, 2-Dichloroethylene
(*c*) *trans*-1, 2-Dichloroethylene
(*d*) None of these compounds. ***(IIT 1987)***

46. *n*-Propyl bromide on treatment with ethanolic potassium hydroxide produces
(*a*) Propane (*b*) Propene
(*c*) Propyne (*d*) Propanol. ***(IIT 1987)***

47. Chlorination of toluene in the presence of light and heat followed by treatment with aqueous NaOH gives
(*a*) *o*-Cresol
(*b*) *p*-Cresol
(*c*) 2, 4-Dihydroxytoluene
(*d*) Benzoic acid. ***(IIT 1990)***

48. 1-Chlorobutane on reaction with alcoholic potash gives
(*a*) 1-Butene (*b*) 1-Butanol
(*c*) 2-Butene (*d*) 2-Butanol. ***(IIT 1991)***

49. The products of reaction of alcoholic silver nitrite with ethyl bromide are
(*a*) Ethane (*b*) Ethene
(*c*) Nitroethane (*d*) Ethyl nitrite. ***(IIT 1991)***

50. Butanenitrite may be prepared by heating
(*a*) Propyl alcohol with KCN
(*b*) Butyl chloride with KCN
(*c*) Butyl alcohol with KCN
(*d*) Propyl chloride with KCN. ***(IIT 1992)***

51. Arrange the following compounds in order of increasing dipole moment : Toluene (I), *m*-dichlorobenzene (II), *o*-dichlorobenzene (III), *p*-dichlorobenzene (IV)
(*a*) I < IV < II < III (*b*) IV < I < II < III
(*c*) IV < I < III < II (*d*) IV < II < I < III.
(IIT 1996)

52. In the following groups :
—OAc (I), —OMe (II),
—OSO_2Me (III), —OSO_2CF_3 (IV)
the order of leaving group ability is
(*a*) I > II > III > IV (*b*) IV > III > I > II
(*c*) III > II > I > IV (*d*) II > III > IV > I.
(IIT 1997)

53. Benzyl chloride ($C_6H_5CH_2Cl$) can be prepared from toluene by chlorination with
SO_2Cl_2 (I), $SOCl_2$ (II), Cl_2 (III), NaOCl (IV)
(*a*) I, II and III (*b*) I and III
(*c*) II and III (*d*) All the four. ***(IIT 1998)***

54. Which of the following will react with water ?
(*a*) $CHCl_3$ (*b*) $CCl_3 . CHO$
(*c*) CCl_4 (*d*) $ClCH_2 . CH_2Cl$.
(IIT 1998)

55. A solution of (+)-2-chloro-2-phenylethane in toluene racemises slowly in the presence of small amount of $SbCl_5$, due to the formation of

(*a*) Carbanion (*b*) Carbene
(*c*) Free radical (*d*) Carbocation. ***(IIT 1999)***

56. The order of reactivities of the following alkyl halides of a S_N2 reaction is

(*a*) RF > RCl > RBr > RI
(*b*) RF > RBr > RCl > RI
(*c*) RCl > RBr > RF > RI
(*d*) RI > RBr > RCl > RF. ***(IIT Screening Test 2000)***

57. Identify the set of reagent/reaction conditions 'X' and 'Y' in the following set of transformations :

$$CH_3—CH_2—CH_2Br \xrightarrow{X} \text{Product} \xrightarrow{Y} CH_3—\underset{\underset{Br}{|}}{CH}—CH_3$$

(*a*) X = dilute aqueous NaOH, 20°C ;
Y = HBr/acetic acid, 20°C
(*b*) X = concentrated alcoholic NaOH, 80°C ;
Y = HBr/acetic acid, 20°C
(*c*) X = dilute aqueous NaOH, 20°C ;
Y = $Br_2 / CHCl_3$, 0°C
(*d*) X = concentrated alcoholic NaOH, 80°C ;
Y = $Br_2 / CHCl_3$, 0°C. ***(IIT Screening Test 2002)***

58. Among the following, the molecule with the highest dipole moment is :

(*a*) CH_3Cl (*b*) CH_2Cl_2
(*c*) $CHCl_3$ (*d*) CCl_4.
(IIT Screening Test 2003)

59. 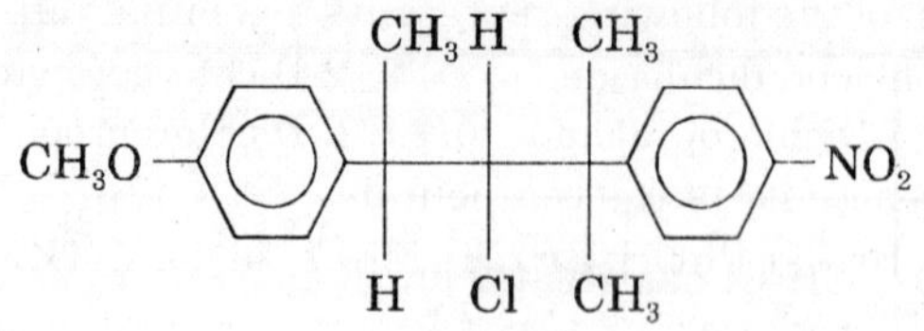

on hydrolysis in aqueous acetone gives,

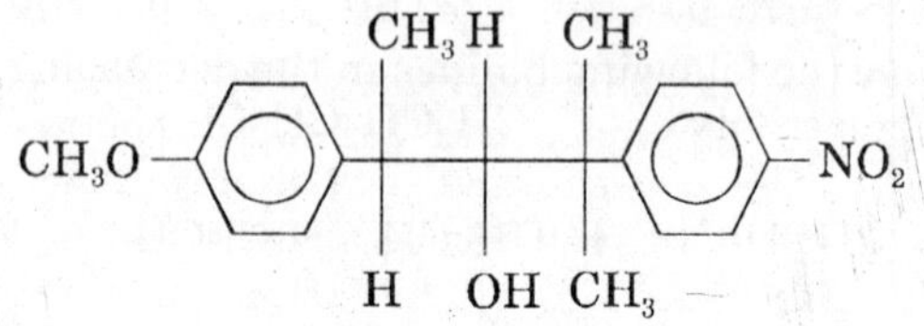 (K)

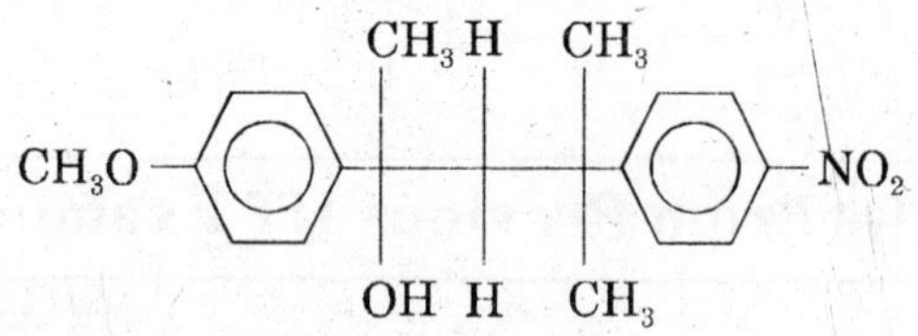 (L)

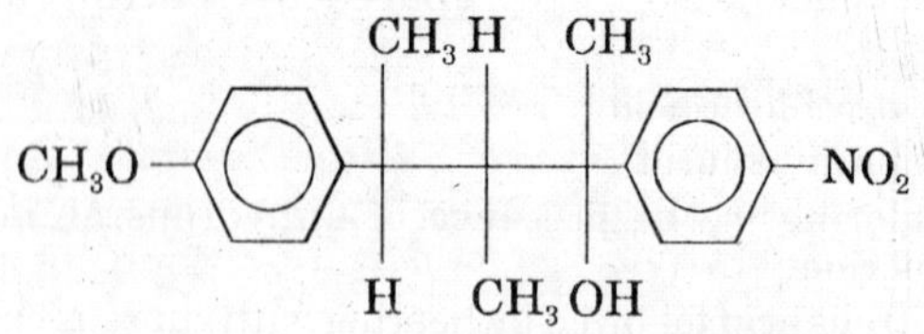 (M)

(*a*) K and L (*b*) only K
(*c*) K and M (*d*) only M.
(IIT Screening Test 2005)

ANSWERS

1. (*a*)	**2.** (*c*)	**3.** (*c*)	**4.** (*c*)	**5.** (*d*)	**6.** (*c*)	**7.** (*b*)	**8.** (*c*)	**9.** (*d*)
10. (*c*)	**11.** (*d*)	**12.** (*d*)	**13.** (*a*)	**14.** (*d*)	**15.** (*d*)	**16.** (*c*)	**17.** (*d*)	**18.** (*c*)
19. (*d*)	**20.** (*c*)	**21.** (*d*)	**22.** (*a*)	**23.** (*c*)	**24.** (*d*)	**25.** (*d*)	**26.** (*a*)	**27.** (*b*)
28. (*d*)	**29.** (*a*)	**30.** (*a*)	**31.** (*c*)	**32.** (*a*)	**33.** (*c*)	**34.** (*a*)	**35.** (*b*)	**36.** (*b*)
37. (*c*)	**38.** (*b*)	**39.** (*d*)	**40.** (*c*)	**41.** (*d*)	**42.** (*b*)	**43.** (*b*)	**44.** (*d*)	**45.** (*c*)
46. (*b*).	**47.** (*d*)	**48.** (*a*)	**49.** (*c, d*)	**50.** (*d*)	**51.** (*b*)	**52.** (*b*)	**53.** (*b*)	**54.** (*b*)
55. (*d*)	**56.** (*d*)	**57.** (*b*)	**58.** (*a*)	**59.** (*a*).				

HINTS/SOLUTIONS

Q. No.	Ans.	Hints & Solutions
1.	(*a*)	CH_3Cl has greater dipole moment than CH_3F because C—Cl bond length is longer than C—F bond length $\mu = e \times d$.
2.	(*c*)	The six dibromoderivatives of butane are : 1, 1 dibromobutane, 1, 2-dibromobutane, 1, 3-dibromobutane, 1, 3-bromobutane, 2, 2-dibromobutane, 2, 3-dibromobutane.
3.	(*c*)	Fluorination of benzene is maximum exothermic because C—F bond is very strong.
4.	(*c*)	CH_3O^- has highest nucleophilicity. Nucleophilicity roughly parallels basicity while comparing nucleophiles that have the same attacking atom.
5.	(*d*)	$CH_3^- > NH_2^- > {}^-OH > F^-$ In going from left to right across a period nucleophilicity decreases.
6.	(*c*)	Carbylamine reaction.
7.	(*b*)	Aryl and vinyl halides donot undergo subsititution readily. On the other hand allyl halides undergo substitution easily.
8.	(*c*)	Gattermann reaction involves reaction of diazonium salt with copper powder in the presence of HCl (or HBr) while Sandmeyer reaction involves reaction of diazonium salt with CuCl (or CuBr).
9.	(*d*)	$C_2H_5Br + AgCN \longrightarrow \mathbf{C_2H_5NC} + AgBr.$
10.	(*c*)	$CCl_3\text{–}CHO + NaOH \longrightarrow \mathbf{CHCl_3} + HCOONa.$
11.	(*d*)	—Cl group is ortho and para directing group.
12.	(*d*)	Vinyl chloride is less reactive towards nucleophilic substitution reactions because C—Cl bond in it has partial double bond character due to resonance.
13.	(*a*)	$CH_3CH_2CH_2Br + 2H \xrightarrow{LiAlH_4} CH_3CH_2CH_3 + HBr.$
14.	(*d*)	$2C_2H_5I + Ag_2O \longrightarrow \mathbf{C_2H_5\text{—}O\text{—}C_2H_5} + 2AgI.$
15.	(*d*)	$CH_3Cl + KCN \longrightarrow \underset{\text{Acetonitrile}}{CH_3CN} + KCl$
16.	(*c*)	β-Elimination obeys Saytzeff's rule.
17.	(*d*)	Vinyl chloride, $CH_2 = CH\text{—}Cl$
18.	(*c*)	$2CHCl_3 + 6\ Ag \longrightarrow HC \equiv HC + 6\ AgCl.$
19.	(*d*)	Vinyl chloride has partial double bond character in C—Cl bond, hence it does not release Cl^- ion.
20.	(*c*)	Halogen atom at benzylic position can be replaced easily.
21.	(*d*)	$CH_3CH_2CHCl_2 \xrightarrow{aq.\ KOH} [CH_3CH_2CH(OH)_2] \xrightarrow{-H_2O} CH_3CH_2CHO$ $\xrightarrow{aq.\ KOH} CH_3C \equiv CH \xrightarrow[\text{Ammoniacal}]{CuCl} \underset{\text{red ppt.}}{CH_3C \equiv CCu}$
22.	(*a*)	When NH_3 is in excess only one of the H atom of NH_3 is replaced by alkyl group (because alkyl halide is limited).
23.	(*c*)	CH_3CN on partial hydrolysis yields CH_3CONH_2.
24.	(*d*)	Benzylic hydrogen is easy to replace because the reaction proceeds through more stable benzylic free radical as intermediate. 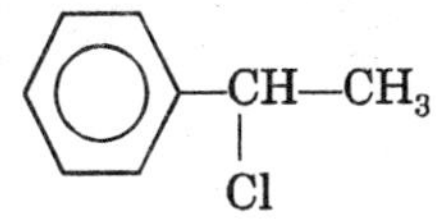 Substitution takes place at benzylic position.
25.	(*d*)	$CHI_3 \xrightarrow{OH^-} [CH(OH)_3] \xrightarrow{-H_2O} HCOOH \xrightarrow{NaOH} HCOONa.$
26.	(*a*)	$3° > 2° > 1°$.
27.	(*b*)	Williamson synthesis involves reaction of alkyl halides with sodium alkoxides.
28.	(*d*)	—Cl group is ortho-para directing and deactivating.
29.	(*a*)	C—I bond is weaker than C—Br, is weaker than C—Cl and C—F.
30.	(*a*)	Reaction through $S_N{}^2$ mechanism leads to racemization whereas reaction through $S_N{}^2$ mechanism leads to inversion of configuration.
31.	(*c*)	Stronger nucleophiles react through $S_N{}^2$ at faster rate. $^-OCH_3$ is the strongest nucleophile while H_2O is the weakest, among the given species.
32.	(*a*)	Primary amines give carbylamine reaction.
33.	(*c*)	N-Bromosuccinimide results in benzylic substitution. 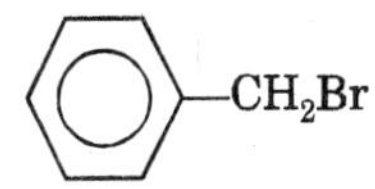
34.	(*a*)	Boiling points of alkyl halides are in the order : RI > RBr > RCl > RF.

Q. No.	Ans.	Hints & Solutions
35.	**(*b*)**	Elimination takes place in *anti* fashion.
36.	**(*b*)**	Bromides and iodides are heavier than water.
37.	**(*c*)**	In Williamson synthesis new C—O bond is formed.
38.	**(*b*)**	The two —Cl atoms in 2, 2-dichloropropane are on the same carbon atom.
39.	**(*d*)**	The reactivity is in the order : Benzylic > alkyl > aryl halides Out of methyl chloride and methyl bromide, methyl bromide is more reactive.
40.	**(*c*)**	The order of reactivity through S_N1 is 3° > 2° > 1°. The compound II is allylic as well as 2° and hence is most reactive.
41.	**(*d*)**	CH_3— group is ortho and para directing group.
42.	**(*b*)**	Self explanatory question.
43.	**(*b*)**	Self explanatory question.
44.	**(*d*)**	Sandmeyer's reaction.
45.	**(*c*)**	*trans*-1, 2-Dichloroethylene has zero dipole moment because dipoles of the two C—Cl bonds are oppositely directed and hence cancel each other.
46.	**(*b*)**	Dehydrohalogenation or β-elimination reaction.
47.	**(*d*)**	$C_6H_5—CH_3 + 3Cl_2 \xrightarrow{\text{light}}$ $C_6H_5CCl_3 \xrightarrow[\text{Hydrolysis}]{\text{NaOH}} C_6H_5COOH$
48.	**(*a*)**	β-Elimination $CH_3CH_2CH_2CH_2Cl \xrightarrow[-\text{HCl}]{\text{alc. KOH}}$ $CH_3CH_2CH = CH_2$ 1-Butene
49.	**(*c, d*)**	Both (*c*) and (*d*) are formed, however, nitroethane is the major product.
50.	**(*d*)**	$CH_3CH_2CH_2–Cl + KCN \longrightarrow$ $CH_3CH_2CH_2CN + KCl$ Butane nitrile

Q. No.	Ans.	Hints & Solutions
51.	**(*b*)**	*p*-Dichlorobenzene has zero dipole moment because the two C—Cl bonds are oppositely directed. *o*-Dichlorobenzene (III) is more polar than *m*-Dichlorobenzene (II) because the angle between the two dipoles is less in it. Toluene (I) has very small dipole moment which is less than (II). Hence, the order of dipole moments is IV < I < II < III.
52.	**(*b*)**	Weaker bases are better leaving groups. The basic strength of the given groups is in the order : $-OMe$ (II) > $-OAc$ (I) > $-OSO_2Me$ (III) > $-OSO_2CF_3$(IV).
53.	**(*b*)**	Sulphuryl chloride (SO_2Cl_2) as well as Cl_2 on reaction with toluene brings about benzylic chlorination.
54.	**(*b*)**	$CCl_3 . CHO + H_2O \longrightarrow CCl_3 - CH(OH)_2$
55.	**(*d*)**	2-Chloro-2-phenylethane ionizes slowly to form carbocation. The carbocation is then attacked equally from both sides forming both the enantiomers resulting in racemization.
56.	**(*d*)**	The alkyl halide having weaker C—X bond has greater reactivity. The C—X bond strength is in the order : C—I < C—Br < C—Cl < C—F Hence, the reactivity is in the order R—I > R—Br > R—Cl > R—F.
57.	**(*b*)**	The first steps corresponds to dehydrohalogenation while the second step involves addition of HBr according to Markownikoff's rule.
58.	**(*a*)**	Dipole moments are in the order $CH_3Cl > CH_2Cl_2 > CHCl_3 > CCl_4$.
59.	**(*a*)**	The reaction follows partly S_N1 and partly S_N2 mechanism. S_N2 mechanism gives K as the product while S_N1 mechanism gives L as the product. The product L is formed through rearrangement of the intermediate 2° carbocation to more stable benzylic carbocation through 1, 2-hydride shift.

3.7. ASSERTION-REASON QUESTIONS

In each of the following questions two statements are given as Assertion A and Reason R. Examine the statements carefully and answer the questions according to the instructions given below :

- *Mark **a** if both **A** and **R** are correct and **R** is the correct reason of **A**.*
- *Mark **b** if both **A** and **R** are correct and **R** is not the correct reason of **A**.*
- *Mark **c** if **A** is correct and **R** is wrong.*
- *Mark **d** if **A** is wrong and **R** is correct.*
- *Mark **e** if both **A** and **R** are wrong.*

1. ***Assertion A.*** *n*-Butyl chloride has higher boiling point than *tert*-butyl chloride.

Reason R. C—Cl bond in *n*-butyl chloride is more polar than in *tert*-butyl chloride.

(*a*)	(*b*)	(*c*)	(*d*)	(*e*)

2. ***Assertion A.*** *iso*-Butyl bromide is a secondary halide.

Reason R. *neo*-Pentyl chloride is a primary halide.

(*a*)	(*b*)	(*c*)	(*d*)	(*e*)

3. ***Assertion A.*** Alkyl halides undergo nucleophilic substitution reactions readily.

Reason R. Halide ion is a very weak base and hence can be easily replaced by the attacking nucleophile.

(*a*)	(*b*)	(*c*)	(*d*)	(*e*)

4. ***Assertion A.*** C—X bond in aryl halides is less polar than the C—X bond in alkyl halides.

Reason R. sp^2 hybridized carbon is more electronegative than sp^3 hybridized carbon atom.

(*a*)	(*b*)	(*c*)	(*d*)	(*e*)

5. ***Assertion A.*** Alkyl chlorides are more reactive than alkyl bromides.

Reason R. C—Cl bond is more polar than C—Br bond.

(*a*)	(*b*)	(*c*)	(*d*)	(*e*)

6. ***Assertion A.*** Allyl chloride undergoes nucleophilic substitution much more readily than vinyl chloride.

Reason R. C—Cl bond in vinyl chloride has partial double bond character due to resonance.

(*a*)	(*b*)	(*c*)	(*d*)	(*e*)

7. ***Assertion A.*** The order of reactivity of alkyl halides in undergoing β-elimination is 3° > 2° > 1°.

Reason R. β-Elimination can follow either E_1 or E_2 mechanism.

(*a*)	(*b*)	(*c*)	(*d*)	(*e*)

8. ***Assertion A.*** HCl on reaction with *iso*-butylene yields *iso*-butyl chloride as the major product.

Reason R. Addition of HCl to *iso*-butylene takes place according to Markownikoff's rule.

(*a*)	(*b*)	(*c*)	(*d*)	(*e*)

9. ***Assertion A.*** Propene can be converted to allyl bromide by reacting with NBS.

Reason R. NBS is a specific reagent for allylic bromination.

(*a*)	(*b*)	(*c*)	(*d*)	(*e*)

10. ***Assertion A.*** Reaction of aromatic diazonium salts with copper powder in the presence of HCl or HBr is known as Sandmeyer's reaction.

Reason R. Sandmeyer's reaction is a good reaction for preparing aryl chlorides and aryl bromides.

(*a*)	(*b*)	(*c*)	(*d*)	(*e*)

11. ***Assertion A.*** Addition of halogen acids to alkenes takes place via free radicals as intermediates.

Reason R. Addition of halogen acids to alkenes obeys Markownikoff's rule.

(*a*)	(*b*)	(*c*)	(*d*)	(*e*)

12. ***Assertion A.*** Benzyl chloride undergoes nucleophilic substitution much more readily than methyl chloride.

Reason R. The intermediate formed during substitution in benzyl chloride is stabilized by resonance and hence is quite stable.

(*a*)	(*b*)	(*c*)	(*d*)	(*e*)

13. ***Assertion A.*** Among the isomeric dichlorobenzene, *o*-dichlorobenzene has the highest melting point.

Reason R. *o*-Dichlorobenzene is most polar among the isomeric dichlorobenzene.

(*a*)	(*b*)	(*c*)	(*d*)	(*e*)

14. ***Assertion A.*** HBr adds to propene more readily than to ethene.

Reason R. Addition of HBr to propene proceeds via more stable intermediate carbocation as compared to ethene.

(*a*)	(*b*)	(*c*)	(*d*)	(*e*)

15. ***Assertion A.*** Chloroform is lighter than water.

Reason R. Chloroform is immiscible with water and forms a separate layer.

(*a*)	(*b*)	(*c*)	(*d*)	(*e*)

16. ***Assertion A.*** Vinyl chloride and vinyl bromide can be distinguished by silver nitrate test.

Reason R. Vinyl chloride gives while precipitate with $AgNO_3$ while vinyl bromide gives yellow precipitate.

(*a*)	(*b*)	(*c*)	(*d*)	(*e*)

17. ***Assertion A.*** Substitution of alkyl halides through S_N^2 mechanism results in inversion of configuration.

Reason R. Substitution of alkyl halides through S_N^2 mechanism proceeds through carbocation as intermediate.

(*a*)	(*b*)	(*c*)	(*d*)	(*e*)

18. ***Assertion A.*** Benzene hexachloride (B.H.C.) is used as an insecticide.

Reason R. IUPAC name of B.H.C. is hexachlorobenzene.

(*a*)	(*b*)	(*c*)	(*d*)	(*e*)

19. ***Assertion A.*** Alcohols cannot be converted into alkyl bromides by reacting with NaBr.

Reason R. Br^- being very weak base cannot displace strong base OH^-.

(*a*)	(*b*)	(*c*)	(*d*)	(*e*)

20. ***Assertion A.*** Ethylene dichloride is a vicinal dihalide.

Reason R. Ethylene dichloride on reaction with sodamide yield ethyne.

(*a*)	(*b*)	(*c*)	(*d*)	(*e*)

ANSWERS

1. (*c*) **2.** (*d*) **3.** (*a*) **4.** (*a*) **5.** (*d*) **6.** (*a*) **7.** (*b*) **8.** (*d*) **9.** (*a*)
10. (*d*) **11.** (*d*) **12.** (*a*) **13.** (*d*) **14.** (*a*) **15.** (*d*) **16.** (*e*) **17.** (*c*) **18.** (*c*)
19. (*a*) **20.** (*b*).

HINTS/SOLUTIONS

Q. No.	Ans.	Hints & Solutions
1.	(*c*)	Statement ***R*** is wrong. C—Cl bond in *tert*-butyl chloride is more polar.
2.	(*d*)	Statement ***A*** is wrong. *iso*-Butyl bromide is a primary halide.
3.	(*a*)	Self explanatory statements.
4.	(*a*)	Self explanatory statements.
5.	(*d*)	Statements ***A*** is wrong. Alkyl chloride are less reactive than alkyl bromides because C—Cl bond is stronger than C—Br bond.

Q. No.	Ans.	Hints & Solutions
6.	(*a*)	Both the statements are correct and the statement ***R*** is the explanation of ***A***.
7.	(*b*)	Both the statements are correct but ***R*** is not the explanation of ***A***.
8.	(*d*)	Statement ***A*** is wrong. HCl adds to *iso*-butylene to form *tert*-butyl chloride in accordance with Markownikoff's rule.
9.	(*a*)	Self explanatory statements. NBS (N-bromosuccinimide) brings about substitution at allylic positiom.

Q. No.	Ans.	Hints & Solutions
10.	(*d*)	Statement ***A*** is wrong. Reaction of aromatic diazonium salts with copper powder in the presence of HCl or HBr is known as **Gattermann reaction** and not Sandmeyer reaction.
11.	(*d*)	Addition of halogen acids to alkenes takes place *via* **carbocations** as intermediates. Hence, the statement ***A*** is wrong.
12.	(*a*)	Both the statements are correct. Benzyl carbocation which is formed as intermediate is stabilized by resonance.
13.	(*d*)	Statements ***A*** is wrong. Among the isomeric dichlorobenzene, para-dichlorobenzene is most symmetrical and hence packs closely in solid state and has highest melting point.

Q. No.	Ans.	Hints & Solutions
14.	(*a*)	Self explanatory statements.
15.	(*d*)	Statement ***A*** is wrong. Chloroform is heavier than water.
16.	(*e*)	The statement ***R*** is wrong. Both of them do not give any precipitate with $AgNO_3$.
17.	(*c*)	Statement ***R*** is wrong. S_N2 mechanism proceeds via transition state.
18.	(*c*)	Statement ***R*** is wrong. IUPAC name of B.H.C. is 1, 2, 3, 4, 5, 6-hexachlorocyclohexane.
19.	(*a*)	Self explanatory question.
20.	(*b*)	Both the statements are correct but are not inter related.

3.8. SELF EVALUATION TEST

Choose the correct answer from the four alternatives given in each of the following questions :

1. The number of isomeric alkyl halides possible for $C_5H_{11}Cl$ is
(*a*) 5 (*b*) 4
(*c*) 7 (*d*) 8.

2. Which of the following alkyl halides can exhibit enantiomerism ?
(*a*) *n*-Butyl chloride (*b*) *iso*-Butyl chloride
(*c*) *sec*-Butyl chloride (*d*) *tert*-Butyl chloride.

3. The best reagent for converting alcohols to alkyl chlorides is
(*a*) PCl_5 (*b*) PCl_3
(*c*) $HCl/ZnCl_2$ (*d*) $SOCl_2$.

4. The reaction,

$$CH_3COOAg + Br_2 \xrightarrow{CCl_4} CH_3Br + CO_2 + AgBr$$

is known as
(*a*) Finkelstein reaction (*b*) Frankland reaction
(*c*) Hunsdiecker reaction (*d*) Sandmeyer reaction.

5. '*Westrosol*' is
(*a*) 1, 1, 2, 2-Tetrachloroethane
(*b*) 1, 1, 2, 2-Tetrachloroethene
(*c*) Trichloroethene
(*d*) 1, 1, 2-Trichloroethane.

6. *Iso*-Butyl chloride on heating with alcoholic solution of KOH yields ...... as the major product.
(*a*) 1-Butene (*b*) 2-Methylpropene
(*c*) 2-Methyl-1-propanol (*d*) 2-Butene.

7. Which of the following compounds can exhibit enantiomerism ?
(*a*) Propadiene (*b*) 1, 3-Dichloropropadiene
(*c*) 1-Chloropropadiene (*d*) 1, 1-Dichloropropadiene.

8. Which of the following halides is not a primary halide ?
(*a*) *iso*-Pentyl chloride (*b*) *neo*-Pentyl chloride
(*c*) *iso*-Butyl chloride (*d*) *iso*-Propyl chloride.

9. $C_2H_5Br \xrightarrow{\text{alc. KCN}} X \xrightarrow{\text{dil. HCl}} Y$
In the above sequence, the compound Y is
(*a*) 1-Chloropropane (*b*) 2-Chloropropane
(*c*) Propanoic acid (*d*) Propanamide.

10. C—X bond energy is maximum in
(*a*) CH_3F (*b*) CH_3Cl
(*c*) CH_3Br (*d*) CH_3I.

11. When C_2H_5Br is treated with limited amount of alc. NH_3, the major product obtained is
(*a*) Ethylamine
(*b*) Diethylamine
(*c*) Triethylamine
(*d*) Tetraethylammonium bromide.

12. When *tert*-butyl bromide is treated with sodium methoxide in methanol, the major product obtained is
(*a*) 2-Methoxy-2-methylpropane
(*b*) 1-Methoxy-2-methylpropane
(*c*) 2-Methylpropene
(*d*) Dimethyl ether.

13. 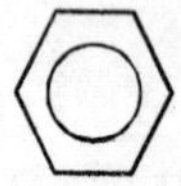 $+ CH_3CH_2CH_2Cl \xrightarrow{AlCl_3} X + HCl$

The compound X (the major product) is
(*a*) *n*-Propyl benzene (*b*) *iso*-Propyl benzene
(*c*) *n*-Hexane (*d*) Diphenyl.

14. Which of the following factors does not favour S_N1 mechanism ?
(*a*) Strong nucleophile
(*b*) Polar solvent
(*c*) Low conc. of nucleophile
(*d*) 3° Halide.

15. Which of the following nucleophiles would react through S_N2 mechanism most readily ?
(*a*) $:\bar{O}H$ (*b*) $H_2\ddot{O}$
(*c*) $\ddot{N}H_3$ (*d*) $:\bar{O}R$.

16. 2-Bromo-3, 3-dimethylbutane on reaction with aqueous KOH yields ...... as the major product. The missing item is
(*a*) 2, 3, 3-Trimethyl-1-propanol
(*b*) 2, 2-Dimethyl-3-butanol
(*c*) 2, 3-Dimethyl-2-butanol
(*d*) None of these.

17. Reaction of alkyl halides towards dehydrohalogenation reaction is
(*a*) 1° > 2° > 3° (*b*) 3° > 1° > 2°
(*c*) 3° > 2° > 1° (*d*) 2° > 1° > 3°.

18. Which of the following halides would undergo nucleophilic substitution most readily ?
(*a*) 1-Chloro-1-butene (*b*) 2-Chloro-1-butene
(*c*) 3-Chloro-1-butene (*d*) 4-Chloro-1-butene.

19. C—Cl bond has partial double bond character in
(*a*) Vinyl chloride (*b*) Chlorobenzene
(*c*) Both (*d*) None.

20. Which of the following is most reactive towards nucleophilic substitution ?
(*a*) Chlorobenzene
(*b*) 2, 4, 6-Trinitrochlorobenzene
(*c*) 2, 4-Dinitrochlorobenzene
(*d*) 2-Nitrochlorobenzene.

21. Iodobenzene can be converted into diphenyl by
(*a*) Wurtz reaction (*b*) Wurtz Fittig reaction
(*c*) Ullmann reaction (*d*) Frankland reaction.

22. C—Cl bond is least polar in
(*a*) Chloroethene (*b*) 1-Chloropropane
(*c*) 2-Chloropropane (*d*) 3-Chloro-1-propene.

23. *iso*-Propylidene dichloride on alkaline hydrolysis yields
(*a*) 1-Propanol (*b*) 2-Propanol
(*c*) Propanol (*d*) Propanone.

24. Ethylene dichloride and ethylidene dichloride yield same product on reaction with
(*a*) Zn dust (*b*) Alcoholic KOH
(*c*) Aqueous KOH (*d*) None of these.

25. Bromobenzene and chlorobenzene can be distinguished by
(*a*) Silver nitrate test (*b*) Lassaigne's test
(*c*) Both (*d*) None.

26. Which of the following alkyl chlorides yields two isomeric products on further substitution with one chlorine atom ?
(*a*) 1-Chloropropane (*b*) 2-Chloropropane
(*c*) 1-Chlorobutane (*d*) 2-Chlorobutane.

27. $CHI_3 \xrightarrow[\Delta]{Ag} X \xrightarrow{+2HCl} Y$

The compound Y is
(*a*) Chloroform (*b*) Ethyl chloride
(*c*) Ethylene dichloride (*d*) Ethylidene dichloride.

28. *neo*-Pentyl chloride on dehydrohalogenation (using low conc. of base) yields mainly
(*a*) 2-Methyl but-2-ene (*b*) 2-Methyl but-1-ene
(*c*) 3-Methyl but-1-ene (*d*) 2-Pentene.

29. During chlorination of benzene using Cl_2 in the presence of $FeCl_3$, the attacking species is
(*a*) Cl^- (*b*) Cl^+
(*c*) Cl_2 (*d*) $FeCl_4^-$.

30. Chlorobenzene when heated with ...... in the presence of conc. H_2SO_4 gives D.D.T. The missing item is
(*a*) Acetone (*b*) Ethanal
(*c*) Chloroform (*d*) Chloral.

31. Chloroform is used for testing the presence of
(*a*) Primary amines (*b*) Secondary amines
(*c*) Tertiary amine (*d*) Nitro compounds.

32. 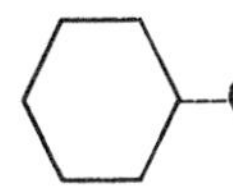 —OH $\xrightarrow{PBr_3}$ X $\xrightarrow{NH_3}$ Y

The compound Y in the above sequence is
(*a*) Cyclohexene
(*b*) Cyclohexylamine
(*c*) 2-Bromocyclohexylamine
(*d*) 4-Bromocyclohexylamine.

33. How many enantiomeric pairs can be obtained by monobromination of *iso*-pentane ?
(*a*) 3 (*b*) 1
(*c*) 2 (*d*) 4.

34. Which of the following represents the correct order of densities ?
(*a*) RI > RBr > RCl > H_2O (*b*) RI > RBr > H_2O > RCl
(*c*) RI > H_2O > RBr > RCl (*d*) RCl > RBr > RI > H_2O.

35. Which of the following represents the correct order of densitities ?
(*a*) $CCl_4 > CHCl_3 > CH_2Cl_2 > CH_3Cl > H_2O$
(*b*) $CCl_4 > CHCl_3 > CH_2Cl_2 > H_2O > CH_3Cl$
(*c*) $CCl_4 > CHCl_3 > H_2O > CH_2Cl_2 > CH_3Cl$
(*d*) $H_2O > CH_3Cl > CH_3Cl_2 > CHCl_3 > CCl_4$.

36. Which of the following halides can be made in good yield by free radical halogenation of parent hydrocarbon ?
(*a*) $CH_3CH_2CH_2Cl$ (*b*) $CH_3-C(CH_3)_2-CH_2Cl$
(*c*) $CH_3CH_2CH(Cl)CH_3$ (*d*) $CH_3CH_2CH_2CH_2Cl$.

37. $CH_3CH=CH_2 \xrightarrow{Cl_2\,(h\nu)} X \xrightarrow[\text{acetone}]{NaI} Y$

The compound Y in the above sequence is
(*a*) 1, 2-Dichloro-3-iodopropane
(*b*) 1-Chloro-2-iodopropane
(*c*) 3-Iodopropene
(*d*) 1, 2-Diodopropane.

38. S_N2 reactions are
(*a*) Stereospecific (*b*) Stereoselective
(*c*) Both (*d*) None.

39. Which of the following alkyl halides is likely to undergo β-elimination most readily ?

(*a*) Ethyl bromide (*b*) *n*-Propylbromide

(*c*) *iso*-Propyl bromide (*d*) *tert*-Butyl bromide.

40. Which of the following compounds would not give ppt. with $AgNO_3$?

(*a*) $CH_3CH = CHCl$ (*b*) $CH_2 = CH - CH_2Cl$

(*c*) 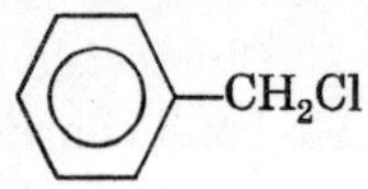 (*d*) $(CH_3)_3CCl$.

ANSWERS

1. (*d*)	**2.** (*c*)	**3.** (*d*)	**4.** (*c*)	**5.** (*c*)	**6.** (*b*)	**7.** (*b*)	**8.** (*d*)	**9.** (*c*)
10. (*a*)	**11.** (*d*)	**12.** (*c*)	**13.** (*b*)	**14.** (*a*)	**15.** (*d*)	**16.** (*c*)	**17.** (*c*)	**18.** (*c*)
19. (*c*)	**20.** (*b*)	**21.** (*c*)	**22.** (*a*)	**23.** (*d*)	**24.** (*b*)	**25.** (*b*)	**26.** (*b*)	**27.** (*d*)
28. (*a*)	**29.** (*b*)	**30.** (*d*)	**31.** (*a*)	**32.** (*a*)	**33.** (*b*)	**34.** (*b*)	**35.** (*b*)	**36.** (*b*)
37. (*c*)	**38.** (*c*)	**39.** (*d*)	**40.** (*a*).					

HINTS/SOLUTIONS

Q. No.	Ans.	Hints & Solutions
1.	(*d*)	The eight isomers are : 1-Chloropentane, 2-chloropentane, 3-chloropentane, 1-chloro-3-methylbutane, 2-chloro-3-methylbutane, 2-chloro-2-methyl-butane, 1-chloro-2-methylbutane, 1-chloro-2, 2-dimethylpropane.
2.	(*c*)	$CH_3 - \overset{*}{C}H(Cl) - CH_2 - CH_3$ *sec*-Butyl chloride contains an asymmetric carbon atom and hence is chiral.
3.	(*d*)	$R\text{-}OH + SOCl_2 \longrightarrow R\text{—}Cl + SO_2 + HCl$ Here the side products SO_2 and HCl are in gaseous state (volatile) and hence can be easily removed.
4.	(*c*)	Hunsdiecker reaction.
5.	(*c*)	$(H)(Cl)C = C(Cl)(Cl)$ Westrosol
6.	(*b*)	$(CH_3)_2CH\text{—}CH_2Cl + KOH\ (alc.) \longrightarrow (CH_3)_2C = CH_2 + KCl + H_2O$
7.	(*b*)	1, 3-Dichloropropadiene is a substituted allene. It is not superimposable on its mirror image.
8.	(*d*)	*iso*-Propyl chloride is a secondary halide.
9.	(*c*)	$C_2H_5Br \xrightarrow{\text{alc. KCN}} C_2H_5CN \xrightarrow{\text{dil. HCl}} CH_3CH_2COOH.$
10.	(*a*)	C-F bond energy is maximum due to smallest size of F among the halogens.
11.	(*d*)	In the presence of excess C_2H_5Br (limited NH_3), all the H atoms of NH_3 are replaced by alkyl groups and further alkyl substituted ammonium salt is formed.
12.	(*c*)	3° Halides on treatment with base undergo elimination readily.

Q. No.	Ans.	Hints & Solutions
13.	(*b*)	$CH_3CH_2CH_2Cl + AlCl_3 \longrightarrow CH_3CH_2\overset{+}{C}H_2 + AlCl_4^-$ $\xrightarrow{\text{1, 2-Hydride shift}}$ $CH_3\overset{+}{C}H - CH_3$ *iso*-Propyl carbocation
14.	(*a*)	Strong nucleophile favours S_N2.
15.	(*d*)	Strong bases react through S_N2 mechanism. $:\bar{O}R$ is the strongest base among the given bases.
16.	(*c*)	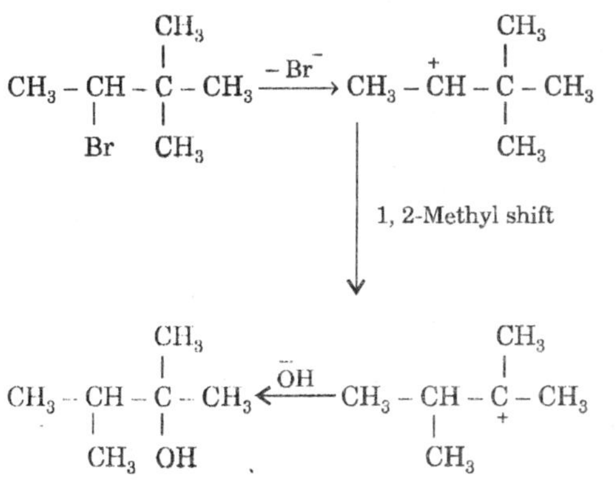 2, 3-Dimethyl-2-butanol
17.	(*c*)	Dehydrohalogenation follows the order 3° > 2° > 1°.
18.	(*c*)	3-Chloro-1-butene is an allylic halide.
19.	(*c*)	C—Cl bond has partial double bond character in vinyl chloride as well as in chlorobenzene due to resonance.
20.	(*b*)	Presence of electron withdrawing groups on ring makes the substitution easier by stabilizing the intermediate carbanion.
21.	(*c*)	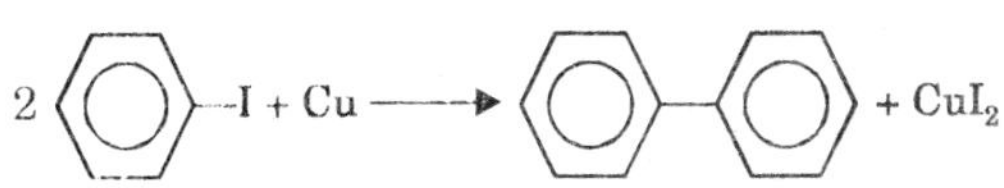 Diphenyl
22.	(*a*)	C—Cl bond in chloroethene is least polar because in vinyl halides electron withdrawing inductive effect of halogen atom is opposed by electron releasing mesomeric effect.
23.	(*d*)	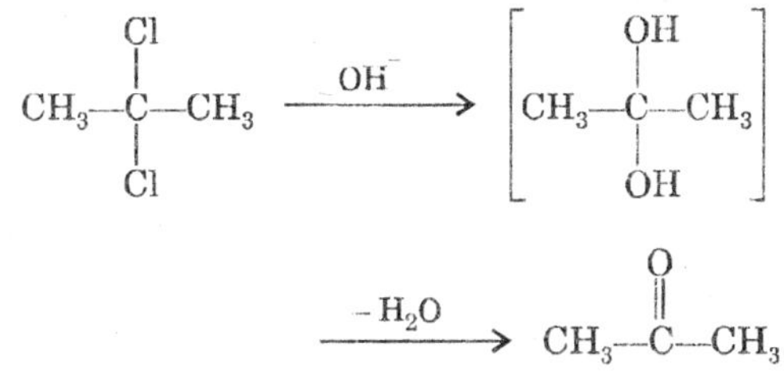
24.	(*b*)	On reaction with alc. KOH both ethylene dichloride and ethylidene dichloride yield ethyne.
25.	(*b*)	In Lassaigne's test bromobenzene would yield light yellow ppt. which are partially soluble in NH_4OH whereas chlorobenzene would yield white ppt. which are soluble in NH_4OH.
26.	(*b*)	2-Chloropropane on further substitution with chlorine atom would yield 1, 2-dichloropropane and 2, 2-dichloropropane.
27.	(*d*)	$CHI_3 \xrightarrow[\Delta]{Ag} HC \equiv CH \xrightarrow{+2HCl} CH_3 - CHCl_2$
28.	(*a*)	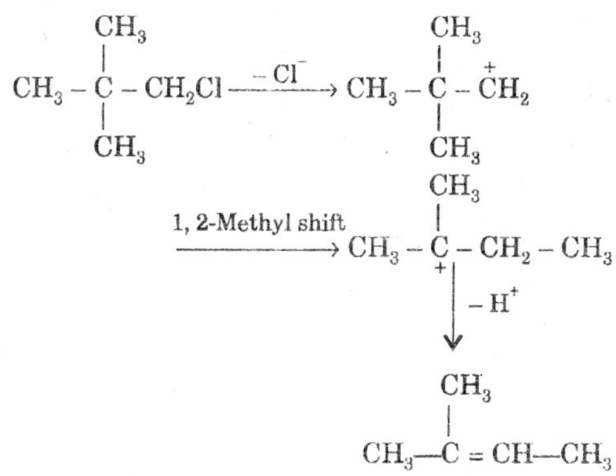 2-Methyl but-2-ene
29.	(*b*)	$FeCl_3 + Cl_2 \longrightarrow FeCl_4^- + Cl^+$
30.	(*d*)	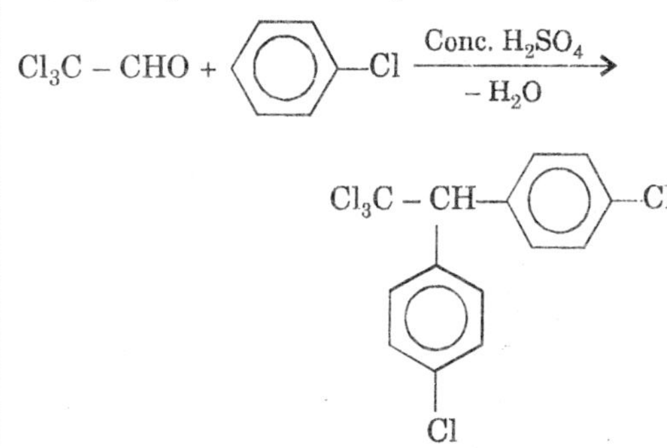 D.D.T.
31.	(*a*)	Carbylamine test for primary amines.

Q. No.	Ans.	Hints & Solutions
32.	(*a*)	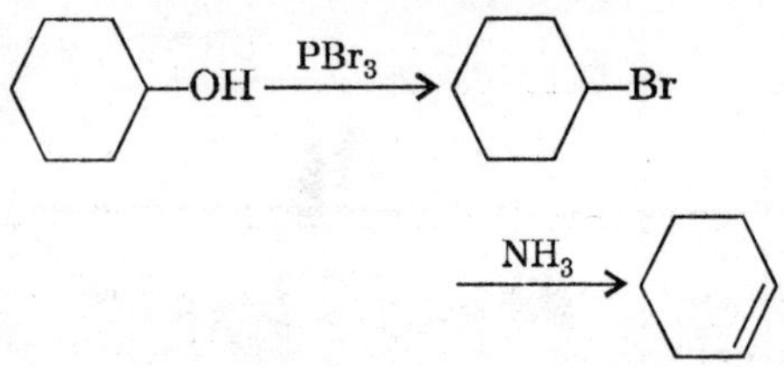 Base (NH_3) brings about dehydrohalogenation of bromocyclohexane.
33.	(*b*)	The one enantiomeric pair is *d* and *l*-2-bromo-3-methyl butane.
34.	(*b*)	Alkyl iodides and bromides are heavier than water while alkyl chlroides are lighter than water.
35.	(*b*)	Polychlorides are heavier than water whereas monochlorides are lighter than water.
36.	**(*b*)**	Monochlorination of *neo*-pentane yields only one product $(CH_3)_3C\ CH_2Cl$. So it can be obtained in good yield.
37.	**(*c*)**	$CH_3-CH=CH_2 \xrightarrow{Cl_2\,(hv)} ClCH_2CH=CH_2 \xrightarrow[\text{acetone}]{NaI} ICH_2CH=CH_2$ The first step is allylic substitution and the second step is Finkelstein reaction.
38.	**(*c*)**	S_N^2 reactions are stereospecific as well as stereoselective.
39.	**(*d*)**	β-Elimination of alkyl halides follows the order $3° > 2° > 1°$. *tert*-Butyl bromide is a 3° alkyl halide.
40.	**(*a*)**	$CH_3CH=CHCl$ is a vinylic halide and hence would not undergo substitution easily.

4 ALCOHOLS, PHENOLS AND ETHERS

4.1. CONCEPTS IN BRIEF

INTRODUCTION

- Alcohols and phenols are the compounds containing one or more hydroxyl groups (—OH). The alcohols contain the —OH group attached to alkyl group whereas in phenols, the —OH group is attached to aromatic ring.

(*a*) **Alcohols**

CH_3CH_2OH	$\underset{OH}{CH_2}—\underset{OH}{CH_2}$	$\underset{OH}{CH_2}—\underset{OH}{CH}—\underset{OH}{CH_2}$
Ethanol or Ethyl alcohol (*Monohydric*)	1, 2-Ethanediol or Ethylene glycol (*Dihydric*)	1, 2, 3-Propanetriol or Glycerol (*Trihydric*)

(*b*) **Phenols**

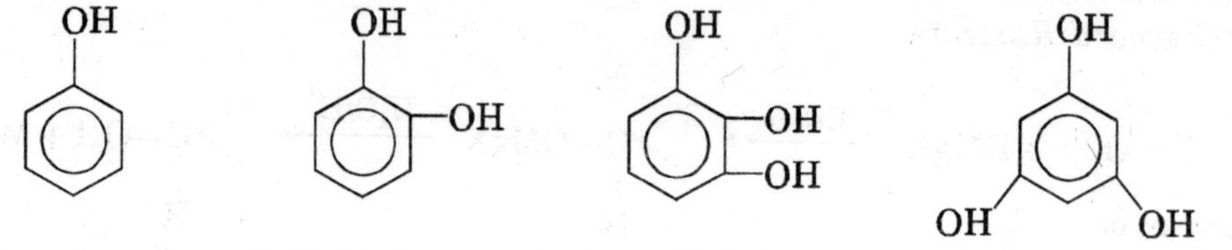

Hydroxybenzene or Phenol (*Monohydric*)	1, 2-Dihydroxy-benzene or Catechol (*Dihydric*)	1, 2, 3-Trihydroxy-benzene or Pyrogallol (*Trihydric*)	1, 3, 5-Trihydroxy-benzene or Phloroglucinol

- Alcohols containing more than one —OH group linked to the same carbon atom are highly unstable and lose water to form more stable aldehydes or ketones.

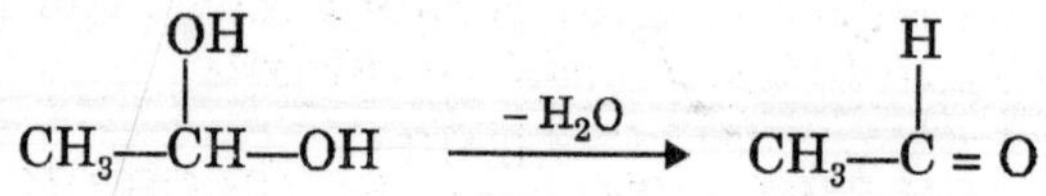

unstable — Ethanal

- Monohydric alcohols are classified as ***primary (1°), secondary (2°) or tertiary (3°)*** alcohols depending upon whether the hydroxyl group is attached to a primary, secondary or tertiary carbon atom.

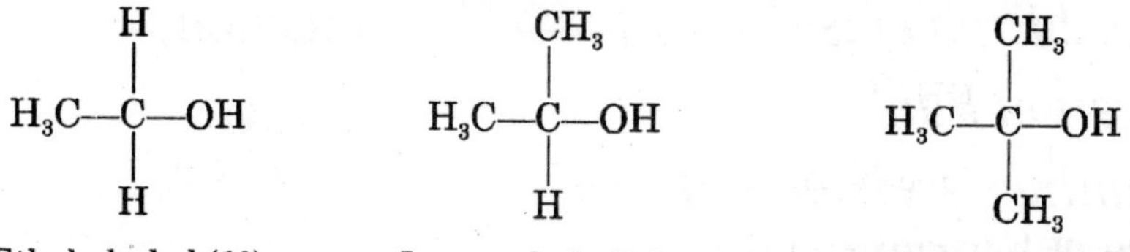

Ethyl alcohol (1°) — *Iso*propyl alcohol (2°) — *tert*-Butyl alcohol (3°)

ALCOHOLS

METHODS OF PREPARATION OF ALCOHOLS

1. From Alkyl Halides

Alkyl halides when boiled with aqueous solution of KOH or moist silver oxide furnish alcohols.

$$C_2H_5Br + KOH\,(aq) \longrightarrow C_2H_5OH + KBr$$

Primary alkyl halides give good yield of alcohols. However, tertiary alkyl halides mainly undergo dehydrohalogenation resulting into formation of alkenes.

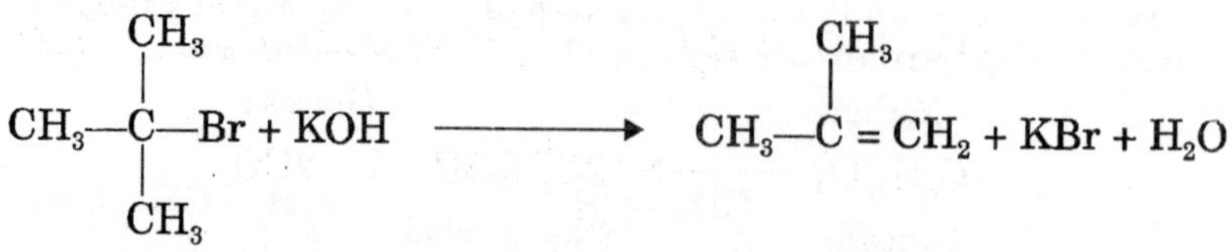

tert-Butyl bromide — *iso*-Butylene (*Major product*)

2. From Alkenes

(*i*) By hydration (*ii*) By hydroboration-oxidation (*iii*) By oxymercuration-demercuration.

These reactions have been discussed in detail in *Chapter 2*.

3. From Aldehydes and Ketones

(i) By using Grignard Reagents

$$\gt C = O + RMgX \xrightarrow{\text{Ether}} \gt \underset{\displaystyle R}{\underset{|}{C}}—OMgX \xrightarrow{H_2O/H^+} \gt \underset{\displaystyle R}{\underset{|}{C}}—OH + Mg\begin{matrix} X \\ OH \end{matrix}$$

Aldehyde or ketone — Alcohol

In this reaction :

—methanal yields 1° alcohols

—other aldehydes yield 2° alcohols

—ketones yield 3° alcohols

(ii) By Reduction. Aldehydes on reduction give primary alcohols and ketones give secondary alcohols. The reducing agents used are $LiAlH_4$, $NaBH_4$ *or* H_2 *with* Ni *catalyst.*

$$\underset{\text{Ethanal}}{CH_3CHO} + 2H \xrightarrow{LiAlH_4} \underset{\text{Ethanol}}{CH_3CH_2OH}$$

$$\underset{\text{Propanone}}{\begin{matrix} CH_3 \\ CH_3 \end{matrix} \gt C = O} + 2H \xrightarrow{LiAlH_4} \underset{\text{Propan-2-ol}}{CH_3\underset{\displaystyle OH}{\underset{|}{C}}HCH_3}$$

4. By Fermentation of Carbohydrates

(i) Fermentation of Sucrose

$$\underset{\text{Sucrose}}{C_{12}H_{22}O_{11}} + H_2O \xrightarrow{\text{Invertase}} \underset{\text{Glucose}}{C_6H_{12}O_6} + \underset{\text{Fructose}}{C_6H_{12}O_6}$$

$$\underset{\text{Glucose or fructose}}{C_6H_{12}O_6} \xrightarrow{\text{Zymase}} 2C_2H_5OH + 2CO_2$$

(ii) Fermentation of Starchy Materials

$$\underset{\text{Starch}}{2(C_6H_{10}O_5)_n} + nH_2O \xrightarrow{\text{Diastase}} \underset{\text{Maltose}}{nC_{12}H_{22}O_{11}}$$

$$\underset{\text{Maltose}}{C_{12}H_{22}O_{11}} + H_2O \xrightarrow{\text{Maltase}} \underset{\text{Glucose}}{2C_6H_{12}O_6}$$

$$\underset{\text{Glucose}}{C_6H_{12}O_6} \longrightarrow \underset{\text{Ethyl alcohol}}{2C_2H_5OH} + 2CO_2$$

5. Oxo process

Alkenes react with carbon monoxide and hydrogen in the presence of $[Co(CO)_4]_2$ as catalyst and at high temperature and pressure to yield aldehydes. Catalytic hydrogenation of aldehydes give primary alcohols.

$$\underset{\text{Ethene}}{CH_2 = CH_2} + CO + H_2 \xrightarrow{[Co(CO)_4]_2} \underset{\text{Propanal}}{CH_3CH_2CHO} \xrightarrow{H_2/Ni} \underset{\text{Propan-1-ol}}{CH_3CH_2CH_2OH}$$

Some more reactions which can be employed for the preparation of alcohols are given under :

- ***Reduction of Carboxylic Acids***

$$RCOOH + 4[H] \xrightarrow{LiAlH_4} RCH_2OH + H_2O$$

- ***Reduction of Esters***

$$RCOOR' + 4[H] \xrightarrow{LiAlH_4} RCH_2OH + R'OH$$

- ***Reaction of Grignard Reagents with Esters***

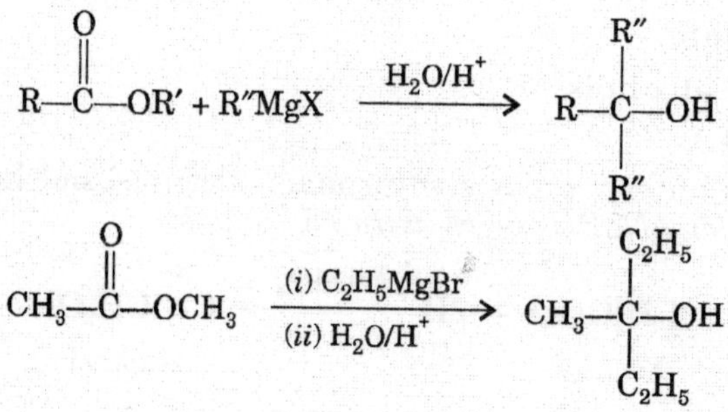

3° Alcohol

- ***From Oxiranes***

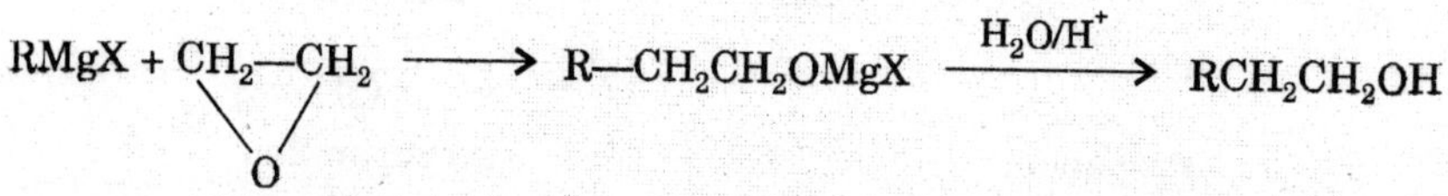

PHYSICAL PROPERTIES OF ALCOHOLS

1. The lower alcohols are colourless liquids having a distinct alcoholic smell and a burning taste. The higher members are colourless, tasteless waxy solids.
2. **Boiling Points**
   - Since alcohols contain a hydrogen attached to strongly electronegative oxygen, there are hydrogen bonds between alcohol molecules. As a result, *the boiling points of alcohols are much higher than those of hydrocarbons, ethers or alkyl halides of comparable molecular masses.*
   - Among the isomeric alcohols the boiling points follow the order :

     primary alcohols > secondary alcohol > tertiary alcohol.
   - With the increase in molecular mass the boiling points increase regularly.
3. **Solubility**
   - The first three members are completely miscible with water. The solubility rapidly decreases with increase in molecular mass. The higher members are almost insoluble in water but are soluble in organic solvents like benzene, ether, etc.
   - *The solubility of lower alcohols is due to the existence of hydrogen bonds between water and polar O—H group of alcohol molecules.*
   - *The solubility of alcohols in water decreases with increase in molecular mass because with increase in molecular mass, the non-polar alkyl group becomes predominant.*

   Among the isomeric alcohols the solubility increases with branching of chain.

CHEMICAL PROPERTIES OF ALCOHOLS

In alcohols, —OH group is the functional group. The reactions of alcohols can be classified into three types :

1. *Reactions involving the cleavage of oxygen-hydrogen bond.*
2. *Reactions involving the cleavage of carbon-oxygen bond.*
3. *Reactions involving both the alkyl and the hydroxyl groups.*

I. Reactions Involving the Cleavage of Oxygen-Hydrogen (R—O┆H) bond.

$$C_2H_5OH \xrightarrow{Na} C_2H_5ONa + \frac{1}{2}H_2 \quad \text{(Sod. ethoxide)}$$

$$C_2H_5OH \xrightarrow{NaH} C_2H_5ONa + H_2$$

$$C_2H_5OH \xrightarrow{RMgX} R\text{—}H + Mg\begin{matrix} X \\ OC_2H_5 \end{matrix}$$

$$C_2H_5OH \xrightarrow[\text{Esterification}]{CH_3COOH/H^+} CH_3COOC_2H_5 + H_2O \quad \text{(Ethyl acetate)}$$

$$C_2H_5OH \xrightarrow[\text{Acylation}]{CH_3COCl} CH_3COOC_2H_5 + HCl \quad \text{(Ethyl acetate)}$$

$$C_2H_5OH \xrightarrow[\text{Acylation}]{(CH_3CO)_2O} CH_3COOC_2H_5 + CH_3COOH \quad \text{(Ethyl acetate)}$$

- The first three reactions illustrate the reactions of alcohols as acids.
- ***Alcohols behave as acids*** because of the presence of polar O—H group. Due to greater electronegativity of oxygen atom the shared pair between O and H is drawn towards the oxygen atom helping in release of H^+ ion.
- ***Alcohols behave as very weak acids.***
- ***The order of acidic strength of different types of alcohols is : 1° > 2° > 3°.***

II. Reactions Involving Cleavage of Carbon-Oxygen (R—C┆—OH) Bond.

1. Reaction with Hydrogen Halides

$$C_2H_5OH \xrightarrow{HCl/ZnCl_2} C_2H_5Cl + H_2O$$

$$C_2H_5OH \xrightarrow[(NaBr + \text{conc. } H_2SO_4)]{HBr} C_2H_5Br + H_2O$$

$$C_2H_5OH \xrightarrow[(KI + H_3PO_4)]{HI} C_2H_5I + H_2O$$

The **order of reactivity of various halogen acids is : HI > HBr > HCl.**

2. Reactions with Phosphorus Halides

$$C_2H_5OH \xrightarrow{PCl_5} C_2H_5Cl + POCl_3 + HCl$$

$$C_2H_5OH \xrightarrow{PCl_3} C_2H_5Cl + H_3PO_3$$

$$C_2H_5OH \xrightarrow[(P + Br_2)]{PBr_3} C_2H_5Br + H_3PO_3$$

$$C_2H_5OH \xrightarrow[(P + I_2)]{PI_3} C_2H_5I + H_3PO_3$$

3. Reactions with Thionyl Chloride

$$\underset{\text{Alcohol}}{ROH} + \underset{\text{Thionyl chloride}}{SOCl_2} \xrightarrow{\text{Pyridine}} \underset{\text{Alkyl chloride}}{RCl} + SO_2\uparrow + HCl\uparrow$$

The **order of reactivity** of alcohols towards this type of reactions is : *tertiary* > *secondary* > *primary.*

4. Reaction with Ammonia

$$C_2H_5OH + NH_3 \xrightarrow[\Delta]{Al_2O_3 \text{ or } ZnCl_2} C_2H_5NH_2 + H_2O$$

III. Reactions Involving Alkyl group as well as Hydroxyl group

1. Acidic Dehydration

$$\underset{\text{Ethanol}}{CH_3CH_2OH} \xrightarrow[443\text{ K}]{\text{Conc. } H_2SO_4} \underset{\text{Ethene}}{H-\underset{H}{\underset{|}{C}}=\underset{H}{\underset{|}{C}}-H} + H_2O$$

$$\underset{\text{Ethanol}}{CH_3CH_2OH} \xrightarrow{Al_2O_3,\ 620\text{ K}} \underset{\text{Ethene}}{CH_2=CH_2} + H_2O$$

- The ease of dehydration of alcohols follows the order

tertiary alcohol > secondary alcohol > primary alcohol.

- The reaction between ethanol and sulphuric acid produces different products under different reaction conditions :

(*i*) $$\underset{\text{Ethyl alcohol}}{C_2H_5OH} + H_2SO_4 \xrightarrow{383\text{ K}} \underset{\text{Ethyl hydrogen sulphate}}{C_2H_5HSO_4} + H_2O$$

(*ii*) $$2C_2H_5OH + H_2SO_4 \xrightarrow[\text{reduced/pressure}]{\text{Heat}} (C_2H_5)_2SO_4 + 2H_2O$$

(*iii*) $$\underset{\text{Excess}}{2C_2H_5OH} \xrightarrow[410\text{ K}]{\text{Conc. } H_2SO_4} \underset{\text{Ethoxyethane}}{C_2H_5OC_2H_5} + H_2O$$

2. Oxidation

Alcohols undergo oxidation with oxidising agents (in neutral or acidic or alkaline medium) such as chromium trioxide (CrO_3), potassium dichromate, potassium permanganate and nitric acid.

- A ***primary alcohol*** on oxidation with acidified potassium dichromate first gives an aldehyde which on further oxidation gives an acid with the *same number of carbon atoms* as the original alcohol.

$$\underset{\text{Ethanol}}{CH_3-\overset{H}{\overset{|}{\underset{H}{\underset{|}{C}}}}-OH} + [O] \xrightarrow[-H_2O]{K_2Cr_2O_7/H^+} \underset{\text{Ethanal}}{CH_3-\overset{H}{\overset{|}{C}}=O} \xrightarrow{+[O]} \underset{\text{Ethanoic acid}}{CH_3-\overset{O}{\overset{||}{C}}-OH}$$

- A ***secondary alcohol*** on oxidation first gives a ketone with the same number of carbon atoms as the original alcohol.

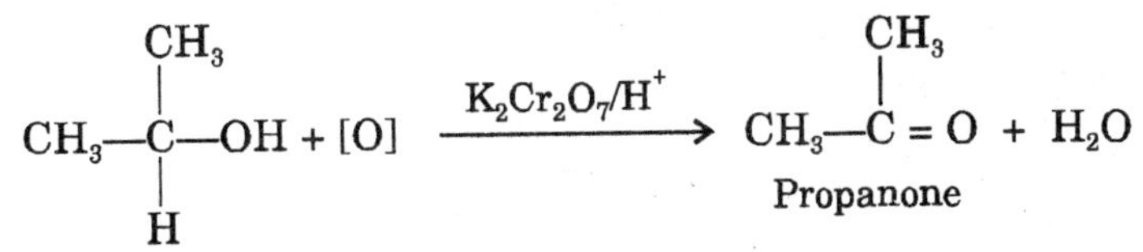

Upon prolonged treatment with the oxidizing agent they get oxidized to acids with lesser number of carbon atoms than the alcohol.

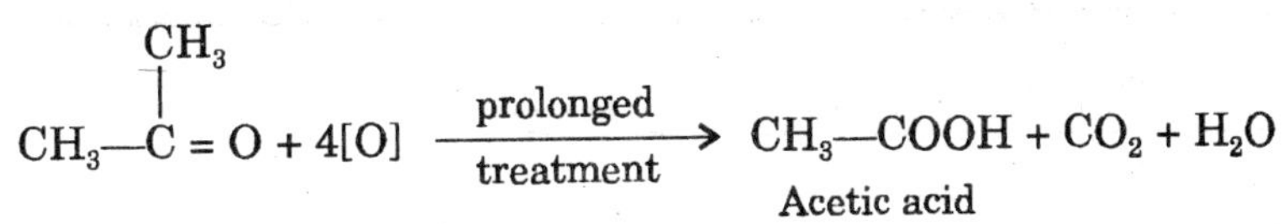

- ***Tertiary alcohols*** are not oxidised with mild oxidizing agents in aqueous, alkaline or under neutral conditions. However, when oxidation is carried out under acidic conditions, tertiary alcohols first undergo dehydration to form alkene. The alkene formed is then oxidized to ketone which is finally oxidised to carboxylic acid with lesser number of carbon atoms than the starting alcohol.

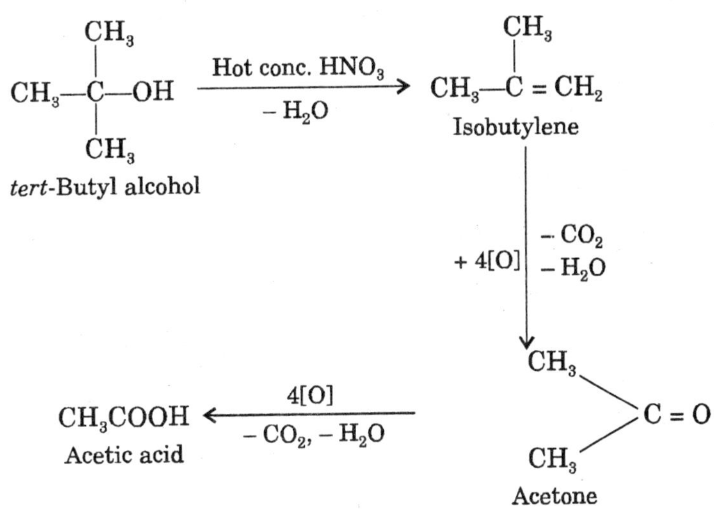

3. **Dehydrogenation** (*Reaction with hot reduced copper*)

- A **primary alcohol** *is dehydrogenated to aldehyde.*

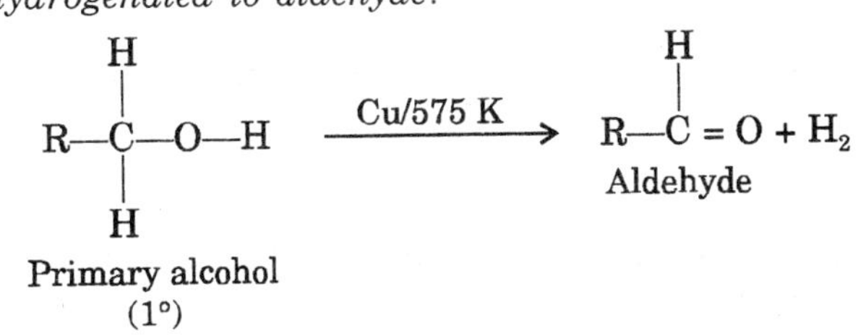

- A **secondary alcohol** *is dehydrogenated to ketone.*

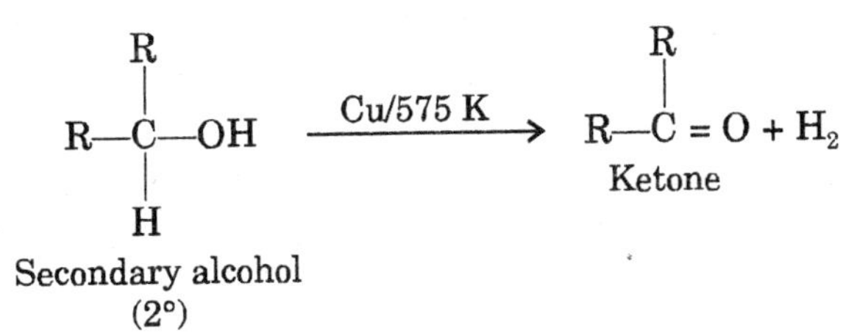

- **Tertiary alcohol** *gets dehydrated to form an alkene.*

$$\underset{\substack{tert\text{-Butyl alcohol}\\(3^\circ)}}{CH_3-\overset{\overset{CH_3}{|}}{\underset{\underset{CH_3}{|}}{C}}-OH} \xrightarrow{Cu/575\ K} \underset{\text{2-Methylpropene}}{CH_3-\overset{\overset{CH_3}{|}}{C}=CH_2} + H_2O$$

METHODS OF PREPARATION OF EHTYLENE GLYCOL

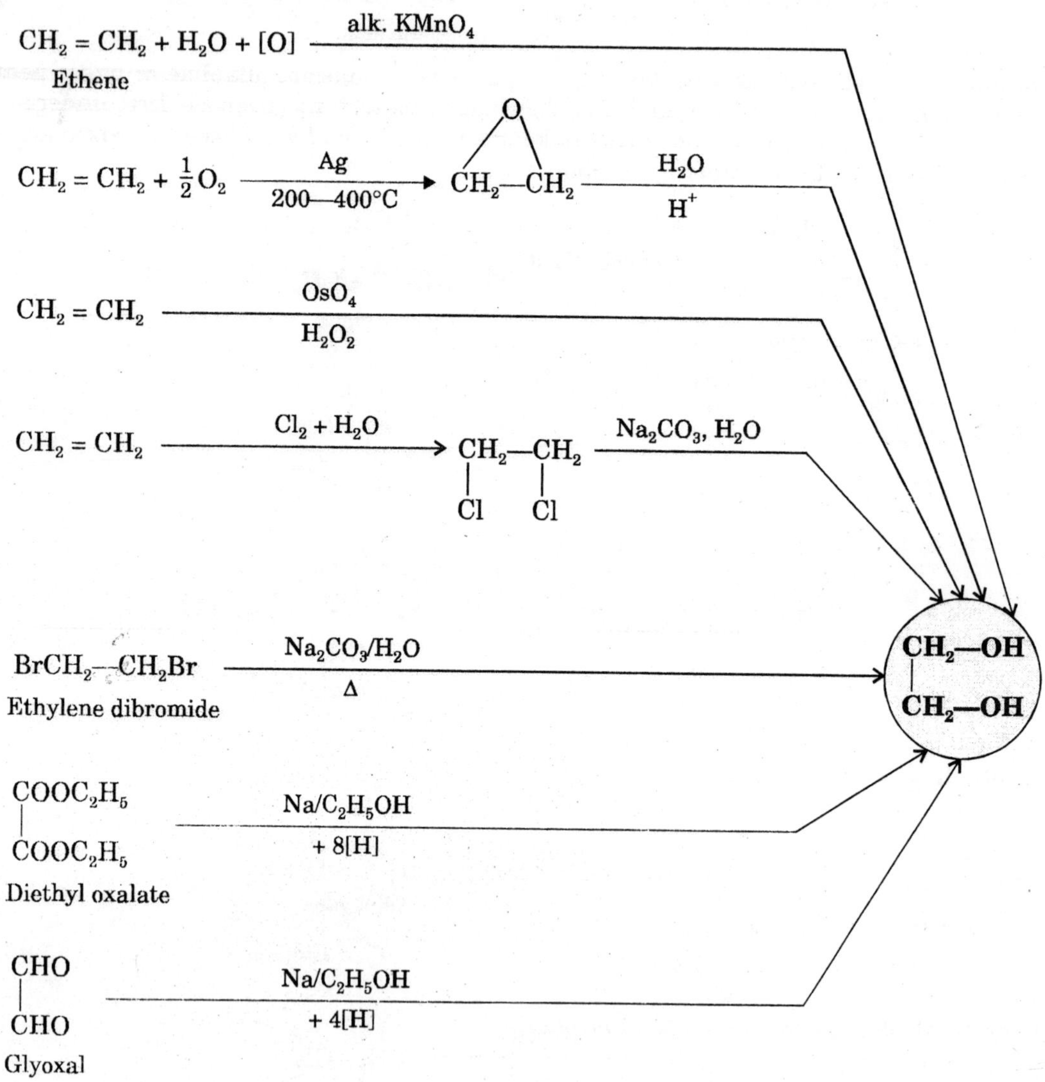

REACTIONS OF ETHYLENE GLYCOL

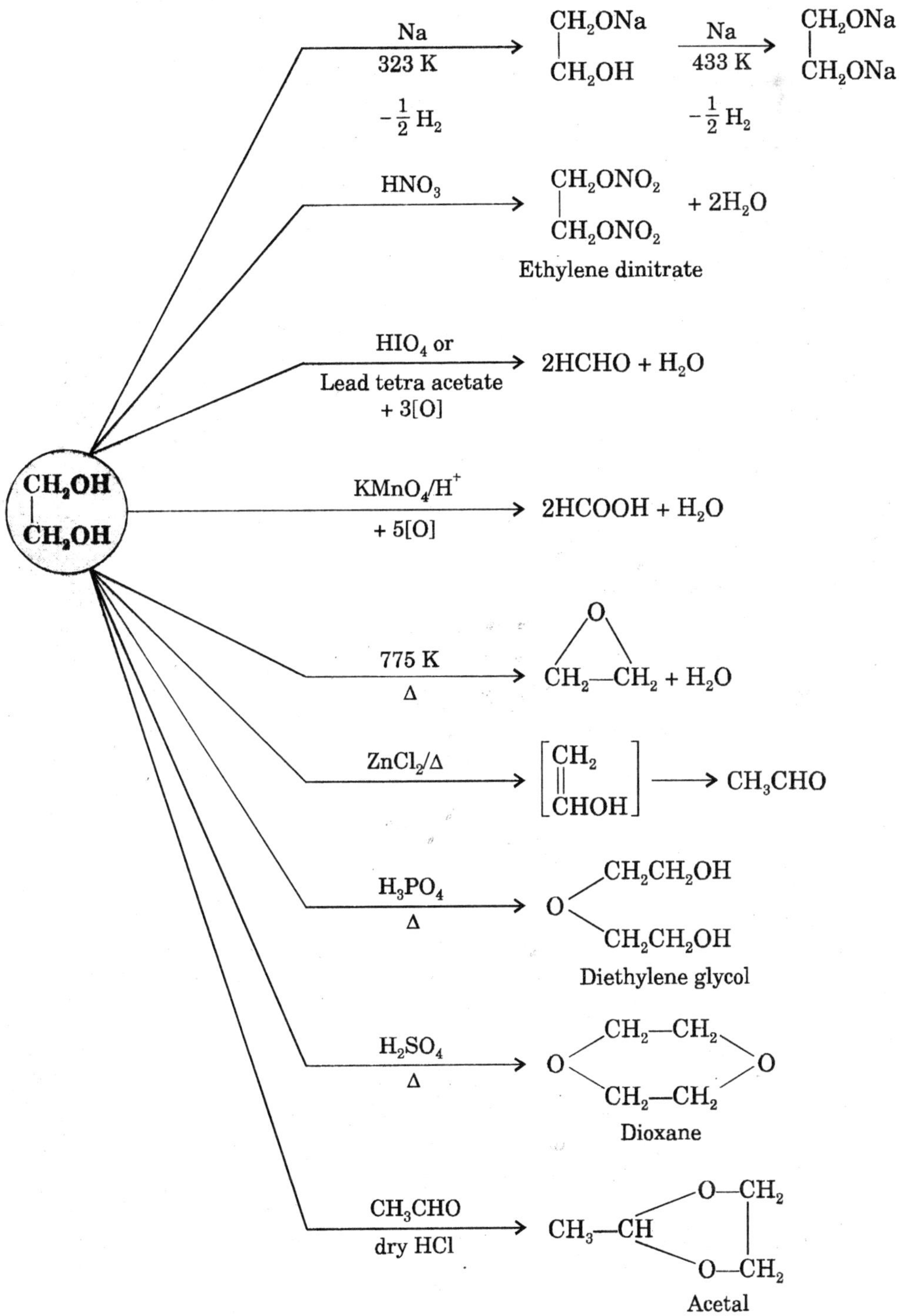

PREPARATION OF GLYCEROL

1. **From oils and Fats.** (*i*) *By hydrolysis.* Oils and fats on hydrolysis with alkalies give glycerol and sodium salts of higher fatty acids (soap). The alkaline hydrolysis of oils and fats is called **saponification.**

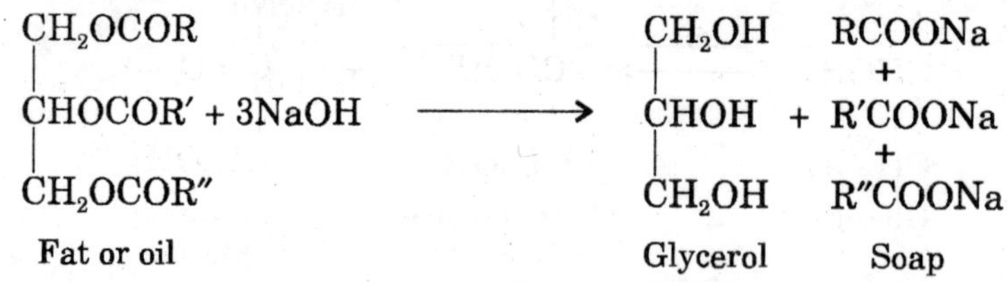

(*ii*) *By trans-esterification*

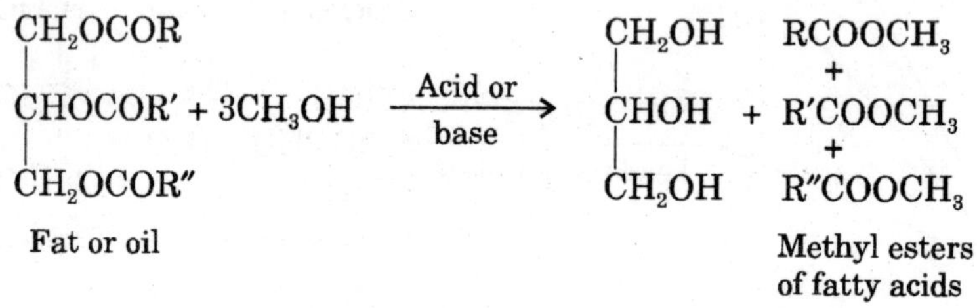

2. **From Propylene**

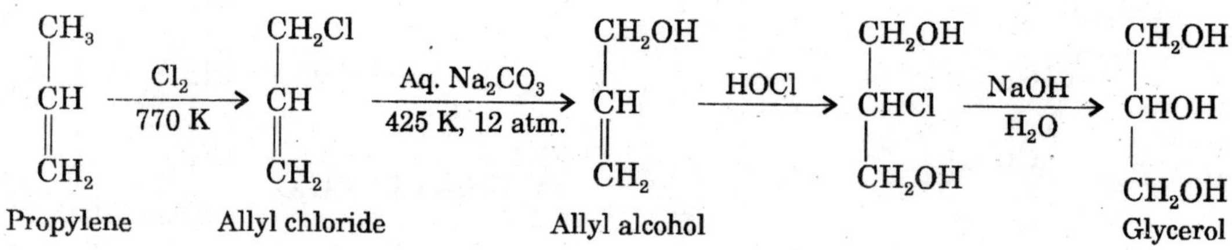

REACTIONS OF GLYCEROL

1. **Nitration.** Glycerol reacts with mixture of conc. HNO_3 and conc. H_2SO_4 at 283-298 K to form glyceryl trinitrate.

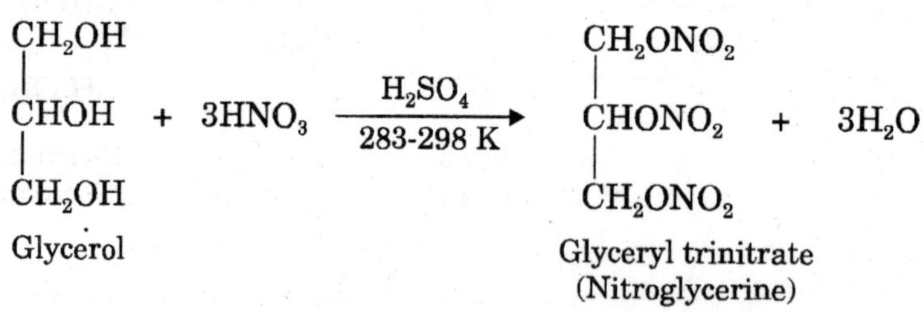

2. **Reaction with Hydrogen Iodide.** When glycerol is warmed with small amount of hydrogen iodide allyl iodide is formed. However, on heating with excess of hydrogen iodide the product obtained is isopropyl iodide.

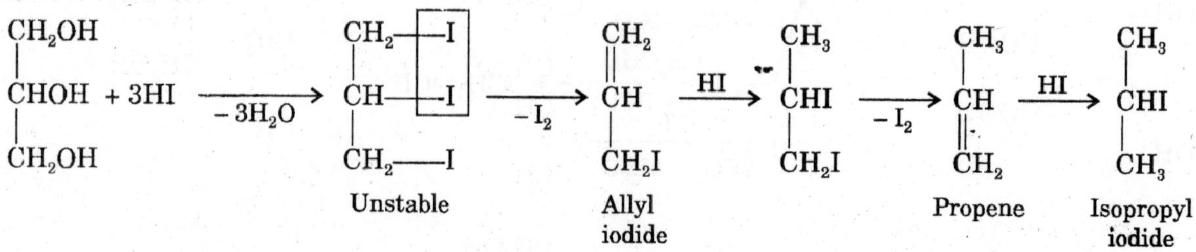

3. Reaction with Fenton's Reagent. Glycerol is oxidised with mild oxidizing agents such as Fenton's reagent ($FeSO_4$ + H_2O_2). On oxidation, a mixture of glyceraldehyde and dihydroxy acetone is obtained. The mixture is called as **glycerose.**

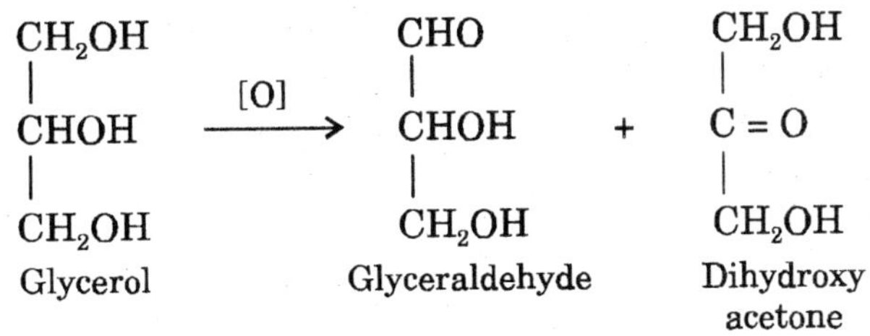

Oxidation with dil. HNO_3 gives a mixture of glyceric acid and tartronic acid

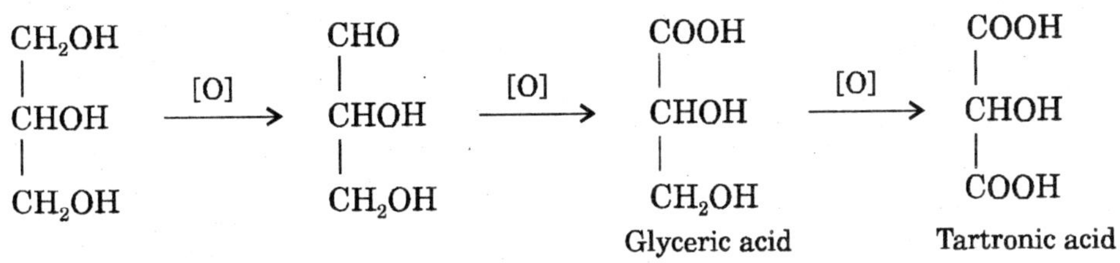

4. Dehydration

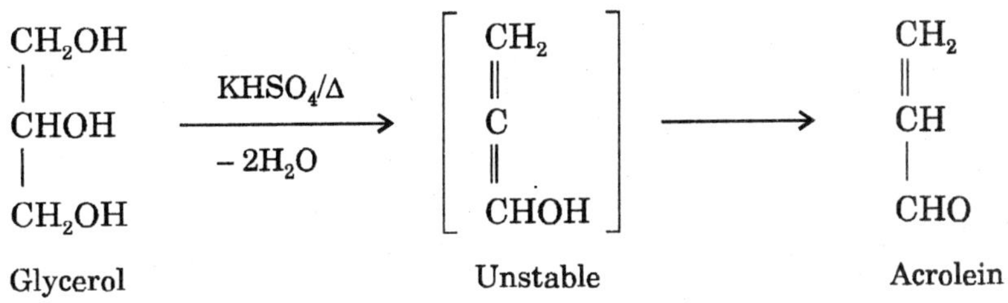

5. Reaction with Sodium

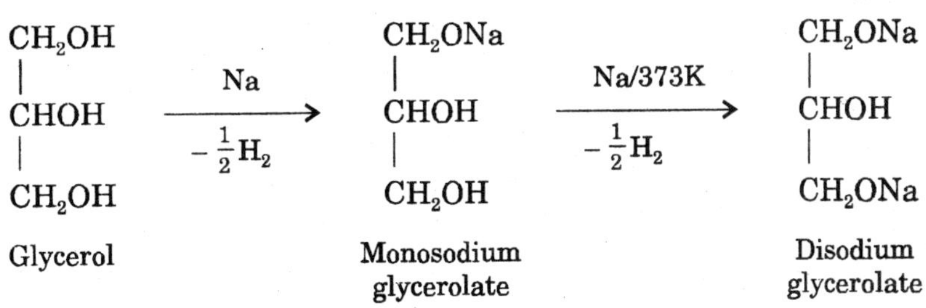

6. Reaction with Oxalic Acid

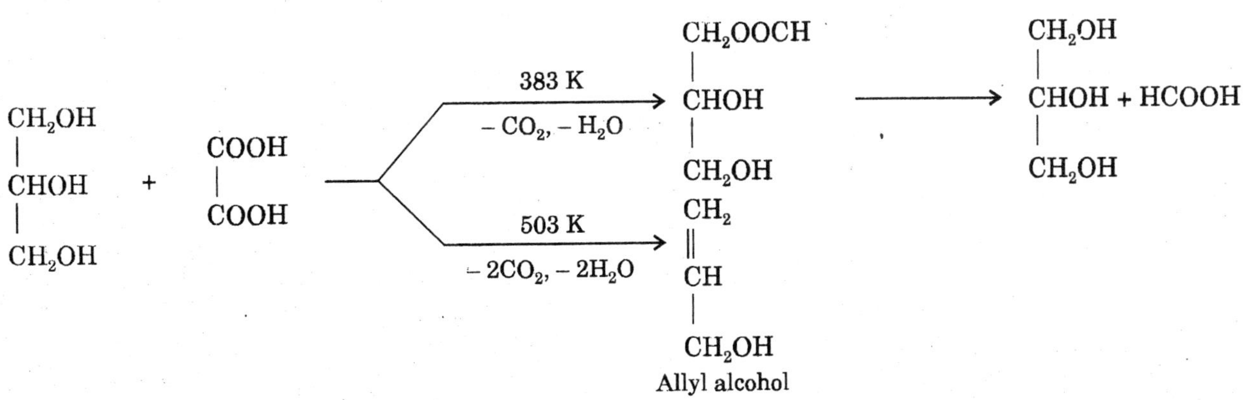

PHENOLS

PREPARATION OF PHENOLS

1. Alkali Fusion of Sulphonates

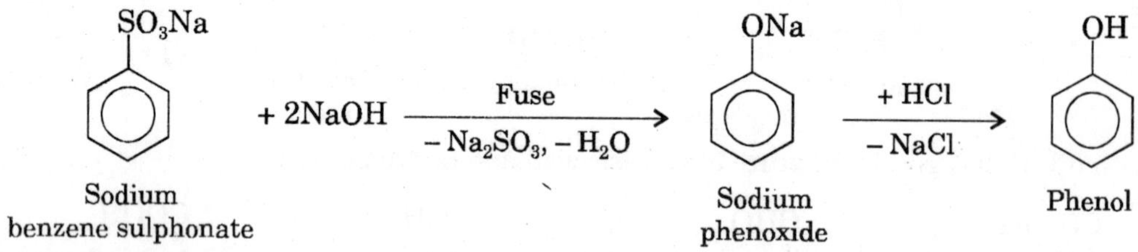

2. From Diazonium Salts

In the laboratory phenols are prepared by hydrolysis of diazonium salts with water or dilute acids.

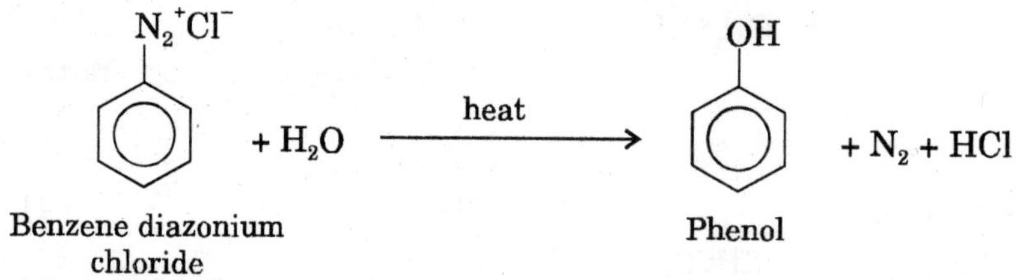

3. From Aryl Halides (*Dow's process*)

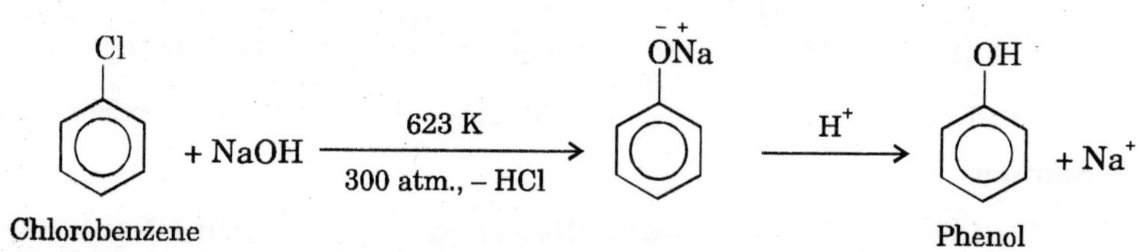

4. Commercial Preparation of Phenol (*From Cumene*)

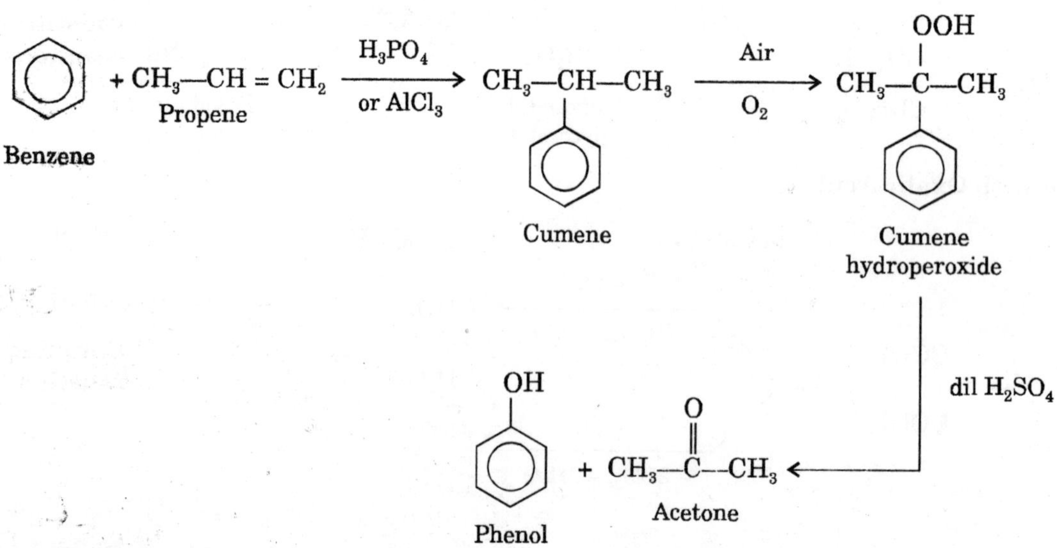

CHEMICAL PROPERTIES OF PHENOLS

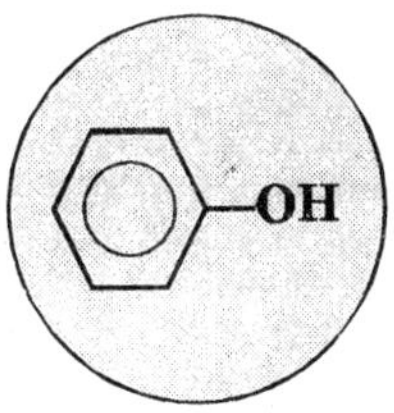

As Acid

- $\xrightarrow{+\ Na} C_6H_5ONa + H_2$
- $\xrightarrow{+\ NaOH} C_6H_5ONa + H_2O$

Acylation

- $\xrightarrow[-\ HCl]{+\ CH_3COCl/Pyridine} C_6H_5COOCH_3$
- $\xrightarrow[-\ CH_3COOH]{+\ (CH_3CO)_2O/H_2SO_4} C_6H_5COOCH_3$

Benzoylation (*Schotten Baumann reaction*)

- $\xrightarrow[aq.\ NaOH]{+\ C_6H_5COCl} C_6H_5COOC_6H_5$

- $\xrightarrow{+\ NH_3/ZnCl_2,\ Heat} C_6H_5—NH_2 + H_2O$
- $\xrightarrow{Zn/\Delta} C_6H_6 + ZnO$
- $\xrightarrow{NaOH} C_6H_5ONa \xrightarrow{RX} C_6H_5—O—R$
- $\xrightarrow{NaOH} C_6H_5ONa \xrightarrow{(CH_3)_2SO_4} C_6H_5OCH_3$

Electrophilic substitution reactions

- $\xrightarrow{Br_2/(aq)}$ 2, 4, 6-Tribromophenol
- $\xrightarrow[273\ K]{Br_2/CS_2}$ *Ortho* and *para*-bromophenol
- $\xrightarrow{Fuming\ HNO_3}$ 2, 4, 6-Trinitrophenol (*Picric acid*)
- $\xrightarrow[293\ K]{dil.\ HNO_3}$ *Ortho* and *para*-nitrophenol
- $\xrightarrow[anhyd.\ AlCl_3]{CH_3Cl}$ *Ortho* and *para*-cresol

Coupling reaction

- $\xrightarrow[273\text{-}278\ K,\ pH \approx 10]{C_6H_5N_2^+Cl^-}$

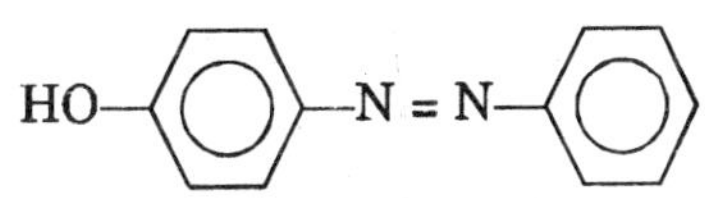

(An azo dye)

Reimer Tiemann reaction

- $\xrightarrow[(ii)\ H_3O^+,\ 340\ K]{(i)\ CHCl_3,\ aq.\ NaOH}$

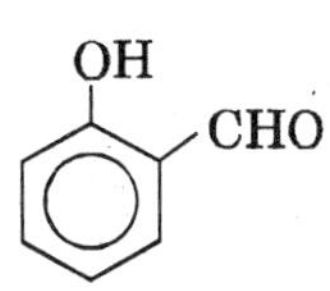

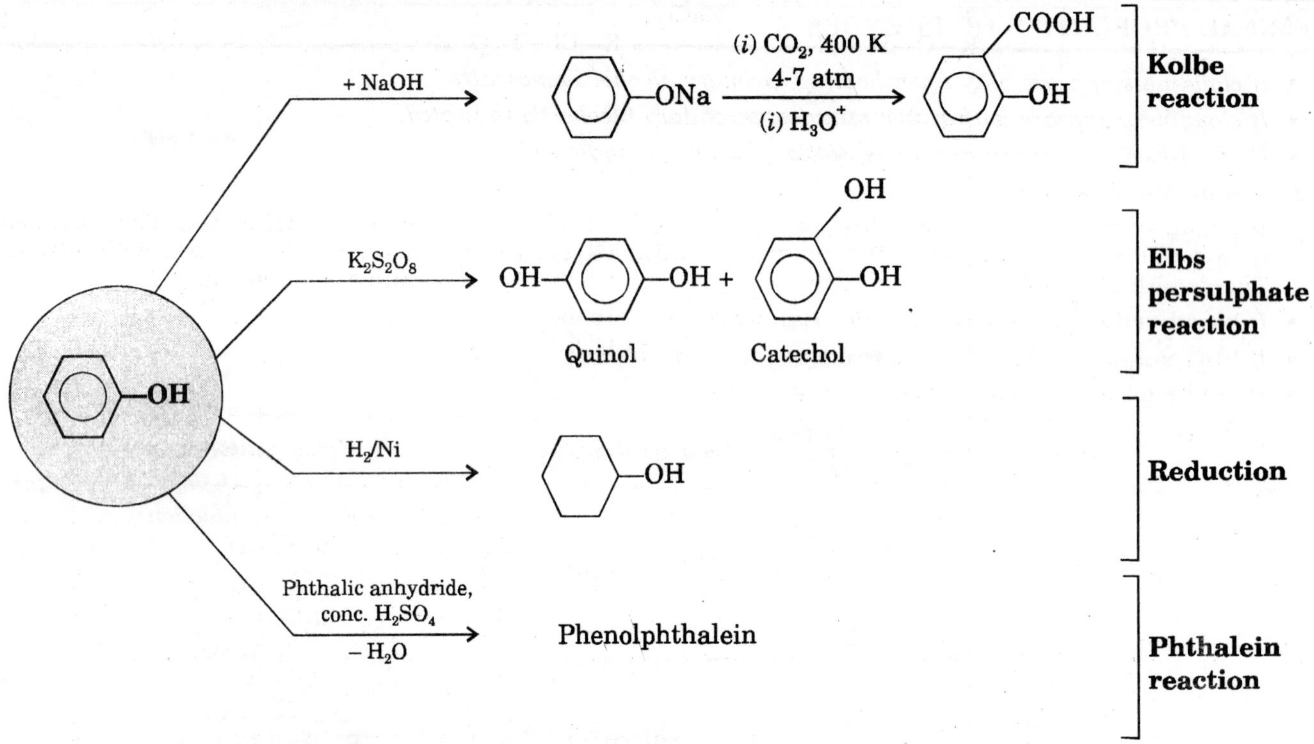

MECHANISM OF REIMER TIEMANN REACTION

$$OH^- + CHCl_3 \rightleftharpoons H_2O + :\bar{C}Cl_3 \longrightarrow Cl^- + :CCl_2$$

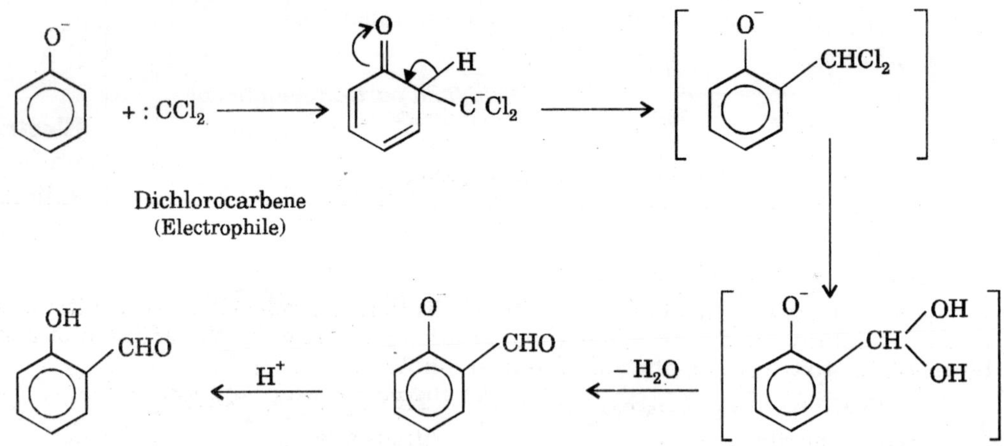

- The reaction involves electrophilic attack of dichlorocarbene (: CCl_2) on the benzene ring.
- *Ortho* product is the major product.

ANALYSIS OF ALCOHOLS AND PHENOLS

1. **Lucas Test**

   In this test, the alcohol is treated with Lucas reagent which is equimolar mixture of HCl and $ZnCl_2$. Alcohols are soluble in Lucas reagent and form a clear solution. On reaction, alkyl chlorides are formed which being insoluble result in cloudiness in the solution.

$$R{-}O{-}H + H{-}Cl \xrightarrow{ZnCl_2} R{-}Cl + H_2O$$

- *If cloudiness appears only upon heating, primary alcohol is indicated.*
- *If cloudiness appears within five minutes, secondary alcohol is indicated.*
- *If cloudiness appears immediately, tertiary alcohol is indicated.*

2. Victor Meyer's Test

In this test, the given alcohol is first converted into its iodide by treating it with HI or P/I_2. The alkyl iodide is then distilled with silver nitrite to convert it into nitroalkane. The nitroalkane is then treated with nitrous acid. The resulting solution is made alkaline with KOH and colour of the solution is observed.

- *If* **blood red** *colour appears, it indicates a* **primary alcohol.**
- *If* **blue colour** *appears, it indicates a* **secondary alcohol.**
- *If* **colourless,** *it indicates a* **tertiary alcohol.**

Primary alcohol	*Secondary alcohol*	*Tertiary alcohol*
$R{-}CH_2OH$ $\downarrow P/I_2$ $R{-}CH_2I$ Alkyl iodide $\downarrow AgNO_2$ $R{-}CH_2{-}NO_2$ $\downarrow HNO_2$ $R{-}C(=N{-}OH){-}NO_2$ Nitrolic acid $\downarrow KOH$ **Blood red colour**	R_2CHOH $\downarrow P/I_2$ R_2CHI Alkyl iodide $\downarrow AgNO_2$ $R_2CH{-}NO_2$ $\downarrow HNO_2$ $R_2C(N{=}O){-}NO_2$ Pseudonitrol $\downarrow KOH$ **Blue colour**	$R_3C{-}OH$ $\downarrow P/I_2$ $R_3C{-}I$ $\downarrow AgNO_2$ $R_3C{-}NO_2$ $\downarrow HNO_2$ No action $\downarrow KOH$ **Colourless**

3. Libermann's Test

Most of the phenols give this test. Red or brown colouration is produced when a nitrite is added to phenol dissolved in conc. sulphuric acid. The colour changes to blue or green by the addition of a strong alkali. **This test can be used to distinguish phenols from alcohols.**

$$\underset{\text{Sodium nitrite}}{2NaNO_2} + H_2SO_4 \longrightarrow \underset{\text{Nitrous acid}}{2HNO_2} + Na_2SO_4$$

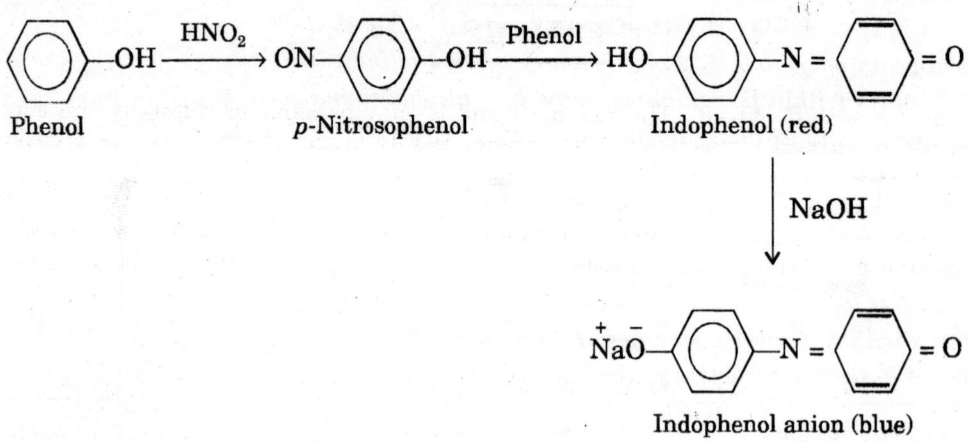

4. Iodoform Test

This tests is used for the identification of alcohols and carbonyl compounds containing $CH_3\overset{\overset{O}{\|}}{C}—$ or $CH_3\overset{\overset{OH}{|}}{C}H—$ group. *It is observed that when the compounds containing such groups are warmed with an aqueous solution of Na_2CO_3 or NaOH and iodide, a yellow precipitate of iodoform is obtained.*

$$2NaOH + I_2 \longrightarrow NaI + NaOI + H_2O$$

$$\mathbf{CH_3—\underset{\underset{OH}{|}}{\overset{\overset{H}{|}}{C}}—R} + NaOI \longrightarrow CH_3—\underset{\underset{O}{\|}}{C}—R + NaI + H_2O$$

$$\mathbf{CH_3—\underset{\underset{O}{\|}}{C}—R} + 3NaOI \longrightarrow Cl_3—\underset{\underset{O}{\|}}{C}—R + 3NaOH$$

$$CI_3—\underset{\underset{O}{\|}}{C}—R + NaOH \longrightarrow \underset{\text{Iodoform}}{CHI_3} + \underset{\text{Sod. salt of acid}}{R—\underset{\underset{O}{\|}}{C}—ONa}$$

Here, R can be H atom or some alkyl group.

5. Periodic Acid Test

This test is given by compounds containing hydroxyl groups on vicinal carbon atoms. The compound is treated with periodic acid (HIO_4). The vicinal diols undergo cleavage and form carbonyl compounds whereas HIO_4 is reduced to iodic acid (HIO_3). Addition of silver nitrate solution gives a precipitate of silver iodate ($AgIO_3$).

$$—\underset{\underset{OH}{|}}{\overset{|}{C}}——\underset{\underset{OH}{|}}{\overset{|}{C}}— + HIO_4 \longrightarrow -\overset{|}{C} = O + O = \overset{|}{C}— + H_2O + HIO_3$$

$$AgNO_3 + HIO_3 \longrightarrow \underset{\text{White ppt.}}{AgIO_3} + HNO_3$$

ETHERS

Ethers have general formula R—O—R′. The groups R and R′ may be same or different and accordingly ethers are known as **simple** or **mixed** respectively. The general IUPAC name of ethers is **alkoxyalkane.**

CH_3-O-CH_3
Methoxymethane
(Dimethyl ether)

$CH_3-O-C_2H_5$
Methoxyethane
(Ethylmethyl ether)

$CH_3CH_2CH_2-OC_2H_5$
1-Ethoxypropane
(Ethyl *n*-propyl ether)

$CH_3=CH-OC_2H_5$
Ethoxyethene
(Ethyl vinyl ether)

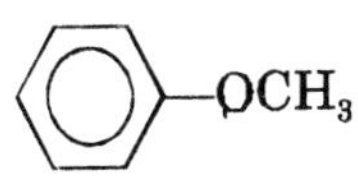
Methoxybenzene
(Anisole)

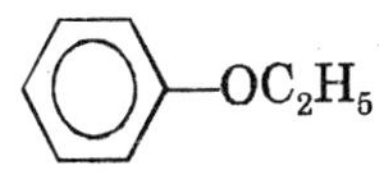
Ethoxybenzene
(Phenetole)

METHODS OF PREPARATION

1. By Williamson Synthesis

$$\underset{\text{Alkyl halide}}{R-X} + \underset{\text{Sod. alkoxide}}{NaOR'} \xrightarrow{\text{ether}} \underset{\text{Ether}}{R-O-R'} + NaX$$

The alkyl halide used in this reaction should preferably be primary because in case of secondary and tertiary alkyl halides the elimination reaction predominates resulting in the formation of alkene as the major product.

$$\underset{(1°\text{ Halide})}{C_2H_5Br} + NaO-\overset{CH_3}{\underset{CH_3}{C}}-CH_3 \longrightarrow \underset{\substack{(\textit{Major product})\\(\textit{ether})}}{C_2H_5-O-\overset{CH_3}{\underset{CH_3}{C}}-CH_3} + NaBr$$

$$\underset{(3°\text{ Halide})}{CH-\overset{CH_3}{\underset{CH_3}{C}}-Br} + NaOC_2H_5 \longrightarrow \underset{\substack{(\textit{Major product})\\(\textit{alkene})}}{CH_3-\overset{CH_3}{C}=CH_2} + NaBr + C_2H_5OH$$

2. By Dehydration of Alcohols

$$\underset{\text{Excess}}{2C_2H_5OH} \xrightarrow[\text{413 K}]{\text{conc. } H_2SO_4} \underset{\text{Diethyl ether}}{C_2H_5-O-C_2H_5} + H_2O$$

$$2C_2H_5OH \xrightarrow[\text{Pressure}]{Al_2O_3/520\text{ K}} C_2H_5-O-C_2H_5 + H_2O$$

3. By Action of Diazomethane on Alcohols. This method is used to prepare **methyl ethers** only.

$$CH_3CH_2OH + CH_2N_2 \xrightarrow{HBF_4} CH_3CH_2{-}OCH_3 + N_2$$

4. By Heating Alkyl halides with dry Silver oxide

$$2C_2H_5I + Ag_2O \xrightarrow{\Delta} C_2H_5{-}O{-}C_2H_5 + 2AgI$$

PHYSICAL PROPERTIES OF ETHERS

1. *Dimethyl ether and ethyl methyl ether are* **gases** *at ordinary temperatures.* The other lower homologues are colourless liquids having typical *'ether'* smell.
2. *The* **boiling points** *of ethers are much lower as compared to isomeric alcohols. Unlike alcohols ether molecules are not associated by hydrogen bonds. The interparticle forces existing in their liquid states are weak* **dipole-dipole forces.**
3. *The* **solubility** *of ethers in water is comparable to those of corresponding alcohols.* For example, *dimethyl ether and ethyl alcohol are both appreciably soluble in water. The solubility of ethers is due to the capability of their molecules to form hydrogen bond with water molecules.*
4. Ethers are polar molecules and have definite dipole moment. This is due to the **bent** structure of their molecules and polar C—O bonds.
5. Ethers are lighter than water. The density of ethers shows a regular increase with increase in molecular mass.

CHEMICAL PROPERTIES OF ETHERS

Ethers are relatively inert compounds and do not react with alkalies, reducing agents, oxidising agents and reactive metals. ***Because of their inertness ethers are frequently used as solvents.*** Ethers, however, undergo cleavage under acidic conditions. Some common reactions of ethers are given below :

1. Cleavage by Acids

- On reaction with conc. HBr or conc. HI ethers undergo cleavage to form an alcohol or phenol and an alkyl halide.

$$C_2H_5 - O - C_2H_5 + HX \longrightarrow C_2H_5X + C_2H_5OH$$

- The **order of reactivity** of halogen acids is

$$HI > HBr > HCl$$

In case the acid is used in excess, alkyl halide is formed as the product

$$C_2H_5 - O - C_2H_5 \quad + \quad \underset{\text{Excess}}{2HX} \longrightarrow C_2H_5X + H_2O.$$

- **Point of cleavage.** (*i*) In case of mixed ethers having both the alkyl groups primary or secondary, the smaller alkyl group gets converted into alkyl halide predominantly.

$$CH_3 - O - C_2H_5 + HI \longrightarrow CH_3I + C_2H_5OH$$

(*ii*) In case of mixed ethers having a tertiary alkyl group, the point of cleavage is such that the tertiary alkyl halide is formed as the major product

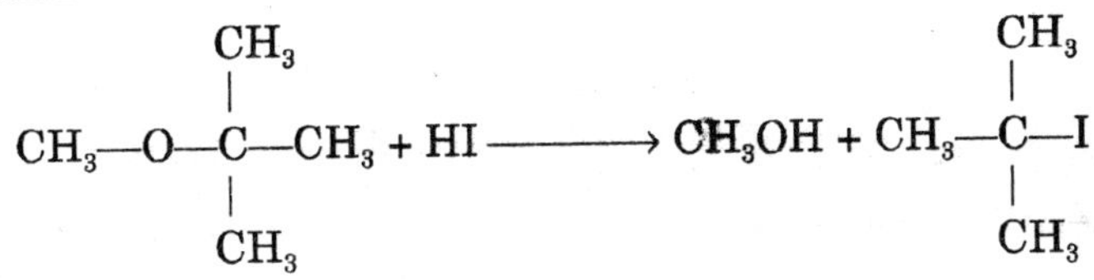

(*iii*) In case of alkyl aryl ethers, the acidic cleavage results in phenol and alkyl halide.

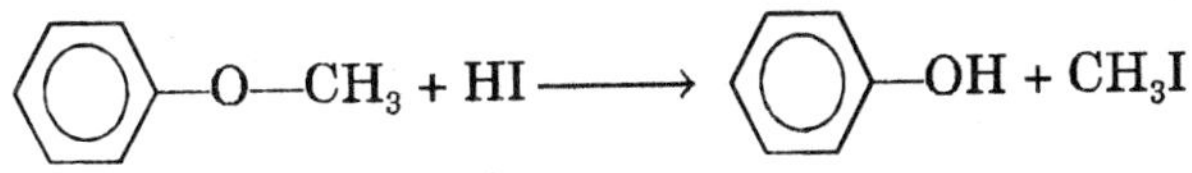

2. Hydrolysis

$$R—O—R' + H_2O \xrightarrow{H_2SO_4} R—OH + R'—OH$$

3. Reaction with conc. H_2SO_4

$$R - O - R' + H_2SO_4 \longrightarrow \underset{\text{Alcohol}}{R - OH} + \underset{\text{Alkyl hydrogen sulphate}}{R'HSO_4}$$

4. Halogenation

- On treatment with halogens in the presence of sunlight, ethers are readily halogenated to form polyhalogenated ethers.

$$\underset{\text{Ethyl ether}}{C_2H_5 - O - C_2H_5} \xrightarrow{Cl_2/h\nu} \underset{\text{Perchlorodiethyl ether}}{C_2Cl_5 - O - C_2Cl_5}$$

- If the reaction is carried out in the absence of sunlight, halogenation takes place at the α-position only.

$$CH_3 - CH_2 - O - CH_2 - CH_3 \xrightarrow{Cl_2} \underset{\alpha,\alpha\text{-Dichlorodiethyl ether}}{CH_3 - CHCl - O - CHCl - CH_3}$$

5. Formation of Peroxides

When stored for long periods, ethers react with oxygen in the presence of light to form peroxides.

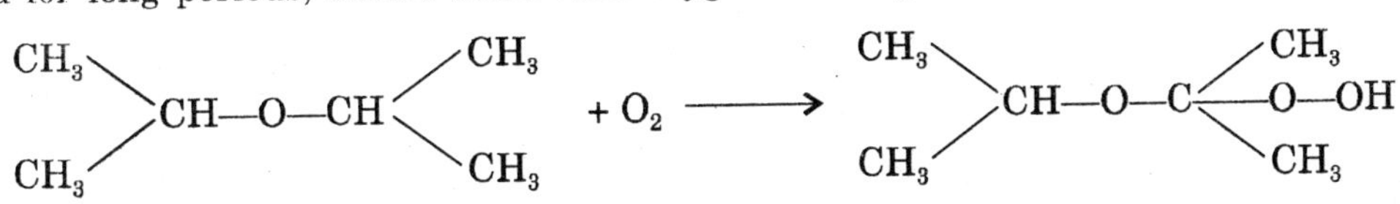

Peroxide

The ether peroxides are unstable compounds and are likely to explode when heated. The presence of peroxides in a ether sample can be detected by shaking the ether with an aqueous solution of ferrous ammonium sulphate and potassium thiocyanate (KSCN). The solution turns red if the peroxide is present

$$Fe^{2+} \xrightarrow{\text{Peroxide}} Fe^{3+} \xrightarrow{SCN^-} \underset{\text{Blood red}}{Fe(SCN)_3}$$

6. Electrophilic Substitution Reactions in Aryl Ethers

Alkoxy group is moderately activating and *ortho-para* directing. Therefore, electrophilic substitution in aromatic ethers takes place mainly at *ortho* and *para* positions.

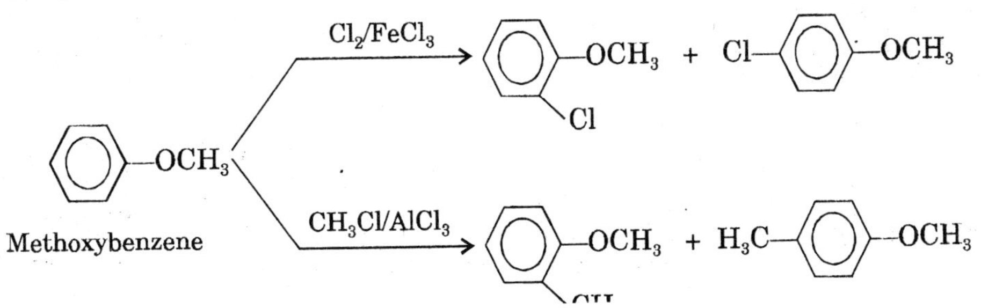

4.2. IMPORTANT FACTS AND TERMS

Add to Knowledge

- For C_4H_9OH, four isomeric alcohols are possible, out of which one (2-butanol) can exhibit enantiomerism.
- Alcohols are functional isomers of ethers. For example, ethanol and dimethyl ether are functional isomers.
- In **carbinol system** of naming alcohols, methyl alcohol is named carbinol while all other alcohols are named as derivatives of carbinol.

  For example,

  CH_3CH_2OH is named methyl carbinol.
- Solubility of alcohols in water decreases with increase in size of the alkyl group, and increases with increase in branching of alkyl group.
- Alcohols are weaker acids than water while phenol is stronger acid than water

  Acidic character : Phenol > H_2O > R—OH.
- The order of reactivity of alcohols in the reactions involving cleavage of oxygen-hydrogen bond is 1° > 2° > 3°.
- The order of reactivity of alcohols in the reactions involving cleavage of carbon-oxygen bond is 3° > 2° > 1°.
- Reaction of Grignard reagents with ethylene oxide (oxirane) followed by hydrolysis yields a primary alcohol having two carbon atoms more than the carbon atoms in Grignard reagent.
- Alcohols are lighter than water.
- Alcohols cannot be dried over anhydrous calcium chloride because they form solid derivatives with it. For example, $CaCl_2.4C_2H_5OH$.
- Methanol is known as **Wood spirit** because it can be obtained by destructive distillation of wood.
- **Pyroligneous Acid.** It is one of the products of destructive distillation of wood. It contains acetic acid (10%), methanol (2.5%), acetone (0.5%) and remaining water.
- Ethanol is known as **grain alcohol.**
- Simple fractional distillation of aqueous solution of ethyl alcohol yield an azeotropic mixture containing 95.87% ethanol and 4.13% water. This is known as **rectified spirit.**
- **Absolute Alcohol.** 100% ethanol is known as absolute alcohol.
- **Methylated spirit or Denatured spirit.** Ethyl alcohol is made unfit for drinking by adding 5-10% methanol. Such alcohol is known as methylated spirit or denatured spirit.
- **Power Alcohol.** Ethyl alcohol when used for the purpose of power generation in internal combustion engines is known as power alcohol. Ethyl alcohol is mixed with petrol in the ratio 1 : 4 for this purpose.
- **Fusel Oil.** It is a mixture of primary alcohols. It mainly contains isopentyl alcohol with smaller amounts of *n*-propyl alcohol, *iso*-butyl alcohol and 2-methyl-1-butanol (*active* amyl alcohol). *It is obtained as the fermentation product of starch along with ethanol.*
- **Proof Spirit.** An alcohol-water mixture containing 57.1% ethyl alcohol (by volume). A sample having higher percentage of alcohol is expressed as **over proof (O.P.)** and the one having lower percentage is expressed as **under proof (U.P.)**. A 10 O.P. sample of alcohol means that 100 ml of the sample contains as much alcohol as is in the 110 ml of proof spirit.
- **Rubbing Alcohol.** *iso*-propyl alcohol is known as rubbing alcohol.
- **Carbolic Acid.** Phenol is also known as carbolic acid.
- **Picric Acid.** 2, 4, 6-Trinitrophenol.
- Phenol on reaction with phthalic anhydride in the presence of conc. H_2SO_4 gives phenolphthalein.
- In Reimer Tiemann reaction, the attacking electrophile is dichlorocarbene (: CCl_2).
- Sulphonation of phenol at room temperature (288-293 K) yields 2-hydroxybenzenesulphonic acid while sulphonation at 373 K yields 4-hydroxybenzenesulphonic acid as the major product.
- **Aspirin** is 2-acetoxybenzoic acid (acetyl salicylic acid).
- **Fries Rearrangement.** Phenyl esters on heating with anhydrous $AlCl_3$ in the presence of CS_2 as solvent undergo rearrangement and yield ortho- and para-hydroxyketones. The reaction is known as Fries rearrangement.

  At low temperature (~ 330 K) para product is major while at high temperature (~ 430 K) the ortho product predominates.

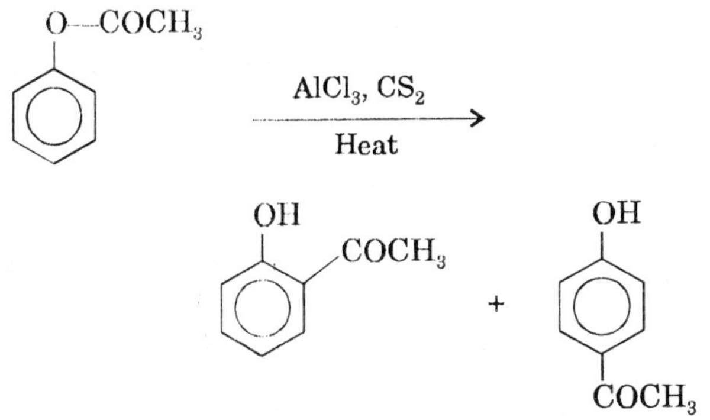

o-Hydroxyaceto-phenone *p*-Hydroxyaceto-phenone

- **Bouveault-Blanc Reduction.** Reduction of carbonyl compounds and esters using sodium and ethyl alcohol.
- **Fenton's Reagent.** An aqueous solution of H_2O_2 and $FeSO_4$. It is a mild oxidizing agent. Fenton's reagent oxidizes glycerol to glyceraldehyde and dihydroxy-acetone.
- Dehydration of glycerol with $KHSO_4$ yields aerolein.
- Benzene ring in phenol is highly activated towards electrophilic substitution reactions due to electron releasing resonance effect of —OH group.
- Alcohols and carbonyl compounds containing $CH_3—CH(OH)—$ or $CH_3—C(=O)—$ group give yellow ppt. of iodoform on reaction with NaOH and I_2.
- **Williamson Synthesis.** Reaction of alkyl halides with sodium alkoxides or sodium phenoxides resulting in the formation of ethers.
- **Anisole on cleavage with HI give phenol and methyl iodide.** Phenyl-oxygen bond due to partial double bond character is stronger and hence difficult to cleave.

4.3. Solved Problems

Enrich your Skill

Example 1. *Propene, $CH_3—CH = CH_2$ can be converted to 1-propanol by oxidation. Which set of reagents among the following is ideal to effect the conversion ?*

(a) *Alkaline $KMnO_4$*

(b) *B_2H_6 and alkaline H_2O_2*

(c) *O_3/Zn dust*

(d) *OsO_4/CH_4, Cl_2.*

Solution. Hydroboration-oxidation converts propene into 1-propanol.

Hence, the answer is ***(b)***.

Example 2. *A compound X with molecular formula C_3H_8O can be oxidized to a compound Y with the molecular formula $C_3H_6O_2$. X is most likely to be a*

(a) *Primary alcohol* (b) *sec-alcohol*

(c) *Aldehyde* (d) *Ketone.*

Solution. Since the compound Y contains two oxygen atoms, it is likely to be a carboxylic acid (CH_3CH_2COOH). The formula of compound X corresponds to that of some alcohol ($C_nH_{2n+1}OH$). Primary alcohols on oxidation give carboxylic acids.

Hence, the correct answer is ***(a)***.

Example 3. *Amongst the following phenols which is most acidic ?*

(a) *Picric acid* (b) *2-Nitrophenol*

(c) *2, 4-Dinitrophenol* (d) *3-Nitrophenol.*

Solution. Greater the number of nitro group on the ring, greater is the acidic strength of the phenol. Picric acid is 2, 4, 6-trinitrophenol.

Hence, ***(a)*** is the strongest acid.

Example 4. *The compound which can form intramolecular hydrogen bond is*

(a) *m-Hydroxybenzaldehyde*

(b) *Salicyladehyde*

(c) *Benzaldehyde*

(d) *p-Hydroxybenzaldehyde.*

Solution. Salicylaldehyde is *o*-hydroxybenzaldehyde. Intramolecular hydrogen bonding is possible in it

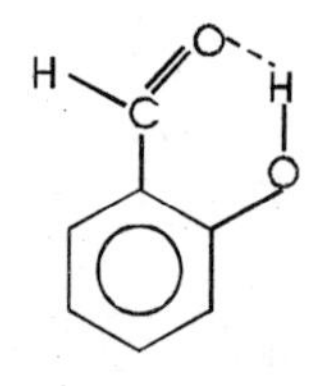

Hence, the correct choice is ***(b)***.

Example 5. *What is the end product in the following sequence of operations ?*

$$\text{Phenol} \xrightarrow{NaOH} A \xrightarrow[140°C]{CO_2} B \xrightarrow{HCl} C$$

(a) *Salicylaldehyde* (b) *Benzoic acid*

(c) *Chlorobenzene* (d) *Salicylic acid.*

Solution. The above sequence of reaction represents Kolbe's reaction. The end product in this sequence is salicylic acid.

Hence, the correct answer is ***(d)***.

Example 6. *Anisole with HNO_3 and conc. H_2SO_4 gives*

(a) *Phenol* (b) *Nitrobenzene*

(c) *o- and p-Nitroanisole* (d) *o-Nitroanisole.*

Solution. Anisole is methoxybenzene. Methoxy group is highly activating and *ortho-para* directing group.

Hence, the correct answer is ***(c)***.

Example 7. *HBr reacts fastest with*

(a) *2-Methylpropan-2-ol* (b) *Propan-1-ol*

(c) *Propan-2-ol* (d) *2-Methylpropan-1-ol.*

Solution. The order of reactivity of alcohols with halogen acids is 3° > 2° > 1°. 2-Methylpropan-2-ol is 3° alcohol.

Hence, the correct answer is ***(a)***.

Example 8. *Bromine water reacts with phenol to form*

(a) *o-Bromophenol*

(b) *p-Bromophenol*

(c) *2, 3, 4-Tribromophenol*

(d) *2, 4, 6-Tribromophenol.*

Solution. Benzene ring in phenol is highly activated at ortho and para positions. Bromine in the presence of polar solvent such as water substitutes all the ortho and para positions in phenol.

Hence, the product is 2, 4, 6-tribromophenol, so, answer is **(*d*)**.

Example 9. *Which of the following compounds does not react with metallic sodium ?*

(*a*) $CH_3CH(OH)CH_3$ (*b*) $CH_3—O—CH_3$

(*c*) CH_3COOH (*d*) C_2H_5OH.

Solution. $CH_3—O—CH_3$ does not contain any acidic proton and hence it would not react with sodium.

Hence, the correct answer is **(*b*)**.

Example 10. *Phenol on reduction with H_2 in presence of nickel catalyst gives*

(*a*) *Benzene* (*b*) *Toluene*

(*c*) *Cyclohexane* (*d*) *Cyclohexanol.*

Solution. Catalytic hydrogenation reduces the aromatic ring in phenol.

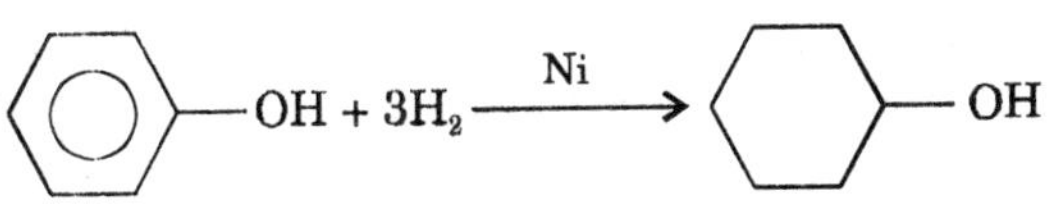

Cyclohexanol

Hence, the correct answer is **(*d*)**.

Example 11. *Phenol, p-Methylphenol, m-Nitrophenol and p-Nitrophenol follows order of increasing acidic strength*

(*a*) *Phenol, p-Methylphenol, p-Nitro phenol, m-Nitrophenol*

(*b*) *p-Methylphenol, Phenol, m-Nitrophenol, p-Nitro phenol*

(*c*) *p-Methylphenol, m-Nitrophenol, Phenol, p-Nitro phenol*

(*d*) *m-Nitrophenol, p-Nitrophenol, Phenol and p-Methylphenol.*

Solution. $—NO_2$ group is electron withdrawing group. Hence, *p*-nitrophenol and *m*-nitrophenol would be stronger acids than the other two. *p*-Nitrophenol is the strongest acid because electron withdrawing effect through resonance is more at para (and ortho) position than at meta position.

Hence, the correct answer is **(*b*)**.

Example 12. *In its reaction with sodium, 1 mol of a compound X gives 1 mol of H_2. Which one of the following compounds might be X ?*

(*a*) $CH_3CH_2CH = CH_2$

(*b*) $CH_3COOCH_2CH_3$

(*c*) $CH_3CH_2CH_2OH$

(*d*) $CH_2OHCH_2CH_2CH_2OH$.

Solution. $R—OH + Na \longrightarrow RONa + \frac{1}{2} H_2$

One mole of monohydric alcohol gives $\frac{1}{2}$ mole of H_2. Since one mole of the given compound is giving 1 mole H_2 it indicates, the compound contains 2 acidic hydrogen atoms per molecule and hence must be a dihydric alcohol.

Hence, the correct answer is **(*d*)**.

Example 13. *When 3, 3-Dimethyl-2 butanol is heated with H_2SO_4, the major product obtained is*

(*a*) *3, 3-Dimethyl-1-butene*

(*b*) *2, 3-Dimethyl-2-butene*

(*c*) *2, 3-Dimethyl-1-butene*

(*d*) *cis and trans-isomers of* (*b*).

Solution.

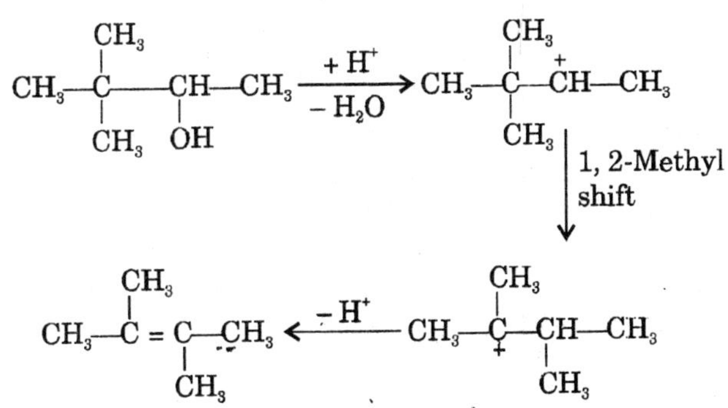

2, 3-Dimethyl-2-butene

Hence, the correct answer is **(*b*)**.

4.4. QUESTION BANK

Develop Confidence

Nomenclature, Isomerism and Classification

1. How many primary alcohols are possible for C_4H_9OH ?
(a) 1 (b) 2
(c) 3 (d) 4.
2. An isomer of ethanol is
(a) Methanol (b) Ethanal
(c) Ethoxy ethane (d) Methoxy methane.
3. Which of the following is a primary alcohol ?
(a) *neo*-Pentyl alcohol (b) *sec*-Pentyl alcohol
(c) *tert*-Pentyl alcohol (d) None of these.
4. *n*-Propyl alcohol—*iso*-propyl alcohol is an example of
(a) Chain isomerism (b) Position isomerism
(c) Metamerism (d) Tautomerism.
5. How many isomers are possible for an aromatic compound having molecular formula C_7H_8O ?
(a) 3 (b) 4
(c) 5 (d) 6.
6. Which of the following acids does not contain —COOH group ?
(a) Picric acid (b) Lactic acid
(c) Citric acid (d) Tartaric acid.
7. Carbolic acid is
(a) Phenol (b) Phenyl benzoate
(c) Phenyl acetate (d) Salol.
8. Vinyl carbinol is
(a) $HO—CH_2—CH = CH_2$
(b) $CH_3C(OH) = CH_2$
(c) $CH_3—CH = CH—OH$
(d) $CH_3—C(CH_2OH) = CH_2$
9. How many isomers of $C_5H_{11}OH$ will be secondary alcohols ?
(a) 5 (b) 4
(c) 2 (d) 3.
10. IUPAC name of glycerol is
(a) 1, 2, 3-Propanol
(b) Tripropanol
(c) Propane-1, 2, 3-triol
(d) 1, 2, 3-Trihydroxy propane.
11. Which of the following is a trihydric phenol ?
(a) Resorcinol (b) *p*-Cresol
(c) Phloroglucinol (d) Catechol.
12. 2-Phenylethanol is an isomer of
(a) Benzyl alcohol (b) Anisole
(c) Phenetole (d) Cresol.

Preparation of Alcohols

13. If phenyl magnesium bromide and acetaldehyde are the reactants, the product formed after hydrolysis would be
(a) Benzyl alcohol (b) 1-Phenylethanol
(c) 2-Phenylethanol (d) Acetone.
14. Propylene is subjected to hydroboration oxidation reaction. The product formed would be
(a) Propanal (b) 1-Propanol
(c) 2-Propanol (d) Propanone.
15. Alcohols can be obtained from all methods except
(a) Hydroboration-oxidation
(b) Oxymercuration-demercuration
(c) Reduction of aldehyde/ketones with Zn—Hg/HCl
(d) Fermentation of starch.
16. The most appropriate reagent to convert
$$RCOOEt \longrightarrow RCH_2OH \text{ is}$$
(a) $LiAlH_4$ (b) $NaBH_4$
(c) H_2/Pd-C (d) Li/NH_3 (*liq*).
17. Propene, $CH_3CH = CH_2$ can be converted into 1-propanol by oxidation. Indicate which set of reagents amongst the following is ideal to effect the above conversion ?
(a) $KMnO_4$ (alkaline)
(b) Osmium tertroxide (OsO_4/CH_2Cl_2)
(c) B_2H_6 and alk. H_2O_2
(d) O_3/Zn.
18. Conversion of cyclohexene to cyclohexanol can be conveniently achieved by
(a) NaOH—H_2O
(b) Br_2—H_2O
(c) Hydroboration—oxidation
(d) Hydroboration—hydrolysis.

19. Which of the following compounds on reaction with CH_3MgBr will give a tertiary alcohol ?

(*a*) C_6H_5CHO (*b*) $C_2H_5CO_2CH_3$

(*c*) C_2H_5COOH (*d*) 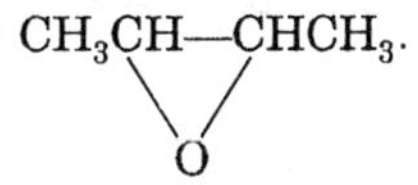

20. Ethyl acetate reacts with CH_3MgBr to form

(*a*) Secondary alcohol

(*b*) Tertiary alcohol

(*c*) Primary alcohol and an acid

(*d*) An acid.

21. The reaction of 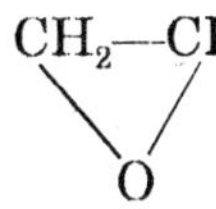 with RMgX leads to the formation of

(*a*) RCHOHR (*b*) $RCHOHCH_3$

(*c*) R_2CHCH_2OH (*d*) RCH_2CH_2OH.

22. The reaction of Grignard reagent with formaldehyde followed by acidification gives

(*a*) An aldehyde (*b*) A primary alcohol

(*c*) A carboxylic acid (*d*) A ketone.

23. Which set of the following reagents (*a* to *d*) would you select to convert $C_6H_5COCH_3$ (acetophenone) to the following alcohol ?

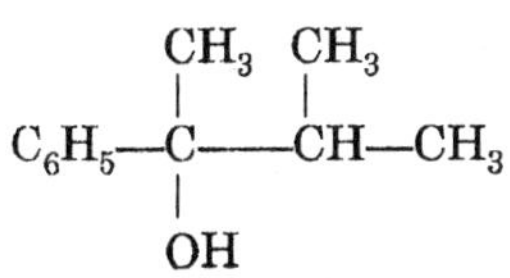

(*a*) $CH_3CH_2CH_2MgBr$ and hydrolysis

(*b*) $CH_3CH(Br)CH_3$. $AlCl_3$

(*c*) $(CH_3)_2CHMgBr$ and acid hydrolysis

(*d*) $CH_3CHCHCH_3$, Zn.

24. 1-Phenylethanol can be prepared by the reaction of benzaldehyde with

(*a*) Methyl bromide

(*b*) Ethyl iodide and magnesium

(*c*) Methyl iodide and magnesium

(*d*) Methyl bromide and aluminium bromide.

25. The enzyme that converts sucrose to glucose and fructose is

(*a*) Zymase (*b*) Diastase

(*c*) Invertase (*d*) Maltase.

26. The enzyme that converts starch into maltose is

(*a*) Maltase (*b*) Zymase

(*c*) Diastase (*d*) Invertase.

27. $CH_2 = CH_2 + CO + H_2 \xrightarrow{[Co(CO)_4]_2} X \xrightarrow{H_2/Ni} Y$

The compounds X and Y are respectively

(*a*) Propanone and propanal

(*b*) Propanal and 1-propanol

(*c*) Propanone and 2-propanol

(*d*) 1-Propanol and 2-propanol.

28. Which of the following methods cannot be used for the preparation of primary alcohols ?

(*a*) Reduction of aldehydes with LAH

(*b*) Reaction of RMgX with ethylene oxide, followed by hydrolysis

(*c*) Reduction of esters with LAH

(*d*) Reaction of esters with excess Grignard reagent, followed by hydrolysis.

29. 2-Phenylethanol may be prepared by the reaction of phenylmagnesium bromide with

(*a*) HCHO (*b*) CH_3CHO

(*c*) CH_3COCH_3 (*d*) 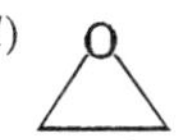

Physical Properties of Alcohols and Phenols

30. Which of the following alcohols is the least soluble in water ?

(*a*) Ethanol (*b*) 1-Propanol

(*c*) 1-Butanol (*d*) 1-Pentanol.

31. Which of the following alcohols is most soluble in water ?

(*a*) *n*-Butyl alcohol (*b*) *iso*-Butyl alcohol

(*c*) *sec*-Butyl alcohol (*d*) *tert*-Butyl alcohol.

32. Which of the following alcohols is least soluble in water ?

(*a*) CH_3OH (*b*) C_2H_5OH

(*c*) n-C_3H_7OH (*d*) n-C_4H_9OH.

33. Which of the following compounds should be least soluble in water ?

(*a*) Phenol (*b*) Ethanol

(*c*) Benzoic acid (*d*) Chlorobenzene.

34. Which of the following is not characteristic of alcohols ?

(*a*) They are lighter than water

(*b*) Their boiling points increase fairly uniformly with increasing molecular mass.

(*c*) Lower members are insoluble in water but solubility regularly increases with molecular mass (*d*) Lower members have pleasant smell and burning taste, while higher members are odourless and tasteless.

35. Which of the following compounds has the highest boiling point ?

(*a*) C_2H_5OH (Mol. mass = 46)
(*b*) CH_3OCH_3 (Mol. mass = 46)
(*c*) CH_3Cl (Mol. mass = 50.5)
(*d*) $CH_3CH_2CH_3$ (Mol. mass = 44).

Chemical Properties of Alcohols

36. Which of the following is most acidic ?

(*a*) CH_3OH
(*b*) C_2H_5OH
(*c*) $CH_3CH_2CH_2OH$
(*d*) H_2O.

37. Which of the following is the strongest base ?

(*a*) *tert*-Butoxide (*b*) Ethoxide
(*c*) *iso*-Propoxide (*d*) Methoxide.

38. 3, 3-Dimethyl-2-butanol, on reaction with conc. H_2SO_4 at 443 K will give ...... as the major product.

(*a*) 3, 3-Dimethyl-1-butene
(*b*) 2, 3-Dimethyl-2-butene
(*c*) 2, 2-Dimethyl-2-butene
(*d*) 2, 2-Dimethyl-1-butene.

39. The compound which gives the most stable carbocation is

(*a*) CH_3CH_2OH (*b*) 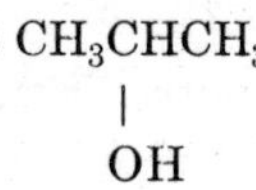

(*c*) 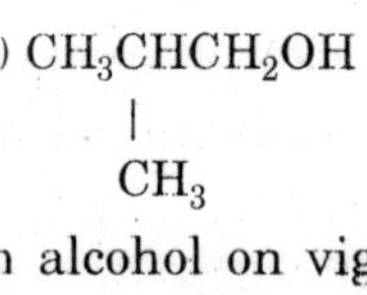 (*d*) 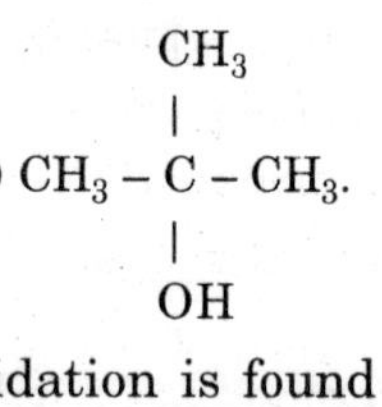

40. An alcohol on vigorous oxidation is found to give ethanoic acid and propanoic acid. The alcohol may be

(*a*) 1-Pentanol (*b*) 2-Pentanol
(*c*) 1-Butanol (*d*) 2-Butanol.

41. Which of the following is the strongest base ?

(*a*) OH^- (*b*) OR^-
(*c*) CH_3^- (*d*) NH_2^-.

42. 3, 3-Dimethyl-2-butanol on reaction with HCl yields mainly

(*a*) 2-Chloro-2, 3-dimethylbutane
(*b*) 1-Chloro-2, 3-dimethylbutane
(*c*) 2-Chloro-3, 3-dimethylbutane
(*d*) 1-Chloro-3, 3-dimethylbutane.

43. The reagent used for dehydration of an alcohol is

(*a*) Phosphorus pentachloride
(*b*) Calcium chloride
(*c*) Aluminium oxide
(*d*) Sodium chloride.

44. The alcohol which does not give a stable compound on dehydration is

(*a*) Ethyl alcohol (*b*) Methyl alcohol
(*c*) *n*-Propyl alcohol (*d*) *n*-Butyl alcohol.

45. The product of the following reaction,

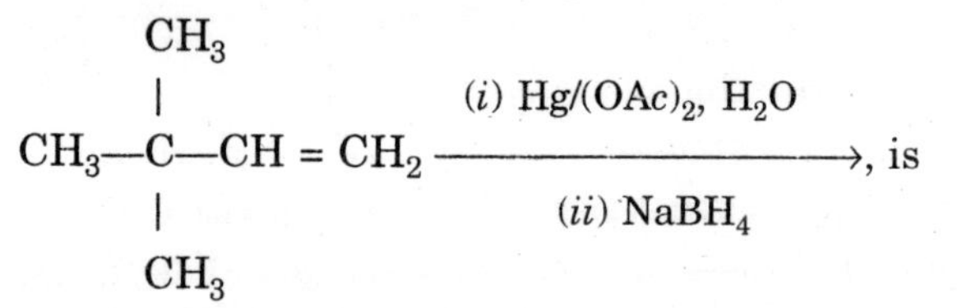

(*a*) 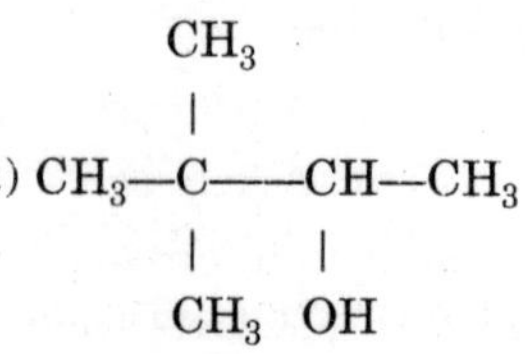

(*b*) 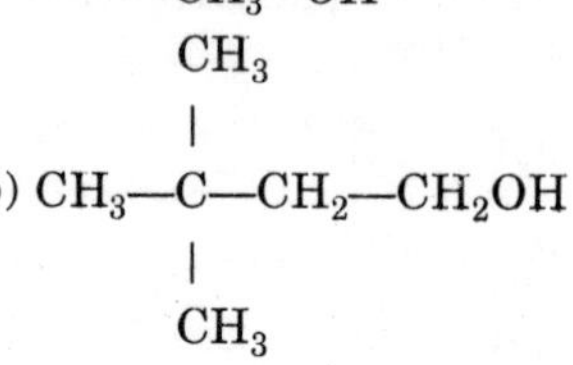

(*c*) 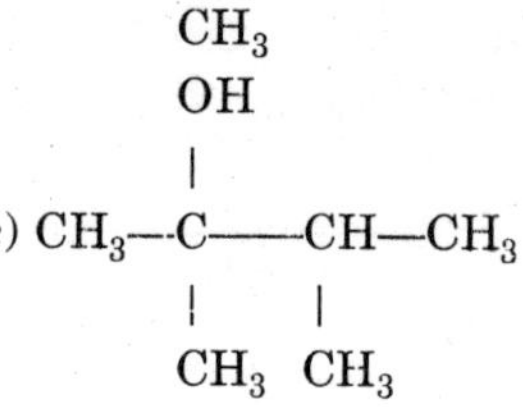

(*d*) 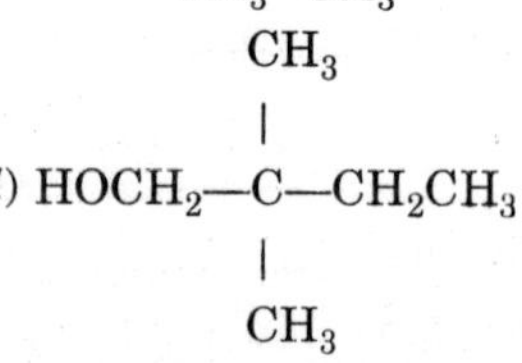

46. When wine is put in air, it becomes sour due to

(*a*) Bacteria
(*b*) Oxidation of C_2H_5OH to CH_3COOH
(*c*) Virus
(*d*) Formic acid formation.

47. When passed over heated copper at 575 K an organic compound was found to give an olefin. The organic compound is most likely to be a/an

(a) Primary alcohol (b) Secondary alcohol
(c) Tertiary alcohol (d) Alkane.

48. Best reagent for preparing an alkyl chloride from an alcohol is
(a) $SOCl_2$ (b) HCl
(c) PCl_3 (d) PCl_5.

49. Which one of the following alcohols can be most easily dehydrated ?
(a) C_2H_5OH (b) $CH_3-\underset{\substack{| \\ CH_3}}{CH}-OH$
(c) $CH_3-\overset{\substack{CH_3 \\ |}}{\underset{\substack{| \\ CH_3}}{C}}-OH$ (d) $CH_3-CH_2-CH_2-OH$.

50. Isopropyl alcohol on treatment with excess hot conc. H_2SO_4 gives a compound X. X on treatment with HBr gives Y. Y when treated with sodium using ether as solvent gives
(a) n-Hexane (b) 2, 3-Dimethylbutane
(c) 2-Methylpentane (d) 3-Methylpentane.

51. When the vapours of ethyl alcohol are passed over red hot copper (300°C), then product formed is
(a) Ethylene (b) Acetaldehyde
(c) Acetic acid (d) None of these.

52. A compound is soluble in conc. H_2SO_4. It does not decolourise bromine in carbon tetrachloride but is oxidised by chromic anhydride in aqueous sulphuric acid within two seconds, turning orange solution to blue, green and then opaque. The original compound is
(a) a primary alcohol (b) a tertiary alcohol
(c) an alkane (d) an ether.

53. Which one of the following on oxidation gives a ketone ?
(a) Primary alcohol (b) Secondary alcohol
(c) Tertiary alcohol (d) All of these.

54. When ethyl alcohol is distilled with conc. H_2SO_4 under reduced pressure, the product is
(a) Ethyl hydrogen sulphate
(b) Ethylene
(c) Diethyl sulphate
(d) Diethyl ether.

55. When phenol is treated with conc. H_2SO_4 at 373 K, the major product is
(a) o-Hydroxy benzenesulphonic acid
(b) p-Hydroxy benzenesulphonic acid
(c) A mixture of o-and p-hydroxy benzenesulphonic acid
(d) m-Hydroxy benzenesulphonic acid.

56. This compound which is most acidic among the following is
(a) p-Chlorophenol (b) p-Nitrophenol
(c) p-Aminophenol (d) p-Cresol.

57. Which one of the following compounds would not be easily oxidised by $K_2Cr_2O_7$ in dil. H_2SO_4 ?
(a) CH_3OH (b) $(CH_3)_3COH$
(c) CH_3CH_2OH (d) CH_3CHO.

58. What is formed when a primary alcohol undergoes catalytic dehydrogenation ?
(a) Aldehyde (b) Ketone
(c) Alkene (d) Acid.

59. An organic compound A reacts with sodium metal and forms B. On heating with conc. H_2SO_4, A gives diethyl ether. A and B are respectively
(a) C_2H_5OH and C_2H_5ONa
(b) C_3H_7OH and C_3H_7ONa
(c) CH_3OH and CH_3ONa
(d) C_4H_9OH and C_4H_9ONa.

60. An alcohol on oxidation is found to give CH_3COOH and CH_3CH_2COOH. The structure of the alcohol is
(a) $CH_3CH_2CH_2OH$ (b) $(CH_3)_2$ $C(OH)CH_2CH_3$
(c) $CH_3(CH_2)_3CH_2OH$ (d) $CH_3CHOHCH_2CH_3$.

61. Oxymercuration-demercuraiton reaction of 1-methylcyclohexene gives
(a) cis-2-Methylcyclohexanol
(b) trans-2-Methylcyclohexanol
(c) 1-Methylcyclohexanol
(d) Mixture of cis- and trans-2-methylcyclohexanol.

62. The order of reactivity of alcohols with sodium metal is
(a) 3° > 2° > 1° (b) 1° > 2° > 3°
(c) 2° > 3° > 1° (d) 3° < 2° > 1°.

63. The intermediate species involved in the acid catalysed dehydration of alcohol is
(a) Free radical (b) Carbocation
(c) Carbanion (d) Carbene.

Chemical Properties of Phenols

64. Which of the following is the strongest acid ?

(a) 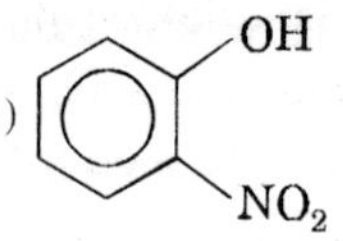

(b) 

(c) 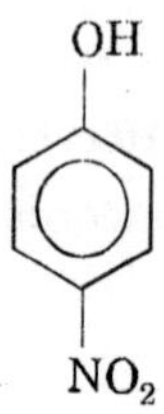

(d) 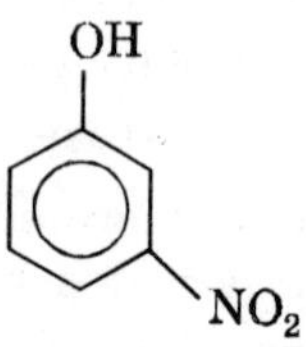

65. The final product obtained by the catalytic hydrogenation of phenol is

(a) Benzene (b) Cyclohexanol
(c) Cyclohexane (d) Benzyl alcohol.

66. Phenol on heating with alc. KOH and chloroform undergoes

(a) Kolbe reaction
(b) Rosenmund reaction
(c) Reimer Tiemann reaction
(d) Cannizzaro reaction.

67. Which of the following compounds is generated by the reaction of phenol and benzenediazonium chloride

(a) Azobenzene
(b) Aniline
(c) Chlorobenzene
(d) *p*-Hydroxyazobenzene.

68. The reaction of CO_2 with sodium phenoxide at 400 K results in

(a) Benzoic acid (b) Sodium benzoate
(c) Sodium salicylate (d) Salicylaldehyde.

69. Which of the following fact(s) explain(s) as to why *p*-nitrophenol is more acidic than phenol ?

I. —I-Effect of nitro group
II. Greater resonance effect of *p*-nitrophenoxy group
III. Steric effect of bulky nitro group

Select the correct answer using the codes given below :

Codes :

(a) I and II (b) I and III
(c) II and III (d) II alone.

70. Phenol on reaction with Br_2 in non-polar aprotric solvent furnishes ...... as the major product.

(a) 2, 4, 6-Tribromophenol (b) *p*-Bromophenol
(c) *o*-Bromophenol (d) *m*-Bromophenol.

71. Electrophilic substitution in phenol takes place at

(a) *ortho* and *para*-positions
(b) *meta*-position
(c) *ortho*-position
(d) *para*-position.

72. Which of the following will not be soluble in sodium bicarbonate ?

(a) 2, 4, 6-Trinitrophenol
(b) Benzoic acid
(c) *o*-Nitrophenol
(d) Benzenesulphonic acid.

73. Which one of the following compounds will be most readily attacked by an electrophile ?

(a) Chlorobenzene (b) Benzene
(c) Phenol (d) Toluene.

74. The following reaction is known as

$$\text{Phenol} \xrightarrow[H^+]{CHCl_3/NaOH} \text{Salicylaldehyde}$$

(a) Gattermann aldehyde synthesis
(b) Duff reaction
(c) Perkin reaction
(d) Reimer-Tiemann reaction.

75. What amount of bromine will be required to convert 2g of phenol into 2, 4, 6-tribromophenol ?

(a) 4.0 (b) 6.0
(c) 10.22 (d) 20.44.

76. The strongest acid among the following aromatic compounds is

(a) *p*-Chlorophenol (b) *p*-Cresol
(c) *p*-Nitrophenol (d) *p*-Aminophenol.

77. Reimer-Tiemann reaction involves

(a) Carbonium ion intermediate
(b) Carbene intermediate
(c) Carbanion intermediate
(d) Free radical intermediate.

78. What is the major product obtained when phenol is treated with chloroform and aqueous alkali ?

(a) 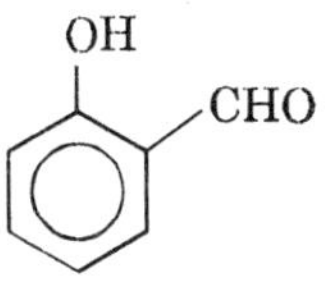

(b) 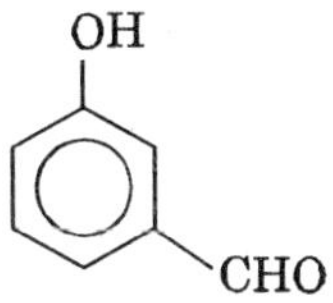

(c) 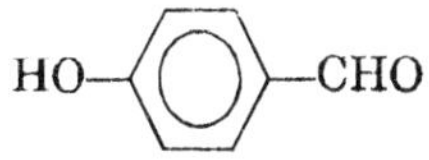

(d) 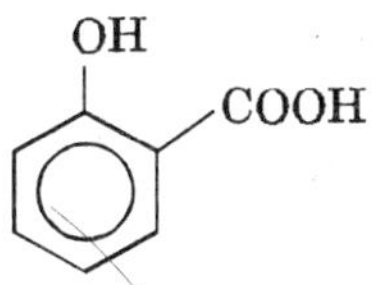

79. Phenol on treatment with conc. HNO_3 gives

(a) o-Nitrophenol (b) p-Nitrophenol

(c) o- and p-Nitrophenol (d) 2, 4, 6-Trinitrophenol.

80. Br_2 dissolved in CS_2 reacts with phenol at 273 K to give ...... as the major product.

(a) o-Bromophenol (b) m-Bromophenol

(c) p-Bromophenol (d) 2, 4, 6-Tribromophenol.

81. Salol is

(a) Acetyl salicylic acid (b) Phenol benzoate

(c) Acetyl salicylate (d) Phenol salicylate.

82. When phenol is heated with NH_3 at high pressure in the presence of $ZnCl_2$ it gives

(a) Benzamide (b) Chlorobenzene

(c) Aniline hydrochloride (d) Aniline.

83. Upon heating phthalic anhydride with phenol in the presence of a little of conc. H_2SO_4 ...... is formed.

(a) Phthalic acid (b) Phenyl acetate

(c) Phenolphthalein (d) Phenyl phthalate.

84. In the following compounds,

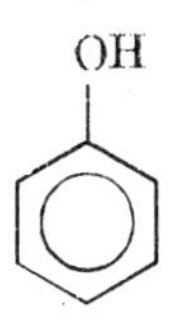

(I)

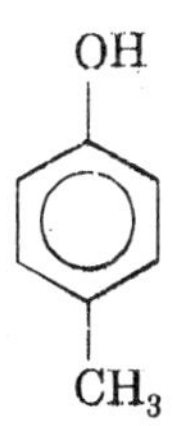

(II)

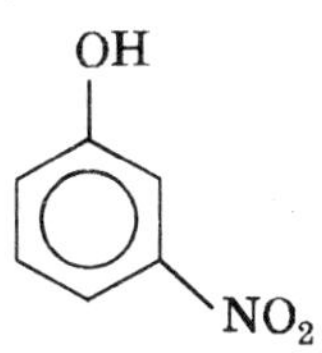

(III)

(IV)

the order of acidity is :

(a) III > IV > I > II (b) I > IV > III > II

(c) II > I > III > IV (d) IV > III > I > II.

85. Identify the product Z in the following sequence of reactions.

$$\text{Phenol} \xrightarrow{\text{NaOH}} X \xrightarrow[\text{4-7 atm, 410 K}]{CO_2} Y \xrightarrow{H_3O^+} Z$$

(a) Aspirin (b) Salicylaldehyde

(c) Benzoic acid (d) Salicylic acid.

86. How many moles of Br_2 are used in converting phenol to 2, 4, 6-tribromophenol ?

(a) 3 (b) 6

(c) 1.5 (d) 4.5.

87. Aspirin is also known as

(a) Phenyl salicylate (b) Acetyl salicylate

(c) Methyl salicylic acid (d) Acetyl salicylic acid.

Methanol and Ethanol

88. Methanol is industrially prepared by

(a) Oxidation of CH_4 by steam at 900°C

(b) Reduction of HCHO using $LiAlH_4$

(c) Reaction HCHO with a solution of NaOH

(d) Reduction of CO using H_2 and $ZnO—Cr_2O_3$.

89. Which of the following is known as '*Wood alcohol*' ?

(a) CH_3OH (b) C_2H_5OH

(c) $CH_3CH_2CH_2OH$ (d) $CH_3-\underset{\mid\atop OH}{CH}-CH_3$.

90. Which of the following is known as '*grain alcohol*' ?

(a) CH_3OH (b) $CH_3-CH_2-CH_2-OH$

(c) CH_3-CH_2-OH (d) $H_3-\underset{\mid\atop OH}{CH}-CH_3$.

91. During the fermentation of mollases the enzyme that converts glucose to alcohol is

(a) Maltase (b) Invertase

(c) Zymase (d) Oxidase.

92. Denatured spirit is a mixture of ethyl alcohol and

(a) Methanol and Toluene

(b) Methanol and pyridine

(c) Methanol and acetic acid

(d) None of the above is correct.

93. Tincture of iodine consists of a solution of

(a) I_2 in ethyl alcohol (b) I_2 in water

(c) I_2 in KI solution (d) CHI_3 in alohol.

94. Rectified spirit is a mixture of

(a) 95% C_2H_5OH and 5% H_2O

(b) 94% C_2H_5OH and 6% H_2O

(c) 95.87% C_2H_5OH and 4.13% H_2O
(d) 94.47% C_2H_5OH and 5.53% H_2O.

95. Power alochol is
(a) An alcohol of 95% purity
(b) Absolute alcohol
(c) Rectified spirit
(d) A mixture of petrol hydrocarbons and ethanol.

96. Proof spirit contains about
(a) 40% alcohol by volume
(b) 60% alcohol by volume
(c) 25% alcohol by volume
(d) 10% alcohol by volume.

97. Fermentation is a/an
(a) Endothermic reaction (b) Exothermic reaction
(c) Reversible reaction (d) None of these.

98. Which of the following enzymes converts starch into maltose ?
(a) Zymase (b) Maltase
(c) Diastase (d) Invertase.

99. Absolute alcohol can be obtained from rectified spirit by
(a) Fractional distillation (b) Azeotropic distillation
(c) Vacuum distillation (d) Steam distillation.

Ethylene Glycol and Glycerol

100. In the reaction sequence

$$CH_2 = CH_2 \xrightarrow[\text{acid}]{\text{Hypochlorous}} X \xrightarrow{Y} \begin{array}{l} CH_2OH \\ | \\ CH_2OH \end{array}$$

X and Y respectively are :

(a) 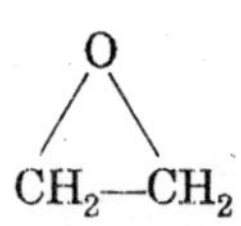 and heat
(b) CH_3CH_2Cl and NaOH
(c) CH_3CH_2OH and H_2SO_4
(d) CH_3ClCH_2OH and $NaHCO_3$.

101. An alcohol produced during manufacture of soap is
(a) Ethanol (b) Ethylene glycol
(c) Glycerol (d) Methanol.

102. In the presence of dry HCl, ethylene glycol reacts with acetaldehyde to yield
(a) An ester (b) A ketal
(c) An acetal (d) An acid.

103. Glycerol is obtained on commercial scale from
(a) Fats (b) Propylene
(c) Both (d) None.

104. Glycerol can be purified by
(a) Steam distillaiton (b) Simple distillation
(c) Vacuum distillation (d) Fractional distillation.

105. Ethylene glycol when heated in the presence of anhydrous $ZnCl_2$ yields
(a) Ethanol (b) Ethylene oxide
(c) Dioxane (d) Diethylene glycol.

106. Ethylene glycol on reaction with alkaline $KMnO_4$ yields
(a) Ethanal (b) Glyoxal
(c) Oxalic acid (d) Acrolein.

107. Glycerol on reaction with excess HI yields
(a) Allyl iodide (b) *n*-Propyl iodide
(c) Propene (d) *iso*-Propyl iodide.

108. On heating ethylene glycol at 775 K, the product obtained is
(a) Oxirane (b) Acetaldehyde
(c) Glyceraldehyde (d) Dioxane.

109. When glycerol is heated with $KHSO_4$, the product formed is
(a) Propane (b) Propene
(c) Allene (d) Acrolein.

110. When ethylene glycol is heated with conc. H_2SO_4, the product formed is
(a) Ethylene oxide (b) Dioxane
(c) Ethanal (d) Diethylene glycol.

111. When glycerol is added to water, which of the following is observed ?
(a) The temperature of water increases
(b) Water evaporates faster
(c) Freezing point of water is lowered
(d) The viscosity of water decreases.

112. Which of the following is *not formed* when glycerol reacts with HI ?
(a) $CH_2 = CH—CH_2I$
(b) $CH_2OH—CHI—CH_2OH$
(c) $CH_3—CH = CH_2$
(d) $CH_3—CHI—CH_3$.

113. Dynamite is prepared by mixing nitroglycerine with
(a) Saw dust and $NH_4\ NO_3$
(b) Cellulose nitrate

(c) Cellulose nitrate and vaseline
(d) T.N.T.

114. Glycerol on heating with oxalic acid at 110°C gives
(a) Ethanol (b) Ethanal
(c) Ether (d) Formic acid.

115. Ethylene reacts with 1% cold alkaline $KMnO_4$ to form
(a) Oxalic acid (b) Ethylene glycol
(c) Ethyl alcohol (d) HCHO.

116. When ethylene glycol is heated with acidified potassium permanganate, the main organic compound obtained is
(a) Oxalic acid (b) Glyoxal
(c) Formic acid (d) Acetaldehyde.

Chemical Tests for Alcohols and Phenols

117. Methanol and ethanol can be distinguished by
(a) Lucas test (b) Iodoform test
(c) Victor Meyer's test (d) All of these.

118. Which of the following compounds reacts fastest with Lucas reagent ?
(a) 1-Butanol (b) 2-Butanol
(c) 2-Methyl-2-propanol (d) 2-Methyl-1-propanol.

119. Which of the following alcohols will give iodoform test ?
(a) Methanol (b) 1-Butanol
(c) 1-Propanol (d) 2-Butanol.

120. Lucas reagent is equimolar solution of
(a) Conc. HNO_3 and conc. H_2SO_4
(b) Conc. HNO_3 and $ZnCl_2$
(c) Conc. HCl and $ZnCl_2$
(d) Conc. of H_2SO_4 and $ZnCl_2$.

121. The reagent, anhydrous zinc chloride and concentrated hydrochloric acid, is used to identify the type of
(a) Nitroalkanes (b) Alkylamines
(c) Alkyl halides (d) Alcohols.

122. For which one of the following pairs of compounds can the members be distinguished by means of iodoform test ?
(a) CH_3OH and $CH_3CH_2CH_2OH$
(b) CH_3CH_2OH and $CH_3CHOHCH_3$
(c) $CH_3CH_2CH_2OH$ and $CH_3CHOHCH_3$
(d) $C_6H_5CH_2OH$ and CH_3OH.

123. Phenol can be distinguished from ethyl alcohol by all reagents except
(a) NaOH (b) $FeCl_3$
(c) Br_2/H_2O (d) Na.

124. The vicinal diols can be detected by the reaction of the compounds with
(a) NaOH and I_2 (b) Conc. H_2SO_4
(c) HIO_4 (d) Acidified $KMnO_4$.

125. The compound which does not react with Lucas reagent at room temperature is
(a) Ethanol
(b) Butan-2-ol
(c) Propan-2-ol
(d) 2-Methylpropan-2-ol.

126. Which of the following gives positive iodoform test ?
(a) CH_3OH (b) $(CH_3)_2CHOH$
(c) $(CH_3)_3COH$ (d) $n—C_3H_7OH$.

Ethers

127. Which of the following is a simple ether ?
(a) $C_2H_5OCH_3$ (b) CH_3OCH_3
(c) $C_6H_5OCH_3$ (d) All are simple ethers.

128. Which of the following compounds will show metamerism ?
(a) $CH_3COC_2H_5$ (b) $C_2H_5SC_2H_5$
(c) CH_3OCH_3 (d) $CH_3OC_2H_5$.

129. Saturated ethers are isomeric with
(a) Aldehydes
(b) Ketones
(c) Alcohols
(d) Carboxylic acids.

130. IUPAC name of ethyl vinyl ether is
(a) Ethoxyethene
(b) 3-Ethoxypropene
(c) 1-Ethoxyprop-2-ene
(d) 2-Ethoxypropene.

131. How many isomeric ethers are possible for the molecular formula $C_4H_{10}O$?
(a) 2 (b) 3
(c) 4 (d) 5.

132. IUPAC name of *tert*-butyl ethyl ether is
(a) 2-Ethoxy-2-methylpropane
(b) 2-Ethyl-2-methoxypropane

(c) 2-Ethoxy-2, 2-dimethylethane

(d) 1-Ethoxy-1, 1-dimethylethane.

133. Ether which is liquid at room temperature is

(a) $C_2H_5OCH_3$ (b) CH_3OCH_3

(c) $C_2H_5OC_2H_5$ (d) None.

134. The interparticle forces in ethers are

(a) H-bonds

(b) Covalent bonds

(c) Dipole-dipole forces

(d) Ion-dipole forces.

135. C—O—C angle would be maximum in

(a) CH_3-O-CH_3

(b) $CH_3-O-C_2H_5$

(c) $C_2H_5-O-C_2H_5$

(d) $(CH_3)_2CH-O-CH(CH_3)_2$.

136. Williamson synthesis involves

(a) Electrophilic substitution

(b) Nucleophilic substitution

(c) Electrophilic addition

(d) Nucleophilic addition.

137. Phenol on treatment with dimethyl sulphate in the presence of NaOH gives

(a) 2-Methylphenol

(b) 4-Methylphenol

(c) Phenetole

(d) Anisole.

138. Ether is obtained from ethyl alcohol

(a) By treating with H_2SO_4 at 413 K

(b) By treating with H_2SO_4 at 474 K

(c) By treating with H_2SO_4 at 383 K

(d) By treating with H_2SO_4 at 273 K.

139. Methyl ethers can be prepared by reaction of alcohols with

(a) Alkyl halides (b) Diazomethane

(c) Grignard reagent (d) None of these.

140. Reaction between sodium ethoxide and bromoethane yields

(a) Methyl ethyl ether (b) Dimethyl ether

(c) Diethyl ether (d) Propane.

141. Which one is formed when sodium phenoxide is heated with ethyl iodide ?

(a) Phenetole (b) Ethyl phenyl alcohol

(c) Phenol (d) None of these.

142. Tetrahydrofuran (T.H.F.) is treated with excess of HBr at 100°C. The product obtained is

(a) 4-Bromo-1-butanol

(b) 1, 4-Butanediol

(c) 1, 4-Dibromobutane

(d) 1, 3-Butadiene.

143. Anisole reacts with Br_2 in the presence of CS_2 as solvent to give

(a) 2, 4, 6-Tribromoanisole

(b) 2-Bromoanisole

(c) 4-Bromoanisole

(d) A mixture of 2-bromoanisole and 4-bromoanisole.

144. 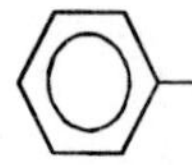 $-OH + CH_2N_2 \xrightarrow{HBF_4}$

The main product in the above reaction is

(a) o-Cresol (b) p-Cresol

(c) Anisole (d) Phenetole.

145. An aromatic ether is not cleaved by HI even at 525 K. The compound is

(a) $C_6H_5OCH_3$ (b) $C_6H_5OC_6H_5$

(c) $C_6H_5OC_3H_7$ (d) Tetrahydrofuran.

146. Which of the following reactions will not result in the formation of anisole ?

(a) Phenol + dimethyl sulphate in presence of a base

(b) Sodium phenoxide is treated with methyl iodide

(c) Reaction of diazomethane with phenol

(d) Reaction of methylmagnesium iodide with phenol.

147. *tert*-Butyl methyl ether on heating with HI of one molar concentration gives

(a) $CH_3OH + (CH_3)_3CI$

(b) $CH_3I + (CH_3)_3COH$

(c) $CH_3I + (CH_3)_3CI$

(d) None of these.

148. On boiling with concentrated hydrobromic acid, ethyl phenyl ether will yield

(a) Phenol and ethyl bromide

(b) Bromobenzene and ethanol

(c) Phenol and ethane

(d) Bromobenzene and ethane.

149. Which of the following reactions does not yield an alkyl halide ?

(a) Diethyl ether + Cl_2

(b) Diethyl ether + HI

(c) Diethyl ether + PCl_5

(d) Diethyl ether + RCOCl.

150. Product C in the reaction,

$$C_2H_5Br \xrightarrow[\text{(aqueous)}]{NaOH} A \xrightarrow{Na} B \xrightarrow{CH_3I} C \text{ will be}$$

(*a*) Propane (*b*) Ethyl iodide
(*c*) Ethane (*d*) Ethyl methyl ether.

151. An organic compound of molecular formula $C_4H_{10}O$ does not react with sodium. With excess of HI, it gives only one type of alkyl halide. The compound is
(*a*) Ethoxyethane (*b*) 2-Methoxypropane
(*c*) 1-Methoxypropane (*d*) 1-Butanol.

152. Which of the following statements is not correct ?
(*a*) Ethers dissolve in conc. H_2SO_4 due to their basic nature and formation of oxonium ion.
(*b*) The peroxides from old samples of ethers can be destroyed by shaking with KI solution.
(*c*) Ethers are heavier than water
(*d*) Ethers are immiscible with water.

Miscellaneous

153. Which will undergo Friedel-Crafts alkylation reaction ?

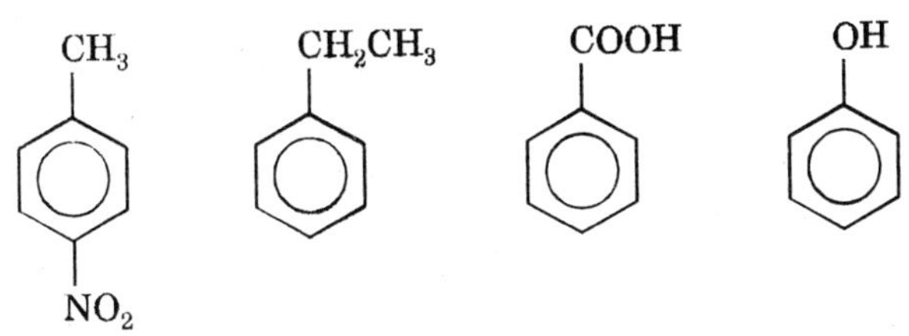

(I) (II) (IIII) (IV)

(*a*) I, II and IV (*b*) I and III
(*c*) II and IV (*d*) I and II.

154. *trans*-Cyclohexane-1, 2-diol can be obtained by the reaction of cyclohexene with
(*a*) $KMnO_4$ (*b*) OsO_4
(*c*) Peroxyformic acid (*d*) SeO_2.

155. Osmium tetroxide is a reagent used for
(*a*) Hydroxylation of acetylenes
(*b*) Hydroxylation of olefins to give *cis*-diols
(*c*) Hydroxylation of olefins to form *trans*-diols
(*d*) Hydroxylation of carbonyl compounds.

156. Ethanol and dimethyl ether form a pair of functional isomers. The boiling point of ethanol is higher than that of dimethyl ether, due to the presence of
(*a*) H-bonding in ethanol
(*b*) H-bonding in dimethyl ether
(*c*) CH_3 group in ethanol
(*d*) CH_3 group in dimethyl ether.

157. Correct increasing order of acidity is as follows :
(*a*) Phenol, H_2O, H_2CO_3
(*b*) H_2O, H_2CO_3, Phenol
(*c*) Phenol, H_2CO_3, H_2O
(*d*) H_2O, Phenol, H_2CO_3.

ANSWERS

1. (*b*)	**2.** (*d*)	**3.** (*a*)	**4.** (*b*)	**5.** (*c*)	**6.** (*a*)	**7.** (*a*)	**8.** (*a*)	**9.** (*d*)
10. (*c*)	**11.** (*c*)	**12.** (*c*)	**13.** (*b*)	**14.** (*b*)	**15.** (*c*)	**16.** (*a*)	**17.** (*c*)	**18.** (*c*)
19. (*b*)	**20.** (*b*)	**21.** (*d*)	**22.** (*b*)	**23.** (*c*)	**24.** (*c*)	**25.** (*c*)	**26.** (*c*)	**27.** (*b*)
28. (*d*)	**29.** (*d*)	**30.** (*d*)	**31.** (*d*)	**32.** (*d*)	**33.** (*d*)	**34.** (*c*)	**35.** (*a*)	**36.** (*d*)
37. (*a*)	**38.** (*b*)	**39.** (*d*)	**40.** (*b*)	**41.** (*c*)	**42.** (*a*)	**43.** (*c*)	**44.** (*b*)	**45.** (*a*)
46. (*b*)	**47.** (*c*)	**48.** (*a*)	**49.** (*c*)	**50.** (*b*)	**51.** (*b*)	**52.** (*a*)	**53.** (*b*)	**54.** (*c*)
55. (*b*)	**56.** (*b*)	**57.** (*b*)	**58.** (*a*)	**59.** (*a*)	**60.** (*d*)	**61.** (*c*)	**62.** (*b*)	**63.** (*b*)
64. (*c*)	**65.** (*b*)	**66.** (*c*)	**67.** (*d*)	**68.** (*c*)	**69.** (*a*)	**70.** (*b*)	**71.** (*a*)	**72.** (*c*)
73. (*c*)	**74.** (*d*)	**75.** (*c*)	**76.** (*c*)	**77.** (*b*)	**78.** (*a*)	**79.** (*d*)	**80.** (*c*)	**81.** (*d*)
82. (*d*)	**83.** (*c*)	**84.** (*d*)	**85.** (*d*)	**86.** (*a*)	**87.** (*d*)	**88.** (*d*)	**89.** (*a*)	**90.** (*c*)
91. (*c*)	**92.** (*b*)	**93.** (*a*)	**94.** (*c*)	**95.** (*d*)	**96.** (*b*)	**97.** (*b*)	**98.** (*c*)	**99.** (*b*)
100. (*d*)	**101.** (*c*)	**102.** (*c*)	**103.** (*c*)	**104.** (*c*)	**105.** (*a*)	**106.** (*c*)	**107.** (*d*)	**108.** (*a*)

109. (*d*)	**110.** (*b*)	**111.** (*c*)	**112.** (*b*)	**113.** (*a*)	**114.** (*d*)	**115.** (*b*)	**116.** (*a*)	**117.** (*b*)
118. (*c*)	**119.** (*d*)	**120.** (*c*)	**121.** (*d*)	**122.** (*c*)	**123.** (*d*)	**124.** (*c*)	**125.** (*a*)	**126.** (*b*)
127. (*b*)	**128.** (*b*)	**129.** (*c*)	**130.** (*a*)	**131.** (*b*)	**132.** (*a*)	**133.** (*c*)	**134.** (*c*)	**135.** (*d*)
136. (*b*)	**137.** (*d*)	**138.** (*a*)	**139.** (*b*)	**140.** (*c*)	**141.** (*a*)	**142.** (*c*)	**143.** (*d*)	**144.** (*c*)
145. (*b*)	**146.** (*d*)	**147.** (*a*)	**148.** (*a*)	**149.** (*a*)	**150.** (*d*)	**151.** (*a*)	**152.** (*c*)	**153.** (*c*)
154. (*c*)	**155.** (*b*)	**156.** (*a*)	**157.** (*d*).					

HINTS/SOLUTIONS

Q. No.	Ans.	Hints & Solutions
1.	**(*b*)**	*n*-Butyl alcohol and *iso*-butyl alcohol.
2.	**(*d*)**	$CH_3CH_2OH(C_2H_6O)$ Ethanol, CH_3OCH_3 (C_2H_6O) Methoxymethane
3.	**(*a*)**	$CH_3-\underset{CH_3}{\overset{CH_3}{\overset{\vert}{\underset{\vert}{C}}}}-CH_2OH$, *neo*-Pentyl alcohol.
4.	**(*b*)**	$CH_3CH_2CH_2OH$, $CH_3\underset{OH}{\underset{\vert}{C}H}CH_3$ Position of —OH is different.
5.	**(*c*)**	*o*, *m* and *p*-cresols, anisole and benzyl alcohol.
6.	**(*a*)**	Picric acid is 2, 4, 6-trinitrophenol.
7.	**(*a*)**	Phenol is also known as carbolic acid.
8.	**(*a*)**	In carbinol system, CH_3OH is known as carbinol. $CH_2=CH-\underset{H}{\overset{H}{\overset{\vert}{\underset{\vert}{C}}}}-OH$ In the given compound one of the H atoms of carbinol has been replaced by vinyl group (CH_2 = CH—).
9.	**(*d*)**	(*i*) $CH_3CH_2CH_2\underset{OH}{\underset{\vert}{C}H}CH_3$ (*ii*) $CH_3CH_2\underset{OH}{\underset{\vert}{C}H}CH_2CH_3$ (*iii*) $CH_3-\underset{CH_3}{\underset{\vert}{C}H}-\underset{OH}{\underset{\vert}{C}H}-CH_3$
10.	**(*c*)**	1, 2, 3-Propanetriol or Propane-1, 2, 3-triol.
11.	**(*c*)**	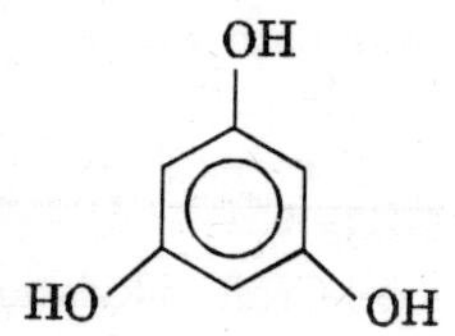 Phloroglucinol
12.	**(*c*)**	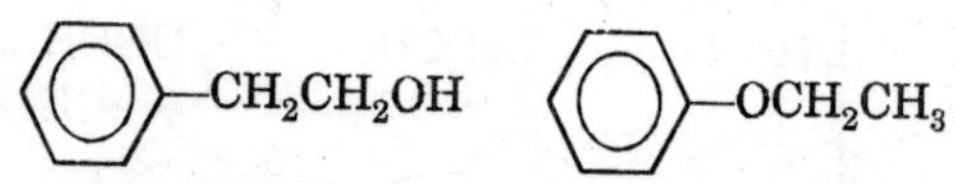 2-Phenylethanol Phenetole
13.	**(*b*)**	$C_6H_5MgBr + CH_3CHO \longrightarrow C_6H_5-\underset{CH_3}{\underset{\vert}{C}H}-OMgBr \xrightarrow{H_2O} C_6H_5-\underset{CH_3}{\underset{\vert}{C}H}-OH$
14.	**(*b*)**	Hydroboration-oxidation of an alkene yields an alcohol formed through anti-Markownikoff addition of H_2O molecule.
15.	**(*c*)**	Reduction of aldehydes and ketones with Zn(Hg)/HCl yields alkanes.
16.	**(*a*)**	$LiAlH_4$ reduces esters to alcohols.
17.	**(*c*)**	Hydroboration-oxidation reaction.
18.	**(*c*)**	Hydroboration-oxidation converts an alkene to alcohol.
19.	**(*b*)**	Esters (except formates) on reaction with Grignard reagents yield tertiary alcohols. $C_2H_5COOCH_3 \xrightarrow[(ii)\ H_2O/H^+]{(i)\ CH_3MgBr} C_2H_5-\underset{CH_3}{\overset{CH_3}{\overset{\vert}{\underset{\vert}{C}}}}-OH$
20.	**(*b*)**	See Hint 19.

Q. No.	Ans.	Hints & Solutions
21.	(*d*)	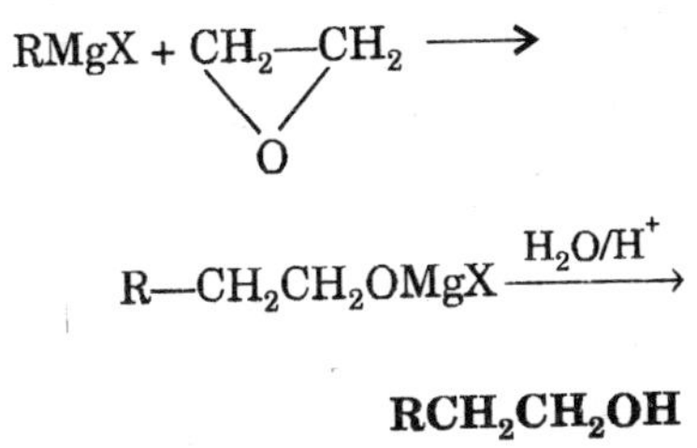
22.	(*b*)	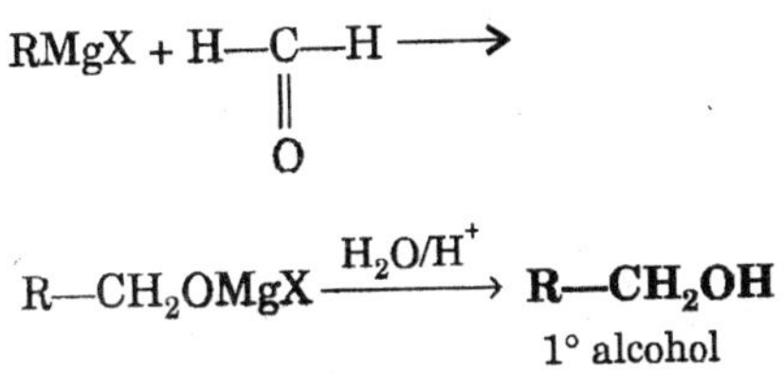
23.	(*c*)	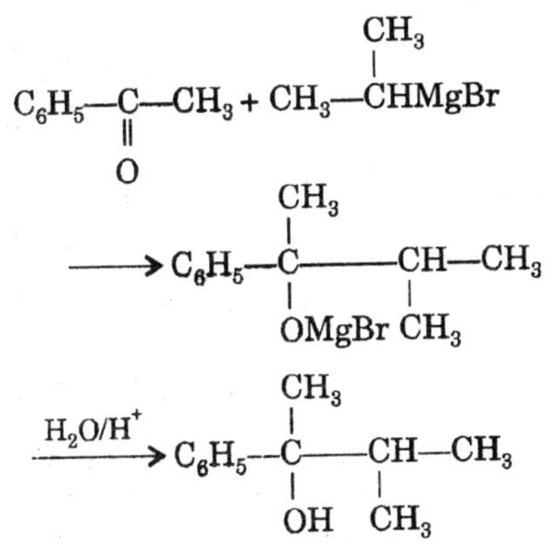
24.	(*c*)	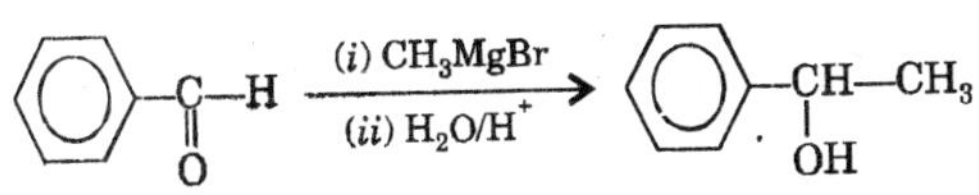
25.	(*c*)	Invertase.
26.	(*c*)	Diastase.
27.	(*b*)	Oxo process. X is propanal and Y is 1-propanol.
28.	(*d*)	Reaction of esters with Grignard reagents yield 3° alcohols.
29.	(*d*)	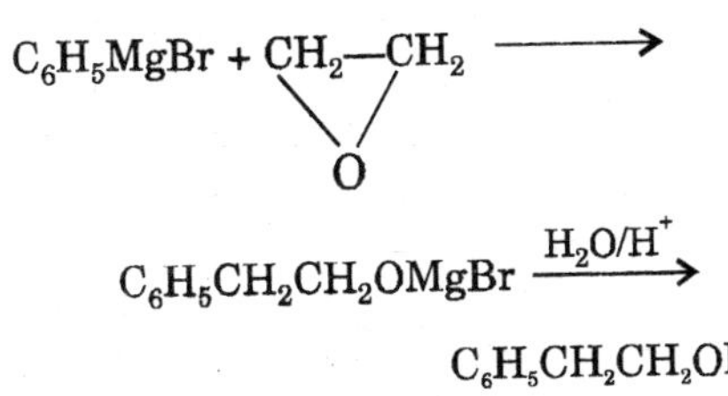
30.	(*d*)	Solubility of alcohols in water decreases with increase in molecular mass.
31.	(*d*)	Solubility of alcohols in water increases with increase in branching in alkyl group.
32.	(*d*)	Alkyl halides are not soluble in water because they cannot form hydrogen bonds with water.
33.	(*d*)	Chlorobenzene cannot form H-bonds with water.
34.	(*c*)	Solubility of alcohols in water decreases with increase in molecular mass of the alcohol.
35.	(*a*)	Among the given compounds, alcohols have the highest boiling point due to intermolecular H-bonding.
36.	(*d*)	Alcohols are weaker acids than water because their conjugate base (alkoxide ion, ^-OR) is less stable than ^-OH due to electron releasing effect of alkyl.
37.	(*a*)	*tert*-butoxide is the conjugate base of *tert*-butyl alcohol which is weakest acid among ethanol, 2-propanol, methanol and *tert*-butyl alcohol.
38.	(*b*)	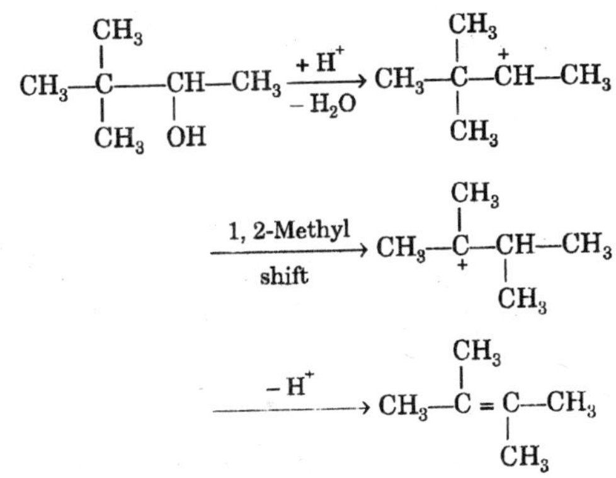
39.	(*d*)	The order of stability of carbocations is 3° > 2° > 1°. The choice (*d*) would give 3° carbocation.
40.	(*b*)	On vigorous oxidation C—C bond undergoes cleavage.
41.	(*c*)	$^-{:}CH_3$ is the conjugate base of the weakest acid (CH_4).
42.	(*a*)	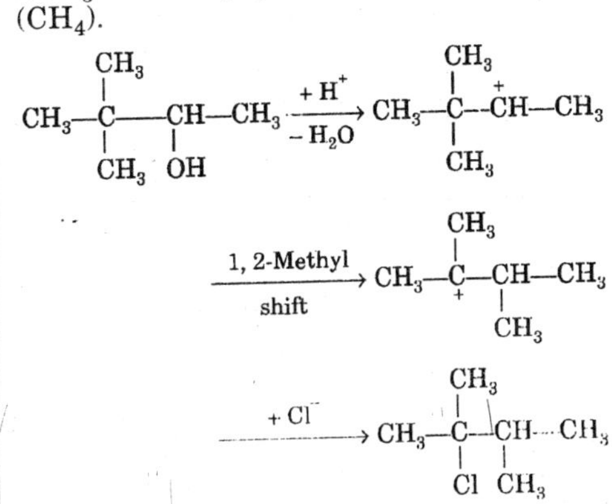
43.	(*c*)	Al_2O_3 is used for dehydration of alcohols.

Q. No.	Ans.	Hints & Solutions
44.	(b)	Dehydration of CH_3OH would yield methylene which is a reactive unstable species $CH_3OH \longrightarrow :CH_2 + H_2O$
45.	(a)	Oxymercuration-demercuration follows Markownikoff's rule. No rearrangement of intermediates takes place.
46.	(b)	Self explanatory question.
47.	(c)	Tertiary alcohols undergo dehydration when heated with reduced copper.
48.	(a)	In this case the side products are gaseous.
49.	(c)	Ease of dehydration is 3° > 2° > 1°.
50.	(b)	$CH_3-CH(OH)-CH_3 \xrightarrow[\text{Heat}]{\text{Conc. } H_2SO_4}$ $CH_3CH=CH_2 \xrightarrow{HBr}$ $CH_3-CH(Br)-CH_3 \xrightarrow[\text{ether}]{Na}$ $CH_3-CH(CH_3)-CH(CH_3)-CH_3$
51.	(b)	Dehydrogenation reaction. Primary alcohols give aldehydes on dehydrogenation. $CH_3CH_2OH \xrightarrow[-H_2]{Cu/575\ K} CH_3CHO$
52.	(a)	Self explanatory question.
53.	(b)	Secondary alcohols on oxidation yield ketones.
54.	(c)	$C_2H_5OH \xrightarrow[\text{Distil under reduced pressure}]{\text{Conc. } H_2SO_4} (C_2H_5)_2SO_4 + H_2O$
55.	(b)	At low temperature (288K) ortho product is major product while at 373K para product is the major product.
56.	(b)	$-NO_2$ group is a very strong electron withdrawing group. It stabilizes the corresponding phenoxide ion by dispersal of negative charge and increases the acid strength.
57.	(b)	Tertiary alcohols do not undergo oxidation under mild conditions.
58.	(a)	Primary alcohols on dehydrogenation yield aldehydes.
59.	(a)	$C_2H_5OH + Na \longrightarrow \mathbf{C_2H_5ONa} + \frac{1}{2}H_2$ $2C_2H_5OH \xrightarrow[\Delta]{\text{Conc. } H_2SO_4} \mathbf{C_2H_5-O-C_2H_5} + H_2O$
60.	(d)	$\mathbf{CH_3-CH(OH)-CH_2CH_3} \xrightarrow{[O]}$ $CH_3-C(=O)-CH_2CH_3 \longrightarrow$ $CH_3-C(=O)-OH + HO-C(=O)-CH_2CH_3$
61.	(c)	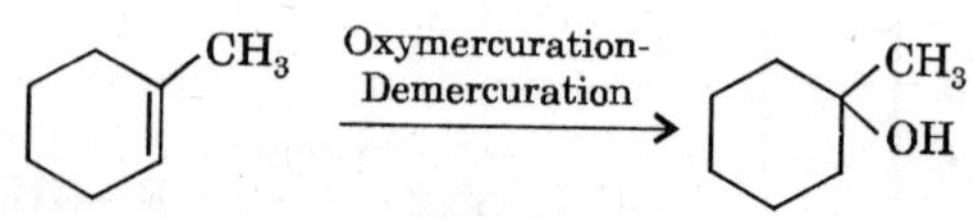 The alcohol formed is the one formed by addition of water molecule according to Markownikoff's rule without rearrangement of the intermediate.
62.	(b)	In all reactions involving cleavage of O—H bond in alcohols, the order of reactivity is **1° > 2° > 3°.**
63.	(b)	$CH_3CH_2OH \overset{H^+}{\rightleftharpoons}$ $CH_3CH_2OH \overset{-H_2O}{\rightleftharpoons} \mathbf{CH_3\overset{+}{C}H_2}$
64.	(c)	$-NO_2$ group is strong electron with drawing group while Cl is a weak electron withdrawing group. Hence, nitrophenols are stronger acids. Among the various isomeric nitrophenols acidic character is in the order para > ortho > meta
65.	(b)	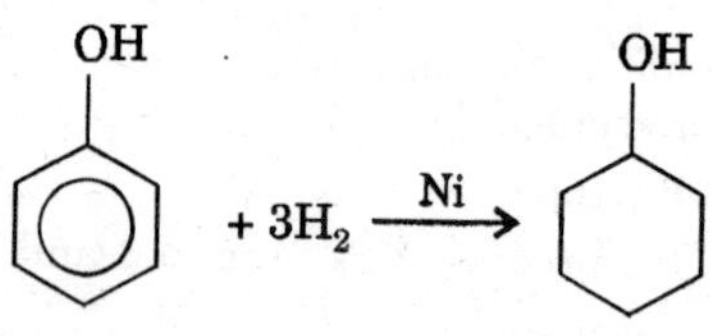
66.	(c)	Reimer Tiemann reaction.
67.	(d)	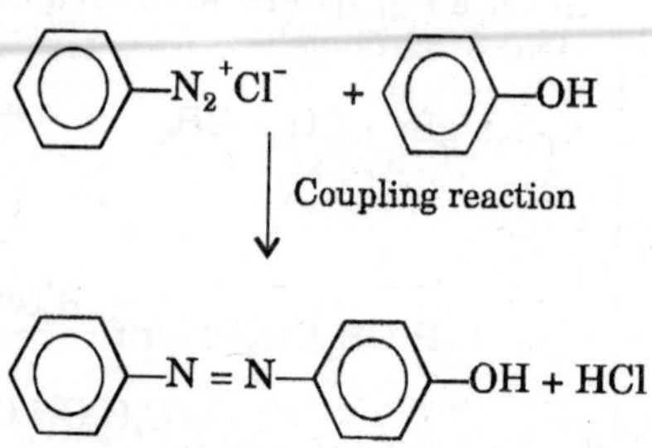
68.	(c)	Kolbe's reaction.
69.	(a)	Self explanatory question.

Q. No.	Ans.	Hints & Solutions
70.	(b)	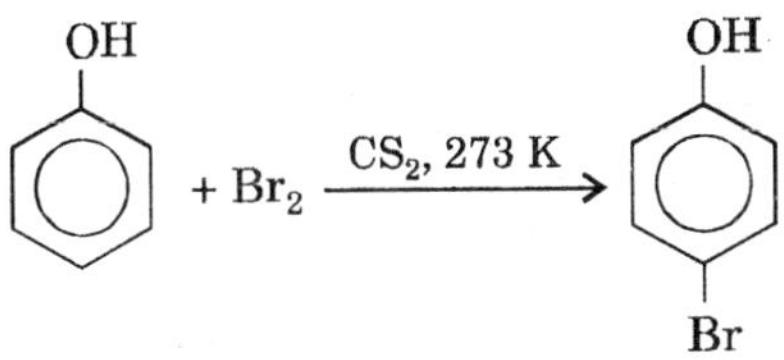 (93%)
71.	(a)	—OH group is ortho-para directing group.
72.	(c)	o-Nitrophenol is the weakest acid among the given compounds.
73.	(c)	—OH group is highly activating group.
74.	(d)	Reimer Tiemann reaction.
75.	(c)	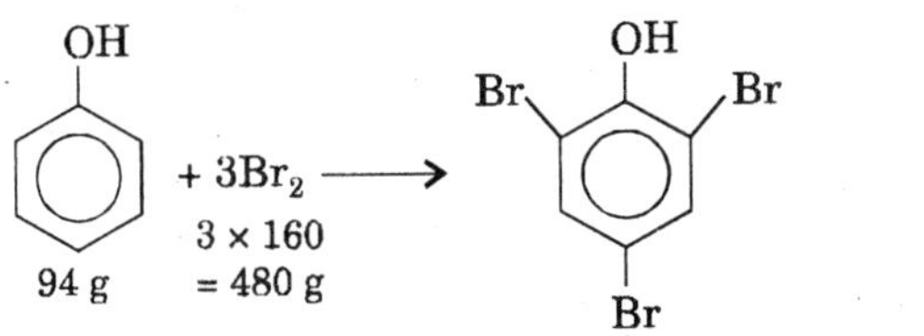 It is clear that amount of Br_2 needed is nearly five times the mass of phenol.
76.	(c)	$—NO_2$ group is strong electron withdrawing group, Cl is mildly electron withdrawing group while $—NH_2$ and $—CH_3$ are electron releasing groups. Stronger the electron withdrawing nature of the group attached to phenolic ring, greater is the acidic character of the phenol.
77.	(b)	$CHCl_3 \xrightarrow[-H_2O]{OH^-} :CCl_3^- \xrightarrow[-Cl^-]{} :CCl_2$ Dichlorocarbene
78.	(a)	Reimer Tiemann reaction.
79.	(d)	When phenol is treated with conc. HNO_3, all the ortho and para positions to the —OH group get substituted by $—NO_2$ groups because —OH group is highly activating and ortho-para directing.
80.	(c)	p-Bromophenol.
81.	(d)	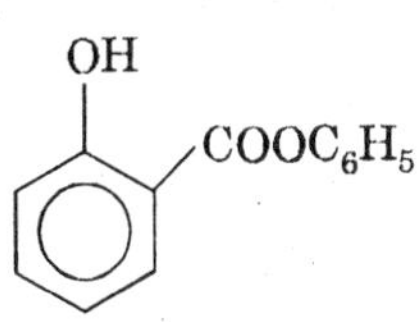 Salol
82.	(d)	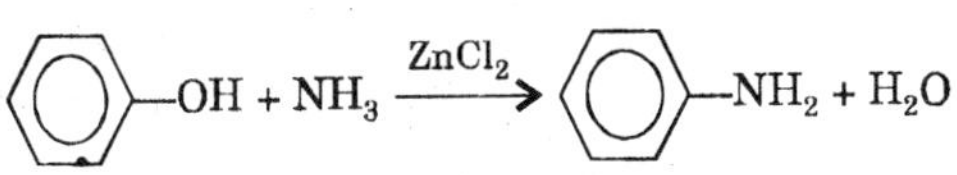
83.	(c)	Phenolphthalein is the product.
84.	(d)	$—NO_2$ is electron withdrawing group. It increases the acid strength of phenol by stabilizing the corresponding phenoxide ion. $—NO_2$ group at p-position stabilizes the phenoxide ion more through resonance. Hence, IV is stronger acid than III which in turn is strong acid than phenol. II is least acidic due to electron releasing effect of methyl group.

Q. No.	Ans.	Hints & Solutions
85.	(d)	Kolbe's reaction. X is sodium phenoxide Y is sodium salicylate Z is salicylic acid.
86.	(a)	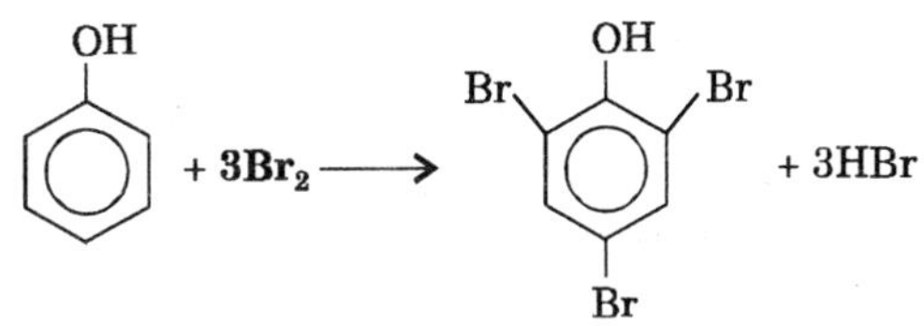
87.	(d)	Aspirin is acetyl salicylic acid. 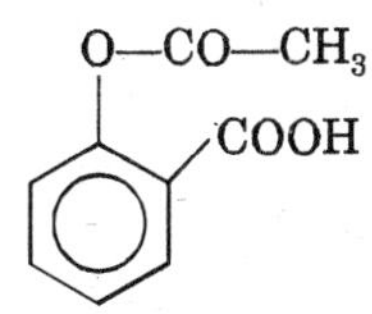 Aspirin
88.	(d)	Self explanatory question.
89.	(a)	**CH_3OH** is known as ***wood alcohol*** because it is one of the products of destructive distillation of wood.
90.	(c)	**CH_3CH_2OH** is known as ***grain alcohol*** because it can be obtained by fermentation of starchy grains.
91.	(c)	Zymase. Factual question.
92.	(b)	Methanol and pyridine.
93.	(a)	Tincture of iodine is solution of iodine in ethyl alcohol.
94.	(c)	Factual question. Ethanol and water in the ratio **95.87 : 4.13** forms a constant boiling azeotrope.
95.	(d)	When ethanol is mixed with petrol, it can be used as fuel in internal combustion engines. This mixture is known as power alcohol.
96.	(b)	Proof spirit contains 57.1% ethanol by volume.
97.	(b)	Fermentation is an exothermic process.
98.	(c)	Diastase (an enzyme) converts starch into maltose.
99.	(b)	Absolute alcohol is prepared by azeotropic distillation of rectified spirit by adding benzene.

Q. No.	Ans.	Hints & Solutions
100.	(d)	$CH_2=CH \xrightarrow{+HOCl} CH_2(Cl)—CH_2(OH)$ $\xrightarrow{NaHCO_3} CH_2(OH)—CH_2(OH)$
101.	(c)	$CH_2OCOR—CHOCOR—CH_2OCOR + 3NaOH \longrightarrow CH_2OH—CHOH—CH_2OH + 3RCOONa$ **Glycerol** Soap
102.	(c)	$CH_3—CHO + CH_2—OH / CH_2—OH \xrightarrow{HCl}$ 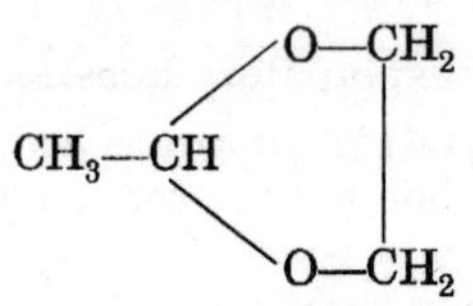 $+ H_2O$ **Acetal**
103.	(c)	Glycerol is obtained on commercial scale from hydrolysis of fats and also from propylene.
104.	(c)	Glycerol has very high boiling point. Hence, it is distilled under reduced pressure.
105.	(a)	Refer to reaction of ethylene glycol.
106.	(c)	$CH_2OH—CH_2OH + 4[O] \longrightarrow COOH—COOH + 2H_2O$ Oxalic acid
107.	(d)	Glycerol on reaction with excess HI gives *iso*-propyl iodide.
108.	(a)	$CH_2(OH)—CH_2(OH) \xrightarrow{775\ K}$ 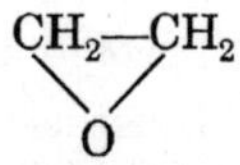 $+ H_2O$ Oxirane
109.	(d)	Glycerol undergoes dehydration in the presence of $KHSO_4$ and gives acrolein ($CH_2=CH—CHO$).
110.	(b)	$2\ CH_2OH—CH_2OH \xrightarrow{Conc.\ H_2SO_4}$ 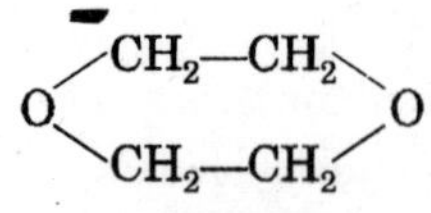 $+ 2H_2O$ Dioxane
111.	(c)	Glycerol is a non volatile solute. It lowers the freezing point of water.
112.	(b)	$HOCH_2—CHI—CH_2OH$ is not formed as any of the products of reactions of HI with glycerol.
113.	(a)	Factual question.
114.	(d)	See reactions of glycerol with oxalic acid.
115.	(b)	$CH_2=CH_2 + H_2O + [O] \longrightarrow CH_2(OH)—CH_2(OH)$
116.	(a)	See Hint 106.
117.	(b)	Ethanol gives positive iodoform test while methanol does not.
118.	(c)	The order of reactivity of various alcohols with Lucas reagent is 3° > 2° > 1°. 2-Methyl-2-propanol is a 3° alcohol.
119.	(d)	$CH_3CH_2CH(OH)CH_3$ contains $CH_3—CH(OH)—$ group.
120.	(c)	Conc. HCl and $ZnCl_2$.
121.	(d)	Lucas reagent is used to identify the type of alcohols, whether 1°, 2° or 3°.
122.	(c)	$CH_3CH_2CH_2OH$ would not give iodoform test while $CH_3CH(OH)CH_3$ would give iodoform test positive.
123.	(d)	Both phenol as well as ethanol react with sodium metal. Hence, it cannot be used for distinguishing between the two.
124.	(c)	Periodic acid test for vicinal diols.
125.	(a)	Primary alcohols react with Lucas reagent only on warming.
126.	(b)	$CH_3—CH(OH)—CH_3$ contains $CH_3—CH(OH)—$ group.
127.	(b)	Ethers (R—O—R) having identical alkyl groups on both the sides of oxygen are known as simple ethers.
128.	(b)	For $C_2H_5—S—C_2H_5$, other isomeric compounds having different distribution of carbon atoms about S atom are possible.

Q. No.	Ans.	Hints & Solutions
		$CH_3—S—CH_2CH_2CH_3$ and $CH_3—S—CH(CH_3)—CH_3$ These are metamers of $C_2H_5—S—C_2H_5$.
129.	(c)	Ethers and alcohols are functional isomers.
130.	(a)	$CH_2 = CH—OC_2H_5$ Ethoxyethene
131.	(b)	$C_2H_5—O—C_2H_5$ $CH_3—O—CH_2CH_2CH_3$ $CH_3—O—CH(CH_3)—CH_3$
132.	(a)	$CH_3—C(CH_3)_2—O—CH_2CH_3$ 2-Ethoxy-2-methylpropane.
133.	(c)	Ethers with four or more carbon atoms are liquid at room temperature.
134.	(c)	Ether molecules are polar due to angular C—O—C bond.
135.	(d)	Bulkier the alkyl groups in the ether, greater is the C—O—C bond angle due to steric factor.
136.	(b)	Nucleophile (alkoxide ion) substitutes halide in alkyl halide. $R—X + :\bar{O}R' \longrightarrow R—O—R' + X^-$
137.	(d)	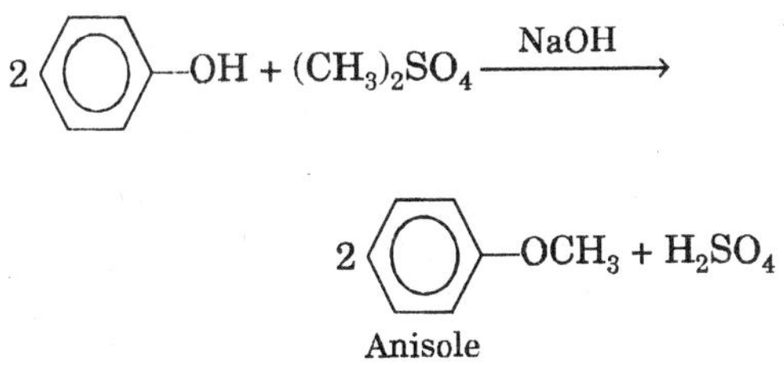
138.	(a)	$2C_2H_5OH \xrightarrow[413\ K]{Conc.\ H_2SO_4} C_2H_5OC_2H_5 + H_2O$
139.	(b)	$R—OH + \mathbf{CH_2N_2} \xrightarrow{HBF_4} R—OCH_3 + N_2$
140.	(c)	$C_2H_5ONa + C_2H_5Br \longrightarrow C_2H_5—O—C_2H_5 + NaBr$ Diethyl ether
141.	(a)	$C_6H_5ONa + C_2H_5I \longrightarrow C_6H_5—OC_2H_5 + NaI$
142.	(c)	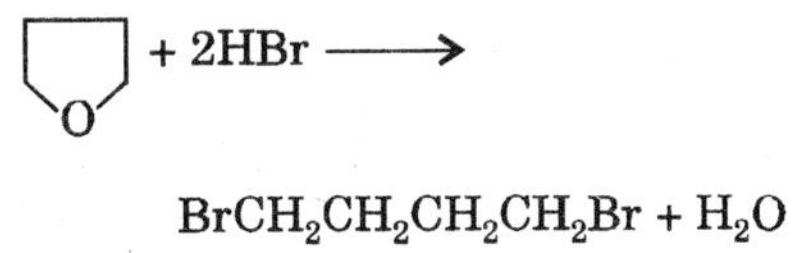
143.	(d)	$—OCH_3$ group is ortho and para directing group.
144.	(c)	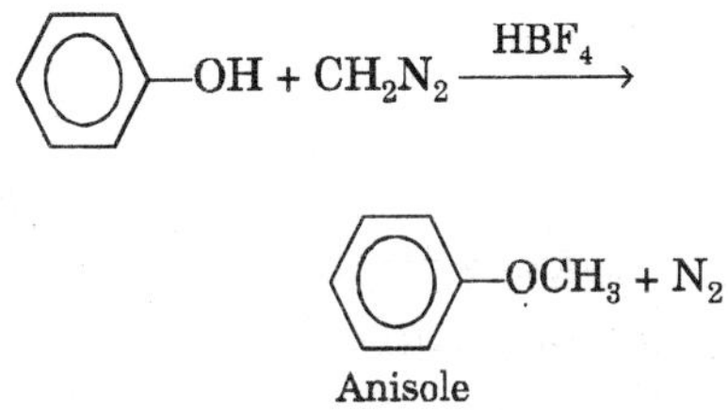
145.	(b)	Phenyl-oxygen bond has double bond character due to resonance and hence is difficult to cleave.
146.	(d)	$C_6H_5OH + CH_3MgI \longrightarrow C_6H_5O—MgI + CH_4$
147.	(a)	$CH_3—C(CH_3)_2—O—CH_3 \xrightarrow{H^+}$ $CH_3—C(CH_3)_2—\overset{\oplus}{O}(H)CH_3 \longrightarrow$ $CH_3—\overset{\oplus}{C}(CH_3)_2 + CH_3OH \xrightarrow{I^-} CH_3—C(CH_3)_2—I$
148.	(a)	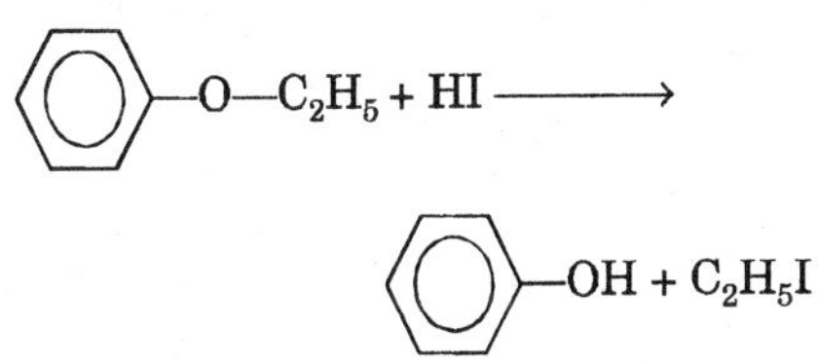 Phenyl-oxygen bond is difficult to cleave due to partial double bond character. Hence, the product of cleavage is phenol and C_2H_5I.

Q. No.	Ans.	Hints & Solutions
149.	(*a*)	$CH_3CH_2—O—CH_2CH_3 + Cl_2 \xrightarrow{h\nu} CH_3—CH(Cl)—O—CH_2—CH_3 + HCl$ H atoms of alkyl are replace by Cl atoms one by one.
150.	(*d*)	$C_2H_5Br \xrightarrow{NaOH} C_2H_5OH \xrightarrow{Na} C_2H_5ONa \xrightarrow{CH_3I} C_2H_5—O—CH_3$ Ethyl methyl ether
151.	(*a*)	$C_2H_5—O—C_2H_5$ (Ethoxyethane) + 2HI (Excess) $\longrightarrow 2C_2H_5I + H_2O$
152.	(*c*)	Ethers are lighter than water.
153.	(*c*)	$—NO_2$ and —COOH groups are electron withdrawing groups and hence deactivate the benzene ring towards electrophilic substitution reactions. $—C_2H_5$ and —OH groups are electron releasing groups. They activate the benzene ring towards electrophilic substitution reactions.
154.	(*c*)	Peroxyformic acid on reaction with cyclohexene followed by hydrolysis gives *trans*-1, 2-diol.
155.	(*b*)	Self explanatory question.
156.	(*a*)	Ethanol has higher boiling point than dimethyl ether due to H-bonding.
157.	(*d*)	Phenol is stronger acid than water but weaker acid than carbonic acid.

4.5. EXAMINER'S VIEWPOINT

Test Your Confidence

1996

1. Which of the following will not be soluble in sodium carbonate solution ?

(a) 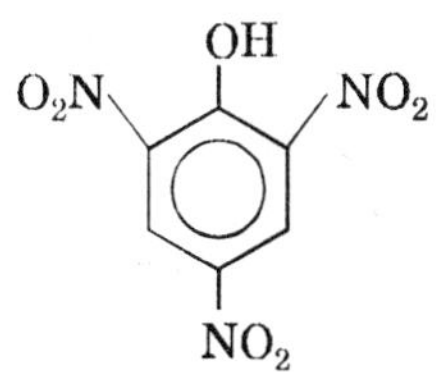

(b) 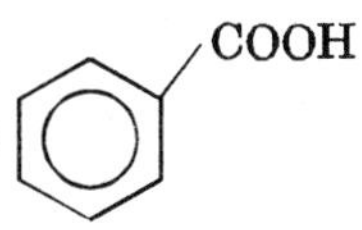

(c) 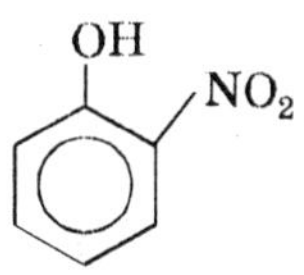

(d) 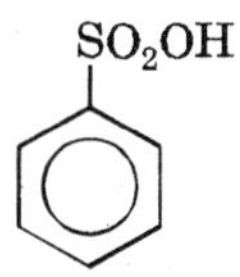

(CEET Kurukshetra)

2. Which of the following reagents will convert propanoic acid to 1-propanol ?
(a) $KMnO_4$ (b) MnO_2
(c) Cr_2O_3 (d) $LiAlH_4$.
(CEET Kurukshetra)

3. The order of reactivity of halogen acids for reaction with C_2H_5OH is
(a) HCl > HBr > HI (b) HI > HBr > HCl
(c) HBr > HI > HCl (d) HBr > HCl > HI.
(MP PMT)

4. The reaction of ethanol with H_2SO_4 does not give
(a) Ethylene
(b) Diethyl ether
(c) Acetylene
(d) Ethyl hydrogen sulphate. **(MP PMT)**

5. Sodium phenoxide reacts with CO_2 at 400 K and 4.7 atm pressure to give
(a) Sodium salicylate (b) Salicylaldehyde
(c) Catechol (d) Benzoic acid. **(MP PMT)**

6. Which of the following does not give yellow ppt. with I_2/NaOH ?
(a) C_2H_5OH (b) CH_3CHO
(c) CH_3COCH_3 (d) HCHO. **(MP PMT)**

7. $R—COOH \longrightarrow R—CH_2OH$. This mode of reduction of an acid to alcohol can be effected by
(a) Zn/HCl
(b) Na-alcohol
(c) Aluminium isopropoxide and isopropyl alcohol
(d) $LiAlH_4$.

8. Identify the final product

$$CH_2 = CH_2 \xrightarrow{O_2,\ Ag} (X) \xrightarrow[\text{Steam}]{473\ K} Y$$

(a) Ethanol (b) Ethanal
(c) Epoxyethane (d) Ethylene glycol.
(CEET)

9. Haloform reaction does not take place with
(a) Acetone (b) 2-Chloropropane
(c) Ethanol (d) Methanol. **(MLNR)**

10. The compound obtained by the reaction of ethene with diborane followed by hydrolysis with alkaline hydrogen peroxide is
(a) Ethanol (b) Propanol
(c) Ethanal (d) Triethyl boride.
(MLNR)

11. An organic compound 'X' on treatment with acidified $K_2Cr_2O_7$ gives a compound 'Y' which reacts with I_2 and sodium carbonate to form triodomethane. The compound 'X' could be
(a) CH_3OH (b) CH_3CHO
(c) $CH_3CH(OH)CH_3$ (d) $CH_3CH_2CH_2OH$.
(CEET)

12. Heating methyl phenyl ether with HI gives
(a) Methanol + Iodobenzene
(b) Methyl iodide + Iodobenzene
(c) Methyl alcohol + Benzyl alcohol
(d) Methyl iodide + Phenol. **(Pb. CET)**

13. Diethyl ether is heated with one mole of HI, which is formed
(a) Ethyl alcohol and ethyl iodide
(b) Ethyl iodide only
(c) Ethyl alcohol only
(d) Ethyl iodide and ethane. **(CPMT)**

14. With oxalic acid, glycerol at 503 K gives
(a) Allyl alcohol (b) Formic acid
(c) Glyceraldehyde (d) None of these.
(Pb. PMT)

15. When benzenediazonium chloride is treated with water, the compound formed is

(a) C_6H_5COOH (b) C_6H_5OH
(c) C_6H_6 (d) $C_6H_5NH_2$. **(CPMT)**

1997

16. A compound is soluble in conc. H_2SO_4. It does not decolourise bromine in CCl_4 but is oxidised by chromic anhydride in aqueous sulphuric acid within two seconds, turning orange solution to blue, green then opaque. The original solution contains
(a) A secondary alcohol (b) An alkene
(c) An ether (d) A primary alcohol. **(KCET)**

17. Which statement is not correct about alcohol ?
(a) Ethyl alcohol is heavier than water
(b) Ethyl alcohol evaporates more quickly than water.
(c) Alcohol with less number of carbon atoms is more soluble in water than alcohol with more number of carbon atoms
(d) Alcohol produces H_2 by reaction with sodium metal. **(AFMC)**

18. 1-Phenylethanol can be prepared by reaction of benzaldehyde with
(a) Methyl iodide and magnesium
(b) Methyl bromide
(c) Methyl bromide and $AlBr_3$
(d) C_2H_5I and Mg. **(CBSE PMT)**

19. In the reaction sequence,

$$CH_2 = CH_2 \xrightarrow[\text{acid}]{\text{hypochlorous}} A \xrightarrow{R} \begin{matrix} CH_2OH \\ | \\ CH_2OH \end{matrix}$$

A and R are respectively
(a) 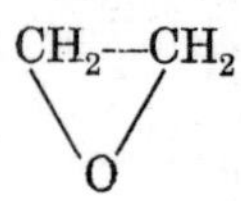 and heat
(b) CH_3CH_2Cl and NaOH
(c) CH_3CH_2OH and H_2SO_4
(d) CH_2Cl . CH_2OH and $NaHCO_3$. **(CBSE PMT)**

20. Which of the following processes is employed to convert alkyl halide into alcohol ?
(a) Addition
(b) Substitution
(c) Dehydrohalogenation
(d) Molecular rearrangement. **(CBSE PMT)**

21. RCH_2CH_2OH can be converted into RCH_2CH_2COOH by the following sequence of steps
(a) PBr_3, KCN, H_3O^+ (b) PBr_3, KCN, H_2/Pt
(c) KCN, H_3O^+ (d) HCN, PBr_3, H_3O^+. **(AIIMS)**

22. In glycerin,
(a) Only one primary —OH group in present
(b) One tertiary —OH group is present
(c) Two secondary —OH groups are present
(d) One secondary —OH group is present. **(AIIMS)**

23. Ethyl alcohol gives ethyl chloride with the help of
(a) $SOCl_2$ (b) NaCl
(c) Cl_2 (d) KCl. **(BHU)**

24. Isopropyl alcohol on oxidation gives
(a) Acetone (b) Ether
(c) Ethylene (d) Acetaldehyde. **(BHU)**

25. Which of the following statements is correct ?
(a) Phenol is less acidic than ethanol
(b) Phenol is more acidic than ethanol
(c) Phenol is more acidic than carbonic acid
(d) Phenol is more acidic than CH_3COOH. **(BHU)**

26. $2R-OH + 2Na \longrightarrow 2RONa + H_2$. In the above reaction the reactivity of different alcohols is
(a) $3° > 2° > 1°$ (b) $3° < 2° < 1°$
(c) $3° < 2° > 1°$ (d) $2° < 1° < 3°$. **(CPMT)**

27. Phenol when heated with concentrated nitric acid gives
(a) Anthranilic acid (b) p-Nitrobenzoic acid
(c) Picric acid (d) Salicyclic acid. **(Pb. CET)**

28. Conversion of ethyl alcohol into acetaldehyde is an example of
(a) Reduction
(b) Molecular rearrangement
(c) Hydrolysis
(d) Oxidation. **(Pb. CET)**

29. The chemical name of Salol is
(a) Acetyl salicylic acid (b) Sodium salicylate
(c) Phenyl salicylate (d) Methyl salicylate.

30. $R—OH + HX \longrightarrow RX + H_2O$
In the above reaction, the reactivity of alcohols is
(a) Tertiary > secondary > primary
(b) Tertiary < secondary < primary
(c) Tertiary > primary > secondary
(d) Secondary > primary > tertiary. **(CPMT)**

31. Zinc powder + ϕ—OH $\longrightarrow$ X. In the above reaction the product X will be
(a) Benzaldehyde (b) Benzene
(c) Anisole (d) Phenyl acetate. **(CPMT)**

32. Which one has highest boiling paint ?
(a) Butan-2-ol (b) Ethane
(c) Butane (d) Pentane. **(CPMT)**

33. The boiling point of glycerol is more than propanal because of
(*a*) Hybridisation (*b*) H-bonding
(*c*) Resonance (*d*) All of these factors.
(CPMT)

34. The heating together of sodium ethoxide and ethyl chloride will give
(*a*) Acetic acid (*b*) Ester
(*c*) Ether (*d*) None of these.
(JIPMER)

35. The alcohol manufactured from water gas is
(*a*) Butanol (*b*) Ethanol
(*c*) Methanol (*d*) Isobutanol. ***(CPMT)***

36. Diethyl ether is prepared by passing vapours of ethyl alcohol over a heated catalyst under high temperature and pressure. The catalyst is
(*a*) SiO_2 (*b*) CuO
(*c*) Al_2O_3 (*d*) Ag_2O. ***(Pb. PMT)***

37. Under basic conditions which are suffers elimination the most :
(*a*) $CH_3CH_2CH_2CH_2Cl$
(*b*) $CH_3CH_2—CH(Cl)—CH_3$
(*c*) $(CH_3)_3CCl$
(*d*) $CH_3—CH(CH_3)—CH_2Cl$ ***(AMU)***

1998

38. What is the product obtained when chlorine reacts with ethyl alcohol in the presence of NaOH ?
(*a*) $CHCl_3$ (*b*) CCl_3CHO
(*c*) C_2H_5Cl (*d*) CH_3Cl. ***(EAM CET)***

39. The reagent used to convert ethanoic acid to ethanol is
(*a*) $LiAlH_4$ (*b*) BH_3
(*c*) PCl_3 (*d*) $K_2Cr_2O_7/H^+$.
(EAM CET)

40. On distilling 2-hydroxy benzoic acid with Zn dust, it gives
(*a*) Phenol (*b*) Benzoic acid
(*c*) Benzaldehyde (*d*) a polymeric compound.
(MP PMT)

41. What amount of bromine will be required to convert 2 g of phenol into 2, 4, 6-Tribromophenol ?
(*a*) 4.00 (*b*) 6.00
(*c*) 10.22 (*d*) 20.44. ***(MP PMT)***

42. Carbolic acid is
(*a*) Aqueous solution of phenol
(*b*) Phenyl benzoate
(*c*) Phenyl acetate
(*d*) Salol. ***(MP PMT)***

43. Which of the following compound is known as *oil of winter green* ?
(*a*) Phenyl benzoate (*b*) Phenyl salicylate
(*c*) Phenyl acetate (*d*) Methyl salicylate.
(MP PMT)

44. Reaction of 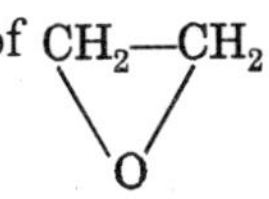 withRMgX followed by hydrolysis will lead to produce

(*a*) RCHOHR (*b*) $RCHOHCH_3$
(*c*) R_2CHCH_2OH (*d*) RCH_2CH_2OH.
(CBSE PMT)

45. Phenol gives ...... colour with neutral $FeCl_3$ solution
(*a*) Violet (*b*) Green
(*c*) Red (*d*) Blue. ***(BITS)***

46. In presence of alkali, C_2H_5OH reacts with Cl_2 to give
(*a*) Ethane (*b*) Ethyl chloride
(*c*) Chloroform (*d*) Pyrene.
(Rajasthan PMT)

47. When phenol is treated with excess bromine water, it gives
(*a*) *m*-Bromophenol
(*b*) *o*- and *p*-Bromophenols
(*c*) 2, 4-Dibromophenol
(*d*) 2, 4, 6-Tribomophenol. ***(Rajasthan PMT)***

48. Glycerol, on heating with oxalic acid at 110°C gives
(*a*) Ethanol (*b*) Methanoic acid
(*c*) Ether (*d*) Acetone.
(Haryana CEET)

49. Which of the following is strongest acid ?

(*a*) 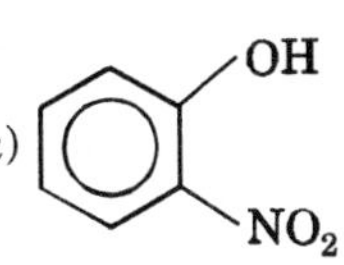

(*b*) 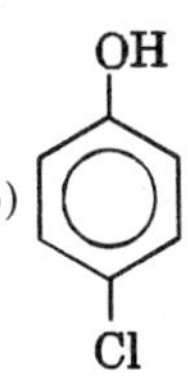

(*c*) 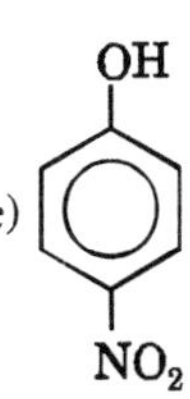

(*d*) 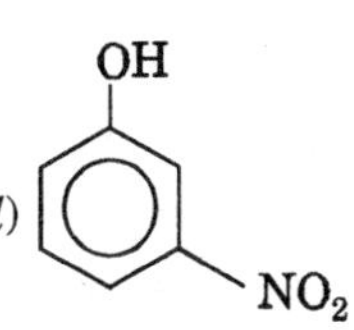

(Haryana CEET)

50. Ethylene reacts with 1% cold alkaline $KMnO_4$ to give
(*a*) Oxalic acid (*b*) Acetone
(*c*) Ethylene glycol (*d*) Formaldehyde.
(Haryana CEET)

51. In the Libermann's nitroso reaction, sequential changes in the colour of phenol occurs as

(*a*) Brown or red ⟶ green ⟶ deep blue

(*b*) Red ⟶ deep blue ⟶ green

(*c*) Red ⟶ green ⟶ white

(*d*) White ⟶ red ⟶ green. ***(AFMC)***

52. An organic compound (A) reacts with sodium metal and forms (B). On heating with conc. H_2SO_4 (A) gives diethyl ether. (A) and (B) are respectively

(*a*) C_2H_5OH and C_2H_5ONa

(*b*) C_3H_7OH and CH_3ONa

(*c*) CH_3OH and CH_3ONa

(*d*) C_4H_9OH and C_4H_9ONa. ***(AFMC)***

53. The reaction of aromatic aryl chloride and phenol in the presence of a base NaOH or pyridine is called

(*a*) Kolbe's reaction

(*b*) Perkin's reaction

(*c*) Sandmeyer's reaction

(*d*) Schotten Baumann reaction. ***(AFMC)***

54. Epichlorohydrin is

(*a*) 3-Chloropropane

(*b*) 3-Chloropropan-1-ol

(*c*) 3-Chloro-1, 2-epoxypropane

(*d*) None of these. ***(AIIMS)***

55. To prepane 3-ethylpentan-3-ol, the reagents needed are

(*a*) $CH_3CH_2MgBr + CH_3COCH_2CH_3$

(*b*) $CH_3MgBr + CH_3CH_2CH_2COCH_2CH_3$

(*c*) $CH_3CH_2MgBr + CH_3CH_2COCH_2CH_3$

(*d*) $CH_3CH_2CH_2MgBr + CH_3COCH_2CH_3$. ***(CET, Chd.)***

56. When wine is put in air, it becomes sour due to

(*a*) Bacteria

(*b*) Oxidation of C_2H_5OH into CH_3COOH

(*c*) Virus

(*d*) Formic acid formation. ***(CET, Chd.)***

57. $(CH_3)_2CHOH \xrightarrow{\text{Mild oxidation}} X \xrightarrow[(ii)\ HOH]{(i)\ CH_3MgBr} Y$

Here Y is

(*a*) *iso*-Butyl alcohol (*b*) *tert*-Butyl alcohol

(*c*) *iso*-Butylene (*d*) *sec*-Butyl alcohol.

(Pb. CET)

58. In order to get Bakelite from phenol which of the following reagents is required ?

(*a*) HCHO (*b*) $CHCl_3$/NaOH

(*c*) CCl_4/NaOH (*d*) HCHO/H^+ or OH^-.

(AIIMS)

59. The product of the following reaction is

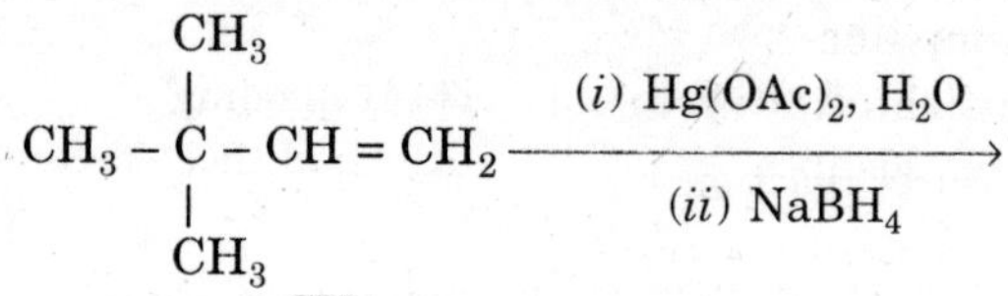

(*a*) 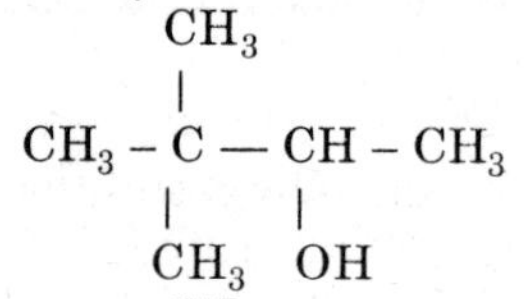

(*b*) 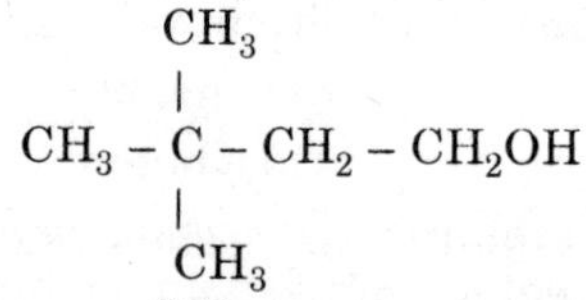

(*c*) 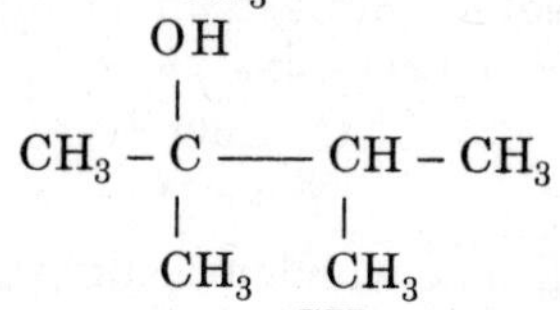

(*d*) 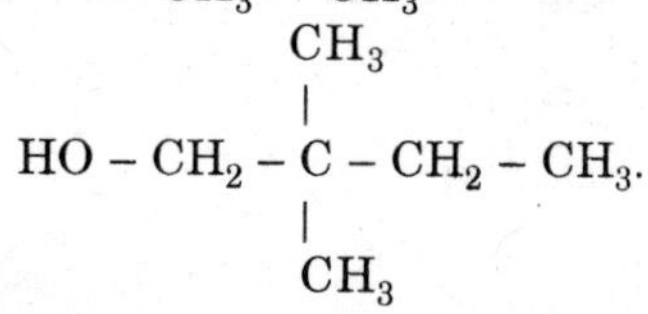

(MP CEE)

1999

60. Which of the following is not true in case of reaction with heated copper at 300° C ?

(*a*) Phenol ⟶ Benzyl alcohol

(*b*) Primary alcohol ⟶ Aldehyde

(*c*) Secondary alcohol ⟶ Ketone

(*d*) Tertiary alcohol ⟶ Olefin. ***(CPMT)***

61. Which of the following will not give iodoform test ?

(*a*) Ethanol (*b*) Ethanal

(*c*) Isopropyl alcohol (*d*) Benzyl alcohol. ***(CPMT)***

62. The correct order of boiling points for primary (1°), secondary (2°) and tertiary (3°) alcohols is

(*a*) 1° > 2° > 3° (*b*) 3° > 2° > 1°

(*c*) 2° > 1° > 3° (*d*) 2° > 3° > 1°. ***(CPMT)***

63. Which of the following has maximum hydrogen bonding ?

(*a*) Ethyl amine (*b*) Ammonia

(*c*) Ethyl alcohol (*d*) Diethyl ether. ***(CPMT)***

64. Which of the following is most acidic ?

(*a*) Phenol (*b*) Benzyl alcohol

(*c*) *m*-Chlorophenol (*d*) Cyclohexanol. ***(CPMT)***

65. Which of the following is the most suitable method for removing the traces of water from ethanol ?

(*a*) Heating with sodium metal

(*b*) Passing dry HCl gas through it

(*c*) Distilling it

(*d*) Reacting with Mg. **(CPMT)**

66. Ethyl chloride is converted into diethyl ether by

(*a*) Wurtz reaction

(*b*) Grignard reaction

(*c*) Perkin's reaction

(*d*) Williamson's synthesis. **(CBSE PMT)**

67. Isopropyl alcohol is obtained by reacting which of the following alkenes with concentrated H_2SO_4 followed by boiling with H_2O ?

(*a*) Ethylene (*b*) Propylene

(*c*) 2-Methylpropene (*d*) Isoprene. **(MP PMT)**

68. The reaction of CH_3MgBr with acetone and hydrolysis of the resulting product gives

(*a*) $CH_3CH_2CH_2OH$ (*b*) $CH_3CH_2CH_2CH_2OH$

(*c*) $(CH_3)_2CHOH$ (*d*) $(CH_3)_3COH$. **(MP PMT)**

69. When diethyl ether is treated with excess of Cl_2 in the presence of sunlight, then the product formed is

(*a*) $CH_3CHCl—O—CH_2CH_3$

(*b*) $CH_3CHCl—O—CHClCH_3$

(*c*) $CCl_3CCl_2—O—CCl_2CCl_3$

(*d*) $CH_3CCl_2—O—CHClCH_3$. **(BHU)**

2000

70. The ionization constant of phenol is higher than that of ethanol because

(*a*) Phenoxide ion is a stronger base than ethoxide ion

(*b*) Phenoxide ion is stabilized through delocalization

(*c*) Phenoxide ion is less stable than ethoxide ion

(*d*) Phenoxide ion is bulkier than ethoxide ion. **(CBSE PMT)**

71. The reagent required to convert propene to 1-propanol is

(*a*) B_2H_6 followed by H_2O_2 / NaOH

(*b*) Conc. H_2SO_4 followed by hydrolysis with boiling water

(*c*) HBr followed by hydrolysis with aqueous KOH

(*d*) $Hg(OCOCH_3)_2$ followed by reduction with $NaBH_4$. **(CBSE PMT)**

72. The order of reactivity of the following towards electrophilic substitution reaction is

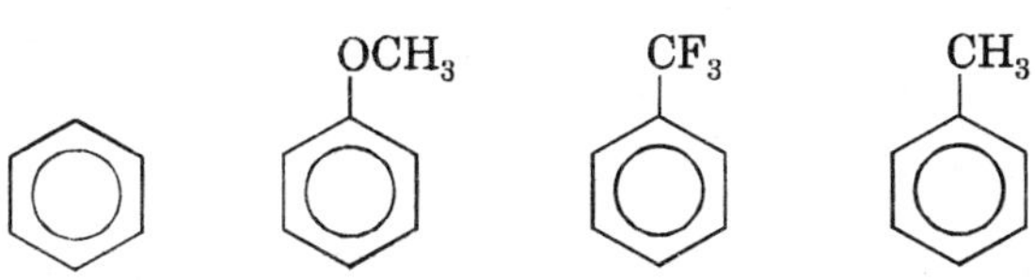

(I) (II) (III) (IV)

(*a*) I > II > III > IV (*b*) II > III > I > IV

(*c*) III > I > IV > II (*d*) II > IV > I > III. **(CBSE PMT)**

73. Which of the following is most acidic ?

(*a*) *p*-Nitrophenol (*b*) *o*-Cresol

(*c*) Phenol (*d*) Anisole. **(Haryana CEET)**

74. Phenol is heated with phthalic anhydride in presence of conc. H_2SO_4. The product gives pink colour with alkali. The product is

(*a*) Phenolphthalein (*b*) Bakelite

(*c*) Salicylic acid (*d*) Fluorescein. **(KCET)**

75. Which of the following is formed as an intermediate in the conversion of ethanol to ethylene in presence of concentrated H_2SO_4 ?

(*a*) $CH_3CH_2OSO_3H$ (*b*) $CH_3CH_2^+$

(*c*) $(CH_3CH_2)_2SO_4$ (*d*) $CH_3CH_2^-$. **(AFMC)**

76. The product of the following reaction is

(*a*) 1-Pentanol (*b*) 2-Pentanol

(*c*) Pentane (*d*) 1, 2-Pentanediol

(*e*) Pentan-2-one. **(Kerala MEE)**

2001

77. The correct acidic order of the following is

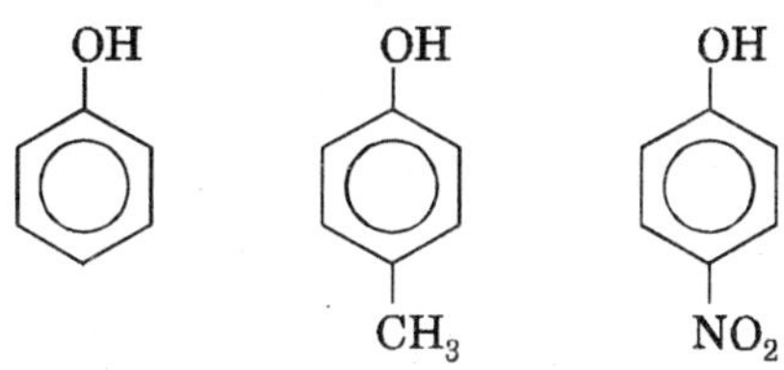

(I) (II) (III)

(*a*) I > II > III (*b*) III > I > III

(*c*) II > III > I (*d*) I > III > II. **(CBSE PMT)**

78. Which of the following is correct ?

(*a*) Reduction of any aldehyde gives secondary alcohol

(*b*) Reaction of vegetable oil with H_2SO_4 gives glycerine

(*c*) Reaction of ethanolic iodine with NaOH gives iodoform

(*d*) Sucrose on reaction with NaCl gives invert sugar. **(CBSE PMT)**

79. Number of metamers represented by molecular formula $C_4H_{10}O$ is

(*a*) 4 (*b*) 3

(*c*) 2 (*d*) 1. **(Tamil Nadu CET)**

80. The reagent used for the preparation of higher ethers from halogenated ethers is

(a) Conc. H_2SO_4 (b) Sodium alkoxide
(c) Dry silver oxide (d) Grignard reagent.
(Tamil Nadu CET)

81. Benzylamine reacts with nitrous acid to form
(a) Azobenzene (b) Benzene
(c) Benzyl alcohol (d) Phenol.
(Tamil Nadu CET)

82. The ether that undergoes electrophilic substitution reaction is
(a) $CH_3OC_2H_5$ (b) $C_6H_5OCH_3$
(c) CH_3OCH_3 (d) $C_2H_5OC_2H_5$. ***(JIPMER)***

83. A compound of the formula $C_4H_{10}O$ reacts with sodium, undergoes oxidation to give a carbonyl compound which does not reduce Tollen's reagent, the original compound is
(a) Diethyl ether (b) *n*-Butyl alcohol
(c) Isobutyl alcohol (d) *sec*-Butyl alcohol.
(JIPMER)

84. Ethanol when reacted with PCl_5 gives A, $POCl_3$ and HCl. A reacts with silver nitrite to form B (major product) and AgCl. A and B respectively are :
(a) C_2H_5Cl and $C_2H_5OC_2H_5$
(b) C_2H_6 and $C_2H_5OC_2H_5$
(c) C_2H_5Cl and $C_2H_5NO_2$
(d) C_2H_6 and $C_2H_5NO_2$. ***(EAMCET Engg.)***

85. 3 moles of ethanol react with one mole of phosphorus tribromide to form 3 moles of bromoethane and one mole of X. Which of the following is X ?
(a) H_3PO_4 (b) H_2PO_2
(c) HPO_3 (d) H_3PO_3.
(EAMCET Engg.)

86. Absolute alcohol (100% alcohol) is prepared by distilling rectified spirit over
(a) Na (b) $CaCl_2$
(c) Mg (d) $Mg(OC_2H_5)_2$.
(EAMCET Engg.)

87. Phenol is heated with $CHCl_3$ and alcoholic KOH when salicylaldehyde is produced. This reaction is known as
(a) Rosenmund's reaction
(b) Reimer-Tiemann reaction
(c) Friedel-Crafts reaction
(d) Sommelet reaction. ***(MP PMT)***

88. Write the IUPAC name of

$$CH_3-\underset{OH}{\overset{H}{\underset{|}{\overset{|}{C}}}}-CH_2-CH_2-CH_2-\underset{Br}{\overset{Br}{\underset{|}{\overset{|}{C}}}}-CH_3$$

(a) 6, 6-Dibromoheptan-2-ol
(b) 2, 2-Dibromoheptan-6-ol
(c) 6, 6-Dibromoheptan-1-ol
(d) None of these. ***(DCE)***

2002

89. When phenol is treated with $CHCl_3$ and NaOH, the product formed is
(a) Benzaldehyde (b) Salicylaldehyde
(c) Salicyclic acid (d) Benzoic acid.
(CBSE PMT)

90. Order of esterification of alcohols is
(a) T > S > P (b) S > T > P
(c) P > S > T (d) None of these.
(Rajasthan PMT)

91. When ether is exposed in air for sometime, an explosive substance produced is
(a) Peroxide (b) TNT
(c) Oxide (d) Super oxide.
(Rajasthan PMT)

92. An organic compound A reacts with methyl magnesium iodide to form an addition product which on hydrolysis forms the compound B. Compound B gives blue colour salt in Victor Meyer's test. The compounds A and B are respectively
(a) Acetaldehyde, tertiary butyl alcohol
(b) Acetaldehyde, ethyl alcohol
(c) Acetaldehyde, isopropyl alcohol
(d) Acetone, isopropyl alcohol. ***(Tamil Nadu)***

93. The compound 'A' when treated with methyl alcohol and few drops of H_2SO_4 gave smell of winter green. The compound 'A' is
(a) Succinic acid (b) Salicylic acid
(c) Tartaric acid (d) Oxalic acid. ***(MP PMT)***

94. In the following reaction,

$$C_2H_5OC_2H_5 + 4\,H \xrightarrow{\text{Red P + HI}} 2\,X + H_2O$$

(a) Ethane (b) Ethylene
(c) Butane (d) Propane. ***(MP PMT)***

95. Lucas test is used for the determination of
(a) Alcohols (b) Phenols
(c) Alkyl halides (d) Aldehyde. ***(AIIMS)***

96. IUPAC name of the compound

$$CH_3-\underset{OH}{\underset{|}{CH}}-CH_2-\underset{CH_3}{\underset{|}{CH}}-CH_3 \text{ is ......}$$

(a) 4-Methylpentan-2-ol
(b) 2-Methylpentanol-4
(c) 4, 4-Dimethyl-butan-2-ol
(d) 4-Methylpentan-2-ol. ***(Orissa JEE)***

97. In the reaction

$$HCHO + CH_3MgI \longrightarrow A \xrightarrow{H_2O} B + Mg(OH)I$$

What are A and B ?

(*a*) CH_3OMgI and CH_3–OH
(*b*) CH_3CH_2OMgI and C_2H_5–O–C_2H_5
(*c*) CH_3CH_2OMgI and CH_3–CH_2–OH
(*d*) CH_3–CH_2–I and CH_3–CH_2–OH. **(*Orissa JEE*)**

98. Acetyl salicylic acid is used as ....

(*a*) Antioxidant (*b*) Analgesic drug
(*c*) Antibiotic drug (*d*) Anesthetics.
(*Orissa JEE*)

2003

99. An'ether is more volatile than an alcohol having the same molecular formula. This is due to

(*a*) Inter-molecular hydrogen bonding in alcohols
(*b*) Dipolar character of ethers
(*c*) Alcohols having resonance structures
(*d*) Inter-molecular hydrogen bonding in ether.
(*AIEE:*)

100. During dehydration of alcohols to alkenes by heating with conc. H_2SO_4, the initiation step is

(*a*) Formation of an ester
(*b*) Protonation of alcohol molecule
(*c*) Formation of carbocation
(*d*) Elimination of water. **(*AIEEE*)**

101. The most reactive nucleophile among the following is

(*a*) CH_3O^- (*b*) $C_6H_5O^-$
(*c*) $(CH_3)_2CHO^-$ (*d*) $(CH_3)_3CO^-$. **(*AIIMS*)**

102. Propan-1-ol can be prepared from propene by

(*a*) H_2O/H_2SO_4
(*b*) $Hg(OAc)_2/H_2O$ followed by $NaBH_4$
(*c*) B_2H_6 followed by H_2O_2
(*d*) CH_3CO_2H/H_2SO_4. **(*AIIMS*)**

2004

103. Among the following compounds which can be dehydrated very easily is

(*a*) $CH_3CH_2CH_2CH_2CH_2OH$
(*b*) $CH_3CH_2CH(CH_3)CH_2CH_2OH$ (i.e. $CH_3CH_2\underset{\large CH_3}{\underset{|}{C}H}CH_2CH_2OH$)
(*c*) $CH_3—CH_2—\overset{\large CH_3}{\overset{|}{\underset{\large OH}{\underset{|}{C}}}}—CH_2—CH_3$

(*d*) $CH_3CH_2CH_2\overset{\large OH}{\overset{|}{C}H}CH_3$. **(*AIEEE*)**

104. The IUPAC name of the compound

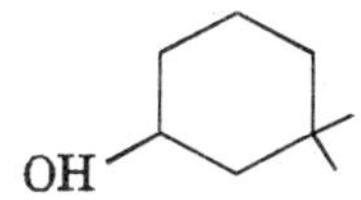

is

(*a*) 3, 3-dimethyl-1-hydroxy cyclohexane
(*b*) 1, 1-dimethyl-3-cyclohexanol
(*c*) 3, 3-dimethyl-1-cyclohexanol
(*d*) 1, 1-dimethyl-3-hydroxy cyclohexane. **(*AIEEE*)**

105. Which one of the following can be oxidised to the corresponding carbonyl compound ?

(*a*) Ortho-nitrophenol
(*b*) Phenol
(*c*) 2-methyl-2-hydroxypropane
(*d*) 2-hydroxypropane. **(*CBSE PMT*)**

106. Which of the following will ***not*** from a yellow precipitate on heating with an alkaline solution of iodine ?

(*a*) $CH_3CH_2CH(OH)CH_3$ (*b*) CH_3OH
(*c*) CH_3CH_2OH (*d*) $CH_3CH(OH)CH_3$.
(*CBSE PMT*)

107. The most suitable reagent for the conversion of $RCH_2OH \rightarrow RCHO$ is :

(*a*) $KMnO_4$
(*b*) $K_2Cr_2O_7$
(*c*) CrO_3
(*d*) PCC (Pyridinium chloro chromate). **(*AIIMS*)**

2005

108. Which one of the following compounds is most acidic ?

(*a*) Cl—CH_2—CH_2—OH (*b*)

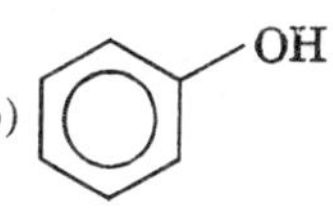

(*c*)

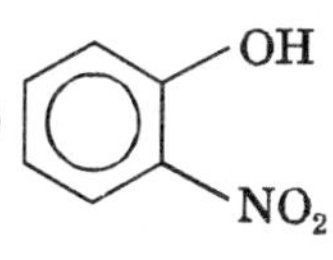

(*d*)

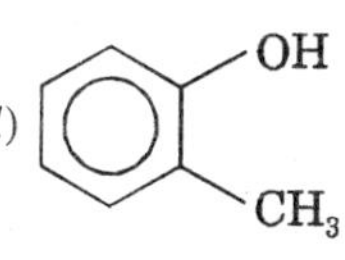

(*CBSE PMT*)

109. Acid catalyzed hydration of alkenes except ethene leads to the formation of

(*a*) Mixture of secondary and tertiary alcohols
(*b*) Mixture of primary and secondary alcohols
(*c*) Secondary or tertiary alcohol
(*d*) Primary alcohol. **(*AIEEE*)**

110. Salicylic acid is prepared from phenol by

(*a*) Reimer Tiemann reaction
(*b*) Kolbe's reaction

(c) Kolbe electrolysis
(d) None of these. **(AFMC)**

111. Lucas test is done for
(a) Alkyl halides (b) Alcohols
(c) Acids (d) Aldehydes. **(AFMC)**

112. Which of the following reactions is correctly represented ?

(a)
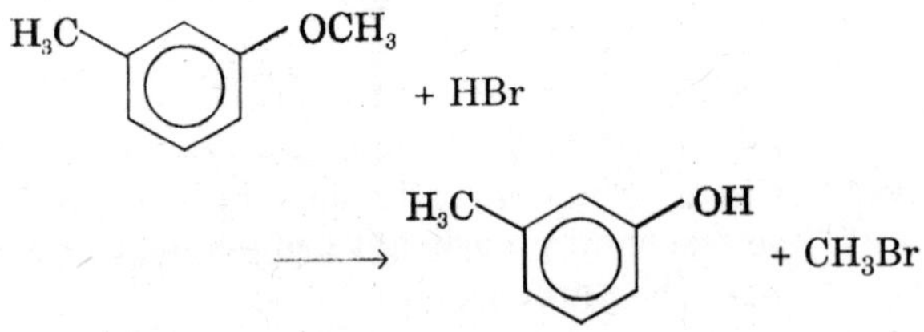

(b)
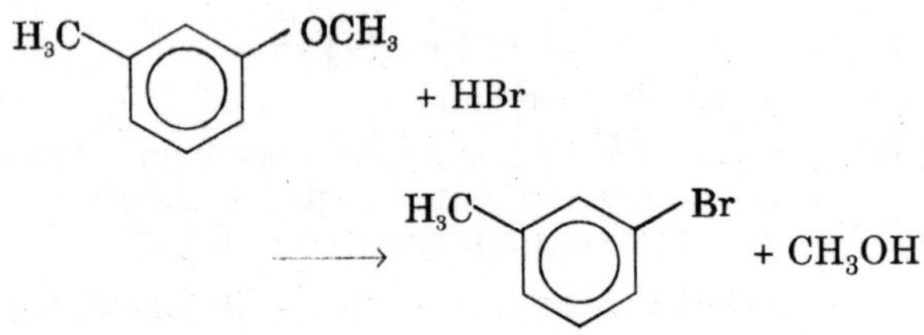

(c)
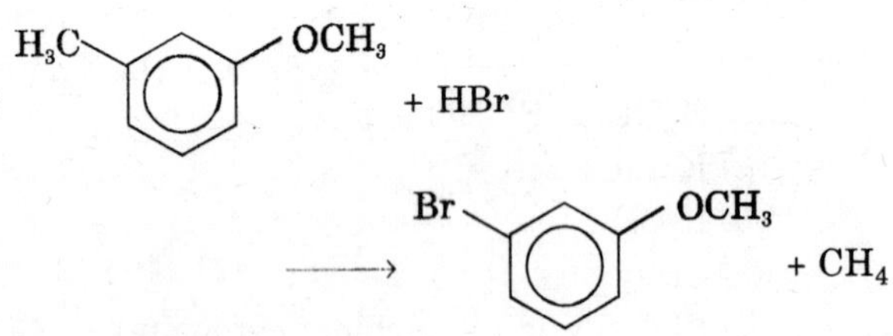

(d)
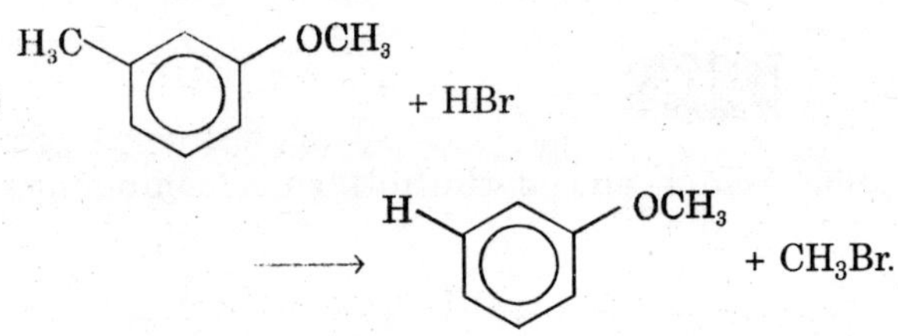

(Orissa JEE)

113. Gycerol is an alcohol which can be classified as
(a) Monohydric (b) Dihydric
(c) Tryhydric (d) Hexahydric.
(J & K CET)

114. Methylphenylether can be obtained by reacting
(a) Phenolate ions with methyl iodide
(b) Methanol and phenol
(c) Bromobenzene and methyl bromide
(d) Methoxide ions and bromobenzene. **(J & K CET)**

115. Reaction of phenol with chloroform/sodium hydroxide to give o-hydroxybenzaldehyde involves the formation of
(a) Dichlorocarbene (b) Trichlorocarbene
(c) Chlorine atoms (d) Chlorone molecules.
(J & K CET)

116.
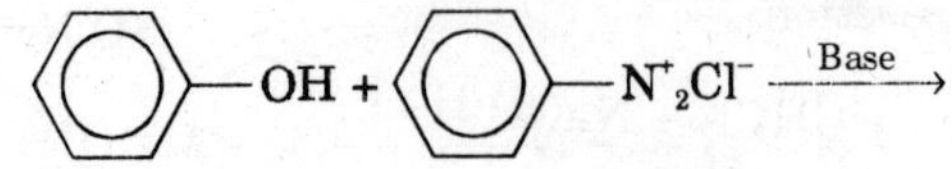

(a)
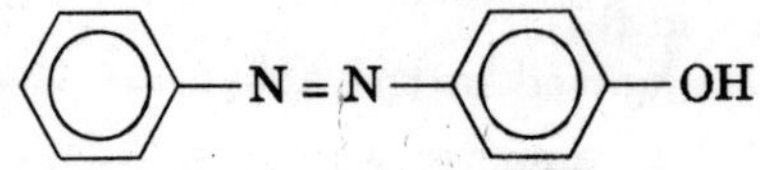

(b)
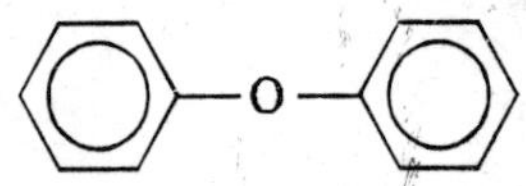

(c)
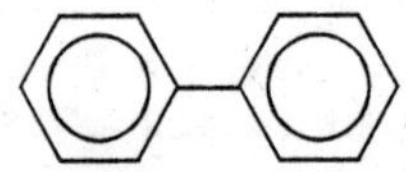

(d)
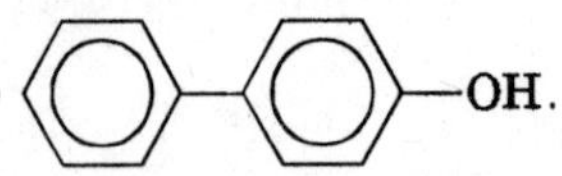
(DPMT)

117. HCHO reacts with CH_3MgI to give
(a) CH_3CH_2OH (b) CH_3OH
(c) CH_3CH_2I (d) CH_3I. **(BHU)**

118. In the reaction,

$$\underset{\text{(Vapour)}}{C_2H_5OH} \xrightarrow[300°C]{Cu} X$$

The molecular formula of X is :
(a) C_4H_6O (b) $C_4H_{10}O$
(c) C_2H_4O (d) C_2H_6. **(EAMCET Engg.)**

119. A compound X undergoes reduction with $LiAlH_4$ to yield Y. When vapours of Y are passed over freshly reduced copper at 300°C. X is formed. What is Y ?
(a) CH_3COCH_3 (b) CH_3CHO
(c) CH_3CH_2OH (d) $CH_3—O—CH_3$.
(EAMCET Medical)

2006

120. The major organic product in the reaction
$CH_3—O—CH(CH_3)_2 + HI \longrightarrow$ Product is
(a) $ICH_2OCH(CH_3)_2$ (b) $CH_3OC(CH_3)_2$ with I attached to C (i.e. $CH_3O\overset{|}{C}(CH_3)_2$—I)
(c) $CH_3I + (CH_3)_2\ CHOH$ (d) $CH_3OH + (CH_3)_2\ CHI$.
(CBSE PMT Screening)

121. Ethylene oxide when treated with Gringard reagent yields
(a) Tertiary alcohol (b) Cyclopropyl alcohol
(c) Primary alcohol (d) Secondary alcohol.
(CBSE PMT Screening)

122. The general molecular formula which represents the homologous series of alkanole is
(a) $C_nH_{2n}O$ (b) $C_nH_{2n+1}O$
(c) $C_nH_{2n+2}O$ (d) $C_nH_{2n}O_2$.
(CBSE PMT Screening)

123. 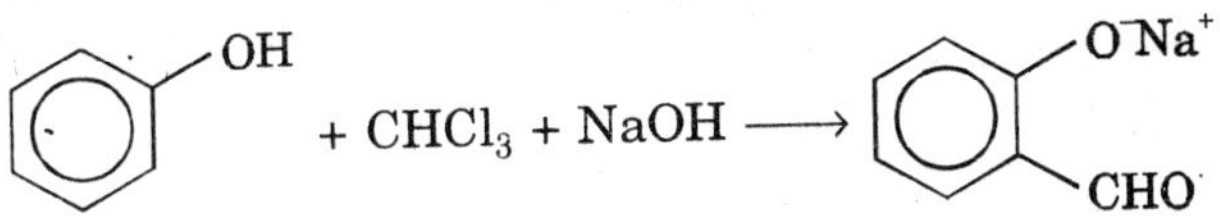

The electrophile involved in the above reaction is

(a) dichloromethyl cation ($\overset{\oplus}{C}HCl_2$)

(b) dichlorocarbene ($:CCl_2$)

(c) trichloromethyl anion ($\overset{\ominus}{C}Cl_3$). **(AIEEE)**

124. The structure of the compound that gives a tribromo derivative on treatment with bromine water is

(a) 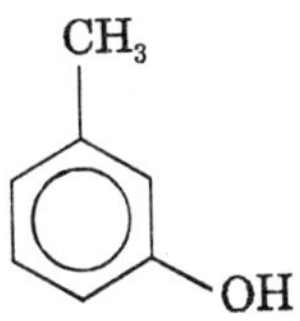

(b) 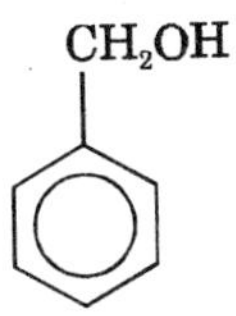

(c) 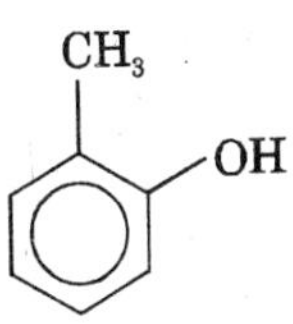

(d)  **(AIEEE)**

125. Among the following the one that gives positive iodoform test upon reaction with I_2 and NaOH is

(a) $CH_3CH_2CH(OH)CH_2CH_3$

(b) $C_6H_5CH_2CH_2OH$

(c) $CH_3-\underset{}{CH}(CH_3)CH_2OH$

(d) $PhCHOHCH_3$. **(AIEEE)**

126. The major product formed in the following reaction

$CH_3CH(Cl)CH_2-CH_2OH \xrightarrow{\text{aq KOH}}$ is

(a) $CH_3CH = CH-CH_2OH$

(b) $CH_2 = CH-CH_2-CH_2OH$

(c) $CH_3-CH-CH_2$ (ring: O—CH_2)

(d) $CH_3-\underset{}{C}OH-CH_2-CH_2OH$. **(AIIMS)**

127. Isopropylbenzene on air oxidation in the presence of dilute acid gives

(a) C_6H_5COOH (b) $C_6H_5COCH_3$

(c) C_6H_5CHO (d) C_6H_5OH. **(AIIMS)**

128. The major product obtained on the monobromination (with $Br_2/FeBr_3$) of the following compound (A) is

(A) = 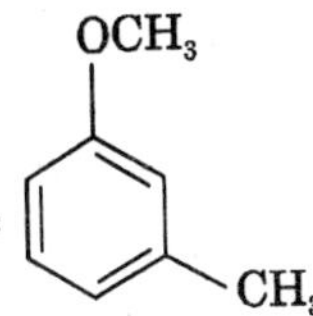

(a) 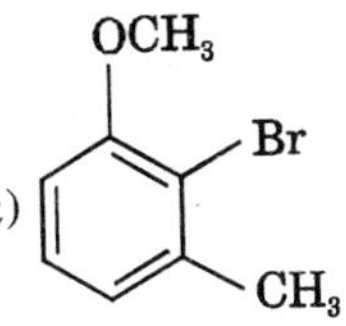

(b) 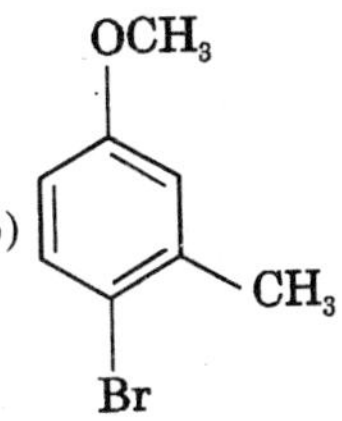

(c) 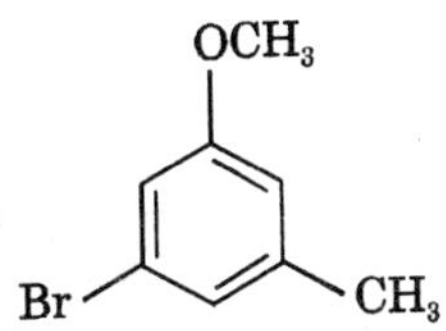

(d) 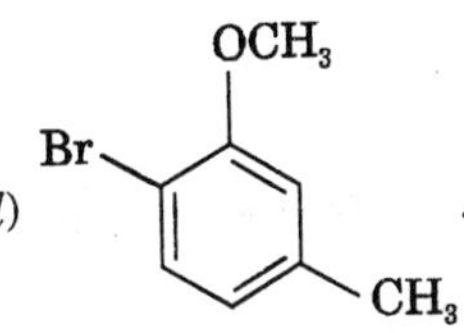 **(AIIMS)**

2007

129. In the reaction :

$CH_3-CH(CH_3)-CH_2-O-CH_2-CH_3 + HI \xrightarrow{\text{Heated}} .....$

Which of the following compounds will be formed ?

(a) $CH_3-CH(CH_3)-CH_2OH + CH_3-CH_2-I$

(b) $CH_3-CH(CH_3)-CH_2-I + CH_3CH_2OH$

(c) $CH_3-CH(CH_3)-CH_3 + CH_3CH_2OH$

(d) $CH_3-CH(CH_3)-CH_2OH + CH_3CH_3$ **(CBSE PMT)**

130. In the following sequence of reactions,

$CH_3CH_2OH \xrightarrow{P + I_2} A \xrightarrow[\text{ether}]{Mg} B \xrightarrow{HCHO} C \xrightarrow{H_2O} D$

the compound 'D' is

(a) n-propyl alcohol (b) Propanal

(c) Butanal (d) n-butyl alcohol. **(AIEEE)**

ANSWERS

1. (*c*)	**2.** (*d*)	**3.** (*b*)	**4.** (*c*)	**5.** (*a*)	**6.** (*d*)	**7.** (*d*)	**8.** (*d*)	**9.** (*d*)
10. (*a*)	**11.** (*c*)	**12.** (*c*)	**13.** (*a*)	**14.** (*a*)	**15.** (*b*)	**16.** (*a*)	**17.** (*a*)	**18.** (*a*)
19. (*d*)	**20.** (*b*)	**21.** (*a*)	**22.** (*d*)	**23.** (*a*)	**24.** (*a*)	**25.** (*b*)	**26.** (*b*)	**27.** (*c*)
28. (*d*)	**29.** (*c*)	**30.** (*a*)	**31.** (*b*)	**32.** (*a*)	**33.** (*b*)	**34.** (*c*)	**35.** (*c*)	**36.** (*c*)
37. (*c*)	**38.** (*a*)	**39.** (*a*)	**40.** (*b*)	**41.** (*c*)	**42.** (*a*)	**43.** (*d*)	**44.** (*d*)	**45.** (*a*)
46. (*c*)	**47.** (*d*)	**48.** (*b*)	**49.** (*c*)	**50.** (*c*)	**51.** (*a*)	**52.** (*a*)	**53.** (*d*)	**54.** (*c*)
55. (*c*)	**56.** (*b*)	**57.** (*b*)	**58.** (*d*)	**59.** (*a*)	**60.** (*a*)	**61.** (*d*)	**62.** (*a*)	**63.** (*c*)
64. (*c*)	**65.** (*d*)	**66.** (*d*)	**67.** (*b*)	**68.** (*d*)	**69.** (*c*)	**70.** (*b*)	**71.** (*a*)	**72.** (*d*)
73. (*a*)	**74.** (*a*)	**75.** (*b*)	**76.** (*a*)	**77.** (*b*)	**78.** (*c*)	**79.** (*b*)	**80.** (*d*)	**81.** (*c*)
82. (*b*)	**83.** (*c*)	**84.** (*c*)	**85.** (*d*)	**86.** (*c*)	**87.** (*b*)	**88.** (*a*)	**89.** (*b*)	**90.** (*c*)
91. (*a*)	**92.** (*c*)	**93.** (*b*)	**94.** (*a*)	**95.** (*a*)	**96.** (*d*)	**97.** (*c*)	**98.** (*b*)	**99.** (*a*)
100. (*b*)	**101.** (*d*)	**102.** (*c*)	**103.** (*c*)	**104.** (*c*)	**105.** (*d*)	**106.** (*b*)	**107.** (*d*)	**108.** (*c*)
109. (*c*)	**110.** (*b*)	**111.** (*b*)	**112.** (*a*)	**113.** (*c*)	**114.** (*a*)	**115.** (*a*)	**116.** (*a*)	**117.** (*a*)
118. (*c*)	**119.** (*c*)	**120.** (*c*)	**121.** (*c*)	**122.** (*c*)	**123.** (*b*)	**124.** (*a*)	**125.** (*d*)	**126.** (*d*)
127. (*d*)	**128.** (*b*)	**129.** (*a*)	**130.** (*a*).					

HINTS/SOLUTIONS

Q. No.	Ans.	Hints & Solutions
1.	**(*c*)**	It is the weaker acid among the given choices.
2.	**(*d*)**	Carboxylic acids are reduced to 1° alcohols with $LiAlH_4$.
3.	**(*b*)**	HI > HBr > HCl.
4.	**(*c*)**	(*a*), (*b*), (*d*) can be obtained by reaction of ethanol with conc. H_2SO_4 under different conditions. **Acetylene** cannot be obtained by reaction of ethanol with H_2SO_4.
5.	**(*a*)**	Kolbe's reaction.
6.	**(*d*)**	HCHO does not contain $CH_3—C(=O)—$ or $CH_3—CH(OH)—$ group.
7.	**(*d*)**	$LiAlH_4$ reduces carboxylic acids to alcohols.
8.	**(*d*)**	X is ethylene oxide and Y is ethylene glycol.
9.	**(*d*)**	Methanol does not contain $CH_3—CH(OH)—$ group.
10.	**(*a*)**	Hydroboration-oxidation reaction.
11.	**(*c*)**	$CH_3—CH(OH)—CH_3$ would be oxidized to dimethyl ketone which can give positive iodoform test.
12.	**(*d*)**	Phenyl-oxygen bond in methyl phenyl ether is difficult to cleave because it has partial double bond character due to resonance. 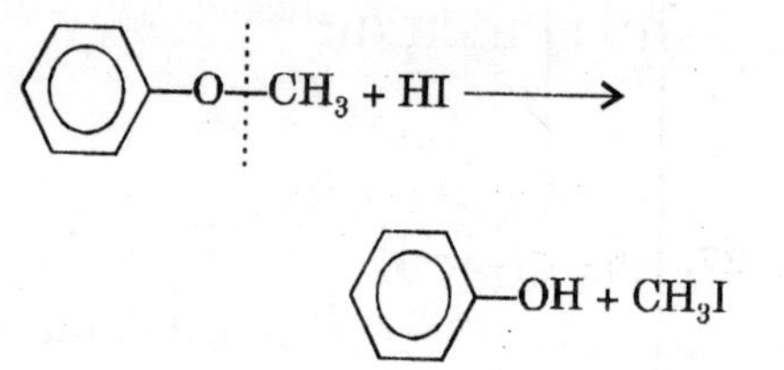
13.	**(*a*)**	$C_2H_5–O–C_2H_5 + HI \longrightarrow C_2H_5OH + C_2H_5I.$
14.	**(*a*)**	Glycerol on reaction with oxalic acid at 503 K gives allyl alcohol.
15.	**(*b*)**	$C_6H_5N_2^+Cl^- + H_2O \longrightarrow C_6H_5OH + N_2 + HCl.$
16.	**(*a*)**	Primary alcohol.
17.	**(*a*)**	Ethyl alcohol is lighter than water.
18.	**(*a*)**	$CH_3I \xrightarrow{Mg} CH_3MgI$ $C_6H_5CHO \xrightarrow[(ii)\ H_2O/H^+]{(i)\ CH_3MgI} C_6H_5—CH(OH)—CH_3$ 1-Phenylethanol

Q. No.	Ans.	Hints & Solutions
19.	(*d*)	$CH_2=CH_2 \xrightarrow{HOCl} CH_2(Cl)\text{—}CH_2(OH) \xrightarrow{NaHCO_3(aq)} CH_2(OH)\text{—}CH_2(OH)$
20.	(*b*)	$R\text{—}X \longrightarrow R\text{—}OH$ In the process X is substituted be —OH.
21.	(*a*)	$RCH_2CH_2OH \xrightarrow{PBr_3} RCH_2CH_2Br \xrightarrow{KCN} RCH_2CH_2CN \xrightarrow{H_2O/H^+} RCH_2CH_2COOH.$
22.	(*d*)	$CH_2(OH)\text{—}\mathbf{CH}(\mathbf{OH})\text{—}CH_2(OH)$ Glycerol or glycerine
23.	(*a*)	$C_2H_5OH + SOCl_2 \longrightarrow C_2H_5Cl + SO_2 + HCl$
24.	(*a*)	$CH_3\text{—}CH(OH)\text{—}CH_3 \xrightarrow{[O]} CH_3\text{—}C(=O)\text{—}CH_3$
25.	(*b*)	Phenol is more acidic than ethanol because phenoxide ion is stabilized by resonance while ethoxide is not.
26.	(*b*)	In the reactions involving cleavage of —O⋮H bond in alcohols the order of reactivity is 1° > 2° > 3°.
27.	(*c*)	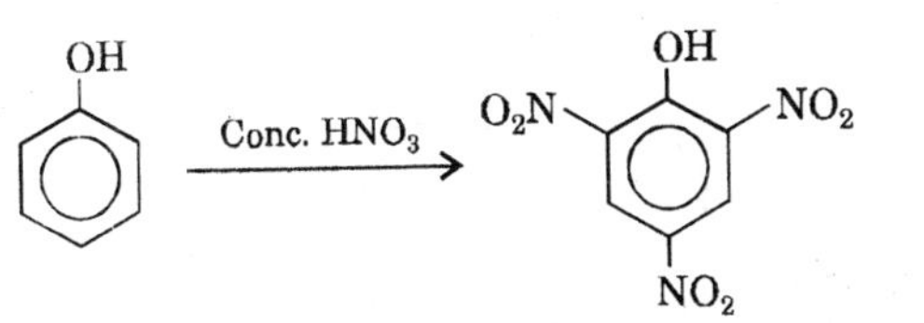
28.	(*d*)	$CH_3CH_2OH \xrightarrow{-H_2} CH_3CHO$ Removal of H_2 is oxidation.
29.	(*c*)	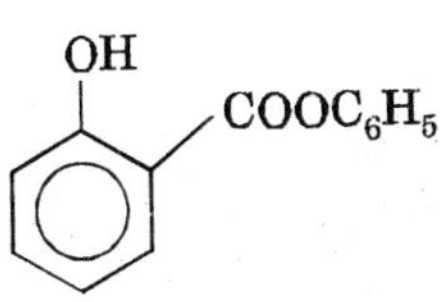 **Salol**
30.	(*a*)	3° > 2° > 1°.
31.	(*b*)	$C_6H_5OH \xrightarrow[\Delta]{Zn} \mathbf{C_6H_6} + ZnO.$
32.	(*a*)	Alcohols have higher boiling point than hydrocarbons.
33.	(*b*)	Glycerol has very strong H-bonds due to three —OH groups.
34.	(*c*)	Williamson synthesis.
35.	(*c*)	Methanol is prepared from CO and H_2.
36.	(*c*)	Self explanatory question.
37.	(*c*)	The reactivity of various alkyl halides towards β-elimination is 3° > 2° > 1°. $(CH_3)_3CCl$ is a 3° halide.
38.	(*a*)	Haloform reaction.
39.	(*a*)	$LiAlH_4$ reduces carboxylic acids to alcohols.
40.	(*b*)	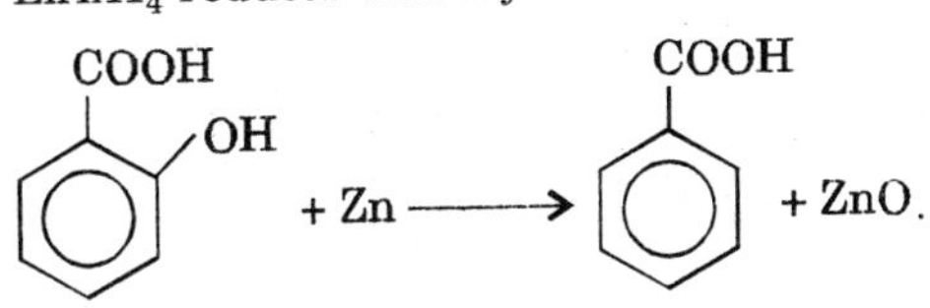
41.	(*c*)	1 mole of phenol (mol. mass = 94) requires 3 moles of Br_2 (mol. mass = 160)
42.	(*a*)	Factual question.
43.	(*d*)	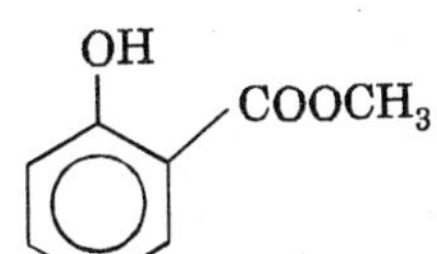 Methyl salicylate (*Oil of winter green*)
44.	(*d*)	$RMgX + CH_2\text{—}CH_2(\text{—}O\text{—}) \longrightarrow R\text{—}CH_2CH_2OMgX \xrightarrow{H_2O/H^+} R\text{—}CH_2CH_2OH$
45.	(*a*)	Violet colour.
46.	(*c*)	Haloform reation.
47.	(*d*)	All the ortho and para positions get substituted by Br atoms because benzene ring in phenol is highly activated.
48.	(*b*)	Methanoic acid.
49.	(*c*)	Electron withdrawing effect of $—NO_2$ group is more at ortho and para positions. Out of the two ***p*-nitrophenol is stronger acid** because in *o*-nitrophenol —OH group is involved in intramolecular H-bonding and hence release of H^+ is relatively difficult.
50.	(*c*)	$CH_2=CH_2 + H_2O + [O] \longrightarrow CH_2(OH)\text{—}CH_2(OH)$
51.	(*a*)	Factual question.
52.	(*a*)	$C_2H_5OH + Na \longrightarrow CH_3CH_2ONa + \frac{1}{2}H_2$

Q. No.	Ans.	Hints & Solutions
53.	(*d*)	Schotten Baumann reaction.
54.	(*c*)	$ClCH_2—CH—CH_2$ (with O bridging the two CH carbons) 3-Chloro-1, 2-epoxypropane (epichlorohydrin)
55.	(*c*)	$C_2H_5MgBr + C_2H_5—CO—C_2H_5 \longrightarrow$ $C_2H_5—C(OMgBr)(C_2H_5)—C_2H_5 \xrightarrow{H_2O/H^+} C_2H_5—C(OH)(C_2H_5)—C_2H_5$
56.	(*b*)	Self explanatory question.
57.	(*b*)	$X = CH_3—CO—CH_3$ $Y = (CH_3)_3COH$, *tert*-Butyl alcohol.
58.	(*d*)	Condensation of phenol with formaldehyde in the presence of acid or alkali gives bakelite.
59.	(*a*)	The reaction involves net addition of H_2O molecule according to Markownikoff's rule without any rearrangement. The reaction is known as oxymercuration-demercuration reaction.
60.	(*a*)	Phenol does not react with hot reduced copper.
61.	(*d*)	Benzyl alcohol ($C_6H_5—CH_2OH$) does not contain any $CH_3—CH(OH)—$ group.
62.	(*a*)	Boiling point decreases with increase in branches in the alkyl group.
63.	(*c*)	Ethyl alcohol.
64.	(*c*)	Phenols are stronger acids than alcohols. Chloro is electron withdrawing group and hence ***m*-chlorophenol** is stronger acid than phenol.
65.	(*d*)	$Mg + 2C_2H_5OH \longrightarrow Mg(OC_2H_5)_2 + H_2$ $Mg(OC_2H_5)_2 + H_2O \longrightarrow Mg(OH)_2 + 2C_2H_5OH.$
66.	(*d*)	$C_2H_5Cl + NaOC_2H_5 \longrightarrow C_2H_5—O—C_2H_5 + NaCl.$
67.	(*b*)	$CH_3CH = CH_2 + H_2SO_4 \longrightarrow$ $CH_3—CH(OSO_3H)—CH_3 \xrightarrow[\Delta]{H_2O} CH_3—CH(OH)—CH_3$
68.	(*d*)	$CH_3—CO—CH_3 + CH_3MgBr \longrightarrow$ $CH_3—C(CH_3)_2—OMgBr \xrightarrow{H_2O/H^+} CH_3—C(CH_3)_2—OH$
69.	(*c*)	$CH_3—CH_2—O—CH_2—CH_3 \xrightarrow[\text{Sunlight}]{Cl_2} CCl_3—CCl_2—O—CCl_2—CCl_3$ Perchlorodiethyl ether All hydrogen atoms are replaced by Cl atoms.
70.	(*b*)	Self explanatory question.
71.	(*a*)	Hydroboration-oxidation yields alcohol with anti Markownikoff addition of water molecule.
72.	(*d*)	$—OCH_3$ and $—CH_3$ are electron releasing groups. Hence, II and IV are more reactive than I. $—OCH_3$ is stronger electron releasing group. Hence II is more reactive than IV.
73.	(*a*)	$—NO_2$ group is electron withdrawing group. It increases the acidic character of phenol. Hence, *p*-nitrophenol is the strongest acid.
74.	(*a*)	Phthalein reaction. Factual question.
75.	(*b*)	$CH_2 = CH_2 + H^+ \longrightarrow CH_3—\overset{+}{C}H_2.$
76.	(*a*)	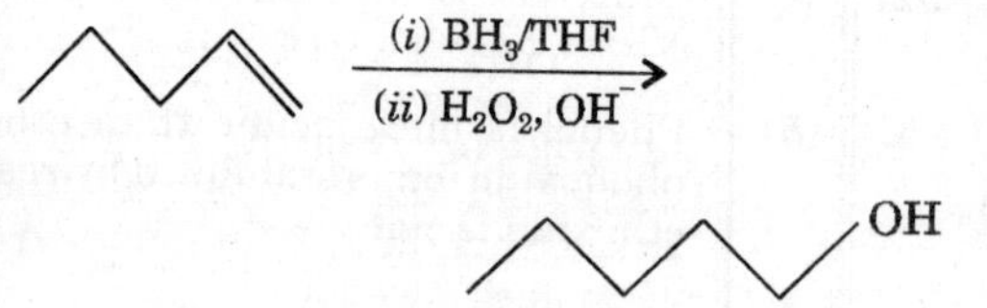 (*i*) BH_3/THF; (*ii*) H_2O_2, OH^- 1-Pentanol Hydroboration-oxidation.
77.	(*b*)	$—NO_2$ group is electron withdrawing group while $—CH_3$ is electron releasing group. Electron withdrawing groups increase the acidic character while electron releasing groups decrease the acidic character of substituted phenol.
78.	(*c*)	CH_3CH_2OH undergoes iodoform reaction with NaOH and I_2, and gives iodoform as product.
79.	(*b*)	The **three** metamers are : $C_2H_5—O—C_2H_5$, $CH_3—O—CH_2CH_2CH_3$ and $CH_3—O—CH(CH_3)—CH_3$
80.	(*d*)	$CH_3—O—CH_2Cl +$ $\mathbf{CH_3MgCl} \longrightarrow CH_3—O—CH_2CH_3 + MgCl_2$
81.	(*c*)	Primary aliphatic and aralkyl amines on reaction with nitrous acid give alcohols.
82.	(*b*)	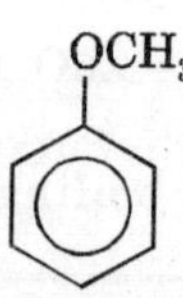 can undergo electrophilic substitution reaction at the benzene ring.

Q. No.	Ans.	Hints & Solutions
83.	(*c*)	The carbonyl compound containing four carbon atoms that does not react with Tollen's reagent is 2-butanone. Therefore, $C_4H_{10}O$ must be 2-butanol.
84.	(*c*)	$C_2H_5OH + PCl_5 \longrightarrow \mathbf{C_2H_5Cl} + POCl_3 + HCl$ $C_2H_5Cl + AgNO_2 \longrightarrow \mathbf{C_2H_5NO_2} + AgCl.$
85.	(*d*)	$3\,C_2H_5OH + PBr_3 \longrightarrow 3\,C_2H_5Br + \mathbf{H_3PO_3}.$
86.	(*c*)	Factual question.
87.	(*b*)	Reimer Tiemann reaction.
88.	(*a*)	Self explanatory question. —OH gets preference over Br while numbering the chain.
89.	(*b*)	Reimer Tiemann reaction.
90.	(*c*)	$1° > 2° > 3°$.
91.	(*a*)	Peroxide is formed. $CH_3CH_2\text{—}O\text{—}CH_2CH_3 + O_2 \longrightarrow CH_3CH_2\text{—}O\text{—}\underset{\displaystyle |}{\overset{\displaystyle OOH}{C}}HCH_3$
92.	(*c*)	Since the compound B gives blue colour in Victor Meyer's test, it is a secondary alcohol. Secondary alcohols are obtained by reaction of Grignard reagents on aldehydes. Hence, compound A is an aldehyde. In choice (*c*), the first compound is aldehyde while second is secondary alcohol.
93.	(*b*)	Oil of winter green is methyl salicylate.
94.	(*a*)	X is ethane.
95.	(*a*)	Lucas test is used for distinguishing between, 1°, 2° and 3° alcohols.
96.	(*d*)	4-Methylpentan-2-ol.
97.	(*c*)	$\begin{matrix}H\\H\end{matrix}\!\!>C=O + CH_3MgI \longrightarrow \underset{\textbf{(A)}}{CH_3\text{—}CH_2\text{—}OMgI}$ $\xrightarrow{H_2O} \underset{\textbf{(B)}}{CH_3CH_2OH} + Mg(OH)I$
98.	(*b*)	Acetyl salicylic acid (*aspirin*) is used to reduce pain and hence is an analgesic.
99.	(*a*)	Alcohols have stronger interparticle forces and hence are less volatile.
100.	(*b*)	$R\text{—}OH + H^+ \longrightarrow R\text{—}\overset{+}{O}H_2.$
101.	(*d*)	*tert*-butoxide is the conjugate base of the weakest acid.
102.	(*c*)	Hydroboration-oxidation yields alcohols from alkenes according to anti-Markownikoff addition of water.
103.	(*c*)	Ease of dehydration of alcohols is $3° > 2° > 1°$. The choice (*c*) is a 3° alcohol.
104.	(*c*)	3, 3-Dimethyl-1-cyclohexanol.
105.	(*d*)	Secondary alcohols on oxidation give ketones with same number of carbon atoms. $2CH_3\text{—}\underset{OH}{\underset{\vert}{C}}H\text{—}CH_3 \xrightarrow{[O]} 2CH_3\text{—}\underset{O}{\underset{\Vert}{C}}\text{—}CH_3 + H_2O$
106.	(*b*)	CH_3OH does not contain $CH_3\text{—}\underset{OH}{\underset{\vert}{C}}H\text{—}$ group. Hence, it does not give positive iodoform test.
107.	(*d*)	PCC converts primary alcohols to aldehydes while others oxidize them to carboxylic acids.
108.	(*c*)	$\text{—}NO_2$ group being electron-withdrawing group stabilizes the phenoxide ion, thus making the corresponding phenol stronger acid.
109.	(*c*)	Secondary or tertiary alcohol is the perferred product.
110.	(*b*)	Kolbe's reaction.
111.	(*b*)	Lucas test is used for distinguishing between 1°, 2 ° and 3° alcohols.
112.	(*a*)	Phenyl ethers on cleavage with HX give phenol and alkyl halides.
113.	(*c*)	Glycerol is a trihydric alcohol Glycerol is 1, 2, 3-propanetriol.
114.	(*a*)	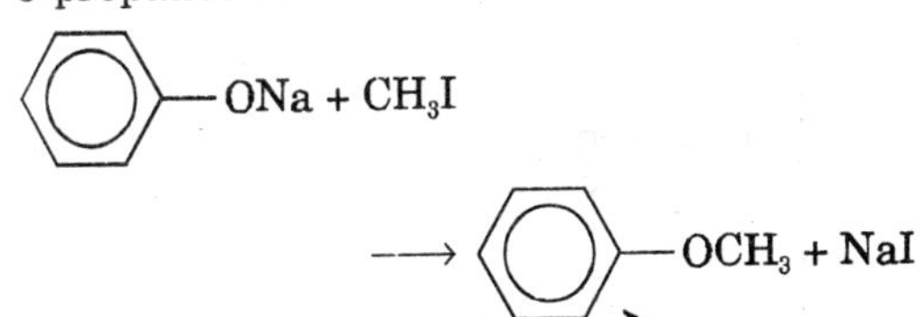
115.	(*a*)	In Reimer Tiemann reaction, dichlorocarbene is formed as intermediate $CHCl_3 + {}^-OH \longrightarrow :CCl_3^- \longrightarrow \underset{\text{Dichlorocarbene}}{:CCl_2} + Cl^-.$
116.	(*a*)	Electrophilic attack of diazonium ion takes place at para position of phenol.
117.	(*a*)	$HCHO + CH_3MgI \longrightarrow CH_3CH_2OMgI$ $\xrightarrow{H_2O/H^+} CH_3CH_2OH.$
118.	(*c*)	CH_3CH_2OH on heating with hot reduced copper at 300°C undergoes dehydrogenation to give ethanol, CH_3CHO.
119.	(*c*)	$\underset{(X)}{CH_3CHO} \xrightarrow{LiAlH_4} \underset{(Y)}{CH_3CH_2OH}$ $\xrightarrow{Cu/300°C} \underset{(Z)}{CH_3CHO}$

Q. No.	*Ans.*	*Hints & Solutions*
120.	**(*c*)**	The reaction proceeds through S_N^2 mechanism, involving attack of I^- ion on less hindered carbon resulting in the formation of CH_3I and $(CH_3)_2$ CHOH 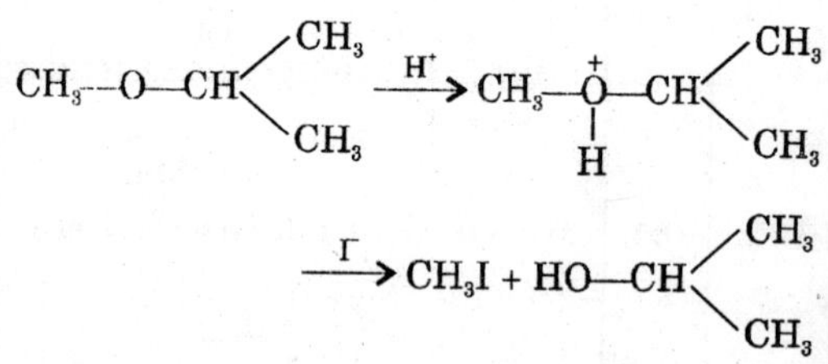
121.	**(*c*)**	RMgX + 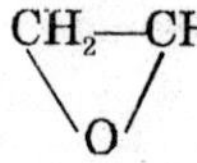 $\longrightarrow$ R—CH_2—CH_2—OMgX $\xrightarrow{H_2O/H^+}$ R—CH_2CH_2OH (1° alcohols)
122.	**(*c*)**	General formula of alkanols is $C_nH_{2n+1}OH$ or $C_nH_{2n+2}O$.
123.	**(*b*)**	In Reimer Tiemann reaction dichlorocarbene is the intermediate electrophile involved $CHCl_3$ + ^-OH $\xrightarrow[-H_2O]{}$: $^-CCl_3$ $\xrightarrow[-Cl^-]{}$: CCl_2.
124.	**(*a*)**	In 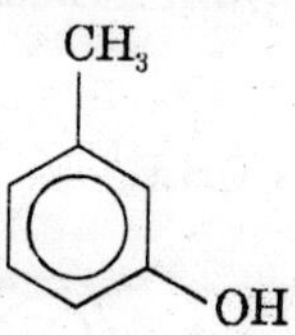 *ortho* and *para* positions with respect to —OH are vacant and there positions would be substituted during reaction with bromine water.
125.	**(*d*)**	Ph—CH(OH)—CH_3 contains CH_3—CH(OH)— group.
126.	**(*d*)**	Aqueous KOH hydrolysis 1° and 2° halides to alcohol.
127.	**(*d*)**	This is the method of preparation of phenol from isopropylbenzene (cumene).
128.	**(*b*)**	Bromination takes place at para position to —OCH_3 because —OCH_3 is stronger activating group than —CH_3.
129.	**(*a*)**	The reaction follows SN^2 mechanism and as a result iodide attacks smaller alkyl group. 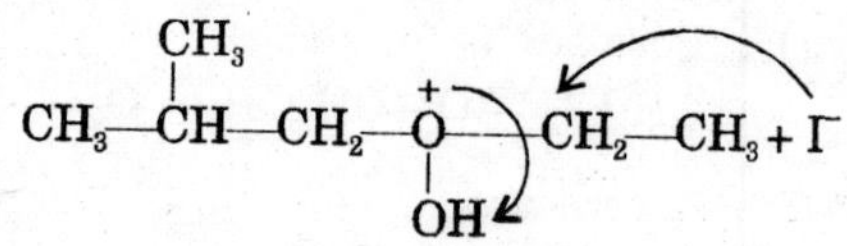
130.	**(*a*)**	$CH_3CH_2OH \xrightarrow{P + I_2} CH_3CH_2I$ (A) $\xrightarrow[\text{ether}]{Mg}$ CH_3CH_2MgI (B) $\xrightarrow{HCHO}$ CH_3CH_2—CH_2OMgI (C) $\xrightarrow{H_2O}$ $CH_3CH_2CH_2OH$ (D)

4.6. MCQs FOR IIT JEE

Brain Twister

A. Multiple Choice Questions With One Correct Answer

1. A compound on treatment with periodic acid yields ethanal and propanal. The compound possibily is

(a)  (b) 

(c) 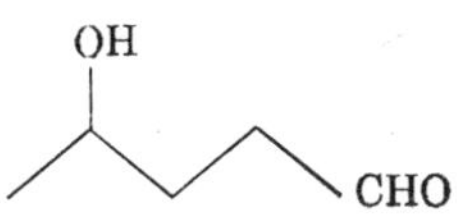 (d) 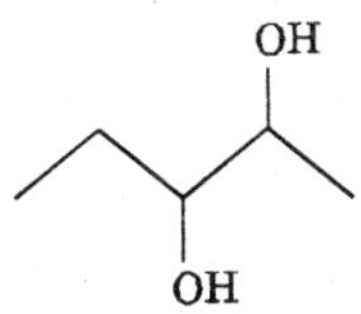

2. Oxymercuration demercuration of 2-methylpropene yields
(a) 2-Methyl-1-propanol (b) 2-Methyl-2-propanol
(c) 2-Methylpropanal (d) 2-Methylpropanoic acid.

3. C—O—C bond in ethers can be cleaved by
(a) $KMnO_4$ (b) $LiAlH_4$
(c) KOH (d) HI.

4. Which of the following does not have H-bonds as the interparticle forces ?
(a) CH_3OCH_3 (b) $CH_3CH_2CH_2OH$
(c) 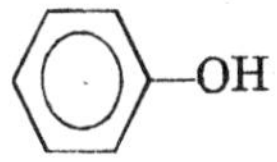 (d) None of these.

5. Which of the following compounds would gives yellow ppt. on reaction with NaOH and I_2 ?
(a) *iso*-Butyl alcohol (b) *sec*-Butyl alcohol
(c) *tert*-Butyl alcohol (d) *n*-Butyl alcohol.

6. Which of the following is the strongest acid ?
(a) Phenol (b) *o*-Nitrophenol
(c) *o*-Chlorophenol (d) *o*-Cresol.

7. 1 mole of ethanol on reaction with sodium would liberate .... of H_2 at STP ?
(a) 22.4 L (b) 44.8 L
(c) 11.2 L (d) 5.6 L.

8. The order of reactivity of alcohols with Luca's reagent is
(a) $3° > 2° > 1°$ (b) $1° > 2° > 3°$
(c) $2° > 3° > 1°$ (d) $3° = 2° = 1°$.

9. Ethyl phenyl ether on reaction with excess HI yields
(a) Ethyl iodide and iodobenzene
(b) Ethyl iodide and phenol
(c) Ethyl alcohol and phenol
(d) Ethyl alcohol and iodobenzene.

10. Which of the following compounds would not react with Grignard reagent ?
(a) Ethanol (b) Ethanal
(c) Propapone (d) Diethyl ether.

11. Ketones can be converted to tertiary alcohols by
(a) Reduction
(b) Oxidation
(c) Reaction with Grignard reagent
(d) None of these.

12. The product obtained from the following sequence of reactions is

$$CH_3C \equiv CH \xrightarrow[\text{dil. } H_2SO_4]{HgSO_4} X \xrightarrow{LiAlH_4} Y$$

(a) 1-Propanol (b) 2-Propanol
(c) Propanone (d) Prop-2-en-1-ol.

13. In the sequence of reactions,

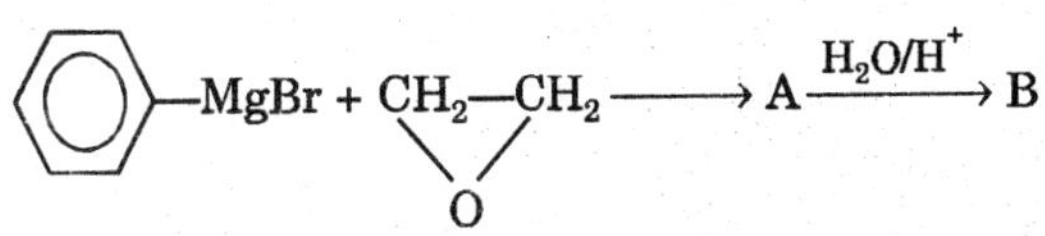

the product B, is
(a) Benzyl alcohol (b) 2-Pheyl ethanol
(c) 1-Phenyl ethanol (d) Quinol.

14. Which of the following is not a stable compound ?
(a) $\underset{OH}{\underset{|}{CH_2}}-CH_2-\underset{OH}{\underset{|}{CH_2}}$ (b) $\underset{OH}{\underset{|}{CH_2}}-\underset{OH}{\underset{|}{CH}}-CH_3$
(c) $CH_3-CH_2-\underset{OH}{\underset{|}{CH}}-OH$ (d) $\underset{OH}{\underset{|}{CH_2}}-\underset{OH}{\underset{|}{CH}}-\underset{OH}{\underset{|}{CH_2}}$

15. How many isomeric secondary alcohols are possible for $C_5H_{11}OH$?
(a) 5 (b) 6
(c) 3 (d) 2.

16. Which of the following would not yield primary alcohol as product ?
(a) Reduction of aldehydes
(b) Reduction of carboxylic acids
(c) Reduction of esters
(d) Reduction of ketone.

17. *iso*-Butylene when subjected to hydroboration-oxidation reaction yields

(*a*) *sec*-Butyl alcohol (*b*) *tert*-Butyl alcohol
(*c*) *iso*-Butyl alcohol (*d*) *n*-Butyl alcohol.

18. $LiAlH_4$ converts ethyl acetate into

(*a*) Acetic acid (*b*) Ethyl alcohol
(*c*) Both (*d*) None.

19. 3-Methyl-1-butene on oxymercuration-demercuration yields .... as the major product.

(*a*) 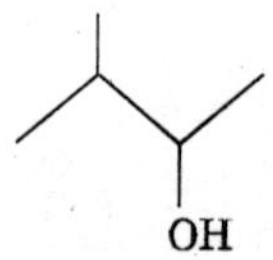 (*b*) 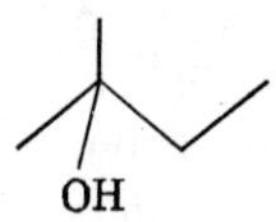

(*c*) 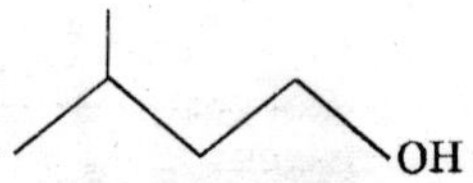 (*d*) 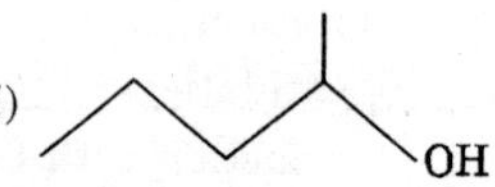

20. Ethyl alcohol is prepared industrially by

(*a*) Hydration of ethylene (*b*) Fermentation of sugars
(*c*) Both (*d*) None.

21. Which of the following will produce a primary alcohol on reaction with CH_3MgI (followed by hydrolysis) ?

(*a*) 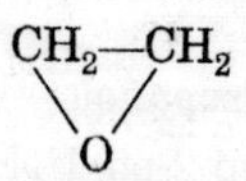 (*b*) CH_3CHO
(*c*) CH_3COCH_3 (*d*) $CH_3CH_2OCOCH_3$.

22. Identify the final product

$$CH_2 = CH_2 \xrightarrow{Cl_2/H_2O} X \xrightarrow{aq.\ KOH} Y$$

(*a*) 2-Chloroethanol (*b*) 2-Chloromethanol
(*c*) 1-Chloroethanol (*d*) Ethylene glycol.

23. IUPAC name of diethyl *n*-propyl carbinol is

(*a*) 3, 3-Diethyl-1-propanol (*b*) 3-Ethyl-3-hexanol
(*c*) 3, 3-Diethyl-2-propanol (*d*) 2-Ethyl-3-heptanol.

24. The IUPAC name of the product obtained by hydroboration-oxidation of 1-methyl cyclohexene is

(*a*) *cis*-2-Methylcyclohexanol
(*b*) Hydroxymethylcyclohexane
(*c*) *trans*-2-Methycyclohexanol
(*d*) 1-Methylcyclohexanol.

25. 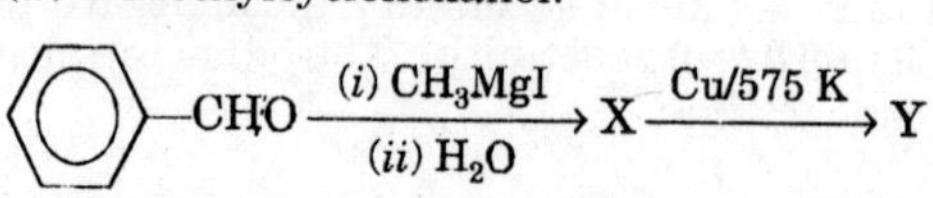

The compound Y in the above sequence is

(*a*) 2-Methyl-2-phenyl-1-propanol
(*b*) 2-Phenyl-2-propanol
(*c*) Acetophenone
(*d*) 2-Methyl-1-phenyl-2-propanol.

26. 1 mole of an alcohol on reaction with sodium liberates 1 moles of H_2. The alcohol may be

(*a*) Ethanol (*b*) Allyl alcohol
(*c*) Ethylene glycol (*d*) Cyclohexanol.

27. Which of the following represents the correct order of acidity ?

(*a*) $H_2O > ROH\ (1°) > ROH\ (2°) > ROH\ (3°) > RC \equiv CH$
(*b*) $H_2O > RC \equiv CH > ROH\ (1°) > ROH\ (2°) > ROH\ (3°)$
(*c*) $ROH\ (1°) > ROH\ (2°) > ROH\ (3°) > H_2O > RC \equiv CH$
(*d*) $ROH\ (1°) > ROH\ (2°) > ROH\ (3°) > RC \equiv CH > H_2O$.

28. Which of the following pairs of compounds cannot be distinguished by iodoform test ?

(*a*) Ethanol and methanol
(*b*) 1-Propanol and 2-propanol
(*c*) 2-Methyl-2-propanol and 2-methyl-1-propanol
(*d*) 1-Butanol and 2-butanol.

29. Cumene hydroproxide on hydrolysis yields

(*a*) Phenol (*b*) Acetone
(*c*) Both (*d*) None of these.

30. In the Reimer-Tiemann reaction the intermediate involved is

(*a*) 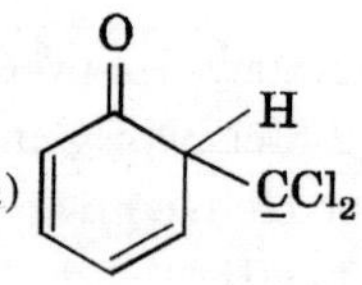 (*b*) 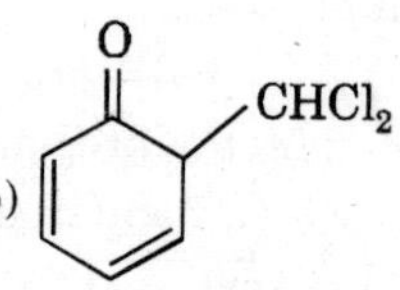

(*c*) 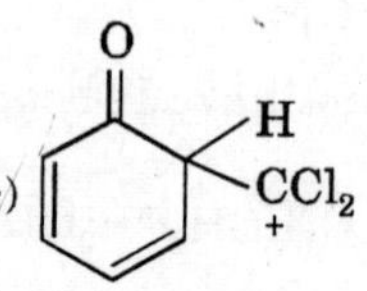 (*d*) 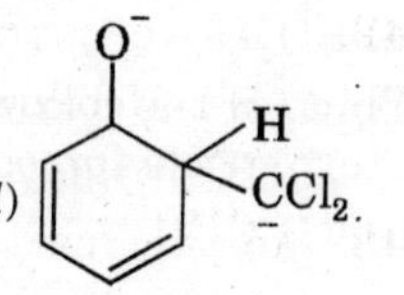

31. Benzyl bromide reacts with alkaline aqueous solution of phenol to give

(*a*) Diphenyl ether (*b*) Diphenylmethane
(*c*) Phenetole (*d*) Benzylphenylether.

32. Which of the following is the strongest acid ?

(*a*) *o*-Chlorophenol (*b*) *o*-Bromophenol
(*c*) *o*-Fluorophenol (*d*) *o*-Iodophenol.

33. *neo*-Pentyl alcohol on reaction with HCl yields

(*a*) *n*-Pentyl chloride (*b*) *iso*-Pentyl chloride
(*c*) *tert*-Pentyl chloride (*d*) *sec*-Pentyl chloride.

34. How many primary isomeric alcohols are possible for $C_5H_{11}OH$?

(*a*) 2 (*b*) 3
(*c*) 4 (*d*) 8.

35. Which of the following compounds is/are soluble in conc. H_2SO_4 ?

(*a*) Ethanol (*b*) Ethyl bromide
(*c*) Both (*d*) None.

36. Which of the following compounds gives positive periodic acid test ?

(*a*) 1, 3-Pentanediol (*b*) 2, 3-Pentanediol
(*c*) 1, 4-Pentanediol (*d*) 2, 4-Pentanediol.

37. Which of the following alcohols would give blue coloration in Victor Meyer's test ?

(*a*) *iso*-Propyl alcohol (*b*) *iso*-Pentyl alcohol
(*c*) *iso*-Butyl alcohol (*d*) *tert*-Pentyl alcohol.

38. Which of the following alcohols on reaction with Luca's reagent produces cloudiness immediately ?

(*a*) 1-Propanol (*b*) 2-Propanol
(*c*) 2-Methyl-2-propanol (*d*) 2-Methyl-1-propanol.

39. Which of the following reactions will yield butan-1-ol ?

(*a*) $CH_3CHO \xrightarrow[Et_2O]{C_2H_5MgBr} \xrightarrow{H_3O^+}$

(*b*) $CH_3CH_2OH \xrightarrow[Et_2O]{C_2H_5MgBr}$

(*c*) 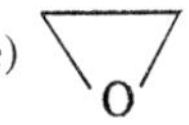 $\xrightarrow[Et_2O]{C_2H_5MgBr} \xrightarrow{H_3O^+}$

(*d*) $CH_3C \equiv N \xrightarrow[Et_2O]{C_2H_5MgBr}$.

40. How many stereoisomers are possible for 2, 3-pentanediol

(*a*) 2 (*b*) 3
(*c*) 4 (*d*) 6.

B. MCQs from Previous IIT Examinations

41. What is Z in the following sequence of reactions ?

$$Z \xrightarrow{PCl_5} X \xrightarrow{\text{Alc. KOH}} Y \xrightarrow[(ii)\ H_2O\ ;\ \text{boil}]{(i)\ \text{Conc. } H_2SO_4} Z$$

(*a*) $CH_3CH_2CH_2OH$ (*b*) $CH_3CHOHCH_3$
(*c*) $(CH_3CH_2)_2CHOH$ (*d*) $CH_3CH = CH_2$.

(IIT 1986)

42. HBr reacts fastest with

(*a*) 2-Methylpropan-1-ol (*b*) 2-Methylpropan-2-ol
(*c*) Propan-2-ol (*d*) Propan-1-ol. ***(IIT 1986)***

43. In the reaction of phenol with $CHCl_3$ and aqueous NaOH at 70°C (343 K), the electrophile attacking the ring is

(*a*) $CHCl_3$ (*b*) $CHCl_2$
(*c*) $:CCl_2$ (*d*) $COCl_2$. ***(IIT 1991)***

44. The products of combustion of an aliphatic thiol (RSH) at 298 K are

(*a*) CO_2 (*g*), H_2O (*g*) and SO_2 (*g*)
(*b*) CO_2 (*g*), H_2O (*l*) and SO_2 (*g*)
(*c*) CO_2 (*l*), H_2O (*l*) and SO_2 (*g*)
(*d*) CO_2 (*g*), H_2O (*l*) and SO_2 (*l*). ***(IIT 1992)***

45. Which one/ones of the following reactions will yield 2-propanol ?

I. $CH_2 = CH—CH_3 + H_2O \xrightarrow{H^+}$

II. $CH_3—CHO \xrightarrow[(ii)\ H_2O]{(i)\ CH_3MgI}$

III. $CH_2O \xrightarrow[(ii)\ H_2O]{(i)\ C_2H_5MgI}$

IV. $CH_2 = CH—CH_3 \xrightarrow{\text{Neutral } KMnO_4}$

Choose the right answer from (*a*), (*b*), (*c*) and (*d*)

(*a*) I and II (*b*) II and III
(*c*) III and I (*d*) II and IV. ***(IIT 1994)***

46. In the following compounds

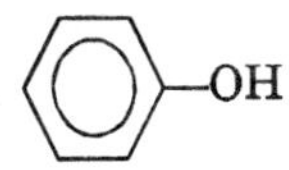

(I)

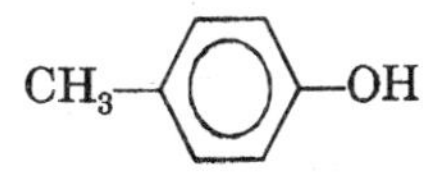

(II)

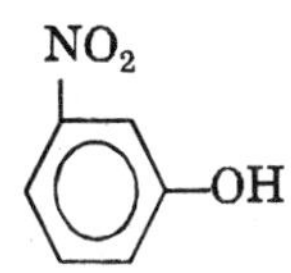

(III)

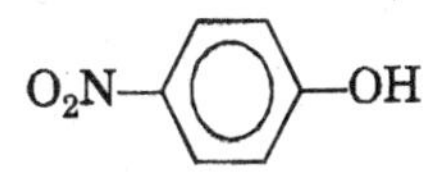

(IV)

The order of acidity is

(*a*) III > IV > I > II (*b*) I > IV > III > II
(*c*) II > I > III > IV (*d*) IV > III > I > II.

(IIT 1996)

47. How many optically active stereoisomers are possible for butan-2, 3-diol ?

(*a*) 1 (*b*) 2
(*c*) 3 (*d*) 4. ***(IIT 1997)***

48. The order of reactivity of the following alcohols towards conc. HCl is

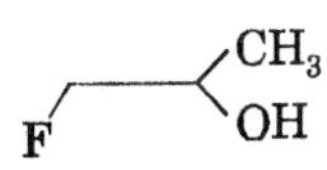

(I)

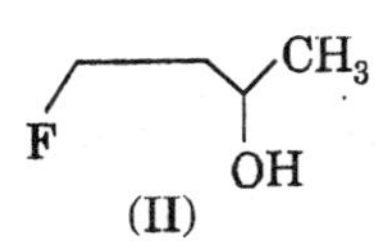

(II)

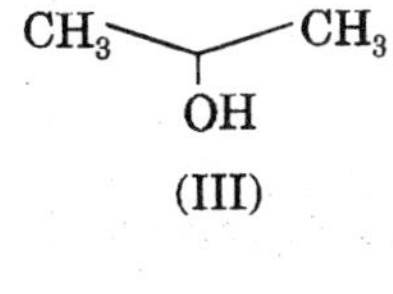

(III)

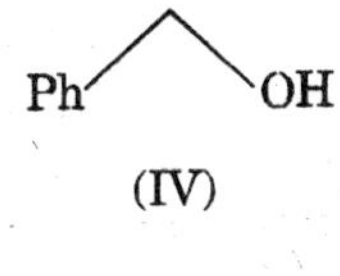

(IV)

(a) I > II > III > IV (b) I > III > II > IV
(c) IV > III > II > I (d) IV > III > I > II.

(IIT 1997)

49. Which of the following will react with water ?
(a) $CHCl_3$ (b) $Cl_3C.CHO$
(c) CCl_4 (d) $ClCH_2 - CH_2 - Cl$.

(IIT 1998)

50. The reaction of

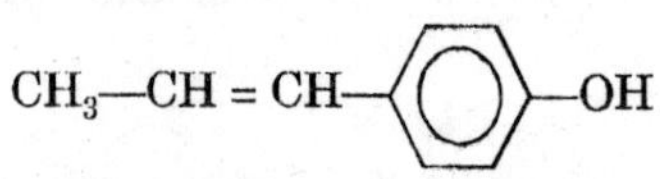

with HBr gives

(a) 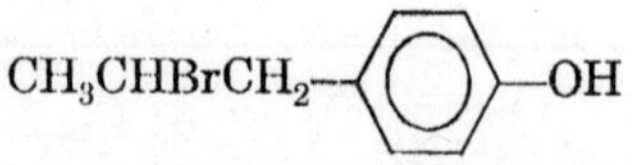

(b) 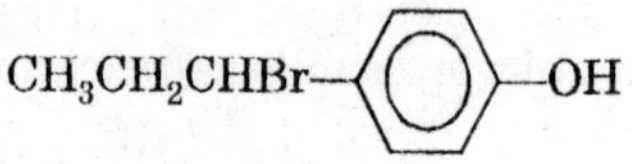

(c) 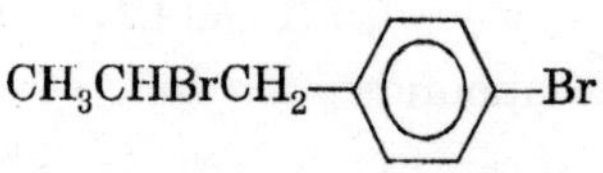

(d) 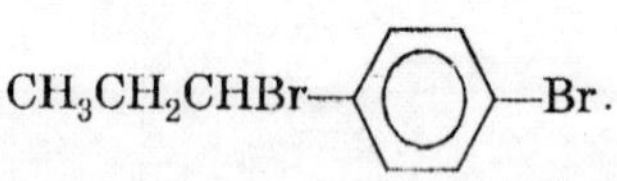 **(IIT 1998)**

51. Among the following compounds, strongest acid is
(a) $HC \equiv CH$ (b) C_6H_6
(c) C_2H_6 (d) CH_3OH. **(IIT 1999)**

52. The ether 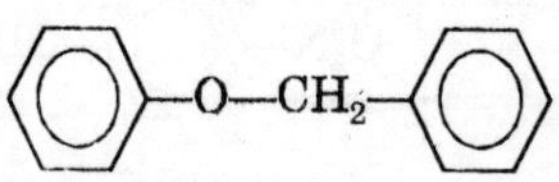 when treated with HI produces

(a) 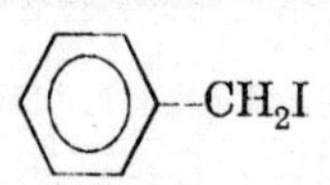 (b) 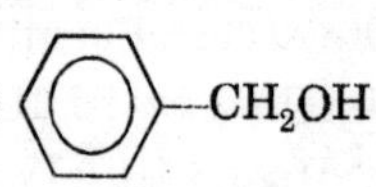

(c) 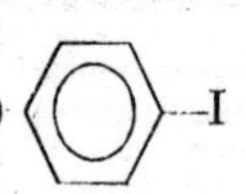 (d) 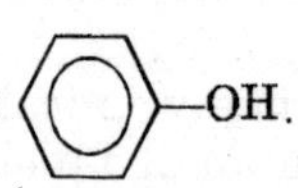 **(IIT 1999)**

53. The most unlikely representation of resonance structures of *p*-nitrophenoxide ion is

(a) 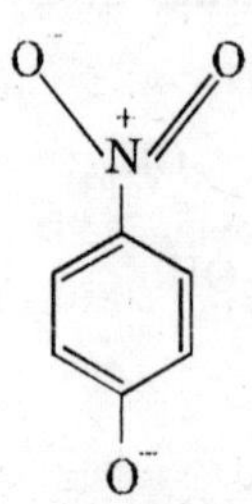 (b) 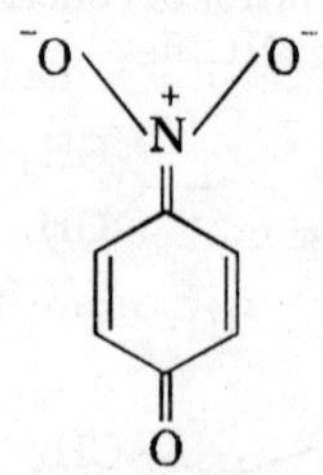

(c) 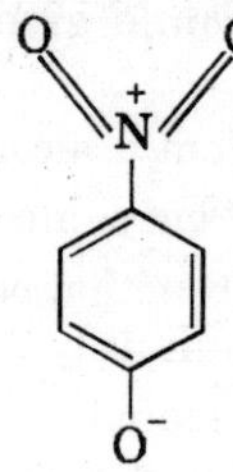 (d) 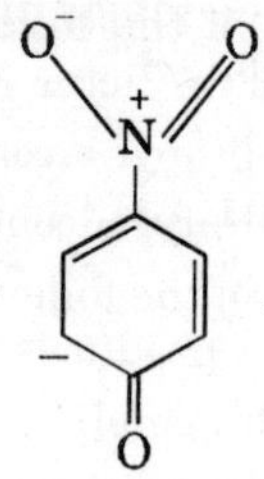 **(IIT 1999)**

54. The enol form of acetone, after treatment with D_2O gives

(a) $CH_3 - \underset{}{\overset{OD}{\overset{|}{C}}} = CH_2$ (b) $CD_3 - \overset{O}{\overset{||}{C}} - CD_3$

(c) $CH_2 - \overset{OH}{\overset{|}{C}} - CH_2D$ (d) $CD_2 = \overset{OD}{\overset{|}{C}} - CD_3$.

(IIT 1999)

55. 1-Propanol and 2-propanol can be distinguished by
(a) Oxidation with alkaline $KMnO_4$ followed by reaction with Fehling solution
(b) Oxidation with acidic dichromate followed by reaction with Fehling solution
(c) Oxidation by heating with copper followed by reaction with Fehling solution
(d) Oxidation with concentrated H_2SO_4 followed by reaction with Fehling solution.

(IIT Screening Test 2002)

56. Ethyl ester $\xrightarrow[\text{excess}]{CH_3MgBr}$ P. The product P will be

(a) 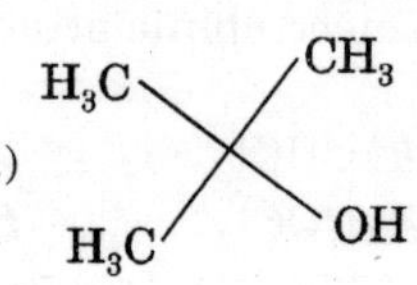 (b) 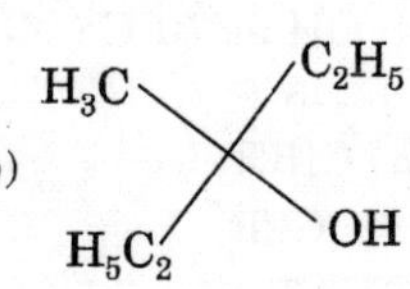

(c) 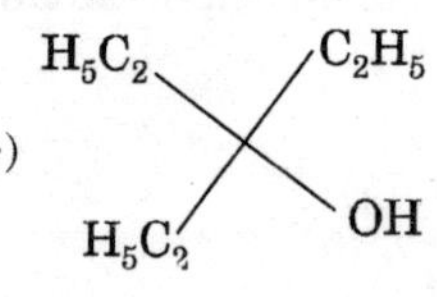 (d) 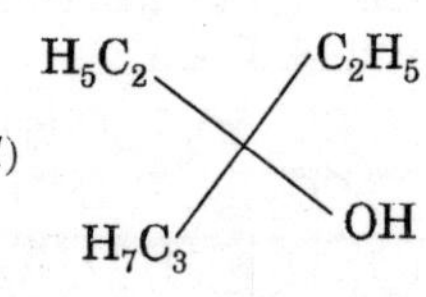

(IIT Screening Test 2003)

57. 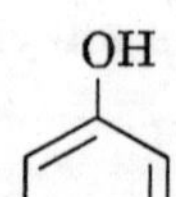 $+ C_2H_5I \xrightarrow[\text{Anhydrous } (C_2H_5OH)]{^-OC_2H_5}$

(a) $C_6H_5OC_2H_5$ (b) $C_2H_5OC_2H_5$
(c) $C_6H_5OC_6H_5$ (d) C_6H_5I.

(IIT Screening Test 2003)

58. The product of acid catalyzed hydration of 2-phenyl propene is
(a) 3-phenyl-2-propanol (b) 1-phenyl-2-propanol
(c) 2-phenyl-2-propanol (d) 2-phenyl-1-propanol

(IIT Screening Test 2004)

59. Identify the major products [P] and [Q] in the following reaction

 $+ CH_3CH_2CH_2Cl \xrightarrow{\text{anhydrous } AlCl_3}$

$$[P] \xrightarrow[\text{(ii) } H_3O^+,\ \Delta]{\text{(i) } O_2} [Q] + PhOH$$

(*a*) 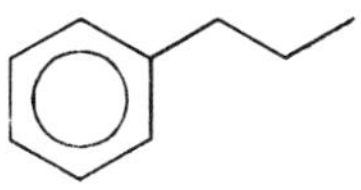 and CH_3COCH_3

(*b*) 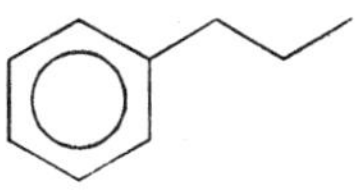 and CH_3CH_2CHO

(*c*) 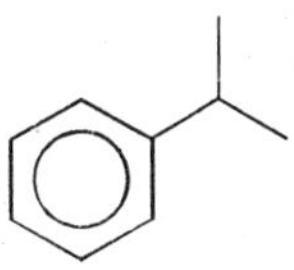 and CH_3COCH_3

(*d*) 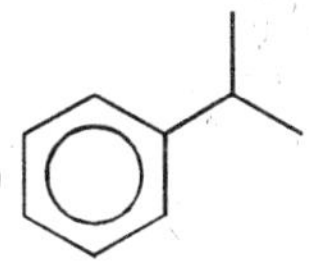 and CH_3CH_2CHO

(*IIT JEE 2006*)

60. Arrange the following compounds according to decreasing boiling points :

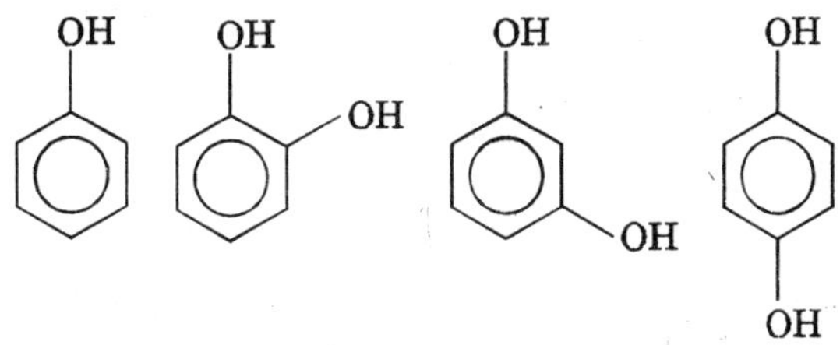

(I) (II) (III) (IV)

(*a*) (IV) > (III) > (II) > (I) (*b*) (III) > (IV) > (II) > (I)
(*c*) (I) > (II) > (III) > (IV) (*d*) (II) > (III) > (I) > (IV)

(*IIT JEE 2006*)

ANSWERS

1. (*d*)	**2.** (*b*)	**3.** (*d*)	**4.** (*a*)	**5.** (*b*)	**6.** (*b*)	**7.** (*c*)	**8.** (*a*)	**9.** (*b*)
10. (*d*)	**11.** (*c*)	**12.** (*b*)	**13.** (*b*)	**14.** (*c*)	**15.** (*c*)	**16.** (*d*)	**17.** (*c*)	**18.** (*b*)
19. (*a*)	**20.** (*c*)	**21.** (*a*)	**22.** (*d*)	**23.** (*b*)	**24.** (*c*)	**25.** (*c*)	**26.** (*c*)	**27.** (*a*)
28. (*c*)	**29.** (*c*)	**30.** (*a*)	**31.** (*d*)	**32.** (*a*)	**33.** (*c*)	**34.** (*c*)	**35.** (*a*)	**36.** (*b*)
37. (*c*)	**38.** (*c*)	**39.** (*c*)	**40.** (*c*)	**41.** (*b*)	**42.** (*b*)	**43.** (*c*)	**44.** (*b*)	**45.** (*a*)
46. (*d*)	**47.** (*b*)	**48.** (*c*)	**49.** (*b*)	**50.** (*b*)	**51.** (*d*)	**52.** (*a, d*)	**53.** (*c*)	**54.** (*d*)
55. (*c*)	**56.** (*a*)	**57.** (*a*)	**58.** (*c*)	**59.** (*c*)	**60.** (*a*).			

HINTS/SOLUTIONS

Q. No.	Ans.	Hints & Solutions
1.	(*d*)	Vicinal diols undergo cleavage of C—C bond on reaction with HIO_4 $$CH_3—\underset{OH}{\underset{\vert}{C}H}—\underset{OH}{\underset{\vert}{C}H}—CH_2—OH \xrightarrow{HIO_4} CH_3CHO + CH_3CH_2CHO$$
2.	(*b*)	Oxymercuration-demercuration brings about addition of water molecule to alkene resulting in the product according to Markownikoff rule.
3.	(*d*)	C—O—C bond in ethers can cleaved by acids.
4.	(*a*)	Ethers have dipole-dipole forces as interparticle foces.
5.	(*b*)	*sec*-Butyl alcohol contains $CH_3—\underset{OH}{\underset{\vert}{C}H}—$ group and hence gives positive iodoform test.
6.	(*b*)	– NO_2 group is strong electron withdrawing group.
7.	(*c*)	$C_2H_5OH + Na \longrightarrow C_2H_5ONa + \frac{1}{2} H_2$ 1 mole ethanol gives 1/2 mole of H_2. Volume of 1/2 mole of H_2 at STP is 11.2 L.

Q. No.	Ans.	Hints & Solutions
8.	(a)	$3° > 2° > 1°$.
9.	(b)	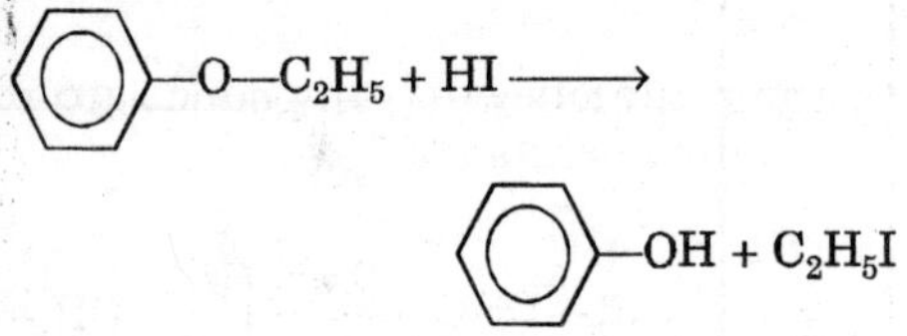 Phenyl-oxygen bond has partial double bond character and hence is difficult to cleave.
10.	(d)	Diethyl ether does not have any active site.
11.	(c)	$R{-}\underset{\underset{O}{\Vert}}{C}{-}R \xrightarrow[(ii)\ H_2O/H^+]{(i)\ R'MgX} R{-}\overset{\overset{R'}{\vert}}{\underset{\underset{OH}{\vert}}{C}}{-}R$
12.	(b)	$CH_3C \equiv CH \xrightarrow[\text{dil. } H_2SO_4]{HgSO_4} CH_3{-}\underset{\underset{O}{\Vert}}{C}{-}CH_3 \xrightarrow{LiAlH_4} CH_3{-}\underset{\underset{OH}{\vert}}{CH}{-}CH_3$
14.	(c)	Two —OH groups on the same carbon are unstable.
15.	(c)	The three secondary alcohols are : 2-pentanol, 3-pentanol and 3-methyl-2-butanol.
16.	(d)	Reduction of ketones yields secondary alcohols as product.
17.	(c)	$(CH_3)_2C = CH_2 \xrightarrow[(ii)\ \text{Oxidation}]{(i)\ \text{Hydroboration}} (CH_3)_2CH{-}CH_2OH$ anti-Markownikoff product is obtained.
18.	(b)	$CH_3CH_2O{-}\underset{}{\overset{\overset{O}{\Vert}}{C}}{-}CH_3 + 4H \xrightarrow{LiAlH_4} 2CH_3CH_2OH.$
19.	(a)	$CH_3-\underset{\underset{CH_3}{\vert}}{CH}-CH = CH_2 \xrightarrow[(ii)\ \text{Demercuration}]{(i)\ \text{Oxymercuration}} CH_3-\underset{\underset{CH_3}{\vert}}{CH}-\underset{\underset{OH}{\vert}}{CH}-CH_3$ Alcohol formed is in accordance with Markownikoff's rule.
20.	(c)	Ethyl alcohol is manafactured by hydration of ethene as well as fermentation of sugars.
21.	(a)	$CH_3MgI + \underset{\diagdown O \diagup}{CH_2{-}CH_2} \longrightarrow CH_3CH_2CH_2OMgI \xrightarrow{H_2O/H^+} CH_3CH_2CH_2OH$ 1° alcohol
22.	(d)	$CH_2 = CH_2 \xrightarrow{Cl_2/H_2O} \underset{\underset{Cl}{\vert}}{CH_2}{-}\underset{\underset{OH}{\vert}}{CH_2} \xrightarrow{\text{aq. KOH}} \underset{\underset{OH}{\vert}}{CH_2}{-}\underset{\underset{OH}{\vert}}{CH_2}$
23.	(b)	$CH_3CH_2CH_2{-}\overset{\overset{C_2H_5}{\vert}}{\underset{\underset{C_2H_5}{\vert}}{C}}{-}OH$, 3-Ethyl-3-hexanol
24.	(c)	*trans*-2-Methylcyclohexanol is the product.
25.	(c)	X = , Y = 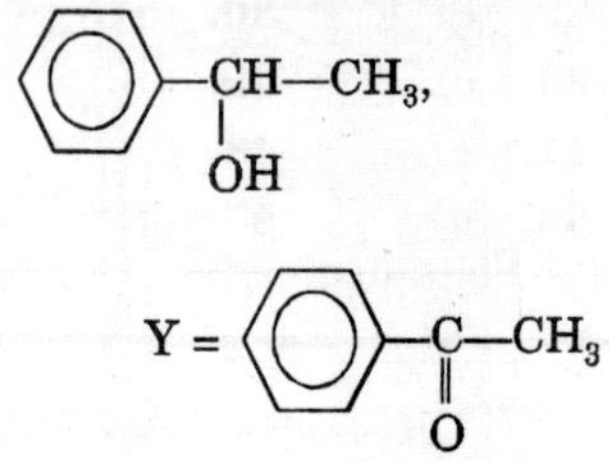
26.	(c)	$\underset{CH_2{-}OH}{\overset{CH_2{-}OH}{\vert}} + 2Na \longrightarrow \underset{CH_2ONa}{\overset{CH_2ONa}{\vert}} + H_2$
27.	(a)	Acidic character of alcohols is $1° > 2° > 3°$. Water is stronger acid than alcohols while acetylene is weaker acid than alcohols.
28.	(c)	None of these would give positive iodoform test.
29.	(c)	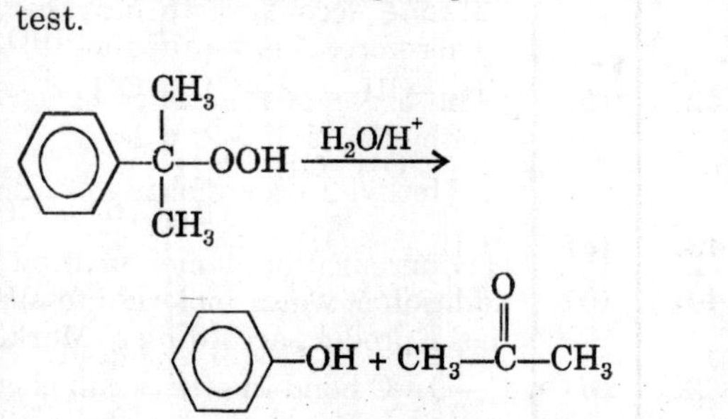

Q. No.	Ans.	Hints & Solutions
30.	(*a*)	See mechanism of Reimer Tiemann reaction.
31.	(*d*)	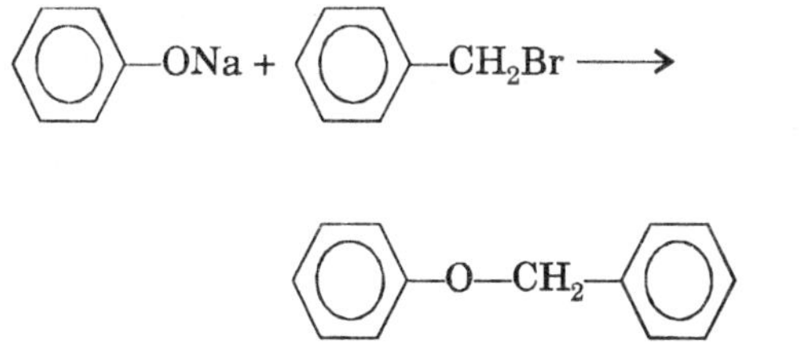 Benzyl phenyl ether
33.	(*c*)	*neo*-Pentyl alcohol on reaction with HCl yields *tert*-pentyl chloride due to rearrangement of the intermediate carbocation.
34.	(*c*)	The four primary alcohols are : 1-pentanol, 3-methyl-1-butanol, 2-methyl-1-butanol and 3, 3-dimethyl-1-propanol.
35.	(*a*)	Alcohols are soluble in conc. H_2SO_4.
36.	(*b*)	2, 3-Pentanediol contains two —OH groups on adjacent carbon atoms.
37.	(*c*)	*iso*-Propyl alcohol is a secondary alcohol. Secondary alcohols give blue colour in Victor Meyer's test.
38.	(*c*)	2-Methyl-2-propanol is a tertiary alcohol.
39.	(*a*)	 $\xrightarrow[Et_2O]{C_2H_5MgBr} CH_3CH_2CH_2CH_2OMgBr$ $\xrightarrow{H_3O^+} CH_3CH_2CH_2CH_2OH.$
40.	(*c*)	2, 3-Pentanediol has two non-equivalent chiral centres $CH_3—\overset{*}{C}H(OH)—\overset{*}{C}H(OH)—CH_2—CH_3$ Hence, there will be $(2)^n$ $[(2)^2]$ stereoisomers.
41.	(*b*)	Z is an alcohol which on reaction with PCl_5 is converted to chloride X. X on reaction with alc. KOH gives alkene Y. Y on reaction with conc. H_2SO_4 and H_2O gives alcohol Z. The alcohol Z is formed by addition of H_2O to alkene according to Markownikoff's rule. Therefore, Z is 2-propanol.
42.	(*b*)	The order of reactivity of various alcohols with HBr is 3° > 2° > 1° 2-Methyl-2-propanol is a 3° alcohol.
43.	(*c*)	$CHCl_3 + OH^- \longrightarrow :CCl_3^- \longrightarrow :\mathbf{CCl_2} + Cl^-$
44.	(*b*)	$CH_3SH + 3O_2 \longrightarrow CO_2 + SO_2 + 2H_2O$ At 298K H_2O will be liquid.
45.	(*a*)	$CH_3CH = CH_2 + H_2O \xrightarrow{H^+} CH_3—CH(OH)—CH_3$ $CH_3—CHO \xrightarrow[(ii)\ H_2O]{(i)\ CH_3MgI} CH_3—CH(OH)—CH_3$ (III) would yield 1-propanol while IV would yield 1, 2-propanediol.
46.	(*d*)	$—NO_2$ group being electron withdrawing group increases the acidic character of phenol while $CH_3—$ group being electron releasing group decreases the acidic character. Effect of group is more when it is at ortho or para position. Hence the order is IV > III > I > II.
47.	(*b*)	Butane-2, 3-diol has three stereoisomers in all, out of which two are optically active while the third (*meso*-form) is optically inactive.
48.	(*c*)	The order of reactivity can be predicted by comparing the relative stability of intermediate carbocation formed. Greater the stability of the intermediate greater is the reactivity. The stability of carbocations formed is in the order. $Ph—\overset{+}{C}H_2 > CH_3—\overset{+}{C}H—CH_3 >$ $F—CH_2—CH_2—\overset{\oplus}{C}H—CH_3 >$ $F—CH_2—\overset{+}{C}H—CH_3$
49.	(*b*)	$Cl_3CCHO + H_2O \longrightarrow Cl_3CCH(OH)_2$ Chloral hydrate
50.	(*b*)	The addition proceeds through more stable benzylic carbocation. The carbocation is further stabilized by electron releasing resonance effect of —OH group. 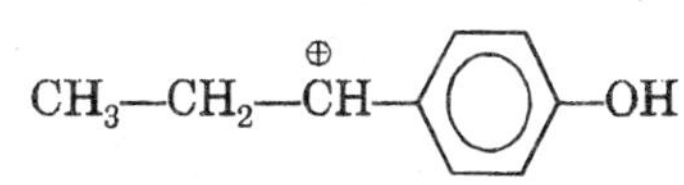
51.	(*d*)	Acidic character is in the order $CH_3OH > HC \equiv CH > C_6H_6 > C_2H_6$

Q. No.	Ans.	Hints & Solutions
52.	(*a*, *d*)	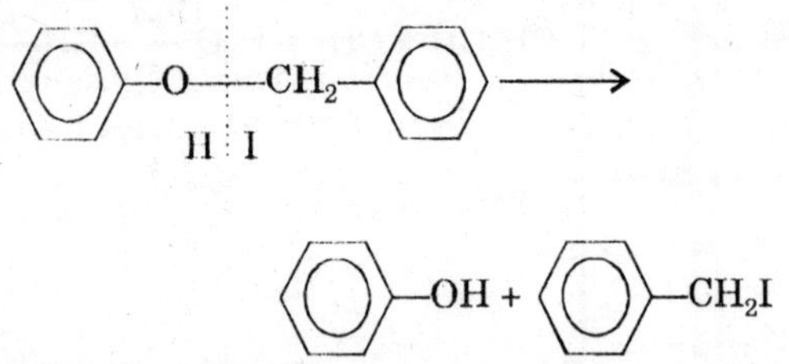
53.	(*c*)	In the given structure (C), there are 10 electrons around nitrogen.
54.	(*d*)	H atoms are substituted one by one by deuterium atoms.
55.	(*c*)	1-Propanol on dehydrogenation gives aldehyde which would give red precipitate with Fehling solution.
56.	(*a*)	An ester reacts with excess Grignard reagent to give a tertiary alcohol in which two alkyl groups come from the Grignard reagent. Thus, the correct choice must contain at least two —CH_3 groups.
57.	(*a*)	$^{-}OC_2H_5$ abstracts a proton from phenol and changes it into phenoxide ion which on reaction with C_2H_5I gives $C_6H_5OC_2H_5$.
58.	(*c*)	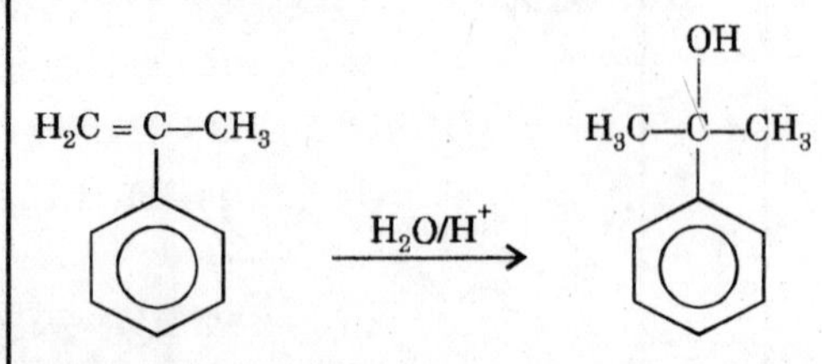 The reaction proceeds through more stable benzylic carbocation.
59.	(*c*)	In the first step *n*-propyl carbocation formed as intermediate rearranges to isopropyl through 1, 2-hydride shift resulting in the subsequent formation of isopropylbenzene (cumene). Cumene on oxidation, followed by hydrolysis gives **acetone** and phenol.
60.	(*a*)	(*iv*) due to its symmetry would have highest boiling point while (I) due its least molecular mass and only one —OH group would have lowest boiling point.

4.7. ASSERTION-REASON QUESTIONS

Quick Revision

In each of the following questions two statements are given as Assertion A and Reason R. Examine the statements carefully and answer the questions according to the instructions given below :

- *Mark* **a** *if both* **A** *and* **R** *are correct and* **R** *is the correct reason of* **A**.
- *Mark* **b** *if both* **A** *and* **R** *are correct and* **R** *is not the correct reason of* **A**.
- *Mark* **c** *if* **A** *is correct and* **R** *is wrong.*
- *Mark* **d** *if* **A** *is wrong and* **R** *is correct.*
- *Mark* **e** *if both* **A** *and* **R** *are wrong.*

1. ***Assertion A.*** Among isomeric butyl alcohols, *sec*-butyl alcohol exhibits enantiomerism.

   Reason R. *sec*-Butyl alcohol has a chiral centre.

(*a*)	(*b*)	(*c*)	(*d*)	(*e*)

2. ***Assertion A.*** 3° alcohols undergo dehydration more readily than 1° alcohols.

   Reason R. 3° alcohols are less acidic than 1° alcohols.

(*a*)	(*b*)	(*c*)	(*d*)	(*e*)

3. ***Assertion A.*** Phenol is stronger acid than alcohols.

   Reason R. Phenol is stabilized by resonance whereas alcohols are not.

(*a*)	(*b*)	(*c*)	(*d*)	(*e*)

4. ***Assertion A.*** Phenols undergo electrophilic substitution at the ring much more readily than aryl halides.

   Reason R. In aryl halides electron density at the ring decreases due to resonance whereas in phenols it increases.

(*a*)	(*b*)	(*c*)	(*d*)	(*e*)

5. ***Assertion A.*** 1-Propanol and 2-propanol can be distinguished by iodoform test.

   Reason R. All secondary alcohols on reaction with NaOH and I_2 yield a yellow precipitate of iodoform.

(*a*)	(*b*)	(*c*)	(*d*)	(*e*)

6. ***Assertion A.*** —OH group in phenols cannot be substitutted easily.

   Reason R. C—O bond in phenols has partial double bond character due to resonance.

(*a*)	(*b*)	(*c*)	(*d*)	(*e*)

7. ***Assertion A.*** Sodium salts of phenols can exist in water whereas sodium salts of alcohols do not exist in water.

   Reason R. Phenol is stronger acid than water whereas alcohol is weaker acid than water.

(*a*)	(*b*)	(*c*)	(*d*)	(*e*)

8. ***Assertion A.*** All ketones on reaction with Grignard reagents, followed by hydrolysis yield tertiary alcohols.

   Reason R. All aldehydes on reaction with Grignard reagents followed by hydrolysis yield secondary alcohols.

(*a*)	(*b*)	(*c*)	(*d*)	(*e*)

9. ***Assertion A.*** Ethanol and ethylene glycol can be distinguished by iodoform test.

   Reason R. Ethylene glycol is a vicinal diol.

(*a*)	(*b*)	(*c*)	(*d*)	(*e*)

10. ***Assertion A.*** *tert*-Butyl alcohol is more soluble in water than *n*-butyl alcohol.

    Reason R. Solubility of an alkyl alcohol in water increases with increase in branching in alkyl group.

(*a*)	(*b*)	(*c*)	(*d*)	(*e*)

11. ***Assertion A.*** Benzyl alcohol turns blue litmus red.

    Reason R. Benzyl alcohol is an isomer of *p*-cresol.

(*a*)	(*b*)	(*c*)	(*d*)	(*e*)

12. ***Assertion A.*** Ethers behave as bases in the presence of mineral acids.

    Reason R. The lone pair on oxygen in ethers can be denoted for sharing with proton from acids.

(a)	(b)	(c)	(d)	(e)

13. ***Assertion A.*** Anisole on reaction with HI gives phenol and CH_3I.

Reason R. Phenyl-oxygen bond is stronger than methyl-oxygen bond in anisole and hence is not cleaved by HI.

(a)	(b)	(c)	(d)	(e)

14. ***Assertion A.*** C—O—C bonds in dimethyl ether are linear.

Reason R. Dimethyl ether is more volatile than ethanol.

(a)	(b)	(c)	(d)	(e)

15. ***Assertion A.*** Glycerol on dehydration with $KHSO_4$ gives acrolein.

Reason R. Acrolein is an α, β-unsaturated aldehyde.

(a)	(b)	(c)	(d)	(e)

16. ***Assertion A.*** Diethyl ether can be dried by passing sodium wire through it.

Reason R. Diethyl ether does not react with sodium but sodium reacts with water and absorbs it.

(a)	(b)	(c)	(d)	(e)

17. ***Assertion A.*** Ethanol is stronger acid than methanol.

Reason R. Ethanol gives positive iodoform test while methanol does not.

(a)	(b)	(c)	(d)	(e)

18. ***Assertion A.*** Phenol is more reactive than benzene towards electrophilic substitution reactions.

Reason R. In the case of phenol, the intermediate carbocation is more resonance stabilized.

(a)	(b)	(c)	(d)	(e)

19. ***Assertion A.*** *p*-Nitrophenol is stronger acid than *o*-nitro phenol.

Reason R. Intramolecular hydrogen bonding in *o*-nitrophenol makes it weaker acid than *p*-nitrophenol.

(a)	(b)	(c)	(d)	(e)

20. ***Assertion A.*** Di-*tert*-butylether cannot be prepared by Williamson synthesis.

Reason R. *tert*-Butyl bromide on treatment with sodium *tert*-butoxide preferentially undergoes elimination to form *iso*-butylene.

(a)	(b)	(c)	(d)	(e)

ANSWERS

1. (*a*)	**2.** (*b*)	**3.** (*b*)	**4.** (*c*)	**5.** (*c*)	**6.** (*a*)	**7.** (*a*)	**8.** (*c*)	**9.** (*b*)
10. (*a*)	**11.** (*d*)	**12.** (*a*)	**13.** (*a*)	**14.** (*d*)	**15.** (*b*)	**16.** (*a*)	**17.** (*d*)	**18.** (*a*)
19. (*a*)	**20.** (*a*).							

HINTS/SOLUTIONS

Q. No.	Ans.	Hints & Solutions
1.	(a)	Self explanatory quesiton.
2.	(b)	The ease of dehydration is 3° > 2° > 1°. The acidic character of alcohols is in the order 1° > 2° > 3°. Hence both the statements are correct but ***R*** is not the explanation of statement ***A***.
3.	(b)	Both the statements are correct but the statement ***R*** is not the explanation ***A***. Phenol is stronger acid than alcohol because the conjugate base of phenol is resonance stabilized while that of alcohol is not.
4.	(c)	The statement ***R*** is wrong. In aryl halides electron density at the ring decreases due to elec-

Q. No.	Ans.	Hints & Solutions
		tron withdrawing inductive effect of halogen atom.
5.	(c)	The statement *R* is wrong. Only those secondary alcohols which have $CH_3-\underset{\displaystyle OH}{\underset{\vert}{CH}}-$ group react with NaOH and I_2 to yield iodoform.
6.	(a)	The statement *R* is the correct explanation of *A*.
7.	(a)	The statement *R* is the correct explanation of *A*.
8.	(c)	The statement *R* is wrong. Formaldehyde on reaction with Grignard reagents yields primary alcohols.
9.	(b)	Ethanol gives positive iodoform test. Ethylene glycol has two —OH groups on adjacent carbon atoms.
10.	(a)	Self explanatory question.
11.	(d)	Statement *A* is wrong. Benzyl alcohol is an aromatic alcohol. It does not turn blue litmus red. Statement **R** is correct.
12.	(a)	$R-\ddot{\underset{\cdot\cdot}{O}}-R + H^+ \longrightarrow R-\underset{\oplus}{\overset{H}{\overset{\vert}{O}}}-R$ Self explanatory question.
13.	(a)	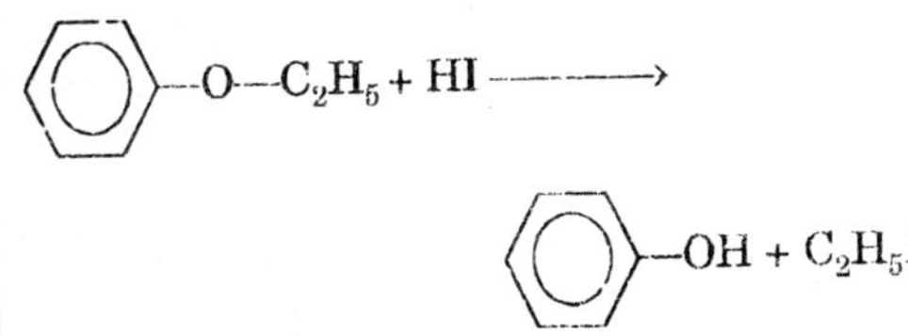 Self explanatory question.
14.	(d)	C—O—C bonds in ethers are bent due to the presence of lone pairs on oxygen. Hence the statement **A** is wrong. Dimethyl ether is more volatile than ethanol because of weaker intermolecular dipole-dipole forces.
15.	(b)	Both the statements are correct, but *R* is not the explanation of *A*.
16.	(a)	Self explanatory correct statements
17.	(d)	Assertion *A* is wrong because methanol is stronger acid than ethanol.
18.	(a)	Self explanatory quesiton. The statement *R* is the correct explanation of *A*.
19.	(a)	The statement *R* is the correct explanation of the statement *A*.
20.	(a)	3° halides undergo elimination readily and hence cannot be used for preparing corresponding ethers by Williamson synthesis.

4.8. SELF EVALUATION TEST

Assess Yourself

Choose the correct answer from the four alternatives given in each of the following questions :

1. Order of reactivity of alcohols towards sodium metal is
(*a*) 3° > 2° > 1° (*b*) 1° > 2° > 3°
(*c*) 3° = 2° = 1° (*d*) 2° > 3° > 1°.

2. Which of the following compounds is oxidised to prepare ethyl methyl ketone ?
(*a*) 2-Butanol (*b*) 2-Propanol
(*c*) 1-Propanol (*d*) 2-Pentanol.

3. Ethylene glycol on oxidation with periodic acid gives
(*a*) Oxalic acid (*b*) Formic acid
(*c*) Formaldehyde (*d*) Glycolic acid.

4. Glycerol on treatment with oxalic acid at 110°C forms
(*a*) Formic acid (*b*) Allyl alcohol
(*c*) CO_2 and CO (*d*) Acrolein.

5. $2ClCH_2CH_2OH \xrightarrow{\text{Conc. } H_2SO_4} X \xrightarrow[\text{Fuse}]{\text{KOH}} Y.$
The compound Y is
(*a*) Ethylene glycol (*b*) Ethanal
(*c*) Divinyl ether (*d*) Vinyl chloride.

6. Which of the following can work as dehydrating agent for alcohols ?
(*a*) Al_2O_3 (*b*) H_2SO_4
(*c*) H_3PO_4 (*d*) All of these.

7. HBr reacts faster with
(*a*) Propan-1-ol (*b*) Propan-2-ol
(*c*) 2-Methylpropan-1-ol (*d*) 2-Methylpropan-2-ol.

8. Fuse oil is a mixture of
(*a*) Ethers (*b*) Alcohols
(*c*) Alcohols and ethers (*d*) Alcohols and ketones.

9. The interparticle forces in ethers are
(*a*) H-bonds (*b*) Dipole-dipole
(*c*) Covalent bonds (*d*) Ion-dipole.

10. Dehydrogenation of *iso*-butyl alcohol yields
(*a*) Butanone (*b*) Butanal
(*c*) 2-Methylpropanal (*d*) 2-Methylpropene.

11. How many isomeric compounds are possible for $C_4H_{10}O$?
(*a*) 4 (*b*) 5
(*c*) 6 (*d*) 7.

12. Which of the following would undergo esterification most readily ?
(*a*) CH_3OH (*b*) CH_3CH_2OH
(*c*) $CH_3CH_2CH_2OH$ (*d*) $CH_3—CH(OH)—CH_3$.

13. An industrial method of preparation of methanol is
(*a*) Catalytic reduction of carbon monoxide in the presence of $ZnO—Cr_2O_3$
(*b*) By reduction of formaldehyde with $LiAlH_4$
(*c*) By reaction of formaldehyde with aq. NaOH
(*d*) By dehydrogenation of methanol with reduced copper at 575 K.

14. Which of the following alcohols is least soluble in water ?
(*a*) CH_3OH (*b*) C_2H_5OH
(*c*) $C_5H_{11}OH$ (*d*) $C_{10}H_{21}OH$.

15. *n*-Propyl alcohol on treatment with excess hot conc. H_2SO_4 gives compound 'X'. 'X' on treatment with HBr in the presence of benzoyl peroxide gives 'Y'. 'Y' when dissolved in ether and treated with sodium gives
(*a*) *n*-Hexane (*b*) 2-Methylpentane
(*c*) 2, 3-Dimethylbutane (*d*) 3-Methylpentane.

16. Which of the following alcohols is most acidic ?
(*a*) CH_3OH (*b*) C_2H_5OH
(*c*) $(CH_3)_2$>CH—OH (*d*) $(CH_3)_3.C.OH$.

17. To prepare butan-2-ol using CH_3MgI, what other chemical would you choose ?
(*a*) Propanal (*b*) Ethanal
(*c*) Ethanol (*d*) Propan-2-ol.

18. When passed over heated copper at 575 K, an organic compound was found to give an olefin. The organic compound is most likely to be a/an
(*a*) Primary alcohol (*b*) Secondary alcohol
(*c*) Tertiary alcohol (*d*) Alkane.

19. Which of the following is not characteristic of alcohols ?
(*a*) Lower alcohols are soluble in water
(*b*) The boiling points of alcohols increase with increase in molecular mass
(*c*) They are lighter than water
(*d*) They are stronger acids than water.

20. 46 g of sodium on complete reaction with methanol would produce ...... of H_2 at N.T.P.

(*a*) 22.4 L (*b*) 11.2 L

(*c*) 44.8 L (*d*) 5.6 L.

21. Which of the following alcohols would give a ketone on reaction with hot reduced copper ?

(*a*) *n*-Butyl alcohol (*b*) *o*-Butyl alcohol

(*c*) *tert*-Butyl alcohol (*d*) *sec*-Butyl alcohol.

22. Which of the following compounds will give a yellow precipitate with iodine and alkali ?

(*a*) 1-Propanol (*b*) 1-Butanol

(*c*) Ethanol (*d*) Methanol.

23. When phenol is treated with bromine in the presence of CS_2 as solvent at 273 K, it gives ...... as the major product.

(*a*) *m*-Bromophenol (*b*) *p*-Bromophenol

(*c*) 2, 4-Dibromophenol (*d*) 2, 4, 6-Tribromophenol.

24. Ethyl alcohol on oxidation with acidified $K_2Cr_2O_7$ gives

(*a*) Ethanal (*b*) Methanal

(*c*) Ethanoic acid (*d*) Methanoic acid.

25. IUPAC name of propargyl alcohol is

(*a*) Prop-2-yn-1-ol (*b*) Prop-1-yn-1-ol

(*c*) Prop-1-yn-2-ol (*d*) Prop-2-en-1-ol.

26. C_2H_5MgBr is treated with CH_3OH. The alkane formed is

(*a*) CH_4 (*b*) C_2H_6

(*c*) $CH_3CH_2CH_3$ (*d*) $CH_3CH_2CH_2CH_3$.

27. Which of the following compounds can exhibit enantiomerism ?

(*a*) 1, 2-Dichloroethane (*b*)- 1, 3-Dichloropropane

(*c*) 1, 2-Dichloropropane (*d*) 1, 1-Dichloropropane.

28. Which of the following would produce a tertiary alcohol on reaction with excess CH_3MgI ?

(*a*) Propanol (*b*) Ethylacetate

(*c*) Ethylene (*d*) Methanal.

29. Which among the following has the highest boiling point ?

(*a*)  (*b*) 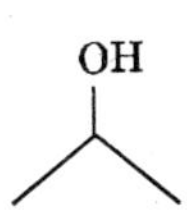

(*c*) 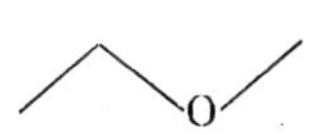 (*d*) 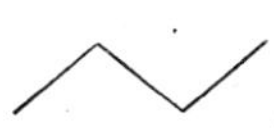

30. The presence of —OH groups on adjacent carbon atoms can be detected by the reaction of the compound with

(*a*) conc. H_2SO_4 (*b*) conc. HNO_3

(*c*) HIO_4 (*d*) acidified $KMnO_4$.

31. Which of the following compounds is most acidic ?

(*a*) 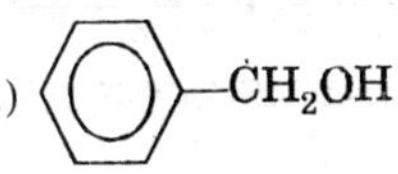 (*b*) 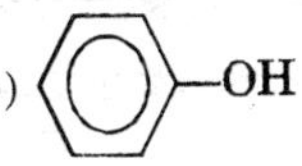

(*c*) CH_3CH_2OH (*d*) $CH_3CH(OH)CH_3$.

32. Acrolein is formed from glycerol by heating with

(*a*) HNO_3 (*b*) $KHSO_4$

(*c*) HI (*d*) Acidified $KMnO_4$.

33. Phenol gives anisole on reaction with

(*a*) Sodium ethoxide

(*b*) Sodium methoxide

(*c*) Diazomethane in the presence of HF

(*d*) Methanol.

34. $CH_3CH = CH_2 \xrightarrow{HCl} X \xrightarrow[\text{Heat}]{\text{Dry } Ag_2O} Y$

The product Y in the above sequence is

(*a*) Di-isopropyl ether (*b*) Di-*n*-propyl ether

(*c*) 2-Propanol (*d*) 1, 2-Epoxypropane.

35. Which of the following isomeric alcohols will produce turbidity first with Lu is reagent ?

(*a*) 1-Pentanol (*b*) 2-Pentanol

(*c*) 3-Methyl-2-butanol (*d*) 2-Methyl-2-butanol.

36. Which one of the following gives red colour in Victor Meyer's test ?

(*a*) *n*-Propyl alcohol (*b*) *iso*-Propyl alcohol

(*c*) *tert*-Butyl alcohol (*d*) *sec*-Butyl alcohol.

37. Which of the following compound has maximum value of Ka ?

(*a*) 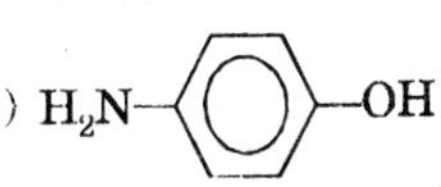 (*b*) 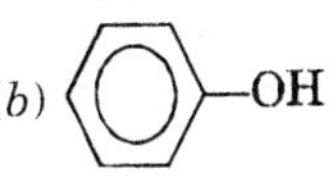

(*c*) 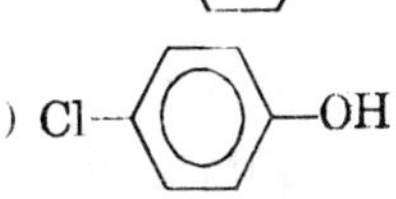 (*d*) 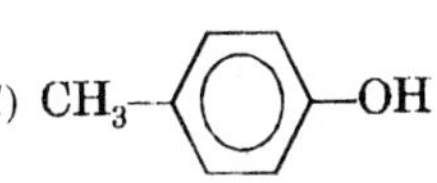

38. '*Rubbing*' alcohol is

(*a*) Methanol (*b*) Ethanol

(*c*) 2-propanol (*d*) Ethylene glycol.

39. Ethers can be cleaved by

(*a*) Alkalies (*b*) Acids

(*c*) Oxidizing agents (*d*) Reducing agents.

40. Propene can be converted to 1-propanol by

(*a*) Oxymercuration-demercuration

(*b*) Acid catalysed hydration

(*c*) Hydroboration oxidation

(*d*) Reaction with ozone.

ANSWERS

1. (*b*)	**2.** (*a*)	**3.** (*c*)	**4.** (*a*)	**5.** (*c*)	**6.** (*d*)	**7.** (*d*)	**8.** (*b*)	**9.** (*b*)
10. (*c*)	**11.** (*d*)	**12.** (*a*)	**13.** (*a*)	**14.** (*d*)	**15.** (*a*)	**16.** (*a*)	**17.** (*a*)	**18.** (*c*)
19. (*d*)	**20.** (*a*)	**21.** (*d*)	**22.** (*c*)	**23.** (*b*)	**24.** (*c*)	**25.** (*a*)	**26.** (*b*)	**27.** (*c*)
28. (*b*)	**29.** (*a*)	**30.** (*c*)	**31.** (*b*)	**32.** (*b*)	**33.** (*c*)	**34.** (*a*)	**35.** (*d*)	**36.** (*a*)
37. (*c*)	**38.** (*c*)	**39.** (*b*)	**40.** (*c*).					

HINTS/SOLUTIONS

Q. No.	*Ans.*	*Hints & Solutions*
1.	**(*b*)**	The order of reactivity of alcohols in the reactions involving cleavage of O—H bond is 1° > 2° > 3°.
2.	**(*a*)**	$CH_3CH_2\underset{\displaystyle OH}{\underset{\vert}{C}}HCH_3 \xrightarrow{[O]} CH_3CH_2—\underset{\displaystyle O}{\underset{\Vert}{C}}—CH_3$ 2-Butanol Ethyl methyl ketone
3.	**(*c*)**	$HOCH_2—CH_2OH + [O] \xrightarrow{HIO_4} 2HCHO + H_2O$
4.	**(*a*)**	Formic acid.
5.	**(*c*)**	$X = ClCH_2CH_2—O—CH_2CH_2Cl$ $Y = CH_2 = CH—O—CH = CH_2$.
6.	**(*d*)**	All these reagents convert alcohols to alkenes.
7.	**(*d*)**	The order of reactivity of alcohols towards HBr is 3° > 2° > 1°. 2 Methyl propan-2-ol is a 3° alcohol.
8.	**(*b*)**	Fusel oil is a mixture of normal alcohols.
9.	**(*b*)**	Due to angular C—O—C bond ethers have net dipole moment.
10.	**(*c*)**	$CH_3—\underset{\displaystyle CH_3}{\underset{\vert}{C}}H—CH_2OH \longrightarrow CH_3—\underset{\displaystyle CH_3}{\underset{\vert}{C}}H—CHO + H_2$
11.	**(*d*)**	For $C_4H_{10}O$, four isomeric alcohols and three isomeric ethers are possible.
12.	**(*a*)**	The order of reactivity of alcohols towards esterification is : CH_3OH > 1° > 2° > 3°.
13.	**(*a*)**	$CO + 2H_2 \xrightarrow{ZnO/Cr_2O_3} CH_3OH$.
14.	**(*d*)**	Solubility of alcohols in water decreases with increase in size of alkyl group.
15.	**(*a*)**	$CH_3CH_2CH_2OH \xrightarrow{Cr\ \ H_2SO_4}$ $CH_3CH = CH_2 \xrightarrow[Peroxide]{HBr}$ $CH_3CH_2CH_2Br \xrightarrow[ether]{Na}$ $CH_3CH_2CH_2CH_2CH_2CH_3$
16.	**(*a*)**	Acid strength among alcohols varies are : CH_3OH > 1° > 2° > 3°.
17.	**(*a*)**	$\mathbf{CH_3CH_2CHO} + CH_3MgI \longrightarrow$ $CH_3CH_2\underset{\displaystyle OMgI}{\underset{\vert}{C}}HCH_3 \xrightarrow{H_2O/H^+}$ $CH_3—CH_2—\underset{\displaystyle OH}{\underset{\vert}{C}}H—CH_3$ Butan-2-ol
18.	**(*c*)**	3° alcohols undergo dehydration to form an alkene, when heated with reduced copper.
19.	**(*d*)**	Alcohols are weaker acids than water.
20.	**(*a*)**	$2Na + 2CH_3OH \longrightarrow 2CH_3ONa + H_2$. 46 g sodium on reaction with methanol produce 1 mole of H_2. Volume of 1 mole H_2 at N.T.P. is 22.4 L.
21.	**(*d*)**	2° alcohols on reaction with hot reduced copper yield ketones.
22.	**(*c*)**	Ethanol contains $CH_3—\underset{\displaystyle OH}{\underset{\vert}{C}}H—$ group and hence can undergo iodoform reaction.

Q. No.	Ans.	Hints & Solutions
23.	(*b*)	In the presence of non-polar solvent such as CS_2 monosubstitution takes place and para-bromophenol is the major product.
24.	(*c*)	Primary alcohols on reaction with acidified $K_2Cr_2O_7$ are oxidised to carboxylic acids with same number of carbon atoms.
25.	(*a*)	$\overset{3}{H{-}C} \equiv \overset{2}{C}{-}\overset{1}{CH_2}OH$ Propargyl alcohol (Prop-2-yn-1-ol).
26.	(*b*)	$C_2H_5MgBr + CH_3OH \longrightarrow C_2H_6 + MgBr(OCH_3)$.
27.	(*c*)	$CH_3{-}\overset{*}{C}H(Cl){-}CH_2Cl$ C-2 is chiral centre.
28.	(*b*)	Esters yield 3° alcohols on reaction with Grignard reagent.
29.	(*a*)	Boiling points of alcohols are higher than that of ethers or alkanes of comparable mass. 1-Propanol has higher boiling point than 2-propanol due to its greater surface area.
30.	(*c*)	Periodic acid test is used to identify vicinal diols.
31.	(*b*)	Phenol is stronger acid than alcohols.
32.	(*b*)	Dehydration of glycerol with $KHSO_4$ yields acrolein.

Q. No.	Ans.	Hints & Solutions
33.	(*c*)	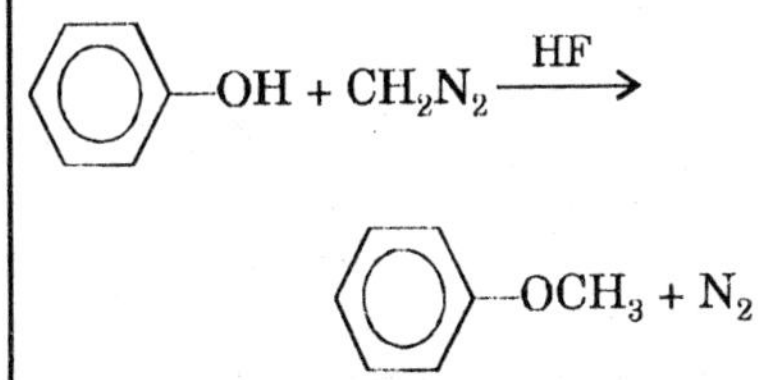
34.	(*a*)	$X = CH_3CH(Cl)CH_3$; $Y = (CH_3)_2CH{-}O{-}CH(CH_3)_2$
35.	(*d*)	Tertiary alcohols produce turbidity fastest.
36.	(*a*)	Primary alcohols give red colour in Victor Meyer's test.
37.	(*c*)	Cl—group is electron withdrawing group. It increases the acid strength of phenol.
38.	(*c*)	*iso*-Propyl alcohol (2-Propanol) is known as '*rubbing alcohol*'.
39.	(*b*)	Ether linkage can be cleaved by acids such as halogen acids, H_2SO_4, etc.
40.	(*c*)	Hydroboration oxidation results in addition of water in anti-Markownikoff fashion.

5 ALDEHYDES AND KETONES

5.1. CONCEPTS IN BRIEF

Develop Logic

GENERAL

- Aldehydes and ketones are first oxidation products of alcohols and carry carbonyl group ($>C = O$) in their molecules.
- Thus functional group of **aldehydes is —CHO** while that for **ketones** is $>\mathbf{C = O}$.
- In **aliphatic aldehdyes,** —CHO group is linked to alkyl group (**R—CHO**) while in **aromatic aldehyde**, —CHO group is linked to aryl group **(Ar—CHO).**
- In **aliphatic ketones**, the CO group is linked to alkyl group **(RCOR)** while in **aromatic ketones,** the CO group is linked to aryl group **(ArCOAr)**
- In **simple ketones**, the alkyl or aryl groups linked to $>C = O$ group are same but in **mixed ketones**, the alkyl or aryl groups linked to $>C = O$ group are different.
- Aldehydes and ketones are functional isomers of each other and can be represented by the formula $\mathbf{C_nH_{2n}O}$.

NOMENCLATURE

- Common names of aldehyde is derived from the corresponding carboxylic acids by replacing the suffix *ic acid* with *aldehyde.*
- Common name of ketones is written by using the names of the alkyl groups followed by the word *ketone.*
- *IUPAC* name of aldehydes is ***alkanals*** while that for ketones is ***alkanones.***

- Aldehydes are **isomeric** with ketones, unsaturated ethers and unsaturated alcohols, cyclic ethers as well as cyclic alcohols.

Illustration

The total number of possible isomers for C_3H_6O are :

CH_3CH_2CHO ;
Propanal

$CH_3\overset{O}{\overset{\|}{C}}CH_3$;
Propanone

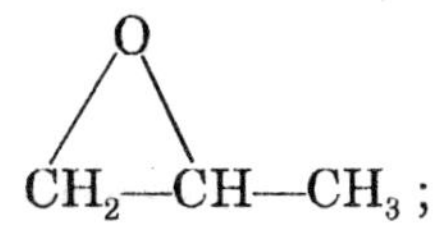
Epoxyethane

$CH_2 = CH—OCH_3$;
Methoxy ethene

$CH_2 = CH—CH_2OH$;
Methoxy ethene

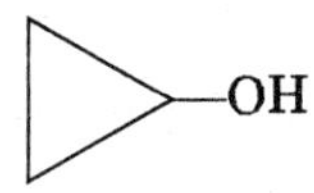
Cyclopropanol

METHODS OF PREPARATION OF ALDEHYDES

1. **From 1° Alcohols (RCH_2OH)**

(*a*) By controlled oxidation using $K_2Cr_2O_7/H_2SO_4$ or K_2MnO_4/H_2SO_4

$$RCH_2OH \xrightarrow{(O)} RCHO + H_2O$$

- Controlled oxidation can be effectively achieved by $(C_5H_5N)_2CrO_3$ in CH_2Cl_2 solvent. The reagent is called **Sarett's Collin's**-reagent which converts $-CH_2OH \longrightarrow -CHO$ group.

(*b*) By dehydrogenation of 1° alcohol using Cu at 575 K.

$$R—CH_2OH \xrightarrow{Cu,\ 575\ K} RCHO + H_2$$

2. **From Acid Chlorides (RCOCl).** By reduction with H_2 in the presence of $Pd/BaSO_4$

$$RCOCl + H_2 \xrightarrow[Xylene]{Pd/BaSO_4} RCHO + HCl$$

- *$BaSO_4$ poisons the catalyst and reduces its efficiency. Thus, it checks the further reduction of aldehydes to alcohols.*
- *This reaction is called **Rosenmund's reaction.***
- *Formaldehyde cannot be prepared by this method because formyl chloride is highly unstable and non-existing.*

3. **From Alkenes**

(*a*) By reducing ozonolysis of alkene of the type RCH = CHR.

$$\underset{\text{(Alkene)}}{R—\overset{H}{\overset{|}{C}} = \overset{H}{\overset{|}{C}}—R} \xrightarrow[(ii)\ H_2O/Zn]{(i)\ O_3} \underset{\text{(Aldehydes)}}{R—\overset{H}{\overset{|}{C}} = O + O = \overset{H}{\overset{|}{C}}—R} + H_2O_2$$

(*b*) **Wacker's process.** It involves the conversion of *terminal alkenes* to aldehydes by oxidation in air in the presence of $PdCl_2/CuCl_2$.

$$\underset{\text{(Alkene)}}{R—CH = CH_2} + \underset{\text{(Air)}}{\tfrac{1}{2}O_2} \xrightarrow[CuCl_2]{PdCl_2} \underset{\text{(Aldehyde)}}{R—CH_2—CHO}$$

(*c*) **Oxo process.** It involves conversion of alkenes to aldehydes with one C atom more than parent alkene.

$$\underset{\text{(Alkene)}}{R—CH = CH_2} + \underset{\text{(Water gas)}}{CO + H_2} \xrightarrow{CO_2(CO)_8} \underset{\text{(Aldehyde)}}{R—CH_2—CH_2—CO}$$

4. From Terminal Gem. Dihalides. By alkaline hydrolysis the terminal gem. dihalides ($—CHCl_2$) get converted into aldehydes

$$RCHCl_2 \xrightarrow{KOH\,(aq)} RCH(OH)_2 \xrightarrow{-H_2O} \underset{\text{(Aldehyde)}}{RCHO}$$

- *This is not a frequently employed method because RCHO in alkaline medium further undergo aldol type condensation.*

5. Hydration of Ethyne

$$\underset{\text{Ethyne}}{CH \equiv CH} + H_2O \xrightarrow[H_2SO_4]{Hg^{2+}} \underset{\text{Ethanal}}{CH_3CHO}$$

- *Only ethanal can be prepared by this method.*

Terminal alkynes can also be converted to aldehydes on ***hydroboronation*** with ****disiamyl borane*** followed by oxidation

$$RC \equiv CH + Si\alpha_2BH \longrightarrow R—CH = CH—BSi\alpha_2 \xrightarrow{H_2O_2/NaOH} \underset{\text{Aldehyde}}{RCH_2CHO}$$

**disiamyl borane is* $\left(CH_3—\overset{CH_3}{\overset{|}{C}H}—\overset{CH_3}{\overset{|}{C}H}—\right)_2 BH$. *It is abbreviated as* $Si\alpha_2BH$.

6. From Nitriles. By Stephen's reduction using $SnCl_2/HCl$.

$$R—C \equiv N \xrightarrow[2H]{SnCl_2/HCl} \underset{\text{Aldimine hydrochloride}}{R—CH = NH\,.\,HCl} \xrightarrow{H_2O} \underset{\text{Aldehyde}}{RCHO} + NH_4Cl$$

7. Decarboxylation of Calcium Salt of fatty Acid. It involves distillation of a mixture of calcium carboxylate and calcium formate

$$(RCOO)_2Ca + Ca(OOCH)_2 \xrightarrow{\text{Dry Distill.}} CaCO_3 + 2RCHO$$

8. From Fatty Acids. Passing vapours of a mixture of formic acid and higher fatty acids over MnO at 575 K.

$$RCOOH + HCOOH \xrightarrow[575\text{ K}]{MnO} RCHO + H_2O + CO_2$$

- *Distillation of formic acid alone under these conditions produces formaldehyde.*

9. From Grignard's Reagents. Acid chlorides and esters can be converted to aldehydes and ketones by reaction with Grignard's reagent.

$$H\overset{O}{\overset{\|}{C}}—OC_2H_5 + \overset{-}{C}H_3\overset{+}{Mg}Br \xrightarrow[H_2O/H^+]{\text{Dry ether}} CH_3CHO + C_2H_5OH$$

$$CH_3—\overset{O}{\overset{\|}{C}}—Cl + CH_3\overset{+}{Mg}Br \xrightarrow[H_2O/H^+]{\text{Dry ether}} CH_3—\overset{O}{\overset{\|}{C}}—CH_3 + HCl$$

10. From Glycol. By oxidation with periodic acid or lead tetra acetate

$$\begin{array}{l} CH_2OH \\ | \\ CH_2OH \end{array} \xrightarrow{HIO_4} 2HCHO + H_2O + HIO_3$$

$$\begin{array}{l} CH_2OH \\ | \\ CH_2OH \end{array} \xrightarrow{(CH_3COO)_4Pb} 2HCHO + 2CH_3COOH + (CH_3COO)_2Pb$$

The above reaction is called **Criegee's reaction.**

11. From Glycerol. By oxidation with Fenton's reagent or periodic acid

$$\begin{array}{l} CH_2OH \\ | \\ CHOH \\ | \\ CH_2OH \end{array} \xrightarrow[\text{Fenton's reagent}]{H_2O_2/FeSO_4} \underbrace{\begin{array}{l} CHO \\ | \\ CHOH \\ | \\ CH_2OH \end{array} + \begin{array}{l} CH_2OH \\ | \\ C=O \\ | \\ CH_2OH \end{array}}_{\text{Glycerose}}$$

$$\begin{array}{l} CH_2OH \\ | \\ CHOH \\ | \\ CH_2OH \end{array} \xrightarrow[\text{Periodic acid}]{HIO_4/\Delta} 2HCHO + HCOOH + 2HIO_3 + H_2O$$

PREPARATION OF AROMATIC ALDEHYDES

1. Oxidation of Toluene

(*a*) Oxidation using chromyl chloride CrO_2Cl_2

$$\underset{\text{Toluene}}{C_6H_5{-}CH_3} \xrightarrow{CrO_2Cl_2} \underset{\text{Complex}}{C_6H_5CH(OCrCl_2OH)_2} \xrightarrow{H_2O} \underset{\text{Benzaldehyde}}{C_6H_5CHO}$$

- *This reaction is called* **Etard's reaction.**
- *If side chain is bigger than —CH_3 group, then terminal C is changed to —CHO group. For example,*

$$C_6H_5CH_2CH_3 \xrightarrow{CrO_2Cl_2} \underset{\text{Phenyl ethanal}}{C_6H_5CH_2CHO}$$

(*b*) Oxidation using CrO_3 is acetic anhydride.

$$C_6H_5{-}CH_3 \xrightarrow[(CH_3CO)_2O]{CrO_3(O)} \underset{\text{Benzal diacetate}}{[C_6H_5CH(OCOCH_3)_2]} \xrightarrow{H_2O} C_6H_5CHO + 2CH_3COOH$$

2. Gattermann Koch Reaction. Involves formylation of aromatic ring with CO.

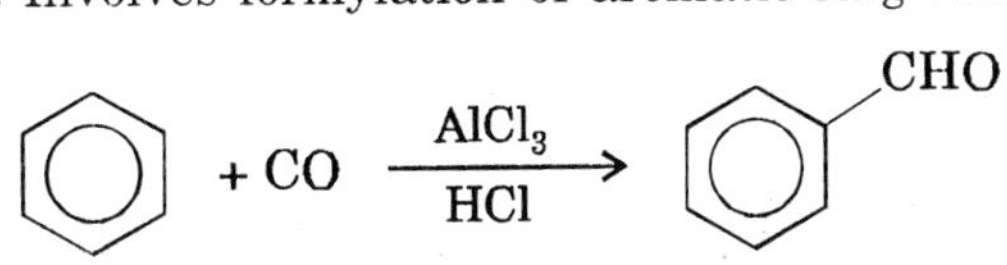

3. Gattermann Aldehyde Synthesis. It involves treatment of benzene or other aromatic compounds having activating group like, —OH, —OR, etc. with a mixture of HCN and HCl in the presence of $AlCl_3$.

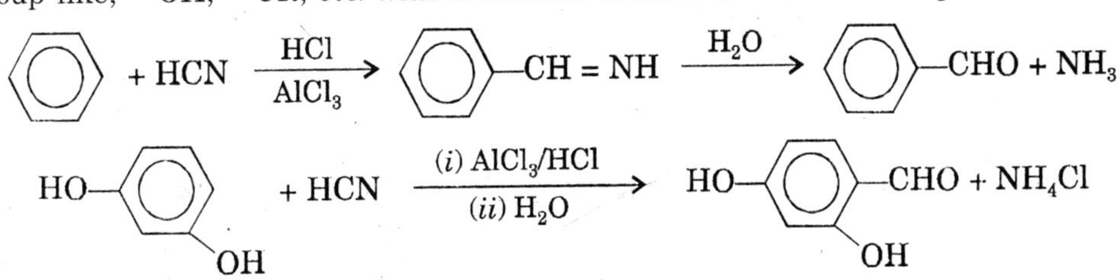

4. **Vilsmeyer Reaction.** It involves the conversion of arene to aromatic aldehyde by using formic acid and 2° amine.

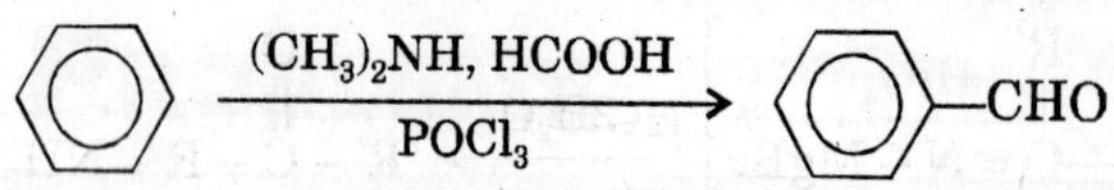

PREPARATION OF KETONES

1. **Oxidation of 2° Alcohols (R_2CHOH)**

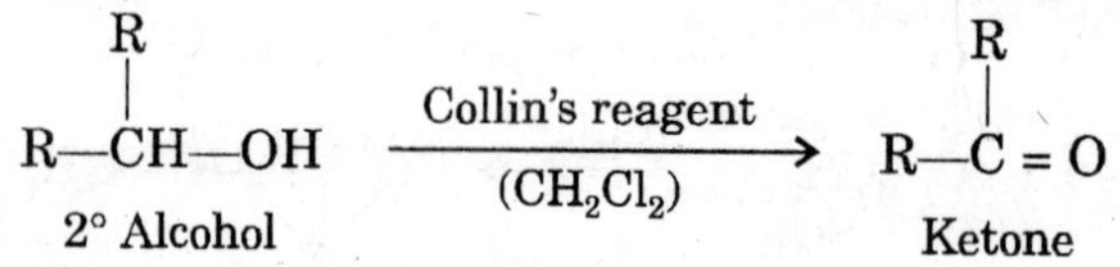

2. **Dehydrogenation of 2° Alcohols**

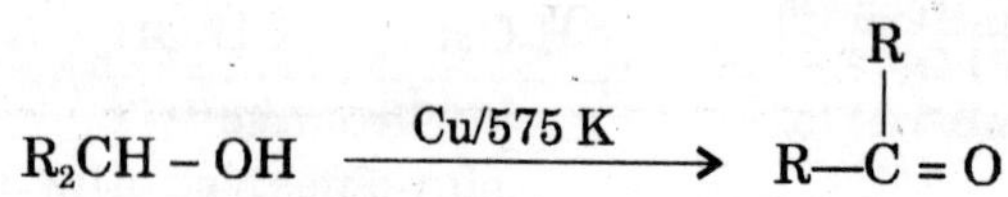

3. **From Acid Chlorides.** By treatment of acid chlorides with dialkyl cadmium (R'_2Cd).

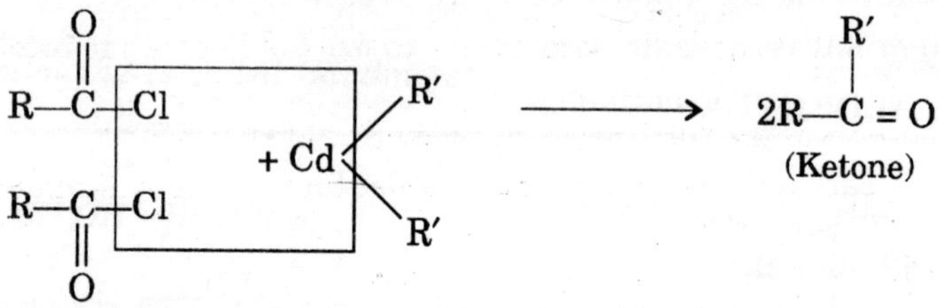

- *The product contains one alkyl group from R_2Cd and the other from RCOCl.*

4. **By Hydration of Higher Terminal Alkynes**

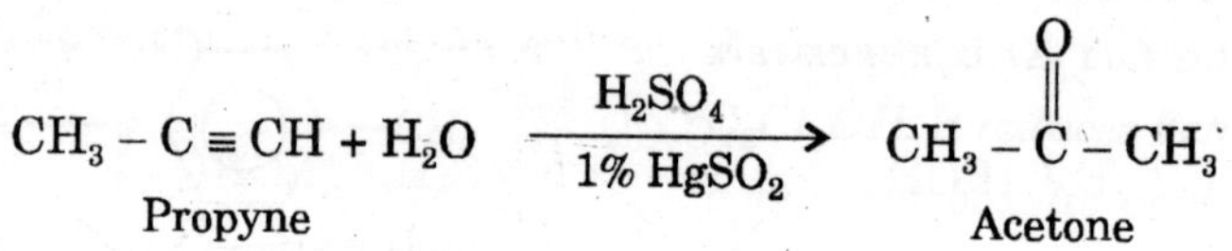

5. **By Reductive Ozonolysis of Alkene of the Type $R_2C = CR_2$**

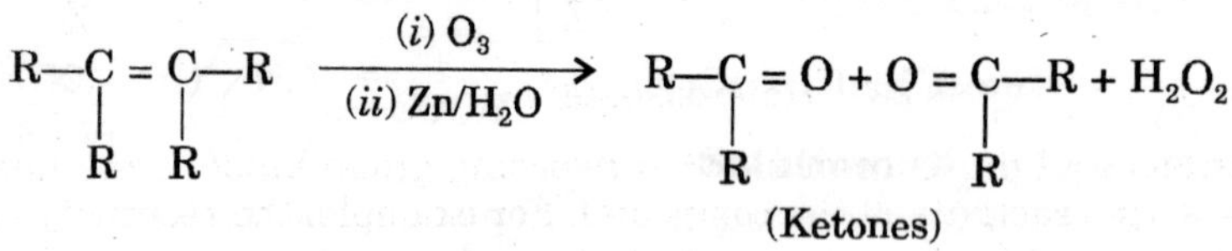

6. **By Distilling the Calcium Salt of Fatty Acids**

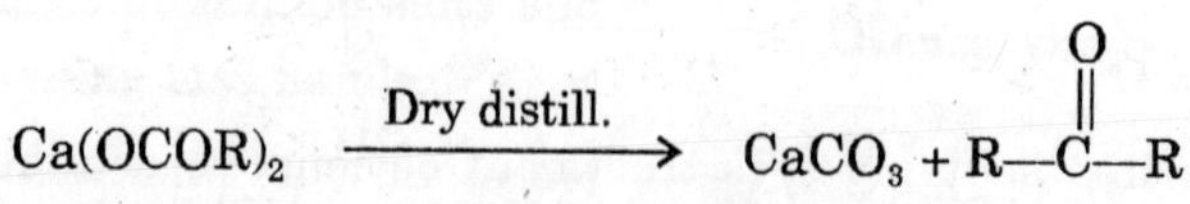

7. **By Alkaline Hydrolysis of Non-terminal Gem. Dehalides**

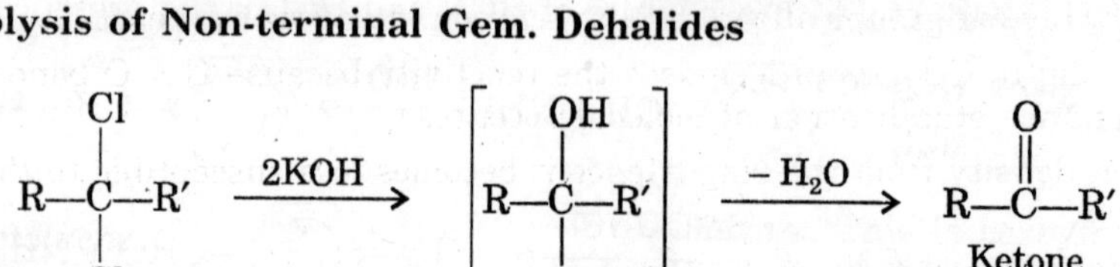

8. From Nitriles. By treatment with Grignard's reagent followed by hydrolysis

$$R'C \equiv N \xrightarrow{RMgX} \left[R-\overset{\displaystyle R'}{\overset{|}{C}} = N - MgBr \right] \xrightarrow{2H_2O} \underset{\text{Ketone}}{R' - \overset{\displaystyle O}{\overset{\|}{C}} - R} + NH_3 + Mg(OH)Br$$

9. Aromatic ketones are generally prepared by Friedel Craft reactions

- **Acylation of benzene**

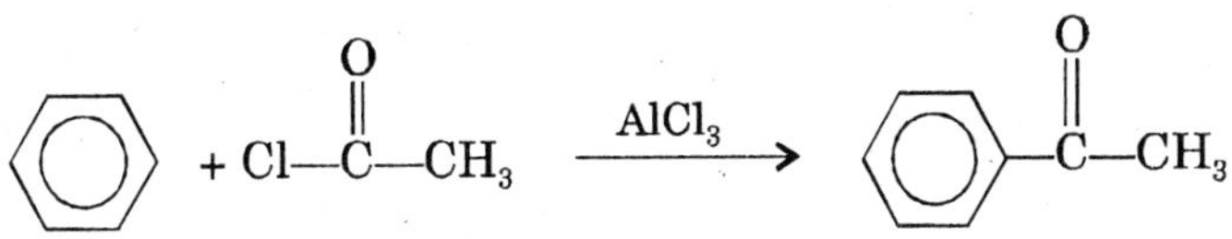

Acetophenone

PROPERTIES OF ALDEHYDES AND KETONES

Physical Properties

- Formaldehyde is gas but other aldehydes are liquids at room temperature.
- Lower ketones are colourless liquids while higher ketones are colourless solids.
- Boiling points of ald./ketones are relatively higher than hydrocarbons but lower than alcohols of comparable molecular mass.
- Boiling points of ketones are more than the corresponding isomeric aldehydes.
- Lower aldehydes/ketones (up to 4 carbons) are soluble in water but higher aldehydes/ketones are almost immiscible with water.

Chemical properties

- Aldehydes and ketones owe their highly reactive nature to the polar carbonyl group $\left(>\overset{\delta+}{C} = \overset{\delta-}{O} \right)$ in their molecules.
- The electron deficient carbon atom of $> C = O$ group readily undergo the nucleophilic attack.

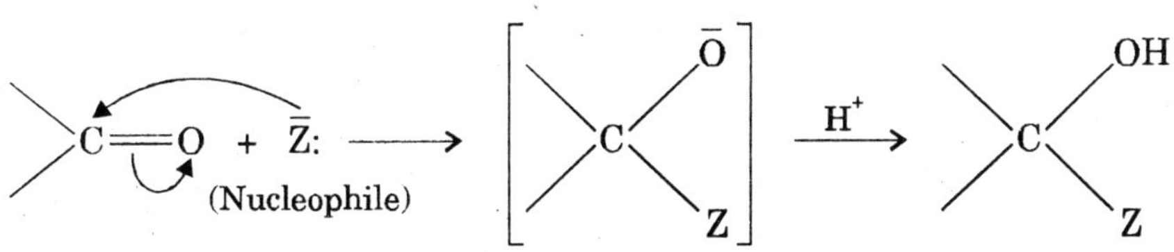

(Intermediate)

- Crowding at the carbonyl carbon and presence of electron releasing group hinder's the nucleophilic attack on the carbonyl carbon and decreases the reactivity of the compound. For example, the reactivity of HCHO, CH_3CHO and $(CH_3)_2CO$ decreases as follows.

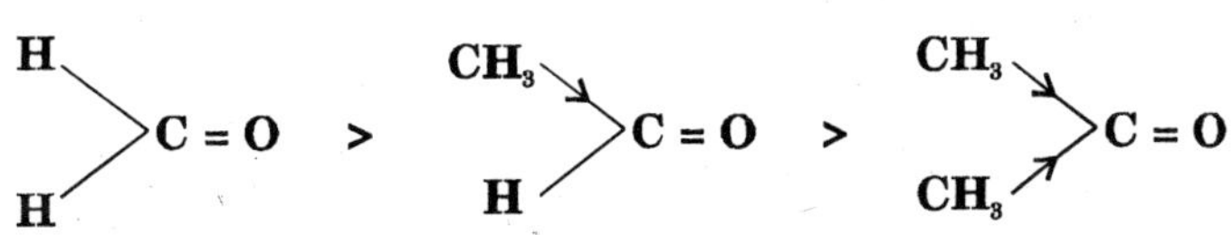

- The presence of electron withdrawing groups at the carbonyl carbon increases the reactivity.
- The attachment of benzene ring to CO group decreases the reactivity because C = O bond comes in conjugation with aromatic ring and resonance stabilisation of molecule occurs.
- CO group withdraws electron density from the ring. Hence it becomes less susceptible to the electrophilic substitution.
- Ring substitution in benzaldehyde occurs at *m*-position.

- **Reactions of Aldehydes and Ketones**

1. Addition across C = O Bond

(*i*) Addition of HCN

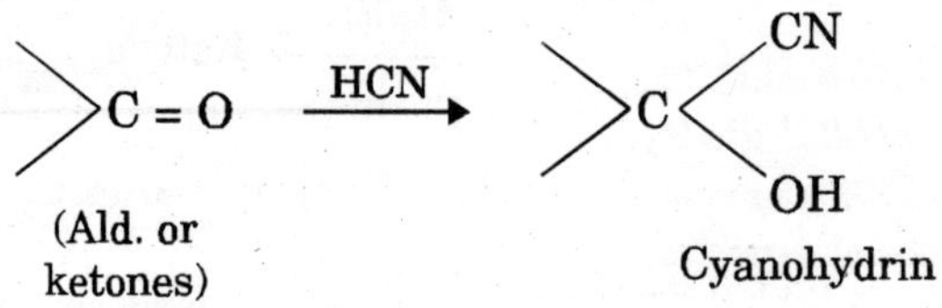

- The reaction is catalysed by acid as well as base.
- Benzaldehyde does not form cyanohydrin but undergoes benzoin condensation in the presence of CN^- ion.
- On further hydrolysis, cyanohydrins give α-hydroxy carboxylic acid. For example

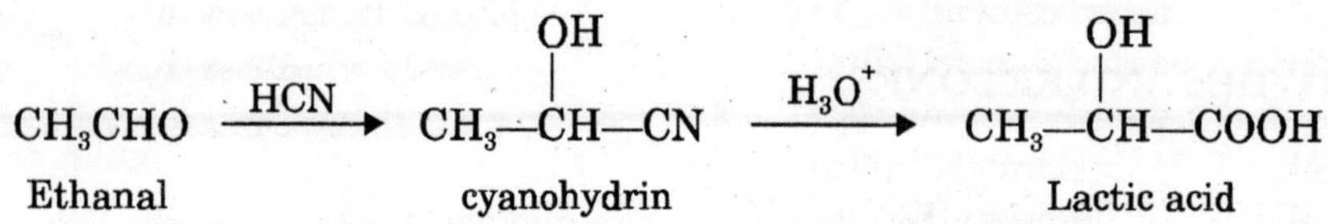

(*ii*) Addition of sodium bisulphite ($NaHSO_3$)

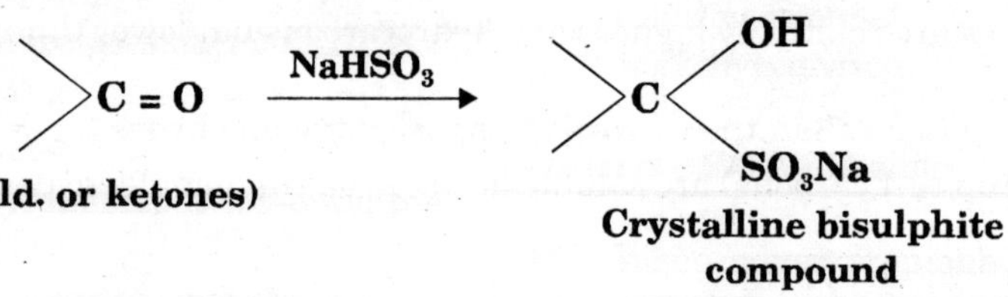

- *Higher ketones do not react with $NaHSO_3$*
- Bisulphite compounds can be decomposed back to aldehyde or ketone by acid or base

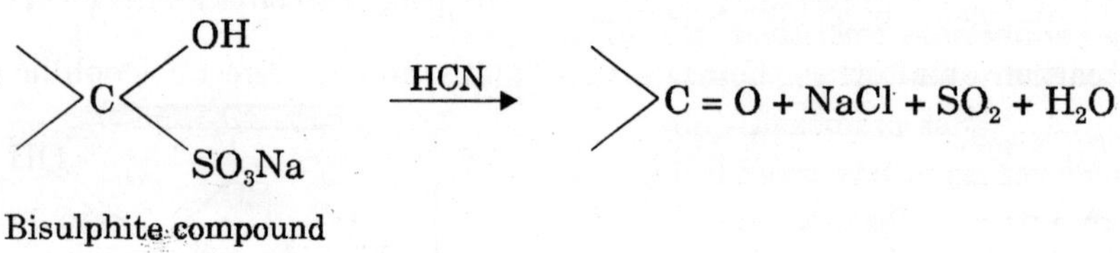

(*iii*) Addition of Grignard's Reagent (RMgX)

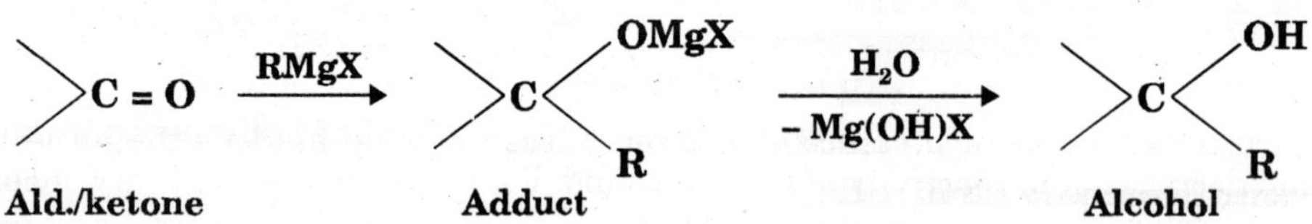

- HCHO gives 1° alcohol ; RCHO gives 2° alcohol and ketone gives 3° alcohol.

2. Replacement of Carbonyl Oxygen by other Groups

(*i*) Reaction with PCl_5 or $SOCl_2$/Py

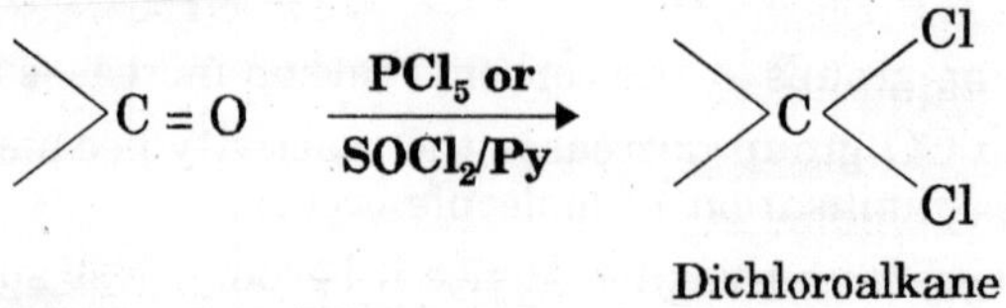

- The products are geminal dihalides.

(ii) Reaction with alcohols

(a) Aldehydes from acetals

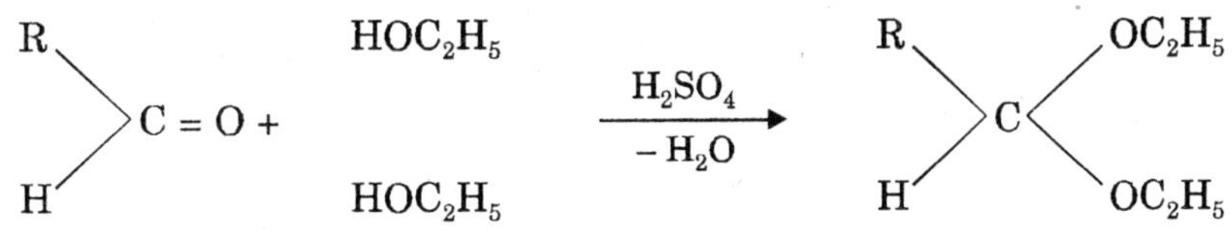

Aldehydes Alcohol Acetals

(b) Ketones react easily with vicinal diols to form cyclic ketals

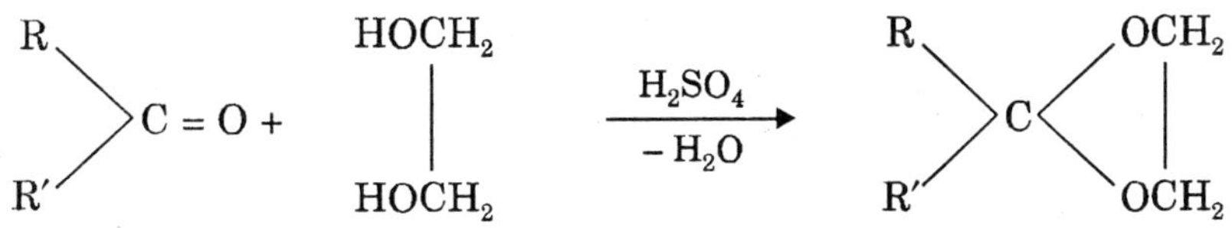

Ketones Cyclic ketals

(iii) Reaction with ammonia derivatives ($H_2H - G$)

$$>C=O + H_2N{-}G \xrightarrow[\text{pH} \approx 5\text{-}6]{H^+} >C=N{-}G$$

(Ald./ketones)

Hydrazine (G is $-NH_2$) gives *hydrazones.*

Phenyl hydrazine (G is $—NH - C_6H_5$) gives *phenyl hydrazone* derivatives.

Hydroxyl amine (G is —OH) gives *oximes.*

Semicarbazide (G is $- NHCONH_2$) gives *semicarbazone derivative.*

2, 4, DNP (G is – NH– 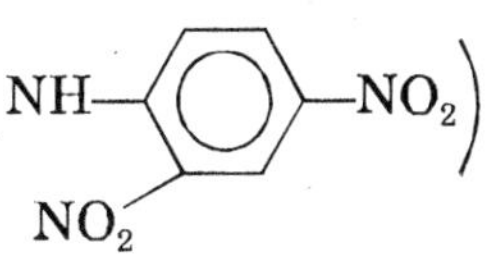 $-NO_2$) gives *2, 4-dinitrophenyl hydrazone* as derivative.

These reactions with ammonia derivatives are acid catalysed because acid facilitates the nucleophilic attack on CO group

$$>C=O \underset{\text{Fast}}{\overset{H^+}{\rightleftharpoons}} \left[>C=\overset{+}{O}H \longleftrightarrow >\overset{+}{C}{-}OH\right]$$

Too highly acidic medium is avoided because NH_2G molecule forms salt $(NH_3G)^+ X^-$ and loses its capacity to act as nucleophile.

Controlled acidic medium is thus required with pH around 5-6.

(iv) Reaction with ammonia (NH_3)

HCHO reacts with NH_3 to give hexamethylene tetramine also called urotropine used as drug in urinary infections.

$$6HCHO + 4NH_3 \longrightarrow (CH_2)_6N_4 + 6H_2O$$

(Urotropine)

Higher aldehydes react to give aldimines.

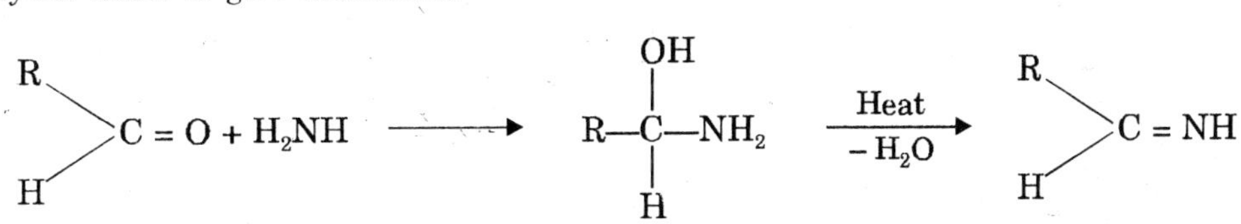

Aldehyde ammonia Aldimine

- Acetone react to give diacetone amine

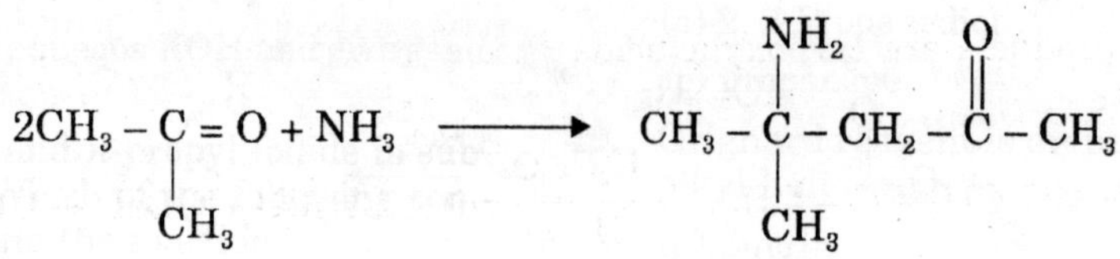

Diacetonamine

- Benzaldehyde reacts to give a product called hydrobenzamide

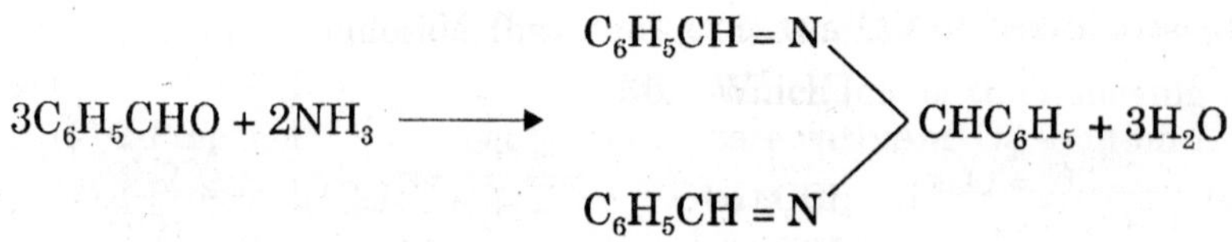

Hydrobenzamide

(v) Reaction with 1° Amines

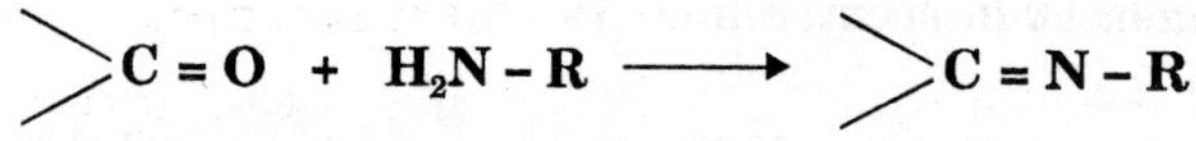

Aldehyde or ketone 1° Amine N-alkylamine (Schiff's base)

- Both aldehyde and ketones give this reaction.

(vi) Wittig reaction. It involves the conversion of >C = O to >C = C< bond by treatment of carbonyl compound with phosphorus **ylide.** Simplest phosphorus **ylide** is methylene triphenylphosphorane.

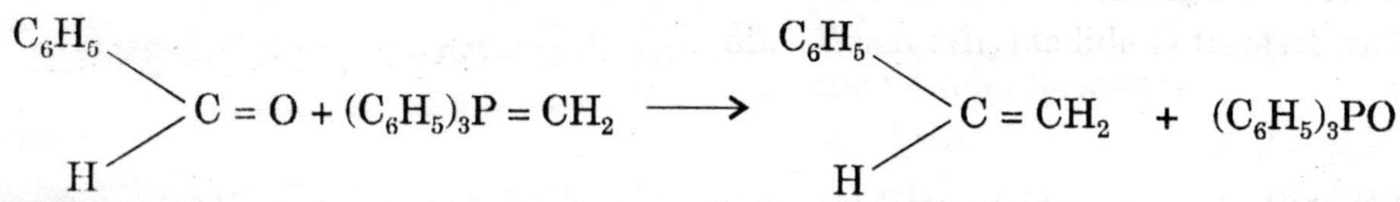

Benzaldehyde Ylide Styrene Triphenyl phosphine oxide

3. Oxidation

- Aldehydes are easily oxidised to carboxylic acids with same number of C atoms

$$\mathbf{RCHO \xrightarrow{(O)} RCOOH}$$

- Not only strong oxidants but even the mild oxidants like Ag^+, Cu^{2+} etc., can oxidise aldehydes.

(a) **Tollen's reagent** is reduced by aldehyde to Ag. This is also called silver mirror test.

$$RCHO + \underset{\text{Tollen's reagent}}{2[Ag(NH_3)_2]OH} \longrightarrow RCOONH_4 + 2Ag + 3NH_3 + H_2O$$

(b) **Fehling solution** is reduced to give reddish ppt. of Cu_2O.

$$RCHO + Cu(OH)_2 + NaOH \longrightarrow RCOONa + \underset{\text{Red ppt.}}{Cu_2O} + 3H_2O$$

- Ketones do not respond to **Tollen's test** or **Fehling solution test.**
- Fehling solution is a mixture of alkaline solution of $CuSO_4$ (*Fehling A*) and alkaline solution of Rochelle's salt *i.e.,* sodium potassium tartrate (*Fehling B*).
- Ketones are resistant to mild oxidation and can be oxidised with strong oxidising agents under drastic conditions.
- Oxidation of ketones involve cleavage of bond between carbonyl C and α-carbon atom.

During oxidative cleavage of mixed ketones, the lower alkyl group is retained by carbonyl group. This is known as **Popoff's Rule.**

The final oxidation products are carboxylic acids

$$CH_3CH_2CH_2-\overset{\overset{O}{\|}}{C}-CH_3 \xrightarrow[H_2SO_4]{K_2Cr_2O_7} CH_3CH_2COOH + CH_3COOH$$

2-Pentanone → Propionic acid + Acetic acid

4. Oxidation with SeO_2 (Selenium dioxide)

SeO_2 oxidises CH_2 part linked to CO group to another CO group.

$$\underset{\text{Ethanal}}{HCH_2CHO} \xrightarrow[CH_3COOH]{SeO_2} \underset{\text{Glyoxal}}{H-\overset{\overset{O}{\|}}{C}-CHO} \;;\; \underset{\text{2-Butanone}}{CH_3\overset{\overset{O}{\|}}{C}-CH_2CH_3} \xrightarrow[AcOH]{SeO_2} \underset{\text{Butanedione}}{CH_3-\overset{\overset{O}{\|}}{C}-\overset{\overset{O}{\|}}{C}-CH_3}$$

5. Oxidation with Sodium Hypohalite

Haloform is formed. The reaction is known as **haloform reaction.**

The reaction is given by aldehydes or ketones having $CH_3-\underset{\underset{O}{\|}}{C}-$ group.

When sodium hypoiodite ($NaOH/I_2$ or Na_2CO_3/I_2) is used as reagent, the product is yellow ppt. of **iodoform**

$$CH_3-\overset{\overset{O}{\|}}{C}-H + 3NaOX \xrightarrow{NaOH/X_2} H-\overset{\overset{O}{\|}}{C}-ONa + CHX_3 + 2NaOH$$

$$CH_3-\overset{\overset{O}{\|}}{C}-R + 3NaOI \xrightarrow{I_2/NaOH} R-\overset{\overset{O}{\|}}{C}-ONa + \underset{\text{Iodoform}}{CHI_3} + 2NaOH$$

6. Reduction

(*a*) **Mild reduction**

This is carried out by reducing agents like H_2/Pt, H_2/Pd, $LiAlH_4$, $NaBH_4$ etc. and converts CO to CHOH.

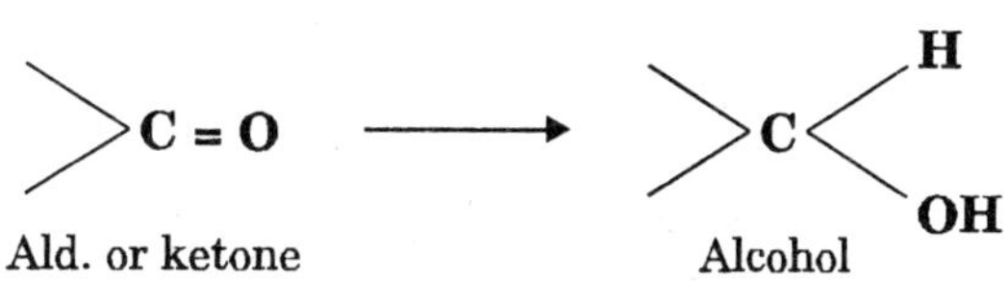

Aldehydes give 1° alcohol and ketones give 2° alcohols.

$NaBH_4$ and $LiAlH_4$ reduces $>C=O$ bond without reducing C = C bonds in the compound. However, $LiAlH_4$ has been found to reduce C = C bond also particularly in α, β unsaturated aldehydes in which β-position carries phenyl group *e.g.*,

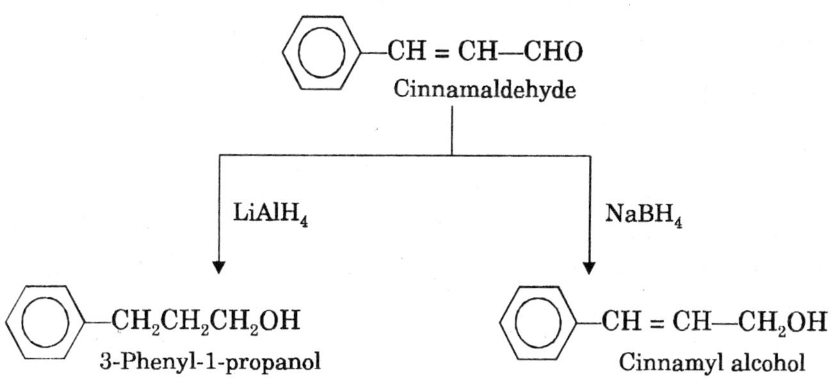

MPV reduction. This is Meervein-Ponndroff-Verley reduction and it involves reduction of $>C=O$ group to $>CHOH$ without affecting the other reducible group in the molecule. The reagent used are *aluminium isopropoxide* and *isopropyl alcohol.*

$$>C=O + (CH_3)_2CHOH \xrightarrow{Al[OCH(CH_3)_2]_3} >CHOH + (CH_3)_2C=O$$

Isopropyl alcohol → Alcohol + Acetone

(*b*) **Strong reduction**

$$>C=O \longrightarrow >C=O$$

Aldehyde/ketone → Alkane

Various reducing agents employed and specific names associated with them are :
Zn-Hg/HCl (*Clemmensen's reduction*) ; NH_2-NH_2/KOH, (470 K) (*Wolf Kishner reduction*), *Na/alcohol HI/red P.*

$$CH_3-\overset{O}{\overset{\|}{C}}-CH_3 \xrightarrow[HCl]{Zn/Hg} CH_3CH_2CH_3 + H_2O$$

Acetone → Propane

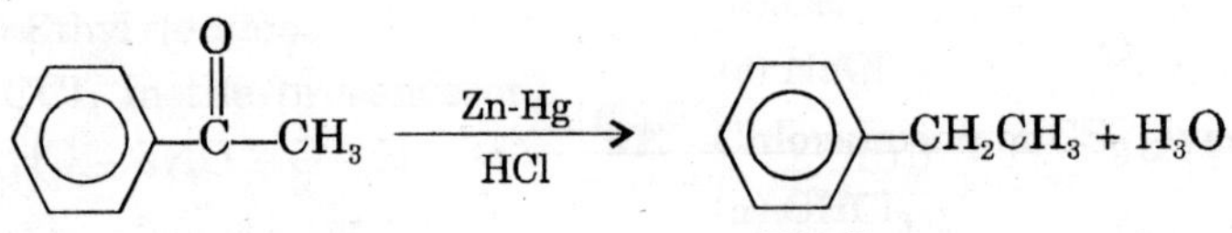

Acetophenone → Ethylbenzene

(*c*) **Bimolecular reduction of acetone**

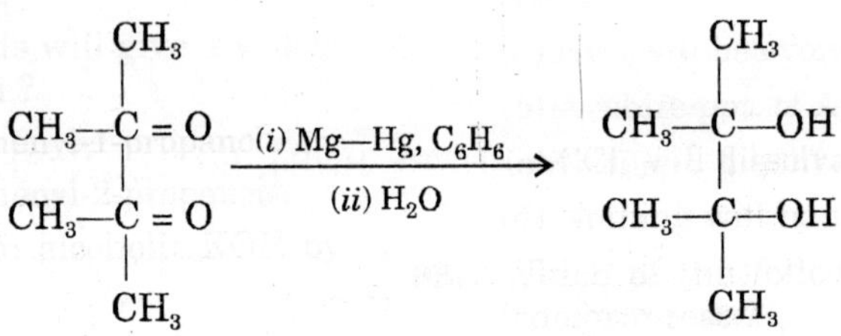

Acetone → Pinacol

7. Condensation Reactions

(*i*) **Aldol condensation.** This is shown by aldehyde/ketones having at least one α-H atom. Here, α-C of one molecules joins with carbonyl carbon of the other :

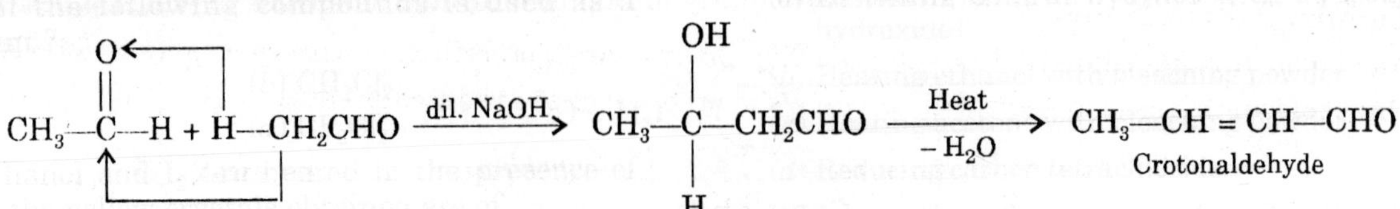

Acetaldehyde → Acetal → Crotonaldehyde

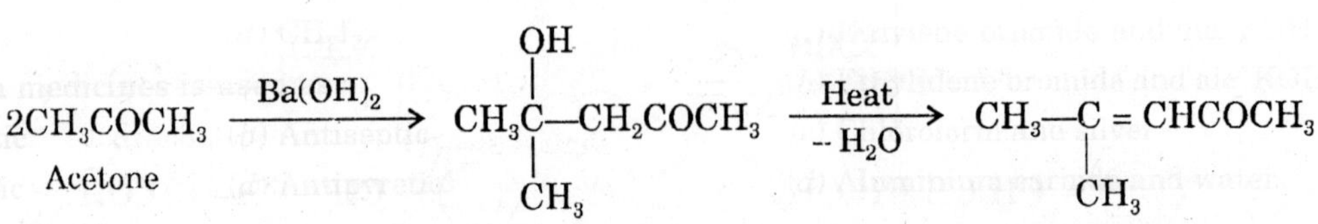

Acetone → Mesityl oxide

(*ii*) **Crossed aldol condensation.** It involves condensation between different molecules or molecules having α-H atom with aldehyde or ketones having no α-H atom.

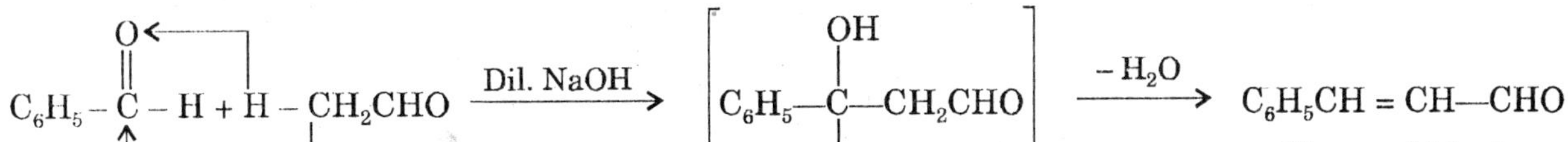

Unstable

Cinnamaldehyde

This type of condensation is called **Claisen Schmidt condensation.**

Here, the α-C of aldehyde having α-H atom joins the molecule with no α-H.

(*iii*) **Internal aldol condensation.** This type of condensation occurs in **diones** and it leads to the formation of cyclic compounds.

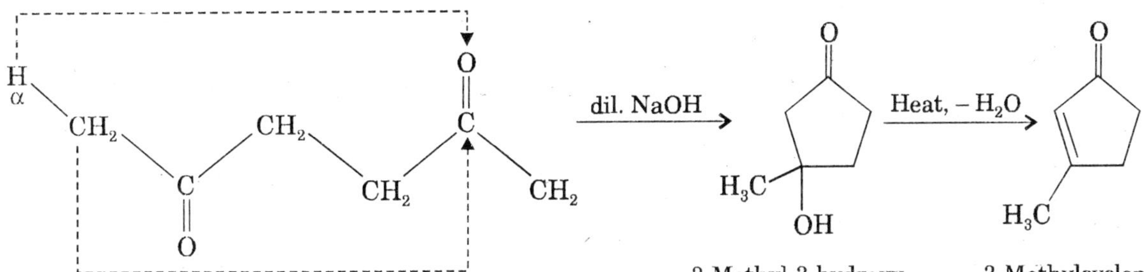

Hexane-2, 5-dione | 3-Methyl-3-hydroxy Cyclopentanone | 3-Methylcyclopent-2-en-1one

(*iv*) **Condensation of ketones in acidic medium**

- Three molecules of acetone condense in the presence of dry HCl to form **phorone**

$$\underset{\displaystyle CH_3-\overset{|}{C}=O}{CH_3} + H_2CH\overset{O}{\overset{||}{C}}-CH_3 + O=\underset{}{\overset{\overset{\displaystyle CH_3}{|}}{C}}-CH_3 \xrightarrow[-2H_2O]{\text{Dry HCl}} CH_3-\overset{\overset{\displaystyle CH_3}{|}}{C}=CH-\overset{\overset{\displaystyle O}{||}}{C}-CH=\overset{\overset{\displaystyle CH_3}{|}}{C}-CH_3$$

Phorone

- On refluxing acetone in conc. H_2SO_4 **mesitylene** is formed.

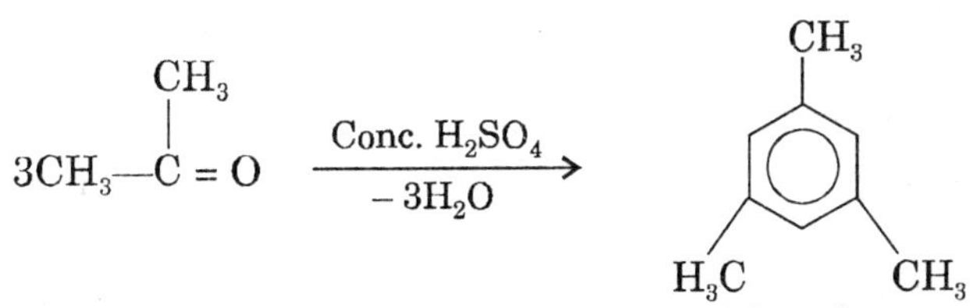

Mesitylene

8. **Cannizzaro's Reaction**

Aldehydes **having no α-H** atom undergo disproportionation on refluxing with conc. alkali, (50% NaOH). One molecule gives oxidised product, while the other gives the reduced product

$$HCHO + HCHO \xrightarrow[NaOH]{50\%} \underset{\text{(Reduction Product)}}{CH_3OH} + \underset{\text{(Oxidation Product)}}{HCOONa}$$

$$C_6H_5CHO + C_6H_5CHO \xrightarrow[NaOH]{50\%} \underset{\text{Benzyl alcohol}}{C_6H_5CH_2OH} + \underset{\text{Sod. benzoate}}{C_6H_5COONa}$$

- **Crossed Cannizzaro's Reaction**

If two different aldehydes with no a-H atom are allowed to react under these conditions, the one which is more susceptible to oxidation gives oxidised product while the other one gives reduced product.

$$C_6H_5CHO + HCHO \xrightarrow{50\% \text{ NaOH}} \underset{\text{Benzylalcohol}}{C_6H_5CH_2OH} + \underset{\text{sod. formate}}{HCOONa}$$

- **Internal Cannizzaro's Reaction**

$$\underset{\text{Glyoxal}}{\begin{matrix} CHO \\ | \\ CHO \end{matrix}} \xrightarrow{50\% \text{ KOH}} \underset{\text{Potassium glycolate}}{\begin{matrix} COOK \\ | \\ CH_2OH \end{matrix}}$$

9. Tischenko Reaction

Condensation between two aldehyde molecules containing α—H atoms in aluminium ethoxide to give **esters**

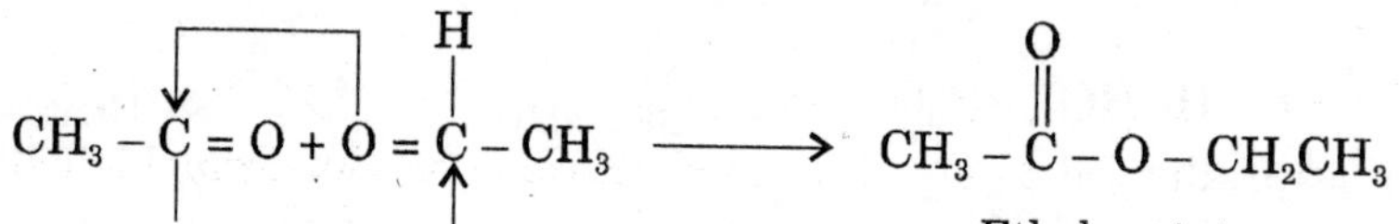

Ethyl acetate

10. Benzoin Condensation. It is a typical reaction of benzaldehyde which undergo self condensation on heating with ethanolic solution of KCN or NaCN to benzoin as product

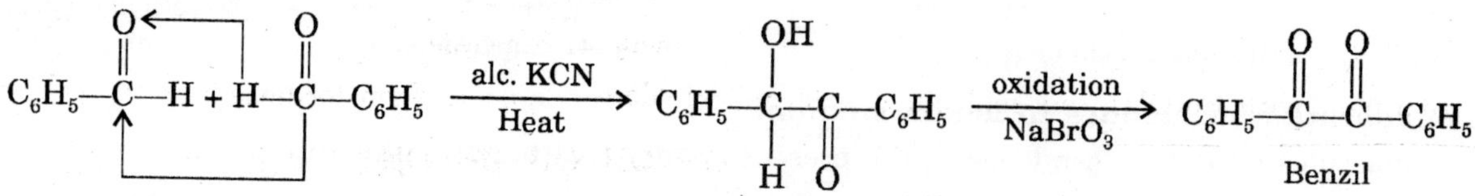

Benzoin Benzil

11. Perkin's Condensation. It involves the reaction between *aromatic aldehyde* and *acid anhydride* along with its corresponding sodium salt to α–β unsaturated acid.

$$C_6H_5CHO + (CH_3CO)_2O \xrightarrow{CH_3COONa} \underset{\text{Cinnamic acid}}{C_6H_5CH = CH - COOH} + CH_3COOH$$

12. Reformatsky Reaction. It involves the reaction of α-bromoester with an aldehyde or ketone in the presence of zinc metal to produce β-hydroxy ester.

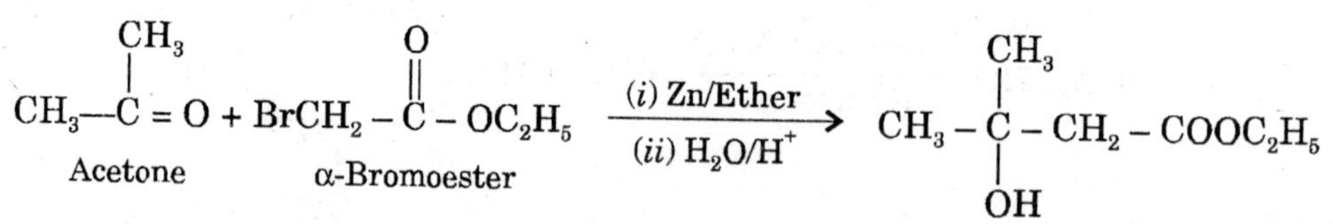

Acetone α-Bromoester Ethyl, β-hydroxy *iso* valerate

13. Knovenageal Reaction. It involves condensation of aldehyde or ketone with malonic acid to form α, β unsaturated acid

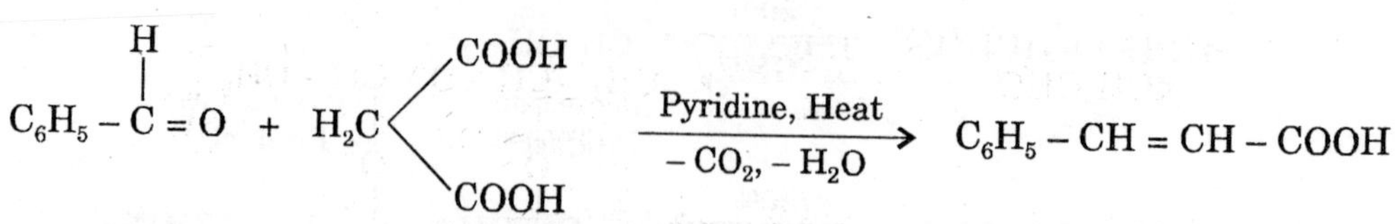

Benzaldehyde Malonic acid Cinnamic acid

$$CH_5-\overset{\overset{\displaystyle H}{|}}{C}=O + H_2C\begin{matrix}\diagup COOH\\ \diagdown COOH\end{matrix} \xrightarrow[-CO_2,\ -H_2O]{\text{Pyridine}} CH_3-CH=CH-COOH$$

Cinnamic acid

14. Reaction with Halogens. This involves substitution of halo group (—Cl or —Br) at the α-carbon atom on heating aldehyde or ketone with halogen in acetic acid solvent

$$\underset{\text{Acetone}}{CH_3COCH_3} + Br_2 \xrightarrow{CH_3COOH} \underset{\alpha\text{-Bromoacetone}}{CH_3COCH_2Br} + HBr$$

$$\underset{\text{Acetophenone}}{C_6H_5COCH_3} + Br_2 \xrightarrow{CH_3COOH} \underset{\text{Phenacyl bromide}}{C_6H_5COCH_2Br} + HBr$$

15. Polymerisation

(a) $nHCHO\,(aq) \xrightarrow{373\ K} (HCHO)_n \,.\, H_2O$

(b) Formaldehyde trimerises if allowed to stand, to form trioxane or metaformaldehyde

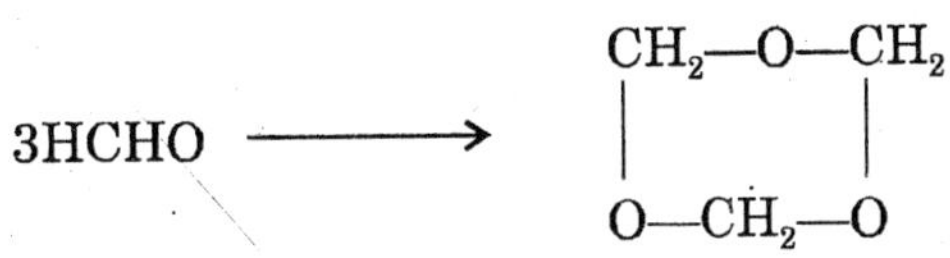

metaformaldehyde

(c) On treatment with $Ca(OH)_2$, six molecules of formaldehyde condense to form formose

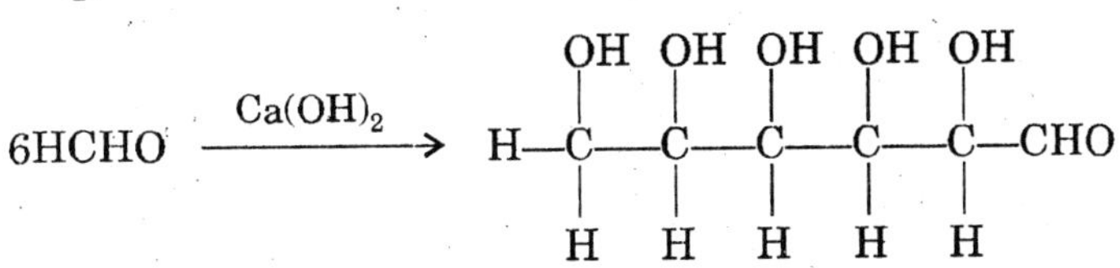

Formose

(d) Acetaldehyde trimerises in conc. H_2SO_4 at room temperature to form paraldehyde

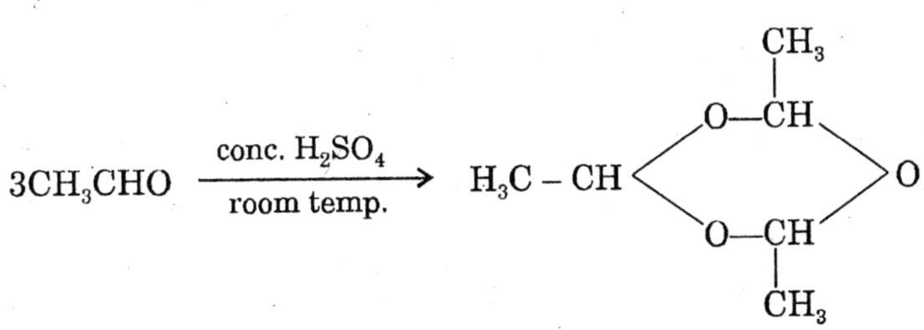

Paraldehyde

(e) Acetaldehyde tetramerise in conc. H_2SO_4 at 273 K to give metaldehyde

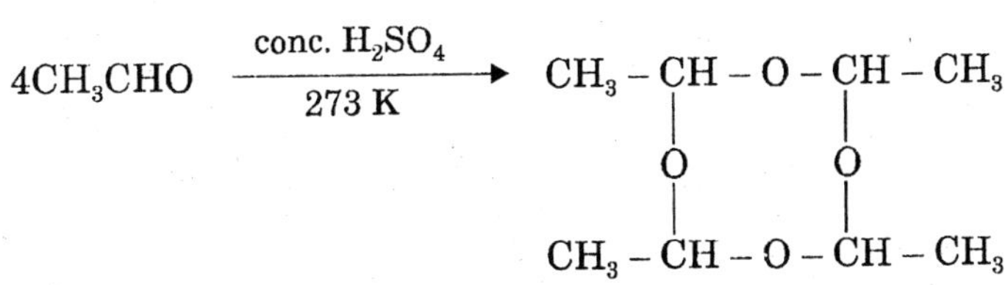

metaldehyde

MECHANISM OF SOME IMPORTANT REACTIONS

- **Addition of HCN to $>C=O$ bond**

$$HCN \rightleftharpoons H^+ + CN^-$$

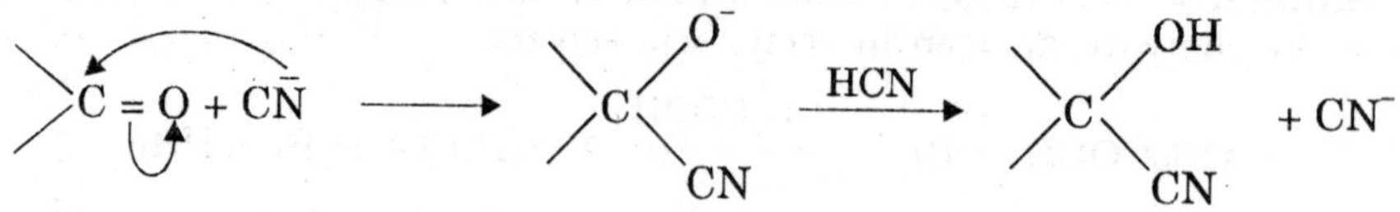

Aldehyde or Ketone

- **Addition of $NaHSO_3$**

$$NaHSO_3 \rightleftharpoons Na^+ + HSO_3^-$$

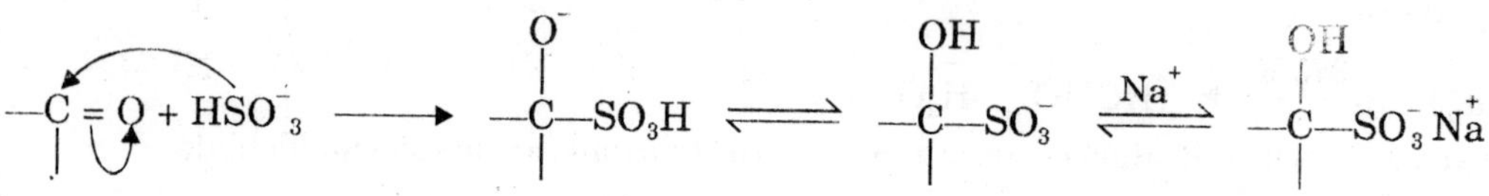

- **Acid catalysed acetal formation**

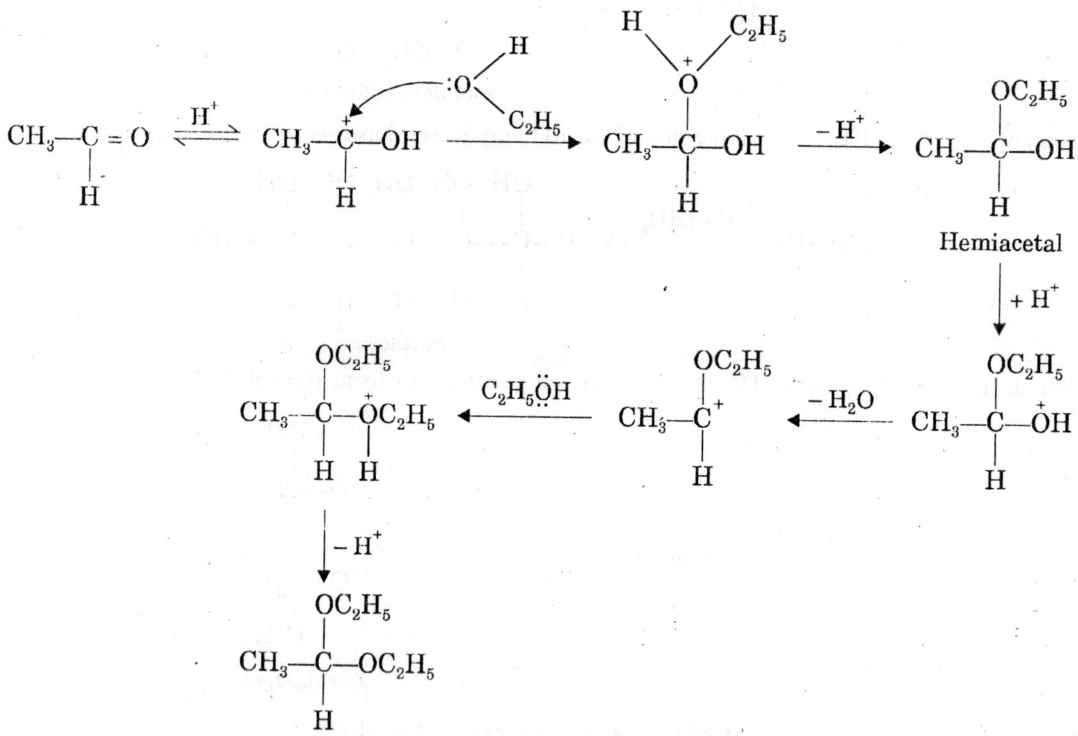

Acetal

- **Aldol condensation.** It is a base catalysed reaction which proceeds through the formation of the carbanion intermediate.

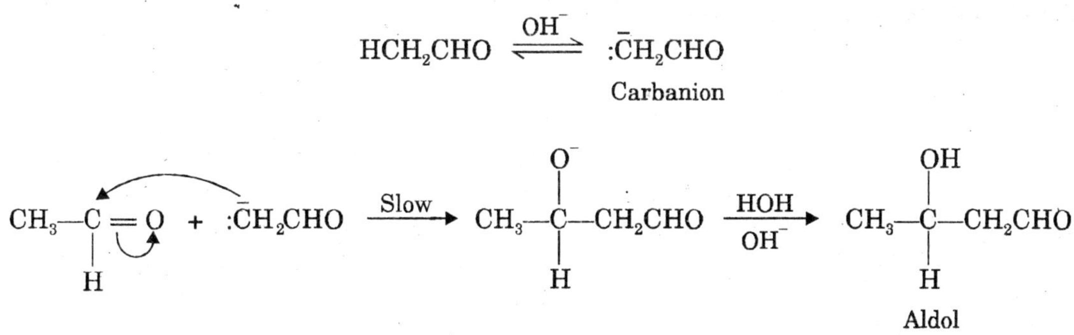

• **Addition of ammonia derivatives**

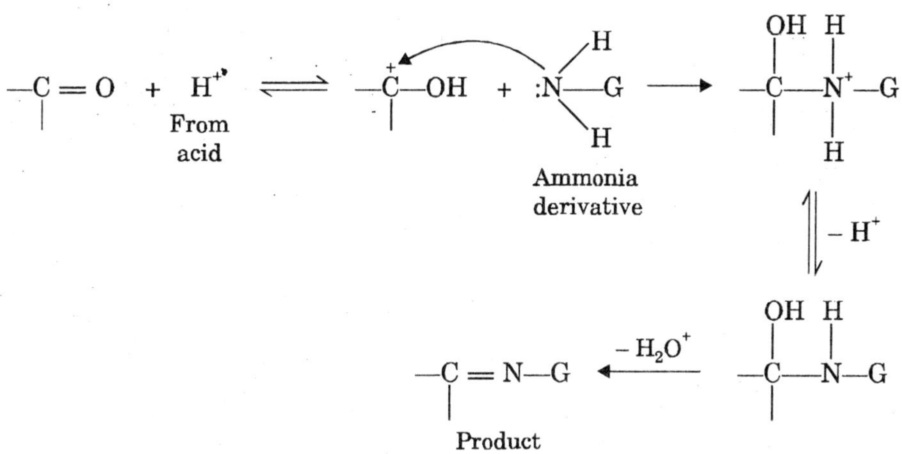

• **Cannizzaro's reaction**

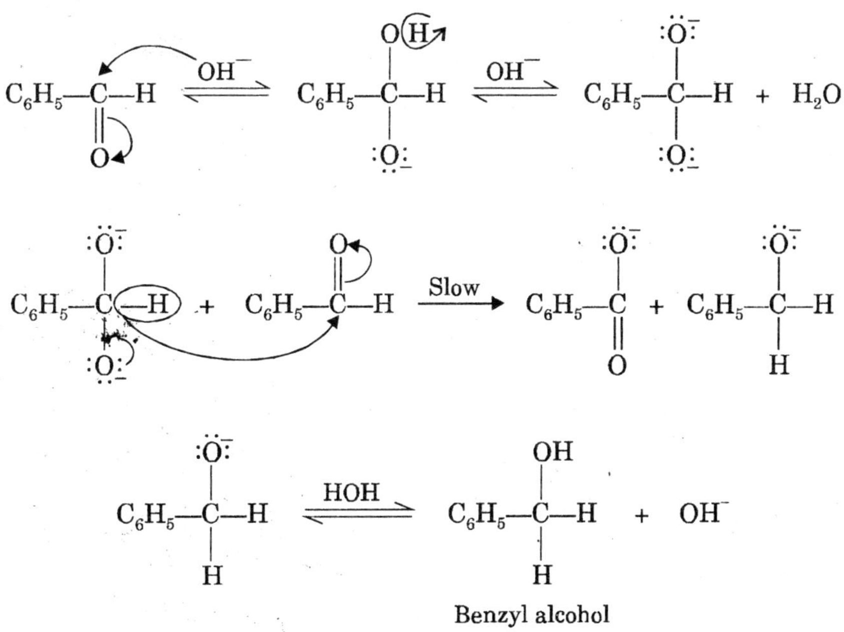

- **Benzoin condensation.** It involves the condensation of benzaldehyde molecules in the presence of : CN^- (KCN), to give benzoin.

$$C_6H_5-\overset{\overset{O}{\|}}{C}-H + C_6H_5-\overset{\overset{O}{\|}}{C}-H \xrightarrow{KCN} C_6H_5\overset{\overset{OH}{|}}{C}H-\overset{\overset{O}{\|}}{C}-C_6H_5$$

Benzoin

The various steps are :

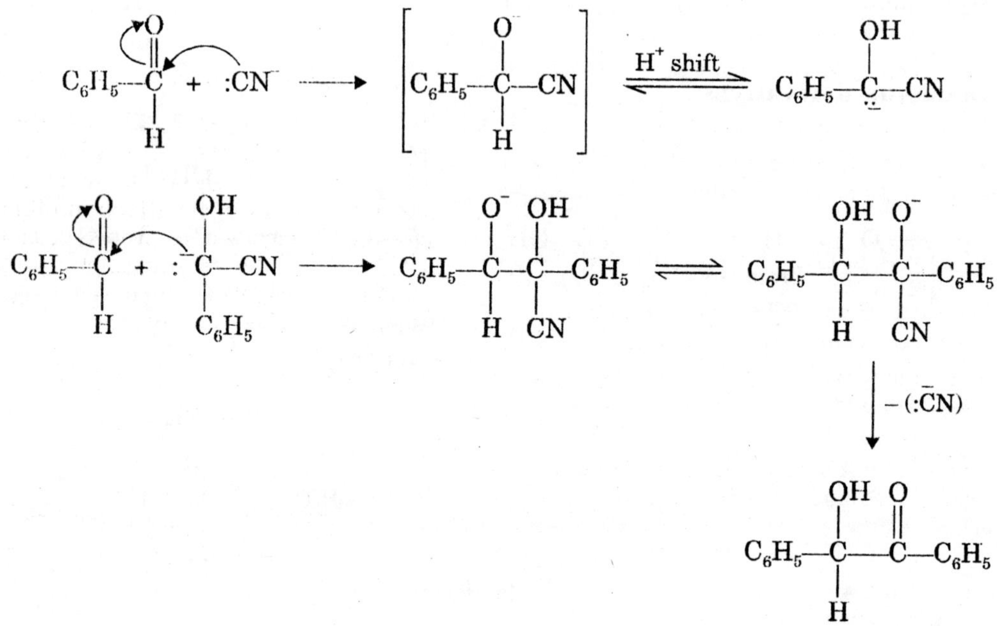

Benzoin

5.2. IMPORTANT FACTS AND TERMS

Add to Knowledge

- Aldehydes and ketones are the first oxidation products of primary and secondary alcohols respectively.
- Aldehydes and ketones are functional isomers of each other.
- Ketones show position isomerism but aldehydes do not, since the aldehyde group being monovalent is always present at the end of the carbon chain.
- Formaldehydes cannot be obtained by Rosenmund's reduction since formyl chloride is unstable at room temperature.
- Ketones cannot be prepared from acid chloride (RCOCl) and Grignard's reagent (RMgX) as the ketone so product react further with Grignard's reagent to produce a 3° alcohol on subsequent hydrolysis.
- In the reaction of toluene with CrO_3, acetic anhydride is used to trap benzaldehyde as benzylidene diacetate to avoid its further oxidation to benzoic acid.
- A mixture of CO + HCl or HCN + HCl can function as formyl chloride and hence has been used for introducing a formyl group (—CHO) into the benzene ring.
- Dialkyl cadmiums are less reactive than Grignard's reagents and hence react with more reactive acid chlorides to form ketones but do not react further with the less *reactive ketones* to form *tert*-alcohols.
- Aldehydes can be prepared by the action of Grignard's reagents on ethyl ortho formate and not on ethyl formate. The reason being that during the reaction of Grignard's reagents with ethyl formate, aldehydes are first formed which react further with Grignard's reagent to form 3° alcohols.
- The common oxidising agents, *i.e.*, $K_2Cr_2O_7/H_2SO_4$ or aqueous/alkaline $KMnO_4$ are not suitable for oxidation of 1° alcohols to aldehydes since the aldehydes produced are more readily oxidised to the corresponding carboxylic acids. To avoid this, a number of specific reagents have been used.

  (*i*) Jones reagents is a solution of $K_2Cr_2O_7 - H_2SO_4$ in aqueous acetone. If, however, excess of reagent is used, aldehydes are further oxidised to acids.

  (*ii*) Collins reagent—a complex of chromium trioxide with two molecules of pyridine, $CrO_3 \cdot 2\ C_5H_5N$.

  (*iii*) Pyridinium chlorochromate (PCC) is a 1 : 1 : 1 complex of CrO_3, C_5H_5N and HCl.
- Alkyl benzene can be oxidised with selenium oxide.

$$C_6H_5-CH_3 \longrightarrow C_6H_5-\overset{\overset{\large O}{\|}}{C}-H + Se + H_2O$$

- Ketones can be oxidised at the α-carbon by using selenium oxide.

$$R-\overset{\overset{\large O}{\|}}{C}-CH_3 + SeO_2 \longrightarrow R-\overset{\overset{\large O}{\|}}{C}-\overset{\overset{\large O}{\|}}{C}-H$$

- **Meerwein-Ponndorf-Verley reduction**

  Meerwein-Ponndorf-Verley (MPV) reduction is reverse of Oppenauer oxidation and involves the reduction of a ketone to 2° alcohol with aluminium isopropoxide in presence of excess of isopropyl alcohol. It also reduces α, β-unsaturated aldehydes or ketones to the corresponding alcohols without reducing the double and triple bonds. For example,

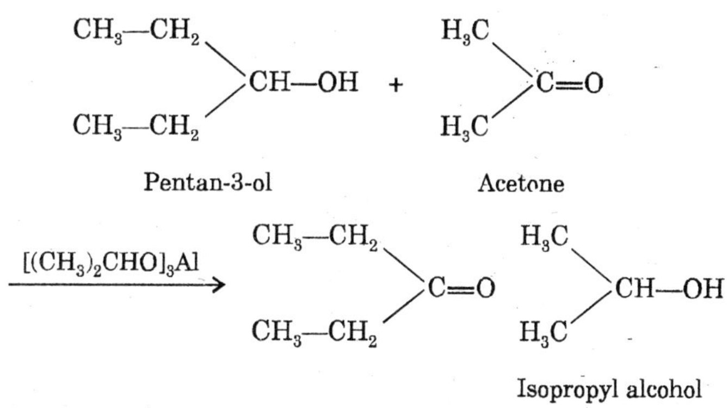

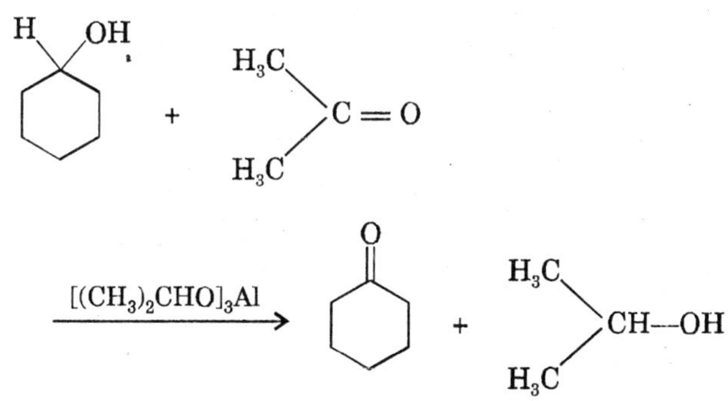

- **Oppenauer oxidation**

  2° Alcohols (saturated or unsaturated) can be conveniently oxidised to ketones (without any danger of being further oxidised to carboxylic acids) by **Oppenauer oxidation.** It involves the treatment of 2° alcohol with aluminium tert-butoxide in presence of excess of acetone.

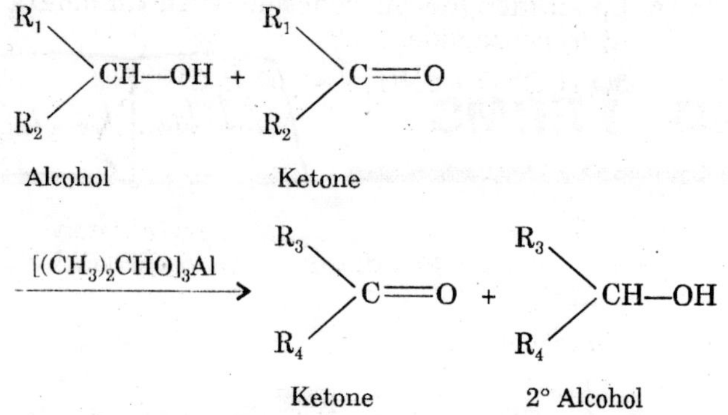

It is reverse of Meerwein-Ponndorf-Verley reduction.

- Aromatic ketones and aliphatic ketones other than methyl ketones, being less reactive, do not react with $NaHSO_3$. Thus, only methyl aliphatic ketones react with $NaHSO_3$.
- Benzaldehyde occurs as glucoside (amygdalins) in bitter almonds and hence it is commonly known as oil of bitter almonds.
- Acetophenone is used in prefumery and in medicine as hypnotic (sleep inducing drug) under the name of hypnone.
- Formalin contains 40% HCHO, 8% CH_3OH and 52% water. It is used as biological preservative.
- The boiling points of individual aldehydes and ketones are higher than those of hydrocarbon of comparable molecular masses but lower than those of corresponding alcohols or carboxylic acids. This is because unlike hydrocarbon, aldehydes and ketones are polar compounds having sufficient dipole-dipole intermolecular attractions. But in contrast to alcohols and carboxylic acids, aldehydes and ketones do not form hydrogen bonds.
- Lower aldehydes and ketones are soluble in water but the solubility decreases as the molecular masses increase.
- In reactions of aldehydes/ketones with ammonia derivatives, the medium should be slightly acidic (pH = 4.5). In too highly acidic medium, ammonia derivative form salt and lose its capacity to act as nucleophile.
- Aldehydes and ketones undergo nucleophilic addition reactions at carbon-oxygen double bond.
- Cyanohydrins on hydrolysis followed by heating form α, β-unsaturated acids, *e.g.*,

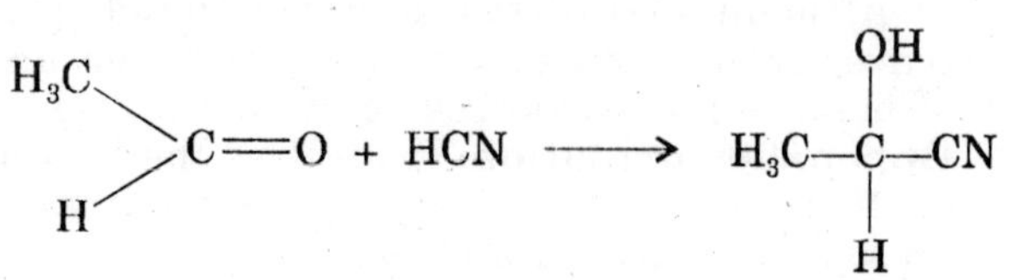

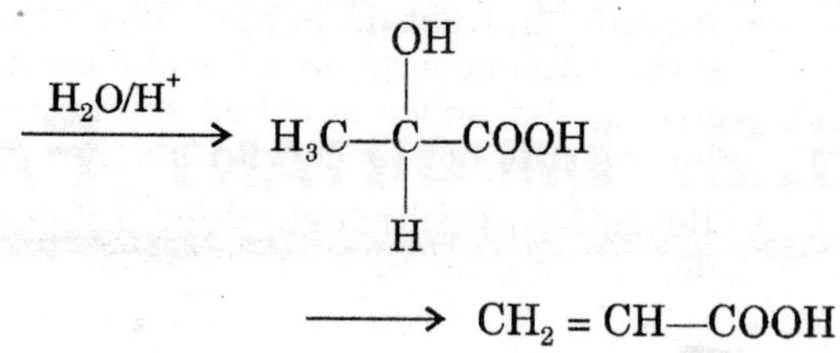

- All aldehydes (including benzaldehyde) and aliphatic methyl ketones react with saturated solution of sodium bisulphite to give insoluble, crystalline bisulphite addition products in good yield. These addition products can be decomposed easily into original aldehyde or ketone with dilute acid or alkali. Amongst aromatic aldehydes and ketones, *benzaldehyde forms $NaHSO_3$ adduct but acetophenone does not.*
- The reaction with sodium bisulphite is used in the separation and identification of aldehydes and ketones.
- The boiling points of aldehydes and ketones are lower than those of the corresponding alcohols due to absence of H-bonding.
- The dipole moments of aldehydes and ketones (2.3 — 2.8 D) are much higher than those of alcohols (1.7 – 1.8 D) or ethers (1.1 – 1.3 D).
- Lower aldehydes and ketones are soluble in water due to H-bonding between the polar carbonyl group and water molecules. However, solubility decreases as the molecular mass increases.
- Aldehydes react with alcohols in presence of dry HCl gas to form acetals *e.g.*,

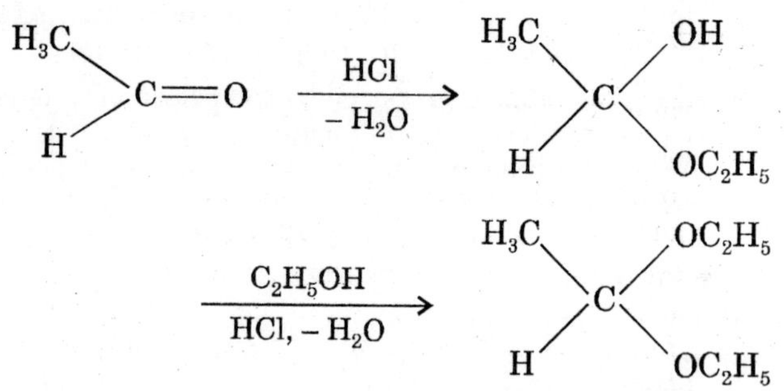

Ketones however do not react with monohydric alcohols, but form cyclic ketals with dihydric alcohols.

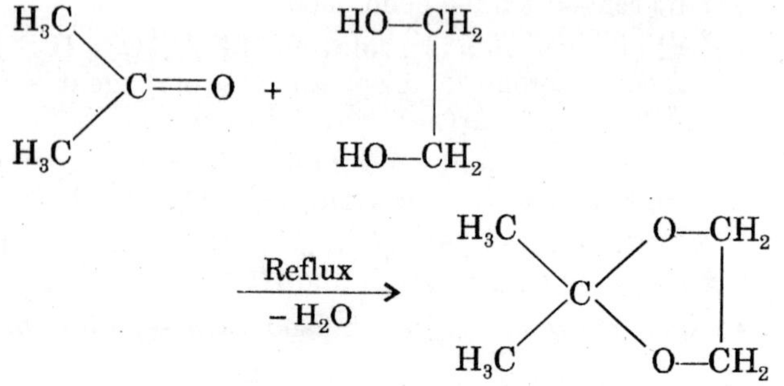

- Acetals and ketals behave like ethers. They are formed under anhydrous acidic conditions and are hydrolyzed under aqueous acidic conditions. They, however, do not react with alkalis.
- Benzaldehyde although reduces Tollen's reagent, H does not reduce Fehling or Benedict solution.
- Due to polarity of the carbonyl group, aldehydes and ketones undergo nucleophilic addition reactions. Reactivity orders are :

  (*i*) $CH_2O > CH_3CHO > (CH_3)_2CO > [(CH_3)_2CH]CO > [(CH_3)_2C]\ CO$ – decreases as *the steric effect and + I-effect of the alkyl groups increases.*

  (*ii*) Due to + *R-effect of the benzene ring, the aromatic aldehydes and ketones are less reactive than the corresponding aliphatic aldehydes and ketones. The order of reactivity decreases as :*

  $C_6H_5CHO > C_6H_5COCH_3 > C_6H_5COC_6H_5$.
- Distinction between aldehydes and ketones can be made by any of the following tests :

  (*a*) Tollen's test (*b*) Fehling solution test (*c*) restoring pink colour of Schiff's reagent.

  These are not shown positive by ketones.
- Aldehydes and ketones having α-hydrogen atom undergo Aldol condensation.
- *Whereas base-catalysed aldol condensation occurs through carbanion intermediate, acid-catalysed aldol condensation occurs through enol form of aldehydes or ketones.*
- Aldehydes which do not contain α-hydrogen's on treatment with conc. alkali undergo disproportionation *i.e.*, self oxidation-reduction to form the corresponding acid and the alcohols. This is called Cannizzaro's reaction.

  For example, 2-Methylpropanal $(CH_3)_2CHCHO$ has an α-hydrogen atom, still it undergoes Cannizzaro's reaction.
- Although the Cannizzaro's reaction is characteristic of aldehydes lacking α-hydrogen atom, certain α-mono alkylated aldehydes also undergo Cannizzaro's reaction.

  $$2(CH_3)_2CHCHO + NaOH\ (Conc.) \longrightarrow (CH_3)_2\ CHCH_2OH + (CH_3)_2\ CHCOO^-Na^+.$$
- All aldehydes (with or without α-hydrogen's) undergo Cannizzaro's reaction on treatment with aluminium ethoxide to form esters—Tischenko reaction.
- Benzaldehyde on reaction with ammonia gives hydrobenzamide.

  $$3C_6H_5CHO + 2NH_3 \longrightarrow \underset{\text{Hydrobenzamide}}{[C_6H_5CH]_3\ N_2} + 3H_2O$$
- 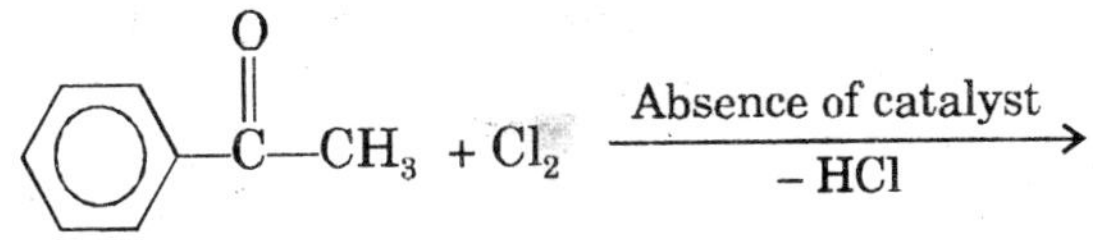

  Acetophenone

  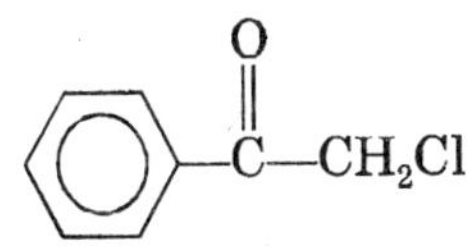

  Phenacyl chloride
  (Lachrymator or Tear gas)
- Two molecules of acetophenone (hypnone) condense in the presence of aluminium tert. butoxide to form Dypnone.

  $$C_6H_5\text{—}\overset{CH_3}{\overset{|}{C}}\text{=O} + H_3C\text{—}\overset{O}{\overset{\|}{C}}\text{—}C_6H_5 \xrightarrow[-H_2O]{\text{Aluminium tert. butoxide}} C_6H_5\text{—}\overset{CH_3}{\overset{|}{C}}\text{=CH—}\overset{O}{\overset{\|}{C}}\text{—}C_6H_5$$
- ***Benzil-benzyllic acid rearrangement.*** The reaction in which Benzil is converted to benzylic acid by treatment with KOH is called as Benzil-benzyllic acid rearrangement.

  $$\underset{\text{Benzil}}{C_6H_5\text{—}\overset{O}{\overset{\|}{C}}\text{—}\overset{O}{\overset{\|}{C}}\text{—}C_6H_5} \xrightarrow[\Delta]{KOH} \underset{\text{Benzillic acid}}{C_6H_5\text{—}\overset{OH}{\overset{|}{\underset{C_6H_5}{\underset{|}{C}}}}\text{—COOH}}$$
- ***Baeyer-Villiger oxidation***

  $$\underset{\text{Acetophenone}}{C_6H_5\text{—}\overset{CH_3}{\overset{|}{C}}\text{=O}} + \xrightarrow{\text{Perbenzoic acid}} \underset{\text{Phenyl acetate}}{C_6H_5\text{—O—}\overset{O}{\overset{\|}{C}}\text{—}CH_3}$$

5.3. SOLVED PROBLEMS

Example 1. *$CH_3CH_2CHO \xrightarrow[\Delta]{NaOH}$ product A. The product A is*

(*a*) $CH_3\,CH_2CH = CH\,CH_2\,CHO$
(*b*) $CH_3\,CH_2\,CH = C(CH_3) - CHO$
(*c*) $CH_3\,CH_2\,CH(OH) - CH_2CH_2\,CHO$
(*d*) $CH_3\,CH_2\,CH(OH) - CH(CH_3) - CHO$.

Solution. During aldol condensation reaction the base abstracts the α-H atom of CH_3CH_2CHO molecule to produce nucleophile.

Hence the correct answer is **(*b*)**.

Example 2. *Caprolactum can be obtained from*

(*a*) Benzaldehyde (*b*) Cyclohexanone
(*c*) Benzophenone (*d*) Adipic acid.

Solution. Caprolactum is

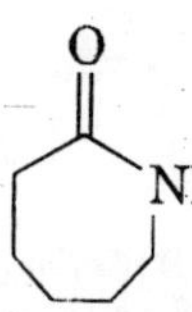

. It is obtained by the reaction of cyclohexanone with NH_2OH followed by heating with conc. H_2SO_4.

Hence the correct answer is **(*b*)**.

Example 3.

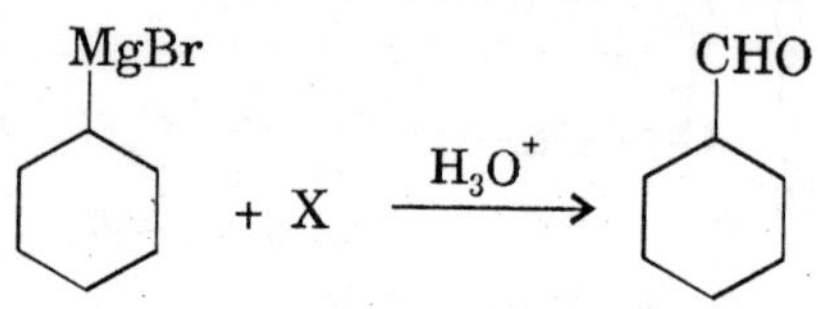

(*a*) CH_3CHO (*b*) HCN
(*c*) CH_3CH_2CHO (*d*) CH_3CN.

Solution. The reaction is

$$HC \equiv N + C_6H_5MgBr \longrightarrow$$

$$\underset{\displaystyle C_6H_5}{H-\overset{|}{C}} = NMgBr \xrightarrow{H_2O/H^+} C_6H_5CHO$$

Only HCN can cause formylation.

Hence the correct answer is **(*b*)**.

Example 4. *Cyclohexanone on reaction with sodium carbonate solution forms*

(*a*)
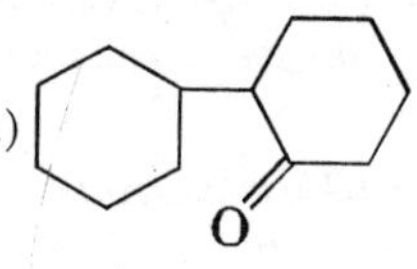

(*b*)
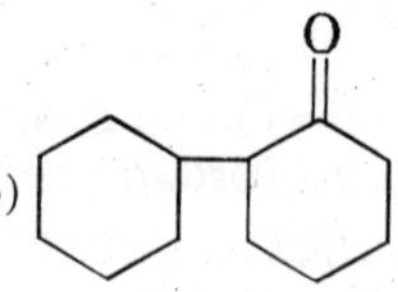

(*c*)
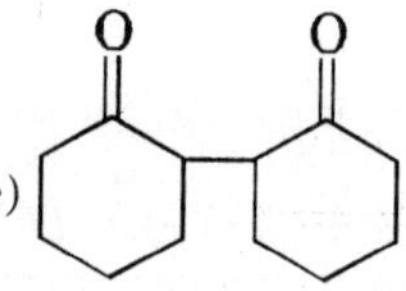

(*d*)
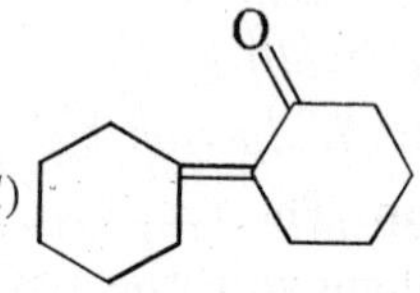

Solution. Cyclohexanone will undergo aldol cendensation in presence of weak basic medium. α-H atom of

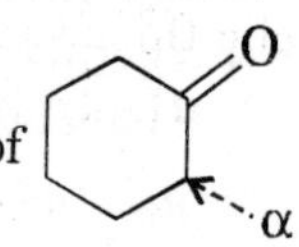

will be abstracted. It will further lose water to form a double bond.

Hence the correct answer is **(*d*)**.

Example 5. *In the reaction,*

$CH_2 = CHCHO \xrightarrow[(ii)\ H_3O^+]{(i)\ MeMgBr}$ A, the principal organic product (A) is

(*a*) $CH_2 = CH—CO—CH_3$
(*b*) $CH_3—CH_2—CH_2\,CHO$
(*c*) $CH_2 = C(Me)—CH_2OH$
(*d*) $CH_3—CH_2—COCH_3$.

Solution. In acrolein $CH_2 = CH - CHO$ there is conjugation between C = C and C = O therefore the reaction yields product as

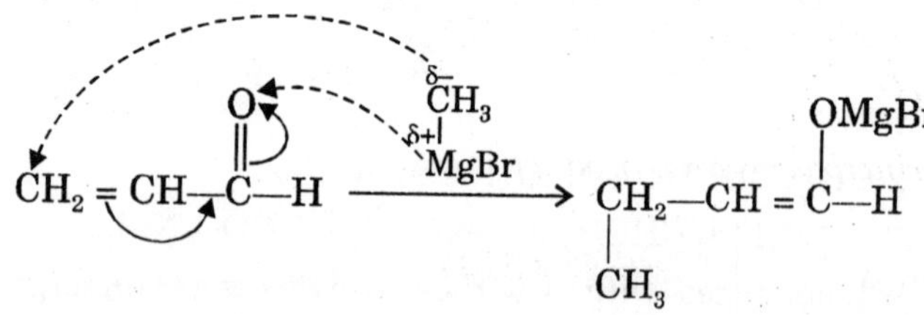

$$\xrightarrow{H_3C^+} CH_3CH_2CH = CHOH$$

$$\xrightleftharpoons{\text{rearrangement}} CH_3CH_2CH_2CHO$$

Hence the correct answer is **(*b*)**.

Example 6. *The transformation,*

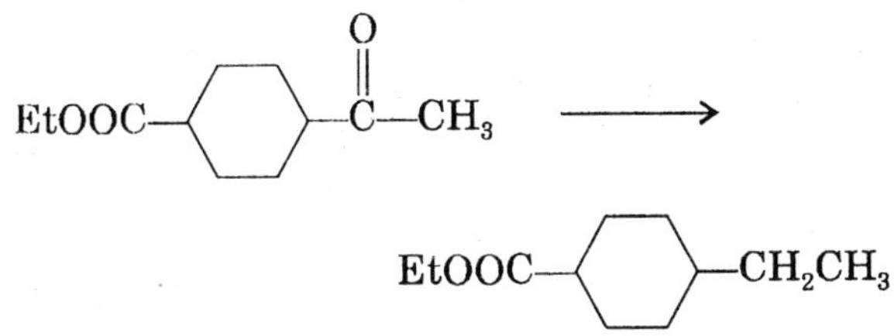

can be best effected by

(*a*) *Lithium aluminium hydride reduction*

(*b*) *Clemmensen's reduction*

(*c*) *Wolff-Kishner reduction*

(*d*) *Reduction by cyclic thio acetal method*

Solution. The reagents *a, b, c* will also reduce the ester group along with the $>C = O$ group. However thio glycols are mild reducing agents. These will add to C = O to form cyclic thioacetal which will get reduced to alkane group.

Hence the answer is **(*d*).**

Example 7. $\begin{matrix} CHO \\ | \\ CHO \end{matrix} \xrightarrow{OH^-} X$. *The product X is*

(*a*) $\begin{matrix} COO^- \\ | \\ COO^- \end{matrix}$ (*b*) $CH_3OH + HCOO^-$

(*c*) $\begin{matrix} CH_2OH \\ | \\ CH_2OH \end{matrix}$ (*d*) $CH_2(OH) - COO^-$.

Solution. It undergoes internal cannizaro's reaction. See concept oriented details.

Hence the correct answer is **(*d*).**

Example 8.

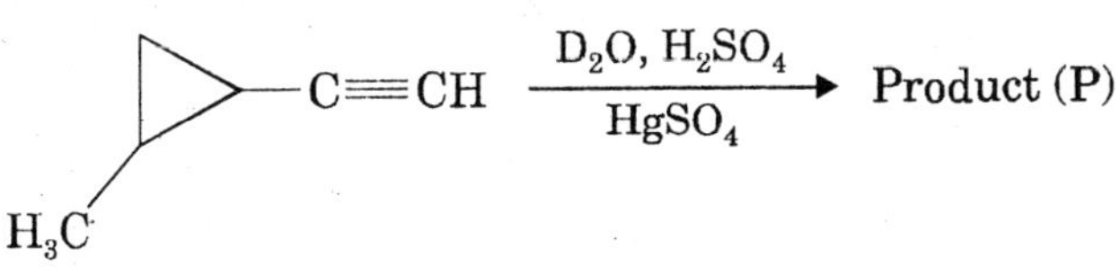

The principal organic product P is

(*a*)

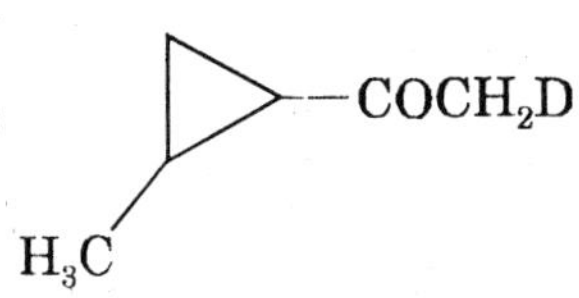

(*b*)

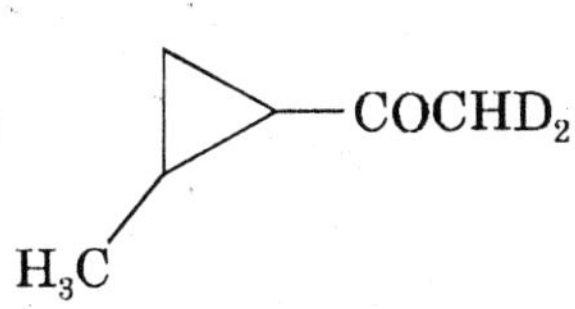

(*c*)

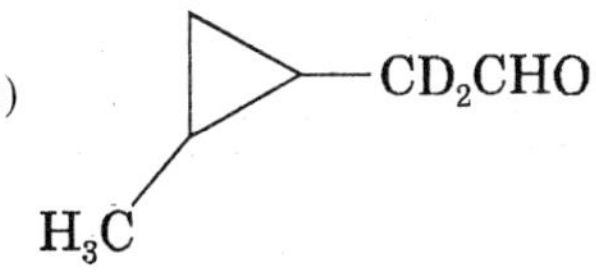

(*d*)

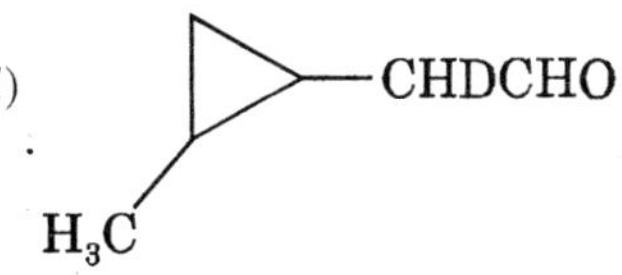

Solution. The addition of D_2O to —C ≡ CH follows Markownikoff rule.

Hence the correct answer is **(*b*).**

Example 9. *Which of the following fails to answer the Iodoform test ?*

(*a*) *Pentan-3-one* (*b*) *Pentan-2-one*

(*c*) *Propanone* (*d*) *Ethanol.*

Solution. Only compounds with $CH_3—\overset{O}{\overset{\|}{C}}—$ or $CH_3CHOH—$ type of grouping will give iodoform test.

Hence the correct answer is **(*a*).**

Example 10. $CH_3—CO—CH_3 \xrightarrow{SeO_2}$ *A. A is*

(*a*) $CHO—CO—CH_3$

(*b*) $CH_3—CO—OCH_3$

(*c*) $CH_3—CO—CH_2OH$

(*d*) $HOCH_2—CO—CH_2OH$.

Solution. SeO_2 can oxidise at the α-carbon. See concept oriented details.

Hence the correct answer is **(*a*).**

Example 11.

$$2D_2-C=O+OH^- \xrightarrow[RX^n]{Cannizaro} \underset{(acid)}{X} + \underset{(alcohol)}{Y} \quad ; X \text{ and } Y \text{ are}$$

(*a*) $D—CO—O^-$, $D_3—C—OH$

(*b*) $D—CO—O^-$, $D_2—CH—OH$

(*c*) $H—CO—O^-$, $D_3—C—OH$

(*d*) *None of these.*

Solution. In Cannizaro's reaction. The α-H atoms should be absent. For further details, see fact file.

Hence the correct answer is **(*a*).**

Example 12. *In the reaction the final product obtained is*

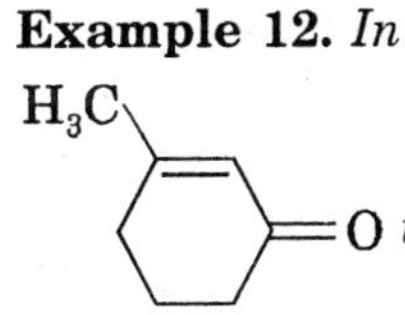

when which of the following is reacted with base

(*a*) $CH_3—CO—CH_2CH_2CH_2—CO—CH_3$

(*b*) $CH_3—CH_2—CH_2—CO—CH_2—CO—CH_3$

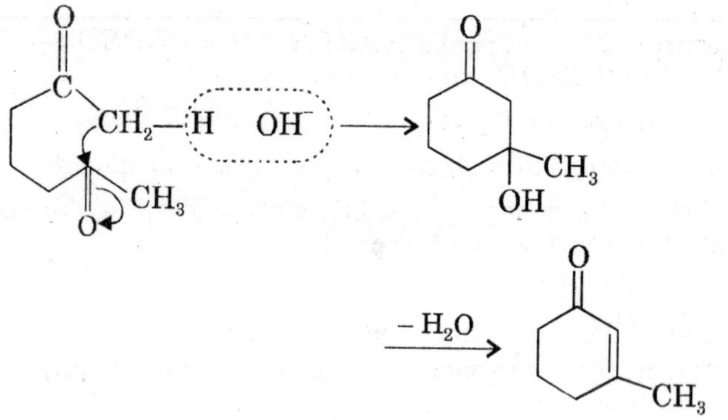

Hence the correct answer is **(a)**.

Example 13. *Which of the following doesn't undergo Cannizzaro's reaction ?*

(a) 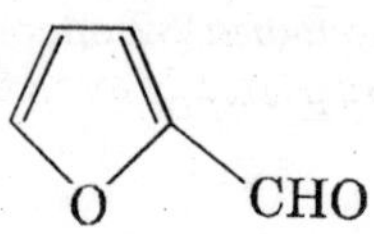 (b) CHO—CHO

(c) Ph—CO—CO—CH_3 (d) $(CH_3)_3C$—CHO.

Solution. Compounds which contain α-H atoms do not undergo Cannizzaro's reaction.

Hence the correct answer is **(c)**.

Example 14. *Ph—CO—CHO* $\xrightarrow{OH^-}$ *X. The compound X is*

(a) $PhCO_2H$ (b) $PhCH_2OH$

(c) $PhCOCH_2OH$ (d) $PhCH(OH)CO_2$.

Solution. It undergoes internal Cannizzaro's reaction.

Hence the correct answer is **(d)**.

Example 15. *Mixture of CH_3CH_2OH and CH_3CHO can be separated by using*

(a) $NaHSO_3$ (b) NH_2OH

(c) HCN (d) NH_2–NH_2.

Solution. Aldehydes and ketones produce crystalline salts with sod. bisulphite whereas alcohols do not. See concept oriented details and fact file.

Hence the correct answer is **(a)**.

Example 16. 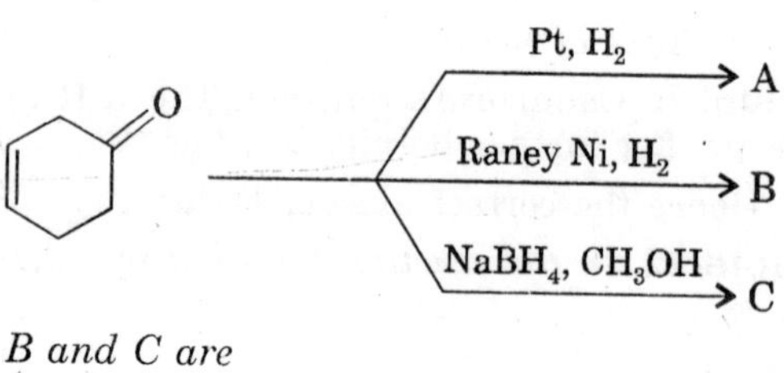

compounds A, B and C are

(a) 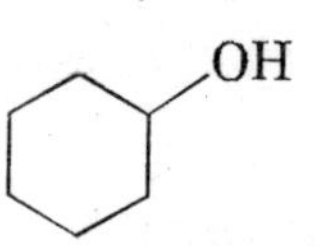 (b) 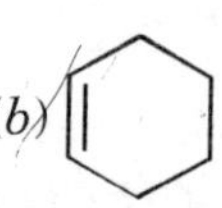

(c) 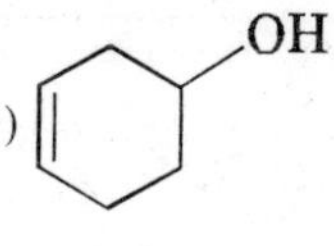

(d) 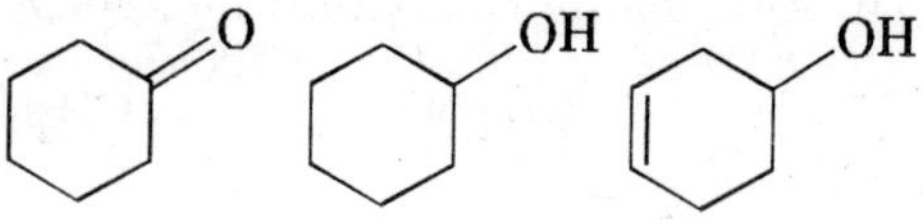

(A) (B) (C)

Solution. Pt, H_2 reduces only double bond.

Raney nickel reduces the double bond and also C = O group to alcohol.

$NaBH_4$ can reduce C = O group to alcoholic group.

Hence the correct answer is **(d)**.

Example 17. *Which of the following gives aldol condensation reaction ?*

(a) $(CH_3)_3CHO$ (b) CCl_3CHO

(c) 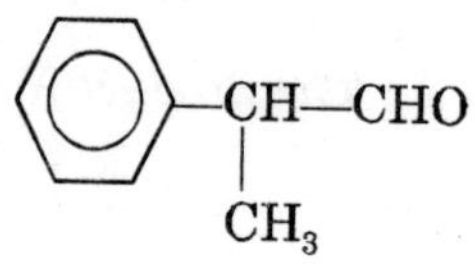 (d) 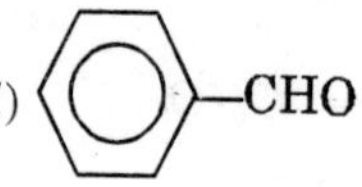

Solution. All carbonyl compounds containing α-H atom can undergo aldol condensation.

Hence the correct answer is (c).

Example 18. *Product of the following reaction is*

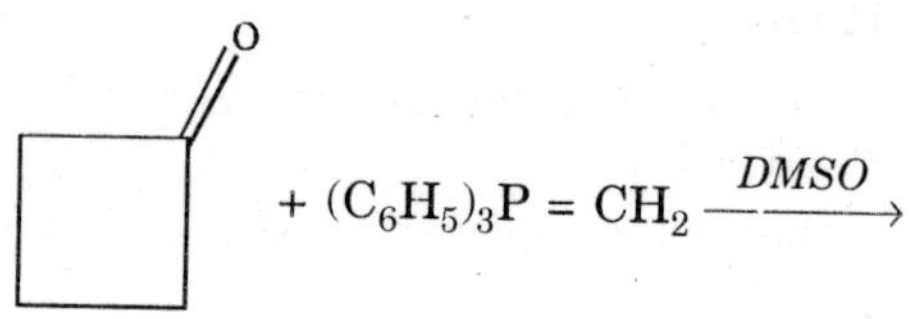

(a) 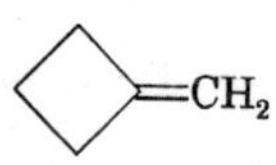 (b) 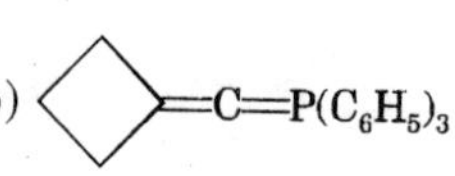

(c) Both (d) None of these.

Solution. It is Wittig reaction. See concept oriented details.

Hence the correct answer is **(a)**.

Example 19. *A carbonyl compound $C_5H_{10}O$ shows Cannizzaro's reaction. What will be its Tischenko reaction product ?*

(a) $(CH_3)_3C$—CO—$OCH_2C(CH_3)_3$

(b) $(CH_3)_3CCOO^-$

(c) $(CH_3)_3C\,CH_2OH$

(d) None.

Solution. Carbonyl compounds which do not contain α-hydrogen undergo Cannizzaro's reaction.

Hence $C_5H_{10}O$ is $(CH_3)_3C—CHO$.

In Tischenko reaction an ester is produced.

Hence the correct answer is **(*a*)**.

Example 20. *An organic compound on treatment with HIO_4 gives HOOC $(CH_2)_4CHO$, the compound is*

(*a*) 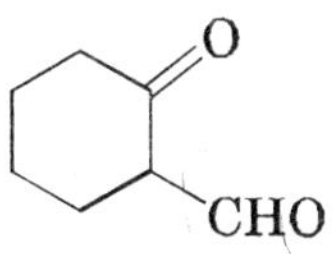 (*b*) 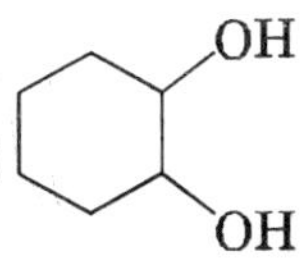

(*c*) 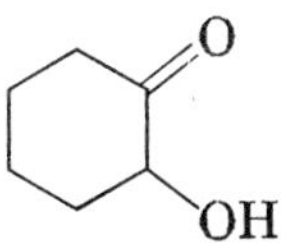 (*d*) 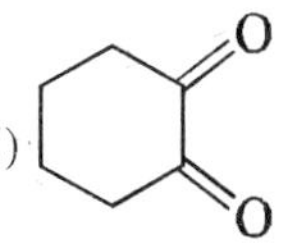

Solution. Periodic acid oxidises polyhydroxy aldehydes or ketones. It also cleaves the carbon-carbon bond if the carbon atoms carry OH or C = O group. As the product contains one COOH and one CHO group it indicates that there is one OH group and the other is C = O group on the two carbon atoms.

$$—OH \xrightarrow{HIO_4} —CHO$$

$$>C=O \xrightarrow{HIO_4} —COOH$$

Hence the correct answer is **(*c*)**.

Example 21. *The conversion $CH_3CH = CHCHO \longrightarrow CH_3CH = CHCOOH$ can be effected by*

(*a*) *alkaline $KMnO_4$* (*b*) *acidic $K_2Cr_2O_7$*

(*c*) *ammonical $AgNO_3$* (*d*) *selenium dioxide.*

Solution. All reagents except amm. $AgNO_3$ *i.e.*, Tollen's reagent can oxidise the C = C also.

Hence the correct answer is **(*c*)**.

Example 22. *The conversion*

$$CH_3CH = CHCHO \longrightarrow CH_3CH = CHCH_2OH$$

can be effected with

(*a*) *Ni/H_2* (*b*) *9 BBN*

(*c*) *Zn-Hg/HCl* (*d*) *none of these.*

Solution. 9 BBN is 9-Borabicyclo [3.3.1] nonane. It can reduce only the carbonyl group and leaves the C = C bond intact.

The structure of 9 BBN is

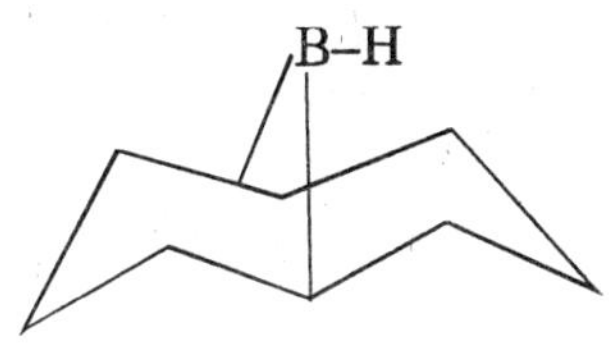

Hence the correct answer is **(*b*)**.

5.4. QUESTION BANK

IUPAC names

1. The IUPAC name of crotonaldehyde is
(*a*) Butanal (*b*) But-2-enal
(*c*) But-1-enal (*d*) None of these.

2. IUPAC name of aldol is
(*a*) α-hydroxybutanal (*b*) 3-Hydroxybutanal
(*c*) 3-Hydroxybutan-2-one (*d*) β-Hydroxypropanal.

3. The IUPAC name of $BrCH_2C(Br)(CH_3)COONa$ is
(*a*) Sodium dibromo isopropionate
(*b*) Sodium 2, 3-dibromo-2-methyl propanoate
(*c*) Dibromo sodium isobutanoate
(*d*) None of the above.

4. Which of the following represents *o*-benzoquinone ?

(*a*) 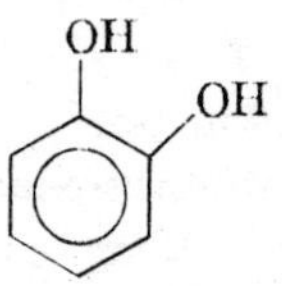

(*b*) 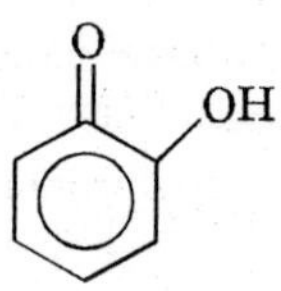

(*c*) 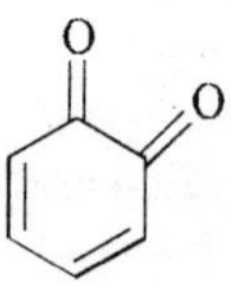

(*d*) 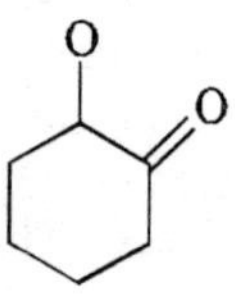

Methods of Preparation

5. Propyne on treatment with aqueous H_2SO_4 in the presence of mercury (II) sulphate gives
(*a*) Propionic acid (*b*) Propanone
(*c*) Propanol-2 (*d*) Propene.

6. What is not true about the following reaction ?

$$RCOCl + H_2 \xrightarrow[BaSO_4]{Pd} RCHO + HCl$$

(*a*) It is Rosenmund's reaction
(*b*) $BaSO_4$ increases the efficiency of the catalyst Pd
(*c*) $BaSO_4$ decreases the efficiency of the catalyst
(*d*) Pd is a heterogeneous catalyst.

7. Propanoyl chloride is subjected to reduction with H_2 in boiling xylene in the presence of Pd supported by $BaSO_4$. The product formed is
(*a*) Acetone (*b*) Propionaldehyde
(*c*) Ethanal (*d*) Propanone.

8. Ethylidene chloride is hydrolysed with aqueous NaOH. The product formed is
(*a*) Ethanol (*b*) Ethylchloride
(*c*) Ethanal (*d*) Ethane.

9. The various of ethyl alcohol are passed over red hot copper at 573 K it forms
(*a*) Methane (*b*) Formaldehyde
(*c*) Acetone (*d*) Acetaldehyde.

10. For obtaining ethyl methyl ketone from acetyl chloride which of the following reagent can be employed ?
(*a*) Grignard reagent
(*b*) Reaction with HI
(*c*) H_2, $Pd/BaSO_4$
(*d*) reaction with $(C_2H_5)_2Cd$ in the presence of dry ether.

11. In order of propare acetone from acetyl chlroide in one step, which of the following reagent will be best ?
(*a*) HI
(*b*) Diethyl cadmium
(*c*) Methyl magnesium bromide
(*d*) Dimethyl cadmium.

12. A dihaloalkane on alkaline hydrolysis produces a ketone with formula C_3H_6O. The dihaloalkane can be
(*a*) 2, 2-Dichloropropane (*b*) 1, 1-Dichloropropane
(*c*) 1, 2-Dichloropropane (*d*) 1, 3-Dichloropropane.

13. Which of the following is Jones' reagent ?
(*a*) Acidified $KMnO_4$
(*b*) CrO_3/CH_3COOH
(*c*) $K_2Cr_2O_7/H_2SO_4$
(*d*) Rosaniline hydrochloride.

14. $$CH \equiv CH \xrightarrow[H_2SO_4]{HgSO_4} X \xrightarrow{LiAlH_4} Y \xrightarrow{P_4/Br_2} Z$$

In the above sequence **Z** is
(*a*) Ethanol (*b*) Ethylediene dibromide
(*c*) Ethylene dibromide (*d*) Ethyl bromide.

15. Which of the following will not form when the mixture of calcium acetate and calcium formate is distilled ?
(*a*) Acetone (*b*) Propanal
(*c*) Ethanal (*d*) Methanal.

Physical Properties

16. The compound having least solubility in water is

(*a*) Methanal (*b*) Acetaldehyde

(*c*) Acetone (*d*) Acetophenone.

17. Which of the following is maximum contributing form of carbonyl group ?

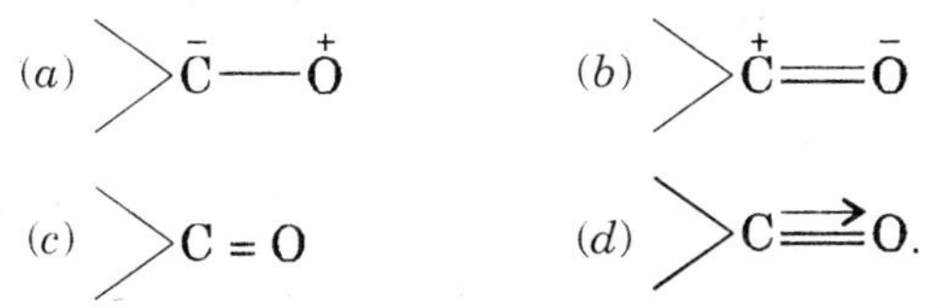

18. Which of the following gives a smell of bitter almonds ?

(*a*) Benzaldehyde (*b*) Benzidene

(*c*) Benzoic acid (*d*) Benzaldichloride.

19. Which of the following is not a resonating form of benzaldehyde ?

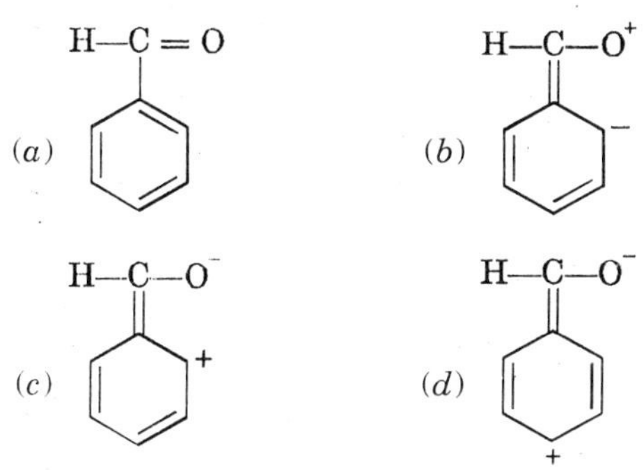

20. Polarisation of acrolein molecule may be correctly written as

(*a*) $\overset{\delta-}{C}H_2 \cdots CH \cdots CH \cdots \overset{\delta-}{O}$

(*b*) $\overset{\delta+}{C}H_2 \cdots CH \cdots CH \cdots \overset{\delta+}{O}$

(*c*) $\overset{\delta+}{C}H_2 \cdots CH \cdots \overset{\delta-}{C} \cdots O$

(*d*) $\overset{\delta+}{C}H_2 \cdots CH \cdots CH \cdots \overset{\delta-}{O}$.

Chemical Properties

21. The rearrangement which can react both with aldehydes and ketones is

(*a*) Fehling solution

(*b*) Schiff's reagent

(*c*) $[Ag(NH_3)_2]$ OH

(*d*) Methyl magnesium bromide.

22. Aldehydes and ketones do not give addition product with

(*a*) HCN (*b*) HCl

(*c*) $NaHSO_3$ (*d*) both with HCN and HCl.

23. In the reaction of $NaBH_4$ with acrylic aldehyde

(*a*) Only CHO group is reduced

(*b*) Only C = C bond gets saturated

(*c*) Both —CHO and C = C bond get reduced

(*d*) Neither C = O, nor C = C bond is reduced.

24. $CH_3CHO \xrightarrow{HCN} X \xrightarrow{H_3O^+} Y$

In the above sequence Y is

(*a*) An aldehyde (*b*) An acid

(*c*) α-Hydroxy acid (*d*) β-Hydroxy acid.

25. Benzaldehyde and malonic acid react to give

(*a*) Ethanal (*b*) Crotonic acid

(*c*) Cinnamaldehyde (*d*) Cinnamic acid.

26. Ethanal is treated with ammonia and adduct formed is warmed. The final product is

(*a*) Acetaldehyde ammonia

(*b*) Acetaldimine

(*c*) Tetramethylene hexamine

(*d*) Ethyl amine.

27. Propanol-2 $\xrightarrow[\text{Oxidation}]{\text{Mild}}$ **X** $\xrightarrow[(ii)\ H_2O]{(i)\ MeMgBr}$ **Y**

In the above sequence of steps **Y** is

(*a*) Butan-2-ol (*b*) Butan-1-ol

(*c*) 2-Methylpropene (*d*) 2-Methylpropan-2-ol.

28. Which of the following ketone will not give yellow precipitate with $NaOH/I_2$?

(*a*) Pentan-2-one (*b*) Acetone

(*c*) Benzophenone (*d*) Hexan-2-one.

29. The adduct of ethyl magnesium iodide and methanal is hydrolysed. The final product is

(*a*) 1-Propanol (*b*) 2-Propanol

(*c*) Ethanol (*d*) 2-Chloro-1-propanol.

30. Which of the following reagent will not react with Ethanal ?

(*a*) HCl (*b*) Cl_2

(*c*) PCl_5 (*d*) $NaHSO_3$ (*aq*).

31. The product formed by the reaction of acetaldehyde with excess of ethanol in the presence of sulphuric acid is

(*a*) $C_2H_5OCH_2CH_2OC_2H_5$ (*b*) A ketal

(*c*) $CH_3CH(OC_2H_5)_2$ (*d*) $CH_3CH(OH)_2$.

32. Which of the following reagent can convert benzaldehyde into a mixture of sodium benzoate and benzyl alcohol ?

(*a*) 50% aq. NaOH

(*b*) H_2/Pd, $BaSO_4$

(*c*) HCN followed by hydrolysis

(*d*) All the above.

33. $C_6H_5 - CH = CHCHO \xrightarrow{X} C_6H_5CH = CHCH_2OH$.

In the above sequence X can be

(*a*) H_2/Ni (*b*) $NaBH_4$

(*c*) $K_2Cr_2O_7/H^+$ (*d*) Both (*a*) and (*b*).

34. When acetaldehyde is treated with hydroxylamine the medium should be

(*a*) Highly basic

(*b*) Highly acidic

(*c*) Moderately acidic

(*d*) Neutral.

35. Which reagent will produce 1, 1-Dichloroethane from Ethanal ?

(*a*) Cl_2 (*b*) $SOCl_2/Py$

(*c*) PCl_5 (*d*) Both (*b*) and (*c*).

36. Which of the following reagents can give UROTROPINE ?

(*a*) Ethanal and Ethylamine

(*b*) Methanal and H_2SO_4

(*c*) Methanal and Ammonia

(*d*) Methanal and 50% NaOH.

37. When pentan-2-one is oxidised using conc. HNO_3, the major products formed will be

(*a*) *n*-Buytric acid and Formic acid

(*b*) Isobutyric acid and Acetic acid

(*c*) Pentanoic acid

(*d*) Ethanoic acid and Propanoic acid.

38. When acetone reacts with chloroform in the presence of KOH, the product formed is

(*a*) Chloropicrin (*b*) Chloretone

(*c*) CCl_4 (*d*) Trichloroacetone.

39. In order to get paraldehyde from ethanal, which of the following reagent is used ?

(*a*) C_2H_5OH (*b*) Conc. H_2SO_4

(*c*) H_2CrO_4 (*d*) All the above.

40. Compound $A(C_5H_{10}O)$ forms a phenyl hydrazone and gives a negative Tollen's test and negative Iodoform test. On reducing A with Zn-Hg/HCl, *n*-pentane is formed. The compound A is

(*a*) 1° alcohol (*b*) An aldehyde

(*c*) Secondary alcohol (*d*) A ketone.

41. Propanal $\xrightarrow[(ii)\ LiAlH_4]{(i)\ HCN/H^+}$ X. Here, **X** is

(*a*) 1-Amino-2-butanol (*b*) 1-Amino butan-1-ol

(*c*) 4-Amino-2-butanol (*d*) 2-Aminobutan-2-ol.

42. Which of the following reaction is used to prepare β-Hydroxyesters from aldehydes ?

(*a*) Wittig reaction (*b*) Claisen condensation

(*c*) Reformatsky reaction (*d*) Michael reaction.

43. Bimolecular reduction of acetone gives

(*a*) Diacetone amine (*b*) Pinacol

(*c*) Chloretone (*d*) Propane.

44. Oxidation of toluene with CrO_3 in the presence of $(CH_3CO)_2O$ gives a product **A** which on treatment with aq. NaOH produce

(*a*) C_6H_5CHO (*b*) C_6H_5COONa

(*c*) $(C_6H_5CO)_2O$ (*d*) $(CH_3CO)_2O$.

45. Identify the final product 'Z' in the following reaction sequence

$$Me_2C = O + HCN \longrightarrow X \xrightarrow{H_3O^+} Y \xrightarrow[\Delta]{H_2SO_4} Z$$

(*a*) $(CH_3)_2C(OH)COOH$

(*b*) $CH_2 = C(CH_3)COOH$

(*c*) $HO - CH_2CH(CH_3)COOH$

(*d*) $CH_3CH = CHCOOH$.

46. *m*-Chlorobenzaldehyde on reaction with conc. KOH at room temperature gives

(*a*) Potassium meta chlorobenzoate and *m*-Hydroxy benzaldehyde.

(*b*) *m*-Hydroxybenzaldehyde and *m*-chlorobenzylalcohol

(*c*) *m*-Chlorobenzyl alcohol and *meta* Hydroxy benzylalcohol

(*d*) Potassium meta chlorobenzoate and *m*-Chlorobenzylalcohol.

47. Benzaldehyde reacts with ammonia to form

(*a*) Benzal amine (*b*) Urotropine

(*c*) Hydrobenzamide (*d*) Benzaldehyde ammonia.

48. Which of the following compounds will react with ethanolic KCN ?

(*a*) Ethane (*b*) Acetone

(*c*) Chlorobenzene (*d*) Benzaldehyde.

49. $CH_3CH = CHCHO$ is oxidised to $CH_3CH = CHCOOH$ by the use of

(*a*) Alkaline $KMnO_4$ (*b*) Selenium dioxide

(*c*) Ammonical $AgNO_3$ (*d*) All of these reagents.

50. One mole of an organic compound is found to require only 0.5 mole of oxygen to produce an acid. Which class of compound does the starting material belong to ?

(*a*) Alcohol (*b*) Aldehyde

(*c*) Ether (*d*) Ketone.

51. Which of the following is formed when cyclohexanone is heated with conc. NaOH ?

(*a*) 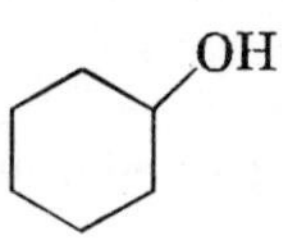

(*b*) 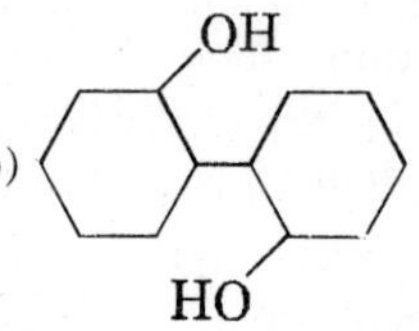

(*c*) 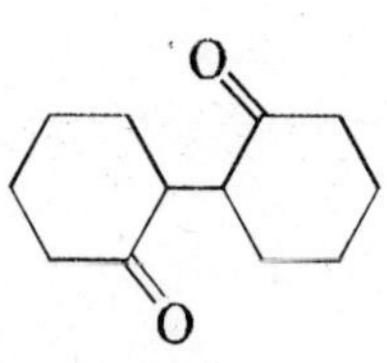

(*d*) 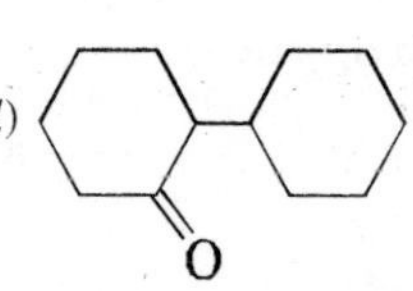

52. 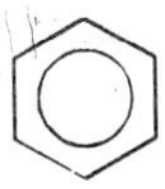 $\xrightarrow[\text{AlCl}_3]{\text{CO + HCl}}$ X + HCl

The compound X in the above sequence is

(*a*) Benzenone (*b*) Carboxybenzene

(*c*) Hydroxybenzene (*d*) Benzene carbaldehyde.

53. Heating of acetone at 950-1030 K results in the formation of

(*a*) Propene and methane

(*b*) Ketene and methane

(*c*) Mixture of ethene and methane

(*d*) Propyne and water.

54. Which reactive intermediate is formed during the condensation reaction between acetaldehyde and formaldehyde ?

(*a*) $:\bar{C}H_2CHO$ (*b*) $\overset{+}{C}H_2CHO$

(*c*) $\overset{+}{C}H_2OH$ (*d*) $:\bar{C}HCHO$.

55. Mesityl oxide is an example of

(*a*) Polymer

(*b*) α, β-unsaturated alcohol

(*c*) α, β-unsaturated ketone

(*d*) α, β-unsaturated acid.

56. Butane-2, 3-dione on reaction with PCl_5 changes its molecular mass by X amu. The value of X is

(*a*) 71 (*b*) 142

(*c*) 110 (*d*) 35.5.

57. What is not true about the oxidation of mixed ketones ?

(*a*) They are resistant to mild oxidation.

(*b*) Oxidation involves cleavage of bond between carbonyl carbon and α-carbon atom.

(*c*) During cleavage carbonyl group remains with higher alkyl group.

(*d*) Oxidative cleavage is governed by Poppof's rule.

58. Aromatic aldehydes react with primary amines to form

(*a*) Amino acids (*b*) Schiff's base

(*c*) Pyrosole (*d*) Pyridine.

59. RCHO is first oxidation product of

(*a*) 1° alcohol (*b*) 2° alcohol

(*c*) alkane (*d*) alkene.

60. Calcium salt of propionic acid is distilled in dry conditions. The product formed will show

(*a*) Fehling test positive

(*b*) Iodoform test negative

(*c*) Victor Meyer test positive

(*d*) Tollen's test positive.

61. **Propyne** $\xrightarrow{\text{conc. } H_2SO_4}$ **X. X** can also be obtained by

(*a*) Acetophenone + conc. H_2SO_4

(*b*) Acetone + conc. H_2SO_4

(*c*) Acetone + conc. NaOH

(*d*) Ethanal + 50% NaOH.

62. In order to get ethanal from calcium carbide which of the following sequence of steps is correct ?

(*a*) (*i*) H_2O (*ii*) $H_2O/HgSO_4/H_2SO_4$

(*b*) (*i*) NH_3 (*ii*) H_2O/H_2SO_4

(*c*) (*i*) H_2O (*ii*) HCN (*iii*) H_2O

(*d*) (*i*) H_2O (*ii*) H_2/Pt (*iii*) $K_2Cr_2O_7/H^+$.

63. What is true about acetaldehyde ?

(*a*) It restores pink colour of Schiff's reagent

(*b*) It gives bakelite on polymerisation with cresol

(*c*) It can be obtained by drastic oxidation of ethanol

(*d*) It gives purple colour with $FeCl_3$.

64. Which of the following change occurs through condensation process ?

(*a*) HCHO $\longrightarrow$ Paraformaldehyde

(*b*) $CH_3COCH_3 \longrightarrow$ Mesityl oxide

(*c*) $CH_3CHO \longrightarrow$ Metaldehyde

(*d*) $CH_2 = CH_2 \longrightarrow$ Polyethene.

65. In order to convert formaldehyde to *n*-butyl alcohol which reagent can be used ?

(*a*) Change cannot be realised

(*b*) (*i*) C_3H_7MgBr (*ii*) H_2O

(*c*) (*i*) $CH_3CH_2CH_2MgBr$ (*ii*) H_2O

(*d*) $K_2Cr_2O_7/H^+$, $LiAlH_4$.

66. $CH \equiv CH \xrightarrow[\text{Zn/}H_2O]{O_3} X \xrightarrow[\text{KOH}]{50\%} Y$.

In the above sequence of reaction Y is

(*a*) Ethanedial (*b*) Glyoxal

(*c*) Pot. glycolate (*d*) Pot. oxalate.

67. $CH_3-\overset{O}{\overset{\|}{C}}-CH_2-CH_2-\overset{O}{\overset{\|}{C}}-CH_3 \xrightarrow[\text{NaOH}]{\text{dil.}} X \xrightarrow{\text{Heat}} Y$.

Here, the final product Y is

(*a*) 3-Methylcyclopent-2-en-1-one

(*b*) 3-Methylcyclopentanol

(*c*) 2, 5-Hexanediol

(*d*) 3-Methylcyclohexanone.

68. Certain carboxyl compound A on reaction with Zn-Hg/HCl give Methylcyclopentane. The compound A could be

(*a*) Cyclopentane carbaldehyde

(*b*) Cyclohexanone

(*c*) 2-Methylcyclopentan-1-one

(*d*) both (*a*) and (*c*).

69. Nitric acid oxidises diphenyl methane to form

(*a*) Acetophenone (*b*) Benzophenone
(*c*) Benzaldehyde (*d*) CO_2 and H_2O.

70. Which of the following will produce dypnone on undergoing aldol type condensation ?

(*a*) Acetophenone (*b*) Benzophenone
(*c*) Acetone (*d*) Ethyl phenyl ketone.

71. Which of the following will produce benzhydrol on treatment with Zn/alc. KOH ?

(*a*) Benzaldehyde (*b*) Acetyl benzene
(*c*) Diphenyl ketone (*d*) Diphenyl ether.

72. The oxime of which of the following species will exist in *syn* and *anti* gemetrical forms ?

(*a*) Benzophenone (*b*) Benzaldehyde
(*c*) Acetone (*d*) Diethyl ketone.

73. Which among the following is least responsive to oxidation ?

(*a*) HCHO (*b*) $C_2H_5COC_2H_5$
(*c*) CH_3CH_2CHO (*d*) C_2H_5CHO.

74. Nitration of acetophenone using nitrating mixture produces mainly

(*a*) *o*-nitro acetophenone (*b*) *p*-nitro acetophenone
(*c*) *m*-nitro acetophenone (*d*) *m*-nitrobenzoic acid.

75. Which of the following will produce acetophenone ?

(*a*) $C_6H_6 + CH_3COCl \xrightarrow{AlCl_3}$

(*b*) $C_6H_5COCl + (CH_3)_2Cd \longrightarrow$

(*c*) $C_6H_5 - C_2H_5 + O_2 \xrightarrow{Mn(CH_3COO)_2}$

(*d*) All of these.

76. The hydrolysis of which of the following will give acetaldehyde ?

(*a*) Acetaldimine (*b*) Ethene
(*c*) Ethylethanoate (*d*) Ethyne.

77. The compound

$$(CH_3)_2\underset{\displaystyle NH_2}{\underset{|}{C}}\,CH_2COCH_3 \text{ is}$$

(*a*) Dimer of acetone (*b*) Acetone amine
(*c*) Amino acid (*d*) Diacetonamine.

78. 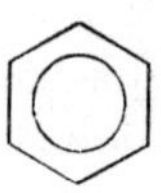 $\xrightarrow[AlCl_3]{CH_3COCl}$ X $\xrightarrow{SeO_2}$ Y. In the given sequence of reactions, Y is

(*a*) Phenyl glyoxal (*b*) Glyoxal
(*c*) Benzoic acid (*d*) Phenyl ethanal.

Naming of reactions/reagents

79. Etard's reaction involves the preparation of benzaldehyde from

(*a*) Toluene (*b*) Ethyl benzene
(*c*) Benzoyl chloride (*d*) Sodium benzoate.

80. The main point of similarity between Etard's reaction and Vilsmeyer reaction is

(*a*) Starting reagent is benzene
(*b*) Final product in both is benzaldehyde
(*c*) Oxidising agent in both is chromyl chloride
(*d*) Both form ketone as product.

81. Which of the following can be used as formylating agent in the presence of $AlCl_3$?

(*a*) HCOOH + HCl (*b*) CO + HCl
(*c*) $CO + Cl_2$ (*d*) $HCl + O_2$.

82. Clemensen's reduction will convert cyclobutanone to

(*a*) Cyclobutanol (*b*) Cyclobutyraldehyde
(*c*) *n*-Butane (*d*) Cyclobutane.

83. Which of the following represent 2, 4-DNP ?

(*a*) $NH_2 - NH - (C_6H_4)_2NO_2$

(*b*) 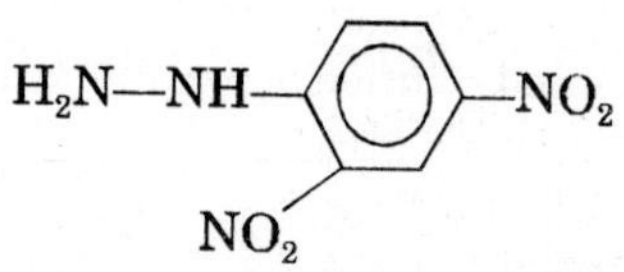

(*c*) 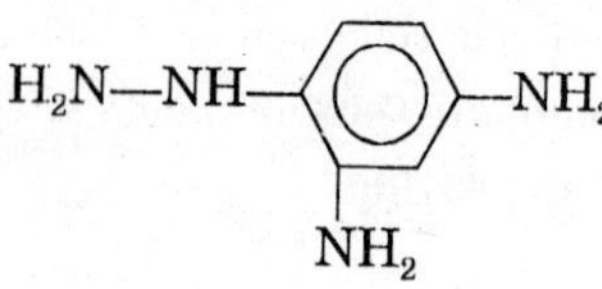

(*d*) 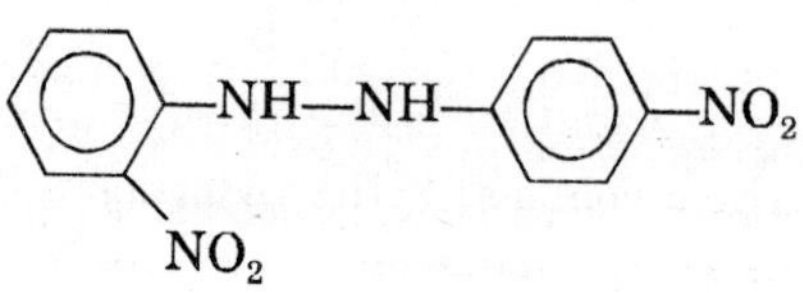

84. Collin's reagent can cause the conversion of

(*a*) $> CO \longrightarrow > CHOH$
(*b*) $> CHO \longrightarrow —COOH$
(*c*) $> CHOH \longrightarrow > CO$
(*d*) $> CHOH \longrightarrow —COOH$.

85. The reaction of acetaldehyde and acetone in the presence of dilute alkali is called

(*a*) Crossed aldol condensation
(*b*) Perkin's condensation
(*c*) Aldol condensation
(*d*) Benzoin condensation.

86. Which of the following aldehyde will undergo Cannizzaro's reaction ?

(*a*) Ethanal (*b*) 2-Methylpropanal
(*c*) 2, 2-Dimethylpropanal (*d*) Phenylacetaldehyde.

87. Which of the following name is associated with the reduction of aldehyde/ketone by using the reagent aluminium isopropoxide ?

(a) Wo...
(b) Meerwe...ONES
(c) Wagner-Mee...
(d) Michael. ...ley

88. Which of the following do... densation ?
(a) Benzene carbaldehyde
(b) p-To...o benzoin con-
(c) Phenylethanal
(d) 4-Methoxy... ...dehyde ...yde.

89. The reaction

$$C_6H_5CHO + CH_3CHO \xrightarrow{NaOH} C_6H_5-CH=CH-CHO + H_2O$$

is known as
(a) Cannizzaro reaction
(b) Aldol condensation
(c) Claisen-Schmidt condensation
(d) Benzoin condensation.

90. Schiff's reagent is
(a) Magenta coloured solution of rosaniline hydrochloride decolourised with H_2SO_3
(b) Magenta solution of rosaniline hydrochloride decolorised with Cl_2
(c) Magenta solution of cobalt chloride solution
(d) Manganese sulphate solution made ammoniacal.

91. Benedict solution provides
(a) Ag^+
(b) Li^+
(c) Cu^{2+}
(d) Ba^{2+}.

92. Oppaneuer oxidation is the reverse process of
(a) Meerwein-Ponndorf-Verley reduction
(b) Clemmensen's reduction
(c) Rosenmund's reduction
(d) Wolff-Kishner reduction.

93. In Stephen's reduction which of the following group is converted to —CHO group.
(a) $-CH_2OH$
(b) -- COCl
(c) – CN
(d) – NC.

94. Which of the following process is helpful is synthesizing acetophenone ?
(a) Cannizzaro's reaction
(b) Wurtz reaction
(c) Friedel Craft's reaction
(d) Rosenmund's reaction.

95. Which of the following combination is employed in Reformatsky reaction ?
(a) Zn/ether
(b) Mg/ether
(c) Li/alcohol
(d) Zn/alcohol.

96. Ylide is one of the reagent used in
(a) Reformatsky reaction
(b) Fittig reaction
(c) Wittig reaction
(d) Perkin's reaction.

97. Reaction of CH_3CHO and aluminium ethoxide is called
(a) Tischenko reaction
(b) Clemmensen's reaction
(c) Perkin's reaction
(d) Etard's reaction.

Distinction

...8. Which of the following reagent can distinguish between acetaldehyde and benzaldehyde ?
(a) NH_2OH
(b) Benedict's solution
(c) ...P
(d) None of these.

...9. The distinction between diethylketone and methyl ethyl ketone can be achieved by using the reagent
(a) Zn-Hg/HCl
(b) Na_2CO_3/I_2
(c) $AgNO_3/NH_4OH$
(d) Fehling soln.

100. Which of the following ketone will not respond to iodoform test ?
(a) 3-Methylbutan-2-one
(b) Ethyl isopropylketone
(c) Methyl phenyl ketone
(d) Dimethyl ketone.

101. Which of the following will not reduce Fehling solution ?
(a) 2-Methylpropanal
(b) Ethanal
(c) Butanal
(d) Acetone.

102. Which of the following reagent can help in separating a mixture of 2-butanone and butanol-2 ?
(a) NaOH
(b) NaCl
(c) $NaHSO_3$
(d) All the above.

103. Which of the following reagent will be able to provide distinction between acetophenone and benzophenone ?
(a) Aqueous $NaHSO_3$
(b) I_2/aqueous Na_2CO_3
(c) Brady's reagent
(d) Benedict solution.

Reactivity

104. Among formaldehyde, acetone and acetaldehyde, the reactivity order is
(a) $HCHO > (CH_3)_2CO > CH_3CHO$
(b) $HCHO > CH_3CHO > (CH_3)_2CO$
(c) $(CH_3)_2C=O > HCHO > CH_3CHO$
(d) Unpredictable.

105. Among the following, the least reactive aldehyde is
(a) HCHO
(b) C_6H_5CHO
(c) CH_3CHO
(d) C_2H_5CHO.

ANSWERS

1. (b)	2. (b)	3. (b)	4. (c)	5. (b)	6. [illegible]	[illegible] (b)	8. (c)	9. (d)
10. (d)	11. (b)	12. (a)	13. (b)	14. (d)	[illegible] (b)	16. (d)	17. (c)	18. (a)
19. (b)	20. (d)	21. (d)	22. (b)	23. (a)	[illegible]	25. (d)	26. (b)	27. (d)
28. (c)	29. (a)	30. (a)	31. (b)	32. (a)	[illegible]	34. (c)	35. (d)	36. (c)
37. (d)	38. (b)	39. (b)	40. (d)	41. (a)	42. (c)	43. (b)	44. (b)	45. (b)
46. (d)	47. (c)	48. (d)	49. (c)	50. [illegible]	51. (d)	52. (d)	53. (b)	54. (a)
55. (c)	56. (c)	57. (c)	58. (b)	59. [illegible]	60. (b)	61. (b)	62. (a)	63. (a)
64. (b)	65. (c)	66. (c)	67. (a)	68. [illegible]	69. (b)	70. (a)	71. (c)	72. (b)
73. (b)	74. (c)	75. (d)	76. (a)	7[illegible] (c)	78. (a)	79. (a)	80. (b)	81. (b)
82. (d)	83. (b)	84. (c)	85. (a)	[illegible]	87. (b)	88. (c)	89. (c)	90. (a)
91. (c)	92. (a)	93. (c)	94. (c)	95. [illegible]	96. (c)	97. (a)	98. (b)	99. (b)
100. (b)	101. (d)	102. (c)	103. [illegible]	10[illegible] (b)	105. (b).			

HINTS/SOLUTIONS

Q. No.	Ans.	Hints & Solutions
1.	(b)	$\overset{4}{CH_3}\overset{3}{CH} = \overset{2}{CH}\overset{1}{CHO}$ is *but-2-enal.*
2.	(b)	Aldol is $CH_3CH(OH)CH_2CHO$. It's IUPAC name is 3-Hydroxybutanal.
3.	(b)	The compound is a salt of propionic acid.
4.	(c)	*o*-Benzoquinone is 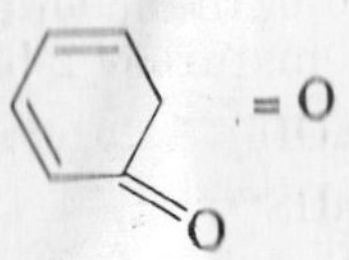
5.	(b)	$CH_3C \equiv CH + HOH \xrightarrow{HgSO_4} CH_3C(OH) = CH_2 \xrightarrow{\text{rearranges}} CH_3COCH_3$ (Acetone)
6.	(b)	$BaSO_4$, poisons the catalyst and decreases its efficiency. It, thus, checks the further reduction of aldehydes to alcohols.
7.	(b)	$CH_3CH_2COCl \xrightarrow[Pd/BaSO_4]{H_2} CH_3CH_2CHO$ (Propionaldehyde)
8.	(c)	$CH_3CHCl_2 \xrightarrow{KOH\,(aq)} CH_3CH(OH)_2$ (unstable) $\xrightarrow{-H_2O} CH_3CHO$ Ethanal
9.	(d)	$CH_3CH_2 \xrightarrow[-H_2]{Cu/575\ K} CH_3CHO.$
10.	(d)	$2CH_3COCl + (C_2H_5)_2Cd \longrightarrow 2CH_3COC_2H_5 + CdCl_2.$
11.	(b)	$2CH_3COCl + (CH_3)_2\,Cd \longrightarrow 2CH_3COCH_3 + CdCl_2.$ (Acetone)
12.	(a)	$CH_3-CCl_2-CH_3 \xrightarrow[-H_2O]{2KOH\,(aq)} CH_3COCH_3 + 2KCl$ (Acetone)
13.	(b)	CrO_3 in acetic acid or H_2CrO_4 in acetone (aq.) is called Jone's reagent.

Q. No.	Ans.	Hints & Solutions
14.	(*d*)	$CH \equiv CH \xrightarrow[H_2SO_4]{HgSO_4} CH_3CHO \xrightarrow{LiAlH_4} CH_3CH_2OH \xrightarrow{P/Br_2} CH_3CH_2Br$ (Ethyl bromide)
15.	(*b*)	$(HCOO)_2Ca \xrightarrow{\Delta} HCHO + CaCO_3$ $(HCOO)_2Ca + Ca(OOCCH_3)_2 \longrightarrow CH_3CHO + CaCO_3$
16.	(*d*)	Acetophenone is least soluble because of its extremely poor tendency to form H-bonds with water molecules.
17.	(*c*)	The structure (*iii*) being neutral is most stable and has maximum contribution to resonance hybrid.
18.	(*a*)	Benzaldehyde has typical smell of bitter almonds.
19.	(*b*)	Oxygen being electronegative cannot carry positive charge.
20.	(*d*)	$CH_2 = CH - CH = O \longleftrightarrow \overset{+}{C}H_2 - CH = CH - \overset{-}{O}$ Resonating forms $\overset{\delta+}{CH_2} \text{=====} CH \text{=====} C \text{=====} \overset{\delta-}{O}$ is hybrid form.
21.	(*d*)	CH_3MgBr can react both with aldehydes as well as ketones.
22.	(*b*)	HCl does not add across C = O bond.
23.	(*a*)	$NaBH_4$ does not reduce C = C bond.
24.	(*c*)	$CH_3CHO \xrightarrow{HCN} CH_3CH(OH)CN \xrightarrow{H_3O^+} CH_3-CH(OH)COOH$ (α-Hydroxy acid)
25.	(*d*)	$C_6H_5-CHO + H_2C(COOH)_2$ (Malonic acid) $\xrightarrow[-H_2O]{-CO_2} C_6H_5CH = CHCOOH$ Cinnamic acid
26.	(*b*)	$CH_3CHO \xrightarrow[(ii)\ Heat]{(i)\ NH_3} CH_3CH = NH + H_2O$ (Acetaldimine)
27.	(*d*)	$CH_3CH(OH)CH_3 \xrightarrow{oxid.} CH_3-\overset{O}{\overset{\Vert}{C}}-CH_3$ (Acetone) (X) $\xrightarrow[(ii)\ H_2O]{(i)\ MeMgBr} CH_3-C(OH)(CH_3)-CH_3$ (Y)
28.	(*c*)	Benzophenone ($C_6H_5COC_6H_5$) does not contain CH_3-CO- grouping. Hence it does not give iodoform test positive.
29.	(*a*)	$HCHO \xrightarrow[(ii)\ HOH]{(i)\ EtMgI} Et.\ CH_2OH + Mg(OH)I$ 1-Propanol
30.	(*a*)	HCl does not add across C = O bond. It give Cl^- ion which is a very weak nucleophile.
31.	(*b*)	Ethanal and C_2H_5OH react in the presence of conc. H_2SO_4 to give acetal.
32.	(*a*)	Cannizzaro's reaction is done in the presence of 50% NaOH.
33.	(*b*)	$NaBH_4$ reduces **> CO** to **> CHOH** but does not reduce C = C.
34.	(*c*)	Moderately acidic medium facilitates the nucleophilic attack on carbonyl carbon, but in highly acidic medium NH_2OH forms salt with the acid and loses its capacity to act as nucleophile.
35.	(*d*)	Both $SOCl_2$ and PCl_5 convert > C = O to > CCl_2.
36.	(*c*)	$6HCHO + 4NH_3 \longrightarrow (CH_2)_6N_4 + 6H_2O$
37.	(*d*)	$CH_3 - CH_2 - COCH_3$ (Pentan-2-one) $\xrightarrow{Conc.\ HNO_3} 2CH_3COOH$ (Acetic acid)
38.	(*b*)	$CH_3 - C(CH_3) = O + HCCl_3 \longrightarrow CH_3 - C(OH)(CH_3) - CCl_3$ (Chloretone)

Q. No.	Ans.	Hints & Solutions
39.	(*b*)	Paraldehyde is formed in the presence of conc. H_2SO_4.
40.	(*d*)	According to the data given, A should be a ketone with CO group carrying no methyl group. Thus, it should be $CH_3CH_2COCH_2CH_3$, *i.e.*, 3-pentanone.
41.	(*a*)	CH_3CH_2CHO (Propanal) $\xrightarrow{HCN}$ $CH_3CH_2CH(OH)-CN$ $\xrightarrow{LiAlH_4}$ $CH_3CH_2CH(OH)CH_2NH_2$ (1-Amino-2-butanol)
42.	(*c*)	Reformatsky reaction is used to prepare β-Hydroxyester by using α-bromoester and aldehydes. $CH_3-CH(Br)-COOC_2H_5 + CH_3-CHO$ (Ethyl, α-Bromo propionate) $\xrightarrow{Zn}$ $(CH_3)(H)C(OH)-CH_2CH(Br)COOC_2H_5$
43.	(*b*)	$CH_3-\overset{O}{\overset{\parallel}{C}}(CH_3) + \overset{O}{\overset{\parallel}{C}}(CH_3)-CH_3 \xrightarrow[H_2O]{Mg-Hg} CH_3-C(OH)(CH_3)-C(OH)(CH_3)-CH_3$
44.	(*b*)	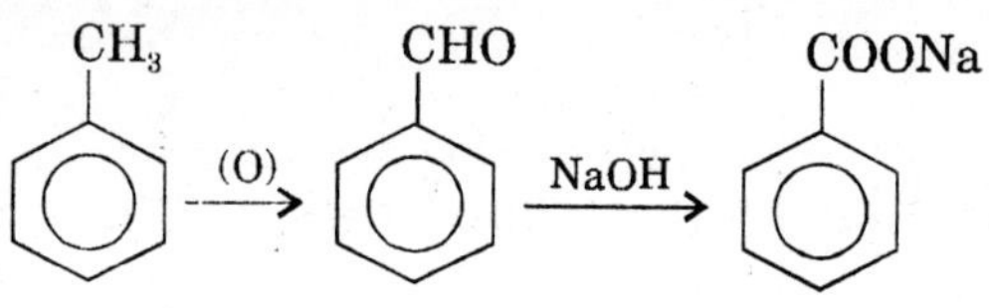
45.	(*b*)	$Me_2C = O \xrightarrow{HCN} Me_2C(OH)-CN$ (X) $\xrightarrow{H_3O^+}$ $Me_2C(OH)-COOH$ (Y) $\xrightarrow[445\ K]{H_2SO_4}$ $CH_2 = C(CH_3)-COOH$ (β-Elimination)
46.	(*d*)	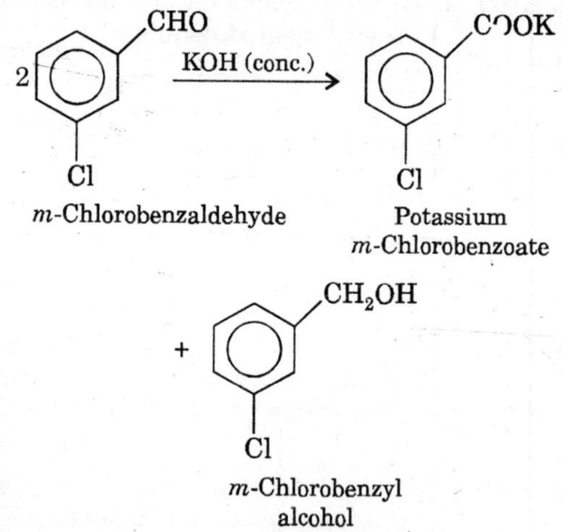
47.	(*c*)	$C_6H_5CHO + NH_3 \longrightarrow$ $C_6H_5CH=N-CH(C_6H_5)-N=CHC_6H_5$ Hydrobenzamide
48.	(*d*)	Benzaldehyde undergoes benzoin condensation.
49.	(*c*)	Ammoniacal $AgNO_3$ (Tollen's reagent) oxidises —CHO group to —COOH group but does not oxidise C = C bond.
50.	(*b*)	RCHO (Aldehyde) $\xrightarrow{1/2\ O_2}$ RCOOH (Acid)
51.	(*d*)	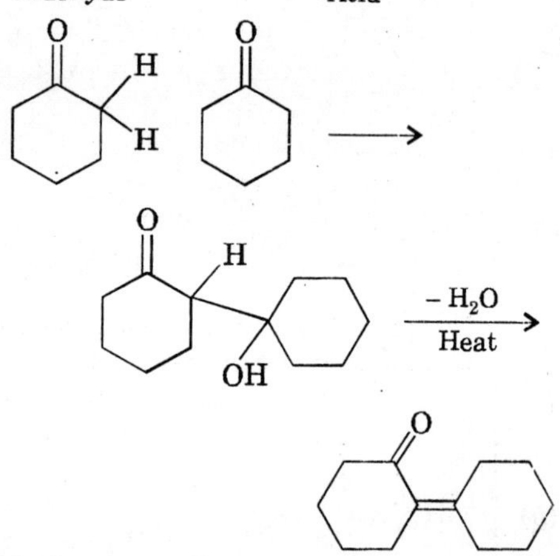 This is an example of aldol type condensation.
52.	(*d*)	This is Gattermann Koch reaction.
53.	(*b*)	$CH_3COCH_3 \xrightarrow{1030\ K} CH_4 + H_2C = C = O$ (Ketene)

Q. No.	Ans.	Hints & Solutions
54.	(a)	See the mechanism of aldol condensation in Concept oriented details.
55.	(c)	Mesityl oxide is $CH_3-\underset{\beta}{\overset{CH_3}{\overset{\vert}{C}}}=\underset{\alpha}{CH}-\overset{O}{\overset{\Vert}{C}}-CH_3$. It is a α, β unsaturated ketone.
56.	(c)	$\underset{\text{(Mol. mass = 86)}}{CH_3-\overset{O}{\overset{\Vert}{C}}-\overset{O}{\overset{\Vert}{C}}-CH_3} \xrightarrow{PCl_5} \underset{\text{(Mol. mass = 196)}}{CH_3-\underset{Cl}{\underset{\vert}{\overset{Cl}{\overset{\vert}{C}}}}-\underset{Cl}{\underset{\vert}{\overset{Cl}{\overset{\vert}{C}}}}-CH_3}$ The change in mol. mass = 196 – 86 = 110.
57.	(c)	According to Popov's rule the $-\overset{O}{\overset{\Vert}{C}}-$ is retained by chain with lesser no. of carbon atoms.
58.	(b)	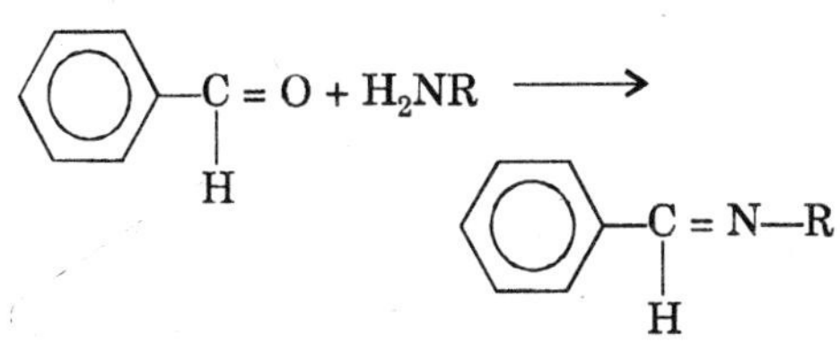 Schiff's base
59.	(a)	$\underset{1^\circ\text{ Alcohol}}{RCH_2OH} \xrightarrow[-H_2O]{(O),\text{ Mild}} \underset{\text{Aldehydes}}{R-\overset{H}{\overset{\vert}{C}}=O}$ The product 3-pentanone will not give any of the tests mentioned in the question.
60.	(b)	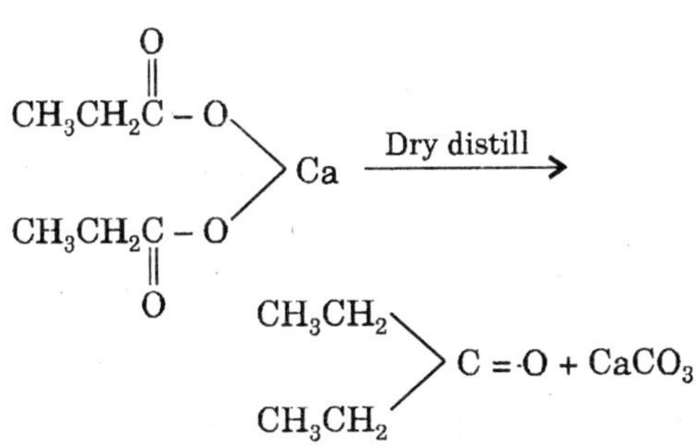 3-Pentanone
61.	(b)	Both propyne and acetone will give trimerised product mesitylene.
62.	(a)	$CaC_2 \xrightarrow{H_2O} Ca(OH)_2 + \underset{\text{(Ethyne)}}{CH\equiv CH}$ $CH\equiv CH \xrightarrow[HgSO_4]{H_2O/H_2SO_4} \underset{\text{(Ethanal)}}{CH_3CHO}$
63.	(a)	Aldehydes restore the pink colour of Schiff's reagent.
64.	(b)	Condensation process involves a loss of H_2O, NH_3 or ROH molecule.
65.	(b)	$CH_2=O \xrightarrow[(ii)\ H_2O]{(i)\ C_3H_7MgBr} H_2C\begin{matrix} \nearrow C_3H_7 \\ \searrow OH \end{matrix}$ *n*-butyl alcohol
66.	(c)	$CH\equiv CH \xrightarrow[(ii)\ H_2O/Zn]{(i)\ O_3} \underset{\text{Glyoxal}}{\begin{matrix} HC=O \\ \vert \\ HC=O \end{matrix}}$ Glyoxal $\xrightarrow[\text{Internal Cannizzaro reaction}]{50\%\ KOH}$ $\begin{matrix} CH_2OH \\ \vert \\ COOK \end{matrix}$ Potassium glycolate
67.	(a)	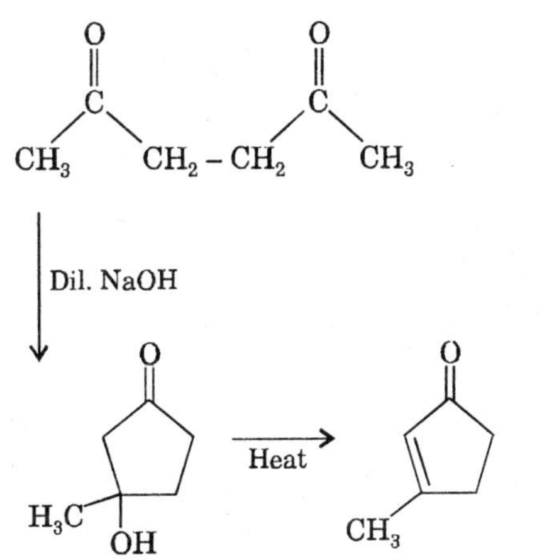 3-Methylcyclopent-2-en-1-one

Q. No.	Ans.	Hints & Solutions
68.	(*d*)	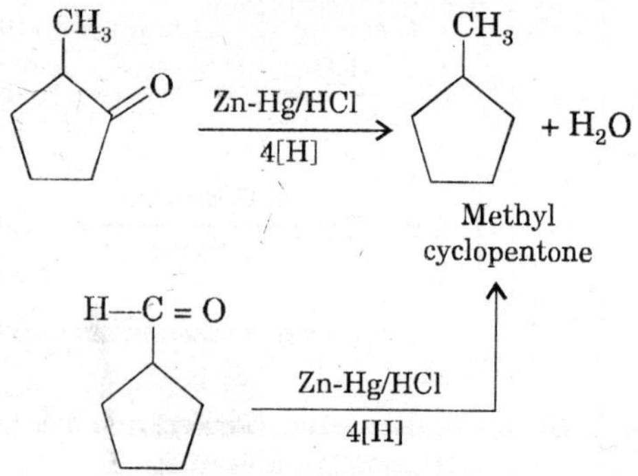
69.	(*b*)	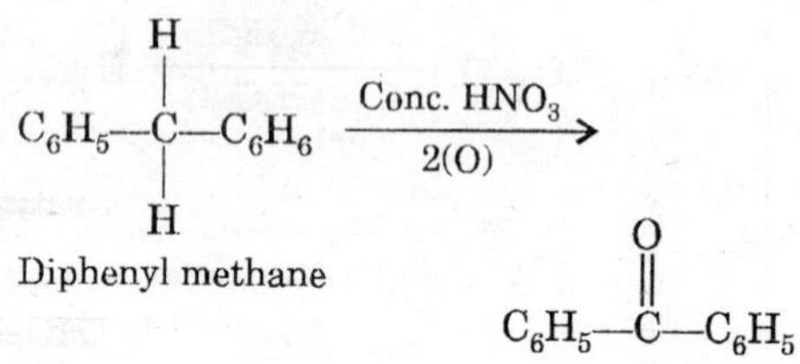
70.	(*a*)	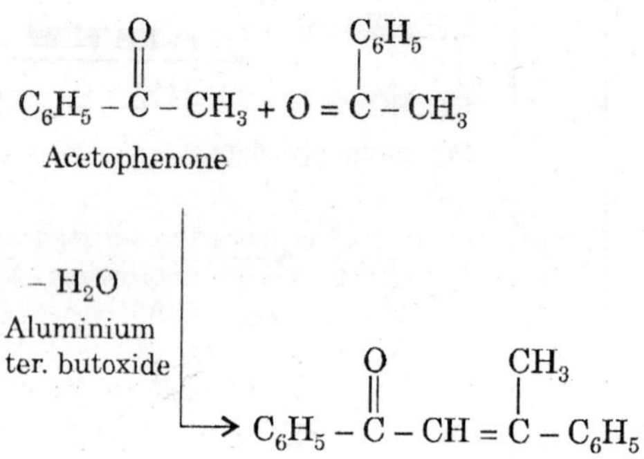
71.	(*c*)	$C_6H_5\overset{O}{\overset{\Vert}{C}}-C_6H_5 \xrightarrow[2[H]]{Zn/alcohol} C_6H_5-\underset{}{\overset{OH}{\overset{\vert}{CH}}}-C_6H_5$ Benzhydrol
72.	(*b*)	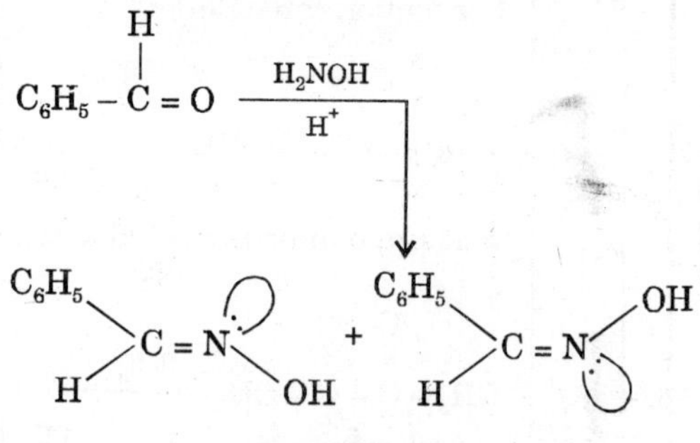 Anti and Syn forms
73.	(*b*)	Ketones are less prone to oxidation. These can be oxidised in drastic conditions only.
74.	(*c*)	—$COCH_3$ group is electron withdrawing. Hence it is *m*-directing.
75.	(*d*)	All the given reactions will produce acetophenone.
76.	(*a*)	$CH_3CH = NH_2 \xrightarrow[H^+]{HOH} CH_3CHO + NH_4^+$
77.	(*d*)	See reaction of ammonia and acetone.
78.	(*a*)	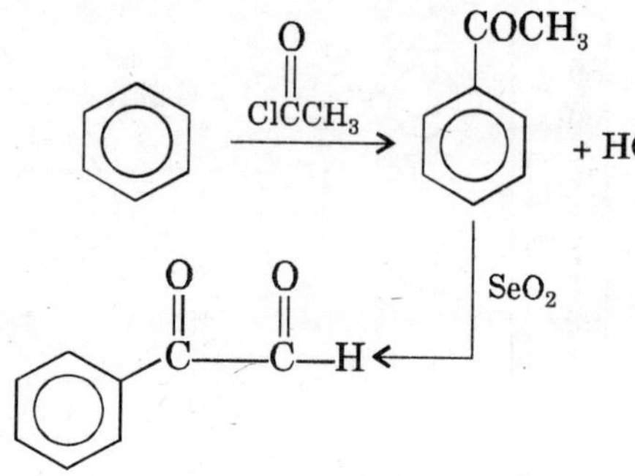 Phenyl glyoxal
79.	(*a*)	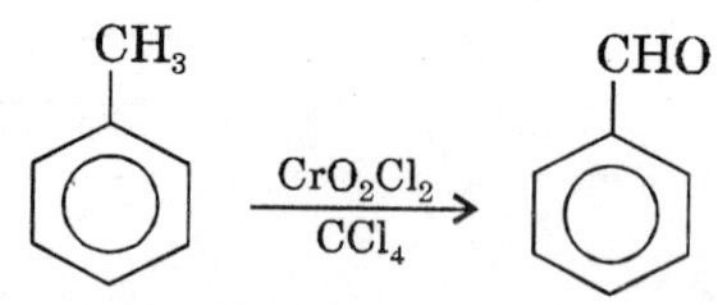
80.	(*b*)	The final product in both cases is benzaldehyde 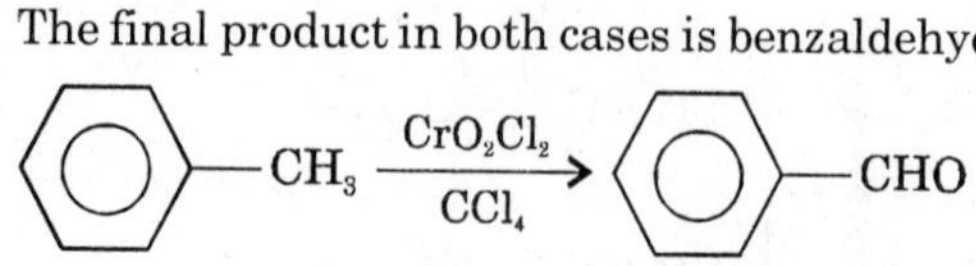 (Etard's reaction) 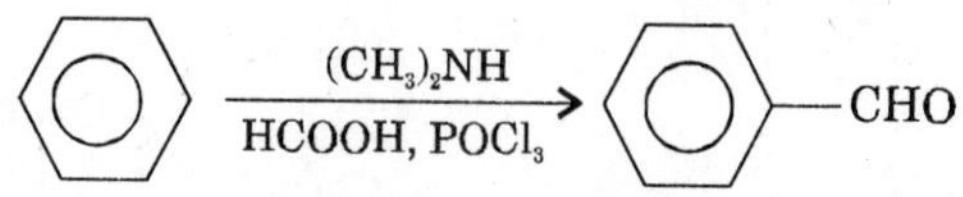 (Vilsmeyer reaction)

Q. No.	Ans.	Hints & Solutions
81.	(*b*)	CO + HCl in presence of $AlCl_3$ acts as formylating agent 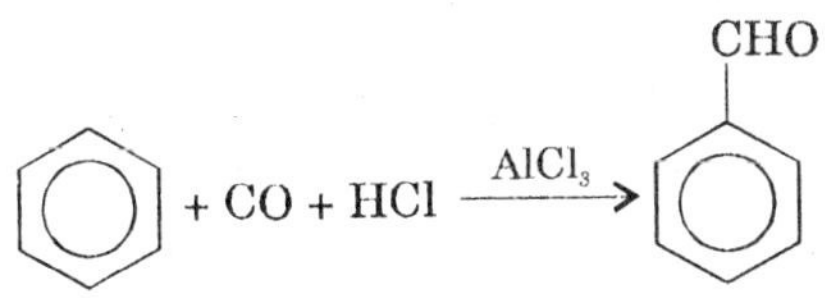
82.	(*d*)	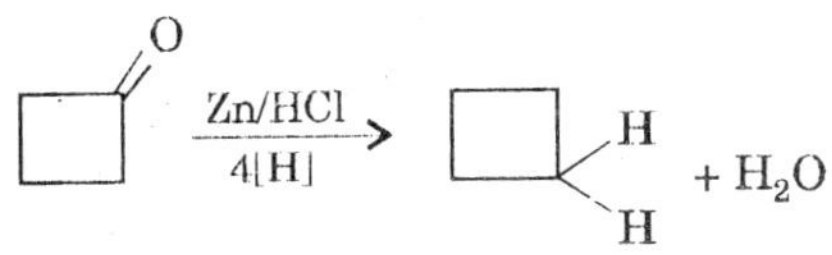 (Cyclobutane)
83.	(*b*)	It is 2, 4-dinitro phenyl hydrazine.
84.	(*c*)	Collin's reagent converts >CHO to >CO.
85.	(*a*)	When aldehyde and ketone are condensed it is called crossed aldol condensation.
86.	(*c*)	2, 2-dimethylpropanal does not have α-H atom.
87.	(*b*)	It is M.P.V. reduction.
88.	(*c*)	Phenylethanal ($C_6H_5CH_2CHO$) does not have —CHO group directly attached to phenyl group, *i.e.,* it is not a aromatic aldehyde.
89.	(*c*)	The reaction is Claisen-Schmidt condensation.
90.	(*a*)	It is *para* rosaniline hydrochloride decolourised with sulphurous acid.
91.	(*c*)	It is aq $CuSO_4$ containing citrate ions.
92.	(*a*)	It is reverse of M.P.V. reduction.
93.	(*c*)	Stephen's reduction converts alkyl cyanides to aldehydes. $R—CN \xrightarrow{SnCl_2/HCl} R—CH = NH_2 \xrightarrow{H_2O^+} RCHO$
94.	(*c*)	Friedel Craft's reaction is quite helpful in synthesis of acetophenone by using benzene and CH_3COCl in the presence of anhydrous $AlCl_3$.
95.	(*a*)	Reformatsky reaction is done in Zn/ether.
96.	(*c*)	$\underset{\text{ylide}}{(C_6H_5)_3P = CH_2} + CH_3CHO \longrightarrow CH_3CH = CH_2 + (C_6H_5)_3P = O.$
97.	(*a*)	$2CH_3CHO \xrightarrow{Al(OC_2H_5)_3} CH_3COOC_2H_5.$ This is called Tischenko reaction.
98.	(*b*)	Benedict's solution react with aldehydes but not with ketones.
99.	(*b*)	Methyl ethyle ketone ($CH_3COC_2H_5$) will give iodoform test positive but diethyl ketone will not.
100.	(*b*)	Ethyl isopropyl ketone [$C_2H_5COCH(CH_3)$] will not give iodoform test positive.
101.	(*d*)	Ketones do not give Fehling test positive.
102.	(*c*)	$NaHSO_3$ react only with carbonyl compounds (aldehydes/ketones) hence it can help in the separation.
103.	(*b*)	$CH_3\overset{O}{\overset{\Vert}{C}} - C_6H_5$ (acetophenone) will show iodoform test positive but $C_6H_5 - \overset{O}{\overset{\Vert}{C}} - C_6H_5$ will not give iodoform test positive.
104.	(*b*)	In going from formaldehyde to ketone, the + I effect as well as crowding at the carbonyl carbon increases which causes decrease in the tendency of nucleophilic attack.
105.	(*b*)	Benzaldehyde is least reactive because of resonance stabilisation of its molecule.

5.5. EXAMINER'S VIEWPOINT

Test Your Confidence

1996

1. Aromatic aldehydes in the presence of cyanide ion as catalyst, are converted into *acyloins*. This reaction is called
(*a*) Perkin's reaction (*b*) Cannizzaro's reaction
(*c*) Benzoin condensation (*d*) Claisen condensation. ***(CEE Delhi)***

2. Para aldehyde is a trimer of
(*a*) Formaldehyde (*b*) Acetaldehyde
(*c*) Benzaldehyde (*d*) Propionaldehyde. ***(AFMC, MP PMT)***

3. The product acetal is produced by reacting alcohol in the presence of dry HCl with
(*a*) Acetaldehyde (*b*) Ketone
(*c*) Carboxylic acid (*d*) Ether. ***(MP PMT)***

4. Acetone is easily oxidised with
(*a*) Tollen's reagent
(*b*) Fehling solution
(*c*) Acidic dichromate solution
(*d*) Benedict's solution. ***(MP PET)***

5. An organic compound X on treatment with acidified $K_2Cr_2O_7$ gives compound Y which reacts with iodine and sodium carbonate to form tri iodomethane. The compound X can be
(*a*) CH_3OH (*b*) $CH_3\overset{\overset{O}{||}}{C}CH_3$
(*c*) CH_3CHO (*d*) $CH_3CH(OH)CH_3$. ***(CEET Kurukshetra)***

6. Aldol condensation will not occur in
(*a*) HCHO (*b*) CH_3CH_2CHO
(*c*) CH_3COCH_3 (*d*) CH_3CHO. ***(CBSE PMT)***

7. Oxidation of toluene to benzaldehyde by the use of chromyl chloride is called
(*a*) Wurtz reaction (*b*) Etard's reaction
(*c*) Fittig reaction (*d*) Rosenmund's reaction. ***(CBSE PMT)***

1997

8. Which of the following compounds does not react with $NaHSO_3$?
(*a*) C_6H_5CHO (*b*) $CH_3COC_6H_5$
(*c*) Acetone (*d*) Acetaldehyde. ***(Karnatake CEE)***

9. Which of the following is a mixed ketone ?
(*a*) Benzophenone (*b*) Acetophenone
(*c*) Dibenzyl (*d*) Di isopropyl ketone. ***(AFMC)***

10. Reductive ozonolysis of benzene produces
(*a*) Acetone (*b*) Maleic anhydride
(*c*) Phthalic anhydride (*d*) Glyoxal. ***(CPMT)***

11. Concentrated sodium hydroxide and benzaldehyde react to produce
(*a*) Benzyl alcohol (*b*) Hydrobenzamide
(*c*) Cinnamic acid (*d*) Benzophenone. ***(CPMT)***

12. Which of the following reaction is expected to give readily a hydrocarbon product in good yields ?
(*a*) $RCOOK \xrightarrow{\text{Electrolysis}}$
(*b*) $RCOOAg \xrightarrow{I_2}$
(*c*) $CH_3CH_3 \xrightarrow[h\nu]{Cl_2}$
(*d*) $(CH_3)_2CCl_2 \xrightarrow{C_2H_5OH}$ ***(CBSE PMT)***

13. Ketones (R_1 COR_2) ; $R_1 = R_2$ = alkyl group, can be obtained in one step by
(*a*) Hydrolysis of esters
(*b*) Oxidation of primary alcohols
(*c*) Oxidation of secondary alcohols
(*d*) Reaction of acid halides and alcohols. ***(CBSE PMT)***

14. Formaldehyde + $NH_3 \longrightarrow$ Y. The product Y is
(*a*) Hexamethylene tetramine
(*b*) Formamide
(*c*) Paraformaldehyde
(*d*) Methanol. ***(CPMT Med.)***

15. In a reaction involving ring substitution of C_6H_5Y, the major product is *meta* isomer. The group Y can be
(*a*) —NH_2 (*b*) —COOH
(*c*) —CH_3 (*d*) —Cl. ***(AIIMS)***

16. Which of the following reaction can produce R—CO—Ar ?
(*a*) ArCOCl + H—Ar $\xrightarrow{AlCl_3}$
(*b*) COCl + RMgX $\longrightarrow$

(c) $R + CrO_3 \longrightarrow$

(d) $RCOCl + Ar—H \xrightarrow{AlCl_3}$ **(AIIMS)**

17. The reaction of benzoyl chloride with $H_2/Pd—BaSO_4$ produces
(a) Benzyl alcohol (b) Benzaldehyde
(c) Benzoic acid (d) Benzoyl cyanide. **(AIIMS)**

18. Acetaldehyde cannot exhibit
(a) Iodoform test (b) Lucas test
(c) Benedict's test (d) Tollen's test. **(AIIMS)**

1998

19. Which of the following is a method of converting a ketone to hydrogen with same number of carbon atoms ?
(a) Clemmensen's reduction
(b) Reimer-Tiemann reaction
(c) Cannizzaro's reaction
(d) Rosenmund's reaction. **(MP PMT, AFMC)**

20. Which of the following aldehyde is most reactive towards nucleophilic addition reaction ?
(a) HCHO (b) CH_3CHO
(c) C_2H_5CHO (d) C_6H_5CHO. **(CEET Kurukshetra)**

21. The oxidation of which of the following will give ethyl methyl ketone ?
(a) Propan-2-ol (b) Butan-1-ol
(c) Butan-2-ol (d) 2-Methylbutane-2-ol. **(CEET Kurukshetra)**

22. The reaction involving the preparation of salicylaldehyde from phenol using chloroform and NaOH is called
(a) Gattermann reaction
(b) Perkin's reaction
(c) Duff's reaction
(d) Reimer-Tiemann reaction. **(CEET Kurukshetra)**

23. $C_6H_5CHO + HCN \longrightarrow C_6H_5CH(CN)OH$
The product would be
(a) Racemate
(b) Optically active
(c) A meso compound
(d) Mixture of diastereomers. **(Pb. PMT)**

24. Reduction of cyclohex-2-enone with aluminium isopropoxide in excess of isopropyl alcohol will produce
(a) Cyclohexanol (b) Cyclohex-2-enol
(c) Cyclohexanone (d) Benzene. **(Delhi PMT)**

25. If acetyl chloride is reduced in the presence of $BaSO_4$ and Pd, then
(a) CH_3CHO is formed (b) CH_3CH_2OH is formed
(c) CH_3COOH is formed (d) CH_3COCH_3 is formed. **(AFMC)**

26. Acetaldehyde when treated with dil. NaOH gives
(a) CH_3CH_2OH (b) CH_3COOH
(c) $CH_3—CH(OH)—CH_2CHO$ (d) $CH_3—CH_3$. **(EAMCET)**

27. C_2H_5CHO and $(CH_3)_2CO$ can be distinguished by testing with
(a) Phenyl hydrazine (b) Hydroxylamine
(c) Fehling solution (d) Sodium bisulphite.

28. The IUPAC name of acetone is
(a) Ethanal (b) Propanone
(c) Butanone (d) Propanal. **(BITS)**

29. Formaldehyde when reacted with methyl magnesium bromide followed by hydrolysis gives
(a) C_2H_5OH (b) CH_3COOH
(c) HCHO (d) CH_3CHO. **(BHU)**

30. In which reaction aromatic aldehyde is treated with acid anhydride in presence of corresponding salt of the acid to give unsaturated acid ?
(a) Friedel-Crafts reaction (b) Perkin's reaction
(c) Wurtz reaction (d) None of these. **(BHU)**

1999

31. When formaldehyde and KOH are heated we get,
(a) Acetylene (b) Methane
(c) Methyl alcohol (d) Ethyl formate. **(MP CET)**

32. Which of the following reagent react differently with HCHO, CH_3CHO and CH_3COCH_3 ?
(a) HCN (b) $NH_2—NH_2$
(c) NH_2OH (d) NH_3. **(MP CET)**

2000

33. Clemmensen's reduction of ketones is carried out in
(a) H_2 with Pd catalyst (b) Glycol with KOH
(c) $LiAlH_4$ in water (d) Zn-Hg with HCl. **(BHU)**

34. Propanal on treatment with dilute sodium hydroxide forms
(a) $CH_3CH_2CH_2CH_2CH_2CHO$
(b) $CH_3CH_2CH(OH)CH_2CH_2CHO$
(c) $CH_3CH_2CH(OH)CH(CH_3)CHO$
(d) CH_3CH_2COONa. **(Kerala MEE)**

35. The product of following reaction

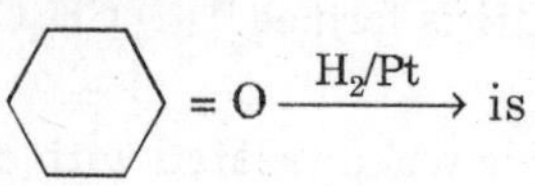

(a) 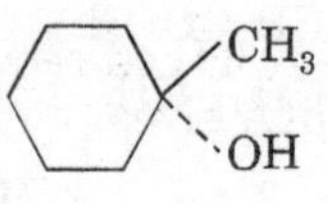 (b) 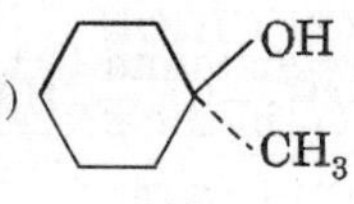

(c) 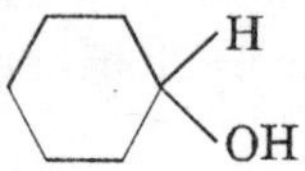 (d) 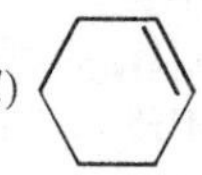

(Kerala EEE)

36. Which of the following products is formed when benzaldehyde is treated with CH_3MgBr and the addition product so obtained is subjected to acid hydrolysis ?

(a) Secondary alcohol (b) A primary alcohol
(c) Phenol (d) *tert*-Butyl alcohol.

(Haryana CEET)

37. The addition of HCN to carbonyl compounds is an example of

(a) Nucleophilic substitution
(b) Electrophilic addition
(c) Nucleophilic addition
(d) Electrophilic substitution. ***(Haryana CEET)***

38. $(CH_3)_2CO \xrightarrow[(HCl)]{NaCN} A \xrightarrow[\Delta]{H_3^+O} B$

In the above sequence of reactions A and B are

(a) $(CH_3)_2C(OH)CN$, $(CH_3)_2C(OH)COOH$
(b) $(CH_3)_2C(OH)CN$, $(CH_3)_2C(OH)_2$
(c) $(CH_3)_2C(OH)CN$, $(CH_3)_2CHCOOH$
(d) $(CH_3)_2C(OH)CN$, $(CH_3)_2C = O$. ***(CPMT)***

39. Reduction of $> C = O$ to CH_2 can be carried out with

(a) Catalytic reduction (b) Na/C_2H_5OH
(c) Wolff-Kishner reduction (d) $LiAlH_4$. ***(DCE)***

40. From which of the following tertiary butyl alcohol is obtained by the action of methyl magnesium iodide ?

(a) HCHO (b) CH_3CHO
(c) CH_3COCH_3 (d) CO_2. ***(MP CET)***

2001

41. Identify the product Y in the sequence

$CH_3CHO + CH_3MgI \xrightarrow{Ether} X \xrightarrow{H_2O/H^+} Y$

(a) CH_3OH (b) CH_3CH_2OH
(c) $(CH_3)_2CHOH$ (d) $(CH_3)_2COH$
(e) CH_3CH_2CHO. ***(Kerala PMT)***

42. Which of the following compounds containing carbonyl group will give coloured crystalline compound with

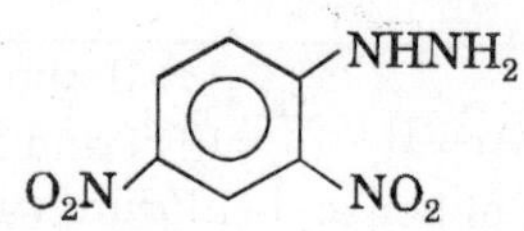

(a) CH_3COCl (b) CH_3COCH_3
(c) $CH_3CO(OC_2H_5)$ (d) CH_3CONH_2
(e) $HO(C_6H_4)COOH$. ***(Kerala PMT)***

43. Which of the following organic compounds exhibits positive Fehling test as well as Iodoform test ?

(a) Methanal (b) Ethanol
(c) Propanone (d) Ethanal.

(Karnataka CET)

44. Rosenmund's reduction of an acyl chloride gives

(a) An aldehyde (b) An alcohol
(c) An ester (d) A hydrocarbon
(e) An alkyl halide. ***(Kerala CET)***

45. Rearrangement of an oxime to an amide in the presence of strong acid is called

(a) Curtius rearrangement
(b) Fries rearrangement
(c) Beckman rearrangement
(d) Aldol condensation
(e) Sandmeyer reaction. ***(Kerala CET)***

46. Chromyl chloride and toluene react to produce

(a) *p*-chlorotoluene (b) Benzaldehyde
(c) Benzyl chloride (d) Benzoic acid. ***(AFMC)***

47. The most reactive compound towards formation of cyanohydrin on treatment with KCN followed by acidification is

(a) Benzaldehyde
(b) *p*-Nitrobonzaldehyde
(c) Phenylacetaldehyde
(d) *p*-Hydroxybenzaldehyde. ***(GATE)***

48. Aldol condensation will not be observed in

(a) Chloral (b) Phenyl acetaldehyde
(c) Hexanal (d) Ethanal. ***(GATE)***

49. Benzophenone can be converted into benzene using

(a) Fused alkali
(b) Anhydrous $AlCl_3$
(c) Sodium amalgam in water
(d) Acidified dichromate. ***(Tamil Nadu CET)***

50. Which of the following statements regarding chemical properties of acetophenone are wrong ?

I. It is reduced to methyl phenyl carbinol by sodium and ethanol.
II. It is oxidised to benzoic acid with acidified $KMnO_4$
III. It does not undergo electrophilic substitution like nitration at meta position
IV. It does not undergo iodoform reaction with iodine and alkali.

(*a*) I and II (*b*) II and IV
(*c*) III and IV (*d*) I and II.
(Tamil Nadu CET)

51. Predict the product 'B' in the sequence of reaction

$$HC \equiv CH \xrightarrow[HgSO_4]{30\%\ H_2SO_4} A \xrightarrow{NaOH} B$$

(*a*) CH_3COONa (*b*) CH_3COOH
(*c*) CH_3CHO (*d*) $CH_3-CH(OH)-CH_2CHO$.
(CBSE PMT)

52. Which of the following react with NaOH to produce an acid and an alcohol ?
(*a*) HCHO (*b*) CH_3COOH
(*c*) CH_3CH_2COOH (*d*) C_6H_5COOH.
(MGIMS Wardha)

2002

53. Which is not true about acetophenone ?
(*a*) Reacts to form 2, 4-dinitrophenyl hydrazine
(*b*) Reacts with Tollen's reagent to form silver mirror
(*c*) Reacts with I_2 / NaOH to form iodoform
(*d*) On oxidation with alkaline $KMnO_4$ followed by hydrolysis gives benzoic acid. ***(Manipal)***

54. The product formed by the reaction of chlorine with benzaldehyde in the absence of a catalyst is
(*a*) Chlorobenzene (*b*) Benzyl chloride
(*c*) Benzoyl chloride (*d*) *o*-Chlorobenzaldehye.
(Tamil Nadu CET)

55. Which of the following reagents is used to distinguish acetone and acetophenone ?
(*a*) $NaHSO_3$ (*b*) Grignard reagent
(*c*) Na_2SO_4 (*d*) NH_4Cl.
(Rajasthan PMT)

56. An organic compound 'A' has the molecular formula C_3H_6O, it undergoes iodoform test. When saturated with dil. HCl it gives 'B' of molecular formula $C_9H_{14}O$. A and B, respectively are
(*a*) Propanal and mesitylene
(*b*) Propanone and mesityl oxide
(*c*) Propanone and 2, 6-dimethyl-2, 5-heptadien-4-one.
(*d*) Propanone and mesitylene oxide. ***(Tamil Nadu)***

57. The end product in the following sequence of reaction is

$$HC \equiv CH \xrightarrow[20\%/H_2SO_4]{1\%\ HgSO_4} A \xrightarrow{CH_3MgX} B \xrightarrow{[O]}$$

(*a*) Acetic acid (*b*) Isopropyl alcohol
(*c*) Acetone (*d*) Ethanol. ***(Bihar CEE)***

2003

58. The compound $CH_3-C(CH_3) = CH-CH_3$ on reaction with $NaIO_4$ in the presence of $KMnO_4$ gives
(*a*) CH_3COCH_3
(*b*) $CH_3COCH_3 + CH_3COOH$
(*c*) $CH_3COCH_3 + CH_3CHO$
(*d*) $CH_3CHO + CO_2$. ***(CBSE PMT)***

59. When *m*-chloro benzaldehyde is treated with 50% KOH solution ; the product(s) obtained is(are)

(*a*) 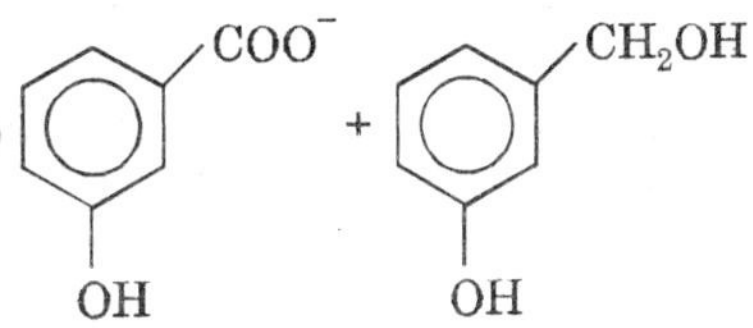

(*b*) 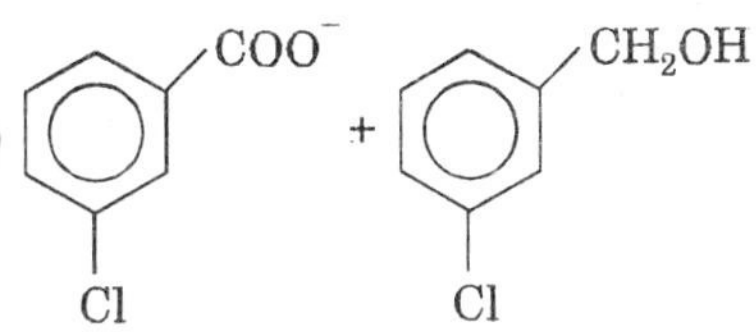

(*c*) 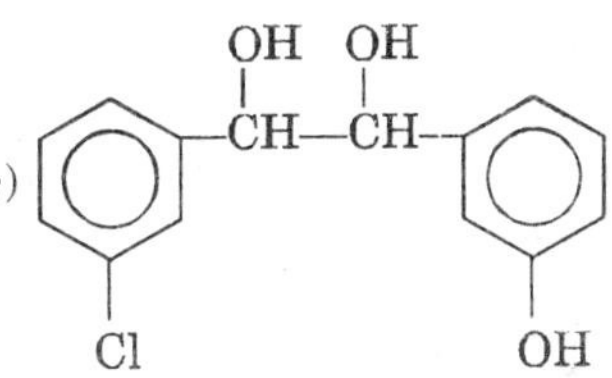

(*d*) 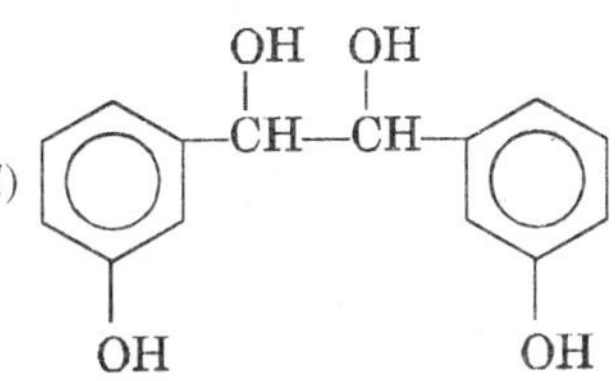
(CBSE PMT)

2004

60. The reagent used for the seperation of acetaldehyde from acetophenone is :
(*a*) $NaHSO_3$ (*b*) $C_6H_5NHNH_2$
(*c*) NH_2OH (*d*) $NaOH-I_2$.
(AIIMS)

61. Which one of the following can be oxidised to the corresponding carbonyl compounds ?
(*a*) 2-hydroxy propane (*b*) Ortho nitrophenol
(*c*) Phenol
(*d*) 2-methyl-2-hydroxy propane. ***(CBSE PMT)***

62. Which one of the following is reduced with Zn and HCl to give the corresponding hydrocarbon ?
(*a*) Ethyl acetate (*b*) Acetic acid
(*c*) Acetamide (*d*) Butan-2-one ***(AIEEE)***

63. Which one of the following undergo reaction with 50% NaOH solution to give the corresponding alcohol and acid ?

(*a*) Phenol (*b*) Benzaldehyde
(*c*) Butanal (*d*) Benzoic acid.

(*AIEEE*)

2005

64. The best reagent to convert pent-3-en-2-ol into pent-3-en-2-one is

(*a*) Pyridiniom chloro chromate
(*b*) Chromic anhydrids in glacial acetic acid
(*c*) Acidic dichromate
(*d*) Acidic permanganate. **(*AIEEE*)**

65. Which of the following will form two isomers with semi carbazide ?

(*a*) Benzaldehyde (*b*) Acetone
(*c*) Benzoquinone (*d*) Benzophenone.

(*Orissa JEE*)

66. Benzyl alcohol and sodium benzoate is obtained by the action of sodium hydroxide on benzaldehyde. This reaction is known as

(*a*) Perkin's reaction (*b*) Cannizzaro's reaction
(*c*) Sandmeyer's reaction (*d*) Claisen condensation.

(*Karnataka CET*)

67. $CH_3—CHO + HCN \rightarrow A$.

Compound *A* on hydrolysis gives

(*a*) $CH_3—CH_2—COOH$ (*b*) $CH_3—CH_2—CH_2—NH_2$
(*c*) $CH_3—CO—COOH$ (*d*) $CH_3CO—CH = NOH$
(*e*) $CH_3—\underset{\underset{OH}{|}}{CH}—COOH$ **(*Kerala PET*)**

68. Which of the following does not undergo Cannizzaro's reaction ?

(*a*) Benzaldehyde
(*b*) 2-methylpropanal
(*c*) *p*-methoxybenzaldehyde
(*d*) 2, 2-dimethylpropanal
(*e*) Formaldehyde. **(*Kerala PET*)**

2006

69. Nucleophilic addition reaction will be most favoured in

(*a*) CH_3CHO

(*b*) $CH_3—CH_2—CH_2\overset{\overset{O}{||}}{C}—CH_3$

(*c*) $(CH_3)_2C = O$

(*d*) CH_3CH_2CHO. **(*CBSE PMT*)**

70. A carbonyl compound reacts with hydrogen cyanide to form cyanohydrin which on hydrolysis forms a racemic mixture of α-hydroxy acid. The carbonyl compound is

(*a*) Formaldehyde
(*b*) Acetaldehyde
(*c*) Acetone
(*d*) Diethyl ketone. **(*CBSE PMT*)**

71. The increasing order of the rate of HCN addition to compounds from A–D is

(A) HCHO (B) CH_3COCH_3
(C) $PhCOCH_3$ (D) PhCOPh

(*a*) A < B < C < D
(*b*) D < B < C < A
(*c*) D < C < B < A
(*d*) C < D < B < A. **(*AIEEE*)**

72. Self-condensation of two moles of ethyl acetate in presence of sodium ethoxide yields

(*a*) Ethyl propionate (*b*) Ethyl butyrate
(*c*) Acetoacetic ester (*d*) Methyl acetoacetate.

(*CBSE PMT*)

73. The increasing order of the rate of HCN addition to compound, A—D is

A. HCHO B. CH_3COCH_3
C. $PhCOCH_3$ D. PhCOPh

(*a*) A < B < C < D (*b*) D < B < C < A
(*c*) D < C < B < A (*d*) C < D < B < A. **(*AIEEE*)**

74. With which of the following reagents, carbonyl compound shows addition cum elimination reaction ?

(*a*) PCl_5 (*b*) Brady's reagent
(*c*) HCN (*d*) all of these. **(*DCE*)**

75. Identify the product Y in the following reaction sequence

$$\begin{matrix} CH_2—CH_2—COO \\ | \\ CH_2—CH_2—COO \end{matrix} \rangle Ca \xrightarrow{\text{heat}} X \xrightarrow[\text{HCl, heat}]{\text{Zn-Hg}} Y$$

(*a*) Pentane (*b*) Cyclobutane
(*c*) Cyclopentane (*d*) Cyclopentanone.

(*KCET*)

76. The diketone $CH_3—\overset{\overset{O}{||}}{C}—(CH_2)_2—\overset{\overset{O}{||}}{C}—CH_3$ on intra-molecular aldol condensation gives the final product

(*a*) 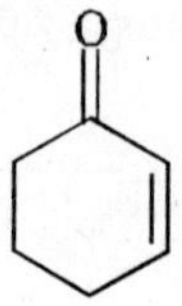

(*b*) 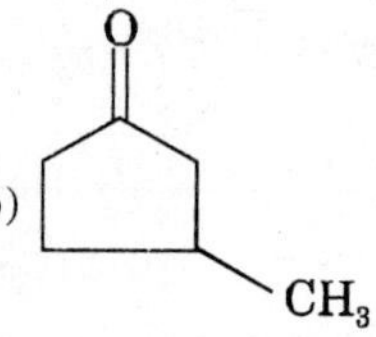

(c) 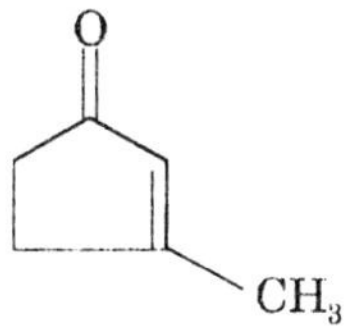 (d) 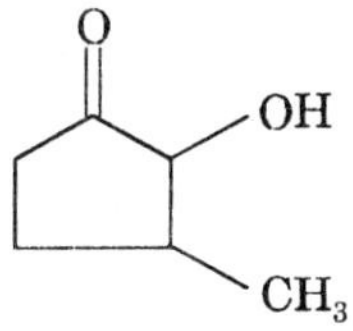

(e) 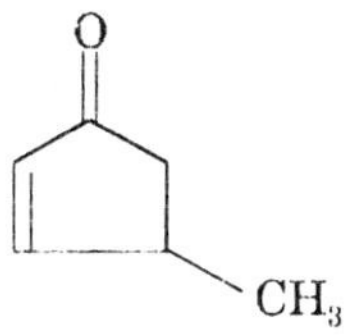 ***(Kerala MEE)***

77. When a mixture of calcium benzoate and calcium acetate is dry distilled, the resulting compound is

(a) Acetophenone (b) Benzaldehyde

(c) Benzophenone (d) Acetaldehyde. ***(KCET)***

78. An acid reacts with an isotopically labelled methanol to produce

(a) Methyl acetate having all the labelled oxygen

(b) Water having all the labelled oxygen

(c) Both methyl acetate and water contain isotopic oxygen

(d) No esterification. ***(HP PMT)***

79. The following reaction is

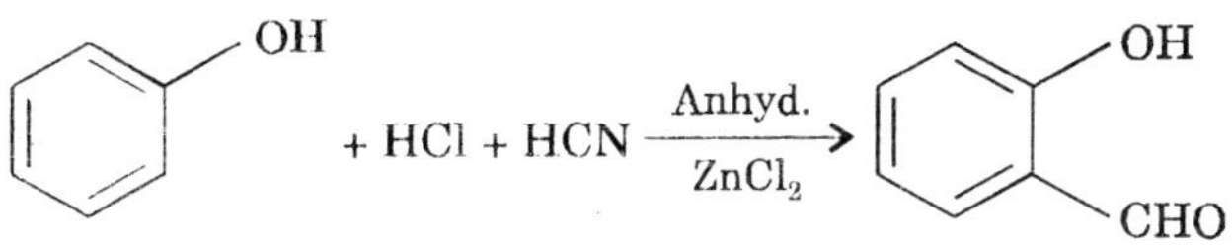

(a) Perkin reaction

(b) Gattermann reaction

(c) Kolbe's reaction

(d) Gattermann aldehyde synthesis. ***(CPMT)***

80. Acetone on distillation with conc. H_2SO_4 forms

(a) Phorone (b) Acrolein

(c) Mesitylene (d) Mesityl oxide. ***(CPMT)***

2007

81. The product in Aldol condensation is

(a) An alpha, beta unsaturated ester

(b) A beta-hydroxy acid

(c) A beta hydroxy aldehyde or beta-hydroxy ketone

(d) An alpha hydroxy aldehyde or ketone.

(CBSE AIPMT Screening)

82. Reduction of aldehydes and ketones into hydrocarbons using zinc amalgam and conc. HCl is called

(a) Wolf-Kishner reaction (b) Clemmensen reduction

(c) Cope reduction (d) Dow reduction.

(CBSE AIPMT Screening)

83. Which of the following on treatment with 50% aq. NaOH yields the corresponding alcohol and acid ?

(a) CH_3COCH_3 (b) $C_6H_5CH_2CHO$

(c) C_6H_5CHO (d) $CH_3CH_2CH_2CHO$.

(CBSE PMT Screening)

ANSWERS

1. (c)	**2.** (b)	**3.** (a)	**4.** (c)	**5.** (d)	**6.** (a)	**7.** (b)	**8.** (b)	**9.** (b)
10. (d)	**11.** (a)	**12.** (a)	**13.** (c)	**14.** (a)	**15.** (b)	**16.** (d)	**17.** (b)	**18.** (b)
19. (a)	**20.** (a)	**21.** (c)	**22.** (d)	**23.** (a)	**24.** (b)	**25.** (a)	**26.** (c)	**27.** (c)
28. (b)	**29.** (a)	**30.** (b)	**31.** (c)	**32.** (d)	**33.** (d)	**34.** (c)	**35.** (c)	**36.** (a)
37. (c)	**38.** (a)	**39.** (c)	**40.** (c)	**41.** (c)	**42.** (b)	**43.** (d)	**44.** (a)	**45.** (c)
46. (b)	**47.** (b)	**48.** (a)	**49.** (a)	**50.** (c)	**51.** (d)	**52.** (a)	**53.** (b)	**54.** (c)
55. (a)	**56.** (c)	**57.** (c)	**58.** (b)	**59.** (b)	**60.** (a)	**61.** (a)	**62.** (d)	**63.** (b)
64. (a)	**65.** (a)	**66.** (b)	**67.** (e)	**68.** (b)	**69.** (a)	**70.** (b)	**71.** (c)	**72.** (c)
73. (c)	**74.** (b)	**75.** (c)	**76.** (c)	**77.** (a)	**78.** (a)	**79.** (d)	**80.** (c)	**81.** (c)
82. (b)	**83.** (c).							

HINTS/SOLUTIONS

Q. No.	Ans.	Hints & Solutions
1.	(*c*)	$2C_6H_5-\overset{O}{\overset{\parallel}{C}}-H \xrightarrow{CN^-} C_6H_5-\underset{H}{\underset{\vert}{\overset{O^-}{\overset{\vert}{C}}}}-\overset{O}{\overset{\parallel}{C}}-C_6H_5$ (Acyloins)
2.	(*b*)	$3CH_3CHO \longrightarrow (CH_3CHO)_3$
3.	(*a*)	Aldehyde react with alcohols to produce Acetals whereas ketones produce ketals.
4.	(*c*)	Acetone can be oxidised by strong oxidising agents.
5.	(*d*)	$CH_3-\overset{OH}{\overset{\vert}{CH}}-CH_3 \xrightarrow[(O)]{Oxid.} CH_3-\overset{O}{\overset{\parallel}{C}}-CH_3 \xrightarrow{I_2/Na_2CO_3} CH_3COONa + CHI_3$
6.	(*a*)	The presence of α-hydrogen atoms is necessary for aldol condensation reaction. HCHO has no α-H atom.
7.	(*b*)	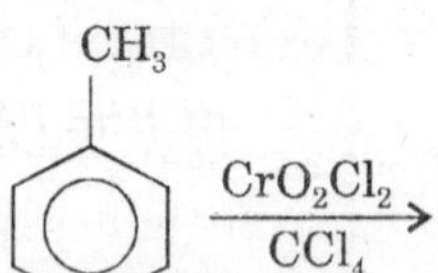
8.	(*b*)	Methyl ketone like $ArCOCH_3$ do not react with $NaHSO_3$. See concept oriented detail.
9.	(*b*)	In mixed betones, $R-\overset{O}{\overset{\parallel}{C}}-R'$, both R and R′ groups are different.
10.	(*d*)	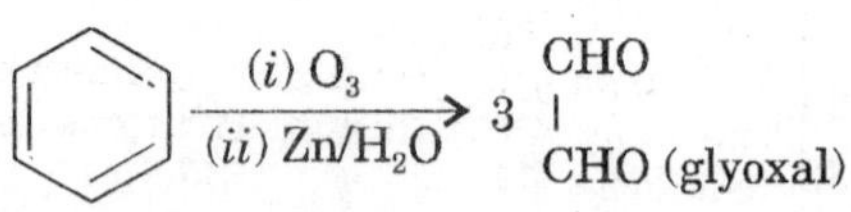
11.	(*a*)	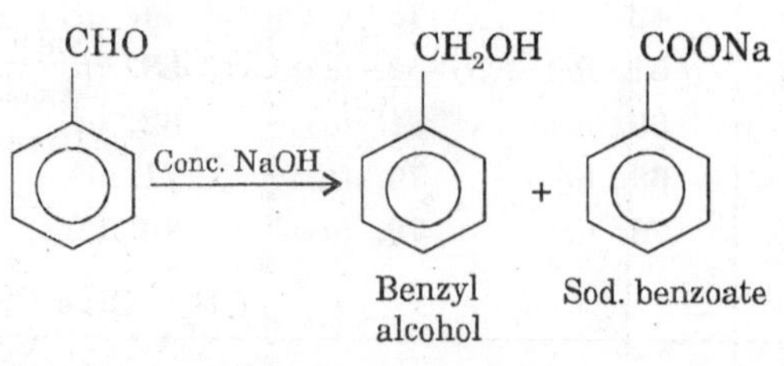
12.	(*a*)	It is Kolbe's electrolytic reaction and produces alkanes $RCOOK \xrightarrow[H_2O]{elec.} R-R + CO_2 + KOH$
13.	(*c*)	2° alcohols on oxidation produce ketones.
14.	(*a*)	$6HCHO + 4NH_3 \longrightarrow (CH_2)_6N_4$ urotropine
15.	(*b*)	Electron withdrawing groups favour meta substitution.
16.	(*d*)	It is Friedal Craft acytation reaction.
17.	(*b*)	This is Rosenmund's reduction.
18.	(*b*)	Lucas test is given by alcohols and not ald. or ketones.
19.	(*a*)	Clemmensen's reduction converts $>C=O$ group to $>CH_2$. Hence it converts ketone to alkane.
20.	(*a*)	Reactivity of HCHO towards nucleophilic attack is highest.
21.	(*c*)	$CH_3CH_2\underset{OH}{\underset{\vert}{C}H}CH_3 \xrightarrow{(O)} CH_3CH_2\underset{O}{\underset{\parallel}{C}}CH_3$ Ethyl methyl ketone
22.	(*d*)	It is known as Reimer-Tiemann reaction.
23.	(*a*)	As the nucleophile can attack from both sides it produces equal amount of *d* and *l* isomers and hence a recemic mixture.
24.	(*b*)	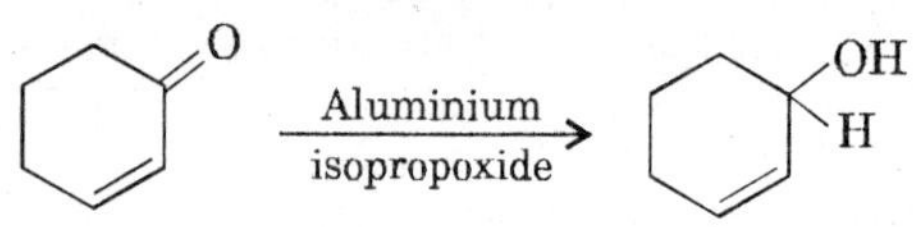
25.	(*a*)	$CH_3COCl \xrightarrow[Pd/B[illegible]_4]{H_2} CH_3CHO$
26.	(*c*)	$CH_3CHO \xrightarrow{dil.\ NaOH} CH_3-\overset{OH}{\overset{\vert}{CH}}-CH_2CHO$
27.	(*c*)	Ketones cannot be oxidised by mild oxidising agents like Fehling solution.
28.	(*b*)	It is propan-2-one.
29.	(*a*)	$HCHO + CH_3MgBr \xrightarrow{H_2O} CH_3CH_2OH + Mg\begin{matrix} Br \\ OH \end{matrix}$

Q. No.	Ans.	Hints & Solutions
30.	(b)	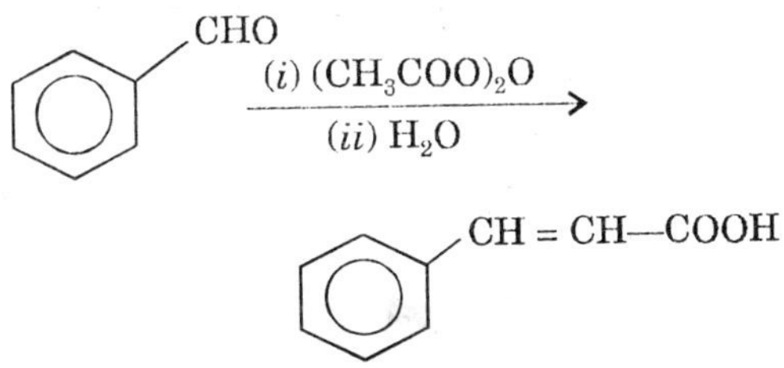 Cinnamic acid
31.	(c)	$HCHO \xrightarrow{KOH} HCOONa + CH_3OH$
32.	(d)	Reactions of different aldehydes/ketones with ammonia produce different product. See concept oriented details.
33.	(d)	$R-\overset{O}{\overset{\parallel}{C}}-R \xrightarrow[HCl]{Zn\text{-}Hg} R-CH_2-R$
34.	(c)	It undergoes aldol condensation. Abstraction of α-hydrogen occurs (see solved problems).
35.	(c)	$>C=O$ group is reduced to —OH group.
36.	(a)	$C_6H_5-\overset{O}{\overset{\parallel}{C}}-H + CH_3MgBr \rightarrow C_6H_5-\overset{OMgBr}{\overset{\vert}{C}H}-CH_3 \xrightarrow{H_2O/H^+} C_6H_5-\overset{OH}{\overset{\vert}{C}H}-CH_3$ (2° alcohol)
37.	(c)	$>C=O$ compounds undergo nucleophilic addition reactions.
38.	(a)	$CH_3-\overset{O}{\overset{\parallel}{C}}-CH_3 \xrightarrow{HCN} CH_3-\underset{CN}{\underset{\vert}{\overset{OH}{\overset{\vert}{C}}}}-CH_3$ (A) $\xrightarrow{H_3O^+} CH_3-\underset{COOH}{\underset{\vert}{\overset{OH}{\overset{\vert}{C}}}}-CH_3$ (B)
39.	(c)	In Wolff-Kishner reduction NH_2NH_2/KOH is used.
40.	(c)	Only ketones can produce tertiary alchols.
41.	(c)	$CH_3-\overset{O}{\overset{\parallel}{C}}-H \xrightarrow{CH_3MgI} CH_3-\overset{OMgI}{\overset{\vert}{C}H}-CH_3 \xrightarrow{H_3O^+} CH_3-\overset{OH}{\overset{\vert}{C}H}-CH_3$
42.	(b)	2, 4-dinitrophenyl hydrazine produces coloured crystalline compound with aldehydes and ketones.
43.	(d)	Aldehydes with 3α H atoms can undergo reaction with Fehling solution as well as iodoform test.
44.	(a)	$-\overset{O}{\overset{\parallel}{C}}-Cl \xrightarrow[Pd\text{-}BaSO_4]{H_2} -CHO$
45.	(c)	$CH_3\overset{CH_3}{\overset{\vert}{C}}=N-OH$ (Oxime) $\xrightarrow{Conc.\ H_2SO_4} CH_3-NH-\overset{O}{\overset{\parallel}{C}}-CH_3$ (amide)
46.	(b)	CrO_2Cl_2 is an oxidising agent, see concept oriented details.
47.	(b)	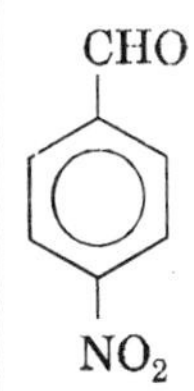 The NO_2 group is electron with drawing group and increases the reactivity.
48.	(a)	Chloral do not have α-hydrogen atoms (CCl_3CHO) which are essential for aldol condensation.
49.	(a)	$C_6H_5-\overset{O}{\overset{\parallel}{C}}-C_6H_5$.
50.	(c)	$C_6H_5-\overset{O}{\overset{\parallel}{C}}-CH_3$. The incoming group goes to meta position in the ring. As it has 3α-H atoms it can form iodoform.
51.	(d)	$HC \equiv CH \xrightarrow[HgSO_4]{H_2SO_4} CH_3CHO \xrightarrow[\text{Aldol Condensation}]{NaOH} CH_3-\overset{OH}{\overset{\vert}{C}H}-CH_2CHO$
52.	(a)	It undergoes Cannizzaro's reaction due to absence of α-hydrogen atoms.
53.	(b)	Ketones do not get oxidised with mild oxidising agents.

Q. No.	Ans.	Hints & Solutions
54.	(c)	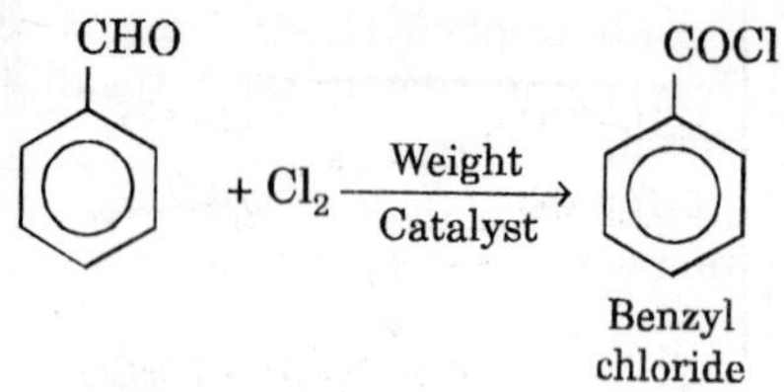
55.	(a)	Acetophenone does not react with $NaHSO_3$ to produce white ppt.
56.	(c)	Iodoform test is given by compound with $CH_3-\overset{O}{\overset{\Vert}{C}}-$ type of grouping. 3 moles of propane condense in the presence of HCl to produce 2, 6-dimethyl-2, 5-heptadiene-4-one. $CH_3-\underset{CH_3}{\underset{\vert}{C}}=CH_2-\overset{O}{\overset{\Vert}{C}}-CH=\underset{CH_3}{\underset{\vert}{C}}-CH_3$
57.	(c)	$HC=CH \xrightarrow[H_2SO_4]{HgSO_4} CH_3CHO \xrightarrow{CH_3MgX}$ $CH_3-\overset{OMgX}{\overset{\vert}{CH}}-CH_3 \xrightarrow{H_3O^+} CH_3-\overset{O}{\overset{\Vert}{C}}-CH_3$ $\xrightarrow{(O)} CH_3-\overset{OH}{\overset{\vert}{CH}}-CH_3$
58.	(b)	Sod. periodate oxidises the double bond giving products.
59.	(b)	In 50% KOH Cannizzaro's reaction takes place, where one mole of —CHO group gets reduced and the other gets oxidised.
60.	(a)	Acetaldehyde forms a crystalline solid with $NaHSO_3$ while acetophenone does not.
61.	(a)	$CH_3-\overset{OH}{\overset{\vert}{CH}}-CH_3 \xrightarrow{[O]} CH_3-\overset{O}{\overset{\Vert}{C}}-CH_3$
62.	(d)	Aldehydes/Ketones get reduced to hydrocarbons with Zn-Hg/HCl (clemmensen reaction).
63.	(b)	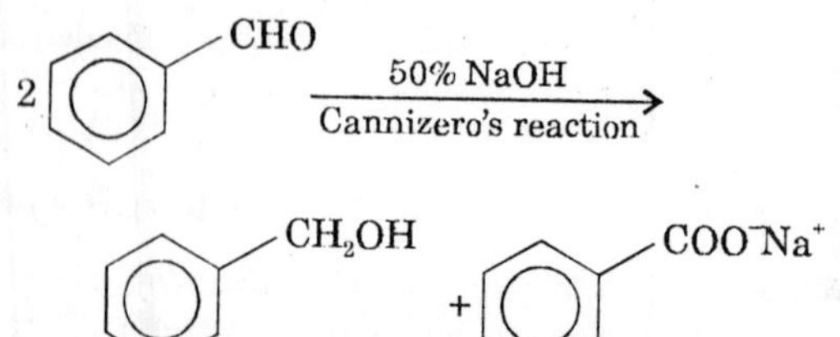

Q. No.	Ans.	Hints & Solutions
64.	(a)	Pyridinium chlorochromate only oxidises —OH group to > = 0 and does not reduce double (=) bond.
65.	(a)	$C_6H_5CH=O+H_2NNHCONH_2$ $\rightarrow C_6H_5CH=NNHCONH_2$ semicarbazone The product shows *E* and *Z* configuration. 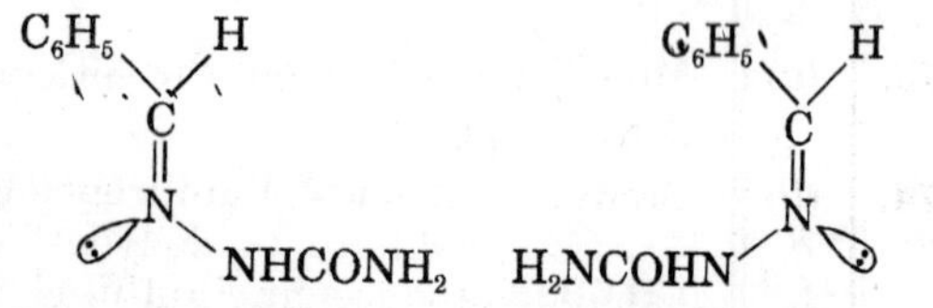*E*-Configuration *Z*-Configuration
66.	(b)	Benzaldehyde on treatment with 50% aqueous or ethanolic alkali solution undergoes Cannizzaro's reaction like HCHO (no α-hydrogen atom) *i.e.*, one molecule is oxidised and one is reduced with the formation of benzoic acid and benzyl alcohol respectively. $2C_6H_5CHO \xrightarrow{NaOH} C_6H_5CH_2OH+C_6H_5COONa$
67.	(e)	$CH_3CHO-HCN \longrightarrow CH_3-\overset{OH}{\overset{\vert}{CH}}-CN \xrightarrow{H_2O}$ $CH_3-\overset{OH}{\overset{\vert}{CH}}-COOH$
68.	(b)	Aldehydes containing no α-H atoms on heating with concentrated alkali solution (50%) undergo Cannizzaro's reaction. In this reaction one molecule is oxidised at the expense of other which is reduced to alcohol.
69.	(a)	Lesser is the steric Hinderance more is the reactivity towards nucleophilic addition.
70.	(b)	$\underset{\text{(Acetaldehyde)}}{CH_3-\overset{O}{\overset{\Vert}{C}}-H}+HCN \longrightarrow CH_3-\underset{H}{\underset{\vert}{\overset{OH}{\overset{\vert}{C}}}}-CN$ $\downarrow H_2O/H^+$ $CH_3-\underset{H}{\underset{\vert}{\overset{OH}{\overset{\vert}{C}}}}-COOH$ Lactic acid Two isomers of lactic acid which are formed which are non super imposible mirror images of each other.
71.	(c)	Lesser the hinderance more is the reactivity.

Q. No.	Ans.	Hints & Solutions
72.	(c)	$2CH_3COOC_2H_5 \xrightarrow{C_2H_5\bar{O}Na^+} CH_3-\overset{O}{\overset{\parallel}{C}}-CH_2-\overset{O}{\overset{\parallel}{C}}-OC_2H_5$
73.	(c)	More is the bulk about the $-\overset{O}{\overset{\parallel}{C}}-$ group lesser is the reactivity.
74.	(b)	Brady's reagent is 2, 4-dinitrophenyl hydrazine.
75.	(c)	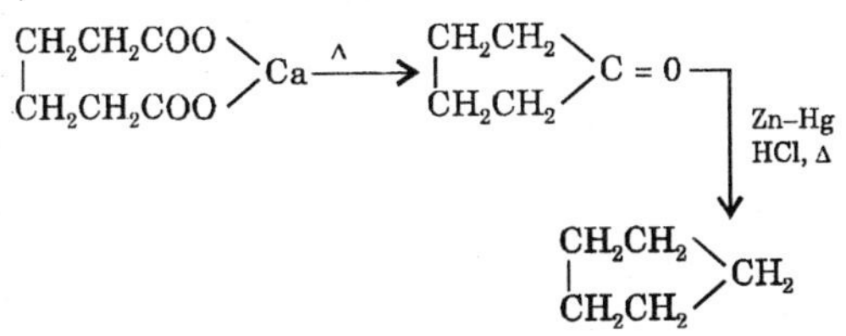
76.	(c)	Intramolecular condensation results in the formation of cyclic 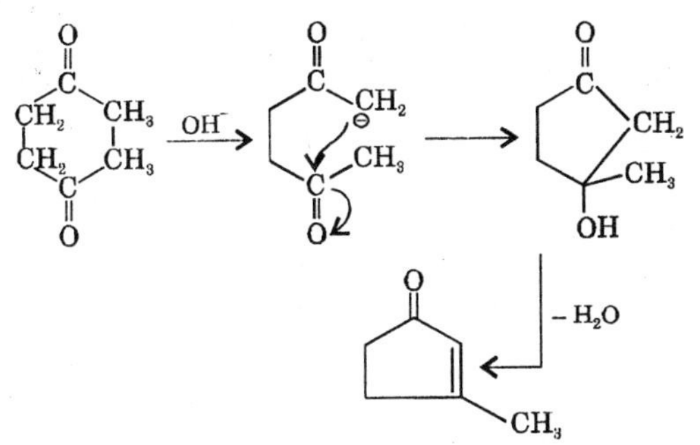
77.	(a)	$(C_6H_5COO)_2Ca + Ca(OOCCH_3)_2 \xrightarrow{\Delta} 2C_6H_5COCH_3 + 2CaCO_3$
78.	(a)	$CH_3COOH + CH_3\overset{*}{O}H \longrightarrow CH_3CO\overset{*}{O}CH_3 + H_2O$
79.	(d)	It is Gattermann aldehyde synthesis.
80.	(c)	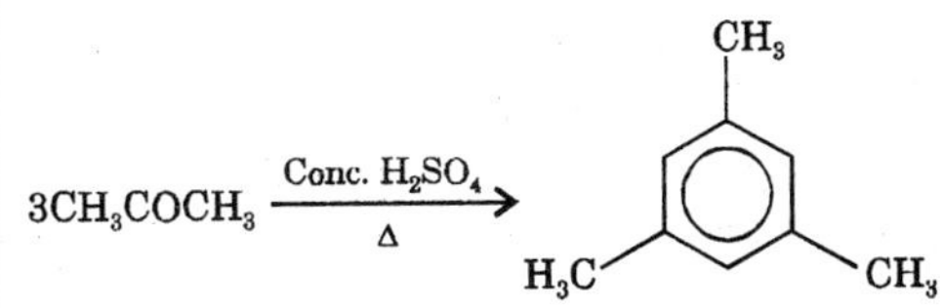
81.	(c)	$2CH_3CHO \xrightarrow{OH^-} CH_3-\underset{}{\overset{OH}{\overset{\mid}{C}H}}-CH_2CHO$ or $2CH_3-\overset{O}{\overset{\parallel}{C}}-CH_3 \xrightarrow{OH^-} CH_3-\overset{OH}{\overset{\mid}{C}H}-CH_2-\overset{O}{\overset{\parallel}{C}}-CH_3$
82.	(b)	It is clemmensen's reduction reaction.
83.	(c)	Benzaldehyde has no α-hydrogen atom and hence it can undergo Cannizaro's reaction.

5.6. MCQs FOR IIT JEE

Brain Twister

A. Multiple Choice Questions With One Correct Answer

1. Which of the following is produced when formaldehyde and ammonia reacts ?
(*a*) Formaline (*b*) Formaldimine
(*c*) Urotropine (*d*) Hexamethylene triamine.

2. What is formed when methanal reacts with ethanal in the presence of dilute caustic soda solution ?
(*a*) β-Hydroxy propionaldehyde
(*b*) 2-Hydroxy ethanal
(*c*) Methanoic acid and CO_2
(*d*) Propanone.

3. Ethanal + Aluminium ethoxide → X. The product X in the above reaction is
(*a*) Aluminium acetate (*b*) Ethanoic anhydride
(*c*) Ethylacetate (*d*) Propanoic acid.

4. Which of the following reagent cannot be used for converting

$$>C—OH \longrightarrow >C—OCOCH_3 \ ?$$

(*a*) CH_3COOH/H_2SO_4 (*b*) CH_3OCH_3
(*c*) $(CH_3CO)_2O$ (*d*) CH_3COCl.

5. A substance A containing three carbon atoms gives white crystalline precipitate with sodium bisulphite solution but does not give red precipitate with Fehling solution. A on treatment with $NH_2—NH_2/KOH$ will yield
(*a*) Propene (*b*) Propane
(*c*) Cyclopropane (*d*) Propionic acid.

6. Acetone $\xrightarrow{\approx 950\text{ K}}$. The products of the above reaction are
(*a*) Ethanal and methane
(*b*) Methane and ethane
(*c*) Ethane and acetic anhydride
(*d*) Methane and ketene.

7. *Phenacyl chloride,* which is an important *lachrymator* (tear gas) and is also called
(*a*) α-Chloroacetophenone (*b*) Allyl chloride
(*c*) Benzyl chloride (*d*) Benzoyl chloride.

8. Which of the following possesses highest value of dipole moment ?
(*a*) 1, Butene (*b*) Butanal
(*c*) Trans 2-Butene (*d*) 2-Butanone.

9. The product of chemical reaction between benzaldehyde and alcoholic KCN is commonly called
(*a*) Aldol (*b*) Acetal
(*c*) Phorone (*d*) Benzoin.

10. Which of the following reagent can convert

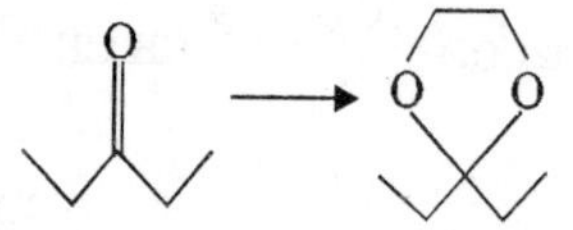

(*a*) $RMgX/H_2O$ (*b*) $(CH_2OH)_2$
(*c*) ROR′ (*d*) RH/H_2O.

11. Which process will give benzaldehyde ?
(*a*) Ethyl formate and phenyl magnesium bromide
(*b*) Toluene and chromyl chloride
(*c*) Benzene and HCN in $ZnCl_2$ followed by hydrolysis
(*d*) Benzene and ethanoyl chloride

12. A certain colourless liquid Q does not respond to Tollen's test but it reacts with semicarbazide to form crystalline derivative. Q can be
(*a*) $CH_3(CH_2)_2CHO$ (*b*) $CH_3CH_2COCH_3$
(*c*) $CH_3(CH_2)_2OC_2H_5$ (*d*) $CH_3CH_2\overset{\overset{O}{\|}}{C}OC_2H_5$.

13. A certain compound Y has a formula C_3H_6O. It combines with hydroxylamine to form two compounds which are geometrical isomers of each other. Y is
(*a*) CH_3CHO (*b*) CH_3CH_2CHO
(*c*) CH_3COCH_3 (*d*) $CH_2 = CHCH_2OH$.

14. Which of the following compound will give yellow precipitate with I_2 / Na_2CO_3 (*aq*) but does not respond to Cannizzaro's reaction ?
(*a*) $(C_2H_5)_2CO$ (*b*) CH_3CHO
(*c*) CH_2O (*d*) $C_2H_5CH_2OH$.

15. Vinyl alcohol and ethanal are
(*a*) Metamers (*b*) Tautomers
(*c*) Position isomers (*d*) Chain isomers.

16. Benzaldichloride $\xrightarrow{KOH\,(aq)}$ B $\xrightarrow[(CH_3CO)_2O]{CH_3COONa}$ C.

In the above sequence, C is

(a) Benzoic acid (b) Phenyl acetate
(c) Methyl benzoate (d) Cinnamic acid.

17. The most appropriate reagent for converting pentan-2-one to butanoic acid is
(a) sodium hypochlorite (b) O_2
(c) H_2O_2 (d) $KMnO_4 / H_2SO_4$.

18. $C_3H_8O \xrightarrow[K_2Cr_2O_7/H^+]{(O)} C_3H_6O \xrightarrow{I_2/NaOH} CHI_3$.
In this sequence the starting compound is
(a) 2-propanol (b) Propanal
(c) Propanone (d) Ethyl methyl ether.

19. Which among the following gives positive iodoform test as well as positive Fehling test ?
(a) Propanal (b) Ethanal
(c) Propanone (d) Acetophenone.

20. The number of isomeric ketones with formula $C_6H_{12}O$ are
(a) Six (b) Two
(c) Five (d) Four.

21. Cannizzaro reaction involves
(a) Oxidation of aldehydes
(b) Oxidation as well as reduction of aldehyde molecule
(c) Reduction of aldehyde molecule
(d) Rearrangement in aldehyde molecule.

22. Which of the following process represent *Oxo process* ?
(a) $CH_2 = CH_2 + H_2O_2 \xrightarrow[H_2O]{PdCl_2} CH_3CHO$
(b) $R—CH = CH_2 + CO + H_2 \xrightarrow{Co_2(CO)8} RCH_2CH_2CHO$
(c) $R—CH = CH—R \xrightarrow[(ii)\, H_2O/Zn]{(i)\, O_3} 2RCHO$
(d) $R—CH_2OH \xrightarrow{(O)} R—COOH$.

23. Which of the following aldehyde contains α-C atom but does not have any α-H atom.
(a) Propionaldehyde (b) Benzaldehyde
(c) Isobutyraldehyde (d) Formaldehyde.

24. $CH_3COCH_3 + BrCH_2COOC_2H_5 \xrightarrow[(ii)\, H_3O^+]{(i)\, Zn/Ether} (CH_3)_2C(OH)CH_2COOC_2H_5$
The above reaction is called
(a) Perkin's reaction (b) HVZ reaction
(c) Reformatsky reaction (d) Cannizzaro's reaction.

25. Which of the following compound will not undergo Cannizzaro's reaction ?
(a) Benzaldehyde (b) 2, 2-Dimethyl propanal
(c) Formaldehyde (d) Phenylethanal.

26. Condensation of acetone in conc. H_2SO_4 and trimerisation of propyne by passing through hot copper tube give some compound A. The compound A is
(a) Phorone (b) Mesitylene
(c) Mesityl oxide (d) Mixture of (a) and (c).

27. A compound X has molecular formula C_2Cl_3OH. It reduces Fehling solution and on oxidation it gives monocarboxylic acid B. X can also be formed by the action of Cl_2 on ethanol. X is
(a) Chloromethane (b) Chloroform
(c) Chloroacetic acid (d) Chloral.

28. Hydrobenzamide is produced by the action of
(a) Benzaldehyde and ammonia
(b) Hydrolysis of Benzamide
(c) Hydrogen on Benzamide
(d) HNO_2 on Benzamide.

29. Which of the following is capable of acting as hypnotic ?
(a) Acetaldehyde (b) Metaldehyde
(c) Paraldehyde (d) Formaldehyde.

30. In which of the following process acetone is one of the final product ?
(a) Ozonolysis of ethyne
(b) Oxidation of 2-butene with $KMnO_4 / H_2SO_4$
(c) Oxidation followed by hydrolysis of cumene
(d) Dehydrogenation of 1-propanal.

31. Which of the following will show disproportionation when treated with 5% aqueous NaOH ?
(a) Benzyl alcohol (b) Ethanal
(c) Phenyl ethanal (d) *m*-nitro benzaldehyde.

32. Which atoms of propionaldehyde molecule assume sp^2-hybrid state ?
(a) C atom of CH_3 group (b) α-C atom of the compound
(c) O atom (d) None of the atoms.

33. Aldehydes/ketones give addition reactions with all except
(a) C_2H_5OH (b) $KCN/Ba(CN)_2$
(c) HCN (d) $NaHSO_3$.

34. Which reagent can convert acetone to propane ?
(a) $NaBH_4$ (b) Na/alcohol
(c) $Al(OC_2H_5)_3$ (d) H_2/Pt.

35. Which of the following do not give Cannizzaro's reaction ?
(a) $(CH_3)_2CHCHO$ (b) HCHO
(c) CCl_3CH_2CHO (d) $(CH_3)_3CCHO$.

36. Which of the following is more reactive than acetone towards reaction with CH_3MgBr ?

(a) Benzophenone (b) Ethyl methyl ketone
(c) Propionaldehyde (d) Cyclohexanone.

37. What is true about acetone ?
(a) It restores the pink colour of Schiff's reagent
(b) It is very easily oxidised by $AgNO_3/NH_3$
(c) It is used for the preparation of the hypnotic, chloretone
(d) It forms pinacol on treatment with Zinc and water.

38. What is not true about acetaldehyde ?
(a) It is very easily oxidised
(b) It forms α-hydroxy nitrile on treatment with HCN
(c) It does not react with Schiff's reagent
(d) It can be obtained by the hydration of ethyne.

39. Which of the following is not isomeric with propionaldehyde ?
(a) 1, 2-epoxypropane (b) Acetone
(c) Propan-2-ol (d) Allyl alcohol.

40. Which reagents can convert isopropyl alcohol to acetone ?
(a) Collin's reagent (b) $CaOCl_2/H_2O$
(c) Zn-Hg/HCl (d) $KOH/NH_2—NH_2$.

41. Which of the following reagents will be able to cause the conversion of cyclopentanone to cyclopentanol ?
(a) Zn-Hg/HCl (b) H_2/Pt
(c) $NaBH_4$ (d) Collin's reagent.

42. Fehling solution cannot be used for the detection of
(a) Ethanol (b) Ethanal
(c) Propanal (d) Propanone.

43. Which of the following molecules can exhibit tautomerism ?
(a) Cyclohexanone (b) Vinyl chloride
(c) Benzophenone (d) Ethanoic acid.

44. Haloform reaction is not given by
(a) $CH_3COCH_2CH_3$ (b) $C_6H_5COC_2H_5$
(c) $CH_3CH(OH)C_2H_5$ (d) $CH_3COC_6H_5$.

45. Which of the following reagent will be able to oxidise ethanal ?
(a) Benedict's solution
(b) Cu / 575 K
(c) Aluminium isopropoxide
(d) NH_2NH_2/KOH.

46. Which of the following will not give aldehyde as the product ?
(a) Reductive ozonolysis of alkanes
(b) Stephen's reduction of nitriles
(c) Hydration of acetylene
(d) Treatment of Grignard's reagent and ethylformate.

47. Which of the following statements about benzaldehyde is true ?
(a) It does not reduce Tollen's reagent
(b) Undergoes aldol condensation
(c) Undergoes Cannizzaro reaction
(d) Does not form an addition compound with sodium hydrogen sulphite.

48. Base-catalysed aldol condensation occurs with
(a) Chloral
(b) Benzaldehyde
(c) 2-Methylpropionaldehyde
(d) 2, 2-Dimethylpropionaldehyde.

49. Which of the following compounds will give a yellow precipitate with iodine and alkali ?
(a) n-Propyl alcohol (b) Acetophenone
(c) Methyl acetate (d) Acetamide.

50. Keto-enol tautomerism is observed in
(a) C_6H_5CHO (b) $C_6H_5COC_6H_5$
(c) $C_6H_5COOC_6H_5$ (d) $C_6H_5COCH_2COCH_3$.

51. Which of the following is an example of aldol condensation ?

(a) $2CH_3CHO \xrightarrow{\text{dil. NaOH}} CH_3CHOHCH_2CHO$

(b) $2CH_3COCH_3 \xrightarrow{\text{dil. NaOH}}$ 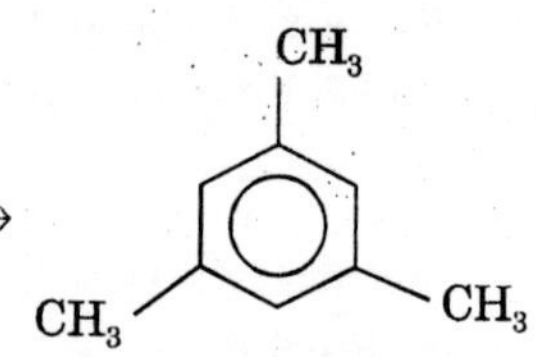

(c) $2HCHO \xrightarrow{\text{dil. NaOH}} CH_3OH + HCOOH$

(d) $C_6H_5CHO + HCHO \xrightarrow{\text{dil. NaOH}} C_6H_5CH(OH)CHO$

52. Under Wolff-Kishner reduction conditions the conversions which can be brought about is
(a) Benzophenone into toluene
(b) Benzaldehyde into benzyl alcohol
(c) Cyclohexanone into cyclohexane
(d) Cyclohexanone into cyclohexanol.

53. Tautomerism is exhibited by all except

(a) 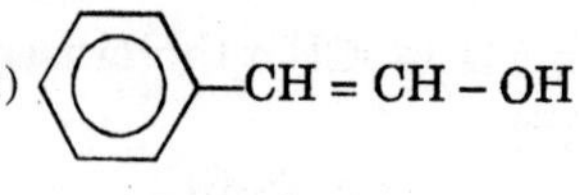

(b) 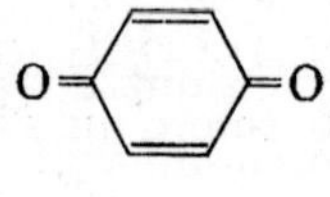

(c) 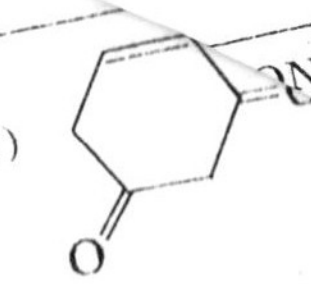

(d) 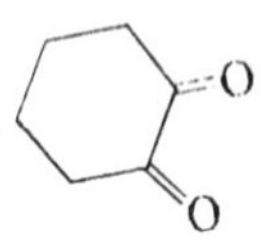

54. A new carbon-carbon bond formation is poss...
(a) Cannizzaro reaction
(b) Friedel-Crafts reaction
(c) Clemmensen reduction
(d) M.P.V. reaction

55. Which of the following will not undergo aldol condensation ?
(a) Acetaldehyde
(b) Propionaldehyde
(c) Benzaldehyde
(d) Trideuteroacetaldehyde.

...m Previous IIT Examinations

56. The reagen... ...ldehydes and k... ...nes react is
(a) Fehling solution (b) Grignard's reagent
(c) Schiff's reagent (d) Tollen's reagent.
(IIT 1981)

57. A compound that gives positive iodoform test is
(a) Pentan-1-ol (b) Petnan-2-on
(c) Pentan-3-one (d) Pentanal. **(IIT 1982)**

58. When acetaldehyde is heated with Fehling solution, it gives a precipitate of
(a) Cu (b) CuO
(c) Cu_2O (d) Cu + CuO + Cu_2O.
(IIT 1983)

59. When propyne is treated with aqueous H_2SO_4 in presence of $HgSO_4$, the major product obtained is
(a) Propanal
(b) n-Propyl hydrogen sulphate
(c) Acetone
(d) Propanol. **(IIT 1983)**

60. The Cannizzaro reaction is not given by
(a) Trimethylacetaldehyde
(b) Acetaldehyde
(c) Benzaldehyde
(d) Formaldehyde. **(IIT 1983)**

61. Which of the following compounds is oxidised to prepare ethyl methyl ketone ?
(a) 2-Propanol (b) 1-Butanol
(c) 2-Butanol (d) t-Butyl alcohol.
(IIT 1987)

62. Benzene reacts with CH_3COCl in the presence of anhydrous $AlCl_3$ to give
(a) $C_6H_5CH_3$ (b) C_6H_5Cl
(c) $C_6H_5O_2Cl$ (d) $C_6H_5COCH_3$. **(IIT 1988)**

63. Polarization of electrons in acrolein may be written as
(a) $\overset{\delta-}{CH_2}$ — CH — $\overset{\delta+}{CH}$ = O
(b) C... = CH — CH = $\overset{\delta+}{O}$
(c) $\overset{\delta+}{CH_2}$ = CH — CH = ...
(d) $\overset{\delta+}{CH_2}$ = CH — CH = $\overset{\delta-}{O}$. **(IIT 1988)**

64. The formation of cyanohydrin from a ketone is an example of
(a) Electrophilic addition
(b) Nucleophilic addition
(c) Nucleophilic substitution
(d) Electrophilic substitution. **(IIT 1990)**

65. m-Chlorobenzaldehyde on reaction with conc. KOH at room temperature gives
(a) Potassium m-Chlorobenzoate and m-hydroxybenzaldehyde
(b) m-Hydroxybenzaldehyde and m-chlorobenzyl alcohol
(c) m-Chlorobenzyl alcohol and m-hydroxybenzyl alcohol
(d) Potassium m-Chlorobenzoate and m-chlorobenzyl alcohol. **(IIT 1991)**

66. Hydrogenation of benzoyl chloride in the presence of Pd on $BaSO_4$ gives
(a) Benzyl alcohol (b) Benzaldehyde
(c) Benzoic acid (d) Phenol. **(IIT 1992)**

67. An organic compound, C_3H_6O does not give a precipitate with 2, 4-dinitrophenylhydrazine reagent and does not react with metallic sodium. It could be
(a) $CH_3—CH_2—CHO$
(b) $CH_2 = CH—CH_2OH$
(c) $CH_3—CO—CH_3$
(d) $CH_2 = CH—O—CH_3$. **(IIT 1993)**

68. Which of the following reactions cannot be used for the reduction of

$$R_2CO \longrightarrow R_2CH_2$$

(a) Clemmensen reduction
(b) Wolff-Kishner reduction
(c) Wurtz reaction
(d) HI/red P at 200°C. ***(Roorkee Screening Test 1993)***

69. Under Wolff-Kishner reduction conditions, the conversions which may be brought about is
(a) Benzaldehyde into Benzyl alcohol
(b) Cyclohexanol into Cyclohexane
(c) Cyclohexanone into Cyclohexanol
(d) Benzophenone into Diphenylmethane. ***(II…***

70. $CH_3-\overset{O}{\overset{\|}{C}}-CH_3 + [O] \xrightarrow{SeO_2} P + S + H…$
Here P is
(a) $CH_3-\overset{O}{\overset{\|}{C}}-\overset{O}{\overset{\|}{C}}-H$ (b) $CH_3-\overset{O}{\overset{\|}{C}}-…H_3$
(c) $CH_3-\overset{O}{\overset{\|}{C}}-CHOH$ (d) None of the above.
(IIT 1995)

71. The compound that will not give iodoform or treatment with alkali and iodine is
(a) Acetone (b) Ethanol
(c) Diethyl ketone (d) Isopropyl alcohol.
(IIT 1995)

72. In the Cannizzaro reaction given below,
$2Ph-CHO \xrightarrow{OH^-} Ph-CH_2OH + PhCO_2^-$, the slowest step is
(a) The attack of —OH at the carbonyl group
(b) The transfer of hydride ion to the carbonyl group
(c) The abstraction of proton from the carboxylic acid
(d) The deprotonation of $Ph-CH_2OH$. ***(IIT 1996)***

73. In a Cannizzaro reaction, the intermediate that will be best hydride donor is

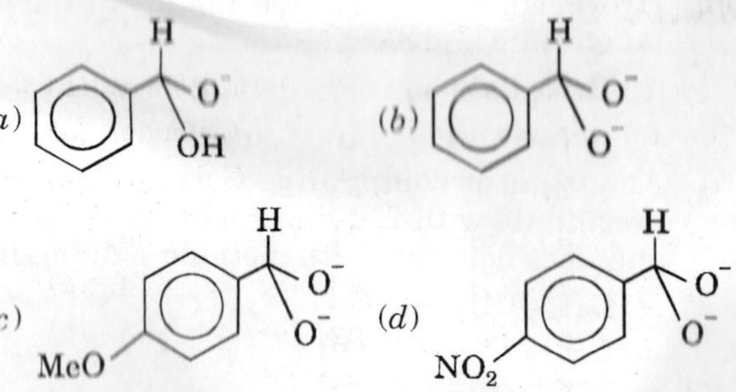

(IIT 1997)

74. $CH_3CHO + H_2NOH \longrightarrow CH_3-CH=N-OH$
The above reaction occurs at
(a) pH = 1 (b) pH = 4.5
(c) Any value of pH (d) pH = 12. ***(IIT 1997)***

75. Among the f… compounds, which will react with aceto… product containing $> C = N -$?
(a) $C_6H…$ (b) $(CH_3)_3N$
(c) … (d) $C_6H_5NHNH_2$. ***(IIT 1998)***

76. … the following will react with water ?
(a) $…Cl_4$ (b) Cl_3CCHO
(c) … (d) $ClCH_2CH_2Cl$.

The product(s) obtained *via* oxymercuration ($HgSO_4 + H_2SO_4$) of 1-butyne would be
(a) $CH_3CH_2COCH_3$
(b) $CH_3CH_2CH_2CHO$
(c) $CH_3CH_2CHO + HCHO$
(d) $CH_3CH_2COOH + HC…$ ***(…T 1999)***

78. The enol form of … treatment with D_2O, gives
(a) $CH_3-\overset{|}{C}=CH_2$ (b) $CD_3-\overset{O}{\overset{\|}{C}}-CD_3$
(c) $CH_2=\overset{OH}{\overset{|}{C}}-CH_2D$ (d) $CD_2=\overset{OD}{\overset{|}{C}}-CD_3$.
(IIT 1999)

79. The appropriate reagent for the transformation

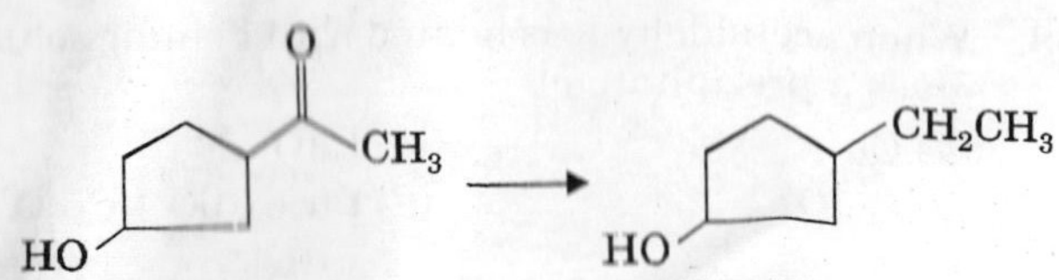

(a) Zn(Hg), HCl (b) NH_2NH_2, OH^-
(c) H_2/Ni (d) $NaBH_4$.
(IIT Screening Test 2000)

80. Which of the following has the most acidic hydrogen ?
(a) 3-hexanone (b) 2, 4-hexanedione
(c) 2, 5-hexanedione (d) 2, 3-hexanedione.
(IIT Screening Test 2000)

81. Which of the following will be most readily dehydrated in acidic conditions ?

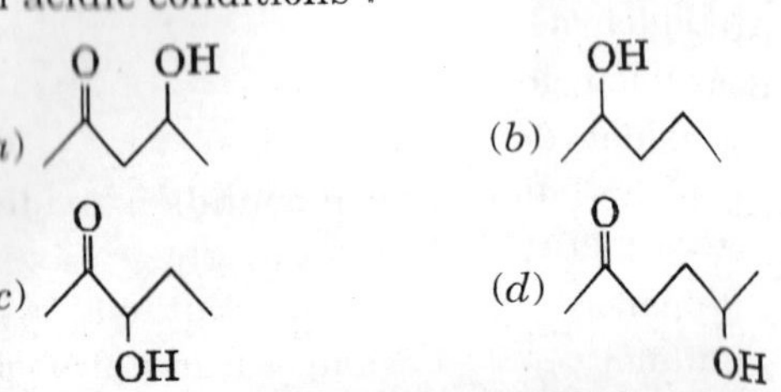

(IIT Screening Test 2000)

82. A mixture of benzaldehyde and formaldehyde on heating with aqueous NaOH solution gives
(a) Benzyl alcohol and sodium formate
(b) Sodium benzoate and methyl alcohol
(c) Sodium benzoate and sodium formate
(d) Benzyl alcohol and methyl alcohol.
(IIT Screening Test 2001)

83. Compound 'A' (molecular formula C_3H_8O) is treated with acidified potassium dichromate to form a product 'B' (molecular formula C_3H_6O). 'B' forms a shining silver mirror on warming with ammoniacal silver nitrate. 'B' when treated with an aqueous solution of $H_2NCONHNH_2$. HCl and sodium acetate gives a product 'C'. Identify the structure of 'C'

(*a*) $CH_3CH_2CH_2 = NNHCONH_2$

(*b*) $CH_3—\underset{CH_3}{\underset{|}{C}H} = NNHCONH_2$

(*c*) $CH_3—\underset{CH_3}{\underset{|}{C}H} = NCONHNH_2$

(*d*) $CH_3CH_2CH—NCONHNH_2$.

(*IIT Screening Test 2002*)

84. 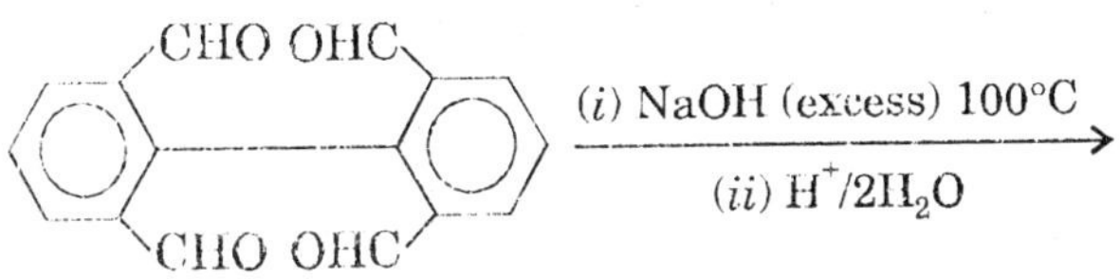

(*a*) 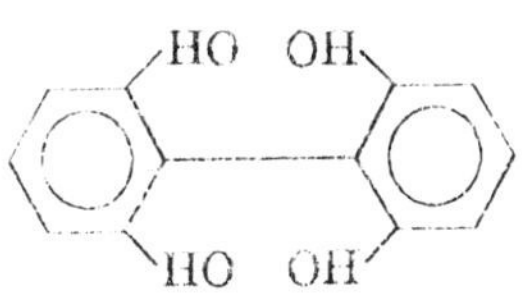

(*b*) 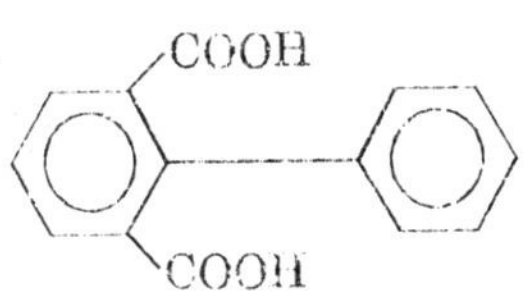

(*c*) 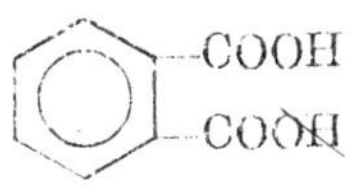

(*d*) 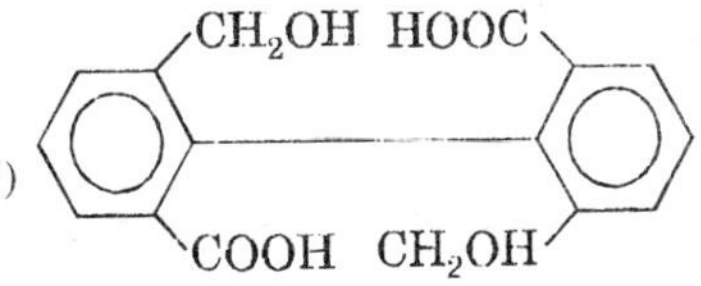

(*IIT Screening Test 2003*)

85. The correct order of reactivity of PhMgBr with

$\underset{(I)}{Ph—\overset{O}{\overset{\|}{C}}—Ph}$ $\quad \underset{(II)}{CH_3—\overset{O}{\overset{\|}{C}}—H}$ $\quad \underset{(III)}{CH_3—\overset{O}{\overset{\|}{C}}—CH_3}$ is

(*a*) I > II > III (*b*) III > II > I

(*c*) II > III > I (*d*) I > III > II.

(*IIT Screening Test 2004*)

86. 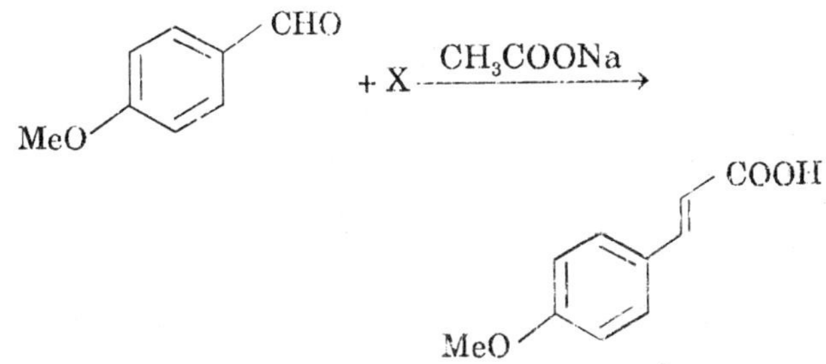

what is X ?

(*a*) CH_3COOH (*b*) $BrCH_2COOH$

(*c*) $(CH_3CO)_2O$ (*d*) CHO—COOH.

(*IIT Screening Test 2005*)

87. But-2-one can be converted to propanoic acid by which of the following

(*a*) NaOH/NaI, H^+ (*b*) Fehling solution

(*c*) NaOH, I_2/H^+ (*d*) Tollen's reagent.

(*IIT Screening Test 2005*)

ANSWERS

1. (*c*)	2. (*c*)	3. (*c*)	4. (*b*)	5. (*b*)	6. (*d*)	7. (*a*)	8. (*d*)	9. (*d*)
10. (*b*)	11. (*b*)	12. (*b*)	13. (*b*)	14. (*b*)	15. (*b*)	16. (*d*)	17. (*a*)	18. (*a*)
19. (*b*)	20. (*a*)	21. (*b*)	22. (*b*)	23. (*b*)	24. (*c*)	25. (*d*)	26. (*b*)	27. (*d*)
28. (*a*)	29. (*c*)	30. (*c*)	31. (*d*)	32. (*c*)	33. (*b*)	34. (*b*)	35. (*a*)	36. (*c*)
37. (*c*)	38. (*c*)	39. (*c*)	40. (*a*)	41. (*c*)	42. (*a*)	43. (*a*)	44. (*b*)	45. (*a*)
46. (*d*)	47. (*c*)	48. (*c*)	49. (*b*)	50. (*d*)	51. (*a*)	52. (*c*)	53. (*b*)	54. (*b*)
55. (*c*)	56. (*b*)	57. (*b*)	58. (*c*)	59. (*c*)	60. (*b*)	61. (*c*)	62. (*d*)	63. (*d*)
64. (*b*)	65. (*d*)	66. (*b*)	67. (*d*)	68. (*c*)	69. (*d*)	70. (*a*)	71. (*c*)	72. (*b*)
73. (*c*)	74. (*b*)	75. (*d*)	76. (*b*)	77. (*a*)	78. (*b*)	79. (*b*)	80. (*b*)	81. (*a*)
82. (*a*)	83. (*a*)	84. (*d*)	85. (*c*)	86. (*c*)	87. (*c*).			

HINTS/SOLUTIONS

Q. No.	Ans.	Hints & Solutions
1.	(c)	$6HCHO + 4NH_3 \longrightarrow (CH_2)_6N_4 + 6H_2O$ (Urotropine)
2.	(a)	$CH_3CHO + HCHO \xrightarrow{OH^-} H_2C(OH)\text{—}CH_2CHO$ (β-Hydroxypropionaldehyde)
3.	(c)	$CH_3CHO + CH_3CHO \xrightarrow{Al(OC_2H_5)_3}$ $CH_3COOH + C_2H_5OH \xrightarrow{-H_2O} CH_3\overset{O}{\overset{\Vert}{C}}\text{—}OC_2H_5$ (Ethyl acetate)
4.	(b)	This is an acylation reaction and CH_3OCH_3 is not an acylating agent.
5.	(b)	A is acetone $CH_3COCH_3 \xrightarrow{NH_2\text{—}NH_2/KOH} CH_3CH_2CH_3 + H_2O$ (Propane)
6.	(d)	$CH_3COCH_3 \xrightarrow{950\text{-}1030\ K} CH_4 + CH_2 = C = O$ (Methane, Ketene)
7.	(a)	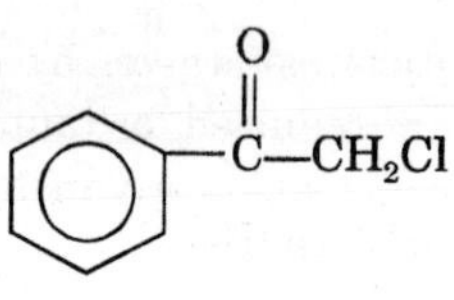 is phenacyl chloride.
8.	(d)	2-butanone has highest dipole moment.
9.	(d)	Benzaldehyde undergoes benzoin condensation when treated with alcoholic KCN solution.
10.	(b)	This is an example of cyclic ketal formed by diethyl ketone and ethylene glycol.
11.	(b)	It is Etard's reaction.
12.	(b)	Q must be ketone.
13.	(b)	Y must be aldehyde with three carbon atoms. Hence it is H_3CH_2CHO 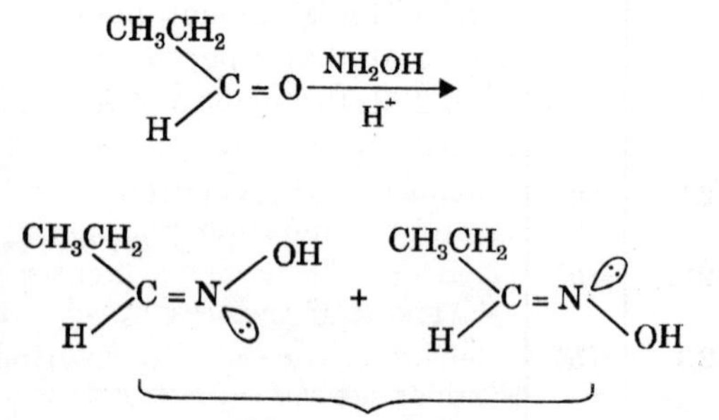 Geometrical isomers
14.	(b)	Among the given compounds, CH_3CHO responds to iodoform test but does not give cannizzaro's reaction because of the presence of α-H atom in its molecule.
15.	(b)	$CH_2 = CHOH$ and $CH_3\underset{O}{\underset{\Vert}{C}}H$ are tautomers.
16.	(d)	$C_6H_5CHCl_2 \xrightarrow{KOH} [C_6H_5CH(OH)_2] \xrightarrow{-H_2O} \underset{(B)}{C_6H_5CHO} \xrightarrow[(CH_3CO)_2O]{CH_3COONa} C_6H_5CH = CHCOOH$ (Cinnamic acid (B))
17.	(a)	$CH_3-\overset{O}{\overset{\Vert}{C}}-CH_2CH_2CH_3 \xrightarrow{NaOCl} CHCl_3 + CH_3CH_2CH_2COONa \xrightarrow{H_3O^+} CH_3CH_2CH_2COOH$
18.	(a)	C_3H_8O is alcohol while C_3H_6O is ketone which shows iodoform test positive. The reaction sequence is $CH_3\overset{OH}{\overset{\vert}{C}}HCH_3 \xrightarrow{(O)} CH_3\overset{O}{\overset{\Vert}{C}}CH_3 \xrightarrow{I_2/NaOH} CHI_3 + CH_3COONa.$

Q. No.	Ans.	Hints & Solutions
19.	(*b*)	Cr... Fehling...
20.	(*a*)	The vario... $C_6H_{12}O$ are ...th Iodoform test and ...ones of formula (*i*) Hexan-2-one (*ii*) Hexan-3-one (*iii*) 3-Methyl-2-pentanone (*iv*) 4-Methyl-2-pentanone (*v*) 2-Methylpentan-3-one (*vi*) 3, 3-Dimethylpentan-2-one.
21.	(*b*)	Involves both oxidation and reduction. It is disproportionation reaction.
22.	(*b*)	Oxo process involves, heating alkene with CO and H_2 in the presence of carbonyl catalyst.
23.	(*b*)	Benzaldehyde does have α-carbon atom but it does not contain an α-H atom.
24.	(*c*)	Reaction of ketones with α-bromo esters is called Reformatsky by reaction.
25.	(*d*)	$C_6H_5CH_2CHO$ contains α-C atom. Hence, it will not undergo Cannizzaro's reaction.
26.	(*b*)	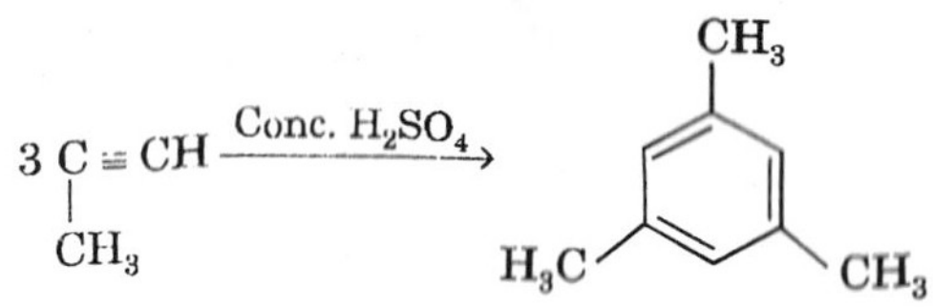 Mesitylene
27.	(*d*)	The compound corresponding to above data is $Cl_3C—CHO$ *i.e.*, chloral.
28.	(*a*)	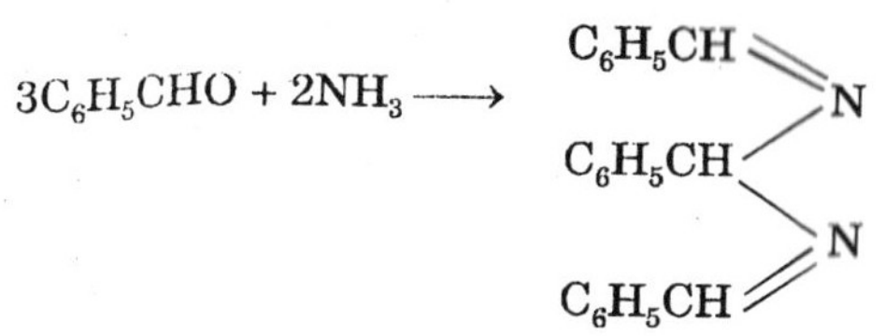 (hydro benzamide)
29.	(*c*)	Paraldehydo acts as an hypriotic to induce deep sleep.
30.	(*c*)	$\phi—CH(CH_3)—CH_3 \xrightarrow{[O]} \phi—C(CH_3)_2—COOH$ $CH_3—C(=O)—CH_3 \leftarrow$
31.	(*d*)	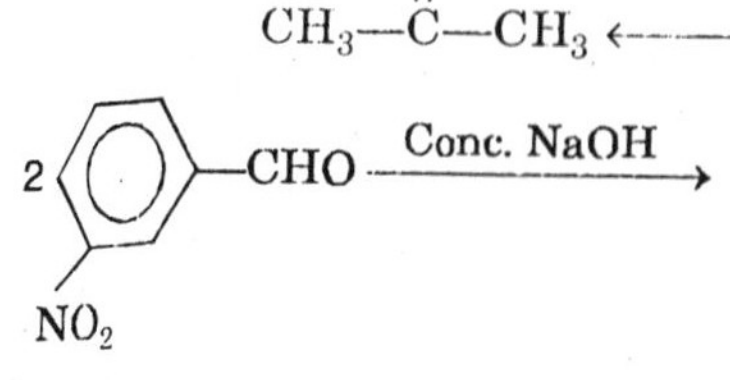 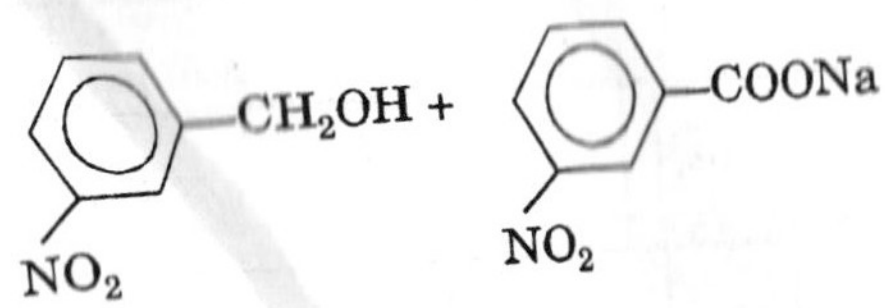
32.	(*c*)	$\overset{\beta}{C}H_3\overset{\alpha}{C}H_2CHO$. The oxygen atom of CHO group is sp^2 hybridised.
33.	(*b*)	O_2 and KCN cannot directly attack the aldehydes and ketones.
34.	(*b*)	Both Zn—Hg/HCl and Na/alcohol are strong reducing agents.
35.	(*a*)	Cannizaro's reaction occurs only in compounds where α-H atom is absent.
36.	[illegible]	More is the bulky group attached to > C = O group lesser is the reactivity.
37.	(*c*)	$CH_3COCH_3 + CHCl_3 \longrightarrow CH_3—C(OH)(CH_3)—CCl_3$ Chloretone
38.	(*c*)	Acetaldehyde can restore the pink colour of Schiff's reagent.
39.	(*c*)	Isomeric compounds should have same molecular formula.
40.	(*a*)	$CH_3—CH(OH)—CH_3 \xrightarrow[\text{reagent}]{\text{Collins}} CH_3COCH_3$
41.	(*c*)	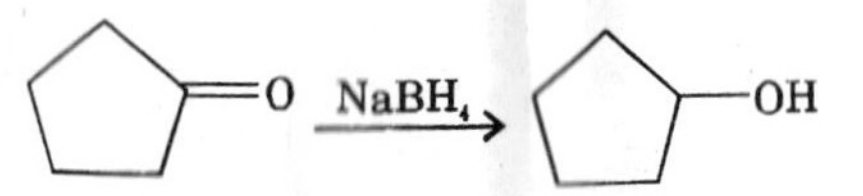
42.	(*a*)	Fehling solution test is given by aldehydes only.
43.	(*a*)	α-H atoms are required for tautomerism.
44.	(*b*)	Haloform reaction is given by compounds containing $CH_3CHOH—$ or $CH_3—\overset{O}{\overset{\Vert}{C}}—$ type of grouping.
45.	(*a*)	Aluminium isopropoxide can bring about Tischenko reaction whereas hydrazine and KOH is a reducing agent.
46.	(*d*)	$CH_3—C \equiv CH \xrightarrow[Hg^{+2}]{H_2O/H^+} CH_3—\overset{O}{\overset{\Vert}{C}}-CH_3$
47.	(*c*)	It does not contain α-H atoms and hence can undergo Cannizzaro's reaction.
48.	(*c*)	α-H atoms are required to cause aldol condensation.
49.	(*b*)	Iodoform test is given by $CH_3CHOH—$ and $CH_3—\overset{O}{\overset{\Vert}{C}}—$ type of groupings.

Q. No.	Ans.	Hints & Solutions
50.	(d)	α-H atoms are required for tautomerism.
51.	(a)	Acetaldehyde condenses in the presence of alkali in aldol condensation.
52.	(c)	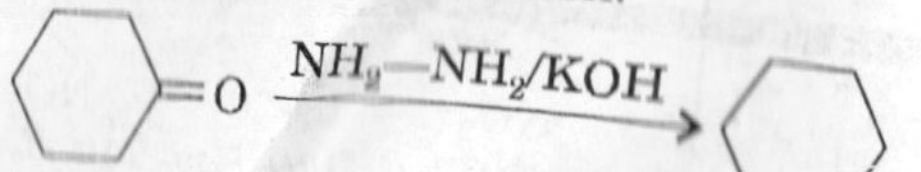
53.	(b)	Benzoquinone will not show tautomerism as it has no α-H atoms.
54.	(b)	Cannizaro's reaction is disproportionat... reaction and no new carbon atom... introduced. Similarly in clemmensen re...tion.
55.	(c)	Benzaldehyde do not have α-H re...uired fo... aldol condensation.
56.	(b)	Only aldehydes react with Fehling ...olution, Schiff's reagent and Tollen's reagent.
57.	(b)	It has $CH_3-\overset{O}{\overset{\parallel}{C}}-$ type of grouping.
58.	(c)	$CH_3CHO + 2CuO \longrightarrow CH_3COOH + \underset{\text{red ppt.}}{Cu_2O}$
59.	(c)	$CH_3-C\equiv CH \xrightarrow[Hg^{+2}]{H_2O,\ H^+} CH_3-\overset{O}{\overset{\parallel}{C}}-CH_3$
60.	(b)	It contains α-H atoms.
61.	(c)	$CH_3-\overset{OH}{\overset{\mid}{CH}}-CH_2CH_3 \xrightarrow{[O]} CH_3-\overset{O}{\overset{\parallel}{C}}-CH_2CH_3$
62.	(d)	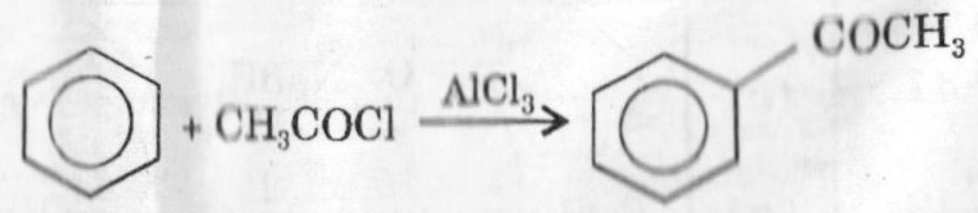
63.	(d)	The charges must be seperated and more electronegative atom should get a negative charge.
64.	(b)	$>C=O$ compounds undergo nucleophilic addition reactions.
65.	(d)	*m*-chloro benzaldehyde undergoes Cannizzaro's reaction.
66.	(b)	It is Rosenmund's reduction.
67.	(d)	Ethers and alcohols do not react with 2, 4 DNP. However alcohols react with Na metal.
68.	(c)	Wurtz reaction is used for preparing higher alkane from alkyl halides.
69.	(d)	On reduction with NH_2-NH_2/KOH $>C=O$ group reduces to $>CH_2$.
70.	(a)	SeO_2 oxidises at α-carbon.
71.	(c)	Iodoform test is given by the compounds with $CH_3-\overset{O}{\overset{\parallel}{C}}-$ type of grouping.

OBJECTI...

Q. No.	Ans.	
		$\xrightarrow[\text{fast}]{} C_6H_5-\overset{OH}{\overset{\mid}{\underset{H}{\underset{\mid}{C}}}}-O^-$
72.	(b)	$\xrightarrow[\text{slow}]{C_6H_5CHO} C_6H_5-COOH + C_6H_5CH_2OH$
	(c)	It is because MeO group in para position will help in the reaction.
74.	(b)	A pH of ≃ 4.5 is required (moderately acidic).
75.	(d)	Phenyl hydrazine, $C_6H_5NHNH_2$, with free NH_2 group can undergo condensation reaction.
76.	(b)	CCl_3CHO due to three –I chlorine atoms can add H_2O to form $CCl_3CHO.H_2O$.
77.	(a)	The addition of water occurs according to Markownikoff's rule.
78.	(b)	$CH_3-\overset{OH}{\overset{\mid}{C}}=CH_2 + D_2O \longrightarrow CH_3-\overset{OH}{\overset{\mid}{\underset{OD}{\underset{\mid}{C}}}}-CH_2D$ $\xrightarrow{-DOH} CH_3-\overset{O}{\overset{\parallel}{C}}-CH_2D \xrightarrow{D_2O} CH_3-\overset{OH}{\overset{\mid}{\underset{OD}{\underset{\mid}{C}}}}-CHD_2$ $\Updownarrow$ $CH_3-\overset{OH}{\overset{\mid}{C}}=CHD$ Thus, all the H atoms or the alkyl groups shall be replaced by D atoms.
79.	(b)	It is Wolff-Kishner reduction.
80.	(b)	In 2, 4-hexanedrone, $CH_3-\overset{O}{\overset{\parallel}{C}}-CH_2-\overset{O}{\overset{\parallel}{C}}-CH_2CH_3$ the C–H bond of CH_2 is highly acidic due to keto-enol tautomerism.
81.	(a)	Dehydration is easy if it results information of conjugated carbonyl compound.
82.	(a)	It is Cannizzaro's reaction.
83.	(a)	Compound A is propanal ∴ It gives silver mirror test. The rest is addition product with semicarbazide.
84.	(d)	It is internal Cannizzaro's reaction (see concepts in brief).
85.	(c)	In nucleophillic addition reaction, lesser is the steric hinderance more is the reactivity.
86.	(c)	It is Perkins reaction.
87.	(c)	$CH_3-\overset{OH}{\overset{\mid}{C}}-CH_2CH_3 \xrightarrow[\text{2 odoform } rx^n]{NaOH/I_2}$ $CHI_3 + CH_3CH_2COOH$

5.7. ASSERTION-REASON QUESTIONS

Quick Revision

In each of the following questions two statements are given as Assertion A and Reason R. Examine the statements carefully and answer the questions according to the instructions given below :

- *Mark **a** if both **A** and **R** are correct and **R** is the correct reason of **A**.*
- *Mark **b** if both **A** and **R** are correct and **R** is not the correct reason of **A**.*
- *Mark **c** if **A** is correct and **R** is wrong.*
- *Mark **d** if **A** is wrong and **R** is correct.*
- *Mark **e** if both **A** and **R** are wrong.*

1. ***Assertion A.*** Higher ketones do not react with sodium bisulphite.

   Reason R. Crowding around carbonyl carbon cause steric hinderance to nucleophilic attack.

(*a*)	(*b*)	(*c*)	(*d*)	(*e*)

2. ***Assertion A.*** Halogen acids add across C = C bond react easily but fail to give addition product with $>C = O$.

   Reason R. C = O bond is less reactive due to its polar nature.

(*a*)	(*b*)	(*c*)	(*d*)	(*e*)

3. ***Assertion A.*** RCOCl fails to give ketone with RMgX.

   Reason R. RCOCl give ketone with R_2Cd.

(*a*)	(*b*)	(*c*)	(*d*)	(*e*)

4. ***Assertion A.*** Aldehyes/ketones do not give nucleophilic substitution reactions.

   Reason R. Aldehydes and ketones are functional isomers of each other.

(*a*)	(*b*)	(*c*)	(*d*)	(*e*)

5. ***Assertion A.*** Chloral is more reactive than acetaldehyde.

   Reason R. Chloro groups increase the magnitude of positive charge on carbonyl carbon atom.

(*a*)	(*b*)	(*c*)	(*d*)	(*e*)

6. ***Assertion A.*** *p*-nitrobenzaldehyde is more reactive than benzaldehyde.

   Reason R. Benzaldehyde is less reactive than acetone.

(*a*)	(*b*)	(*c*)	(*d*)	(*e*)

7. ***Assertion A.*** The medium should not be highly acidic during addition of hydroxyl amine to aldehydes.

   Reason R. In highly acidic medium the hydroxyl amine forms salt and loses its capacity to act as nucleophile.

(*a*)	(*b*)	(*c*)	(*d*)	(*e*)

8. ***Assertion A.*** Ketones have boiling points higher than corresponding isomeric aldehydes.

   Reason R. Ketones are relatively more polar as compared to aldehydes with same molecular mass.

(*a*)	(*b*)	(*c*)	(*d*)	(*e*)

9. ***Assertion A.*** Formyl cannot be prepared by Rosenmund's reaction.

Reason R. Formyl chloride is highly unstable and does not exist at room temperature.

(a)	(b)	(c)	(d)	(e)

10. ***Assertion A.*** Keto-form of phenol is more stable than phenol itself.

Reason R. In phenol, the lone pair of O atom is delocalised over the ring.

(a)	(b)	(c)	(d)	(e)

11. ***Assertion A.*** Addition of HCN across $>C=O$ bond produces cyanohydrin.

Reason R. Addition of HCN across $>C=O$ proceed through the attack of CN^- ion on the carbonyl group.

(a)	(b)	(c)	(d)	(e)

12. ***Assertion A.*** Both acetone and benzaldehyde are less reactive to nucleophilic attack as compared to acetaldehyde.

Reason R. Both acetone and benzaldehyde are resonance stabilized.

(a)	(b)	(c)	(d)	(e)

13. ***Assertion A.*** Aldehydes can be easily prepared by the reduction of carboxylic acids with $LiAlH_4$.

Reason R. In going from —COOH to —CHO group oxidation number of C decreases.

(a)	(b)	(c)	(d)	(e)

14. ***Assertion A.*** Formaldehyde can be obtained by oxidation of glycol with HIO_4.

Reason R. Glycol contains two primary alcoholic groups.

(a)	(b)	(c)	(d)	(e)

15. ***Assertion A.*** Diisopropyl ketone is more reactive than acetone.

Reason R. Acidic medium makes the carbonyl group more susceptible to nucleophilic attack.

(a)	(b)	(c)	(d)	(e)

ANSWERS

1. (a)	**2.** (c)	**3.** (b)	**4.** (b)	**5.** (a)	**6.** (b)	**7.** (a)	**8.** (a)	**9.** (a)
10. (d)	**11.** (b)	**12.** (c)	**13.** (d)	**14.** (b)	**15.** (d).			

5.8. SELF EVALUATION TEST

Choose the correct answer from the four alternatives given in each of the following questions :

1. Which of the following is called Rochelle's salt ?
(*a*) Potassium tartrate
(*b*) Sodium hydrogen phosphate
(*c*) Sodium potassium tartrate
(*d*) Sodium ammonium tartrate.

2. Glyoxal $\xrightarrow{\text{NaOH}}$ ...... gives mainly
(*a*) $2CH_3OH$ (*b*) $H_2COH—COO^- Na^+$
(*c*) $CH_3OH + HCOOH$ (*d*) $Na_2C_2O_4$.

3. Schiff's base and Schiff's reagent are
(*a*) Diastereomers (*b*) Anomers
(*c*) Structural isomers (*d*) Different compounds.

4. Pinacolone is
(*a*) 1-phenyl-2-propanone
(*b*) 3, 3-Dimethyl-2-butanone
(*c*) 2, 3-dimethyl-2, 3-butanediol
(*d*) 2, 3-Dimethyl-but-2-ene.

5. Aldehydes and ketones react with α-bromoesters in the presence of metallic zinc to form hydroxy acids. The reaction is known as
(*a*) Claisen condensation (*b*) Reformatsky reaction
(*c*) Knoevenagel reaction (*d*) Ziesel's reaction.

6. Chloretone belongs to which of the following category ?
(*a*) Simple aldehyde (*b*) β-chloroketone
(*c*) β-chloroester (*d*) Trichloro alcohol.

7. $C_6H_5COCH_3 + Br_2 \xrightarrow{CCl_4}$ ....
The product of this reaction is mainly
(*a*) *p*-isomer of $Br—C_6H_4COCH_3$
(*b*) *o*-isomer of $Br—C_6H_4COCH_3$
(*c*) $C_6H_5COCH_2Br$
(*d*) C_6H_5COBr.

8. 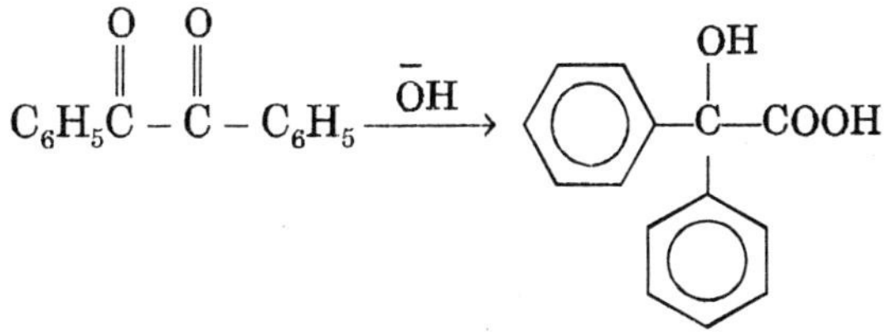

The above reaction is known as
(*a*) Backmann rearrangement
(*b*) Benzoin condensation
(*c*) Internal aldol condensation
(*d*) Benzilic acid rearrangement.

9. 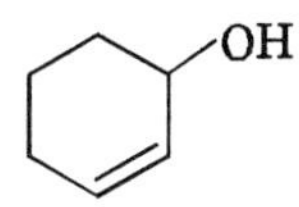 $\xrightarrow[\text{Acetone}]{[(CH_3)_3CO]_3Al}$ X. The product of this reaction is

(*a*) 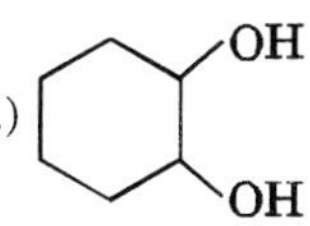 (*b*) 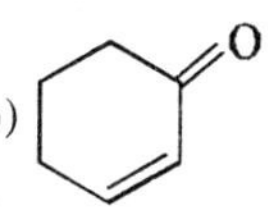

(*c*) 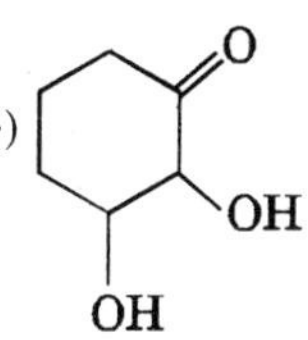 (*d*) 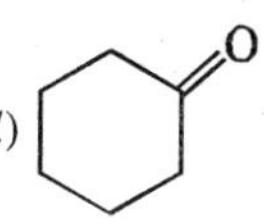

10. $CH_3CHO \xrightarrow[\text{oxidation}]{SeO_2}$ .... The product of the reaction is
(*a*) $CO_2 + H_2O$ (*b*) OHC—CHO
(*c*) $(COOH)_2$ (*d*) $(CH_3)_2CO$.

11. Which reagent is most effective in converting **but-2-enal** to **but-2-en-1-ol ?**
(*a*) $K_2Cr_2O_7/H_2SO_4$ (*b*) OsO_4
(*c*) H_2/Pt (*d*) $NaBH_4$.

12. Wacker synthesis involve the preparation of aldehyde/ketone from
(*a*) Alkenes (*b*) Alkane
(*c*) Alkyl halides (*d*) Isocyanides.

13. Which of the following reagents produce diacetone alcohol ?
(*a*) $(CH_3)_2CO + EtOH$ (*b*) $(CH_3)_2CO$ + conc. H_2SO_4
(*c*) $(CH_3)_2CO + Ba(OH)_2$ (*d*) $(CH_3)_2CO + LiAlH_4$.

14. Gas **A** is produced by dropping calcium carbide in water, is bubbled through very dilute H_2SO_4 containing mercury (II) sulphate to form **B**.
B + X ⟶ *Ethylidene dichloride.* **X** can be
(*a*) Cl_2 (*b*) $SOCl_2$/Pyridine
(*c*) $SbCl_5$ (*d*) $CHCl_3$.

15. The product formed during the reaction of CCl_3CHO and H_2O is

(a) Glyoxal (b) Chloral hydrate
(c) Trichloroacetic acid (d) HOOC—CHO.

16. Hypnone is a commercial name of
(a) Methyl phenyl ketone (b) Acetone
(c) Benzophenone (d) Diethyl ketone.

17. Certain ketone **A** (capable of showing iodoform test positive) on reduction gives **B. B** on acidic dehydration at 443 K gives **C. C**, on reductive ozonolysis gives only MeCHO. **A** and **C** in the above sequence are respectively.
(a) CH_3COCH_3 and $CH_3CH = CH_2$
(b) $CH_3COC_2H_5$, $CH_3CH_2CH = CH_2$
(c) $CH_3COC_2H_5$ and $CH_3CH = CHCH_3$
(d) $CH_3COC_2H_5$, $CH_3CH_2 = CH(OH)CH_3$.

18. $\mathbf{3CH_3COCH_3} \xrightarrow{\text{dry HCl gas}}$ **X.** The main product X in this reaction is
(a) Mesitylene (b) Acetylene
(c) Phorone (d) Dypnone.

19. Which of the following reagent can convert acetone to 1, 3, 5-trimethylbenzene ?
(a) dilute $Ba(OH)_2$ (b) conc. H_2SO_4
(c) conc. NaOH (d) Cl_2/CCl_4.

20. Cyclohexanone is treated with methylene triphenyl phosphorane, $(C_6H_5)_3P = CH_2$. The product formed is

(a) 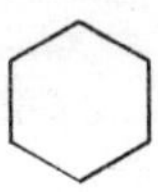

(b) 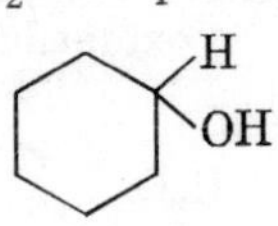

(c) 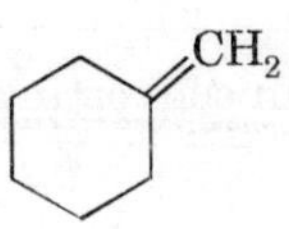

(d) 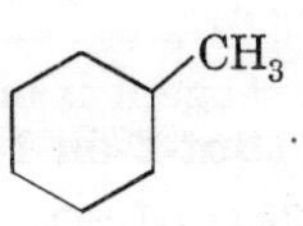

In each of the following questions, three statements are given. Mark the correct answer as per following instructons :
(a) **If statement I and II are correct**
(b) **if statement II and III are correct**
(c) **if statement I and III are correct**
(d) **if only statement III is correct.**

21. I. Aldehydes are functional isomers of ketones
II. Ketones are less polar than the corresponding isomeric aldehydes.
III. Tertiary alcohols cannot be formed by reduction of ketones.
(a) (b)
(c) (d)

22. I. Benzaldehyde does not react with Benedict solution.
II. Benzaldehyde is less reactive than acetaldehyde.
III. Cannizzaro's reaction is not shown by benzaldehyde.
(a) (b)
(c) (d)

23. I. Aldol condensation proceeds through carbocation intermediate
II. Aldol condensation is acid catalysed reaction.
III. Aldol condensation is not given by benzaldehyde.
(a) (b)
(c) (d)

24. I. The high reactivity of formaldehyde is due to the absence of electron releasing groups at the carbonyl carbon
II. Formaldehyde and ethanal react with NH_3 to form different products
III. Acetone cannot give aldol type condensation.
(a) (b)
(c) (d)

25. I. Paraldehyde is a tetramer of acetaldehyde.
II. Dry distillation of barium acetate salt produces acetone.
III. Metaldehyde is a tetramer of ethanal.
(a) (b)
(c) (d)

26. Aldehydes can be conveniently separated from alcohols by treating with
(a) Na_2SO_4 (b) NaCN
(c) $NaHSO_3$ (d) Schiff's reagent.

27. Which of the following compound after treatment with $LiAlH_4$ will give Lucas test readily ?
(a) $CH_3CH_2OCH_2CH_3$ (b) $CH_3CH_2COOCH_3$
(c) CH_3CH_2CHO (d) $CH_3CH_2COCH_3$.

28. Iodoform test is not shown by
(a) 2-Butanone (b) *Sec*-butyl alcohol
(c) Propanone (d) Pentan-3-ol.

29. In what respects C = C and C = O bonds differ ?
(a) Hybrid state of carbons (b) Number of π and σ bonds
(c) Polarity of bonds (d) Planar disposition.

30. 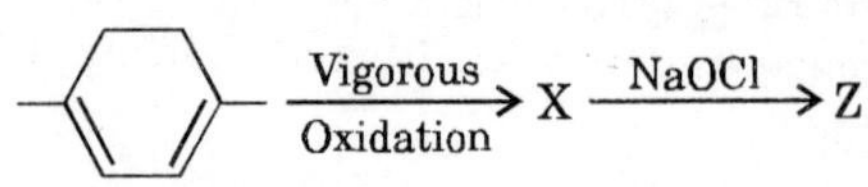

In the above sequence, Z is

(a) 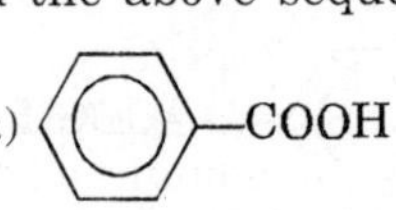

(b) 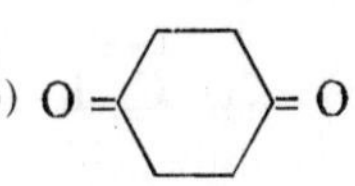

(c) 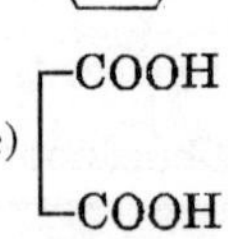

(d) 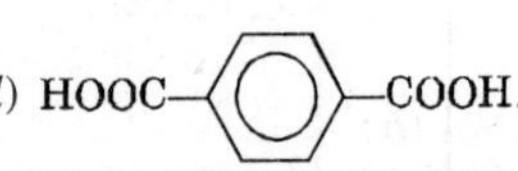

31. Which of the following compounds will react with enanolic solution of KCN ?
(*a*) Chlorobenzene (*b*) Benzaldehyde
(*c*) Ethyl alcohol (*d*) Propanone.

32. Benzaldehyde and PCl_5 react to give
(*a*) Chlorobenzene (*b*) Benzoyl chloride
(*c*) Benzal chloride (*d*) Benzyl chloride.

33. Which of the following is/are **dione**
(*a*) Acetal (*b*) Acetyl acetone
(*c*) Diacetyl (*d*) Acetyl benzene.

34. Which has lowest reactivity towards nucleophilic attack ?
(*a*) Butanone (*b*) Ethanal
(*c*) Propanal (*d*) Propanone.

35. Alkaline hydrolysis of 2, 2-dichloropropene gives
(*a*) Alcohol (*b*) A diol
(*c*) A ketone (*d*) An ester.

36. In the Cannizzaro's reaction involving benzaldehyde the slowest step is
(*a*) Attack of OH^- ion at the carbonyl carbon
(*b*) The transfer of hydride ion to the carbonyl group
(*c*) The abstraction of proton from carboxylic acid
(*d*) The deprotonation of $PhCH_2OH$.

37. What is not true for formaldehyde ?
(*a*) Its aqueous solution on heating to dryness gives paraldehyde
(*b*) Its 40% aqueous solution is called formaline
(*c*) Formaline is used as preservative for anatomical specimens
(*d*) It is one of the reagent which produces urotropine.

38. Dry distillation of calcium propionate gives
(*a*) Acetic acid (*b*) Ethane
(*c*) 3-Pentanone (*d*) Acetone.

39. The number of isomeric aldehydes of the formula $C_5H_{10}O$ is
(*a*) 4 (*b*) 5
(*c*) 3 (*d*) 2.

40. Which of the following gives iodoform test positive but not Tollen's test ?
(*a*) Diethyl ketone (*b*) Acetaldehyde
(*c*) 2-propanol (*d*) Acetophenone.

ANSWERS

1. (*c*)	**2.** (*b*)	**3.** (*d*)	**4.** (*b*)	**5.** (*b*)	**6.** (*d*)	**7.** (*c*)	**8.** (*d*)	**9.** (*b*)
10. (*b*)	**11.** (*d*)	**12.** (*a*)	**13.** (*c*)	**14.** (*b*)	**15.** (*b*)	**16.** (*a*)	**17.** (*b*)	**18.** (*c*)
19. (*b*)	**20.** (*c*)	**21.** (*c*)	**22.** (*a*)	**23.** (*d*)	**24.** (*a*)	**25.** (*b*)	**26.** (*c*)	**27.** (*d*)
28. (*d*)	**29.** (*c*)	**30.** (*c*)	**31.** (*b*)	**32.** (*c*)	**33.** (*b*)	**34.** (*a*)	**35.** (*c*)	**36.** (*b*)
37. (*a*)	**38.** (*c*)	**39.** (*a*)	**40.** (*c*)					

HINTS/SOLUTIONS

Q. No.	*Ans.*	*Hints & Solutions*
1.	(*c*)	Rochelle's salt is sodium potassium tartrate.
2.	(*b*)	Glyoxal undergoes internal Cannizzaro's reaction $$\begin{array}{l} H{-}C=O \\ \quad\;\, \vert \\ H{-}C=O \end{array} \xrightarrow{NaOH} \begin{array}{l} CH_2OH \\ \;\vert \\ COONa \end{array}$$
3.	(*d*)	Schiff's reagent is rosaniline hydrochloride while Schiff's base in N-alkyldimine a product of reaction between aldehydes or ketones with 1° amines.
4.	(*b*)	Pinacolone is 3, 3-Dimethyl-2-butanone.
5.	(*b*)	The reaction between aldehydes/ketones and α-bromoesters in the presence of Zn to give β-hydroxyester is called Reformatsky reaction. $$R_2C=O + BrCH_2COOC_2H_5 \xrightarrow[(ii)\ H_3O^+]{(i)\ Zn\ (Benzene)} \begin{array}{c} OH \\ \vert \\ R-C-CH_2COOC_2H_5 \\ \vert \\ R \end{array}$$ β-hydroxy ester

Q. No.	Ans.	Hints & Solutions
6.	(*d*)	Chloretone is $CH_3-\underset{CCl_3}{\overset{CCl_3}{C}}-OH$, a trichloro alcohol.
7.	(*c*)	$C_6H_5COCH_3 + Br_2 \xrightarrow{CCl_4} \underset{\text{Phenacyl bromide}}{C_6H_5COCH_2Br} + HBr$
8.	(*d*)	The given reaction in Benzillic acid rearrangement which involves the treatment of benzil with strong base (NaOH) to form sodium salt of benzillic acid. $\underset{\text{Benzil}}{C_6H_5-\overset{O}{\overset{\|\|}{C}}-\overset{O}{\overset{\|\|}{C}}-C_6H_5} \xrightarrow{NaOH}$ $\left[C_6H_5-\underset{C_6H_5}{\overset{O}{\overset{\|}{C}}}-\overset{O}{\overset{\|\|}{C}}-OH\right] \longrightarrow \underset{\text{Sodium salt of benzillic acid}}{C_6H_5-\underset{C_6H_5}{\overset{O}{\overset{\|}{C}}}-\overset{O}{\overset{\|\|}{C}}-ONa}$
9.	(*b*)	The given process is oppenaur oxidation which causes oxidation of C—OH group to C = O group without oxidising C = C bond.
10.	(*b*)	Refer to SeO_2 oxidation.
11.	(*d*)	$\underset{\text{But-2-enal}}{CH_3—CH = CH—CHO} \xrightarrow{NaBH_4} \underset{\text{But-2-en-1-ol}}{CH_3—CH = CH—CH_2OH}$
12.	(*a*)	$CH_2 = CH_2 + PdCl_2 + H_2O \xrightarrow{CuCl_2} CH_3CHO + Pd + 2HCl$
13.	(*c*)	$CH_3-\overset{O}{\overset{\|\|}{C}}-CH_3 + CH_3-\overset{O}{\overset{\|\|}{C}}-CH_3 \xrightarrow{Ba(OH)_2} \underset{\text{Diacetone alcohol}}{CH_3-\underset{CH_3}{\overset{OH}{C}}-CH_2-\overset{O}{\overset{\|\|}{C}}-CH_3}$
14.	(*b*)	$CaC_2 + H_2O \longrightarrow Ca(OH)_2 + \underset{(A)}{CH \equiv CH}$ $CH \equiv CH \xrightarrow[H_2SO_4]{Hg^{2+}} \underset{(B)}{CH_3CHO} \xrightarrow[Py]{SOCl_2} CH_3CHCl_2.$
15.	(*b*)	$CCl_3 . CHO + H_2O \longrightarrow \underset{\text{Chloral hydrate}}{Cl_3C—CH(OH)_2}$
16.	(*a*)	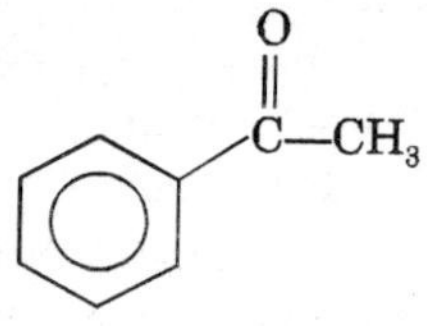 is known as hypnone.
17.	(*b*)	$CH_3COC_2H_5 \xrightarrow[LiAlH_4]{\text{Reduction}} \underset{(B)}{CH_3\overset{OH}{C}HCH_2CH_3} \xrightarrow{H_2SO_4} \underset{(C)}{CH_3CH = CHCH_3}$ $\xrightarrow[(ii)\ Zn/H_2O]{(i)\ O_3} CH_3CHO + CH_3CHO$
18.	(*c*)	$CH_3—\overset{CH_3}{C} = O + H_2CH—\overset{O}{\overset{\|\|}{C}}—CHH_2 + O = \overset{CH_3}{C}—CH_3 \xrightarrow{\text{dry HCl}} \underset{\text{Phorone}}{CH_3—\overset{CH_3}{C} = CH—\overset{O}{\overset{\|\|}{C}}—CH = \overset{CH_3}{C}—CH_3}$
19.	(*b*)	$3CH_3COCH_3 \xrightarrow[H_2SO_4]{\text{Conc.}}$ 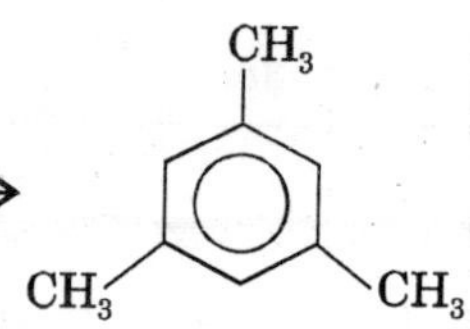
20.	(*c*)	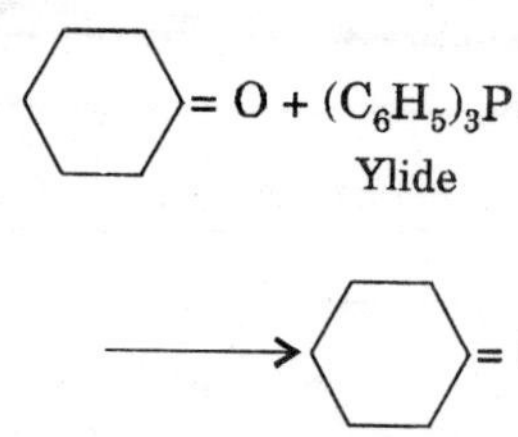 $= O + (C_6H_5)_3P = CH_2$ (Ylide) $\longrightarrow$ $= CH_2 + (C_6H_5)P = O$ Methylenecyclohexene The reaction is called is called **Wittig reaction.**
21.	(*c*)	I and III are both correct.
22.	(*a*)	Both statement I and II are correct.
23.	(*d*)	Statement III is correct.
24.	(*a*)	I and II are correct.
25.	(*b*)	Statement II and III are correct.
26.	(*c*)	Separation of carbonyl compounds from non-

Q. No.	Ans.	Hints & Solutions
26.	(c)	Separation of carbonyl compounds from non-carbonyl compounds can be achieved by sodium bisulphite.
27.	(d)	Reduction of both CH_3COCH_3 and $CH_3CH_2COCH_3$ give secondary alcohols which will respond to Lucas test.
28.	(d)	It does not have $CH_3-\underset{}{\overset{OH}{\overset{\mid}{C}H}}-$ type of grouping.
29.	(c)	Bond lengths and polarities of C = C and C = O bonds are different.
30.	(c)	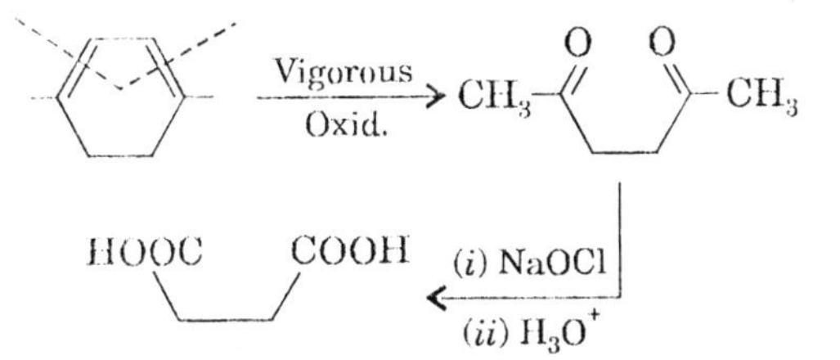
31.	(b)	Benzaldehyde reacts with KCN to give benzoin condensation and C_2H_5Cl also reacts with KCN.
32.	(c)	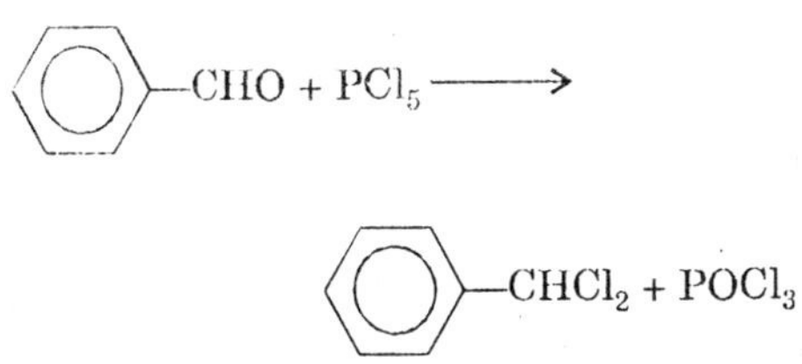 Benzaldichloride
33.	(b)	Diacetyl is $H_3C-\overset{O}{\overset{\parallel}{C}}-\overset{O}{\overset{\parallel}{C}}-CH_3$ and acetyl acetone is $CH_3-\overset{O}{\overset{\parallel}{C}}-CH_2-\overset{O}{\overset{\parallel}{C}}-CH_3$.
34.	(a)	$CH_3-\overset{O}{\overset{\parallel}{C}}-CH_2CH_3$ has two bulky alkyl groups surrounding >C = O and hence show steric hinderance to attack by nucleophile.
35.	(c)	$CH_3-\underset{Cl}{\underset{\mid}{\overset{Cl}{\overset{\mid}{C}}}}-CH_3 \xrightarrow{NaOH(aq)} CH_3-\underset{OH}{\underset{\mid}{\overset{OH}{\overset{\mid}{C}}}}-CH_3 \xrightarrow{-H_2O} \underset{\text{Acetone}}{CH_3-\overset{O}{\overset{\parallel}{C}}-CH_3}$
36.	(b)	The second step is slowest. See concept oriented details or answer to Q. 72 of IIT section.
37.	(a)	Formaldehyde will form paraform.
38.	(c)	$(CH_3-CH_2\overset{O}{\overset{\parallel}{C}}-O)_2Ca \xrightarrow{\text{Dry Distill}} CaCO_3 + \underset{\text{Diethyl ketone}}{CH_3CH_2COCH_2CH_3}$
39.	(a)	$CH_3CH_2CH_2CH_2CHO$, $CH_3-\overset{CH_3}{\overset{\mid}{C}H}-CH_2CHO$, $CH_3-\underset{CHO}{\underset{\mid}{\overset{CH_3}{\overset{\mid}{C}}}}-CH_3$, $CH_3-CH_2\underset{CHO}{\underset{\mid}{C}H}-CH_3$
40.	(c)	Iodoform test is given by $CH_3-\overset{O}{\overset{\parallel}{C}}-$ or CH_3CHOH- type of group whereas Tollen's reagent can only oxidise aldehydes.

6 CARBOXYLIC ACIDS AND THEIR DERIVATIVES

6.1. CONCEPTS IN BRIEF

Develop Logic

CARBOXYLIC ACIDS

- Carboxylic acids are organic compounds containing —COOH functional group. This functional group is supposed to be made of carboxyl group (> C = O) and hydroxyl (—OH) groups.
- If —COOH group is linked to alkyl chain they are called *aliphatic acids* **(R—COOH).** On the other hand if —COOH group is directly linked to aryl group, they are called *aromatic acids* **(Ar—COOH).**
- General formula of monocarboxylic acids is $\mathbf{C_nH_{2n+1}COOH}$ where $n = 0$ for the first member.
- Higher monocarboxylic acids are called **Fatty acids.**

NOMENCLATURE

- **Common** or **trivial** names are derived from the source of individual acids.
- **IUPAC names** are obtained by replacing terminal **e** of corresponding alkanes by the suffix **oic-acid.**

METHODS OF PREPARATION

1. **By Oxidation of Primary Alcohols.** Oxidation can be achieved by $KMnO_4/H_2SO_4$ or $K_2Cr_2O_7/H_2SO_4$.

$$\underset{\text{1° Alcohol}}{R-CH_2OH} \xrightarrow[K_2Cr_2O_7/H_2SO_4]{(O)} \underset{\text{Carboxylic acid}}{R-COOH}$$

2. By Oxidation of Aldehydes

$$\underset{\text{Aldehyde}}{\text{R—CHO}} \xrightarrow[\text{KMnO}_4/\text{H}^+]{(\text{O})} \underset{\text{Carboxylic acid}}{\text{R—COOH}}$$

- In the above methods the number of carbon atoms in the acid obtained and those in the parent alcohol or aldehyde is same.

3. By Haloform Reaction

- This method involves oxidation of methyl ketones with hypohalite ion, XO^- ($NaOH + X_2$)

$$\underset{\text{(Ketone)}}{\text{R—}\overset{\text{O}}{\overset{\|}{\text{C}}}\text{—CH}_3} + 3\text{NaOI} \xrightarrow{\text{NaOH/I}_2} \underset{\text{(Salt of acid)}}{\text{R—}\overset{\text{O}}{\overset{\|}{\text{C}}}\text{—ONa}} + \text{CHI}_3 + 2\text{NaOH}$$

$$\text{R—}\overset{\text{O}}{\overset{\|}{\text{C}}}\text{—ONa} \xrightarrow[-\text{NaCl}]{\text{HCl}} \underset{\text{(Acid)}}{\text{R—}\overset{\text{O}}{\overset{\|}{\text{C}}}\text{—OH}}$$

- In this method the acid produced has one C atom less than the parent ketone.

4. By Hydrolysis of Cyanides and other Acid Derivatives

$$\underset{\text{Alkyl cyanides}}{\text{R—CN}} + 2\text{H}_2\text{O} + \text{HCl} \longrightarrow \underset{\text{Acid}}{\text{R—COOH}} + \text{NH}_4\text{Cl}$$

$$\underset{\text{Alkyl chlorides}}{\text{R—COCl}} + \text{HOH} \longrightarrow \underset{\text{Acid}}{\text{R—COOH}} + \text{HCl}$$

$$\underset{\text{Ester}}{\text{R—COOR}'} + \text{HOH} \xrightarrow{\text{HCl}} \underset{\text{Acid}}{\text{RCOOH}} + \text{ROH}$$

$$\underset{\text{Acid anhydride}}{\text{RCO—O—COR}} + \text{HOH} \longrightarrow \underset{\text{Acid}}{2\text{RCOOH}}$$

$$\underset{\text{Amides}}{\text{R—CONH}_2} + \text{HOH} \xrightarrow{\text{HCl}} \underset{\text{Acid}}{\text{R—COOH}} + \text{NH}_4\text{Cl}$$

5. By Carbonation of Grignard's Reagent

$$\text{RMgX} + \text{O}=\text{C}=\text{O} \longrightarrow \text{R—}\overset{\text{O}}{\overset{\|}{\text{C}}}\text{—OMgX} \xrightarrow{\text{H}_2\text{O}} \text{R—}\overset{\text{O}}{\overset{\|}{\text{C}}}\text{—OH} + \text{Mg(OH)X}$$

This method gives carboxylic acid with *one carbon atom more than the starting Grignard's reagent.*

6. From Olefins

(*a*) By oxidative cleavage of ethylenic bond in symmetrical alkenes of the type R—CH = CHR by using $KMnO_4/H^+$

$$\text{R—CH}=\text{CH—R} \xrightarrow[\text{KMnO}_4/\text{H}_2\text{SO}_4]{(\text{O})} 2\text{RCOOH}$$

- In case the alkene is unsymmetrical, mixture of acids will be formed.

(*b*) By heating olefin with CO and steam under pressure at 570 K to 675 K in the presence of catalyst.

$$\underset{\text{Ethylene}}{\text{CH}_2=\text{CH}_2} + \text{CO} + \text{H}_2\text{O} \xrightarrow[\text{H}_2\text{PO}_4]{570\text{—}675\text{ K}} \underset{\text{Propanoic acid}}{\text{CH}_3\text{CH}_2\text{COOH}}$$

$$\underset{\text{Propene}}{CH_3CH = CH_3} + CO + H_2O \xrightarrow[H_2PO_4]{570\text{—}675\ K} \underset{iso\text{ Butyric acid}}{CH_3-\underset{}{\overset{CH_3}{\overset{|}{C}}}H-COOH}$$

- This reaction is known as KOCH REACTION.

7. Decarboxylation of Alkyl Substituted Malonic Acid

$$\underset{\text{Alkyl malonic acid}}{R\text{—}CH(COOH)_2} \xrightarrow{Heat} \underset{\text{Carboxylic acid}}{RCH_2COOH + CO_2}$$

- If malonic acid is used, the product is acetic acid.

8. From Sodium Alkoxide and Carbon Monoxide

Heating of sodium alkoxide with CO under pressure yields sodium salt of fatty acids which on subsequent acidification gives acid.

$$\underset{\substack{\text{Sodium}\\\text{alkoxide}}}{RONa} + CO \xrightarrow[\text{Pressure}]{\text{high}} \underset{\substack{\text{Sodium salt}\\\text{of acid}}}{RCOONa} \xrightarrow{HCl} \underset{\text{Fatty acid}}{RCOOH} + NaCl$$

9. From alkyl benzenes. Benzoic acid can be prepared by oxidation of alkyl benzenes with alkaline potassium permanganate, chromic anhydride or conc. nitric acid. The ***alkyl side chain gets oxidised*** to **—COOH** group irrespective of the size of the chain. Many aromatic acids are obtained industrially by this method.

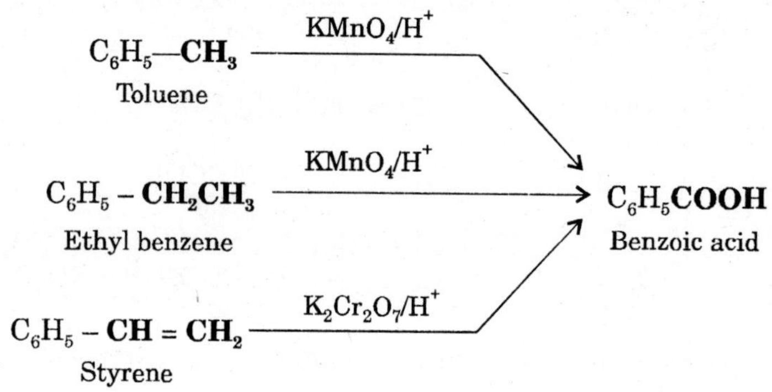

PHYSICAL PROPERTIES OF CARBOXYLIC ACIDS

1. **Physical State.** The first three aliphatic acids are colourless liquids with pungent smell. The next six are oily liquids with an odour of rancid butter while the higher members are colourless, odourless waxy solids. Benzoic acid is crystalline solid.
2. **Solubility.** The first four aliphatic members are soluble in water due to their ability of intermolecular hydrogen bonding with water molecules. With increasing size of the alkyl group, the non-polar part of the molecule predominates thereby decreasing thee solubility in water. The higher members are practically insoluble in water.
3. **Boiling Points.** Carboxylic acids have high boiling points due to intermolecular H-bonding which result in the formation of ***Cyclic dimers.***

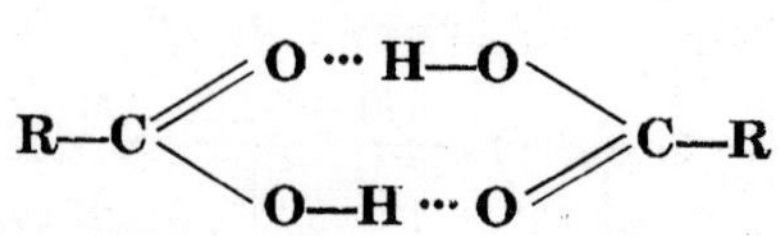

4. **Melting Points.** The melting points of first ten *n*-monocarboxylic acids follow ***saw tooth rule,*** *i.e.,* they show alteration from one member to the next. The means that increase in the *m.p.* from odd carbon acid to an even carbon acid is relatively higher than that from even carbon acid to the odd carbon acid.

5. The melting/boiling points of aromatic acids are relatively higher than the corresponding aliphatic acids of comparable molecular masses.

ACIDIC NATURE OF CARBOXYLIC ACIDS

The acidic nature of carboxylic acids can be explained by resonating structures of carboxylic acid and the corresponding conjugate base, *i.e.*, the carboxylate ion.

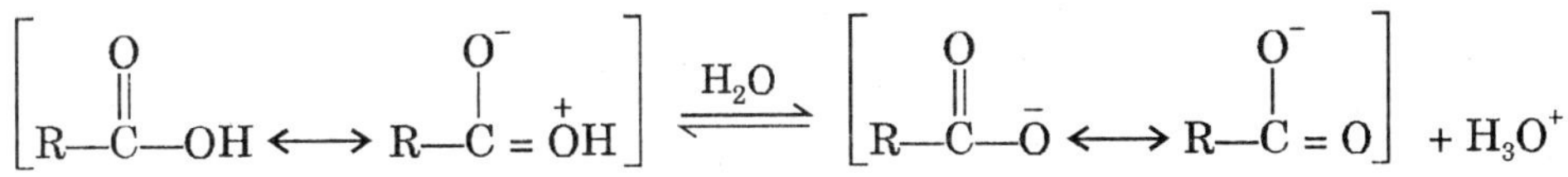

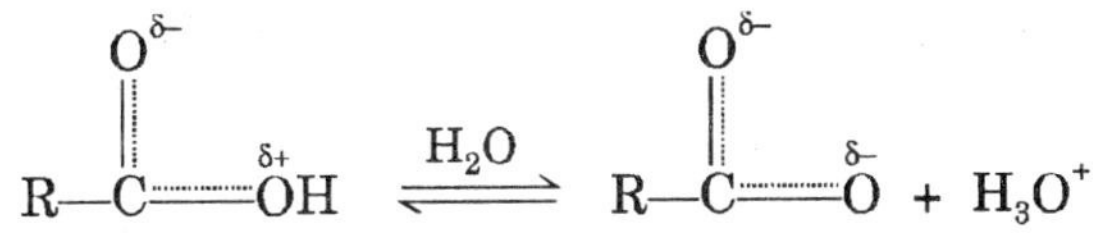

(Acid molecule) (Carboxylate ion)

- Carboxylate ion is relatively more stabilised (due to equivalent structures) than the carboxylic acid molecule. Thus, acid has a tendency to form carboxylate ion by losing a proton to the base.
- Any substituent which further **stabilises the carboxylate** ion **increases** the acid strength and that which destabilises the carboxylate ion decreases the acid strength.
- The acid strength is expressed by dissociation constant (K_a) or pK_a (= – log K_a).
- Larger the K_a or lower the pK_a of the acid, stronger will be the acid itself.
- The **electron attracting substituents** on carboxyl carbon or α-carbon, increases the acid strength. For example, acid strength increases with the increase in the electron attracting power of the substituent.

$$CH_3COOH < ClCH_2COOH < BrCH_2COOH < FCH_2COOH < NCCH_2COOH < O_2NCH_2COOH$$

ACID STRENGTH INCREASES →

- Acid strength also increases with the increase in the number of electron withdrawing substituents.

$$CH_3COOH < ClCH_2COOH < Cl_2CHCOOH < Cl_3CCOOH$$

ACID STRENGTH INCREASES →

- The presence of electron releasing substituents cause + I effect and destablise the carboxylate ion. This results in the decrease of acid strength.

$$HCOOH > CH_3COOH > CH_3CH_2COOH > (CH_3)_2CHCOOH > (CH_3)_3C\,.\,COOH$$

ACID STRENGTH DECREASES →

- Benzoic acid (k_a = 6.3 × 10^{-5}) is slightly stronger acid than CH_3COOH (k_a = 1.8 × 10^{-5}), because in benzoic acid —COOH group is attached to sp^2 hybrid carbon which is relatively more electronegative than sp^3 hybrid carbon.
- Among aromatic acids, the presence of electron withdrawing groups at *p*-position increases the acid strength whereas presence of electron releasing groups at *p*-position decrease acid strength

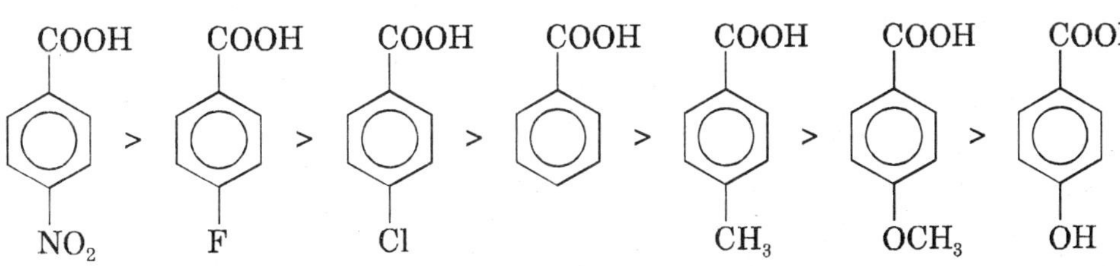

ACID STRENGTH DECREASES →

- The presence of electron releasing or electron attracting group at ortho position w.r.t. —COOH group cause slight increase in acid strength. This is called **ortho effect.**

DISSOCIATION CONSTANTS K_a OF SOME ACIDS

1. Monocarboxylic Acids

Acid	K_a	*Acid*	K_a
HCOOH	17.7×10^{-5}	$BrCH_2COOH$	125×10^{-5}
CH_3COOH	1.75×10^{-5}	ICH_2COOH	67×10^{-5}
$CH_3CH_2CH_2COOH$	1.52×10^{-5}	FCH_2COOH	260×10^{-5}
$C_6H_5CH_2COOH$	4.9×10^{-5}	$Cl_2CHCOOH$	5530×10^{-5}
Cl—CH_2COOH	136×10^{-5}	$Cl_3CHCOOH$	23200×10^{-5}

2. Substituted Benzoic Acids

C_6H_5COOH ; $K_a = 6.3 \times 10^{-5}$					
Substituent	K_a	*Substituent*	K_a	*Substituent*	K_a
p-NO_2	36×10^{-5}	*o*-NO_2	670×10^{-5}	*m*-NO_2	32×10^{-5}
p-Cl	10.3×10^{-5}	*o*-Cl	120×10^{-5}	*m*-Cl	15.1×10^{5}
p-CH_3	4.2×10^{-5}	*o*-CH_3	12.4×10^{-5}	*m*-CH_3	5.4×10^{-5}
p-OH	2.6×10^{-5}	*o*-OH	105×10^{-5}	*m*-OH	8.3×10^{-5}
p-OCH_3	3.3×10^{-5}	*o*-OCH_3	8.2×10^{-5}	*m*-OCH_3	8.2×10^{-5}

3. Dicarboxylic acids

Name	*Formula*	K_{a_1}	K_{a_2}
Oxalic acid	HOOC—COOH	5400×10^{-5}	5.2×10^{-5}
Malonic acid	HOOC—CH_2—COOH	140×10^{-5}	0.20×10^{-5}
Succinic acid	HOOC—CH_2—CH_2—COOH	6.4×10^{-5}	0.23×10^{-5}
Maleic acid	*cis*-HOOC—CH = CH—COOH	1000×10^{-5}	0.055×10^{-5}
Fumeric acid	*trans*-HOOC—CH = CH—COOH	96×10^{-5}	4.1×10^{-5}
Phthalic acid	1, 2 $C_6H_4(COOH)_2$	110×10^{-5}	0.4×10^{-5}

CHEMICAL PROPERTIES OF CARBOXYLIC ACIDS

Reactions of carboxylic acids can be discussed under four heads :

(i) Reactions due to H atom of —COOH group

(ii) Reactions due to —OH part

(iii) Reactions due to —COOH group

(iv) Reactions due to R—part.

1. Reactions due to H Atom of —COOH Group

$$R\text{—}\overset{O}{\overset{\|}{C}}\text{—}O\text{—}\mathbf{H} \begin{cases} \xrightarrow{Na} R\text{—}CO\overset{-}{O}\overset{+}{Na} + \frac{1}{2}H_2 \\ \xrightarrow{NaOH} R\text{—}CO\overset{-}{O}\overset{+}{Na} + H_2O \\ \xrightarrow{NaHCO_3} R\text{—}CO\overset{-}{O}\overset{+}{Na} + CO_2 + H_2O \\ \xrightarrow{Na_2CO_3} RCO\overset{-}{O}\overset{+}{Na} + CO_2 + H_2O \\ \xrightarrow{NH_3} RCO\overset{-}{O}(\overset{+}{N}H_4) \end{cases}$$

These reactions justify the acidic nature of carboxylic acids.

2. Reactions due to —OH Part

These reactions involve the formation of acid derivatives as listed below :

$R\text{—}\overset{O}{\overset{\|}{C}}\text{—}\mathbf{OH}$ Carboxylic acid

Amide Formation

$$\xrightarrow{NH_3} RCOONH_4 \xrightarrow[\text{Heat}]{-H_2O} \underset{\text{Amide}}{R\text{—}\overset{O}{\overset{\|}{C}}\text{—}NH_4}$$

Acid Chloride Formation. Acid chlorides can be prepared by reaction with PCl_5 or $SOCl_2$

$$\xrightarrow[\text{Phosphorus pentachloride}]{\mathbf{PCl_5}} \underset{\text{Acid chloride}}{R\text{—}\overset{O}{\overset{\|}{C}}\text{—}Cl} + POCl_3 + HCl$$

$$\xrightarrow[\text{Thionyl chloride}]{\mathbf{SOCl_2}} \underset{\text{Acid chloride}}{R\text{—}\overset{O}{\overset{\|}{C}}\text{—}Cl} + SO_2 + HCl$$

Ester Formation. Esters are formed by reaction with alcohols

$$\xrightarrow[\text{Alcohol}]{\mathbf{HOR'}\ \text{Conc. } H_2SO_4} \underset{\text{Ester}}{R\text{—}\overset{O}{\overset{\|}{C}}\text{—}OR'} + H_2O$$

- *The reactivity of esterification decreases with the increase in crowding of alkyl groups around —OH among alcohol and around —COOH among acids.* For example,

$$CH_3OH > C_2H_5OH > (CH_3)_2CHOH > (CH_3)_3COH$$

Also $HCOOH > CH_3COOH > (CH_3)_2CHCOOH > (CH_3)_3C - COOH$.

- *The reactivity of aromatic acids towards esterification decreases with increase in the number of alkyl substituents at ortho position.*

Carboxylic acid — **Anhydride Formation.** Acid anhydrides are formed by heating the acid with P_2O_5. Two molecules of acid participate in reaction

$$\underset{\text{Acid}}{HO\overset{O}{\overset{\|}{C}}\text{—}R} \xrightarrow[-H_2O]{P_2O_5} \underset{\text{Acid anhydride}}{R\text{—}\overset{O}{\overset{\|}{C}}\text{—}O\text{—}\overset{O}{\overset{\|}{C}}\text{—}R}$$

3. Reactions Involving Complete —COOH Group

(*i*) **Soda lime decarboxylation.** Sodium or potassium salts of carboxylic acids on heating with soda-lime (NaOH and CaO) give *alkanes with one carbon less than the parent acids.*

$$\underset{\text{Sodium salt of acid}}{RCOONa} + NaOH \xrightarrow{CaO} \underset{\text{Alkane}}{R—H} + Na_2CO_3$$

- *The reaction can be employed for descending down the series i.e., decreasing one carbon in a chain.*
- If sodium salt has n carbon atoms, the alkane will have $(n - 1)$ carbon atoms.

(*ii*) **Kolbe's electrolytic decarboxylation.** It involves electrolysis of aqueous solution of potassium carboxylate

$$2RCOOK + 2H_2O \xrightarrow{\text{Electrolysis}} \underset{\text{Alkane}}{R-R} + 2CO_2 + H_2 + 2KOH$$

- If there are n carbon atoms in potassium salt, the alkane produced will have $2(n - 1)$ carbon atoms
- Alkane and CO_2 are obtained at anode while H_2 gas is obtained at cathode.

(*iii*) **Dry distillation of calcium salts of carboxylic acids.** Calcium salts of carboxylic acids on heating given aldehydes or ketones.

- *When calcium formate is distilled, formaldehyde is formed.*

$$\underset{\text{Calcium formate}}{Ca(OCOH)_2} \xrightarrow{\Delta} CaCO_3 + \underset{\text{Formaldehyde}}{HCHO}$$

- *When calcium formate is distilled with calcium acetate, acetaldehyde is formed.*

$$\underset{\text{Cal. formate}}{Ca(OCOH)_2} + \underset{\text{Cal. acetate}}{Ca(OCOCH_3)} \xrightarrow{\Delta} 2CaCO_3 + \underset{\text{Acetaldehyde}}{CH_3CHO}$$

- *When calcium acetate is distilled alone, acetone is formed.*

$$Ca(OCOCH_3)_2 \xrightarrow{\text{Distill}} CaCO_3 + \underset{\text{Acetone}}{CH_3\overset{\overset{O}{\|}}{C}CH_3}$$

(*iv*) **Reduction**

- *Partial reduction.* Carboxylic acids are reduced to primary alcohols with lithium aluminium hydride ($LiAlH_4$), or with hydrogen (H_2) in the presence of copper chromite ($CuCr_2O_4$) catalyst (also called **Adkin's Catalyst**). Here, **—COOH** group is reduced to **– CH_2OH.**

$$RCOOH + H_2 \xrightarrow{CuCr_2O_4} \underset{\text{Alcohol}}{RCH_2OH}$$

- *Complete reduction.* Carboxylic acids are reduced to alkanes on reaction with HI in the presence of phosphorus at about 500 K. Here, **—COOH** group is reduced to **– CH_3.**

$$\underset{\text{Acid}}{RCOOH} + 6HI \xrightarrow[500\text{ K}]{P} \underset{\text{Alkane}}{RCH_3} + 3H_2O + 3I_2$$

- The alkane produced has same number of carbon atoms as in parent acid.
- **Reaction with hydrazoic acid (HN_3)**

It involves reaction of carboxylic acid with HN_3 in the presence of sulphuric acid to give amines. The reaction is known as **Schmidt reaction.**

$$R—\overset{\overset{O}{\|}}{C}—OH + \underset{\text{Hydrazoic acid}}{HN_3} \xrightarrow{H_2SO_4} \underset{1^\circ\text{ Amine}}{R-NH_2} + CO_2 + N_2$$

- The HN_3 being highly toxic is prepared *in situ* by adding NaN_3 to H_2SO_4 slowly.
- The reaction proceeds through the formation of *isocyanate intermediate.*

4. Reaction Involving Alkyl Group of Carboxylic Acids

(*i*) **Oxidation.** When acids are treated with mild oxidising agents such as H_2O_2, the alkyl group is oxidised at the β-position. For example :

$$\underset{n\text{-Butyric acid}}{CH_3\overset{\beta}{C}H_2\overset{\alpha}{C}H_2COOH} \xrightarrow[H_2O_2]{[O]} \underset{\beta\text{-Hydroxy butyric acid}}{CH_3\overset{OH}{\overset{|}{C}}HCH_2COOH}$$

(*ii*) **Hell Volhard Zelinksy (HVZ) Reaction (a-*Halogenation of aliphatic acids*).** When carboxylic acids are treated with Cl_2 or Br_2 in the presence of red phosphorus, the a-hydrogen atoms of carboxylic acids are replaced by chlorine or bromine. The reaction does not stop at monosubstitution but continues till all the a-hydrogens are replaced. For example :

$$\underset{\text{Acetic acid}}{CH_3COOH} \xrightarrow[P]{Cl_2} \underset{\text{Chloroacetic acid}}{CH_2ClCOOH} \xrightarrow[P]{Cl_2} \underset{\text{Dichloroacetic acid}}{CHCl_2COOH} \xrightarrow[P]{Cl_2} \underset{\text{Trichloroacetic acid}}{CCl_3COOH}$$

The halogen atom of the α-halo acid can be further replaced by number of other groups, like $—NH_2$, —OH, —CN, etc., by suitable reactions.

5. Ring Substitution Reactions in Benzoic Acid

Carboxylic group is *electron withdrawing group.* It decreases the electron density of benzene ring and hence, deactivates it towards electrophilic substitution reactions. Substitution in benzoic acid takes place mainly at *meta* position. Some examples are given below :

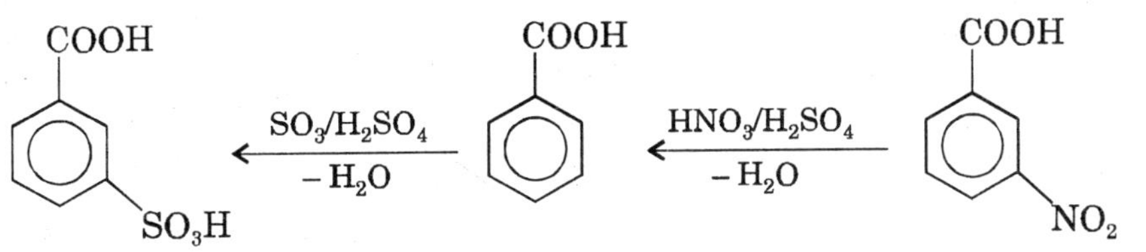

3-Carboxybenzene sulphonic acid | Benzoic acid | 3-Nitrobenzoic acid

DICARBOXYLIC ACIDS

- General formula is $C_nH_{2n}\,(COOH)_2$, where n = 0, 1, 2, ...
- According to IUPAC system they are called **Alkanedioic acids.**
- The common and IUPAC names of some common dicarboxylic acids are :

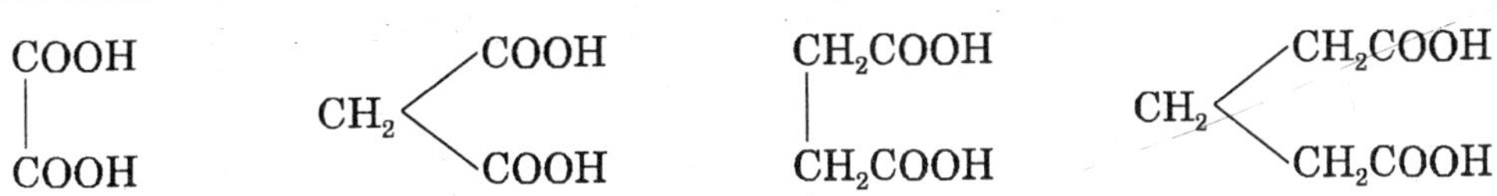

Oxalic acid	Malonic acid	Succinic acid	Gluteric acid
Ethanedioic acid	1, 3-Propane dioic acid	1, 4-Butanedioic acid	1, 5-Pentanedioic acid

PREPARATION AND PROPERTIES OF OXALIC ACID

Preparation

1. **By oxidation of vic dihydric alcohols, hydroxy aldehydes, hydroxy acids dialdehydes and aldehydic acids.**

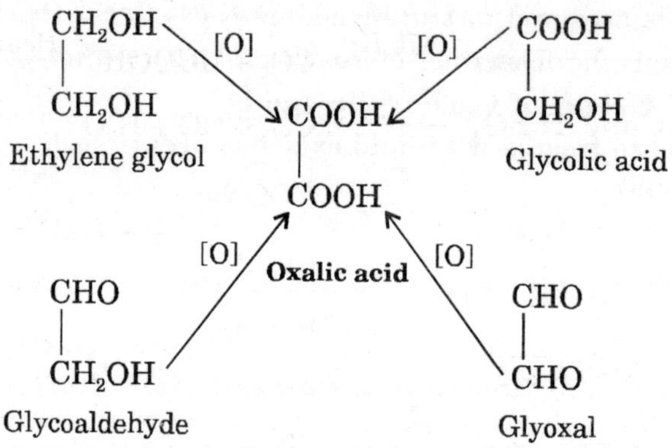

2. By hydrolysis of alkanedinitrile

$$\underset{\text{Cyanogen}}{\begin{array}{c} CN \\ | \\ CN \end{array}} \xrightarrow{4H_2O/HCl} \underset{\text{Oxalic acid}}{\begin{array}{c} COOH \\ | \\ COOH \end{array}}$$

3 By heating of cane sugar with conc. HNO_3 in the presence of V_2O_5

$$\underset{\text{Cane sugar}}{C_{12}H_{22}O_{11}} + 18[O] \xrightarrow[V_2O_5]{HNO_3} 6 \underset{\text{Oxalic acid}}{\begin{array}{c} COOH \\ | \\ COOH \end{array}} + 5H_2O$$

4. By heating sodium or potassium in current of CO_2

$$2Na + 2CO_2 \xrightarrow{633\ K} \begin{array}{c} COONa \\ | \\ COONa \end{array} \xrightarrow{HCl} \underset{\text{Oxalic acid}}{\begin{array}{c} COOH \\ | \\ COOH \end{array}} + NaCl$$

5. By heating sodium formate

$$2NaOH + 2CO \xrightarrow[6\text{-}10\ atm]{583\ K} \underset{\text{Sod. formate}}{2HCOONa} \xrightarrow{660\ K} H_2 + \begin{array}{c} COONa \\ | \\ COONa \end{array} \xrightarrow{HCl} \begin{array}{c} COOH \\ | \\ COOH \end{array} + NaCl$$

Properties

- Colourless crystalline substance with m.p. 374.1 K and b.p. 462.5 K.
- Contains 2 molecules of hydration per acid molecule ($H_2C_2O_4 . 2H_2O$)
- Loses water of crystallisation on heating to 373 K.

$$\begin{array}{c} COOH \\ | \\ COOH \end{array} 2H_2O \xrightarrow{373\ K} \begin{array}{c} COOH \\ | \\ COOH \end{array} + 2H_2O$$

- Solute in water but insoluble in ether.

- Some important chemical properties are listed below :

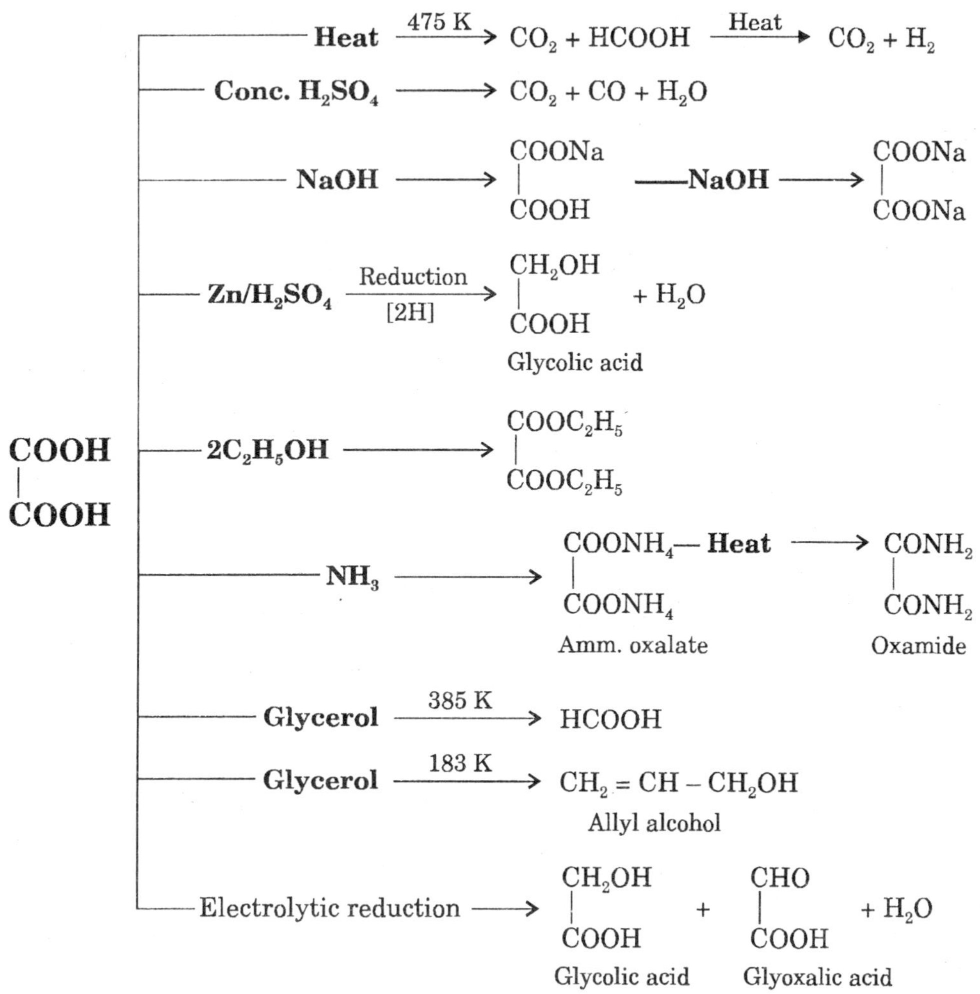

PREPARATION AND PROPERTIES OF MALONIC ACID $\left(H_2C{<}^{COOH}_{COOH}\right)$

Preparation

1. **By oxidation of malic acid with chromic acid**

$$\underset{\text{Malic acid}}{\begin{matrix} CH_2COOH \\ | \\ CH(OH)COOH \end{matrix}} + 2O \xrightarrow{CrO_3} \underset{\text{Malonic acid}}{CH_2{<}^{COOH}_{COOH}}$$

2. **By heating potassium chloro acetate with aqueous KCN followed by hydrolysis**

$$CH_2{<}^{Cl}_{COOK} + KCN \longrightarrow CH_2{<}^{CN}_{COOK} \xrightarrow{H_3O^+} CH_2{<}^{COOH}_{COOH}$$

Properties

- Colourless crystaline solid has m.p. = 408.6 K.
- Highly soluble in water and ethanol.

- Some chemical properties are listed below :

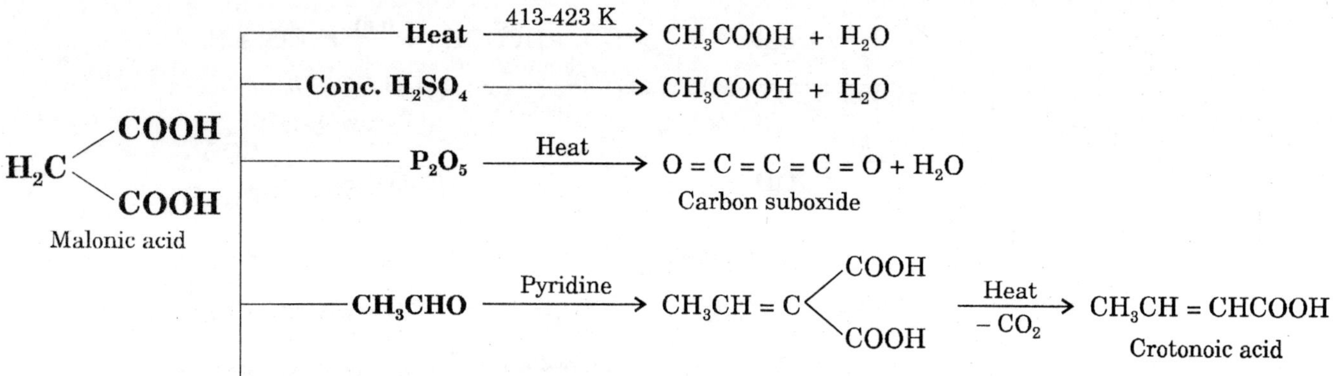

- This reaction is **Knovenagel reaction.**

Like oxalic acid it also forms, esters, amides and acid chlorides.

Blanc's Rule. This deals with effect of heat on various dicarboxylic acids.

(*a*) Dicarboxylic acids having two COOH groups on the same C atom give monocarboxylic acids on heating.

(*b*) 1, 4 and 1, 5-dicarboxylic acids give on heating cyclic anhydrides

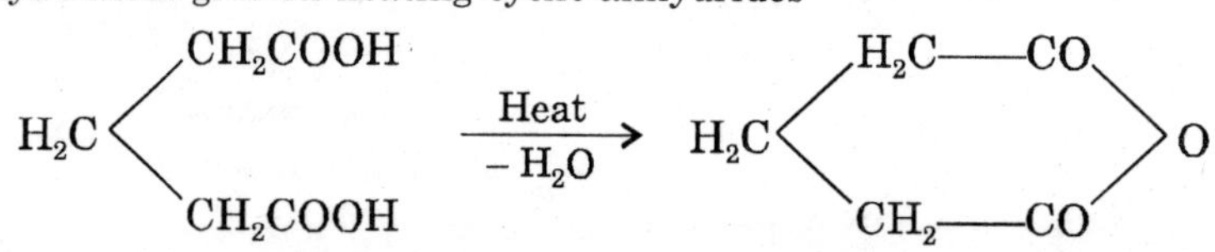

Gluteric acid Gluteric anhydride

(*c*) 1, 6 and 1, 7 dicarboxylic acids on heating with acetic anhydrides give cyclic ketones

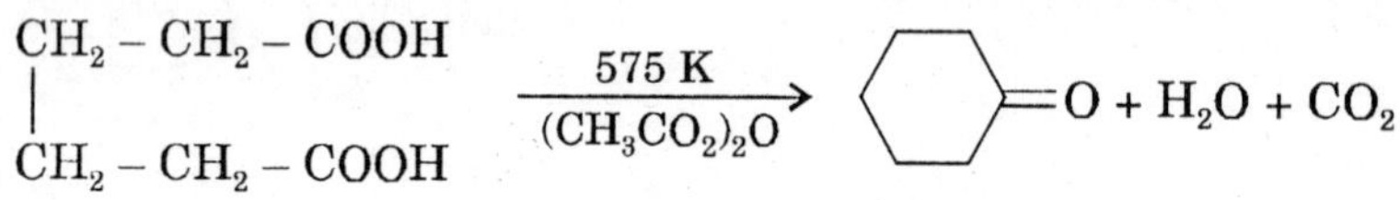

Adipic acid Cyclohexanone

SOME IMPORTANT SUBSTITUTED ACIDS

These are the compounds formed by replacing one or more H atoms in the alkyl group of carboxylic acids.

LACTIC ACID (CH_3—CH(OH)—COOH)

- It is α-hydroxy propionic acid or 2-chloropropanoic acid.
- It is found in sour milk.
- On large scale it is prepared by fermentation of molasses, cane sugar, glucose or milk sugar by the enzymes **Bacillus acid lactiti.**

$$\underset{\text{Lactose}}{C_{12}H_{22}O_{11}} + H_2O \longrightarrow \underset{\text{Lactic acid}}{4CH_3CH(OH)COOH}$$

- Due to its chiral structure, it is optically active compound.

- Some chemical properties are

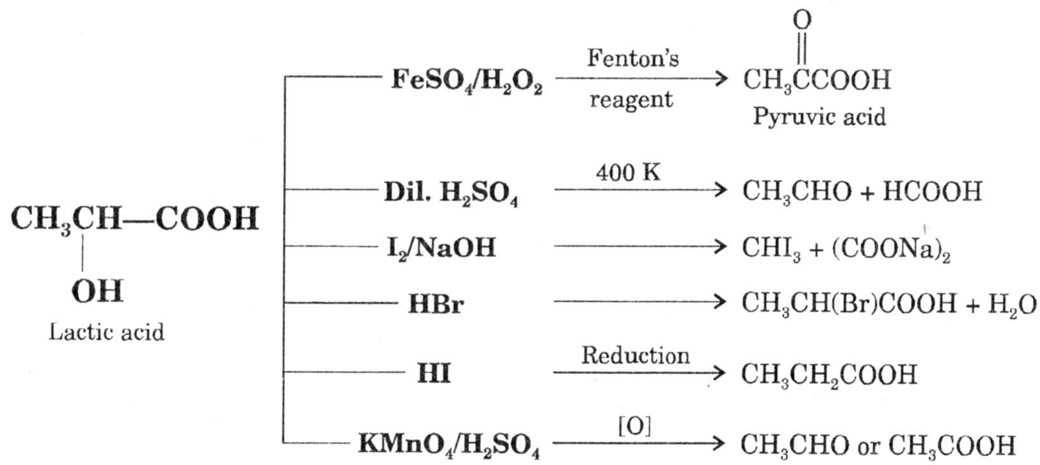

PREPARATION AND PROPERTIES OF TARTARIC ACID

- It is 2, 3-dihydroxy-1, 4-butanedioic acid.
- Present as its potassium salt in tamarind, grapes, plums and berries.
- Naturally occurring tartaric acid is (*d*) tartaric acid only.

Preparation

1. **From argol (cream of tartar) by treatment with $Ca(OH)_2$**

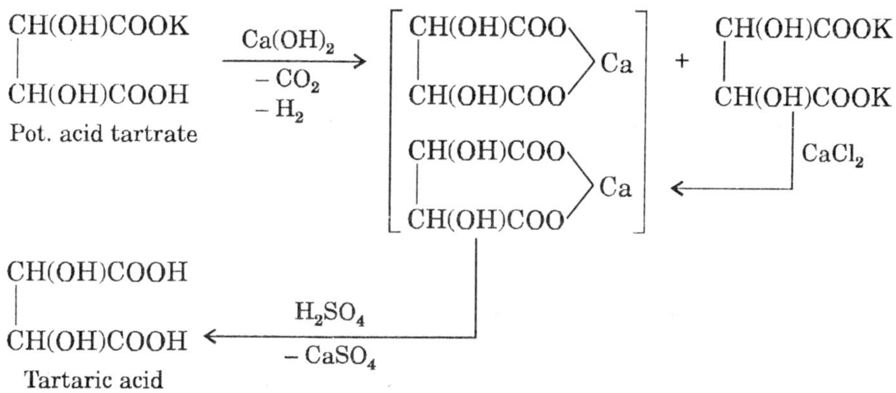

2. **By hydrolysis of glyoxal cyanohydrin**

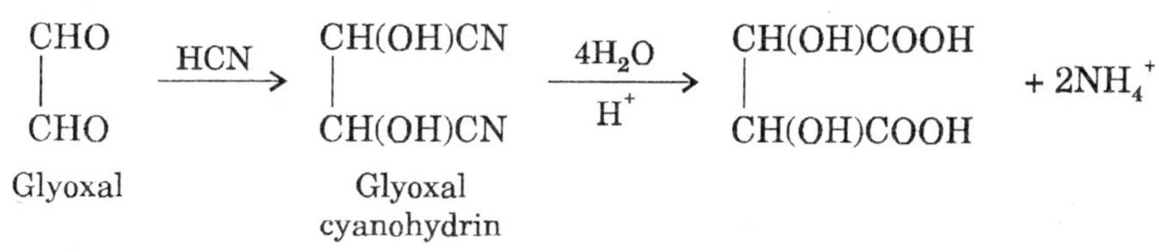

3. **By oxidation of fumeric acid with $KMnO_4$**

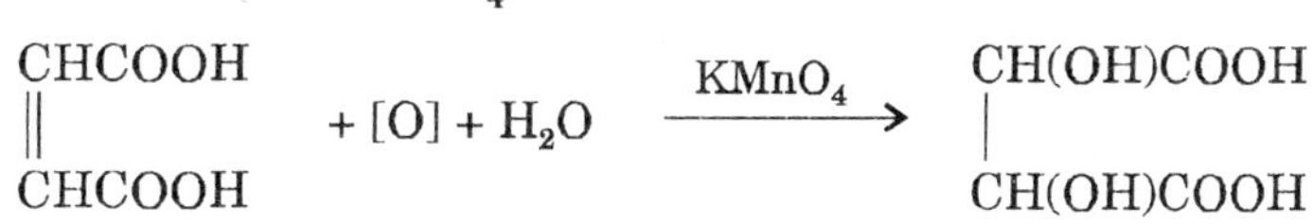

Properties

- It is dibasic acid and forms two series of salts.
- Some important salts of tartaric acid are :

CH(OH)COOH | CH(OH)COOK	CH(OH)COONa | CH(OH)COOK	CH(OH)COO(SbO) | CH(OH)COOK
Pot. hydrogen tartrate	Sod. pot. tartrate (Rochell's salt)	Pot. antimony tartrate (Tataremetic)

- Some important reactions of tartaric acid are listed below :

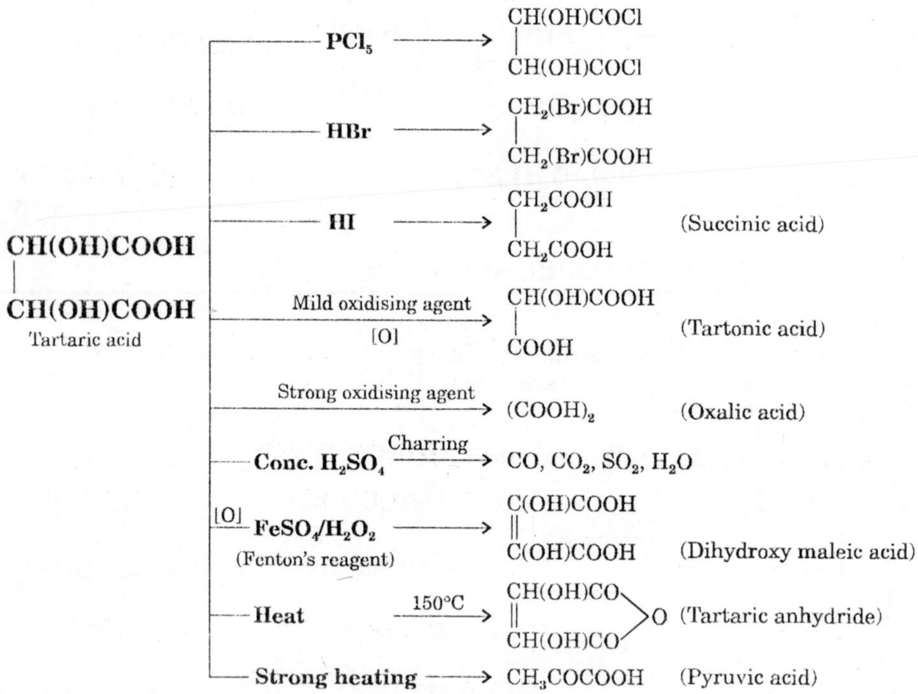

PREPARATION AND PROPERTIES OF CITRIC ACID

- It is an essential constituent of citrus fruits.
- Lemon juice contains about 10% of citric acid.
- Its IUPAC name is **2-Hydroxypropane-1, 2, 3-tricarboxylic acid.**
- It is prepared by fermentation of molasses in the presence of **citromyces pfeferiames** or **aspergillus niger** or from lemon juice.

• Synthesis from Glycerol

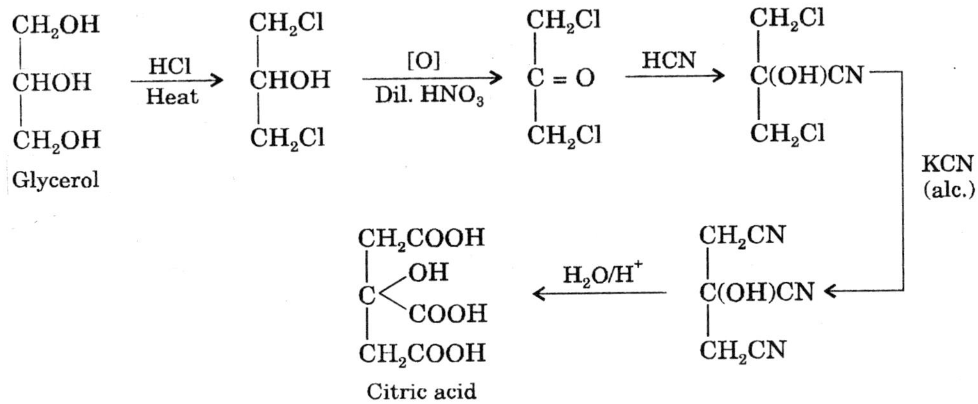

• Chemical reactions of citric acid are listed below :

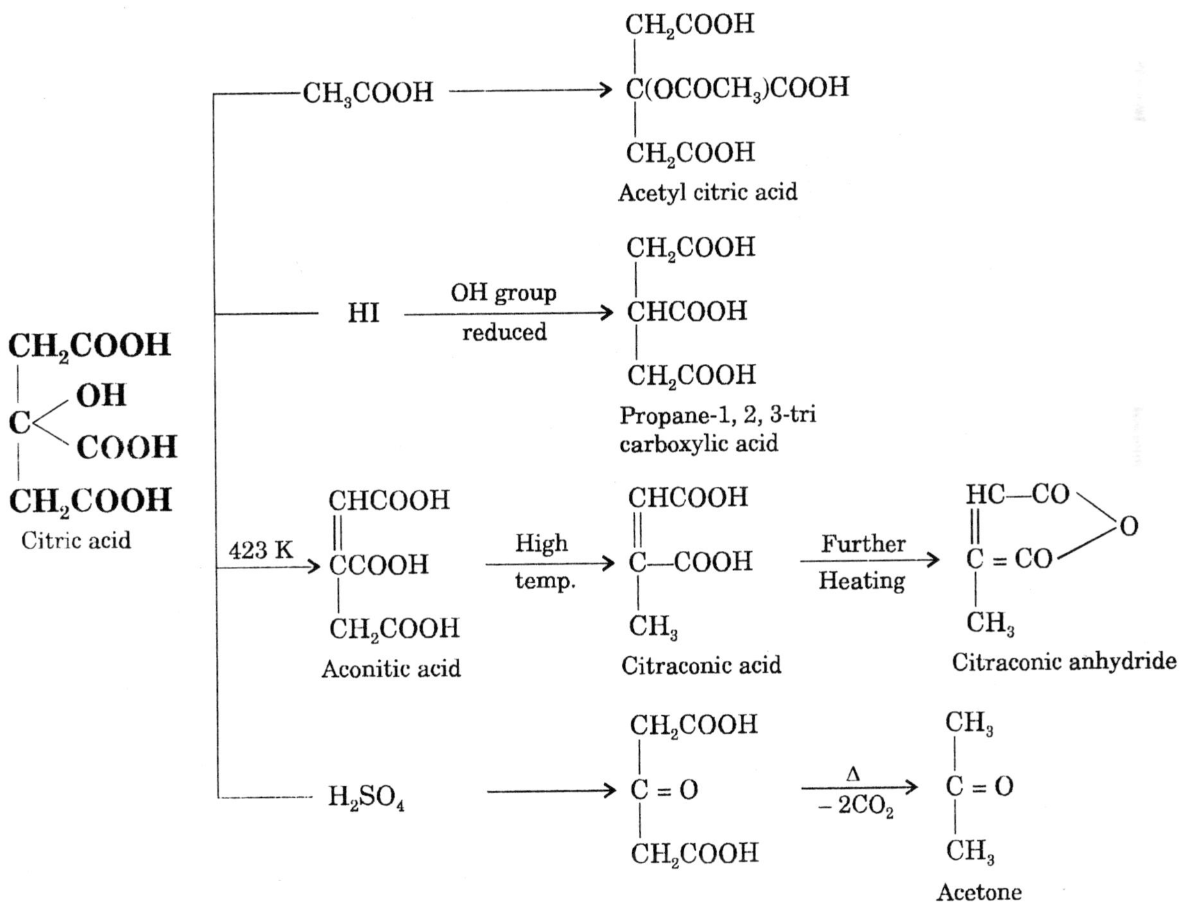

ACID DERIVATIVES

The important functional derivatives of acid are

$R-\overset{\overset{O}{\|}}{C}-\mathbf{Cl}$	$R-\overset{\overset{O}{\|}}{C}-\mathbf{O}-\overset{\overset{O}{\|}}{C}-R$	$R-\overset{\overset{O}{\|}}{C}-\mathbf{OR'}$	$R-\overset{\overset{O}{\|}}{C}-\mathbf{NH_2}$
Acid chlorides Alkanoyl chlorides	Acid anhydrides Alkanoic anhydrides	Esters Alkyl alkanoate	Amides Alkanamides

ACID CHLORIDES (RCOCl)

- According to IUPAC system they are called **alkanoyl chlorides**

Preparation

1. **From carboxylic acids.** By reaction of PCl_5 or $SOCl_2$ on carboxylic acids.

$$R—COOH + PCl_5 \longrightarrow RCOCl + POCl_3 + HCl$$
$$R—COOH + SOCl_2 \longrightarrow RCOCl + SO_2 + HCl$$

2. **From the salts of carboxylic acids.** By the action of sodium or calcium salt of acids with PCl_3 or SO_2Cl_2 or $POCl_3$.

$$3R—COONa + PCl_3 \longrightarrow 3RCOCl + Na_3PO_3$$
$$R—COONa + \underset{\text{Phosphoryl chloride}}{POCl_3} \longrightarrow 3RCOCl + NaCl + \underset{\text{Sodium metaphosphite}}{NaPO_3}$$
$$(R—COO)_2Ca + \underset{\text{Sulphuryl chloride}}{SO_2Cl_2} \longrightarrow 2RCOCl + CaSO_4$$

Physical Properties

1. Physical state, smell, etc. The lower acyl chlorides are colourless liquids while the higher members are colourless low melting solids with pungent and irritating odours. *They fume in air due to the formation of hydrochloride acid by hydrolysis.*

2. Solubility. They are readily soluble in most of the organic solvents such as ether, benzene, chloroform, etc.

3. Boiling points. Unlike carboxylic acids, acyl chlorides do not contain acidic hydrogen. Hence, they do not form *intermolecular hydrogen bonds.* Therefore, their boiling points are lower than those of their parent acids. For example, acetic acid boils at 391 K while acetyl chloride boils at 326 K. Similarly, benzoic acid boils at 523 K while benzoyl chloride boils at 470 K.

Chemical Properties

Acid chloride is a very reactive class of compound. The important chemical reactions are being summarised as follows :

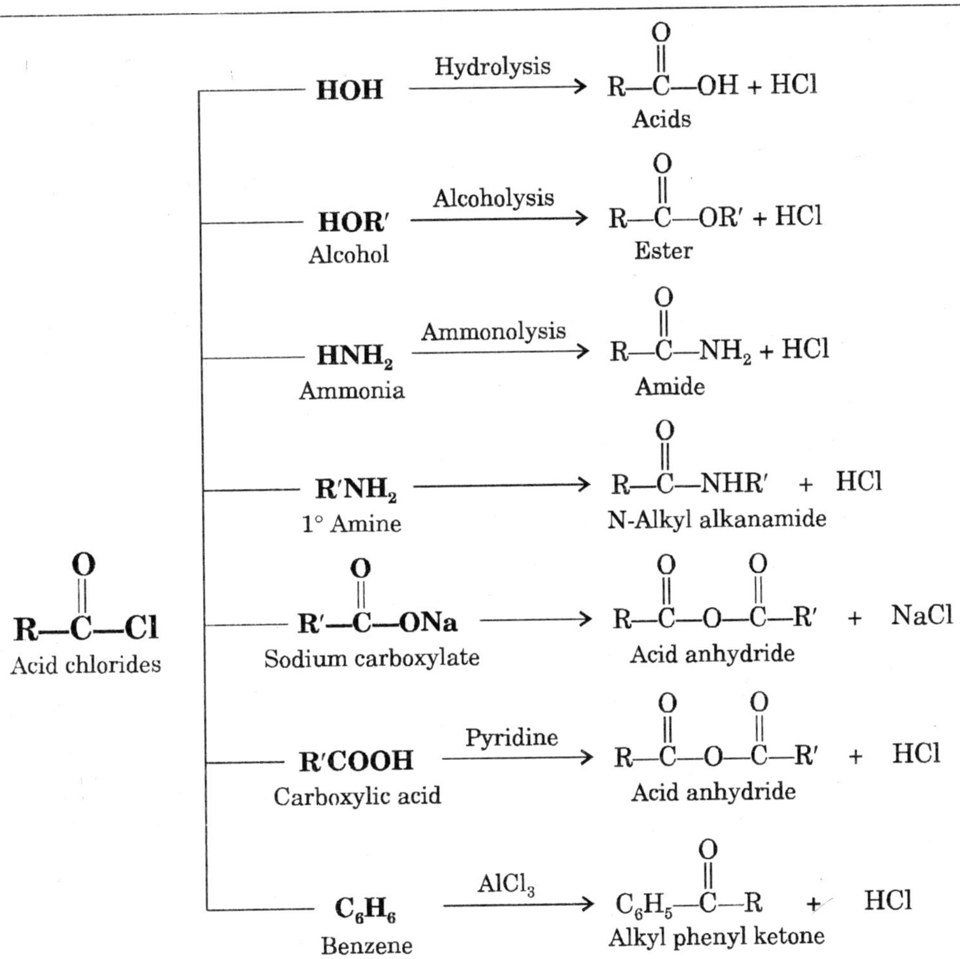

• *All the above reactions are called acylation reactions.*

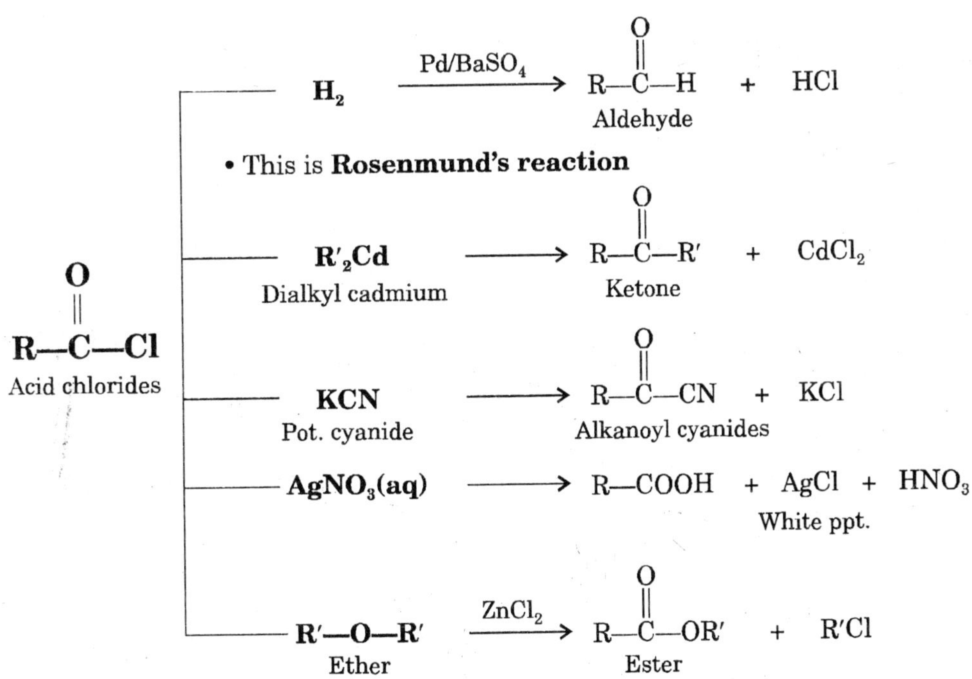

ACID ANHYDRIDES (RCO—O—COR)

Preparation

1. By heating carboxylic acids with P_2O_5

$$2R\text{—}COOH \xrightarrow{P_2O_5} R\text{—}\overset{\overset{O}{\|}}{C}\text{—}O\text{—}\overset{\overset{O}{\|}}{C}\text{—}R$$

By this method only symmetrical anhydrides are produced.

2. By treatment of carboxylic acid with acid chloride. Carboxylic acids are treated with acid chlorides in the presence of pyridine.

$$R\text{—}\overset{\overset{O}{\|}}{C}\text{—}OH + R'\text{—}\overset{\overset{O}{\|}}{C}\text{—}Cl \xrightarrow{\text{Pyridine}} R\text{—}\overset{\overset{O}{\|}}{C}\text{—}O\text{—}\overset{\overset{O}{\|}}{C}\text{—}R' + HCl$$

By this method, both symmetrical as well as unsymmetrical anhydrides can be produced.

3. By passing ketene into glacial acetic acid

$$\underset{\text{Ketone}}{CH_2 = C = O} + \underset{\text{Acetic acid}}{CH_3COOH} \longrightarrow \underset{\text{Acetic anhydride}}{CH_3\text{—}\overset{\overset{O}{\|}}{C}\text{—}O\text{—}\overset{\overset{O}{\|}}{C}\text{—}CH_3}$$

4. By reaction of acetylene and glacial acetic acid

$$\begin{matrix}CH\\ \|\!|\\ CH\end{matrix} + 2CH_3COOH \xrightarrow[Ba^{2+}]{H_2SO_4} \underset{\text{Ethylidene diacetate}}{CH_3-CH(OCOCH_3)_2} \xrightarrow[\text{—}CH_3CHO]{\text{Distill}} \underset{\text{Acetic anhydride}}{(CH_3CO)_2O}$$

Chemical properties of Acid anhydrides are listed below :

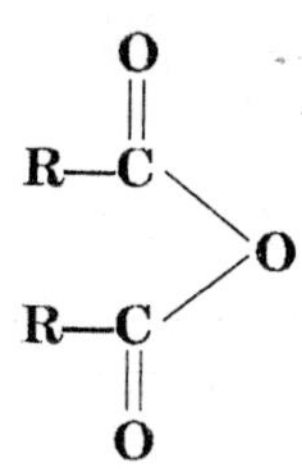

Acid anhydride

- $HOH \xrightarrow{\text{Hydrolysis}} \underset{\text{Carboxylic acid}}{2RCOOH}$
- $R'OH \xrightarrow{\text{Alcoholysis}} \underset{\text{Ester}}{RCOOR'} + RCOOH$
- $NH_3 \xrightarrow{\text{Ammonolysis}} \underset{\text{Amide}}{RCONH_2} + RCOOH$
- $R'NH_2 \longrightarrow \underset{\text{Substituted amide}}{RCONH\text{—}R'} + RCOOH$
- $C_6H_6 \xrightarrow{AlCl_3} \underset{\text{Alkyl phenyl ketone}}{C_6H_5COR} + RCOOH$
- $(i)\ LiAlH_4,\ (ii)\ H_3O^+ \longrightarrow \underset{\text{Alcohol}}{2RCH_2OH}$
- $HCl\ (\text{dry}) \longrightarrow \underset{\text{Alkanoyl chloride}}{RCOCl} + RCOOH$
- $PCl_5 \longrightarrow \underset{\text{Alkanoyl chloride}}{2RCOCl} + POCl_3$

ESTERS (R—COOR)

- These are the derivative of carboxylic acids in which the —OH part of carboxylic acids has been replaced by alkoxy part (OR′). They are named by writing the name of *alkyl* or *aryl* group of OR′ part of the molecule before the name of the parent acid and replacing the suffix **ic** of IUPAC or common name of the corresponding acid by **ate.**

Preparation

1. Acylation. Esters are prepared by the *acylation of alcohols or phenols.* The acylating agents can be any of the following :

Carboxylic acid / H_2SO_4 or *Acyl chloride* or *Acid anhydride*

$$\underset{\text{Alcohol}}{\mathrm{R'OH}} \xrightarrow[\mathrm{H_2SO_4}]{\mathrm{RCOOH}} \mathrm{R{-}\overset{\overset{\large O}{\|}}{C}{-}OR'} + \mathrm{H_2O}$$

$$\underset{\text{Alcohol}}{\mathrm{R'OH}} \xrightarrow{\mathrm{RCOCl}} \mathrm{R{-}\overset{\overset{\large O}{\|}}{C}{-}OR'} + \mathrm{HCl}$$

$$\underset{\text{Alcohol}}{\mathrm{R'OH}} \xrightarrow{\mathrm{RCO{-}O{-}COR'}} \mathrm{R{-}\overset{\overset{\large O}{\|}}{C}{-}OR'} + \mathrm{RCOOH}$$

2. By reaction of acids with diazomethane. Acids on treating with an ethereal solution of diazomethane yield **methyl esters.**

$$\underset{\text{Acid}}{\mathrm{RCOOH}} + \underset{\text{Diazo methane}}{\mathrm{CH_2N_2}} \longrightarrow \underset{\text{Ester}}{\mathrm{RCOOCH_3 + N_2}}$$

3. By Tischenko reaction. When aldehydes containing α-hydrogen atoms are treated with aluminium ethoxide. They undergo condensation to produce esters.

$$\underset{\text{Ethanal}}{\mathrm{CH_3CHO}} + \mathrm{O = CHCH_3} \xrightarrow{\mathrm{Al(OC_2H_5)_2}} \underset{\text{Ethyl acetate}}{\mathrm{CH_3{-}\overset{\overset{\large O}{\|}}{C} - OCH_2CH_3}}$$

Physical Properties

1. Esters are colourless, liquids with a characteristic fruity odour. The odour of the most of the flowers and fruits is due to esters present in them. The characteristic tastes and odours of different esters find applications in the manufacture of artificial flavouring and perfuming agents.

2. Esters are sparingly soluble in water but are quite miscible in organic solvents like alcohols and ethers. In fact, most of the esters are themselves very good solvents for plastics and nitrocellulose.

3. The boiling points of esters are always less than the corresponding carboxylic acids because esters do not form hydrogen bonds.

Chemical Properties

Some important chemical properties of esters are being given as follows :

Reactions of ester $R{-}\overset{O}{\overset{\|}{C}}{-}OR'$ (Ester):

$$R{-}\overset{O}{\overset{\|}{C}}{-}OR' + \textbf{HOH} \xrightarrow[\text{Acidic Hydrolysis}]{H_2SO_4} \underset{\text{Acid}}{R{-}\overset{O}{\overset{\|}{C}}{-}OH} + \underset{\text{Alcohol}}{R'OH}$$

$$R{-}\overset{O}{\overset{\|}{C}}{-}OR' + \textbf{NaOH (aq)} \xrightarrow[\text{Basic Hydrolysis}]{} \underset{\text{Salt of acid}}{R{-}\overset{O}{\overset{\|}{C}}{-}OH} + \underset{\text{Alcohol}}{R'OH}$$

- It may be noted that acidic hydrolysis of esters is reversible whereas basic hydrolysis of esters is irreversible and is also called **saponification.**

$$R{-}\overset{O}{\overset{\|}{C}}{-}OR' + \underset{\text{Ammonia}}{\textbf{HNH}_2} \xrightarrow{\text{Ammonolysis}} \underset{\text{Amide}}{R{-}\overset{O}{\overset{\|}{C}}{-}NH_2} + \underset{\text{Alcohol}}{R'OH}$$

$$R{-}\overset{O}{\overset{\|}{C}}{-}OR' + \underset{\text{Phosphorus pentachloride}}{\textbf{PCl}_5} \longrightarrow \underset{\text{Acid chloride}}{R{-}\overset{O}{\overset{\|}{C}}{-}Cl} + R'Cl + POCl_3$$

$$R{-}\overset{O}{\overset{\|}{C}}{-}OR' + \underset{\text{Thionyl chloride}}{\textbf{SOCl}_2} \longrightarrow \underset{\text{Acid chloride}}{R{-}\overset{O}{\overset{\|}{C}}{-}Cl} + R'Cl + SO_2$$

$$R{-}\overset{O}{\overset{\|}{C}}{-}OR' + \underset{\text{Alcohol}}{\textbf{R''OH}} \xrightarrow[\text{Base}]{\text{Acid or}} \underset{\text{Ester}}{R{-}\overset{O}{\overset{\|}{C}}{-}OR''} + R'OH$$

- In this reaction, the alkoxy part of ester and alcohols are exchanged. This is called **transesterification reaction.**

$$R{-}\overset{O}{\overset{\|}{C}}{-}OR' \xrightarrow[\text{(i) LiAlH}_4\text{, (ii) H}_2\text{O or Na/alcohol}]{4[H]} \underset{\text{Alcohols}}{RCH_2OH + R'OH}$$

- Reduction with Na/ethanol reagent is called ***Bouveault Blanc reduction.***

$$R{-}\overset{O}{\overset{\|}{C}}{-}OR' + \underset{\text{(Grignard's reagent)}}{\textbf{R''MgX}} \longrightarrow \left[R{-}\overset{OMgX}{\underset{R''}{C}}{-}OR'\right] \xrightarrow[-\,R''OMgX]{HOH} \underset{\text{Ketone}}{R{-}\overset{R''}{C}{=}O} \xrightarrow[(ii)\ H_2O]{(i)\ R'MgX} \underset{\text{Alcohol}}{R{-}\overset{R'}{\underset{R'}{C}}{-}OH}$$

In this reaction,

- ***Alkyl formates*** given ***secondary alcohols*** and ***esters of other acids*** give ***tertiary alcohols.***

Claisen condensation. This reaction is given by esters having α-H atom. Such esters on treatment with strong bases like $NaOC_2H_5$ undergo self-condensation to form β-*keto esters. Here α-C atom of one ester molecule attaches itself to carbonyl carbon of the other.*

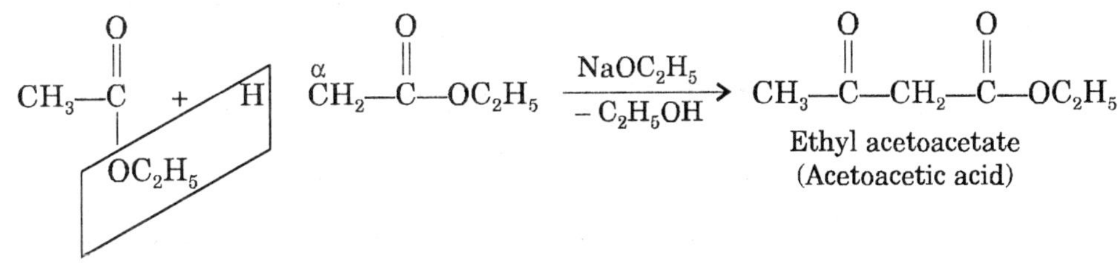

Ethyl acetoacetate
(Acetoacetic acid)

AMIDES $\left(R-\overset{\overset{\large O}{\|}}{C}-NH_2\right)$

- They may be regarded as the derivatives of carboxylic acids in which —OH part of acids in replaced by $-NH_2$. They are named by replacing suffix ***ic acid*** of common and ***oic acid*** of IUPAC name of corresponding carboxylic acids.

Preparation

1. From acid chlorides or acid anhydrides. Amides are generally prepared by reaction of *ammonia* or *amines* with *acid chlorides* or *acid anhydrides*

$$RCOCl + NH_3 \longrightarrow \underset{\text{Amide}}{RCONH_2} + HCl$$

$$\underset{\text{Acid anhydride}}{(RCO)_2O} + 2NH_3 \longrightarrow RCOONH_4 + \underset{\text{Amide}}{RCONH_2}$$

2. From ammonium salts of carboxylic acids. Amides may also be prepared by the decomposition of ammonium salts of carboxylic acids.

$$\underset{\text{Amm. Salt}}{RCOONH_4} \xrightarrow{\text{Heat}} \underset{\text{Amide}}{RCONH_2} + H_2O$$

3. From nitriles. Alkanenitriles on partial hydrolysis produce amides.

$$\underset{\text{Alkane nitrile}}{R-C\equiv N} \xrightarrow[H_2O_2/OH^-]{\text{Conc. HCl or}} \underset{\text{Amide}}{R\overset{\overset{O}{\|}}{C}NH_2}$$

Physical Properties

All amides except formamide, are crystalline solids at room temperature. They have relatively high melting and boiling points. It is due to association of amide molecules by intermolecular hydrogen bonds.

Chemical Properties

Amphoteric nature. Amides can act as very feeble acids as well as base. The amphoteric character can be explained on the basis of following resonating structures of amides.

$$\left[\underset{\text{I}}{R-\overset{\overset{:\ddot{O}}{\|}}{C}-\ddot{N}H_2} \longleftrightarrow \underset{\text{II}}{R-\overset{\overset{:\ddot{O}:^-}{|}}{C}=\overset{+}{N}H_2}\right]$$

The ***basic character*** is acounted on the basis of lone pair on N atom. However, the withdrawal of electron density towards carboxyl group greatly decreases the electron donating capacity of amides.

The ***acidic character*** can be explained on the basis of structure II which shows development of positive charge on NH_2 group. This causes the tendency of release of proton in very strong basic media.

The chemical reactions of amides have been summed up as follows :

Amide $R{-}\overset{O}{\overset{\|}{C}}{-}NH_2$ reacts with:

$$R{-}\overset{O}{\overset{\|}{C}}{-}NH_2 + HOH \xrightarrow[\text{Acidic Hydrolysis}]{HCl} R{-}\overset{O}{\overset{\|}{C}}{-}OH + NH_4Cl$$

$$R{-}\overset{O}{\overset{\|}{C}}{-}NH_2 + NaOH\,(aq) \xrightarrow{\text{Basic Hydrolysis}} \underset{\text{Sod. salt}}{R{-}\overset{O}{\overset{\|}{C}}{-}ONa} + NH_3$$

$$R{-}\overset{O}{\overset{\|}{C}}{-}NH_2 \xrightarrow[4[H]]{\text{Reduction}} \underset{1^\circ\text{ Amine}}{R{-}CH_2NH_2} + H_2O$$

(reagent: (*i*) $LiAlH_4$, (*ii*) H_3O or Na/alcohol)

- Here, the primary amine formed has same number of C atoms as in parent amide.

$$R{-}\overset{O}{\overset{\|}{C}}{-}NH_2 + P_2O_5 \xrightarrow{\text{Heat}} \underset{\text{Alkane nitrile}}{R{-}CN} + H_2O$$

$$R{-}\overset{O}{\overset{\|}{C}}{-}NH_2 + PCl_5 \text{ or } SOCl_2/P_y \longrightarrow R{-}CN + H_2O$$

$$R{-}\overset{O}{\overset{\|}{C}}{-}NH_2 + \underset{\text{Nitrous acid}}{HONO} \xrightarrow{NaNO_2/HCl} R{-}\overset{O}{\overset{\|}{C}}{-}OH + N_2 + H_2O$$

$$R{-}\overset{O}{\overset{\|}{C}}{-}NH_2 + HCl \longrightarrow \underset{\text{Amide hydrochloride}}{R{-}\overset{O}{\overset{\|}{C}}\overset{+}{N}H_3Cl^-}$$

- This reaction depicts feeble basic nature of amides.

$$R{-}\overset{O}{\overset{\|}{C}}{-}NH_2 + Na \longrightarrow \underset{\text{Sodium salt of amide}}{R{-}\overset{O}{\overset{\|}{C}}{-}\overset{-}{N}H\overset{+}{Na}} + \frac{1}{2}H_2$$

$$R{-}\overset{O}{\overset{\|}{C}}{-}NH_2 + HgO \longrightarrow (R{-}COONH)_2Hg + H_2O$$

- The above two reactions show acidic nature of amides.

$$R{-}\overset{O}{\overset{\|}{C}}{-}NH_2 + 4KOH + Br_2 \longrightarrow \underset{\text{Primary amine}}{R{-}NH_2} + 2KBr + K_2CO_3 + 2H_2O$$

- This reaction is called **Hoffmann's bromamide reaction**. Here, the primary amine formed has one carbon less than the parent amide.
- This reaction is used in step down synthesis.

PREPARATION AND PROPERTIES OF UREA ($H_2N\overset{O}{\overset{\|}{C}}NH_2$)

- It is also called **carbamide** or **aminomethanamide.**
- Urea is the first organic compound synthesized in the laboratory by Wholer by heating ammonium cyanate (1828).

$$\underset{\text{Amm. cyanate}}{NH_4CNO} \xrightarrow[\text{Rearrangement}]{\Delta} \underset{\text{Urea}}{H_2N\text{—}\overset{\overset{\displaystyle O}{\|}}{C}\text{—}NH_2}$$

- Urea is present in the urine as urea mononitrate, $H_2NCONH_2 \cdot HNO_3$

Preparation

1. **By reaction of phosgene and ammonia**

$$\underset{\text{Phosgene}}{Cl\text{—}\overset{\overset{\displaystyle O}{\|}}{C}\text{—}Cl} + 2NH_3 \longrightarrow \underset{\text{Urea}}{H_2N\text{—}\overset{\overset{\displaystyle O}{\|}}{C}\text{—}NH_2} + 2HCl$$

2. **By carbonation of ammonia**

$$O=C=O + H\text{—}NH_2 \underset{150\text{ atm}}{\overset{190°C}{\rightleftharpoons}} \underset{\text{Carbamic acid}}{O=C\begin{matrix} OH \\ NH_2 \end{matrix}} \xrightarrow{NH_3} \underset{\text{Amm. carbamate}}{O=C\begin{matrix} ONH_2 \\ NH_2 \end{matrix}} \rightleftharpoons \underset{\text{Urea}}{O=C\begin{matrix} NH_2 \\ NH_2 \end{matrix}} + H_2O$$

Properties

- White crystalline solid with no smell, m.p. = 405 K.
- It is one of the best nitrogenous fertilizer with 46.6% nitrogen.
- Its aqueous solution is neutral.
- It is a monoacid base.
- Some of the chemical properties are summarised below :

Urea, $O=C\begin{matrix} NH_2 \\ NH_2 \end{matrix}$:

- **Heat**: $\xrightarrow[-NH_3]{\text{Above 410 K}} \underset{\text{Biuret}}{H_2N\text{—}\overset{\overset{\displaystyle O}{\|}}{C}\text{—}NH\text{—}\overset{\overset{\displaystyle O}{\|}}{C}\text{—}NH_2}$

  Ž On adding a drop of $CuSO_4$ to an alkaline solution of biuret, violet appears. This is called **biuret test.** This test is given by the compounds containing —CO—NH— group *i.e., peptide linkage.*

- **$H_2SO_4 + SO_3$** (Fuming sulphuric acid) $\longrightarrow \underset{\text{Sulphomic acid}}{2NH_2\text{—}SO_3H} + CO_2$

- **$HO\text{—}C_2H_5$** (Ethanol) $\xrightarrow{\text{Heat}} \underset{\text{Urethane}}{NH_2\text{—}CONHC_2H_5} + H_2O$

- **$H_2N\text{—}NH_2$** (Hydrazine) $\xrightarrow{373\text{ K}} \underset{\text{Semi carbazide}}{H_2N\text{—}CONH\text{—}NH_2} + NH_3$

- **$CH_2\begin{matrix} COOH \\ COOH \end{matrix}$** (Malonic acid) $\xrightarrow[-H_2O]{POCl_3} \underset{\text{Barbituric acid}}{O=C\begin{matrix} NH\text{—}CO \\ NH\text{—}CO \end{matrix}CH_2}$

THE REACTIVITY OF ACID DERIVATIVES TOWARDS ACYL SUBSTITUTION

The reactivity of various functional derivatives of acids towards acyl substitution is

$$R{-}\overset{\overset{O}{\|}}{C}{-}Cl > R{-}\overset{\overset{O}{\|}}{C}{-}OCOR > R{-}\overset{\overset{O}{\|}}{C}{-}OR' > R{-}\overset{\overset{O}{\|}}{C}{-}NH_2$$

REACTIVITY DECREASES →

This can be explained as follows :

- **Basicity of leaving group (L).** Weaker the basic character of the leaving group, more will be the ease with which the bond can be ruptured and thus, more is the reactivity. Among the acid derivatives the basic character of leaving group decreases as $-\bar{N}H_2 > -\bar{O}R' > RCOO^- > Cl^-$. Hence the reactivity increases in the same order.
- **Resonance effect.** All acid derivatives show resonance as follows :

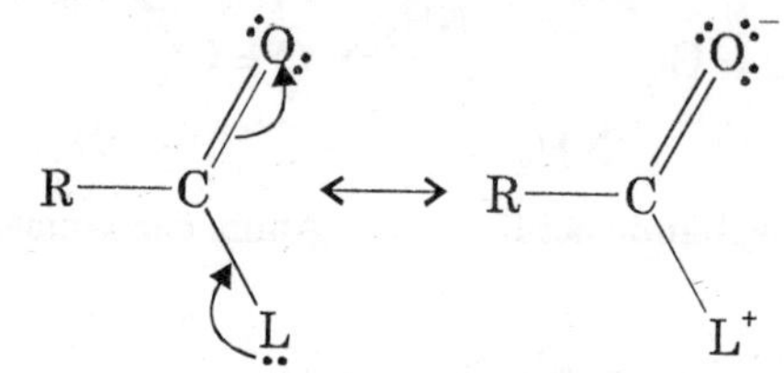

I II

L = Leaving group

It is evident that C—L bond attains double bond character. Now higher is the electronegativity of L, lesser will be the stability of structure II and larger will be its reactivity. For example Cl^- is most electronegative group. Hence acid chlorides have most reactivity.

MECHANISM OF SOME IMPORTANT REACTIONS

- **Esterification.** This is an acid catalysed reaction

$$R{-}\overset{\overset{O}{\|}}{C}{-}OH + H{-}O{-}R' \xrightarrow{H_2SO_4} \underset{\text{Ester}}{R{-}\overset{\overset{O}{\|}}{C}{-}OR'} + H_2O$$

Mechanism. $H_2SO_4 \longrightarrow H^+ + HSO_4^-$

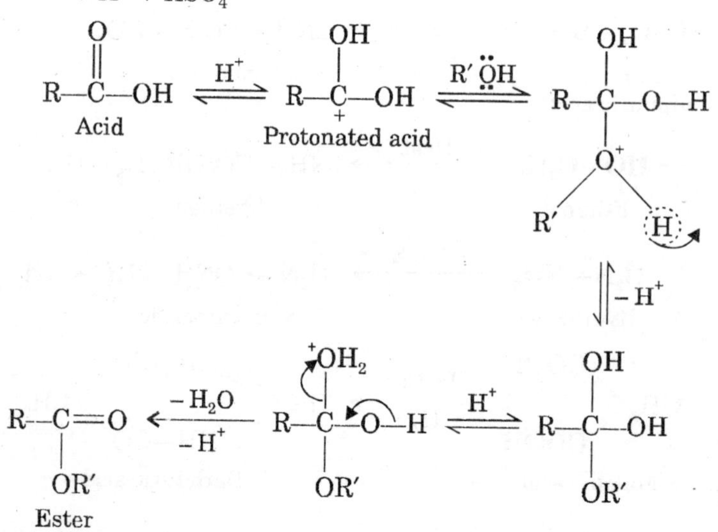

- **Base Catalysed Hydrolysis of Esters (Saponification)**

  The reaction is initiated by the nucleophilic attack of a base (OH^-) on carbonyl carbon.

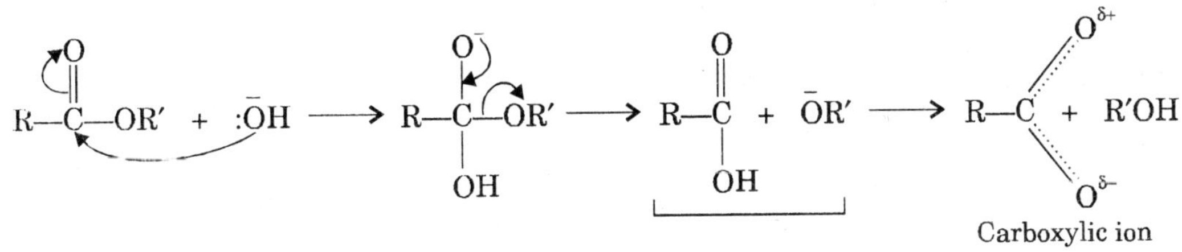

- **Hoffmann's Bromamide Reaction**

  The reaction involves a conversion of 1° amide into 1° amine

$$R-CONH_2 \xrightarrow{KOH/Br_2} RNH_2$$

Various steps are

$$2KOH + Br_2 \longrightarrow KBr + K^+(OBr)^- + H_2O$$

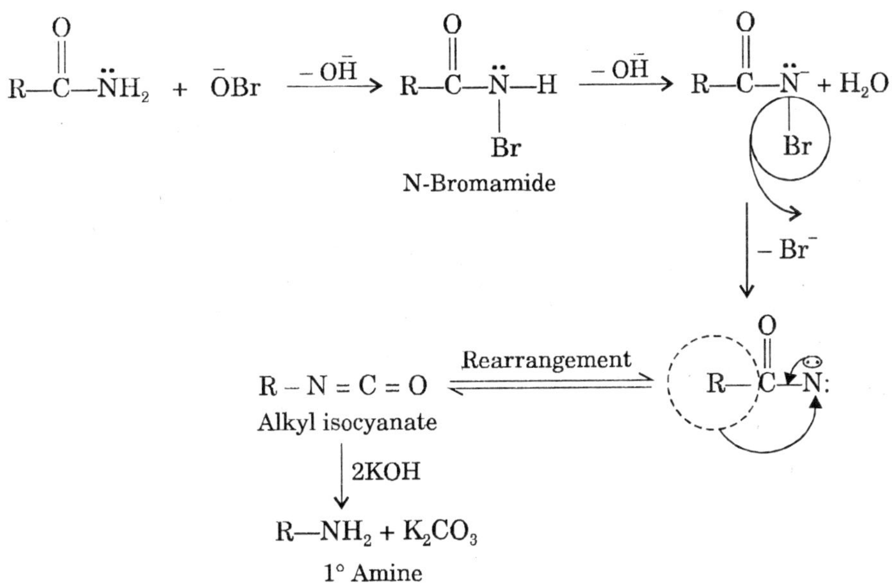

6.2. IMPORTANT FACTS AND TERMS

Add to Knowledge

- In the oxidation of alkyl benzene with alkaline $KMnO_4$, benzoic acid is always formed irrespective of chain length.
- Methyl alcohol reacts with CO in presence of rhodium to form acetic acid.

$$CH_3OH + CO \xrightarrow{\text{Rhodium Catalyst}} CH_3\text{—}COOH$$

- Vinegar is 6-10% solution of acetic acid. In Quick Vinegar process, it is prepared by oxidation of wort (fermented liquor obtained by fermentation of molasses, starch etc.) by means of bacteria, *Mycoderma aceti* or *Acetobactor aceti*.

$$CH_3CH_2OH + O_2 \xrightarrow{\text{Acetobacter aceti}} CH_3COOH + H_2O$$

- The melting point of carboxylic adds show alternation effect or oscillation effect, *i.e.*, melting point of even number acid is always higher than the next lower and higher odd number acid. This is because even number acids are more symmetrical and fit well into the crystal lattice and hence more energy is required to break.
- Formic acid was first obtained by the distillation of red ants and hence the name (*Latin* : *Formica* means ants).
- Below 16.6°C pure acetic acid is converted into ice like solid and hence is called glacial acetic acid.
- Malic acid is found in apples, grapes etc.
- Carboxylic acids are stronger acids as compared to phenols or alcohols. Carboxylic acids are stronger acids than phenols since carboxylate ion is better stabilized by resonance than phenoxide ion.
- Acid strength increases by the presence of a group with – I effect in the alkyl group.
- Acid strength decreases by the attachment of a group with + I effect in the alkyl part.

(a) Decreasing order of acid strength is given below :

(*i*) $HCOOH > CH_3\text{—}COOH > (CH_3)_2\text{—}CH\text{—}COOH > (CH_3)_3C\text{—}COOH$

(Acidity decreases as the + I-effect of the alkyl group increases).

(*ii*) $O_2NCH_2COOH > FCH_2COOH > ClCH_2COOH > BrCH_2COOH$

(Acidity decreases as the – I-effect of the group or halogen decreases).

(*iii*) $CH_3CH_2 - C(Cl)_2 - COOH > CH_3 - CH(Cl)CH(Cl)COOH > ClCH_2 - CH(Cl)CH_2COOH$

(Acidic strength decreases as the distance of the halogen from the carboxyl group increases.)

(*iv*) $F_3C\text{—}COOH > Cl_3C\text{—}COOH > Br_3C\text{—}COOH$

(Acidic strength decreases as the electro negativity of halogen atoms decreases).

(b) Effect of substitutents on the acid strength of aromatic acids.

(*i*) Electron-donating substituents like —R, —OH, —NH_2, etc. tend to decrease while electron-withdrawing substituents tend to increase the acid strength of substituted benzoic acids relative to benzoic acid ; the effect of these substituents being more pronounced at *p*-than at *m*-positions.

(*ii*) *o*-substituted benzoic acids are usually stronger acids than benzoic acid regardless of the nature of the substituent-ortho effect.

In the light of the above facts, the acid strength decreases as :

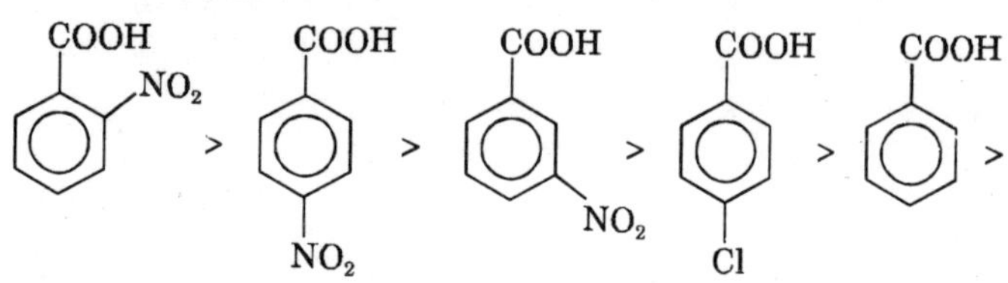

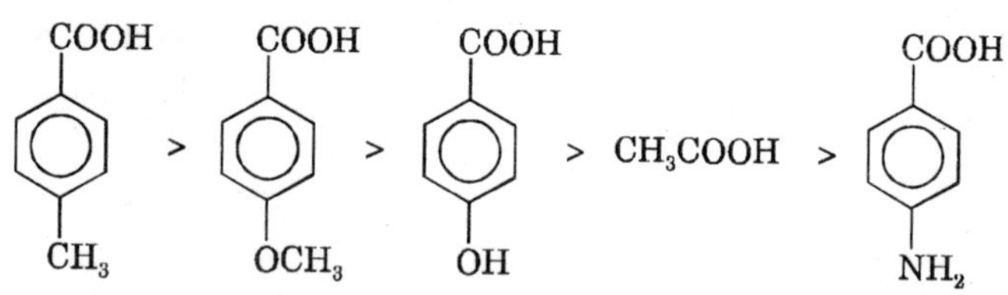

- Carboxylic acids do not give reactions of carbonyl-group as this group enters into resonance with the lone pairs of —OH group.
- $CuO.CuCr_2O_4$ (Copper chromite) is known as Adkin's catalyst.
- $CH_3\text{-}CH_2\text{—}COOH \xrightarrow{Br_2/\text{Red } P_4} CH_3\text{—}CH(Br)\text{—}COOH.$

The resulting bromo acid can be converted to a variety of compounds by using different reagents :

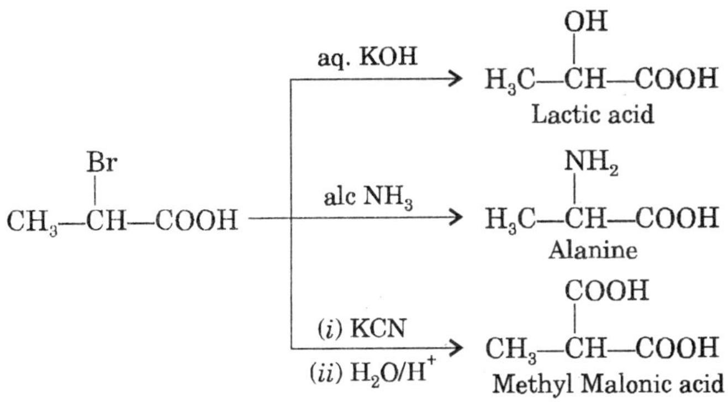

- Carboxylic acids form stronger H-bonds and hence their boiling points are much higher than those of alcohols with comparable molecular masses. Most of the carboxylic acids in the solid, liquid or even in the vapour state exists as cyclic dimers.
- Carboxylic acids do not give the characteristic reactions of the carbonyl group. The reason being that due to resonance, the double bond character of the C = O bond in carboxylic acids is greatly reduced as compared to that in aldehydes and ketones. Consequently, the carbonyl group in carboxylic acids is not a true carbonyl group and hence does not give some of the characteristic reactions of the carbonyl group such as oxime, hydrazones and semicarbazone formation.
- Formic acid acts as reducing agent due to —CHO group. It reduces Tollen's reagent, Fehling solution, mercuric chloride and $KMnO_4$.

$$HCOOH + Ag_2O \longrightarrow CO_2 + H_2O + \underset{\text{(Silver mirror)}}{2Ag}$$

$$2CuO \longrightarrow CO_2 + H_2O + \underset{\text{(Red ppt.)}}{Cu_2O}$$

$$2HCOOH + HgCl_2 \longrightarrow 2CO_2 + 4HCl + \underset{\text{(Black)}}{2Hg}$$

$$5HCOOH + 2KMnO_4 + 3H_2SO_4 \longrightarrow K_2SO_4 + 2MnSO_4 + 5CO_2 + 8H_2O$$

- Sodium formate on decarboxylation with soda lime gives H_2I. The boiling points of acid chlorides and esters are lower than those of corresponding acids because of absence of hydrogen bonding in their molecules.
- The boiling and melting points of amides are higher than those of corresponding acids from which they are derived due to strong intermolecular hydrogen bonding in their molecules.
- The boiling points of acid anhydrides are higher than the acids from which they have been derived because of stronger van der Waal's forces of intermolecular attraction owing to the comparatively large size of their molecules.
- The reactivity of various acid derivatives towards acyl substitution is

  $RCOCl > (RCO)_2O > RCOOR' > RCONH_2$
- Claisen condensation

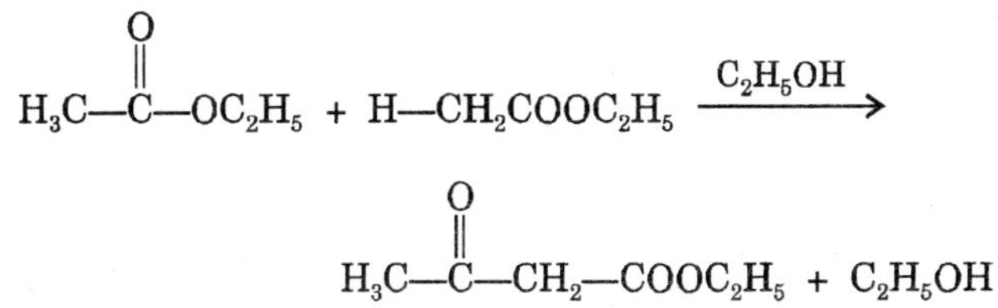

- Formamide is a liquid while all other amides are colourless crystalline solids at room temperature.
- Urea is hydrolysed in the presence of enzyme *urease* (present in soil and soyabeans) forming CO_2 and NH_3. Enzymatic hydrolysis of urea is a quantitative reaction and hence used for estimating urea by measuring the amount of ammonia liberated.
- Smell of ammonia observed in public urinals is due to hydrolysis of urea of urine by urease present in atmosphere.
- Rochelle salt (sod. pot. tartarate) is used as complexing agent in Fehling solution.
- Magnesium citrate is used as an antacid.
- Succinic acid was prepared by the distillation of amber (*Latin* : *succinium* means *amber*).
- **Blanc rule.** According to this rule, dicarboxylic acids 1, 4-and 1, 5-dicarboxylic acids on heating or on distillation with acetic anhydride from cyclic anhydrides while 1, 6-and 1, 7-dicarboxylic acids on distillation with acetic acid anhydride form cyclic ketones.
- Malonic acid on heating with P_2O_5 forms carbon suboxide.

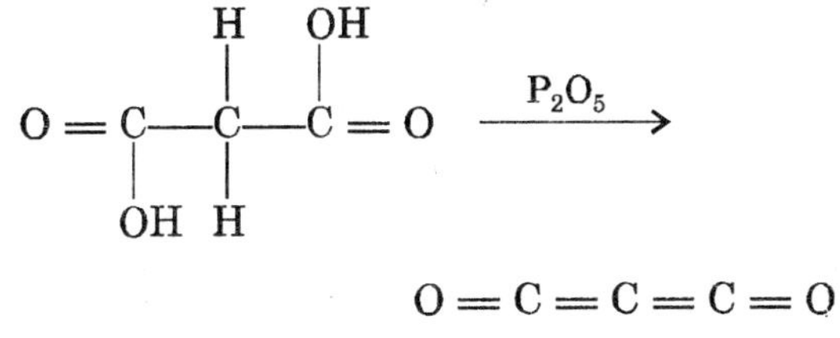

- Tartaremetic is used to produce nausea and vomitting in the treatment of poisoning.

- Cream of tartar (pot. hydrogen tartarate) is used in baking powder.
- The boiling points of acid derivatives follow the order :

  $RCONH_2 > (RCO)_2O > RCOOH > RCOCl$ as explained below :

  (*i*) The boiling points of acid chloride and esters are lower than those of their parent acids due to absence of H-bonding in their molecules.

  (*ii*) The boiling points of acid anhydrides are higher than those of the acids from which they are derived because of stronger van der Waal's forces of attraction owing to the larger size of their molecules.

  (*iii*) The melting points and boiling points of acid amides are much higher than those of acids from which they are derived due to strong intermolecular H-bonding even though their molecular masses are almost identical.
- The reactivity of acid derivatives towards nucleophilic acyl substitution reactions decreases in the order :

  $RCOCl > (RCO)_2O > RCOOR' > RCONH_2$, as the basicity of the leaving group increases *i.e.*, $Cl^- < RCOO^- < RO^- < NH_2^-$.
- Aromatic acid chlorides are less reactive than aliphatic acid chlorides primarily due to greater electron-donating effect of the benzene ring over alkyl group which tends to reduce the electron deficiency of aromatic acyl carbon.
- The reaction of aromatic acid chlorides with alcohols, phenols and amines (1° or 2°) to form the corresponding esters or amides is called Schotten-Baumann reaction.
- During esterification reaction, the reactivity of the acid or the alcohol decreases according as the size of the acid or the alcohol increases. For example :

  (*i*) Reactivity of alcohols : $CH_3OH > CH_3CH_2OH > (CH_3)_2CHOH > (CH_3)_3COH$

  (*ii*) Reactivity of acids : $HCOOH > CH_3COOH > (CH_3)_2CHCOOH > (CH_3)_3CCOOH$
- Pthalic acid and succinic acid on heating lose a molecule of water to form their corresponding anhydrides which when heated with ammonia to form phthalimide and succinimide respectively.
- Phthalimide and succinimide on treatment with Br_2-KOH, undergo Hoffmann bromamide reaction to form anthranilic acid and β-aminopropionic acid respectively.
- 2% Urea acts as a mono acidic base and reacts with conc. HNO_3 and saturated solution of oxalic acid to form urea nitrate and urea oxalate respectively.

6.3. SOLVED PROBLEMS

Enrich your Skill

Example 1. *Alkaline hydrolysis of CH_3CN forms*

(*a*) CH_3COOH (*b*) CH_3COO^-

(*c*) $HCOO^-$ (*d*) $HCOOH$.

Solution. In alkaline solution the cyanide hydrolysis to produce $CH_3COO^- Na^+$.

Hence the answer is **(*b*).**

Example 2.

$$\underset{\text{(monobasic acid)}}{X} \xrightarrow[NH_3]{\Delta} Y \xrightarrow[KBrO]{\Delta} Z \xrightarrow[0°C]{NaNO_2/HCl,} (CH_3)_3C—OH,$$

X is :

(*a*) $(CH_3)_3C—COOH$ (*b*) $(CH_3)_3C—CH_2OH$

(*c*) $(CH_3)_3C—CONH_2$ (*d*) $(CH_3)_2CH—COOH$.

Solution. The sequence of reaction is

$$(CH_3)_3C—COOH \xrightarrow{NH_3,\ \Delta} (CH_3)_3C—CONH_2$$

$$\xrightarrow{KBrO,\ \Delta} (CH_3)_3C—NH_2 \xrightarrow{HNO_2} (CH_3)_3C—OH.$$

Hence answer is **(*a*).**

Example 3. *Which is correct alternate ?*

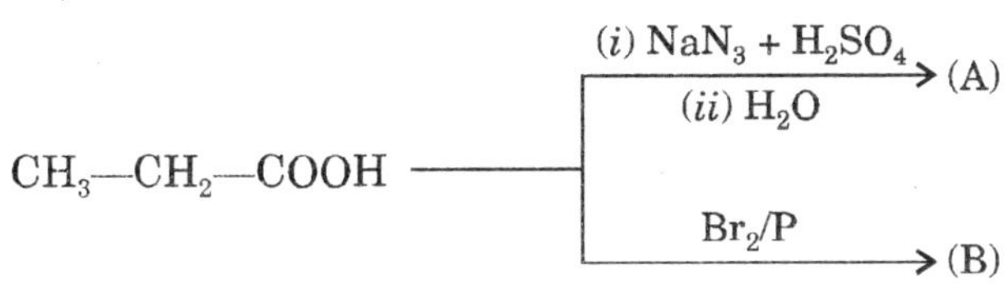

(*a*) $CH_3CH_2NH_2$ (*A*) ; $CH_3CH(Br)COOH$ (*B*)

(*b*) $CH_3CH_2CONH_2$ (*A*) ; CH_3CH_2COBr (*B*)

(*c*) $CH_3CH_2NH_2$ (*A*) ; CH_3CH_2COBr (*B*)

(*d*) *None is correct.*

Solution. NaN_3 and H_2SO_4 produces N_3H (hydrazoic acid). This reacts with acid to produce azide which rearrange to give isocyanate whose hydrolysis gives amine. It is curtius reaction or rearrangement

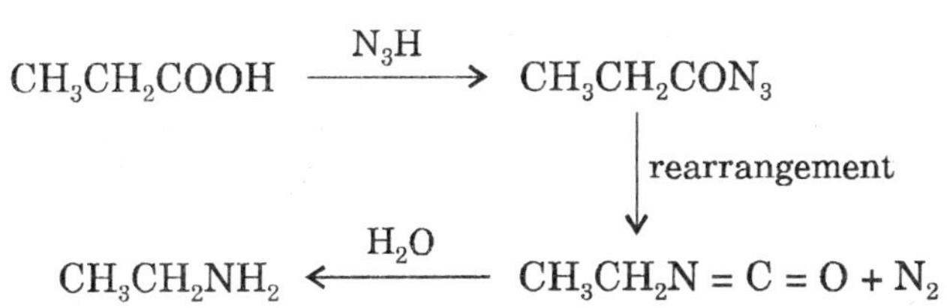

Hence the answer is **(*a*).**

Example 4. *Maximum dehydration to form α, β unsaturated acid takes place in*

(*a*) *Lactic acid*

(*b*) *Glycolic acid*

(*c*) *3-hydroxy propanoic acid*

(*d*) *Alanine.*

Solution. The OH group on the β-carbon can readily undergo dehydration as compared to OH group on the α-carbon in acids.

Hence answer is **(*c*).**

Example 5. $CH_3—CO—\overset{*}{O}H + CH_3CH_2OH \xrightarrow{H^+,\ \Delta}$ *ester (A) + Water (B). A and B are*

(*a*) $CH_3—CO—OC_2H_5,\ H_2\overset{*}{O}$

(*b*) $CH_3—CO—\overset{*}{O}C_2H_5,\ H_2.$

(*c*) *Both are correct*

(*d*) *None is correct.*

Solution. In esterification reaction the —OH bond of acids cleave.

Hence the answer is **(*a*).**

Example 6. $CH_3—CO—Cl + Nu^- \rightarrow CH_3—CO—Nu + Cl^-$

Reactivity order of different nucleophiles (NH_2^-, CH_3COO^-, OH^-) is in order :

(*a*) $NH_2^- < CH_3COO^- < OH^-$

(*b*) $CH_3COO^- < OH^- < NH_2^-$

(*c*) $NH_2^- < OH^- < CH_3COO^-$

(*d*) $CH_3COO^- < NH_2^- < OH^-$.

Solution. More basic is the nucleophile greater is its reactivity.

Hence the answer is **(*b*).**

Example 7. *The compounds A and B in the reaction given below are :*

$$H_3C—\overset{\overset{\displaystyle O}{\|}}{C}—C_2H_5 \xrightarrow{C_6H_5CO_3H} A \xrightarrow{\text{Tischenko reaction}} B$$

(*a*) $CH_3COOC_2H_5,\ CH_3CHO$

(*b*) $CH_3COOC_2H_5,\ CH_3CH_2CHO$

(*c*) $CH_3CH_2COOCH_3 > CH_3CH_2CHO$

(*d*) *None is correct.*

Solution. Ketones react with peracids to form esters. (Baeyer-Villiger reaction).

In Tischenko reaction, 2 moles of acetaldehyde will yield the same ester.

Hence answer is **(*a*)**.

Example 8. *Ease of esterification of following acids with* CH_3OH *:*

$HCOOH$ (I), CH_3COOH (II), CH_3CH_2COOH (III)

(*a*) III < II < I (*b*) I < II < III

(*c*) II < I < III (*d*) *equal.*

Solution. The esterification reaction involves the cleavage of — OH bond of acids. More is the + I effect, ease is with which this bond cleaves.

Hence the answer is **(*a*)**.

Example 9. *Ease of esterification of following alcohols with HCOOH :*

CH_3CH_2OH (I), $(CH_3)_2CHOH$ (II), $(CH_3)_3COH$ (III)

(*a*) I < II < III (*b*) III < II < I

(*c*) II < I < III (*d*) *equal.*

Solution. Alcohols involve the cleavage of O—H bond of alcohols. Greater the + I effect, less is the ease with which O—H bond cleaves.

Hence the answer is **(*b*)**.

Example 10. $CH_3COOH + HCOOH \xrightarrow{MnO,\ 300°C} ?$

Main product of this reaction is

(*a*) CH_3CHO (*b*) CH_3COCH_3

(*c*) $HCHO$ (*d*) $(CH_3CO)_2O$.

Solution. On heating in the presence of MnO at 300°C, decarboxylation occurs producing an aldehyde.

Hence answer is **(*a*)**. See concept oriented details.

Example 11. $CH_3CH_2ONa \xrightarrow[(ii)\ H^+]{(i)\ CO,\ \Delta} A$;

$$CH_2 = CH_2 + CO + H_2O \xrightarrow[400°C]{H_3PO_4} B$$

A and B are

(*a*) CH_3CH_2COOH *in both cases*

(*b*) CH_3CH_2CHO *in both cases*

(*c*) CH_3CH_2COOH, CH_3CHO

(*d*) CH_3CHO, CH_3COOH.

Solution. Heating alkoxides with CO produces an acid. In the second case the reaction is called oxo process. See concept oriented details.

Hence answer is **(*a*)**.

Example 12. $R—\overset{\overset{O}{\|}}{C}—R' \xrightarrow[C_6H_5CO_3H]{H_2SO_5 \text{ or}} A$. *Product A in the reaction and the name of the reaction are*

(*a*) R—CO—O—R′, *Baeyer-Villiger oxidation*

(*b*) RCOOH, *Baeyer oxidation*

(*c*) RCOOH, R′OH, *trans-esterification*

(*d*) *None is correct.*

Solution. Reaction of ketones with peracids produces esters.

Hence the answer is **(*a*)**.

Example 13. *In the compounds*

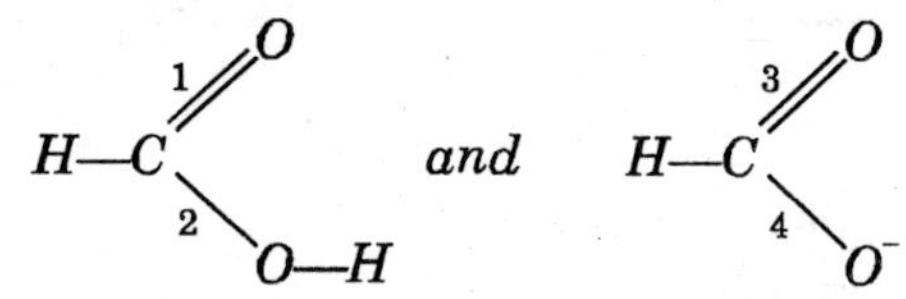

(I) (II)

(C—O) bond lengths designated by 1, 2, 3 and 4 are in order

(*a*) 1 = 3 < 2 = 4 (*b*) 1 < 2 < 3 = 4

(*c*) 1 < 3 = 4 < 2 (*d*) *all are equal.*

Solution. The two C—O bond lengths in formic acid are different but same in the case of formate ion. For details see concept oriented details.

Hence answer is **(*c*)**.

Example 14. $R—CO—NH_2 + KOH + Br_2 \rightarrow$ *Intermediates of this reaction are except*

(*a*) R—N = C = O (*b*) R—CO—N̈ :

(*c*) R—CO—NH—Br (*d*) Both (*a*) and (*c*).

Solution. It is Hoffmann's Bromamide reaction. See summary for details.

The answer is **(*b*)**.

Example 15. $CH_3CHO \xrightarrow{Al(OC_2H_5)_3} A \xrightarrow{C_2H_5ONa} B$,

A and B are

(*a*) $CH_3COOC_2H_5$, $CH_3COCH_2COOC_2H_5$

(*b*) CH_3COOH, $CH_3COOC_2H_5$

(*c*) CH_3COOH, $C_2H_5OH^-$

(*d*) *All are incorrect.*

Solution. The correct answer is **(*a*)**.

Example 16. *Which will show Hoffmann degradation reaction ?*

(*a*) NH_2CONH_2 (*b*) CH_3CONH_2

(*c*) $CH_3—CO—CH_2NH_2$ (*d*) $H—CO—NH_2$.

Solution. Only amides give this reaction.

Hence answer is **(*b*)**.

Example 17. *End product of the following sequence of reactions is*

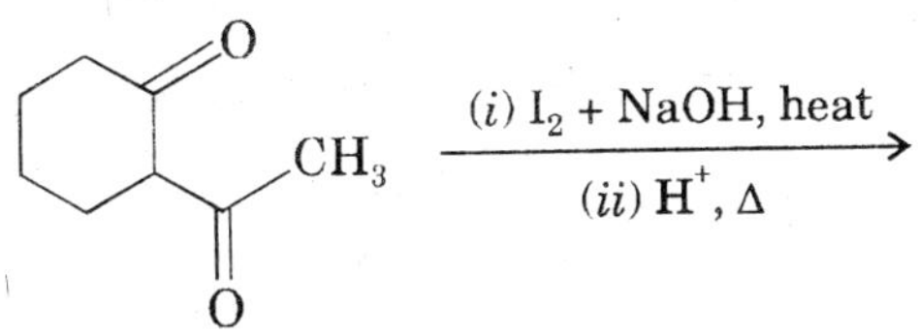

(*a*) *Yellow ppt. of* CHI_3, 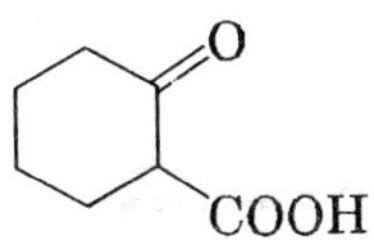

(*b*) *yellow ppt. of* CHI_3, 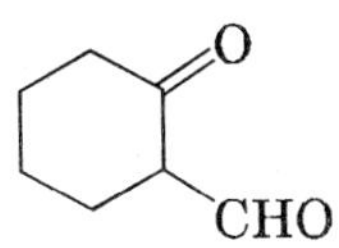

(*c*) *Yellow ppt. of* CHI_3, 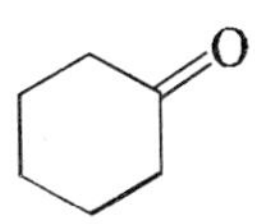

(*d*) *yellow ppt. of* CHI_3, 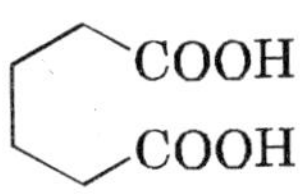

Solution. Compounds with $CH_3-\overset{\overset{\large O}{\|}}{C}-$ type of groupings produce iodoform on reaction with I_2 and NaOH. On heating the resulting acid (β-keto acid) loses CO_2.

Hence the answer is **(*c*)**.

Example 18. *Peroxy acetic acid* (CH_3CO_3H) *is a weaker acid than acetic acid* (CH_3CO_2H) *since*

(*a*) *Negative charge in* CH_3COO^- *can't be delocalized into the carbonyl group*

(*b*) CH_3 *group in* CH_3CO_3H *shows I-effect*

(*c*) *Both are correct*

(*d*) *None is correct.*

Solution. Answer is **(*a*)**.

Example 19. *For the following acids,*
$(CH_3)_3CCH_2CO_2H$ *(I)*, $(CH_3)_3NCH_2CO_2H$ *(II)*, CH_3COOH *(III)*; pK_a *values will be in order :*

(*a*) $I > II > III$ (*b*) $I < II < III$

(*c*) $I > III > II$ (*d*) $II > III > I$.

Solution. $pK_a = -\log k_a$, lower the value of pK_a more stronger is an acid.

Hence answer is **(*c*)**.

Example 20. *Strongest acid out of*

CH_3CO_2H, $CH_2 = CHCO_2H$, 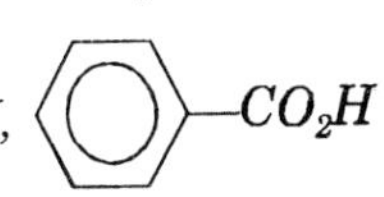

(*a*) CH_3CO_2H (*b*) $CH_2 = CHCO_2H$

(*c*) 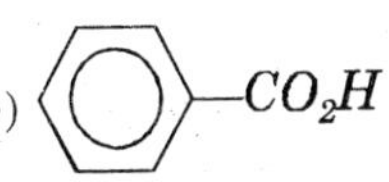 (*d*) *equal* K_a.

Solution. Benzoic acid gets resonance stabilised after losing H^+ ion. Hence it is the strongest acid.

Hence the answer is **(*c*)**.

- Carboxylic acids do not give reactions of carbonyl-group as this group enters into resonance.

6.4. QUESTION BANK

Develop Confidence

IUPAC names of Isomerism

1. The formula AcOAc represents
   (a) Ester (b) Ether
   (c) Carboxylic acids (d) Acid anhydride.
2. The IUPAC name of the acid which was originally obtained by the distillation of red ants is
   (a) Acetic acid (b) Palmitic acid
   (c) Methanoic acid (d) Ethanoic acid.
3. Which of the following is not a derivative of carboxylic acids ?
   (a) Alkanoyl chloride (b) Alkanamide
   (c) Alkyl alkanoate (d) Alkanals.
4. The IUPAC name of $BrCH_2C(Br)(CH_3)COONa$ is
   (a) Sodium dibromo isopropionate
   (b) Sodium, 2, 3-dibromo-2-methylpropanoate
   (c) Dibromo sodiumisobutanoate
   (d) None of the above.
5. 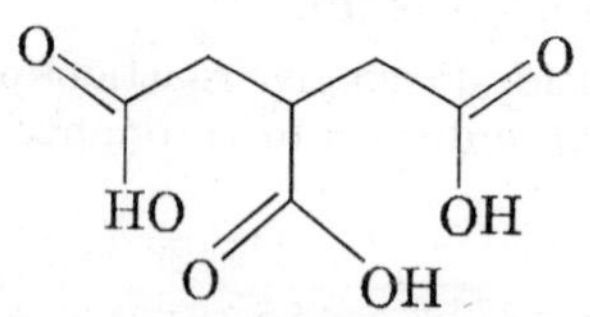
   The IUPAC name of the above acid is
   (a) 1, 2, 3-propane tricarboxylic acid
   (b) 3-hydroxy-1, 5-pentanedioic acid
   (c) 3-hydroxy-3-keto-1, 5-pentanedioic acid
   (d) 1, 3, 5-trihydroxypentane-1, 5-dione.
6. Which of the following is α-hydroxy acid ?
   (a) Stearic acid (b) Lactic acid
   (c) Picric acid (d) Isobutyric acid.
7. The IUPAC name of 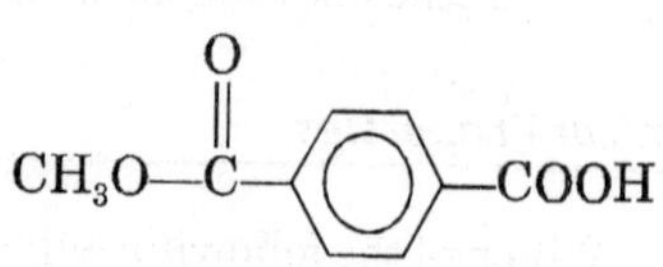 is
   (a) 4-Methoxybenzoic acid
   (b) 4-Carboxy-methylbenzoate
   (c) 4-Methoxycarbonyl-benzoic acid
   (d) None of these.
8. Which acid derivative is a functional isomer of carboxylic acids ?
   (a) Acid chlorides (b) Esters
   (c) Acid amides (d) Acid anhydrides.
9. Which of following acid can show optical isomerism ?
   (a) 2, 2-Dimethylpropanoic acid
   (b) 2-Methylpropanoic acid
   (c) 2-Methylbutanoic acid
   (d) Ethanoic acid.
10. Which of the following is p-anisic acid ?
    (a) 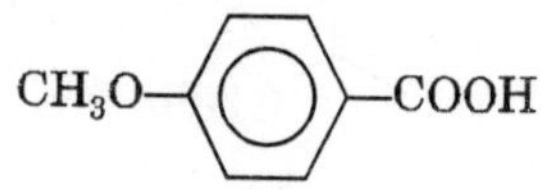
    (b) 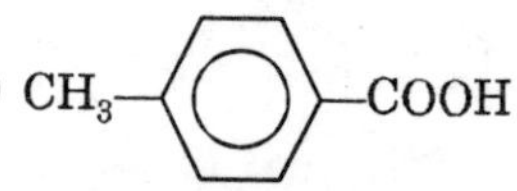
    (c) 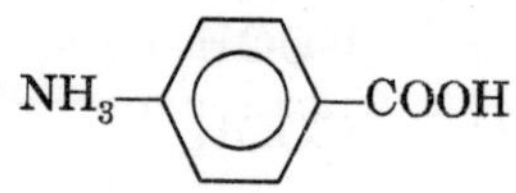
    (d) 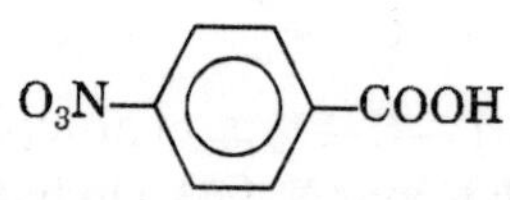
11. The number of isomeric carboxylic acids and esters possible for the formula $C_4H_8O_2$ are respectively
    (a) 2 + 2 (b) 3 + 2
    (c) 2 + 4 (d) 4 + 2.

Methods of Preparation

12. Acid chlorides react with Grignard's reagent to form
    (a) Alcohol (b) Esters
    (c) Aldehydes (d) Ketones.
13. In order to prepare propanal from propanoyl chloride which reagent is suitable ?
    (a) LAH (b) $(CH_3)_2Cd$
    (c) $H_2 - Pd/BaSO_4$ (d) SO_2Cl_2.
14. Benzene can be produced by soda lime decarboxylation of
    (a) Phenol (b) Sodium phenate
    (c) Sodium benzoate (d) Sodium salicylate.
15. Which acid is produced when ethyl benzene is subjected to $KMnO_4$ oxidation ?
    (a) Benzoic acid (b) Phenylacetic acid
    (c) Ethanoic acid (d) Phthalic acid.

16. 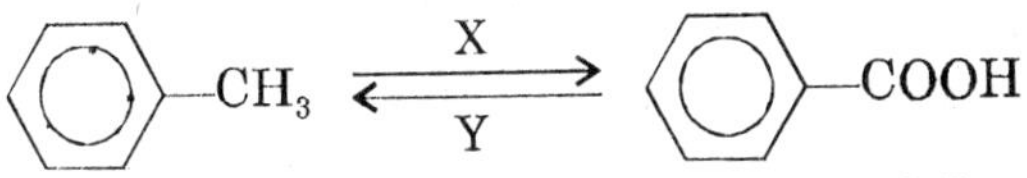

In the above sequence of reaction X and Y are respectively

(a) H_2/Pt, Br_2 (b) $KMnO_4$, H_2/Pt
(c) $KMnO_4$ (aq), HI/P (d) $NH_2—NH_2$/KOH, HI/P.

17. $CH_3Cl \xrightarrow[\text{Ether}]{Mg} X \xrightarrow{CH_3COCl}$ ......

The end product in the above reaction will be

(a) Acetone (b) Ethanol
(c) Tertiary butyl alcohol (d) Isopropyl alcohol.

18. Sulphuryl chloride and calcium propionate react to give

(a) Propyl hydrogen sulphate
(b) Propyl acetate
(c) Propionyl chloride
(d) Propanoic anhydride.

19. Acetaldehyde $\xrightarrow{HCN}$ **X** $\xrightarrow[H_2O/H^+]{}$ **Y** $\xrightarrow{\text{Heat}}$ **Z.**

In the above sequence, the end product Z is

(a) Prop-2-enoic acid (b) But-2-enoic acid
(c) Lactic acid (d) Tartaric acid.

20. $CH_3COC_2H_5 \xrightarrow{NaOH/I_2}$ X $\xrightarrow{Ag}$ Y(g)

Yellow ppt.

In this sequence, the gaseous product Y is

(a) Ethyne (b) Ethane
(c) Ethene (d) Propane.

21. Which of the following cannot produce acetyl chloride ?

(a) $(CH_3COO)_2Ca$ and SO_2Cl_2
(b) CH_3COOH and $SOCl_2$
(c) CH_3COONa and $POCl_3$
(d) CH_3COOH and HCl.

Physical Properties

22. Which of the following has highest solubility in water ?

(a) n-Butyric acid (b) Propionic acid
(c) Isobutyric acid (d) Acetic acid.

23. In which of the following, the carbon-oxygen bonds are of equal lengths ?

(a) CH_3COO^- (b) CH_3COOH
(c) $CH_3COOC_2H_5$ (d) In all.

24. Carboxylic acids do not give the characteristic reaction of

(a) alkyl part (R—) (b) —OH group
(c) —C = O group (d) —COOH group.

25. The hybrid structure of carboxylate ion can be best represented as

(a) 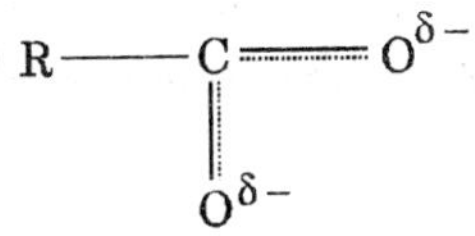 (b) 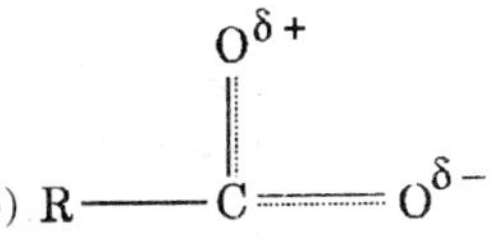

(c) 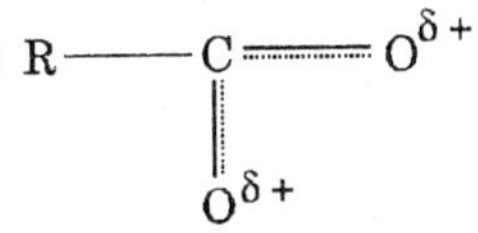 (d) 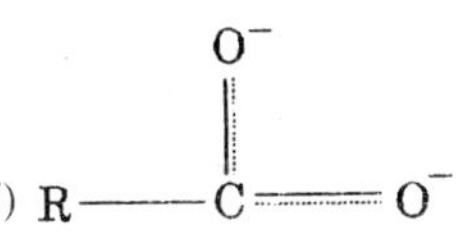

26. Acetic acid shows a molecular mass of 120 when dissolved in benzene. This is because of

(a) Solvation process
(b) Intermolecular H-bonding
(c) Presence of H atom at α-C
(d) Dimerisation.

27. What is not true about cinnamic acid ?

(a) It produces brisk effervescence with $NaHCO_3$
(b) It contains —COOH and —OH groups
(c) It decolourises bromine solution
(d) It decolourises alkaline $KMnO_4$.

28. Esters are quite soluble in

(a) Alcohol
(b) Benzene
(c) Both alcohol and benzene
(d) Water.

29. The compound with a formula H_2NCH_2COOH behave as

(a) Strong acid (b) Strong base
(c) Amphoteric substance (d) Strong reducing agent.

30. —OH group in ethanol is almost neutral but the same is acidic in AcOH because

(a) In ethanol —OH is linked to electron releasing C_2H_5— group
(b) In AcOH, C = O group causes electron withdrawing effect
(c) Resonance cause the development of partial positive charge on O atom of O—H group in AcOH.
(d) All the above reasons are correct.

Chemical Properties

31. Which of the following will not be able to produce acetyl chloride by its reaction with acetic acid ?

(a) PCl_5 (b) $SOCl_2$/Py
(c) PCl_3 (d) Cl_2.

32. Propionic acid $\xrightarrow{PCl_5}$ X $\xrightarrow[AlCl_2]{C_6H_6}$ Y.

In the above sequence Y is

(*a*) Phenylethylketone (*b*) Phenylacetate
(*c*) Ethyl benzene (*d*) Ethyl benzoate.

33. Treatment of tartaric acid with Fenton's reagent gives mainly

(*a*) Pyruvic acid (*b*) Maleic acid
(*c*) Tartaric anhydride (*d*) Dihydroxymaleic acid.

34. Which combination produces oxalyl chloride ?

(*a*) $(COOH)_2 + Cl_2$ (*b*) $(COOH)_2 + KCl$
(*c*) $(COOH)_2 + SOCl_2$ (*d*) $(COOH)_2 + ZnCl_2$.

35. Which reagent can convert α-chloroacetic acid to 2-chloroethanoyl chloride ?

(*a*) HCl (*b*) PCl_5
(*c*) Cl_2 (*d*) $MnCl_2$.

36. Chloroacetic acid is allowed to react with moist Ag_2O (Ag_2O/H_2O), the main product is

(*a*) Silver lactate
(*b*) α-hydroxyethanoic acid
(*c*) Silver pyruvate
(*d*) The reaction does not occur.

37. The mixture of formic acid and acetic acid vapours are passed over heated manganous oxide at 575 K. The main product is

(*a*) Ethyl ehtanoate (*b*) Methyl formate
(*c*) Acetone (*d*) Acetaldehyde.

38. $$C_2H_5MgBr \xrightarrow[(ii)\ H_2O]{(i)\ CO_2} C_nH_{2n+1}COOH.$$

The value of *n* is

(*a*) 1 (*b*) 2
(*c*) 3 (*d*) 0.

39. During esterification of carboxylic acid with alcohol which bond of carboxylic acid undergoes cleavage ?

(*a*) C—C (*b*) C = O
(*c*) C---O (*d*) O—H.

40. *Acetone* and *Sodium hypochlorite* react to give

(*a*) Monochloroacetone (*b*) Dichloroacetone
(*c*) Chloroform (*d*) Chloretone.

41. A colourless gas evolved by the reaction of propanamide with $NaNO_2/HCl$ is

(*a*) Dinitrogen (*b*) Ammonia
(*c*) Dioxygen (*d*) Propane.

42. Acetamide produce 1° amine with

(*a*) NaOH (*b*) HCl
(*c*) $NaOH/Br_2$ (*d*) HgO.

43. RCOCl reacts with sodium salt of hydrazoic acid to produce

(*a*) Acylazide (*b*) 1° Amine
(*c*) Urea (*d*) 2° Amine.

44. Heating of sodium formate with soda lime produces

(*a*) Dihydrogen (*b*) Ethane
(*c*) Ethyne (*d*) Methane.

45. The families of reactant and the corresponding product in Hoffmann's bromamide reaction are respectively

(*a*) Amide, 1° amine (*b*) Amine, 2° amine
(*c*) Amine, amide (*d*) 1 mide, 1 mine.

46. A concentrated solution of urea gives white precipitate with

(*a*) HNO_3 (conc.) (*b*) HNO_2
(*c*) $AgNO_3$ (*d*) NaOH.

47. Chemical reaction of any amine with acid chloride is termed as

(*a*) Esterification (*b*) Condensation
(*c*) Acylation (*d*) Saponification.

48. Acetyl chloride can be reduced to acetaldehyde by

(*a*) $LiAlH_4$ (*b*) $H_2/Pd - BaSO_4$
(*c*) H_2/Ni (*d*) Na/C_2H_5OH.

49. Formation of acetophenone by the reaction of acetylchloride and benzene is an example of

(*a*) Aldol condensation (*b*) Perkin's reaction
(*c*) Claisen's condensation (*d*) Friedel-Craft's reaction.

50. A compound **A** with formula $C_4H_{10}O_3$ on acylation with acetic anhydride give another compound with molecular mass 190. The number of hydroxyl group in the compound **A** are

(*a*) Two (*b*) One
(*c*) Three (*d*) Unpredictable.

51. $$C_2H_2 \xrightarrow[HgSO_4]{CH_3COOH\ (excess)} \mathbf{A} \xrightarrow{Distill} \mathbf{B + C}.$$

In the above sequence B and C are

(*a*) Ethanal, Ethanol
(*b*) Ethanal, Ethanoic anhydride
(*c*) Acetic anhydride, Ethanol
(*d*) Acetic acid.

52. A halogen compound A on hydrolysis with dilute alkali followed by acidification gives acetic acid. The compound A is

(*a*) CH_3CCl_3 (*b*) CH_3CHCl_2
(*c*) $ClCH_2CH_2Cl$ (*d*) $ClCH_2CHCl_2$.

53. N-Bromosuccimide is allowed to react with α-amino propionic acid. The main products are

(*a*) CH_3CHO, CO_2/NH_3 (*b*) $CH_3CH(Br)COOH$
(*c*) $CH_3CH_2CONH_2$ (*d*) $CH_3CH(NH)_2CONH_2$.

54. The alkanes produced by the decarboxylation of sodium propionate and that by reduction of propionic acid with HI/P are respectively

(*a*) Propane and propane (*b*) Ethane and propane
(*c*) Propane and ethane (*d*) Ethane and ethane.

55. Which of the following reagents will produce acetophenone from benzene ?

(a) CH_3COOH/H_2SO_4 (b) $(CH_3CO)_2O$
(c) CH_3COCl (d) Both (b) and (c).

56. The calcium salt of which of the following acid on dry distillation produces 2, 4-dimethyl pentane-3-one ?
(a) Propionic acid (b) Isobutyric acid
(c) Butyric acid (d) Adipic acid.

57. $\mathbf{X} \xleftarrow[P_2O_5]{\text{Heat}}$ **Malonic acid** $\xrightarrow[423\text{ K}]{\text{Heat}} \mathbf{Y}$.
X and Y in the above sequence are respectively
(a) C_3O_2, CH_3COOH (b) CO_2, CH_3COOH
(c) CO_2, CO_2 (d) CH_3CHO, CH_3COOH.

58. Certain ester 'A' is boiled with KOH. The product is cooled and acidified with conc. HCl to get white crystalline precipitate. The ester is likely to be
(a) Ethylacetate (b) Methylbenzoate
(c) Isopropyl acetate (d) Methyl formate.

59. Acetyl chloride and phenol react to produce
(a) Phenetole (b) Phenoxy chloride
(c) Phenyl acetate (d) Methylbenzoate.

60. Certain hydrocarbon C_6H_{12} decolourises bromine solution and yields n-hexane on hydrogenation. Oxidation of C_6H_{12} gives two different monobasic carboxylic acids other than formic acid. The hydrocabron is
(a) Hex-2-ene (b) Cyclohexane
(c) Hex-3-ene (d) Hex-1-ene.

61. Which of the following acid will not produce hydrocarbon on soda lime decarboxylation process ?
(a) Salicylic acid
(b) Benzoic acid
(c) Cyclopropane carboxylic acid
(d) Isobutyric acid.

62. A mixture of colourless organic compound and $NaNO_2$ produces brisk effervescence when dilute HCl is added to it. The organic compound could be
(a) Glucose (b) Urea
(c) Benzoic acid (d) Ethanol.

63. Which is not a reacting species in Schotten Baumann reaction ?
(a) Phenol (b) Aniline
(c) Benzoyal chloride (d) Benzaldehyde.

64. RCOOH and Etheral solution of diazomethane react to produce
(a) RNH_2 (b) RCOOR
(c) $RCOCH_3$ (d) $RCOOCH_3$.

65. 0.122 g of which of the follwoing acid will require 20 ml of M/20 KOH solution for neutralisation.
(a) C_6H_5COOH (b) CH_3COOH
(c) $ClCH_2COOH$ (d) $(CH_3)_3CHCOOH$.

66. $X \xrightarrow[\text{Hydrolysis}]{\text{Alkaline}} Y \xrightarrow{\text{HCl (dil.)}} Z$
$\text{Acetonitrile} \xrightarrow{H_3O^+} Z$

In the above sequence X is
(a) $ClCH_2CH_2Cl$ (b) CH_3CHCl_2
(c) $CH_2Cl \cdot CHCl_2$ (d) CH_3CCl_3.

67. $\mathbf{X} \xrightarrow{\text{NaOH}} \mathbf{Y + CHCl_3, Y} \xrightarrow{H_3O^+}$ **ethanoic acid.**
Here, X is
(a) Cl_3CCOCH_3 (b) $Cl_3CCOC_2H_5$
(c) $Cl_2CHCOCH_3$ (d) $Cl_2CHCOC_2H_5$.

68. Which alkane cannot be obtained from propionic acid by using methods like Kolbe's decarboxylation, soda lime decarboxylation or reduction with HI ?
(a) Ethane (b) Butane
(c) Methane (d) Propane.

69. Phosgene and ammonia react to give
(a) Carbonic acid (b) Carbolic acid
(c) Carbamic acid (d) Carbamide.

70. Which acid can be prepared by Reimer Tiemann reaction ?
(a) Phthalic acid (b) Salicylic acid
(c) Cinnamic acid (d) Benzoic acid.

71. Which of the following ester cannot undergo Claisen self condensation ?
(a) $CH_3(CH_2)_3COOC_2H_5$ (b) $C_6H_5COOCH_2CH_3$
(c) $CH_3CH_2COOC_6H_5$ (d) $C_6H_5CH_2COOC_2H_5$.

72. Salicylic acid on treatment with fuming HNO_3 produces
(a) Phenol (b) Picric acid
(c) Toluic acid (d) Benzoic acid.

73. p-tolyl chloride and sodium p-toluate react to produce
(a) p-toluic anhydride (b) Toluic acid
(c) Benzoic anhydride (d) p-toludene.

74. **Sod. salt of phthalamic acid** $\xrightarrow{Br_2/NaOH} \mathbf{X} \xrightarrow{H_3O^+} \mathbf{Y}$. Here, Y is
(a) Anthranilic acid (b) Phthalic acid
(c) Terephthalic acid (d) o-phenylene diamine.

Acidic Strength

75. Among ethanol (I), acetic acid (II), phenol (III) and benzoic acid (IV), the correct order of increasing acid strength is

(a) I < II < III < IV (b) I < III < II < IV
(c) I < III < IV < II (d) III < IV < I < II.

76. Which of the following acid has highest tendency to ionise in aqueous solution ?
(a) HCOOH (b) CH_3COOH
(c) FCH_2COOH (d) $BrCH_2COOH$.

77. NH_2CONH_2 is
(a) Monoacid base (b) Monobasic acid
(c) Dibasic acid (d) Diacid base.

78. Among α-nitro acetic acid (1), α-fluoroacetic acid (2) α-bromoacetic acid (3), α-cyanoacetic acid (4), the correct order of increasing acid strength is
(a) 3 < 2 < 4 < 1 (b) 1 < 2 < 3 < 4
(c) 2 < 1 < 3 < 4 (d) 4 < 1 < 2 < 3.

79. Which acid has lowest value of pK_a ?
(a) p-Methoxybenzoic acid (b) p-Chlorobenzoic acid
(c) p-Aminobenzoic acid (d) p-Toluic acid.

80. The weakest acid among the following is
(a) Water (b) Ethyne
(c) Phenol (d) Ethanoic acid.

81. Which of following anion is a strongest base ?
(a) $C_6H_5COO^-$ (b) $HCOO^-$
(c) CH_3COO^- (d) $(CH_3)_2CHCOO^-$.

82. Among the acids given below :
CH_3CH_2COOH (X) ; $CH_2 = CHCOOH$ (Y) and
$CH \equiv C\text{-}COOH$ (Z).
The correct order of increasing acid strength is
(a) X < Y < Z (b) X < Z < Y
(c) Y < X < Z (d) Z < Y < X.

83. Me_3SiCH_2COOH is weaker acid as compared to Me_3CCH_2COOH because
(a) C is less electronegative than Si
(b) Ionisation energy of C is greater than Si
(c) Si is more electropositive then C
(d) Si does not form covalent bonds with carbon.

Reactivity

84. What name is associated with the reverse of acidic esterification ?
(a) Alcoholysis (b) Hydrolysis
(c) Neutralisation (d) Trans esterification.

85. In general, the order of reactivity towards aryl substitution among acid derivatives is
(a) $RCOCl > RCONH_2 > RCOOR' > (RCO)_2O$
(b) $RCOCl > (RCO)_2O > RCOOR' > RCONH_2$
(c) $(RCO)_2O > RCOCl > RCONH_2 > RCOOR'$
(d) $RCOOR' > RCOCl > (RCO)_2O > RCONH_2$.

86. In esterification reaction, the role of conc. H_2SO_4 is
(a) To act as dehydrating agent
(b) To act as hydrolytic agent
(c) To act as catalyst
(d) To act as catalyst as well as dehydrating agent.

87. Given three acids

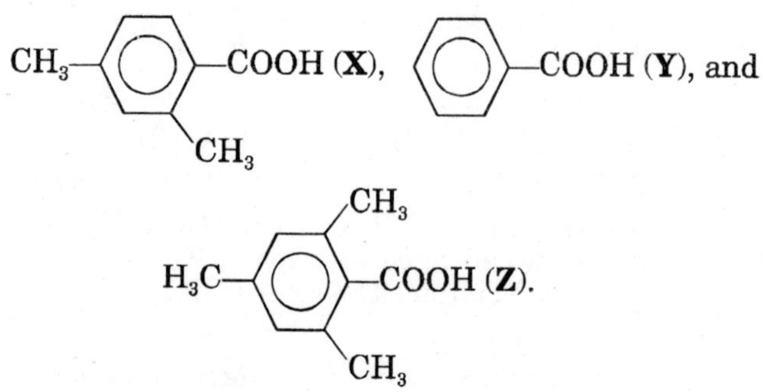

The correct order of ease of acid catalysed esterification is
(a) X > Z > Y (b) Y > X > Z
(c) Z > X > Y (d) Y > Z > X.

88. Rate of esterification is highest for the acid
(a) $CH_3(CH_2)_2COOH$ (b) $(CH_3)_2CHCOOH$
(c) CH_3COOH (d) $Cl(CH_2)_4COOH$.

Distinction

89. Which of the following can provide distinction between formic acid and acetic acid ?
(a) Litmus solution
(b) Alcoholic $AgNO_3$
(c) Sodium bicarbonate
(d) Diammine silver (I) hydroxide solution.

90. Which of the following acid produces violet colour with $FeCl_3$ solution ?
(a) Benzoic acid (b) Acetic acid
(c) Picric acid (d) Oxalic acid.

91. Which of the following compound will give purple colour with $FeCl_3$ solution ?
(a) Acetic acid (b) Salicylic acid
(c) Propionic acid (d) α-hydroxypropionic acid.

General

92. Which of the following is present in Vinegar ?
(a) Acetic acid (b) Citric acid
(c) Tartaric acid (d) Lactic acid.

93. Which resonating structure accounts for the inability of carboxylic acids to react with $NaHSO_3$ like aldehydes ?

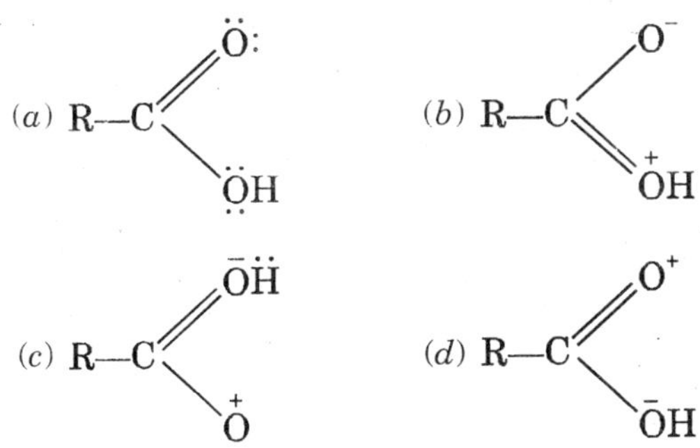

94. Which among the following is a constituent of baking powder ?

(a) Citric acid (b) Cream of tartar
(c) Glycerol (d) Lactic acid.

95. Which of the following is used to remove ink and rust stains on cloth ?

(a) Oxalic acid (b) Ethanol
(c) Kerosene oil (d) Olive oil.

96. Which of the following is used in synthetic lemonade ?

(a) Tartaric acid (b) Acetic acid
(c) Citric acid (d) Oxalic acid.

97. A detergent is a

(a) cleansing agent (b) drug
(c) catalyst (d) soap.

98. Which of the following is responsible for the rancidity of fats ?

(a) Hydrogenation
(b) Hydrolysis
(c) Oxidation and hydrolysis
(d) Dehydration.

99. Tamrind contains

(a) (+) Tartaric acid (b) (–) Tartaric acid
(c) ± Tartaric acid (d) none of these.

100. Which of the folloiwng acid is capable of producing three series of salts ?

(a) Oxalic acid (b) Citric acid
(c) Tartaric acid (d) Salicylic acid.

ANSWERS

1. (d)	2. (c)	3. (d)	4. (b)	5. (a)	6. (b)	7. (c)	8. (b)	9. (c)
10. (a)	11. (c)	12. (a)	13. (c)	14. (c)	15. (a)	16. (c)	17. (c)	18. (c)
19. (a)	20. (a)	21. (d)	22. (d)	23. (a)	24. (c)	25. (a)	26. (d)	27. (b)
28. (c)	29. (c)	30. (d)	31. (d)	32. (a)	33. (d)	34. (c)	35. (b)	36. (b)
37. (d)	38. (b)	39. (c)	40. (c)	41. (a)	42. (c)	43. (a)	44. (a)	45. (a)
46. (a)	47. (c)	48. (b)	49. (d)	50. (a)	51. (b)	52. (a)	53. (a)	54. (b)
55. (d)	56. (b)	57. (a)	58. (b)	59. (c)	60. (a)	61. (a)	62. (b)	63. (d)
64. (d)	65. (a)	66. (d)	67. (a)	68. (c)	69. (d)	70. (b)	71. (b)	72. (b)
73. (a)	74. (a)	75. (b)	76. (c)	77. (a)	78. (a)	79. (b)	80. (b)	81. (d)
82. (a)	83. (c)	84. (b)	85. (b)	86. (d)	87. (b)	88. (c)	89. (d)	90. (c)
91. (b)	92. (a)	93. (b)	94. (b)	95. (a)	96. (c)	97. (a)	98. (c)	99. (a)
100. (b).								

HINTS/SOLUTIONS

Q. No.	Ans.	Hints & Solutions
1.	(d)	AcOAc is $R-\overset{O}{\overset{\|}{C}}-O-\overset{O}{\overset{\|}{C}}-R$. Thus, it represents acid anhydrides.
2.	(c)	It is formic acid (*formica* = red ants).
3.	(d)	Alkalnals are not functional derivatives of carboxylic acids.
4.	(b)	Following the rules, the given compound is sodium, 2, 3-dibromo-2-methylpropanoate.
5.	(a)	The given compound is 1, 2, 3-propane tricarboxylic acid.

Q. No.	Ans.	Hints & Solutions
6.	(*b*)	Lactic acid is α-hydroxy propionic acid.
7.	(*c*)	The correct IUPAC name is 4-methoxycarbonyl benzoic acid.
8.	(*b*)	Acids and esters are functional isomers of each other.
9.	(*c*)	2-Methylbutanoic acid has chiral structure hence, it shows optical activity.
10.	(*a*)	*p*-anisic acid has CH_3O group in para position to COOH group.
11.	(*c*)	The acids are $CH_3CH_2CH_2COOH$ and $CH_3CH(CH_3)COOH$ The esters are $HCOOCH_2CH_2CH_3$, $HCOOCH(CH_3)_2$ $CH_3COOC_2H_5$, $CH_3CH_2COOCH_3$.
12.	(*a*)	$R{-}COCl + R'MgX \longrightarrow RCOR' + MgXCl$ $RCOR' + R'MgX \longrightarrow R{-}C(R')_2{-}OMgX$ $\xrightarrow{H_2O} R{-}C(R')_2{-}OH + Mg(OH)X.$ (Alcohol)
13.	(*c*)	$CH_3CH_2COCl + H_2 \xrightarrow{Pd/BaSO_4} CH_3CH_2CHO + HCl$ (Propanal)
14.	(*c*)	$C_6H_5COONa \xrightarrow[CaO]{NaOH} C_6H_6 + Na_2CO_3.$
15.	(*a*)	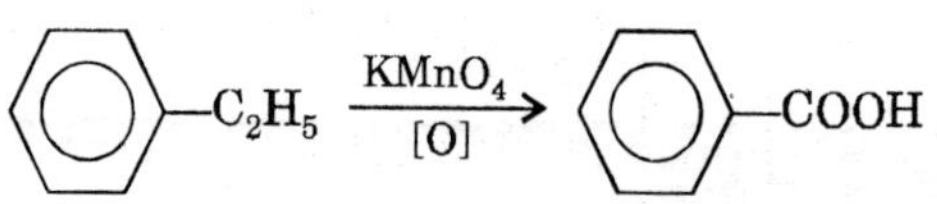 Benzoic acid
16.	(*c*)	$KMnO_4$ converts – CH_3 in toluene to —COOH while HI reduces COOH group to —CH_3.
17.	(*c*)	$CH_3Cl \xrightarrow[Ether]{Mg} CH_3MgBr \xrightarrow{CH_3COCl} (CH_3)_3COH$ Tert. butyl alcohol
18.	(*c*)	$(CH_3CH_2COO)_2Ca + SO_2Cl_2 \longrightarrow CaSO_4 + (CH_3CO)_2O.$
19.	(*a*)	$CH_3CHO \xrightarrow{HCN} CH_3CH(OH)CN \xrightarrow{H_3O^+} CH_3CH(OH)COOH \xrightarrow{H_3O^+} CH_2 = CHCOOH$ Acrylic acid
20.	(*a*)	$CH_3COC_2H_5 \xrightarrow{NaOI} CHI_3 + C_2H_5COONa$ (Yellow) $2CHI_3 + Ag \longrightarrow CH \equiv CH + 6AgI.$ (Y)
21.	(*d*)	HCl can not convert —COOH group to —COCl.
22.	(*d*)	Due to H bonding.
23.	(*a*)	Carboxylic ion is hybrid of two equivalent resonating forms. Hence C—O bonds are of equal length.
24.	(*c*)	—COOH group exhibits resonance as 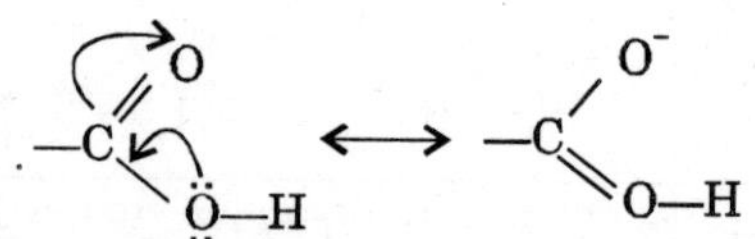 Thus, there is no complete >C = O character.
25.	(*a*)	Salicylic acid contains phenolic groups as well as —COOH group.
26.	(*d*)	It is due to dimerisation.
27.	(*b*)	Cinnamic acid (C_6H_5 – CH = CH – COOH) does not contain —OH group.
28.	(*c*)	R—COOR compounds have a slightly polar character and hence soluble in alcohol. Due to larger hydrocarbon part these are also soluble in benzene.
29.	(*c*)	It is amphoteric because it can neutralise strong acid as well as strong base.
30.	(*d*)	The OH group in ROH does not ionise whereas in —COOH, due to resonance stabilisation of carboxyl group, H^+ ions are released with ease.
31.	(*d*)	Cl_2 brings about substitution at α-C atom.
32.	(*a*)	$CH_3CH_2COOH \xrightarrow{PCl_5} CH_3CH_2COCl \xrightarrow[AlCl_3]{C_6H_6} C_6H_5CO\,C_2H_5 + HCl$ Ethylphenyl ketone
33.	(*d*)	$HO{-}CH{-}COOH \mid HO{-}CH{-}COOH \xrightarrow[(O)]{FeSO_4/H_2O_2} HO{-}C{-}COOH \parallel HO{-}C{-}COOH.$ Tartaric acid → Dihydroxy maleic acid

Q. No.	Ans.	Hints & Solutions
34.	(c)	$\begin{matrix}COOH\\ \vert\\ COOH\end{matrix} + SOCl_2 \xrightarrow{Py} \begin{matrix}COCl\\ \vert\\ COCl\end{matrix} + SO_2$ Oxalyl chloride
35.	(b)	$ClCH_2COOH + PCl_5 \longrightarrow ClCH_2COCl$
36.	(b)	$ClCH_2COOH \xrightarrow[H_2O]{Ag_2O} HOCH_2COOH$ α-Hydroxyethanoic acid
37.	(d)	$CH_3COOH + HCOOH \xrightarrow{MnO} CH_3CHO + CO + H_2O$
38.	(b)	$C_2H_5MgBr \xrightarrow[(ii)\ H_2O]{(i)\ CO_2} C_2H_5COOH.$ Thus, value of $n = 2$.
39.	(c)	$R—\overset{O}{\overset{\Vert}{C}}\boxed{—O—H + H—}O—R' \longrightarrow R—\overset{O}{\overset{\Vert}{C}}—OR' + H_2O$ Thus, it involves the cleavage of C—O bond of carboxylic acid molecule.
40.	(c)	$CH_3\overset{O}{\overset{\Vert}{C}}CH_3 \xrightarrow{NaOCl} CHCl_3 + CH_3COONa.$
41.	(a)	$CH_3CH_2CONH_2 + HNO_2 \longrightarrow C_2H_5COOH + N_2\uparrow.$
42.	(c)	It refers to Hoffmann's bromamide reaction.
43.	(a)	$RCOCl + NaN_3 \longrightarrow R-\overset{O}{\overset{\Vert}{C}}-N_3 + NaCl$ Acyl azide
44.	(a)	$HCOONa + NaOH \xrightarrow{CaO} Na_2CO_3 + H_2.$
45.	(a)	Hoffmann's bromamide reaction involves conversion of amides into 1° amines.
46.	(a)	Urea is a feeble monoacidic base ($K_b = 1.5 \times 10^{-14}$) which forms crystalline white precipitate of ureanitrate with conc. HNO_3 $(H_2N)_2CO + HNO_3 \longrightarrow O = C(NH_2) . HNO_3.$ Ureanitrate
47.	(c)	The reaction is known as acylation.
48.	(b)	This is in accordance with Rosenmund's reaction.
49.	(d)	It is Friedel craft's acylation reaction.

Q. No.	Ans.	Hints & Solutions
50.	(a)	$—OH \xrightarrow[+CH_3CO^-]{-H} -OCOCH_3$ The change in molecular mass for acylation of one —OH group = 42 amu. The change in molecular mass for acylation of $C_4H_{10}O_3 = 190 - 106 = 84$. ∴ The number of —OH groups acylated = 84/42 = 2.
51.	(b)	$CH \equiv CH \xrightarrow[Hg^{2+}]{2CH_3COOH} CH_3CH(OCOCH_3)$ Ethylidene diacetate (A) $\xrightarrow{Distill} CH_3CHO + (CH_3CO)_2O$ (B) (C)
52.	(a)	$\underset{(A)}{CH_3CCl_3} \xrightarrow{NaOH\ (aq)} CH_3\overset{OH}{\underset{OH}{C}}—OH \xrightarrow[NaOH]{-2H_2O} CH_3COONa \xrightarrow[-NaCl]{HCl} CH_3COOH.$
53.	(a)	NBS causes decarboxylation of α-amino acids in aqueous medium. $CH_3CH(NH_2)COOH \xrightarrow{NBS} CH_3CHO + NH_3 + CO_2.$
54.	(b)	$CH_3CH_2COONa \xrightarrow[CaO]{NaOH} CH_3CH_3.$ $CH_3CH_2COOH \xrightarrow{HI/P} CH_3CH_2CH_3.$
55.	(c)	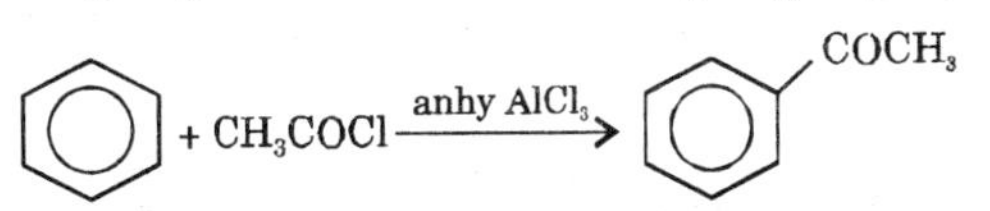
56.	(b)	$2[(CH_3)_2CHCOO]_2Ca \xrightarrow{Dry\ distill} (CH_3)_2CH\overset{O}{\overset{\Vert}{C}}CH(CH_3)_2 + CaCO_3.$
57.	(a)	$CH_2(COOH)_2 \xrightarrow{423\ K} \underset{(Y)}{CH_3COOH} + CO_2$ $CH_2(COOH)_2 \xrightarrow{P_2O_5} \underset{(X)}{O=C=C=C=O} + 2H_2O.$
58.	(b)	$C_6H_5COOCH_3 \xrightarrow{KOH} C_6H_5COOK + CH_3OH$

Q. No.	Ans.	Hints & Solutions
		$C_6H_5COOK \xrightarrow{HCl} \underset{(ppt.)}{C_6H_5COOH} + KCl$ Other Esters will produce acids which are not solid at room temperature.
59.	(*d*)	$C_6H_5OH + ClCOCH_3 \longrightarrow C_6H_5O-\overset{O}{\overset{\parallel}{C}}-CH_3$ Phenyl acetate
60.	(*a*)	Formation of different mono carboxylic acids other than HCOOH indicates that the double bond in C_6H_{12} is not terminal. Hence it should be hex-2-ene $CH_3CH_2CH_2CH=CH-CH_3 \xrightarrow{oxid.} CH_3CH_2CH_2COOH + CH_3COOH$
61.	(*a*)	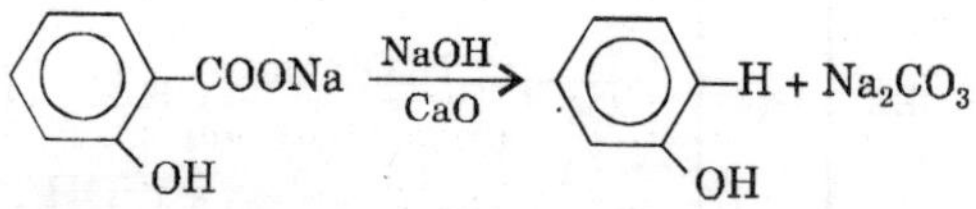 Sodium salicylate Phenol
62.	(*b*)	$(NH_2)_2CO + 2HNO_2 \longrightarrow 2H_2O + 2N_2 + H_2CO_3$.
63.	(*d*)	Scholten Bauman reaction involves benzoylation of phenol or aniline.
64.	(*d*)	$R—COOH + CH_2N_2 \xrightarrow{Ether} \underset{Ester}{R—COOCH_3} + N_2$.
65.	(*a*)	Moles of KOH required $= 20 \times 10^{-3} \times \frac{1}{20} = 10^{-3}$ 10^{-3} mol KOH neutralise acid $= \frac{0.122}{GMM}$ moles 1 mol KOH neutralise acid $= \frac{0.122}{GMM \times 10^{-3}}$ moles Since 1 mol KOH should neutralise 1 mol of acid. $\therefore \frac{0.122 \times 10^3}{GMM} = 1$ or $GMM = 0.122 \times 10^3 = 122$ g. Thus, the acid should be C_6H_5COOH with GMM = 122 g.
66.	(*d*)	$\underset{(X)}{CH_3CCl_3} \xrightarrow{(i)\ NaOH} \underset{(Y)}{CH_3COONa} \xrightarrow{HCl} CH_3COOH$ $CH_3CN \xrightarrow{H_3O^+} CH_3COOH$ (Z)

Q. No.	Ans.	Hints & Solutions
67.	(*a*)	$\underset{(X)}{Cl_3C-\overset{O}{\overset{\parallel}{C}}-CH_3} \xrightarrow{NaOH} CHCl_3 + \underset{(Y)}{CH_3\overset{O}{\overset{\parallel}{C}}-ONa}$ $\underset{(Y)}{CH_3COONa} \xrightarrow{H_3O^+} CH_3COOH + NaCl$
68.	(*c*)	Different methods can yield propane, ethane or butane as follows , $CH_3CH_2COOH \xrightarrow[P]{HI} \underset{Propane}{CH_3CH_2CH_3}$ $CH_3CH_2COONa \xrightarrow[CaO]{NaOH} \underset{Ethane}{CH_3CH_3}$ $2CH_3CH_2COOK + 2H_2O \xrightarrow{Electrolysis} \underset{Butane}{CH_3CH_2—CH_2CH_3} + 2CO_2 + 2KOH$.
69.	(*d*)	$Cl-\overset{O}{\overset{\parallel}{C}}-Cl + 2NH_3 \xrightarrow{-HCl} H_2N-\overset{O}{\overset{\parallel}{C}}-NH_2$. Carbamide or urea
70.	(*b*)	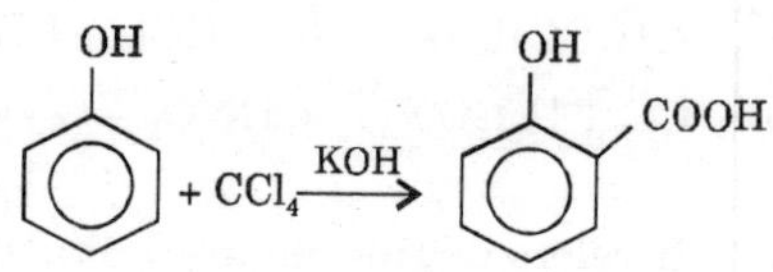 (Salicyclic acid) It is Reimer Tiemann reaction.
71.	(*b*)	$C_6H_5COOC_2H_5$ does not have α-H atom. Hence it cannot undergo self Claisen condensation.
72.	(*b*)	Fuming HNO_3 brings out decarboxylation of —COOH group and at the same time nitration of the ring to give picric acid.
73.	(*a*)	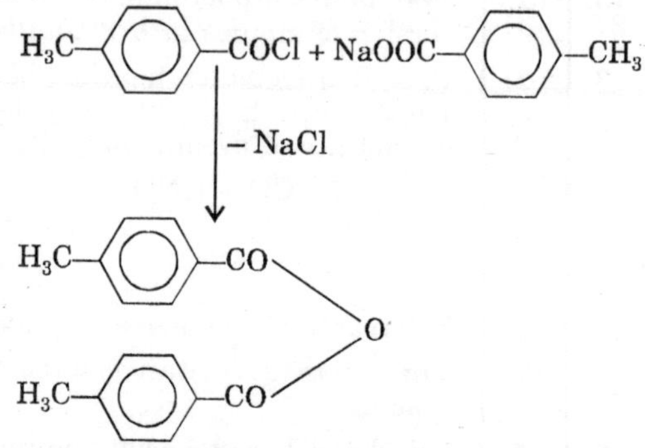 *p*-toluic anhydride

Q. No.	Ans.	Hints & Solutions
74.	(*a*)	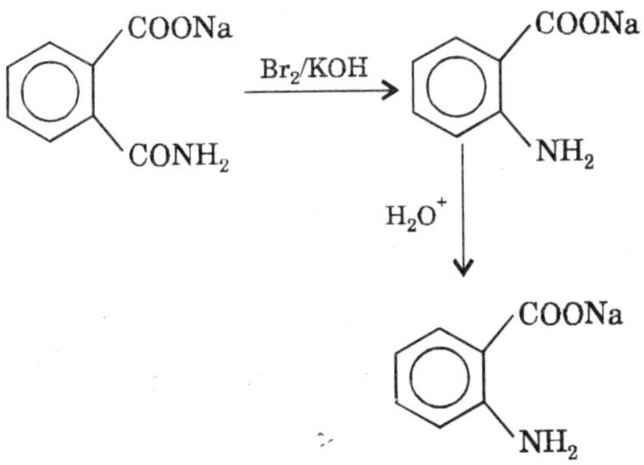 anthranilic acid
75.	(*b*)	Benzoic acid is stronger acid as compared to acetic acid due to—I effect of C_6H_5-group. Acetic acid is stronger acid than phenol and phenol is more acidic than ethanol.
76.	(*c*)	F has maximum electron withdrawing effect. Hence among the given acids FCH_2COOH is strongest.
77.	(*a*)	It is monoacid base.
78.	(*a*)	Electron withdrawing effect decreases as — NO_2 > —CN > —F > —Br. Hence, the acid strength is 3 < 2 < 4 < 1.
79.	(*b*)	*p*-Chlorobenzoic acid is strongest among the given acids.
80.	(*b*)	Ethyne is the weakest acid.
81.	(*d*)	$(CH_3)_2CHOOH$ is a weakest acid. Hence its conjugate pair $(CH_3)_2CHCOO^-$ is a strongest base among the given species.
82.	(*a*)	*sp*-hybrid C is more electronegative than sp^2 hybrid C which is more electronegative than sp^3-hybrid C. As the hybrid state of α-C changes from sp^3 to *sp*, the acid strength tend to increase.
83.	(*c*)	Si is less electro negative than C.
84.	(*b*)	Hydrolysis is reverse of esterification.
85.	(*b*)	See the reactivity of acid derivatives.
86.	(*d*)	See mechanism of esterification.
87.	(*b*)	Acid catalysed esterification involves the nucleophilic attack of alcohol molecule on the protonated acid molecule $(R{-}\underset{\displaystyle OH}{\underset{\vert}{C}}{=}\overset{+}{O}H \leftrightarrow R{-}\underset{\displaystyle OH}{\underset{\vert}{\overset{+}{C}}}{-}OH)$. The crowding of electron releasing alkyl groups at the α-C atom makes the acid less susceptible to nucleophilic attack. Hence correct order of esterification is Y > X > Z.
88.	(*c*)	It is greatest for the acid having lesser +I effect of the alkyl group.
89.	(*d*)	Tollen's reagent is reduced by formic acid. Formic acid is the only caboxylic acid that can reduce Tollen's reagent other acids can't do it.
90.	(*c*)	Picric acid is trinitro phenol and this gives purple colour with $FeCl_3$.
91.	(*b*)	Compounds with enolic group $\left({=}\overset{\displaystyle OH}{\overset{\vert}{C}}{-}\right)$ produce characteristic colour with $FeCl_3$.
92.	(*a*)	Vinegar contains 6-10% of acetic acid.
93.	(*b*)	—COOH group does not have a free >C = O group due to resonance stabilisation of COOH group.
94.	(*b*)	$\underset{\displaystyle CH(OH)COOH}{\overset{\displaystyle CH(OH)COOH}{\vert}}$
95.	(*a*)	Oxalic acid is used to remove ink and rust stains.
96.	(*c*)	Citric acid is used in synthetic lemonade.
97.	(*a*)	It is cleansing agent.
98.	(*c*)	Oxidation and hydrolysis of fats result in the development of foul smell due to formation of large number of organic species. This is referred to as rancidity.
99.	(*a*)	Tamrind has (+) tartaric acid.
100.	(*b*)	Citric acid is a tribasic acid. Hence, it can produce three series of salts.

6.5. EXAMINER'S VIEWPOINT

Test Your Confidence

1996

1. Lactic acid on heating with conc. H_2SO_4 gives
   (*a*) Acetic acid (*b*) Propionic acid
   (*c*) Acrylic acid (*d*) Formic acid. **(*MP PET*)**
2. The refluxing of $(CH_3)_2NCOCH_3$ with acid gives
   (*a*) $2CH_3NH_2 + CH_3COOH$ (*b*) $2CH_3OH + CH_3CONH_2$
   (*c*) $(CH_3)_2NH + CH_3COOH$ (*d*) $(CH_3)_2NCOOH + CH_4$. **(*CEET Kurukshetra*)**
3. On strong heating ammonium cyanate gives
   (*a*) Acetamide (*b*) Methyl cyanide
   (*c*) Urea (*d*) Formamide. **(*MLNR*)**
4. The following reaction
   $$2CH_3COOC_2H_5 \xrightarrow{C_2H_5O^-} CH_3COCH_2COOC_2H_5 \text{ is}$$
   called
   (*a*) Perkin's reaction
   (*b*) Tischenko reaction
   (*c*) Claisen condensation
   (*d*) Schotten Baismen reaction. **(*CEET Kurukshetra*)**

1997

5. Which of the following reaction is expected to give readily a hydrocarbon product in good yield ?
   (*a*) $RCOOK \xrightarrow{\text{Electrolysis}}$
   (*b*) $RCOOAg \xrightarrow{I_2}$
   (*c*) $CH_3CH_3 \xrightarrow{Cl_2,\, h\nu}$
   (*d*) $(CH_3)_2CCl_2 \xrightarrow{C_2H_5OH}$ **(*CBSE PMT*)**
6. Formic acid when treated with conc. H_2SO_4 gives
   (*a*) $(COOH)_2$ (*b*) CH_3COOH
   (*c*) C_2H_5OH (*d*) CO. **(*Pb. PMT*)**
7. Which of the following salt can be produced by the reaction of CO and caustic soda ?
   (*a*) CH_3COONa (*b*) $Na_2C_2O_4$
   (*c*) HCOONa (*d*) CH_3ONa. **(*CPMT*)**
8. In a reaction involving ring substitution of C_2H_5Y, the major product is meta isomer. The group Y can be
   (*a*) $—CH_3$ (*b*) —OH
   (*c*) —Cl (*d*) —COOH. **(*AIIMS*)**
9. Which test is not shown positive by acetaldehyde ?
   (*a*) Lucas test (*b*) Tollen's test
   (*c*) Fehling test (*d*) Schiff's reagent test. **(*AIIMS*)**

1998

10. The acid showing salt like character in aqueous solution is
    (*a*) Acetic acid (*b*) Benzoic acid
    (*c*) Formic acid (*d*) α-aminoacetic acid. **(*MP PMT*)**
11. Which of the following reaction is suitable for preparing α-chloroacetic acid ?
    (*a*) Hell Volhard Zelinsky reaction
    (*b*) Stephan's reaction
    (*c*) Perkin's reaction
    (*d*) None of these. **(*MP PMT*)**
12. Lacrymator or tear gas is
    (*a*) C_6H_5COCl (*b*) $C_6H_5OC_6H_5$
    (*c*) $C_6H_5COCH_2Cl$ (*d*) $C_6H_5COCH_3$. **(*Delhi PMT*)**
13. The order of decreasing ease of reaction with ammonia is
    (*a*) Anhydrides, ester, ether
    (*b*) Anhydrides, ethers, esters
    (*c*) Ethers, anhydrides, esters
    (*d*) Esters, ethers, anhydrides. **(*PMT Pb.*)**
14. The principal product formed in the reaction
    $$CH_2 = CH(CH_2)_8COOH + HBr \xrightarrow{\text{Peroxide}} \text{is}$$
    (*a*) $CH_3CH(Br)(CH_2)_8\ COOH$
    (*b*) $CH_2 = CH(CH_2)_8\ COBr$
    (*c*) $CH_2Br(CH_2)_9\ COOH$
    (*d*) $CH_2 = CH(CH_2)_7CH(Br)COOH$. **(*Pb. PMT*)**
15. As ester (A) with molecular formula $C_9H_{10}O_2$ was treated with excess of CH_3MgBr and the compound so formed was treated with conc. H_2SO_4 to form olefin (B). Ozonolysis of B gave ketone with formula C_8H_8O which shows iodoform test positive. The structure of A is

(*a*) $C_6H_5COOC_2H_5$
(*b*) $CH_3OCH_2COC_6H_5$
(*c*) $CH_3CO—C_6H_5—COCH_3$
(*d*) $C_6H_5COOC_6H_5$. **(*CBSE PMT*)**

16. Which one of the following is strongest acid among the following ?
(*a*) CF_3COOH (*b*) CBr_3COOH
(*c*) CH_3COOH (*d*) CCl_3COOH. **(*BHU*)**

17. Acetic acid reacts with PCl_5 to form
(*a*) $CH_2ClCOOH$ (*b*) $CHCl_2COOH$
(*c*) CH_3COCl (*d*) Both (*a*) and (*c*). **(*EAMCET*)**

1999

18. Lower carboxylic acid are soluble in water due to
(*a*) Lower molecular mass (*b*) Hydrogen bonding
(*c*) Dissociation into ions (*d*) Easy hydrolysis. **(*MP CET*)**

19. Ethyl acetate reacts with CH_3MgBr to form
(*a*) Secondary alcohol
(*b*) Tertiary alcohol
(*c*) Primary alcohol and acid
(*d*) Acid. **(*MP CET*)**

20. A colourless liquid at room temperature reacts with soda lime to form sodium salt of carboxylic acid and ammonia gas. The liquid is
(*a*) Propamamide (*b*) Propanoic acid
(*c*) Formamide (*d*) Methyl ethanoate. **(*Roorkee Screening*)**

2000

21. Identify the product C in the series

$$CH_3CN \xrightarrow{Na/C_2H_5OH} A \xrightarrow{HNO_2} B \xrightarrow{\text{Tollen's reagent}} C$$

(*a*) CH_3COOH (*b*) CH_3CH_2NHOH
(*c*) CH_3CONH_2 (*d*) CH_3CHO. **(*MP CET*)**

22. The product of the reaction,

$$CH_3COOH \xrightarrow{P_2O_5} \text{.......} \text{ is}$$

(*a*) CO and H_2O (*b*) Formic acid
(*c*) Ethanoic anhydride (*d*) Ethyl ethanoate. **(*JIPMER*)**

23. What will happen if $LiAlH_4$ is added to an ester ?
(*a*) Two units of alcohol are obtained
(*b*) One unit of alcohol and one unit of acid is obtained
(*c*) Two units of acids are obtained
(*d*) None of these. **(*CBSE PMT*)**

24. Which reagent will bring about the conversion of carboxylic acids into esters ?
(*a*) C_2H_5OH (*b*) Dry HCl + C_2H_5OH
(*c*) $LiAlH_4$ (*d*) $Al(OC_2H_5)_3$. **(*CBSE PMT*)**

25. A fruity smell is produced by the reaction of C_2H_5OH with
(*a*) PCl_5 (*b*) CH_3COCH_3
(*c*) CH_3COOH (*d*) none of these. **(*AFMC*)**

26. Which of the following orders of relative strengths of acids is correct ?
(*a*) $FCH_2COOH > ClCH_2COOH > BrCH_2COOH$
(*b*) $ClCH_2COOH > BrCH_2COOH > FCH_2COOH$
(*c*) $BrCH_2COOH > ClCH_2COOH > FCH_2COOH$
(*d*) $ClCH_2COOH > FCH_2COOH > BrCH_2COOH$. **(*AIIMS*)**

27. When acetamide is treated with NaOBr, the product formed is
(*a*) CH_3CN (*b*) $CH_3CH_2NH_2$
(*c*) CH_3NH_2 (*d*) None of the above. **(*Haryana CEET*)**

28. Hydrolysis reaction of fats by caustic soda is known as
(*a*) Acetylation (*b*) Carboxylation
(*c*) Esterification (*d*) Saponification. **(*Kerala CET*)**

29. The fatty acid which shows reducing property is
(*a*) Acetic acid (*b*) Ethanoic acid
(*c*) Oxalic acid (*d*) Formic acid. **(*Kerala CET*)**

2001

30. Saponification of ethyl benzoate with caustic soda as alkali, gives
(*a*) Benzyl alcohol, ehtanoic acid
(*b*) Sodium benzoate, ethanol
(*c*) Benzoic acid, sodium ethoxide
(*d*) Phenol, ethanoic acid
(*e*) Sodium benzoxide, ethanoic acid. **(*Kerala PMT*)**

31. Which of the following reagents is used to distinguish between methanoic acid and ethanoic acid ?
(*a*) Amm. silver nitrate solution
(*b*) Neutral ferric chloride
(*c*) Sodium hydroxide solution
(*d*) Sodium carbonate solution
(*e*) Phenolphthalein. **(*Kerala CET*)**

32. Hydrolysis of an ester gives a carboxylic acid which on Kolbe's electrolysis yields ethane. The ester is

(*a*) Ethyl methonoate (*b*) Methyl ethanoate

(*c*) Propylamine (*d*) Ethylamine.

(Manipal PMT)

33. When propanamide reacts with Br_2 and NaOH then which of the following compounds is formed ?

(*a*) Ethyl alcohol (*b*) Propyl alcohol

(*c*) Propyl amine (*d*) Ethylamine.

(Manipal)

34. HCOOH reacts with conc. H_2SO_4 to produce

(*a*) CO (*b*) CO_2

(*c*) NO (*c*) NO_2. ***(Manipal)***

35. The reaction

$$RCH_2CH_2COOH \xrightarrow[Br_2]{\text{Red P}} R\text{—}CH_2\text{—}\underset{\underset{Br}{|}}{CH}\text{—}COOH$$

is called

(*a*) Reimer-Tiemann reaction

(*b*) Hell-Volhard Zelinsky reaction

(*c*) Cannizzaro reaction

(*d*) Sandmeyer reaction. ***(NSE)***

36. Which of the following is the weakest acid ?

(*a*) 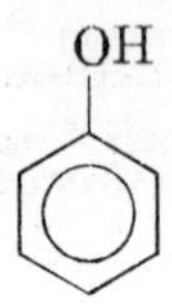 (*b*) CH_3COOH

(*c*) HCOOH (*d*) 

(CPMT)

37. Which alkene is formed from the following ylide carbonyl pair ?

$CH_3CH_2CH_2CH = PPh_3$ + 2-Butanone

(*a*) 3-Methyl-3-heptene (*b*) 4-Methyl-3-heptene

(*c*) 5-Methyl-3-heptene (*d*) 1-Methyl-5-methane.

(Manipal)

38. Consider the following statements :

Acetophenone can be prepared by

(I) Oxidation of 1-phenylethanol

(II) Reaction of benzaldehyde with methyl magnesium bromide.

(III) Friedel craft's reaction of benzene with acetyl chloride

(IV) Distillation of calcium benzoate.

Which of the above statements are correct ?

(*a*) I and II (*b*) I and IV

(*c*) I and III (*d*) III and IV. ***(SCRAE)***

39. Which one of the following pairs is not correctly matched ?

(*a*) $>C=O \xrightarrow{\text{Clemmensen's reduction}} >CH_2$

(*b*) $>C=O \xrightarrow{\text{Wolf-Kishner reduction}} >CHOH$

(*c*) $–COCl \xrightarrow{\text{Rosenmund's reduction}} CHO$

(*d*) $–C \equiv N \xrightarrow{\text{Stephan reduction}} CHO$. ***(SCRAE)***

40. Which of the following gives aldol condensation reaction ?

(*a*) C_6H_5OH (*b*) $C_6H_5\text{—}\overset{\overset{O}{||}}{C}\text{—}C_6H_5$

(*c*) $CH_3CH_2\text{—}\overset{\overset{O}{||}}{C}\text{—}CH_3$ (*d*) $(CH_3)_3C\text{—}\overset{\overset{O}{||}}{C}\text{—}CH_3$.

(CPMT)

41. Carboxylic acids are more acidic than phenol and alcohol because of

(*a*) Intermolecular hydrogen bonding

(*b*) Formation of dimers

(*c*) Highly acidic hydrogen

(*d*) Resonance stabilization of their conjugate base.

(Tamil Nadu)

42. In the mechanism of Hoffman's reaction, which intermediate rearranges to alkyl isocyanate ?

(*a*) Bromamide (*b*) Methene

(*c*) Nitrene (*d*) Amide. ***(Tamil Nadu)***

43. Oxalic acid when reduced with zinc and H_2SO_4 gives

(*a*) Glyoxallic acid (*b*) Glyoxal

(*c*) Glycollic acid (*d*) Glycol. ***(Tamil Nadu)***

44. Lactic acid on oxidation by alkaline potassium permanganate gives

(*a*) Tartaric acid (*b*) Pyruvic acid

(*c*) Cinnamic acid (*d*) Propionic acid.

(Tamil Nadu)

2002

45. Ethyl acetate is obtained when methyl magnesium iodide reacts with

(*a*) Ethyl formate (*b*) Ethyl chloroformate

(*c*) Acetyl chloride (*d*) Carbon dioxide

(Tamil Nadu CET)

46. Glacial acetic acid is obtained by
(*a*) Distilling vinegar
(*b*) Crystallizing, separating and melting acetic acid
(*c*) Treating vinegar with dehydrating agent
(*d*) Chemically separating acetic acid.
(Karnataka CET)

47. Dimerisation in carboxylic acid is due to
(*a*) Ionic bond
(*b*) Covalent bond
(*c*) Coordinate bond
(*d*) Intermolecular hydrogen bond. ***(Karnataka CET)***

48. In the following reaction, product P is

$$R-\overset{\overset{\large O}{||}}{C}-Cl \xrightarrow[Pd-BaSO_4]{H_2} P$$

(*a*) $R\ CH_2OH$ (*b*) $R\ COOH$
(*c*) $R\ CHO$ (*d*) $R\ CH_3$. ***(CBSE PMT)***

49. 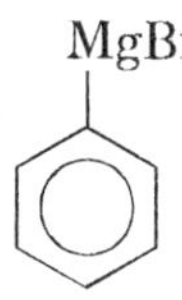 $\xrightarrow[(ii)\ H_3\overset{+}{O}]{(i)\ CO_2}$ P

In the above reaction, product P is

(*a*) 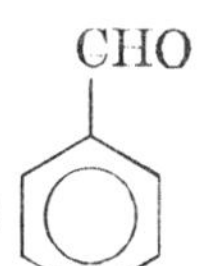 (*b*) 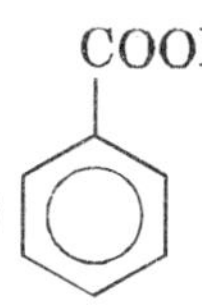

(*c*) 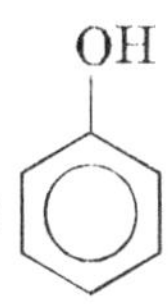 (*d*) $C_6H_5-\overset{\overset{O}{||}}{C}-C_6H_5$

(CBSE PMT)

50. $CH_3CH_2COOH \xrightarrow[Red\ P]{Cl_2} A \xrightarrow{Alc.\ KOH} B.$

What is B ?
(*a*) CH_3CH_2COCl (*b*) CH_3CH_2CHO
(*c*) $CH_2 = CHCOOH$ (*d*) $ClCH_2CH_2COOH$.
(AIEEE)

51. On vigorous oxidation by permanganate solution $(CH_3)_2C = CH-CH_2CH_3$ gives

(*a*) $CH_3-\underset{\underset{CH_3}{|}}{\overset{\overset{OH}{|}}{C}}-\overset{\overset{OH}{|}}{CH}-CH_2CH_3$

(*b*) $\begin{matrix}CH_3\\CH_3\end{matrix}\!\!>CHCO_2H + CH_3CH_2COOH$

(*c*) $\begin{matrix}CH_3\\CH_3\end{matrix}\!\!>CHOH + CH_3CH_2CH_2OH$

(*d*) $\begin{matrix}CH_3\\CH_3\end{matrix}\!\!>C = O + CH_3CH_2COOH$ ***(AIEEE)***

52. The reaction

$$CH_3CH = CH_2 \xrightarrow[H^+]{CO + H_2O} CH_3-\underset{\underset{COOH}{|}}{CH}-CH_3$$

is known as
(*a*) Wurtz reaction
(*b*) Koch reaction
(*c*) Clemmensen's reduction
(*d*) Kolbe's reaction. ***(MP MPT)***

53. By aerial oxidation, which one of the following gives phthalic acid.
(*a*) Naphthalene (*b*) Benzene
(*c*) Mesitylene (*d*) Toluene.
(Tamil Nadu CET)

54. Identify the wrong statement's from the following
(*a*) Salicylic acid is a monobasic acid
(*b*) Methyl salicylate is an ester
(*c*) Salicylic acid gives violet colour with neutral ferric chloride as well as brisk effervescence with sodium bicarbonate
(*d*) Methyl salicylate does not occur in natural oils.
(Tamil Nadu CET)

2003

55. The correct order of increasing basic nature for bases $\ddot{N}H_3$, $CH_3\ddot{N}H_2$ and $(CH_3)_2\ddot{N}H$ is
(*a*) $CH_3NH_2 < NH_3 < (CH_3)_2NH$
(*b*) $(CH_3)_2NH < NH_3 < CH_3NH_2$
(*c*) $NH_3 < CH_3NH_2 < (CH_3)_2NH$
(*d*) $CH_3NH_2 < (CH_3)_2NH < NH_3$. ***(AIEEE)***

56. Ethyl isocyanide on acidic hydrolysis generates
(*a*) Ethylamine salt and methanoic acid
(*b*) Propanoic acid and ammonium salt
(*c*) Ethanoic acid and ammonium salt
(*d*) Methyl amine salt and ethanoic acid. ***(AIEEE)***

57. The reaction of chloroform with alcoholic KOH and *p*-toludine forms

(*a*) 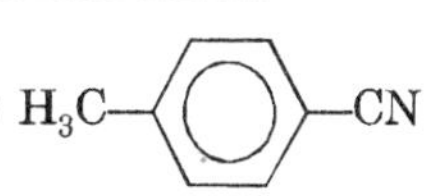 (*b*) 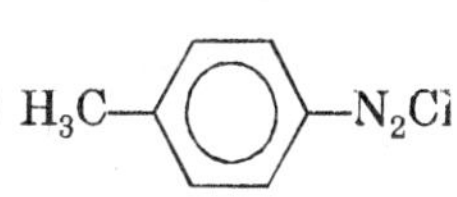

(*c*) 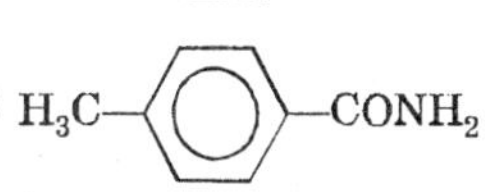 (*d*) 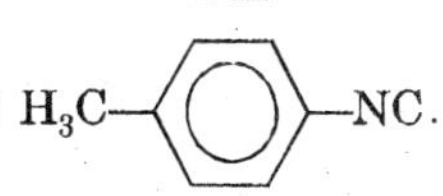

(AIEEE)

58. $R-\overset{O}{\overset{\|}{C}}-R' \xrightarrow{HCN/KCN} A \xrightarrow{B} RR'C(OH)(CH_2NH_2)$

In the above sequence of reactions, A and B are

(*a*) A is $RR'C(CN)(OH)$, B is $LiAlH_4$

(*b*) A is $RR'C(OH)(COOH)$, B is NH_3

(*c*) A is $RR'C(CN)(OH)$, B is H_3O^+

(*d*) A is $RR'CH_2CN$, B is NaOH. **(*CBSE PMT*)**

59. 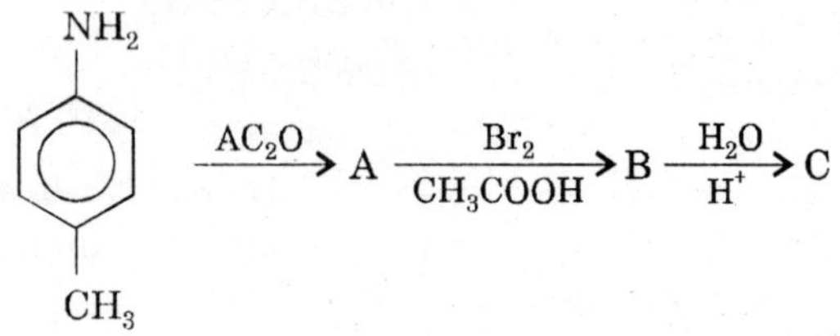

The final product 'C' in the above reaction is

(*a*) 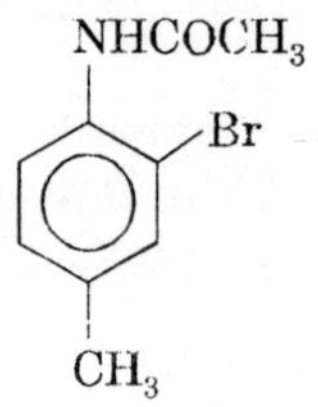 (*b*) 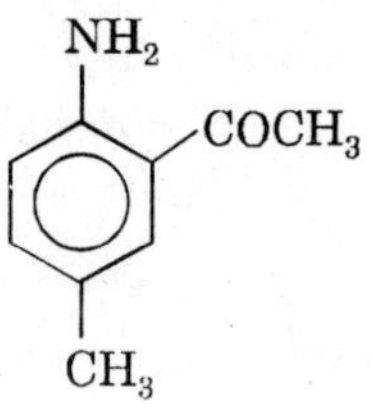

(*c*) 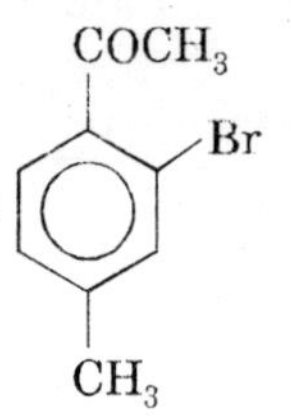 (*d*) 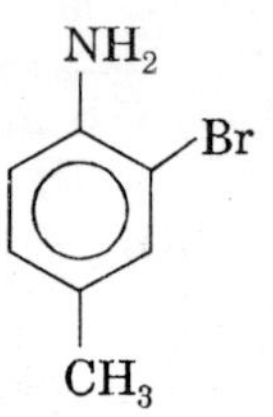

(*CBSE PMT*)

60. Among the following the weakest base is

(*a*) $C_6H_5CH_2NH_2$ (*b*) $C_6H_5CH_2NHCH_3$

(*c*) $O_2NCH_2NH_2$ (*d*) CH_3NHCHO. **(*AIIMS*)**

61. Identify the product Z in the following reaction scheme

$C_6H_5NH_2 \xrightarrow{AC_2O} X \xrightarrow{Br_2/CCl_4} Y \xrightarrow{HOH} Z$

(*a*) *p*-bromoaniline (*b*) *p*-bromoacetophenone

(*c*) *p*-bromoacetanilide (*d*) *o*-bromoacetophenone

(*e*) *o*-bromoacetanilide. **(*Kerala PMT*)**

2004

62. Toluic acid on reaction with Br_2 + Fe gives

(*a*) 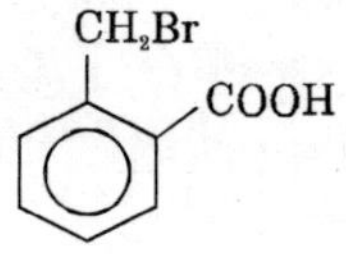 (*b*) 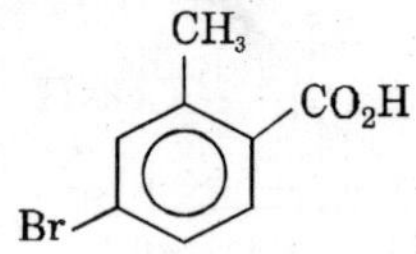

(*c*) 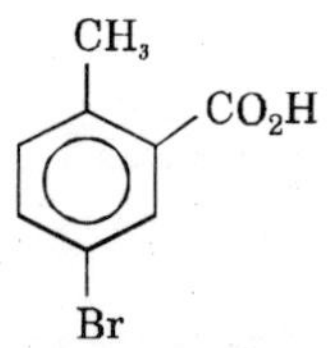 (*d*) 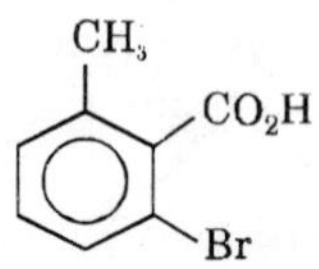 **(*AIIMS*)**

63. Melting points are normally the highest for

(*a*) Tertiary amides (*b*) Secondary amides

(*c*) Primary amides (*d*) Amines. **(*AIIMS*)**

64. The OH group of an alcohol or the —COOH group of a carboxylic acid can be replaced by —Cl using

(*a*) PCl_5 (*b*) HOCl

(*c*) Cl (*d*) HCl. **(*CBSE PMT*)**

65. The rate of the reaction

$$R-C(=O)-Z + \overset{\ominus}{Nu} \longrightarrow R-C(=O)-Nu + \overset{\ominus}{Z}$$

is fastest when Z is

(*a*) Cl (*b*) NH_2

(*c*) OC_2H_5 (*d*) $OCOCH_3$ **(*AIEEE*)**

66. Consider the acidity of the carboxylic acids

(*a*) PhCOOH (*b*) *o*—$NO_2C_6H_4COOH$

(*c*) *p*-$NO_2C_6H_4COOH$ (*d*) *m*-$NO_2C_6H_4COOH$

which is the correct order ?

(*a*) $a > b > c > d$ (*b*) $b > d > c > a$

(*c*) $b > d > a > c$ (*d*) $b > c > d > a$.

(*AIEEE*)

67. On mixing ethyl acetate with aq NaCl, the composition of the resultant solution is

(*a*) $CH_3COOC_2H_5 + NaCl$

(*b*) $CH_3COONa + C_2H_5OH$

(*c*) $CH_3COCl + C_2H_5—OH + NaOH$

(*d*) $CH_3Cl + C_2H_5COONa$. **(*AIEEE*)**

68. Acetyl bromide reacts with excess of CH_3MgI followed by treatment with a saturated solution of NH_4Cl gives

(*a*) Acetone (*b*) Acetamide

(*c*) 2-methyl-2-propanol (*d*) Acetyl iodide.

(*AIEEE*)

2005

69. Among the following acids which has the lowest pKa value ?
(a) CH_3CH_2COOH (b) $(CH_3)_2CH—COOH$
(c) HCOOH (d) CH_3COOH. **(AIEEE)**

70. The reaction,

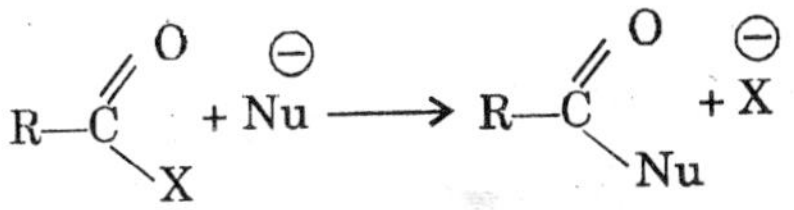

is fastest when X is
(a) OCOR (b) OC_2H_5
(c) NH_2 (d) Cl. **(AIEEE)**

71. Acetic acid will be obtained on oxidation of
(a) Ethanol (b) Propanal
(c) Methanal (d) Glyoxal. **(J & K CET)**

72. A carboxylic acid is converted into its anhydride using
(a) Thionyl chloride (b) Sulphur chloride
(c) Sulphuric acid (d) Phosphorus pentaoxide. **(J & K CET)**

73. CH_3COOCH_3 + excess PhMgBr

$\rightarrow$ product $\xrightarrow{H^+}$ X

The product X is
(a) 1, 1-diphenylethanol
(b) 1, 1-diphenylmethanol
(c) Methyl phenylethanol
(d) Methyl phenylketone. **(Orissa JEE)**

74. Order of hydrolysis for the following
(I) RCOCl (II) RCOOR
(III) $RCONH_2$ (IV) $(RCO)_2O$.
(a) I > IV > II > III (b) I > II > III > IV
(c) I > III > II > IV (d) IV > III > II > I. **(DPMT)**

2006

75. In a set of reactions propionic acid yielded a compound D.

$$CH_3CH_2COOH \xrightarrow{SOCl_2} B \xrightarrow{NH_3} C \xrightarrow[Br_2]{KOH} D$$

The structure of D would be
(a) $CH_3CH_2NH_2$ (b) $CH_3CH_2CH_2NH_2$
(c) $CH_3CH_2CONH_2$ (d) $CH_3CH_2NHCH_3$. **(CBSE PMT)**

76. Self condensation of two moles of ethyl acetate in presence of sodium ethoxide yields
(a) Ethyl propionate
(b) Ethyl butyrate
(c) Acetoacetic ester
(d) Methyl acetoacetate. **(CBSE PMT)**

77. The correct order of increasing acid strength of the compounds
(a) CH_3CO_2H (b) $MeOCH_2CO_2H$
(c) CF_3CO_2H (d) 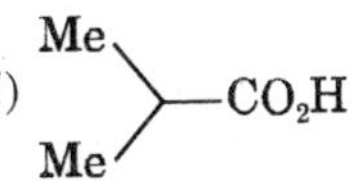

is
(a) b < d < a < c (b) d < a < c < b
(c) d < a < b < c (d) a < d < c < b. **(AIEEE)**

78. Hydroxylamine reacts with
(a) CH_3COCl (b) CH_3COCH_3
(c) $CH_3COOC_2H_5$ (d) CH_3CONH_2. **(BHU Mains)**

79. Which has highest pK_a value ?

(a) 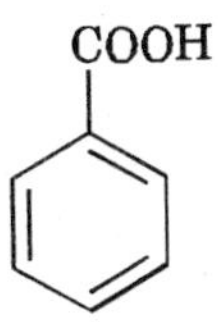

(b) 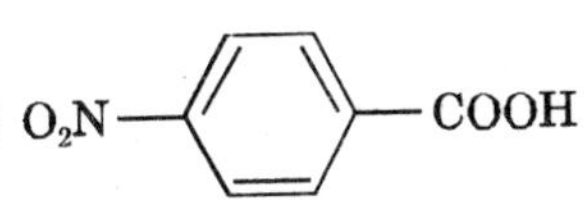

(c) 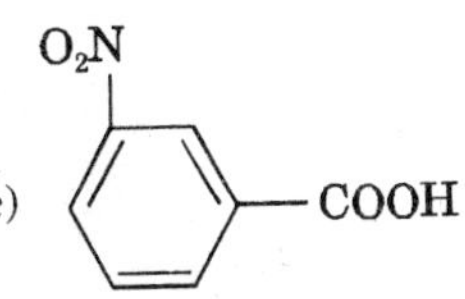

(d) 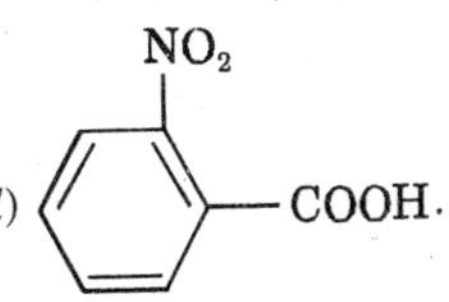 **(DCE)**

80. Which of the following does not give benzoic acid on hydrolysis is ?
(a) Phenyl cyanide (b) Benzoyl chloride
(c) Benzyl chloride (d) Methyl benzoate. **(KCET)**

81. Consider the following compounds
(A) C_6H_5COCl

(B) 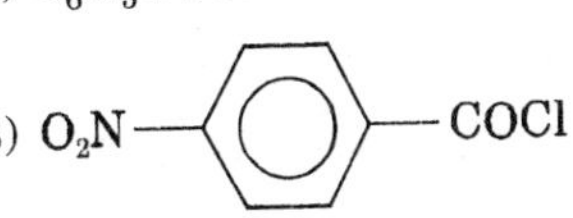

(C) 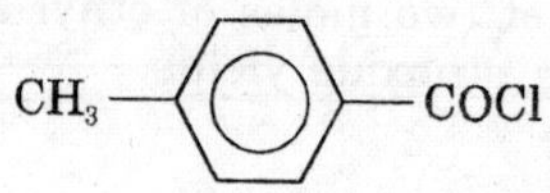

(D) 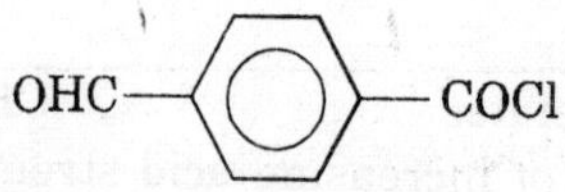

The correct decreasing order of their reactivity to wards hydrolysis is

(a) $b > d > a > c$ (b) $b > d > c > a$

(c) $a > b > c > d$ (d) $d > b > a > c$.

(CBSE AIPMT Screening)

82. Which of the following presents the correct order of the acidity in the given compounds ?

(a) $FCH_2COOH > ClCH_2COOH > BrCH_2COOH > CH_3COOH$

(b) $CH_3COOH > BrCH_2COOH > ClCH_2COOH > FCH_2COOH$

(c) $FCH_2COOH > CH_3COOH > BrCH_2COOH > ClCH_2COOH$

(d) $BrCH_2COOH > ClCH_2COOH > BrCH_2COOH > CH_3COOH$.

(CBSE AIPMT Screening)

83. The compound formed as a result of oxidation of ethyl benzene by $KMnO_4$ is

(a) Benzo phenone (b) Aceto phenone

(c) Benzoic acid (d) Benzyl alcohol. ***(AIEEE)***

ANSWERS

1. (c)	**2.** (c)	**3.** (c)	**4.** (c)	**5.** (a)	**6.** (d)	**7.** (c)	**8.** (d)	**9.** (a)
10. (d)	**11.** (a)	**12.** (c)	**13.** (a)	**14.** (c)	**15.** (a)	**16.** (a)	**17.** (c)	**18.** (c)
19. (b)	**20.** (c)	**21.** (a)	**22.** (c)	**23.** (a)	**24.** (b)	**25.** (c)	**26.** (a)	**27.** (c)
28. (d)	**29.** (d)	**30.** (b)	**31.** (a)	**32.** (b)	**33.** (d)	**34.** (a)	**35.** (b)	**36.** (a)
37. (a)	**38.** (c)	**39.** (b)	**40.** (c)	**41.** (d)	**42.** (c)	**43.** (c)	**44.** (b)	**45.** (b)
46. (b)	**47.** (d)	**48.** (c)	**49.** (b)	**50.** (c)	**51.** (d)	**52.** (b)	**53.** (a)	**54.** (d)
55. (c)	**56.** (a)	**57.** (d)	**58.** (a)	**59.** (d)	**60.** (d)	**61.** (a)	**62.** (c)	**63.** (c)
64. (a)	**65.** (a)	**66.** (d)	**67.** (a)	**68.** (c)	**69.** (c)	**70.** (d)	**71.** (a)	**72.** (d)
73. (a)	**74.** (a)	**75.** (a)	**76.** (c)	**77.** (c)	**78.** (b)	**79.** (a)	**80.** (c)	**81.** (a)
82. (a)	**83.** (c).							

HINTS/SOLUTIONS

Q. No.	Ans.	Hints & Solutions
1.	(c)	$CH_3CH(OH)—COOH \xrightarrow{\text{Conc. } H_2SO_4} CH_2 = CH—COOH + H_2O$. Acrylic acid
2.	(c)	$CH_3CON(CH_3)_2 \xrightarrow{H^+} CH_3COOH + (CH_3)_2NH$
3.	(c)	$NH_4CNO \xrightarrow{\text{Heat}} H_2NCONH_2$. (urea)

Q. No.	Ans.	Hints & Solutions
4.	(c)	It is Claisen in condensation.
5.	(a)	Refers to Kolbe's electrolytic decarboxylation.
6.	(d)	$HCOOH \xrightarrow{H_2SO_4} H_2O + CO$.
7.	(c)	$CO + NaOH \longrightarrow HCOONa$.
8.	(d)	Among the given groups only —COOH is meta-directing.
9.	(a)	Lucas test is meant for alcohols.
10.	(d)	α-amino acids exist in the form Zwitter ion which is also referred to as internal salt structure.

Q. No.	Ans.	Hints & Solutions
		$NH_2—CH_2—COOH \rightleftharpoons H_3\overset{+}{N}—CH_2—COO^-$ α-Amino acid Zwitter ion
11.	(a)	HVZ reaction causes substitution of halo groups at α-C atom of carboxylic acids.
12.	(c)	It contains phen acyl chloride.
13.	(a)	The reactivity of acid derivatives follows the order : anhydride > ester > ethers.
14.	(c)	Addition of bromine across the C = C occurs against Markownikoff rule in the presence of peroxide.
15.	(a)	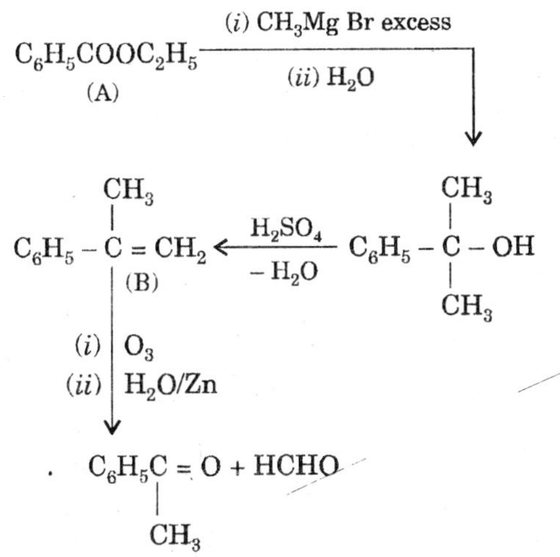
16.	(a)	CF_3COOH has a strong –I group of F.
17.	(c)	It undergoes HVZ reaction.
18.	(c)	Acetic acid reacts in the presence of PCl_5 to form acid chloride.
19.	(b)	$CH_3COOC_2H_5 \xrightarrow[(ii)\ H_2O]{(i)\ CH_3MgBr} CH_3—\underset{CH_3}{\overset{CH_3}{\overset{\vert}{\underset{\vert}{C}}}}—OH$ 3° alcohol
20.	(c)	$HCONH_2 \xrightarrow{NaOH} HCOONa + NH_3$ Propanamide will also give similar products, but it is not a liquid at room temperature, it is solid. Hence correct answer is $HCONH_2$.
21.	(a)	$\underset{(A)}{CH_3CN} \xrightarrow{Na/EtOH} CH_3CH_2NH_2 \xrightarrow{HNO_2} CH_3CH_2OH \xrightarrow{HNO_2} CH_3CHO \xrightarrow{[Ag(NH_3)_2]OH} CH_3COOH$
22.	(c)	P_2O_5 is a dehydrating agent.
23.	(a)	Choice itself is the answer.
24.	(b)	It is Fischer-Spierer reaction.
25.	(c)	Esters have fruity smell.
26.	(a)	The electron with drawing effect of Halogen atoms is in the order I < Br < Cl < F and hence the answer.
27.	(c)	It is Hoffmann's Bromamide reaction.
28.	(d)	Choice itself is the answer.
29.	(d)	$H—\overset{O}{\overset{\Vert}{C}}—OH$ has H atom which is easily oxidisable.
30.	(b)	See concept oriented details.
31.	(a)	HCOOH has $H—\overset{O}{\overset{\Vert}{C}}—$ type of grouping and hence reacts with Tollen's reagent (amm. $AgNO_3$).
32.	(b)	CH_3COOH or Kolbe's electrolysis produces ethane ∴ The ester is methyl ethanoate.
33.	(d)	It is Hoffmann's Bromamide reaction.
34.	(a)	H_2SO_4 dehydrates HCOOH to form CO.
35.	(b)	It is H.V.Z. reaction.
36.	(a)	Its dissociation constant is less as compound to carboxylic acids.
37.	(a)	It is Wittig's reaction.
38.	(c)	It can be obtained by reaction with CH_3COCl/anhy $AlCl_3$ and oxidation of alcohols.
39.	(b)	Wolff Kishner reduction changes >C = O to >CH_2.
40.	(c)	For aldol condensation α-H atoms is required to be present.
41.	(d)	—COOH group undergoes resonance stabilisation and help in the easy release of H^+ ions.
42.	(c)	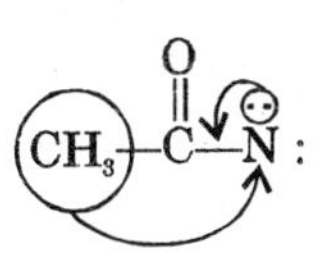 Nitrene rearranges to isocyanate.
43.	(c)	$\underset{COOH}{\overset{COOH}{\vert}} \xrightarrow{Zn/H_2SO_4} \underset{COOH}{\overset{CH_2OH}{\vert}}$
44.	(b)	OH group of pyruvic acid is oxidised to >C = O group.

Q. No.	Ans.	Hints & Solutions
45.	(b)	$H—\overset{O}{\overset{\parallel}{C}}—OCH_2Cl + CH_3MgI \longrightarrow H—\overset{O}{\overset{\parallel}{C}}—OCH_2CH_3 + Mg\langle^{I}_{Cl}$
46.	(b)	Glacial acetic acid is 100% pure acetic acid.
47.	(d)	It is due to H-bonding 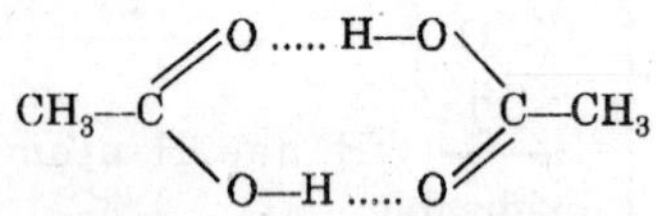
48.	(c)	It is Rosenmund's reduction.
49.	(b)	It is Kolbe's reaction.
50.	(c)	CH_3CH_2COOH C $CH_2CH_3 \xrightarrow{Cl_2/P_4}$ $CH_3\overset{OH}{\overset{\vert}{C}H}$ COOH $\xrightarrow{\text{alc KOH}}$ $CH_2 = CH$ COOH
51.	(d)	Cleavage occurs at the double bond along with oxidation.
52.	(b)	Alkenes undergo carbonylation in Koch reaction.
53.	(a)	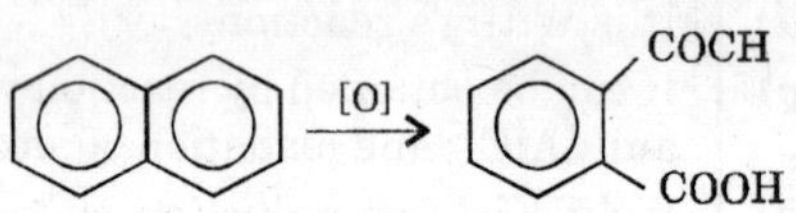
54.	(d)	Methyl oalicylate is also present in natural oils.
55.	(c)	More is the no. of + I groups greater is the basic character.
56.	(a)	$C_2H_5N \equiv C + H_2O \xrightarrow{H^+} C_2H_5NH_2 + HCOOH$
57.	(d)	It is an example of carbylamine reaction. See concepts in brief.
58.	(a)	HCN adds to C = O group to give cyano hydrins. CN group reduces to CH_2NH_2 group with $LiAlH_4$.
59.	(d)	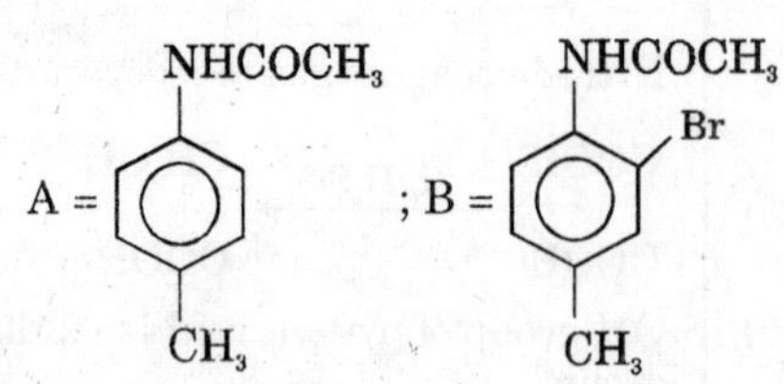
		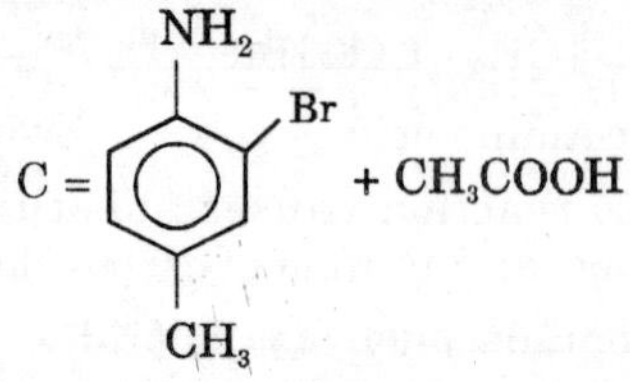
60.	(d)	Because in amides the lone pair on nitrogen is in conjugation with C O group.
61.	(a)	. $C_6H_5NH_2 \xrightarrow{Ac_2O} C_6H_5NH—\overset{O}{\overset{\parallel}{C}}—CH_3$ 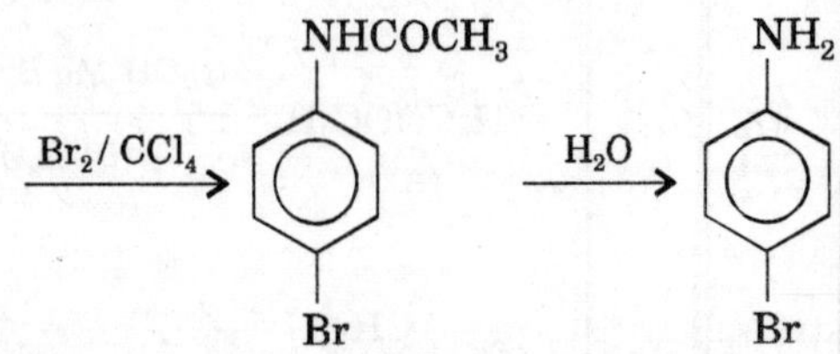
62.	(c)	Bromination occurs and in coming group occupies *para* position w.r.t. electron releasing group.
63.	(c)	1° amides have high m.p. due to extensive H-bonding.
64.	(a)	PCl_5 is the best chlorination agents for OH group of alcohol Cr—COOH group of acids.
65.	(a)	Cl^- is the best leaving group among the given options.
66.	(d)	*o*-nitrobenzoic acid is the strongest due to electron with drawing NO_2 group the order of various substituted derivatives is $o > p > m$.
67.	(a)	$CH_3COOC_2H_5 \xrightarrow{\text{aq. NaCl}}$ no reaction.
68.	(c)	$CH_3COBr \xrightarrow[(ii)\ \text{Satd } NH_4Cl]{(i)\ CH_3MgI} CH_3—\overset{CH_3}{\overset{\vert}{\underset{CH_3}{\underset{\vert}{C}}}}—OH$ 2-methyl-2-propanol
69.	(c)	pKa = – log Ka *i.e.*, lower the value of Ka more is pKa. Since a strong acid has larger Ka value ∴ its pKa will be least.
70.	(d)	Cl is good leaving group and hence the reaction would be fastest.
71.	(a)	$C_2H_5OH \xrightarrow{[O]} CH_3COOH$
72.	(d)	$\begin{matrix} RCOOH \\ RCOOH \end{matrix} \xrightarrow{P_2O_5} \begin{matrix} RCO \\ RCO \end{matrix}\rangle O + H_2O$ P_2O_5 acts as a dehydrating agent.

Q. No.	Ans.	Hints & Solutions
73.	(a)	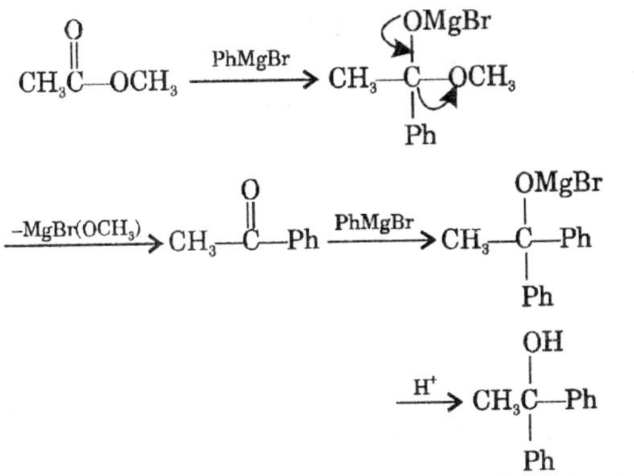 1, 1-diphenylethanol
74.	(a)	Of all the acid derivatives, acid chlorides, *i.e.*, CH_3COCl is most reactive. The order of reactivity of acid derivatives decreases in the following order : $RCOCl > (RCO)_2O > RCOOR > RCONH_2$.
75.	(a)	$CH_3CH_2COOH \xrightarrow{SOCl_2} CH_3CH_2COCl\ (B)$ $CH_3CH_2COCl \xrightarrow{NH_3} CH_3CH_2CONH_2\ (C)$ $CH_3CH_2CONH_2\ (C) \xrightarrow[Br_2]{KOH} CH_3CH_2NH_2$
76.	(c)	$2CH_3-\overset{O}{\overset{\Vert}{C}}-OC_2H_5 \xrightarrow{C_2H_5ONa} CH_3-\overset{O}{\overset{\Vert}{C}}-CH_2-\overset{O}{\overset{\Vert}{C}}-OC_2H_5$ Acetoacetic ester

Q. No.	Ans.	Hints & Solutions
77.	(c)	The acidic strength depends upon inductive effect. –I groups increase the acidic strength while +I groups decrease the acidic strength.
78.	(b)	$CH_3-\overset{O}{\overset{\Vert}{C}}-CH_3 + NH_2OH \longrightarrow (CH_3)_2C=N-OH$ Oxime
79.	(a)	The strongest acid will have lowest pK_a value and vice-versa.
80.	(c)	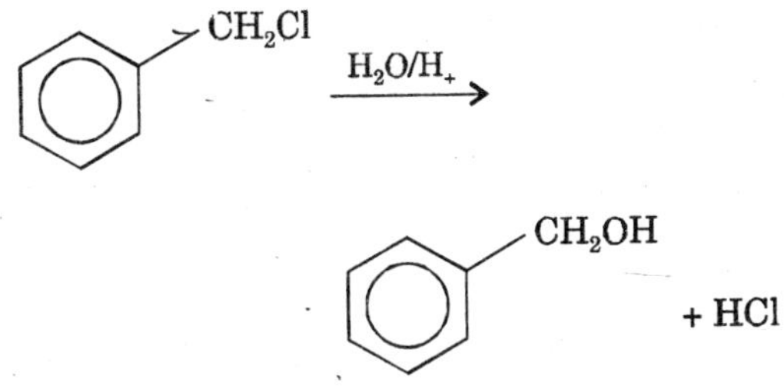
81.	(a)	The attack of the nucleophile on the carbonyl carbon is the rate determining step and hence the order.
82.	(a)	More is the –I effect of the group greater is the acidic character.
83.	(c)	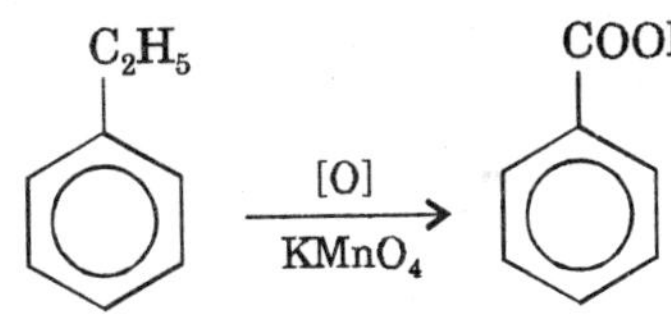

6.6. MCQs FOR IIT JEE

Brain Twister

A. Multiple Choice Questions With One Correct Answer

1. The name of the family associated with general formula $(RO)_2CR_2$ is
   (*a*) Acid anhydride (*b*) Ketones
   (*c*) Acetals (*d*) Hemiacetals.
2. The main product formed when propionic acid vapours are passed over heated manganese oxide at 575 K is
   (*a*) Propionic anhydride (*b*) Propanal
   (*c*) 3-pentanone (*d*) Propanone.
3. Which of the following will be produced when ethanoic acid and diazomethane are allowed to react ?
   (*a*) Ethyl acetate (*b*) Methyl methanoate
   (*c*) Methyl ethanoate (*d*) Ethanoyl cyanide.
4. Which compound will be formed when ethylacetate and excess of ethyl magnesium bromide allowed to react and product is hydrolysed ?
   (*a*) 3-Ethyl-3-pentanol (*b*) 3-Methyl-3-pentanol
   (*c*) Hexan-3-ol (*d*) 2-Methyl-2-propanol.
5. Which of the following reagent will distinguish between two functional isomers of formula $C_2H_4O_2$?
   (*a*) $NaHCO_3$ (*b*) $CuSO_4$
   (*c*) C_2H_5Cl (*d*) All the above.
6. $\mathbf{C_6H_5COCl \xrightarrow{NH_3} X \xrightarrow{Y} C_6H_5CN}$. In this sequence Y is
   (*a*) KCN (*b*) $Ba(CN)_2$
   (*c*) P_2O_5 (*d*) P_4.
7. HVZ reaction can be used to convert propionic acid to
   (*a*) Malonic acid (*b*) Stearic acid
   (*c*) Lactic acid (*d*) Succinnic acid.
8. Which of the following will liberate ethane as one of the product ?
   (*a*) Soda lime decarboxylation of sodium methanoate
   (*b*) Electrolysis of aqueous solution of sodium methanoate
   (*c*) Reduction of ethanoic acid with HI/P
   (*d*) Treatment of ethanoic acid with H_2/Pd.
9. Claisen condensation involves the treatment of
   (*a*) Ethyl acetate and ethanol
   (*b*) Ethyl acetate and sodium ethoxide
   (*c*) Ethanol and peroxy acetic acid
   (*d*) Ethyl acetate and Grignard's reagent.
10. Which of the following substance will not give silver mirror with Tollen's reagent ?
    (*a*) Ethanal (*b*) Methanoic acid
    (*c*) Propanone (*d*) Propanal.
11. In which of the following structure the carbon-oxygen bond lengths are equal ?
    (*a*) $CH_3-\overset{\overset{O}{||}}{C}-OH$ (*b*) $CH_3-\overset{\overset{O}{||}}{C}-O^-$
    (*c*) $C_6H_5-\overset{\overset{O}{||}}{C}-OH$ (*d*) $CH_3-\overset{\overset{O}{||}}{C}-OCH_3$.
12. $\mathbf{BrCH_2CH_2CH_2Br \xrightarrow{Alc.\ NaCN} X \xrightarrow{H_3O^+} Y}$. The product Y in the above sequence is
    (*a*) Gluteric acid (*b*) 1, 3-propanedioic acid
    (*c*) Succinic acid (*d*) Melonic acid.
13. In order to convert $\mathbf{C_6H_5CONH_2}$ to $\mathbf{C_6H_5-CH_3}$, which of the following sequence of steps is correct ?
    (*a*) Dil. NaOH ; CaO/NaOH ; $CH_3Cl/AlCl_3$
    (*b*) $NaNO_2/HCl$; NH_3 ; NaOH
    (*c*) KOH/Br_2 ; $K_2Cr_2O_7/H^+$; $CH_3Cl/AlCl_3$
    (*d*) $CH_3Cl/AlCl_3$; CaO/NaOH ; Dil. NaOH.
14. When ketene is passed through glacial acetic acid, which of the following product is formed ?
    (*a*) Ethyl acetate (*b*) Ethanoic anhydride
    (*c*) Propanoic acid (*d*) Pyruvic acid.
15. Which of the following aromatic acid has lowest value of K_a ?
    (*a*) *p*-nitrobenzoic acid (*b*) *p*-methoxybenzoic acid
    (*c*) *p*-chlorobenzoic acid (*d*) Benzoic acid.
16. $\mathbf{R-COCH_3 \xrightarrow{X_2/OH^-} CHX_3 + Carboxylate\ ion \xrightarrow{H^+} Carboxylic\ Acid}$
    In the above sequence, the carboxylic acid obtained is
    (*a*) CH_3COOH (*b*) HCOOH
    (*c*) RCOOH (*d*) RCH_2COOH.
17. The increasing order of acid strength is correct in
    (*a*) Phenol, Acetic acid, Ethanol, Chloroacetic acid
    (*b*) Acetic acid, Ethanol, Phenol, Chloroacetic acid

(c) Ethanol, Phenol, Acetic acid, Chloroacetic acid
(d) Chloroacetic acid, Acetic acid, Ethanol, Phenol.

18. Which of the following is present in Vinegar ?
(a) Acetic acid (b) Citric acid
(c) Tartaric acid (d) Lactic acid.

19. Acid chlorides react with Grignard's reagent to form
(a) Alcohols (b) Esters
(c) Aldehydes (d) Ketones.

20. Which of the following is strongest reducing agent ?
(a) Benzoic acid (b) Acetic acid
(c) Isobutyric acid (d) Methanoic acid.

21. The percentage of acetic acid in pyroligneous acid is approximately
(a) 2% (b) 10%
(c) 20% (d) 15%.

22. Fermentation of ethanol to ethanoic acid is achieved by
(a) Bacterium *Mycoderma aceti*
(b) Acetic invertase
(c) Diastase
(d) Zymase.

23. Which of the following is insoluble in acetic acid ?
(a) Calcium oxalate (b) Calcium carbonate
(c) Quicklime (d) Calcium hydroxide.

24.
$$\begin{array}{l} CH_2COOH \\ | \\ C(OH)COOH + H_2SO_4 \text{ Conc.} \xrightarrow[\text{(Fuming)}]{} ? \\ | \\ CH_2COOH \end{array}$$
The main product in the above reaction is
(a) Acetone (b) Acetone dicarboxylic acid
(c) Dihydroxy acetone (d) Citranoic acid.

25. $CH_3COOH + Cl_2 \xrightarrow{\text{Red P}}$ .... In this reaction acetic acid gets converted into
(a) CH_3COCl (b) $CHCl_3$
(c) $Cl_3CH(OH)_2$ (d) Cl_3CCOOH.

26. When adipic acid is heated strongly what is formed ?

(a) 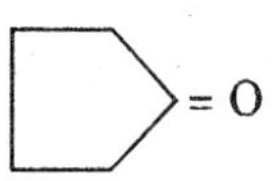 (b) 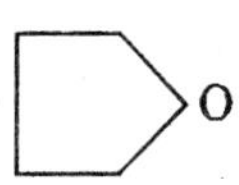

(c) $\begin{array}{l} CH_2—CH_2COOH \\ | \\ CH_2—CH_2COOH \end{array}$ (d) 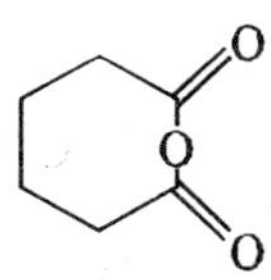

27. Ammonium acetate on heating gives
(a) N_2 gas (b) Ammonia gas
(c) Acetic acid vapours (d) Ethanamide.

28. Alkali salt of palmitic acid is known as
(a) An epoxide (b) A soap
(c) An alkaloid (d) A lipid.

29. Acetamide and nitrous acid react to form
(a) CH_3NH_2 and N_2 (b) CH_3OH and NH_3
(c) CH_3COOH and N_2 (d) CO_2 CH_3NH_2.

30. The inorganic compound which on heating produces organic compound is
(a) Sodamide (b) Acetamide
(c) Ammonium cyanate (d) Ammonium phosphate.

31. Which among the following reagents will justify the acidic nature of amides ?
(a) HgO (b) NHO_2
(c) Br_2/KOH (d) HCl.

32. The number of moles of ethanoyl chloride which produces 10.2 g of ethanoic anhydride on reaction with sodium acetate is
(a) 0.51 mol (b) 1.02 mol
(c) 0.10 mol (d) 10.2 mol.

33. Which acid derivative react with hydroxyl amine to produce hydroxamic acid ?
(a) Esters (b) Acid chloride
(c) Both (a) and (b) (d) None of these.

34. Which of the following is not a hydroxy acid ?
(a) Lactic acid (b) Tartaric acid
(c) Citric acid (d) Succinic acid.

35. Which of the following is not a dicarboxylic acid ?
(a) Oxalic acid (b) Tartaric acid
(c) Malonic acid (d) Toluic acid.

36. Distinction between oxalic acid, malonic acid and succinic acid can be achieved by
(a) Heating (b) Acidified $KMnO_4$
(c) Br_2 water (d) PCl_5.

37. Soda lime decarboxylation of which of the following will produce 1, 1, 2, 2-tetrabromethane ?
(a) $Br_3C . COOH$ (b) $CH_2Br . CHBrCOOH$
(c) CH_2BrCBr_2COOH (d) $CHBr_2CBr_2COOH$.

38. Which of the following are generally used in the preparation of artificial flavours ?
(a) Lower alcohols
(b) Esters of unsaturated fatty acids
(c) Esters of lower fatty acids
(d) Esters of higher fatty acids.

39. Which of the following reagents can provide distinction between acetic acid and formic acid ?
(a) Na (b) $NaNH_2$
(c) dil. acidic $KMnO_4$ (d) 2, 4 DNP.

40. Which one of the following is expected to be highly ionised in aqueous solution ?

(a) $CH_2ClCH_2CH_2COOH$ (b) $CH_3CHClCH_2COOH$
(c) $CH_3CH_2CHClCOOH$ (d) $CH_3CH_2CCl_2COOH$.

41. Amides contain C = O group, yet they do not give characteristic reactions of C = O group because
(a) They dimerise
(b) Of resonance
(c) They possess cyclic structure
(d) Of attached alkyl group.

42. Almost all amides exist in
(a) Liquid state (b) Gaseous state
(c) Solid state (d) Plasma state.

43. Which of the following acid does not form anhydride when treated with P_2O_5 at elevated temperature ?
(a) Acetic acid (b) Propanoic acid
(c) Oxalic acid (d) Benzoic acid.

44. Which of the following sodium salt on treatment with soda lime does not produce alkane ?
(a) Sodium ethanoate (b) Sodium isobutyrate
(c) Sodium propanoate (d) Sodium methanoate.

45. Reduction of esters with sodium and alcohol is referred to as
(a) Bouvealt-Blanc reaction
(b) Mendius reaction
(c) Clemmensen's reduction
(d) MPV reduction.

46. The end product C in the following reaction is

$$CH_3OH \xrightarrow{PCl_5} A \xrightarrow{KCN} B \xrightarrow{H_3O^+} C.$$

(a) Methanoic acid (b) Ethanoic acid
(c) Methanamide (d) Ethanamide.

47. Which of following will reduce carboxylic acid to 1° alcohol ?
(a) $H_2/BaSO_4$ (b) HI
(c) $LiAlH_4$ (d) Na/C_2H_5OH.

48. The pK_a of which acid is lowest ?
(a) C_6H_5COOH (b) CH_3COOH
(c) CH_3CH_2COOH (d) CH_2FCOOH.

49. Which acid is formed when sucrose is heated with conc. HNO_3 ?
(a) Gluconic acid (b) Ascorbic acid
(c) Oxalic acid (d) Citric acid.

50. Bees wax is
(a) Myricyl palmitate
(b) Myricyl stearate
(c) Mixture of higher hydrocarbon
(d) Myricyloleate.

51. A dicarboxylic acid has four (CH_2) groups in its molecule.
(a) Adipic acid (b) Pivalic acid
(c) Pimelic acid (d) Gluteric acid.

52. $KMnO_4$ oxidation of which of the following will produces benzoic acid ?
(a) Ethyl benzene (b) Toluene
(c) Aniline (d) Anisole.

53. Which of the following pair are not isomers of each other ?
(a) Cyclopropane, carboxylic acid, crotonic acid
(b) Methylacetate and propanoic acid
(c) Ethyl acetate and crotonic acid
(d) Acetic acid and hydroxyethanal.

54. Which of the following groups present at para position w.r.t. —COOH in substituted benzoic acids enhance the acid strength ?
(a) $—NO_2$ (b) $—OCH_3$
(c) —OH (d) $—C(CH_3)_3$.

55. Which of the following will produce tartaric acid ?
(a) Hydrolysis of glyoxal
(b) Oxidation of fumaric acid by $KMnO_4$
(c) Reduction of malonic acid
(d) Addition of water to succinic acid.

56. The chemical reaction between alcohol and esters in acidic medium
(a) Involve exchange of alkoxy parts
(b) Are called esterification
(c) Involve exchange of their alkyl parts
(d) Are called self esterification.

57. Which carboxylic acid will show HVZ reaction ?
(a) Formic acid
(b) Trichloroethanoic acid
(c) Phenyl acetic acid
(d) 2, 2-dimethyl propanoic acid.

58. An ester on treatment with excess of ethyl magnesium iodide and subsequent hydrolysis produces 3-pentanol. The ester could be
(a) Methyl acetate (b) Methylmethanoate
(c) Ethyl acetate (d) Ethyl propionate.

59. Propanamide $\xrightarrow{H_2/Pt}$ X

Propanamide $\xrightarrow{Br_2/KOH}$ Y

In the above sequence of reactions
(a) X and Y are functional isomers
(b) X and Y are optically active amines
(c) X and Y are homologoues
(d) Y is bromo derivative of X.

60. Which of the following reagents will be able to produce propionylchloride from propionic acid ?

(a) anhy $ZnCl_2$ (b) Cl_2/P_4
(c) HCl (d) $SOCl_2$.

61. Which chemical reaction can be referred to as acylation process ?
(a) EtOH + AcOEt (b) $EtNH_2$ + Aldehyde
(c) $EtNH_2 + (CH_3CO)_2O$ (d) EtCOOHEt + Et_3N.

62. In the Hoffmann's bromamide reaction, the intermediate involved are
(a) RCOBr (b) R—N = C = O
(c) R—NHBr (d) R—$CONBr_2$.

63. Which of the following acids do not contain —COOH group ?
(a) Acrylic acid (b) Prussic acid
(c) Acetic acid (d) Salicylic acid.

64. Which of the following species donot show resonance stabilisation ?
(a) CH_3COOH (b) $C_6H_5COO^-$ ion
(c) $CH_2 = CH—CH_2O^-$ (d) C_6H_5OH.

65. Which of the following produces acetic acid ?
(a) Hydrolysis of pyruvic acid
(b) Oxidation of but-2-ene with hot $KMnO_4$
(c) Dehydration of ethanamide
(d) Hydrogenation of acrylic acid.

66. The acids which do not contain a —COOH group are
(a) Propionic acid (b) Picric acid
(c) Carbonic acid (d) Lactic acid.

67. Which of the following statements is not true about HCOOH ?
(a) It is a stronger acid than CH_3COOH
(b) It forms formyl chloride with PCl_5
(c) It gives CO and H_2O on heating with conc. H_2SO_4
(d) It reduces Tollen's reagent.

68. Which of the following on oxidation with alkaline $KMnO_4$ followed by acidification with HCl gives benzoic acid ?
(a) p-Cresol (b) Ethylbenzene
(c) o-Xylene (d) p-Xylene.

69. Acetophenone is obtained when
(a) Benzoyl chloride is treated with anhy $AlCl_3$
(b) Acetyl chloride is treated with dimethylcadmium
(c) Acetyl chloride is treated with benzene in presence of anhydrous $AlCl_3$
(d) Benzoyl chloride is reduced with H_2 in presence of Lindlar's catalyst.

70. Which of the following on reduction with $LiAlH_4$ will not give ethyl alcohol ?
(a) $(CH_3CO)_2O$ (b) CH_3COCl
(c) CH_3CONH_2 (d) $CH_3COOC_2H_5$.

B MCQs From Previous IIT Examinations

71. The pK_a of acetylsalicylic acid (aspirin) is 3.5. The pH of gastric juice in human stomach is about 2-3 and pH in the small intestine is about 8. Aspirin will be
(a) Unionized in the small intestine and in the stomach
(b) Completely ionized in the stomach and almost unionized in the small intestine
(c) Ionised in the stomach and almost unionized in the small intestine
(d) Ionised in the small intestine and almost unionized in the stomach. ***(IIT 1988)***

72. Chlorination of toluene in presence of light and heat followed by treatment with aqueous NaOH gives
(a) o-Cresol (b) p-Cresol
(c) 2, 4-Dihydroxytoluene (d) Benzoic acid. ***(IIT 1990)***

73. Reaction of ethyl formate with excess of CH_3MgI followed by hydrolysis gives
(a) n-Propyl alcohol (b) Ethanal
(c) Propanal (d) Isopropyl alcohol. ***(IIT 1992)***

74. Phenol and benzoic acid may be distinguished by their reaction with
(a) Aqueous NaOH (b) Aqueous $NaHCO_3$
(c) Neutral $FeCl_3$ (d) Aqueous NH_3. ***(IIT 1992)***

75. The major product of nitration of Benzoic acid is
(a) 3-Nitrobenzoic acid (b) 4-Nitrobenzoic acid
(c) 2-Nitrobenzoic acid (d) 2, 4-Dinitrobenzoic acid. ***(IIT 1993)***

76. Which of the following carboxylic acids undergoes decarboxylation easily ?
(a) C_6H_5—CO—CH_2—COOH
(b) C_6H_5—CO—COOH
(c) C_6H_5—CH(OH)—COOH
(d) C_6H_5—CH(NH_2)—COOH. ***(IIT 1995)***

77. The molecular weight of benzoic acid in benzene as determined by depression in freezing point method corresponds to
(a) Ionization of benzoic acid
(b) Dimerization of benzoic acid

(*c*) Trimerization of benzoic acid
(*d*) Solution of benzoic acid. **(*IIT 1996*)**

78. $CH_3CH_2COOH \xrightarrow[\text{Red P}]{Br_2} X \xrightarrow{NH_3(alc.)} Y.$

Y in the above reaction is

(*a*) Lactic acid (*b*) Ethylamine
(*c*) Propylamine (*d*) Alanine.

(*IIT 1996*)

79. The correct statement about the compounds A, B and C is

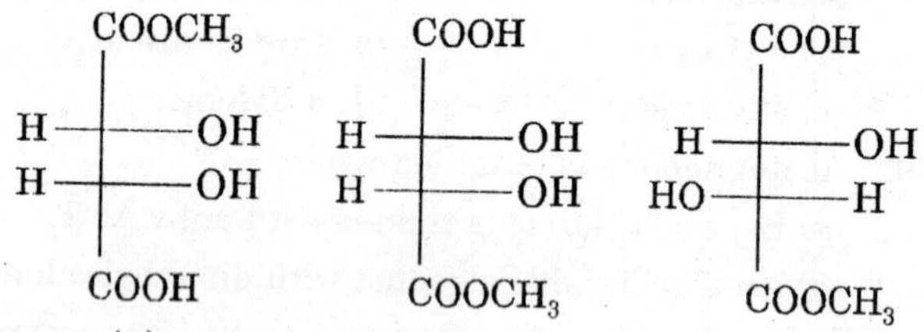

(*a*) (*b*) (*c*)

(*a*) *a* and *b* are identical
(*b*) *a* and *b* are diastereomers
(*c*) *a* and *c* are enantiomers
(*d*) *a* and *b* are enantiomers. **(*IIT 1997*)**

80. Among the given compounds, the most susceptible to nucleophilic attack at the carbonyl group is

(*a*) MeCOCl (*b*) MeCHO
(*c*) MeCOOMe (*d*) MeCOOCOMe.

(*IIT 1997*)

81. When propionic acid is treated with aqueous sodium bicarbonate CO_2 is liberated. The 'C' of CO_2 comes from

(*a*) Methyl group (*b*) Carboxylic acid group
(*c*) Methylene group (*d*) Bicarbonate. **(*IIT 1999*)**

82. Benzoyl chloride is prepared from benzoic acid by

(*a*) Cl_2, *hν* (*b*) NaOCl
(*c*) $SOCl_2$ (*d*) Cl_2, H_2O.

(*IIT Screening Test 2000*)

83. Which of the following acids has the smallest dissociation constant ?

(*a*) $CH_3CHFCOOH$ (*b*) FCH_2CH_2COOH
(*c*) $BrCH_2CH_2COOH$ (*d*) $CH_3CHBrCOOH$.

(*IIT Screening Test 2002*)

84. Identify the correct order of boiling points of the following compounds :

$CH_3CH_2CH_2CH_2OH$, (1) $CH_3CH_2CH_2CHO$, (2)
$CH_3CH_2CH_2COOH$ (3)

(*a*) 1 > 2 > 3 (*b*) 3 > 1 > 2
(*c*) 1 > 3 > 2 (*d*) 3 > 2 > 1.

(*IIT Screening 2002*)

85. The IUPAC name of C_6H_5COCl is

(*a*) Chlorobenzyl ketone
(*b*) Benzene chloroketone
(*c*) Benzenecarbonyl chloride
(*d*) Chlorophenyl ketone. **(*IIT 2006*)**

86. Which of the following reactants on reaction with concentrated NaOH followed by acidification gives the following lactone as the only product ?

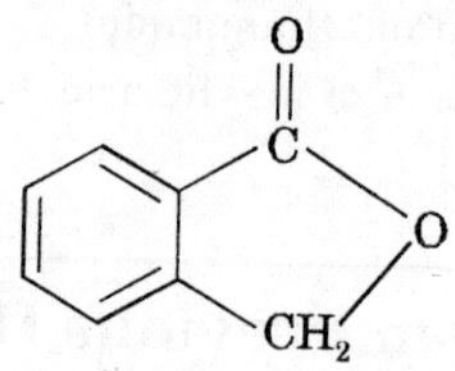

(*a*) 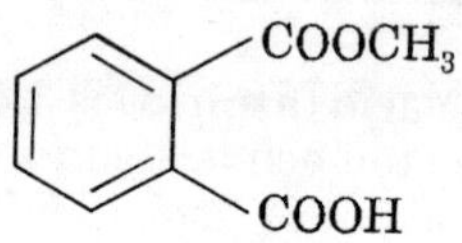

(*b*) 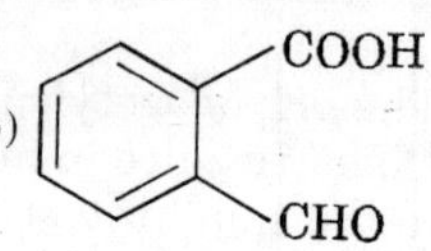

(*c*) 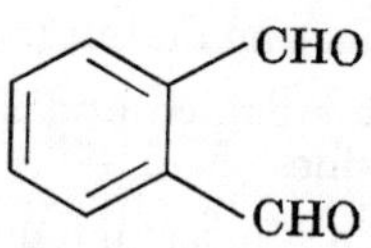

(*d*) 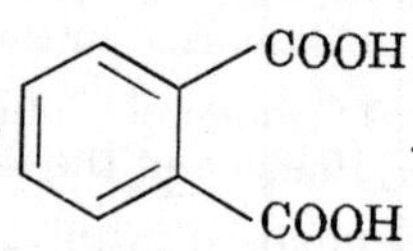

(*IIT 2006*)

ANSWERS

1. (*c*)	**2.** (*c*)	**3.** (*c*)	**4.** (*a*)	**5.** (*a*)	**6.** (*c*)	**7.** (*c*)	**8.** (*c*)	**9.** (*b*)
10. (*c*)	**11.** (*b*)	**12.** (*a*)	**13.** (*a*)	**14.** (*b*)	**15.** (*b*)	**16.** (*c*)	**17.** (*c*)	**18.** (*a*)
19. (*a*)	**20.** (*d*)	**21.** (*b*)	**22.** (*a*)	**23.** (*a*)	**24.** (*b*)	**25.** (*d*)	**26.** (*a*)	**27.** (*d*)
28. (*b*)	**29.** (*c*)	**30.** (*c*)	**31.** (*a*)	**32.** (*c*)	**33.** (*b*)	**34.** (*d*)	**35.** (*d*)	**36.** (*a*)
37. (*d*)	**38.** (*c*)	**39.** (*c*)	**40.** (*d*)	**41.** (*b*)	**42.** (*c*)	**43.** (*c*)	**44.** (*d*)	**45.** (*a*)
46. (*b*)	**47.** (*c*)	**48.** (*d*)	**49.** (*c*)	**50.** (*a*)	**51.** (*a*)	**52.** (*b*)	**53.** (*c*)	**54.** (*a*)
55. (*b*)	**56.** (*a*)	**57.** (*c*)	**58.** (*b*)	**59.** (*c*)	**60.** (*d*)	**61.** (*c*)	**62.** (*b*)	**63.** (*b*)

64. (*c*)	**65.** (*b*)	**66.** (*b*)	**67.** (*b*)	**68.** (*b*)	**69.** (*c*)	**70.** (*c*)	**71.** (*d*)	**72.** (*d*)
73. (*d*)	**74.** (*b*)	**75.** (*a*)	**76.** (*a*)	**77.** (*b*)	**78.** (*d*)	**79.** (*d*)	**80.** (*a*)	**81.** (*d*)
82. (*c*)	**83.** (*c*)	**84.** (*b*)	**85.** (*c*)	**86.** (*c*).				

HINTS/SOLUTIONS

Q. No.	Ans.	Hints & Solutions
1.	(*c*)	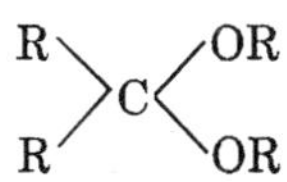 is formula of acetals.
2.	(*c*)	$CH_3CH_2\overset{O}{\overset{\Vert}{C}}—OH + HO\overset{O}{\overset{\Vert}{C}}—CH_2CH_3 \longrightarrow CH_3CH_2\overset{O}{\overset{\Vert}{C}}CH_2CH_3 + CO_2 + H_2O$. Pentanone-3
3.	(*c*)	$CH_3COOH + CH_2N_2 \longrightarrow CH_3\overset{O}{\overset{\Vert}{C}}—OCH_3 + N_2$ Methyl ethanoate
4.	(*a*)	$CH_3\overset{O}{\overset{\Vert}{C}}—OC_2H_5 + 2C_2H_5MgBr \xrightarrow[H_2O]{} CH_3—\underset{C_2H_5}{\overset{C_2H_5}{C}}—OH + Mg(OH)Br + Mg(OC_2H_5)Br.$ 3-Methylpentan-3-ol
5.	(*a*)	The functional isomers are CH_3COOH and $HCOOCH_3$. Out of these CH_3COOH will neutralise $NaHCO_3$ giving efferversence due to evolution of CO_2.
6.	(*c*)	$C_6H_5COCl \xrightarrow[-HCl]{NH_3} C_6H_5CONH_2 \xrightarrow[Heat]{P_2O_5} C_6H_5CN.$ Hence, Y is P_2O_5 or it can be PCl_5 also.
7.	(*c*)	$CH_3CH_2COOH \xrightarrow[Cl_2/P]{HVZ} CH_3\underset{Cl}{CH}COOH \xrightarrow{KOH\ (aq.)} CH_3\underset{OH}{CH}COOH$ Lactic acid

Q. No.	Ans.	Hints & Solutions
8.	(*c*)	$CH_3COOH \xrightarrow{HI/P} CH_3CH_3.$ Ethane
9.	(*b*)	$2CH_3—\overset{O}{\overset{\Vert}{C}}—OC_2H_5 \xrightarrow{Na^+O^-C_2H_5} CH_3—\overset{O}{\overset{\Vert}{C}}—CH_2—\overset{O}{\overset{\Vert}{C}}—OC_2H_5$ This is called Claisen condensation.
10.	(*c*)	Ketones do not given silver mirror test.
11.	(*b*)	In CH_3COO^- ion the carbon oxygen bonds are of equal length due to equivalent resonating forms.
12.	(*a*)	$BrCH_2CH_2CH_2Br \xrightarrow{Alc.\ NaCN} NC—CH_2CH_2CH_2CN \xrightarrow{H_3O^+} HOOC—CH_2CH_2CH_2—COOH$ Gluteric acid
13.	(*a*)	$C_6H_5CONH_2 \xrightarrow{Dil.\ NaOH} C_6H_5COONa \xrightarrow[CaO]{NaOH} C_6H_6 \xrightarrow[CH_3Cl]{AlCl_3} C_6H_5 - CH_3 + HCl.$ Toluene
14.	(*b*)	See methods of preparation of acid anhydrides.
15.	(*b*)	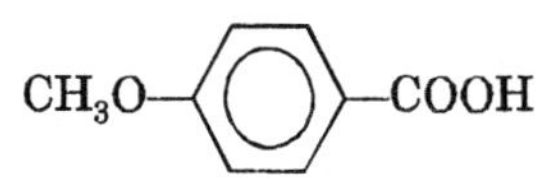 is weakest among the given acids. Hence it has lowest value of K_a.
16.	(*c*)	$RCOHH_3 \xrightarrow{X_2/OH^-} CHX_3 + RCOO^- \xrightarrow{H^+} RCOOH.$
17.	(*c*)	$C_2H_5OH < C_6H_5OH < CH_3COOH < ClCH_2COOH.$

Q. No.	Ans.	Hints & Solutions
18.	(*a*)	$R{-}\overset{O}{\overset{\parallel}{C}}{-}Cl + R'MgX \longrightarrow MgXCl + R\overset{O}{\overset{\parallel}{C}}R'$ $R{-}\overset{O}{\overset{\parallel}{C}}{-}R' + R'MgX \xrightarrow[H_3O^+]{} R{-}\overset{OH}{\overset{\parallel}{C}}{-}R'$ with R' below (3° alcohol).
19.	(*a*)	$CH_3{-}\overset{O}{\overset{\parallel}{C}}{-}Cl + RMgX \longrightarrow CH_3{-}\overset{O}{\overset{\parallel}{C}}{-}R + Mg\begin{matrix}OR\\X\end{matrix}$ $CH_3{-}\overset{OH}{\overset{\mid}{C}}{-}R$ (R below) $\xleftarrow[(ii)\ H_2O]{(i)\ RMgX}$
20.	(*d*)	Among the carboxylic acids, HCOOH is strongest reducing agent which can reduce Tollen's reagent as well.
21.	(*b*)	It contains ≃ 10% acetic acid.
22.	(*a*)	$C_2H_5OH \xrightarrow{\text{Acetobacter}} CH_3COOH$ Bacterium mycoderma Acetic is also known as Acetobacter.
23.	(*a*)	Acetic acid will dissolve carbonate, hydroxide and oxide of calcium but calcium oxalate is insoluble.
24.	(*b*)	$HO{-}C(CH_2COOH)_2{-}COOH \xrightarrow[\text{Fuming}]{\text{conc. } H_2SO_4}$ $CH_2COOH{-}C{=}O{-}CH_2COOH$ (i.e. $O{=}C(CH_2COOH)_2$) $+ CO + H_2O$.
25.	(*d*)	It is HVZ reaction.
26.	(*a*)	$\begin{matrix}CH_2{-}CH_2{-}COOH\\ \mid \\ CH_2{-}CH_2{-}COOH\end{matrix} \xrightarrow[575\ K]{\text{Heat}}$ Adipic acid 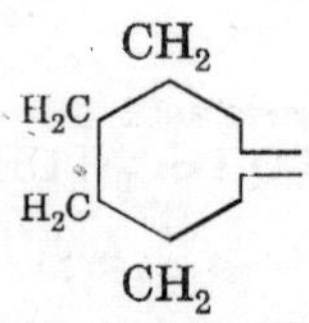 $= O + CO_2 + H_2O$.
27.	(*d*)	$CH_3COONH_4 \xrightarrow{\text{Heat}} CH_3CONH_2 + H_2O$. Acetamide
28.	(*b*)	Sod salt of higher fatty acids like palmitic acid are known as soaps.
29.	(*c*)	$CH_3CONH_2 + HNO_2 \longrightarrow CH_3COOH + N_2 + H_2O$.
30.	(*c*)	Ammonium cyanate of heating gives urea, an organic compound.
31.	(*a*)	HgO reacts with acetamide to form mercury salt.
32.	(*c*)	$\underset{(1\ mol)}{CH_3COCl} + NaOCOCH_3 \longrightarrow \underset{(102\ g)}{(CH_3CO)_2O} + NaCl$. 10.2 g of ethanoic anhydride is produced from acetylchloride = 0.1 mol.
33.	(*b*)	$CH_3{-}\overset{O}{\overset{\parallel}{C}}{-}Cl + NH_2OH \longrightarrow CH_3{-}\overset{O}{\overset{\parallel}{C}}{-}NHO$ Hydro oxamic acid.
34.	(*d*)	Succinic acid is $\begin{matrix}CH_2COOH\\ \mid \\ CH_2COOH\end{matrix}$
35.	(*d*)	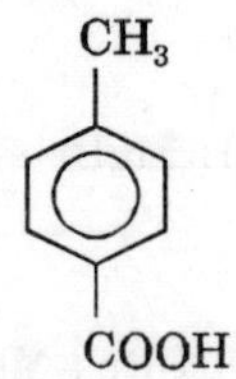 is known as toluic acid.
36.	(*a*)	$\begin{matrix}COOH\\ \mid \\ COOH\end{matrix} \xrightarrow{\text{Heat}} HCOOH + CO_2$ Oxalic acid $H_2C\begin{matrix}COOH\\COOH\end{matrix} \xrightarrow{\text{Heat}} CH_3COOH + CO_2$ Malonic acid $\begin{matrix}CH_2{-}COOH\\ \mid \\ CH_2{-}COOH\end{matrix} \xrightarrow{\text{Heat}} \begin{matrix}CH_2{-}CO\\ \mid \\ CH_2{-}CO\end{matrix}\!\!>\!O$ Succinic acid — Succinic anhydride
37.	(*d*)	$CHBr_2CBr_2COOH \xrightarrow[(ii)\ NaOH/CaO]{(i)\ Na}$ $CHBr_2CHBr_2 + Na_2CO_3$ 1, 1, 2, 2-Tetrabromoethane

Q. No.	Ans.	Hints & Solutions
38.	(c)	Esters of lower fatty acids have a pleasant fragrance.
39.	(c)	Dilute acidic $KMnO_4$ oxidises formic acid but does not oxidise acetic acid.
40.	(d)	$CH_3CH_2CCl_2COOH$ is strongest acid among the given compounds due to the presence of 2-electron withdrawing Cl atoms at the α-position.
41.	(b)	C = O does not undergo nucleophilic attack easily due to adjacent NH_2 group $\left[\begin{matrix}R{-}C=O\\ \vert\\ NH_2\end{matrix} \longleftrightarrow \begin{matrix}R{-}C=O^-\\ \Vert\\ {}^+NH_2\end{matrix}\right]$
42.	(c)	Amides have H-bonding and therefore exist as solids.
43.	(c)	Oxalic acid cannot dehydrate to form an anhydride.
44.	(c)	$HCOONa \xrightarrow[CaO]{NaOH} Na_2CO_3 + H_2$
45.	(c)	It is Blanc reaction. $CH_3COOC_2H_5 \xrightarrow[EtOH]{Na} CH_3CH_2C_2H_5$
46.	(b)	$CH_3OH \xrightarrow{PCl_5} CH_3Cl \xrightarrow{KCN} CH_3CN \xrightarrow{H_3O^+} CH_3COOH$
47.	(c)	CH_3FCOOH is strongest among the given compounds and thus has lowest pH.
48.	(d)	CH_2FCOOH is the strongest among the given compounds and thus has lowest pH.
49.	(c)	$C_{12}H_{22}O_{11} \xrightarrow[(O)]{HNO_3} \begin{matrix}COOH\\ \vert\\ COOH\end{matrix} + CO_2 + H_2O.$
50.	(a)	Bees wax is an ester.
51.	(a)	Adipic acid is $HOOC{-}(CH_2)_4{-}COOH$.
52.	(b)	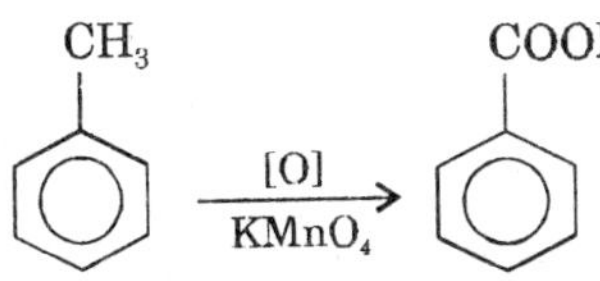
53.	(c)	Isomers should have same formula.
54.	(a)	The electron withdrawing group at para position stabilises the COO^- ion and hence increases the acidic strength.
55.	(b)	$\begin{matrix}CHCOOH\\ \vert\\ CHCOOH\end{matrix} \xrightarrow[KMnO_4/KOH]{[O]/H_2O} \begin{matrix}CH(OH)COOH\\ \vert\\ CH(OH)COOH\end{matrix}$
56.	(a)	$R{-}\overset{O}{\overset{\Vert}{C}}{-}OR' + ROH \longrightarrow R{-}\overset{O}{\overset{\Vert}{C}}{-}OR + R'OH$
57.	(c)	The acid containing α-Hydrogen atoms will show HVZ reaction.
58.	(b)	$H{-}\overset{O}{\overset{\Vert}{C}}{-}OCH_3 \xrightarrow[H_2O]{C_2H_5MgI} CH_3CH_2CHO \xrightarrow[H_2O]{C_2H_5MgI} CH_3CH_2\overset{OH}{\overset{\vert}{C}}HCH_2CH_3$
59.	(c)	$CH_3CH_2CHNH_2 \xrightarrow{H_2/Pt} CH_3CH_2CH_2NH_2(X)$; $CH_3CH_2CHNH_2 \xrightarrow{Br_2/KOH} CH_3CH_2NH_2(Y)$. As X and Y differ by CH_2 group, these are homologues.
60.	(d)	$CH_3CH_2COOH \xrightarrow{SOCl_2} CH_3CH_2CoCl + SO_2 + HCl$
61.	(c)	Addition of CH_3CO-group is acylation.
62.	(b)	$R{-}\overset{O}{\overset{\Vert}{C}}{-}\ddot{N}: \xrightarrow{rearrange} R{-}N=C=O.$
63.	(b)	HCN is known as prussic acid.
64.	(c)	$CH_2 = CH{-}CH_2{-}O:$ has got no conjugation.
65.	(b)	$CH_3CH = CHCH_3 + [O] \xrightarrow{hot\ KMnO_4} 2CH_3COOH$
66.	(b)	picric acid is 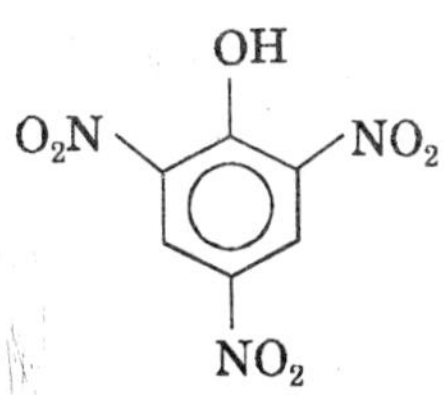 (trinitro phenol)

Q. No.	Ans.	Hints & Solutions
67.	(*b*)	Formyl chloride is highly unstable and is therefore not formed in the reaction.
68.	(*b*)	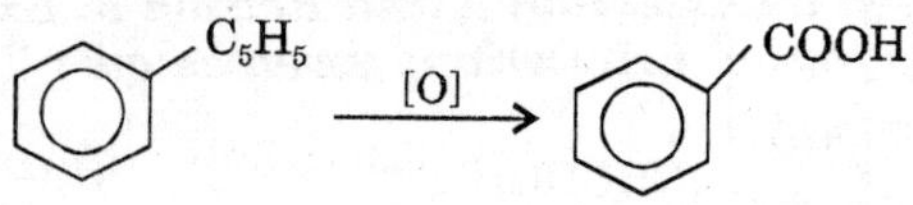
69.	(*c*)	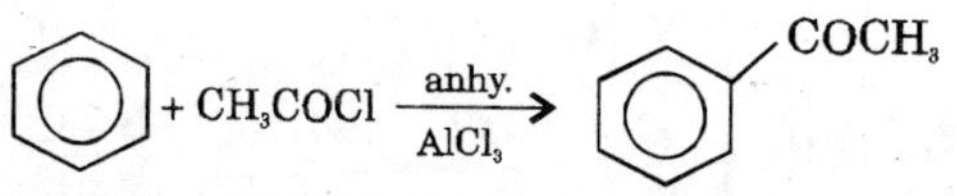
70.	(*c*)	$CH_3CONH_2 \xrightarrow{\text{red}} CH_3CH_2NH_2$
71.	(*d*)	$pK_a = -\log K_a$ or $= \log \frac{1}{K_a}$ The value is suggestive that aspirin will ionise in small intestine and almost unionised in stomach.
72.	(*d*)	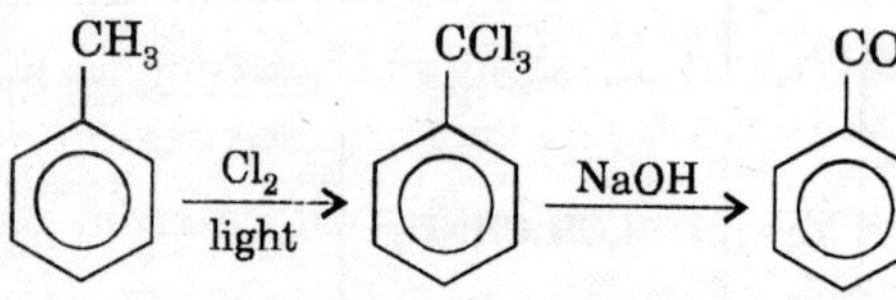
73.	(*d*)	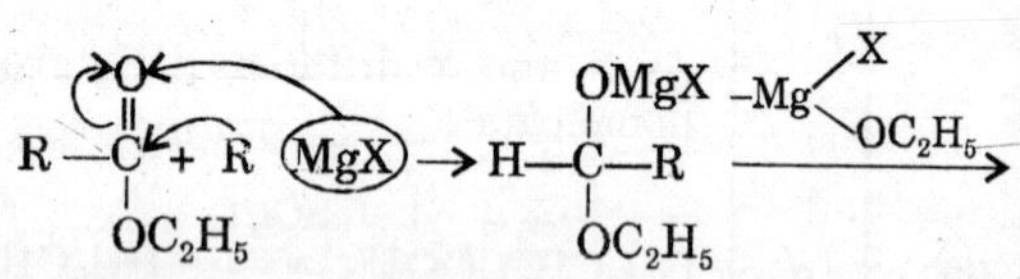 $CH_3CHO \xrightarrow[H_2O]{CH_3MgI} CH_3-CH(OH)-CH_3$
74.	(*b*)	Phenol do not decompose $NaHCO_3$ $\therefore$ K_a (phenol) $= 10^{-8} < K_a (H_2CO_3) = 10^{-7}$.
75.	(*a*)	—COOH group is meta directing.
76.	(*a*)	β-keto acids can readily under go decarboxylations.

Q. No.	Ans.	Hints & Solutions
77.	(*b*)	$(C_6H_5COOH)_2$ is formed due to inter molecular H-bonding.
78.	(*d*)	$CH_3CH_2COOH \xrightarrow{Br_2} CH_3-CH(Br)-COOH \xrightarrow{NH_3} CH_3-CH(NH_3)-COOH$ alanine
79.	(*c*)	A pair of compounds that are not super imposable mirror images are called enantiones.
80.	(*d*)	In MeCHO, the presence of C = O group facilitates the attack by nucleophile.
81.	(*d*)	$—COOH + NaHCO_3 \longrightarrow —COO^-Na^+ + H_2O + CO_2$.
82.	(*c*)	$C_6H_5COOH + SOCl_2 \longrightarrow C_6H_5COCl + SO_2 + HCl$. Thronyl chloride is a good chlorinating agent.
83.	(*c*)	The electron with drawing Br group is pittached at β-carbon and therefore the with drawing effect (– I) of Br is considerably less.
84.	(*b*)	b.p. of acids > alcohols > carbonyl compounds.
85.	(*c*)	The correct IUPAC name is benzene carbonyl chloride.
86.	(*c*)	On reaction with conc. NaOH it undergoes internal cannizaro's reaction which in the presence of acids forms a lactone. 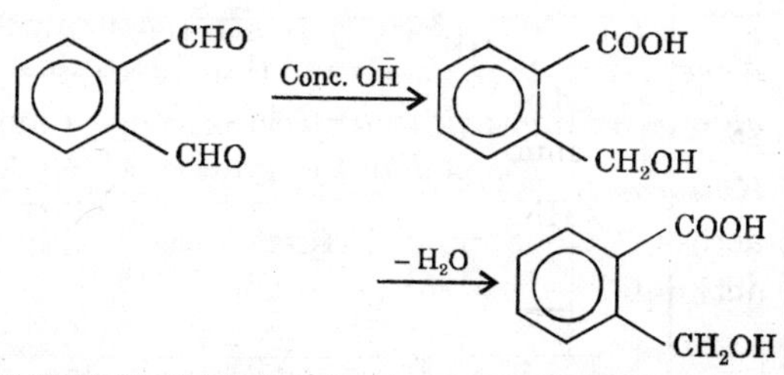

6.7. ASSERTION-REASON QUESTIONS

Quick Revision

In each of the following questions two statements are given as Assertion A and Reason R. Examine the statements carefully and answer the questions according to the instructions given below :

- *Mark **a** if both **A** and **R** are correct and **R** is the correct reason of **A**.*
- *Mark **b** if both **A** and **R** are correct and **R** is not the correct reason of **A**.*
- *Mark **c** if **A** is correct and **R** is wrong.*
- *Mark **d** if **A** is wrong and **R** is correct.*
- *Mark **e** if both **A** and **R** are wrong.*

1. ***Assertion A.*** Acetic anhydride is preferred to acetyl chloride for acylation reactions.

   Reason R. Chemical reaction in case of acetyl chloride is more vigorous and is difficult to control.

(*a*)	(*b*)	(*c*)	(*d*)	(*e*)

2. ***Assertion A.*** Amides are amphoteric in nature.

   Reason R. Amides are associated through H-bonding.

(*a*)	(*b*)	(*c*)	(*d*)	(*e*)

3. ***Assertion A.*** Boiling points of esters are higher than corresponding isomeric carboxylic acid.

   Reason R. Acid molecule dimerise in non-aqueous solutions.

(*a*)	(*b*)	(*c*)	(*d*)	(*e*)

4. ***Assertion A.*** $> C = O$ group is present both in aldehydes and acid derivatives.

   Reason R. Aldehydes give nucleophilic addition across $C = O$ bond but esters donot exhibit such reactions.

(*a*)	(*b*)	(*c*)	(*d*)	(*e*)

5. ***Assertion A.*** Acid anhydrides and esters are functional isomers of each other.

   Reason R. Both acid anhydrides and esters hydrolyse in acidic medium to give carboxylic acids.

(*a*)	(*b*)	(*c*)	(*d*)	(*e*)

6. ***Assertion A.*** Alkylated amides and acylated amines may be represented by same formula.

   Reason R. Amides are most preferred acylating agents among acid derivatives.

(*a*)	(*b*)	(*c*)	(*d*)	(*e*)

7. ***Assertion A.*** Distillation of calcium carboxylate and sulphuryl chloride can produce acid anhydrides.

   Reason R. Distillation of calcium carboxylate produces ketones.

(*a*)	(*b*)	(*c*)	(*d*)	(*e*)

8. ***Assertion A.*** Benzamide and methyl benzoate are derivatives of benzoic acid.

   Reason R. Benzamide is less easily hydrolysed as compared to methyl benzoate.

(*a*)	(*b*)	(*c*)	(*d*)	(*e*)

9. ***Assertion A.*** Benzoic acid does not undergo Friedel-Craft's reaction.

   Reason R. —COOH group deactivates the benzene ring by its electron withdrawing nature.

(*a*)	(*b*)	(*c*)	(*d*)	(*e*)

10. ***Assertion A.*** Acetic acid is more stronger acid as compared to phenol.

    Reason R. Resonance stabilisation of acetate ion is more than that of phenoxide ion.

(*a*)	(*b*)	(*c*)	(*d*)	(*e*)

11. ***Assertion A.*** Acetic acid is less soluble in water than benzoic acid.

Reason R. Acetic acid dissociates in aqueous solution while it dimerises in CCl_4.

(*a*)	(*b*)	(*c*)	(*d*)	(*e*)

12. ***Assertion A.*** *p*-Fluorobenzoic acid is weaker acid as compared to *p*-chlorobenzoic acid.

Reason R. Fluoroacetic acid is strongest acid as compared to chloroacetic acid.

(*a*)	(*b*)	(*c*)	(*d*)	(*e*)

13. ***Assertion A.*** Acetic acid does contain C = O group but does not react with $NaHSO_3$ like acetaldehyde.

Reason R. Both acetic acid and benzoic acid can decompose $NaHCO_3$.

(*a*)	(*b*)	(*c*)	(*d*)	(*e*)

14. ***Assertion A.*** *m*-Nitrobenzoic acid is less stronger acid as compared to *p*-nitrobenzoic acid.

Reason R. Acetic acid is stronger acid than benzoic acid but formic acid is still stronger than both.

(*a*)	(*b*)	(*c*)	(*d*)	(*e*)

15. ***Assertion A.*** Claisen condensation involves the self condensation of ester molecules having α-hydrogen.

Reason R. Claisen condensation occurs in the presence of a strong base such as sodium ethoxide.

(*a*)	(*b*)	(*c*)	(*d*)	(*e*)

ANSWERS

1. (*a*)	**2.** (*b*)	**3.** (*d*)	**4.** (*b*)	**5.** (*d*)	**6.** (*b*)	**7.** (*d*)	**8.** (*b*)	**9.** (*a*)
10. (*a*)	**11.** (*d*)	**12.** (*b*)	**13.** (*b*)	**14.** (*a*)	**15.** (*b*).			

6.8. SELF EVALUATION TEST

Assess Yourself

Choose the correct answer from the four alternatives given in each of the following questions :

1. Which of the following is known as oil of winter-green ?
   (*a*) Methyl salicylate (*b*) Phenyl salicylate
   (*c*) Ethyl salicylate (*d*) Ethyl,*p*-hydroxybenzoate.

2. Which is formed as a product in the following reaction ?
   $$\text{Methanol} + \text{CO} \xrightarrow{\text{Rh}} \ldots\ldots$$
   (*a*) Ethanol (*b*) Ethanoic acid
   (*c*) Methylmethanoate (*d*) Methanal.

3. $H_2NCONH_2 \xrightarrow{\text{Heat}}$ **X. X** in the above reaction is
   (*a*) $NH_3 + CO_2$ (*b*) urotropine
   (*c*) $H_2N-\overset{\overset{O}{\|}}{C}-NH-\overset{\overset{O}{\|}}{C}-NH_2$ (*d*) Urethane.

4. Which of the following acid does not contain carboxylic group ?
   (*a*) Styphinic acid (*b*) Oxalic acid
   (*c*) Pivalic acid (*d*) Salicylic acid.

5. What is the formed when ethanoic acid and hydrazoic acid are allowed to react in the presence of H_2SO_4 ?
   (*a*) Methanamine (*b*) Ethanamide
   (*c*) Ethanamine (*d*) Oxamide.

6. **Lactic acid** $\xrightarrow{X}$ **Lactile chloride.**
   The reagent X can be
   (*a*) PCl_5 (*b*) Cl_2
   (*c*) NaCl (*d*) ICl.

7. Formic acid reduces Tollen's reagent because
   (*a*) It is strongest carboxylic acid
   (*b*) It does carry CHO group in its molecule
   (*c*) Oxidation number of C in it is + 2
   (*d*) It is weakest carboxylic acid.

8. **Silver acetate +** $I_2 \xrightarrow{CS_2}$ .... The main product formed in the reaction is
   (*a*) CH_3I (*b*) CHI_3
   (*c*) CH_3COI (*d*) CH_3COOCH_3.

9. Chemically banana oil is
   (*a*) Mixture of hydrocarbons obtained from banana leaves
   (*b*) Mixture of compounds having 3-centre bonds in their molecules
   (*c*) Isoamyl acetate
   (*d*) Isooctyl butyrate.

10. $NH_2CONH_2 + NH_2—NH_2 \longrightarrow X + NH_3$.
    In the above reaction X is mainly
    (*a*) Phenyl hydrazine (*b*) Semicarbazide
    (*c*) Acetyl urea (*d*) Biuret.

11. What is formed when diethyl ether react at about 383 K with boron trifluoride ?
    (*a*) Ethyl acetate (*b*) $(C_2H_5)_2CO_3$
    (*c*) Et OH (*d*) $[(C_2H_5)_2O]BF_3$.

12. Which dicarboxylic acid gives AcOH on strong heating ?
    (*a*) Ethanedioic acid (*b*) Malonic acid
    (*c*) Gluteric acid (*d*) Butane-1, 4-dioic acid.

13. $CH_3COCl \xrightarrow{X} Y \xrightarrow[H_2SO_4]{C_2H_5OH}$ **Acetal.**
    In the sequence X is
    (*a*) H_2/Pt (*b*) $H_2/Pd - BaSO_4$
    (*c*) $Cr_2O_7^{2-}/H^+$ (*d*) Zn-Hg/HCl.

14. Silver salt of which following acid on treatment with bromine will produce 2-bromopropane
    (*a*) Propanoic acid (*b*) Butyric acid
    (*c*) Isobutyric acid (*d*) Gluteric acid.

15. Methanoic acid is manufactured synthetically by the reaction of CO and
    (*a*) NaOH (*b*) dil. HCl
    (*c*) conc. H_2SO_4 (*d*) $POCl_3$.

16. Which of the following acid undergoes decarboxylation most easily ?
    (*a*) $C_6H_5COCH_2COOH$ (*b*) $C_6H_5CO\ COOH$
    (*c*) $C_6H_5\underset{\underset{OH}{|}}{CH}\ COOH$ (*d*) $C_6H_5\underset{\underset{NH_2}{|}}{CH}—COOH$

17. 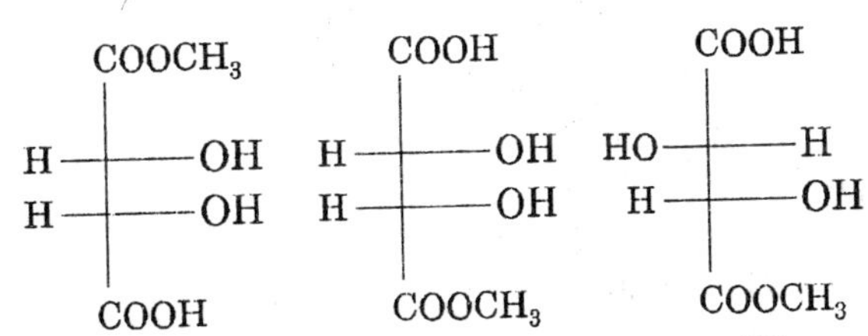

    (A) (B) (C)

What is correct about A, B, C ?

(*a*) A and C are diastereomers

(*b*) A and B are identical

(*c*) A and B are meso structures

(*d*) A and C are enantiomers.

18. $CH_3COCH_2CH_2COOC_2H_5 \xrightarrow{NaBH_4}$ .... X.

The product X is

(*a*) $CH_3CH_2CH_2CH_2COOC_2H_5$

(*b*) $CH_3COCH_2CH_2CH_2OH + C_2H_5OH$

(*c*) $CH_3CH(OH)—CH_2CH_2CH_2OH + C_2H_5OH$

(*d*) $CH_3CH(OH)CH_2CH_2COOC_2H_5$

19. Which of the following is most acidic ?

(*a*) $CH_3CH_2CH(Cl)COOH$ (*b*) $CH_3CH(Cl)CH_2COOH$

(*c*) $ClCH_2CH_2CH_2COOH$ (*d*) $(CH_3)_2CH$ COOH.

20. The hydrolysis of which of the following takes the longest time ?

(*a*) CH_3COCl (*b*) $CH_3COOC_2H_5$

(*c*) $CH_3COOCOCH_3$ (*d*) CH_3CONH_2.

In each of the following questions, three statements are given. Mark the correct answer as per following instructons :

(*a*) **if only II is correct** (*b*) **if all are wrong**

(*c*) **if II and III is correct** (*d*) **if I and II are correct.**

21. I. Dissolution of acetic acid in benzene lowers its molecular mass.

II. The two C—O bonds in acetic acid molecule have different bond lengths.

III. Lactic acid is β-hydroxy propionic acid.

(*a*) (*b*) (*c*) (*d*)

22. I. Hydrolysis of prussic acid produces formic acid.

II. Oxidation of benzoyl chloride with $KMnO_4$ produces benzoic acid.

III. Hydrogenation of benzoyl chloride in the presence of Pt produces benzaldehyde.

(*a*) (*b*) (*c*) (*d*)

23. I. Acetone gives tertiary butyl alcohol on reduction.

II. Acetone gives acetic acid on drastic oxidation.

III. Acetone undergoes bimolecular reduction with Mg—Hg/H_2O.

(*a*) (*b*) (*c*) (*d*)

24. I. Oxidation of cane sugar with conc. HNO_3 yield oxalic acid.

II. Fenton's reagent is $H_2SO_4/FeSO_4$.

III. Carboxylic acids dimerise in water.

(*a*) (*b*) (*c*) (*d*)

25. I. Carboxylate ion is a hybrid of two non-equivalent structures.

II. The weaker the base, the better the leaving group.

III. Acetyl chloride is most effective acylating agent among the acid derivatives.

(*a*) (*b*) (*c*) (*d*)

26. I. HO—C(=NH)—NH_2 is imide form of urea.

II. Conc. H_2SO_4 and formic acid react to form carbon monoxide gas

III. pK_a of chloroacetic acid is higher than that of benzoic acid.

(*a*) (*b*) (*c*) (*d*)

27. I. Tartaric acid does not contain —COOH group at all

II. Crotonic acid does not decolourise bromine disolved in CS_2.

III. Alkane with same number of C atoms cannot be produced from carboxylic acid.

(*a*) (*b*) (*c*) (*d*)

28. Hydrolysis of which of the following does not give acetic acid

(*a*) Ethyl ethanoate (*b*) Ethyl cyanide

(*c*) Ethanamide (*d*) Isopropylethanoate.

29. Isopropyl acetate on treatment with $LiAlH_4$ does not produce

(*a*) C_2H_5OH (*b*) $CH_3CH(OH)CH_3$

(*c*) CH_3OH (*d*) Both *a* and *b*.

30. $KMnO_4$ oxidation of which of the following will not produce benzoic acid ?

(*a*) Toluene (*b*) Benzene

(*c*) Ethylbenzene (*d*) Cumene.

31. Silver acetate and 2-bromopropane will react to produce

(*a*) Methyl ethanoate (*b*) $CH_3—COO—C_2H_5$

(*c*) 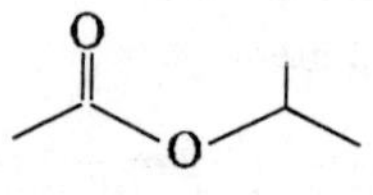 (*d*) 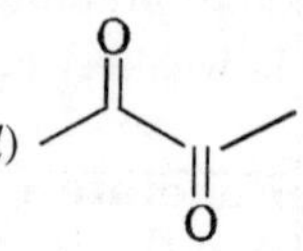

32. The starting materials in Hoffmann's bromamide reaction is/are

(*a*) RCONHBr (*b*) RC ≡ N

(*c*) Br_2/KOH (*d*) Br_2/CCl_4.

33. Cyanohydrin of compound X gives lactic acid, the reaction of X with sodium ethoxide will give

(*a*) CH_3COOH (*b*) CH_3COOCH_3

(*c*) $CH_3COOC_2H_5$ (*d*) $CH_3OC_2H_5$.

34. Which of the following reaction will give cyclohexane ?

(a) 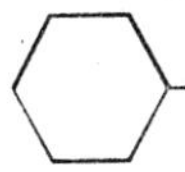 $-COOH \xrightarrow[(ii)\ NaOH/CaO]{(i)\ Na}$

(b) 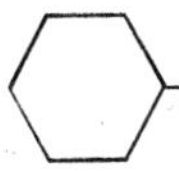 $-CHO + Zn-Hg/HCl \longrightarrow$

(c) 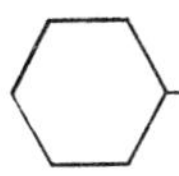 $-COOH + HI/(P) \longrightarrow$

(d) 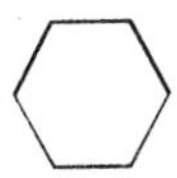 $= O + H_2/Pt \longrightarrow$

35. Sour taste of lemon is due to

(a) Ascorbic acid (b) Citric acid

(c) Lactic acid (d) Acetic acid.

36. Which of the following reagent can distinguish between formic acid and formaldehyde ?

(a) $NaHCO_3$ (b) Tollen's reagent

(c) Fehling solution (d) Ferric chloride.

37. During acid catalysed esterification, the attacking nucleophile is

(a) $H_2\ddot{O}$: (b) $R\ddot{O}H$

(c) : CH_2 (d) : OH^-.

38. **Isopropylchloride** $\xrightarrow{\text{KCN (alc)}}$ **X** $\xrightarrow{H_3O^+}$ **Y.**

In the given sequence Y is

(a) Propanoic acid (b) Butanoic acid

(c) Isobutyric acid (d) Valeric acid.

39. Which of the following acid can decolourise $KMnO_4$ solution ?

(a) Acrylic acid (b) Acetic acid

(c) Butyric acid (d) Stearic acid.

40. Which of the following are resonating structures of acetamide ?

(I) $CH_3-\overset{\overset{O}{\parallel}}{C}-NH_2$ (II) $CH_2 = \overset{\overset{O}{\parallel}}{C}-NH_2$

(III) $CH_3-\overset{\overset{OH}{|}}{C} = NH$ (IV) $CH_3-\overset{\overset{O^-}{|}}{C} = \overset{+}{N}H_2$.

(a) I and IV (b) II and IV

(c) II, III, IV (d) I, III, II.

ANSWERS

1. (a)	**2.** (b)	**3.** (c)	**4.** (a)	**5.** (a)	**6.** (a)	**7.** (b)	**8.** (d)	**9.** (c)
10. (b)	**11.** (d)	**12.** (b)	**13.** (b)	**14.** (c)	**15.** (a)	**16.** (a)	**17.** (a)	**18.** (d)
19. (a)	**20.** (d)	**21.** (a)	**22.** (d)	**23.** (c)	**24.** (d)	**25.** (c)	**26.** (d)	**27.** (b)
28. (b)	**29.** (c)	**30.** (b)	**31.** (c)	**32.** (c)	**33.** (c)	**34.** (a)	**35.** (b)	**36.** (a)
37. (b)	**38.** (c)	**39.** (a)	**40.** (a).					

HINTS/SOLUTIONS

Q. No.	Ans.	Hints & Solutions
1.	(a)	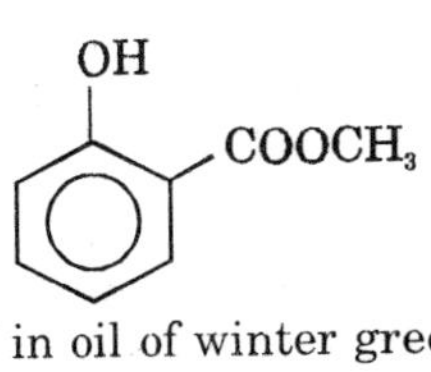 methyl salicylate is used in oil of winter green.
2.	(b)	$CH_3OH + CO \xrightarrow{Rh} CH_3COOH$.
3.	(c)	$2H_2NCONH_2 \xrightarrow[303\ K]{Heat}$ $H_2N-CO-NH-\overset{\overset{O}{\parallel}}{C}-NH_2$ Biuret

Q. No.	Ans.	Hints & Solutions
4.	(a)	Styphinic acid does not contain —COOH group. It is 2, 4, 6-trinitro resorcinol 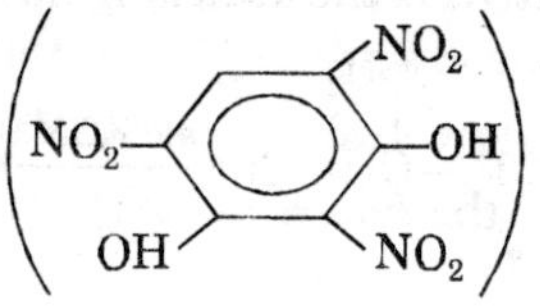
5.	(a)	$CH_3COOH + HN_3 \xrightarrow{H_3O^+} CH_3NH_2 + N_2 + CO_2$. Methanamine
6.	(a)	$CH_3CH(OH) - COOH \xrightarrow{PCl_5} CH_3 - CH(OH) - COCl$
7.	(b)	$H—\overset{O}{\overset{\parallel}{C}}—OH$ has $H—\overset{O}{\overset{\parallel}{C}}—$ group and hence reduces Tollen's reagent.
8.	(d)	$CH_3COOAg + I_2 \xrightarrow{Cs_2} CH_3I + AgI$. $CH_3I + AgOCOCH_3 \longrightarrow CH_3OCOCH_3$. Methylacetate
9.	(c)	Isoamylacetate is also called banana oil.
10.	(b)	$NH_2CONH_2 + NH_2—NH_2 \longrightarrow H_2N—NHCONH_2 + NH_3$. Semi carbazide
11.	(d)	ether due to lone pair acts as donor whereas BF_3 is a Lewis acid, therefore an adduct is formed.
12.	(b)	$CH_2(COOH)_2 \xrightarrow[- CO_2]{Heat} CH_3—COOH + CO_2$ Acetic acid
13.	(b)	$CH_3COCl \xrightarrow{H_2/Pd - BaSO_4} CH_3CHO \xrightarrow{2C_2H_5OH} CH_3CH(OC_2H_5)_2$. Acetal
14.	(c)	$CH_3 - CH(CH_3) - COOAg + Br_2 \xrightarrow[CCl_4]{H_2SO_4} CH_3CH(CH_3) - Br + CO_2 + AgBr$. 2-bromopropane
15.	(a)	CO and NaOH react to give sodium formate which on hydrolysis give methanoic acid.
16.	(a)	$C_6H_5—\overset{O}{\overset{\parallel}{C}}—CH_2—COOH$ is easily decarboxylated *i.e.*, β-Keto acids can easily decarboxylate.
17.	(c)	A and C are diastereomers.
18.	(d)	$NaBH_4$ reduces $> C = O$ group to $> CHOH$ but does not reduce $—COOC_2H_5$ group.
19.	(a)	In $CH_3CH_2CH_2CH(Cl)COOH$, Cl-group which is electron withdrawing group is nearest to —COOH group. Hence its acid strength is maximum.
20.	(d)	CH_3CONH_2 is least reactive towards acyl substitution.
21.	(a)	Only statement II is correct.
22.	(d)	Both I and II are correct.
23.	(c)	II and III are correct.
24.	(d)	Statement I and II are correct.
25.	(c)	Statement II and III are correct.
26.	(d)	I and II are correct.
27.	(b)	All statements are wrong.
28.	(b)	Ethyl cyanide (C_2H_5CN) on hydrolysis produces propanoic acid.
29.	(c)	$CH_3 - \overset{O}{\overset{\parallel}{C}} - O - CH(CH_3) - CH_3 \xrightarrow[Red.]{LiAlH_4} CH_3CH_2OH + (CH_3)_2CHOH$.
30.	(b)	Except benzene, all others will oxidise to produce benzoic acid. The side chain gets oxidised to —COOH by $KMnO_4$.
31.	(c)	$CH_3 - CH(CH_3) - Br + AgO\overset{O}{\overset{\parallel}{C}} - CH_3 \longrightarrow CH_3 - CH(CH_3) - O - \overset{O}{\overset{\parallel}{C}} - CH_3 + AgBr$. Isopropyl acetate

Q. No.	Ans.	Hints & Solutions
32.	(c)	$CH_3-\overset{O}{\overset{\Vert}{C}}-NH_2 \xrightarrow{Br_2/KOH} CH_3NH_2$
33.	(c)	X is CH_3CHO whose Cyanohydrin ($CH_3CH(OH)CN$) produces lactic acid on hydrolysis. $2CH_3CHO \xrightarrow{NaOC_2H_5} CH_3-\overset{O}{\overset{\Vert}{C}}-OC_2H_5$. This reaction is Tischenko reaction.
34.	(a)	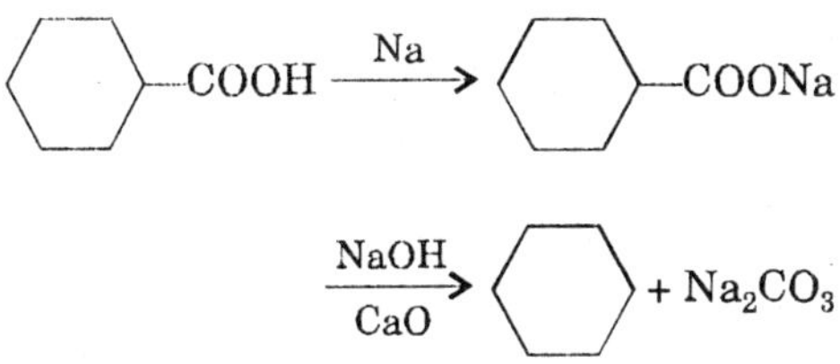
35.	(b)	It is due to citric acid $CH_2COOH-C(OH)(COOH)-CH_2COOH$
36.	(a)	$NaHCO_3$ gives effervescene with formic acid but does not react with HCHO.
37.	(b)	ROH is the attacking nucleophile.
38.	(c)	$CH_3-\underset{}{CH}(CH_3)-Cl \xrightarrow{alc\ KCN} CH_3-CH(CH_3)-CN \xrightarrow{H_2O/H^+} CH_3-CH(CH_3)-COOH$
39.	(a)	$CH_2=CH-COOH$ (acrylic acid) contains a double bond which get oxidised by alk. $KMnO_4$.
40.	(a)	$CH_3-\overset{O}{\overset{\Vert}{C}}-\ddot{N}H_2 \leftrightarrow CH_3-\overset{O^{\ominus}}{\overset{\vert}{C}}=\overset{\oplus}{N}H_2$.

7 ORGANIC COMPOUNDS WITH FUNCTIONAL GROUPS CONTAINING NITROGEN

7.1. CONCEPTS IN BRIEF

IMPORTANT FUNCTIONAL GROUPS CONTAINING NITROGEN

Some important functional groups containing nitrogen element are :

$-C \equiv N$	$-N \stackrel{\rightarrow}{=} C$	$-N \genfrac{}{}{0pt}{}{\diagup\!\!\!= O}{\searrow O}$	$-O-N=O$	$-NH_2$
(Cyano)	(Isocyano)	(Nitro)	(-Nitrito)	(Primary amino)
$-\overset{\vert}{N}H$	$=NH$	$-\underset{\vert}{\overset{\vert}{N}}$	$-\overset{+}{N} \equiv N$	$-C=N=O$
(sec. amino)	(Imino)	(ter. amino)	(Diazo)	(Cyanate)

CYANIDES OR CYANO COMPOUNDS

- These are the compounds which contain **cyano ($-C \equiv N$)** group in their molecules.
- They are classified as alkyl or aryl cyanides.

$$\mathbf{R-C \equiv N} \quad ; \quad \mathbf{Ar-C \equiv N}$$

(Alkyl cyanide) ; *(Aryl cyanides)*

- In common system cyanides are named either by using suffix ***cyanide*** after the name of alkyl or aryl group or by adding the suffix ***onitrile*** in place of **ic acid** of the common name of the corresponding acid produced by the hydrolysis of cyanide compound.
- In IUPAC system, alkyl cyanides are named as ***alkanenitriles*** whereas aryl cyanides as ***arenecarbonitrile.***

Some example are :

CH_3CH_2CN
Propanenitrile
(Ethyl cyanide)

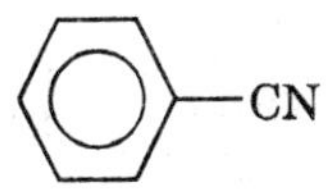

Benzenecarbonitrile
(Benzonitrile)

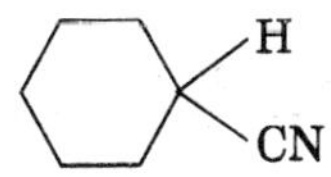

Cyclohexanecarbonitrile

ORBITAL PICTURE OF —CN GROUP

- Both C and N atoms assume *sp*-hybrid states.
- There is one *sp-sp* sigma bond and two $p_\pi - p_\pi$ bonds between C and N atoms.
- The lone pair of N is present in *sp*-hybrid orbital as shown.

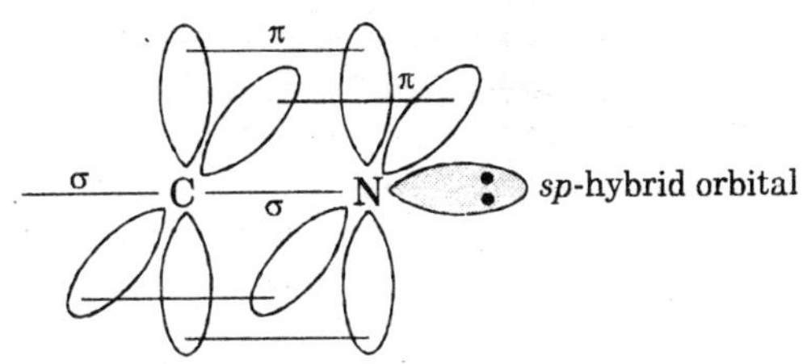

METHODS OF PREPARATION OF CYANIDES

Some of the methods of preparation of **cyanides** are :

1. **From alkyl halides.** Alkyl halides on treatment with *ethanolic solution of sodium cyanide* or *potassium cyanide* give alkyl cyanides.

$$\underset{\text{Alkyl halide}}{RX} + KCN(alc.) \longrightarrow \underset{\text{Alkyl cyanide}}{RCN} + \overset{+\,-}{KX}$$

- In fact, CN^- ion is an ambident nucleophile which can link the alkyl chain through carbon as well as through N atom. KCN is considerably ionic and CN^- ion exists freely. It prefers to form C—C bond rather than C—N bond because it is relatively more stronger. Hence major product formed in this reaction is **R—CN.**
- Some amount of **RNC** is also formed during this reaction.
- The method is not suitable for preparing aryl halides because of their less reactivity towards nucleophilic substitution.

2. **By Dehydration of Primary Amides.** The dehydration of primary amides can be achieved by *phosphorus pentoxide (P_2O_5) or thionyl chloride ($SOCl_2$).*

$$\underset{1^\circ\ \text{Amide}}{R\text{—}\overset{\overset{\displaystyle O}{\|}}{C}\text{—}\overset{\overset{\displaystyle H}{|}}{N}\text{—}H} \xrightarrow[\text{Heat}]{P_2O_5} \underset{\text{Alkyl cyanide}}{R\text{—}C \equiv N} + H_2O$$

3. **By Dehydration of Ammonium Carboxylates.** Ammonium carboxylate on heating with phosphorus pentoxide also yield alkyl cyanides

$$\underset{\text{Ammonium carboxylate}}{RCOONH_4} \xrightarrow[\text{Heat}]{P_2O_5} \underset{\text{Alkanenitrile}}{RCN} + 2H_2O$$

4. **By Dehydration of Aldoximes.** Dehydration of aldoximes with P_2O_5 or *acetic anhydride* also gives cyanides. For example,

$$\underset{\text{Acetaldoxime}}{CH_3CH = NOH} \xrightarrow{(CH_3CO)_2O} \underset{\text{Ethanenitrile}}{CH_3C \equiv N} + H_2O$$

5. **From Grignard's Reagent.** This method is particularly useful for preparing tertiary alkyl cyanides. It involves the treatment of the Grignard's reagent with *cyanogen chloride* (Cl—CN).

$$RMgX + Cl{-}CN \xrightarrow{\text{Ether}} RCN + Mg(X)Cl$$

6. **From Aryl Diazonium Salt.** It involves the treatment of *benzene diazonium salt with a mixture of copper (I) cyanide and sodium or potassium cyanide.*

$$\underset{\text{Benzenediazonium chloride}}{C_6H_5\overset{+}{N_2}\overset{-}{Cl}} \xrightarrow[\text{KCN}]{\text{CuCN}} \underset{\text{Phenl cyanide}}{C_6H_5CN} + N_2 + CuCl$$

- This reaction is a special case of *Sandmeyer's reaction.*
- This method is applicable to the preparation of aryl cyanides only.

PROPERTIES OF CYANIDES

- **Physical state and smell.** The lower members of the family are colourless liquids whereas the higher members are crystalline solids. They are stable compounds with *pleasant smell.*
- **Boiling points.** Due to the presence of polar $-C \equiv N$ group in their molecules, they have high dipole moment and consequently, possess high melting and boiling points.
- **Solubility.** The cyanides with low molecular mass are soluble in water. *This is due to the tendency of their* molecules to form hydrogen bonds with water molecules. However, with the increase in molecular mass, the bulk of the non-polar portion (*i.e.*, R—) increases and consequently, solubility in water decreases. However, the cyanides are fairly soluble in organic solvents.

Summary of Chemical Reactions

Some of the important chemical reactions of nitriles are summed up as follows :

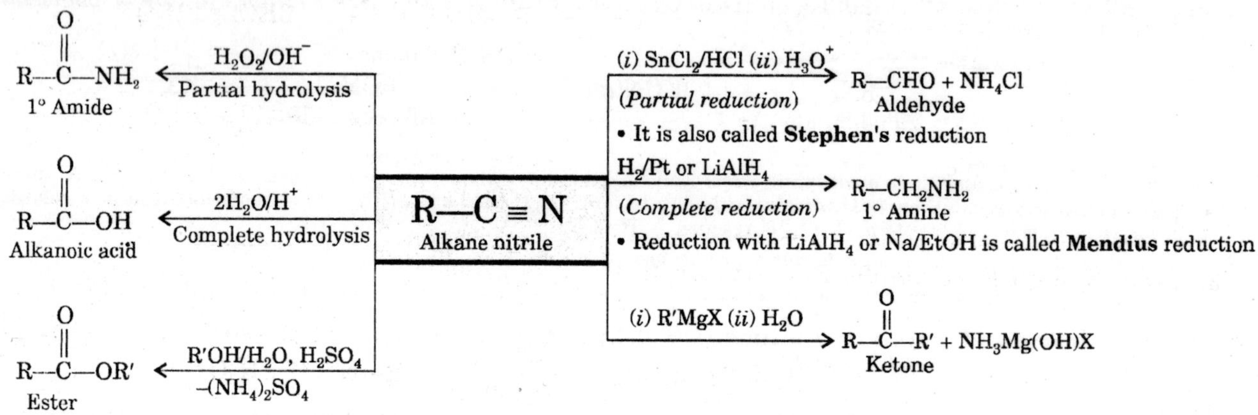

ISOCYANIDES

These are the compounds which contain ***isocyano*** **(—N ⇉ C)** group in their molecules.

- They are structural isomers of cyanides.
- They can be classified as alkyl as well as aryl isocyanides

R—N ⇉ C (*Alkyl isocyanides*) ; **Ar—N ⇉ C** (*Aryl isocyanides*)

- They are also called **carbylamines.**

ORBITAL PICTURE OF R—N $\overset{\rightarrow}{=}$ C

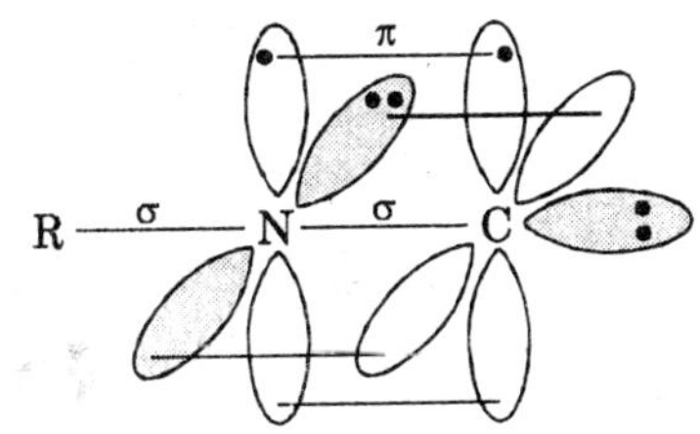

- Both N and C atoms of isocyanide group assume *sp* hybrid state.
- There is one *sp-sp*σ bond between N and C atoms and one p_π-p_π bond. Another p_π-p_π bond is formed by overlap of *p*-orbital of N containing lone pair and empty *p*-orbital of *sp*-hybrid carbon. This is called $p_\pi - p_\pi$ back bonding.

PREPARATION OF ISOCYANIDES

Some of the methods of preparation of isocyanides are :

1. **From Alkyl Halides.** Alkyl halides, on heating with ethanolic solution of AgCN (silver cyanide) give alkyl isocyanides as major product.

$$R + Br + AgCN \xrightarrow[\text{Heat}]{C_2H_5OH} \underset{\text{Alkyl isocyanide}}{R{-}N \overset{\rightarrow}{=} C} + AgBr$$

AgCN is considerably covalent in nature. Hence in this reaction the nucleophilic attack on the alkyl halide molecule takes place mostly through N atom thereby forming R—N $\overset{\rightarrow}{=}$ C as major product.

- The minor product during this reaction is **alkyl cyanide.**

2. **From Primary Amines.** Both aromatic as well as aliphatic amines on treatment with $CHCl_3$ in the presence of potassium hydroxide give carbylamines.

$$\underset{1^\circ \text{ Amine}}{R{-}NH_2} + CHCl_3 + 3KOH \longrightarrow \underset{\text{Alkyl carbylamine}}{R{-}N \overset{\rightarrow}{=} C} + 3KCl + 3H_2O$$

$$\underset{\text{Aniline}}{C_6H_5NH_2} + CHCl_3 + 3KOH \longrightarrow \underset{\text{Phenyl isocyanide}}{C_6H_5N \overset{\rightarrow}{=} C} + 3KCl + 3H_2O$$

- This reaction is known as ***carbylamine reaction.***
- This reaction is a test for primary amino ($-NH_2$) group.

3. **From N-alkyl Formamide.** By reaction with $POCl_3$ in pyridine

$$\underset{\text{N-Alkyl formamides}}{R{-}NH{-}\overset{\overset{\displaystyle O}{\|}}{C}{-}H} \xrightarrow[\text{(Pyridine)}]{POCl_3} \underset{\text{Alkyl isocyanide}}{R{-}N \overset{\rightarrow}{=} C} + H_2O$$

PROPERTIES OF ISOCYANIDES

- They are colourless liquids with pungent irritating smell.
- They are relatively less polar than corresponding R—CN. Thus, their m.p. or b.p. are lower than cyanides.
- They are less soluble in water due to their inability to form H-bonds with water molecules.

Summary of chemical reactions of isocyanides

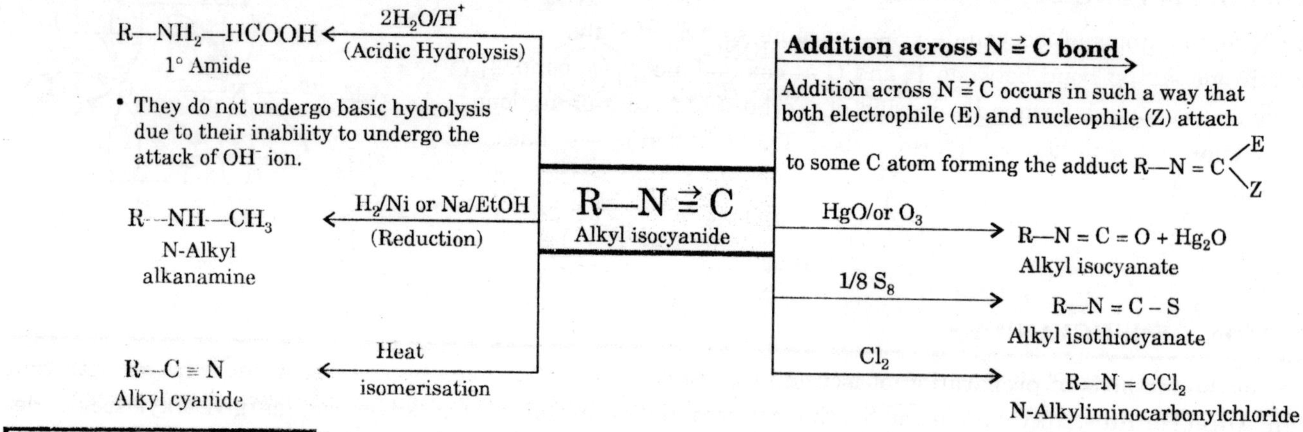

NITRO COMPOUNDS

These are the compounds which contain **nitro** functional group **(—N(→O) = O)** in their molecules.

- They can be classified as alkyl and aryl nitro compounds.

$$\mathbf{R—NO_2} \quad ; \quad \mathbf{Ar—NO_2}$$

(Nitroalkanes) *(Nitroarenes)*

- Aliphatic nitrocompounds are further classified as primary, secondary and tertiary nitro compounds

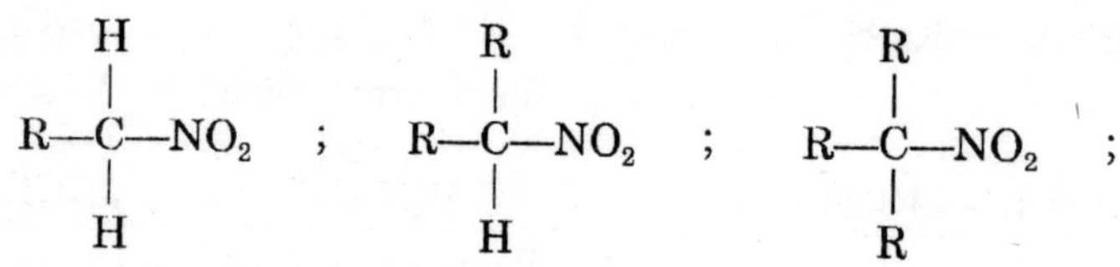

Primary (1°) Secondary (2°) Tertiary (3°)

- The compounds containing NO_2 group linked through O atom are called ***alkyl nitrites*** **(R—O—N = O).**
- Alkyl nitrites and nitroalkanes are *functional isomers* of each other.

PREPARATION OF NITRO COMPOUNDS

1. **From Alkyl Halides.** By treatment with alcoholic solution of silver nitrite

$$R—I + AgNO_2 \longrightarrow R—NO_2 + AgI$$

- Minor product in this reaction is alkyl nitrite.
- Treatment of halides with KNO_2 gives R—O—N = O (alkyl nitrite) as major product.
- Even $NaNO_2$ or KNO_2 also give good yield of nitro compounds if the reaction is carried out in solvents like DMF or DMSO.
- This method is not suitable for preparing aromatic nitro compounds, due to low reactivity of aryl halides.

2. **Direct Nitration**

- Nitroarenes can be prepared by treating arenes with nitrating mixtures (conc. H_2SO_4 and conc. HNO_3)

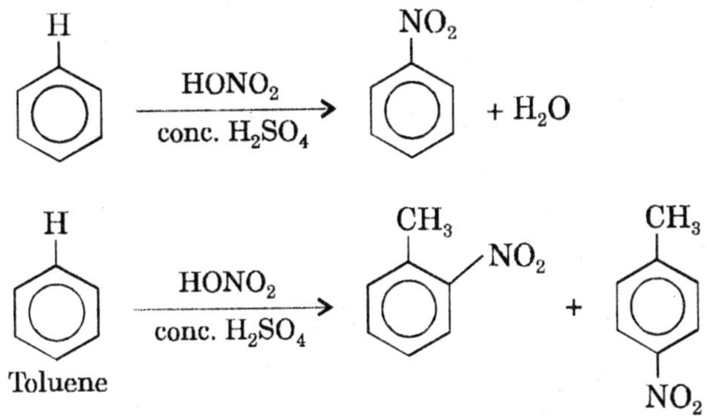

- The function of conc. H_2SO_4 acid in nitrating mixture is not to act as dehydrating agent but it reacts with conc. HNO_3 to produce attacking electrophile nitronium ion (NO_2^+). (See mechanism of nitration of arene page ......)
- The other nitrating agents which can be used for nitration of aromatic compounds are :
  [HNO_3 + $(CH_3CO)_2O$] *mixture, acetylnitrate* (CH_3COONO_2), *nitronium perchlorate* ($NO_2^+ClO_4^-$) and *nitronium fluoborate* ($NO_2^+BF_4^-$)
- The method of direct nitration cannot be applied to alkanes because :
  (*a*) Alkanes lower than C_6H_{14} can only be nitrated by **vapour phase nitration** technique

$$CH_4(g) + HNO_3(g) \xrightarrow[\text{Red hot Si tube}]{675\text{ K}} CH_3NO_2 + H_2O$$

  (*b*) At high temperature, undesirable products are obtained due to cleavage of C—C bonds.

$$\underset{\text{Propane}}{CH_3CH_2CH_3} \xrightarrow[675\text{ K}]{HNO_3} \left[\underset{25\%}{CH_3(CH_2)_2NO_2} + \underset{40\%}{(CH_3)_2CHNO_2} + \underset{10\%}{CH_3CH_2NO_2} + \underset{25\%}{CH_3NO_2}\right]$$

PROPERTIES OF NITRO COMPOUNDS

- Lower nitroalkanes are colourless liquids with pleasant smell. Aromatic nitro compounds like nitrobenzene have pale yellow colour. Nitrobenzene, particularly has a smell of bitter almonds.
- Nitrobenzene is also called **oil of mirbane.**
- Nitroalkanes are sparingly soluble in water (lower members are relatively more soluble than the higher members). Aromatic nitro compounds are not soluble in water at all. They are soluble in organic solvents.
- Nitro compounds have abnormally high boiling points because of their highly polar nature.

Summary of the Chemical Reactions of Nitro Compounds

1. **Reduction.** Reduction is one of the important property of nitro compounds.

The various reduction stages of nitro group are given below :

$$\underset{\text{Nitro}}{-NO_2} \xrightarrow[-H_2O]{+2H} \underset{\text{Nitroso}}{-N=O} \xrightarrow[-H_2O]{+2H} \underset{\text{Substituted hydroxylamine}}{-NHOH} \xrightarrow[-H_2O]{+2H} \underset{p\text{-Amine}}{-NH_2}$$

The final product, however, depends upon the nature of reducing agent as well as the pH of the medium as described below :

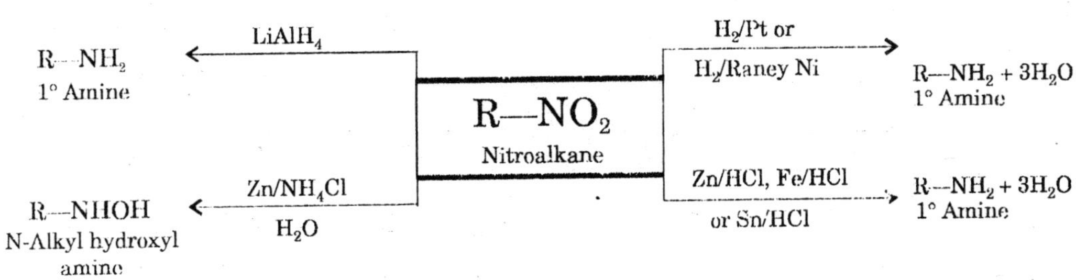

The reduction products of nitrobenzene under different conditions have been summarized below :

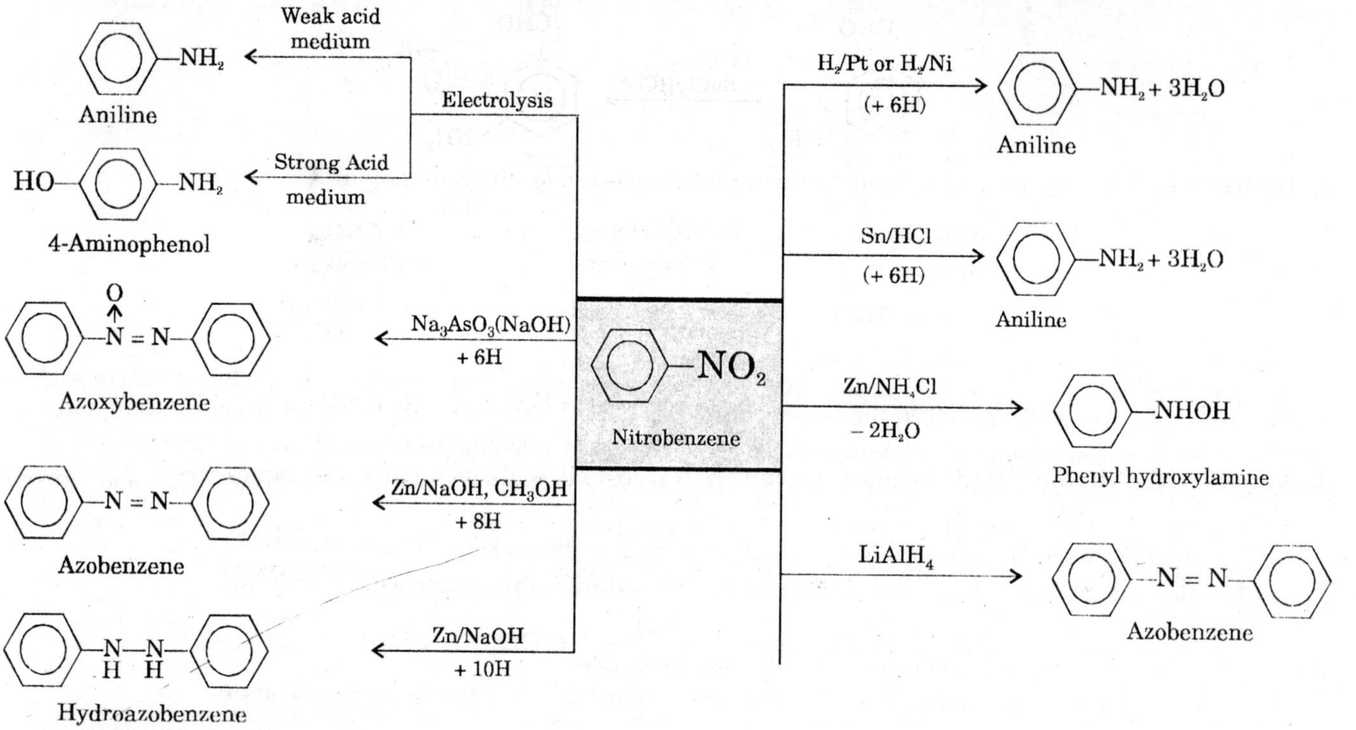

- Conversion of 1° or 2° nitroalkanes to aldehydes or ketones respectively can be done by using ***McMurry reagent*** *(Aqueous* $TiCl_3$*)*. The reaction probably occur through the formation of *imine* intermediates.

$$>CH{-}NO_2 \xrightarrow{\text{Reduction}} >C{=}NH \xrightarrow{H_2O} >C{=}O$$

1° or 2° nitroalkane (Imine) Ald or ketone

- **Selective reduction**

If more than one nitro groups are linked to the aromatic ring, then ***selective reduction*** *of one of the* $-NO_2$ *group can be achieved without affecting the other nitro groups by using very mild reducing agent such as ammonium sulphide,* $(NH_4)_2S$ *or sodium polysulphide* (Na_2S_x)*.*

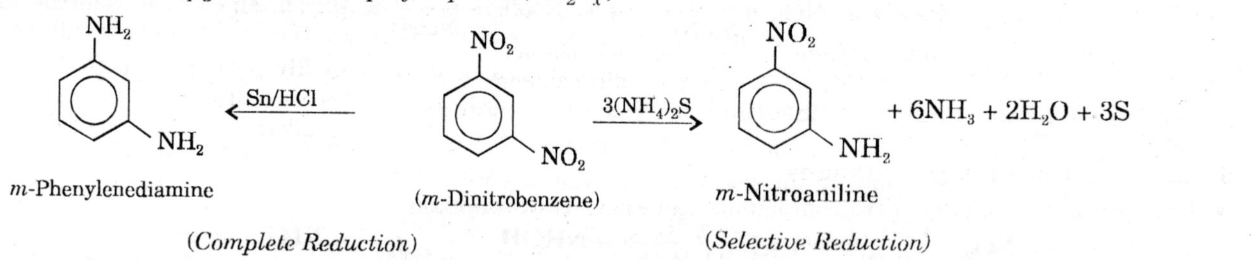

- Selective reduction can also be achieved by the use of NH_4HS or $SnCl_2/HCl$. The action of these reagents is highly specific as indicated by following example, NH_4HS reduces NO_2 *para* to $-CH_3$ group while $SnCl_2/HCl$ reduces NO_2 group *ortho* to $-CH_3$ group.

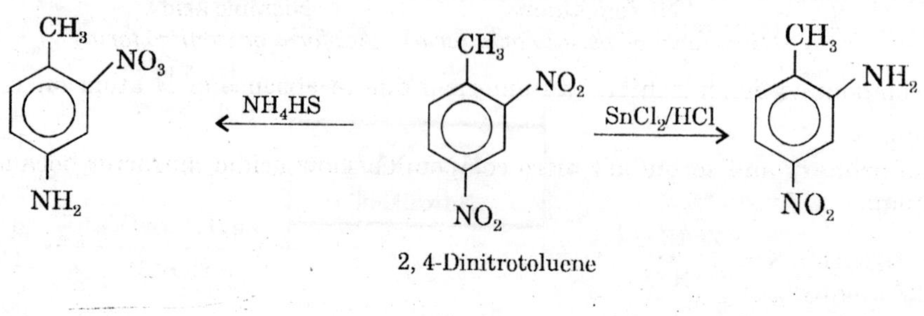

2, 4-Dinitrotoluene

- It may be noted that $SnCl_2/HCl$ is not able to reduce —CHO but it reduces —NO_2 group.

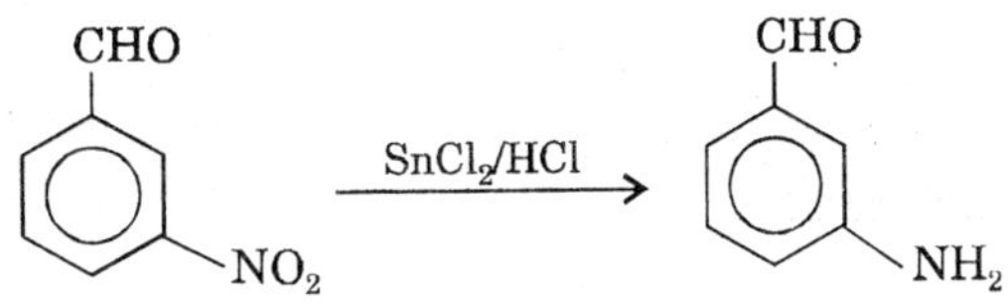

2. Hydrolysis. Primary, secondary and tertiary nitroalkanes give different products

RCH_2NO_2 (1° Nitroalkane)	R_2CHNO_2 (2° Nitroalkane)	R_3CNO_2 (3° Nitroalkane)
↓ HCl/H_2O, Boil	↓ HCl/H_2O, Boil	↓ HCl/H_2O, Boil
$RCOOH + NH_2OH.HCl$	$R_2C = O + N_2O + H_2O$ (Ketone)	No reaction

3. Action with Nitrous Acid. Primary, secondary, tertiary nitro alkanes react differently.

RCH_2NO_2 (1° Nitroalkane)	R_2CHNO_2 (2° Nitroalkane)	R_3CNO_2 (3° Nitroalkane)
↓ HONO, $-H_2O$	↓ HONO, $-H_2O$	↓ HONO
$R-C(NO_2)=NOH$ (Nitrolic acid)	$R_2C(NO)(NO_2)$ (Pseudonitrol)	No reaction

4. Substitution in Alkyl Group (Halogenation). Primary and secondary nitroalkanes on treatment with Cl_2 or Br_2 in the presence of NaOH give *halonitroalkanes*. The α-H atom of nitroalkanes are sucessively replaced by halogen atoms.

$$\underset{p\text{-nitroalkane}}{R—CH_2—NO_2} \xrightarrow[NaOH]{Br_2} \underset{\text{Monobromo nitro alkane}}{R—CH(Br)NO_2} \xrightarrow[NaOH]{Br_2} \underset{\text{Dibromonitro alkane}}{R—CBr_2—NO_2}$$

5. Tautomerism in Nitro Alkanes

- Primary and secondary nitro compounds can show **tautomerism.**

$$\underset{\substack{\text{Nitromethane} \\ (\textit{Nitro form or pseudo acid form})}}{H_2C(H)—N(\rightarrow O) = O} \rightleftharpoons \underset{\substack{\text{Nitronic acid} \\ (\textit{Aci-form or isonitro form})}}{H_2C = N(\rightarrow O)—OH}$$

- Tertiary nitro compounds donot exhibit tautomerism due to absence of H atom on α carbon atom.

6. Acidic Nature

- The α-H atom of primary and secondary nitro compounds show acidic character because of electron withdrawing effect of NO_2 group.

- When sodium salts of nitronic acids (*i.e.*, conjugate base of aliphatic nitro compounds) is treated with 50% H_2SO_4 at room temperature aldehydes or ketones are formed. This is known as **Nefs carbonyl synthesis.**

$$2\ \overset{R}{\underset{H}{}}\!\!>CH-NO_2 \xrightarrow{NaOH} 2R-CH=N{<}^{O^-Na^+}_{O} \xrightarrow[H_2SO_4]{50\%} 2RCHO + N_2O + 2NaHSO_4 + H_2O$$

1° Nitro compound — Sodium Salt — Aldehyde

7. Ring substitution in nitroarenes

- Nitro group **deactivates** the ring and hence makes it less reactive towards electrophilic ring substitution.
- NO_2 group makes *ortho* and *para* positions electron deficient because of resonance.
  Thus, electrophilic ring substitution occurs at metaposion.

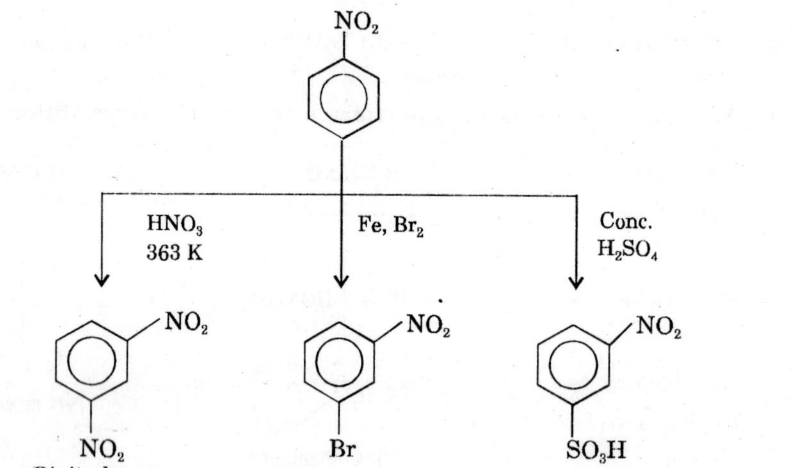

m-Dinitrobenzene — *m*-Bromonitrobenzene — *m*-Nitrobenzenesulphonic acid

- Nitrobenzene does not undergo Friedel Craft's reaction due to deactivation of the ring.
- Due to inert nature of nitrobenzene, it is used as solvent in many organic reactions.
- **Nucleophilic ring substitution** in nitrobenzene occurs at the *ortho* and *para* positions because these are relatively electron deficient positions.

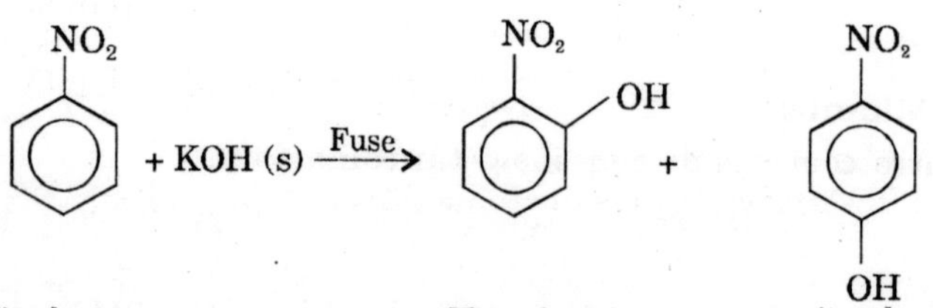

Nitrobenzene — *o*-Nitrophenol — *p*-nitrophenol

- **Nitro group also** influences the reactions of other substituents present in the ring. For example *chlorobenzene* does not undergoes nucleophilic substition easily but *4-Nitrochloro benzene* undergoes alkaline hydrolysis easily. Picryl chloride, (2, 4, 6-trinitrochlorobenzene) on the other hand hydrolyse even at room temperature.

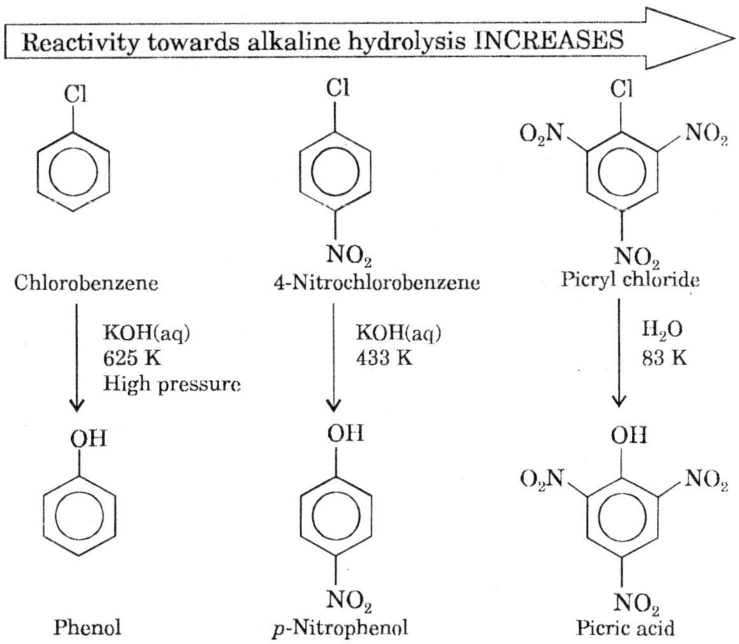

ETHYL NITRITE

Preparation

- *By adding **HCl** or **H_2SO_4** to aqueous solution of sodium nitrite and ethanol.*

$$NaNO_2 + HCl \longrightarrow NaCl + HNO_2$$
$$C_2H_5OH + HNO_2 \longrightarrow C_2H_5ONO + H_2O$$

- *By action of N_2O_3 on C_2H_5OH*

$$C_2H_5OH + N_2O_3 \longrightarrow C_2H_5{-}ON = O + H_2O$$

- *By action of KNO_2 on C_2H_5I*

$$C_2H_5I + KNO_2 \longrightarrow C_2H_5O{-}N = O + KI$$

- Hydrolysis and reduction of ethyl nitrite are given below

$$\underset{\text{Ethyl nitrite}}{C_2H_5{-}O{-}N = O} \xrightarrow[\text{Reduction}]{Sn/HCl} C_2H_5OH + NH_2OH$$
$$\underset{\text{Ethyl nitrite}}{C_2H_5{-}O{-}N = O} \xrightarrow[\text{Hydrolysis}]{H_2O/H^+} C_2H_5OH + HNO_2$$

- Since hydrolysis produces HNO_2. It can be called ester of nitrous acid.
- It dialates blood vessels and lowers blood pressure. Its 4% solution in alcohol is known as sweet spirit of nitre and in used as **diuretic.**

AMINES

- Amines are regarded as alkyl or aryl derivatives of ammonia.
- They are obtained by replacement of one or more H atoms of ammonia by alkyl or aryl groups.
- They can be classified into primary, secondary or tertiary amines depending upon whether the N atom is primary, secondary or tertiary.

$-NH_2$	$-\overset{\vert}{N}H$	$-\overset{\vert}{\underset{\vert}{N}}$
1° Amino group	2° Amino group	3° Amino group

- N atom in amines adopts sp^3 hybrid state and thus they have **pyramidal** structure. However, the angle around the N atom depends upon the number and nature of alkyl groups. For example, in trimethylamine the angle around N atom is 108°.

- In common system amines are **named** as **alkyl amines** or **aryl amines.** In IUPAC system they are named as **alkanamines** or **arenamines.**

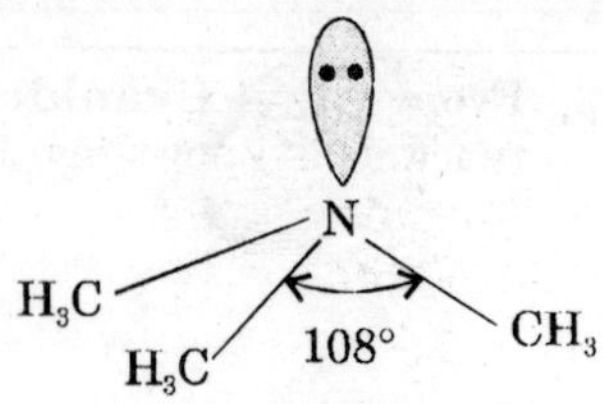

Pyramidal structure of Amines

Illustrations

$CH_3CH_2NH_2$	$CH_3NHC_2H_5$	$(CH_3)N(CH_3)—CH_3$
Ethyl amine or (Ethanamine)	Ethyl methyl amine or (N-methylethanamine)	Trimethylamine or (N, N-Dimethylmethanamine)

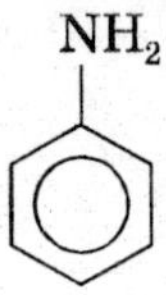

Aniline (Benzenamine)

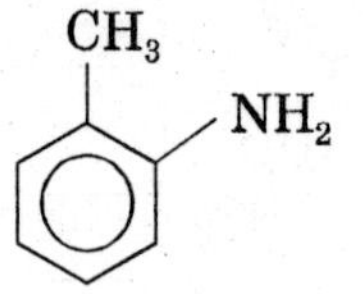

o-Toluidine or 2-Methylaniline (2-Methylbenzenamine)

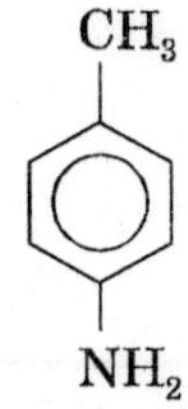

(*p*-Toluidine) (4-Methylbenzenamine)

METHODS OF PREPARATION

1. From Alkyl Halides

(*a*)

$$RX \xrightarrow[-HX]{NH_3} \underset{1° \text{ Amine}}{RNH_2} \xrightarrow[-HX]{RX} \underset{2° \text{ Amine}}{R_2NH} \xrightarrow[-HX]{RX} \underset{3° \text{ Amine}}{R_3N} \xrightarrow{RX} \underset{\text{Quaternary salt}}{[R_4N]^+X^-}$$

- *The above process is called* ***Hoffmann's ammonolysis*** *reaction.*
- *Ammonolysis of alkyl halides gives mixture of amines the composition of which depends upon the molar ratio of starting material.*
- *A good yield of 1° Amine may be obtained by using NH_3 in large excess.*
- *Order of reactivity of Alkyl halides is RI > RBr > RCl.*
- *Ammonolysis is not suitable for preparation of aryl amines because of low reactivity of aryl halides towards nucleophilic replacement.*

(*b*) **Gabriel's Phthalimide Synthesis.** This method is used for preparing **pure primary aliphatic amines** from alkyl halides. The method involves the following steps :

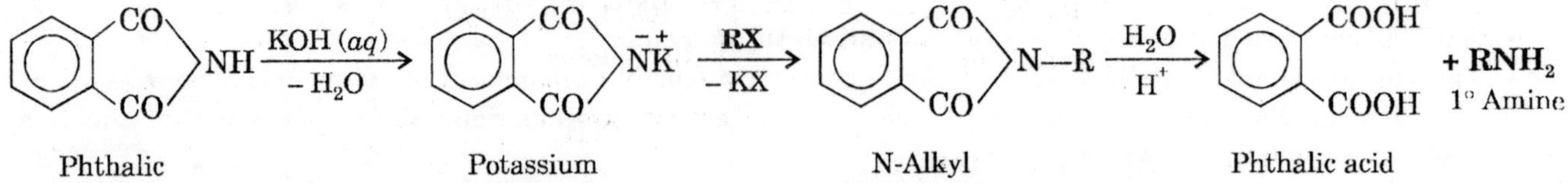

Phthalic anhydride — Potassium Phthalimide — N-Alkyl Phthalimide — Phthalic acid

- The reaction cannot be used to prepare 2° and 3° Amines.
- Aromatic amines cannot be prepared by this method.

2. From Alkyl Cyanides. This method involves the reduction of R-CN with H_2/Ni or $LiAlH_4$ or Na/C_2H_5OH. The reaction is known as ***Mendius reaction.***

$$\underset{\text{Alkyl cyanides}}{R—C \equiv N} \xrightarrow[4H]{Na/C_2H_5OH} \underset{1^\circ \text{ Amine}}{RCH_2NH_2}$$

- *The reduction of isocyanide by similar reagents gives* **secondary amines.**

$$\underset{\text{Alkyl isocyanide}}{R—N \rightleftarrows C} \xrightarrow[4H]{Na/C_2H_5OH} \underset{2^\circ \text{ Amine}}{R-\overset{\overset{\large H}{|}}{N}-CH_3}$$

3. From Aldoximes. Reduction of aldoximes or ketoxime with $LiAlH_4$ or Na/C_2H_5OH also gives primary amines.

$$\underset{\text{Aldoxime}}{R—CH = NOH} \xrightarrow[4H]{Na/C_2H_5OH} \underset{1^\circ \text{ Amine}}{R—CH_2—NH_2} + H_2O$$

4. From Amides

(*i*) **Reduction** of amides with $LiAlH_4$ or Na/C_2H_5OH produce 1° amine with ***same number of carbon atoms*** as in parent amide

$$\underset{\text{Amide}}{R—CO—NH_2} \xrightarrow{LiAlH_4} \underset{1^\circ \text{ Amine}}{R—CH_2NH_2}$$

- *B_2H_6 can also be used in THF (tetrahydrofurane) to reduce primary, secondary and tertiary amides to corresponding amines.*

$$\underset{3^\circ \text{ Amide}}{RCO—N\begin{matrix} R'' \\ R' \end{matrix}} \xrightarrow[THF]{B_2H_6} \underset{3^\circ \text{ Amine}}{R—CH_2—\overset{\overset{\large R''}{|}}{N}—R'}$$

(*ii*) **Hoffmann's Bromamide reaction.** In this case amides gives amines with ***one carbon atom less than the parent amides.***

$$\underset{\text{Amide}}{R—\overset{\overset{\large O}{||}}{C}—NH_2} \xrightarrow{KOH/Br_2} \underset{1^\circ \text{ Amine}}{R—NH_2} + K_2CO_3 + KBr + H_2O$$

5. By Reductive Amination of Aldehydes or Ketones. Aldehydes and ketones react with NH_3 to form imines which can be reduced by catalytic hydrogenation to primary amines.

$$\underset{\text{Aldehyde}}{R—\overset{\overset{\large H}{|}}{C} = O} + NH_3 \xrightarrow[-H_2O]{Heat} \underset{\text{Imine}}{R—\overset{\overset{\large H}{|}}{C} = NH} \xrightarrow{H_2/Ni} \underset{1^\circ \text{ Amine}}{R—CH_2NH_2}$$

6. From Nitro Compounds. This involves the reduction of NO_2 group either by catalytic hydrogenation (H_2/Pt or V) or by chemical means using Fe/HCl or Sn/HCl.

$$\underset{\text{Nitro alkane}}{R—NO_2} \xrightarrow[Fe/HCl\ (6H)]{3H_2/Ni \text{ or}} \underset{\text{Alkanamine}}{RNH_2} + 2H_2O$$

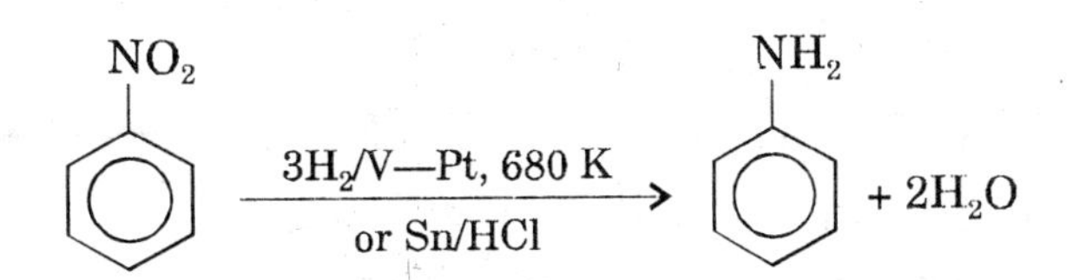

Nitrobenzene Aniline

7. From Isocyanides and Isocyanates

(i) Acidic hydrolysis of isocyanides give 1° amine

$$\underset{\text{Alkyl isocyanide}}{R—N \rightleftarrows C} + HOH \xrightarrow{HCl} \underset{1^\circ\ \text{Amine}}{RNH_2} + HCOOH$$

(ii) Basic hydrolysis of isocynates give 1° amine

$$\underset{\text{Alkyl isocyanate}}{R—N = C = O} + 2KOH \longrightarrow \underset{1^\circ\ \text{Amine}}{RNH_2} + K_2CO_3$$

8. From Grignard's Reagent. It involves the treatment of Grignard's reagent with chloramine ($Cl—NH_2$).

$$R—MgX + Cl—NH_2 \longrightarrow RNH_2 + MgXCl$$

- *The method can be used to prepare tertiary alkyl amine* (R_3CNH_2).

9. From Amino Acids. Decarboxylation of amino acids also yields, amines

$$\underset{\alpha\text{-Amino acid}}{R—\underset{|}{\overset{NH_2}{CH}}—COOH} \xrightarrow[\text{Distill}]{Ba(OH)_2} \underset{1^\circ\ \text{Amine}}{R\overset{NH_2}{\overset{|}{CH_2}}} + CO_2$$

10. By Use of Schmidt Reaction. The reaction involves the treatment of carboxylic acids with hydrazoic acid (HN_3) in conc. H_2SO_4

$$\underset{\text{Carboxylic acid}}{R—COOH} + HN_3 \xrightarrow{\text{Conc. } H_2SO_4} \underset{1^\circ\ \text{Amine}}{R—NH_2} + N_2 + CO_2$$

11. By Use of Curtius Reaction. It involves the treatment of acid chlorides with sodium azide (NaN_3) to form acid azide which is subsequently converted into amine.

$$\underset{\text{Acid chloride}}{R—\overset{O}{\overset{\|}{C}}—Cl} \xrightarrow{NaN_3} \underset{\text{Acidazide}}{R—\overset{O}{\overset{\|}{C}}—N_3} \xrightarrow[\text{Heat}]{C_2H_5OH} \underset{\text{Alkyl isocyanate}}{R—N = C = O}$$

$$\underset{\text{Alkyl isocyanate}}{R—N = C = O} \xrightarrow{C_2H_5OH} \underset{\text{Urethane}}{R—NH—\overset{O}{\overset{\|}{C}}—C_2H_5}$$

$$\underset{1^\circ\ \text{Amine}}{RNH_2} + C_2H_5OH + CO_3^{2-} \xleftarrow[HOH]{OH^-} \underset{\text{Urethane}}{R—NH—\overset{O}{\overset{\|}{C}}—C_2H_5}$$

PROPERTIES OF AMINES

1. Physical state and smell

- Among aliphatic amines, the lower members are gases while higher members are liquids.
- Among arylamines the lower members are liquids while higher members are solids.
- Methyl amine and ethyl amine have ammoniacal smell but higher amines have fishy smell.
- Most of the aromatic amines are colourless in pure form. However, they become coloured on keeping, due to oxidation in air. Aromatic amines are generally toxic. They are easily absorbed through the skin.

2. Boiling points

- All amines, except the tertiary amines are capable of forming intermolecular hydrogen bonds due to the presence of polar N—H bonds. As a result, they have higher boiling points than non-polar compounds.
- Among isomeric amines boiling points decrease as **1° > 2° > 3°.**

3. Solubility

- Aliphatic amines of lower molecular mass are soluble in water. With increase in molecular mass, the solubility in water decreases, the borderline being reached at about six carbon atoms.

- The higher amines are more soluble in less polar solvents such as ethers, alcohols and benzene.
- The solubility of lower aliphatic amines in water is due to their capability to form hydrogen bonds with water molecules.
- Among isomeric amines the solubility decreases as **1° amine > 2° amine > 3° amine.**

SUMMARY OF CHEMICAL REACTIONS OF AMINES

Chemical reactions of amines are summarized below :

Reagents	**AMINES**			
	RNH_2 (1° Amine)	**R_2NH** (2° Amine)	**R_3N** (3° Aniline)	**$C_6H_5NH_2$** (1° Aniline)
1. Basic Nature				
H_2O Water	$R\overset{+}{N}H_3\overset{-}{O}H$ Mono alkyl amm. hydroxide	$R_2\overset{+}{N}H_2\overset{-}{O}H$ Dialkyl amm. hydroxide	$R_3\overset{+}{N}H\overset{-}{O}H$ Trialkyl amm. hydroxide	No reaction with H_2O
HCl	$R\overset{+}{N}H_3Cl^-$ Mono alkyl amm. chloride	$R_2\overset{+}{N}H_2Cl^-$ Dialkyl amm. chloride	$R_3\overset{+}{N}HCl^-$ Tri alkyl amm. chloride	$C_6H_5\overset{+}{N}H_3\overset{-}{Cl}$ Aniline hydrochloride
2. Alkylation of Amines CH_3Br *(Methyl bromide)*	$R—NHCH_3$ N-Methyl alkan-amine	R_2NCH_3 Tertiary amine	$[R_3NCH_3]^+Br^-$ Quaternary salt	$C_6H_5NHCH_3$ N-Methyl aniline
3. Acylation of Amines CH_3COCl *(Acetyl chloride)* or $(CH_3CO)_2O$ *(Acetic anhydride)*	$R—NHCOCH_3$ N-Alkyl ethanamide	$R_2N—COCH_3$ N, N-Dialkyl ethanamide	-	$C_6H_5\overset{H}{\overset{\vert}{N}}COCH_3$ Acetanilide
4. Benzoylation of Amines C_6H_5COCl *(Benzoyl chloride)*	$R—NHCOC_6H_5$ N-Methyl benzamide	$R_2N—COC_6H_5$ N, N-Dialkyl benzamide	—	$C_6H_5—\overset{H}{\overset{\vert}{N}}COC_6H_5$ N-Phenyl benzamide
5. Carbyl Amine Reaction $CHCl_3/KOH$ *(Chloroform)*	$R—N \overset{\rightarrow}{=} C$ Alkyl isocyanide	—	—	$C_6H_5—N \overset{\rightarrow}{=} C$ Phenyl isocyanide

6. Reaction with Nitrous Acid HNO_2 Nitrous acid is prepared in *situ* by using $NaNO_2$/HCl	R—OH + N_2 + H_2O Alcohol **In this reaction alkene is side product*	R_2N—NO Dialkyl nitrosoamine	$R_3\overset{+}{N}HNO_2^-$ Trialkyl amm. nitrite	At 273—278 K $C_6H_5N_2Cl$ Benzene diazonium chloride At 323 K $C_6H_5OH + N_2 + H_2O$ Phenol
7. Formation of Schiff's Bases $>C=O$ (Aldehyde or Ketone)	$>C=NR$ N-Alkylaldimine or N-Alkylketimine	—	—	$>C=N—C_6H_5$ Schiff's bases
8. Reaction with Phenyl Isocyanate $C_6H_5—N=C=O$	$C_6H_5—NH—\overset{O}{\overset{\|\|}{C}}—NHR$ Substituted urea	$C_6H_5NH—\overset{O}{\overset{\|\|}{C}}—NR_2$ Substituted urea	—	Same as 1° amine
9. Reaction with Nitrosyl Chloride NOCl	RCl + N_2 + H_2O	R_2N'—NO + HCl Oily compound	—	
10. Reaction with $COCl_2$ (Phosgene) $COCl_2$	$(RNH)_2C=O$ Substituted urea	$(R_2N)_2C=O$ Substituted urea	—	$C_6H_5—N=C=O$ Phenyl isocyanate
11. Reaction with Hoffmann's Reagent (Diethyloxalate) $COOC_2H_5$ | $COOC_2H_5$	CONHR | CONHR N, N-Dialkyl oxamide (solid)	$CONR_2$ | $COOC_2H_5$ N, N-Dialkyl oxamic ester (liquid)	—	Same as 1° Amine
12. Reaction with Grignard's Reagent R′MgX	R′H + RNHMgBr Alkane	R′H + R_2H—MgBr	No reaction	Same as 1° Amine

13. Reaction with Carbon Disulphide (CS_2)

- This reaction is given by 1° Amines only and is known as **Hoffmann's mustred oil reaction.**

$$RNH_2 + CS_2 \xrightarrow[\text{Heat}]{HgCl_2} \underset{\text{Alkyl isothiocyanate}}{R{-}N = C = S} + 2HCl + HgS$$

- Aniline reacts in a different way to produce N, N′-Diphenyl thiourea initially which on reaction with HNO_3 gives phenylisothiocyanate

$$\underset{\text{Aniline}}{2C_6H_5NH_2} + CS_2 + 2KOH \xrightarrow{-K_2S,\ -2H_2O} \underset{\text{N, N'-Diphenyl thiourea}}{(C_6H_5NH)_2\,C = S} \xrightarrow[HCl]{HNO_3 \text{ or}} \underset{\text{Pheyl isothiocyanate}}{C_6H_5N = C = S} + C_6H_5NH_3^+Cl^-$$

14. Oxidation. The products of oxidation depends upon the nature of oxidising agent as well as amine as discussed below.

(*i*) **Oxidation of 1° Amines.** The oxidation products of various 1° amines with $KMnO_4$ and Caro's acid (H_2SO_5) are given below :

Amine	*$KMnO_4$*	*H_2SO_5 (Caro's acid)*
1° Amine		
$R{-}CH_2{-}NH_2$	$\underset{\text{Aldimine}}{R{-}CH = NH} \xrightarrow{H_3O^+} \underset{\text{Aldehyde}}{R{-}CHO} + NH_3$	$R{-}CH_2NOH + \underset{\text{Aldoxime}}{RCH = NOH}$ + $\underset{\text{Hydroxamic acid}}{R{-}C(OH){=}NOH}$
$R_2CH{-}NH_2$	$\underset{\text{Kitimine}}{R_2C = NH} \xrightarrow{H_3O^+} \underset{\text{Ketone}}{R_2C = O} + NH_3$	$\underset{\text{Ketoxime}}{R_2C = NOH}$
$R_3C{-}NH_2$	$R_3C{-}NO_2$ Nitroalkane	$R_3C{-}NO$ Nitrosoalkane
2° Amine		
$2R_2NH$	$R_2N{-}NR_2 + H_2O$ Tetraalkylhydrazine	$R_2N{-}OH$ N, N-Dialkyl hydroxyl amine
3° Amine		
R_3N	No reaction	$R_3N \rightarrow O$ Amine oxide

Amines also undergo oxidation in air to give N-oxides

$$C_2H_5NH_2 \xrightarrow[\text{air}]{[O]} \underset{\text{Ethylamine oxide}}{C_2H_5\overset{O\uparrow}{N}{-}H_2}$$

Aniline can be readily oxidised by $Na_2Cr_2O_7/H_2SO_4$ to give *p*-benzoquinone

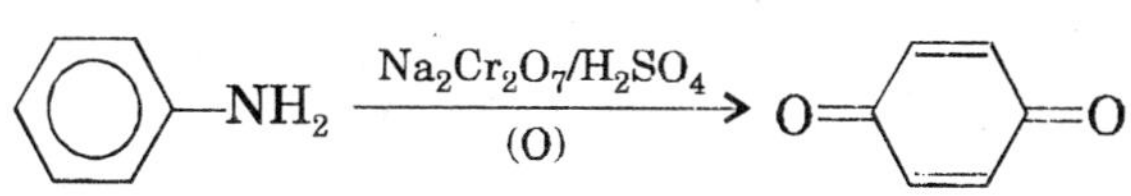

p-benzoquinone

15. Reaction with Metal Ions

Lower aliphatic amines form co-ordination complexes with metal ions like Ag^+ and Cu^{2+}.

$$\underset{\text{Silver chloride}}{AgCl} + \underset{\text{Ethyl amine}}{2C_2H_5\ddot{N}H_2} \longrightarrow \underset{\text{Bis (ethylamine) silver(I) chloride (soluble)}}{[Ag(C_2H_5NH_2)_2]^+ Cl^-}$$

$$\underset{\text{Copper (II) sulphate}}{CuSO_4} + 4C_2H_5\ddot{N}H_2 \longrightarrow \underset{\text{Tetrakis (ethylamine) copper (II) sulphate (Deep blue)}}{[Cu(C_2H_5NH_2)_4]SO_4}$$

16. Ring Substitution Reactions in Aniline

The ring substitution reactions are electrophilic substitution reactions :

Benzene ring in aniline is highly activated due to the delocalisation of lone pair of nitrogen on the ring which results in increase in the electron density on the ring and hence facilitates the electrophilic attack. The substitution mainly takes place at *ortho* and *para* positions.

- **Bromination.** Benzene ring in aniline is so activated that on reaction with aqueous bromine all the *ortho* and *para* positions get substituted resulting in the formation of 2, 4, 6-tri-bromoaniline.

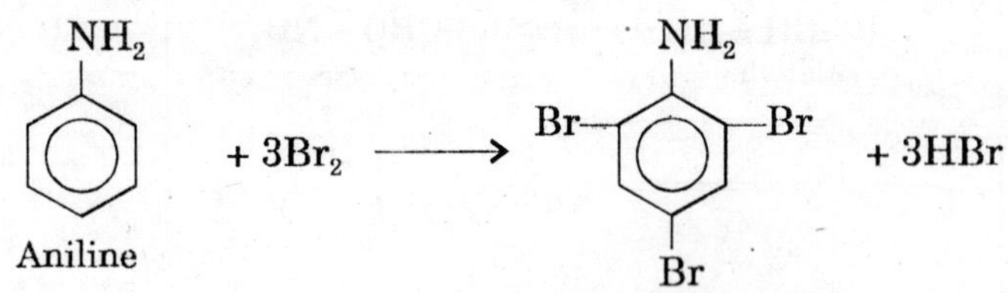

2, 4, 6-Tribromoaniline

- For getting **monobromo compound,** —NH_2 is acylated to reduce its activating effect, before nitration.

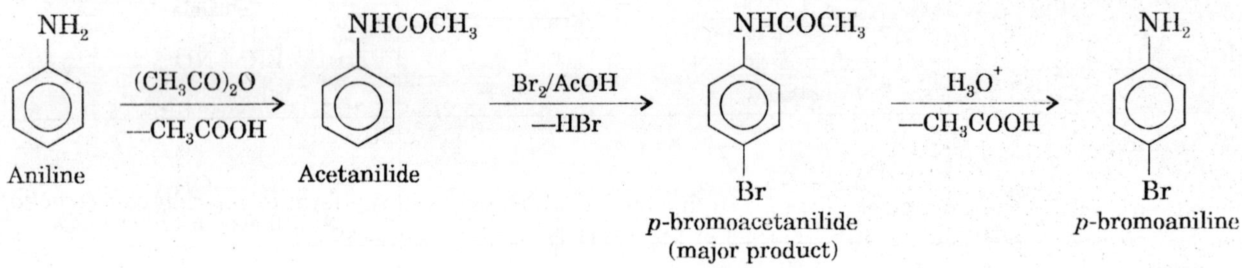

- **Purpose of acylation.** In acyl derivative of aniline, the electron pair is withdrawn towards carbonyl group, thus, decreasing the reactivity.

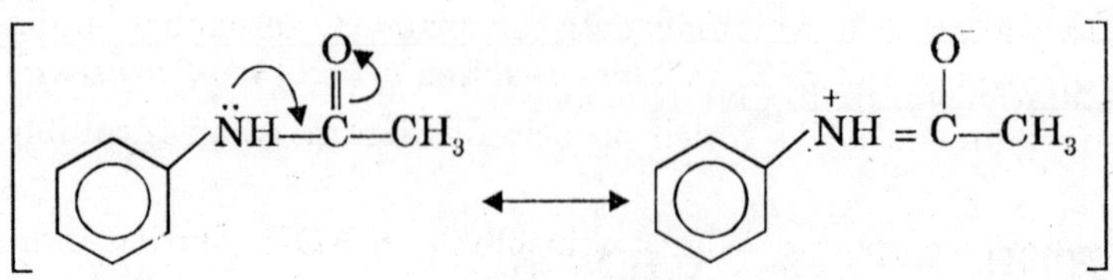

- **Nitration.** Nitration of aniline is also carried out after protecting the —NH_2 group by its acylation.

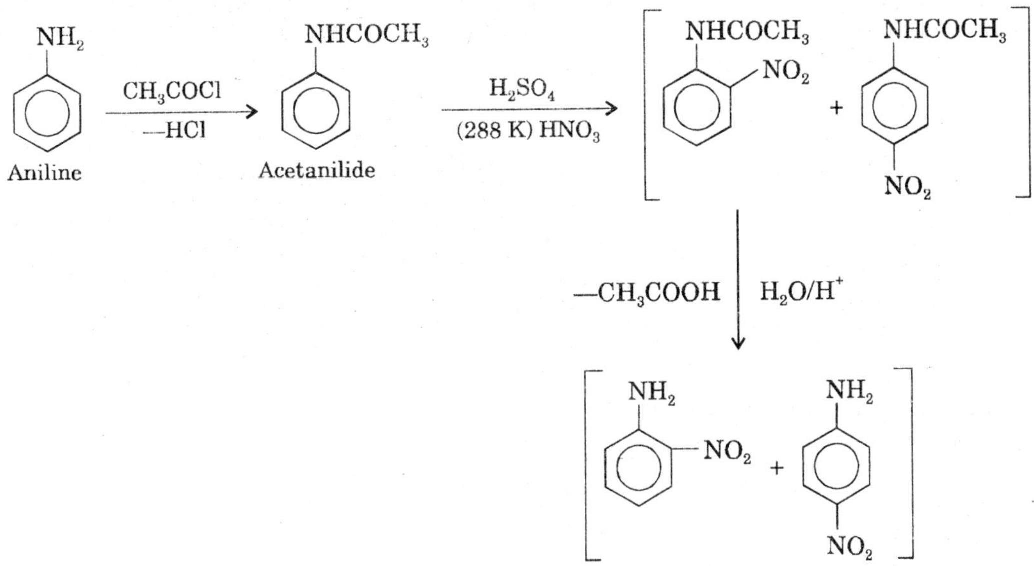

- Nitration without acylation leads to the oxidation of highly activated benzene ring forming black tarry oxidation products.
- **Sulphonation.** Aniline reacts with conc. H_2SO_4 to form the salt anilinium hydrogen sulphate which on '*baking*' at 455 – 475 K gives sulphanilic acid.

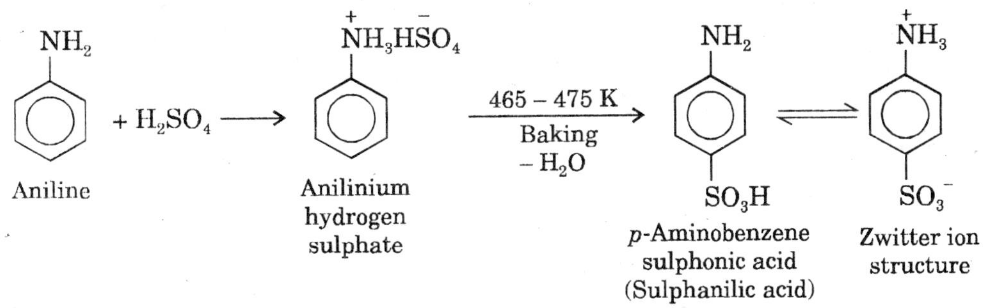

Sulphanilic acid exists as **zwitter ion**, *i.e.*, a dipolar ion which exists in the form of *internal salt structure*. Such ion has positive as well as negative charge within same molecular structure.

ANALYSIS OF AMINES

1. **Hinsberg Test.** This test helps to distinguish between primary, secondary and tertiary amines. The amine is treated with ***benzene sulphonyl chloride in the presence of excess of aqueous potassium hydroxide.***
   (*i*) A clear solution in aqueous KOH which on acidification gives an insoluble material indicates **primary amine.**
   (*ii*) A precipitate (insoluble compound) which is insoluble in KOH solution indicates **secondary amine.**
   (*iii*) Tertiary amines do not react with benzene sulphonyl chloride. An insoluble compound in KOH solution which dissolves *by the addition of acid indicates* **tertiary amine.**

$$\underset{1^\circ \text{ Amine}}{R{-}NH_2} \xrightarrow[-HCl]{C_6H_5SO_2Cl} \underset{\substack{\text{N-Alkyl benzene}\\ \text{sulphonamide}}}{C_6H_5SO_2NHR} \overset{KOH}{\rightleftharpoons} C_6H_5SO_2\overset{-}{N}R\overset{+}{K} \text{ (soluble)}$$

$$\underset{2^\circ \text{ Amine}}{R_2NH} \xrightarrow[-HCl]{C_6H_5SO_2Cl} \underset{\substack{\text{N, N-Dialkyl benzene}\\ \text{sulphonamide}}}{C_6H_5SO_2NR_2} \quad \textit{(insoluble in KOH)}$$

$$\underset{3^\circ \text{ Amine}}{R_3N} \xrightarrow{C_6H_5SO_2Cl} \text{No reaction}$$

2. **Carbylamine Test.** This test is employed to identify *primary amines* (aliphatic as well as aromatic). In this test, the compound is warmed with chloroform in the presence of alcoholic solution of potassium hydroxide. The primary amines yield iso-cyanides which have an extremely unpleasant smell.

$$RNH_2 + CHCl_3 + 3KOH \longrightarrow \underset{\substack{\text{Alkyl}\\ \text{isocyanide}}}{RNC} + 3KCl + 3H_2O$$

$$ArNH_2 + CHCl_3 + 3KOH \longrightarrow \underset{\substack{\text{Aryl}\\ \text{isocyanide}}}{ArNC} + 3KCl + 3H_2O$$

3. The distinction between aromatic 1° amine and aliphatic 1° amine can be made by coupling reaction with benzene diazonium chloride. Aromatic 1° amines form coloured azo compounds with benzene diazonium chloride while aliphatic 1° amines do not form such compounds.

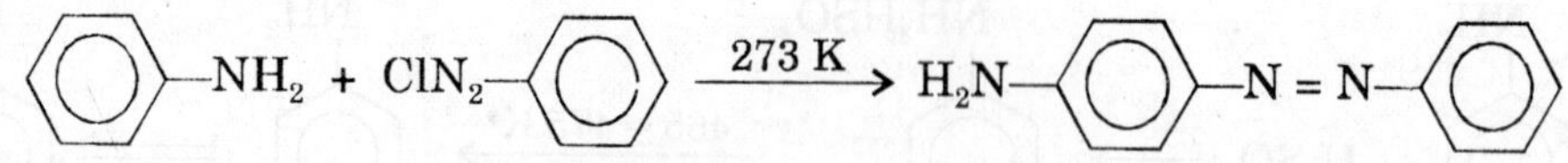

Aromatic amine — Benzene diazonium chloride — Azo compound (a dye)

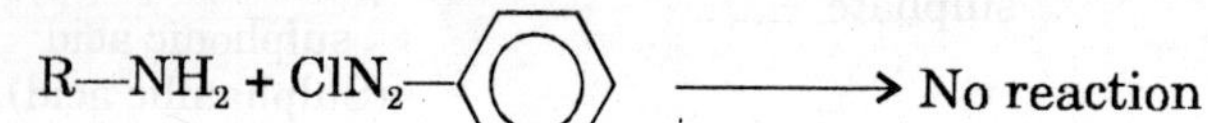

Aliphatic amine

SEPERATION OF 1°, 2°, 3° AMINES

There are different methods for separation of 1°, 2°, 3° amines. Two of the important methods are : **Hoffman method** and **Hinsberg Method** which are being described here.

1. Hoffman Method

Mixture of RNH_2, R_2NH, R_3N

Treat the mixture with diethyl oxalate

- RNH_2 forms $(CONHR)_2$
- R_2NH forms $CONR_2$ | $COOC_2H_5$
- R_3N does not react

↓

Distillation of Resulting mixture

→ *Distillate*: **R_3N** (3° Amine)

→ *Residual mixture*: Contain $(CONHR)_2$ as solid and oxamic ester as liquid

Filtration

→ *Filtrate*: $CONR_2$ | $COOC_2H_5$ (oxamic ester) —Boil with KOH→ COOK | COOK **+ R_2NH** (2° Amine)

→ *Residue*: CONHR | CONHR (oxamide) —KOH(aq), Boil→ COOK | COOK **+ $2RNH_2$** (1° Amine)

2. Hinsberg Method

Mixture of RNH_2, R_2NH, R_3N

Treat the mixture with Hinsberg reagent ($C_6H_5SO_2Cl$) followed by addition of KOH to make it alkaline

- RNH_2 forms $(C_6H_5SO_2NR)K$ which is soluble
- R_2NH forms $C_6H_5SO_2NR_2$ which remain in soluble
- R_3N remains as such

↓

Distil the Resulting mixture

Distillate
R_3N
3° Amine

Residual mixture
$(C_6H_5SO_2NR_2 + C_6H_5SO_2NRK)$

Filter

Residue
$(C_6H_5SO_2NR_2)$
$\downarrow H_3O^+$
$R_2NH + C_6H_5SO_2OH$
2° Amine

Filtrate
$(C_6H_5SO_2NR^-K^+)$
$\downarrow$ (*i*) H_3O^+ (*ii*) H_2O
$C_6H_5SO_2OH + RNH_2$
1° Amine

BASICITY OF AMINES

Amines are derivative of **ammonia.** They exhibit basic character due to the presence of lone pair of electrons on the N-atom. Some important features of their basic character are summarized as follows :

- Basicity of amines is expressed by their basicity constant K_n or by pK_b (= – log K_b). ***The larger value of K_b or lower value of pK_b correspond to higher basic strength of the corresponding amine.***
- **Alkylamines are stronger bases than ammonia.** Alkyl groups by their electron releasing effect, concentrate electron density on nitrogen and hence, make the lone pair of nitrogen more easily available for sharing with acids. Moreover, electron releasing effect of alkyl groups stabilizes the alkyl ammonium ion formed

$$R \rightarrow \ddot{N}H_2 + H_2O \rightleftharpoons R\text{—}\overset{\oplus}{N}H_3 + \overset{-}{O}H$$

Basic Strength of Primary, Secondary and Tertiary Alkylamines

Among primary, secondary and tertiary aliphatic amines, the electron releasing effect is maximum in tertiary amines and minimum in primary amines. Thus, basic strength should increase as

$$\underset{1°}{R\text{—}\ddot{N}H_2} < \underset{2°}{R_2\ddot{N}H} < \underset{3°}{R_3\ddot{N}}$$

The observed basic strength of 1°, 2° and 3° alkyl amines have been found to follow the expected order in *gas phase* or in *non-aqueous solvents.* However, the basic strength of 1°, 2° and 3° amines does not follow the expected order in aqueous solution. For example, the basic strengths of 1°, 2°, 3° methyl and ethyl amines in aqueous medium is as follows :

	$(CH_3)_2NH$	>	CH_3NH_2	>	$(CH_3)_3N$
	2°		1°		3°
pK_b	(3.29)		(3.38)		(4.22)
	$(C_2H_5)_2NH$	>	$(C_2H_5)_3N$	>	$C_2H_5NH_2$
	2°		3°		1°
pK_b	(3.0)		(3.20)		(3.27)

In fact, the basic strength in aqueous solutions not only depends on electron releasing effect but also depends upon ***steric effect*** and ***hydration effect.***

(*a*) **Steric effect** refers to the crowding of alkyl groups around N atom which hinders the attack of proton on the amine molecule and thus, decreases its basic strength. Since crowding of alkyl groups around N atom increases from primary to tertiary amines. Consequently basic strength of amines should decrease from 1° > 2° > 3°.

(*b*) **Hydration effect** refers to the stabilisation of the protonated amine by the water molecules. The water molecules form H-bonds with the protonated amine and release energy called hydrogen energy.

Now greater the extent of H-bonding in protonated amine more will be its stabilisation and consequently, greater will be the basic strength of the corresponding amine. The hydration due to H-bonding is maximum in monoalkyl ammonium ion (protonated cation of 1° amine), it is less in dialkyl ammonium ion and still less in trialkyl ammonium ion. Therefore, basic strength should decrease from 1° > 2° > 3°.

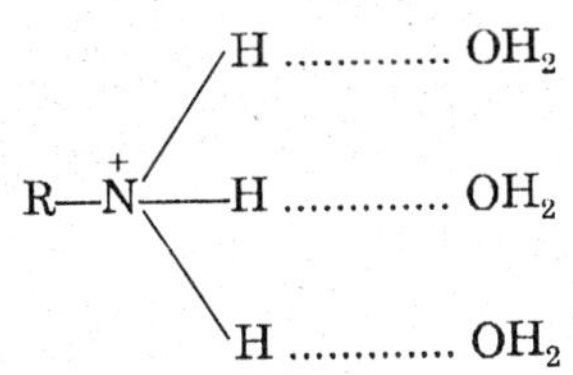

Monoalkyl ammonium ion

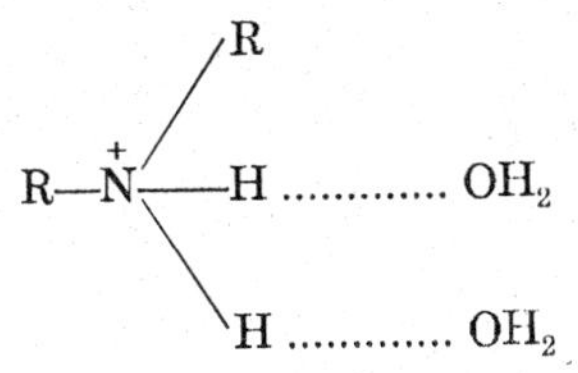

Dialkyl ammonium ion

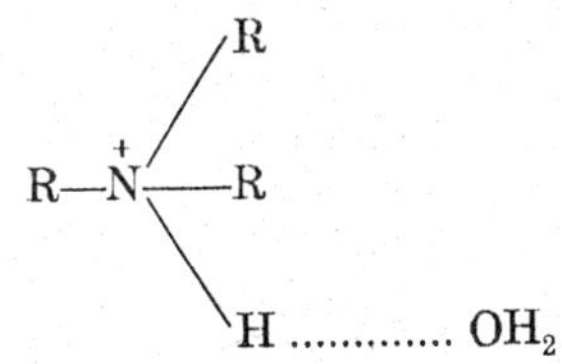

Trialkyl ammonium ion

The extent of hydration and stabilisation DECREASES

The resultant of all the factors, namely + I *effect, Steric effect* and *Hydration effect* cause 2° amine to become most basic.

- **Aromatic amines such as aniline are less basic than ammonia.** This can be explained in terms of electron withdrawing effect of aromatic ring due to resonance. For example, Aniline is resonance hybrid of the following contributing structures :

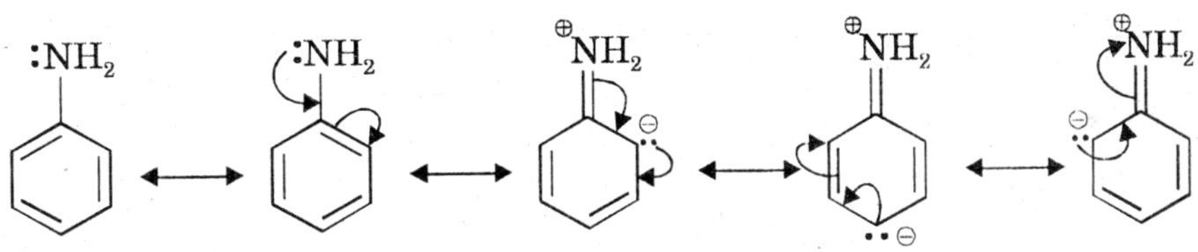

I II III IV V

It is quite evident from the above resonating structures that the lone pair of electrons is withdrawn away from it and is being partially shared with benzene ring. Therefore, long pair of nitrogen in aniline is not fully available for sharing with acids whereas in ammonia it is fully available for sharing. Hence, ***aniline is weaker base than aliphatic amines and ammonia.***

- **Effect of substituents on basic strength of aromatic amines.** The presence of **electron releasing groups (ERG)** increase the basic strength. On the other hand presence of **electron withdrawing groups (EWG)** on the ring (particularly at *p*-position) decrease the basic strength.

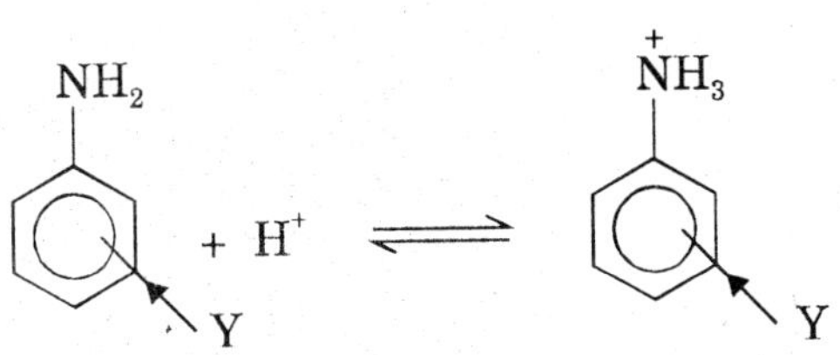

Y : releases electrons ; stabilises cation, increases basic strength

Y = $—NH_2$ or $—OH$ or $—OCH_3$ or $—CH_3$

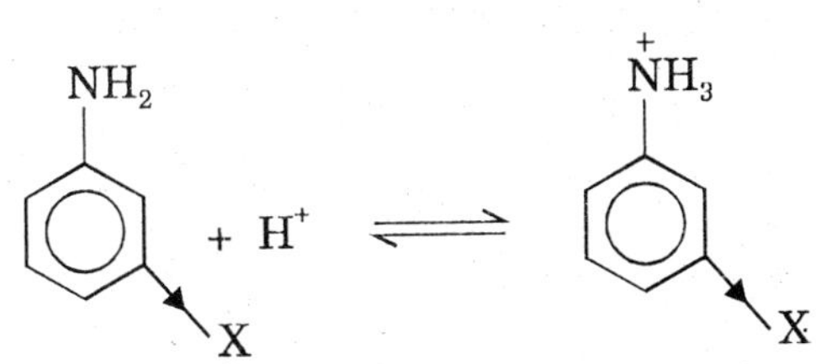

X : withdraws electrons ; destabilises cation ; decreases basic strength

Y = $—\overset{+}{N}H_3$ or $—NO_2$ or $—CHO$ or $—COOH$

- The *base weakening effect* of **EWG** and *base strengthening effect* of **ERG** is more pronounced at ***p*-position** in comparison to that at ***m*-position.**
- **Ortho effect.** The presence of electron releasing or electron attracting group at the *ortho* position cause decrease of basic strength. However, the effect of electron attracting group has more pronounced effect in decreasing the basic strength.

The overall basic strength of *ortho, meta* and *para* substituted anilines depends on *electron donating or electron withdrawing* **resonance** effect as well as **inductive** effect.

ORDER OF BASIC STRENGTHS OF SOME AROMATIC AMINES

Nitro anilines **pk_b**	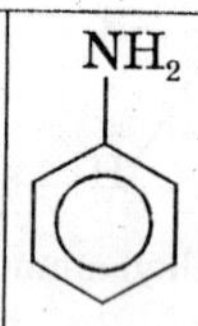 9.376	>	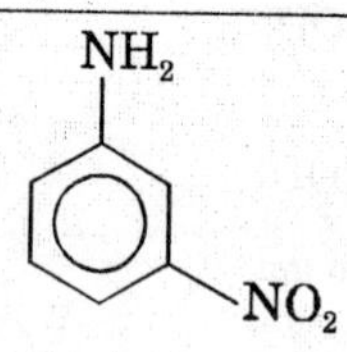 11.53	>	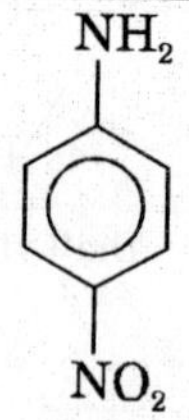 13	>	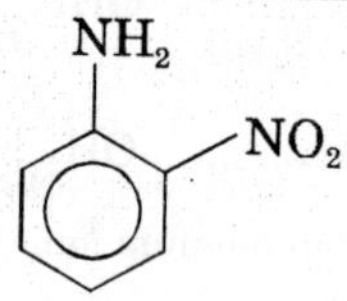 14.22	**$-NO_2$** has **(–R)** and **(–I)** effect.
Toluenes **pk_b**	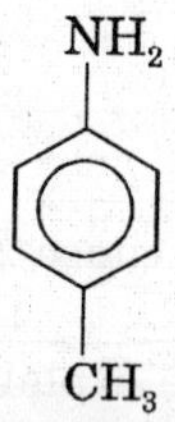 8.92	>	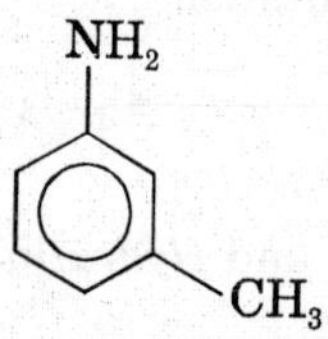 9.30	>	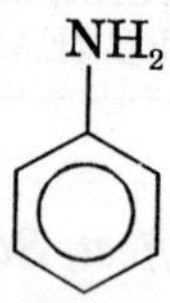 9.376	>	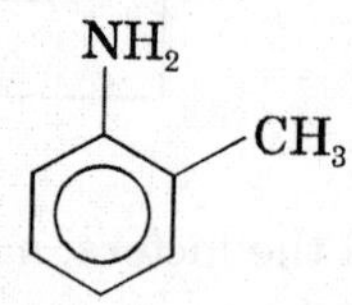 9.52	**$-CH_3$** has **(+I)** as well as **(+R)** effect.
Anisidines **pk_b**	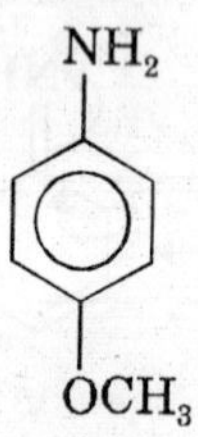 8.69	>	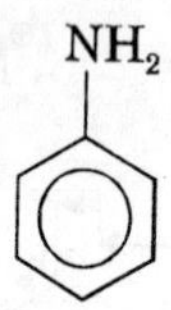 9.376	>	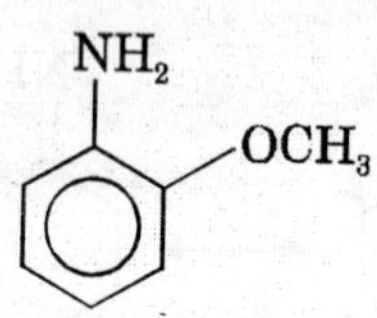 9.52	>	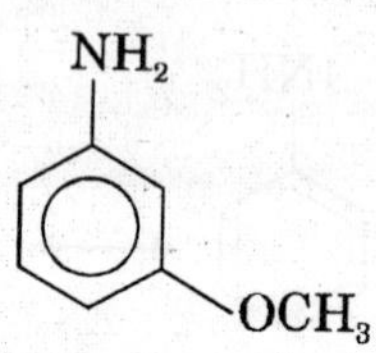 9.69	**$-OCH_3$** has strong **(+R)** and weak **(–I)** effect.
Aminophenols **pk_b**	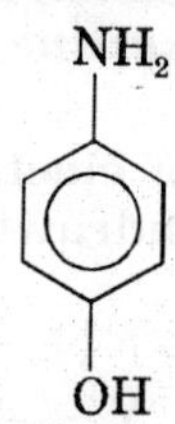 8.5	>	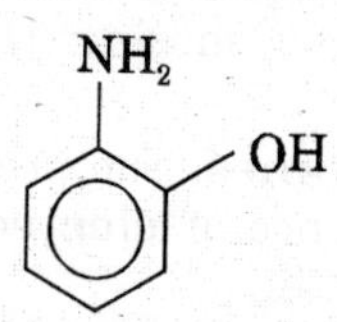 9.28	>	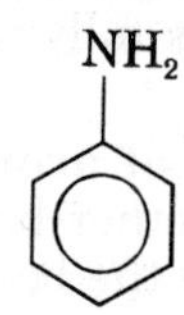 9.376	>	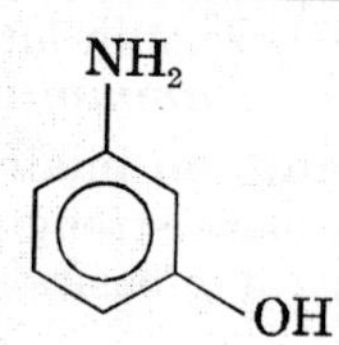 9.80	**–OH** has strong **(+R)** ; weak **(–I)** effect. Orthohydroxy anilinium ion is stabilised by intramolecular H-Bonding
Phenylenedi-amines **pk_b**	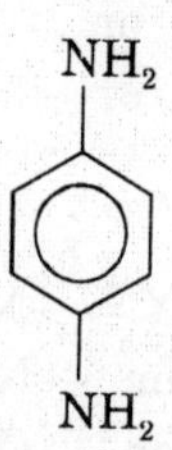 7.85	>	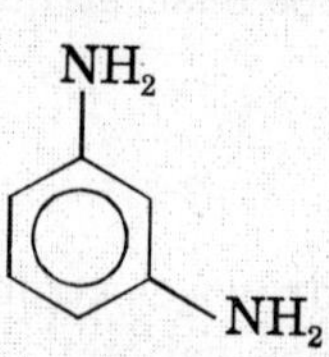 9.0	>	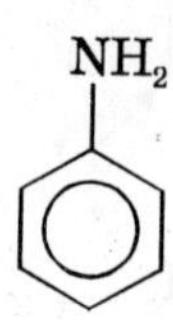 9.376	>	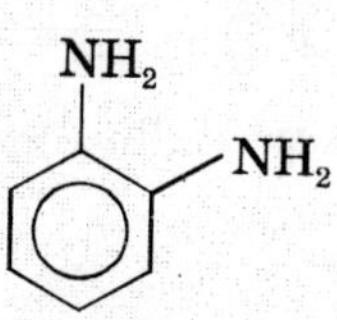 9.52	**$-NH_2$** has strong **(+R)** ; very weak **(+I)** effect.

Chloroanilines pk_b	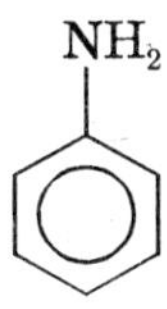 9.376 >	 10.0 >	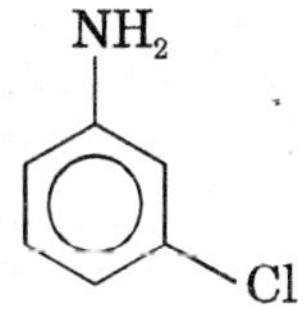 10.5 >	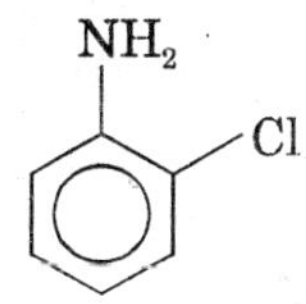 11.3	**-Cl** has weak **(-R)** and strong **(-I)** effect.
N-substituted alkyl anilines pk_b	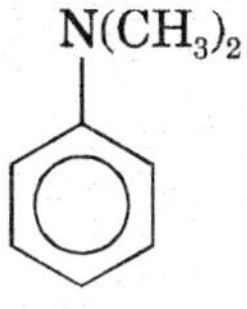 8.93 >	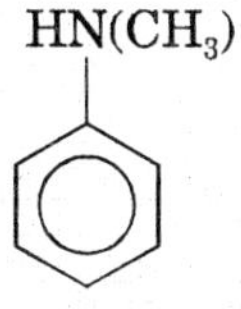 9.1 >	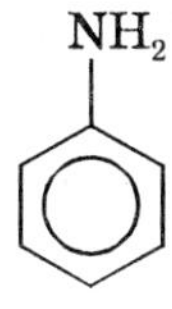 9.3		$-CH_3$ is **ERG.**
N-substituted phenyl anilines pk_b	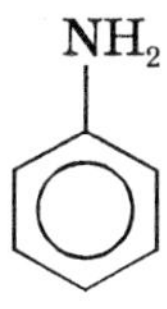 9.376 >	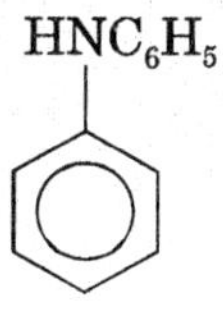 13.22 >	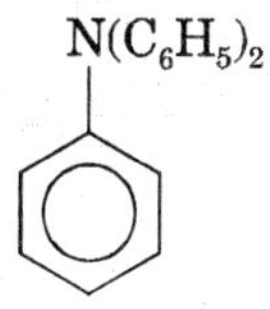 —		$-C_6H_5$ is **EWG.**

DIAZONIUM SALTS

The diazonium salts have general formula $\mathbf{ArN_2^+X^-}$, where X^- may be anion like Cl^-, Br^-, I^-, HSO_4^-, etc., and $\overset{+}{N_2}$ group ($-N = N$) is called diazonium group. They are named by adding a suffix diazonium to the name of the parent aromatic compound followed by the name of the anion. Some examples are :

$C_6H_5N_2^+Cl^-$	$C_6H_5N_2^+HSO_4^-$
Benzene diazonium chloride	Benzene diazonium hydrogen sulphate

Preparation. They are prepared by the reaction of aromatic primary amine with $NaNO_2/HCl$, *i.e.,* (HNO_2) at 273—280 K. The reaction is called **diazotisation.**

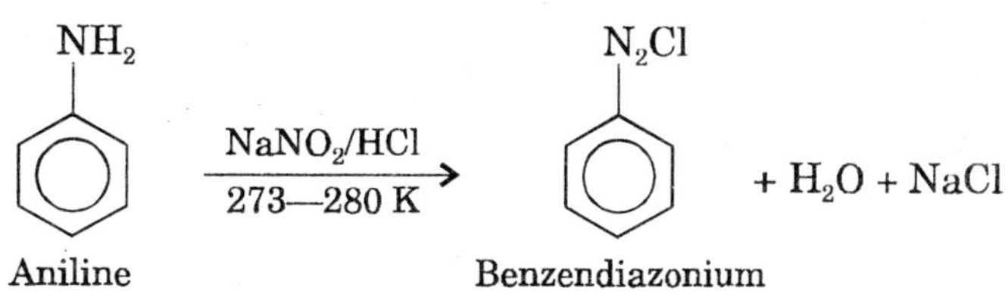

Aniline Benzendiazonium chloride

Properties. Diazonium salts are

- generally colourless crystalline solids
- readily soluble in water
- unstable and explode is dry state, hence, they are generally used in solution state.
- Some important chemical properties have been summed up as follows :

1. Coupling Reactions

Arenediazonium salts reacts with electron rich aromatic compounds like phenol, aniline to form brightly coloured azo compounds.

- Coupling generally occurs at the para position.
- If para position is occupied then coupling occurs at *ortho* position.
- Coupling tendency is enhanced if electron donating group is present at the *p*-position of $N_2^+Cl^-$ group.

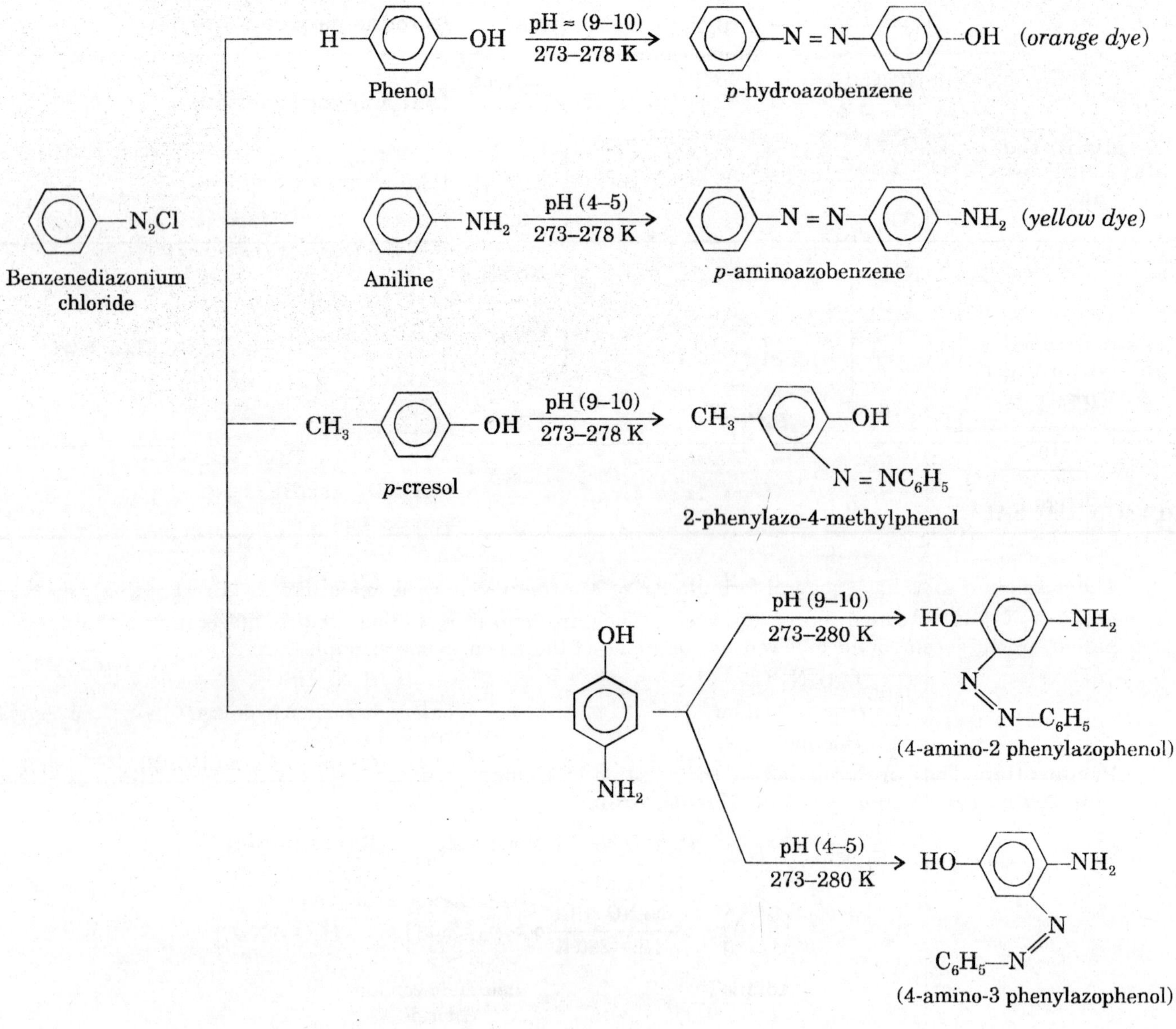

2. Replacement Reactions

In these reactions, the diazonium group ($-\overset{+}{N_2}$) is replaced by other substituents.

$C_6H_5\overset{+}{N_2}\overset{-}{Cl}$ (Benzene diazonium chloride)

$$C_6H_5N_2Cl \xrightarrow{CuCl/HCl} C_6H_5\mathbf{Cl} + N_2 + Cl^-$$ Chlorobenzene **(Sandmeyer reaction)**

$$C_6H_5N_2Cl \xrightarrow{CuBr/HBr} C_6H_5\mathbf{Br} + N_2 + CuCl$$ Bromobenzene **(Sandmeyer reaction)**

$$C_6H_5N_2Cl \xrightarrow{CuCN/KCN} C_6H_5\mathbf{CN} + N_2 + CuCl$$ Benzonitrile **(Sandmeyer reaction)**

$$C_6H_5N_2Cl \xrightarrow{Cu/HCl} C_6H_5\mathbf{Cl}$$ **(Gattermann reaction)**

$$C_6H_5N_2Cl \xrightarrow{Cu/HBr} C_6H_5\mathbf{Br}$$ **(Gattermann reaction)**

$$C_6H_5N_2Cl \xrightarrow{KI,\ Heat} C_6H_5\mathbf{I} + KCl + N_2$$ Iodobenzene

$$C_6H_5N_2Cl \xrightarrow[-HCl]{HBF_4} C_6H_5N_2^+ (BF_4)^- \xrightarrow{Heat} C_6H_5F + N_2 + BF_3$$ Benzenediazonium tetra fluoborate; Fluorobenzene **(Balz-Schiemann reaction)**

$$C_6H_5N_2Cl \xrightarrow[-HCl]{HBF_4} C_6H_5N_2^+ (BF_4)^- \xrightarrow{NaNO_2} C_6H_5\mathbf{NO_2} + NaBF_4 + N_2$$ Nitrobenzene

$$C_6H_5N_2Cl \xrightarrow[(ii)\ H_2O/Cu^+]{(i)\ H_3PO_2} C_6H_6 + N_2 + H_3PO_3 + HCl$$ Benzene **(Deamination)**

$$C_6H_5N_2Cl \xrightarrow[Boiling\ H_2SO_4]{H_2O} C_6H_5\mathbf{OH} + N_2 + HCl$$ Phenol

$$C_6H_5N_2Cl \xrightarrow{C_6H_6/NaOH} C_6H_5\text{—}\mathbf{C_6H_5} + N_2 + HCl + H_2O$$ Biphenyl **(Gomberg reaction)**

$$C_6H_5N_2Cl \xrightarrow[CuSO_4]{Na_3AsO_3} [C_6H_5AsO_3Na_2] + NaCl + N_2$$ **(Bart reaction)**

$$[C_6H_5AsO_3Na_2] \xrightarrow{2\ HCl} C_6H_5AsO_3H_2 + 2NaCl$$ Phenylarsonic acid

Reduction:

$$C_6H_5N_2Cl \xrightarrow{SnCl_2/HCl} C_6H_5\text{—}NH\text{—}NH_2$$ Phenylhydrazine

$$C_6H_5N_2Cl \xrightarrow{Zn/HCl} [C_6H_5NH\text{—}NH_2] \xrightarrow{Zn/HCl} C_6H_5NH_2 + NH_3$$ Aniline

MECHANISM OF SOME IMPORTANT REACTIONS

1. Nitration of Arenes

(*i*) $HO-NO_2 + H_2SO_4 \longrightarrow H_2\overset{+}{O}-NO_2 + H\overset{-}{S}O_4 \longrightarrow H_2O + \overset{+}{N}O_2$ (Nitronium ion)

(*ii*) 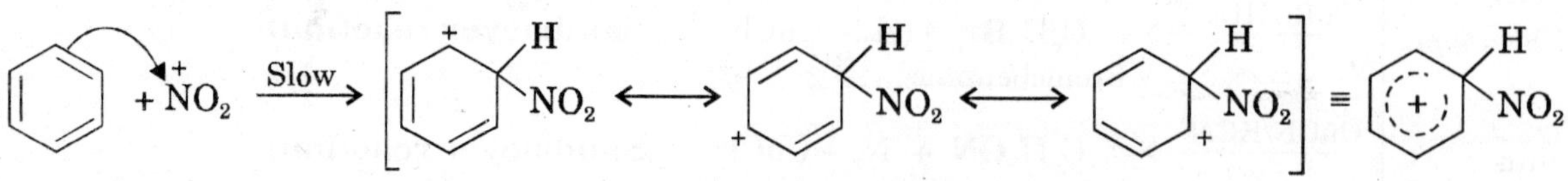

This step is rate determining step.

(*iii*) 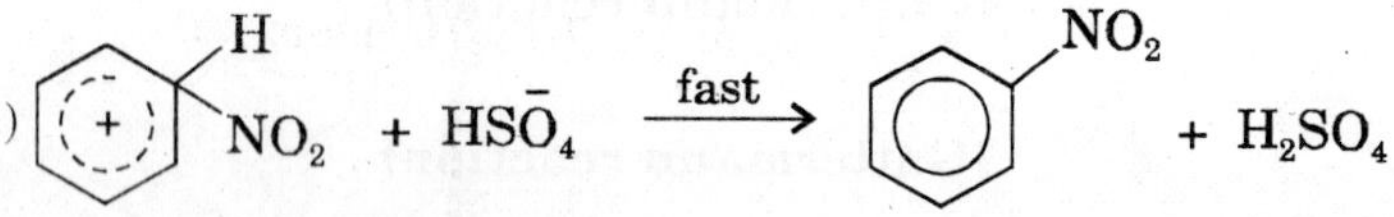

2. Mechanism of Diazotisation

(*i*) $O=N-OH + \overset{+}{H} \rightleftharpoons O=N-\overset{+}{O}H_2 \longrightarrow \overset{+}{N}=O + H_2O$

Nitrous acid — Nitrosonium ion

(*ii*) 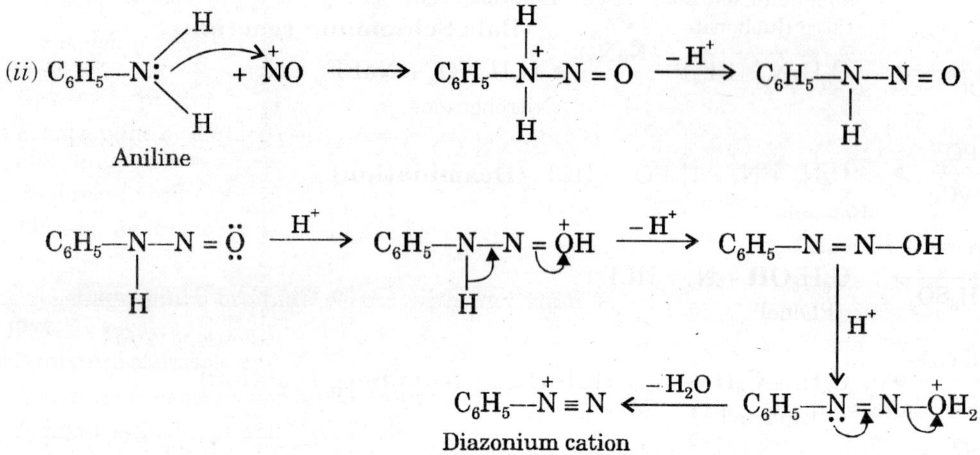

3. Mechanism of Coupling Reaction

Coupling reaction is a typical example of electrophilic substitution reaction in which diazonium ion acts as electrophile while phenol or aniline act as nucleophile.

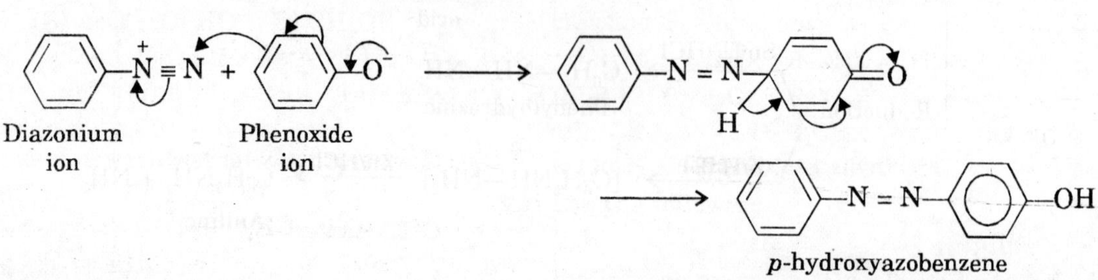

4. Exhaustive Methylation and Hoffmann Elimination. The reaction of amines (1°, 2° or 3°) with excess of an alkyl halide to form quaternary ammonium salts is called **exhaustive alkylation.** However, *if the alkyl halide used is methyl iodide, the reaction is known as* **exhaustic methylation.**

When the quaternary ammonium halide obtained above is treated with moist Ag_2O, silver halide gets precipitated and the corresponding quaternary ammonium hydroxide is produced. When this quaternary ammonium hydroxide is heated to 400 K or above, a β-hydrogen *w.r.t.* N-atom is abstracted by OH^- ion to form an alkene. *This pyrolysis of quaternary ammonium hydroxide to produce alkenes is called* **Hofmann elimination.** For example,

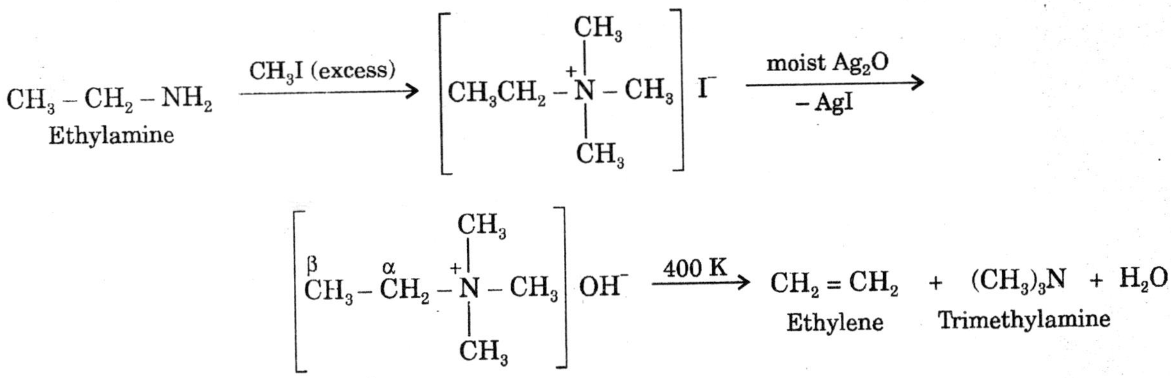

It may be noted that **Hofmann elimination** *usually gives least substituted alkene in contrast to dehydrohalogenation of alkyl halides which invariably gives most highly substituted alkene.* **(Saytzeff rule).** For example,

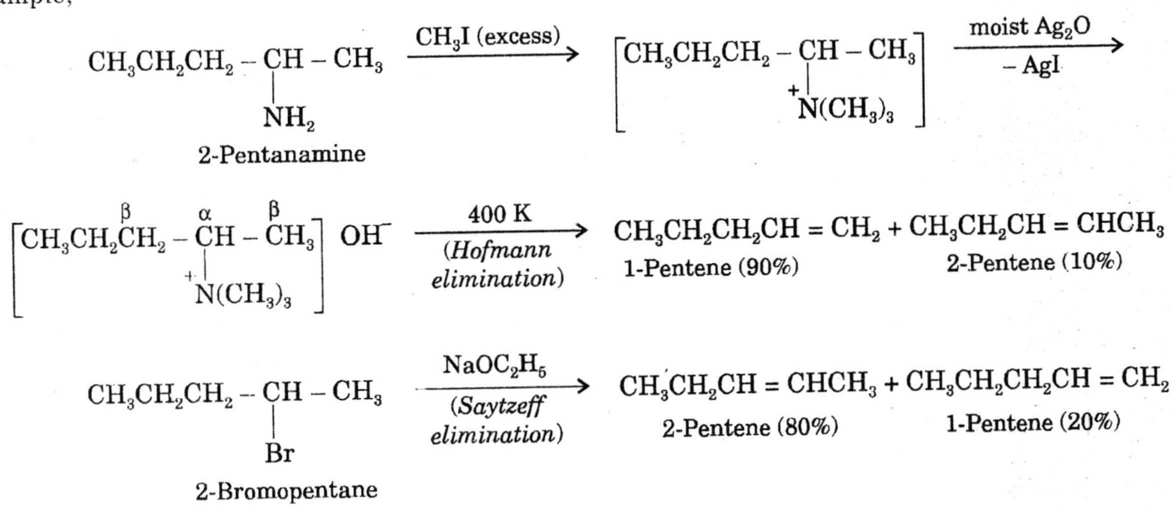

5. **Cope Elimination.** Tertiary amine oxides in which at least one of the alkyl groups contains a β-hydrogen on heating to 423—473 K, eliminate a molecule of dimethylhydroxylamine to yield an alkene. For example,

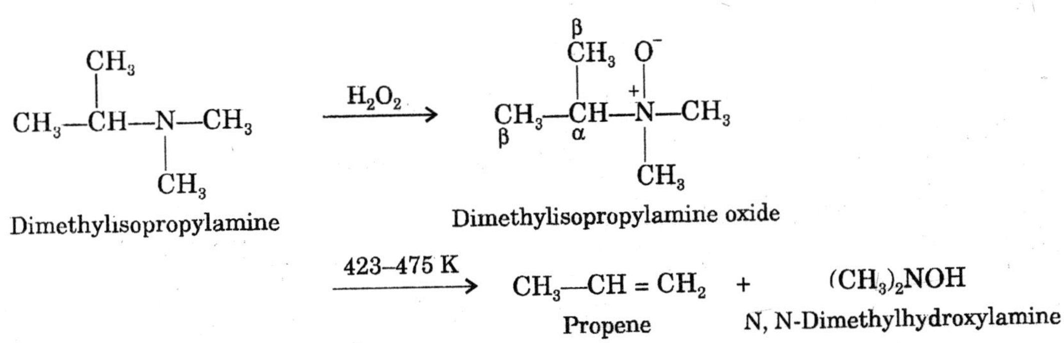

This reaction is called **cope elimination.** *Like Hofmann elimination, cope elimination is also used to determine the structure of amines and for the preparation of alkenes from tertiary amines.*

6. Mechanism of Curtius Reaction. It involves the decomposition of *acylazide* in inert solvent like $CHCl_3$, C_6H_6 to form *isocyanate* which can react with water or alcohol to form 1° amine or urethane respectively.

$$\underset{\text{Acyl azide}}{R{-}CON_3} \xrightarrow[C_6H_6]{\text{Heat}} \underset{\text{Isocyanate}}{R{-}N=C=O} \xrightarrow{H_2O} \underset{\text{1° Amine}}{RNH_2} + CO_2$$

$$R{-}N=C=O \xrightarrow{ROH} \underset{\text{Urethane}}{RNHCOOR}$$

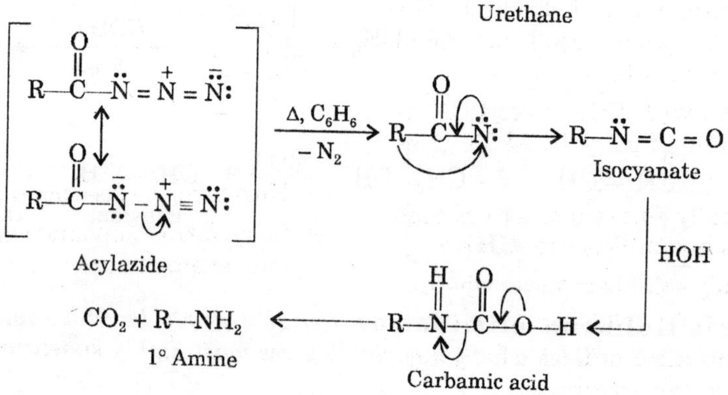

7. Mechanism of Schmidt Reation. In involves the reaction of carboxylic acid and hydrazoic acid in conc. H_2SO_4 to form RNH_2.

$$RCOOH + N_3H \xrightarrow{H_2SO_4} \underset{\text{1° Amine}}{R{-}NH_2} + CO_2 + N_2$$

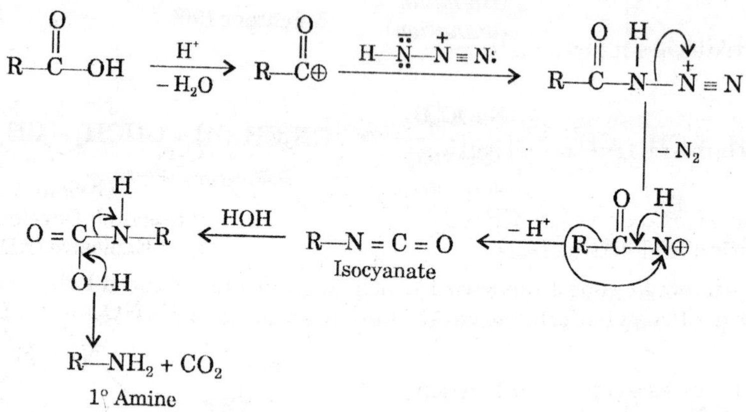

7.2. IMPORTANT FACTS AND TERMS

Add to Knowledge

- R—CN and RNC are functional isomers of each other.
- Carbon and N atoms of —CN group are *sp* hybridised. In isocyanide group also carbon is *sp* hybridised. The isocyano (—NC) group contains one σ bond, one π bond and one *p*π-*p*π back bond from N to C.
- Treatment of halides (RX′) with KCN gives RCN as major product while reaction of RX with AgCN gives R—NC as major product.
- Alkyl cyanides have generally pleasant order whereas alkyl isocyanides have highly unpleasant smell.
- Alkyl cyanides are generally soluble in water whereas alkyl isocyanides are generally insoluble in water.
- Due to electron deficient nature of C of alkyl isocyanides they give some addition reactions.
- Cyanide (—CN) or isocyano (—N—C) group when attached to aryl group act as *meta* directing groups during electrophilic ring substitution.
- Alkyl cyanides are relatively more polar than corresponding alkyl isocyanides.
- Nitroalkanes are formed in excellent yield when 1° or 2° alkyl halides (bromides or iodides) are treated with silver nitrite.
- Nitroalkanes and alkylnitrites are functional isomers of each other.
- NO_2 group highly electron withdrawing and is *meta* directing during electrophilic ring substitution reaction.
- In nitrating mixture (conc. H_2SO_4 and conc. $CHNO_3$), HNO_3 undergoes protonation and acts as base.
- Reduction of nitro alkanes in acidic conditions give 1° amine whereas reduction of alkyl nitrite give 1° alcohols.
- Nitroalkanes are almost twice as acidic as aldehydes or ketones are.
- Due to two strong electron withdrawing groups in *m*-Dinitrobenzene it cannot be further nitrated to form tri-nitrobenzene.
- Nitrobenzene also undergoes nucleophilic substitution reaction at *o* and *p*-positions. However yields are low.

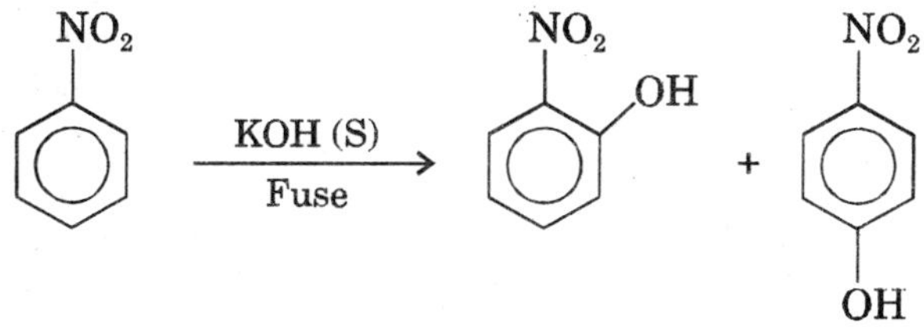

- Nitrobenzene does not undergo Friedel Craft reaction.
- Some nitro compounds are highly explosive. The common examples are :

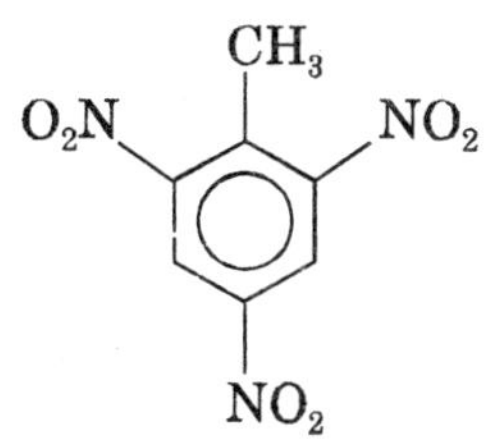

Trinitrotoluene (TNT)

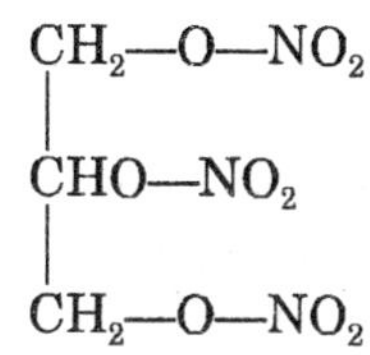

Nitroglycerine (Nobel's oil)

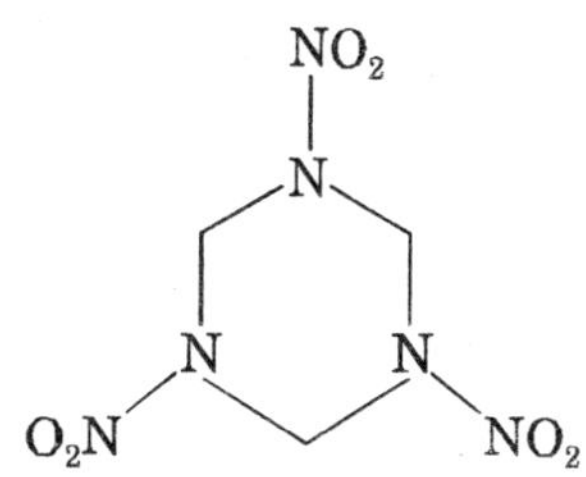

Cyclonite Research Development Explosive (RDX)

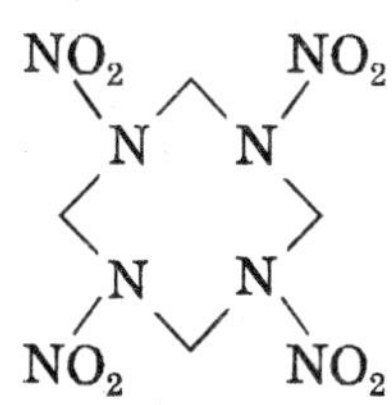

Her Majesty's Explosive (HMX)

- A mixture of TNT (20%) and NH_4NO_3 (80%) is called **amatol** used in coal mining.
- A mixture of TNT (30%), NH_4NO_3 (47%), aluminium powder (22%) and charcoal (1%) is known as **aluminonal** and is used for blasting purposes.

- A mixture of RDX, TNT and Al powder is used in torpedo heads.
- A mixture of trinitroglycerine, dinitroglycerine absorbed over Kieselguhr is known as **dynamite.**
- Amines are polar compounds and are capable of forming intermolecular hydrogen bonding except in case of amines which have no hydrogen atom. Hence amines have higher boiling points than the non-polar hydrocarbons of comparable molecular mass.
- The boiling points of amines are lower than those of alcohols and carboxylic acids of comparable molecular masses because hydrogen bonds in amines are weaker due to lower electronegativity of nitrogen in comparison with oxygen. As such the intermolecular H-bonds in amines are weaker than those in alcohols and carboxylic acids.
- Among aniline, ammonia and aliphatic amines the order of basic strength $C_6H_5NH_2 < NH_3 < RNH_2$.
- Among ethyl amines the order of basic strength is :

  (*i*) $(C_2H_5)_3N < C_2H_5NH_2 < (C_2H_5)_2NH$ ***in aqueous solution.***

  (*ii*) $C_2H_5NH_2 < (C_2H_5)_2NH < (C_2H_5)_3N$ ***in acetic acid or chlorobenzene solvents.***

  In aqueous solution, the basic strength of alkyl amines also depend upon the nature of the alkyl group or groups. For example, in case of methyl amines the order is :

  $(CH_3)_3N < CH_3NH_2 < (CH_3)_2NH$

  and in case of *n*-propyl amines the order is :

  $CH_3CH_2CH_2NH_2 < (CH_3CH_2CH_2)_3N < (CH_3CH_2CH_2)_2NH.$
- In general, basic nature decreases within the homologous series with increase in molecular mass.
- Among *p*-substituted anilines the order of basic strength is as under :

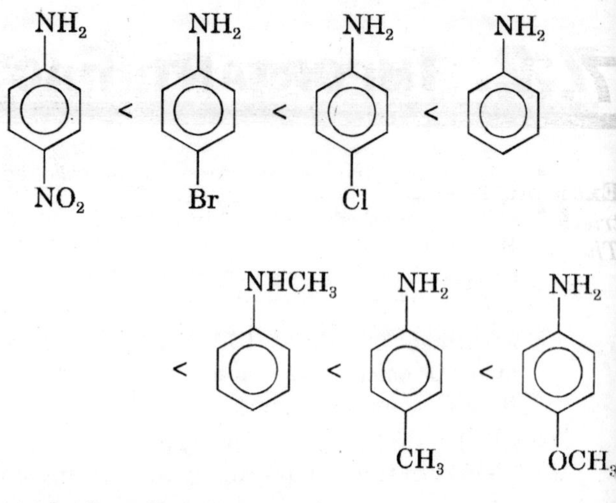

- *o*-substituted amines are usually weaker bases than aniline irrespective of the nature of substituent (whether electron releasing or withdrawing group). This is called *ortho-effect* and is probably due to steric and electronic factors.
- Amines have chiral N-atom but cannot be resolved into enantiomeric forms because of rapid flipping of one enantiomeric form into the other.
- Nitrosoamines are carcinogens (cancer causing agents).
- Sulphanilic acid exist as Zwitter ion which is amphoteric in nature.
- Hydrazobenzene, Azobenzene, Azoxybenzene can be converted into one antoher as follows

$$C_6H_5\text{—NH—NH—}C_6H_5$$
$$\text{Zn/NaOH} \uparrow\downarrow NaIO_3$$
$$C_6H_5\text{—NH—NH—}C_6H_5$$
$$\text{Fe} \uparrow\downarrow CH_3COOOH$$
$$C_6H_5\text{—NH—NH—}C_6H_5$$
$$\downarrow$$
$$O$$

- Hydrazobenzene, Azobenzene and azoxybenzene, all on reduction with Sn/HCl give aniline as final product.

7.3. Solved Problems

Enrich your Skill

Example 1. *Mixture of ethanamine, diethyl amine and triethyl amine is treated with diethyl oxalate and filtered. The residue contains*

(*a*) $(C_2H_5)_3N$

(*b*) *Mixture of* $(C_2H_5)_3N$ *and* $(C_2H_5)_2NH$

(*c*) $CONHC_2H_5$
|
$CONHC_2H_5$

(*d*) $CON(C_2H_5)_2$
|
$COOC_2H_5$.

Solution. Ethanamine reacts with $COOC_2H_5$ | $COOC_2H_5$ to form $CONHC_2H_5$ | $CONHC_2H_5$ (N, N=diethyl oxamide which is solid and appears as residue after filteration. $(C_2H_5)NH$ forms N, N-diethyl oxamic ester which is liquid and remains in the filtrate along with R_3N.

Hence correct choice is (*c*).

Example 2. 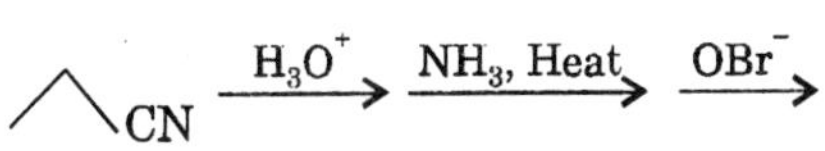

A. The product A of the above reaction is

(*a*) 

(*b*) 

(*c*) 

(*d*) 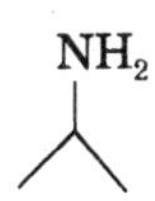

Solution. The sequence of reaction is

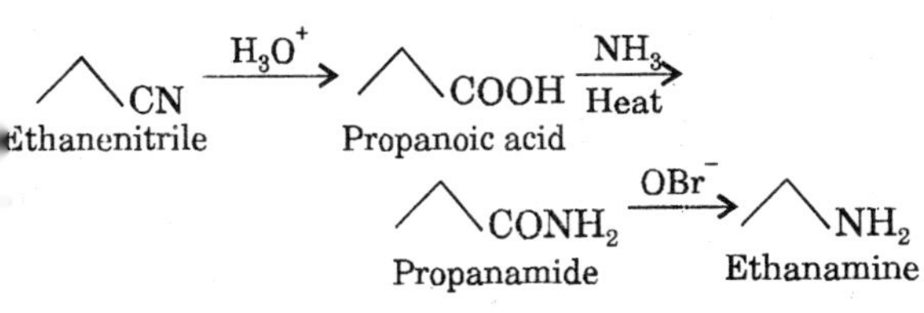

Hence, correct answer is (*b*).

Example 3. *In order to convert ethanoic acid to glycine which sequence of steps as given below is correct ?*

(*a*) (*i*) NH_3 (*ii*) P_2O_5/*Heat*

(*b*) (*i*) NH_3 (*ii*) *heat* (*iii*) OBr^-

(*c*) (*i*) P/Br_2 (*ii*) NH_3 (*iii*) H_3O^+

(*d*) (*i*) NH_3 (*ii*) *Heat* (*iii*) *NaOH*.

Solution. The conversion of acetic acid to glycine is as given below

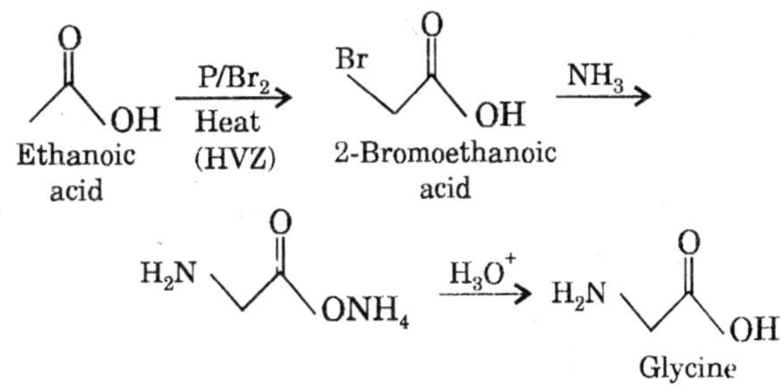

Hence the correct choice is (*c*).

Example 4. *When cyclohexanone reacts with methanamine the final product obtained is*

(*a*) 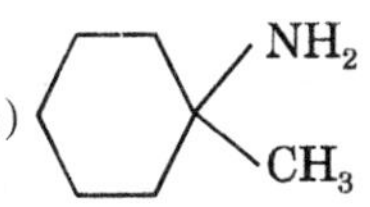

(*b*) 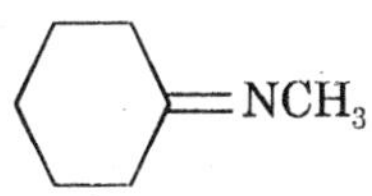

(*c*) 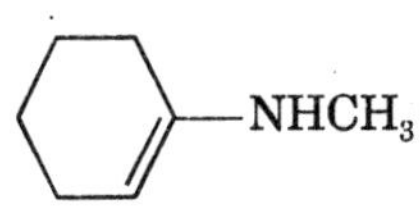

(*d*) 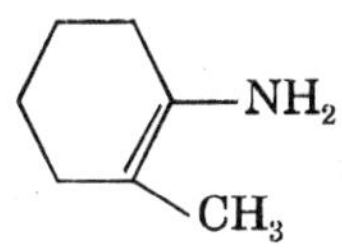

Solution.

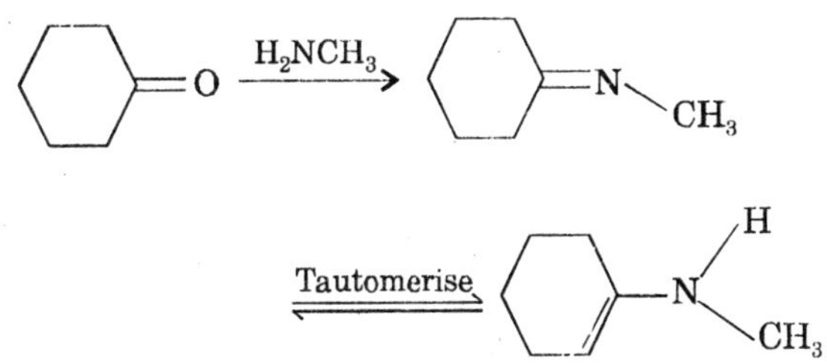

Hence correct choice is (*c*).

Example 5. 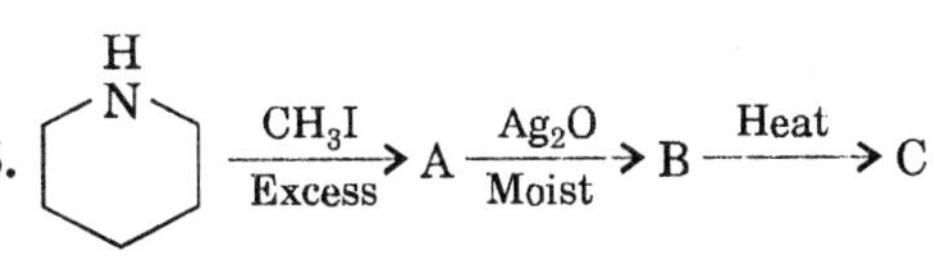

The compound C in the above sequence is

(*a*) 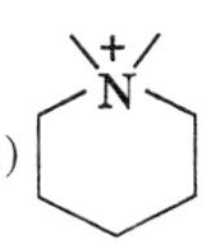

(*b*) 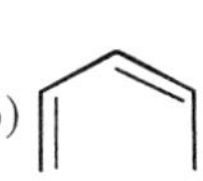

(*c*) 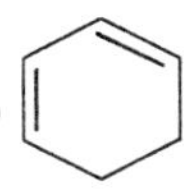

(*d*) 

Solution. The sequence of reactions is

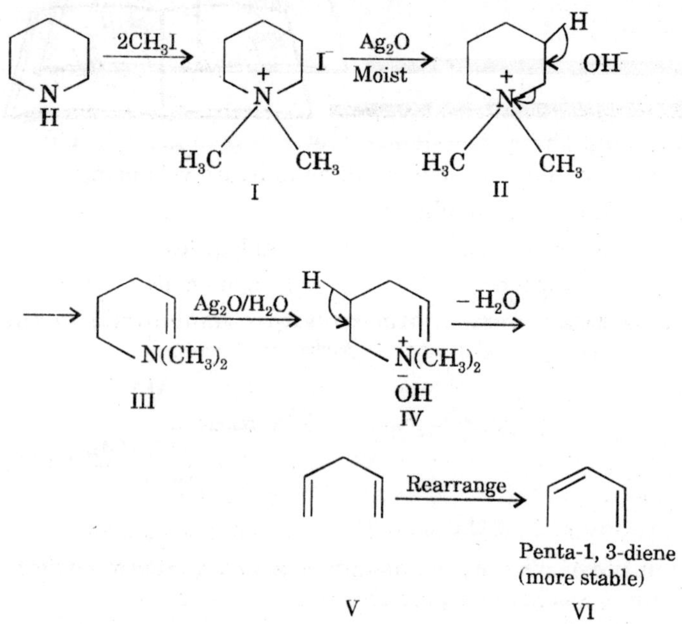

Thus correct choice is ***(b)***.

Example 6. *A student adopted the following path for converting propan-1-ol into an amine*

$$\text{Alcohol} \xrightarrow[H_2SO_4]{K_2Cr_2O_7} \xrightarrow{SOCl_2} \xrightarrow{NH_3} \xrightarrow{\overline{O}Br} \text{Amine.}$$

The amine he obtained is

(a) propan-1-amine *(b) propan-2-amine*
(c) N-Methyl ethanamine *(d) Ethanamine.*

Solution. The sequence of reaction is

$$\underset{\text{Propan-1-ol}}{CH_3CH_2CH_2OH} \xrightarrow[(O)]{K_2Cr_2O_7/H^+} \underset{\text{Propanoic acid}}{CH_3CH_2COOH}$$

$$\xrightarrow[Py]{SOCl_2} \underset{\text{Propanoyl chloride}}{CH_3CH_2COCl} \xrightarrow{NH_3(alc.)} \underset{\text{Ethanamine}}{CH_3CH_2NH_2}$$

$$\xrightarrow[(Br_2/KOH)]{OBr^-} \underset{\text{Propanamide}}{CH_3CH_2CONH_2}$$

Hence the correct choice is ***(d)***.

Example 7. *n-Propyl amine is allowed o react with $NaNO_2$/HCl. Which of the following is not possible to be formed ?*

(a) $CH_3CH_2CH_2OH$ *(b) $CH_3CH = CH_2$*
(c) $CH_3NHC_2H_5$ *(d) $CH_3CH(OH)CH_3$.*

Solution. The reaction proceeds through the formation of carbocation intermediate there can be variety of different products (I → VI) formed as a result of rearrangement of carbocation as shown below.

$$CH_3CH_2CH_2NH_2 \xrightarrow[NaNO_2/HCl]{HNO_2} \underset{\text{unstable diazonium salt}}{CH_3CH_2CH_2\overset{+}{N}_2Cl^-}$$

$$\xrightarrow{-N_2} \underset{\text{carbocation intermediate}}{CH_3CH_2\overset{+}{C}H_2}$$

$$CH_3CH_2\overset{+}{C}H_2 \xrightarrow{H_2O/SN^2} CH_3CH_2CH_2OH \text{ (I)}$$

$$CH_3CH_2\overset{+}{C}H_2 \xrightarrow{Cl^-(SN)^2} CH_3CH_2CH_2Cl \text{ (II)}$$

$$CH_3CH_2\overset{+}{C}H_2 \xrightarrow{1,2\ H^-\ \text{shift}} \underset{2^\circ\ \text{carbocation}}{CH_3\overset{+}{C}HCH_3}$$

$$CH_3\overset{+}{C}HCH_3 \xrightarrow{H_2O/SN^2} CH_3CH(OH)CH_3 \text{ (III)}$$

$$CH_3\overset{+}{C}HCH_3 \xrightarrow{Cl^-(SN^2)} CH_3CH(Cl)CH_3 \text{ (IV)}$$

The carbocation can also form elimination product (*v*) and (*vi*) involving loss of H^+ ions.

$CH_3CH = CH_2$ (*v*) and

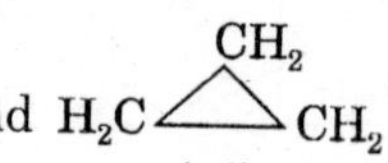

(*vi*)

There is no possibility of formation of $CH_3NHC_2H_5$. Hence correct choice is ***(c)***.

Example 8. *Nitrobenzene is subjected to following sequence of steps, the final product in the process is*

(i) $LiAlH_4$ (ii) H_2, Ni, 15 atm, 215 K

(a) $C_6H_5N = NC_6H_5$ *(b) Azoxybenzene*
(c) m-nitroaniline *(d) Aniline.*

Solution. The reactions involved are

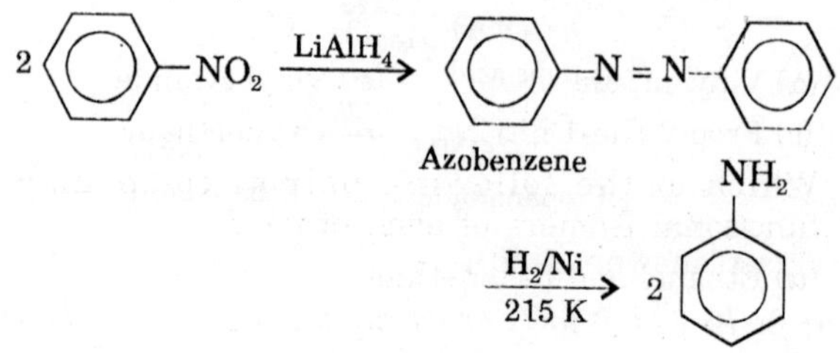

The intermediate reduction product of azobenzene can also be hydroazobenzene. Among the given choices the correct choice is ***(d)***.

7.4. QUESTION BANK

Cyanides and Isocyanides

1. The IUPAC name of 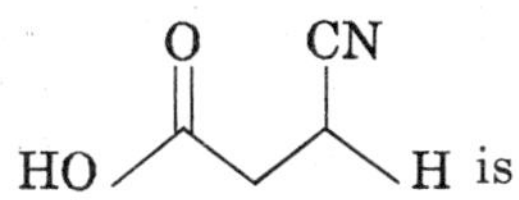 is
   (*a*) Butanenitrolic acid (*b*) Carboxypropanenitrile
   (*c*) 3-Cyanopropanoic acid (*d*) 4-Cyanobutanoic acid.
2. The IUPAC name of the compound
   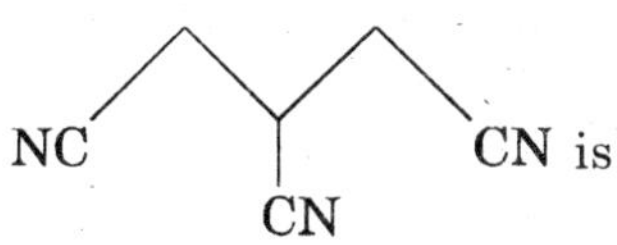 is
   (*a*) 1, 2, 3-Propanetrinitrite
   (*b*) 1, 2, 3-Propanetricarbonitrile
   (*c*) α, β-Dicyano butyronitrile
   (*d*) 1, 2, 3-tricyanopropane.
3. The IUPAC name of the compound
   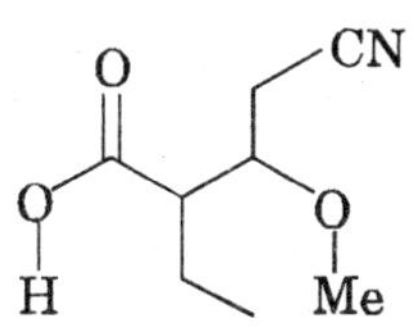 is
   (*a*) Methyl-2-cyanobutanoate
   (*b*) 4-Cyano-2-ethyl-3-methoxybutanoic acid
   (*c*) 3-Cyano-1-hydroxy-1-methoxybutanal
   (*d*) 4-Cyano-2-keto-3-methoxypentan-1-ol.
4. The IUPAC name of 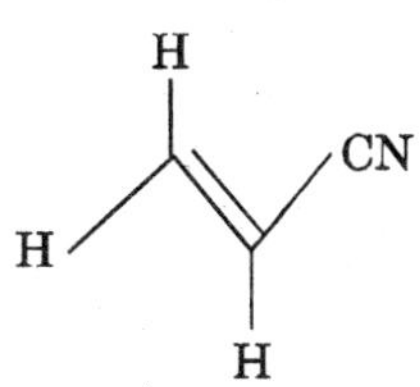 is
   (*a*) Vinyl nitrile (*b*) Vinyl cyanide
   (*c*) Prop-2-ene-1-nitrile (*d*) Cyanoethane.
5. Which of the following pair of compounds are functional isomers of each other ?
   (*a*) Ethanol and Ethanethiol
   (*b*)  and 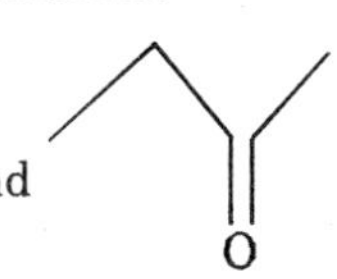
   (*c*) Propenenitrile and vinyl cyanide
   (*d*) Propenenitrile and ethyl carbylamine.
6. An alkyl cyanide on alkaline hydrolysis using H_2O_2/OH^- produces
   (*a*) Alkyl amine (*b*) Aldehydes
   (*c*) Ketones (*d*) Alkanamide.
7. Ethanol on treatment with ethanenitrile in the presence of dilute sulphuric will yield
   (*a*) Ethyl ethanoate (*b*) Butyraldehyde
   (*c*) Methyl propanoate (*d*) 2-butanone.
8. Alkyl isocyanide on reduction with Zn—Hg/HCl gives
   (*a*) 1° Amine (*b*) 3° Amine
   (*c*) N-alkylalkanamine (*d*) N-Methylalkanamine.
9. Which of the following reagent on treatment with benzenamine produces benzene isonitrile ?
   (*a*) Dichloromethane
   (*b*) Carbon tetrachloride
   (*c*) Trichloromethane and KOH
   (*d*) CH_3Cl/KOH.
10. Methyl carbylamine is treated with ozone, the product formed is
    (*a*) Formic acid
    (*b*) Formic acid and methyl amine
    (*c*) Methyl isocyanate
    (*d*) Formaldimine.
11. In order to produce ethyl isothiocyanate, which of the following reagents are required ?
    (*a*) Ethyl amine and sulphur
    (*b*) Ethanamide and sulphur
    (*c*) Ethyl carbylamine and sulphur
    (*d*) Ethyl cyanide and sulphur.
12. The hybrid states of nitrogen, carbon 1 and carbon 2 in ethanenitrile are
    (*a*) sp^3, sp^2, sp (*b*) sp, sp, sp^3
    (*c*) sp^2, sp, sp (*d*) sp, sp^2, sp.
13. Which reactants are involved in Mendius reaction ?
    (*a*) RCN and $LiAlH_4$ (*b*) RCN and H_2O/HCl
    (*c*) RCN and Zn/NH_4Cl (*d*) RCN and $SnCl_4$/HCl.
14. The number of isomeric aromatic compounds possible for the formula C_7H_9N are
    (*a*) 2 (*b*) 3
    (*c*) 4 (*d*) 5.
15. Which combination will give methyl isocyanate ?
    (*a*) $CH_3Cl + (O)$ (*b*) $CH_3NH_2 + CH_2Cl_2$
    (*c*) $CH_3NH_2 + COCl_2$ (*d*) $C_2H_5OH + COCl$.

16. Treatment of 2-bromopropane with ethanolic solution of KCN results in the formation of 2-methylpropanenitrile and also propene. Which of following step results the formation of propene ?

(*a*) 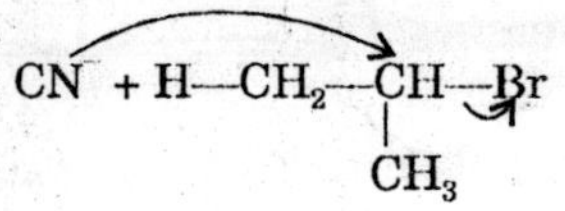

(*b*) 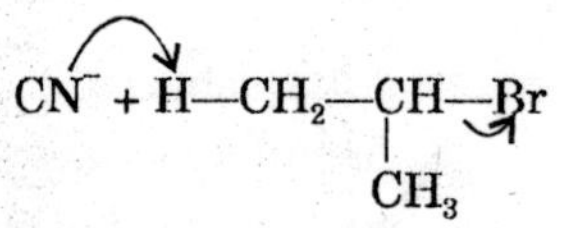

(*c*) 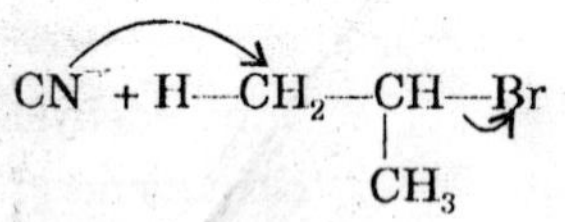

(*d*) 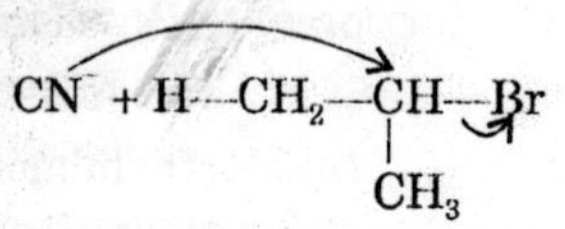

17. Which of the following sequence will convert methyl cyanide to ethyl isocyanide ?

(*a*) H_3O^+, LAH, P/I_2, AgCN

(*b*) LAH, $NaNO_2/HCl$, KCN

(*c*) H_3O^+, NH_3, P_2O_5

(*d*) None of the above.

18. CH_3CN: does not act as a base inspite of presence of lone pair of electrons on N atom because

(*a*) Lone pair of N is present in *sp* hybrid orbital and thus its availability for protonation decreases

(*b*) Electronegativity of C is more than that of nitrogen

(*c*) – CH_3 group causes – I effect

(*d*) – CN group causes + I effect.

19. Carbylamine reaction can help us to test which of the following functional group in the organic molecule ?

(*a*) —NH_2 (*b*) = NH

(*c*) —NH
|
(*d*) —OH.

20. Stephan's reduction will convert  CN to

(*a*) Ethanal (*b*) EtOH

(*c*) MeCOMe (*d*) EtCHO.

21. Cyanide ion is

(*a*) Electrophile (*b*) Ambiphile

(*c*) Ambident nucleophile (*d*) Zwitter ion.

22. $\text{MeBr} \xrightarrow{\text{AgCN}} \text{A} \xrightarrow[\text{(Predominant)}]{\text{Reduction}} \text{B}.$

In the above sequence B is

(*a*) $MeNH_2$ (*b*) $EtNH_2$

(*c*) MeNHMe (*d*) $EtNO_2$.

23. 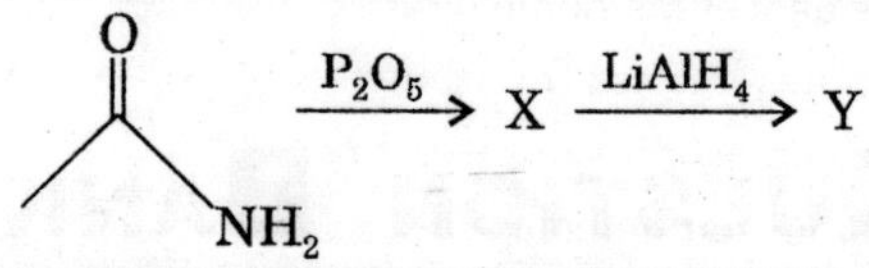

Y in the above sequence is

(*a*) Ethylamine (*b*) methylamine

(*c*) Acetonitrile (*d*) Ammonium acetate.

24. Benzonitrile on reaction with MeMgBr followed by acidic hydrolysis gives

(*a*) Benzophenone (*b*) Acetophenone

(*c*) Bromobenzene (*d*) Benzoic acid.

25. Which of the following statements is *not* correct ?

(*a*) Alkyl isocyanides have bad odours while akyl cyanides have pleasant odours

(*b*) Alkyl cyanides are not as poisonous as KCN

(*c*) Alkyl cyanides have lower boiling points than the corresponding alkyl isocyanides

(*d*) Acetonitrile is soluble in water but methylcarbylamine is not.

26. Alkyl isocyanides when heated with mercuric oxide or sulphur powder give addition products of the type. These reactions suggest that carbon atom of the isocyanide group is

(*a*) Electron deficient

(*b*) Electron rich

(*c*) Has strong affinity for oxygen or sulphur

(*d*) None of the above.

27. Methyl isocyanate (MIC), the poisonous gas which caused Bhopal tragedy in 1984, on alkaline hydrolysis gives

(*a*) Methylamine (*b*) Methylcarbamic acid

(*c*) Ethylamine (*d*) Methyl cyanide.

Nitro Compounds

28. Which of the following pair does not represent functional isomers of each other ?

(*a*) $CH_3CH_2NO_2$ and CH_3CH_2ONO

(*b*) C_2H_5CHO and CH_3COCH_3

(*c*) $CH_3CH_2CH_2NH_2$ and $(CH_3)_2CHNH_2$

(*d*) $CH_3CH_2CH_2NH_2$ and $(CH_3)_2N$.

29. The compounds

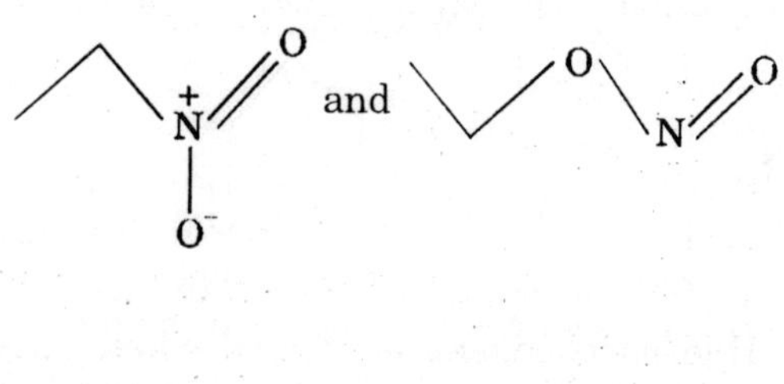

are :

(a) Isomers as well as homologues
(b) Homologues but not the isomers
(c) Neither homologues nor isomers
(d) Isomers but not homologues.

30. The main product of reduction of nitrobenzene with lithium aluminium hydride is
(a) Azoxybenzene (b) Aniline
(c) Azobenzene (d) Diazonium salt.

31. Which attacking species is generated from nitrating mixture during the nitration of phenol ?
(a) NO_2^- (b) NO_2^+
(c) NO_2 free radical (d) N_2O_3.

32. Which of the following treatment will yield azoxybenzene from nitrobenzene ?
(a) electrolysis in acidic conditions
(b) NaOH/Zn, CH_3OH
(c) Sn/HCl
(d) Na_3AsO_3/NaOH.

33. The major product of the reaction of ethyl bromide and alcoholic solution of silver nitrite is
(a) ethyl cyanide (b) nitroethane
(c) ethene (d) ethyl nitrite.

34. Which of the following is not a nitrating agent for aromatic hydrocarbons
(a) Conc. HNO_3 + Conc. H_2SO_4
(b) Nitronium fluoborate
(c) $HNO_2 + (CH_3CO)_2O$
(d) Nitronium perchlorate.

35. Which of the following is called *sweet spirit of nitre* ?
(a) CH_3CH_2ONO (b) $C_6H_5NO_2$
(c) C_6H_5Cl (d) CH_3NO_2.

36. Which of the following will give red colour with NaOH ?

(a) 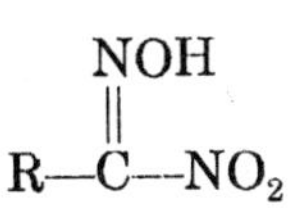 (b) 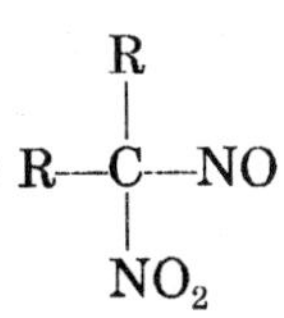

(c) $R_3C—NO_2$ (d) R_3CONO.

37. Tertiary nitroalkane cannot tautomerise because
(a) Their tautomeric forms are highly unstable
(b) They do not contain any multiple bond
(c) They do not have labile H atom
(d) They are not basic in nature.

38. Which among the following is least reactive to nitration ?
(a) Benzene (b) Nitrobenzene
(c) Chlorobenzene (d) Aniline.

39. Hydroazobenzene is formed when nitrobenzene is reduced with
(a) Zn/HCl (b) Zn/NaOH
(c) Zn/NaOH and CH_3OH (d) Zn/NH_4Cl.

40. When 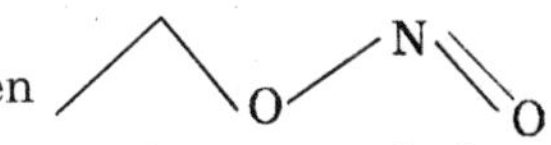 is reduced with Sn/HCl, the pair of compounds formed are
(a) Ethanol, hydroxylamine hydrochloride
(b) Ethanol, ammonium hydroxide
(c) Ethanal, NH_2OH
(d) $C_2H_5NH_2$, H_2O.

41. Which one among the following when boiled with hydrochloric acid gives hydroxylamine hydrochloride ?
(a) Any nitroarene
(b) Any tertiary nitro compound
(c) Any 2° nitroalkane
(d) Any 1° nitroalkane.

42. In order to prepare 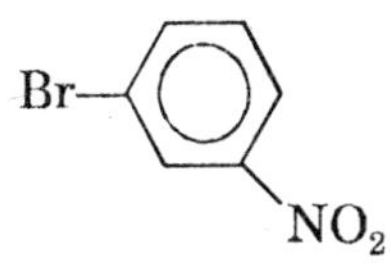
(a) m-nitrophenol is treated with P/Br_2
(b) $C_6H_5NO_2$ is treated with Br_2 in presence of Fe
(c) C_6H_5Br is treated with a mixture of conc. H_2SO_4/conc. HNO_3
(d) p-hydroxy sulphonic acid is treated with HNO_3.

43. $CH_3NO_2 \xrightarrow[\text{Excess}]{Cl_2/NaOH} X$
X can also be prepared by the reaction of
(a) CH_3NH_2 and HNO_2 (b) Chloroform and HNO_3
(c) CH_2Cl_2 and HNO_2 (d) Acetone and HNO_3.

44. In which of the following compound the nitration will occur at the *meta* position

(a) 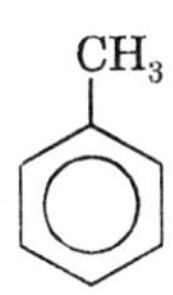 (b) 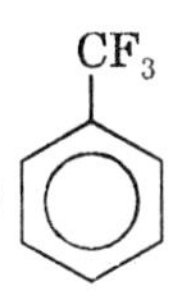

(c) 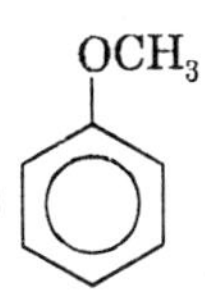 (d) 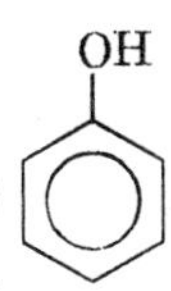.

45. Which of the following does not reduce Tollen's reagent ?
(a) HCOOH (b) $C_6H_5NO_2$
(c) C_6H_5NHOH (d) MeCHO.

46. Which reactions among the following donot give the product as indicated ?

(I) 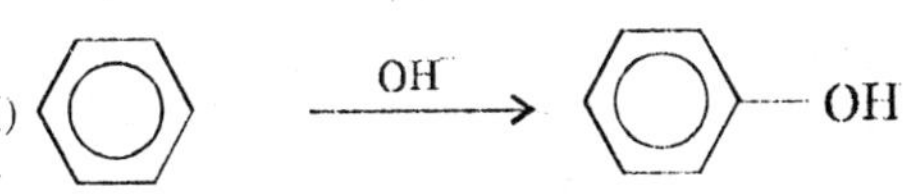

(II) 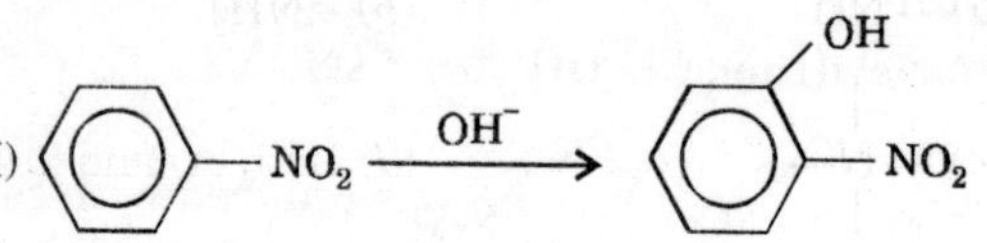

(III) 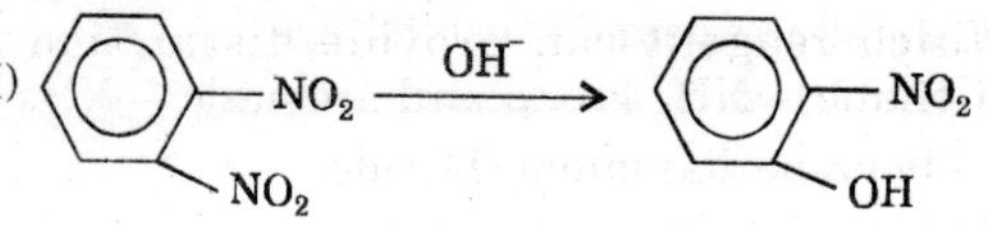

(IV) 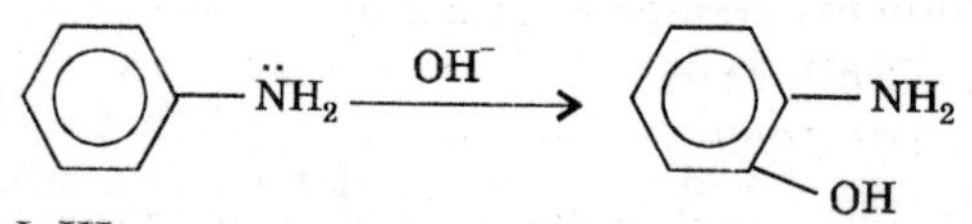

(*a*) I, III (*b*) II, III
(*c*) I, IV (*d*) III, IV.

47. 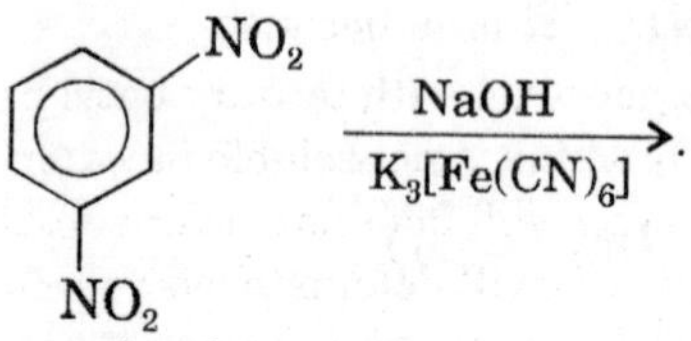 The product of above reaction is

(*a*) 3-Nitrophenol (*b*) 1, 3-cyanobenzene
(*c*) 1, 3-Benzenediol (*d*) 2, 4-Dinitrophenol.

48. Which name is not associated with

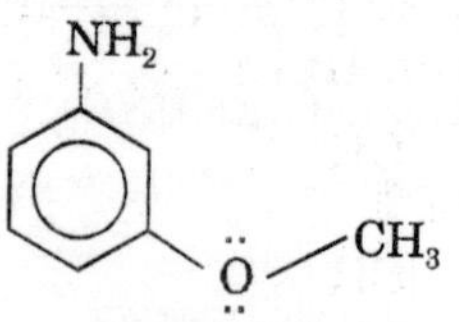

(*a*) *m*-Anisidine (*b*) 3-Amino anisole
(*c*) 3-Methoxyaniline (*d*) 3-Methoxybenzenamine.

49. Which of the following reaction does not yield amine ?

(*a*) $R—X + NH_3 \longrightarrow$

(*b*) $RCH = NOH \xrightarrow{Na/C_2H_5OH}$

(*c*) $R—CN \xrightarrow{H_2O/H^+}$

(*d*) $RCONH_2 \xrightarrow{LiAlH_4}$

50. The number of structural isomers of C_3H_9N are

(*a*) one (*b*) two
(*c*) three (*d*) four.

Amines

51. Which among the following has highest melting point ?

(*a*) Aniline (*b*) *o*-Toludine
(*c*) *p*-Toludine (*d*) *m*-Toludine.

52. $CH_3\ddot{N}H_2$ is more basic than $CH_3\ddot{O}H$. The correct reason for this is that

(*a*) Hyperconjugative resonance of $CH_3\overset{+}{N}H_3$ is better than that of $CH_3\overset{+}{O}H_2$

(*b*) Resonance for $CH_3\overset{+}{O}H_2$ is poorer than that for $CH_3NH_3^+$

(*c*) Electronegativity of oxygen is higher than that of nitrogen.

(*d*) The statement itself is wrong because $CH_3\ddot{O}H$ is stronger base than $CH_3\ddot{N}H_2$ because oxygen has two non-bonded electron pairs.

53. Which one of the following amines is most basic in aqueous medium ?

(*a*) CH_3NH_2 (*b*) $(CH_3)_2NH$
(*c*) $(CH_3)_3N$ (*d*) $(CH_3)_4NBr$.

54. Which of the following has lowest value of pk_b ?

(*a*) $CH_3CH_2\ddot{N}H_2$ (*b*) $CH_3CH_2\ddot{O}H$
(*c*) $CH_2CH_2CO\ddot{N}H_2$ (*d*) $\ddot{N}H_3$.

55. The major product of the reaction given below is

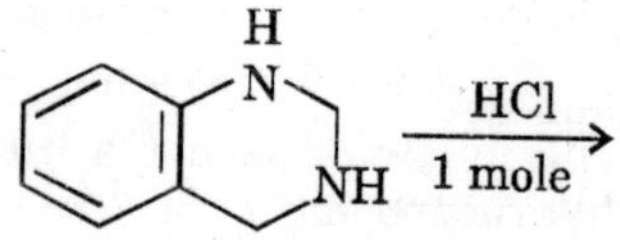

(*a*) 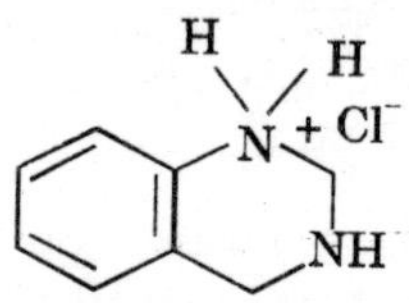 (*b*) 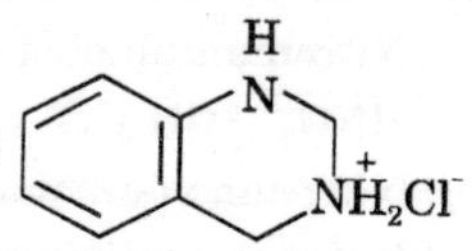

(*c*) 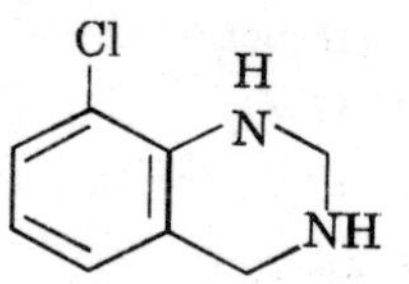 (*d*) 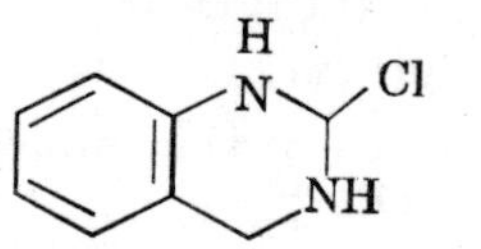

56. The precipitate formed by mixing silver nitrate and sodium chloride solution disappear on adding ethyl amine. It is due to the formation of

(*a*) C_2H_5Cl (*b*) $C_2H_5NO_3$
(*c*) $[Ag(C_2H_5NH_2)_2]Cl$ (*d*) CH_3COCl.

57. When diethylamine is shaken with cold solution of $NaNO_2/HCl$, the product formed will be

(*a*) $(C_2H_5)_2NNO$ (*b*) $C_2H_5NO_2$
(*c*) $(C_2H_5)_2\overset{+}{N}H_2NO_2^-$ (*d*) C_2H_5OH.

58. Schiff's bases are produced by the reaction of aniline with

(*a*) Alcohols (*b*) Aldehydes
(*c*) Amides (*d*) Alkyl halides.

59. Which of the following on reaction with chloroform and KOH produces carbylamine ?

(a) RNH_2 (b) R_2NH
(c) R_3N (d) All the above.

60. Treatment of ethyl amine and methyl magnesium halide produces a gaseous substance X. X is
(a) Ethane (b) Ammonia
(c) Methane (d) Propane.

61. The oxidation of aniline by $K_2Cr_2O_7/H_2SO_4$ produces mainly
(a) Nitrobenzene (b) Benzoic acid
(c) p-benzoquinone (d) Benzamide.

62. 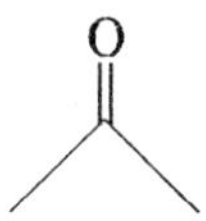 $\xrightarrow{NH_2OH}$ X $\xrightarrow{LiAlH_4}$ Y. In the above sequence Y is
(a) Tertiary amine (b) Secondary amine
(c) Primary amine (d) 2-Nitropropane.

63. Chemical formula of Hinsberg's reagent is
(a) $C_6H_5SO_2Cl$ (b) C_6H_5COCl
(c) C_6H_5COOH (d) $[C_6H_5NH(CH_3)_2]Cl$.

64. The most basic compound among the following is
(a) Benzylamine (b) Aniline
(c) Acetanilide (d) p-Nitroaniline.

65. Which of the following reagent can make distinction between primary and sec. amines ?
(a) NH_3 (b) $NaNO_2/HCl$
(c) HCl (d) all the above.

66. Oxidation of aniline with manganese dioxide and sulphuric acid produces

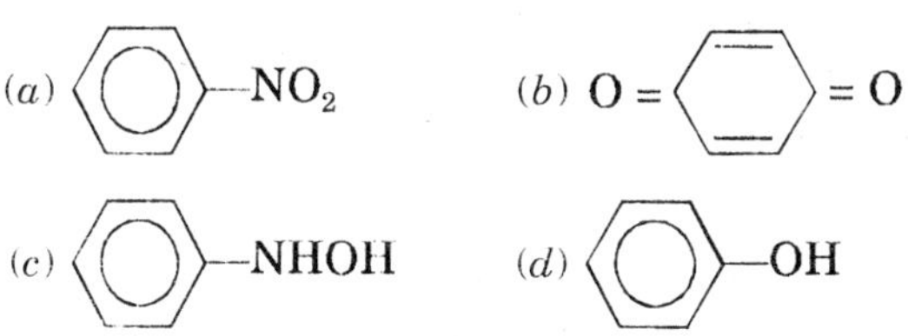

67. Treatment of aniline with bromine water produces
(a) 2, 4, 6-tribromoaniline
(b) Mixture of ortho and para bromoaniline
(c) Bromobenzene
(d) N-bromoaniline.

68. Which of the following will respond to carbylamine test ?
(a) N-Methylaniline (b) p-toludine
(c) Phenyl-n-butyl amine (d) N, N-Diethyl aniline.

69. An amine Y reacts with benzene sulphonyl chloride and the product, thus, formed is soluble in KOH. The correct representation of the functional group of amine is
(a) —NH (b) $-NH_2$
(c) $-\overset{|}{\underset{|}{N}}-$ (d) Both (a) and (b).

70. Which reagent can provide distinction between aliphatic $-NH_2$ group and aromatic $-NH_2$ group ?
(a) Benzene diazonium chloride
(b) Benzene sulphonyl chloride
(c) $CHCl_3/KOH$
(d) CH_3COCl.

71. Tertiary butyl amine cannot be prepared by the following reaction

$(CH_3)_3C—Br + \ddot{N}H_3$. This is because
(a) Ammonia does not react with tertiary butyl bromide
(b) Tertiary butyl bromide is not soluble in water
(c) Elimination process will dominate over substitution
(d) Substitution process will dominate over elimination.

72. Benzoyl chloride reacts with benzenamine in the presence of a base to form
(a) Benzamide (b) Hydrobenzamide
(c) N-Phenylethanamide (d) Benzanilide.

73. Which of the following will give crystalline precipitate on treatment with HCl ?
(a) $CH_2{=}CH_2$ (b) $CH{\equiv}CH$
(c) Ethanol (d) Ethanamine.

74. Which of the following amine will not react with diethyl oxalate ?
(a) $(CH_3)_2CH—NH_2$ (b) $C_2H_5NH_2$
(c) $(C_2H_5)_3N$ (d) $(C_2H_5)_3C—NH_2$.

75. In the neutral resonating structure of aniline, the number of σ and π bonds are respectively
(a) 12σ, 3π (b) 14σ, 3π
(c) 6σ, 6π (d) 3σ, 3π.

76. Which of the following will give white precipitate with an aqueous solution of ethyl amine ?
(a) $[Ag(NH_3)_2]^+$ (b) Iron (II) sulphate
(c) Zinc sulphate (d) Cupric sulphate.

77. Which of the following will not give dinitrogen gas with nitrous acid ?
(a) Propanamine (b) Methanamine
(c) Trimethylamine (d) Isopropylamine.

78. Acetone is treated with hydroxyl amine and the derivative so formed is subjected to catalytic hydrogenation. The final product is
(a) Isopropylamine (b) N-methylethanamine
(c) CH_4 and ethanamine (d) Propanamine.

79. Ethyl amine is subjected to $KMnO_4$ oxidation followed by hydrolysis. The final product is

(*a*) Propanal (*b*) Ethanal
(*c*) Ethanolic acid (*d*) CO_2, H_2O.

80. Which of the following process will give tetramethylhydrazine ?

(*a*) $NH_2—NH_2 + CH_2Cl_2$ (*b*) $NH_2—\overset{+}{N}H_3\overset{-}{Cl} + CH_4$
(*c*) $(CH_3)_2NH + H_2SO_5$ (*d*) $(CH_3)_2NH + KMnO_4$.

81. An organic molecule belonging to the category of Schiff's bases is subjected to reduction. The product formed could be

(*a*) Aniline (*b*) 2°-amine
(*c*) 1°-amine (*d*) 3°-amine.

82. 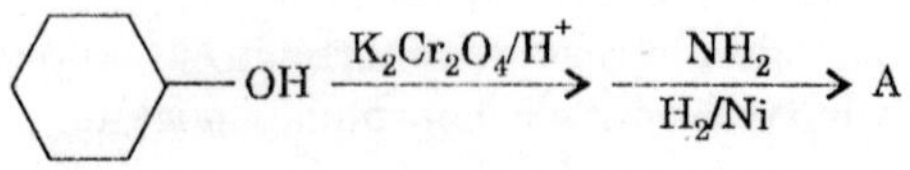

The product A in the above sequence is

(*a*) 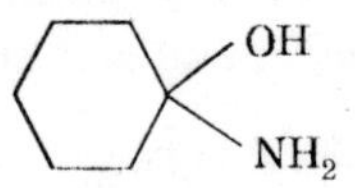 (*b*) 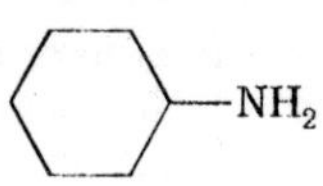

(*c*) 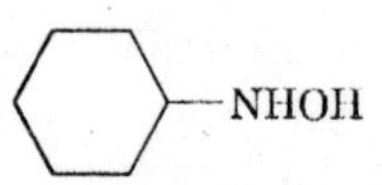 (*d*) 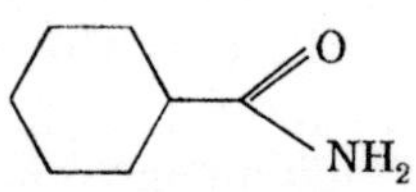

83. The number of 1° amines with molar mass 73 g mol^{-1} and pertaining to formula $C_nH_{2n+3}N$ is

(*a*) 4 (*b*) 3
(*c*) 2 (*d*) only one.

84. Which of the following reagent can distinguish between 1° amine and 2° amine ?

(*a*) $Br_2/NaOH$ (*b*) $NaNO_2/HCl$
(*c*) NH_3 (*d*) $HClO_4$.

85. 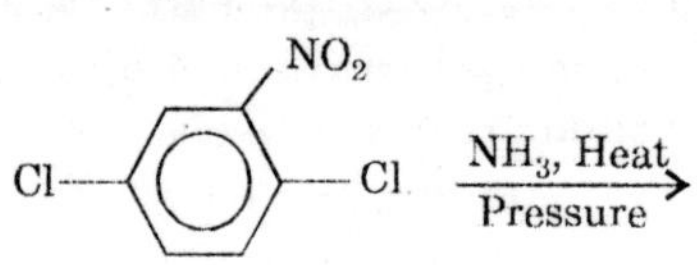

The main product of above reaction is

(*a*) 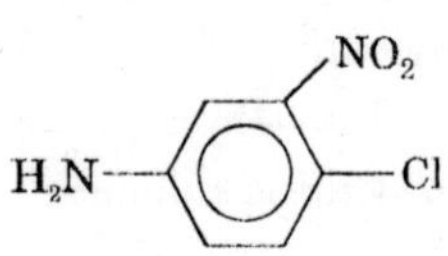 (*b*) 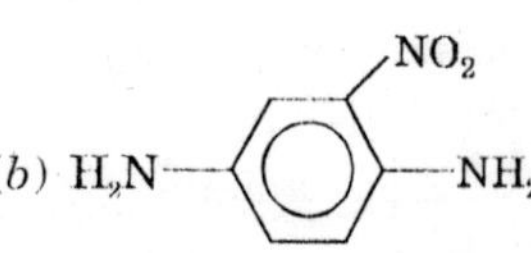

(*c*) 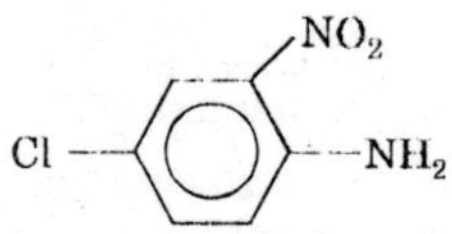 (*d*) 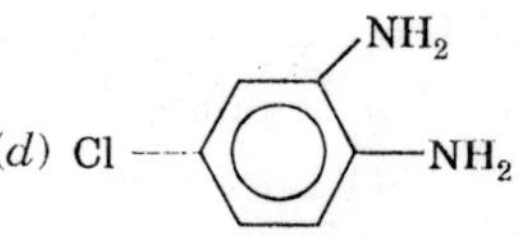

86. When aqueous solutions of ethylamine and iron (III) chloride are mixed,

(*a*) A colourless solution is formed
(*b*) Brown precipitate is formed
(*c*) Violet coloured solution is produced
(*d*) Black precipitate is formed.

87. 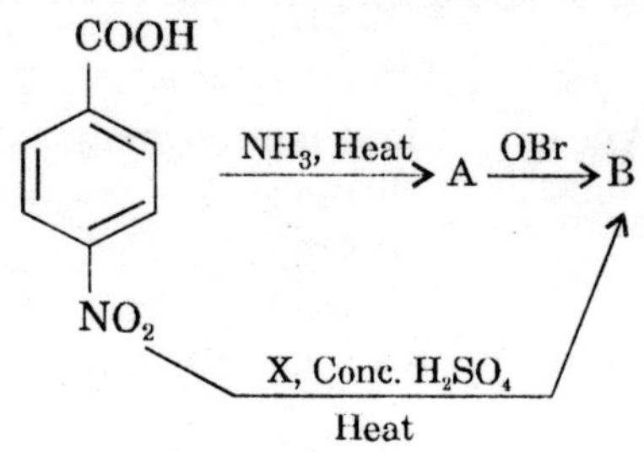

In the above sequence of reaction. X is

(*a*) $NH_2—NH_2$ (*b*) HN_3
(*c*) NH_2OH (*d*) KNH_2.

88. Which of the following is not correctly named ?

(*a*) $NH_2—(CH_2)_3NH_2$;
1, 3-Diaminopropane or trimethylene diamine
(*b*) $C_6H_5—N(CH_3)_2$; N-Phenyldimethylamine
(*c*) $CH_3—NH—(CH_2)_2NH—CH_3$; 2, 5-Diazohexane
(*d*) 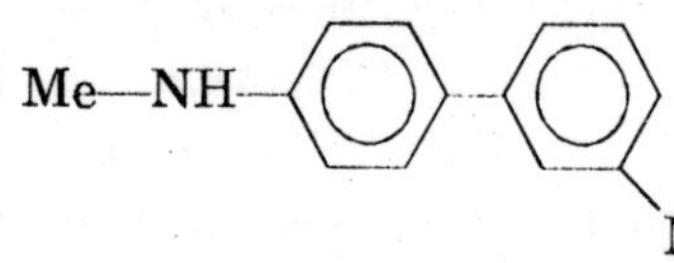
3, 4-bis (N-Methylamino) biphenyl

89. Certain primary amino compound on treatment with X gave alkyl chloride, N_2 and water. X can be

(*a*) CS_2 (*b*) $COCl_2$
(*c*) NOCl (*d*) HNO_2.

90. The correct increasing order of basic strength among *p*-nitroaniline (I) ; *p*-bromoaniline (II) ; *p*-toludine (III) is

(*a*) I < II < III (*b*) III < II < I
(*c*) II < I < III (*d*) III < I < II.

91. Among Dimethylaniline (I), Aniline (II) and Methylaniline (III) the increasing order of basic strength

(*a*) II < III < I (*b*) I < III < II
(*c*) II < I < III (*d*) I < II < III.

92. The order of increasing basic strength among *m*-toludine (I), *p*-toludine (II) and *o*-toludine (III) is

(*a*) III < II < I (*b*) II < III < I
(*c*) III < I < II (*d*) II < I < III.

93. Among *p*-phenylenediamine (I), *p*-anisidine (II) and *p*-toludine (III) the correct order of decreasing basic strength is

(*a*) I > II > III (*b*) II > III > I
(*c*) I > III > II (*d*) III > II > I.

94. Among *o*-nitroaniline (I), *p*-nitroaniline (II), *m*-nitroaniline (III), the order of increasing basic strength is

(*a*) I < III < II (*b*) II < I < III
(*c*) III < I < II (*d*) I < II < III.

95. The conjugate acid and conjugate base of aniline are respectively

(*a*) $C_6H_5NH_2$, $C_6H_5\overset{+}{N}H_3$ (*b*) $C_6H_5\overset{+}{N}H_3$, $C_6H_5\ddot{N}H$

(*c*) $C_6H_5\ddot{N}H$, $C_6H_5\overset{+}{N}H_3$ (*d*) $(C_2H_5)_2NH$, $C_6H_4(NH_2)_2$.

96. ***Isopropylamine*** $\xrightarrow{KMnO_4}$ ***X*** $\xrightarrow{H_3O^+}$ ***Y***. In the above sequence X and Y are respectively

(*a*) Acetaldimine, Ethanal (*b*) Ethanal, Ketimine

(*c*) Ketimine, Acetone (*d*) Acetone, Propan-2-ol.

97. Aniline reacts with bromine in aqueous medium to form

(*a*) 2, 4, 6-tribromoaniline (*b*) 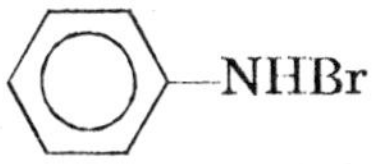

(*c*) N, N-dibromoaniline (*d*) *m*-bromoaniline.

98. Among the given bases, the one with lowest value of pk_b is

(*a*) Propanamide

(*b*) *n*-Propylamine

(*c*) 2-Chloro-1-propanamine

(*d*) *m*-bromoaniline.

99. 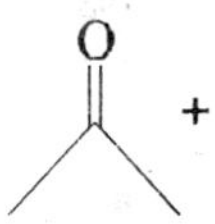 + $HCOONH_4$ $\xrightarrow{475\ K}$ $X + H_2O + CO_2$

The product X in the above reaction is a/an

(*a*) Isopropyl amine (*b*) 2° amine

(*c*) 3° amine (*d*) 1° amide.

100. $(CH_3)_3CCH_2NH_2 \xrightarrow{NaNO_2/HCl}$ A (*a major product*)

A in the above reaction is

(*a*) $(CH_3)_3CCH_2OH$ (*b*) Et—C(CH$_3$)(OH)—CH$_3$

(*c*) $(CH_3)_2CH—CH=CH_2$ (*d*) $(CH_3)_3CH_2\overset{+}{N}_2Cl$.

Diazonium Salts

101. During coupling reaction of benzene diazonium chloride and aniline the pH of reaction medium should be approximately

(*a*) 1-2 (*b*) 9-10

(*c*) 4-5 (*d*) 7-8.

102. Which of the following reagent can be used to convert benzene diazonium chloride to benzene ?

(*a*) Phosphorus acid (*b*) Phosphoric acid

(*c*) Hypophosphoric acid (*d*) Metaphosphoric acid.

103. Benzene diazonium chloride on heating with very dilute H_2SO_4 gives

(*a*) Benzene sulphonic acid

(*b*) Benzene sulphonyl chloride

(*c*) Benzene

(*d*) Phenol.

104. During preparation of arenediazonium salts the excess of nitrous acid, if any, is removed by adding

(*a*) Aqueous Na_2CO_3 (*b*) Aqueous NaOH

(*c*) Aqueous NH_2CONH_2 (*d*) NH_4OH.

105. 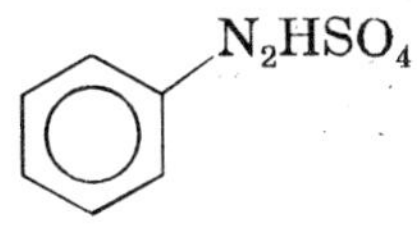 $\xrightarrow[\text{Warm}]{\text{MeOH}}$

The product of above reaction is

(*a*) Anisole (*b*) Ahenetole

(*c*) Benzenol (*d*) Phenyl methanol.

106. $C_6H_5N_2Cl \xrightarrow[\text{Zn/HCl}]{\text{Reduction}}$ X. The product X is

(*a*) Benzene (*b*) Benzamine

(*c*) Hydrazobenzene (*d*) Benzamide.

107. Which of the following diazonium salt is most stable ?

(*a*) *p*-methoxybenzene diazonium chloride

(*b*) 2, 4, 6-trinitrobenzene diazonium chloride

(*c*) *p*-nitrobenzene diazonium chloride

(*d*) 2, 4-dinitrobenzene diazonium chloride.

108. Which of the following molecule can exhibit *syn* and *anti* isomers

(*a*) $CH_3CH=O$ (*b*) CH_3COCH_3

(*c*) $C_6H_5N=NC_6H_5$ (*d*) $(CH_3)_2C=NOH$.

109. During diazotisation of benzenamine with sodium nitrite and hydrochloric acid the excess of hydrochloric acid is used primarily to

(*a*) Check the hydrolysis of ϕ-OH

(*b*) Ensure a stochiometric amount of nitrous acid

(*c*) Check the conc. of free aniline

(*d*) Neutralise any base formed during reaction.

110. Which of the following aryl amines undergoes diazotisation most readily ?

(*a*) 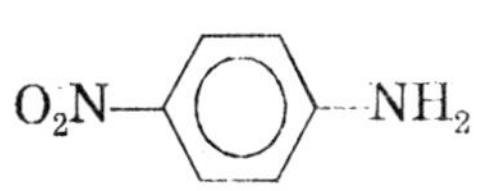 (*b*) 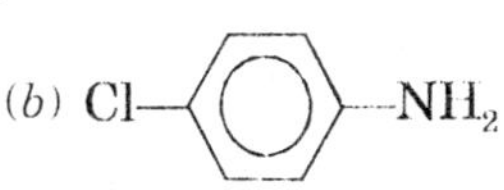

(*c*) 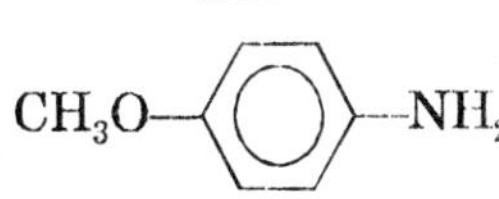 (*d*) 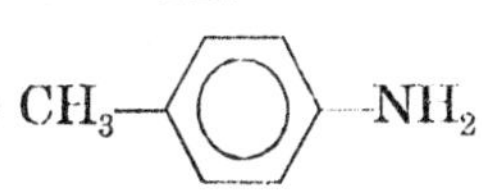

111. Benzene diazonium chloride does not form a coloured dye with

(*a*) Benzenol (*b*) β-Naphthol

(*c*) Aniline (*d*) Benzyl alcohol.

112. 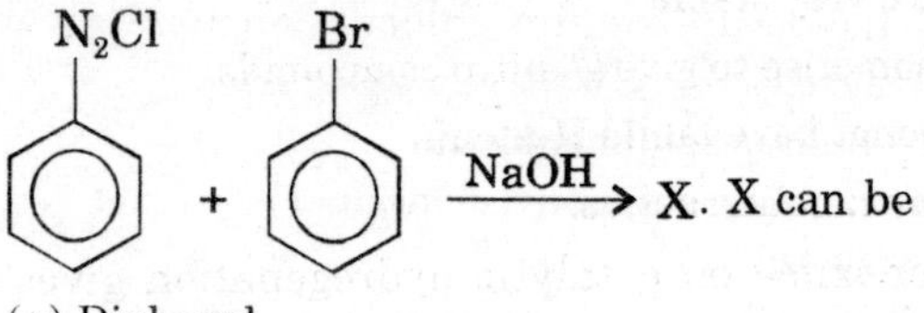 X can be

(a) Diphenyl

(b) 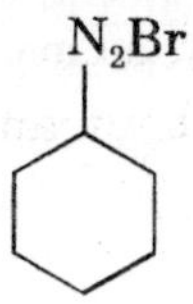

(c) 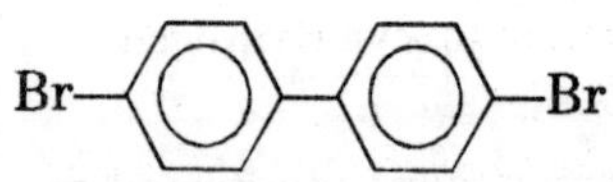

(d) 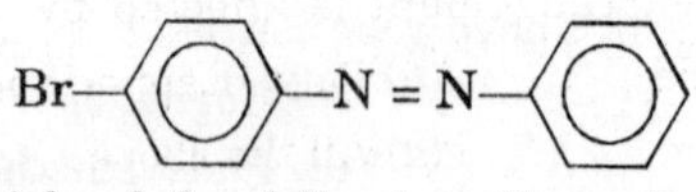

113. Which of the following diazonium salt when boiled with dil. H_2SO_4 gives corresponding phenol most readily

(a) 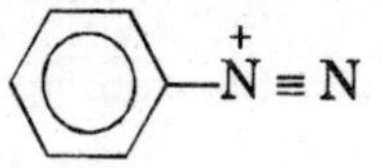 (b) 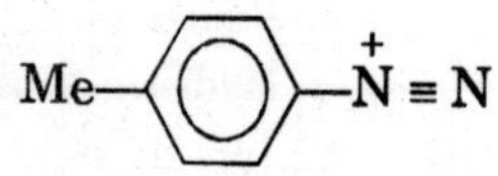

(c) 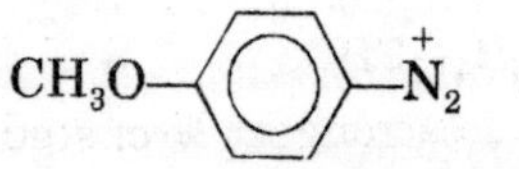 (d) 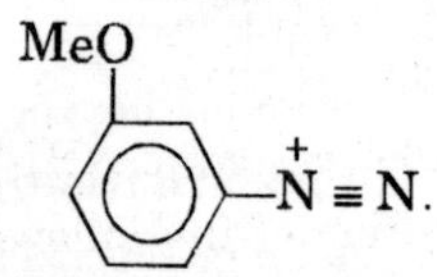

114. Which of the following reagents cannot convert $C_6H_5N_2Cl$ to $C_6H_5NH—NH_2$

(a) Zn/HCl (b) Na_2SO_3

(c) $NaHSO_3$ (d) $SnCl_2$/HCl.

115. In order to prepare benzonitrile from $C_6H_5\overset{+}{N}_2\overset{-}{Cl}$ the reagent required is

(a) KCN in presence of zinc powder

(b) CuCN dissolved in KCN

(c) $Ba(CN)_2$ and HCN

(d) Aqueous HCN.

Miscellaneous Questions

116. Sodalime decarboxylation of sodium glycinate produces

(a) Glycine (b) Ethanoic acid

(c) Methylamine (d) Alanin.

117. In the following compounds,

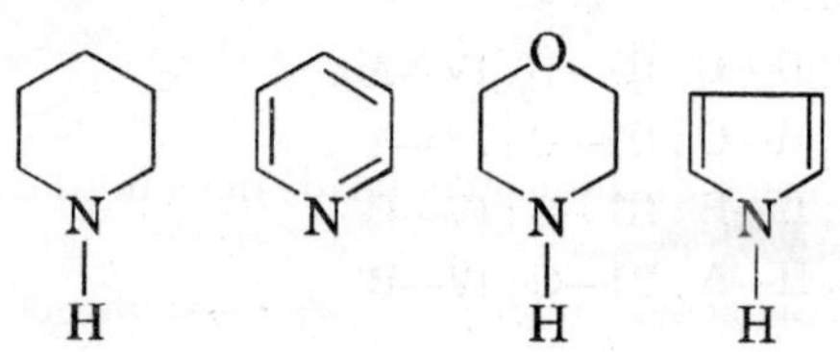

I II III IV

The order of basicity as :

(a) IV > I > III > II (b) III > I > IV > II

(c) II > I > III > IV (d) I > III > II > IV.

118. Among the ions given below, the correct order of increasing reactivity towards azo coupling is

I. 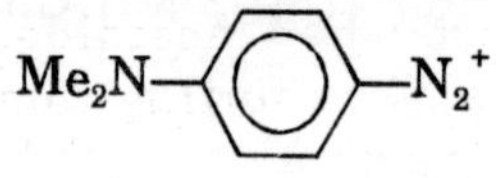 II. 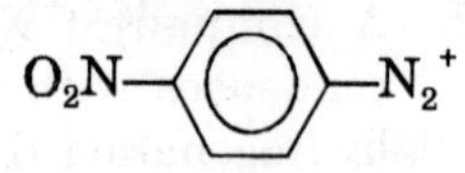

III. 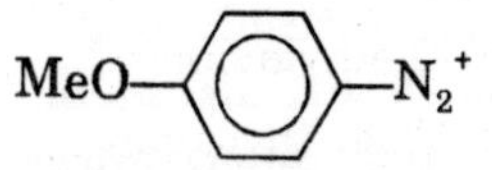 IV. 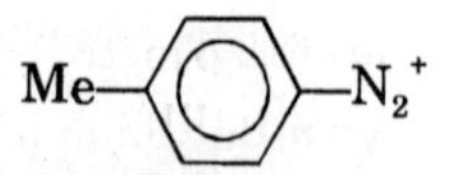

(a) I < II < IV < III (b) III < I < IV < II

(c) I < III < IV < II (d) III < I < II < IV.

119. Among the statements given about the coupling reaction between 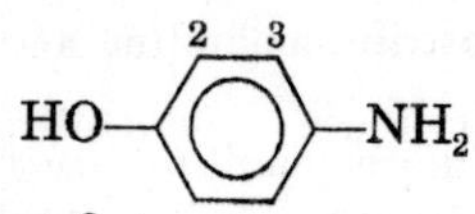 and $C_6H_5\overset{+}{N}_2\overset{-}{Cl}$, which are correct ?

I. coupling occurs at position 2 in alkaline medium

II. coupling does not occur at all

III. coupling occurs at position 3 in acidic medium

IV. coupling occurs at position 3 both in acidic as well as in alkaline medium

(a) I, II (b) I, III

(c) I, IV (d) only IV.

120. The product of Hoffmann elimination of 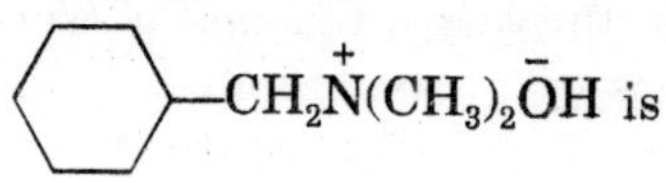 is

(a) 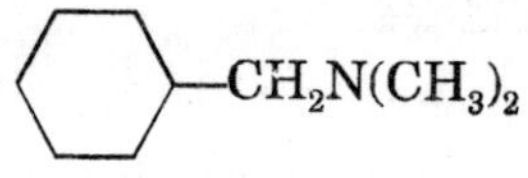

(b) 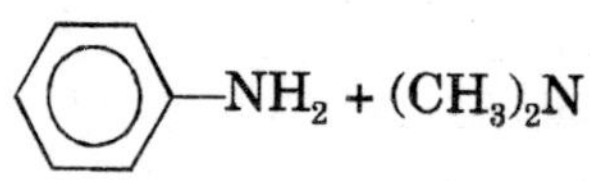

(c) 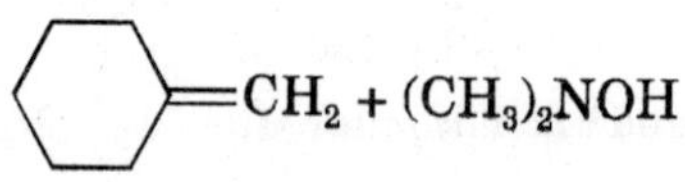

(d) 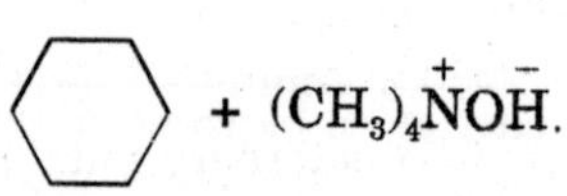

121. 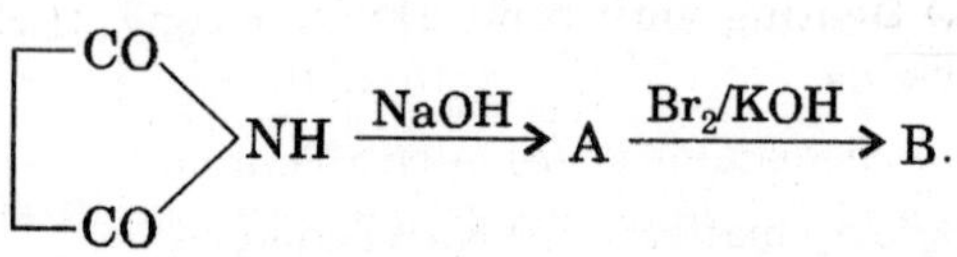

In the above reaction B is mainly

(a) β-alanine (b) α-alanine

(c) Ethylenediamine (d) γ-aminobutyric acid.

122. The most reactive compound among the following towards electrophilic substitution of the ring is

(*a*) $C_6H_5NO_2$ (*b*) $C_6H_5NH_2$

(*c*) $C_6H_5\overset{+}{N}H_3Cl^-$ (*d*) $C_6H_5NHCOCH_3$.

123. A compound X is formed when acetaldehyde condenses with ethylamine. X on further catalytic hydrogenation (H_2/Ni) will yield

(*a*) Ethyl methyl amine (*b*) Diethylamine

(*c*) *n*-Butyl amine (*d*) Trimethyl amine.

124. An organic compound **A** reacts with nitrous acid to form N-methyl-N-nitrosoethanamine. **A** can be obtained by reduction of

(*a*) Propanenitrile (*b*) Acrylonitrile

(*c*) Methylisocyanate (*d*) Ethylisocyanide.

125. Ammonium salt of benzoic acid is heated strongly with P_2O_5 and the product so formed is reduced and then treated with $NaNO_2$/HCl at low temperature. The final compound formed is

(*a*) Benzene diazonium chloride

(*b*) Benzyl alcohol

(*c*) Phenol

(*d*) Nitrosobenzene.

126. Certain nitrogeneous compound with mol. mass (180) shows an increase in its molecular mass to 348, after treatment with acetyl chloride. The number of possible $—NH_2$ groups in the molecule is

(*a*) 5 (*b*) 4

(*c*) 3 (*d*) 6.

127. For carbylamine reaction, we require hot alcoholic KOH along with

(*a*) Any 1° amine and chloroform

(*b*) Chloroform and silver powder

(*c*) Primary amine and an alkyl halide

(*d*) N-alkylamine and trichloromethane.

128. C_3H_9N cannot represent

(*a*) 1° Amine (*b*) 2° Amine

(*c*) 3° Amine (*d*) quaternary salt.

129. Aniline on heating with conc. HNO_3 + conc. H_2SO_4 mixture yields

(*a*) *o*- and *p*-Nitroanilines (*b*) *m*-Nitroaniline

(*c*) A black tarry matter (*d*) No reaction.

130. Tertiary nitro compounds cannot show tautomerism because they

(*a*) Are very stable

(*b*) Isomerise to give 2° nitro compounds

(*c*) Donot have labile H-atom

(*d*) Are highly reactive.

131. Acetonoxime on catalytic hydrogenation gives

(*a*) 1-Propanamine (*b*) Ethyl methyl amine

(*c*) CH_4 and ethanamine (*d*) Isopropyl amine.

132. Which statement is not correct about aniline ?

(*a*) It is less basic than ethyl amine

(*b*) It can be steam distilled

(*c*) It reacts with Sodium to give Hydrogen

(*d*) It is soluble in water.

133. A Beckmann rearrangement is affected by

(*a*) Sulphuric acid (*b*) Polyphosphoric acid

(*c*) PCl_5 (*d*) All the above.

134. Compound **A** has a molecular formula C_7H_7NO on treatment with Br_2/KOH, **A** gives an amine **B** which respond to carbylamine test. **B** on diazotisation and coupling with phenol give an azo dye. **A** can be

(*a*) $C_6H_5CONHCH_3$

(*b*) $C_6H_5CONH_2$

(*c*) C_6H_5NO

(*d*) *o*, *m* or *p*-$NH_2—C_6H_4—CHO$.

135. Which of the following is strongest Bronsted acid ?

(*a*) 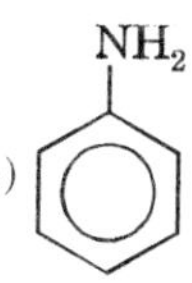

(*b*) 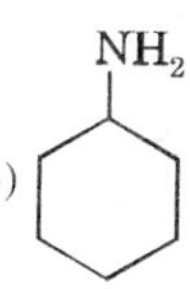

(*c*) 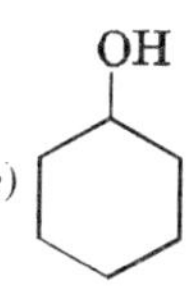

(*d*) 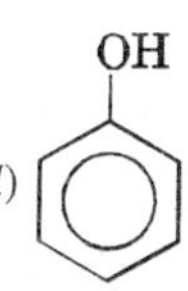

136. Match the compounds in List I with their nature from List II, as seen in aqueous medium from the combinations shown:

List I	List II
I. Acetamide	A. Acidic
II. Benzonitrile	B. Basic
III. Triethylamine	C. Neutral
IV. Phenol	

(*a*) I—C ; II—C ; III—B ; IV—A

(*b*) I—B ; II—C ; III—C ; IV—A

(*c*) I—C ; II—B ; III—B ; IV—C

(*d*) I—A ; II—A ; III—C ; IV—B.

ANSWERS

1. (*c*)	**2.** (*b*)	**3.** (*b*)	**4.** (*c*)	**5.** (*d*)	**6.** (*d*)	**7.** (*a*)	**8.** (*d*)	**9.** (*c*)
10. (*c*)	**11.** (*c*)	**12.** (*b*)	**13.** (*a*)	**14.** (*d*)	**15.** (*c*)	**16.** (*b*)	**17.** (*a*)	**18.** (*a*)
19. (*a*)	**20.** (*d*)	**21.** (*c*)	**22.** (*c*)	**23.** (*a*)	**24.** (*b*)	**25.** (*c*)	**26.** (*a*)	**27.** (*a*)
28. (*c*)	**29.** (*d*)	**30.** (*c*)	**31.** (*b*)	**32.** (*d*)	**33.** (*b*)	**34.** (*c*)	**35.** (*a*)	**36.** (*a*)
37. (*c*)	**38.** (*b*)	**39.** (*b*)	**40.** (*a*)	**41.** (*d*)	**42.** (*b*)	**43.** (*b*)	**44.** (*b*)	**45.** (*b*)
46. (*c*)	**47.** (*d*)	**48.** (*b*)	**49.** (*c*)	**50.** (*d*)	**51.** (*c*)	**52.** (*c*)	**53.** (*b*)	**54.** (*a*)
55. (*b*)	**56.** (*c*)	**57.** (*a*)	**58.** (*b*)	**59.** (*a*)	**60.** (*c*)	**61.** (*c*)	**62.** (*c*)	**63.** (*a*)
64. (*a*)	**65.** (*b*)	**66.** (*b*)	**67.** (*a*)	**68.** (*b*)	**69.** (*b*)	**70.** (*a*)	**71.** (*c*)	**72.** (*d*)
73. (*d*)	**74.** (*c*)	**75.** (*b*)	**76.** (*c*)	**77.** (*c*)	**78.** (*a*)	**79.** (*b*)	**80.** (*c*)	**81.** (*b*)
82. (*d*)	**83.** (*a*)	**84.** (*b*)	**85.** (*b*)	**86.** (*b*)	**87.** (*d*)	**88.** (*b*)	**89.** (*c*)	**90.** (*a*)
91. (*a*)	**92.** (*c*)	**93.** (*a*)	**94.** (*a*)	**95.** (*b*)	**96.** (*c*)	**97.** (*a*)	**98.** (*b*)	**99.** (*a*)
100. (*b*)	**101.** (*c*)	**102.** (*c*)	**103.** (*d*)	**104.** (*c*)	**105.** (*a*)	**106.** (*b*)	**107.** (*a*)	**108.** (*c*)
109. (*c*)	**110.** (*c*)	**111.** (*d*)	**112.** (*d*)	**113.** (*d*)	**114.** (*a*)	**115.** (*b*)	**116.** (*c*)	**117.** (*d*)
118. (*c*)	**119.** (*b*)	**120.** (*c*)	**121.** (*a*)	**122.** (*b*)	**123.** (*b*)	**124.** (*a*)	**125.** (*b*)	**126.** (*b*)
127. (*a*)	**128.** (*d*)	**129.** (*c*)	**130.** (*c*)	**131.** (*d*)	**132.** (*d*)	**133.** (*a*)	**134.** (*b*)	**135.** (*d*)
136. (*a*).								

HINTS/SOLUTIONS

Q. No.	Ans.	Hints & Solutions
1.	**(c)**	The name is 3-cyanopropanoic acid.
2.	**(b)**	IUPAC name of the compound is 1, 2, 3-propanetricarbonitrile.
3.	**(b)**	The correct IUPAC name is 4-Cyano-2-ethyl-3-methoxybutanoic acid.
4.	**(c)**	IUPAC name is prop-2-ene-1-nitrile.
5.	**(d)**	CH_3CH_2CN and $CH_3CH_2—N \overset{\rightarrow}{=} C$ functional isomers as well as tautomers.
6.	**(d)**	$RC \equiv N \xrightarrow{H_2O_2/OH^-} RCONH_2$
7.	**(a)**	$CH_3CN + C_2H_5OH \xrightarrow{H_2O/H^+} CH_3COOC_2H_5$
8.	**(d)**	$R—N \overset{\rightarrow}{=} C \xrightarrow[4[H]]{Zn\text{-}Hg/HCl} R—NH—CH_3$ N-Methylalkanamine
9.	**(c)**	$C_6H_5—NH_2 \xrightarrow{CHCl_3/KOH} C_6H_5—N \overset{\rightarrow}{=} C$ Benzenamine → Benzene isonitrile
10.	**(c)**	$CH_3NC + O_3 \longrightarrow CH_3N = C = O + O_2$.
11.	**(c)**	$C_2H_5—N \overset{\rightarrow}{=} C + S \longrightarrow C_2H_5N = C = S$. (Ethyl isothiocyanate)
12.	**(b)**	$\overset{sp^3}{CH_3}—\overset{sp}{C} \equiv \overset{sp}{N}$.
13.	**(a)**	Mendius reaction involves the reduction of RCN by $LiAlH_4$.
14.	**(d)**	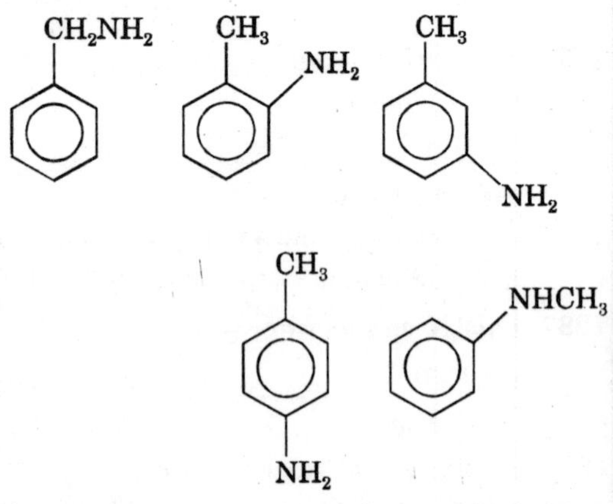
15.	**(c)**	$CH_3NH_2 + COCl_2 \longrightarrow CH_3—N = C = O + 2HCl$ Methyl isocyanate
16.	**(b)**	The formation of propene results due to β-elimination process which is represented in 'b'.

Q. No.	Ans.	Hints & Solutions
17.	(a)	$CH_3CN \xrightarrow{H_3O^+} CH_3COOH \xrightarrow{LAH}$ $CH_3CH_2OH \xrightarrow{P/I_2} CH_3CH_2I$ $\xrightarrow{AgCN} CH_3CH_2NC$
18.	(a)	Choice '*a*' is correct explanation of the inability of CH_3CN to act as base.
19.	(a)	Carbylamine reaction is given by 1° amine.
20.	(d)	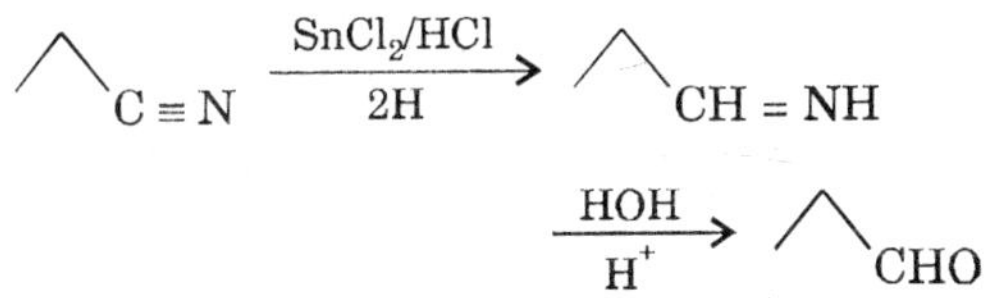
21.	(c)	CN^- ion is an ambident nucleophile because it has two nucleophilic sites.
22.	(c)	$MeBr \xrightarrow{AgCN} MeNC \xrightarrow{Red.} MeNHCH_3$
23.	(a)	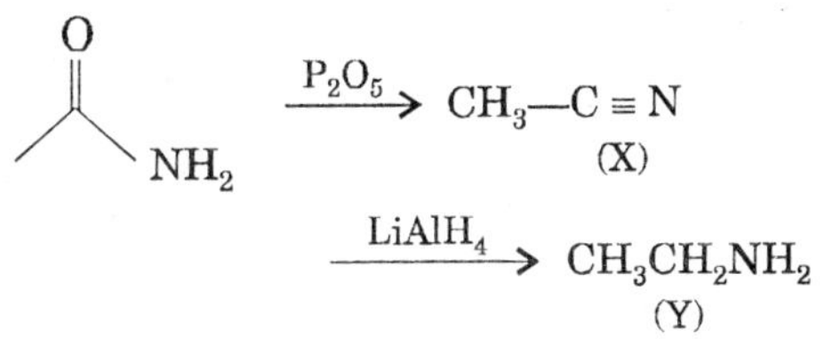
24.	(b)	$C_6H_5CN \xrightarrow[(ii)\ H_2O]{(i)\ MgBr} C_6H_5—\underset{\underset{O}{\|}}{C}—CH_3$ (Acetophenone)
25.	(c)	Alkyl cyanides have higher boiling points than corresponding alkyl isocyanides. Hence statement (*c*) is wrong.
26.	(a)	$R—\overset{+}{N} \equiv \overset{-}{C}: \longleftrightarrow R—\ddot{N} = C:$ (I) (II) In structure II, there are only six electrons and therefore, it is electron deficient.
27.	(a)	$CH_3—N = C = O + H_2O \longrightarrow (CH_3—NHCOOH) \longrightarrow CH_3NH_2 + CO_2$
28.	(c)	The pair of isomers given in choice (*c*) are position isomers.
29.	(d)	The given pair of species are isomers of each other and isomers cannot be homologues.
30.	(c)	$2C_6H_5—NO_2 \xrightarrow{LiAlH_4} C_6H_5—N = N—C_6H_5$
31.	(b)	Nitrating mixture generates NO_2^+ electrophile.
32.	(d)	Nitrobenzene produces azoxybenzene on treatment with $Na_3AsO_3/NaOH$.
33.	(b)	$C_2H_5Br + AgNO_2\ (alc.) \longrightarrow C_2H_5NO_2 + C_2H_5O—N = O$ (Main) (Minor)
34.	(c)	$HNO_2/(CH_3CO)_2O$ is not a nitrating agent.
35.	(a)	Ethylnitrite is called sweet spirit of nitre.
36.	(a)	'*a*' is nitrolic acid which gives red colour with NaOH.
37.	(c)	There is no labile H atom in tertiary nitro compounds.
38.	(b)	Among the given species nitrobenzene is least reactive due to electron withdrawing effect of $—NO_2$ group.
39.	(b)	Reduction of nitrobenzene with Zn/NaOH gives Hydroazobenzene.
40.	(a)	$C_2H_5—O—N = O \xrightarrow[{[H]}]{Sn/HCl} C_2H_5OH + NH_2OH + HCl$
41.	(d)	$RCH_2NO_2 + H_2O + HCl \xrightarrow{\Delta} NH_2OH \cdot HCl + RCOOH$
42.	(b)	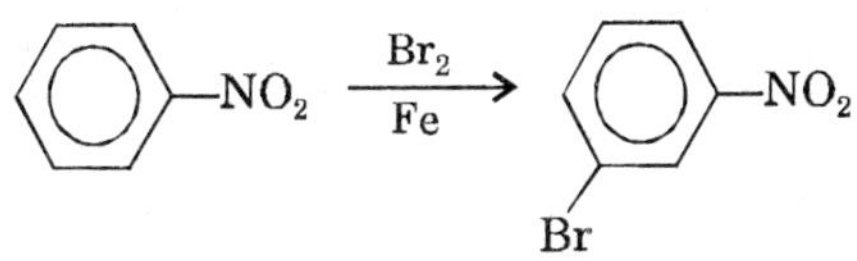
43.	(b)	$CH_3NO_2 + 3Cl_2 \xrightarrow{NaOH} CCl_3NO_2 + 3HCl$ (Chloropicrin) $CHCl_3 + HNO_3 \longrightarrow CCl_3.NO_2 + H_2O$.
44.	(b)	In 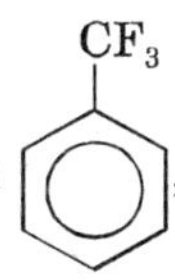, the electron with drawing effect of fluoro groups cause electron deficiency at *ortho* and *para* positions. Hence electrophitic attack occurs at *meta* position.
45.	(b)	$C_6H_5NO_2$ does not reduce Tollen's reagent.
46.	(c)	Reactions I, IV cannot be realised as such because benzene ring or aniline donot give nucleo philic ring substitutions.
47.	(d)	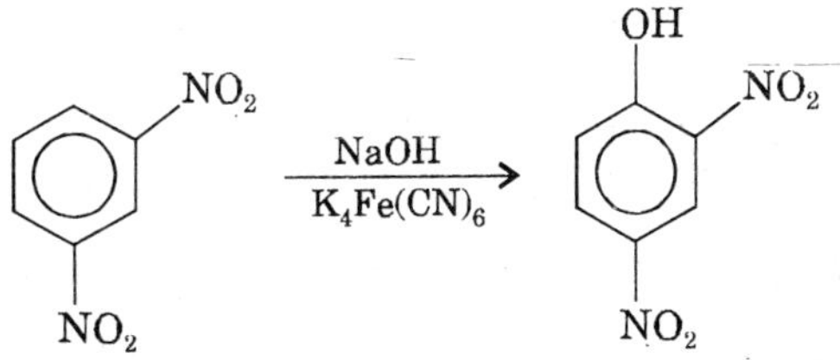

Q. No.	Ans.	Hints & Solutions
		This reaction involves nucleophilic substitution of OH^- at *o* or *p*-position w.r.t. NO_2 group.
48.	(*b*)	The name **3-aminoanisole** is wrong as it goes against the rules. All other names are acceptable.
49.	(*c*)	R-CN on hydrolysis gives R—COOH.
50.	(*d*)	Various structural isomers of C_3H_9N are : $CH_3CH_2CH_2NH_2$ (1-Propanamine) ; $CH_3CH(CH_3)—NH_2$ (2-Propanamine) ; $CH_3—CH_2—N(H)—CH_3$ (N-Methyl ethanamine) ; $CH_3—N(CH_3)—CH_3$ (N, N-Dimethyl methanamine)
51.	(*c*)	Toludines have higher melting points in general as compared to aniline. Among toludines, the *p*-toludine is most symmetrical and thus fits into crystal lattice most easily. Therefore, it has highest melting point. *p*-toludine is solid at room temperature whereas all others are liquids.
52.	(*c*)	The higher electronegativity of oxygen does not allow it to act as better donor of electron pair as compared to nitrogen atom.
53.	(*b*)	Among the given species $(CH_3)_2NH$ is most basic. (See discussion on basicity of amines in concepts in brief).
54.	(*a*)	Lowest pk_b refers to largest k_b and hence greatest basic strength. Among the given species $CH_3CH_2NH_2$ is most basic and thus has lowest pk_b.
55.	(*b*)	In 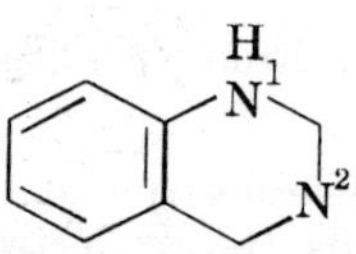, the lone pair of the N atom adjecent to the ring is delocalised over the ring due to conjugation thus making it poor donor. On the other hand, the nitrogen atom (marked as 2) is better donor of electron pair. Thus the product formed is 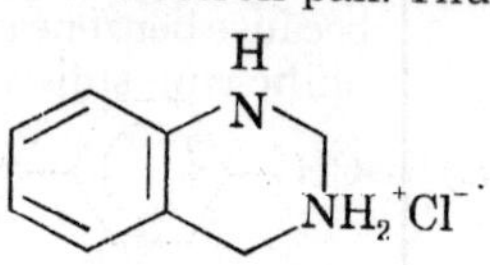
56.	(*c*)	$AgCl + 2C_2H_5NH_2 \longrightarrow [Ag(C_2H_5NH_2)_2]Cl$ (Bis-(ethylamine) silver (I) chloride (soluble complex).
57.	(*a*)	$(C_2H_5)_2NH + HONO \longrightarrow (C_2H_5)_2N—NO + H_2O$ (Nitrosoamine)
58.	(*b*)	Aldehydes and aniline react to give Schiff's bases.
59.	(*a*)	Only 1° amines give carbylamine reaction.
60.	(*c*)	$CH_3CH_2NH_2 + CH_3MgBr \longrightarrow CH_4\uparrow + (CH_3CH_2NH)MgBr.$
61.	(*c*)	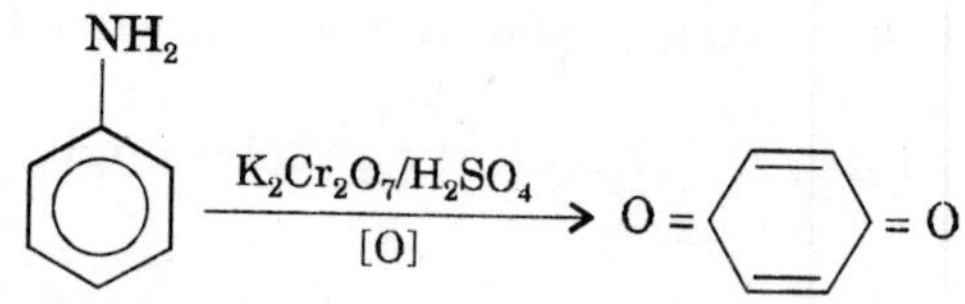 *p*-benzoquinone
62.	(*c*)	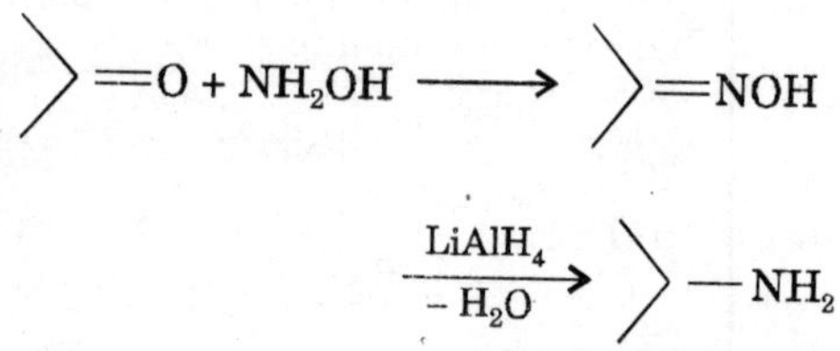 2-Propanamine (1° Amine)
63.	(*a*)	Chemical formula of Hinsberg's reagent is $C_6H_5SO_2Cl$.
64.	(*a*)	In benzyl amine, the $—NH_2$ group is more like aliphatic amino group but in other amines there is a delocalisation of lone pair of electrons of N atom on the ring. Hence benzyl amine is most basic among the given structures.
65.	(*b*)	$NaNO_2$/HCl gives HNO_2 which gives different products with primary and secondary amines.
66.	(*b*)	See reactions of aniline.
67.	(*a*)	Refer reactions of aniline.
68.	(*b*)	Among the given species *p*-toludine is the only 1° amine. Hence, it responds to carbylamine test.
69.	(*b*)	The given observations refer to primary amino group.
70.	(*a*)	Benzenediazonium salt forms a dye with aromatic—NH_2 by coupling reaction but does not form any such compound with aliphatic NH_2 group.
71.	(*c*)	Major product will be alkene due to dominance of β-elimination process.

Q. No.	Ans.	Hints & Solutions
72.	(d)	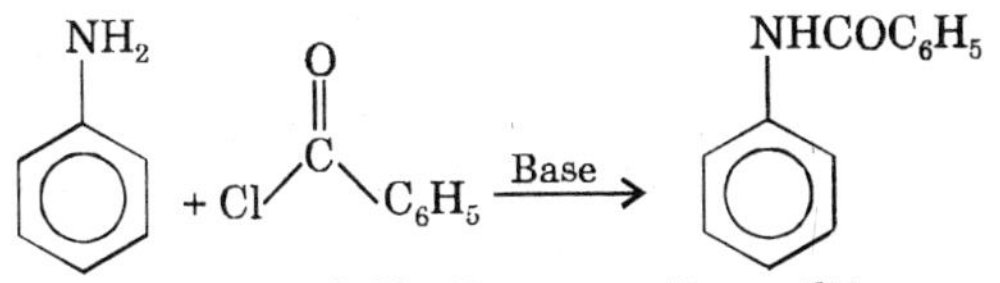 Aniline Benzoyl chloride Benzanilide
73.	(d)	$C_2H_5NH_2 + HCl \longrightarrow C_2H_5\overset{+}{N}H_3Cl^-$.
74.	(c)	Tertiary amines do not react with diethyloxalate.
75.	(b)	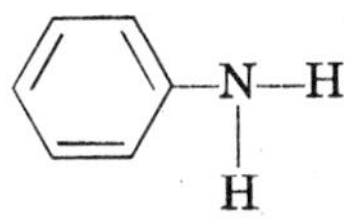 ; Evidently, σ bonds = 14, π bonds = 3.
76.	(c)	Aqueous solutions of $C_2H_5NH_2$ contains $C_2H_5\overset{+}{N}H_3$ and $O\overset{-}{H}$ ions, which will react with zinc ions (Zn^{2+}) to form white precipitate of $Zn(OH)_2$.
77.	(c)	Only primary amines (– NH_2) gives N_2 with HNO_2.
78.	(a)	$CH_3—\overset{CH_3}{\overset{\vert}{C}} = O + H_2\ddot{N}OH \longrightarrow CH_3—\overset{CH_3}{\overset{\vert}{C}} = NOH$ Acetone Oxime $\xrightarrow{\text{Reduction}} CH_3—\overset{CH_3}{\overset{\vert}{C}}HNH_2 + H_2O$ (Isopropylamine)
79.	(b)	$CH_3NH_2 \xrightarrow[KMnO_4]{(O)} CH_3CH = NH \xrightarrow{H_3O^+} CH_3CHO$ Aldimine Ethanal
80.	(c)	H_2SO_5 oxidises $(CH_3)_2NH$ to form $(CH_3)_2N—N(CH_3)_2$.
81.	(b)	$R—CH = N—R + [H] \xrightarrow{\text{Redn.}} R_2CH_2—NH—R$ Schiff's base sec. Amine
82.	(d)	$(CH_3)_3C—NH_2$ contains $—NH_2$ group. Hence it is primary amine.
83.	(a)	$CH_3CH_2CH_2CH_2NH_2$; $(CH_3)_2CHCH_2NH_2$; $(CH_3)_3C—NH_2$; $CH_3\underset{NH_2}{\underset{\vert}{C}}HC_2H_5$.
84.	(b)	HNO_2 (produced from $NaNO_2$/HCl) reacts differently with 1° amine and 2° amines.
85.	(c)	Only the Cl^- group present ortho to NO_2 gets activated and is replaced by NH_2 group. The – Cl group present at meta position w.r.t. NO_2 is not activated for nucleophilic substitution.
86.	(b)	$[C_2H_5\overset{+}{N}H_3]OH^- + FeCl_3 \longrightarrow 3[C_2H_5\overset{+}{N}H_3]Cl^- + Fe(OH)_3$. Brown ppt.
87.	(b)	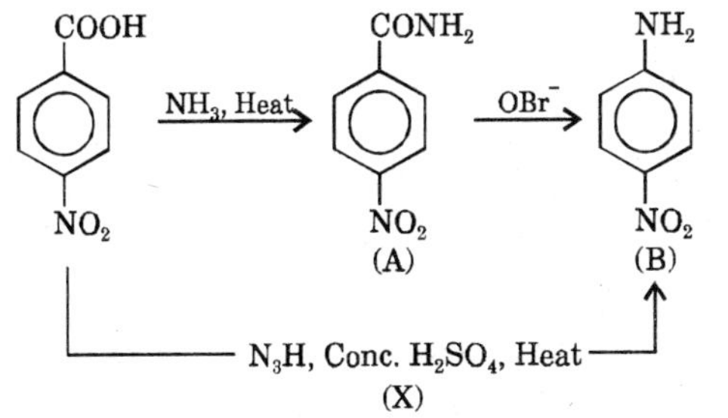
88.	(b)	The correct name of (b) is N, N-Dimethyl benzeneamine.
89.	(c)	$RNH_2 + NOCl \longrightarrow R—Cl + N_2 + H_2O$.
90.	(a)	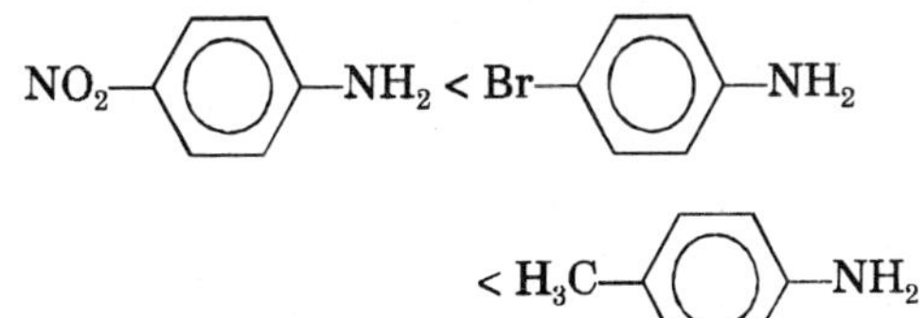 NO_2— is electron withdrawing group. While CH_3— group is electron repelling. Br group has – I effect but + R effect.
91.	(a)	As the number of alkyl group on the N atom increases, the basic strength in general also increases among the alkyl substituted anilines.
92.	(c)	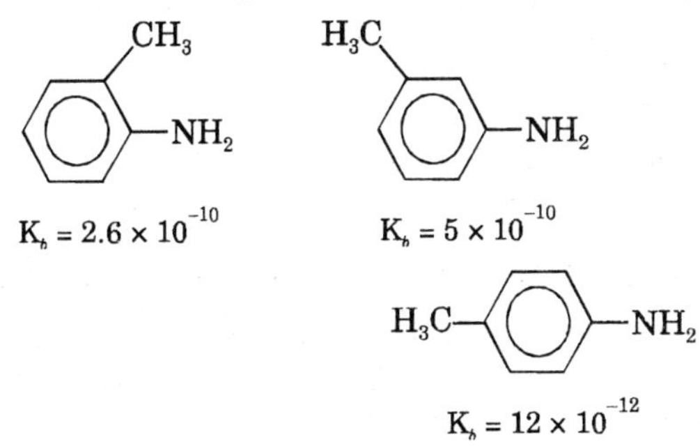 $K_b = 2.6 \times 10^{-10}$ $K_b = 5 \times 10^{-10}$ $K_b = 12 \times 10^{-12}$
93.	(a)	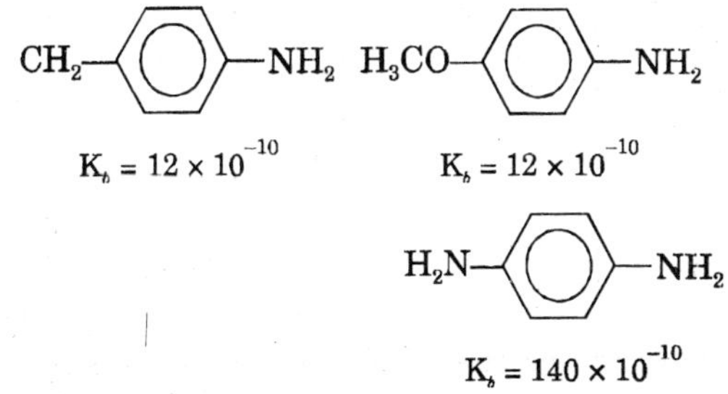 $K_b = 12 \times 10^{-10}$ $K_b = 12 \times 10^{-10}$ $K_b = 140 \times 10^{-10}$

Q. No.	Ans.	Hints & Solutions
94.	(a)	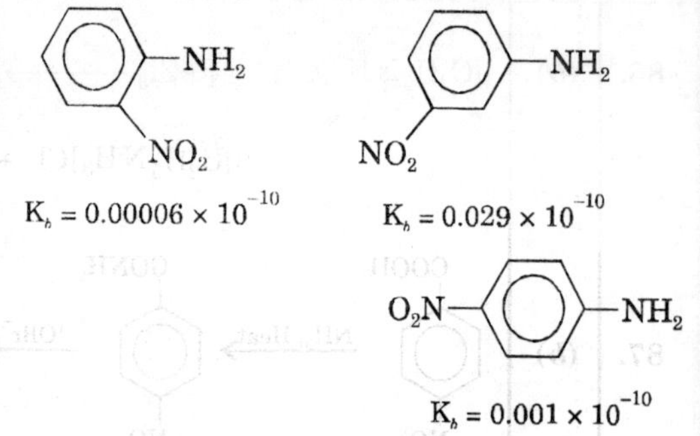
95.	(b)	$C_6H_5\overset{+}{N}\overset{-}{H} \xleftarrow{-H^+} C_6H_5NH_2 \xrightarrow{+H^+} C_6H_5\overset{+}{N}H_3$ Conjugate base of aniline; Aniline; Conjugate acid of aniline
96.	(c)	$CH_3-CH(CH_3)-NH_2 \xrightarrow{KMnO_4} CH_3-C(CH_3)=NH$ (Ketimine) $\xrightarrow{H_3O^+} CH_3-C(CH_3)=O$ (Ketone)
97.	(a)	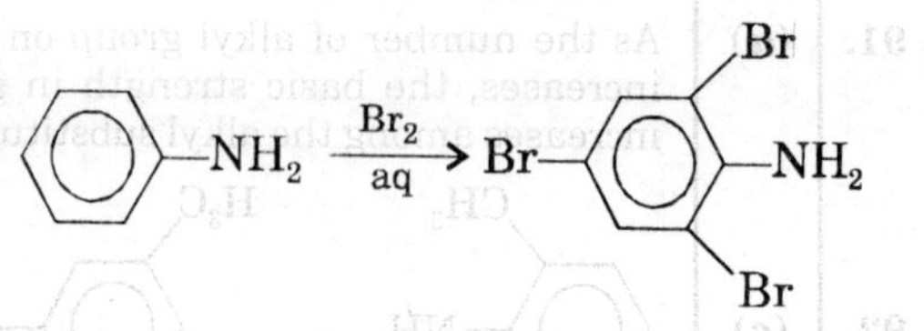
98.	(b)	$CH_3CH_2CH_2-\ddot{N}H_2$ is least basic among these.
99.	(a)	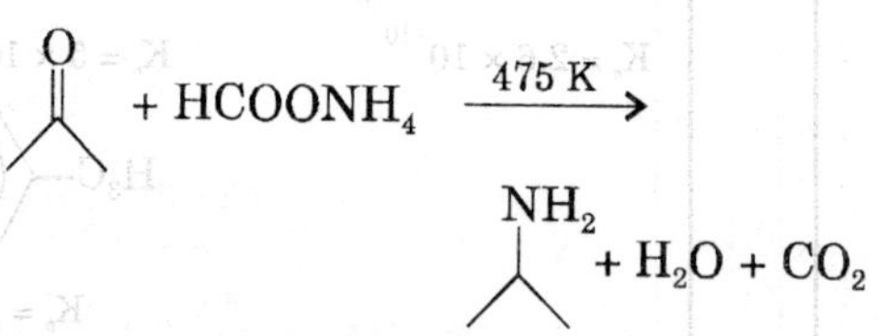 The reaction is called Leuckart reaction.
100.	(b)	The carbocation formed as intermediate rearranges to give a more stable tertiary carbocation. This gives $CH_3-C(CH_3)(OH)-CH_2CH_3$ as major product.
101.	(c)	Medium should be slightly acidic with pH around 4-5.
102.	(c)	$C_6H_5\overset{+}{N}_2\overset{-}{Cl} \xrightarrow[H_3O^+]{H_3PO_2} C_6H_6 + N_2 + HCl$.
103.	(d)	Diazonium group is replaced by —OH. Thus, phenol is product.
104.	(c)	Urea reacts with nitrous acid and produces N_2, CO_2 and H_2O $NH_2CONH_2 + HNO_2 \longrightarrow 2N_2 + CO_2 + H_2O.$ Thus it doesnot interfere with the diazotisation reaction.
105.	(a)	$C_6H_5N_2HSO_4 + MeOH \xrightarrow{warm} C_6H_5-OCH_3 + H_2SO_4 + N_2$ (Anisole)
106.	(b)	$C_6H_5N_2Cl \xrightarrow{Zn/HCl} C_6H_5NH-NH_2 \xrightarrow{Zn/HCl} C_6H_5NH_2 + NH_4Cl$
107.	(a)	*p*-methoxybenzene diazonium salt is most stable due to dispersal of positive charge because of resonance.
108.	(c)	$C_6H_5N = NC_6H_5$ exhibit isomeric forms as 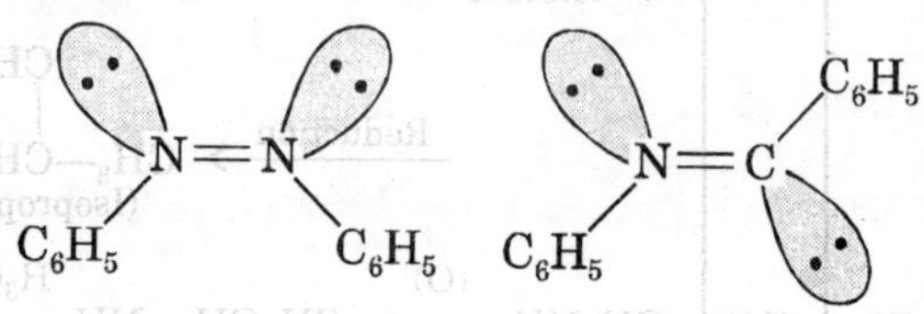 syn anti
109.	(c)	It checks the concentration of free aniline which otherwise may undergo coupling reaction.
110.	(c)	The diazonium salt formed is stablised by powerful + R effect of—OCH_3 group.
111.	(d)	Benzyl alcohol does not have phenolic group. Hence it does not give coupling reaction with benzenediazonium chloride.
112.	(d)	The product of reaction is 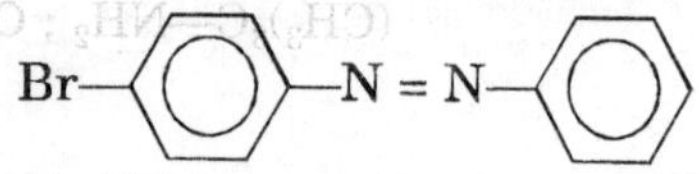
113.	(d)	MeO-group at meta position causes electron releasing effect which tends to weaken $Ar - N_2^+$ bond there, by fecilitating the decomposition.
114.	(a)	Zn/HCl being stronger reducing agent further converts $C_6H_5NH–NH_2$ to $C_6H_5NH_2$.

Q. No.	Ans.	Hints & Solutions
115.	(*b*)	Conversion of $C_6H_5N_2Cl$ to C_6H_5CN is achieved by Sandmeyer reaction using CuCN dissolved in KCN.
116.	(*c*)	$\underset{\mid \atop NH_2}{CH_2COONa} \xrightarrow[CaO]{NaOH} CH_3NH_2 + Na_2CO_3$
117.	(*d*)	In I and III lone pair of N is in sp^3 hybrid orbital but in III lone pair is less easily available due to electronegative O atom present in the ring skeleton. Hence III is less basic than I. II and IV have lone pair of N in sp^2 hybrid orbital. Thus, they are less basic than I and III. In IV, the lone pair is delocalised over the entire cyclic structure (Huckel rule). Hence it is least basic. Thus, the order of basic strength is I > III > II > IV.
118.	(*c*)	Lesser the stability of diazonium salt is, more will be its reactivity. The stability of the given diazonium ions decreases as the electron donating tendency of the substituent at para position decreases. It is as I > III > IV > II. Consequently, reactivity order is I < III < IV < II.
119.	(*b*)	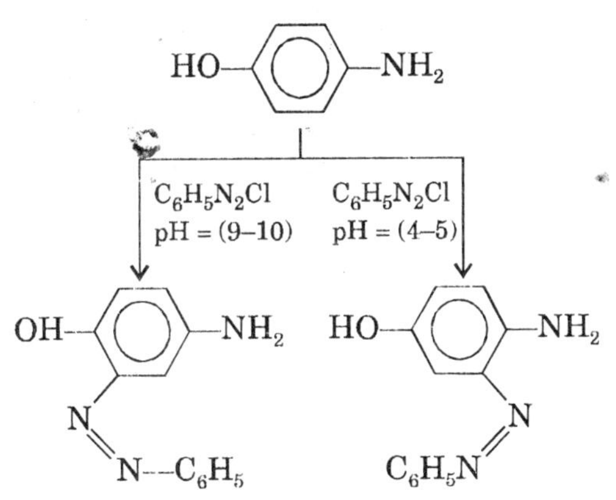
120.	(*c*)	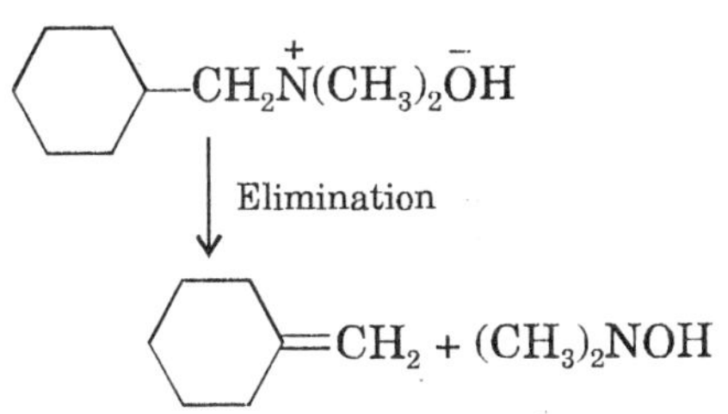
121.	(*a*)	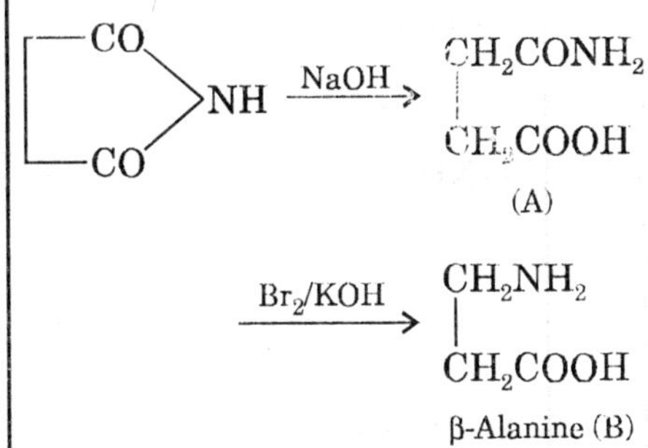
122.	(*b*)	Among the given species $C_6H_5NH_2$ is most reactive because NH_2 group activates the ring but other substituents deactivate the ring by virtue of their electron withdrawing effect.
123.	(*b*)	$CH_3CHO + H_2N—C_2H_5 \xrightarrow{-H_2O} \underset{(X)}{CH_3CH=NC_2H_5} \xrightarrow{H_2/Pt} \underset{\text{Diethyl amine}}{CH_3CH_2—\overset{H}{\overset{\mid}{N}}—C_2H_5}$
124.	(*d*)	$C_2H_5N \rightleftharpoons C \xrightarrow{\text{Reduction}} C_2H_5NH—CH_3 \xrightarrow{HNO_2} C_2H_5—N(CH_3)(NO)$ N-methyl-N-nitrosoamine
125.	(*b*)	$C_6H_5COONH_4 \xrightarrow[\text{Heat}]{P_2O_5} C_6H_5-C\equiv N \xrightarrow[LiAlH_4]{\text{Reduction}} \underset{\text{Benzyl amine}}{C_6H_5—CH_2NH_2} \xrightarrow{HNO_2} \underset{\text{Benzylalcohol}}{C_6H_5CH_2OH} + N_2 + H_2O$

Q. No.	Ans.	Hints & Solutions
126.	(*b*)	During acylation of $—NH_2$ group, there is increase in the molecular mass by 42 units for each acyl ($CH_3CO—$) group added $—NH_2 + Cl—\overset{\overset{O}{\parallel}}{C}—CH_3 \xrightarrow[+\,CH_3CO]{-\,H} —NH—\overset{\overset{O}{\parallel}}{C}—CH_3$ Net increase in mol. mass after acylation $= 348 - 180 = 168$ $\therefore$ Number of $CH_3\overset{\overset{O}{\parallel}}{C}—$groups added $= \frac{168}{42} = 4.$ Hence the compound has 4 ($—NH_2$) groups.
127.	(*a*)	Carbylamine reaction involves 1° amine KOH and $CHCl_3$.
128.	(*d*)	Quaternary salt does not correspond to formula C_3H_9N.
129.	(*c*)	HNO_3 being oxidising agent oxidises most of aniline to give black tarry oxidation products with only very small nitrated products.
130.	(*c*)	No labile H atom (*i.e.*, α-H atom) is present in 3° nitro compounds.
131.	(*d*)	$\underset{\text{Acetonoxime}}{(CH_3)_2C = NOH} \xrightarrow{H_2/Pt} \underset{\text{(Isopropyl amine)}}{(CH_3)_2CHNH_2} + H_2O$
132.	(*d*)	Aniline is not soluble in water.
133.	(*a*)	Beckmann rearrangement involves the conversion of oximes into substituted amides in acidic medium $\underset{\text{Oxime}}{(C_6H_5)_2C = NOH} \xrightarrow{H_2SO_4} \underset{\text{Benzanilide}}{C_6H_5—\overset{\overset{O}{\parallel}}{C}—NH—C_6H_5}$
134.	(*b*)	A is benzamide, the sequence of reactions is $\underset{(A)}{C_6H_5CONH_2} \xrightarrow{KOH/Br_2} \underset{(B)}{C_6H_5NH_2} \xrightarrow[KOH]{CHCl_3} C_6H_5NC.$
135.	(*d*)	Among the given species, C_6H_5OH is strongest Bronsted acid because it gives conjugate base $C_6H_5O^-$ which is resonance stabilised.
136.	(*a*)	Acetanilide and benzonitrile are neutral, triethylamine is basic and phenol is acidic. Hence the correct matchings are I—C, II—C, III—B, IV—A.

7.5. EXAMINER'S VIEWPOINT

1996

1. The reaction of nitrobenzene with zinc and alkali results in the formation of

(*a*) Aniline (*b*) Hydrazobenzene
(*c*) Nitrosobenzene (*d*) Phenyl hydroxylamine.

(*Pb. CET*)

2. The reaction,

$$C_6H_5NO_2 + KOH(s) \xrightarrow{\text{Heat}} o\text{-and } p\text{-nitrophenol}$$

is an example of

(*a*) Nucleophilic substitution
(*b*) Electrophilic substitution
(*c*) Free radical substitution
(*d*) Electrophilic addition. **(*Har. CEET*)**

3. Aniline reacts with concentrated HNO_3 to give

(*a*) 

(*b*) 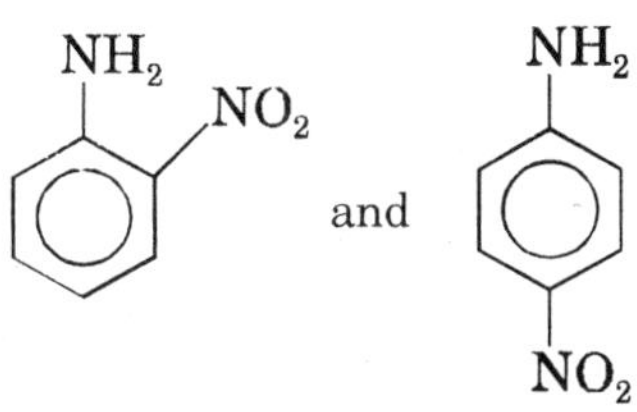

(*c*) 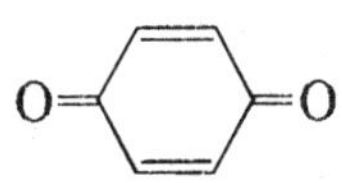

(*d*) 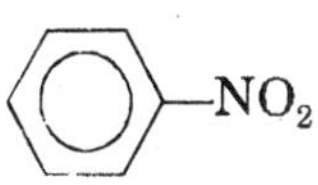 **(*Har. CEET*)**

4. *p*-Nitrotoluene on further nitration gives

(*a*) 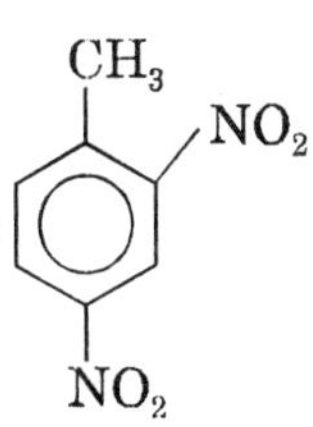

(*b*) 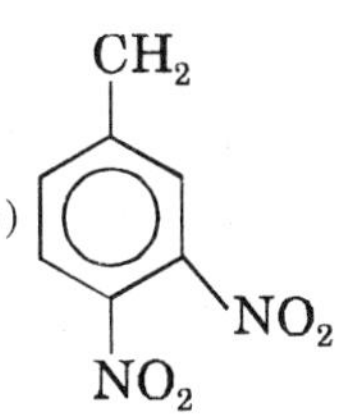

(*c*) 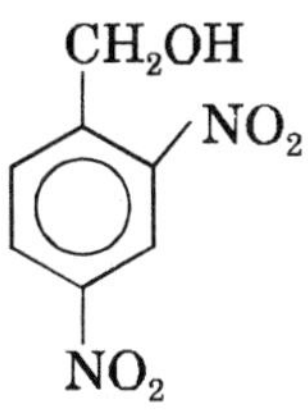

(*d*) 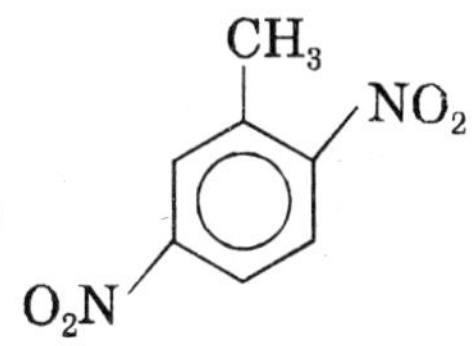

(*CBSE PMT*)

5. Identify X in the sequence given below :

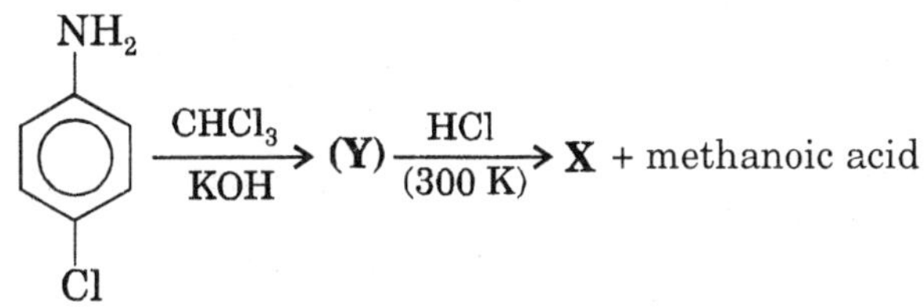

(*a*) 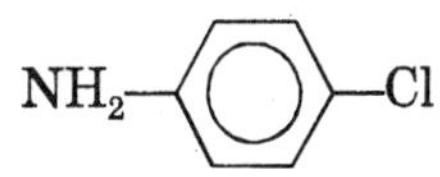

(*b*) 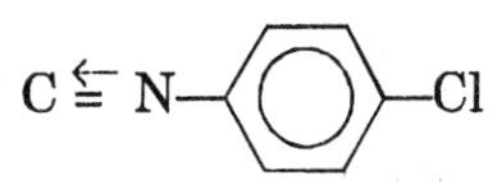

(*c*) 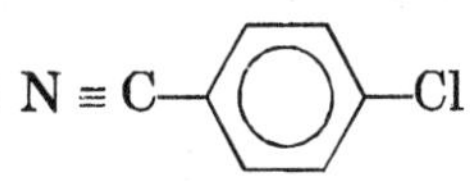

(*d*) 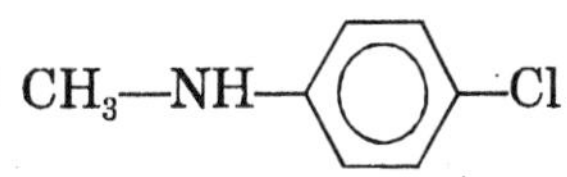 **(*CPMT*)**

6. Aniline reacts with KOH and phosgene to form

(*a*) 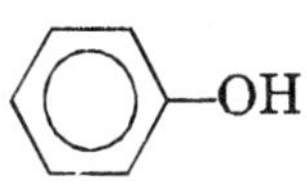

(*b*) 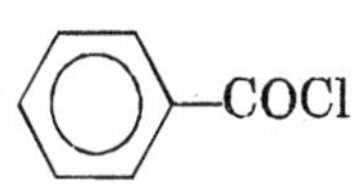

(*c*) 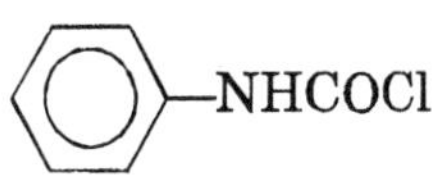

(*d*) 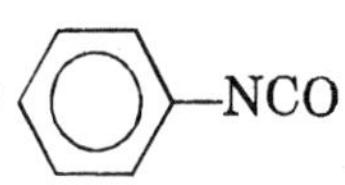

(*Har. CEET*)

1997

7. The compound formed by heating a mixture of 1° amine and chloroform with ethanolic solution of potassium hydroxide (KOH) is

(*a*) An alkyl isocyanide
(*b*) An alkyl isothiocyanate
(*c*) An amide
(*d*) An amide and nitro compound. **(*CBSE PMT*)**

8. The major product (70%—80%) of reaction between *m*-dinitrobenzene with $(NH_4)_2 S_x$ is

(*a*) 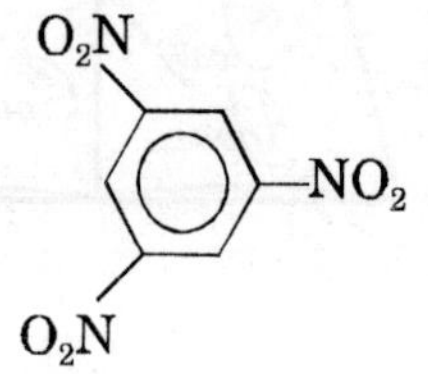

(*b*) 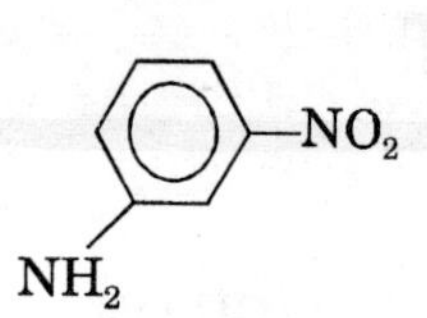

(*c*) 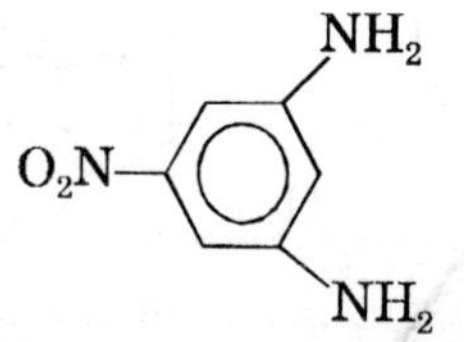

(*d*) 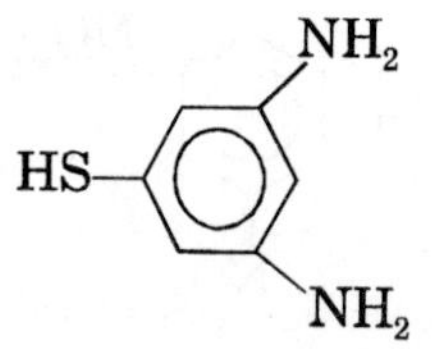

(AIIMS)

9. Which of the following on hydrolysis forms acetic acid ?
(*a*) CH_3CN (*b*) C_2H_5OH
(*c*) $C_2H_5NH_2$ (*d*) CH_3OH. **(BHU)**

10. The colour of *p*-aminoazobenzene is
(*a*) Orange (*b*) Congo red
(*c*) Bismark brown (*d*) Indigo. **(BHU)**

11. Acetaldoxime reacts with P_2O_5 to give
(*a*) Methyl cyanide (*b*) Ethyl cyanate
(*c*) Ethyl cyanide (*d*) Mixture of all these. **(AFMC)**

12. Arrange the following :
CH_3NH_2(I) ; $(CH_3)_2NH$(II) ; $C_6H_5NH_2$(III) and $(CH_3)_3N$(IV) in increasing order of basicity in aqueous medium
(*a*) II < I < IV < III (*b*) III < IV < I < II
(*c*) I < II < III < IV (*d*) II < III < I < IV. **(AFMC)**

13. When *primary amine* is heated with CS_2 in the presence of excess of mercury chloride, it gives *isothiocyanate*. This reaction is called
(*a*) Hoffmann Bromoamide reaction
(*b*) Hoffmann mustard oil reaction
(*c*) Perkin's condensation
(*d*) Hoffmann elimination reaction. **(Har. CPMT, CEET)**

14. If N and S elements are present in organic compound, then during Lassaigne's test both change into
(*a*) Na_2S and NaCN (*b*) NaSCN
(*c*) Na_2SO_3 and Na_2CO_3 (*d*) Na_2S and NaCNO. **(CPMT)**

15. Which of the following is least basic ?

(*a*) 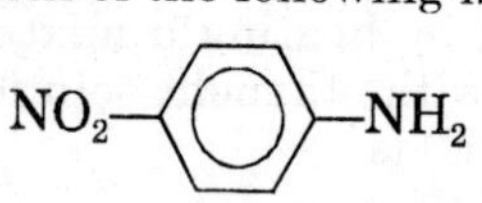

(*b*) 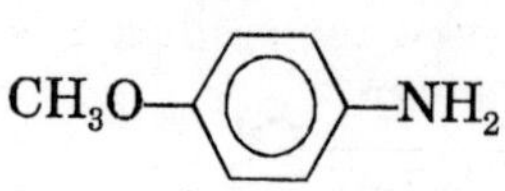

(*c*) 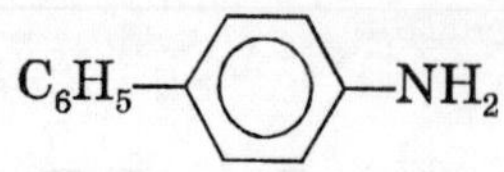

(*d*) All are equally basic. **(CPMT)**

16. The type of isomerism exhibited by $C_6H_5C \equiv N$ and $C_6H_5N \rightleftarrows C$ is
(*a*) Position
(*b*) Functional
(*c*) Enantiomerism
(*d*) Functional as well as tautomerism. **(CPMT)**

17. Which of the following will exhibit tautomerism ?
(*a*) $(CH_3)_2CNO$ (*b*) $(CH_3)_2NH$
(*c*) R_3CNO_2 (*d*) RCH_2NO_2. **(CBSE PMT)**

1998

18. Piperidine is
(*a*) Homocyclic compound
(*b*) Heterocylic aromatic compound
(*c*) Heterocyclic alicyclic compound
(*d*) Acyclic compound. **(Har. CEET)**

19. Which group deactivates the benzene ring but causes electrophilic substitution to occur at ortho and para positions ?
(*a*) $—NH_2$ (*b*) —Cl
(*c*) $—NO_2$ (*d*) $—C_2H_5$. **(Pb. PMT)**

20. Which species act as electrophile during the nitration of arene ?
(*a*) NO_2^- (*b*) NO_2^+
(*c*) NO_2 (*d*) NO_3^-. **(Pb. PMT)**

21. In the diazotisation of aniline the excess of hydrochloric acid is used primarily with sodium nitrite because it
(*a*) Supresses the conc. of free aniline
(*b*) Supresses the hydrolysis to phenol
(*c*) Ensures a stoichiometric amount of nitrous acid
(*d*) Neutralises the base liberated. **(Pb. PMT)**

22. Aniline is reacted with bromine water and the resulting product is treated with an aqueous solution of sodium nitrite in the presence of dilute HCl. The compound so formed is converted into tetrafluoroborate which is subsequently heated to dryness. The final product is
(*a*) *p*-bromofluorobenzene
(*b*) *p*-bromoaniline
(*c*) 2, 4, 6-tribromofluorobenzene
(*d*) 1, 3, 5-tribromobenzene. **(CBSE PMT)**

23. Nitrosoamines ($R_2N—N = O$) are insoluble in water. On heating them with conc. H_2SO_4. They give secondary amines. The reaction is called

(*a*) Liberman nitroso reaction
(*b*) Etard's reaction
(*c*) Fries's reaction
(*d*) Perkin's reaction. **(*AFMC*)**

24. In the presence of acid, the hydrolysis of methyl cyanide gives
(*a*) Methanoic acid (*b*) Ethanoic acid
(*c*) Methylamine (*d*) Methyl alcohol. **(*MP PMT*)**

25. The compound which on reaction with cold nitrous acid gives oily nitrosoamine is
(*a*) Methylamine (*b*) Dimethylamine
(*c*) Trimethylamine (*d*) Triethylamine. **(*MP PMT*)**

26. Which of the following compound gives *dye test* ?
(*a*) Aniline (*b*) Methylamine
(*c*) Diphenyl amine (*d*) Ethylamine. **(*MP PMT*)**

27. Reaction of nitrous acid with aliphatic 1° amine in cold gives
(*a*) A diazonium salt (*b*) An alcohol
(*c*) A nitrile (*d*) A dye. **(*MP PMT*)**

28. Gabriel synthesis is used for the preparation of
(*a*) 1° Aromatic amines (*b*) 1° Aliphatic amines
(*c*) 2° Amines (*d*) Tertiary amines. **(*Har. CEET*)**

29. Hydrolysis of acetonitrile in acidic medium gives
(*a*) CH_3CH_2OH (*b*) CH_3COOH
(*c*) CH_3NC (*d*) CH_3CH_2CHO. **(*EAMCET*)**

30. Reduction of nitrobenzene in the presence of Zn/NH_4Cl gives
(*a*) Azobenzene
(*b*) Aniline
(*c*) N-Phenylhydroxylamine
(*d*) Hydrazobenzene. **(*EAMCET*)**

31. Attacking species in the nitration of toluene in the presence of nitrating mixture
(*a*) SO_3 (*b*) SO_3^+
(*c*) NO_2^- (*d*) NO_2^+ **(*Raj. PMT*)**

32. A compound Z reacts with three mole of CH_3I and gives a product which on hydrolysis gives $[(CH_3)_4N]^+\bar{O}H$. Z is
(*a*) CH_3NH_2 (*b*) $(CH_3)_2NH$
(*c*) $(CH_3)_3N$ (*d*) $(CH_3)_4N^+Cl$. **(*Raj. PMT*)**

33. Which one of the following is a typical example of compound which exists as zwitter ion ?
(*a*) Aniline (*b*) Aminophenol
(*c*) Glycine (*d*) Acetamide. **(*SCRA*)**

34. Identify product D in the following sequence of reactions

$$CH_3Cl \xrightarrow{KCN} A \xrightarrow{H_3O^+} B \xrightarrow{NH_3} C \xrightarrow{\Delta} D$$

(*a*) $CH_3CH_2NH_2$ (*b*) CH_3CN
(*c*) $HCONH_2$ (*d*) CH_3CONH_2. **(*MP PET*)**

1999

35. How many isomeric forms of C_7H_9N contain a benzene ring ?
(*a*) 4 (*b*) 5
(*c*) 6 (*d*) 7. **(*CPMT*)**

36. Which one of the following does not participate in carbylamine reaction ?
(*a*) KOH (*b*) Ethanol
(*c*) $CHCl_3$ (*d*) Aniline. **(*MP CET*)**

37. In the reaction,

$$CH_3CN + 2H \xrightarrow[\text{Ether}]{SnCl_2/HCl} X \xrightarrow[H_2O]{\text{Boiling}} Y.$$

The term Y, here is
(*a*) Acetone (*b*) Ethanamine
(*c*) Acetaldehyde (*d*) Dimethylamine. **(*CBSE PMT*)**

38. Phenyl isocyanides are prepared from which of the following reaction ?
(*a*) Rosenmund's reaction
(*b*) Carbylamine reaction
(*c*) Reimer Tiemann reaction
(*d*) Wurtz reaction. **(*CBSE PMT*)**

39. In the following reaction X is

$$X \xrightarrow{\text{Bromination}} Y \xrightarrow[+ HCl]{NaNO_2} Z \xrightarrow{H_3PO_2} \text{Tribromobenzene}$$

(*a*) Benzoic acid (*b*) Salicylic acid
(*c*) Phenol (*d*) Aniline. **(*CPMT*)**

40. In the reduction of nitrobenzene, which of the following is intermediate ?
(*a*) $C_6H_5N = O$ (*b*) $C_6H_5NH—NH—C_6H_5$
(*c*) $C_6H_5N = NC_6H_5$ (*d*) $C_6H_5N = NH_6H_5$. (with $\downarrow$ O) **(*CPMT*)**

41. In the reaction given below the product Y is

$$C_6H_6 \xrightarrow[H_2SO_4]{HNO_3} X \xrightarrow[FeCl_3]{Cl_2} Y$$

(a) 4-nitrochlorobenzene (b) 1-Nitrochlorobenzene
(c) 3-Nitrochlorobenzene (d) None of the above.

(AIIMS)

42. 'Z' in the following sequence of reactions is

$$C_6H_6 \xrightarrow[\Delta]{HNO_3/H_2SO_4} W \xrightarrow{Zn/HCl} X \xrightarrow[HCl]{NaNO_2} Y \xrightarrow{H_3PO_2/H_2O} Z$$

(a) 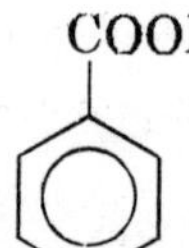

(b) 

(c) 

(d) 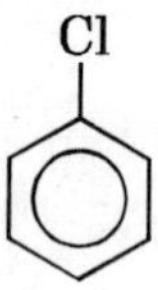

(JIPMER)

43. The reaction of $C_6H_5N_2^+Cl^-$ with CuCl gives
(a) C_6H_5Cl (b) C_6H_6
(c) $C_6H_5 - C_6H_5$ (d) $C_6H_4Cl_2$. **(Pb. CET)**

44. Primary and secondary amines can be distinguished by
(a) Schiff's reagent
(b) Carbylamine reaction
(c) Hoffmann's bromamide reaction
(d) Biuret reaction. **(UP SEAT)**

45. When benzenediazonium chloride in hydrochloric acid reacts with cuprous chloride, then chlorobenzene is formed. The reaction is called
(a) Gattermann reaction (b) Perkin reaction
(c) Etard reaction (d) Sandmeyer reaction.

(UP SEAT)

46. Identify the product C in the series :

$$CH_3CN \xrightarrow{Na/C_2H_5OH} A \xrightarrow{HNO_2} B \xrightarrow{Cu/573\ K} C$$

(a) CH_3COOH (b) CH_3CH_2NHOH
(c) CH_3CONH_2 (d) CH_3CHO. **(MP PET)**

47. The amine that does not react with acetyl chloride is
(a) CH_3NH_2 (b) $(CH_3)_2NH$
(c) $(CH_3)_3N$ (d) None of these.

(MP PET)

48. Nitrobenzene on further excessive nitration gives
(a) *Sym*-Trinitrobenzene (b) *m*-Dinitrobenzene
(c) *p*-Dinitrobenzene (d) All of these.

(Manipal PMT)

2000

49. 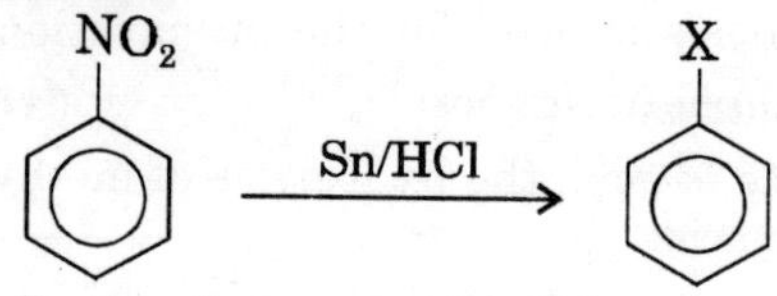

In the above reaction X is
(a) Cl (b) NH_2
(c) $NH_3^+Cl^-$ (d) $N_2^+Cl^-$. **(Kerala CET)**

50. In the sequence given below Z is

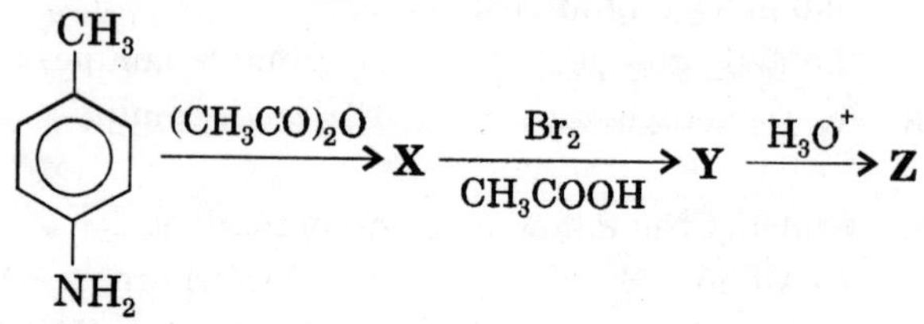

(a) 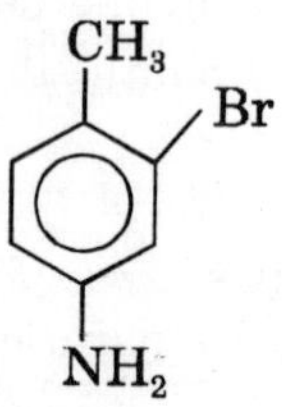

(b) 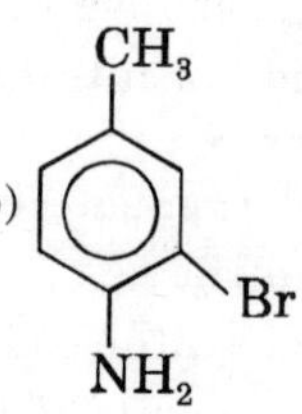

(c) 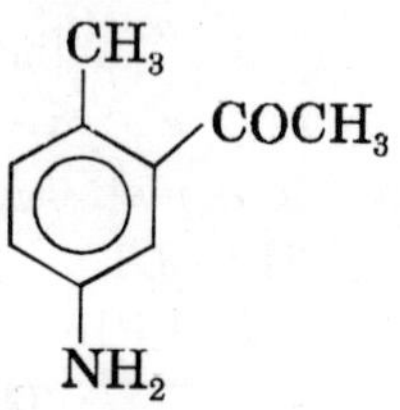

(d) 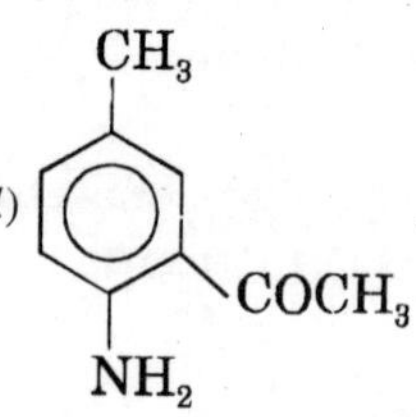

(DPMT)

51. Hinsberg Reagent is
(a) $C_6H_5SO_3H$ (b) C_2H_5NO
(c) $C_6H_5SO_2Cl$ (d) $C_6H_5N_2Cl$. **(MP PMT)**

52. Reaction of aniline with acetyl chloride in the presence of NaOH gives
(a) Acetanilide (b) *p*-chloroaniline
(c) A red dye (d) Aniline hydrochloride.

(BHU)

53. C_3H_9N cannot represent
(a) 1° amine (b) 2° amine
(c) 3° amine (d) quaternary salt. **(BHU)**

54. Compound A on reduction gives B which on further reaction with $CHCl_3$ and alcoholic KOH gives compounds C which on further hydrolysis gives aniline. The compound A is
(a) Nitrobenzene (b) Methylamine
(c) Nitromethane (d) Nitrosobenzene.

(CBSE PMT)

55. Which of the following compounds cannot be identified by carbylamine test ?

(a) $C_6H_5NHC_6H_5$ (b) $CH_3CH_2NH_2$
(c) $CHCl_3$ (d) $C_6H_5NH_2$. ***(SCRA)***

56. Gabriel synthesis is used for the preparation of
(a) Primary aliphatic amines
(b) Secondary amines
(c) Carboxylic acids
(d) Aldehydes
(e) 3° Amines. ***(Kerala MEE)***

57. Which of the following will be most basic ?
(a) Aniline (b) p-methoxy aniline
(c) p-nitroaniline (d) p-methylaniline
(e) Benzylamine. ***(Kerala MEE)***

58. When acetamide is treated with NaOBr, the product formed is
(a) CH_2CN (b) CH_3COBr
(c) CH_3NH_2 (d) CH_3OH. ***(Har. CEET)***

2001

59. Primary nitro compounds reacts with nitrous acid to form nitrolic acid which dissolve in NaOH giving
(a) Yellow solution (b) Blue solution
(c) Colourless solution (d) Red solution.
(Tamil Nadu CET)

60. What is formed when $[(CH_3)_4N]^+ \bar{O}H$ is heated
(a) CH_3NH_2 (b) $C_2H_5NH_2$
(c) $(CH_3)_3N$ (d) $(CH_2)_3N$.
(Manipal PMT)

61. Identify **Y** in the following sequence of reactions

$$C_6H_5NO_2 \xrightarrow{Sn/HCl} X \xrightarrow{C_6H_5COCl} Y + HCl$$

(a) acetanilide (b) Benzanilide
(c) azobenzene (d) Hydroazobenzene.
(EAMCET)

62. The electrophile involved in the nitration of arenes
(a) NO_2 (b) NO_2^+
(c) NO (d) NO_2^-
(e) HNO_3. ***(Kerala MEE)***

63. The unpleasant smell of carbylamine is obtained when chloroform and alcoholic KOH are heated with any
(a) Aliphatic amine (b) Amine
(c) Primary amine (d) Aromatic amine.
(Karnataka CET)

64. Which among the following compounds will give offensive smell when heated with chloroform and alcoholic potash ?
(a) $C_2H_5NH_2$ (b) $(C_2H_5)_2NH$
(c) $(CH_3)_3N$ (d) CH_3CN
(e) $C_6H_5CONH_2$. ***(Kerala CET)***

65. Action of HNO_2 on ethylamine gives
(a) Ethane (b) C_2H_5OH
(c) Ammonia (d) Nitromethane.
(MP PET)

66. Which of the following is most basic in nature ?
(a) NH_3 (b) CH_3NH_2
(c) $(CH_3)_2NH$ (d) $C_6H_5N(CH_3)_2$. ***(AFMC)***

67. Nitrobenzene gives azoxybenzene and hydrazobenzene when reduced
(a) In acidic medium (b) In neutral medium
(c) Electrolytically (d) In alkaline medium.
(NSE)

68. An organic compound 'A' having molecular formula C_2H_3N on reduction gave another compound 'B'. Upon treatment with nitrous acid, 'B' gave ethyl alcohol. On warming with chloroform and alcoholic KOH, B formed an offensive smelling compound 'C'. The compound 'C' is
(a) $CH_3CH_2NH_2$ (b) $CH_3CH_2N \overset{\rightarrow}{=} C$
(c) $CH_3C \equiv N$ (d) CH_3CH_2OH. ***(NSE)***

2002

69. When a primary amine reacts with chloroform in ethanolic KOH, then the product is a/an
(a) Isocyanide (b) Aldehyde
(c) Cyanide (d) Alcohol. ***(AIEEE)***

70. 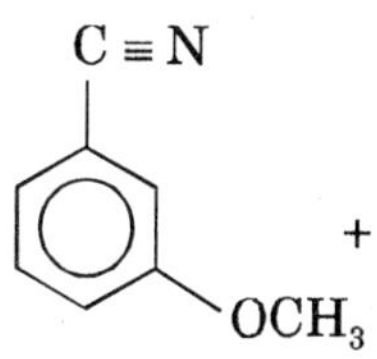

$+ CH_3MgBr \longrightarrow Q \xrightarrow{H_3O^+} P$

The product 'P' in the above reaction is

(a) 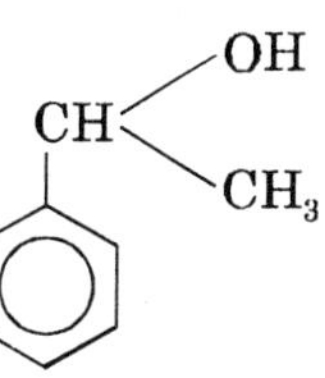

(b) 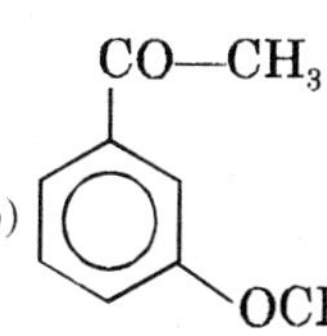

(c) 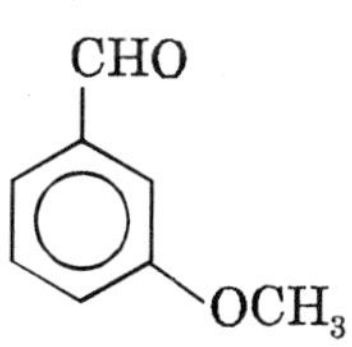

(d) 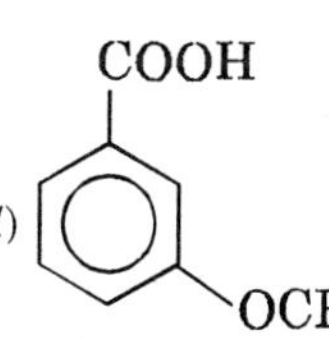

(CBSE PMT)

71. $CH_3CH_2Cl \xrightarrow{NaCN} X \xrightarrow{Ni/H_2} Y \xrightarrow{\text{acetic anhydride}} Z$

Z in the above reaction sequence is
(*a*) $CH_3CH_2CH_2NHCOCH_3$
(*b*) $CH_3CH_2CH_2NH_2$
(*c*) $CH_3CH_2CH_2CONHCH_3$
(*d*) $CH_3CH_2CH_2CONHCONH_3$. ***(CBSE PMT)***

72. Which is most basic ?
(*a*) $C_6H_5NH_2$ (*b*) $(C_6H_5)_2NH$
(*c*) CH_3NH_2 (*d*) $(CH_3)_2NH$. ***(Manipal PMT)***

73. Which of the following reagent can be used to convert benzenediazonium chloride into benzene ?
(*a*) CH_3OH (*b*) H_3PO_2
(*c*) $Br_2—H_2O$ (*d*) $LiAlH_4$. ***(BVP Pune)***

74. Nitrobenzene on electrolytic reduction in strongly acidic medium gives
(*a*) Aniline (*b*) *p*-aminophenol
(*c*) *m*-nitroaniline (*d*) Nitrosobenzene. ***(Tamil Nadu CET)***

75. The reagent that reacts with nitromethane to form methylhydroxylamine is
(*a*) Zn/HCl (*b*) Zn/NH_4Cl
(*c*) Zn/NaOH (*d*) Sn/HCl. ***(MP PET)***

76. A compound 'A' when treated with HNO_3 (in presence of H_2SO_4) gives compound 'B', which is then reduced with Sn and HCl to aniline. The compound 'A' is
(*a*) Toluene (*b*) Benzene
(*c*) Ethane (*d*) Acetamide. ***(MP PET)***

77. Reaction of aniline with benzaldehyde is
(*a*) Substitution (*b*) Addition
(*c*) Condensation (*d*) Polymerization. ***(KCET)***

78. The indicator that is obtained by coupling the diazonium salt of sulphanilic acid with N, N-dimethylaniline is
(*a*) Phenanthroline (*b*) Methyl orange
(*c*) Methyl red (*d*) Phenolphthalein
(*e*) Indigo. ***(Kerala CEE)***

2003

79. The correct order of increasing basic nature for bases $\ddot{N}H_3$, $CH_3\ddot{N}H_2$ and $(CH_3)_2\ddot{N}H$ is
(*a*) $CH_3NH_2 < NH_3 < (CH_3)_2NH$
(*b*) $(CH_3)_2NH < NH_3 < CH_3NH_2$
(*c*) $NH_3 < CH_3NH_2 < (CH_3)_2NH$
(*d*) $CH_3NH_2 < (CH_3)_2NH < NH_3$. ***(AIEEE)***

80. Ethyl isocyanide on acidic hydrolysis generates
(*a*) Ethylamine salt and methanoic acid
(*b*) Propanoic acid and ammonium salt
(*c*) Ethanoic acid and ammonium salt
(*d*) Methyl amine salt and ethanoic acid. ***(AIEEE)***

81. The reaction of chloroform with alcoholic KOH and *p*-toluidine forms
(*a*) 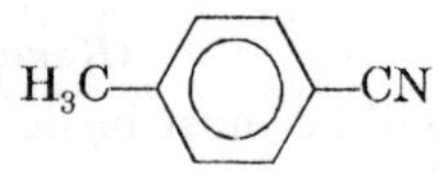 (*b*) 
(*c*) 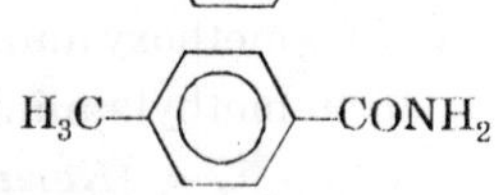 (*d*) 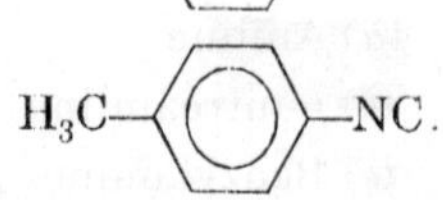 ***(AIEEE)***

82. $R—\overset{O}{\overset{\|}{C}}—R' \xrightarrow{HCN/KCN} A \xrightarrow{B} \underset{R'}{\overset{R}{>}}C\underset{CH_2NH_2}{\overset{OH}{<}}$

In the above sequence of reactions, A and B are respectively
(*a*) A is $RR'C\underset{OH}{\overset{CN}{<}}$, B is $LiAlH_4$
(*b*) A is $RR'C\underset{COOH}{\overset{OH}{<}}$, B is NH_3
(*c*) A is $RR'C\underset{OH}{\overset{CN}{<}}$, B is H_3O^+
(*d*) A is $RR'CH_2CN$, B is NaOH. ***(CBSE PMT)***

83. 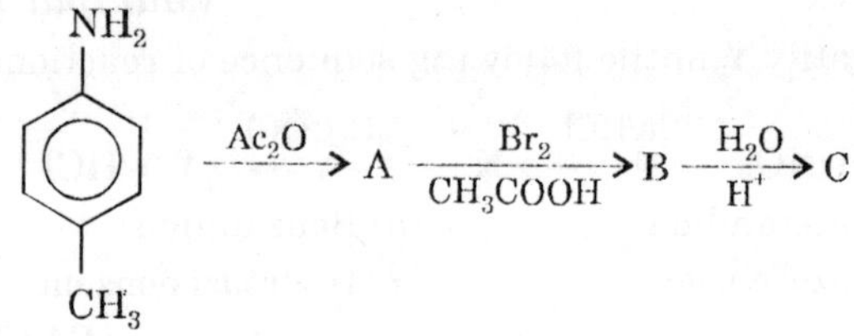

The final product 'C' in the above reaction is
(*a*) 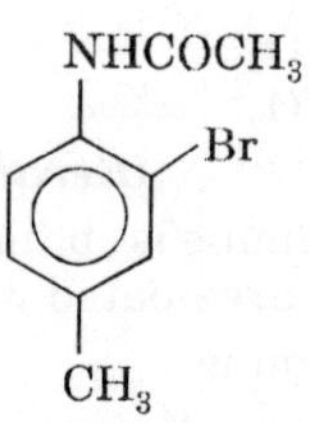 (*b*) 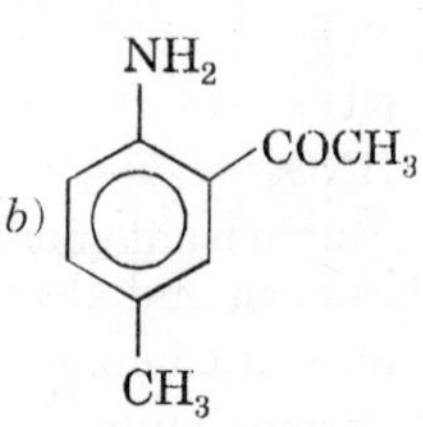
(*c*) 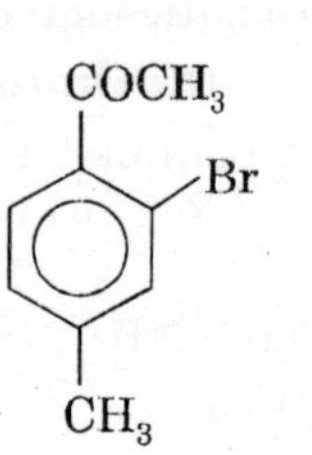 (*d*) 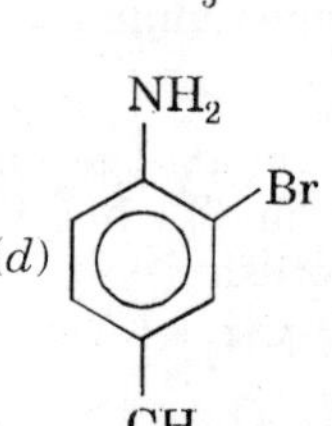
(CBSE PMT)

84. Among the following the weakest base is
(*a*) $C_6H_5CH_2NH_2$ (*b*) $C_6H_5CH_2NHCH_3$
(*c*) $O_2NCH_2NH_2$ (*d*) CH_3NHCHO. ***(AIIMS)***

85. Identify the product Z in the following reaction scheme

$$C_6H_5NH_2 \xrightarrow{AC_2O} X \xrightarrow{Br_2/CCl_4} Y \xrightarrow{HOH} Z$$

(*a*) *p*-bromoaniline (*b*) *p*-bromoacetophenone
(*c*) *p*-bromoacetanilide (*d*) *o*-bromoacetophenone
(*e*) *o*-bromoacetanilide. ***(Kerala PMT)***

2004

86. The strongest base among the following is

(*a*)  (*b*) 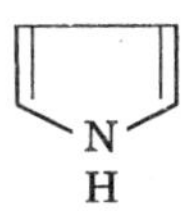

(*c*) 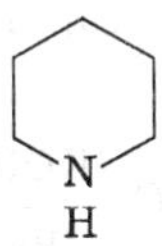 (*d*) 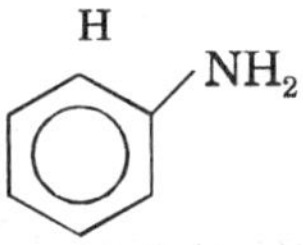

(AIIMS)

87. Aromatic nitriles are not prepared by
(*a*) ArX + KCN (*b*) ArN_2^+ + CuCN
(*c*) $ArCONH_2 + P_2O_5$ (*d*) $ArCONH_2 + SOCl_2$ ***(AIIMS)***

88. Chloropicrin is obtained by the reaction of
(*a*) Steam on carbon tetrachloride
(*b*) Nitric acid on chlorobenzene
(*c*) Chlorine on picric acid
(*d*) Nitric acid on chloroform. ***(CBSE PMT)***

89. Aniline when diazotised in cold and then treated with dimethyl aniline gives a coloured product. Its structure would be

(*a*) 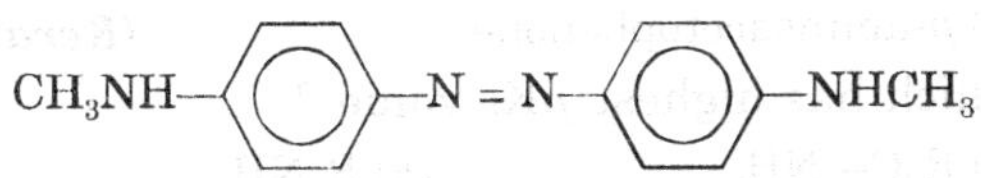

(*b*) 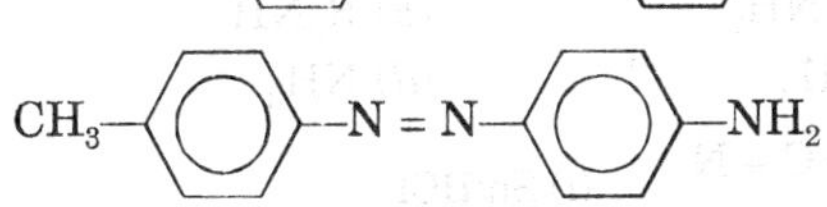

(*c*) 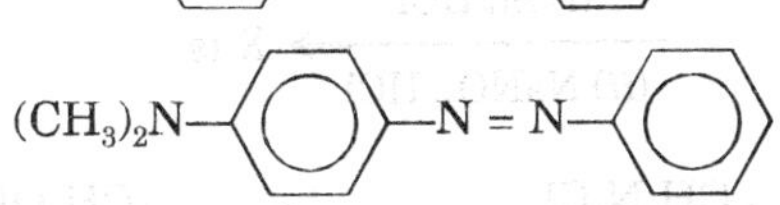

(*d*) 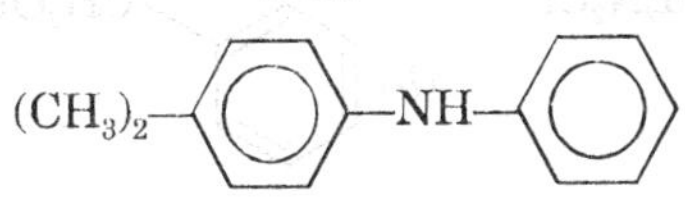

(CBSE PMT)

90. In which case formation of butanenitrile is possible ?
(*a*) C_3H_7Br + KCN (*b*) C_4H_9Br + KCN
(*c*) C_3H_7OH + KCN (*d*) C_4H_9OH + KCN ***(Orissa PMT)***

91. Which of the following is a strongest base ?

(*a*) 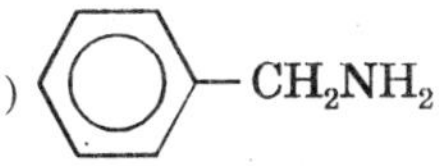 (*b*) 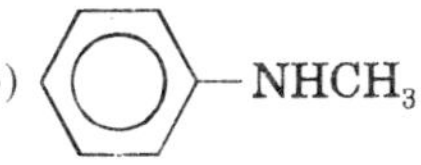

(*c*) 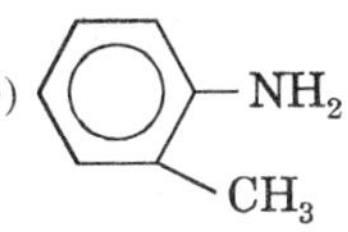 (*d*) 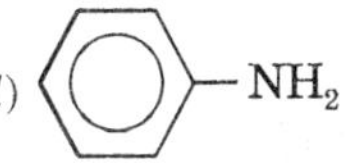

(AIEEE)

2005

92. Electrolytic red reduction of nitrobenzene in weakly acidic medium gives
(*a*) N-phenylhydroxylamine
(*b*) Nitrosobenzene
(*c*) Aniline
(*d*) *p*-Hydroxyaniline ***(CBSE PMT)***

93. Aniline in a set of reactions yielded a product D

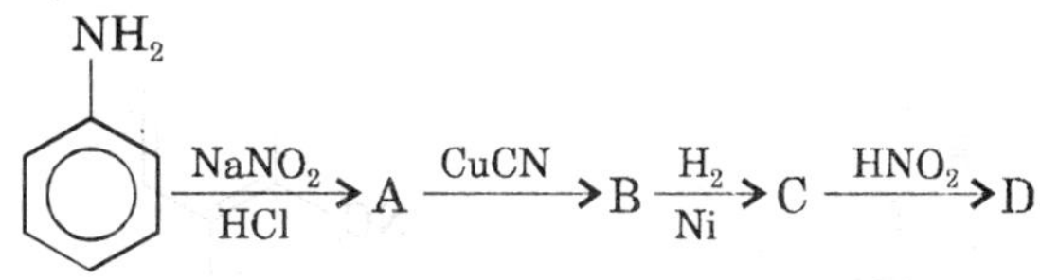

The structure of D is
(*a*) C_6H_5NHOH (*b*) $C_6H_5NHCH_2CH_3$
(*c*) $C_6H_5CH_2NH_2$ (*d*) $C_6H_5CH_2OH$. ***(CBSE PMT)***

94. Which of the following method is neither meant for separation nor for the synthesis of amines
(*a*) Curtius reaction (*b*) Wurtz reaction
(*c*) Hoffmann method (*d*) Hinsberg method. ***(AIEEE)***

95. Amongst the following the most basic compound is
(*a*) *p*-nitroaniline (*b*) acetanilide
(*c*) Aniline (*d*) Benzylamine. ***(AIEEE)***

96. Which of the following gives paracetamol on acetylation ?

(*a*) 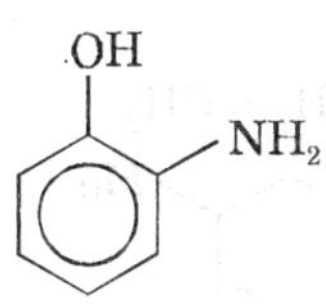 (*b*) 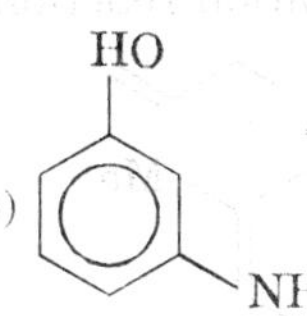

(*c*) 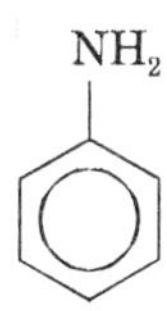 (*d*) 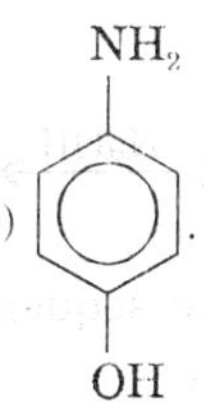

(EAMCET Medical)

97. Which of the following is obtained in carbylamine reaction ?
(*a*) $C_6H_5NH_2$ (*b*) $COCl_2$
(*c*) C_6H_5CN (*d*) C_6H_5NC.
(EAMCET Med.)

2006

98. Which of the following is more basic than aniline ?
(*a*) Benzylamine (*b*) Diphenylamine
(*c*) Triphenylamine (*d*) *p*-Nitroaniline.
(CBSE PMT)

99. 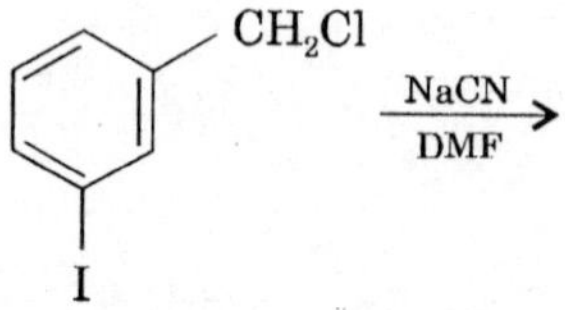

The main product in the above reaction is

(*a*) 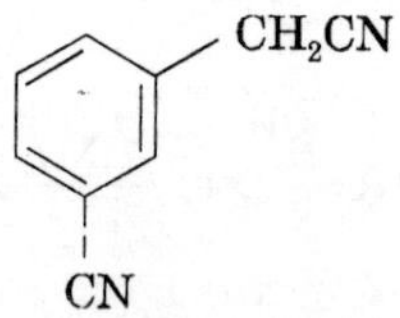

(*b*) 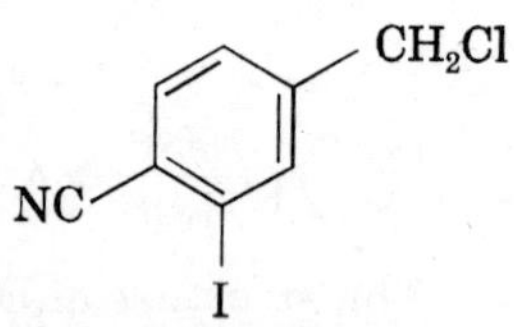

(*c*) 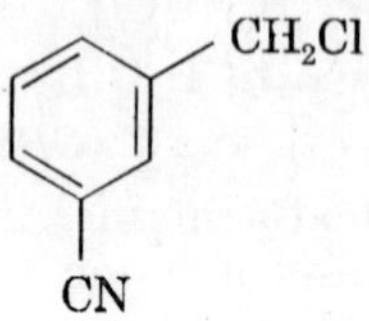

(*d*) 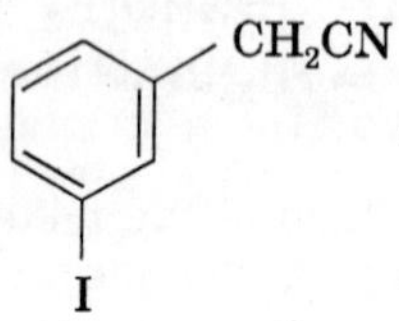

(AIEEE)

100. 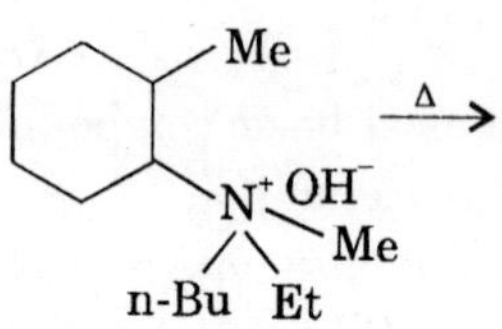

The alkene formed as a major product in the above elimination reaction is

(*a*) 

(*b*) $CH_2 = CH_2$
(*c*) 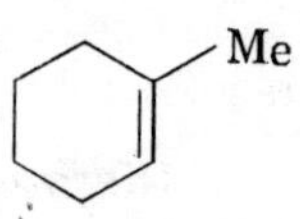

(*d*) 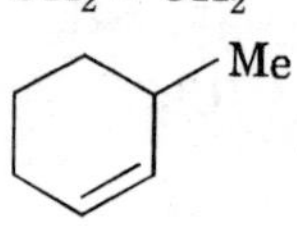

(AIEEE)

101. 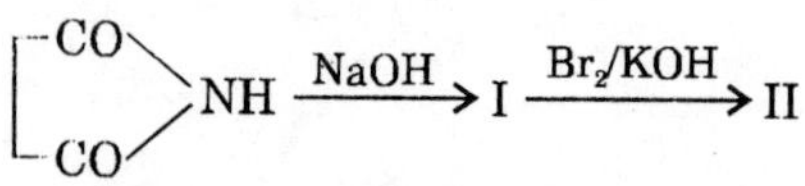

In the above sequence, II is
(*a*) β-alanine (*b*) α-alanine
(*c*) ethylenediamine (*d*) γ-aminobutyric acid.
(BHU Mains)

102. In the following reaction,

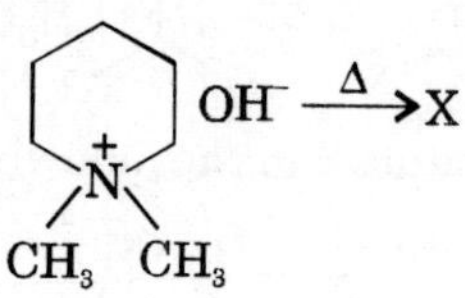

The organic product X is

(*a*) 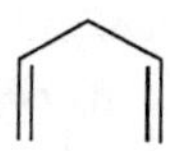
(*b*) 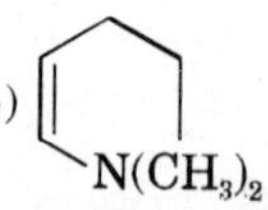

(*c*) 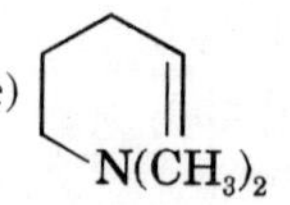

(*d*) 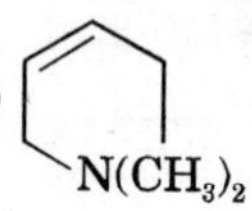

(*e*) 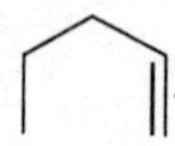
(Kerala MEE)

103. In a set of reactions given below, propanoic acid yielded a compound D.

$$CH_3CH_2COOH \xrightarrow[\text{Py}]{SOCl_2} B \xrightarrow[\text{Heat}]{NH_3} C \xrightarrow[Br_2]{KOH} D$$

The compound D is
(*a*) $CH_3CH_2NH_2$ (*b*) $CH_3CH_2CH_2NH_2$
(*c*) $CH_3CH_2CONH_2$ (*d*) $CH_3CH_2NHCH_3$.
(CBSE PMT)

104. The amine which will not liberate nitrogen on reaction with nitrous acid is
(*a*) Trimethylamine (*b*) Ethylamine
(*c*) *sec*-butylamine (*d*) *t*-butylamine
(*e*) Isopropylamine. ***(Kerala EEE)***

105. Which of the following exists as zwitter ion ?
(*a*) *p*-aminophenol (*b*) Salicylic acid
(*c*) Sulphanilic acid (*d*) Ethanolamine
(*e*) *p*-aminoacetophenone. ***(Kerala EEE)***

106. Which has highest pK_b value ?
(*a*) $R_3C—NH_2$ (*b*) R_2NH
(*c*) $R—NH_2$ (*d*) NH_3. ***(DCE)***

107. 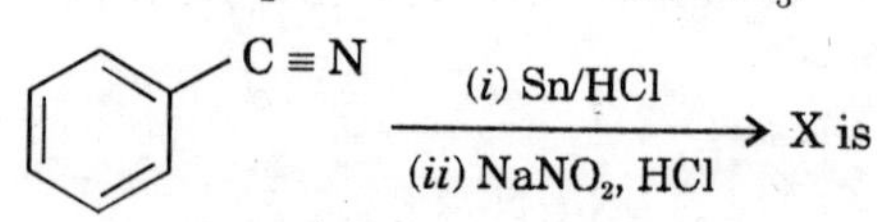

(*a*) 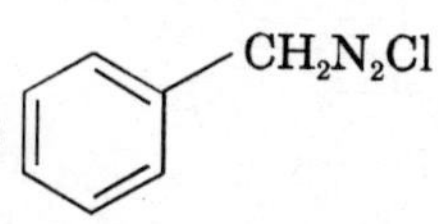

(*b*) 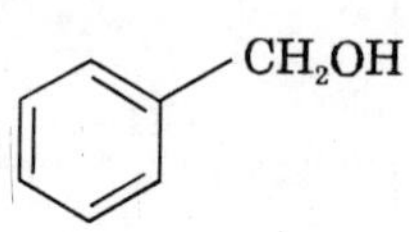

(*c*) 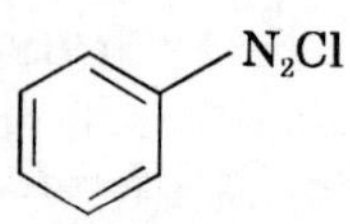

(*d*) 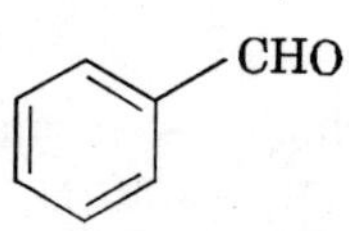

(DCE)

108. 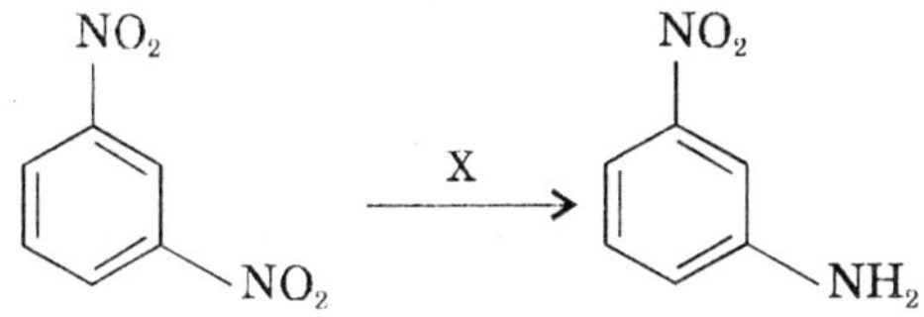 X is

(*a*) Na_2S (*b*) Sn/HCl
(*c*) $LiAlH_4$ (*d*) all of these. **(*DCE*)**

2007

109. Which one of the following on reduction with lithium aluminium hydride yields a secondary amine
(*a*) Methyl cyanide (*b*) Nitroethane
(*c*) Methyl isocyanide (*d*) Acetamide.
(*CBSE PMT*)

110. Which of the following is a strongest base in aqueous solution ?
(*a*) Dimethyl amine (*b*) Methyl amine
(*c*) Trimethyl amine (*d*) Aniline. **(*AIEEE*)**

111. In a chemiacl reaction,
$CH_3CH_2NH_2 + CHCl_3 + 3KOH \longrightarrow (A) + (B) + 3H_2O$
A and B are respectively
(*a*) C_2H_5NC and K_2CO_3 (*b*) C_2H_5NC and KCl
(*c*) C_2H_5CN and KCl (*d*) $CH_3CH_2CONH_2$ and KCl.
(*AIEEE*)

112. The presence of nitro group on benzene ring
(*a*) Deactivates the ring towards nucleophilic substitution
(*b*) Deactivates the ring towards electrophilic substitution
(*c*) Activates the ring towards electrophilic substitution
(*d*) Renders to ring basic. **(*AIEEE*)**

113. Which of the following amine can be directly oxidised to the corresponding nitro compound by potassium permenganate ?
(*a*) CH_3NH_2 (*b*) $(CH_3)_2CHNH_2$
(*c*) $(CH_3)_2NH$ (*d*) $C_6H_5NH_2$
(*e*) $(CH_3)_3C$—NH_2. **(*Kerala PMT*)**

114. Treatment of cyclobutyl methylamine with nitrous acid does not give

(*a*) 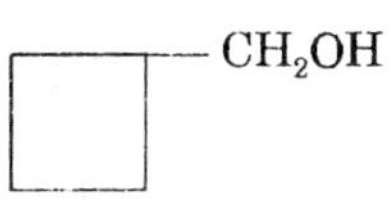

(*b*) 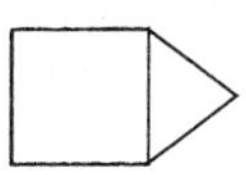

(*c*) 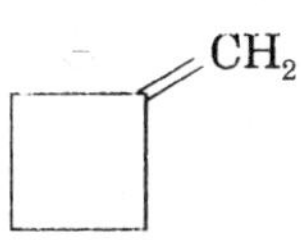

(*d*) 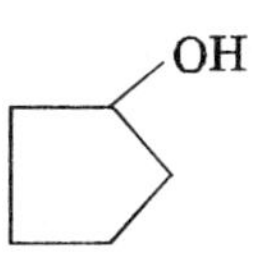

(*e*) 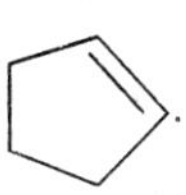

(*Kerala PMT*)

115. In the following sequence of reactions predict the products X and Y

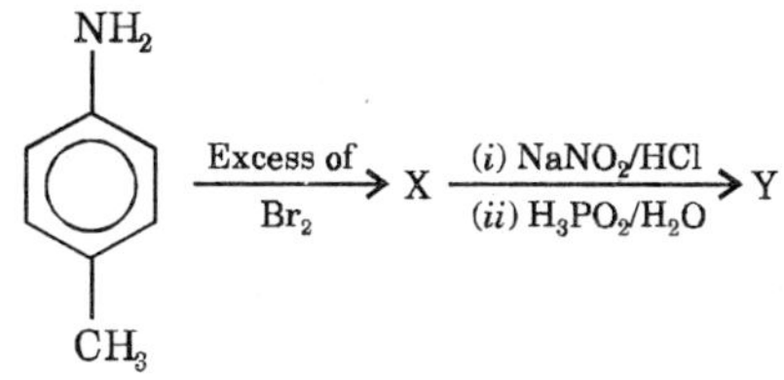

(*a*) 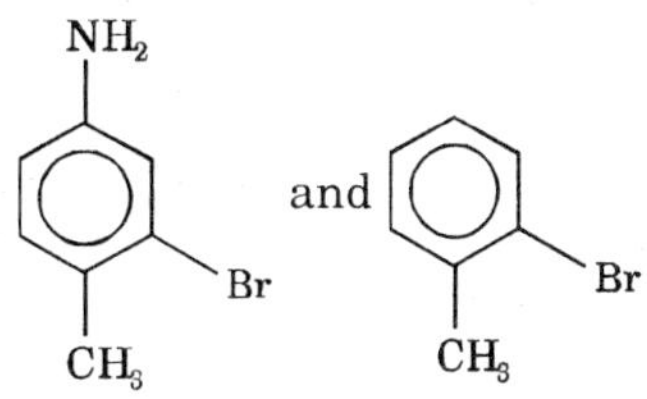

(*b*) 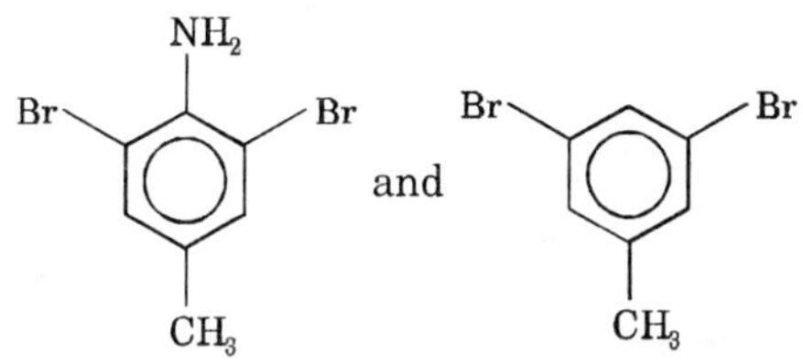

(*c*) 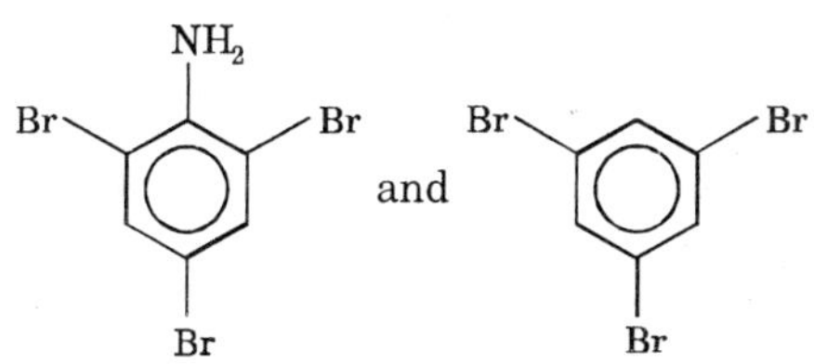

(*d*) 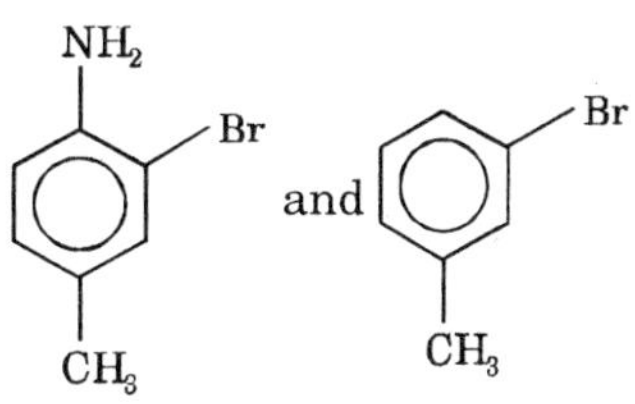

(*e*) 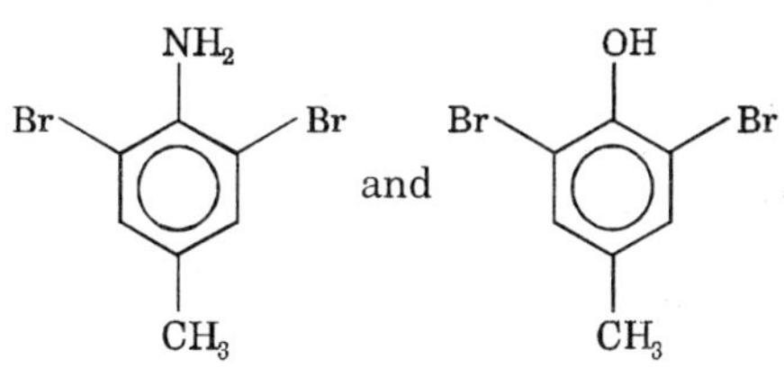

(*Kerala PMT*)

ANSWERS

1. (*b*)	**2.** (*a*)	**3.** (*c*)	**4.** (*a*)	**5.** (*a*)	**6.** (*d*)	**7.** (*a*)	**8.** (*b*)	**9.** (*a*)
10. (*a*)	**11.** (*a*)	**12.** (*b*)	**13.** (*b*)	**14.** (*b*)	**15.** (*a*)	**16.** (*d*)	**17.** (*d*)	**18.** (*c*)
19. (*b*)	**20.** (*b*)	**21.** (*a*)	**22.** (*c*)	**23.** (*a*)	**24.** (*b*)	**25.** (*b*)	**26.** (*a*)	**27.** (*b*)
28. (*b*)	**29.** (*b*)	**30.** (*c*)	**31.** (*d*)	**32.** (*a*)	**33.** (*c*)	**34.** (*d*)	**35.** (*b*)	**36.** (*b*)
37. (*c*)	**38.** (*b*)	**39.** (*d*)	**40.** (*a*)	**41.** (*c*)	**42.** (*b*)	**43.** (*a*)	**44.** (*b*)	**45.** (*d*)
46. (*d*)	**47.** (*c*)	**48.** (*b*)	**49.** (*c*)	**50.** (*d*)	**51.** (*c*)	**52.** (*a*)	**53.** (*d*)	**54.** (*a*)
55. (*a*)	**56.** (*a*)	**57.** (*e*)	**58.** (*c*)	**59.** (*d*)	**60.** (*c*)	**61.** (*b*)	**62.** (*b*)	**63.** (*c*)
64. (*a*)	**65.** (*b*)	**66** (*c*)	**67.** (*d*)	**68.** (*b*)	**69.** (*a*)	**70.** (*b*)	**71.** (*a*)	**72.** (*d*)
73. (*b*)	**74.** (*b*)	**75.** (*b*)	**76.** (*b*)	**77** (*c*)	**78.** (*b*)	**79.** (*c*)	**80.** (*a*)	**81.** (*d*)
82. (*a*)	**83.** (*d*)	**84.** (*d*)	**85.** (*a*)	**86.** (*c*)	**87.** (*a*)	**88.** (*d*)	**89.** (*c*)	**90.** (*a*)
91. (*a*)	**92.** (*c*)	**93.** (*d*)	**94.** (*b*)	**95.** (*d*)	**96.** (*d*)	**97.** (*d*)	**98.** (*a*)	**99.** (*d*)
100. (*b*)	**101.** (*a*)	**102.** (*c*)	**103.** (*a*)	**104.** (*a*)	**105.** (*c*)	**106.** (*d*)	**107.** (*b*)	**108.** (*a*)
109. (*c*)	**110.** (*a*)	**111.** (*b*)	**112.** (*b*)	**113.** (*e*)	**114.** (*b*)	**115.** (*b*).		

HINTS/SOLUTIONS

Q. No.	Ans.	Hints & Solutions
1.	(*b*)	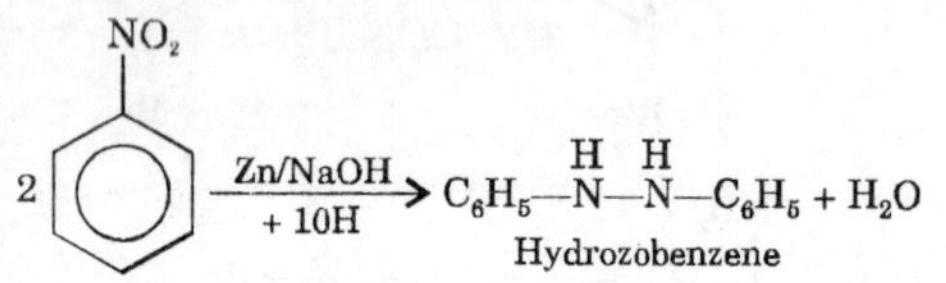
2.	(*a*)	NO_2 group is electron withdrawing and thus creates electron deficiency at *o* and *p*-positions where the incoming nucleophite (OH^-) ion attacks.
3.	(*c*)	Conc. HNO_3 oxidises aniline to *p*-benzoquinone.
4.	(*a*)	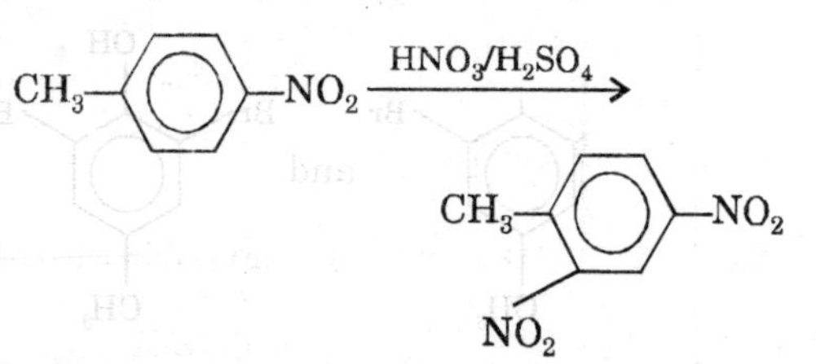 —NO_2 group is *m*-directing while CH_3-group is *o*-directing.
5.	(*a*)	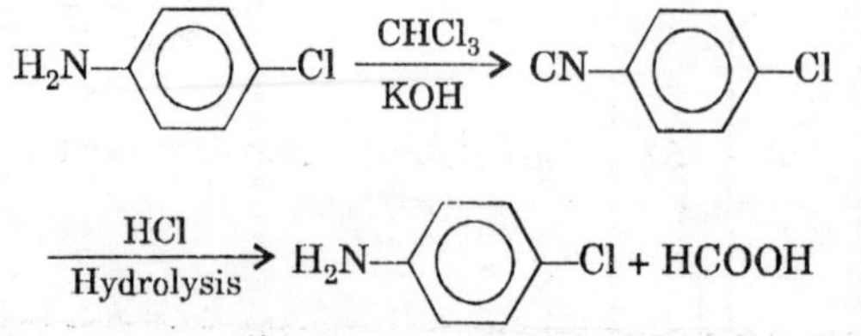

Q. No.	Ans.	Hints & Solutions
6.	(*d*)	$C_6H_5NH_2 + COCl_2 \xrightarrow{\text{Base}} C_6H_5NCO + 2HCl$
7.	(*a*)	It is a carbylamine reaction and product is alkyl or aryl isocyanides.
8.	(*b*)	$(NH_4)_2S_x$ reduces one —NO_2 to —NH_2 group. Hence the product is 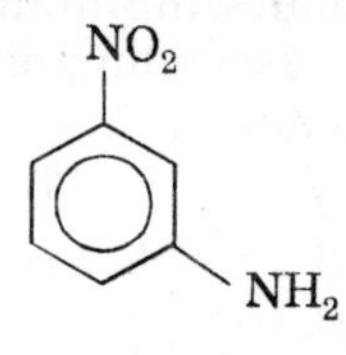
9.	(*a*)	$CH_3CN \xrightarrow{H_2O/H^+} CH_3COOH.$
10.	(*a*)	*p*-aminoazobenzene is an orange coloured dye.
11.	(*a*)	$CH_3CH = NOH \xrightarrow{P_2O_5} CH_3C \equiv N + H_2O.$
12.	(*b*)	See basic nature of amines in Concepts in Brief.
13.	(*b*)	The given reaction is called Hoffmann's mustard oil reaction.
14.	(*b*)	N and S during fusion with sodium metal change into NaCNS.
15.	(*a*)	It is due to electron withdrawing nature of NO_2 group.
16.	(*d*)	They refer to functional as well as tautomerism.

Q. No.	Ans.	Hints & Solutions
17.	(*d*)	The tautomeric forms of $RCH_2—NO_2$ are $$\underset{\text{(Nitro form)}}{R—CH_2—\overset{O}{\overset{\|\|}{N}} \rightarrow O} \rightleftharpoons \underset{\text{(Aci form)}}{R—CH = \overset{OH}{\overset{\|}{N}} \rightarrow O}$$
18.	(*c*)	Piperdiene is 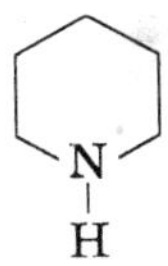
19.	(*b*)	—Cl group has – I effect. Yet it gives *o* and *p* substitution due to delocalisation of *p*-electrons on the ring.
20.	(*b*)	During nitration, NO_2^+ ion is liberated by reaction between H_2SO_4 and HNO_3 and acts as electrophile.
21.	(*a*)	Conceptual question. During diazotisation, approximately, 3 mol of acid is used per mol of amine. The break up is as follows. One mole is used to dissolve the amine, one mole librates HNO_2 from $NaNO_2$ and one mole maintains the proper acid medium which prevents the coupling of free aniline, *i.e.*, undiazotised aniline to form azo dye. Thus, choice (*a*) is most appropriate.
22.	(*c*)	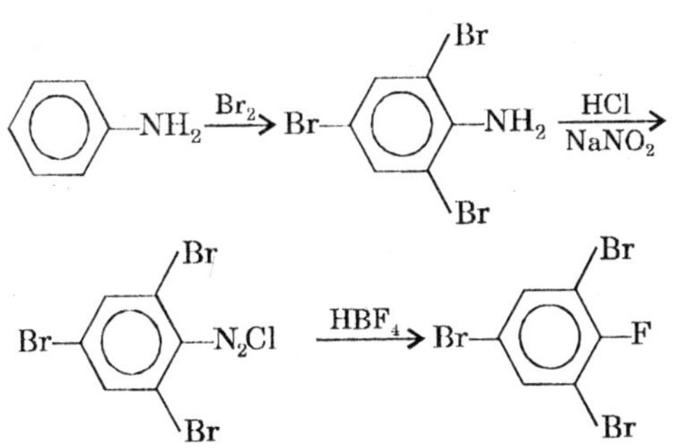
23.	(*a*)	This reaction refers to Libermann's nitroso reaction.
24.	(*c*)	Methyl cyanide hydrolyses in acidic medium to form acetic acid.
25.	(*b*)	$(CH_3)_2NH + HONO \longrightarrow \underset{\text{Nitrosoamine}}{(CH_3)_2N—NO}$.
26.	(*a*)	Dye test is given by 1° aromatic amines.
27.	(*b*)	$R—NH_2 + HNO_2 \longrightarrow R—OH + N_2 + H_2O$. Mainly alcohol is produced.
28.	(*b*)	Gabriel synthesis is used for the preparaton of 1° amine.
29.	(*b*)	Acetonitrile is CH_3CN which on hydrolysis gives acetic acid.
30.	(*c*)	$C_6H_5NO_2 \xrightarrow[4H]{Zn/(NH_4Cl)} C_6H_5NHOH + H_2O$.
31.	(*d*)	Attacking species is NO_2^+ (nitronium ion).
32.	(*a*)	$CH_3NH_2 + 3CH_3I \longrightarrow [(CH_3)_4N]^+I^- \xrightarrow[KOH\ (aq)]{Hydrolysis} [(CH_3)_4N]^+OH^-$
33.	(*c*)	$\underset{\text{glycine}}{NH_2CH_2COOH} \longrightarrow \underset{\text{(Zwitter ion)}}{\overset{+}{N}H_3CH_2COO^-}$
34.	(*d*)	$CH_3Cl \xrightarrow{KCN} CH_3CN \xrightarrow{H_3O^+} CH_3COOH \xrightarrow{NH_3} CH_3COONH_4 \xrightarrow[-H_2O]{Heat} CH_3CONH_4$
35.	(*b*)	The various isomeric forms are : 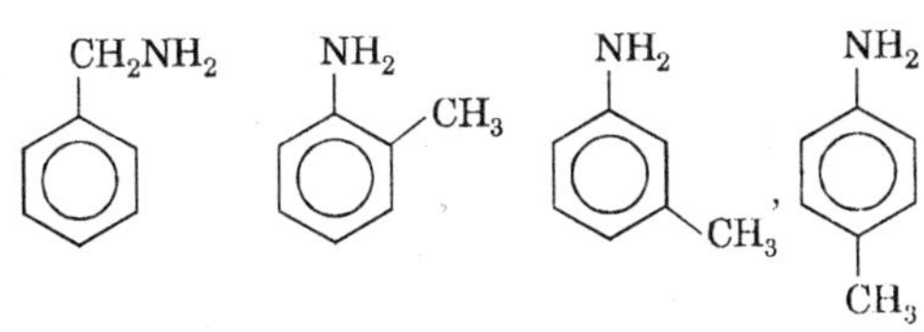 and 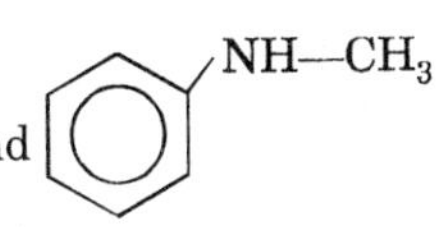
36.	(*b*)	Ethanol is not required in carbylamine reaction.
37.	(*c*)	$CH_3CN + 2H \xrightarrow[Ether]{SnCl_2/HCl} CH_3CH = NH \xrightarrow[H_2O]{Boil} \underset{\text{Acetaldehyde}}{CH_3CHO}$
38.	(*b*)	Carbylamine reaction is used in the preparation of isocyanides.

Q. No.	Ans.	Hints & Solutions
39.	(*d*)	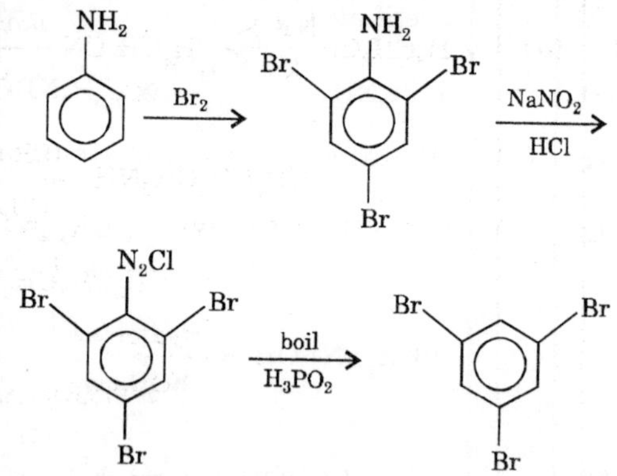
40.	(*a*)	C_6H_5NO is one of the intermediate formed during reduction of nitrobenzene.
41.	(*c*)	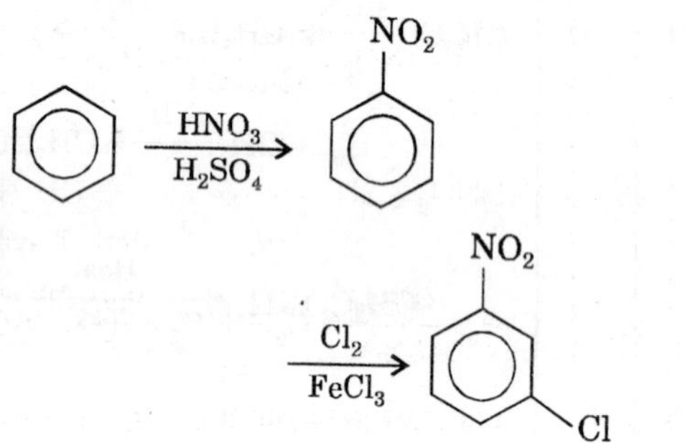
42.	(*b*)	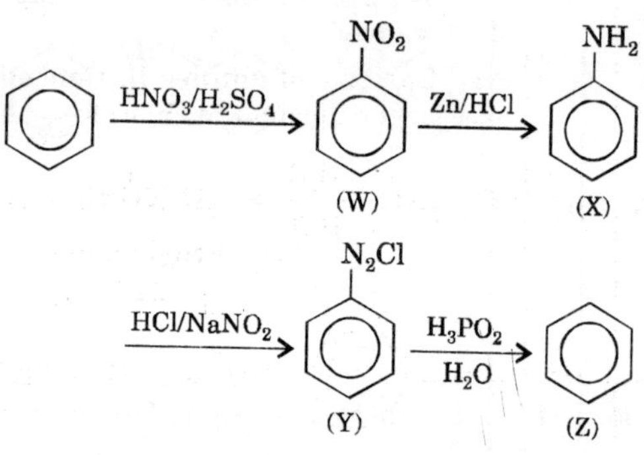
43.	(*a*)	$C_6H_5N_2Cl \xrightarrow{CuCl} C_6H_5Cl + N_2$
44.	(*b*)	1° amine gives carbylamine reaction but 2° amine donot.
45.	(*d*)	The reaction is called Sandmeyer's reaction.
46.	(*d*)	$CH_3CN \xrightarrow{Na/EtOH} CH_3CH_2NH_2 \xrightarrow{HNO_2} CH_3CH_2OH \xrightarrow{Cu/575\ K} CH_3CHO$ Ethanal
47.	(*c*)	$(CH_3)_3N$ does not undergo acylation due to absence of H on N atom.
48.	(*b*)	Once dinitrobenzene is formed further nitration does not occur due to strong deactivating effect of two NO_2 groups.
49.	(*c*)	$C_6H_5NO_2 \xrightarrow{Sn/HCl} C_6H_5\overset{+}{N}H_3\overset{-}{Cl}$ salt formation occurs after the reduction due to acidic medium.
50.	(*d*)	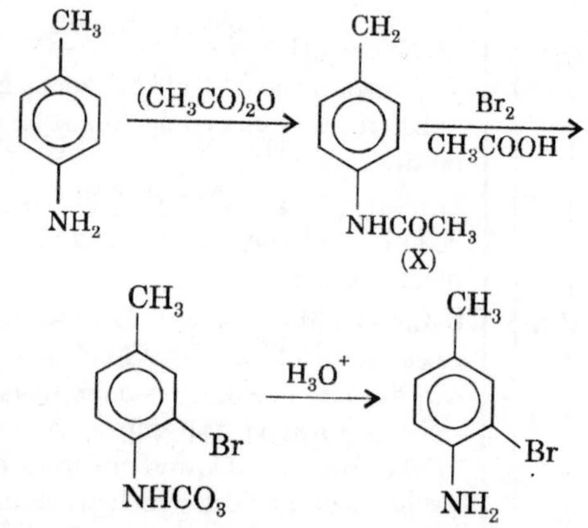
51.	(*c*)	Hinsberg's reagent is $C_6H_5SO_2Cl$.
52.	(*a*)	Aniline reacts with acetyl chloride to give acetanilide.
53.	(*d*)	C_3H_9N cannot represent a quaternary salt.
54.	(*a*)	$C_6H_5NO_2 \xrightarrow{Redn.} C_6H_5NH_2 \xrightarrow[CHCl_3]{KOH} C_6H_5NC.$
55.	(*a*)	Carbylamine reaction is not given by secondary amines.
56.	(*a*)	Gaberiel's synthesis is used for the preparation of 1° amine.
57.	(*e*)	Benzylamine is aralkylamine and hence it is more basic than corresponding arylamines.
58.	(*c*)	$CH_3CONH_2 \xrightarrow{NaOBr} CH_3NH_2$. This is Hoffmann bromamide reaction.
59.	(*d*)	Nitrolic acid dissolves in NaOH to give red coloured solution.
60.	(*c*)	$(CH_3)_4N^+OH^- \xrightarrow{\Delta} (CH_3)_3N + CH_3OH$
61.	(*b*)	$C_6H_5NO_2 \xrightarrow{Sn/HCl} \underset{(X)}{C_6H_5NH_2} \xrightarrow[-HCl]{C_6H_5COCl} \underset{\text{Benzanilide (Y)}}{C_6H_5NHCOC_6H_5}$

Q. No.	Ans.	Hints & Solutions
62.	(*b*)	Nitration of arenes is initiated by electrophilic attack of nitronium ion (NO_2^+).
63.	(*c*)	1° Amine give carbylamine reaction.
64.	(*a*)	$C_2H_5NH_2$ being 1° amine gives carbylamine reaction.
65.	(*b*)	$C_2H_5NH_2 + HNO_2 \longrightarrow C_2H_5OH + N_2 + H_2O$.
66.	(*c*)	Among the given species dimethylamine is most basic.
67.	(*d*)	$C_6H_5NO_2 + 6H \xrightarrow{\text{Glucose + NaOH}} C_6H_5-\overset{\overset{O}{\uparrow}}{N}=N-C_6H_5 + 3H_2O$ Azoxybenzene
68.	(*b*)	$\underset{(A)}{CH_3C\equiv N} \xrightarrow{\text{Reduction}} \underset{(B)}{CH_3CH_2NH_2} \xrightarrow{HNO_2} \underset{\text{Ethanol}}{CH_3CH_2OH}$ $\underset{(B)}{CH_3CH_2NH_2} \xrightarrow{CHCl_3/\,KOH} \underset{(C)}{CH_3CH_2NC}$
69.	(*a*)	The product of this reaction is isocyanide.
70.	(*b*)	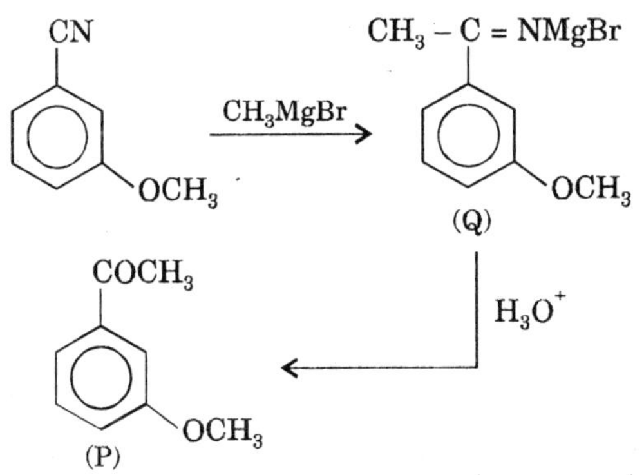
71.	(*a*)	$CH_3CH_2Cl \xrightarrow{NaCN} \underset{(X)}{CH_3CH_2CN} \xrightarrow{\textit{Raney}\ Ni/H_2}$ $\underset{(Y)}{CH_3CH_2CH_2NH_2} \xrightarrow[-CH_3COOH]{(CH_3CO)_2O}$ $CH_3CH_2CH_2NH-COCH_3$ *n*-Propylacetamide (Z)
72.	(*d*)	Refer discussion on basicity of amines.
73.	(*b*)	H_3PO_2 converts benzenediazonium chloride into benzene.
74.	(*b*)	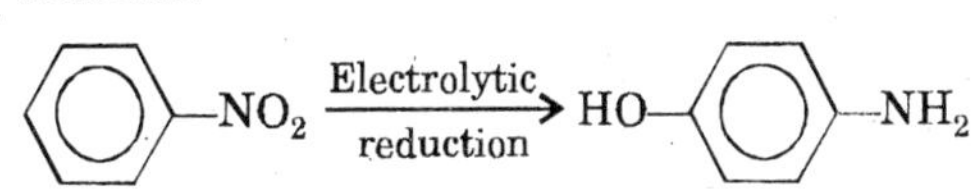
75.	(*b*)	$CH_3NO_2 + 2H \xrightarrow{Zn/NH_4Cl} CH_3NHOH$ | H_2/Ni Methyl hydroxylamine
76.	(*b*)	$\underset{(A)}{C_6H_6} \xrightarrow[H_2SO_4]{HNO_3} \underset{(B)}{C_6H_5NO_2} \xrightarrow{Sn/HCl} \underset{(C)}{C_6H_5NH_2}$
77.	(*c*)	This reaction is condensation reaction.
78.	(*b*)	See coupling reactions. The product will be methyl orange.
79.	(*c*)	See Basicity of amines in Concepts in Brief of benzene diazonium chloride.
80.	(*a*)	$C_2H_5NC \xrightarrow[HCl]{H_2O} C_2H_5\overset{+}{N}H_3Cl^- + HCOOH$ Ethyl amine salt
81.	(*d*)	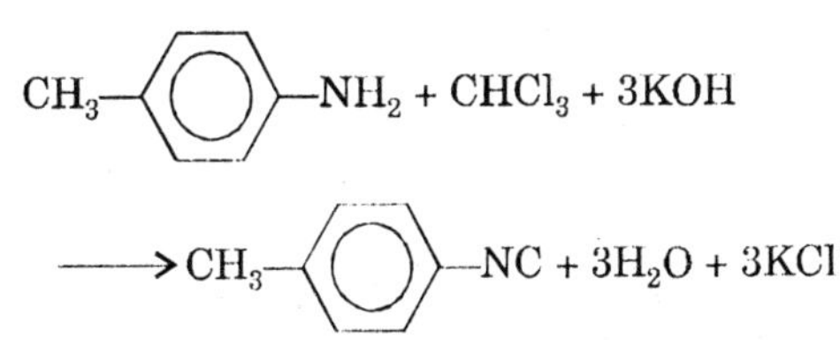
82.	(*a*)	$\begin{matrix}R\\R'\end{matrix}\!\!>C=O \xrightarrow[KCN]{HCN} \begin{matrix}R\\R'\end{matrix}\!\!>C<\!\!\begin{matrix}CN\\OH\end{matrix}$ $\xrightarrow[LAH]{\text{Reduction}} \begin{matrix}R\\R'\end{matrix}\!\!>C<\!\!\begin{matrix}CH_2NH_2\\OH\end{matrix}$

Q. No.	Ans.	Hints & Solutions
83.	(*d*)	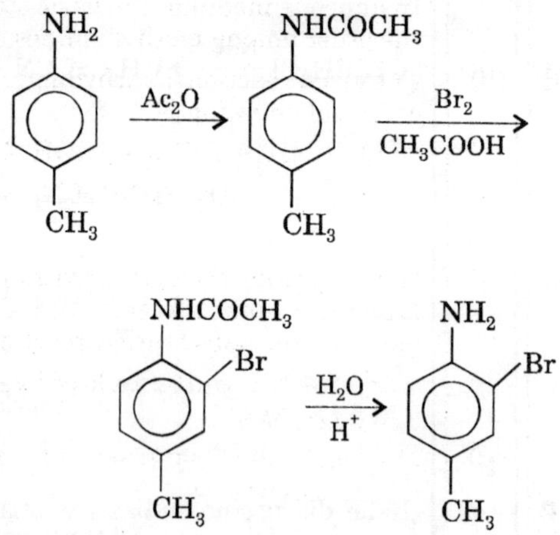
84.	(*d*)	Electron withdrawing effect of C = O group causes base weakening effect in CH_3NHCHO.
85.	(*a*)	Refer bromination of Aniline.
86.	(*c*)	The strongest base among the given species is 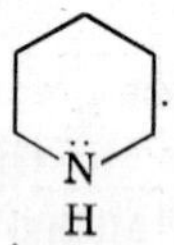. In other species the lone pair of N atom is not easily available for protonation because of their delocalisation on ring.
87.	(*a*)	Aryl halide does not react with KCN directly to produce ArCN.
88.	(*d*)	$Cl_3C—H + HONO_2 \longrightarrow \underset{\text{Chloropicrin}}{Cl_3C—NO_2} + H_2O$
89.	(*c*)	$C_6H_5NH_2$ gets converted to benzenediazonium salt on diazotisation which further reacts with dimethyl aniline as 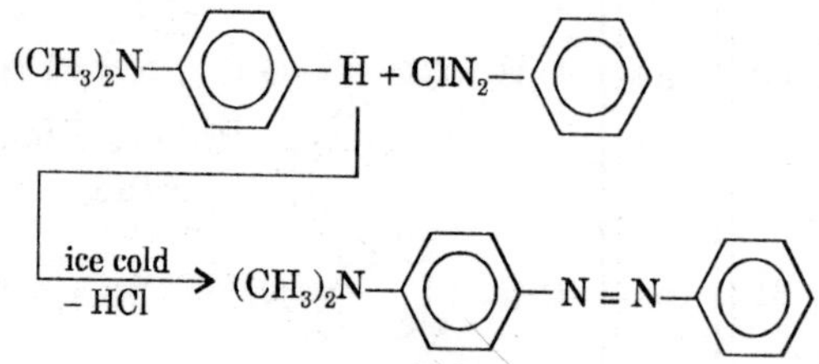 (A dye)
90.	(*a*)	$C_3H_7Br + KCN(alc.) \longrightarrow \underset{\text{Butanenitrile}}{C_3H_7CN} + KBr$
91.	(*a*)	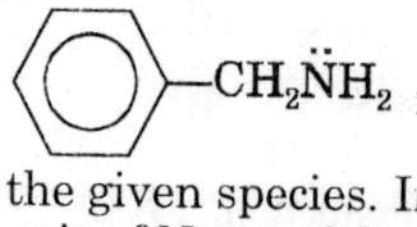 is a strongest base among the given species. In all other species, the lone pair of N gets delocalised over the ring and its availability to a proton among decreases.
92.	(*c*)	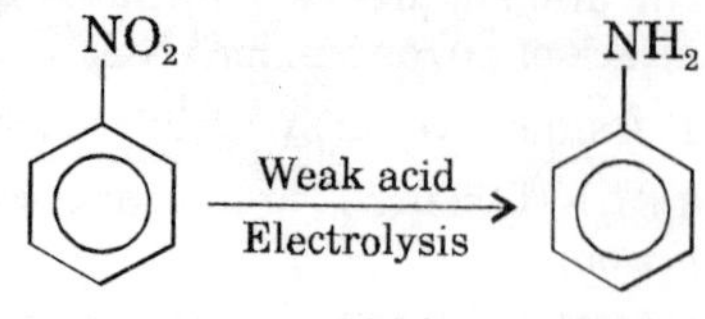 Aniline
93.	(*d*)	$C_6H_5NH_2 \xrightarrow[HCl]{NaNO_2} \underset{(A)}{C_6H_5N_2Cl} \xrightarrow{CuCN} \underset{(B)}{C_6H_5CN}$ $\xrightarrow{H_2/Ni} \underset{(C)}{C_6H_5CH_2NH} \xrightarrow{HNO_2} \underset{(D)}{C_6H_5CH_2OH}$
94.	(*b*)	Wurtz reaction is neither used for synthesis nor for the separation of amines.
95.	(*d*)	Among the given compounds, $C_6H_5CH_2NH_2$ is most basic because in this the pair of electron on nitrogen is most easily available.
96.	(*d*)	Paracetamol is 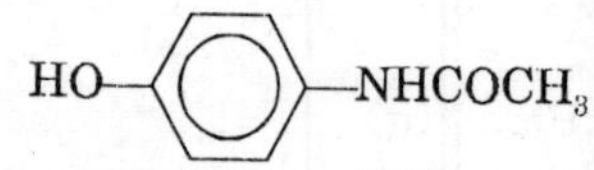 which is obtained by acetylation of—NH_2 group of *p*-aminophenol.
97.	(*d*)	Carbylamine reaction involves the production of phenylisocyanide.
98.	(*a*)	Among the given amines only benzyl amine is more basic than aniline. It is due to the absence of conjugation of lone pair of N atom.
99.	(*d*)	NaCN replaces chloro group from the side-chain following SN^2 mechanism. Iodo group is directly bonded to aromatic system and is therefore not replaced under these conditions.
100.	(*b*)	The heating involves Hoffmann elimination giving least substituted alkene along with 3° amine. The products are $CH_2 = CH_2$ and 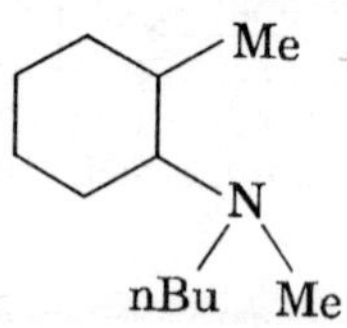
101.	(*a*)	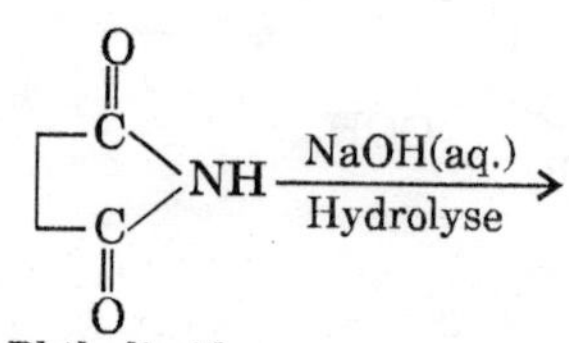 Phthalimide $HOOC—CH_2—CH_2—CONH_2 \xrightarrow{Br_2/KOH} \underset{\beta\text{-Alanine}}{HOOC—CH_2—CH_2—NH_2}$

Q. No.	Ans.	Hints & Solutions
102.	(c)	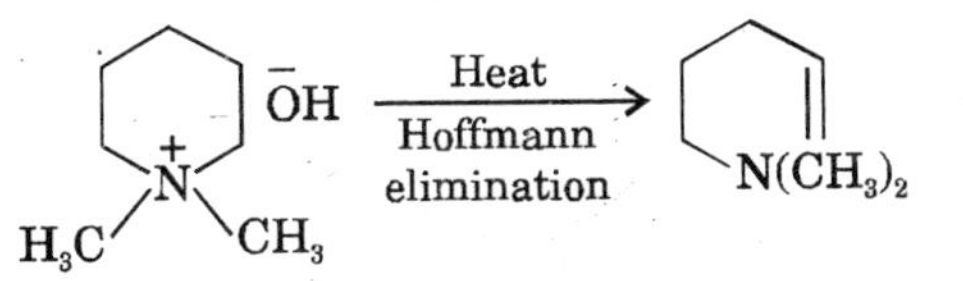
103.	(a)	$CH_3CH_2COOH \xrightarrow[Py]{SOCl_2} CH_3CH_2COCl \xrightarrow[Heat]{NH_3} CH_3CH_2CONH_2 \xrightarrow[Br_2]{KOH} CH_3CH_2NH_2$
104.	(a)	Only 1° Amine on reaction with HNO_2 give N_2 gas, 2° and 3° amines donot liberate N_2 gas by reaction with HNO_2
105.	(c)	Sulphanilic acid exists as internal salt structure, 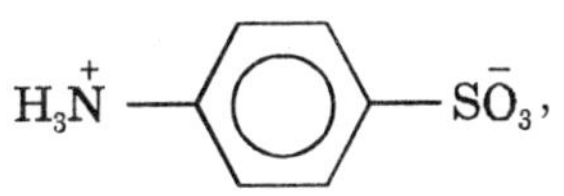 also called Zwitter ion.
106.	(d)	NH_3 is weakest base among these. Hence, it has highest pK_b value.
107.	(b)	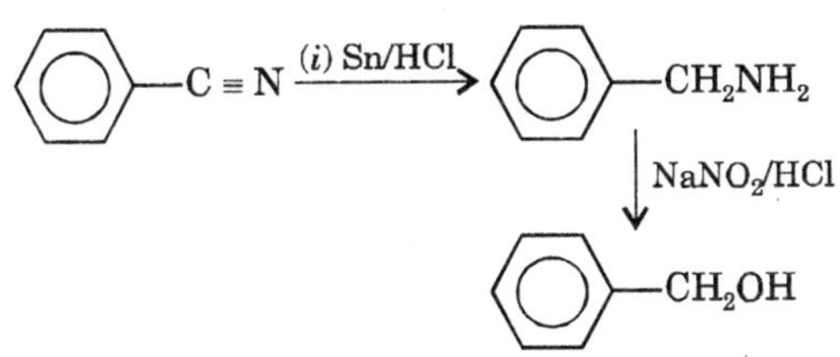
108.	(a)	Selective reduction is carried out by using Na_2S when one of the NO_2 group is reduced to —NH_2.
109.	(c)	$CH_3N \rightleftarrows C \xrightarrow{LiAlH_4} CH_3NH—CH_3$. 2° Amine
110.	(a)	Dimethyl amine is a strongest base in aqueous medium because of resultant of + I effect, crowding effect and hydration effect.
111.	(b)	In aqueous medium, the basic strength follows the order among methyl amines as 2° > 1° > 3°. The given reaction is carbylamine reaction. The chemical equation is $CH_3CH_2NH_2 + CHCl_3 + 3KOH \longrightarrow \underset{(A)}{CH_3CH_2NC} + \underset{(B)}{3KCl} + 3H_2O$
112.	(b)	NO_2— group is electron withdrawing group. Hence it deactivates, the ring towards electrophilic substitution reactions.
113.	(e)	$(CH_3)_3C—NH_2$ can be oxidised by $KMnO_4$ to form $(CH_3)_3C—NO_2$.
114.	(b)	During the course of reaction of 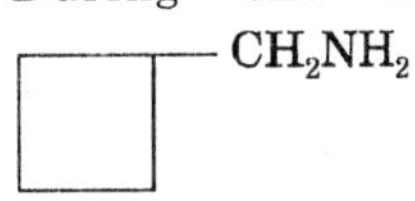 and HNO_2, the intermediate carbocation formed is 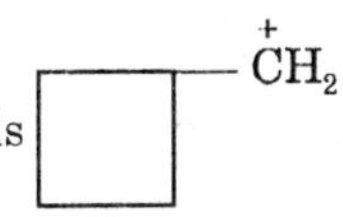. It can rearrange and form all the products following substitution or elimination processes as given in choices *a*, *c*, *d*, *e* but 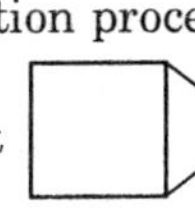 cannot be formed.
115.	(b)	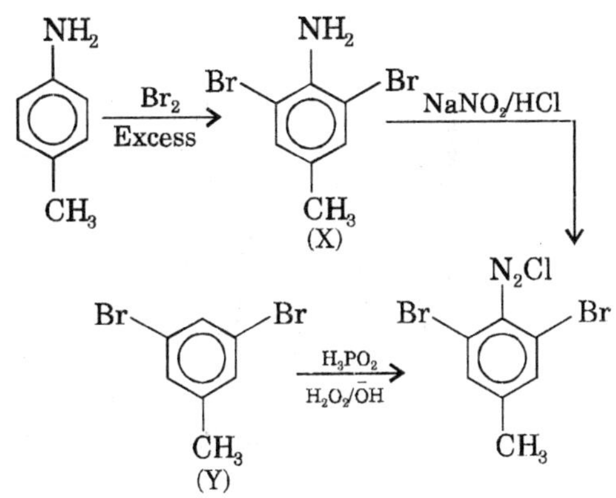

7.6. MCQs FOR IIT JEE

Brain Twister

A. Multiple Choice Questions With One Correct Answer

1. *p*-toluidine is treated with an ice cold solution of sodium nitrite and hydrochloric acid. The crystalline ppt. so formed are boiled with water. The resulting product is

(*a*) Toluic acid (*b*) Phenol
(*c*) *p*-cresol (*d*) *p*-nitro toluene.

2. Match the compounds,

(A) 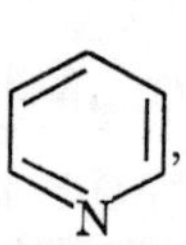, (B) 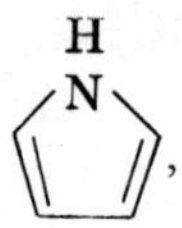, (C) 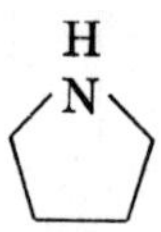

with values of dissociation constants, (1) $K_b \sim 10^{-4}$ (2) $K_b \sim 10^{-3}$ and (3) $K_b \sim 10^{-9}$. The correct matching are

(*a*) A – 1, B – 2, C – 3 (*b*) A – 3, B – 1, C – 2
(*c*) A – 3, B – 2, C – 1 (*d*) A – 2, B – 3, C – 1.

3. Among the compounds given below which one has five hydrogen atoms, sp^2 hybridised nitrogen atom and a planar structure

(*a*)  (*b*) 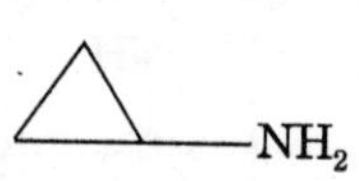

(*c*) 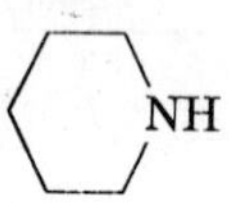 (*d*) 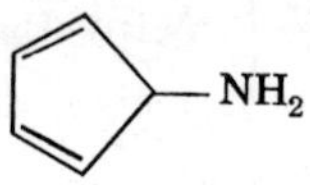

4. In order to prepare 1° amine from 1° alcohol (RCH_2OH) a student offers three routes

RCH_2OH
- 1. (*i*) P/Br_2, (*ii*) KCN, (*iii*) H_2/Ni
- 2. (*i*) Cu/Δ, (*ii*) NH_3/Heat (*iii*) H_2/Ni
- 3. (*i*) $KMnO_4$, (*ii*) $SOCl_2$ (*iii*) NH_3/Heat (*iv*) $\bar{O}Br$

which path should be followed so that the number of carbons in amine and alcohol remain same ?

(*a*) 1and 3 (*b*) only 2
(*c*) only 3 (*d*) 2 and 3.

5. The correct route for the change given in the box is

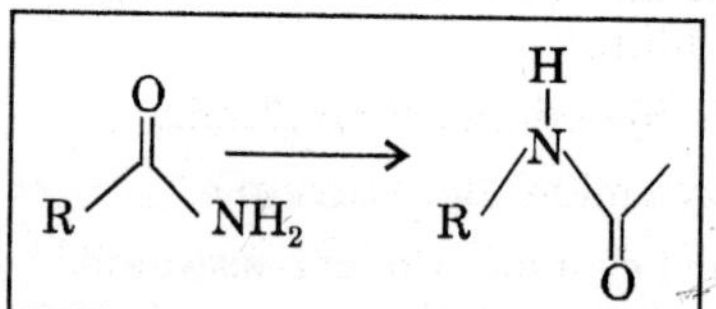

(*a*) (*i*) Br_2/KOH, (*ii*) H_2/Pt, (*iii*) CH_3COOH
(*b*) (*i*) P_2O_5/Heat, (*ii*) H_2/Pt, (*iii*) Ac_2O
(*c*) (*i*) Ac_2O, (*ii*) H_2/Pt, (*iii*) Br_2/KOH
(*d*) (*i*) Br_2/KOH, (*ii*) CH_3COCl.

6. $R-N \overset{\rightarrow}{=} C + HgO \longrightarrow$ **A** + Hg_2O. In the above sequence A is

(*a*) 1° Amine (*b*) 2° Amide
(*c*) Alkylcyanate (*d*) Alkyl isocyanate.

7. *Nitromethane* $\xrightarrow{575\text{ K}}$ A + B + C. In this reaction, the products are

(*a*) $CH_4 + N_2 + CO$ (*b*) $CO_2 + \frac{3}{2}H_2 + \frac{1}{2}N_2$
(*c*) $C_2H_6 + N_2 + H_2$ (*d*) $CO_2 + H_2 + N_2$.

8. Aniline + $COCl_2 \longrightarrow$? The products of the reaction are

(*a*) Phenyl cyanate (*b*) Phenyl isocyanate
(*c*) Phenyl cyanide (*d*) Phenyl isocyanide.

9. $C_2H_5NH_2 + NOCl \longrightarrow H_2O + A + B$

A and B in the above reaction are

(*a*) N_2 and C_2H_5Cl (*b*) NH_3 and C_2H_5OCl
(*c*) $(C_2H_5)_2N$ and N_2 (*d*) C_2H_4 and N_2.

10. A student plans to prepare *cyclohexylamine* from *cyclohexanol* by following the route (*i*) PBr_3 (*ii*) NH_3. But he did not get desired yield. What side products he could have got ?

(*a*) 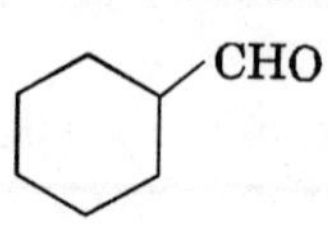 (*b*) 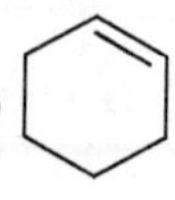

(*c*) 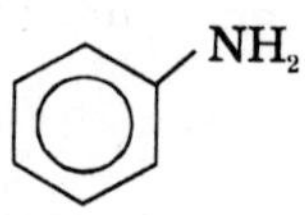 (*d*) 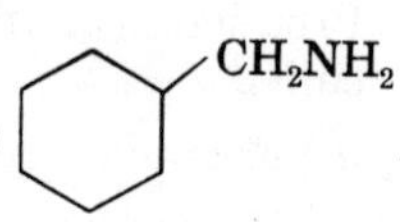

11. Which of the following amine can exhibit optical activity ?

(*a*) $CH_3N—H$ with C_2H_5 on N (*b*) CH_3CHNH_2 with C_2H_5 on CH

(c) $(CH_3)_2CNH_2$ with C_2H_5 on C   (d) $CH_3—N—C_3H_7$ with C_2H_5 on N

12. 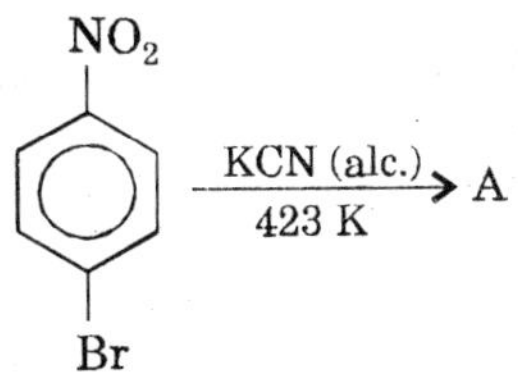

The product A in the above reaction is

(a) 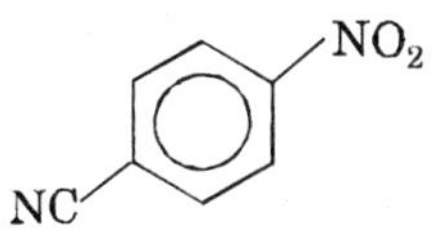 (b) 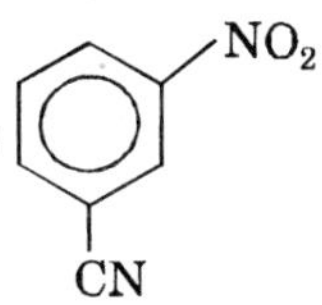

(c) 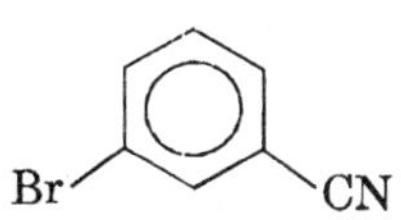 (d) 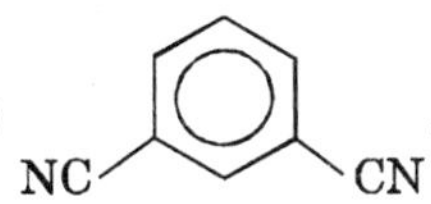

13. Which of the following does not produce salt with hydrochloric acid ?

(a) Ethylamine (b) Aniline

(c) Acetamide (d) Phenol.

14. What is true about Aniline ?

(a) p-electrons of N atom are delocalised over the ring

(b) It is a stronger base than benzyl amine

(c) It is a strong base than cyclohexyl amine

(d) It is a weaker base than benzamide.

15. Amatol is one of the blasting material which is obtained by mixing TNT with

(a) NH_4Cl (b) NH_4Br

(c) NH_4NO_3 (d) $(NH_4)_2SO_4$.

16. Reduction of benzenediazonium chloride with sodium sulphite produces

(a) 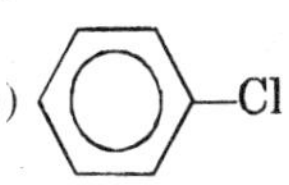 (b) 

(c) $Ph—NH—\overset{+}{N}H_3\overset{-}{Cl}$ (d) 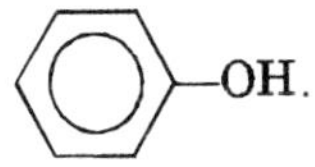

17. Production of amines by ammonia and alkylhalides is called

(a) Frankland reaction

(b) Hoffmann ammonolysis

(c) Hoffmann's mustard oil reaction

(d) Hoffmann's bromamide reaction.

18. The formula of dimethyldiazene is

(a) $CH_3NH—NH—CH_3$ (b) $CH_3N = NCH_3$

(c) $(CH_3)_2C = NH$ (d) $(CH_3)_2N—N(CH_3)_2$.

19. Which of the following cannot show tautomerism ?

(a) 2-Methyl-2-nitropropane

(b) 2-Nitropropane

(c) 1-Nitropropane

(d) Vinyl alcohol.

20. The structural formula of the compound **A**, $(C_6H_{11}N)$ which is optically active, dissolves in dilute aqueous HCl and also produces N_2 on reaction with nitrous acid is

(a) 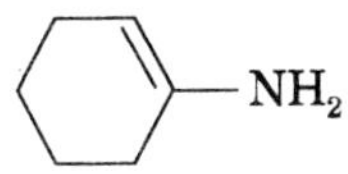 (b) 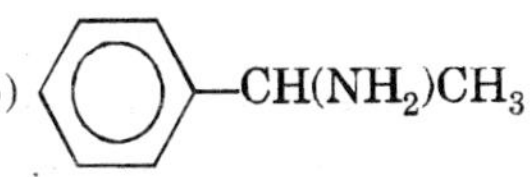

(c) 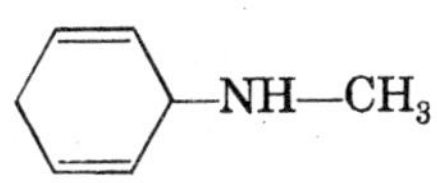 (d) 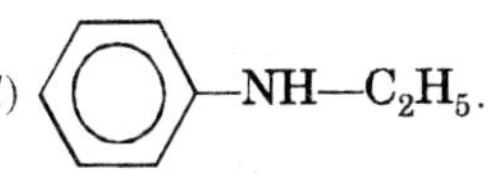

21. Which pair contains reagents each of which can convert acetamide into primary amine ?

(a) PCl_5, H_2/Pt (b) P_2O_5, Br_2/KOH

(c) H_2/Pt, P_2O_5 (d) Br_2/KOH, H_2/Pt.

22. Which of the following pair of amines will give carbylamine reaction ?

(a) Dimethylaniline, ethyl amine

(b) Dimethylamine, Methanamine

(c) p-toluidine, aniline

(d) Aniline, Dimethylamine.

23. Which of the following amines can undergo chemical reaction with C_2H_5Cl ?

(a) Methylamine

(b) Dimethylamine

(c) Tertralkyl ammonium chloride

(d) Both (a) and (b).

24. Which of the following reaction will not give primary amine ?

(a) $CH_3CN \xrightarrow{LiAlH_4}$ (b) $CH_3NC \xrightarrow{LiAlH_4}$

(c) $CH_3NO_2 \xrightarrow{Sn/HCl}$ (d) $C_2H_5CN \xrightarrow{LiAlH_4}$

25. The products of reaction of alcoholic silver nitrite with ethyl bromide are

(a) Ethane, Ethene (b) Ethene, Ethyl nitrite

(c) Ethyl nitrite, CH_3NO_2 (d) Nitroethane, C_2H_5–ONO.

26. Which of the following pair of compounds are isomers of o-toluidine ?

(a) N-Methylaniline, Benzylamine

(b) Benzylamine, Benzenamine

(c) Phenyl cyanide, Phenyl isocyanide

(d) N, N-Dimethyl aniline, N-Methylaniline.

27. Benzene diazonium chloride forms orange red dye with

(*a*) Phenol and β-Naphthol (*b*) *p*-cresol and ethanol

(*c*) β-Naphthol and ethanol (*d*) Ethanol and phenol.

28 Which of the following is less reactive towards electrophilic substitution as compared to benzene ?

(*a*) Nitrobenzene (*b*) Aniline

(*c*) Phenol (*d*) Toluene.

29. Which of the following pair of compounds are less basic than $C_2H_5NH_2$?

(*a*) Pyridine and Et_2 NH_2

(*b*) Diethylamine and $C_6H_5NH_2$

(*c*) Aniline and pyridine

(*d*) Ammonia and Et_2 NH.

30. Ammonia and phthalic acid react to produce

(*a*) Phthalamide (*b*) *o*-Amino benzoic acid

(*c*) *o*-phenylenediamine (*d*) Phthalimide.

31. The oxidation of which of the following compounds will produce *p*-benzoquinone.

(*a*) Terephthalic acid (*b*) *p*-Xylene

(*c*) Aniline (*d*) Nitrobenzene.

32. $\underset{\textbf{(A)}}{\text{Hydrazobenzene}} \xrightarrow{NaIO_3} \textbf{(X)} \xrightarrow{CH_3CO_3H} \textbf{(Y)}.$

Both **X** and **Y** on reduction with Sn/HCl give **Z**. Which of the following does not represent X, Y or Z.

(*a*) Azobenzene (*b*) Phenol

(*c*) Aniline (*d*) Azoxybenzene.

33. The intermediate stages during the reduction of nitrobenzene to aniline are

(*a*) 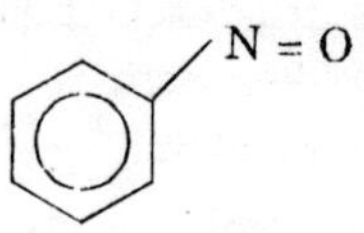

(*b*) 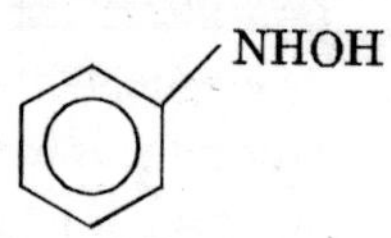

(*c*) $C_6H_5NH—NHC_6H_5$ (*d*) both '*a*' and '*b*'.

34. Aniline produces Schiff's bases on reaction with

(*a*) Ethyl bromide (*b*) Benzyl alcohol

(*c*) Ethynoyl chloride (*d*) Acetone.

35. Which processes given below result in the formation of *p*-nitroaniline.

I. Sulphanilic acid $\xrightarrow{HNO_3}$

II. Benzenesulphonic acid $\xrightarrow{HNO_3/H_2SO_4}$

III. Aniline $\xrightarrow{HNO_3}$

IV. Aniline $\xrightarrow[(iii)\ \text{Hydrolysis}]{(i)\ (CH_3CO)_2O,\ (ii)\ HNO_3/H_2SO_4}$

(*a*) only I (*b*) II, III

(*c*) I, IV (*d*) III, IV.

36. Which of the following will result in formation of phenol ?

(*a*) $C_6H_5N_2Cl \xrightarrow{\text{Warm}}$

(*b*) $C_6H_5CN \xrightarrow[(ii)\ HNO_2]{(i)\ \text{Reduction}}$

(*c*) $C_6H_5NH_2 \xrightarrow[50°C]{NaNO_2/HCl}$

(*d*) $C_6H_6 + NaOH \xrightarrow[\text{Heat}]{}$

37. What is true about *p*-anisidine ?

(*a*) It consists $—NH_2$ group as well as $—CH_3$ group

(*b*) It is a weaker base than aniline

(*c*) It is stronger base than aniline

(*d*) It is amphoteric substance.

38. Which method as listed below, is not suitable for the preparation of secondary amines ?

(*a*) Phthalimide synthesis

(*b*) Reduction of alkylisocyanide

(*c*) Reduction of N-alkylalkanamide

(*d*) Treatment of alkylhalide and 1° amine.

39. Which of the following not an ambident nucleophile ?

(*a*) NO_2^+ (*b*) NO_2^-

(*c*) CN^- (*d*) SCN^-.

40. $\underset{\text{(A)}}{\text{Cyclopropane Carboxylic acid}} \xrightarrow{NH_3,\ \text{Heat}} X \xrightarrow{KOH/Br_2} Y \xrightarrow{CHCl_3/KOH} Z \xrightarrow{H_3O^+}$

The final products of above squence of steps are

(*a*) X and Z (*b*) HCOOH and Y

(*c*) CH_3NH_2 and A (*d*) A and X.

41. What is not true about benzene diazonium chloride ?

(*a*) It is soluble in water

(*b*) Its structure is 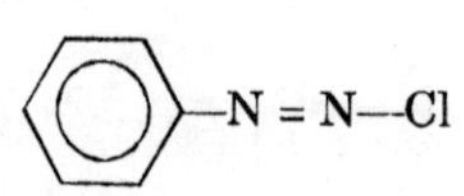

(*c*) Chlorine is present as Cl^- ion.

(*d*) It is one of the reactant in Balz-Schieman reaction.

42. Which amines will produce oily nitrosoamines with nitrous acid ?

(*a*) Aniline

(*b*) Ethanamine

(*c*) N, N-Diethylbenzenamine

(*d*) 2-(N-ethyl) propanamine.

43. An aliphatic amine with molar mass 73 g mol^{-1} on heating with excess of CH_3I gave a quaternary salt. The amine can be

(*I*) N, N-Diethylethanamine

(*II*) N-Methylpropanamine

(*III*) 2-Butanamine

(*IV*) Neopentylamine.

(*a*) II, III (*b*) only II

(*c*) I, III (*d*) only IV.

44. Which of the following will produce isopropyl amine ?

I. $(CH_3)_2CO \xrightarrow{NH_2OH} X \xrightarrow{LiAlH_4}$

II. $CH_3CH_2CHO \xrightarrow[Heat]{NH_3} X \xrightarrow{LiAlH_4}$

III. $(CH_3)_2CHOH + PCl_5 \longrightarrow X \xrightarrow{NH_3}$

IV. $CH_3CH_2CH_2NH_2 \xrightarrow{Heat}$

(*a*) I, II (*b*) II, III

(*c*) I, III (*d*) IV only.

45. Which of the following amine is less basic than aniline ?

(*a*) 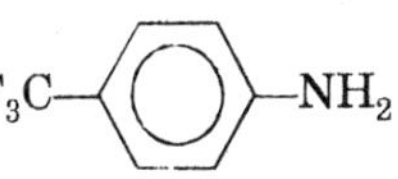

(*b*) $C_2H_5NH_2$

(*c*) 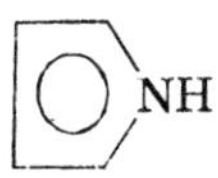

(*d*) Both '*a*' and '*c*'.

46. Which process can convert ammonium acetate into ethanamide ?

(*a*) Heating

(*b*) Heating in the presence of P_2O_5

(*c*) Heating in glacial acetic acid

(*d*) Heating with Br_2/KOH.

47. Which of the following represent electrophilic substitution ?

I. 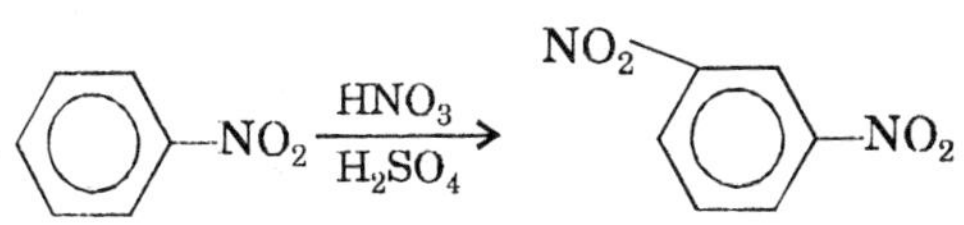

II. 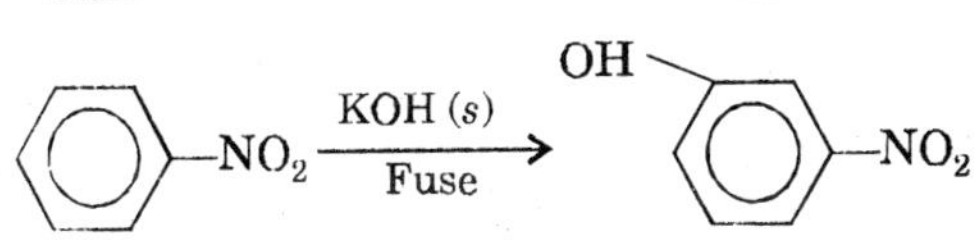

III. 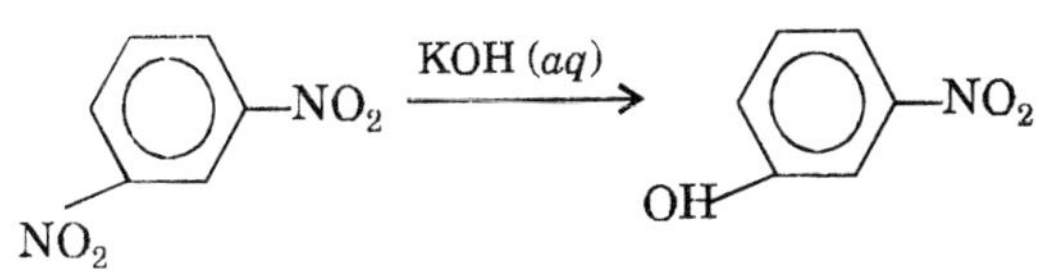

IV. 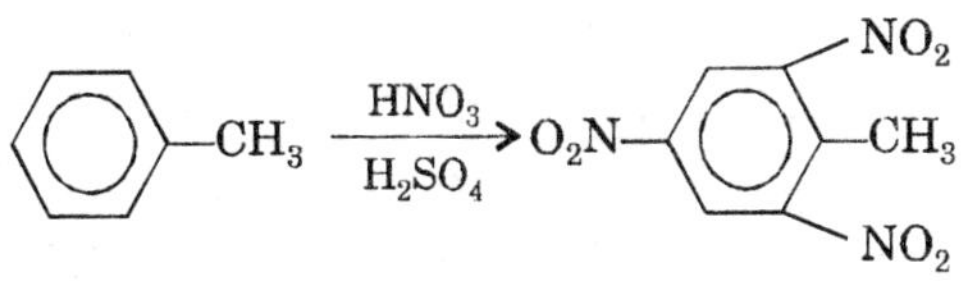

(*a*) I, II (*b*) I, II, III

(*c*) III, IV (*d*) I, IV.

48. An organic compound A contains C, H, N and reacts with dilute HCl to form formic acid as one of the product. The compound A can be

(*a*) CH_3NC (*b*) CH_3CN

(*c*) $C_2H_5NO_2$ (*d*) C_2H_5CN.

49. Among the compounds given below which has lowest value of pK_a

(*a*) 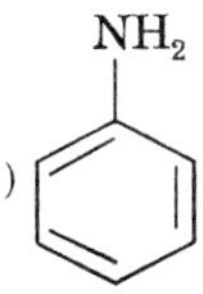

(*b*) 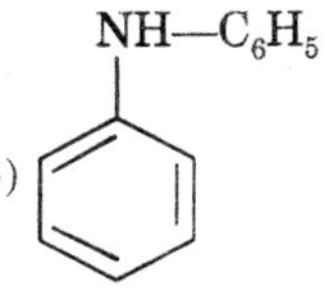

(*c*) 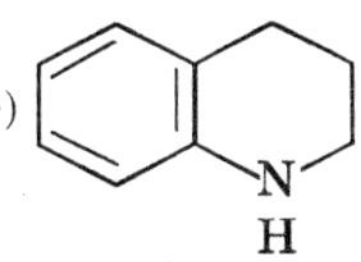

(*d*) 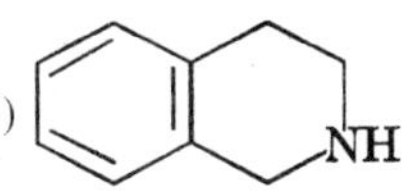

50. Which of the following on reaction with ethanal will produce *1-nitroprop-1-ene* ?

(*a*) CH_3NO_2

(*b*) CH_3NH_2

(*c*) $HCHO + HNO_3$

(*d*) Aldol condensation followed by treatment with HNO_3

51. A compound X is an antipyretic and is obtained by the following sequence of steps from *p*-nitrophenol (*i*) NaOH/C_2H_5Br (*ii*) Sn/HCl (*iii*) Ac_2O. Here X is

(*a*) Paracetanol (*b*) Phenacetin

(*c*) Aspirin (*d*) Novalgin.

52. 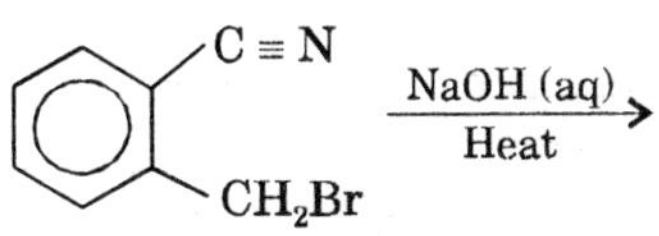

The major product of above reaction is

(*a*) 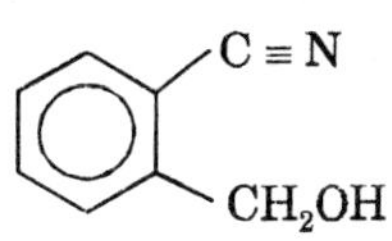

(*b*) 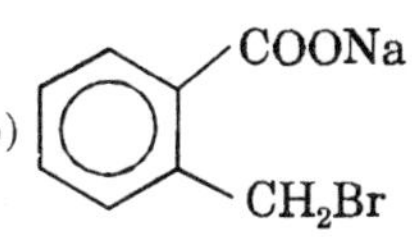

(*c*) 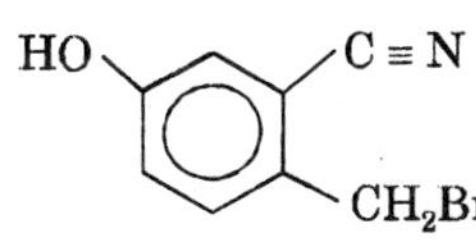

(*d*) 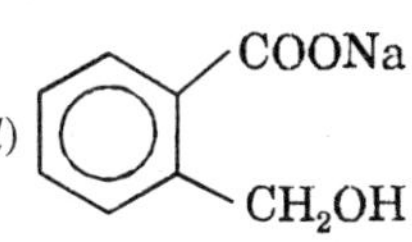

B. MCQs From Previous IIT Examinations

53. IUPAC name for the amine

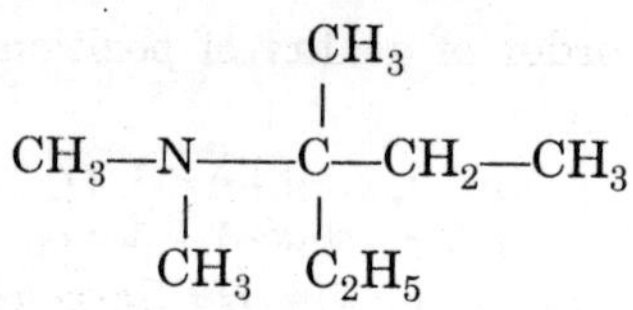

is

(*a*) 3-Dimethylamino-3-methylpentane
(*b*) 3(N, N-Triethyl)-3-aminopentane
(*c*) 3-N, N-Trimethyl pentanamine
(*d*) 3-N, N-Dimethylamino-3-methylpentane. **(*IIT 1991*)**

54. Butanenitrile may be prepared by heating
(*a*) *n*-Propyl alcohol with KCN
(*b*) *n*-Butyl alcohol with KCN
(*c*) *n*-Butyl chloride with KCN
(*d*) *n*-Propyl chloride with KCN. **(*IIT 1992*)**

55. When nitrobenzene is treated with Br_2 in the presence of $FeBr_3$, the major product formed is *m*-bromonitrobenzene. Which of the following statement is/are related to the process ?
(*a*) Electron density on *m*-C is more than that on *ortho* or *para* carbons
(*b*) Loss of aromaticity when Br^+ attacks at the *ortho* or *para* positions and not at *meta* position
(*c*) Intermediate carbocation is formed after initial attack of Br^+ at the *meta* position is least stabilized
(*d*) Easier loss of H^+ to regain aromaticity from *meta* position than from *ortho* and *para* position. **(*IIT 1992*)**

56.

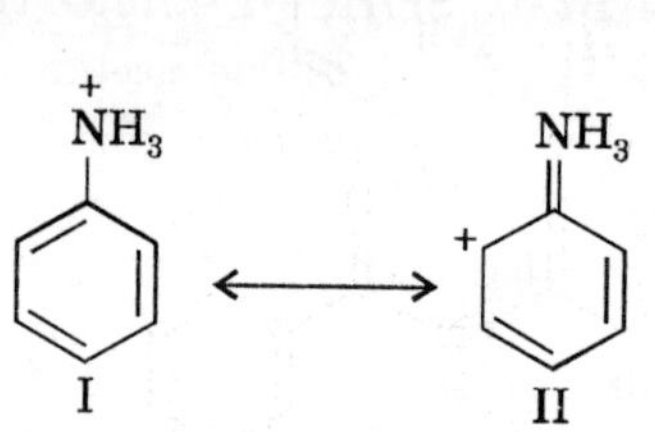

(*a*) II is not an acceptable cannonical structure because carbonium ions are less stable than ammonium ions
(*b*) II is not an acceptable cannonical structure because it is non-aromatic
(*c*) II is not an acceptable cannonical structure because the nitrogen has 10 valence electrons
(*d*) II is an acceptable cannonical structure. **(*IIT 1993*)**

57. Allyl isocyanide contains σ and π bonds, as
(*a*) 9σ and 3π (*b*) 9σ and 9π
(*c*) 3σ and 4π (*d*) 5σ and 7π. **(*IIT JEE 1995*)**

58. Nitrobenzene can be prepared from benzene by using a mixture of conc. HNO_3 and conc. H_2SO_4. In the nitrating mixture, nitric acid acts as a
(*a*) Base (*b*) Acid
(*c*) Reducing agent (*d*) Catalyst. **(*IIT 1997*)**

59. Among following statements on the nitration of aromatic compounds, the false one is
(*a*) The rate of nitration of benzene is almost the same as that of hexadeuterobenzene
(*b*) The rate of nitration of toluene is greater than that of benzene
(*c*) The rate of nitration of benzene is greater than that of hexadeuterobenzene
(*d*) Nitration is an electrophilic substitution reaction. **(*IIT 1997*)**

60. Benzenediazonium chloride on reaction with phenol in weakly basic medium gives
(*a*) Diphenyl ether (*b*) *p*-Hydroxyazobenzene
(*c*) Chlorobenzene (*d*) Benzene. **(*IIT 1998*)**

61. *p*-chloroaniline and anilinium hydrochloride cannot be distinguished by
(*a*) Sandmeyer reaction (*b*) $NaHCO_3$
(*c*) $AgNO_3$ (*d*) Carbylamine test. **(*IIT 1998*)**

62. The most unlikely representation of resonance structures of *p*-nitrophenoxide ion is

(*a*) 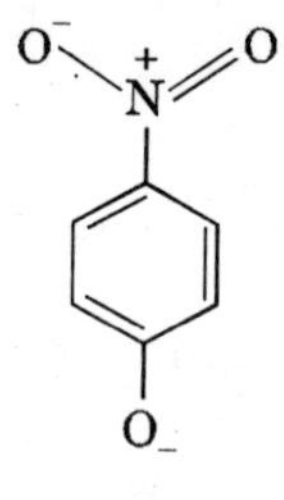

(*b*) 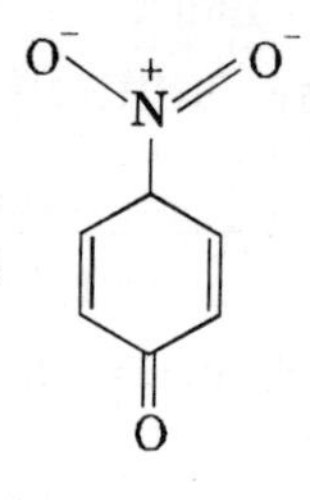

(*c*) 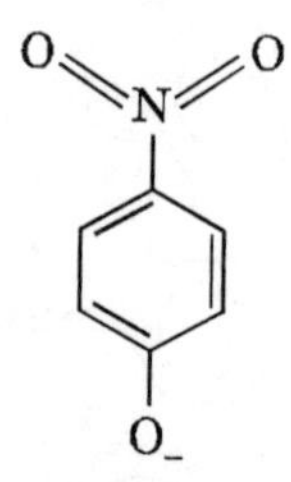

(*d*) 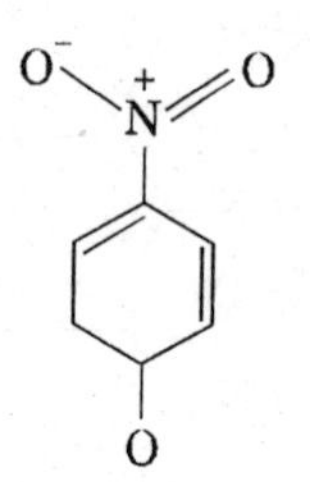

(*IIT 1999*)

63. A positive carbylamine test is shown by
(*a*) N, N-Dimethylaniline
(*b*) 2, 4-Dimethylaniline
(*c*) N-methyl-*o*-methylaniline
(*d*) *p*-methyl benzylamine. **(*IIT 1999*)**

64. Among the following, the strongest base is

(*a*) $C_6H_5NH_2$ (*b*) $p—NO_2—C_6H_4NH_2$
(*c*) $m—NO_2—C_6H_4NH_2$ (*d*) $C_6H_5CH_2NH_2$.

(*IIT Screening Test 2000*)

65. The compound that will react most readily with NaOH to form methanol is

(*a*) $(CH_3)_4N^+I^-$ (*b*) CH_3OCH_3
(*c*) $(CH_3)_3S^+I^-$ (*d*) $(CH_3)_3CCl$.

(*IIT Screening Test 2001*)

66. The correct order of basicities of the following compounds is

$CH_3—C(=NH)—NH_2$ (1) $CH_3CH_2NH_2$ (2)

$(CH_3)_2NH$ (3) $CH_3—C(=O)—NH_2$ (4)

(*a*) 2 > 1 > 3 > 4 (*b*) 1 > 3 > 2 > 4
(*c*) 3 > 1 > 2 > 4 (*d*) 1 > 2 > 3 > 4.

(*IIT Screening Test 2001*)

67. Which of the following represents a given mode of hybridisation sp^2, sp^2, sp, sp ?

(*a*) $H_2C = CH—C \equiv N$ (*b*) $HC \equiv C—C \equiv CH$
(*c*) $H_2C = C = C = CH_2$ (*d*) H_2C=CH–CH=CH_2.

(*IIT Screening Test 2003*)

68. 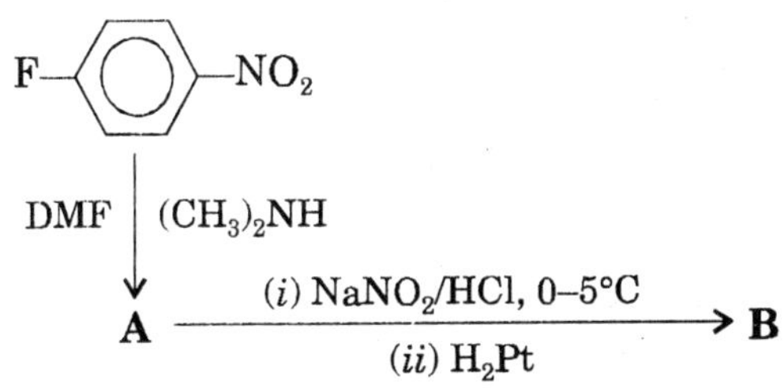

In the above sequence B is

(*a*) 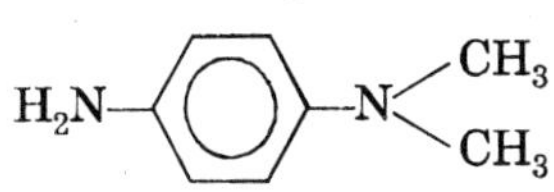

(*b*) 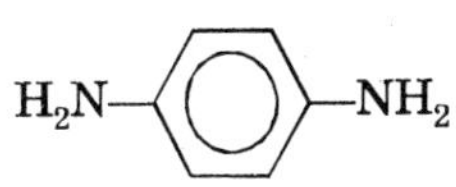

(*c*) 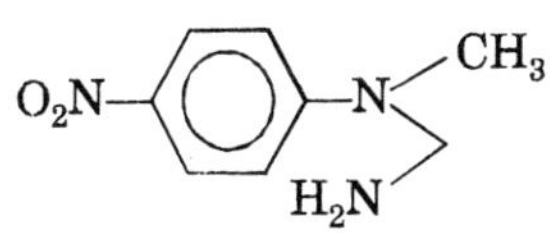

(*d*) 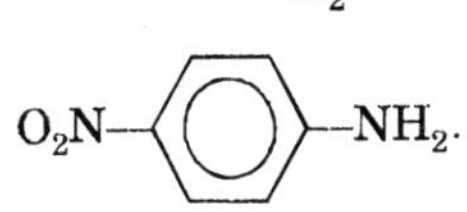 **(*IIT Screening Test 2003*)**

69. In the compound given below

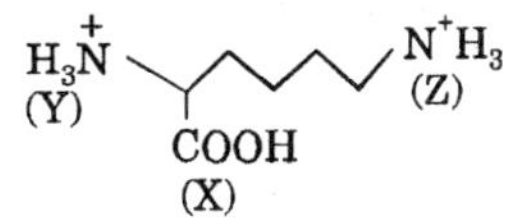

The correct order of acidity of positions (X), (Y), (Z) is

(*a*) Z > X > Y (*b*) X > Y > Z
(*c*) X > Z > Y (*d*) Y > X > Z.

(*IIT Screening Test 2004*)

70. The product formed by reaction of benzamide and $POCl_3$ is

(*a*) Aniline (*b*) Chlorobenzene
(*c*) Benzylamine (*d*) Benzonitrile.

(*IIT Screening Test 2004*)

71. $CH_3NH_2 + CHCl_3 + KOH \longrightarrow$ nitrogen containing compound + KCl + H_2O

Nitrogen containing compound is

(*a*) $CH_3—C \equiv N$ (*b*) $CH_3—NH—CH_3$
(*c*) $CH_3—\overset{-}{N} \equiv \overset{+}{C}$ (*d*) $CH_3—\overset{+}{N} \equiv \overset{-}{C}$.

(*IIT JEE 2006*)

72. In the following reaction

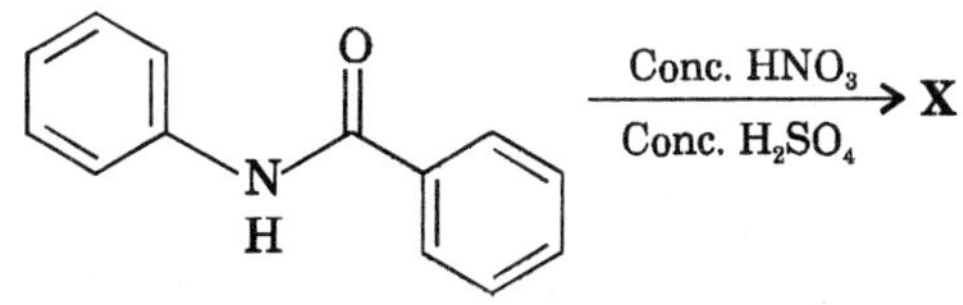

The product X is

(*a*) 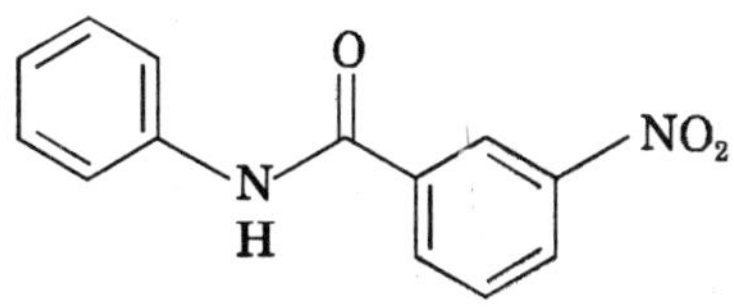

(*b*) 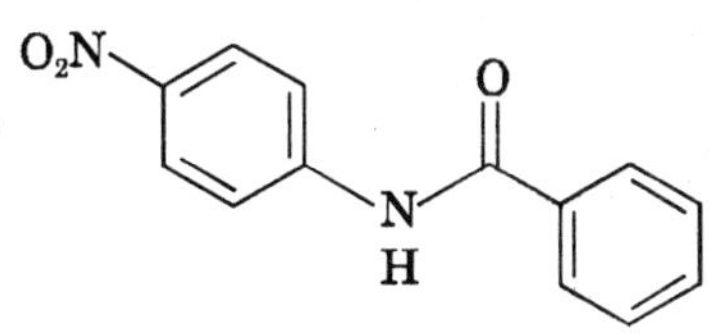

(*c*) 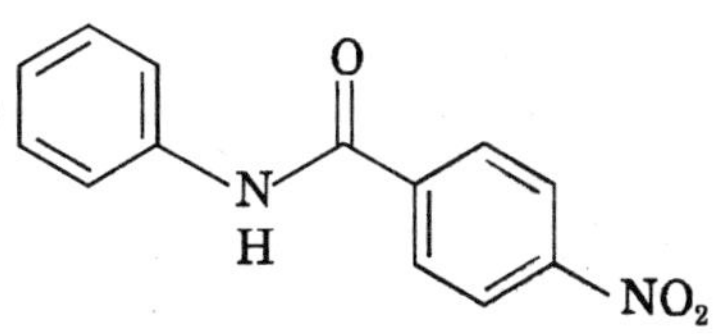

(*d*) 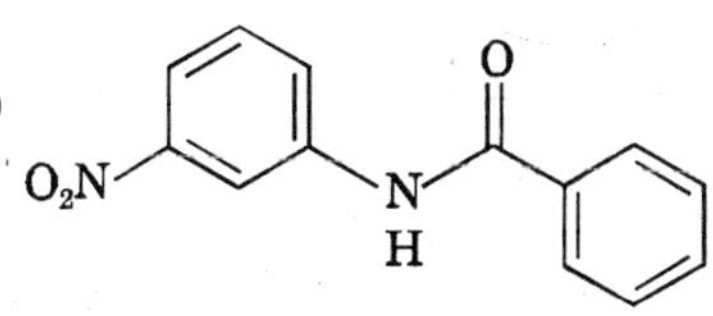

(*IIT JEE 2007*)

ANSWERS

1. (*c*)	**2.** (*b*)	**3.** (*a*)	**4.** (*b*)	**5.** (*d*)	**6.** (*d*)	**7.** (*b*)	**8.** (*b*)	**9.** (*a*)
10. (*b*)	**11.** (*b*)	**12.** (*c*)	**13.** (*d*)	**14.** (*a*)	**15.** (*c*)	**16.** (*c*)	**17.** (*b*)	**18.** (*b*)
19. (*a*)	**20.** (*b*)	**21.** (*d*)	**22.** (*c*)	**23.** (*d*)	**24.** (*b*)	**25.** (*d*)	**26.** (*a*)	**27.** (*a*)
28. (*a*)	**29.** (*c*)	**30.** (*d*)	**31.** (*c*)	**32.** (*b*)	**33.** (*d*)	**34.** (*d*)	**35.** (*c*)	**36.** (*d*)
37. (*c*)	**38.** (*a*)	**39.** (*a*)	**40.** (*b*)	**41.** (*b*)	**42.** (*d*)	**43.** (*a*)	**44.** (*c*)	**45.** (*d*)
46. (*b*)	**47.** (*d*)	**48.** (*a*)	**49.** (*d*)	**50.** (*a*)	**51.** (*b*)	**52.** (*d*)	**53.** (*d*)	**54.** (*d*)
55. (*a*)	**56.** (*b, c*)	**57.** (*a*)	**58.** (*a*)	**59.** (*c*)	**60.** (*b*)	**61.** (*a, b*)	**62.** (*c*)	**63.** (*b, d*)
64. (*d*)	**65.** (*a*)	**66.** (*b*)	**67.** (*a*)	**68.** (*a*)	**69.** (*b*)	**70.** (*d*)	**71.** (*d*)	**72.** (*b*).

HINTS/SOLUTIONS

Q. No.	Ans.	Hints & Solutions
1.	(*c*)	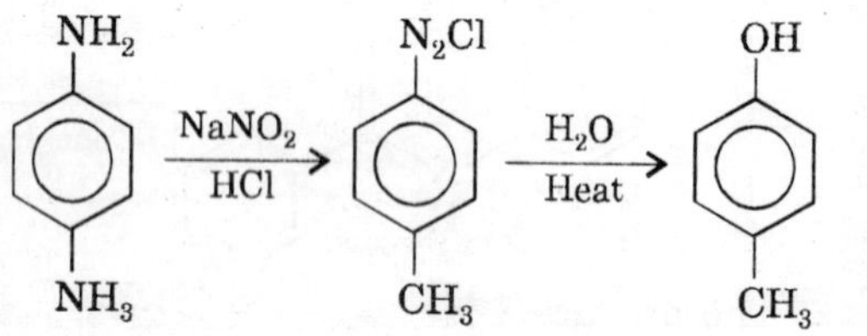 (*p*-cresol)
2.	(*b*)	Among the given compounds, C (pyrollidine) has highest basic strength due to easy availability of lone pair of N. The compound B (pyrole) is least basic because lone pair of N is used up for developing aromatic character and its availability for protanation is considerably decreased. The compound A (pyridine) is moderately basic. Hence the basic order is C > A > B.
3.	(*a*)	Among the given compounds, only *a i.e.*, pyridine satisfies the given requirements.
4.	(*b*)	Route 2 will give 1° amine with same no of C $RCH_2OH \xrightarrow[\text{Heat}]{\text{Cu}} RCHO \xrightarrow[\text{Heat}]{NH_3} RCH = NH \xrightarrow{H_2/Ni} RCH_2NH_2$ Route 1 will give upgraded amine $(RCH_2CH_2NH_2)$ Route 3 will give degraded amine (RNH_2).
5.	(*d*)	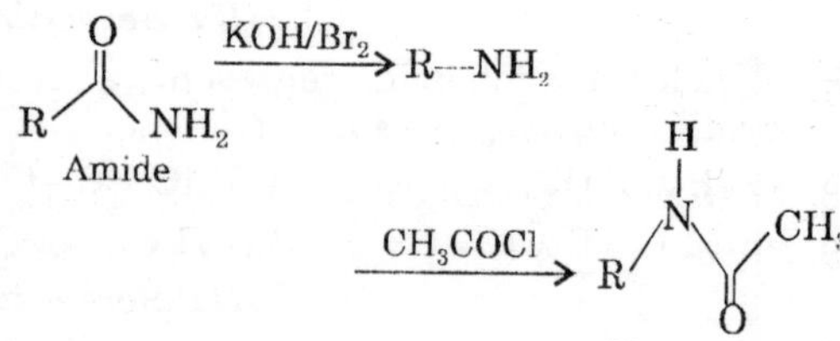 N-Acetylamine
6.	(*d*)	$R—N \overset{\rightarrow}{=} C + 2HgO \longrightarrow R—N = C = O + Hg_2O.$ Alkyl isocyanate
7.	(*b*)	$2CH_3NO_2 \xrightarrow{575\ K} 2CO_2 + 3H_2 + N_2.$
8.	(*b*)	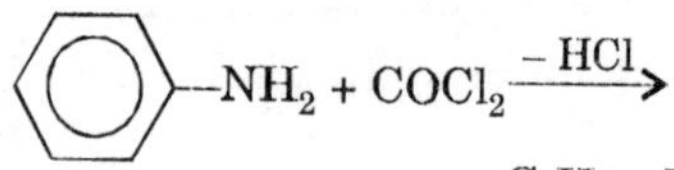 $C_2H_5 - N = C = O$ Phenyl isocyanate
9.	(*a*)	$C_2H_5NH_2 + NOCl \longrightarrow N_2 + H_2O + C_2H_5Cl.$
10.	(*b*)	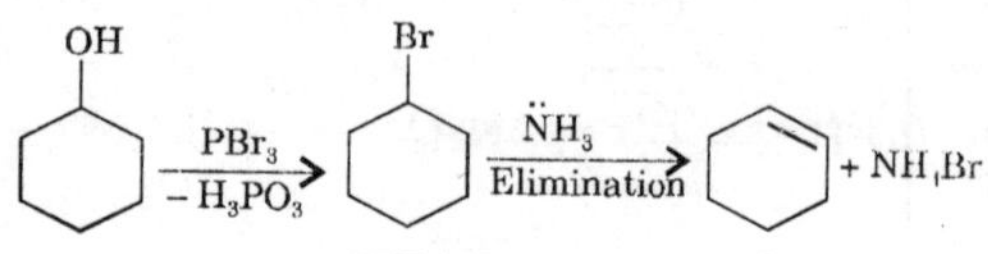 2° Halide. 2° Halide undergoes β-elimination which is a major side product.
11.	(*b*)	$CH_3CH(C_2H_5)—NH_2$ has chiral structure.
12.	(*c*)	When halonitrobenzene is heated with ethanolic KCN at 423 K then NO_2 group leaves the ring and – CN group attaches itself to the

Q. No.	Ans.	Hints & Solutions
		ring at the ortho position w.r.t. to the position occupied by leaving – NO_2 group. The reaction is called **Von Richter reaction.** 
13.	(*d*)	Phenol does not produce salt with mineral acids.
14.	(*a*)	Aniline is a stronger base than benzamide but weaker than benzylamine or cyclohexyl amine. Hence statements *b*, *c*, *d* are wrong. However, '*a*' is correct because resonance causes delocalisation of *p*-electrons of N on the ring.
15.	(*c*)	Amatol is obtained by mixing TNT with NH_4NO_3.
16.	(*c*)	$C_6H_5N_2Cl \xrightarrow[\text{or } SnCl_2/HCl]{NaHSO_3/HCl} C_6H_5NH{-}NH_2HCl$ (Phenyl hydrazine hydrochloride)
17.	(*b*)	The reaction is called Hoffmann ammonolysis reaction.
18.	(*b*)	Dimethylazene is $CH_3N = NCH_3$.
19.	(*a*)	2-Methyl-2-nitropropane does not have hydrogen at the α-position.
20.	(*b*)	For a compound to show optical activity, it should possess chiral structure. Among the given formulae only $C_6H_5CH(NH_2)CH_3$ possesses chiral structure.
21.	(*d*)	The reagent H_2/Pt converts CH_3CONH_2 to $CH_3CH_2NH_2$ whereas Br_2/KOH converts CH_3CONH_2 to CH_3NH_2.
22.	(*c*)	*p*-toluidine and aniline are 1° amines and thus will give carbylamine reaction.
23.	(*d*)	CH_3NH_2 and $(CH_3)_2NH$, both can react with C_2H_5Cl to form secondary and tertiary amines respectively.
24.	(*b*)	The reaction given in choice (*b*) will produce 2° amine.
25.	(*d*)	$C_2H_5Br + AgNO_2 \longrightarrow C_2H_5NO_2 + C_2H_5ONO$. (Major) (Minor)
26.	(*a*)	Benzylamine ($C_6H_5CH_2NH_2$), N-methylaniline ($C_6H_5NH{-}CH_3$) are isomers of *o*-toluidine 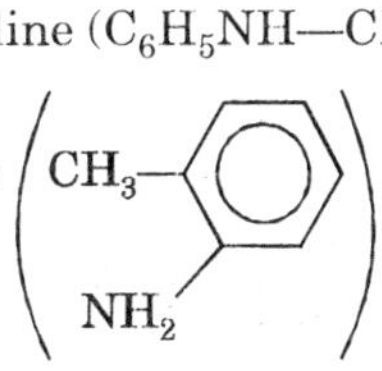
27.	(*a*)	Both phenol and β-naphthol will give orange red dye as a result of coupling reaction.
28.	(*a*)	Nitrobenzene is less reactive because of electron withdrawing effect of —NO_2.
29.	(*c*)	$C_6H_5NH_2$ and C_5H_5N are less basic than $C_2H_5NH_2$.
30.	(*d*)	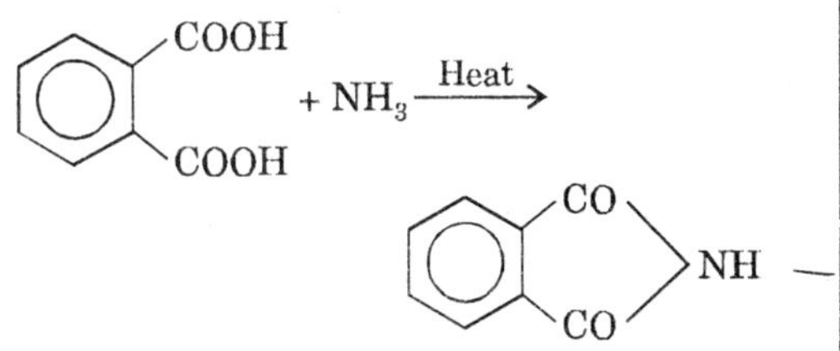 Phthalimide
31.	(*c*)	Aniline ($C_6H_5NH_2$) on oxidation will produce *p*-benzoquinone.
32.	(*b*)	(A) $C_6H_5{-}NH{-}NH{-}C_6H_5$ Hydrozobenzene $\downarrow NaIO_3$ (X) $C_6H_5{-}N = NH{-}C_6H_5$ Azobenzene $\downarrow CH_3CO_3H$ (Y) $C_6H_5 = N(\rightarrow O){-}C_6H_5$ Azoxybenzene; (A), (X), (Y) $\xrightarrow{Sn/HCl} 2C_6H_5NH_2$ (Z) Aniline
33.	(*d*)	Various stages of reduction of NO_2 group are — $NO_2 \longrightarrow {-}NO \longrightarrow {-}NHOH \longrightarrow {-}NH_2$.
34.	(*d*)	Both Benzaldehyde and acetone produce Schiff's bases by reaction with aniline.
35.	(*c*)	Both the roots as given in (*a*) and (*b*) give *p*-nitroaniline as major product.
36.	(*d*)	$C_6H_5N_2Cl \xrightarrow[\text{Warm}]{H_2O} C_6H_5OH + N_2 + HCl$ $C_6H_5NH_2 \xrightarrow[50°C]{NaNO_2/HCl} C_6H_5OH + N_2 + H_2O$
37.	(*c*)	*p*-anisidine is 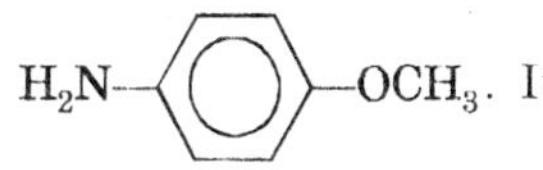. It is a stronger base than 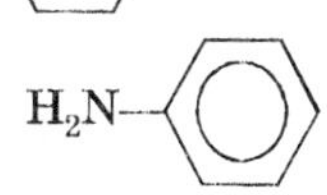 due to electron donating nature of —OCH_3 group.
38.	(*a*)	Secondary amine cannot be prepared by phthalimide synthesis. All other methods given in question can produce 2° amine.
39.	(*a*)	All except $\overset{+}{N}O_2$ are ambident nucleophile.
40.	(*b*)	The chemical reactions involved in above sequence of steps are as follows :

Q. No.	Ans.	Hints & Solutions
		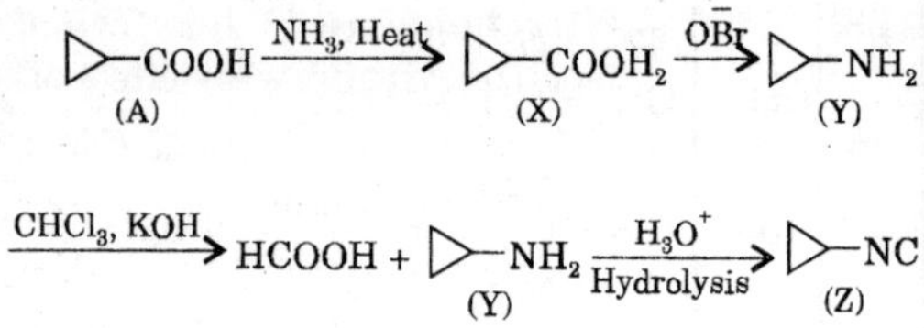 Hence correct choice is **(*b*)**.
41.	**(*b*)**	Benzene diazonium chloride is $[C_6H_5N\equiv N]^+Cl^-$.
42.	**(*d*)**	Nitrosoamine is produced by secondary amines on reaction with HNO_2.
43.	**(*a*)**	Let the amine be $C_nH_{2n+3}N$. The molar mass $= (14n + 17)$ $14n + 17 = 73$ or $n = 4$ Among the given amines only II and III correspond to the given data and also have four carbons.
44.	**(*c*)**	$(CH_3)_2CO \xrightarrow{NH_2OH} (CH_3)_2CH=NOH \xrightarrow{LiAlH_4} (CH_3)_2CHNH_2$ $(CH_3)_2CHOH \xrightarrow{PCl_5} (CH_3)_2CHCl \xrightarrow{NH_3} (CH_3)_2CHNH_2$
45.	**(*d*)**	(*a*) is less basic than aniline due to electron withdrawing effect of —CF_3 group. (*c*) is less basic than aniline because lone pair of N atom is delocalised on the entire ring skeleton.
46.	**(*b*)**	$CH_3COONH_4 \xrightarrow[Heat]{P_2O_5} CH_3CONH_2 + H_2O$
47.	**(*d*)**	Electrophilic substitution is represented by I and IV.
48.	**(*a*)**	CH_3NC gives formic acid as one of the hydrolytic product.
49.	**(*d*)**	In *a*, *b* and *c* the lone pair of N atom gets delocalised over adjecent benzene rings. In '*d*' the lone pair is not delocalised over the ring and is easily available for protonation. Hence basic strenth of '*d*' is highest and thus it has lowest value of pK_b.
50.	**(*a*)**	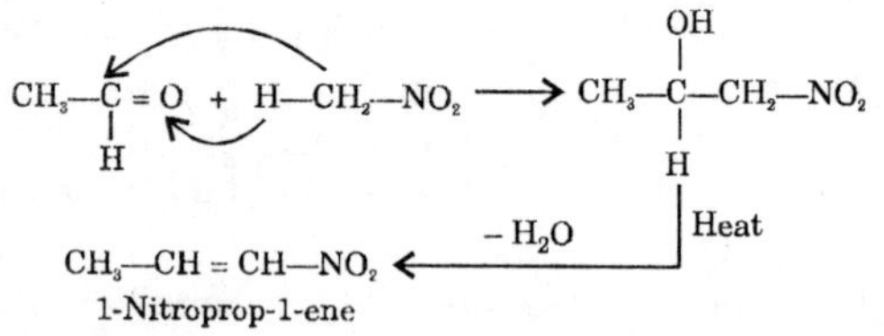

Q. No.	Ans.	Hints & Solutions
51.	**(*b*)**	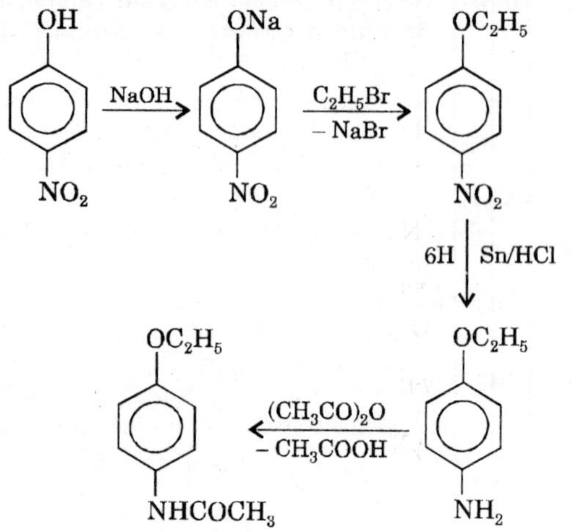 Phenacetin
52.	**(*d*)**	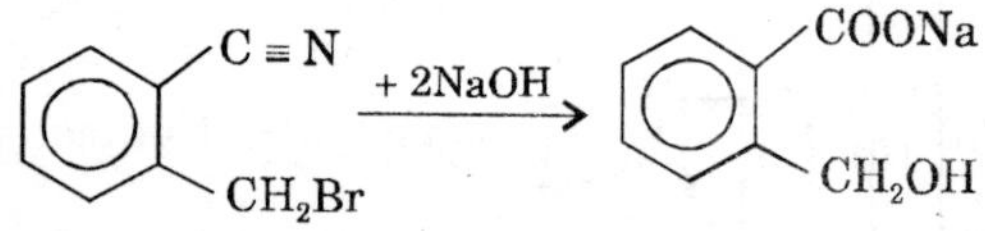 $+ NaBr + NH_3$
53.	**(*d*)**	Correct name of the compound is 3(N, N-Dimethylamino)-3-methylpentane among the given options.
54.	**(*d*)**	$CH_3CH_2CH_2Cl + KCN \longrightarrow CH_3(CH_2)CN + KCl.$ (Butanenitrile)
55.	**(*a*)**	The statement given in (*a*) provides explanation to the question.
56.	**(*b, c*)**	The statements given in choices (*b*) and (*c*) are factual.
57.	**(*a*)**	Allyl isocyanide is $CH_2=CH-CH_2-N \overset{\rightarrow}{=} C$. It has 9σ and 3π bonds.
58.	**(*a*)**	HNO_3 acts as a base and undergoes protonation to generate nitronium ion ($\overset{+}{N}O_2$) which further attacks the ring.
59.	**(*c*)**	In the electrophilic substitution of benzene, the rate determining step is the attack of electrophile on the benzene ring and not the breaking of *C—H bond*. Hence (*c*) is a false statement.
60.	**(*b*)**	$C_6H_5N_2Cl + C_6H_5OH \xrightarrow[medium]{Acidic}$ 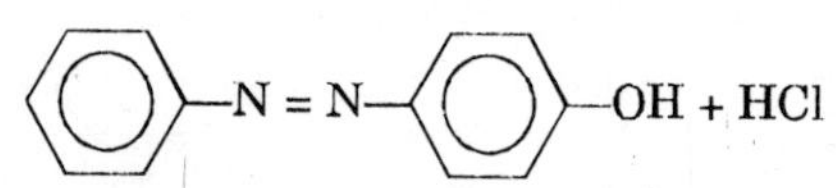 (4-Hydroxy azobenzene)

Q. No.	Ans.	Hints & Solutions
61.	(*a, b*)	Both Sandmeyer reaction and $NaHCO_3$ cannot provide distinction between $C_6H_5NH_2$ and $C_6H_5NH_3^+Cl^-$.
62.	(*c*)	Among the given structures, the most unlikely structure is given in choice '*c*' because in this structure N is shown to pentavalent.
63.	(*b, d*)	2, 4-dimethyl aniline and *p*-methyl benzylamine both being 1° amines can show carbylamine reaction.
64.	(*d*)	Benzylamine is strongest base among the given species.
65.	(*a*)	$(CH_3)_4N^+I^- + NaOH \rightarrow NaI + (CH_3)_3N + CH_3OH$ Due to greater electronegativity of N over S, the positive charge on N atom makes CH_3 group more electron deficient in $(CH_3)_3N^+I^-$. Therefore it gives nucleophilic substitution more readily $H\bar{O} + CH_3—\overset{+}{N}(CH_3)_3 \longrightarrow$ $CH_3OH + (CH_3)_3N$
66.	(*b*)	Protonation of $CH_3—\underset{\underset{NH}{\parallel}}{C}—NH_2$ gives cation which is resonance stabilised. $\left[CH_3—\underset{\underset{NH_2}{\mid}}{\overset{+}{C}}—NH_2 \leftrightarrow CH_3—\underset{\underset{NH_2}{\mid}}{C}=\overset{+}{N}H_2 \leftrightarrow CH_3—\underset{\underset{\overset{+}{N}H_2}{\parallel}}{C}—NH_2\right]$ Hence 1 is more basic than 2 which is 1° amine. Further 2° amine is more basic than 1° amine as well as amide. Amide is least basic among all. Hence correct order of basicity is 1 > 3 > 2 > 4. *i.e.*, option (*b*) is correct.
67.	(*a*)	$\overset{sp^2}{CH_2} = \overset{sp^2}{CH}—\overset{sp}{C} \equiv \overset{sp}{N}$.
68.	(*a*)	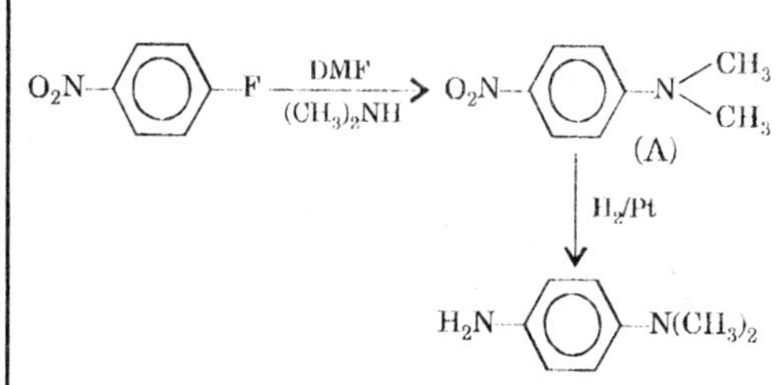 No reaction occurs with $NaNO_2/HCl$ and (A).
69.	(*b*)	The position X is obviously most acidic. Among positions Y and Z, Y is more acidic than Z due to electron withdrawing group (—COOH) in its close proximity. Hence the order of acidity is **X > Y > Z.**
70.	(*b*)	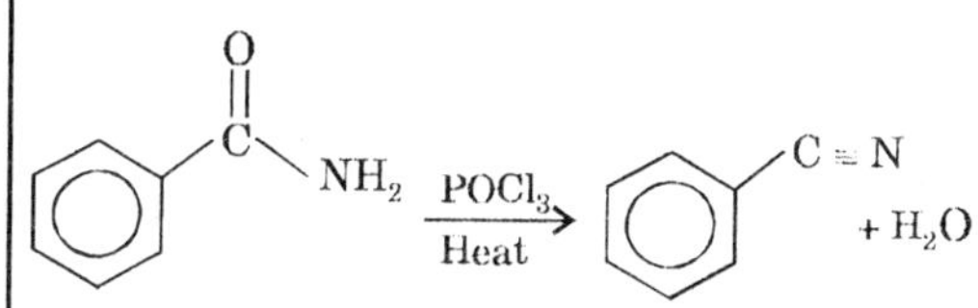
71.	(*d*)	$CH_3NH_2 + CHCl_3 + 3KOH \longrightarrow$ $CH_3—\overset{+}{N} \equiv \bar{C} + 3KCl + 3H_2O$
72.	(*b*)	In benzanilide the lone pair on N atom activates the adjacent benzene ring and will cause the NO_2 group to attack preferably at *p*-position. The ortho position is sterically hindered. Hence the correct choice is (*b*).

7.7. ASSERTION-REASON QUESTIONS

In each of the following questions two statements are given as Assertion A and Reason R. Examine the statements carefully and answer the questions according to the instructions given below :

- *Mark **a** if both **A** and **R** are correct and **R** is the correct reason of **A**.*
- *Mark **b** if both **A** and **R** are correct and **R** is not the correct reason of **A**.*
- *Mark **c** if **A** is correct and **R** is wrong.*
- *Mark **d** if **A** is wrong and **R** is correct.*
- *Mark **e** if both **A** and **R** are wrong.*

1. ***Assertion A.*** Boiling point of trimethyl amine is higher than that of *n*-propyl amine.

   Reason R. H-bonding is more extensive in tertiary amines.

(*a*)	(*b*)	(*c*)	(*d*)	(*e*)

2. ***Assertion A.*** Sulphanilic acid exists as zwitter ion.

   Reason R. Sulphanilic acid is soluble in NaOH but not in HCl.

(*a*)	(*b*)	(*c*)	(*d*)	(*e*)

3. ***Assertion A.*** Urea and thiourea are functional isomers of each other.

   Reason R. The elements present in thiourea and urea are nitrogen, hydrogen and carbon.

(*a*)	(*b*)	(*c*)	(*d*)	(*e*)

4. ***Assertion A.*** $COCl_2$ reacts with RNH_2, and $C_6H_5NH_2$ to form different products.

   Reason R. $COCl_2$ is very good acylating agent.

(*a*)	(*b*)	(*c*)	(*d*)	(*e*)

5. ***Assertion A.*** Methylbromide and sivercyanide react to form methyl isocyanide as major product.

   Reason R. AgCN has sufficient covalent character.

(*a*)	(*b*)	(*c*)	(*d*)	(*e*)

6. ***Assertion A.*** Nitroalkanes and alkylnitrites are functional isomers of each other.

   Reason R. Both alkylnitrites and nitroalkanes give same hydrolytic products.

(*a*)	(*b*)	(*c*)	(*d*)	(*e*)

7. ***Assertion A.*** Acetonitrile is another name of ethanenitrile.

   Reason R. α-H atom of acetonitrile exhibit acidic character.

(*a*)	(*b*)	(*c*)	(*d*)	(*e*)

8. ***Assertion A.*** CN^- is ambident nucleophlies as well as pseudo halide ion.

   Reason R. RCN and R—NC are tautomers of each other.

(*a*)	(*b*)	(*c*)	(*d*)	(*e*)

9. ***Assertion A.*** Chlorination of benzenecarbonitrile produces mainly *m*-chloro derivatives.

   Reason R. —CN group deactivates the benzene ring.

(*a*)	(*b*)	(*c*)	(*d*)	(*e*)

10. ***Assertion A.*** Nitrobenzene cannot be prepared by treatment of chlorobenzene with $AgNO_2$.

    Reason R. In the nitrating mixture H_2SO_4 acts as an oxidising agent.

(*a*)	(*b*)	(*c*)	(*d*)	(*e*)

11. ***Assertion A.*** During nitration of benzene, the slowest step is cleavage of C_6H_5—H bond.

    Reason R. Rate of nitration of benzene and hexadeutrobenzene is same.

(*a*)	(*b*)	(*c*)	(*d*)	(*e*)

12. ***Assertion A.*** Nitrosoalkane is reduction product of R—NHOH.

    Reason R. Boiling point of nitroethane is higher than ethylnitrite.

(*a*)	(*b*)	(*c*)	(*d*)	(*e*)

13. ***Assertion A.*** R—ONO_2 can be considered as ester of nitrous acid.

Reason R. It gives HNO_2 as one of the product on hydrolysis.

(a)	(b)	(c)	(d)	(e)

14. ***Assertion A.*** Before nitration, aniline is converted into acetanilide.

Reason R. Conversion of aniline to acetanilide before nitrating it, is done to protect —NH_2 group from oxidation.

(a)	(b)	(c)	(d)	(e)

15. ***Assertion A.*** Diazo group of benzene diazonium salt can be easily replaced with hydrogen by boiling with water.

Reason R. Alkyl diazonium salts donot exist because of their unstable structures.

(a)	(b)	(c)	(d)	(e)

ANSWERS

1. (e) **2.** (d) **3.** (e) **4.** (c) **5.** (a) **6.** (c) **7.** (b) **8.** (b) **9.** (a)
10. (e) **11.** (c) **12.** (e) **13.** (a) **14.** (a) **15.** (d).

HINTS/SOLUTIONS

Q. No.	Ans.	Hints & Solutions
1.	(e)	Both ***A*** and ***R*** are against the facts.
2.	(d)	Statement ***A*** is correct but ***R*** is wrong because sulphanilic acid is soluble in NaOH as well as in HCl.
3.	(e)	Urea (NH_2CONH_2) and thiourea (NH_2CSNH_2) have different formulae.
4.	(c)	***A*** is correct but ***R*** is wrong because $COCl_2$ is not a good acylating agent
5.	(a)	***R*** is explanation of ***A***.
6.	(c)	Statement ***A*** is correct but ***R*** is wrong because hydrolytic products of alkanenitrite and nitroalkanes are different.
7.	(b)	Both the statements are correct.
8.	(b)	Both ***A*** and ***R*** are correct.
9.	(a)	Statements ***R*** provides explanation of ***A***.
10.	(e)	Both the statements are against the facts.
11.	(c)	Statement ***A*** is wrong because slowest step during nitration is attack of nitronium ion.
12.	(e)	Both ***A*** and ***R*** are wrong.
13.	(a)	***R*** provides an explanation of ***A***.
14.	(a)	***R*** is an explanation of ***A***.
15.	(d)	***A*** is wrong because on warming diazonium salt with water diazo group is replaced by —OH group.

7.8. SELF EVALUATION TEST

Choose the correct answer from the four alternatives given in each of the following questions :

1. The reduction of which of the following compounds will give us the desired product $CH_3NHC_2H_5$.
   I. 2-Nitrobutane II. Ethylisocyanide
   III. N-Methylethanamide.
   (*a*) I only (*b*) II, III
   (*c*) I, II (*d*) I, III.

2. Which one of the following reactions will not give 1° amine as one of the product.
   (*a*) $RCONH_2 \xrightarrow{Br_2/KOH}$ (*b*) $R—NC \xrightarrow{H_3O^+}$
   (*c*) $R—CN \xrightarrow{H_3O^+}$ (*d*) $RCH = NOH \xrightarrow{Na/EtOH}$.

3. Among various amines represented by the molecular formula C_3H_9N, the one which has weakest basic character in aqueous medium is
   (*a*)  (*b*) 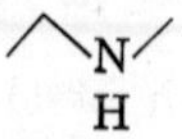
   (*c*) $(CH_3)_3N$ (*d*) 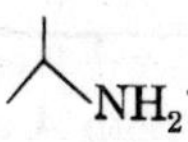

4. The reagents which are used in Hoffmann's Mustard oil reaction are
   (*a*) 1° Amine and Br_2/KOH
   (*b*) 1° Amine and CS_2 in $HgCl_2$
   (*c*) 1° Amine with $CHCl_3/KOH$
   (*d*) 1° Amine and C_6H_5COCl.

5. Among the structures given below which can exist as zwitter ions.
   (*a*) 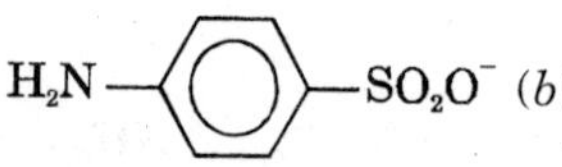 (*b*) 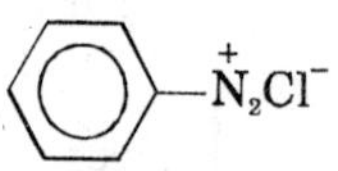
   (*c*) 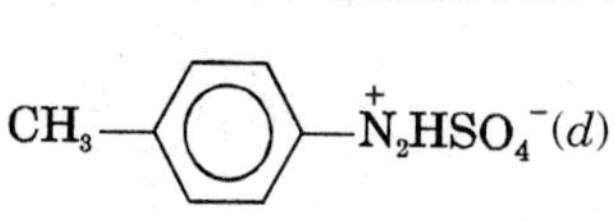 (*d*) 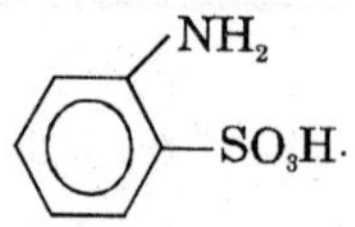

6. The rate controlling step in the nitration process of the benzene molecule is
   (*a*) 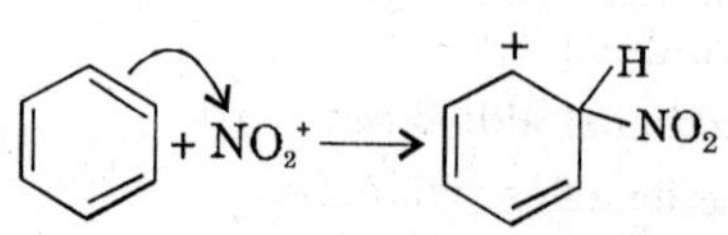
   (*b*) 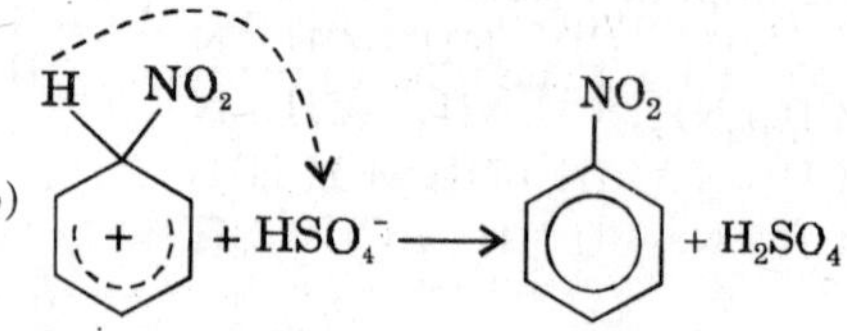
   (*c*) 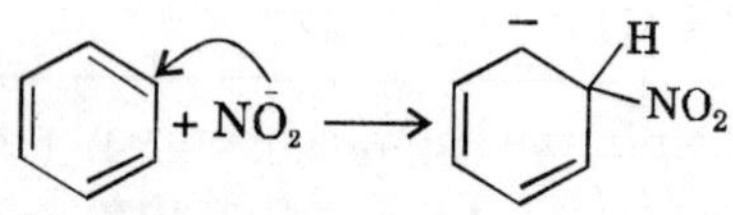
   (*d*) $HNO_3 + H_2SO_4 \longrightarrow NO_2^+ + HSO_4^-$.

7. Which of the following nitroalkanes cannot exhibit tautomerism
   (*a*) CH_3NO_2 (*b*) $CH_3CH_2NO_2$
   (*c*) $(CH_3)_2CHNO_2$ (*d*) $(CH_3)_3C—NO_2$.

8. Which process is most suitable for getting primary amine from alkyl isocyanide ?
   (*a*) Reduction with H_2/Pt (*b*) Reduction with Zn/HCl
   (*c*) Acidic hydrolysis (*d*) None of the above.

9. Acetoxime on reduction with $LiAlH_4$ produces
   (*a*) 1-Propanamine (*b*) 2-Propanamine
   (*c*) Ethanamine (*d*) Ethanenitrile.

10. Acetamide is less basic than aniline because
    (*a*) It is an aliphatic compound
    (*b*) Of greater electron withdrawing effect of adjacent carbonyl group
    (*c*) Its molecular mass is low
    (*d*) Of intermolecular H-bonding.

11. Deep blue colour formed by addition of copper (II) sulphate solution of ethyl amine is due to formation of
    (*a*) Free Cu^{2+} ions in solution
    (*b*) $(NH_4)_2SO_4$
    (*c*) $[Cu(C_2H_5NH_2)_4]^{2+}$ ions
    (*d*) $Cu(OH)_2$.

12. In order to get phenylisothiocyanate from aniline which combination of reagents should be used ?
    (*a*) $CS_2/HgCl_2$ (*b*) $S_8/NaOH$
    (*c*) S_8/CCl_4 (*d*) $(NH_4)_2S_x$.

13. Which of the following reagent reacts only with primary amines ?
    (*a*) $NaNO_2/HCl$ (*b*) CH_3COCl
    (*c*) CH_3MgI (*d*) CH_3CHO.

14. Which of the following is produced when aniline is treated with conc. HNO_3 ?
(*a*) *o* and *p*-nitroaniline (*b*) *m*-nitroaniline
(*c*) Nitrobenzene (*d*) A black tarry material.

15. The hybrid state of nitrogen in diethyl amine is
(*a*) sp^3 (*b*) sp^2
(*c*) *sp* (*d*) sp^3d.

16. Which of the following is correct order of basic strength among aliphatic amines in benzene solution ?
(*a*) $CH_3NH_2 > (CH_3)_2NH > (CH_3)_3N$
(*b*) $(CH_3)_2NH > CH_3NH_2 > (CH_3)_3N$
(*c*) $(CH_3)_3N > (CH_3)_2NH > CH_3NH_2$
(*d*) $(CH_3)_3N > CH_3NH_2 > (CH_3)_2NH$.

17. $C_6H_5COCl \xrightarrow{NH_3} X \xrightarrow{P_2O_5} Y \xrightarrow[H_2]{Ni} Z$. The end product in the above sequence of reactions is
(*a*) Benzoic acid (*b*) Aniline
(*c*) Benzyl amine (*d*) Benzonitrile.

18. One mole of ethylamine when reacts with nitrous acid will produce dinitrogen gas equal to
(*a*) 1 L at S.T.P. (*b*) 22.4 L at S.T.P.
(*c*) 11.2 L at S.T.P. (*d*) unpredictable.

19. Further nitration of nitrobenzene by treatment with nitrating mixture produces
(*a*) 1, 3-Dinitrobenzene
(*b*) 1, 4-Dinitrobenzene
(*c*) 1, 3, 5-Trinitrobenzene
(*d*) 1 : 1 mixture of DNB and TNB

20. Identify S in the following sequence of reactions :

$$C_6H_5N_2Cl \xrightarrow[KCN]{CuCN} X \xrightarrow{LiAlH_4} Y \xrightarrow{HNO_2} S$$

(*a*) Benzoic acid (*b*) Phenyl acetic acid
(*c*) Benzyl alcohol (*d*) Benzamide.

21. Identify the end product in the following sequence :

$$C_2H_5NH_2 \xrightarrow[320\ K]{NaNO_2/HCl} X \xrightarrow{NaOH(aq.)} Y \xrightarrow{C_2H_5I} Z$$

(*a*) 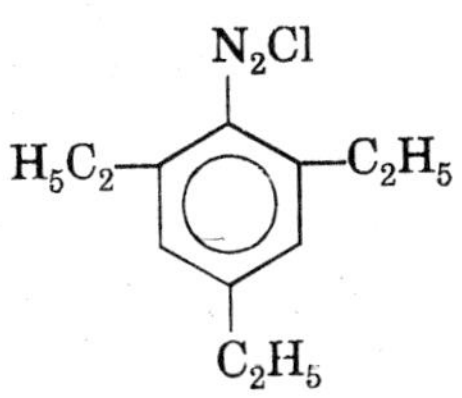 (*b*) 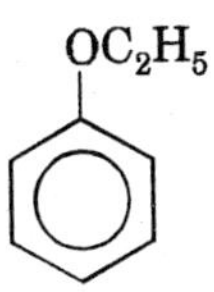

(*c*) 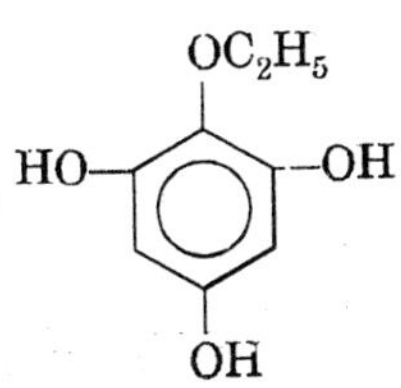 (*d*) 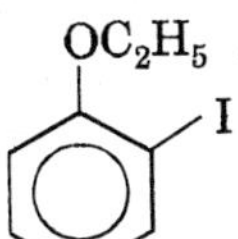

22. Which of the following will not give coloured dye with benzene diazonium chloride ?

(*a*) 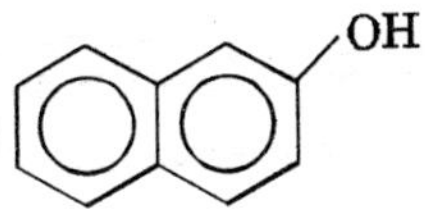 (*b*) 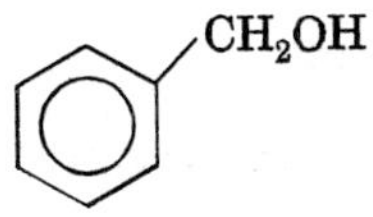

(*c*) 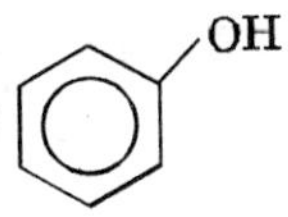 (*d*) 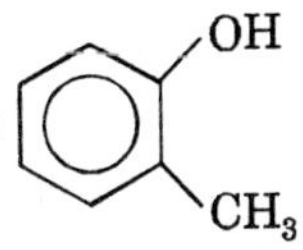

23. *m*-bromobenzamide $\xrightarrow{Br_2/KOH}$ X. The compound X can be
(*a*) *m*-bromoaniline (*b*) 3, 5-Dibromobenzamide
(*c*) *m*-bromobenzyl amine (*d*) *m*-bromophenol.

24. Which of the following species will give methyl orange by reaction with diazonium salt of sodium *p*-aminobenzene sulphonate ?
(*a*) Aniline (*b*) N, N-Dimethylaniline
(*c*) *m*-nitro aniline (*d*) *p*-cresol.

25. The total number of structural isomers of aromatic amines containing carboxylic rings having formula C_7H_9N is
(*a*) 2 (*b*) 3
(*c*) 4 (*d*) 5.

26. Which of the following exists as Zwitter ion structure ?
(*a*) *p*-aminobenzenesulphonic acid
(*b*) *p*-cresol
(*c*) Picric acid
(*d*) Salicylic acid.

27. Which of the following aromatic amine is least basic ?

(*a*)  (*b*) 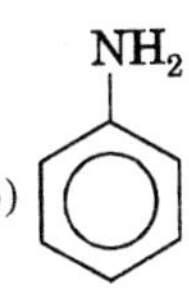

(*c*) 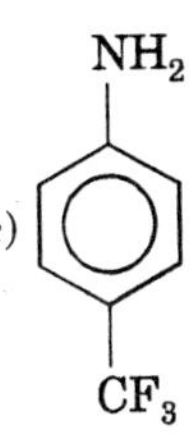 (*d*) 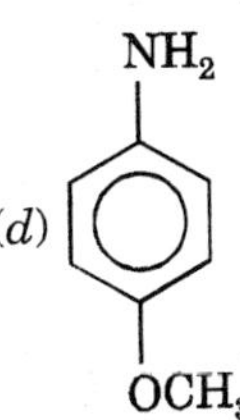

28. Which of the following process will not yield secondary amines ?
(*a*) Reduction of aldoximes
(*b*) Reduction of carbylamines

(c) Reduction of Schiff's bases
(d) Reduction of N-Methylethanamide.

29. Which of the following is intermediate species in Hoffmann's degradation reaction for the preparation of primary amines ?
(a) $RCONBr_2$ (b) R—N = C = O
(c) R—CN (d) R—CH_2NHBr.

30. Which of the following amine will not react with benzoyl chloride ?
(a) Isopropyl amine (b) p-toludine
(c) N, N-Diethyl aniline (d) N-methyl aniline.

31. Which of the following nitrogenous compound is amphoteric ?
(a) Aniline (b) Diethylamine
(c) β-Naphthylamine (d) Ethanamide.

32. Which of following substance is insoluble in water but gives cystalline precipitate with hydrochloric acid ?
(a) Methyl amine (b) Acetamide
(c) Aniline (d) Dimethyl amine.

33. Which of the following substance does not give dinitrogen gas with nitrous acid ?
(a) Acetamide (b) Methyl amine
(c) Diethyl amine (d) Isopropyl amine.

34. The substance which gives benzamide on alkaline hydrolysis and can be obtained from benzamide by treatment with $SOCl_2$ is
(a) Benzonitrile (b) Benzenamine
(c) Benzylamine (d) Benzylnitrile.

35. Which of the following substance does produce a salt with nitrous acid but does not react with $CHCl_3$/KOH combination ?
(a) Dimethylamine (b) Aniline at 273 K
(c) Methyl amine (d) Trimethyl amine.

36. In which of the following species the lone pair of N atom is not delocalised over the benzene ring ?
(a) p-Toluidine
(b) Anilinium hydrogen sulphate
(c) Aniline
(d) p-nitroaniline.

37. Which of the following will not produce aryl halide from benzene diazonium chloride ?
(a) Cu/HCl (b) CuI/HI
(c) CuBr (d) HBF_4, Heat.

38. Which sequence of steps will be able to produce 3, 3′-Dinitrobiphenyl from benzene ?
(a) HNO_3/H_2SO_4, $Cl_2/FeCl_3$, Na/Ether
(b) $Cl_2/FeCl_3$, HNO_3/H_2SO_4, Na/Ether
(c) $Cl_2/FeCl_3$, H_2SO_4, Na/Ether
(d) I_2/HIO_3, $Cl_2/FeCl_3$, $C_6H_5NO_2$.

39. Which of the following is metamer of diethylamine ?
(a) 2-N-Methylaminopropane
(b) 1-N-Methylpropanamine
(c) 2-N-Methylbutanamine
(d) Both (a) and (b).

40. $\text{Propene} \xrightarrow{Cl_2,\ 778\ K} \mathbf{X} \xrightarrow[\text{Excess}]{NH_3\ (alc.)} \mathbf{Y}$ *(Major product)*

The product Y in the above sequence is
(a) Allylamine (b) n-Propylamine
(c) 1, 2-Diaminopropane (d) 1, 1-Diaminopropane.

ANSWERS

1. (b)	2. (c)	3. (c)	4. (b)	5. (d)	6. (a)	7. (d)	8. (c)	9. (c)
10. (b)	11. (c)	12. (a)	13. (d)	14. (d)	15. (a)	16. (b)	17. (c)	18. (b)
19. (a)	20. (c)	21. (b)	22. (b)	23. (a)	24. (b)	25. (a)	26. (a)	27. (c)
28. (a)	29. (b)	30. (c)	31. (d)	32. (c)	33. (c)	34. (a)	35. (d)	36. (b)
37. (b)	38. (a)	39. (d)	40. (a).					

HINTS/SOLUTIONS

Q. No.	Ans.	Hints & Solutions
1.	(*b*)	$C_2H_5NH \xrightarrow{\text{Reduction}} C_2H_5NHCH_3$ Ethylisocynide $CH_3—CONHCH_3 \xrightarrow[H_2/Pt]{\text{Reduction}} CH_3CH_2NH—CH_3$ N-Methylethamide N-Methylethamide
2.	(*c*)	Alkyl cyanides on hydrolysis gives carboxylic acids and not 1° amine.
3.	(*c*)	The basic strength of $(CH_3)_3N$ is lowest. It is due to maximum crowding around N atom and also due to lesser extent of the hydration of protonated amine, $[(CH_3)_3NH]^+$.
4.	(*b*)	$RNH_2 + CS_2 \xrightarrow[\text{Heat}]{HgCl_2} R{=}N{=}C = S + 2HCl + HgS$
5.	(*d*)	*o*-aminosulphonic acid exists as a zwitter ion as 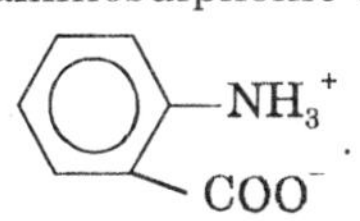
6.	(*a*)	The electrophilic attack of NO_3^+ ion on benzene is slowest step and also the rate controlling step.
7.	(*d*)	3° nitro compounds donot have α-H atom. Hence they cannot exhibit tautomerism.
8.	(*c*)	$R—N \rightleftharpoons C \longrightarrow HCOOH + RNH_2$ 1° Amine
9.	(*c*)	$(CH_3)_2CH = NOH \xrightarrow{LiAlH_4} (CH_3)_2CH—NH_2$ Acetoxime 1° Amine
10.	(*b*)	$\left[CH_3—\overset{O}{\overset{\Vert}{C}}—\ddot{N}H_2 \longrightarrow CH_3—\overset{O^-}{\overset{\vert}{C}} = \overset{+}{N}H_2\right]$
11.	(*c*)	$Cu^{2+} + 4C_2H_5NH_2 \longrightarrow [Cu(C_2H_5NH_2)_4]^{2-}$ Bluish Tetrakis [ethylamine] copper (II) ion (Deep blue complex)
12.	(*a*)	$C_6H_5NH_2 + S = C = S \xrightarrow{HgCl_2} C_6H_5—N = C = S$ Phenyl isothiocynanate
13.	(*d*)	Aldehydes and ketones react only with primary amines and form Schiff's bases.
14.	(*d*)	Direct treatment of aniline (without protection) with nitrating mixture leads to the formation of a black tarry mixture because of many oxidation products.
15.	(*a*)	N is surrounded by 3 bond-pairs and one lone pair. Hence, its, hybrid state should be sp^3.
16.	(*b*)	See basicity of amines in concept in brief.
17.	(*c*)	$C_6H_5COCl \xrightarrow{NH_3} C_6H_5CONH_2 \xrightarrow{P_2O_5} C_6H_5CN \xrightarrow{H_2/Ni} C_6H_5CH_2NH_2$ Benzyl amine
18.	(*b*)	$C_2H_5NH_2 + HONO \longrightarrow C_2H_5OH + N_2 + H_2O$ (1 mol) (1 mol) Thus, 1 mol of $C_2H_5NH_2$ gives out N_2 = 1 mol = 22.4 L at S.T.P.
19.	(*a*)	—NO_2 group is *meta* directing. ∴ *m*-Dinitrobenzene is produced. The presence of two deactivating NO_2 groups in *m*-dinitrobenzene checks further nitration.
20.	(*c*)	$C_2H_5\overset{+}{N_2}\overset{-}{Cl} \xrightarrow{CuCN} C_6H_5CN \xrightarrow{LiAlH_4} C_6H_5CH_2NH_2 \xrightarrow{HNO_2} C_6H_5CH_2OH$ (Benzyl alcohol)
21.	(*b*)	$C_6H_5NH_2 \xrightarrow[50°C]{NaNO_2/HCl} C_6H_5OH \xrightarrow{NaOH} C_6H_5ONa \xrightarrow{C_2H_5I} C_6H_5OC_2H_5$ (X) (Y) Phenetole (Z)
22.	(*b*)	Coloured dyes are produced by the reaction of phenols and benzene Diazonium chloride through coupling reaction. Benzyl alcohol does not contain phenolic group. Hence it does not form coloured compound with diazonium salt.
23.	(*a*)	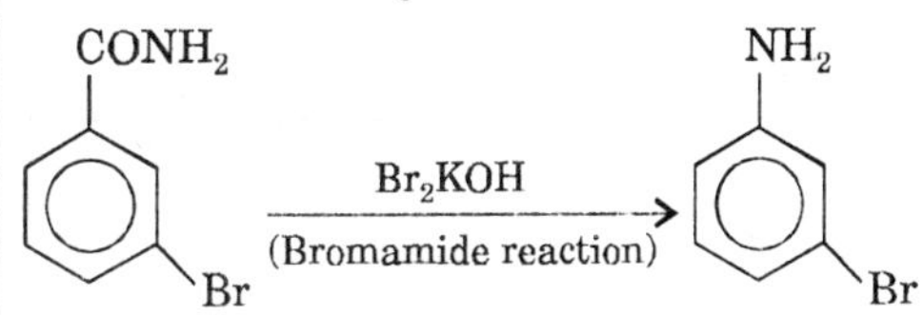 (*m*-Bromobenzamide) (*m*-Bromoaniline) $+ K_2CO_3 + 2KBr + H_2O$

Q. No.	Ans.	Hints & Solutions
24.	(*b*)	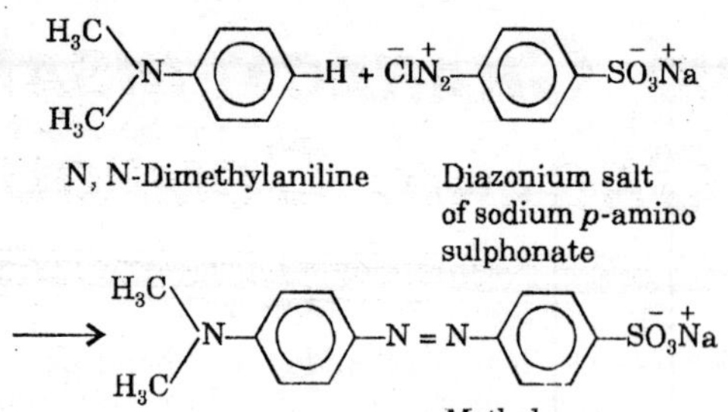 N, N-Dimethylaniline; Diazonium salt of sodium *p*-amino sulphonate; Methyl orange
25.	(*d*)	Various isomeric structures are : 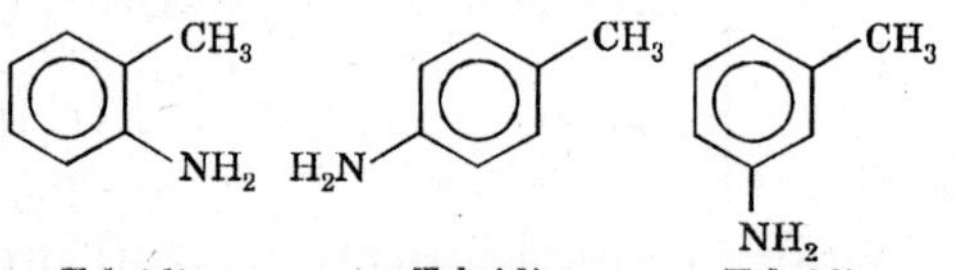 *o*-Toluidine, *p*-Toluidine, *m*-Toluidine 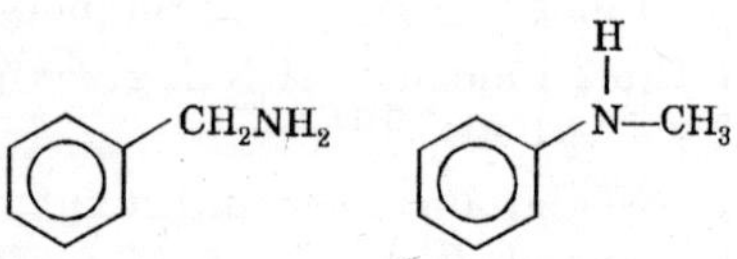 Benzylamine, N-Methylaniline
26.	(*a*)	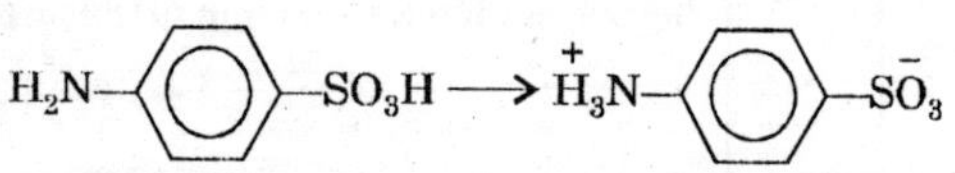 *p*-Aminobenzene sulphonic acid; Zwitter ion or Internal salt structure
27.	(*c*)	—CF_3 group cases withdrawal of electrons creating deficiency of electrons on the N atom, thus, making it least basic among the given species.
28.	(a)	Reduction of aldoxime give 1° Amine. Others will yield secondary amines.
29.	(*b*)	$R—C = N = O$ (alkyl isocyanate) is one of the intermediate of Hoffmann bromamide reaction.
30.	(*c*)	Tert. amines do not react with acetyl chloride or benzoyl chloride.
31.	(*d*)	Amides are amphoteric in nature.
32.	(*c*)	Aniline gives crystalline ppt. of aniline hydrochloride but is not soluble in water.
33.	(*c*)	Sec. amines do not give N_2 gas with HNO_2.
34.	(*a*)	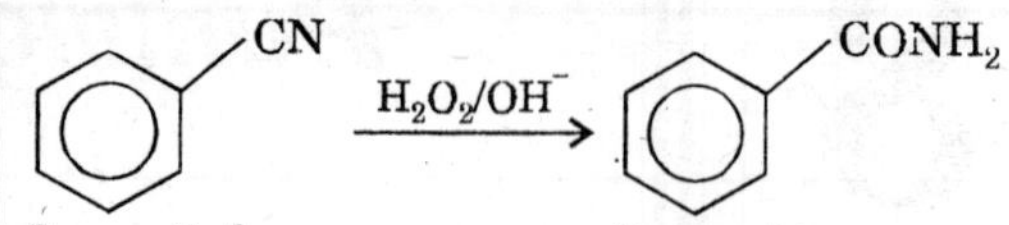 Benzonitrile; Benzamide 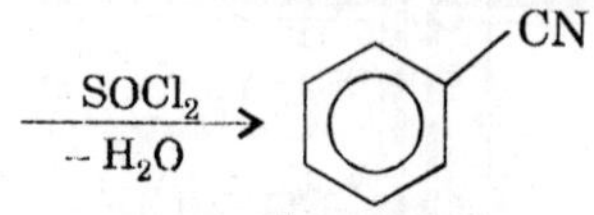 Benzonitrile
35.	(*d*)	Tert. amines produce triakyl ammonium nitrite with HNO_2.
36.	(*b*)	In anilinium ion $(C_6H_5NH_3)^+$ the lone pair is not present on the N atom.
37.	(*b*)	CuI does not produce iodobenzene from benzene diazonium chloride.
38.	(*a*)	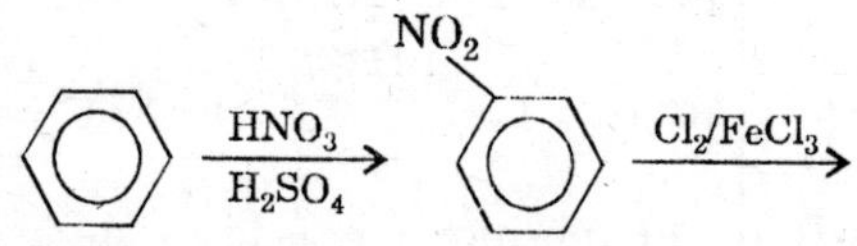 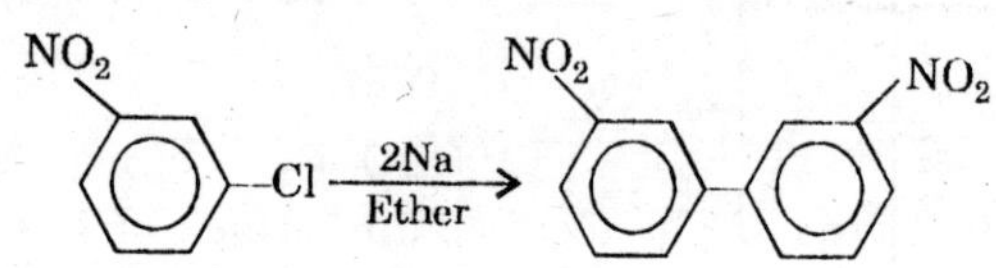 (*Fittig reaction*) 3, 3-Dinitrobiphenyl
39.	(*d*)	$CH_3CH_2—\overset{H}{\overset{\vert}{N}}—CH_2CH_3$ has following metamers $CH_3CH_2CH_2—\overset{H}{\overset{\vert}{N}}—CH_3$ and $CH_3—\overset{CH_3}{\overset{\vert}{CH}}—\overset{H}{\overset{\vert}{N}}—CH_3$.
40.	(*a*)	$CH_2 = CH - CH_3 \xrightarrow[785\ K]{Cl_2(g)} CH_2 = CH - CH_2Cl \xrightarrow[Excess]{NH_3} CH_2 = CH - CH_2NH_2$ Allylamine

8 PURIFICATION AND CHARACTERISATION OF ORGANIC COMPOUNDS

8.1. CONCEPTS IN BRIEF

Develop Logic

PURIFICATION OF ORGANIC COMPOUNDS

Different methods by which organic compounds can be purified are as follows :

1. **Simple Crystallization.** This is the most common method used to purify **organic solids.** It is based upon the fact that whenever a crystal is formed, it tends to leave out the impurities. For crystallization, a suitable solvent with following characteristics is selected :
   (i) *which dissolves more of the substance at higher temperature than at room temperature.*
   (ii) *in which impurities are either insoluble or dissolve to an extent that they remain in solution (in the mother liquor) upon crystallization.*
   (iii) *which is not highly inflammable, and*
   (iv) *which does not react chemically with the compound to be crystallized.* The most commonly used solvents for crystallization are : *water, alcohol, ether, chloroform, carbon tetrachloride, acetone, bezene, petroleum ether* etc.

   Some examples are :
   - A mixture of **benzoic acid** and **naphthalene** can be separated from hot water in which benzoic acid dissolves but naphthalene does not.
   - **Sugar** having impurity of **common salt** can be purified by crystallisation from hot ethanol. Sugar dissolves in hot ethanol but common salt does not.
2. **Fractional Crystallization.** *This is the process of separation of different components of a mixture by repeated crystallizations.* The mixture is dissolved in a solvent in which the two components have *different solubilities.* When a hot saturated solution of this mixture is allowed to cool, the less soluble component crystallises out first while the

more soluble substance remains in solution (*mother liquor*). The mother liquor left after crystallization of the less soluble component is again concentrated and then allowed to cool when the crystals of the more soluble component are obtained. The two components thus separated are recrystallized from the same or different solvent to yield both the components of the mixture in pure form.

3. **Sublimation.** *It is a process of direct conversion of a solid into the gaseous state on heating without passing through the intervening liquid state and vice-versa on cooling.* Only those substances whose vapour pressures become equal to the atmospheric pressure much before their respective melting points can be purified by this method. For example, *camphor, naphthalene, benzoic acid, anthracene, phthalic anhydride, ammonium chloride, iodine* etc.

   This method is particularly useful for the purification of such solids which sublime on heating and are associated with non-volatile impurities.

4. **Simple Distillation.** This method is used for purification of those liquids which do not decompose at their respective boiling points and contain non-volatile impurities. When the given impure liquid is heated to boiling, it gets converted into vapour and the vapours on cooling condense to give back the original liquid in the pure form while the non-volatile impurities remain in the flask. Thus, *the process of converting the liquid into its vapours and condensing the vapours back to give the original liquid in the pure form is called* **distillation.**

   Simple distillation can also be used to separate the mixture of liquids whose boiling points differ by 30 – 40°C. Here, the more volatile component distills over while the less volatile component distills afterwards. Some examples of liquid mixtures that can be separated by simple distillation are :
   - Mixture of C_6H_6 (b.p. 353 K) and $C_6H_5NH_2$ (b.p. 457 K)
   - Mixture of *ether* (b.p. 308 K) and *toluene* (b.p. 384 K)
   - Mixture of *n-Hexane* (b.p. 342 K) and *toluene* (b.p. 384 K).

5. **Fractional Distillation.** If the boiling points of the two liquids of a mixture are very close, *i.e.,* differ by 10—15 K, the separation cannot be achieved by simple distillation method. The reason being that at *the b.p. of the more volatile liquid of the mixture, there will be sufficient vapours of the less volatile liquid as well.* As a result, both the liquids of the mixture will distill together and the separation is not possible. The separation of such a liquid mixture into individual components can, however, be achieved by **fractional distillation** *which involves repeated distillations and condensations.* Fractional distillation is usually carried out using a *fractionating column.* The actual purpose of using a fractionating column is to increase the cooling surface area and to provide hurdles or obstructions to the ascending vapours and descending liquid.

   As a result of distillation and condensation at each point of the fractionating column, the vapours rising up become richer in more volatile component and the liquid falling back into the flask becomes richer in less volatile component. Thus, *the low boiling liquid distills first while the higher boiling liquid distils afterwards.*

   Some examples of mixtures which can be separated by fractional distillation technique are :

- *Acetone* (b.p. 329 K) and *methyl alcohol* (b.p. 338 K)
- Useful fractions such as *gasoline, kerosene oil, diesel oil, lubricating oil,* etc. are separated from *crude petroleum.*

6. **Azeotropic Distillation.** *Azeotropes are constant boiling mixtures.* For example, rectified spirit which contains ethanol and water in the ratio 95.87 : 4.13 by weight boils at 351.1 K (78.1°C). *The components of an azeotropic mixture cannot be separated by fractional distillation since both the components of the mixture will distill together in a fixed proportion by weight at a fixed temperature.* Therefore, a special type of distillation called azeotropic distillation is used in which a third low boiling liquid like benzene is added to the azeotropic mixture.

   This method has been commercially used to prepare *absolute alcohol* from *rectified spirit.* For this purpose, rectified spirit is mixed with a suitable amount of benzene and subjected to fractional distillation. The first fraction which distills at 337.8 K is a *ternary azeotrope* consisting of all the water, some alcohol and benzene (water 7.4%, benzene 74.1%, alcohol 18.5%). The second fraction which distils at 341.2 K is a *binary azeotrope* consisting of all the remaining benzene and some alcohol (benzene 67.6%, alcohol 32.4%). The last fraction which distills at 351 K is absolute alcohol.

7. **Vacuum Distillation or Distillation under Reduced Pressure.** If a liquid decomposes at or below its boiling point, it cannot be purified by simple distillation. Such liquids can, however, be purified by vacuum distillation. The underlying principle is that *a liquid boils at a temperature when its vapour pressure becomes equal to the atmospheric pressure. Obviously, the same liquid would boil at a lower temperature when, the pressure acting on it is reduced.* Since now the liquid boils at the lower temperature, it does not undergo decomposition.

   Glycerol which decomposes at its boiling point (563 K) under atmospheric pressure can be distilled without decomposition at 453 K under 12 mm Hg pressure. Similarly, sugarcane juice is concentrated in sugar industry by evaporation under reduced pressure.

8. **Steam Distillation.** This method is applicable for the separation and purification of those organic compounds (solids or liquids) which (*i*) *are insoluble in water* (*ii*) *are volatile in steam* (*iii*) *possess a high vapour pressure 10—15 mm Hg at 373 K and* (*iv*) *contain non-volatile impurities.*

In this method, a mixture of two immiscible substances, *i.e.,* water and the organic compound is heated. Each would exert its own vapour pressure independently of the other and the mixture would begin to boil at a temperature when the sum of the vapour pressure of the ogranic substance (p_1) and that of water (p_2) becomes equal to that of the atmospheric pressure (p).

$$p = p_1 + p_2$$

Unless the vapour pressure of the organic substance is zero, *the temperature at which the mixture boils will always be lower than the normal boiling points of both the organic substance and water.* In other words, the organic liquid boils at a temperature much lower than its normal b.p. and hence its decomposition is avoided. In this way, *steam distillation can be regarded analogous to distillation under reduced pressure.*

Examples. *Aniline* (b.p. 453 K) can be purified by steam distillation since it boils at a temperature of 371.5 K in presence of steam. Other compounds which can be purified by steam distillation are : *nitrobenzene, bromobenzene, o-nitrophenol, salicylaldehyde, o-hydroxyacetophenone, essential oils* etc.

9. Differential Extraction. This method is used to separate an organic compound present in aqueous solution by shaking it with a suitable organic solvent which is immiscible with water and in which the organic compound is more soluble than in water.

The efficiency of the process, however, depends upon the number of times the extraction is repeated. Thus, *with a given volume of the solvent, larger the number of extractions, more is the amount of the material extracted.*

10. Chromatography. It is the most modern and versatile method used for the separation and purification of organic compounds. It was discovered by the Russian Botanist *Tswett* in 1906. In this technique, *the separation and purification is brought about by the differential movement of the individual components of a mixture through a stationary phase under the influence of a mobile phase.* Depending upon the nature of the **stationary** and the **mobile phase,** the different types of chromatographic techniques commonly used are :

	Type of Chromatography	*Mobile / Stationary Phase*	*Applications*
1.	Adsorption or column chromatography	Liquid/Solid	Large scale separations.
2.	Thin-layer chromatography	Liquid/Solid	Qualitative analysis (identification and characterization of organic compounds).
3.	High performance liquid chromatography	Liquid/Solid	Qualitative and quantitative analysis.
4.	Gas-liquid chromatography (GLC)	Gas/Liquid	Qualitative and quantitative analysis.
5.	Paper or Partition chromatography	Liquid/Liquid	Qualitative and quantitative analysis of polar organic compounds (sugars, α-amino acids and inorganic compounds).

A. Column chromatography. In column chromatography, the *stationary phase* is a solid (*adsorbent*) such as silica gel, alumina, magnesium oxide, starch, charcoal etc. while the *mobile phase* is a pure solvent or a mixture of solvents of varying polarity. The solid adsorbent is made into a slurry with a suitable solvent (preferably non-polar) and packed in a glass column of suitable length. A concentrated solution of the organic mixture to be separated in a suitable solvent (preferably non-polar) is then applied to the top of the column *when the different components of the mixture are adsorbed to different extents depending upon their polarity.* Thereafter, the column is extracted with a suitable solvent called the *eluent.* As the eluent passes down the column, it dissolves the different components. The most weakly adsorbed component is eluted first by the least polar solvent while the more strongly adsorbed component is eluted later by the solvents of higher polarity. In this way, the various components of the mixture can be **separated.**

B. Thin Layer Chromatography (TLC). This is another type of chromatographic technique which involves separation of substances over a thin layer of adsorbent. A thin layer ($\approx$ 0.2 mm thick) of adsorbent (*silica gel* or *alumina*) is spread over a glass plate of suitable size. This is called *thin layer chromatographic plate or* **TLC plate.** A solution of mixture to be separated is applied as a small spot about 2 cm from one end of TLC plate. The plate is then placed in suitable solvent contained in a jar. As the solvent moves up it elutes the components and they also move up the plate but to a different distances depending on their relative degrees of adsorption and separation is affected. The relative adsorption of each component of the mixture is expressed in terms of its **retention factor** denoted as **R_f value.**

$$R_f = \frac{\text{Distance moved by the substance from base line}}{\text{Distance moved by the solvent from base line}}.$$

C. Paper chromatography. This is an example of **partition chromatography.** Here both the stationary phase and mobile phase are liquids. In this technique a strip of a **chromatographic paper** (*which contains water trapped in it that acts as stationary phase*) is spotted at the base by the solution of the mixture and is suspended in suitable solvent. The solvent rises the paper by capillary action and flows over the spot. The paper selectively retains different components according to their different partition in two phases. The paper strip so developed is called **chromatogram.** The spots of separated coloured compounds are visible at different heights from the position of initial spot on the chromatogram.

11. Chemical Methods. Besides above physical methods, a large number of chemical methods can also be used depending upon the nature of the substance to be purified. Some examples are :

- *Purification of commercial benzene* : *Commercial benzene* obtained from coal-tar distillation contains 3—5% *thiophene* as an impurity which can be removed by extraction with conc. H_2SO_4. *This purification is based upon the fact that thiophene undergoes sulphonation much more easily than benzene.* Thus, when commercial benzene is shaken with conc. H_2SO_4 in a separating funnel, thiophene undergoes sulphonation to form thiophene-2-sulphonic acid which dissolves in conc. H_2SO_4 while benzene does not.

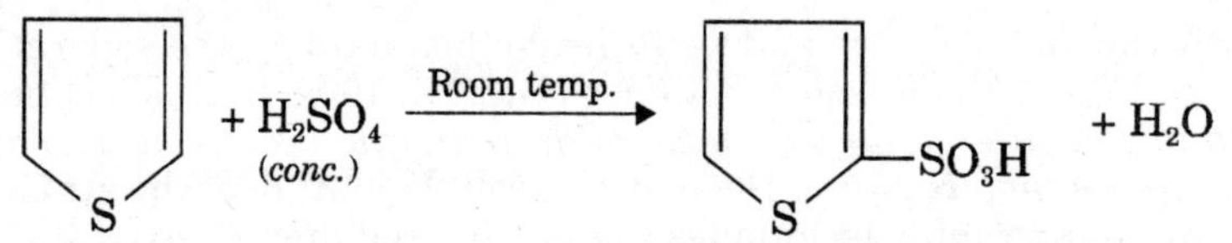

Thiophene → Thiophene-2-sulphonic acid (*Dissolves in conc.* H_2SO_4)

After this treatment, the benzene layer is removed, washed with water to remove unreacted H_2SO_4, dried over anhyd. $CaCl_2$ and then distilled to give pure benzene.

- **Preparation of absolute alcohol from rectified spirit** rectified spirit is kept over a calculated amount of quick lime (CaO) for few hours and is then refluxed. Water present is absorbed by CaO to form $Ca(OH)_2$. During refluxing absolute alcohol is distilled leaving behind $Ca(OH)_2$.
- **Separation of phenols from the carboxylic acids**

Phenols can be separated from carboxylic acids with an aqueous solution of $NaHCO_3$ when carboxylic acids dissolve evolving CO_2 but phenol donot react.

DETECTION OF ELEMENTS IN THE ORGANIC COMPOUNDS

1. Detection of Carbon and Hydrogen

Detection of C and H is carried out by heating the organic compound with dry *cupric oxide* when carbon is oxidized to CO_2 and hydrogen to H_2O . CO_2 is detected by lime water which turns milky while H_2O is detected by anhydrous $CuSO_4$ (*white*) which turns it *blue.*

2. Detection of Nitrogen

Nitrogen is generally detected by the following tests :

- **Soda-lime Test For Nitrogen.** The organic compound is heated with soda-lime when N of the organic compound is converted into NH_3. This test is, however, not reliable since certain compounds like nitro, azo etc. do not evolve NH_3 when heated with soda-lime.
- **Lassaigne's Test for Nitrogen.** The organic compound is fused with a small piece of Na metal when C and N of the organic compound combine to form NaCN. The red hot tube is broken in distilled water, boiled and filtered. The filtrate is called *Lassaigne's extract.* It is then heated with $FeSO_4$ solution in presence of alkali, the solution is cooled and acidified with dil. HCl. If a *green or blue colouration* is obtained, it confirms the presence of N in the organic compound. The chemistry of the test is

$$\text{Na} + \underset{\text{(From organic compound)}}{\text{C + N}} \xrightarrow{\Delta} \text{NaCN}$$

$$2NaCN + FeSO_4 \longrightarrow Fe(CN)_2 + Na_2SO_4$$

$$Fe(CN)_2 + 4NaCN \longrightarrow \underset{\text{Sod. ferrocyanide}}{Na_4[Fe(CN)_6]}$$

$$3Na_4[Fe(CN)_6] + 4Fe^{3+} \longrightarrow \underset{\text{Ferric ferrocyanide}}{Fe_4[Fe(CN)_6]_3} + 2Na^+$$

This test is very delicate and is given by all compounds containing C and N. The compounds such as NH_2NH_2, NH_4Cl, $NaNO_3$ etc. *do not respond to this test since they do not contain carbon.*

Sometimes on acidifying with dil. HCl, *blood red colour* is obtained. This indicates the presence of both N and S together.

$$Na + \underset{\text{(From organic compound)}}{C + N + S} \xrightarrow{\Delta} \underset{\text{Sod. thiocyanate or Sod. sulphocyanide}}{NaCNS}$$

$$Fe^{3+} + 3NaCNS \longrightarrow \underset{\text{Ferric thiocyanate (blood red colour)}}{Fe(CNS)_3} + 3Na^+$$

3. Detection of Sulphur

If **S** is present, Na_2S is formed then during fusion with Na metal. Na_2S in Lassaigne's extract may be tested as follows :

- On adding **sodium nitroprusside** solution to LE, violet colouration is produced.

$$Na_2S + \underset{\text{Sodium nitroprusside}}{Na_2[Fe(CN)_5NO]} \longrightarrow \underset{\text{(violet colour)}}{Na_4[Fe(CN)_5NOS]}$$

- On adding **lead acetate** solution to LE, black ppt. of PbS is formed

$$Na_2S + (CH_3COO)_2Pb \longrightarrow \underset{\text{(black ppt.)}}{PbS} + 2CH_3COONa$$

4. Detection of Halogens

The presence of halogens in organic compound is detected by following tests :

- *Beilstein's test.* A clean copper wire is heated in the Bunsen flame till it does not impart any green colour to the flame. The heated end is dipped in the organic compound and heated again. The appearance of a *green* or *bluish green* flame due to the formation of volatile cupric halides indicates the presence of some halogen in the organic compound.

  Though this test is very sensitive yet it does not confirm the presence of halogens in an organic compound since certain organic compounds like urea, thiourea etc. which do not contain halogens give this test due to the formation of volatile *cupric cyanide.*

- *Lassaigne's test.* When the organic compound is fused with Na metal, the halogens combine with Na to sodium halides :

$$Na + Cl \longrightarrow NaCl\ ; \quad Na + Br \longrightarrow NaBr\ ; \quad Na + I \longrightarrow NaI$$

The presence of these halides is tested with $AgNO_3$ solution.

(*i*) *A white ppt. soluble in NH_4OH and insoluble in dil. HNO_3 indicates Cl.*

$$NaCl + AgNO_3 \longrightarrow \underset{\text{(White ppt.)}}{AgCl} + NaNO_3\ ; AgCl + 2NH_4OH \longrightarrow \underset{\text{(Soluble silver complex)}}{[Ag(NH_3)_2]Cl} + 2H_2O$$

(*ii*) *A pale yellow ppt. partially soluble in ammonia indicates Br and*

(*iii*) *a yellow ppt. insoluble in ammonia indicates I.*

$$NaBr + AgNO_3 \longrightarrow \underset{\text{(Pale yellow ppt.)}}{AgBr} + NaNO_3\ ; NaI + AgNO_3 \longrightarrow \underset{\text{(Yellow ppt.)}}{AgI} + NaNO_3$$

If the organic compound also contains N or S, the sodium extract is first boiled with dil. HNO_3 to decompose any cyanides or sulphides.

$$NaCN + HNO_3 \longrightarrow NaNO_3 + HCN \uparrow \quad ; \quad Na_2S + 2HNO_3 \longrightarrow 2NaNO_3 + H_2S \uparrow$$

If cyanides or sulphides are not decomposed these will form ppt. with $AgNO_3$ solution.

$$NaCN + AgNO_3 \longrightarrow \underset{\text{(White ppt.)}}{AgCN} + NaNO_3 \; ; \quad Na_2S + 2AgNO_3 \longrightarrow 2NaNO_3 + \underset{\text{(Black ppt.)}}{Ag_2S}$$

5. Detection of Phosphorus

Phosphorus is detected by fusing the organic compound with sodium peroxide when phosphorus is converted into sodium phosphate.

$$\underset{\text{(From organic compound)}}{2P} + 5Na_2O_2 \longrightarrow 2Na_3PO_4 + 2Na_2O$$

The fused mass is extracted with H_2O, boiled with conc. HNO_3 and then ammonium molybdate is added. Appearance of *yellow ppt. or colouration* due to the formation of ammonium phosphomolybdate indicates the presence of phosphorus.

$$Na_3PO_4 + 3HNO_3 \xrightarrow{\Delta} H_3PO_4 + 3NaNO_3$$

$$H_3PO_4 + \underset{\text{Amm. molybdate}}{12(NH_4)_2MoO_4} + 21HNO_3 \longrightarrow \underset{\substack{\text{Amm. phosphomolybdate} \\ \text{(Yellow ppt.)}}}{(NH_4)_3PO_4.12MoO_3} + 21NH_4NO_3 + 12H_2O$$

ESTIMATION OF ELEMENTS

Estimation of elements in organic compound refers to the quantitative analysis of various elements in organic compound. It involves determination of %age composition of various elements. Various methods used are

1. Estimation of Carbon and Hydrogen

- **Liebig's method.** A known mass of the organic compound is heated strongly with excess of dry copper oxide in a current of dry air or oxygen (*free from CO_2*) when carbon present in the organic compound is oxidised to CO_2 and hydrogen to H_2O.

$$C + 2CuO \xrightarrow{\Delta} CO_2 + 2Cu \; ; \quad 2H + CuO \xrightarrow{\Delta} H_2O + Cu$$

CO_2 evolved is absorbed in a U-tube containing a conc. solution of KOH or NaOH + CaO mixture. The *increase in the mass of KOH tube gives the mass of CO_2 produced.*

H_2O evolved is absorbed in another U-tube containing anhydrous $CaCl_2$ or magnesium perchlorate. From the mass of CO_2 and H_2O produced, percentage of carbon and hydrogen is calculated as follows :

$$\textbf{\% of carbon} = \frac{12}{44} \times \frac{\text{Mass of } CO_2 \text{ formed}}{\text{Mass of substance taken}} \times 100$$

$$\textbf{\% of hydrogen} = \frac{2}{18} \times \frac{\text{Mass of } H_2O \text{ formed}}{\text{Mass of substance taken}} \times 100.$$

2. Estimation of Nitrogen

- **Dumas method.** *This method is of general applicability and nitrogen in all kinds of organic compounds can be estimated.* In this method, a known mass of the organic compound is heated strongly with excess of copper oxide in an atmosphere of CO_2 and C and H of the organic compound are oxidised to CO_2 and H_2O respectively while nitrogen is set free as N_2 gas. Any oxides of nitrogen, if formed, are reduced back to free nitrogen by passing over a hot reduced copper gauge.

$$C + 2CuO \xrightarrow{\Delta} CO_2 + 2Cu \; ; \quad 2H + CuO \xrightarrow{\Delta} H_2O + Cu$$

$$N + CuO \longrightarrow N_2 + \text{Small amounts of oxides of nitrogen}$$

$$\text{Oxides of nitrogen} + Cu \xrightarrow{\Delta} CuO + N_2.$$

The N_2 gas thus formed is collected over conc. KOH solution in a nitrometer. The volume of N_2 thus collected is noted and then reduced to STP by applying the gas equation, $P_1V_1/T_1 = P_2V_2/T_2$. From the volume of N_2 at STP, the percentage of nitrogen in the organic compound is calculated as follows :

$$\textbf{\% of N} = \frac{28 \times \text{Vol. of } N_2 \text{ (L) at STP} \times 100}{\text{Mass of substance} \times 22.4 \text{ L}}$$

- **Kjeldahl's method.** This is a more convenient and simpler method for estimation of nitrogen and is largely used for estimation of nitrogen in *fertilizers, drugs, food stuffs etc.* in the laboratory. This method is, however, not applicable to compounds containing nitrogen in the ring (*e.g.,* pyridine, quinoline etc.) and compounds containing nitro and *azo* (—N = N—) groups since nitrogen in these compounds is *not completely converted into* $(NH_4)_2SO_4$ *during digestion.*

  A known mass of the organic compound is heated with conc. H_2SO_4 when nitrogen present in the organic compound is quantitatively converted into ammonium sulphate. Ammonium sulphate thus, obtained is boiled with conc. NaOH solution. The NH_3 thus liberated is absorbed in a known excess of standard H_2SO_4 or HCl.

$$\text{N (from organic compound)} + \text{conc. } H_2SO_4 \xrightarrow{\Delta} (NH_4)_2SO_4$$

$$(NH_4)_2SO_4 + 2NaOH \longrightarrow Na_2SO_4 + 2H_2O + 2NH_3 \;; \qquad 2NH_3 + H_2SO_4 \longrightarrow (NH_4)_2SO_4.$$

  The volume of the standard acid left unused is found by titration against standard alkali solution. From the volume of acid used (or the volume of NH_3 evolved), the percentage of nitrogen is calculated as follows :

$$\%\text{ of N} = \frac{1.4 \times (\text{Normality of acid})(\text{Vol. of acid used (mL)})}{\text{Mass of the substance}}$$

or

$$\%\text{ of N} = \frac{1.4 \times (\text{Molarity of acid})(\text{Basicity of acid})(\text{Vol. of acid used (mL)})}{\text{Mass of the substance}}$$

3. Estimation of Halogens

- **Carius method.** A known mass of the organic compound is heated with excess of fuming HNO_3 and a **few crystals of $AgNO_3$ in a sealed tube called** *Carius tube.* Under these conditions, C is oxidised to CO_2 and H to H_2O while halogens are converted into silver halides. The ppt. of silver halide thus formed is filtered, washed, dried and weighed. From the mass of silver halide thus formed, the percentage of halogen can be estimated as follows :

$$\%\text{ of Cl} = \frac{35.5 \times \text{Mass of AgCl formed} \times 100}{143.5 \times \text{Mass of substance taken}}$$

$$\%\text{ of Br} = \frac{80 \times \text{Mass of AgBr formed} \times 100}{188 \times \text{Mass of substance taken}}$$

$$\%\text{ of I} = \frac{127 \times \text{Mass of AgI formed} \times 100}{235 \times \text{Mass of substance taken}}.$$

4. Estimation of Sulphur

- **Carius method.** A known mass of organic compound is heated with fuming HNO_3 in a *Carius tube* when C is oxidised to CO_2 and H to H_2O while S present in the organic compound is oxidised to H_2SO_4. Sulphuric acid thus formed is precipitated as $BaSO_4$ by adding excess of $BaCl_2$ solution. The ppt. of $BaSO_4$ is filtered, washed, dried and weighed. From the mass of $BaSO_4$ thus formed, the percentage of sulphur can be calculated as follows :

$$\%\text{ of S} = \frac{32 \times \text{Mass of } BaSO_4 \text{ formed} \times 100}{233 \times \text{Mass of substance taken}}.$$

5. Estimation of Phosphorus

A known mass of the organic compound is heated with fuming HNO_3 in a *Carius tube* when C is oxidised to CO_2, H to H_2O while P of the organic compound is oxidized to H_3PO_4. Phosphoric acid thus formed is precipitated as magnesium ammonium phosphate by adding *magnesia mixture* (a solution containing $MgCl_2$, NH_4Cl and NH_4OH.

$$MgCl_2 + NH_4Cl + H_3PO_4 \longrightarrow \underset{\text{Mag. ammonium phosphate}}{MgNH_4PO_4} + 3HCl$$

The ppt. of $MgNH_4PO_4$ are filtered, washed, dried and then ignited. The magnesium pyrophosphate thus formed is weighed.

$$2MgNH_4PO_4 \xrightarrow{\Delta} \underset{\text{Mag. pyrophosphate}}{Mg_2P_2O_7} + 2NH_3 + H_2O$$

Knowing the mass of $Mg_2P_2O_7$ obtained, the percentage of phosphorus can be calculated as follows :

$$\textbf{\% of P} = \frac{62 \times \text{Mass of } Mg_2P_2O_7 \text{ formed} \times 100}{222 \times \text{Mass of substance taken}}.$$

6. **Estimation of Oxygen.** There is no direct method for estimation of oxygen. It is obtained by the *difference method.*

$$\textbf{\% of O} = 100 - (\text{Sum of \% of all other elements}).$$

DETERMINATION OF MOLECULAR MASS

Different methods for the determination of molecular masses are as follows :

1. **Victor Meyer's Method**

This method is used for determination of molecular mass of *volatile substances.* In this method, a known mass of the volatile substance (say **wg**) is heated in a *Victor Meyer's* tube and the vapours of the substance thus formed displace an equal volume of air which is collected over water. The volume of air is then reduced to STP by applying gas equation,

$$P_1V_1/T_1 = P_2V_2/T_2.$$

From the volume of the air displaced at STP (V cm^3), the molecular mass of the volatile substance is calculated as follows :

$$\textbf{Mol. Mass} = \frac{W_{(g)} \times 22400}{V_{(cm^3)} \text{ at STP}}.$$

2. **Volumetric Method**

This method is used for determination of molecular masses of acids or bases. A known mass of acid say $w(g)$ is dissolved in water to make the total volume solution as $V_{(L)}$.

A known volume of this solution (say V_1) is then titrated with some standard solution of base whose molarity is known (say M_2). From the volume of acid solution taken (say V_1) and that of base consumed for titration (say V_2), the molarity (M_1) of acid solution is calculated by applying the stoichiomeric equation $\frac{M_1V_1}{x} = \frac{M_2V_2}{y}$. The molecular mass is then calculated by applying the relation :

$$\textbf{Mol mass} = \frac{w(g)}{M_1 \times \text{Volume of acid solution } (V_L)}.$$

3. **Silver salt Method.** This method is applicable to organic acids only. A known mass of organic acid is dissolved in water and excess of $AgNO_3$ sol. is added till precipitation of silver salt is complete. The precipitate is separated, dried, weighed and ignited till it is completely decomposed to form silver. From the mass of silver salt precipitated (***w* g**) and that of silver formed (***x* g**), the molar mass of organic acid can be calculated as follows. If n is the basicity of acid, then

$$\textbf{Mol. mass of acid} = n\left(\frac{108\ W}{x} - 107\right).$$

4. Platinic Chloride method

This method is applicable to organic bases. A known mass of organic base is allowed to react with chloroplatinic acid (H_2PtCl_6) in conc. HCl to form precipitate of platinic chloride. The precipitate is separated, dried, weighed and ignited to get free platinum. From the mass of platinic chloride (***w g***) and that of platinum liberated (***x* g**), the molecular mass of base is calculated as follows. If n is acidity of the base, then

$$\textbf{Molar mass of base} = \frac{n}{2}\left(\frac{195\ \text{W}}{x} - 410\right).$$

5. Mass Spectrometry

It is modern technique of determining molecular mass of the substance. In this method a gaseous substance is bombarded with high speed electron beam. These high speed electrons ionise the molecular of the substance and also decompose the molecules in smaller fragments. Different fragments have different mass to charge ratio (m/e). The spectrometer is so adjusted that bulk of ions produced carry unit positive charge *i.e.*, $e = 1$ so that its mass corresponds to m/e ratio. The positive ions formed in ionisation chamber are accelerated and are also subjected to magnetic field when they deflect into circular paths depending upon their m/e ratio, and finally strike the detector in different regions. The output of mass spectrometer is expressed in the form of a plot of relative intensity of mass of the ion.

The most intense peak in the spectrum (called base peak, and mass corresponding to base peak is molecular mass of substance.

EMPIRICAL AND MOLECULAR FORMULAE

- **Empirical formula** *of a substance gives the simplest whole number ratio between the atoms of the various elements present in one molecule of the substance.* For example, empirical formula of glucose is CH_2O, *i.e.*, for each carbon atom, there are two H-atoms and one oxygen atom, its molecular formula is, however $C_6H_{12}O_6$.
- **Calculation of empirical formula.** The main points are :
  - (*i*) Divide the percentage of each element by its atomic mass. This gives the *relative number of atoms.*
  - (*ii*) Divide the figures obtained in step (*i*) by the lowest one. This gives the *simplest ratio* of the various elements present.
  - (*iii*) If the simplest ratio obtained in step (*ii*) is not a whole number ratio, then multiply all the figures with a suitable integer *i.e.*, 2, 3 etc. to make it *simplest whole number ratio.*
  - (*iv*) Write down the symbols of the various elements side by side with the above numbers at the lower right corner of each. This gives the *empirical or the simplest formula.*
- **Molecular formula** *of a substance gives the actual number of atoms present in one molecule of the substance.*

$$\text{Molecular formula} = n \times \text{Empirical formula}$$

where n is a simple integer 1, 2, 3 ...... etc. given by the equation,

$$n = \frac{\text{Molecular mass of the compound}}{\text{Empirical formula mass of the compound}}$$

where molecular mass of the compound is determined experimentally by any one of the methods discussed above, empirical formula mass is calculated by adding the atomic masses of all the atoms present in the empirical formula.

EUDIOMETRY

Eudiometry is a direct method for determination of molecular formula of gaseous hydrocarbons without determining the percentage composition of various elements in it and without knowing the molecular mass of the hydrocarbon. The various steps involved are :

(*i*) A known volume of the gaseous hydrocarbon is mixed with an excess of oxygen in the eudiometer tube kept in a trough of mercury.

(*ii*) The mixture is exploded by passing an electric spark between the platinum electrodes. As a result, carbon and hydrogen of the hydrocarbon are oxidised to CO_2 and H_2O vapours respectively.

(*iii*) The tube is allowed to cool to room temperature when water vapours condense to give liquid water which has a negligible volume as compared to the volume of water vapours. Thus, the *gaseous* mixture left behind in the eudiometer tube after explosion and cooling consists of only *CO_2 and unused O_2.*

(*iv*) Caustic potash or caustic soda solution is then introduced into the eudiometer tube which absorbs CO_2 completely and only unused O_2 is left behind.

$$2NaOH + CO_2 \longrightarrow Na_2CO_3 + H_2O.$$

The decrease in volume on introducing NaOH or KOH solution gives the volume of CO_2 formed.

Sometimes, the volume of O_2 left unused is found by introducing pyrogallol and noting the decrease in volume.

Calculations. From the volume of CO_2 formed and the total volume of O_2 used, it is possible to calculate the molecular formula of gaseous hydrocarbon with help of the following equation,

$$\underset{1\text{ vol}}{C_xH_y(g)} + \underset{(x+y/4)\text{ vol}}{(x + y/4)\,O_2(g)} \longrightarrow \underset{x\text{ vol}}{x\,CO_2\,(g)} + \underset{y/2\text{ vol}}{y/2\,H_2O(l)}$$

(*negligible volume on condensation*)

From the above equation, it is evident that for one volume of hydrocarbon,

(*i*) $(x + y/4)$ volume of O_2 is used

(*ii*) x volume of CO_2 is produced

(*iii*) $y/2$ volume of H_2O vapours is produced which condense to give liquid H_2O with negligible volume.

(*iv*) contration in volume after explosion and cooling $= [(1 + x + y/4) - x] = 1 + y/4$.

By equating the experimental values with the theoretical values from the above combustion equation, the values of x and y and hence the molecular formula of the gaseous hydrocarbon can be easily determined.

IMPORTANT FORMULAE

1. Estimation of C and H

$$\%\,C = \frac{12 \times W_{(CO_2)} \times 100}{44 \times W_{(substance)}}$$

$$\%\,H = \frac{2 \times W_{(H_2O)} \times 100}{44 \times W_{(substance)}}$$

2. Estimation of Nitrogen

(*a*) **Dumas method**

$$\%\,N = \frac{28 \times V_{(N_2)}\,(\text{in Litre at STP}) \times 100}{22.4 \times W_{(substance)}}$$

(*b*) **Kjeldahl's method**

$$\%\,N = \frac{1.4 \times N_{(acid)}\,V_{(acid\ used)}}{W_{(substance)}}$$

$$= \frac{1.4 \times M_{(acid)} \times \text{Basicity}_{(acid)} \times V_{(acid)}}{W_{(substance)}}$$

3. Estimation of Sulphur

$$\%\,S = \frac{32 \times W_{(BaSO_4)} \times 100}{233 \times W_{(substance)}}$$

4. Estimation of halogens

$$\%\,Cl = \frac{35.5 \times W_{(AgCl)} \times 100}{143.5 \times W_{(substance)}}$$

$$\%\,Br = \frac{80 \times W_{(AgBr)} \times 100}{188 \times W_{(substance)}}$$

$$\%\,I = \frac{127 \times W_{(AgI)} \times 100}{235 \times W_{(substance)}}$$

5. Estimation of Phosphorus

$$\%\,P = \frac{62 \times W_{(Mg_2P_2O_7)} \times 100}{222 \times W_{(substance)}}$$

6. Victor Meyer Method

$$\text{Mol. mass} = \frac{W_{(subs.)} \times 22400}{V_{(mL)}\ \text{at STP (of displaced air)}}$$

7. Silver salt method

$$\text{Mol. mass (acid)} = n\left(\frac{108\,W}{x} - 107\right)$$

n = basicity of acid

W = mass of silver salt

x = mass of silver.

8. Platinic Chloride Method

$$\text{Mol. mass (bases)} = \frac{n}{2}\left(\frac{195\,W}{x} - 410\right)$$

n = acidity of base

W = mass of platinic salt

x = mass of platinum.

8.2. IMPORTANT FACTS AND TERMS

- Mixture of KCl and $KClO_3$ can be separated by fractional crystallisation. KCl is more soluble in hot water whereas $KClO_3$ is less soluble in hot water.
- The process of crystallisation can be made rapid by **seeding.** It involves the addition of a few small crystals of pure substance to the concentrated solution to *induce* crystallisation.
- Azeotropes cannot be separated by fractional distillation.
- Water soluble compounds such as *urea, oxalic acid,* etc., can be separated from water insoluble compounds such as naphthalene, *p*-dichlorobenzene, *m*-dinitrobenzene, etc, by **water**.
- *Steam distillation* can be regarded as analogous to *distillation under reduced pressure.*
- Carbohydrates are insoluble in ether. These are, therefore, separated from ether soluble organic compounds such as *m*-dinitrobenene, aspirin, *p*-dichlorobenzene naphthalene etc. by **ether.**
- Water insoluble carboxylic acids are separated from non-acidic compounds by shaking with an aqueous solution of $\mathbf{NaHCO_3}$.
- Water insoluble amines can be separated from non-basic organic compounds by shaking with a dil. solution of **HCl.**
- Water insoluble phenols can be separated from non-acids compounds by shaking with an aqueous solution of **NaOH.**
- **Mixed melting point** is used to identify whether the two unknown compounds having approx. the same melting point, are identical or not.
- **Beilstein's test** is not given by fluorine since cupric fluoride is not volatile.
- A freshly prepared $FeSO_4$ solution is used in the Lassaigne's test for nitrogen because on keeping $FeSO_4$ solution is oxidised to basic ferric sulphate which cannot be used for detection.
- **Lithium** is not used in Lassaigne's test since it reacts slowly and its compounds are generally covalent. **Potassium** cannot be used since it reacts violently and cannot be handled properly.
- In Kjeldahl's method for estimation of nitrogen, $\mathbf{K_2SO_4}$ is added to raise the boiling point of H_2SO_4 to ensure complete conversion of nitrogen into $(NH_4)_2SO_4$ while $\mathbf{CuSO_4}$ **or mercury** is added to catalyse the above conversion.
- Estimation of sulphur in organic compound is also done by **Messenger's method**. In this method, the organic compound is heated with alkaline $KMnO_4$ solution when sulphur of the organic compound is oxidised to K_2SO_4 which is then estimated as $BaSO_4$.
- For continuous extraction of organic compounds with a minimum amount of organic solvent, **soxhlet extractor** is used. For example, the flavouring agent vanillin is extracted from vanilla beans by this apparatus.
- An alkaline solution of **pyrogallol,** *i.e.,* 1, 2, 3-trihydroxybenzene or benzene-1, 2, 3-triol is used to absorb oxygen.

8.3. SOLVED PROBLEMS

Enrich your Skill

Example 1. *0.60 g of a silver salt of an acidic organic compound on ignition gave 0.36 g of pure silver. If molar mass of the compound is 219 g mol^{-1} the basicity of acid is*

(*a*) 4 (*b*) 3 (*c*) 2 (*d*) 1.

Solution. Mol mass (Acid) $= n\left(\dfrac{108\text{ W}}{x} - 107\right)$

or $n = \dfrac{\text{Mol. mass}}{\left(\dfrac{108\text{ W}}{x} - 107\right)}$, Here W = 0.6 ; x = 0.36

$$= \frac{219}{\dfrac{108 \times 0.6}{0.36} - 107} = \frac{219}{180 - 107}$$

$$= \frac{219}{73} = \mathbf{3.}$$

Hence correct choice is **(*b*)**.

Example 2. *A hydrocarbon has empirical formula CH_2 and its density at STP is 1.25 gL^{-1}. The molecular formula of hydrocarbon is*

(*a*) C_2H_4 (*b*) C_3H_6 (*c*) C_3H_8 (*d*) C_6H_8.

Solution. Molar mass of hydrocarbon

= Mass of 22.4 L of it at STP

= 1.25 × 22.4

= 28 g mol^{-1}

Now the ratio, $\dfrac{\text{Mol. mass}}{\text{Empirical mass}} = \dfrac{28}{14} = 2$

∴ Mol. formula = $(CH_2)_2$ = $\mathbf{C_2H_4}$

Hence correct choice is **(*a*)**.

Example 3. *0.5 g of an organic compound was Kjeldahlised and NH_3 evolved was absorbed in a certain volume of 1 N H_2SO_4. The residual acid required 60 cm^3 of N/2 NaOH. If the % of N in organic compound is 56% then volume of 1 N H_2SO_4 taken was*

(*a*) 40 mL (*b*) 80 mL (*c*) 100 mL (*d*) 50 mL.

Solution. 60 cm^3 of N/2 NaOH ≡ v ml of 1 N H_2SO_4 left unused

or $60 \times \dfrac{1}{2} = v \times 1$ or v = 30 ml

Let volume of acid taken = V ml

Volume of 1 N acid used by NH_3 = (V – 30) mL

$$\% \text{ N} = \frac{1.4 \times \text{N}_{(\text{acid})} \times \text{V}_{(\text{acid})}}{\text{W}_{\text{subst.}}}$$

or $\text{V}_{(\text{acid})} = \dfrac{\% \text{ N} \times \text{W}_{\text{subs.}}}{1.4 \times \text{N}_{(\text{acid})}} = \dfrac{56 \times 0.5}{1.4 \times 1} = 20$ mL

Now V – 30 = 20 or V = 20 + 30 = **50 mL**

Hence correct answer is **(*d*)**.

Example 4. *0.98 of chloroplatinate salt of some mono acid base when ignited left 0.39 g of platinum as residue. The molar mass of the base is*

(*a*) 80 g mol^{-1} (*b*) 160 g mol^{-1}

(*c*) 40 g mol^{-1} (*d*) 240 g mol^{-1}.

Solution. $\text{M}_{(\text{Base})} = \dfrac{n}{2}\left(\dfrac{\text{W} \times 195}{x} - 410\right)$

Here n = 1, W = 0.98 g, x = 0.39 g

$$\text{M}_{\text{Base}} = \frac{1}{2}\left(\frac{0.98 \times 195}{0.39} - 410\right) = \mathbf{40\ g\ mol^{-1}}$$

Hence correct choice is **(*c*)**.

Example 5. *3.75 g of an organic acid is dissolved in water and solution has a volume of 500 mL. 20 ml of this solution required 25 mL of M/5 NaOH. The basicity of acid (whose molar mass is 90 g mol^{-1}) is*

(*a*) 1 (*b*) 4 (*c*) 3 (*d*) 2.

Solution. Let basicity of acid is n

Equation is $H_nX + n\text{NaOH} \longrightarrow Na_nX + nH_2O$

Now $\dfrac{M_1V_1}{1} = \dfrac{M_2V_2}{n}$ or $M_1 = \dfrac{M_2V_2}{nV_1} = \dfrac{25 \times 1}{5 \times 20 \times n}$

Molarity of acid $(M_1) = \dfrac{1}{4n}$

$$\text{M} = \frac{\text{W}_{\text{acid}} \times 1000}{\text{GMM}_{\text{acid}} \times \text{V}_{(\text{ml})}}$$

or $\dfrac{1}{4n} = \dfrac{3.75 \times 1000}{90 \times 500}$ or $n = \dfrac{1 \times 90}{3.75 \times 2 \times 4} = \mathbf{3.}$

Hence correct choic is **(*c*)**.

Example 6. *An organic compound containing C, H and O has 52.2% carbon and 13.04% hydrogen V.D. of the compound is 23. The molecular formula of the compound is* (*a*) C_2H_6O (*b*) C_4H_8O

(*c*) C_3H_8O (*d*) $C_5H_{10}O$.

Solution. % of O = 100 – (52.2 + 13.04) = 34.76

$$C : H : O = \frac{52.2}{12} : \frac{13.04}{1} : \frac{34.76}{16}$$

$$= 4.35 : 13.04 : 2.17$$

$$= 2 : 6 : 1$$

Thus, empirical formula is $\mathbf{C_2H_6O}$

V.D. = 23 ∴ Molar mass = 23 × 2 = 46 g mol^{-1}

$$\frac{\text{Mol. mass}}{\text{Empirical mass}} = \frac{46}{46} = 1$$

∴ Mol. formula is $\mathbf{C_2H_6O}$

Hence correct choice is (***a***).

Example 7. *10 mL of gaseous hydrocarbon was exploded with excess of oxygen in eudiometer and subsequent cooling resulted in contraction of 20 ml in volume. On further shaking gaseous content with conc. NaOH further contraction of 20 mL was noticed. Assuming all measurement under similar conditions. The molecular formula of hydrocarbon is*

(*a*) C_2H_6 (*b*) C_2H_4 (*c*) C_3H_6 (*d*) CH_4.

Solution. Let hydrocarbon be C_xH_y

$$C_xH_y + (x + y/4)O_2 \longrightarrow xCO_2 + y/2\ H_2O(l)$$

$$10 \text{ mL} \qquad 10\left(x + \frac{y}{4}\right) \text{mL} \qquad 10\,x \text{ mL}$$

Contraction in volume after reaction

$$= 10 + 10\left(x + \frac{y}{4}\right) - 10x$$

$$= 10 + \frac{y}{4}$$

Now $10 + \frac{y}{4} = 20$ or $y = 4$

Volume of CO_2 produced = 20 mL

or $10x = 20$ or $x = 2$

The formula is $\mathbf{C_2H_4}$

Hence correct choice is (***b***).

8.4. QUESTION BANK

Develop Confidence

Purification of Organic Compounds

1. Azeotropic mixture of two liquids can be separated by
   (*a*) Simple distillation (*b*) Fractional distillation
   (*c*) Azeotropic distillation (*d*) Steam distillation.
2. From azeotropic mixture of EtOH and H_2O, absolute ethanol can be obtained by
   (*a*) Crystallisation
   (*b*) Chemical method
   (*c*) Distillation under reduced pressure
   (*d*) Steam distillation.
3. Which among the following can be purified by steam distillation ?
   (*a*) Cresol (*b*) Aniline
   (*c*) Benzoic acid (*d*) *p*-Nitrophenol.
4. Steam distillation is used for the extraction of
   (*a*) Essential oils (*b*) Higher alkynes
   (*c*) Mineral oils (*d*) Fatty acids.
5. The separation of an organic compound from its aqueous solution can be achieved by
   (*a*) Distillation (*b*) Steam distillation
   (*c*) Solvent extraction (*d*) Fractional distillation.
6. Chromatographic technique was introduced by
   (*a*) Duma (1803) (*b*) Kjeldahl (1903)
   (*c*) Tswett (1906) (*d*) Lavosier.
7. The separation of the constituents of a mixture by column chromatography depends upon their
   (*a*) Different solubilities
   (*b*) Different boiling points
   (*c*) Different refractive indices
   (*d*) Differential adsorption.
8. In paper chromatography,
   (*a*) the mobile phase is a liquid and the stationary phase is a solid
   (*b*) the mobile phase is a solid and the stationary phase is a liquid
   (*c*) both phases are liquids
   (*d*) both phases are solids.
9. The technique of gas chromatography is suitable for compounds which
   (*a*) Are liquids
   (*b*) Have very low boiling points
   (*c*) Are soluble in water
   (*d*) Vapourize without decomposition.
10. A mixture of camphor and benzoic acid can be separated by which of the following technique ?
    (*a*) Chemical methods
    (*b*) Sublimation
    (*c*) Fractional distillation
    (*d*) Extraction with a solvent.
11. Separation of two substances by fractional crystallisation depends upon their difference in
    (*a*) Densities (*b*) Solubilities
    (*c*) Melting point (*d*) Boiling points.
12. Sugar containing an impurity of common salt can be purified by crystallization from
    (*a*) Benzene (*b*) Ethanol
    (*c*) Petroleum ether (*d*) Water.
13. A mixture of benzoic acid and naphthalene can be separated by crystallization from
    (*a*) Hot water (*b*) Cold water
    (*c*) Benzene (*d*) Ether.
14. When a solid vapourises directly without melting, the process is called
    (*a*) Evaporation (*b*) Sedimentation
    (*c*) Saponification (*d*) Sublimation.
15. Naphthalene can be purified by
    (*a*) Sublimation
    (*b*) Crystallization
    (*c*) Extraction with a solvent
    (*d*) Distillation.
16. Sublimation cannot be used for purification of
    (*a*) Naphthalene (*b*) Camphor
    (*c*) Benzoic acid (*d*) Urea.
17. The boiling point of a compound does not depend upon
    (*a*) Solubility of the compound in water
    (*b*) Hydrogen bonding
    (*c*) Size of molecules
    (*d*) Polarity of the molecule.
18. By decreasing the pressure, the boiling point of a liquid
    (*a*) Increases
    (*b*) Decreases

(*c*) Remains constant

(*d*) May increase or decrease.

19. Simple distillation can be used to separate

(*a*) A mixture of benzene (b.p. 80°C) and thiophene (b.p. 84°C)

(*b*) A mixture of ethanol (b.p. 78.1°C) and water (b.p. 110°C)

(*c*) A mixture of ether (b.p. 35°C) and toluene (b.p. 110°C)

(*d*) None of the above.

20. Two volatile liquids A and B differ in their boiling points by 15°C. The process which can be used to separate them is

(*a*) Fractional distillation

(*b*) Steam distillation

(*c*) Distillation under reduced pressure

(*d*) Simple distillation.

21. A mixture of *acetone* and *methanol* can be separated by

(*a*) Vacuum distillation (*b*) Steam distillation

(*c*) Fractional distillation (*d*) Simple distillation.

22. Refining of petroleum involves the process of

(*a*) Simple distillation

(*b*) Fractional distillation

(*c*) Steam distillation

(*d*) Distillation under reduced pressure.

23. A substance which decomposes below its boiling point can be best purified by

(*a*) Simple distillation

(*b*) Steam distillation

(*c*) Fractional distillation

(*d*) Distillation under reduced pressure.

24. Aniline is insoluble in water and possesses a vapour pressure of 10—15 mm Hg at 373 K. It can be conveniently purified by

(*a*) Sublimation (*b*) Crystallization

(*c*) Steam distillation (*d*) Simple distillation.

25. In steam distillation, the sum of the vapour pressure of the volatile compound and that of water is

(*a*) Equal to atmospheric pressure

(*b*) Less than atmospheric pressure

(*c*) More than atmospheric pressure

(*d*) Exactly half of the atospheric pressure.

26. A mixture of *o-hydroxyacetophenone* and *p-hydroxyacetophenone* can be separated by

(*a*) Sublimation

(*b*) Steam distillation

(*c*) Fractional crystallization

(*d*) Distillation.

Detection of Elements in Organic Compound

27. The formation of violet colour when sodium nitroprusside is added to sodium extract indicates the presence of

(*a*) Nitrogen (*b*) Sulphur

(*c*) Oxygen (*d*) Halogens.

28. During the test of sulphur in organic compound, the purple colour formed by adding sodium nitroprusside to L.E. is due to

(*a*) $Na_4[Fe(CN)_5NO]$ (*b*) $Na_4[Fe(CN)_5NOS]$

(*c*) $Na_3[Fe(CN)_5NS]$ (*d*) $Fe_4[Fe(CN)_6]_3$.

29. For detection of sulphur in an organic compound, the Lassaigne's extract is acdified with dilute acetic acid and lead acetate is then added to it. The black ppt. is obtained which is due to the formation of

(*a*) PbS (*b*) Na_2S

(*c*) $PbSO_4$ (*d*) PbS_2O_3.

30. A compound which does not give a positive test in the Lassaigne's test for nitrogen is

(*a*) Urea (*b*) Hydroxyl amine

(*c*) Glycine (*d*) Phenylhydrazine.

31. Which of the following elements in an organic compound cannot be detected by Lassaigne's test ?

(*a*) N (*b*) S

(*c*) Cl (*d*) H.

32. In Lassaigne's test, the organic compound is fused with a piece of sodium metal in order to

(*a*) Increase the ionization of the compound

(*b*) Decrease the melting point of the compound

(*c*) Increase the reactivity of the compound

(*d*) Convert the covalent compound into a mixture of ionic compounds.

33. Which of the following statements is not applicable to Lassaigne's test ?

(*a*) The elements (N, S and halogens) are tested as their inorganic salts

(*b*) During fusion of the organic compound with sodium, Na_2S, NaCN and NaX are formed

(*c*) It can even be used for the detection of oxygen in the organic compound

(*d*) It is possible to differentiate between the different halognes (chlorine, bromine and iodine).

34. Concentrated nitric acid is added to Lassaigne's extract in detection of halogens to

(*a*) Increase the concentration of nitrate ions

(*b*) Increase the solubility product of AgCl

(*c*) Help in the precipitation of AgCl

(*d*) Decompose Na_2S and NaCN, if formed.

35. When an organic compound containing phosphorus is fused with fusion mixture, it gives

(*a*) Na_2HPO_4 (*b*) Na_3PO_4
(*c*) NaH_2PO_4 (*d*) Na_3PO_3.

36. For the detection of phosphorus, the organic compound after fusion with Na_2O_2 is extracted with water, boiled with HNO_3 and then ammonium molybdate is added to it. A yellow ppt. is obtained which is due to the formation of

(*a*) Ammonium phosphate
(*b*) Ammonium phosphomolybdate
(*c*) Ferric phosphate
(*d*) Disodium ammonium phosphate.

37. The presence of carbon in an organic compound is detected by heating it with

(*a*) Sodium metal to convert it into NaCN
(*b*) CaO to convert it into CO which burns with a blue flame
(*c*) CuO to convert it into CO_2 which turns lime water milky
(*d*) Cu wire to give a bluish-green flame.

38. The water formed during detection of hydrogen is passed over

(*a*) Anhydrous $MgSO_4$ (*b*) Anhydrous $ZnCl_2$
(*c*) Anhydrous $AlCl_3$ (*d*) Anhydrous $CuSO_4$.

39. Copper wire test for halogens is known as

(*a*) Beilstein's test (*b*) Duma's test
(*c*) Liebig's test (*d*) Lassaigne's test.

40. The Beilstein's test is a rapid test used for organic compounds to detect

(*a*) Phosphorus (*b*) Sulphur
(*c*) Halogens (*d*) Nitrogen.

41. Which of the following statements is not applicable to Beilstein test ?

(*a*) Green or bluish-green flame is due to the formation of volatile cupric halides
(*b*) It does nto tell as to which halogen (chlorine, bromine or iodine) is actually present in the organic compound
(*c*) It is a very sensitive test and can be easily performed
(*d*) It is a sure test for the presence of halogens.

42. Amongst the following elements present in an organic compound, the one which has no direct test for detection is

(*a*) Cl (*b*) N
(*c*) S (*d*) O.

43. In the Lassaigne's test for the detection of nitrogen in an organic compound, the blue or green colour is due to the formation of

(*a*) $K_3[Fe(CN)_6]$ (*b*) $K_4[Fe(CN)_6]$
(*c*) $Fe(CN)_6$ (*d*) $Fe_4[Fe(CN)_6]_3$.

44. In the Lassaigne's test, the blood red colouration is due to the formation of

(*a*) $Fe(CNS)_2$ (*b*) NaCNS
(*c*) NH_4CNS (*d*) $Fe(CNS)_3$.

45. For which of the following compound, the Lassaigne's test for N will not be positive ?

(*a*) Nitrobenzene (*b*) Aniline
(*c*) Diazane (*d*) Ethanamide.

Estimation of Elements in Organic Compounds

46. Kjeldahl's method *cannot* be used for the estimation of nitrogen in

(*a*) Pyridine (*b*) Nitro compounds
(*c*) Azo compounds (*d*) All the above three.

47. Which of the following methods is not used to estimate nitrogen in the laboratory ?

(*a*) Kjeldahl's method (*b*) Duma's method
(*c*) Carius method (*d*) Both (*a*) and (*c*).

48. In organic compounds, sulphur is estimated as

(*a*) $BaSO_4$ (*b*) $BaCl_2$
(*c*) $Ba_3(PO_4)_2$ (*d*) H_2SO_4.

49. In organic compounds, phosphorus is estimated as

(*a*) $Mg_2P_2O_7$ (*b*) $Mg(NH_4)PO_4$
(*c*) $Mg_3(PO_4)_2$ (*d*) H_3PO_4.

50. Percentage of hydrogen is maximum in

(*a*) CH_4 (*b*) C_2H_4
(*c*) C_6H_6 (*d*) C_2H_4.

51. In the Carius method for the estimation of sulphur which reagent is taken along with organic compound in Carius tube ?

(*a*) Conc. HNO_3 and Na_2SO_4
(*b*) Conc. H_2SO_4
(*c*) Conc. H_2SO_4 and $NaNO_3$
(*d*) Conc. HNO_3 and barium chloride.

52. Which liquids are taken in Schiff's nitrometer in Duma's method ?

(*a*) Water (*b*) Water and Hg
(*c*) KOH (*aq*) and Hg (*d*) Hg and dil. H_2SO_4.

53. A current of which gas is passed through the combustion tube in Duma's method ?

(*a*) Air (*b*) Pure and dry oxygen
(*c*) An inert gas (*d*) Dry carbon dioxide.

54. In the Liebig's method for the estimation of C and H, the combustion tube is packed with

(*a*) CuO pallets (*b*) Copper turnings
(*c*) Iron filings (*d*) Zinc-copper couple.

55. During estimation of carbon and hydrogen, the CO_2 evolved during combustion is absorbed in

(a) $CaCl_2$ tube
(b) Distilled water
(c) Potash bulbs
(d) Small bulbs carrying freshly cut potassium.

56. In Duma's method the funciton of reduced copper gauze in the combustion tube is

(a) To remove CO_2
(b) To remove SO_2
(c) To absorb NO_2
(d) To reduce oxides of nitrogen of free N_2.

57. Which gas is introduced into the combustion tube in Liebig's method ?

(a) Pure and dry CO_2 (b) Pure and dry N_2
(c) Pure and dry O_2 (d) Pure and dry He.

58. Which reagents are taken in Carius tube along with organic compound during estimation of halogens ?

(a) Conc. HNO_3 + Conc. H_2SO_4
(b) Conc. HNO_3 + $AgNO_3$
(c) Conc. HNO_3 + $Ba(NO_3)_2$
(d) Conc. HNO_3 + Na_2SO_4.

Molecular Mass Determination ; Empirical and Molecular Formula

59. Empirical formula of a compound represents

(a) Actual number of atoms of the various elements
(b) Simplest ratio of atoms
(c) Types of bonds between different atoms
(d) Arrangement of various atoms in the molecule.

60. Empirical formula mass is calculated by

(a) Adding the atomic masses of all the atoms present in the empirical formula
(b) By dividing the molecular mass by two
(c) By multiplying the vapour density by two
(d) By multiplying the equivalent mass by valency.

61. Which of the following statements is not true about the molecular mass ?

(a) It is obtained by adding atomic masses of all the atoms present in the molecule of a substance
(b) It is obtained by multiplying the vapour density of the substance by two
(c) It is obtained by adding the atomic masses of all the atoms present in the empirical formula
(d) Molecular mass is n times the empirical formula weight.

62. Molecular formula of a compound represents

(a) Simplest ratio of atoms of the various atoms present in one molecule
(b) Actual number of atoms of the various elements present in one molecule
(c) Type of bonds present between different atoms
(d) Arrangement of various atoms within the molecule.

63. Chloroplatinic salt method is useful for determining the molecular mass of

(a) Organic acids
(b) Organic bases
(c) Any organic compound
(d) Only sulphonic acids.

64. The percentage of oxygen in organic compound can be determined by

(a) Carius method (b) Liebig's method
(c) Method of difference (d) All of these.

65. Org. compound $\xrightarrow{\text{heat}}$ O_2 + Other gases

$$2C + O_2 \longrightarrow 2CO$$
$$5CO + I_2O_5 \longrightarrow I_2 + 5CO_2$$

The reactions given above form the basis of direct estimation of

(a) N in organic compound
(b) O in organic compound
(c) P in organic compound
(d) I in organic compound.

66. Silver salt method for determining molecular mass is used for

(a) Organic compounds (b) Organic Amines
(c) Organic acids (d) All the above.

Miscellaneous Questions

67. Mass spectrometry is a technique to

(a) Determine % composition of various elements
(b) Know the functional groups of organic compound
(c) Determine molar mass of organic compound
(d) Determine resistance of compound towards ionisation.

68. The compounds $CH_3CH = CH_2$ and cyclohexane will have

(a) Same density
(b) Same percentage composition
(c) Different percentage composition
(d) Different empirical formulae.

69. Which of the following pairs of compounds will give blood red colouration in Lassaigne's test for N ?

(a) $(NH_2)_2C = O$, $C_6H_5SO_3H$
(b) Thiourea, sulphanilic acid
(c) $C_6H_5SO_3H$, $(NH_2)_2C = S$
(d) Sulphanilic acid and alanine

70. In mass spectrogram, the molecular mass corresponds to

(*a*) Base peak
(*b*) Peak with lowest intensity
(*c*) Average of mol. masses pertaining highest and lowest peak
(*d*) Peak having highest *e/m* ratio.

71. The masses of carbon, hydrogen and oxygen in an organic compound are in the ratio 6 : 1 : 8 respectively. Which of the following pairs of compounds correspond to above data ?

(*a*) $CH_2O, C_3H_6O_3$ (*b*) $C_2H_6O_2, CH_2O_2$
(*c*) CH_2O_2, CH_2O (*d*) $C_3H_6O_3, C_2H_6O_2$.

72. Benzene and aniline are both liquids. Their mixture can be separated by

I. Distillation II. Steam distillation
III. Dil. HCl IV. Dil. NaOH.

(*a*) I, II (*b*) II, III
(*c*) I, III (*d*) II, IV.

73. Presence of halogens in organic compounds is tested by

(*a*) Iodoform test (*b*) Millon's test
(*c*) Silver nitrate test (*d*) Soda lime test.

74. An organic compound contains about 52% carbon. It could be

I. Ethanol II. Dimethyl ether
III. Acetic acid IV. Phenol.

(*a*) I, II (*b*) II, III
(*c*) III, IV (*d*) I only.

75. The empirical formula of a compound is CH. To which of the following pairs of hydrocarbon series does it belong ?

(*a*) Alkanes, cycloalkanes (*b*) Alkenes, cycloalkanes
(*c*) Alkynes, dienes (*d*) None of the above.

76. Which of the following can be used as desiccants for absorbing water during Liebig's method for estimation of carbon and hydrogen ?

I. Anhyd. $CaCl_2$ II. Anhyd. Na_2SO_4
III. $MgSO_4.7H_2O$ IV. $Mg(ClO_4)_2$.

(*a*) I, IV (*b*) II, IV
(*c*) II, III (*d*) I, II, IV.

77. Which of the following is not purified by vacuum distillation ?

I. Ethanol II. Ethanal
III. Glycerol IV. Aniline.

(*a*) I, II, IV (*b*) II, III
(*c*) Only IV (*d*) All.

78. Chromatographic technique is used in the separation of

(*a*) All volatile liquids
(*b*) Amino acids
(*c*) Plant pigments
(*d*) Amino acids as well as plant pigments.

79. Mixed melting point of an organic compound is determined to

(*a*) Check the purity of an organic compound
(*b*) Check homology of two compounds
(*c*) Check chemical nature of compounds
(*d*) Check whether the two compounds can be separated or not.

80. Kjeldahl's method cannot be used for estimation of nitrogen in

I. $C_6H_5CONH_2$
II. Pyridine
III. $C_6H_5-N=N-C_6H_5$
IV. $C_6H_5NHCOCH_3$.

(*a*) I, II (*b*) III, IV
(*c*) II, III (*d*) I, III.

81. Two elements X (atomic mass = 75) and Y (atomic mass = 16) combine to give a compound having 75.8% X. The formula of the compound is

(*a*) XY (*b*) X_2Y
(*c*) X_2Y_2 (*d*) X_2Y_3.

82. **A** is a lighter phenol and **B** is an aromatic carboxylic acid. Separation of a mixture of **A** and **B** can be carried out easily by using a solution of

(*a*) Sodium hydroxide (*b*) Sodium sulphate
(*c*) Calcium chloride (*d*) Sodium bicarbonate.

83. The general equation for combustion of C_xH_y is

(*a*) $C_xH_y + \left(\frac{x+y}{4}\right)O_2 \longrightarrow xCO_2 + yH_2O$

(*b*) $C_xH_y + \left(x+\frac{y}{2}\right)O_2 \longrightarrow xCO_2 + \frac{y}{2}H_2O$

(*c*) $C_xH_y + \left(x+\frac{y}{4}\right)O_2 \longrightarrow xCO_2 + \frac{y}{2}H_2O$

(*d*) $C_xH_y + (x+y)\,O_2 \longrightarrow xCO_2 + \frac{y}{2}H_2O$.

84. Lassaigne's test for the detection of nitrogen fails in

(*a*) $NH_2CONHNH_2.HCl$ (*b*) $NH_2NH_2.HCl$
(*c*) NH_2CONH_2 (*d*) $C_6H_5NHNH_2.HCl$.

85. The most suitable method for separation of a 1 : 1 mixture of *ortho* and *para*-nitrophenol is

(a) Sublimation (b) Chromatography
(c) Crystallization (d) Steam distillation.

86. The compound that does not give a blue colour in Lassaigne's test for nitrogen is
(a) Aniline (b) Glycine
(c) Hydrazine (d) Urea.

87. The following is the percentage composition of a compound : Na = 16.08%, C = 4.19%, O = 16.78% and H_2O = 62.95%. Its molecular formula is
(a) Na_2CO_3 (b) $Na_2CO_3 . H_2O$
(c) $Na_2CO_3 . 10H_2O$ (d) $Na_2CO_3 . 5H_2O$.

88. When pieces of human hair are heated strongly with soda-lime, smell of ammonia can be detected. Which one of the following conclusion can be drawn from this observation ?
(a) Ammonia is present in the human hair
(b) Ammonium salt is present in hair
(c) Hair contains amino acids
(d) None of the above.

ANSWERS

1. (c)	**2.** (b)	**3.** (b)	**4.** (a)	**5.** (c)	**6.** (c)	**7.** (d)	**8.** (c)	**9.** (d)
10. (a)	**11.** (b)	**12.** (b)	**13.** (a)	**14.** (b)	**15.** (a)	**16.** (d)	**17.** (a)	**18.** (b)
19. (c)	**20.** (a)	**21.** (c)	**22.** (b)	**23.** (d)	**24.** (c)	**25.** (a)	**26.** (b)	**27.** (b)
28. (b)	**29.** (a)	**30.** (b)	**31.** (d)	**32.** (d)	**33.** (c)	**34.** (d)	**35.** (b)	**36.** (b)
37. (c)	**38.** (d)	**39.** (a)	**40.** (a)	**41.** (d)	**42.** (d)	**43.** (d)	**44.** (d)	**45.** (c)
46. (d)	**47.** (c)	**48.** (a)	**49.** (a)	**50.** (a)	**51.** (d)	**52.** (c)	**53.** (d)	**54.** (a)
55. (c)	**56.** (d)	**57.** (c)	**58.** (b)	**59.** (b)	**60.** (a)	**61.** (c)	**62.** (b)	**63.** (b)
64. (c)	**65.** (b)	**66.** (c)	**67.** (c)	**68.** (b)	**69.** (b)	**70.** (a)	**71.** (a)	**72.** (b)
73. (c)	**74.** (a)	**75.** (d)	**76.** (b)	**77.** (a)	**78.** (d)	**79.** (a)	**80.** (c)	**81.** (d)
82. (d)	**83.** (c)	**84.** (b)	**85.** (d)	**86.** (c)	**87.** (c)	**88.** (c).		

HINTS/SOLUTIONS

Q. No.	Ans.	Hints & Solutions
1.	(c)	Azeotropic mixture of two liquids can be separated indirectly by azeotropic distillation.
2.	(b)	Absolute alcohol can be obtained from azeotropic mixture of EtOH and water either by azeotropic distillation or by chemical methods.
3.	(b)	Only aniline is steam volatile.
4.	(a)	Only essential oils are volatile while others are not. Essential oils are volatile oils isolated from sap and tissues of certain plants and are extensively used in perfumery.
5.	(c)	Organic compounds are usually more soluble in organic solvents than in water and hence can be extracted by solvent extraction with a water immiscible organic solvent such as C_6H_6, $CHCl_3$, CCl_4, ether etc.

Q. No.	Ans.	Hints & Solutions
6.	(c)	Russian Botanist, Tswett in 1906.
7.	(d)	Column chromatography is based upon differential adsorption-desorption of different components of a mixture.
8.	(c)	Paper chromatography is a liquid-liquid partition chromatography in which the water adsorbed or chemically bound to cellulose of paper acts as the stationary phase while the mobile phase is another liquid which is usually a mixture of two or three solvents in which water is one of the components.
9.	(d)	Vapourize without decomposition.
10.	(a)	Chemical method using $NaHCO_3$ solution.
11.	(b)	The basis of separation of constituents by fractional distillation is different solubilities of constituents

Q. No.	Ans.	Hints & Solutions
12.	(*b*)	Sugar dissolves in alcohol but NaCl does not.
13.	(*a*)	Naphthalene does not dissolve in hot water but benzoic acid dissolves.
14.	(*b*)	The statement refers to sublimation process.
15.	(*a*)	Naphthalene sublimes on heating.
16.	(*d*)	All other compounds sublime except urea.
17.	(*a*)	The b.p. of a compound depends upon interparticle bonding, size and polarity of the molecules but does not depend upon its solubility in water.
18.	(*b*)	Boiling point decreases as the pressure acting on it decreases.
19.	(*c*)	Simple distillation can be used only when the b.ps. of the two liquids differ by more than 30°. Further, ethanol and H_2O form H-bonds and hence distills together even though their b.ps. differ by 21.9°C.
20.	(*a*)	Fractional distillation can be used when the b.ps. differ by 10–15°C.
21.	(*c*)	Boiling points of acetone (56°C) and methanol (65°C) differ by 9°C and hence can be separated only by fraction distillation.
22.	(*b*)	Petroleum is refined by fractional distillation.
23.	(*d*)	Distillation under reduced pressure decreases the boiling point and hence the liquid can be distilled without decomposition.
24.	(*c*)	A liquid which is immiscible in water and has a vapour pressure of 10—15 mm of Hg at 373 K can be conveniently purified by steam distillation.
25.	(*a*)	In steam distillation, the mixture begins to boil when the sum of the vapour pressure of the organic liquid and that of water becomes equal to that of the atmosphere.
26.	(*b*)	Because of intramolecular H-bonding (*chelation*) *o*-hydroxyacetophenone exists as discrete molecules, while *p*-hydroxyacetophenone exists as an associated molecules. Consequently, *o*-hydroxyacetophenone is *steam volatile* but *p*-hydroxyacetophenone is not.
27.	(*b*)	Formation of purple colour is due to $Fe_4[Fe(CN)_6]_3$ *i.e.*, ferric ferrocyanide which indicates the presence of nitrogen.
28.	(*b*)	$Na_4[Fe(CN)_5NOS]$.
29.	(*a*)	Lead sulphide, PbS.
30.	(*b*)	NH_2OH does not contain C required for conversion of N of the organic compound into NaCN.
31.	(*d*)	All the three, *i.e.*, N, S and Cl (halogens).
32.	(*d*)	To convert covalent compounds into ionic compounds (NaCN, Na_2S, NaX).
33.	(*c*)	Oxygen cannot be detected by Lassaigne's test.
34.	(*d*)	To decompose Na_2S and NaCN, if formed.
35.	(*b*)	Fusion of organic compound with fusion mixture ($Na_2CO_3 + K_2CO_3$) converts P into Na_3PO_4.
36.	(*b*)	Ammonium phosphomolybdate.
37.	(*c*)	Heating with CuO converts C into CO_2.
38.	(*d*)	Anhyd. $CuSO_4$ (white) which turns blue.
39.	(*a*)	Beilstein's test.
40.	(*a*)	Halogens.
41.	(*d*)	It is not a sure test for the presence of halogens, since pyridine, urea, thiourea etc. also give this test.
42.	(*d*)	Oxygen.
43.	(*d*)	Ferric ferrocyanide, $Fe_4[Fe(CN)_6]_3$.
44.	(*d*)	Ferric thiocyanate or sulphocyanide, $Fe(CNS)_3$.
45.	(*c*)	Diazane ($NH_2–NH_2$) does not give Lassaigne's test because it does not contain carbon.
46.	(*d*)	Kjeldahl's method cannot be used for the estimation of N in nitro, azo compounds and compounds containing N in the ring (*i.e.*, pyridine). It is because of the fact that these compounds are not completely converted into $(NH_4)_2SO_4$ during digestion.
47.	(*c*)	Kjeldahl's method.
49.	(*a*)	Magnesium pyrophosphate, $Mg_2P_2O_7$.
50.	(*a*)	CH_4 has highest ratio of H to C.
51.	(*d*)	The reagent used are conc. HNO_3 which oxidises S to SO_4^{2-} and $BaCl_2$ which forms $BaSO_4$ quantitatively.
52.	(*c*)	N_2 is collected by downward displacement of KOH solution and Hg act as seal.
53.	(*d*)	In Duma's method, a current of pure and dry CO_2 used.
54.	(*a*)	CuO pallets oxidise H to H_2O and C to CO_2.
55.	(*c*)	CO_2 is absorbed in KOH solution contained in potash bulbs.
56.	(*d*)	Reduced copper gauze reduces any oxide of nitrogen formed during combustion to free N_2 gas.
57.	(*c*)	A current of pure and dry stream of O_2 is passed through the combustion tube.

Q. No.	Ans.	Hints & Solutions
58.	(*b*)	The reagents used for estimation of halogens in Carius tube are $AgNO_3$ and conc. HNO_3 along with the organic compound.
59.	(*b*)	Simplest ratio of atoms.
60.	(*a*)	Statement (*a*) is based on facts.
61.	(*c*)	Statement (*c*) is correct.
62.	(*b*)	Actual number of atoms of various elements present in one molecule.
63.	(*b*)	Chloroplatenic method is used for determining molar masses of organic bases.
64.	(*c*)	The estimation of oxygen in organic compound is carried out generally, by method of difference.
65.	(*b*)	These reactions are used for direct estimation of oxygen in the organic compound from the mass of CO_2 produced.
66.	(*c*)	Factual question.
67.	(*c*)	Mass spectrometry is generally used to determine molecular masses.
68.	(*b*)	The compounds have same molecular formulae, hence their percentage composition is also same.
69.	(*b*)	Both thiourea and sulphanilic acid contain S and also N in their molecules.
70.	(*a*)	Base peak corresponds to highest *m/e* ratio for fragment M^+. *e* being unity, the ratio refers to mol. mass.
71.	(*a*)	Both CH_2O and $C_3H_6O_3$ have same empirical formula.
72.	(*b*)	The mixture of C_6H_6 and $C_6H_5NH_2$ can be separated by steam distillation or by using HCl.
73.	(*c*)	Halogens are tested by silver nitrate test with Lassaigne's extract.
74.	(*a*)	Both ethanol and dimethyl ether have same molecular formula.

Q. No.	Ans.	Hints & Solutions
75.	(*d*)	CH does not represent any homologous series.
76.	(*b*)	Desiccants used are anhydrous Na_2SO_4 or $Mg(ClO_4)_2$
77.	(*a*)	Vacuum distillation cannot be used for purification of C_2H_5OH, CH_3CHO and $C_6H_5NH_2$.
78.	(*d*)	Both, plant pigments and aminoacids are separated more effectively by chromotographic techniques.
79.	(*a*)	The concept of mixed m.p. is employed to check purity of organic compound.
80.	(*c*)	Kjeldahe's method cannot be used for estimation of N in pyridine and acetanilide.
81.	(*d*)	$x : y = \frac{75.8}{75} \times \frac{16}{24.2} = 2 : 3$ ∴ The empirical formula is X_2Y_3.
82.	(*d*)	Carboxylic acids dissolve in aqueous solution of $NaHCO_3$ but phenols do not.
83.	(*c*)	The balanced equation is $C_xH_y + \left(x + \frac{y}{4}\right)O_2 \longrightarrow xCO_2 + \frac{y}{2}H_2O.$
84.	(*b*)	$NH_2NH_2.HCl$ since it does not contain C which is needed for the formation of NaCN.
85.	(*d*)	*o*-Nitrophenol because of chelation exists as monomer and hence is steam volatile.
86.	(*c*)	Hydrazine since it does not contain C.
87.	(*c*)	$Na : C : O : H_2O$ $= \frac{16.08}{23} : \frac{4.19}{12} : \frac{16.78}{16} : \frac{62.95}{18}$ $= 2 : 1 : 3 : 10$ ∴ Empirical formula = $Na_2CO_3 . 10H_2O$.
88.	(*c*)	Hair contains amino acids which upon fusion with soda-lime (NaOH + CaO) evolve NH_3.

8.5. NUMERICAL BANK

Practice with Formulae

1. 0.2 g of an organic compound on Kjeldahl's analysis gave ammonia enough to just neutralize 20 cm^3 of 0.1 N H_2SO_4. The percentage of nitrogen in the compound is

   (*a*) 14 (*b*) 20
   (*c*) 70 (*d*) 24.

2. 0.28 g of a nitrogenous compound was kjeldahilised to produce 0.17 g of NH_3. The percentage of nitrogen in the organic compound is

   (*a*) 5 (*b*) 25
   (*c*) 50 (*d*) 40.

3. 0.216 g of an organic compound gave 28.8 cm^3 of moist nitrogen measured at 288 K and 732.7 mm pressure. The percentage of nitrogen in the substance is approximately (aq. tension at 288 K = 12.7 mm)

   (*a*) 15 (*b*) 20
   (*c*) 25 (*d*) 30.

4. 0.10 g of an organic compound containing phosphorus gave 0.222 g of $Mg_2P_2O_7$. Then the percentage of phosphorus in it is

   (*a*) 6.2 (*b*) 62
   (*c*) 3.1 (*d*) 31.

5. An organic compound contains 29.27% carbon, 5.69% hydrogen and 65.04% bromine. Its empirical formula is

   (*a*) C_3H_7Br (*b*) C_3H_6Br
   (*c*) C_3H_3Br (*d*) $C_2H_4Br_2$.

6. The simplest formula of a compound containing 50% of element X (at. mass = 10) and 50% of element Y (at. wt. = 20) is

   (*a*) XY (*b*) XY_2
   (*c*) X_2Y (*d*) X_2Y_2.

7. Which of the following compound has empirical formula CH_2O but molar mass equal to 60 g mol^{-1} ?

   (*a*) CH_3CHO (*b*) Ethanol
   (*c*) CH_3OH (*d*) Methyl formate.

8. A compound with empirical formula CH_2O has a vapour density of 90. Its molecular formula is

   (*a*) $C_2H_4O_2$ (*b*) CH_2O
   (*c*) $C_3H_6O_3$ (*d*) $C_6H_{12}O_6$.

9. The empirical formula of a compound containing six atoms of oxygen is CH_2O. Its molecular formula should be

   (*a*) CH_2O_6 (*b*) CH_2O
   (*c*) $CH_{12}O_6$ (*d*) $C_6H_{12}O_6$.

10. The empirical formula of a compound is CH_2. The mass of one mole of the compound is 42 g. The probable structural formula is

    (*a*) $CH_3CH_2CH_3$ (*b*) $CH_3-CH=CH_2$
    (*c*) $CH_2=CH-CH=CH_2$ (*d*) $CH_3-C\equiv CH$.

11. The silver salt of a monobasic acid on ignition gave 60% of Ag. The molecular mass of the acid is

    (*a*) 37 (*b*) 33
    (*c*) 73 (*d*) 77.

12. An organic compound with carbon = 40% and hydrogen = 6.6% will have an empirical formula

    (*a*) CH_2 (*b*) CH_2O
    (*c*) $C_2H_4O_2$ (*d*) C_2H_4.

13. 32 g of organic compound contains 12 g carbon, 4 g hydrogen and rest oxygen. Its simple formula is

    (*a*) CH_2O (*b*) $C_2H_4O_2$
    (*c*) CH_4O (*d*) $C_2H_8O_2$.

14. An organic compound contains 10% C, 0.84% H and 89.2% Cl, its empirical formula is

    (*a*) CH_2Cl_2 (*b*) $CHCl_3$
    (*c*) $C_2H_4Cl_2$ (*d*) $C_2H_3Cl_3$.

15. The chloroplatinate of a monoacid organic base contains 39% of Pt. The molecular mass of the base is

    (*a*) 73 (*b*) 59
    (*c*) 65 (*d*) 45.

16. 10 ml of a gaseous hydrocarbon were mixed with 100 ml of oxygen and exploded in a eudiometer tube. The volume of the residual gases was 95 ml of which 20 ml was absorbed by caustic soda and the remainder by pyrogallol. The molecular formula of the gas is

    (*a*) C_2H_2 (*b*) C_2H_4
    (*c*) C_3H_6 (*d*) C_4H_8.

17. 9.0 ml of a mixture of methane and ethylene was exploded with 30 ml (excess) of oxygen. After cooling, the volume was 21.0 ml. Further treatment with caustic potash reduced the volume to 7.0 ml. The amounts of methane and ethylene in the mixture respectively are

    (*a*) 5 ml of CH_4 + 5 ml of C_2H_4
    (*b*) 6 ml of CH_4 + 4 ml of C_2H_4
    (*c*) 3 ml of CH_4 + 7 ml of C_2H_4
    (*d*) 4 ml of CH_4 + 6 ml of C_2H_4.

18. Certain hydrocarbon contains 8 g of carbon for each gram of hydrogen. The empirical formula of the compound is

(*a*) C_2H_3 (*b*) CH_2
(*c*) C_3H_4 (*d*) C_4H_3.

19. The sulphur content of an amino acid *cystine* is 26.7%. Given that each molecule of cystine contain two S atoms, what will be molar mass of cystine ?

(*a*) 100 g mol^{-1} (*b*) 170 g mol^{-1}
(*c*) 239.7 g mol^{-1} (*d*) 200.8 g mol^{-1}.

20. Silver salt of certain organic acid with atomic ratio C : H : O as 2 : 3 : 2 contains 65.06% of silver. If each molecule of acid has two ionisable H atoms, the formula of the acid is

(*a*) $C_3H_6O_3$ (*b*) $C_4H_6O_4$
(*c*) $C_2H_3O_2$ (*d*) $C_3H_6O_2$.

21. 0.1092 g of certain dibasic organic acid neutralised 21 mL of decinormal solution of NaOH. The molar mass of acid is

(*a*) 75 g mol^{-1} (*b*) 52 g mol^{-1}
(*c*) 208 g mol^{-1} (*d*) 104 g mol^{-1}.

22. Tyrosine, (an amino acid) is one of the constituent of certain protein and is present to the extent of 0.22% by mass. If molar mass of tyrosine is 181 g mol^{-1}. The minimum molecular mass of protein is

(*a*) 7×10^4 U (*b*) 82273 U
(*c*) 92200 U (*d*) Above 10^5 U.

23. A certain organic compound contains 13.46% of nitrogen by mass. X g of the compound produced 50 cm^3 of nitrogen gas at 27°C and 720 mm pressure in Duma's method. The value of X is

(*a*) 0.9 g (*b*) 0.8 g
(*c*) 0.6 g (*d*) 0.4 g.

24. Certain organic compound contains carbon atoms twice the number of nitrogen atoms and hydrogen atoms seven times the number of nitrogen atom. The compound is monoacid base and 0.45 g of the compound neutralises 0.01 mol of HCl. The compound is

(*a*) $(CH_3)_2NH$ (*b*) $C_2H_5NH_2$
(*c*) Either (*a*) or (*b*) (*d*) None.

25. One gram of X–N_2Cl on strong heating gave 160 mL of N_2 gas at STP, what could X be

(*a*) C_6H_5– (*b*) CH_3–
(*c*) C_2H_5– (*d*) C_6H_{13}–.

26. A Grignard's reagent RMgBr gave a gaseous hydrocarbon (RH) on boiling with water. 14.0 L of RH at S.T.P. conditions is found to weigh 10 g. The correct formula of Grignard's reagent is

(*a*) CH_3CH_2MgBr (*b*) C_6H_5MgBr
(*c*) CH_3MgBr (*d*) $(CH_3)_2CHMgBr$.

27. 7.3 g of $XONH_2$ on heating with excess of NaOH gave ammonia gas which requires 200 mL of 0.5 M hydrochloric acid for exact neutralisation. The correct formula of organic compound is

(*a*) CH_3CONH_2 (*b*) $C_2H_5CONH_2$
(*c*) $C_6H_5CONH_2$ (*d*) $HCONH_2$.

28. Which of the following organic compounds will contain 23.728% of nitrogen by mass

I. CH_3CONH_2 II. CH_3CH_2NO
III. $CH_3CH = NOH$ IV. $C_2H_5NO_2$

(*a*) IV (*b*) III, IV
(*c*) I, IV (*d*) I, II, III.

29. Complete combustion of a sample of hydrocarbon gives 0.66 g of CO_2 and 0.36 g of H_2O. The empirical formula of the compound is

(*a*) C_3H_8
(*b*) C_3H_4
(*c*) CH_2
(*d*) Unpredictable due to insufficient data.

30. A dibasic acid containing C, H and O was found to contain, C = 26.7% and H = 2.2%. The vapour density of its diethyl ester was found to be 73. What is the molecular formula of the acid ?

(*a*) CH_2O_2 (*b*) $C_2H_2O_4$
(*c*) $C_3H_3O_4$ (*d*) $C_4H_4O_4$.

31. The percentage composition of a compound is C = 90% and H = 10%. Which of the following hydrocarbon corresponds to above data ?

(*a*) C_6H_{10} (*b*) $C_{15}H_{30}$
(*c*) $C_{15}H_{20}$ (*d*) $C_{15}H_{12}$.

32. Two hydrocarbons A and B, both on analysis yields the same percentage composition by mass

C = (12/13) × 100% and H = (1/13) × 100%.

A decolourises Br_2-water but B does not. Identify A and B

(*a*) A = C_2H_2, B = C_6H_6 (*b*) A = C_6H_6, B = C_2H_2
(*c*) A = C_2H_4, B = C_2H_6 (*d*) A = C_2H_2, B = C_3H_8.

33. Complete combustion of 0.959 g of compound X gives 2.63 g of CO_2 and 1.28 g of H_2O. The lowest molecular mass X can have is

(*a*) 43 (*b*) 86
(*c*) 129 (*d*) 172.

34. Each mole of haemoglobin contains four mol of iron. If the percentage of iron in haemoglobin is 0.35 by mass. The molar mass of haemoglobin is

(*a*) 6.4×10^4 g mol^{-1} (*b*) 5.6×10^4 g mol^{-1}
(*c*) 6.4×10^5 g mol^{-1} (*d*) 5.6×10^5 g mol^{-1}.

35. 1.55 g of certain dihydric alcohol on treatment with excess of sodium metal produced 560 mL of dihydrogen gas at STP. The molar mass of alcohol is

(*a*) 80 g mol^{-1} (*b*) 124 g mol^{-1}
(*c*) 31 g mol^{-1} (*d*) 62 g mol^{-1}.

36. A certain compound contains six O atoms per molecule. If the mass percentage of C and hydrogen in the compound is 32% and 4% respectively. The molecular formula is

(*a*) $C_3H_2O_6$ (*b*) $C_4H_3O_6$
(*c*) $C_6H_{12}O_6$ (*d*) $C_5(H_2O)_6$.

37. An organic liquid has atomic ratio C : H : N : S as 2 : 3 : 1 : 1. Each mol of this basic liquid can be neutralised by 2 mol of HCl. Also 1.02 g of chloroplatinate of this base, on ignition gave 0.4 g of platinum. The ratio of molecular mass to empirical mass of liquid is

(*a*) 1 (*b*) 2
(*c*) 3 (*d*) 4.

38. 6.0 g of urea was heated strongly with excess of caustic soda solution to liberate NH_3 gas. What volume of semimolar HCl is required to neutralise ammonia evolved ?

(*a*) 200 mL (*b*) 400 mL
(*c*) 450 mL (*d*) 500 mL.

ANSWERS

1. (*a*)	**2.** (*c*)	**3.** (*a*)	**4.** (*b*)	**5.** (*a*)	**6.** (*c*)	**7.** (*d*)	**8.** (*d*)	**9.** (*d*)
10. (*b*)	**11.** (*c*)	**12.** (*b*)	**13.** (*c*)	**14.** (*b*)	**15.** (*d*)	**16.** (*a*)	**17.** (*b*)	**18.** (*a*)
19. (*c*)	**20.** (*b*)	**21.** (*d*)	**22.** (*b*)	**23.** (*d*)	**24.** (*c*)	**25.** (*a*)	**26.** (*c*)	**27.** (*b*)
28. (*d*)	**29.** (*a*)	**30.** (*c*)	**31.** (*b*)	**32.** (*a*)	**33.** (*a*)	**34.** (*a*)	**35.** (*d*)	**36.** (*b*)
37. (*a*)	**38.** (*b*).							

HINTS/SOLUTIONS

Q. No.	Ans.	Hints & Solutions
1.	**(*a*)**	% of N $= \frac{1.4 \times 0.1 \times 20}{0.2} = 14.$
2.	**(*c*)**	% of N $= \frac{14}{17} \times \frac{0.17}{0.28} \times 100 = 50.$
3.	**(*a*)**	Volume of N_2 at STP $= \frac{28.8\,(732.7 - 12.7)}{288} = \frac{273}{760} = 25.86\ cm^3$ % of N $= \frac{28}{22400} \times \frac{25.86}{0.216} \times 100$ $= 14.96 = 15$ (approx.)
4.	**(*b*)**	% of P $= \frac{62}{222} \times \frac{0.222}{0.1} \times 100 = 62.$
5.	**(*a*)**	C : H : Br $= \frac{29.27}{12} : \frac{5.69}{1} : \frac{65.04}{80}$ $= 2.43 : 5.69 : 0.813 = 3 : 7 : 1$ or E.F. $= C_3H_7Br$.

Q. No.	Ans.	Hints & Solutions
6.	**(*c*)**	X : Y $= \frac{50}{10} : \frac{50}{20} = 2 : 1$ or E.F. $= X_2Y$.
7.	**(*d*)**	E.F. mass of $CH_2O = 12 \times 1 + 2 \times 1 + 1 \times 16 = 30$ Mol. mass = 60 (*Given*) $\therefore$ M.F. = E.F. $\times \frac{60}{30} = (CH_2O) \times 2$ $C_2H_4O_2 \equiv CH_3COOH$.
8.	**(*d*)**	E.F. mass of CH_2O = 30 Mol. mass = 2 × V.D. = 2 × 90 = 180 $\therefore$ M.F. = E.F. $\times \frac{180}{30} = (CH_2O)_6 = C_6H_{12}O_6$.
9.	**(*d*)**	M.F. $= 6 \times [CH_2O] = C_6H_{12}O_6$.
10.	**(*b*)**	E.F. mass of $CH_2 = 12 \times 1 + 1 \times 2 = 14$ Mol. mass = 42 $\therefore$ M.F. $= \frac{42}{12} \times$ E.F. $= 4 \times CH_2 = C_3H_6$ Thus, the compound is $CH_3 - CH = CH_2$.

Q. No.	Ans.	Hints & Solutions
11.	(c)	Mol. mass of acid $= n\left(\frac{108\,W}{x} - 107\right)$ $= 1\left(\frac{108 \times 100}{60} - 107\right)$ $= 180 - 107 = 73\ u$.
12.	(b)	% O = 100 – 40 – 6.6 = 53.4. $\therefore$ C : H : O $= \frac{40}{12} : \frac{6.6}{1} : \frac{53.4}{16} = 1 : 2 : 1$ or E.F. = CH_2O.
13.	(c)	O = 32 – 12 – 4 = 16 g $\therefore$ C : H : O $= \frac{12}{12} : \frac{4}{1} : \frac{16}{16} = 1 : 4 : 1$ or E.F. = CH_4O.
14.	(b)	C : H : Cl $= \frac{10}{12} : \frac{0.84}{1} : \frac{89.2}{35.5} = 1 : 1 : 3$ or E.F. = $CHCl_3$.
15.	(d)	Molecular mass of the base $= \frac{1}{2}\left(\frac{100}{39} \times 195 - 410\right)$ $= \frac{1}{2}(500 - 410) = 45\ u$.
16.	(a)	Vol. after explosion and cooling *i.e., vol. of CO_2 formed + unused O_2 = 95 ml.* Vol. after introducing KOH solution *i.e., vol. of CO_2 formed* = 20 ml Vol. absorbed by alkaline pyrogallol, *i.e., vol. of unused* O_2 = 95 – 20 = 75 ml. $\therefore$ Vol. of O_2 used = 100 – 75 = 25 ml. Applying the combustion equation, $C_xH_y + (x + y/4)\,O_2 \longrightarrow xCO_2 + y/2\ H_2O$ 10 ml, 10 (x + y/4) ml, 10x ml, negligible volume. Equating theoretical and experimented values, $10x = 20$ and $10(x + y/4) = 25$ $\therefore$ $x = 2$ and $10(2 + y/4) = 25$ or $y = 2$ Thus the hydrocarbon is C_2H_2.

Q. No.	Ans.	Hints & Solutions
17.	(b)	Let the vol. of $CH_4 = x$ ml $\therefore$ Vol. of $C_2H_4 = (100 - x)$ ml. Equations for combustion of CH_4 and C_2H_4 respectively are : (i) $CH_2 + 2O_2 \longrightarrow CO_2 + 2H_2O$ — x ml, 2x ml, x ml, negligible volume (ii) $C_2H_4 + 3O_2 \longrightarrow 2CO_2 + 2H_2O$ — (10 – x)ml, 3(10 – x)ml, 2(10 – x)ml, negligible vol. Vol. after contraction and cooling *i.e., Vol. of CO_2 produces + O_2 left unused* = 21 ml Vol. of O_2 left unused = 7 ml $\therefore$ Vol. of CO_2 produced = 21 – 7 = 14 ml ...(iii) But from Eq. (i) and (ii), total vol. of CO_2 produced = x ml + 2(10 – x)ml ...(iv) Equating Eq. (iii) and (iv), we have $x + 2(10 - x) = 14$ or $x = 6$ $\therefore$ *Vol. of CH_4 = 6 ml and that of* C_2H_4 = 10 – 6 = 4 ml.
18.	(a)	No. of g-atoms of C present for each g-atom of H $= \frac{8}{12} = \frac{2}{3}$ g-atom No. of g-atoms of H for 1 g atom of C $= \frac{3}{2} = 1.5$ $\therefore$ Formula is $CH_{1.5}$ or C_2H_3.
19.	(c)	26.7 g of S is present in Cystine = 100 g 2 × 32 g of S is present in Cystine $= \frac{100 \times 64}{26.7} = 239.7$ g.
20.	(b)	Empirical formula is $C_2H_3O_2$ Mol. mass = Basicity $\left(\frac{108\,W}{x} - 107\right)$ $= 2\left(\frac{108 \times 100}{65.06} - 107\right) = 118.0$ $\frac{\text{Mol. mass}}{\text{Empirical mass}} = \frac{118}{59} = 2$ $\therefore$ Mol. formula $= (C_2H_3O_2)_2 = C_4H_6O_4$.

Q. No.	Ans.	Hints & Solutions
21.	(d)	$H_2X + 2NaOH \longrightarrow Na_2X + 2H_2O$ M. mol of NaOH consumed = 21 × 0.1 = 2.1 M. mol of acid = $\frac{1}{2}$ (M. mol of base) $= \frac{1}{2} \times 2.1 = 1.05$ Mol of acid = 1.05×10^{-3} Now $1.05 \times 10^{-3} = \frac{0.1092}{GMM}$ or $GMM_{(acid)} = \frac{0.1092}{1.05 \times 10^{-3}} = 104 \text{ g mol}^{-1}$.
22.	(b)	0.22 g of tyrosin is present in protein = 100 g 181 g of tyrosin is present in protein $= \frac{100 \times 181}{0.22} = 82272.7$ g or ≈ 82273 g.
23.	(d)	Vol. of N_2 at STP = $\frac{50 \times 720 \times 273}{300 \times 760} = 43.10$ mL $W_{comp} = \frac{28 \times V_{ml} \text{ at STP} \times 100}{22400 \times \% \text{ age}}$ $= \frac{28 \times 43.10 \times 100}{22400 \times 13.46} = 0.400$ g.
24.	(c)	0.45 g of base ≡ 0.01 mol of HCl ∴ $\frac{0.45}{GMM} = 0.01$ mol or Mol. mass of compound = $\frac{0.45}{0.01}$ $= 45 \text{ g mol}^{-1}$ Now as per given conditions, atomic ratio C : H : N is 2 : 7 : 1 Empirical formula is C_2H_7N $\frac{\text{Mol. mass}}{\text{Empirical mass}} = \frac{45}{45} = 1$ Thus, both the formulae $(CH_3)_2NH$ as well as $C_2H_5NH_2$ correspond to above data.
25.	(a)	N_2 produced by 1 g of XN_2Cl = 160 ml $= \frac{160}{22400}$ mol 1 mol of N_2 will be produced from salt $= \frac{22400}{160}$ g = 140 g ∴ Molar mass of XN_2Cl = 140 g mol^{-1} Molar mass of X = 140 – 28 – 35.5 = 76.5 g mol^{-1} This pertains to C_6H_5 – group Hence X is C_6H_5.
26.	(c)	14.0 L of RH at STP weighs = 10 g 22.4 L of RH at STP weighs = $\frac{10 \times 22.4}{14}$ g = 16 Thus RH must be CH_4 and R-must be CH_3 – Hence Grignard's reagent is CH_3MgBr.
27.	(b)	Millimoles of HCl required by NH_3 = 200 × 0.5 = 100 Mol of HCl required by NH_3 = $100 \times 10^{-3} = 0.1$ Mol of NH_3 produced = 0.1 Mol of $XCONH_2$ taken = 0.1 If M is molar mass of $XCONH_2$ $\frac{7.3}{M} = 0.1$ or $M = \frac{7.3}{0.1} = 73 \text{ g mol}^{-1}$ The formula corresponding to molar mass of 73 g mol^{-1} is $C_2H_5CONH_2$.
28.	(d)	Structures I, II, III have same molecular formula, C_2H_5NO and thus, they have same % of N by mass.
29.	(a)	Let the mass of substance be W g $\% C = \frac{12 \times 0.66 \times 100}{44 \times W} = 18/W$ $\% H = \frac{2 \times 0.36 \times 100}{18 \times W} = 4/W$ C : H ratio = $\frac{18}{12 W} : \frac{4}{1 W} = \frac{3}{2} : 4 = 3 : 8$ Hence formula is C_3H_8.
30.	(c)	C : H : O = $\frac{26.7}{12} : \frac{2.2}{1} : \frac{71.1}{16}$ = 1 : 1 : 2 Empirical formula is CHO_2 Mol. mass of diethyl ester = 2 × 73 = 146 *u* Formula mass of ethyl group (– C_2H_5) = 29 *u* In diethyl ester two ethyl group must have replaced the H atoms of dibasic acid The mol. mass of acid = 146 – (2 × 29) + 2 = 90 *u* ∴ Molecular formula of acid $= \frac{90}{45} \times (CHO_2) = C_2H_2O_4$.

Q. No.	Ans.	Hints & Solutions
31.	(b)	C : H ratio = $\frac{90}{12} : \frac{10}{1}$ = 7.5 : 10 = 1 : 1.33 or 3 : 4 The empirical formula is C_3H_4 Among the given species $C_{15}H_{20}$ also has empirical formula C_3H_4.
32.	(a)	C : H = $\frac{12 \times 100}{13 \times 12} : \frac{1 \times 100}{13 \times 1}$ = 1 : 1 ∴ EF = CH Since A decolourises bromine water, but B does not, therefore, A = C_2H_2 and B = C_6H_6.
33.	(a)	% C = $\frac{12}{44} \times \frac{2.63}{0.858} \times 100 = 83.59$ % H = $\frac{2}{18} \times \frac{1.28}{0.858} \times 100 = 16.57$ Now, C : H = $\frac{83.59}{12} : \frac{16.57}{1}$ = 6.96 : 16.57 = 1 : 2.38 or 3 : 8 ∴ E.F. = C_3H_8 E.F. mass = 3 × 12 + 8 × 1 = 44 Lowest mol. mass of a compound is equal to its empirical formula mass = 44 The only figure close to 44 is 43. Therefore option (a) is correct.
34.	(a)	Number of iron atoms per molecule = 4 Mass of iron per molecule = 4 × 56 = 224 *u* 0.35 parts by mass of iron is present in haemoglobin = 100 parts 224 parts by mass of iron is present in haemoglobin = $\frac{100 \times 224}{0.35}$ = 64000 or 6.4×10^4 g mol^{-1}

Q. No.	Ans.	Hints & Solutions
35.	(d)	Let the alcohol be $X(OH)_2$ $X(OH)_2 + 2Na \longrightarrow X(ONa)_2 + H_2$ 1 mol 1 mol H_2 produced = 560 mL of STP = $\frac{560}{22400}$ = 0.025 mol Mass of dihydric alcohol that produces 0.25 mol of alcohol = 1.55 g Mass of alcohol which produces (1 mol) of alcohol = $\frac{1.55}{0.025}$ = 62 g mol^{-1}.
36.	(b)	% mass of O = 100 – 32 – 4 = 64 Atomic ratio, C : H : O is $\frac{32}{12} : \frac{2}{1} : \frac{64}{16}$ or 2.66 : 2 : 4 or 1.33 : 1 : 2 or 4 : 3 : 6 Hence the formula = $\mathbf{C_4H_3O_6}$.
37.	(a)	Empirical formula is C_2H_3NS Acidity of base = 2 Molar mass = $\frac{2}{2}\left(\frac{1.02 \times 195}{0.4} - 410\right)$ = 87 g mol^{-1} $\frac{\text{Molar mass}}{\text{Empirical mass}} = \frac{87}{87} = 1.$
38.	(b)	$(NH_2)_2C = O \xrightarrow{2\,NaOH} Na_2CO_3 + 2NH_3$ 60 g 2 mol 6 g of urea liberated NH_3 = 0.2 mol HCl required to neutralise NH_3 = 0.2 ml Now, V × $\frac{1}{2}$ = 0.2 or V = 0.4 L = 0.4 × 10^3 mL = 400 mL.

8.6. EXAMINER'S VIEWPOINT

Test Your Confidence

1996

1. 0.24 g of a volatile liquid on vaporization gives 45 ml of vapours at STP. What will be the vapour density of the substance ? (Density of H_2 = 0.089 gL^{-1})
(*a*) 95.39 (*b*) 39.95
(*c*) 99.53 (*d*) 59.93. **(CBSE PMT)**

2. Liebig's method is used for the estimation of
(*a*) Nitrogen (*b*) Sulphur
(*c*) Carbon and hydrogen (*d*) Halogens. **(CMEET)**

3. The percentage of sulphur in an organic compound whose 0.32 g produces 0.233 g of $BaSO_4$ (At. wt. of Ba = 137, S = 32) is
(*a*) 1.0 (*b*) 10.0
(*c*) 23.5 (*d*) 32.1. **(Pb. CET)**

4. A hydrocarbon has C = 85.72% and rest being hydrogen. The hydrocarbon is
(*a*) C_2H_4 (*b*) C_2H_6
(*c*) C_2H_2 (*d*) CH_4. **(MP PMT)**

5. For the separation of two immiscible liquids which method is used ?
(*a*) Chromatography (*b*) Fractionating column
(*c*) Fractional distillation (*d*) Separating funnel. **(CPMT)**

6. A mixture of benzene and chloroform is separated by
(*a*) Sublimation (*b*) Separation funnel
(*c*) Crystallization (*d*) Fractional distillation. **(AFMC)**

1997

7. Nitrogen in an organic compound can be estimated by
(*a*) Kjeldahl's method only (*b*) Duma's method only
(*c*) Both the methods (*d*) None of these methods. **(Pb. CET)**

8. In Lassaigne's test for nitrogen, the blue colour is due to the formation of
(*a*) Potassium ferricyanide (*b*) Sodium cyanide
(*c*) Sodium ferrocyanide (*d*) Ferri-ferrocyanide. **(Pb. CET)**

9. Aniline is usually purified by
(*a*) Chromatographic technique
(*b*) Steam distillation
(*c*) By addition of oxalic acid
(*d*) Fractional crystallization. **(JIPMER)**

10. Impure glycerine is purified by
(*a*) Steam distillation (*b*) Simple distillation
(*c*) Vaccum distillation (*d*) None of the above. **(Pb. PMT)**

1998

11. In Kjeldahl's method, nitrogen present is estimated as
(*a*) N_2 (*b*) NO
(*c*) NH_3 (*d*) NO_2. **(Har. CEET)**

12. If 0.2 gram of an organic compound containing carbon, hydrogen and oxygen on combustion, yielded 0.147 gram carbon dioxide and 0.12 gram water. What will be the content of oxygen in the substance ?
(*a*) 73.29% (*b*) 78.45%
(*c*) 83.23% (*d*) 89.50%. **(AFMC)**

13. If a compound on analysis was found to contain C = 18.5%, H = 1.55%, Cl = 55.04% and O = 24.81%, then its empirical formula is
(*a*) CHClO (*b*) CH_2ClO
(*c*) C_2H_5OCl (*d*) $ClCH_2O$. **(AIIMS)**

14. The equivalent mass of an acid is equal to
(*a*) Molecular mass × acidity
(*b*) Molecular mass × basicity
(*c*) Molecular mass/basicity
(*d*) Molecular mass/acidity. **(AIIMS)**

15. A compound with empirical formula CH_2O has a vapour density of 30. Its molecular formula is
(*a*) $C_2H_2O_2$ (*b*) $C_2H_4O_2$
(*c*) $C_3H_6O_3$ (*d*) $C_6H_{12}O_6$. **(AIIMS)**

16. The latest technique used for purification of organic compounds is
(*a*) Chromatography (*b*) Vacuum distillation
(*c*) Fractional distillation (*d*) Crystallization.

17. Empirical formula of a hydrocarbon containing 80% carbon and 20% hydrogen is
(*a*) CH (*b*) CH_2
(*c*) CH_3 (*d*) CH_4. **(MP CEE)**

18. Molecular mass of a volatile substance may be obtained by
(*a*) Kjeldahl's method (*b*) Duma's method
(*c*) Victor-Meyer's method (*d*) Liebig's method. **(Pb. CET)**

1999

19. An organic compound with C = 40% and H = 6.7% will have the empirical formula
(a) CH_4 (b) CH_2O
(c) $C_3H_6O_3$ (d) $C_2H_4O_2$. ***(MP CEE)***

20. An organic compound contains C = 40%, H = 13.33% and N = 46.67%. Its empirical formula would be
(a) CHN (b) C_2H_2N
(c) CH_4N (d) C_3H_7N. ***(CBSE PMT)***

21. Empirical formula of a compound is CH_2O. It its molecular mass is 180, then the molecular formula of the compound is
(a) $C_6H_{12}O_6$ (b) $C_5H_{10}O_5$
(c) $C_3H_6O_3$ (d) $C_3H_8O_4$. ***(AFMC, AIIMS)***

22. The Beilstein test for organic compounds is used to detect
(a) Nitrogen (b) Sulphur
(c) Carbon (d) Halogens. ***(AIIMS)***

23. Which of the following is the best scientific method to test presence of water in a liquid ?
(a) Smell
(b) Taste
(c) Use of litmus paper
(d) Use of anhydrous copper sulphate. ***(AIIMS)***

24. Which of the following has molecular mass of 92 *u* ?
(a) Toluene (b) Benzene
(c) Methylene (d) Propene. ***(Manipal)***

2000

23. Which of the following compounds does not show Lassaigne's test for nitrogen ?
(a) Urea (b) Hydrazine
(c) Phenylhydrazine (d) Azobenzene. ***(DCE)***

26. 0.1914 g of an organic acid is dissolved in about 20 ml of water. 0.25 ml of 0.12 N NaOH is required for the complete neutralization of the acid solution. The equivalent mass of the acid is
(a) 65.0 (b) 64.0
(c) 63.8 (d) 62.5. ***(MP CEE)***

27. An organic compound contains 49.3% carbon, 6.84% hydrogen and its vapour density is 73. Molecular formula of the compound is
(a) $C_3H_5O_2$ (b) $C_6H_{10}O_4$
(c) $C_3H_{10}O_2$ (d) $C_4H_{10}O_2$. ***(MP CEE)***

28. The empirical formula of an acid is CH_2O_2, the probable molecular formula of the simplest acid may be
(a) CH_2O (b) CH_2O_2
(c) $C_2H_4O_2$ (d) $C_3H_6O_4$. ***(AFMC)***

29. In paper chromotagraphy,
(a) Moving phase is liquid and stationary phase is solid
(b) Moving phase is liquid and stationary phase is liquid
(c) Moving phase is solid and stationary phase is solid
(d) Moving phase is solid and stationary phase is liquid. ***(CPMT)***

2001

30. In a hydrocarbon, mass ratio of hydrogen and carbon is 1 : 3, the empirical formula of hydrocarbon is
(a) CH (b) CH_2
(c) CH_3 (d) C_2H_4
(e) CH_4. ***(Kerala PMT)***

31. 0.765 g of an acid gives 0.535 g of CO_2 and 0.14 g of H_2O. Then the ratio of the percentage of carbon and hydrogen is
(a) 19 : 2 (b) 18 : 11
(c) 20 : 17 (d) 1 : 7. ***(CPMT)***

32. In *peroxidase anhydrase enzyme* the percentage of selenium (Se) by mass is 0.5%. If at. mass of Se is 78.4 U, then minimum mol. mass of peroxidase anhydrase enzyme is
(a) 1.568×10^4 u (b) 1.568×10^3 u
(c) 1.568 u (d) 2.136×10^4 u. ***(CBSE PMT)***

2002

33. The empirical formula of compound is CH_2. One mol of a compound has a mass of 42 g. Its molecular formula is
(a) CH_2 (b) C_2H_4
(c) C_3H_6 (d) C_3H_8. ***(MP PMT)***

34. In a compound, C, H, N atoms are present in the ratio 9 : 1 : 3.5 by mass. If molecular mass of the compound is 108, then molecular formula of the compound is
(a) $C_2H_6N_2$ (b) C_3H_4N
(c) $C_6H_8N_2$ (d) $C_9H_{12}N_3$. ***(AIEEE)***

2004

35. The compound formed in the positive test for nitrogen with the Lassaigne's solution of an organic compounds is
(a) $Na_4[Fe(CN)_5NOS]$ (b) $Na_3]Fe(CN)_6]$
(c) $Fe(CN)_3$ (d) $Fe_4[Fe(CN)_6]_3$. ***(AIEEE)***

36. Ammonia evolved for the treatment 0.30 g of an organic compound for the estimation of nitrogen was passed in 100 mL of 0.1 M sulphuric acid. The excess of acid required 20 mL of 0.5 M sodium hydroxide solution for complete neutralisation. The organic compound is

(*a*) thiourea (*b*) benzamide
(*c*) urea (*d*) acetamide **(*AIEEE*)**

2005

37. The best method used for the separation of naphthalene and benzoic acid from their mixture is

(*a*) Distillation (*b*) Sublimation
(*c*) Chromatography (*d*) Crystallisation.

(*CBSE PMT*)

38. Four gram of hydrocarbon on complete combustion gave 12 g of CO_2. The emperical formula of the hydrocarbon is

(*a*) CH_3 (*b*) C_4H_9
(*c*) CH (*d*) C_3H_8. **(*EAMCET Med.*)**

39. The best method for the separation of naphthalene and benzoic acid from their mixture is :

(*a*) Distillation (*b*) Sublimation
(*c*) Chromatography (*d*) Crystallisation.

(*CBSE PMT*)

40. An organic compound having molar mass 60 g mol^{-1} found to contain C = 20%, H = 6.67% and N = 46.67%, rest beeing oxygen. On heating it gives NH_3 along with solid residue. The solid residue gives violet colour with alkaline copper sulphate solution. The compound is

(*a*) $CH_3CH_2CONH_2$ (*b*) $(NH_2)_2CO$
(*c*) CH_3CONH_2 (*d*) CH_3NCO. **(*AIEEE*)**

41. For the purification, isolation and separation of organic compounds. The latest technique followed is

(*a*) Chromatography
(*b*) Steam distillation
(*c*) Fractional crystallisation
(*d*) Sublimation
(*e*) Vaccum distillation. **(*Kerala PMT*)**

42. An organic compound weighing 0.31 g gave 0.444 g of magnesium pyrophosphate in the estimation of phosphorous by carius method. The % of P in the compound is

(*a*) 20 (*b*) 60
(*c*) 15 (*d*) 30
(*e*) 40. **(*Kerala PMT*)**

ANSWERS

1. (*d*)	**2.** (*c*)	**3.** (*b*)	**4.** (*a*)	**5.** (*d*)	**6.** (*d*)	**7.** (*c*)	**8.** (*d*)	**9.** (*b*)
10. (*c*)	**11.** (*c*)	**12.** (*a*)	**13.** (*a*)	**14.** (*c*)	**15.** (*b*)	**16.** (*a*)	**17.** (*c*)	**18.** (*c*)
19. (*b*)	**20.** (*c*)	**21.** (*a*)	**22.** (*d*)	**23.** (*d*)	**24.** (*a*)	**25.** (*b*)	**26.** (*c*)	**27.** (*c*)
28. (*b*)	**29.** (*b*)	**30.** (*e*)	**31.** (*a*)	**32.** (*a*)	**33.** (*c*)	**34.** (*c*)	**35.** (*d*)	**36.** (*c*)
37. (*d*)	**38.** (*d*)	**39.** (*d*)	**40.** (*b*)	**41.** (*a*)	**42.** (*e*).			

HINTS/SOLUTIONS

Q. No.	Ans.	Hints & Solutions
1.	**(*d*)**	$\text{V.D.} = \dfrac{\text{Mass of 45 ml vapours at STP}}{\text{Mass of 45 ml of } H_2 \text{ at STP}}$ $= \dfrac{0.24}{45 \times 0.000089} = 59.93.$ ($\because$ Density of $H_2 = 0.089 \text{ g L}^{-1} = 0.000089$ g/ml)
2.	**(*c*)**	Carbon and hydrogen.
3.	**(*b*)**	$\% \text{ of S} = \dfrac{32}{233} \times \dfrac{0.233}{0.32} \times 100 = 10.$
4.	**(*a*)**	$C : H = \dfrac{85.72}{12} : \dfrac{14.28}{1} = 1 : 2$ $\therefore$ Empirical formula is CH_2 and its molecular formula would be a simple multiple of CH_2 *i.e.*, $CH_2 \times 2 = C_2H_4$.

Q. No.	Ans.	Hints & Solutions
5.	(*d*)	Separating funnel since the two liquids are immiscible.
6.	(*d*)	Distillation particularly fractional distillation because the b.p. of benzene (80°C) and chloroform (61.5°C) are close.
7.	(*c*)	Both Kjeldahl's and Duma's method.
8.	(*d*)	Ferric ferrocyanide.
9.	(*b*)	Steam distillation.
10.	(*c*)	Glycerine is generally purified by vaccum distillation.
11.	(*c*)	NH_3.
12.	(*a*)	$\% C = \frac{12}{44} \times \frac{0.147}{0.2} \times 100 = 20.045$ $\% H = \frac{2}{18} \times \frac{0.12}{0.2} \times 100 = 6.666$ $\therefore \quad \% O = (100 - 20.045 - 0.6666)$ $= 73.289 = 73.29$ (approx.)
13.	(*a*)	$C : H : Cl : O$ $= \frac{18.5}{12} : \frac{1.55}{1} : \frac{55.04}{35.5} : \frac{24.81}{16}$ $= 1 : 1 : 1 : 1$ $\therefore$ EF = CHClO.
14.	(*c*)	Eq. mass = Mol. mass/basicity.
15.	(*b*)	Mol mass = 2 × V.D. = 2 × 30 = 60 $\therefore \quad M.F. = \frac{\text{Mol. mass}}{\text{E.F. mass}} \times E.F. = \frac{60}{30} \times CH_2O$ $= C_2H_4O_2$.
16.	(*a*)	Chromatography.
17.	(*c*)	$C : H = \frac{80}{12} : \frac{20}{1} = 1 : 3$ $\therefore$ E.F. = CH_3.
18.	(*c*)	Victor-Meyer's method.
19.	(*b*)	$\% O = 100 - 40 - 6.7 = 53.33$ $C : H : O = \frac{40}{12} : \frac{6.7}{1} : \frac{53.33}{16}$ $\therefore \quad = 1 : 2 : 1$ $\therefore$ E.F. = CH_2O.
20.	(*c*)	$C : H : N = \frac{40}{12} : \frac{13.33}{1} : \frac{46.67}{14} = 1 : 4 : 1$ $\therefore$ E.F. = CH_4N.
21.	(*a*)	E.F. mass = 1 × 12 + 2 × 1 + 16 × 1 = 30 $M.F. = \frac{\text{Mol. wt.}}{\text{E.F. wt.}} \times E.F.$ $= \frac{180}{30} \times CH_2O = C_6H_{12}O_6$.
22.	(*d*)	Halogens.
23.	(*d*)	Anhyd. $CuSO_4$ turns blue in presence of water. $\underset{(White)}{CuSO_4} + 5H_2O \longrightarrow \underset{(Blue)}{CuSO_4 . 5H_2O}$
24.	(*a*)	Mol. mass of toluene ($C_6H_5CH_3$). $= 12 \times 7 + 8 \times 1 = 92$.
25.	(*b*)	Hydrazine (NH_2NH_2) does not contain C and hence on fusion with Na metal, it cannot form NaCN. Therefore hydrazine does not show Lassaigne's test.
26.	(*c*)	25 ml of 0.12 N NaOH ≡ 0.1914 g of acid $\therefore$ 100 ml of 1 N NaOH $= \frac{0.1914 \times 1000}{25 \times 0.12} = 63.8$ $\therefore$ Eq. mass of the acid = 63.8.
27.	(*c*)	$\% O = 100 - 49.3 - 6.84 = 43.86$ Now $C : H : O = \frac{49.3}{12} : \frac{6.84}{1} : \frac{43.86}{16}$ $= 1.49 : 2.49 : 1 = 3 : 5 : 2$ $\therefore$ Empirical Formula = $C_3H_5O_2$.
28.	(*b*)	Since an acid contain COOH group, therefore, simplest acid containing this group is HCOOH, *i.e.*, CH_2O_2.
29.	(*b*)	In paper chromatography both moving as well as stationary phases are liquids.
30.	(*e*)	C : H mol ratio is $\frac{3}{12} : \frac{1}{1}$ or 1 : 4 Hence empirical formula is CH_4.
31.	(*a*)	$\% C = \frac{0.535 \times 12 \times 100}{44 \times 0.765} = 19\%$ $\% H = \frac{0.14 \times 2 \times 100}{18 \times 0.765} = 2.03\%$ $\therefore \quad \frac{\% C}{\% H}$ is 19 : 2.
32.	(*a*)	$\%$ of Se $= \frac{\text{At. mass of Se} \times 100}{\text{Mini. mol. mass}}$ $\therefore$ Minimum mol. mass $= \frac{\text{At. mass of Se} \times 100}{\% \text{ of Se}}$ $= \frac{7.84 \times 100}{0.5} = 1.568 \times 10^4\ u$.
33.	(*c*)	$\frac{\text{Mol. mass}}{\text{Empirical mass}} = \frac{42}{14} = 3$ $\therefore$ Mol. formula = $(CH_2)_3 = C_3H_6$.

Q. No.	Ans.	Hints & Solutions
34.	**(*c*)**	Mass ratio C : H : N 9 : 1 : 3.5 At. ratio C : H : N $\frac{9}{12} : \frac{1}{1} : \frac{3.5}{14}$ or 0.75 : 1 : 0.25 or 3 : 4 : 1 Hence Emp. formula is C_3H_4N $\frac{\text{Mol. mass}}{\text{Emp. mass}} = \frac{108}{54} = 2$ ∴ Mol. formula is $(C_3H_4N)_2$ or $C_6H_8N_2$.
35.	**(*d*)**	Lassaigne's extract contains N as NaCN which reacts with $FeSO_4$ to form $Na_4[Fe(CN)_6]$ in solution. This reacts with Fe^{3+} ions of $FeCl_3$ solution in acid medium to form $Fe_4[Fe(CN)_6]_4$ a blue coloured compound.
36.	**(*c*)**	0.1 M H_2SO_4 used by NH_3 = 50 mL % of N = $\frac{1.4 \times 0.1 \times 2 \times 50}{0.3}$ = 46.6% Among the given species only urea supports the above data. Hence, the correct choice is (*c*).
37.	**(*d*)**	The mixture of naphthalene and benzoic acid is separated by crystallisation from hot water. Benzoic acid dissolves in hot water whereas naphthalene is insoluble.
38.	**(*d*)**	% of C = $\frac{12 \times 12 \times 100}{44 \times 4}$ = 81.8 % of H = 100 – 81.8 = 18.2 Relative number of atoms = C = $\frac{81.8}{12}$ = 6.81 Simplest atomic ratio C : H = $\frac{6.81}{6.81} : \frac{18.2}{6.81}$ = 1 : 2.67 or 3 : 8 Empirical formula of hydrocarbon C_3H_8.
39.	**(*d*)**	Napthalene and benzoic acid mixture can be separated by crystallisation from hot water.
40.	**(*b*)**	From the %age data, the empirical formula comes out to the N_2H_4CO.
41.	**(*a*)**	Latest technique of separation and isolation is chromatography.
42.	**(*d*)**	% of P = $\frac{62 \times W_{(Mg_2P_2O_7)} \times 100}{222 \times W_{(compound)}}$ = $\frac{62 \times 0.444 \times 100}{222 \times 0.31}$ = 40

8.7. ASSERTION-REASON QUESTIONS

Quick Revision

In each of the following questions two statements are given as Assertion A and Reason R. Examine the statements carefully and answer the questions according to the instructions given below :

- *Mark **a** if both **A** and **R** are correct and **R** is the correct reason of **A**.*
- *Mark **b** if both **A** and **R** are correct and **R** is not the correct reason of **A**.*
- *Mark **c** if **A** is correct and **R** is wrong.*
- *Mark **d** if **A** is wrong and **R** is correct.*
- *Mark **e** if both **A** and **R** are wrong.*

1. ***Assertion A.*** Carbon and hydrogen elements are estimated by Leibig's method.

   Reason R. All the organic compounds contain C and H elements as essential constituents.

(a)	(b)	(c)	(d)	(e)

2. ***Assertion A.*** Melting and boiling points are significant criteria of purity of organic compounds.

   Reason R. Homologues have same boiling points.

(a)	(b)	(c)	(d)	(e)

3. ***Assertion A.*** Beilstein test is used to detect halogens in organic compounds.

   Reason R. Beilstein test is not always dependable.

(a)	(b)	(c)	(d)	(e)

4. ***Assertion A.*** Victor Meyer method can be used to determine the molecular mass of all types of substance.

   Reason R. In Victor Meyer method, the vapours of the substance are finally collected in the graduated tube.

(a)	(b)	(c)	(d)	(e)

5. ***Assertion A.*** Column chromatography is based on the principle of adsorption.

   Reason R. In paper chromatrography the mobile phase as well as stationary phase is constituted by liquid.

(a)	(b)	(c)	(d)	(e)

6. ***Assertion A.*** Lithium can be used in place of sodium in the Lassaigne's test.

   Reason R. The compounds of Li are generally ionic in nature.

(a)	(b)	(c)	(d)	(e)

7. ***Assertion A.*** Aqueous solution of NaOH can bring about the separation of water insoluble phenols from non acidic compounds.

   Reason R. Water insoluble phenols do not react with NaOH.

(a)	(b)	(c)	(d)	(e)

8. ***Assertion A.*** In Kjeldahl's method, a little K_2SO_4 is added to the contents while heating the organic compound with conc. H_2SO_4.

   Reason R. K_2SO_4 raises the boiling point of H_2SO_4 to ensure complete conversion of N into $(NH_4)_2SO_4$.

(a)	(b)	(c)	(d)	(e)

9. ***Assertion A.*** N_2 produced in Duma's method is collected over dilute aqueous solution of KOH.

   Reason R. KOH solution makes it free from CO_2 and H_2O.

(a)	(b)	(c)	(d)	(e)

10. ***Assertion A.*** The compounds having same empirical formula must have same properties.

    Reason R. Molecular formula is whole number multiple of empirical formula.

(a)	(b)	(c)	(d)	(e)

11. ***Assertion A.*** Essential oils can be purified by steam distillation.

    Reason R. Essential oils are insoluble in water but volatile in steam.

(a)	(b)	(c)	(d)	(e)

12. ***Assertion A.*** Diazonium salts do not show Lassaigne's test for nitrogen.

    Reason R. Diazonium salts do not contain nitrogen at all.

(a)	(b)	(c)	(d)	(e)

ANSWERS

1. (*c*) **2.** (*c*) **3.** (*b*) **4.** (*e*) **5.** (*b*) **6.** (*e*) **7.** (*c*) **8.** (*a*) **9.** (*a*)
10. (*d*) **11.** (*a*) **12.** (*a*).

HINTS/SOLUTIONS

Q. No.	Ans.	Hints & Solutions
1.	**(*c*)**	Statement ***R*** is wrong. There may be organic compounds which do not contain H.
2.	**(*c*)**	Statement ***R*** is wrong because homologues do not have same boiling points.
3.	**(*b*)**	Both the statements are factual.
4.	**(*e*)**	Both the statements are against the facts.
5.	**(*b*)**	Both statements are factual
6.	**(*e*)**	Both statement are against the facts.
7.	**(*c*)**	Statement ***R*** is wrong.
8.	**(*a*)**	***R*** provides explanation of ***A***.
9.	**(*a*)**	***R*** is an explanation of ***A***.
10.	**(*d*)**	***A*** is wrong. Different compounds can have same properties.

8.8. SELF EVALUATION TEST

Assess Yourself

Choose the correct answer from the four alternatives given in each of the following questions :

1. 0.50 g of an organic compound was Kjeldahlised and the NH_3 evolved was absorbed in a 50 mL of 0.5 M H_2SO_4. The residual acid required 60 cm^3 of NaOH. The % age N in the compound is
(*a*) 56% (*b*) 46%
(*c*) 39.2% (*d*) 29.2%.

2. 0.8 g of the silver salt of a tribasic acid on ignition gave 0.40 g of pure silver. The molar of the acid is
(*a*) 109 g mol^{-1} (*b*) 280 g mol^{-1}
(*c*) 327 g mol^{-1} (*d*) 218 g mol^{-1}.

3. 0.532 g of the chloroplatinate of an organic base (mol. mass 244 *u*) gave 0.195 g of Pt. on ignition. Then the acidity of the base is
(*a*) 1 (*b*) 2
(*c*) 3 (*d*) 4.

4. A hydrogen having empirical formula as CH_2 has a density of 1.25 g/L. The molecular formula of the hydrocarbon is
(*a*) C_2H_2 (*b*) C_2H_4
(*c*) C_2H_6 (*d*) C_3H_8.

5. The percentage composition of a compound is C = 90% and H = 10%. A plausible formula of the compound is
(*a*) C_8H_{10} (*b*) $C_{15}H_{30}$
(*c*) $C_{15}H_{20}$ (*d*) $C_{15}H_{12}$.

6. Complete combustion of a sample of a hydrocarbon gives 0.66 g of CO_2 and 0.36 g of H_2O. The empirical formula of the compound is
(*a*) CH_2 (*b*) C_3H_4
(*c*) C_3H_8 (*d*) C_6H_8.

7. A dibasic acid containing C, H and O was found to contain C = 26.7% and H = 2.2%. The vapour density of its diethyl ester was found to be 73. What is the molecular formula of the acid ?
(*a*) CH_2O_2 (*b*) $C_2H_2O_4$
(*c*) $C_3H_3O_4$ (*d*) $C_4H_4O_4$.

8. 20.0 cm^3 of a mixture of CO, CH_4 and He gas was exploded with excess of O_2 in a eudiometer tube at room temperature. After cooling the volume contraction was found to be 13.0 cm^3. A further contraction of 14.0 cm^3 occurred when the residual gas was treated with KOH solution. The composition of the mixture is
(*a*) CO = 10 cm^3, CH_4 = 6 cm^3 and He = 4 cm^3
(*b*) CO = 10 cm^3, CH_4 = 4 cm^3 and He = 6 cm^3
(*c*) CO = 6 cm^3, CH_4 = 10 cm^3 and He = 4 cm^3
(*d*) CO = 4 cm^3, CH_4 = 6 cm^3 and He = 10 cm^3.

9. Sublimation cannot be used for purification of
(*a*) Urea (*b*) Camphor
(*c*) Benzoic acid (*d*) Naphthalene.

10. Thiophene can be removed from commercial benzene by
(*a*) Shaking it with a mixture of conc. HNO_3/H_2SO_4
(*b*) Shaking it with conc. H_2SO_4
(*c*) Steam distillation
(*d*) Fractional distillation.

11. An organic compound from its aqueous solution is to be extracted with 100 ml. of $CHCl_3$. The amount of the organic compound extracted is maximum when
(*a*) 100 ml of $CHCl_3$ is used in one lot
(*b*) Two 50 ml portions of $CHCl_3$
(*c*) Four 25 ml portions of $CHCl_3$
(*d*) Ten 10 ml portions of $CHCl_3$.

12. Which of the following compounds will give Lassaigne's test for nitrogen ?
(*a*) NH_2NH_2 (*b*) NH_4Cl
(*c*) NaCN (*d*) $NaNO_3$.

13. Beilstein's test is shown by
(*a*) Halogens (*b*) Pyridine
(*c*) Thiourea (*d*) All the above.

14. In Kjeldahl's method, during digestion, the nitrogen of the organic compound is converted into
(*a*) NH_4Cl (*b*) $(NH_4)_2SO_4$
(*c*) NH_4NO_3 (*d*) NH_3.

15. The molecular mass of an organic compound which contains only one nitrogen atom can be
(*a*) 73 (*b*) 76
(*c*) 146 (*d*) 152.

16. An organic compound having carbon, hydrogen and sulphur contains 4% of sulphur. The minimum molecular weight of the compound is
(*a*) 200 (*b*) 400
(*c*) 600 (*d*) 800.

17. If the percentage of nitrogen in an organic compound is 12.5% then how much of the organic compound should be taken so as to produce 50 ml of N_2 at 300 K and 715 mm pressure (Aq. tension = 15 mm) ?
(*a*) 0.419 g (*b*) 0.149 g
(*c*) 0.914 g (*d*) 0.941 g.

18. Which of the following is used as dessicant for absorbing water in Liebig's method for estimation of carbon and hydrogen ?

(*a*) $MgSO_4.7H_2O$ (*b*) $CuSO_4.5H_2O$
(*c*) Anhydrous $CaCl_2$ (*d*) KOH.

19. In the estimation of nitrogen, the Kjeldahlised liquid is heated with

(*a*) Conc. HNO_3 (*b*) Conc. Na_2SO_4
(*c*) Conc. HCl (*d*) Conc. NaOH.

20. For the estimation of sulphur by Carius method, the organic compound is heated in Carius tube with

(*a*) $BaCl_2/HCl$ (*b*) $BaCl_2/H_2SO_4$
(*c*) AgCl/HCl (*d*) $CaCl_2/HCl$.

21. Which pair of compounds give purple colouration with sodium nitroprusside ?

(*a*) Toluic acid, benzoic acid

(*b*) 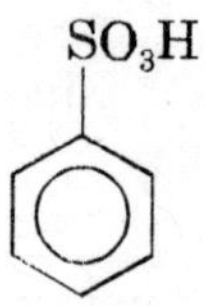 and 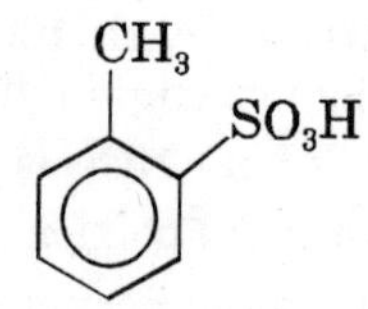

(*c*) 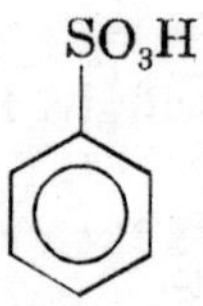 and 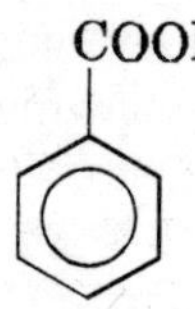

(*d*) None of the above.

22. For which of the following compound, steam-distillation can be used for purification ?

(*a*) *p*-hydroxyphenol (*b*) Phenol
(*c*) Salicylaldehyde (*d*) All of these.

23. For which of the following pairs of compound, Kjeldahl's method cannot be employed for the estimation of nitrogen

(*a*) 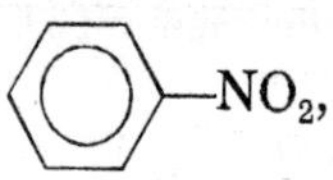 

(*b*) 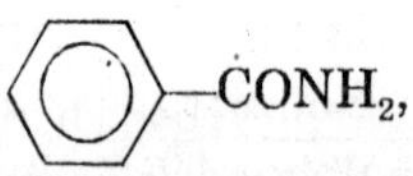 

(*c*) Benzamide, acetamide
(*d*) Nitrobenzene, benzamide.

24. Haemoglobin contains 0.33% of iron by mass. The molecular mass of haemoglobin is approximately 67200. The number of iron atoms [at. mass of Fe = 56] present is one molecule of haemoglobin is

(*a*) 6 (*b*) 1
(*c*) 4 (*d*) 2.

25. 6.0 g of organic compound on heating with NaOH gave NH_3 which neutralised 200 ml of 1 M HCl. The percentage of N in organic compound is

(*a*) 46.67% (*b*) 35.5%
(*c*) 56.6% (*d*) 67%.

26. Complete combustion of 0.858 g of compound X gives 2.577 g of CO_2 and 1.398 g of H_2O. The lowest molecular mass that X can have is

(*a*) 44 (*b*) 22
(*c*) 129 (*d*) 172.

27. A gaseous hydrocarbon having empirical formula CH_2 has a density of 1.25 g L^{-1} at STP. The molecular formula of the hydrocarbon is

(*a*) C_2H_2 (*b*) C_2H_4
(*c*) C_2H_6 (*d*) C_3H_8.

28. Two volatile liquids A and B differ in their boiling points by 15 K. Which process can be used to cause their separation ?

(*a*) Steam distillation
(*b*) Fractional distillation
(*c*) Simple distillation
(*d*) Distillation under reduced pressure.

29. Which method can be used for the purification of naphthalene ?

(*a*) Sublimation
(*b*) Distillation
(*c*) Steam distillation
(*d*) Distillation under reduced pressure.

30. An organic compound contains carbon = 12.76%, Hydrogen = 2.13% and rest being bromine. The vapour density of the compound is 95. Its molecular formula is

(*a*) CH_3Br (*b*) $C_2H_4Br_2$
(*c*) $C_3H_6Br_3$ (*d*) C_4H_4Br.

31. Two elements A (at. mass = 12) and B (at. mass = 16) combine to form a compound with mass percentage A being 27.3%. The formula of the compound is

(*a*) A_2B_2 (*b*) AB
(*c*) A_2B (*d*) AB_2.

32. For a given organic liquid, the best criteria to judge its purity is its

(*a*) colour (*b*) mixed boiling point
(*c*) odour (*d*) none of these.

33. 15 ml of gaseous hydrocarbon required 45 ml of oxygen for complete combustion and 30 ml of CO_2 is formed. The formula of hydrocarbon is

(*a*) C_3H_6 (*b*) C_2H_4
(*c*) C_2H_6 (*d*) C_2H_2.

34. Molecular mass of monobasic acid A is 60. Which among the following represents its empirical fomula ?

(*a*) $C_2H_4O_2$ (*b*) CH_2O
(*c*) $C_3H_6O_3$ (*d*) $C_2H_4O_3$.

35. Certain organic compound contains 10% of sulphur in it. Which of the following represent the minimum molecular mass of the compound

(*a*) 320 (*b*) 160

(*c*) 400 (*d*) 200.

36. During the estimation of phosphorus, in the organic compound, which of the following represent final precipitate ?

(*a*) $Mg_3(PO_4)_2$

(*b*) Magnesium pyrophosphate

(*c*) Magnesium hydrogen phosphate

(*d*) P_4O_{10}.

37. Which of the following compound does not correspond to the following percentage data ? (C = 85.45% and H = 14.55 %)

(*a*) C_2H_4 (*b*) C_6H_{12}

(*c*) C_2H_6 (*d*) C_4H_8.

38. In the estimation of sulphur in the organic compound, fuming nitric acid converts sulphur into

(*a*) SO_2 (*b*) SO_4^{2-}

(*c*) SO_3^{2-} (*d*) S^{2-}.

39. 60 g of organic compound on analysis gave C = 24 g ; H = 4 g and O = 32 g. The empirical formula of the compound is and

(*a*) CH_2O (*b*) C_2H_2O

(*c*) CH_2O_2 (*d*) CH_3O.

40. 0.75 g of silver salt of dibasic organic acid on ignition left 0.45 g of metallic silver. The molar mass of acid is

(*a*) 73.0 g mol^{-1} (*b*) 108 g mol^{-1}

(*c*) 60 g mol^{-1} (*d*) 146 g mol^{-1}.

ANSWERS

1. (*c*)	**2.** (*c*)	**3.** (*d*)	**4.** (*b*)	**5.** (*c*)	**6.** (*c*)	**7.** (*b*)	**8.** (*b*)	**9.** (*a*)
10. (*b*)	**11.** (*d*)	**12.** (*c*)	**13.** (*a*)	**14.** (*b*)	**15.** (*a*)	**16.** (*d*)	**17.** (*a*)	**18.** (*c*)
19. (*d*)	**20.** (*a*)	**21.** (*d*)	**22.** (*c*)	**23.** (*a*)	**24.** (*c*)	**25.** (*a*)	**26.** (*a*)	**27.** (*b*)
28. (*b*)	**29.** (*a*)	**30.** (*b*)	**31.** (*d*)	**32.** (*b*)	**33.** (*b*)	**34.** (*b*)	**35.** (*a*)	**36.** (*b*)
37. (*c*)	**38.** (*b*)	**39.** (*a*)	**40.** (*d*).					

HINTS/SOLUTIONS

Q. No.	Ans.	Hints & Solutions
1.	(*c*)	$H_2SO_4 + 2NaOH \longrightarrow Na_2SO_4 + H_2O$ $\therefore \frac{M_1V_1}{1} = \frac{M_2V_2}{2}$ or $V_1 = \frac{M_2V_2}{2M_1}$ $= \frac{0.6 \times 60}{2 \times 0.5} = 36$ mL Volume of H_2SO_4 used = 50 – 36 = 14 mL $\%N = \frac{1.4 \times V_{acid} \times \text{basicity} \times M_{acid}}{W_{subst.}}$ $= \frac{1.4 \times 14 \times 2 \times 0.5}{0.5} = 39.2\%$
2.	(*c*)	Molar mass $= 3\left(\frac{108 \times 0.8}{0.4} - 107\right) = 327$ g mol^{-1}.
3.	(*d*)	Let acidity of base be n Mol mass $= \frac{n}{2}\left(\frac{195\text{ W}}{x} - 410\right)$ or $n = \frac{2 \times \text{mol. mass}}{(195\text{ W}/x - 410)}$ $= \frac{2 \times 264}{(195 \times 0.532/0.195 - 410)}$ $= \frac{2 \times 244}{122} = 4.$
4.	(*b*)	1 Mole of hydrocarbon = Mass of 22.4 litres at STP = 1.25 × 22.4 = 28 $\therefore$ Mol. formula $= CH_2 \times \frac{28}{14} = C_2H_4$.

Q. No.	Ans.	Hints & Solutions
5.	(c)	$C : H = \frac{90}{12} : \frac{10}{1} = 3 : 4$ $\therefore$ Empirical formula = C_3H_4. Since M.F. is n times E.F., therefore, the M.F. of the hydrocarbon = $C_{15}H_{20}$.
6.	(c)	Let the mass of the substance taken = w $\therefore \% C = \frac{12}{44} \times \frac{0.66}{w} \times 100 = \frac{18}{w}$ and $\% H = \frac{2}{18} \times \frac{0.36}{w} \times 100 = \frac{4}{w}$ Now, $C : H = \frac{18}{12w} : \frac{4}{1w} = 3 : 8$ Hence E.F. = C_3H_8.
7.	(b)	$C : H : O = \frac{26.7}{12} : \frac{2.2}{1} : \frac{71.1}{16} = 1 : 1 : 2$ $\therefore$ E.F. of the compound = CHO_2 and E.F. mass = 12 + 1 + 32 = 45 Mol. mass of diethyl ester = 2 × 73 = 146 Mol. mass of the ethyl group = 29 Since the acid is dibasic, therefore, it must have two ethyl groups. Thus, Mol. of the acid = 146 – 2 × 29 + 2 = 90 $\therefore$ M.F. of the dibasic acid $= \frac{90}{45} \times CHO_2 = C_2H_2O_4$.
8.	(b)	Let the vol. of CO in the mixture = x cm^3 and that of CH_4 = y cm^3 $\therefore$ Vol. of He in the mixture = 20 – $(x + y)$ cm^3 Contraction in volume due to explosion and cooling = 13.0 cm^3 and contraction in volume due to adsorption in KOH = 14.0 cm^3 = vol. of CO_2 produced. The chemical equations representing combustion of CO and CH_4 respectively are : $2CO + O_2 \longrightarrow 2CO_2$ ...(i) x cm^3 $x/2$ cm^3 x cm^3 $CH_4 + 2O_2 \longrightarrow CO_2 + 2H_2O$ ...(ii) y cm^3 $2y$ cm^3 y cm^3 negligible volume Contraction in volume due to reaction (i) $x + \frac{x}{2} - 2 = \frac{x}{2}$ cm^3 Contraction in volume due to reaction (ii) $y + 2y - y = 2y$ cm^3 $\therefore$ Total contraction in volume $= \left(\frac{x}{2} + 2y\right)$ cm^3 But $\frac{x}{2} + 2y = 13$ (Given) ...(iii) Total volume of CO_2 produced in reaction (i) and (ii) = $(x + y)$ cm^3 But $x + y = 14$ (Given) ...(iv) Solving (iii) and (iv) for x and y, we have $x = 10.0$ cm^3 and $y = 4.0$ cm^3 Thus, volume of CO = 10 cm^3, of CH_4 = 4 cm^3 and of He = 20 – (10 + 4) = 6 cm^3. and of He.
9.	(a)	Only urea does not sublime while naphthalene, camphor and benzoic acid do.
10.	(b)	By shaking it with conc. H_2SO_4.
11.	(d)	For a given amount of the solvent, larger the number of extractions, more is the compound extracted.
12.	(c)	During fusion with Na metal, the organic compound is converted into NaCN and if NaCN is the given compound, it will respond to Lassaigne's test.
13.	(d)	The bluish-green flame obtained during Beilstein test is due to the formation of volatile cupric halides. Pyridine and thiourea also give this test because they form volatile cupric cyanide which also give bluish-green flame.
14.	(b)	$(NH_4)_2SO_4$.
15.	(a)	All compounds containing an odd number of N atoms (*i.e.*, 1, 3, 5, ... etc.) have odd masses and those which contain even number of N atoms (*i.e.*, 2, 4, 6, ... etc.) have even masses. This is called *nitrogen rule*. Thus, option (a) is correct.
16.	(d)	The minimum mol. mass must contain at least one S atom $\% S = \frac{\text{Mass of one S atom}}{\text{Min. mol. mass}} \times 100$ or $4 = \frac{32}{\text{Min. mol. mass}} \times 100$ or Min. mol. mass = $\frac{32}{4} \times 100 = 800$.
17.	(a)	Let the mass of organic compound taken = w $\therefore$ Mass of N present in 100 g of the organic compound

Q. No.	Ans.	Hints & Solutions
		$= \frac{12.5\,w}{100} \times g$ But 22400 ml of N_2 at STP, weighs = 28 g $\therefore$ Vol. of $\frac{12.5 \times w}{100}$ g of N_2 at STP $= \frac{22400}{28} \times \frac{12.5}{100} \times w = 100\,w$ ml. Applying gas equation, $P_1 = 760$ mm ; $P_2 = 715 - 15 = 700$ mm $V_1 = 100\,w$ ml ; $V_2 = 50$ ml $T_1 = 273$ K ; $T_2 = 300$ K $\therefore \frac{760 \times 100w}{273} = \frac{700 \times 50}{300}$ or $w = \frac{700 \times 50}{300} \times \frac{273}{760 \times 100} = 0.419$ g.
18.	(*c*)	Among the given species, dessicants used in Liebig's method is anhydrous $CaCl_2$.
19.	(*d*)	Kjeldahlised liquid contains $(NH_4)_2SO_4$. Hence it is heated with NaOH to liberate ammonia.
20.	(*a*)	$BaCl_2$ causes the precipitation of sulphur as $BaSO_4$.
21.	(*d*)	Purple colour is given by sodium sulphide and sodium nitroprusside. Hence none of the given pairs will give purple colouration directly.
22.	(*c*)	Among the given species only salicylaldehyde can be purified by steam distillation.
23.	(*a*)	Among the given pairs, the Kjeldahl's method cannot be employed to nitrobenzene and pyridine.
24.	(*c*)	Total mass of Fe in haemoglobin $= \frac{0.33}{100} \times 67200 = 221.76$ At. mass of Fe = 56 Total no. of Fe atoms per molecule of haemoglobin $= \frac{221.76}{56} = 3.96$ Thus no. of Fe atoms = 4.
25.	(*a*)	M. mol of HCl = 200 × 1 = 200 M. mol of NH_3 = 200 Mass of $NH_3 = 200 \times 10^{-3} \times 17$ g Mass of N $= 200 \times 10^{-3} \times 14$ g % of N $= \frac{200 \times 10^{-3} \times 14 \times 100}{6} = 46.67\%$.
26.	(*a*)	$\% C = \frac{12}{44} \times \frac{2.577}{0.858} \times 100 = 81.9$ $\% H = \frac{2}{18} \times \frac{1.398}{0.858} \times 100 = 18.1$ Now, $C : H = \frac{81.9}{12} : \frac{18.1}{1} = 6.825 : 18.1$ $= 1 : 2.65$ or 3 : 8 $\therefore$ Empirical formula = C_3H_8 E.F. mass = 3 × 12 + 8 × 1 = 44 Lowest mol. mass of a compound is equal to its empirical formula mass = 44.
27.	(*b*)	Mol. mas of hydrocarbon = 1.25×22.4 g mol^{-1} = 28 g mol^{-1} E.F. = mass = 14 $\therefore n = \frac{28}{14} = 2.33 \approx 2$ $\therefore$ Molecular Formula = C_2H_4.
28.	(*b*)	Fractional distillation will cause the separation.
29.	(*a*)	Purification of naphthalene can be caused by sublimation.
30.	(*b*)	% of Br = 100 – (2.13 + 12.76) = 85.11% $C : H : Br = \frac{12.76}{12} : \frac{2.13}{1} : \frac{85.11}{80}$ = 1.06 : 2.13 : 1.06 = 1 : 2 : 1 Empirical formula is CH_2Br $n = \frac{\text{Mol. mass}}{\text{Empirical mass}} = \frac{92 \times 2}{95} = 2$ Mol. formula is $(CH_2Br)_2$ or $C_2H_4Br_2$.
31.	(*d*)	$A = 27.3 ; \frac{27.3}{12} = 2.275 ; \frac{2.275}{2.275} = 1$ $B = 72.4 ; \frac{72.4}{16} = 4.5 ; \frac{4.5}{2.275} = 2$ Thus E.F. is AB_2.
32.	(*b*)	The given sample of liquid is mixed with purified liquid and its boiling point is determined. If the mixture has sharp and fixed boiling point as that of pure liquid. It implies that the sample liquid is pure.
33.	(*b*)	The given volume ratio of HC : O_2 : CO_2 is 1 : 3 : 2 which is justified only by C_2H_4 among the given species $C_2H_4 + 3O_2 \longrightarrow 2CO_2 + 2H_2O$ 1 vol. 3 vol. 2 vol.

Q. No.	Ans.	Hints & Solutions
34.	(*b*)	A has mol. mass = 60 but it is not the simplest formula. The simplest formula of A is CH_2O. Hence choice (*b*) is correct.
35.	(*a*)	If S is 10 the mol. mass is 100 If S is 32 parts the mol. mass $= \frac{100}{10} \times 32 = 320.$
36.	(*b*)	Final precipitate is $Mg_2P_2O_7$.
37.	(*c*)	C_2H_6 does not correspond to above data. The other species have same empirical formulae CH_2.
38.	(*b*)	Fuming nitric acid oxidises sulphur to SO_4^{2-} ion during carius method.

Q. No.	Ans.	Hints & Solutions
39.	(*a*)	$C = 24\ g = \frac{24}{12} = 2$ g-atom ; $H = 4g = \frac{4}{1} = 4$ g atom ; O = 32 g = 2 g-atom Hence atomic ratio C : H : O is 2 : 4 : 2 or 1 : 2 : 1 Thus, empirical formula is CH_2O.
40.	(*d*)	Molar mass $= n\left(\frac{108\ W}{x} - 107\right)$ $= 2\left(\frac{108 \times 0.75}{0.450} - 107\right)$ $= 2 \times 73 = 146\ g\ mol^{-1}$

9 BIOMOLECULES

9.1. CONCEPTS IN BRIEF

Develop Logic

INTRODUCTION

- **Biochemistry.** The branch of science that deals with the study of the chemical composition and structure of living organisms and also various changes taking place within them.
- **Biomolecules.** The complex organic molecules which form the basis of life, *i.e.,* which build up living organisms and are also required for their growth and maintenance. For example, carbohydrates, proteins, lipids, vitamins, etc.

CARBOHYDRATES

- **Carbohydrates.** Carbohydrates are defined as the polyhydroxy aldehydes and polyhydroxy ketones or the other compounds which can be hydrolysed to them.
- **Types of carbohydrates.** Depending upon their behaviour towards hydrolysis, carbohydrates are divided into following types :

1. ***Monosaccharides.*** *These are simple carbohydrates which cannot be hydrolysed to still simpler carbohydrates.* Glucose and fructose are common examples.
2. ***Oligosaccharides.*** *These are the carbohydrates which on hydrolysis give two to nine units of monosaccharides.*

Disaccharides give two units of monosaccharides on hydrolysis. Common examples are sucrose and maltose. Both have molecular formula $C_{12}H_{22}O_{11}$.

Trisaccharides give three units of monosaccharides on hydrolysis. Raffinose, $C_{18}H_{32}O_{16}$ is a common example.

Tetrasaccharides give four units of monosaccharides on hydrolysis. Stachyose, $C_{24}H_{42}O_{21}$ is a common example.

3. ***Polysaccharides.*** *These are the carbohydrates which are polymeric molecules and can be hydrolysed to give large number of monosaccharide units.* The commonly occurring polysaccharides have the general formula $(C_6H_{10}O_5)_n$. The common examples are starch, glycogen and cellulose.

- **Sugars and Non-sugars.** Carbohydrates which are sweet in taste are collectively called **sugars** while those which are not sweet are called **non-sugars.** *Monosaccharides and Disaccharides are* **sugars** *but polysaccharides are* **non-sugars.**

The relative degree of sweetness of some common sugars is given below in tabular form :

Sugar	Lactose	Maltose	Galactose	Glucose	Sucrose	Fructose
Relative Sweetness	16	32	32	74	100	173

- **Structures of Some Monosaccharides**

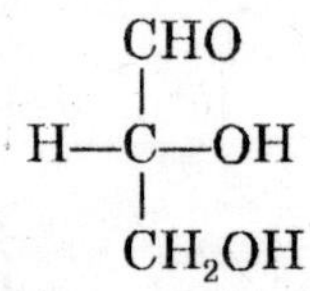

Glyceraldehyde
(Aldotriose)

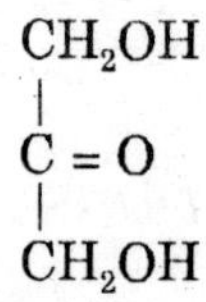

Dihydroxyacetone
(Ketotriose)

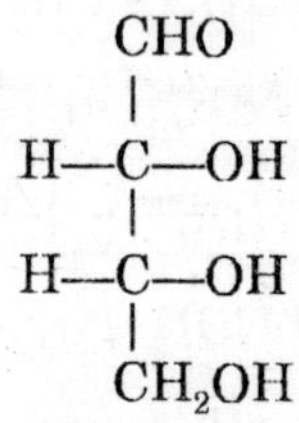

Erythrose
(Aldotetrose)

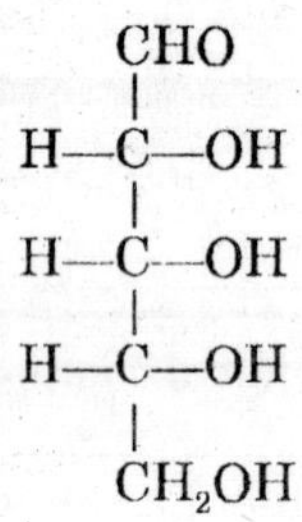

Ribose
(Aldopentose)

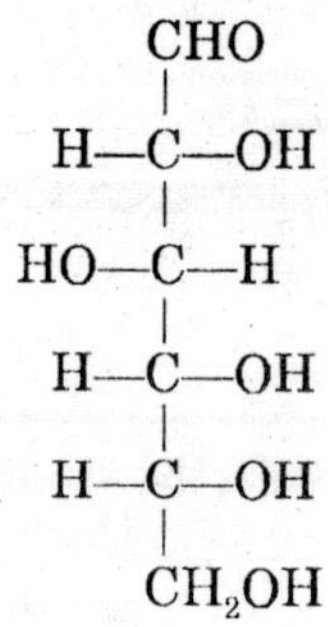

Glucose
(Aldohexose)

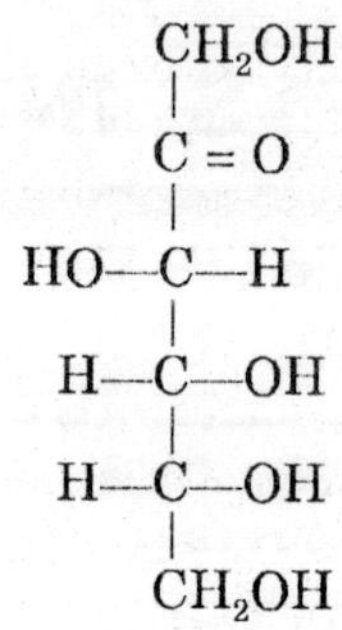

Fructose
(Ketohexose)

- The simplest monosaccharide has three carbon atoms.
- The various monosaccharide units are linked to each other in a polysaccharide through **glycoside linkage** which is an **ether linkage.**
- The sugars which can reduce Tollen's reagent and Fehling's solution (or Benedict's solution) are known as **reducing sugars.** All monosaccharides and most of the disaccharides are reducing sugars.
- Sucrose is a non-reducing sugar.
- α-D-Glucose and β-D-Glucose are **anomers.** Anomers are the diastereomers which have different orientation of hydroxyl groups at C-1.
- Monosaccharides containing aldehyde group are called **aldoses** while those containing ketone group are called **ketoses.** Glucose is an aldohexose while fructose is a ketohexose.
- **Epimers.** These are the diastereomeric aldoses which differ in orientation of hydroxyl groups at C-2. For example, D-glucose and D-mannose.
- The letter D before the name of monosaccharides does not refer to dextrorotatory, but represents the configuration as derived from glyceraldehyde. Glyceraldehyde can be represented by two enantiomeric as shown below :

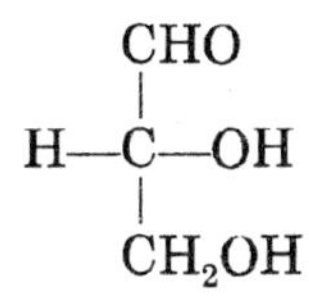

D-configuration

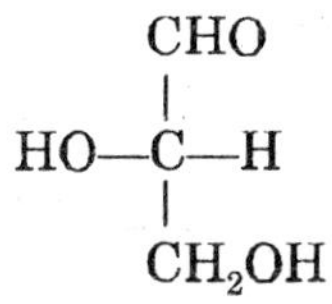

L-configuration

By convention a monosaccharide is given a **D-configuration** *if hydroxyl group attached to the carbon adjacent to $-CH_2OH$ is on right hand side.* On the other hand, it is assigned **L-configuration** *if hydroxyl group attached to the carbon adjacent to $-CH_2OH$ is on left hand side.*

- **Mutarotation.** The two forms of glucose *i.e.,* α-D-glucose and β-D-glucose possess different specific rotations. Specific rotation of α-D-glucose is + 112° whereas that of β-D-glucose is + 19°. When either form of glucose is dissolved in water, the specific rotation attains an equilibrium value which is + 52°. *The change in specific rotation of either form of glucose in aqueous solution to that of equilibrium mixture is called* **mutarotation.**
- **Hydrolysis products of some carbohydrates**

  Sucrose ⟶ Glucose + Fructose

  Maltose ⟶ Glucose + Glucose

  Lactose ⟶ Glucose + Galactose

  Starch ⟶ Glucose

  Cellulose ⟶ Glucose
- Starch is formed by polymerisation of α-D-glucose.
- In cellulose, the monomer units are β-D-glucose.

INVERSION OF SUGAR

Sucrose is dextro rotatory with specific rotation + 66.5°C. On hydrolysis in the presence of HCl or enzymes invertase, it produces mixture of D(+) glucose (specific rotation + 52.7°) and D(–) fructose (specific rotation = – 92.4°). Since specific rotation of D(–) fructose is greater than D(+) glucose, the resulting solution becomes laevorotatory. As hydrolysis produces change in optical nature from dextro rotatory to laevorotatory, the process is called **inversion of sugar.** The mixture of D(+) gluose and D(–) fructose obtained as a result of hydrolysis is called **INVERT SUGAR.**

- Gums and pectins are also examples of polysaccharides.

REACTIONS OF GLUCOSE

- Glucose reacts with acetyl chloride or acetic anhydride to form penta-acetyl derivative.
- Glucose reacts with CH_3OH in the presence of HCl gas forming α-methyl glucoside and β-methyl glucoside.
- Glucose, on oxidation with mild oxidizing agent, such as Br_2 water gives **gluconic acid.**
- Glucose reacts with Tollen's reagent as well as Fehling solution and is oxidized to **gluconic acid.**
- Glucose on oxidation with HNO_3 gives **glutaric acid.**
- Glucose reacts with phenylhydrazine in equimolar proportion to form phenyl hydrazone however, with excess phenyl hydrazine it forms **glucosazone.** In this reaction one molecule of glucose reacts with three molecules of phenylhydrazine.
- Glucose on treatment with dilute sodium hydroxide solution undergoes a reversible isomerization and is converted into a mixture of D-glucose, D-fructose and D-mannose. This reaction is known as **Lobry de Bruyn van-Ekenstein** rearrangement.
- Glucose in aqueous solution is reduced with sodium amalgam to hexahydric alcohol **sorbitol.**
- Glucose on prolonged heating with HI forms *n*-hexane.

PROTEINS

- The monomeric units in polypeptides and proteins are α-**amino acids.**
- Proteins contain about 21 different types of amino acids, out of which 10 are essential amino acids.

- Amino acids possess zwitter ion structure.

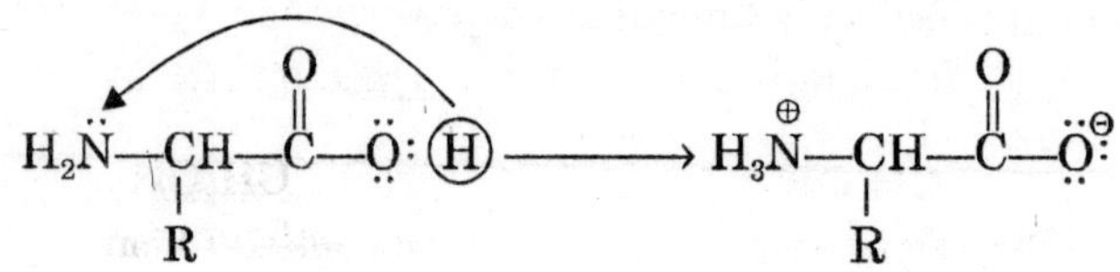

Zwitter ion

ISOELECTRIC POINT OF α-AMINO ACIDS OR PROTEINS

α-Amino acids and proteins have zwitter ion structure. At low pH they behave as cations and migrate towards cathode whereas at high pH, they behave as anions and migrate towards anode. At some definite intermediate pH, which depends on nature of protein, the positive and negative charge are exactly balanced *i.e.,* there is no net charge on the protein molecule. The molecules of proteins do not migrate in an electric field in this state. In this condition the protein is said to be at its **isoelectric point.** At isoelectric point the protein has least solubility in water, least viscosity and least osmotic pressure.

- In proteins and polypeptide various amino acids are linked through **peptide linkage** which is an amide linkage.
- **Simple proteins.** *These proteins on hydrolysis give mixture of α-amino acids only.*
- **Conjugated proteins.** *The proteins on hydrolysis give a non-protein part in addition to α-amino acids.* This non-protein part is called **prosthetic group.**
- **Derived proteins.** These proteins are obtained by partial hydrolysis of simple or conjugated proteins.
- **Primary Structure of Proteins.** It refers to the sequence in which amino acids are arranged in protein.
- **Secondary Structure of Proteins.** It refers to the arrangement of polypeptide chains which arises as a result of hydrogen bonding. Two common types of secondary structures are α-**Helix structure** and β-**Pleated sheet structure.**
- **Tertiary Structure of Proteins.** It refers to the definite geometric pattern in which the entire protein molecule folds up to produce a specific shape. Two important tertiary structures are : **fibrous structure** and **globular structure.**
- **Denaturation of Proteins.** The process (such as heating, treatment with acids, etc.) that brings about changes in the physical as well as biological properties of the proteins is called denaturation. Denaturation changes the secondary and tertiary structures of proteins but has no effect on the primary structure.

ENZYMES

- **Enzymes are a group of globular proteins which act as catalysts in biological processes.**

Table 9.1. Some Common Enzymes

Enzymes	*Reaction catalysed*
Amylase	Starch $\longrightarrow$ Glucose
Maltase	Maltose $\longrightarrow$ Glucose
Lactase	Lactose $\longrightarrow$ Glucose + Galatose
Invertase	Sucrose $\longrightarrow$ Glucose + Fructose
Urease	Urea $\longrightarrow$ $CO_2 + NH_3$
Carbonic anhydrase	$H_2CO_3 \longrightarrow CO_2 + H_2O$
Pepsin	Proteins $\longrightarrow$ Amino acids
Trypsin	Proteins $\longrightarrow$ Amino acids
Nucleases	DNA, RNA $\longrightarrow$ Nucleotides
DNA Polymerase	Deoxynucleotide triphosphates $\longrightarrow$ DNA

An enzyme molecule may contain a non-protein component known as **prosthetic group**. The prosthetic group which is covalently bonded with an enzyme is known as *cofactor*. On the other hand a prosthetic group which gets attached to the enzyme at the time of reaction is known as *coenzyme*.

NUCLEIC ACIDS

- There are two types of nucleic acids :
  (*i*) *Deoxyribonucleic acid* (DNA) (*ii*) *Ribonucleic acid* (RNA)
- The monomeric units in nucleic acids are known as **nucleotides.**
- Each nucleotide consists of three parts : a phosphate group, a five carbon sugar and a nitrogen containing heterocyclic base.
- The bases found in nucleotides are derivatives of purine and pyrimidine. The bases derived from purine are ***aldenine*** and ***guanine*** whereas bases derived from pyrimidine are ***thymine, uracil*** and ***cytosine.***
- The bases present in RNA nucleotides are : adenine, guanine, uracil and cytosine.
  The bases present in DNA are : adenine, guanine, thymine and cytosine.
- The **sugars** present in DNA is 2-deoxyribose and in RNA is ribose.

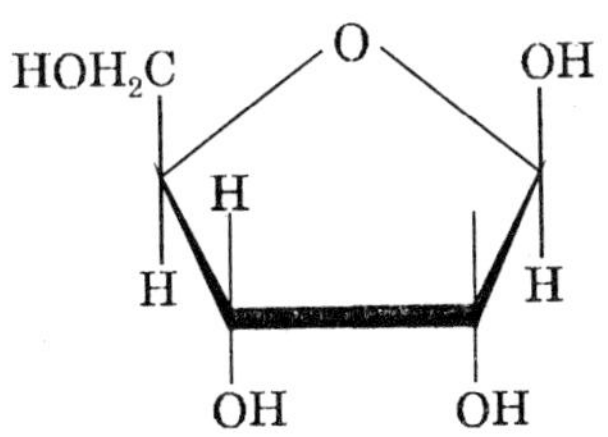

D-Ribose

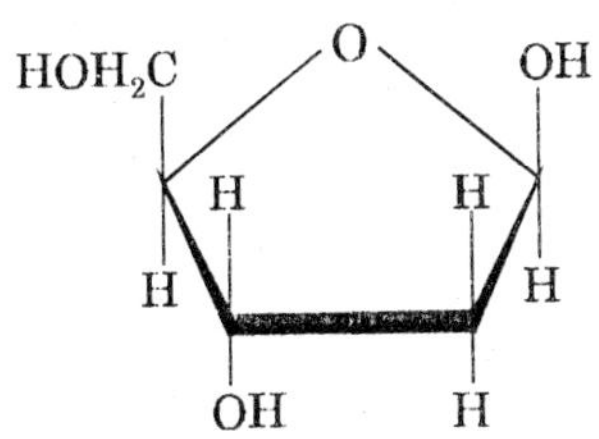

D-2-deoxyribose

- It was observed by **E. Chargaff** that the base composition in DNA varied from one species to the other. However, it was found that the amount of adenine (A) was equal to that of thymine (T) and the amount of guanine (G) was equal to that of cytosine (C). Thus, the total amount of purines (A + G) was found to be equal to the total amount of pyrimidines (C + T). However, the ratio A + T/C + G varies from one species to the other and is characteristic for a particular species. This ratio is 1 : 52 in man.
- The **double helix** structure of DNA was proposed by **Watson and Crick.**
  Hydrogen bonds between bases on the two strands of DNA is highly specific. *Thymine pairs with adenine while cytosine pairs with guanine.*
- On heating, the two strands of DNA separate from each other. This process is known as melting. The temperature at which the two strands of DNA separate completely is known as melting temperature (T_m) of DNA.
- The major function of RNA is protein synthesis while that of DNA is transference of hereditary character.

LIPIDS

- **Lipids** are those constituents of plants and animals tissue that are insoluble in water but are soluble in solvents of low polarity.
- There are **three types of lipids :**
  (*i*) *Simple lipids* or *glycerides.* These are esters of glycerol with higher fatty acids.
  (*ii*) *Waxes.* These are esters of higher fatty acids with long chain fatty alcohols. For example, been wax is ceryl myristate, $C_{13}H_{27}COOC_{26}H_{53}$.
  (*iii*) *Phospholipids.* These are the esters of glycerol. In these two —OH groups of glycerol are esterified with higher fatty acids while third —OH group is esterified with some derivative of phosphoric acid.
  For example, Lecithin. **Phospholipids form the structures of cell membranes.**
- Glycerides on hydrolysis yield glycerol and fatty acids.

METABOLISM

- **Anabolism.** The set of chemical reactions which lead to the synthesis of complex organic molecules in the living systems.
- **Catabolism.** The series of chemical reactions which involve breakdown of complex organic molecules in the living systems.
- **Metabolism.** Anabolism and catabolism together constitute metabolism.
- **ATP (Adensine triphosphate).** ATP is a nucleotide containing adenine (as base), ribose (as sugar) and three interlinked phosphate units. ***ATP molecule contains two high energy phosphate bonds.*** During hydrolysis of ATP, energy is released.

$$\textbf{Adenosine triphosphate (ATP)} \xrightarrow[\text{Energy released} = 30.93 \text{ kJ mol}^{-1}]{\text{(Hydrolysis)}} \textbf{Adenosine diphosphate (ADP)} \xrightarrow[\text{Energy released} = 28.42 \text{ kJ mol}^{-1}]{\text{(Hydrolysis)}} \textbf{Adenosine monophosphate (AMP)}$$

- **Digestion.** *The process of conversion of complex organic molecules in the food into simpler molecules with the help of enzymes is called digestion.* In mammals, digestion mainly occurs in mouth, stomach and small intestine. Digestion products of carbohydrates, fats and proteins has been summarized, below :

$$\textbf{Starch} \xrightarrow[\text{(Amylase)}]{\text{Ptyalin}} \text{Maltose} \xrightarrow{\text{Maltase}} \text{Glucose}$$

$$\textbf{Proteins} \xrightarrow{\text{Pepsin}} \text{Polypeptides} \xrightarrow[\text{(Peptidases)}]{\text{Trypsin/Chymotrpysin}} \alpha\text{-Amino acids}$$

$$\textbf{Fats} \xrightarrow{\text{Bile juices}} \text{Emulsified fat} \xrightarrow{\text{Lipases}} \text{Fatty acids + Glycerol}$$

- **Glycolysis.** The first stage of oxidation of glucose is known as glycolysis. This process does not require oxygen and is ***anaerobic process.*** During this process one molecule of glucose breaks into two molecules of pyruvic acid and there is a net gain of two molecules of ATP for each molecule of glucose.
- **The second stage** of oxidation of glucose is ***aerobic.***
- The complete metabolic conversion of 1 molecule of glucose into CO_2 and H_2O is accompanied by conversion of 38 molecules of ADP to ATP.
- **Lipid metabolism**

$$\textbf{Lipids} \xrightarrow{\text{hydrolysis}} \textbf{Fatty acids + Glycerol} \xrightarrow{\text{oxidation}} \textbf{Acetyl CoA} \xrightarrow[\textit{cycle}]{\text{citric acid}} \textbf{ATP}$$

- Metabolic conversion of one molecule of palmitic acid (a C-16 fatty acid) produces 130 molecules of ATP.

BLOOD

- An average person has about 6.8 litres of blood which is about 6-10% of the body weight.
- pH of blood is about 7.4.
- Haemoglobin is a globular protein. It is made up of ***four polypeptide chains*** which are arranged in tetrahedral manner. Each chain is associated with a non protein part called ***haeme.***

HORMONES

- **Hormones** *are the chemical substances which are produced by ductless glands in the body.* Hormones act as **chemical messengers.**
- Some examples of ductless (**endocrine**) **glands** are : thyroid, pituitary, adrenals, pancreas, testes and ovaries.
- Hormones are divided into three types :
  (*i*) Steroids (*ii*) Proteins or polypeptides (*iii*) Amines.

Table 9.2. Some Typical Hormones

Hormone	*Source*	*Chemical Name*	*Function*
1. Thyroxin	Thyroid	Amino acid	Stimulates metabolism.
2. Adrenaline	Adrenal medula	Amine	Increases pulse rate and blood pressure, releases glucose from glycogen and fatty acids from fats.
3. Insulin	Pancreas	Peptide	Decreases blood glucose.
4. Glucogon	Pancreas	Peptide	Increases blood glucose.
5. Testosterone	Testes	Steroid	Controls normal functioning of male sex organs.
6. Estrone and Estradiol	Ovary	Steroid	Controls normal functioning of female sex organs.
7. Progesterone	Ovary	Steroid	Prepares uterus for pregnancy, controls menstrual cycle.
8. Cortisone	Adrenal cortex	Steroid	Metabolism of water, mineral salts, fats, proteins and carbohydrates.

ITAMINS

- ***The organic compounds other than carbohydrates, proteins and fats that are necessary to maintain normal health, growth and nutrition are called vitamins.***
- The vitamins are complex organic molecules, however, for the sake of simplicity, they are represented by letters such as A, B, C, D, E, K.
- Vitamins are broadly classified into two types, *water soluble vitamins* and *oil soluble vitmains.* ***Vitamins A, D, E and K are oil soluble whereas vitamins B and C are water soluble.*** The various vitamins and the diseases caused by their deficiencies are listed in Table 9.3.

Table 9.3. Vitamins and the Diseases caused by their Deficiency

Chemical Nature	*Deficiency Diseases*
Vitamin A (*Carotenoids* or *Axerophytol* or *Retinol*). Soluble in oils and fats, but insoluble in water.	Night blindness, Xerophthalmia (cornea becomes opaque), drying of skin.
Vitamin B_1 (*Thiamine*). Soluble in water, destroyed by heat.	Beriberi, loss of appetite.
Vitamin B_2 (*Riboflavin*). Soluble in water. Stable to heat, destroyed by light.	Cracked lips, sore tongue and skin disorders.
Vitamins B_6 (*Pyridoxine*).	Nervous disturbances and convulsions (*pernious anaemia*).
Vitamin B_{12} (*Cyano cobalamin*). Soluble in water and contains cobalt, red crystalline.	A serious type of anaemia.

Vitamin C (*Ascorbic Acid,* $C_6H_8O_6$). Soluble in water, destroyed by cooking and exposure to air.	Scurvy, dental caries, pyorrhea, anaemia.
Vitamin D (*Calciferol*), mixture of 4 complex compounds containing C, H and O. Soluble in fats and oils but insoluble in water. Stable towards heat and oxidation. This vitamin regulates the absorption of calcium and phosphate in intestine.	Infantile rickets ; deformation of bones and teeth.
Vitamin E (*Tocopherol*). Mixture of 3 complex substances containing C, H and O. Soluble in fats and oils but insoluble in water. Stable to heat and oxidation.	Loss of sexual power and degeneration of muscle fibres in animals.
Vitamin H (*Biotin*). Neither soluble in water nor in fat.	Dermatitis, loss of hair and paralysis.
Coenzyme Q_{10}. This is present in chloroplasts of green plants and mitochondria of animals.	Low order of immunity of body against many diseases.
Vitamin K, mixture of two complex substances containing C, H and O. Soluble in fats but insoluble in water. Stable to heat and oxidation.	Tendency to haemorrhage and impaired clotting of blood.

DISEASES

1. **Viral diseases.** Some examples are : *chicken pox, small pox, measles, poliomyelities, influenza, AIDS, viral heptatitis, rabies and mumps*, etc.
2. **Bacterial diseases.** The different bacterial diseases are *cholera, typhoid, leprosy, meningitis, tuberculosis*, etc.
3. **Venereal diseases.** The most common of these are *syphilis* and *gonorrhoea.*
4. **Protozoal diseases.** A few diseases belonging to this category are *amoebic dysentery, malaria* etc.

9.2. IMPORTANT FACTS AND TERMS

Add to Knowledge

- Fructose is the sweetest sugar.
- Starch and glycogen are condensation polymers of α-D-glucose.
- Cellulose is condensation polymer of β-D-glucose.
- Monosaccharides which differ in configuration at C_1 in aldoses and at C_2 in ketoses are known as **anomers.** For example, α-D-glucose and β-D-glucose represent a pair of anomers.
- Monosaccharides which differ in configuration at a carbon atom other than anomeric carbon are known as **epimers.** For example, glucose and mannose are epimers.
- All monosaccharides can reduce Tollen's reagent and Fehling's solution and are thus reducing sugars.
- Sucrose is a non-reducing sugar.
- **Mutarotation.** Change of specific rotation of a carbohyrate to an equilibrium value when dissolved in water is known as muta rotation. All reducing sugars exhibit mutarotation.
- **Inversion of sucrose.** Hydrolysis of sucrose into (+)-glucose and (–)-fructose is referred to as inversion of sucrose.
- Different monomeric units in polysaccharides are united through **glycoside linkage.**
- All α-amino acids except glycine are optically active.
- All naturally occurring carbohydrates have D-configuration while all naturally occurring α-amino acids have L-configuration.
- Proteins and α-amino acids have zwitter ion structures.
- Different α-amino acids are linked to each other in proteins through **peptide linkage**.
- **α-Helix** and **β-pleated sheet** structures are the secondary structures of proteins.
- The disease **phenylketone urea** is caused by the deficiency of enzyme ***phenyl alanine hydroxylase*** while the disease **albinism** is caused by the deficiency of enzyme ***tyrosinase.***
- Enzyme **streptokinase** is used to dissolve blood clots when a person suffers from a heart attack.
- Insulin is composed of two polypeptide chains containing a total 51 amino acids. The two chains are linked by two disulphide bridges.
- **Nucleotides.** The monomeric units in nucleic -acids.
- **Nucleoside.** The sugar-base pair in nucleic acids.
- **Sugars** present in nucleic acids are : **Ribose** in RNA and **2-deoxy ribose** in DNA.
- *Bases in nucleic acids are*

  Pyrimidines : Uracil, thymine and cytosine

  Purines : Adenine, guanine.

  Uracil is present only in RNA while thymine is present only in DNA.
- Complimentary base pairs in DNA are thymine-adenine, cytosine-guanine.
- For a particular species the ratio A + T/C + G in DNA is constant. For man this ratio is 1.52.
- The **double helix structure** of DNA was proposed by **Watson and Crick.**
- **Waxes** are esters of long chain fatty acids with long chain alcohols.
- **Phospholipids** are the major constituents of cell membranes.
- **Lecithin** in an example of a phospholipid.
- The complete metabolic conversion of 1 molecule of glucose into CO_2 and H_2O is accompanied by conversion of 38 molecules of ADP into ATP.
- **Glycolysis** is an anaerobic process.
- Metabolic conversion of 1 molecule of palmitic acid produces 130 molecules of ATP.

9.3. QUESTION BANK

Develop Confidence

Carbohydrates

1. Which of the following is an example of aldohexose ?
(a) Ribose (b) Fructose
(c) Sucrose (d) Glucose.

2. Glycoside linkage is
(a) An amide linkage (b) An ether linkage
(c) An ester linkage (d) None of these.

3. Which of the following is not sweet in taste ?
(a) Lactose (b) Maltose
(c) Galactose (d) Starch.

4. A pyranose ring consists of a skeleton of
(a) 5 carbon atoms and one oxygen atom
(b) 6 carbon atoms
(c) 6 carbon atoms and one oxygen atom
(d) 4 carbon atoms and one oxygen atom.

5. Sucrose on hydrolysis yields a mixture which is
(a) Optically inactive (b) Dextrorotatory
(c) Laevorotatory (d) Racemic.

6. The sweetest carbohydrate is
(a) Sucrose (b) Glucose
(c) Fructose (d) Saccharine.

7. Which of the following is a non reducing sugar ?
(a) Glyceraldehyde (b) Glucose
(c) Fructose (d) Sucrose.

8. Which of the following is not a monosaccharide ?
(a) Glucose (b) Fructose
(c) Glyceraldehyde (d) Sucrose.

9. Hydrolysis of sucrose into (+) glucose and (–) fructose is known as
(a) Muta rotation (b) Inversion
(c) Pyrolysis (d) None of these.

10. Lactose on hydrolysis yields
(a) Glucose (b) Fructose
(c) Glucose and fructose (d) Glucose and galactose.

11. Which of the following carbohydrates gives a dark blue colour with iodine ?
(a) Glucose (b) Cellulose
(c) Starch (d) Sucrose.

12. The reagent which forms crystalline osazone derivative when treated with glucose is
(a) Fehling solution (b) Phenylhydrazine
(c) Benedict solution (d) Hydroxylamine.

13. The letter D in D-glucose signifies
(a) Dextrorotatory (b) Mode of synthesis
(c) Its configuration (d) Its diamagnetic nature.

14. Which disaccharide is present in milk ?
(a) Maltose (b) Galactose
(c) Sucrose (d) Lactose.

15. Which of the following is a ketohexose ?
(a) Fructose (b) Maltose
(c) Glucose (d) Ribose.

16. The linkage that holds monosaccharide units together in a polysaccharide is called
(a) Peptide linkage (b) Glycoside linkage
(c) Hydroxide linkage (d) Nucleoside linkage.

17. Which of the following is the simplest carbohydrate ?
(a) Glucose (b) Maltose
(c) Sucrose (d) Glyceraldehyde.

18. Starch is polymer of
(a) α-D-Glucose
(b) β-D-Glucose
(c) α-D-Glucose and β-D-Glucose
(d) α-D-Fructose.

19. Cellulose on hydrolysis yields
(a) β-D-Fructose (b) α-D-Glucose
(c) β-D-Glucose (d) α-D-Fructose.

20. A sucrose molecule on hydrolysis yields
(a) Glucose (b) Fructose
(c) Glucose and fructose (d) Glucose and galactose.

21. Maltose on hydrolysis yields
(a) Fructose (b) Glucose
(c) Galactose (d) Glucose + Fructose.

22. Digestion of carbohydrates mainly takes place in
(a) Stomach (b) Small intestine
(c) Large intestine (d) Mouth.

23. Glucose when treated with CH_3OH in presence of dry HCl gas gives α- and β-methylglucosides because it contains
(a) An aldheydic group (b) A —CH_2OH group
(c) A ring structure (d) Five —OH group.

24. Ribose is an example of

(*a*) Polysaccharide (*b*) Disaccharide

(*c*) Hexose (*d*) Pentose.

25. Which of the following statements about glucose is not correct ?

(*a*) It contains one —CHO group

(*b*) It contains one primary —OH group

(*c*) It contains four secondary —OH group

(*d*) It contains six —OH groups.

26. Ribose and 2-deoxyribose are

(*a*) Isomers (*b*) Diastereomers

(*c*) Enantiomers (*d*) Aldopentoses.

27. Glucose on oxidation with bromine water gives

(*a*) Glyceric acid (*b*) Gluconic acid

(*c*) Saccharic acid (*d*) Pyruvic acid.

28. Glucose reacts with CH_3OH in the presence of dry HCl to give

(*a*) α-Methyl glucoside (*b*) β-Methyl glucoside

(*c*) Both of these (*d*) None of these.

29. Which of the following compounds can reduce Tollen's reagent ?

(I) Sucrose, (II) Glucose,

(III) Fructose, (IV) Maltose.

(*a*) I, II, III and IV (*b*) II

(*c*) II and III (*d*) II, III and IV.

30. Glucose and fructose are

(*a*) Isomers of each other

(*b*) Homologues of each other

(*c*) Anomers of each other

(*d*) Enantiomers of each other.

31. Invert sugar is

(*a*) A variety of cane sugar

(*b*) Optically inactive form of sugar

(*c*) Mixture of glucose and fructose in equimolar proportion

(*d*) Mixture of glucose and galactose.

32. Fructose has

(*a*) Bitter taste (*b*) Sweet taste

(*c*) Tasteless (*d*) Nothing certain.

33. The constituent units of sucrose are

(*a*) Lactose and glucose (*b*) Glucose and fructose

(*c*) Galactose and glucose (*d*) Glucose and maltose.

34. When hemiacetal reacts with alcohol, the product is

(*a*) Dihemiacetal (*b*) Alcohol

(*c*) Peptide (*d*) Acetal.

35. Amylopectin is a polymer of

(*a*) β-D-glucose (*b*) α-D-glucose

(*c*) β-D-fructose (*d*) α-D-fructose.

36. Glucose cannot be classified as

(*a*) Hexose (*b*) Carbohydrate

(*c*) Aldose (*d*) Oligosaccharide.

37. Glucose gives silver mirror with Tollen's reagent. It shows the presence of a/an

(*a*) Alcoholic group (*b*) Keto group

(*c*) Aldehydic group (*d*) Acidic group.

38. Which carbohydrate is essential constituent of plant cells ?

(*a*) Starch (*b*) Sucrose

(*c*) Glucose (*d*) Cellulose.

39. Which of the following can be used for detection of traces of iodine ?

(*a*) Glucose in aqueous solution

(*b*) Starch in aqueous solution

(*c*) Cellulose in alcoholic solution

(*d*) Cellulose in aqueous solution.

40. Calorific vlaues of carbohydrates (I), fats (II) and proteins (III) are in the order :

(*a*) I > II > III (*b*) III > II > I

(*c*) II > I > III (*d*) II > III > I.

41. Which of the following pairs form the same osazone ?

(*a*) Glucose and fructose

(*b*) Glucose and galactose

(*c*) Glucose and arabinose

(*d*) Lactose and maltose.

42. Which of the following compounds is formed by condensation of α-D-glucose units ?

(*a*) Sucrose (*b*) Matlose

(*c*) Lactose (*d*) Cellulose.

43. α-D-Glucose and β-D-glucose are

(*a*) Epimers (*b*) Structural isomers

(*c*) Enantiomers (*d*) Diastereomers.

44. Which of the following is example ketohexose.

(*a*) Fructose (*b*) Ribose

(*c*) Glucose (*d*) Lactose.

45. On heating glucose with Fehling's solution, we get a precipitate whose colour is

(*a*) Yellow (*b*) Red

(*c*) Black (*d*) White.

α-Amino Acids and Proteins

46. Which of the following is an example of zwitter ion ?
(a) Alanine (b) Glycine hydrochloride
(c) Both (d) None.

47. The main structural feature of proteins is
(a) Peptide linkage (b) Glycoside linkage
(c) Ether linkage (d) All the three.

48. Which of the following is not proteinous ?
(a) Wool (b) Hair
(c) Nail (d) DNA.

49. The amino acids are the end products of the digestion of
(a) Lipids (b) Fats
(c) Proteins (d) Enzymes.

50. Which of the following α-amino acids is not optically active ?
(a) Alanine (b) Glycine
(c) Phenylalanine (d) All are optically active.

51. The name of the dipeptide

$$\underset{\displaystyle CH_3}{H_2N\underset{|}{C}HCONHCH_2COOH}$$ is

(a) Glycylglycine (b) Glycylalanine
(c) Glycine alanine (d) Alanylglycine.

52. The force of attraction between the neighbouring peptide chains is
(a) Van der Waal's force (b) Covalent bond
(c) Hydrogen bond (d) Peptide linkage.

53. β-Pleated structure of proteins is
(a) Primary structure (b) Secondary structure
(c) Tertiary structure (d) Quaternary structure.

54. Denaturation of protein does not have any effect on its
(a) Secondary structure (b) Tertiary structure
(c) Quaternary structure (d) Primary structure.

55. The proteins with a prosthetic group are known as
(a) Complex proteins (b) Conjugated proteins
(c) Secondary proteins (d) Essential proteins.

56. Peptides on hydrolysis give
(a) Ammonia (b) Amines
(c) Amino acids (d) Hydroxy acids.

57. Which of the following is a test for proteins ?
(a) Molisch's test (b) Beilstein test
(c) Biuret test (d) Benedict's test.

58. Enzymes are
(a) Steroids (b) Proteins
(c) Fatty acids (d) Nucleic acids.

59. The destruction of the biological nature and activity of proteins by heat or chemical agent is called
(a) Dehydration (b) Denaturation
(c) Denitrogenation (d) Deammination.

60. The non-proteinous substances which certain enzymes require for their activity are called
(a) Catalysts (b) Inhibitors
(c) Co-enzymes (d) Epimers.

61. An example of zwitter ion is
(a) Glycine hydrochloride (b) Ammonium acetate
(c) Aniline (d) Alanine.

62. At isoelectric point, the amino acid has
(a) Least viscosity
(b) Maximum surface tension
(c) Maximum solubility
(d) All of these.

63. Biuret test is used for the detection of
(a) Sugars (b) Proteins
(c) Fats (d) Starch.

64. α-Amino acids behave as crystalline ionic solids and have high melting point due to the presence of
(a) $—NH_2$ group
(b) —COOH group
(c) Both $—NH_2$ and —COOH group
(d) None of these.

65. In an α-amino acid, the basic group is
(a) $—NH_2$ (b) —COOH
(c) $—NH_3^+$ (d) $—COO^-$.

66. In an α-amino acid, the acidic group is
(a) —COOH (b) $—NH_2$
(c) $—NH_3^+$ (d) $—COO^-$.

67. Which of the following is an example of globular protein ?
(a) Myosin (b) Collagen
(c) Keratin (d) Haemoglobin.

68. Which of following is an example of fibrous protein ?
(a) Insulin (b) Haemoglobin
(c) Fibroin (d) Glucogen.

69. The bond that determines the secondary structure of a protein is
(a) Co-ordinate bond (b) Covalent bond
(c) Hydrogen bond (d) Ionic bond.

70. Amino acids are formed on hydrolysis of
(a) Fats (b) Proteins
(c) Nucleic acids (d) Carbohydrates.

71. Amino acids are least soluble in water
(*a*) at pH = 7 (*b*) at pH > 7
(*c*) at pH < 7 (*d*) At isoelectric point.

72. α-helix refers to
(*a*) Primary structure of proteins
(*b*) Secondary structure of proteins
(*c*) Tertiary structure of proteins
(*d*) Quaternary structure of proteins.

73. Cheese is an example of
(*a*) Denatured protein (*b*) Derived protein
(*c*) Conjugated protein (*d*) Globular protein.

74. On heating with conc. HNO_3, proteins give yellow colour. This test is called
(*a*) Oxidizing test (*b*) Xanthoproteic test
(*c*) Biuret test (*d*) Acid base test.

75. Which of the following α-amino acids is optically inactive ?
(*a*) Alanine (*b*) Glycine
(*c*) Phenylalanine (*d*) Lysine.

76. The prosthetic group in glycoprotein is
(*a*) Carbohydrate (*b*) Fat
(*c*) Vitamin (*d*) Nucleic acid.

Enzymes

77. The disease phenyl ketone urea is caused by the deficiency of
(*a*) Acetophenone
(*b*) Phenyl acetone
(*c*) Tyrosinase
(*d*) Phenylalanine hydroxylase.

78. Which of the following enzymes helps in digestion of proteins ?
(*a*) Invertase (*b*) Trypsin
(*c*) Tyrosinase (*d*) Urease.

79. Enzymes are
(*a*) Complex nitrogeneous substances produced in living cells
(*b*) Steroids
(*c*) Living organisms
(*d*) Dead organisms.

80. Insulin is secreted by
(*a*) Pancreas (*b*) Stomach
(*c*) Thyroid (*d*) Adernal medula.

81. The enzyme that converts starch into maltose is called
(*a*) Zymase (*b*) Maltase
(*c*) Invertase (*d*) Amylase.

82. Enzyme pepsin hydrolyses
(*a*) Fats to fatty acids
(*b*) Proteins to α-amino acids
(*c*) Starch to glucose
(*d*) Glucose to ethyl alcohol.

83. Which of the following enzymes hydrolyses triglycerides to fatty acids and glycerol ?
(*a*) Amylase (*b*) Maltase
(*c*) Lipases (*d*) Pepsin.

Nucleic Acids

84. Pyrimidine bases present in RNA are
(*a*) Adenine and guanine (*b*) Thymine and uracil
(*c*) Uracil and cytosine (*d*) Thymine and cytosine.

85. Adenosine monophosphate (AMP) is
(*a*) A nucleotide (*b*) A nucleoside
(*c*) An enzyme (*d*) A nucleic acid.

86. In adenosine diphosphate (ADP), the sugar present is
(*a*) D-Ribose (*b*) D-Glucose
(*c*) D-Fructose (*d*) D-2-Deoxyribose.

87. In AMP the sequence is
(*a*) Sugar-base-phosphate (*b*) Sugar-phosphate-base
(*c*) Phosphate-sugar-base (*d*) Phosphate-base-sugar.

88. Which one of the following is not present in RNA ?
(*a*) Uracil (*b*) Thymine
(*c*) Ribose (*d*) Phosphate.

89. In nucleic acids, the sequence is
(*a*) Phosphate-Base-Sugar
(*b*) Sugar-Phosphate-Base
(*c*) Phosphate-Sugar-Base
(*d*) Base-Phosphate-Sugar.

90. Which of the following bases is not present in DNA ?
(*a*) Thymine (*b*) Uracil
(*c*) Adenine (*d*) Guanine.

91. The sequence of bases on *m*-RNA molecule, synthesized on the GCATATGGA strand of DNA is
(*a*) CGUAUACCU (*b*) CGTATACCT
(*c*) TACGCGATTC (*d*) ATCGCGTTC.

92. Which of the following bases is not present in RNA ?
(*a*) Thymine (*b*) Cytosine
(*c*) Uracil (*d*) Adenine.

93. Corresponding to guanine in DNA, the complementary base in RNA is
(*a*) Cytosine (*b*) Thymine
(*c*) Uracil (*d*) Adenine.

94. Major function of DNA is
(*a*) To control metabolism
(*b*) To catalyse biochemical reactions
(*c*) To control synthesis of proteins
(*d*) To transfer genetic information from one generation to the next.

95. Which of the following statements about RNA is *not correct* ?
(*a*) It has a single strand
(*b*) It does not undergo replication
(*c*) It does not contain any pyridimine base
(*d*) It controls the synthesis of proteins.

96. The pair of bases in DNA are held together by
(*a*) Covalent bonds
(*b*) Hydrogen bonds
(*c*) Dipole-dipole interactions
(*d*) Van der Waal's attraction.

97. A nucleotide is made up of
(*a*) A base and sugar
(*b*) A base and phosphoric acid unit
(*c*) A sugar and phosphoric acid unit
(*d*) A sugar, a base and a phosphoric acid unit.

98. Which of the following is a pyrimidine base ?
(*a*) Adenine (*b*) Guanine
(*c*) Uracil (*d*) None of these.

99. The relation between nucleotide triplets and the amino acids is called
(*a*) Transcription (*b*) Duplication
(*c*) Genetic code (*d*) Gene.

100. Nucleic acids are polymers of
(*a*) Nucleotides (*b*) Nucleosides
(*c*) Nuclei of heavy metals (*d*) Proteins.

101. DNA has 2-deoxyribose, base and a third component known as
(*a*) Ribose (*b*) Adenine
(*c*) Phosphoric acid (*d*) Guanine.

102. The RNA/s which take part in the synthesis of proteins is/are
(*a*) *m*-RNA (*b*) *r*-RNA
(*c*) *t*-RNA (*d*) All the three above.

103. Guanine is an example of
(*a*) Nucleotide (*b*) Nucloside
(*c*) Purine bases (*d*) Pyrimidine base.

104. Which of the following is not present in nucleotides ?
(*a*) Adenine (*b*) Thyroxine
(*c*) Guanine (*d*) Cytosine.

105. Which of the following sets of bases is present both in DNA and RNA ?
(*a*) Adenine, uracil, thymine
(*b*) Adenine, guanine, cytosine
(*c*) Adenine, guanine, uracil
(*d*) Adenine, guanine, thymine.

106. Which of the following compounds is responsible for the transmission of heredity characters ?
(*a*) RNA (*b*) DNA
(*c*) ATP (*d*) ADP.

107. Which of the following is derived from purine ?
(*a*) Uracil (*b*) Thymine
(*c*) Guanine (*d*) Cyosine.

108. The chemical change in DNA molecule that could lead to synthesis of protein with an altered amino acid sequence is called
(*a*) Replication (*b*) Lipid formation
(*c*) Cellular membrane (*d*) Mutation.

Lipids

109. Oils and fats are esters of higher fatty acids with
(*a*) Ethanol (*b*) Glycol
(*c*) Glycerol (*d*) Methanol.

110. Stearic acid, palmitic acid and linolenic acid are
(*a*) α-Amino acids (*b*) Fatty acids
(*c*) Nucleic acids (*d*) α-Hydroxy acids.

111. $C_nH_{2n}O_2$ is the formula of
(*a*) Carbohydrates
(*b*) Proteins
(*c*) Unsaturated fatty acids
(*d*) Saturated fatty acids.

112. Waxes are esters of higher fatty acids with long chain
(*a*) Monohydric alcohols (*b*) Dihydric alcohols
(*c*) Trihydric alcohols (*d*) All the three above.

113. The calorific values of carbohydrates, fats and proteins vary in the order
(*a*) Fats > Carbohydrates > Proteins
(*b*) Fats > Proteins > Carbohydrates
(*c*) Carbohydrates > Fats > Proteins
(*d*) Carbohydrates > Proteins > Fats.

114. Which of the following is an ester ?
(*a*) Kerosene oil (*b*) Soap
(*c*) Bees wax (*d*) Peptoses.

115. Which of the following is an example of PUFA ?
(*a*) Oleic acid (*b*) Lauric acid
(*c*) Stearic acid (*d*) Linolenic acid.

116. Lecithin is an example of

(a) A simple lipid (b) Waxes
(c) Phospholipids (d) None of these.

117. The chief constituents of cell membranes are

(a) Simple triglycerides (b) Waxes
(c) Phospholipids (d) Proteins.

118. Which of the following fatty acids is an unsaturated fatty acids ?

(a) Lauric acid (b) Oleic acid
(c) Stearic acid (d) Palmitic acid.

119. The most concentrated source of energy in the human body is

(a) Fats (b) Sugars
(c) Proteins (d) Nuclei acids.

Digestion and Metabolism

120. Pyruvic acid has the formula

(a) $CH_3COCOOH$
(b) $CH_2 = CH - COOH$
(c) $CH_2CHCOOH$ (with OH on the central carbon)
(d) $OH - CH_2 - COOH$.

121. Metabolic conversion of one mole of glucose to CO_2 and H_2O is accompanied by conversion of ...... mole of ADP to ATP.

(a) 18 (b) 19
(c) 2 (d) 38.

122. In which of the following changes maximum energy is released ?

(a) ATP $\rightarrow$ ADP (b) ADP $\rightarrow$ ATP
(c) ADP $\rightarrow$ AMP (d) AMP $\rightarrow$ ATP.

123. The enzyme present in saliva is

(a) Pepsin (b) Ptyalin
(c) Trypsin (d) Lipases.

124. The substances which act as emulsifiers in lipid metabolism are

(a) Bile juices (b) Fatty acids
(c) Amino acids (d) Glycerol.

125. The digestion of fats in the intestines is aided by

(a) Diffusion (b) Protection
(c) Peptization (d) Emulsification.

126. Last product of protein digestion is

(a) polypeptides (b) DNA
(c) amino acids (d) peptones.

127. Which of the following enzymes hydrolyse triglycerides to fatty acids and glycerol ?

(a) Amylase (b) Maltase
(c) Lipases (d) Pepsin.

128. Enzyme trypsin converts

(a) Proteins into α-amino acids
(b) Starches into sugar
(c) Glucose into glycogen
(d) α-Amino acids into proteins.

129. Which of the following is a molecular disease ?

(a) Allergy (b) Cancer
(c) German measels (d) Sickel cell anamia.

Blood

130. Transport of molecular oxygen from lungs to tissues and CO_2 from tissues to lungs is accomplished by

(a) Leucocytes (b) Erythrocytes
(c) Thrombocytes (d) Fibrinogen.

131. Reaction of O_2 and CO with haemoglobin gives

(a) Only O_2-haeme complex
(b) Only CO-haeme complex
(c) Both (a) and (b), but CO-haeme complex is more stable
(d) Both (a) and (b), but O_2-haeme complex is more stable.

132. The protein responsible for blood-clotting is

(a) Fibrinogen (b) Albumin
(c) Insulin (d) Fibroin.

133. The protein which transports oxygen in the blood stream is

(a) Albumin (b) Collagen
(c) Fibroin (d) Haemoglobin.

134. The pH of the blood in human beings is maintained by the buffer :

(a) HCO^-_3/H_2CO_3 (b) CH_3COO^-/CH_3COOH
(c) $H_3N^+—CHR—COO^-$ (d) NH_3/NH_4^+

Hormones

135. Which of the following is a steroid hormone ?

(a) Insulin (b) Thyroxine
(c) Adrenaline (d) Progesterone.

136. The chemical messengers produced in the duct less glands are

(a) Vitamins (b) Lipids
(c) Hormones (d) Antibodies.

137. Which of the following is an amine hormone ?

(a) Insulin (b) Adrenaline
(c) Estrone (d) Cortisone.

138. Which of the following hormones contains iodine ?
(*a*) Adrenaline (*b*) Thyroxine
(*c*) Estrone (*d*) Progesterone.

139. Which of the following is a male sex hormone ?
(*a*) Estrone (*b*) Progesterone
(*c*) Testosterone (*d*) Cortisone.

140. Which of the following hormones is produced by pancreas ?
(*a*) Adrenaline (*b*) Thyroxine
(*c*) Cortisone (*d*) Glucogen.

141. The first hormone to be isolated and also the first to be synthesized in the laboratory was
(*a*) Insulin (*b*) Testosterone
(*c*) Cortisone (*d*) Adrenaline.

142. Insulin is secreted by
(*a*) Stomach (*b*) Pancreas
(*c*) Adernal cortex (*d*) Adernal medulla.

Vitamins

143. Cod liver oil is rich in
(*a*) Vitamin C (*b*) Vitamin A
(*c*) Vitamin B_6 (*d*) Vitamin B_{12}.

144. Niacin is vitamin
(*a*) B_1 (*b*) B_2
(*c*) B_{12} (*d*) B_4.

145. The vitamin which contains cobalt is
(*a*) Vitamin B_6 (*b*) Vitamin A
(*c*) Vitamin B_{12} (*d*) Vitamin E.

146. An example of water soluble vitamin is
(*a*) Vitamin A (*b*) Vitamin C
(*c*) Vitamin D (*d*) Vitamin E.

147. The best source of vitamin C is
(*a*) Cod liver oil (*b*) Egg yolk
(*c*) Citrus fruits (*d*) Fish liver oil.

148. Disease scurvy is caused by the deficiency of vitamin
(*a*) A (*b*) B_6
(*c*) C (*d*) D.

149. Vitamin B_1 is known as
(*a*) Ascorbic acid (*b*) Carotenoids
(*c*) Thiamine (*d*) Pyridoxine.

150. Which of the following vitamins is oil soluble ?
(*a*) A (*b*) B_6
(*c*) B_{12} (*d*) B_1.

151. Which of the following statements about vitamin B-12 is incorrect ?
(*a*) It has a cobalt atom
(*b*) It also occurs in plants
(*c*) It is also present in rain water
(*d*) It is needed for human body in very small amounts.

152. Which of the following vitamins is water soluble ?
(*a*) K (*b*) E
(*c*) D (*d*) B_1.

153. Deficiency of vitamin A leads to a disease known as
(*a*) Scurvy (*b*) Night blindness
(*c*) Beriberi (*d*) Rickets.

154. The deficiency of vitamin K causes
(*a*) Haemorrhage
(*b*) Lenghening time of blood clotting
(*c*) Inflammation of tung
(*d*) Both (*a*) and (*b*).

155. Vitamin A is also known as
(*a*) Thiamine (*b*) Pyridoxine
(*c*) Carotenoids (*d*) Tocopherol.

156. Ascorbic acid is
(*a*) Vitamin K (*b*) Vitamin E
(*c*) Vitamin B (*d*) Vitamin C.

157. Citrus fruits are rich in vitamin
(*a*) A (*b*) C
(*c*) B (*d*) K.

158. Vitamin E is also called
(*a*) Cyanocobalamin (*b*) Tocopherol
(*c*) Lactoflavin (*d*) Ascorbic acid.

159. Which of the following if taken excessively can accumulate in body and cause toxicity ?
(*a*) Vitamin C (*b*) Vitamin D
(*c*) Vitamin B_2 (*d*) None of these.

160. Vitamin B_6 is known as
(*a*) Pyridoxine (*b*) Thiamine
(*c*) Tocopherol (*d*) Riboflavin.

161. The vitamin which is water soluble and antioxidant is
(*a*) Vitamin B_6 (*b*) Vitamin B_{12}
(*c*) Vitamin C (*d*) Vitamin E.

162. Vitamin D is also known as
(*a*) Growth vitamin (*b*) Sunshine vitamin
(*c*) Reproductive vitamin (*d*) Ascorbic acid.

163. Deficiency of vitamin E causes
(*a*) Night blindness (*b*) Loss of fertility
(*c*) Scurvy (*d*) Impaired clotting.

164. Milk contains vitamins
(*a*) A, D and E (*b*) A, B_{12} and D
(*c*) C, D and K (*d*) B_1, B_6 and D.

165. Prolonged deficiency of nicotinic acid (niacin) in human diet leads to

(*a*) Beri-beri (*b*) Pellagra

(*c*) Scurvy (*d*) Anaemia.

Diseases

166. Which of the following is a venereal disease ?

(*a*) Syphilis (*b*) Malaria

(*c*) Meningitis (*d*) Cholera.

167. Which of the following is not a viral disease ?

(*a*) Small pox (*b*) Rabies

(*c*) AIDS (*d*) Leprosy.

168. Which of the following is not a bacteria disease ?

(*a*) Malaria (*b*) Typhoid

(*c*) Tuberculosis (*d*) Cholera.

169. Which of the following is a protozoal disease ?

(*a*) Mumps (*b*) Measels

(*c*) Syphilis (*d*) Malaria.

170. Most viruses are composed of

(*a*) Proteins

(*b*) Proteins and nucleic acids

(*c*) Cellulose and fat

(*d*) Fats and proteins.

Miscellaneous

171. A certain compound gives negative test with ninhydrin and positive test with Benedict's solution. The compound is

(*a*) A protein (*b*) A monosaccharide

(*c*) A lipid (*d*) An amino acid.

172. Insulin is

(*a*) A protein (*b*) A hormone

(*c*) Both of these (*d*) None of these.

173. Relation between amino acids and proteins is similar to the one present between

(*a*) Nucleotides and nucleic aids

(*b*) Nucleosides and nucleic acids

(*c*) RNA and DNA

(*d*) Glucose and fructose.

174. Among these the latest discovery in cytology is that of

(*a*) Respiration (*b*) Genetic code

(*c*) Enzyme (*d*) Nucleic acids.

175. Which of the following compounds is not a vitamin ?

(*a*) Niacin (*b*) Riboflavin

(*c*) Thiamine (*d*) Guanine.

176. Which of the following statements is **not** correct ?

(*a*) Glycine is the simplest α-amino acid

(*b*) Glycine is the only a-amino acid which does not show optical activity

(*c*) Most naturally occurring amino acids have L-configuration.

(*d*) At isoelectric point amino acids have maximum solubility in water.

177. Which of the following statements is **not** correct ?

(*a*) Insulin is made up of two polypeptide chains

(*b*) The number of amino acid units in insulin is 51.

(*c*) The two polypeptide chains in insulin are cross linked at three places by disulphide bridges.

(*d*) The two polypeptide chains contain 21 and 30 amino acid units respectively.

178. Which of the following structures represents purine ?

(*a*) 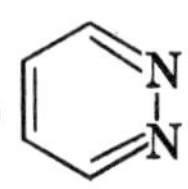

(*b*) 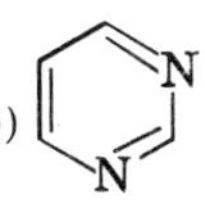

(*c*) 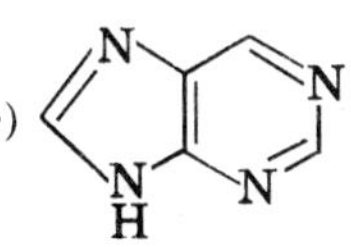

(*d*) 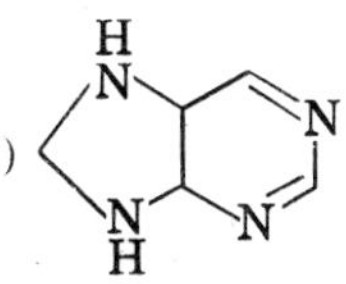

179. Which of the following represents adenine ?

(*a*) 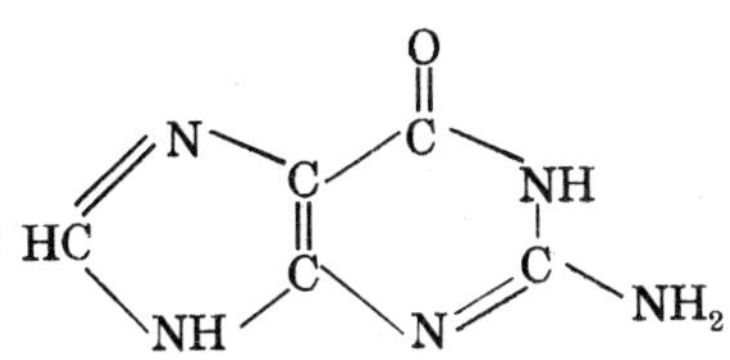

(*b*) 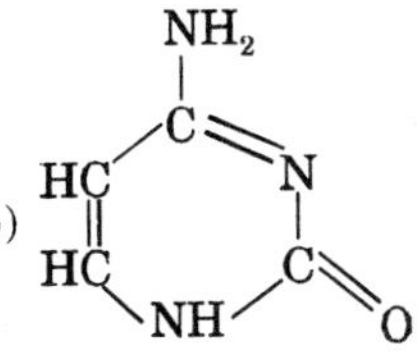

(*c*) 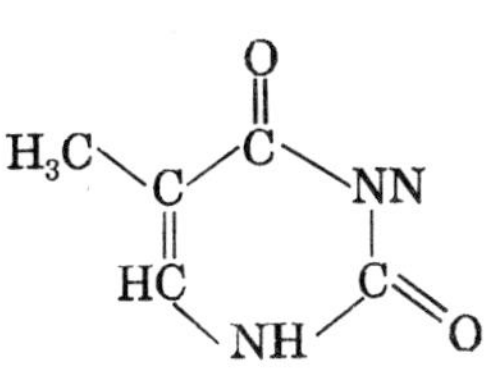

(*d*) 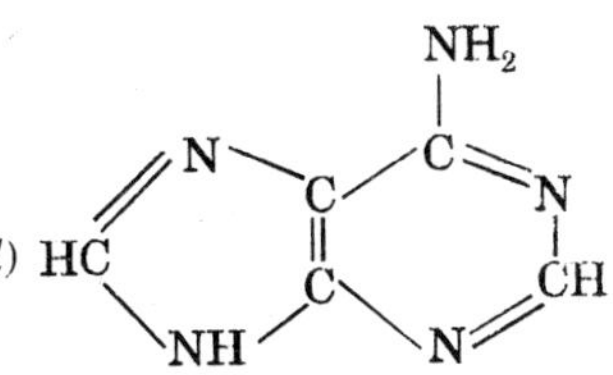

180. Which of the following statements is **not** correct ?

(*a*) The structure of the sugar present in RNA is

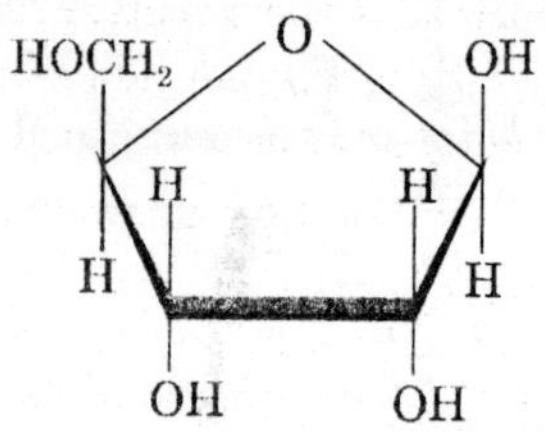

(*b*) The structure of the sugar present in DNA is

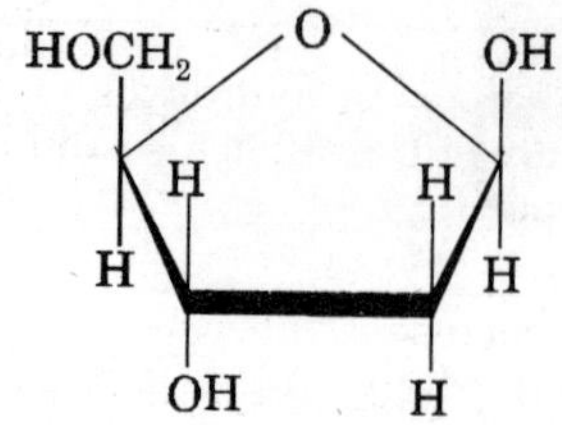

(*c*) The pyrimidine base uracil is not present in RNA but is present in DNA

(*d*) The purine bases present in nucleic acids are adenine and guanine.

ANSWERS

1. (*d*)	**2.** (*b*)	**3.** (*d*)	**4.** (*a*)	**5.** (*c*)	**6.** (*c*)	**7.** (*d*)	**8.** (*d*)	**9.** (*b*)
10. (*d*)	**11.** (*c*)	**12.** (*b*)	**13.** (*c*)	**14.** (*d*)	**15.** (*a*)	**16.** (*b*)	**17.** (*d*)	**18.** (*a*)
19. (*c*)	**20.** (*c*)	**21.** (*b*)	**22.** (*b*)	**23.** (*c*)	**24.** (*d*)	**25.** (*d*)	**26.** (*d*)	**27.** (*b*)
28. (*c*)	**29.** (*d*)	**30.** (*a*)	**31.** (*c*)	**32.** (*b*)	**33.** (*b*)	**34.** (*d*)	**35.** (*b*)	**36.** (*d*)
37. (*c*)	**38.** (*d*)	**39.** (*b*)	**40.** (*c*)	**41.** (*a*)	**42.** (*c*)	**43.** (*d*)	**44.** (*a*)	**45.** (*b*)
46. (*a*)	**47.** (*a*)	**48.** (*d*)	**49.** (*c*)	**50.** (*b*)	**51.** (*d*)	**52.** (*c*)	**53.** (*b*)	**54.** (*d*)
55. (*b*)	**56.** (*c*)	**57.** (*c*)	**58.** (*b*)	**59.** (*b*)	**60.** (*c*)	**61.** (*d*)	**62.** (*a*)	**63.** (*b*)
64. (*d*)	**65.** (*d*)	**66.** (*c*)	**67.** (*d*)	**68.** (*c*)	**69.** (*c*)	**70.** (*b*)	**71.** (*d*)	**72.** (*b*)
73. (*a*)	**74.** (*b*)	**75.** (*b*)	**76.** (*a*)	**77.** (*d*)	**78.** (*b*)	**79.** (*a*)	**80.** (*a*)	**81.** (*d*)
82. (*b*)	**83.** (*c*)	**84.** (*c*)	**85.** (*a*)	**86.** (*a*)	**87.** (*c*)	**88.** (*b*)	**89.** (*c*)	**90.** (*b*)
91. (*a*)	**92.** (*a*)	**93.** (*a*)	**94.** (*d*)	**95.** (*c*)	**96.** (*b*)	**97.** (*d*)	**98.** (*c*)	**99.** (*c*)
100. (*a*)	**101.** (*c*)	**102.** (*d*)	**103.** (*c*)	**104.** (*b*)	**105.** (*b*)	**106.** (*b*)	**107.** (*c*)	**108.** (*d*)
109. (*c*)	**110.** (*b*)	**111.** (*d*)	**112.** (*a*)	**113.** (*a*)	**114.** (*c*)	**115.** (*d*)	**116.** (*c*)	**117.** (*c*)
118. (*b*)	**119.** (*a*)	**120.** (*a*)	**121.** (*d*)	**122.** (*a*)	**123.** (*b*)	**124.** (*a*)	**125.** (*d*)	**126.** (*c*)
127. (*c*)	**128.** (*a*)	**129.** (*d*)	**130.** (*b*)	**131.** (*c*)	**132.** (*a*)	**133.** (*d*)	**134.** (*a*)	**135.** (*d*)
136. (*c*)	**137.** (*b*)	**138.** (*b*)	**139.** (*c*)	**140.** (*d*)	**141.** (*a*)	**142.** (*b*)	**143.** (*b*)	**144.** (*d*)
145. (*c*)	**146.** (*b*)	**147.** (*c*)	**148.** (*c*)	**149.** (*c*)	**150.** (*a*)	**151.** (*b*)	**152.** (*d*)	**153.** (*b*)
154. (*d*)	**155.** (*c*)	**156.** (*d*)	**157.** (*b*)	**158.** (*b*)	**159.** (*b*)	**160.** (*a*)	**161.** (*c*)	**162.** (*b*)
163. (*b*)	**164.** (*a*)	**165.** (*b*)	**166.** (*a*)	**167.** (*d*)	**168.** (*a*)	**169.** (*d*)	**170.** (*b*)	**171.** (*b*)
172. (*c*)	**173.** (*a*)	**174.** (*b*)	**175.** (*d*)	**176.** (*d*)	**177.** (*c*)	**178.** (*c*)	**179.** (*d*)	**180.** (*c*).

HINTS/SOLUTIONS

Q. No.	Ans.	Hints & Solutions
1.	(*d*)	Aldohexose is a carbohydrate with six carbon atoms and containing aldehydic group. Glucose satisfies these conditions.
2.	(*b*)	Glycoside linkage is an ether linkage (C—O—C).
3.	(*d*)	All monosaccharides and disaccharides are sweet in taste. Starch is not sweet in taste, it is a polysaccharide.
4.	(*a*)	A pyranose ring is a six membered ring, containing 5 carbon atoms and one oxygen atom.

Q. No.	Ans.	Hints & Solutions
5.	(*c*)	Sucrose on hydrolysis yields equimolar mixture of D-(–)-fructose and D-(+)-glucose. Since specific rotation of D-(–)-fructose is greater than D-(+)-glucose, the mixture is laevorotatory.
6.	(*c*)	Fructose is the sweetest carbohydrate.
7.	(*d*)	All monosaccharides and disaccharides (except sucrose) are reducing sugars.
8.	(*d*)	Sucrose is a disaccharide.
9.	(*b*)	Hydrolysis of sucrose is known as inversion of sucrose.
10.	(*d*)	Lactose on hydrolysis gives glucose and galactose.
11.	(*c*)	Starch forms blue complex with iodine.
12.	(*b*)	Glucose reacts with excess phenylhydrazine to form osazone.
13.	(*c*)	The letter D indicates the configuration of the glucose molecule.
14.	(*d*)	Milk sugar is lactose.
15.	(*a*)	Fructose is a monosaccharide containing six carbon atoms and an aldehyde group.
16.	(*b*)	Monosaccharide units are held together in a polysaccharide through an ether linkage (C—O—C), called glycoside linkage.
17.	(*d*)	Glyceraldehyde, with only three carbon atoms is the simplest carbohydrate.
18.	(*a*)	Starch on hydrolysis gives α-(D)-glucose.
19.	(*c*)	Cellulose on hydrolysis gives β-(D)-glucose.
20.	(*c*)	Sucrose gives glucose and fructose on hydrolysis.
21.	(*b*)	Maltose yields α-(D)-glucose on hydrolysis.
22.	(*b*)	Small intestine.
23.	(*c*)	Due to ring structure two methyl glucosides are possible which are stereoisomeric.
24.	(*d*)	Ribose is an aldopentose.
25.	(*d*)	Glucose contains five —OH groups and one —CHO group.
26.	(*d*)	Ribose and 2-deoxyribose are aldopentoses. They are not isomers hence they cannot be enantiomers or diastereomers.
27.	(*b*)	Bromine water oxidises —CHO group to —COOH.
28.	(*c*)	Glucose reacts with CH_3OH is the presence of dry HCl to form α and β-methyl glucosides.
29.	(*d*)	Sucrose is a non-reducing sugar.
30.	(*a*)	Glucose and fructose have same molecular formula and hence are isomers.
31.	(*c*)	Self explanatory answer.
32.	(*b*)	Fructose is the sweetest carbohydrate.
33.	(*b*)	Sucrose is formed by condensation of a molecule of glucose and fructose.
34.	(*d*)	Self explanatory answer.
35.	(*b*)	Amylopectin is a form of starch and is branched chain polymer of α-(D)-glucose.
36.	(*d*)	Oligosaccharides have 2-9 monosaccharide units while glucose is a monosaccharide.
37.	(*c*)	Aldehydes reduce Tollen's reagent to silver.
38.	(*d*)	Cellulose is the essential constituent of cell membrances of plant cells.
39.	(*b*)	Starch solution gives blue colour with iodine.
40.	(*c*)	Calorific values are in the order : Fats > carbohydrates > proteins.
41.	(*a*)	Glucose and fructose are epimers and hence form same osazone.
42.	(*c*)	Maltose is formed by condensation of two α-D-glucose units.
43.	(*d*)	α-D-glucose and β-D-glucose are stereoisomers which are not mirror image of each other. Hence, these are diastereomers.
44.	(*a*)	Fructose is a ketohexose becuase it contains keto group and six carbon atoms.
45.	(*b*)	Fehling's solution is reduced to cuprous oxide (red ppt.) by glucose.
46.	(*a*)	α-amino acids exist as zwitter ions. $CH_3—\underset{\overset{\vert}{\overset{+}{N}H_3}}{CH}—COO^-$
47.	(*a*)	In proteins α-amino acids are linked through peptide linkage.
48.	(*d*)	Hair, wool and nail are proteins.
49.	(*c*)	Proteins during digestion hydrolyse to amino acids.
50.	(*b*)	Glycine does not contain any chiral centre and hence is optically inactive.
51.	(*d*)	Alanylglycine.
52.	(*c*)	Neighbouring peptide chains are held by hydrogen bonds bewteen —CO— and —NH—.
53.	(*b*)	α-Helix and β-pleated structures are secondary structures of proteins.
54.	(*d*)	Self explanatory answer.
55.	(*b*)	Conjugated proteins have a non-protein group (called prosthetic group) attached to them.
56.	(*c*)	Peptides are formed by condensation of α-amino acids. Therefore, on hydrolysis they yield α-amino acids.
57.	(*c*)	Biuret test is a test for proteins.

Q. No.	Ans.	Hints & Solutions
58.	(*b*)	Enzymes are globular proteins.
59.	(*b*)	Self explanatory answer.
60.	(*c*)	Co-enzymes.
61.	(*d*)	α-amino acids have zwitter ion structure.
62.	(*a*)	Least viscosity.
63.	(*b*)	Biuret test is for proteins.
64.	(*d*)	Due to zwitter ion structure.
65.	(*d*)	α-Amino acids have —COO^- as the basic group and —NH_3^+ as the acidic group.
66.	(*c*)	See Hint 65.
67.	(*d*)	Haemoglobin. Others are fibrous proteins.
68.	(*c*)	Fibroin is a fibrous protein.
69.	(*c*)	Secondary structure of protein is determined by hydrogen bonding between various polypeptide chains.
70.	(*b*)	Proteins yield α-amino acids on hydrolysis.
71.	(*d*)	Isoelectric point.
72.	(*b*)	α-Helix is the secondary structure of proteins.
73.	(*a*)	Cheese is denatured milk protein.
74.	(*b*)	Xanthoproteic test of proteins.
75.	(*b*)	Glycine has no chiral centre and hence is optically inactive.
76.	(*a*)	Carbohydrate is the prosthetic group in glycoproteins.
77.	(*d*)	Phenylalanine hydroxylase.
78.	(*b*)	Trypsin.
79.	(*a*)	Self explanatory answer.
80.	(*a*)	Pancreas.
81.	(*d*)	Amylase.
82.	(*b*)	Pepsin hydrolyses proteins to α-amino acids.
83.	(*c*)	Lipase.
84.	(*c*)	Uracil and cytasine are derivatives of pyrimidine.
85.	(*a*)	Adenosine monophosphate is a nucleotide as it contains phosphate-sugar-base components.
86.	(*a*)	D-Ribose.
87.	(*c*)	In AMP, phosphate and base both are linked to sugar.
88.	(*b*)	The base thymine is not present in RNA.
89.	(*c*)	Phosphate-sugar-base.
90.	(*b*)	The base uracil is not present in DNA.
91.	(*a*)	DNA: Guanine (G), Cytosine (C), Adenine (A), Thymine (T) — *m*-RNA: Cytosine (C), Guanine (G), Uracil (U), Adenine (A).

DNA	*m*-RNA
Guanine (G)	Cytosine (C)
Cytosine (C)	Guanine (G)
Adenine (A)	Uracil (U)
Thymine (T)	Adenine (A).

Q. No.	Ans.	Hints & Solutions
92.	(*a*)	The base thymine is not present in RNA.
93.	(*a*)	See Hint 91.
94.	(*d*)	Self explanatory answer.
95.	(*c*)	RNA contains two pyrimidine bases, cytosine and uracil.
96.	(*b*)	The bases on the two strands of DNA are held by hydrogen bonds.
97.	(*d*)	Sugar-base-phosphoric acid unit are the three components of a nucleotide.
98.	(*c*)	Uracil.
99.	(*c*)	Genetic code.
100.	(*a*)	Monomeric units in nucleic acids are nucleotides.
101.	(*c*)	Phosphoric acid.
102.	(*d*)	The three RNA's which participate in protein synthesis are : **t**ransfer-RNA, **m**essenger-RNA, **r**ibosome-RNA.
103.	(*c*)	Guanine and adenine are purine bases.
104.	(*b*)	Thyroxine is a hormone. It is not present in nucleotides.
105.	(*b*)	Base thymine is not present in RNA and the base uracil is not present in DNA.
106.	(*b*)	DNA.
107.	(*c*)	Guanine is a purine base.
108.	(*d*)	Mutation.
109.	(*c*)	Glycerol. Self explanatory answer.
110.	(*b*)	These are fatty acids with high molecular masses.
111.	(*d*)	Saturated fatty acids have the formula $C_mH_{2m+1}\cdot COOH$. Thus, molecular formula of saturated fatty acids is $C_{m+1}H_{2m+2}O_2$. or $C_nH_{2n}O_2$ (where $n = m + 1$).
112.	(*a*)	Waxes are esters of higher fatty acids with higher monohydric alcohols.
113.	(*a*)	Factual question.
114.	(*c*)	See Hint 112.
115.	(*d*)	Linolenic acid is a polyunsaturated fatty acid (PUFA) because it contains more than one double bonds.
116.	(*c*)	Lecithin is a phospholipid.
117.	(*c*)	Phospholipids are the major constituents of cell membranes.
118.	(*b*)	Oleic acid.
119.	(*a*)	Fats are rich source of energy.
120.	(*a*)	Pyruvic acid is 2-keto propanoic acid, $CH_3CO.COOH$.
121.	(*d*)	Factual question. 38 molecules of ATP.
122.	(*a*)	Maximum energy is released when ATP changes to ADP.

Q. No.	Ans.	Hints & Solutions
123.	(b)	Ptyalin.
124.	(a)	Bile juices emulsify fats before digestion.
125.	(d)	See Hint 124.
126.	(c)	During digestion proteins are finally hydrolysed into amino acids.
127.	(c)	Enzyme lipases.
128.	(a)	Proteins to α-amino acids.
129.	(d)	Sickel cell anaemia is the disease caused due to defect in the molecular structure of haemoglobin.
130.	(b)	Erythrocytes.
131.	(c)	Self explanatory answer.
132.	(a)	Fibrinogen.
133.	(d)	Haemoglobin.
134.	(a)	HCO_3^-/H_2CO_3 buffer maintains pH of blood.
135.	(d)	Progesterone is a steroidal hormone.
136.	(c)	Self explanatory statement.
137.	(b)	Adrenaline.
138.	(b)	Thyroxine contains iodine.
139.	(c)	Testosterone.
140.	(d)	Glucogen.
141.	(a)	Insulin.
142.	(b)	Insulin is secreted by pancreas.
143.	(b)	Vitamin A
144.	(d)	Niacin is vitamin B_4.
145.	(c)	Vitamin B-12.
146.	(b)	Vitamin C is water soluble.
147.	(c)	Citrus fruits are rich in vitamin C.
148.	(c)	Deficiency of vitamin C.
149.	(c)	Vitamin B_1 is known as thiamine.
150.	(a)	Vitamin A is oil soluble while all vitamins of B-complex group are water soluble.
151.	(b)	Factual question.
152.	(d)	Vitamin B_1.
153.	(b)	Deficiency of vitamin A causes night blindness.
154.	(d)	Self explanatory answer.
155.	(c)	Vitamin A is known as carotenoids.
156.	(d)	Vitamin C is known as ascorbic acid.
157.	(b)	Vitamin C.
158.	(b)	Vitamin E is known as tocopherol.
159.	(b)	Oil soluble vitamins such as D, A, E, etc., if taken in excess accumulate in body.
160.	(a)	Vitamin B_6 is known as pyridoxine.
161.	(c)	Vitamin C.
162.	(b)	Vitamin D is known as sunshine vitamin.
163.	(b)	Deficiency of vitamin E causes loss of fertility.
164.	(a)	Milk contains vitamins A, D and E.
165.	(b)	Pellagra.
166.	(a)	Syphilis.
167.	(d)	Leprosy is a bacterial disease.
168.	(a)	Malaria is caused by protozoa.
169.	(d)	Malaria.
170.	(b)	Viruses are composed of proteins and nucleic acids.
171.	(b)	Monosaccharides reduce Benedict's solution.
172.	(c)	Insulin is a protein hormone.
173.	(a)	Proteins are polymers of α-amino acids while nucleic acids are polymers of nucleotides.
174.	(b)	Genetic code.
175.	(d)	Guanine is a purine base.
176.	(d)	At isoelectric point amino acids have least solubility in water.
177.	(c)	The two polypeptide chains in insulin are cross linked at two places by disulphide (S—S) bridges.
178.	(c)	(c) is purine while (b) is pyrimidine.
179.	(d)	(a) is guanine, (b) is cytosine, (c) is thymine and (d) is adenine.
180.	(c)	The base uracil is present in RNA and not in DNA.

9.4. EXAMINER'S VIEWPOINT

Test Your Confidence

1996

1. Sucrose is made up of
   (a) D-Glucose + L-Fructose (b) D-Glucose + D-Fructose
   (c) L-Glucose + L-Fructose (d) L-Glucose + D-Fructose.
2. Match the items in list I with those in list II

List I	List II
I. Pepsin	A. Genetic material
II. Nucleic acid	B. Vitamin
III. Testosterone	C. Sex hormone
IV. Ascorbic acid	D. Anitbiotic
	E. Digestion system

   (a) I-E, II-A, III-C, IV-B (b) I-D, II-A, III-C, IV-B
   (c) I-D, II-A, III-B, IV-C (d) I-E, II-B, III-C, IV-D.
   (EAM CET)
3. The enzyme which converts glucose and fructose into ethyl alcohol is
   (a) Diastase (b) Invertase
   (c) Zymase (d) Maltase. ***(MP PMT)***
4. Protein is a polymer of
   (a) Amino acids (b) Glycols
   (c) α-Hydroxy acids (d) Glucose. ***(MP PMT)***
5. Which of the following is not present in RNA ?
   (a) Uracil (b) Thymine
   (c) Ribose (d) Phosphate. ***(MP PMT)***
6. The bond that determines the secondary structure of proteins is
   (a) Covalent bond (b) Hydrogen bond
   (c) Ionic bond (d) Co-ordinate bond.
   (MP PMT)
7. In nucleic acids, the nucleotides are linked to one another through
   (a) Hydrogen bond (b) Peptide bond
   (c) Glycosidic linkage (d) Phosphate groups.
   (Pb. CET)
8. Enzymes belong to which class of compounds ?
   (a) Polysaccharides
   (b) Polypeptides
   (c) Polynitro heterocyclic compounds
   (d) Hydrocarbons. ***(CBSE Kurukshetra)***
9. Glucose reacts with acetic anhydride to form
   (a) Monoacetate (b) Tetra-acetate
   (c) Penta-acetate (d) Hexa-acetate.
   (CEET Kurukshetra)
10. Which of the following does not reduce Benedict's solution ?
    (a) Glucose (b) Fructose
    (c) Sucrose (d) Aldehyde. ***(CPMT)***
11. The helical structure of protein is stabilized by
    (a) Peptide bonds (b) Dipeptide bonds
    (c) Hydrogen bonds (d) Van der Waals forces.
    (MLNR)

1997

12. Which of the following chemical unit is certainly to be found in an enzyme ?
    (a) 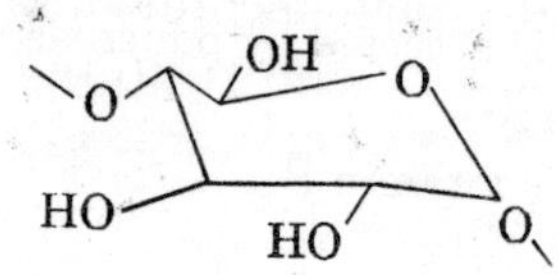

    (b) 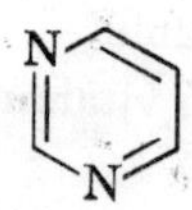

    (c) $\overset{\text{H}}{\underset{|}{}}\ \ \overset{\text{O}}{\underset{\|}{}}$ ...N – C – ...
    (d) 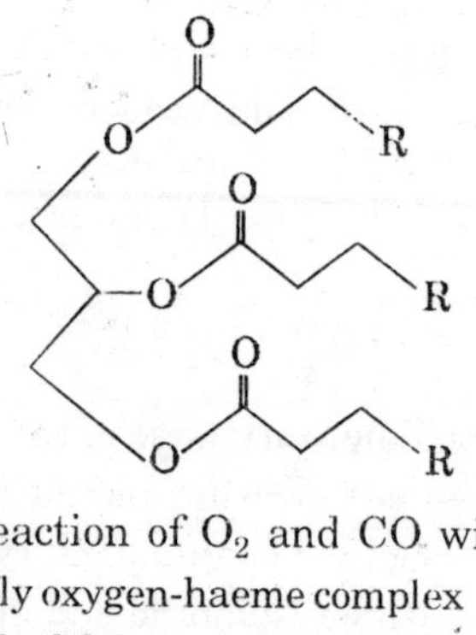

    (CBSE PMT)
13. The reaction of O_2 and CO with haemoglobin gives
    (a) Only oxygen-haeme complex
    (b) Only CO-haeme complex
    (c) Both but oxygen-haeme complex is more stable
    (d) Both but CO-haeme complex is more stable.
    (CBSE PMT)

14. The function of enzymes in the living system is to
(*a*) Transport oxygen
(*b*) Provide immunity
(*c*) Catalyse biochemical reactions
(*d*) Provide energy. **(*CBSE PMT*)**

15. The sweetest sugar is
(*a*) Sucrose (*b*) Fructose
(*c*) Glucose (*d*) Maltose. **(*BHU*)**

16. Which amino acid has no asymmetric carbon atom ?
(*a*) Histidine (*b*) Alanine
(*c*) Glycine (*d*) Phenylalanine. **(*AFMC*)**

17. Molisch test is answered by
(*a*) Glucose (*b*) Sucrose
(*c*) Fructose (*d*) All carbohydrates. **(*KCET*)**

18. The purine base present in RNA is
(*a*) Guanine (*b*) Cytosine
(*c*) Uracil (*d*) Thymine. **(*KCET*)**

19. The deficiency of which of the following vitamins adversely affects eye sight
(*a*) A (*b*) D
(*c*) B_2 (*d*) E. **(*AMU*)**

20. If one strand of DNA has the sequence ATGCTTGA, the sequence in the complimentary strand would be
(*a*) TACGAACT (*b*) TCCGAACT
(*c*) TACGTACT (*d*) TACGTAGT. **(*AMU*)**

21. The hormone which helps in the conversion of glucose into glycogen is
(*a*) Insulin (*b*) Glucogen
(*c*) Adrenaline (*d*) None of these. **(*Pb. CET*)**

22. Oleic, stearic and palmitic acids are
(*a*) Nucleic acids (*b*) Amino acids
(*c*) Fatty acids (*d*) None. **(*CPMT*)**

23. DNA has deoxyribose, a base and the third compound is
(*a*) Phosphoric acid (*b*) Ribose
(*c*) Adenine (*d*) Thymine. **(*DCE*)**

1998

24. In DNA, the complementary bases are
(*a*) Uracil and adenine : cytosine and guanine
(*b*) Adenine and thymine : guanine and cytosine
(*c*) Adenine and thymine : guanine and uracil
(*d*) Adenine and guanine : thymine and cytosine. **(*CBSE PMT*)**

25. Glucose molecule reacts with X number of molecules of phenylhydrazine to yield osazone. The value of X is
(*a*) Three (*b*) Two
(*c*) One (*d*) Four. **(*CBSE PMT*)**

26. The number of molecules of ATP produced in the lipid metabolism of a molecule of palmitic acid is
(*a*) 86 (*b*) 36
(*c*) 56 (*d*) 130. **(*CBSE PMT*)**

27. Vitamin A is called
(*a*) Ascorbic acid (*b*) Retinol
(*c*) Calciferol (*d*) Tocopherol. **(*AFMC*)**

28. Galactose is converted into glucose in
(*a*) Mouth (*b*) Stomach
(*c*) Liver (*d*) Anintestine. **(*AFMC*)**

29. Enzymes in the living systems
(*a*) Provide energy
(*b*) Provide immunity
(*c*) Transport oxygen
(*d*) Catalyse biological processes. **(*BHU*)**

30. Which substance is not present in nucleic acids ?
(*a*) Cytosine (*b*) Adenine
(*c*) Thymine (*d*) Guanidine. **(*MP PMT*)**

31. Which of the following is not an amino acid ?
(*a*) Glycine (*b*) Alanine
(*c*) Histidine (*d*) Benzidine. **(*MP PMT*)**

32. The acid showing salt-like character in aqueous solution is
(*a*) Acetic acid (*b*) Benzoic acid
(*c*) Formic acid (*d*) α-Aminoacetic acid. **(*MP PMT*)**

33. In the conversion of sucrose to glucose and fructose, the enzyme is
(*a*) Zymase (*b*) Lipase
(*c*) Invertase (*d*) Diastase. **(*CPMT*)**

34. The segment of DNA which acts as the instruction manual for the synthesis of the protein is
(*a*) Nucleoside (*b*) Nucleotide
(*c*) Ribose (*d*) Gene. **(*Pb. PMT*)**

35. When amylases catalyse the hydrolysis of starch, the product obtained is chiefly
(*a*) Cellobiose (*b*) Glucose
(*c*) Maltose (*d*) Sucrose. **(*Pb. PMT*)**

36. Which of the following diseases is caused by virus ?
(*a*) Small pox (*b*) Typhoid
(*c*) Sleeping sickness (*d*) Diphtheria. **(*Pb. CET*)**

37. Insulin, a hormone, chemically is
(*a*) Fat (*b*) Steroid
(*c*) Protein (*d*) Carbohydrate. **(*Haryana CEET*)**

38. Peptides on hydrolysis give
(*a*) Amines (*b*) Amino acids
(*c*) Ammonia (*d*) Alcohols. ***(Haryana CEET)***

39. Which of the following enzymes is not useful in the digestion of proteins ?
(*a*) Chymotripisn (*b*) Pepsin
(*c*) Tripsin (*d*) Lipases. ***(KCET)***

40. The double helical structure of DNA was proposed by
(*a*) Watson and Crick (*b*) Meicher
(*c*) Emil Fischer (*d*) Khorana. ***(KCET)***

41. Enzymes are essential as biocatalysis. They function in
(*a*) Aqueous medium, temperature = 30—35°C ; pH = 7
(*b*) Organic medium
(*c*) Aqueous medium under extreme pH conditions
(*d*) None of these. ***(AMU)***

42. Assuming sweetness of the cane sugar to be 10, the sweetness of glucose is about
(*a*) 15 (*b*) 1.5
(*c*) 12.5 (*d*) 7.5. ***(JIPMER)***

1999

43. Which of the following is the sweetest sugar ?
(*a*) Sucrose (*b*) Glucose
(*c*) Fructose (*d*) Maltose. ***(CBSE PMT)***

44. Enzyme in the living systems
(*a*) Provide energy
(*b*) Provide immunity
(*c*) Transport oxygen
(*d*) Catalyse biological reactions. ***(CBSE PMT)***

45. Which of the following gives maximum energy in metabolic processes ?
(*a*) Proteins (*b*) Vitamins
(*c*) Lipids (*d*) Carbohydrates. ***(PM CEE)***

46. The enzyme pepsin hydrolysis
(*a*) Proteins to amino acids
(*b*) Fats to fatty acids
(*c*) Glucose to ethyl alcohol
(*d*) Polysaccharides to monosaccharides. ***(MP CEE)***

47. The commonest disaccharide has the molecular formula
(*a*) $C_{12}H_{22}O_{11}$ (*b*) $C_{10}H_{18}O_9$
(*c*) $C_{10}H_{20}O_{10}$ (*d*) $C_{18}H_{32}O_{18}$. ***(MP CEE)***

48. Amino acids are the building blocks of
(*a*) Carbohydrates (*b*) Vitamins
(*c*) Fats (*d*) Proteins. ***(UP CPMT)***

49. Which carbohydrate is an essential constituent of plant cells ?
(*a*) Starch (*b*) Cellulose
(*c*) Sucrose (*d*) Vitamins. ***(MP CEE)***

50. The structure of DNA is
(*a*) Single helix (*b*) Double helix
(*c*) Triple helix (*d*) Linear. ***(AFMC)***

51. Allergy is caused by the production of .... in the body
(*a*) Hormones (*b*) Histamines
(*c*) Vitamins (*d*) Saliva. ***(JIPMER)***

52. Lack of essential amino acids in the diet leads to the disease
(*a*) Night blindness (*b*) Pernicious anaemia
(*c*) Kwashiorkor (*d*) Sickle cell anaemia. ***(JIPMER)***

53. Which of the following molecule is capable of forming zwitter ion ?
(*a*) CH_3—CHOH—NH_2 (*b*) NH_2 CH_2 COOH
(*a*) CH_3—COOH (*d*) $CCl_3 - NO_2$. ***(Haryana CEET)***

54. Enzyme which breakes down starch to maltose is
(*a*) Maltose (*b*) Zymase
(*c*) Diastase (*d*) Invertase. ***(Pb. CET)***

55. Vitamin C is chemically
(*a*) Ascorbic acid (*b*) Citric acid
(*c*) Aspirin (*d*) Aspartic acid. ***(Pb. CET)***

56. Match List I (name of vitamin) with List II (deficiency result/disease) and select the correct answer using the codes given below the lists :

List I	*List II*
I. Ascorbic acid	A. Beri-beri
II. Retinol	B. Cracked lips
III. Riboflavin	C. Scurvy
IV. Thiamine	D. Night blindness

Codes :
(*a*) I-B, II-A, III-C, IV-D (*b*) I-A, II-B, III-C, IV-D
(*c*) I-D, II-C, III-B, IV-A (*d*) I-C, II-D, III-B, IV-A. ***(SCRAE)***

2000

57. α-D-(+)-glucose and β-D-(+)-glucose are
(*a*) Enantiomers (*b*) Epimers
(*c*) Anomers (*d*) Conformers. ***(CBSE PMT)***

58. Which of the following is correct about enzymes ?
(*a*) It changes the equilibrium constant
(*b*) It increases the activation energy
(*c*) It reduces the activation energy
(*d*) None of these. ***(CBSE PMT)***

59. Which of the following has magnesium ?
(a) Carbonic anhydrase (b) Haemocyanin
(c) Chlorophyll (d) Vitamin-B_{12}.
(CBSE PMT)

60. The hormone which controls the processes of burning of facts, proteins and carbohydrates to liberate energy in the body is
(a) Cortisone (b) Adrenaline
(c) Thyroxine (d) Insulin. ***(CBSE PMT)***

61. Which of the following reagents cannot distinguish between glucose and fructose ?
(a) Tollen's reagent (b) Fehling solution
(c) Benedict's solution (d) All of these. ***(CPMT)***

62. Glucose gives silver mirror test with Tollen's reagent. It shows the presence of
(a) Ketonic group (b) Aldehydic group
(c) Alcoholic group (d) Acidic group. ***(CPMT)***

63. The helical structure of proteins is stabilized by
(a) H-bonding (b) van dar Waal's forces
(c) ionic bond (d) peptide bond.
(Haryana CEET)

64. Which of the following prepares human for emergency ?
(a) Adrenaline (b) Oxytocin
(c) Testosterone (d) Progesterone. ***(Pb. CET)***

65. Amino acids usually exist in the form of zwitter ions. This means that amino acid contains
(a) The basic group $-NH_2$ and the acidic group $-COOH$
(b) The basic $-NH_3^+$ and the acidic group $-COO^-$
(c) The basic group $-COO^-$ and the acidic group $-NH_3^+$
(d) There is no acidic or basic group. ***(KCET)***

66. Which of the following is not a sex hormone ?
(a) Testosterone (b) Estrone
(c) Estradiol (d) Cortisone. ***(MP PMT)***

67. In polysaccharides, the linkage connecting monosaccharides is called
(a) glycoside linkage (b) nucleoside linkage
(c) glycogen linkage (d) peptide linkage.
(MP PMT)

68. Iodine test is shown by
(a) Polypeptides (b) Glycogen
(c) Starch (d) Glucose. ***(BHU)***

69. The best solvent for removing butter stain from cloth is
(a) $CHCl_3$ (b) C_2H_5OH
(c) $C_2H_5OC_2H_5$ (d) H_2O. ***(JIPMER)***

70. Which amino acid is achiral ?
(a) Alanine (b) Valine
(c) Proline (d) Histidine
(e) None of the above. ***(Kerala MEE)***

71. The structural feature which distinguishes proline from other natural α-amino acids ?
(a) It is optically inactive
(b) It contains aromatic group
(c) It is a dicarboxylic acid
(d) It is secondary amine
(e) It contains two amino groups. ***(Kerala MEE)***

2001

72. Which is correct statement ?
(a) Starch is a polymer of α-glucose
(b) Amylose is a component of cellulose
(c) Proteins are compounds of only one type of amino acids
(d) In cyclic structure of fructose, there are four carbon atoms and one oxygen atom. ***(CBSE PMT)***

73. Which statement is *incorrect* about peptide bond ?
(a) C—N bond length in proteins is longer than usual bond length of C—N bond
(b) Spectroscopic analysis shows planar structure $-\underset{\underset{O}{\|}}{C}-NH-$ bond
(c) C—N bond length in proteins is smaller than usual bond length of C—N bond
(d) None of these. ***(CBSE PMT)***

74. An example of water soluble vitamin is
(a) Vitamin D (b) Vitamin E
(c) Vitamin A (d) Vitamin C. ***(AIIMS)***

75. Which substance chars when heated with conc. H_2SO_4
(a) Carbohydrate (b) Hydrocarbon
(c) Fat (d) Protein. ***(AIIMS)***

76. The aldehyde which is formed during photosynthesis in plants is
(a) Methanal (b) Acetaldehyde
(c) Propanal (d) Phenyl methanol.
(Tamil Nadu CET)

77. α-Tocopherol is found in
(a) vegetable oils (b) lemon
(c) peanuts (d) tomatoes.
(Tamil Nadu CET)

78. Scruvy is caused due to the deficiency of vitamin
(a) B_1 (b) C
(c) K (d) D. ***(Manipal PMT)***

79. Violet colour is obtained when dilute $CuSO_4$ is added in alkaline solution of protein. This test is known as
(a) Biuret (b) Xanthoproteic test
(c) Hopkins-cole (d) Millon's test.
(Manipal PMT)

80. A nucleotide consists of
(a) Sugar (b) nitrogen containing base
(c) phosphoric acid (d) all of these.
(Manipal PMT)

81. What happens when H_2SO_4 is treated with sugar ?
(*a*) Oxidation (*b*) Reduction
(*c*) Dehydration (*d*) Hydrolysis. ***(Manipal PMT)***

82. Which polymer is used in making rayon ?
(*a*) Polyester (*b*) Nylon
(*c*) Cellulose (*d*) Starch. ***(MP PET PMT)***

83. Which of the following is not a function of proteins ?
(*a*) Nail formation
(*b*) Skin formation
(*c*) Muscle formation
(*d*) Providing energy for metabolism. ***(MP PET PMT)***

2002

84. Cellulose is a polymer of
(*a*) Glucose (*b*) Fructose
(*c*) Ribose (*d*) Sucrose. ***(CBSE PMT)***

85. Enzymes are made up of
(*a*) Edible proteins
(*b*) Proteins with specific structure
(*c*) Nitrogen containing carbohydrates
(*d*) Carbohydrates. ***(CBSE PMT)***

86. Purine derivative among the following bases is
(*a*) Thymine (*b*) Uracil
(*c*) Guanine (*d*) Cytosine. ***(KCET)***

87. Drying oil invariably contains
(*a*) Linoleic acid (*b*) Lauric acid
(*c*) Stearic acid (*d*) Butyric acid. ***(KCET)***

88. Oils are converted into fats by
(*a*) Hydration (*b*) Decarboxylation
(*c*) Hydrogenation (*d*) Dehydrogenation
(*e*) Hydrogenolysis. ***(Kerala MEE)***

89. α-Helix is found in
(*a*) DNA (*b*) RNA
(*c*) Lipid (*d*) Carbohydrates
(*e*) Proteins. ***(Kerala CEE)***

90. In fructose, the possible optical isomers are
(*a*) 12 (*b*) 8
(*c*) 16 (*d*) 4. ***(Tamil Nadu CET)***

91. The functional group, which is found in amino acid is
(*a*) —COOH group (*b*) $-NH_2$
(*c*) $-CH_2$ (*d*) both (*a*) and (*b*). ***(AIEEE)***

92. RNA is different from DNA because RNA contains
(*a*) Ribose sugar and thymine
(*b*) Ribose sugar and uracil
(*c*) Deoxyribose sugar and thymine
(*d*) Deoxyribose sugar and uracil. ***(AIEEE)***

93. Two vitamins absorbed from intestine along with fats are
(*a*) A, D (*b*) A, B
(*c*) A, C (*d*) D, B. ***(Manipal PMT)***

94. Iodine value is related to
(*a*) Fats and oils (*b*) Alcohols
(*c*) Esters (*d*) Hydrocarbons. ***(MP PET)***

95. A triglyceride can have how many different acyl groups ?
(*a*) 3 (*b*) 2
(*c*) 1 (*d*) 4. ***(DPMT)***

2003

96. Phospholipids are esters of glycerol with
(*a*) Two carboxylic acid residues and one phosphate group
(*b*) One carboxylic acid residue and two phosphate groups
(*c*) Three phosphate groups
(*d*) Three carboxylic acid residues. ***(CBSE PMT)***

97. Vitamin B_{12} contains
(*a*) Co(III) (*b*) Zn(II)
(*c*) Ca(II) (*d*) Fe(II). ***(CBSE PMT)***

98. Complete hydrolysis of cellulose gives
(*a*) L-glucose (*b*) D-fructose
(*c*) D-ribose (*d*) D-glucose. ***(AIEEE)***

99. The reason for double helical structure of DNA is operation of
(*a*) Electrostatic attractions
(*b*) Van der Waal's forces
(*c*) Dipole-dipole interaction
(*d*) Hydrogen bonding. ***(AIEEE)***

100. The structure given below is known as

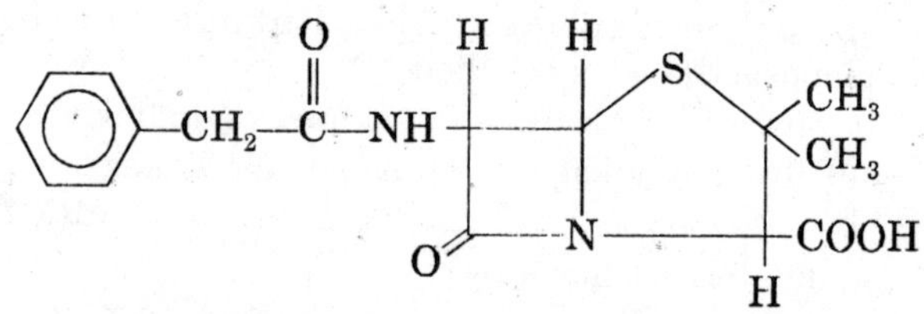

(*a*) Penicillin F (*b*) Penicillin G
(*c*) Penicillin K (*d*) Ampicillin
(*e*) Sulphadiazine. ***(Kerala PMT)***

2004

101. Identify the correct statement regarding enzymes :
(*a*) Enzymes are specific biological catalysts that can normally function at very high temperature (T ~ 1000 K)
(*b*) Enzymes are specific biological catalysts that possess well-defined active sites.

(c) Enzymes are specific biological catalysts that cannot be poisoned

(d) Enzymes are normally heterogeneous catalysts that are very specific in their action. **(AIEEE)**

102. Insulin production and its action in human body are responsible for the level of diabetes. This compound belongs to which of the following categories ?

(a) A co-enzyme (b) An antibiotic

(c) An enzyme (d) A hormone **(AIEEE)**

103. Which base is present in RNA but not in DNA ?

(a) Uracil (b) Thymine

(c) Guanine (d) Cytosine. **(AIEEE)**

104. A sequence of how many nucleotides in messenger RNA makes a codon for an amino acid ?

(a) Four (b) One

(c) Two (d) Three. **(CBSE PMT)**

105. The hormone that helps in the conversion of glucose to glycogen is

(a) Bile acids (b) Adrenaline

(c) Insulin (d) Cortisone. **(CBSE PMT)**

106. The correct statement in respect of protein haemoglobin is that it

(a) Maintains blood sugar level

(b) Acts as an oxygen carrier in the blood

(c) Forms antibodies and offers resistance to diseases

(d) Functions as a catalyst for biological reactions. **(CBSE PMT)**

107. Which one of the following structures represents the peptide chain ?

(a) 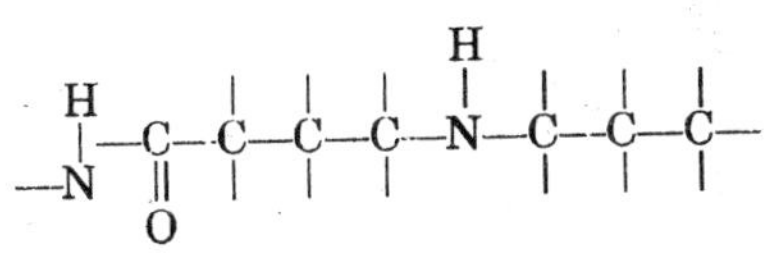

(b) 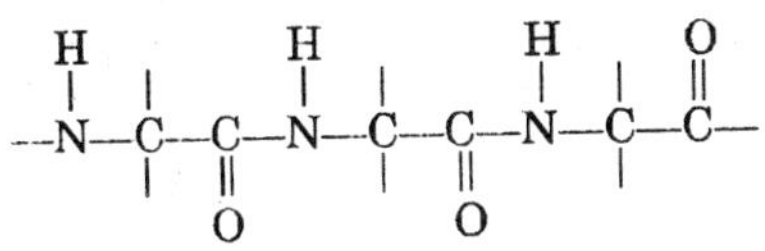

(c) 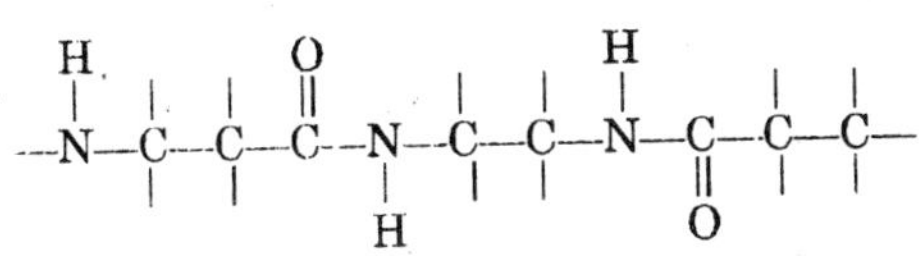

(d) 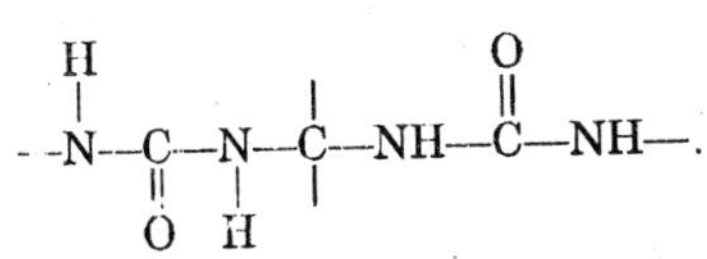 **(CBSE PMT)**

108. Number of chiral carbons in β-D-(+)-glucose is

(a) Six (b) Three

(c) Four (d) Five. **(CBSE PMT)**

109. The helical structure of protein is stabilized by

(a) Hydrogen bonds (b) Ether bonds

(c) Peptide bonds (d) Dipeptide bonds. **(CBSE PMT)**

110. The enzyme which hydrolyses triglycerides to fatty acids and glycerol is called

(a) Lipase (b) Zymase

(c) Pepsin (d) Maltase. **(CBSE PMT)**

111. The nucleic acid base having two possible binding sites is :

(a) Thymine (b) Cytosine

(c) Guanine (d) Adenine. **(AIIMS)**

112. The pair of compounds in which both the compounds give positive test with Tollen's reagent is

(a) Glucose and Sucrose

(b) Fructose and Sucrose

(c) Acetophenone and Hexanal

(d) Glucose and Fructose. **(IIT)**

2005

113. In both DNA and RNA, heterocyclic base and phosphate ester linkages are at :

(a) C_5' and C_1' respectively of the sugar molecule

(b) C_1' and C_5' respectively of the sugar molecule

(c) C_2' and C_5' respectively of the sugar molecule

(d) C_5' and C_2' respectively of the sugar molecule. **(AIEEE)**

114. Which functional group participates in disulphide bond formation in proteins ?

(a) Thioester (b) Thioether

(c) Thiol (d) Thiolactone. **(CBSE PMT)**

115. The cell membranes are mainly composed of

(a) Fats (b) Proteins

(c) Phospholipids (d) Carbohydrates. **(CBSE PMT)**

116. The two forms of D-glucopyranose are called

(a) Enantiomers (b) Epimers

(c) Anomers (d) Diastereomers. **(IIT Screening Test)**

117. Which one of the following biomolecules is insoluble in water ?

(a) α-keratin (b) Haemoglobin

(c) Ribonuclease (d) Adenine. **(AIIMS)**

118. Which one of the following statements is true for protein synthesis (translation) ?

(a) Amino acids are directly recognized by m-RNA

(b) The third base of the codon is less specific

(c) Only one codon codes for an amino acid

(d) Every t-RNA molecule has more than one amino acid attachment. **(AIIMS)**

119. The group linkage present in fats is
(*a*) Peptide linkage (*b*) Ester linkage
(*c*) Glycosidic linkage (*d*) None of these. ***(AFMC)***

120. Which of the following amino acid is optically inactive ?
(*a*) Phenyl alanine (*b*) Glycine
(*c*) Glutamic acid (*d*) Asparagine. ***(BHU)***

121. Glucose has difference from fructose in that it
(*a*) Does not undergo hydrolysis
(*b*) Gives silver mirror with Tollen's reagent
(*c*) Monosaccharide
(*d*) None of these. ***(BHU)***

122. Sucrose molecule is made up of
(*a*) A gluco pyranose and a fructo pyranose
(*b*) A gluco pyranose and a fructo furanose
(*c*) A gluco furanose and a fructo pyranose
(*d*) A gluco furanose and a fructo furanose.
(Karnataka CET)

123. Water insoluble component of starch is
(*a*) Amylopectin (*b*) Amylose
(*c*) Cellulose (*d*) None of these.
(Karnataka CET)

124. An example for a saturated fatty acid, present in nature is
(*a*) Oleic acid (*b*) Linoleic acid
(*c*) Linolenic acid (*d*) Palmitic acid.
(Karnataka CET)

125. A nanopeptide contains .... peptide linkages
(*a*) 10 (*b*) 8
(*c*) 9 (*d*) 18. ***(Karnataka CET)***

126. Which of the following biomolecules contain non-transition metal ion ?
(*a*) Vitamin B_{12} (*b*) Chlorophyll
(*c*) Haemoglobin (*d*) Insulin.
(Karnataka CET)

127. Which of the following is *not* correct ?
(*a*) Chlorophyll is responsible for the synthesis of carbohydrates in plants
(*b*) The compound formed in the addition of oxygen to haemoglobin is called oxyhaemoglobin
(*c*) Acetyl salicylic acid is known as aspirin
(*d*) The metal ion present in Vitamin B_{12} is Mg^{2+}.
(EAMCET Engg.)

2006

128. Which one of the following is a peptide hormone ?
(*a*) Testosterone (*b*) Thyroxin
(*c*) Adrenaline (*d*) Glucagon.
(CBSE PMT Screening)

129. During the process of digestion, the proteins present in food materials are hydrolysed to amino acids. The two enzymes involved in the process

Proteins $\xrightarrow{\text{Enzyme (A)}}$ Polypeptides $\xrightarrow{\text{Enzyme (B)}}$ Aminoacids, are respectively
(*a*) Diastase and Lipase (*b*) Pepsin and Trypsin
(*c*) Invertase and Zymase (*d*) Amylase and Maltase.
(CBSE PMT Screening)

130. The pyrimidine bases present in DNA are
(*a*) Cytosine and adenine (*b*) Cytosine and guanine
(*c*) Cytosine and thymine (*d*) Cytosine and uracil.
(AIEEE)

131. Glucose contains in addition to aldehyde group
(*a*) One secondary OH and four primary OH groups
(*b*) Four secondary OH and one primary OH groups.
(*c*) Two primary OH and three secondary OH groups.
(*d*) Three primary OH and two secondary OH groups.
(Karnataka CET)

132. A distinctive and characteristic functional group of fats is
(*a*) A peptide group (*b*) An ester group
(*c*) An alcoholic group (*d*) A ketonic group.
(Karnataka CET)

133. Insulin regulates the metabolism of
(*a*) Minerals (*b*) Amino acids
(*c*) Glucose (*d*) Vitamins.
(Karnataka CET)

134. At *p*H = 4, glycine exists as
(*a*) $H_3\overset{+}{N}—CH_2—COO^-$ (*b*) $H_3\overset{+}{N}—CH_2—COOH$
(*c*) $H_2N—CH_2—COOH$ (*d*) $H_2N—CH_2—COO$.
(Karnataka CET)

2007

135. Methyl-α-D-glucoside and methyl-β-D-glucoside are
(*a*) Epimers
(*b*) Anomers
(*c*) Enantiomers
(*d*) Conformational diastereomers. ***(AIIMS)***

136. All monosaccharides .... Tollen's reagent
(*a*) Oxidise (*b*) Condense with
(*c*) Reduce (*d*) Add to. ***(J.K. CET)***

137. The secondary structure of a protein refers to
(*a*) Sequence of α-amino acids
(*b*) Fixed configuration of the polypeptide backbone
(*c*) α-helical backbone
(*d*) Hydrophobic interactions. ***(AIEEE)***

138. Which of the following vitamins is water soluble ?
(*a*) Vitamin A (*b*) Vitamin B
(*c*) Vitamin E (*d*) Vitamin K. ***(CBSE PMT)***

139. RNA and DNA are chiral molecules, their chirality is due to
(*a*) D-sugar compound
(*b*) L-sugar component
(*c*) Chiral bases
(*d*) Chiral phosphate ester groups. ***(CBSE PMT)***

ANSWERS

1. (*b*)	**2.** (*a*)	**3.** (*c*)	**4.** (*a*)	**5.** (*b*)	**6.** (*b*)	**7.** (*d*)	**8.** (*b*)	**9.** (*c*)
10. (*c*)	**11.** (*c*)	**12.** (*c*)	**13.** (*d*)	**14.** (*c*)	**15.** (*b*)	**16.** (*c*)	**17.** (*d*)	**18.** (*a*)
19. (*a*)	**20.** (*a*)	**21.** (*a*)	**22.** (*c*)	**23.** (*a*)	**24.** (*b*)	**25.** (*a*)	**26.** (*d*)	**27.** (*b*)
28. (*c*)	**29.** (*d*)	**30.** (*d*)	**31.** (*d*)	**32.** (*d*)	**33.** (*c*)	**34.** (*d*)	**35.** (*b*)	**36.** (*a*)
37. (*c*)	**38.** (*b*)	**39.** (*d*)	**40.** (*a*)	**41.** (*a*)	**42.** (*d*)	**43.** (*c*)	**44.** (*d*)	**45.** (*c*)
46. (*a*)	**47.** (*a*)	**48.** (*d*)	**49.** (*b*)	**50.** (*b*)	**51.** (*b*)	**52.** (*c*)	**53.** (*b*)	**54.** (*c*)
55. (*a*)	**56.** (*d*)	**57.** (*c*)	**58.** (*c*)	**59.** (*c*)	**60.** (*a*)	**61.** (*d*)	**62.** (*b*)	**63.** (*a*)
64. (*a*)	**65.** (*c*)	**66.** (*d*)	**67** (*a*)	**68.** (*c*)	**69.** (*a*)	**70.** (*e*)	**71.** (*d*)	**72.** (*a*)
73. (*c*)	**74.** (*d*)	**75.** (*a*)	**76.** (*a*)	**77.** (*c*)	**78.** (*b*)	**79.** (*a*)	**80.** (*d*)	**81.** (*c*)
82. (*c*)	**83.** (*d*)	**84.** (*a*)	**85.** (*b*)	**86.** (*c*)	**87.** (*a*)	**88.** (*c*)	**89.** (*e*)	**90.** (*b*)
91. (*d*)	**92.** (*b*)	**93.** (*a*)	**94.** (*a*)	**95.** (*a*)	**96.** (*a*)	**97.** (*a*)	**98.** (*d*)	**99.** (*d*)
100. (*b*)	**101.** (*b*)	**102.** (*d*)	**103.** (*a*)	**104.** (*d*)	**105.** (*c*)	**106.** (*b*)	**107.** (*c*)	**108.** (*d*)
109. (*a*)	**110.** (*a*)	**111.** (*a*, *d*)	**112.** (*d*)	**113.** (*b*)	**114.** (*c*)	**115.** (*c*)	**116.** (*c*)	**117.** (*a*)
118. (*b*)	**119.** (*b*)	**120.** (*b*)	**121.** (*d*)	**122.** (*b*)	**123.** (*a*)	**124.** (*d*)	**125.** (*b*)	**126.** (*b*)
127. (*d*)	**128.** (*d*)	**129.** (*b*)	**130.** (*c*)	**131.** (*b*)	**132.** (*b*)	**133.** (*c*)	**134.** (*b*)	**135.** (*b*)
136. (*c*)	**137.** (*b*)	**138.** (*b*)	**139.** (*a*).					

HINTS/SOLUTIONS

Q. No.	Ans.	Hints & Solutions
1.	(*b*)	D-Glucose + D-Fructose.
2.	(*a*)	I-E, II-A, III-C and IV-B is the correct matching.
3.	(*c*)	Enzyme zymase.
4.	(*a*)	Protein is a polymer of α-amino acids.
5.	(*b*)	The base thymine is not present in RNA.
6.	(*b*)	Hydrogen bond.
7.	(*d*)	The various nucleotides are linked through phosphate groups in nucleic acids.
8.	(*b*)	Enzymes are proteins or polypeptides.
9.	(*c*)	There are 5-OH groups in glucose and hence the product pentacetate is formed.
10.	(*c*)	Sucrose is a non reducing sugar.
11.	(*c*)	Hydrogen bonds between —CO— group and —NH— groups of different peptide linkages stabilize the helical structure of protein.
12.	(*c*)	Enzymes being proteins contain $-\underset{\underset{O}{\|}}{C}-\underset{\underset{H}{\mid}}{N}-$ linkage.
13.	(*d*)	Self explanatory answer.
14.	(*c*)	Enzymes are biocatalysts.
15.	(*b*)	The sweetest sugar is fructose.
16.	(*c*)	Glycine (α-amino acetic acid) is the only α-amino acid which is achiral.
17.	(*d*)	All carbohydrates respond to Molisch test.
18.	(*a*)	Purine bases present in RNA are guanine and adenine.
19.	(*a*)	Vitamin A.
20.	(*a*)	Adenine (A) pairs with thymine (T) while guanine (G) pairs with cytosine (C).

Q. No.	Ans.	Hints & Solutions
21.	(a)	Insulin.
22.	(c)	Oleic acid, stearic acid and palmitic acid are fatty acids.
23.	(a)	Phosphoric acid.
24.	(b)	See Hint 20.
25.	(a)	One molecule of glucose reacts with three molecules of phenyl hydrazine to form osazone.
26.	(d)	130 molecules. Factual question.
27.	(b)	Vitamin A is called retinol.
28.	(c)	Galactose is converted into glucose in liver.
29.	(d)	Enzymes are biocatalysts.
30.	(d)	Guanidine is not present in nucleic acids.
31.	(d)	Benzidine is not an amino acid.
32.	(d)	α-Amino acetic acid has zwitter ion stucture.
33.	(c)	Enzyme invertase converts sucrose to glucose and fructose.
34.	(d)	Gene. Self explanatory answer.
35.	(b)	Amylases hydrolyse starch to maltose.
36.	(a)	Small pox is a viral disease.
37.	(c)	Insulin is a protein hormone.
38.	(b)	Peptides on hydrolysis give α-amino acids.
39.	(d)	Enzyme lipases is involved in the digestion of fats.
40.	(a)	Watson and Crick. Factual question.
41.	(a)	Self explanatory answer.
42.	(d)	Sweetness of glucose is about 70 percent that of sucrose.
43.	(c)	Fructose is the sweetest sugar.
44.	(d)	Enzymes catalyse biological reactions.
45.	(c)	Lipids release maximum energy.
46.	(a)	Pepsin hydrolyses protein to α-amino acids.
47.	(a)	Sucrose ($C_{12}H_{22}O_{11}$) is the commonest disaccharide.
48.	(d)	Proteins are formed (by condensation of amino acids).
49.	(b)	Cellulose is present in all plant cells.
50.	(b)	DNA has double helix structure.
51.	(b)	Histamines produce allergy symptoms.
52.	(c)	Kwashiorkor is a protein deficiency disease.
53.	(b)	Amino acids can exist as zwitter ions.
54.	(c)	Enzyme diastase breaks down starch to maltose.
55.	(a)	Vitamin C is ascorbic acid.
56.	(d)	I-C, II-D, III-B, IV-A is the correct matching.
57.	(c)	α-D-glucose and β-D-glucose have different configuration at C-1 and hence are **anomers.**
58.	(c)	Enzymes reduce the activation energy.
59.	(c)	Chlorophyll is a complex of magnesium.
60.	(a)	Cortisone controls metabolism of fats, proteins and carbohydrates.
61.	(d)	None of these reagents can distinguish between glucose and fructose.
62.	(b)	Aldehydic group.
63.	(a)	Helical structure of proteins stabilized by hydrogen bonding.
64.	(a)	Adrenaline.
65.	(c)	In zwitter ion structure the acidic group is NH_3^+ and the basic group is —COO^-.
66.	(d)	Cortisone is not a sex hormone.
67.	(a)	Glycoside linkage.
68.	(c)	Starch gives blue colouration with iodine.
69.	(a)	Fats are soluble in chloroform, $CHCl_3$.
70.	(e)	Only glycine is achiral amino acid.
71.	(a)	Proline is a secondary amine.
72.	(a)	Starch is a polymer of α-(D)-glucose.
73.	(c)	C—N bond in proteins has partial double bond character due to resonance.
74.	(d)	Vitamin C is water soluble.
75.	(a)	Carbohydrates undergo charring when heated with conc. H_2SO_4 due to dehydration.
76.	(a)	Methanal.
77.	(c)	Oil seeds are rich in vitamin E (or α-Tocopherol).
78.	(b)	Deficiency of vitamin C causes scurvy.
79.	(a)	Biuret test.
80.	(d)	A nucleotide is a base-sugar-phosphoric acid unit.
81.	(c)	Sugar undergoes dehydration and gets charred.
82.	(c)	Cellulose is used for making rayon.
83.	(d)	Proteins are not the normal source of energy.
84.	(a)	Cellulose is a polymer of β-(D)-glucose.
85.	(b)	Enzymes are proteins with specific structure.
86.	(c)	Purine bases are **guanine** and adenine.
87.	(a)	Drying oils contain unsaturated fatty acids. Linoleic acid is an unsaturated fatty acids.
88.	(c)	Oils are converted into solid fats by hydrogenation.

Q. No.	Ans.	Hints & Solutions
89.	(*a*)	Proteins have α-helix structure or β-pleated sheet structure as secondary structures.
90.	(*b*)	Fructose has three chiral carbon atoms. Hence, number of optical isomers possible for it is $(2)^3$.
91.	(*d*)	Amino acids contain both —COOH and—NH_2 groups.
92.	(*b*)	RNA contains ribose sugar and uracil base while DNA contains 2-deoxyribose sugar and thymine base.
93.	(*a*)	Vitamins A and D are oil soluble.
94.	(*a*)	Iodine value is used for estimation of unsaturation in oils and fats.
95.	(*a*)	A triglyceride is formed by acylation of all the three —OH groups of glycerol.
96.	(*a*)	Phospholipids are esters of glycerol with two carboxylic acid residues and one phosphate group.
97.	(*a*)	Cobalt (III).
98.	(*d*)	D-Glucose.
99.	(*d*)	Hydrogen bonding.
100.	(*b*)	Penicillin G.
101.	(*b*)	Enzymes are specific biological catalysts that possess well defined active sites.
102.	(*d*)	Insulin is a hormone secreted by pancreas.
103.	(*a*)	Uracil is present in RNA and not in DNA.
104.	(*d*)	The sequence of three nucleotides (triplet) acts as a codon for an amino acid.
105.	(*c*)	Glucose is converted into glycogen by hormone insulin.
106.	(*b*)	Haemoglobin combines with oxygen to form oxy-haemoglobin and thus acts as oxygen carrier in the blood.
107.	(*c*)	A peptide linkage is an amide $\left(-\underset{\underset{O}{\|}}{C}-NH-\right)$ linkage.
108.	(*d*)	β-D-(+) glucose has five chiral carbon atoms. 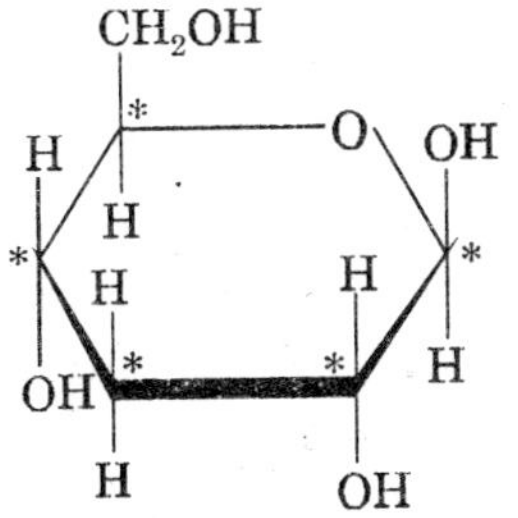
109.	(*a*)	The helical structure of protein is stabilized by hydrogen bonds between C = O and —NH— groups.
110.	(*a*)	Enzyme lipase hydrolyses triglycerides to fatty acids and glycerol.
111.	(*a*)	Thymine has two possible bonding site.
112.	(*d*)	Glucose (an aldehyde) and fructose (an α-hydroxyketone) both give positive test with Tollen's reagent.
113.	(*b*)	Heterocyclic base is attached to C_1' of the sugar while phosphate ester linkage is at C_5' 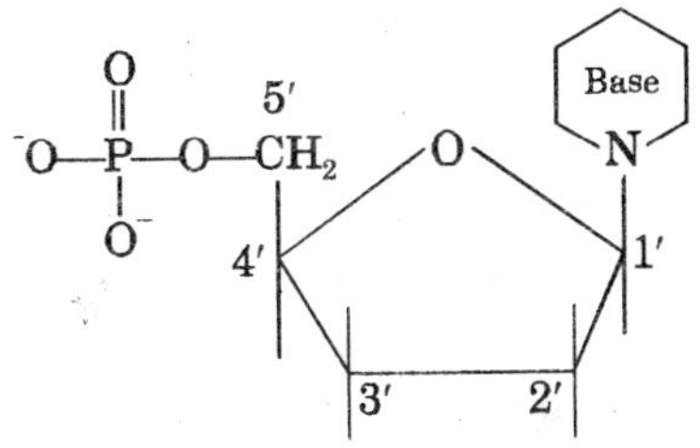
114.	(*c*)	Thiol group.
115.	(*c*)	The cell membranes are mainly composed of phospholipids.
116.	(*c*)	The two forms of D-glucopyranose differ in the orientation of groups around C-1. Such isomers are called anomers.
117.	(*a*)	α-Keratin is insoluble in water.
118.	(*b*)	The third base of codon is less specific. A particular amino acid may be represented by more than one codons. In these codons of a particular amino acid, the first two bases are common while the third base is different.
119.	(*b*)	Fat molecules contain ester linkage.
120.	(*b*)	Glycine is the only α-amino acid which is optically inactive. It does not contain any chiral centre. $H_2N—CH_2—COOH$. Glycine
121.	(*d*)	None of these.
122.	(*b*)	Sucrose molecule contains glucose as tyranose ring and fructose as furanose ring.
123.	(*a*)	Amylobectin is water insoluble whereas amylose is water soluble.
124.	(*d*)	Palmatic acid ($C_{15}H_{31}COOH$) is a saturated fatty acid.
125.	(*b*)	A nanopeptide is formed by condensation of nine amino acids. It contains **eight** peptide linkages.
126.	(*b*)	Chlorophyll contains magnesium which is a non-transition metal.
127.	(*d*)	Vitamin B_{12} contains cobait ion.

Q. No.	Ans.	Hints & Solutions
128.	**(*d*)**	Testosterone is a steroid hormone while thyroxine and drenaline have only one amino acid unit. Glucagon has 29 amino acid units which are joined by peptide links.
129.	(*b*)	The enzyme pepsin hydrolyses proteins to polypeptides while the enzyme tryprin hydrolyses polypeptides to amino acids.
130.	(*c*)	In DNA, pyrimidine bases present are cytosine and thymine.
131.	(*b*)	Glucose molecule contains 4 secondary —OH groups and one primary —OH group.
132.	(*b*)	Fats contain ester group. For example, oils are triesters of glycerol.
133.	(*c*)	Insulin regulates metabolism of glucose.
134.	(*b*)	In acidic medium —NH_2 group changes to 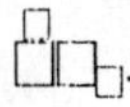.
135.	(*b*)	These are anomers.
136.	(*c*)	All monosaccharides **reduce** Tollen's reagent to silver.
137.	(*b*)	The secondary structure of a protein refers to fixed configuration the polypeptide backbone.
138.	(*b*)	Vitamin B is water soluble.
139.	(*a*)	DNA and RNA are chiral due to chirality in the sugar component (D-2-deoxyribose and D-ribose).

9.5. ASSERTION-REASON QUESTIONS

Quick Revision

In each of the following questions two statements are given as Assertion A and Reason R. Examine the statements carefully and answer the questions according to the instructions given below :

- *Mark **a** if both **A** and **R** are correct and **R** is the correct reason of **A**.*
- *Mark **b** if both **A** and **R** are correct and **R** is not the correct reason of **A**.*
- *Mark **c** if **A** is correct and **R** is wrong.*
- *Mark **d** if **A** is wrong and **R** is correct.*
- *Mark **e** if both **A** and **R** are wrong.*

1. ***Assertion A.*** Hydrolysis of sucrose is known as inversion of sucrose.

Reason R. Sucrose is dextrorotatory whereas the product formed on hydrolysis is laevorotatory.

(a)	(b)	(c)	(d)	(e)

2. ***Assertion A.*** D-(+)-Glucose and L-(–)-fructose are enantiomers.

Reason R. Glucose molecule is mirror image of fructose molecule.

(a)	(b)	(c)	(d)	(e)

3. ***Assertion A.*** DNA molecule has double helix structure.

Reason R. The monomeric units in DNA are known as nucleosides.

(a)	(b)	(c)	(d)	(e)

4. ***Assertion A.*** Fructose is the sweetest sugar.

Reason R. Fructose is a functional isomer of glucose.

(a)	(b)	(c)	(d)	(e)

5. ***Assertion A.*** Starch on hydrolysis yields α-D-glucose.

Reason R. Starch is hydrolysed to glucose by the action of enzymes amlyase and maltase.

(a)	(b)	(c)	(d)	(e)

6. ***Assertion A.*** Enzymes are proteins.

Reason R. Enzymes are highly specific in nature.

(a)	(b)	(c)	(d)	(e)

7. ***Assertion A.*** Purine bases present in DNA are adenine and guagine.

Reason R. The base uracil present in DNA whereas the base thymine is present in RNA.

(a)	(b)	(c)	(d)	(e)

8. ***Assertion A.*** Thymine pairs with adenine whereas cytosine pairs with guanine in DNA molecule.

Reason R. The hydrogen bonding between bases of two strands is highly specific.

(a)	(b)	(c)	(d)	(e)

9. ***Assertion A.*** All monosaccharides are sweet in taste.

Reason R. All monosaccharides have the general formula $C_6H_{12}O_6$.

(a)	(b)	(c)	(d)	(e)

10. ***Assertion A.*** In polysaccharides, monosaccharide units are linked by glycoside linkage.

Reason R. Glycoside linkage is an ether linkage.

(a)	(b)	(c)	(d)	(e)

11. ***Assertion A.*** α-Amino acids have zwitter ion structure.

Reason R. α-Amino acids have least conductivity at isoelectric point.

(a)	(b)	(c)	(d)	(e)

12. ***Assertion A.*** One molecule of glucose on glycolysis yields two molecules of pyruvic acid.

Reason R. Glycolysis is an anaerobic process.

(a)	(b)	(c)	(d)	(e)

13. ***Assertion A.*** Thyroxin is a hormone that stimulates metabolism.

Reason R. Thyroxin contains iodine.

(a)	(b)	(c)	(d)	(e)

14. ***Assertion A.*** The enzyme lipases hydrolyse fats.
Reason R. Digestion of fats take place in small intestines.

(a)	(b)	(c)	(d)	(e)

15. ***Assertion A.*** Digestion of proteins begins in stomach and continues in small intestines.
Reason R. Proteins are hydrolysed to polypeptides by pepsin.

(a)	(b)	(c)	(d)	(e)

ANSWERS

1. (a)	**2.** (e)	**3.** (c)	**4.** (b)	**5.** (b)	**6.** (b)	**7.** (c)	**8.** (a)	**9.** (c)
10. (b)	**11.** (b)	**12.** (b)	**13.** (b)	**14.** (b)	**15.** (b).			

9.6. SELF EVALUATION TEST

Assess Yourself

Choose the correct answer from the four alternatives given in each of the following questions :

1. α-D-Glucose and β-D-glucose are
(*a*) Enantiomers (*b*) Anomers
(*c*) Epimers (*d*) Laevorotatory.

2. Which of the following give(s) silver mirror test with Tollen's reagent ?
(*a*) Glucose (*b*) Fructose
(*c*) Both of these (*d*) None of these.

3. Which of the following is most abundant in the living world ?
(*a*) Starch (*b*) Proteins
(*c*) Glucose (*d*) Cellulose.

4. Gums and pectins are examples of
(*a*) Proteins (*b*) Lipids
(*c*) Enzymes (*d*) Polysaccharides.

5. Glycylalanine and alanylglycine are
(*a*) Stereoisomers (*b*) Structural isomers
(*c*) Diastereomers (*d*) Identical.

6. In nuclei acids, the sequence is
(*a*) Phosphate-Base-Sugar (*b*) Sugar-Phosphate-Base
(*c*) Phosphate-Sugar-Base (*d*) Base-Phosphate-Sugar.

7. Which of the following bases is not present in DNA ?
(*a*) Thymine (*b*) Uracil
(*c*) Adenine (*d*) Guanine.

8. A nucleoside is made up of
(*a*) A base and sugar
(*b*) A base and phosphoric acid
(*c*) A sugar and phosphoric acid
(*d*) A sugar, a base and phosphoric acid.

9. Which of the following bases is a purine ?
(*a*) Thymine (*b*) Uracil
(*c*) Cytosine (*d*) Adenine.

10. The linkage which holds various amino acid units in primary structure of proteins is
(*a*) Glycoside linkage (*b*) Peptide linkage
(*c*) Ionic linkage (*d*) Hydrogen bond.

11. Maltose is made of
(*a*) α-D-glucose (*b*) α and β-D-glucose
(*c*) Glucose and fructose (*d*) Fructose only.

12. Which of the following is an example of zwitter ion ?
(*a*) Aniline (*b*) Alanine hydrochloride
(*c*) Glycine (*d*) All of these.

13. The sequence of bases on *m*-RNA molecule synthesized on the CGATAC strand of DNA is
(*a*) GCUAUG (*b*) GCTATAG
(*c*) GCAUAG (*d*) ATCGCA.

14. The correct order of sweetness of the following compounds is
Sucrose (I), Fructose (II), Glucose (III), Maltose (IV)
(*a*) I > II > III > IV (*b*) II > I > III > IV
(*c*) II > I > IV > III (*d*) IV > I > III > II.

15. Which of the following fatty acids is a saturated fatty acid ?
(*a*) Oleic acid (*b*) Linoleic acid
(*c*) Linolenic acid (*d*) Stearic acid.

16. Which of the following is an example of aldopentose ?
(*a*) Glyceraldehyde (*b*) Ribose
(*c*) Fructose (*d*) Erythrose.

17. Glucose on oxidation with nitric acid yields
(*a*) Glucaric acid (*b*) Gluconic acid
(*c*) Oxalic acid (*d*) Sorbitol.

18. Glucose on reaction with excess phenylhydrazine gives glucosazone. In this reaction one molecule of glucose reacts with .... molecules of phenylhydrazine.
(*a*) Two (*b*) Three
(*c*) Four (*d*) Six.

19. The number of hydrogen bonds possible between a molecule of cytosine and a molecule of guanine in DNA molecule is
(*a*) 1 (*b*) 3
(*c*) 2 (*d*) 4.

20. Digestion of lipids occurs primarily in
(*a*) Stomach (*b*) Small intestine
(*c*) Mouth (*d*) Large intestine.

21. In stomach the proteins are broken down into peptides by enzyme
(*a*) Trypsin (*b*) Pepsin
(*c*) Chymotrypsin (*d*) Amylase.

22. The first stage of oxidation of carbohydrates is known as
(*a*) Glycolysis (*b*) Cellular respiration
(*c*) Kreb's cycle (*d*) Anabolism.

23. The pH of blood is approximately
(*a*) 5 (*b*) 9
(*c*) 6.6 (*d*) 7.4.

24. The energy released during conversion of ATP to ADP is approximately
(*a*) 31 kJ/mole (*b*) 3.0 kJ/mole
(*c*) 28 kJ/mole (*d*) 2.8 kJ/mole.

25. The number of polypeptide chains present in a molecule of haemoglobin is
(*a*) 8 (*b*) 6
(*c*) 4 (*d*) 12.

26. Pyridoxine is the name given to vitamin
(*a*) B_2 (*b*) B_6
(*c*) B_1 (*d*) B_{12}.

27. The number of molecules of ATP produced per molecule of glucose in glycolysis is
(*a*) 38 (*b*) 36
(*c*) 18 (*d*) 2.

28. In myoglobin the number of polypeptide chains associated with one heme unit is
(*a*) Three (*b*) One
(*c*) Two (*d*) Four.

29. Which of the following is not a sugar ?
(*a*) Sucrose (*b*) Glucose
(*c*) Fructose (*d*) All are sugars.

30. During glycolysis one molecule of glucose breaks into
(*a*) Two molecules of acetic acid
(*b*) Three molecules of acetic acid
(*c*) Two molecules of pyruvic acid
(*d*) Three molecules of pyruvic acid.

31. Catabolism refers to
(*a*) Breakdown of complex molecules in living organisms.
(*b*) Synthesis of complex molecules.
(*c*) Both synthesis as well as breakdown of complex molecules.
(*d*) None of the above.

32. Photosynthesis invoves
(*a*) Conversion of chemical energy into radiant energy.
(*b*) Conversion of chemical energy into mechanical energy.
(*c*) Conversion of solar energy into chemical energy.
(*d*) Conversion of mechanical energy into chemical energy.

33. Biochemical reactions which require oxygen are called
(*a*) Metabolic (*b*) Aerobic
(*c*) Anaerobic (*d*) No specific name.

34. Hyperglycemia refers to
(*a*) High blood sugar level
(*b*) Low blood sugar level
(*c*) High conc. of salts in blood
(*d*) Low conc. of salts in blood.

35. Which one of the following vitamins checks night blindness ?
(*a*) A (*b*) B
(*c*) C (*d*) D.

36. Which one of the following is synthesized in our body by sun rays ?
(*a*) Vitamin D (*b*) Vitamin B
(*c*) Vitamin K (*d*) Vitamin A.

37. With which of the following diasease is vitamin K associated ?
(*a*) Rickets
(*b*) Coagulating property of blood
(*c*) Scurvy
(*d*) Sore throat.

38. ATP represents
(*a*) Abnormal temperature and pressure.
(*b*) Atmospheric temperature.
(*c*) Adenosine triphosphate.
(*d*) None of these.

39. The region of plants where photosynthesis occurs is called
(*a*) Mitochondria (*b*) Chloroplast
(*c*) Cytoplasm (*d*) None of the above.

40. In cellular respiration, there is a net gain of ...... per molecule of glucose
(*a*) 38 ATP molecules (*b*) 50 ATP molecules
(*c*) 36 ATP molecules (*d*) 10 ATP molecules.

41. Which name is associated with the chemical substances produced in endocrine glands ?
(*a*) Vitamins (*b*) Antigens
(*c*) Bile acids (*d*) Hormones.

42. During digestion which enzyme converts peptides into amino acids
(*a*) Peptidase (*b*) Invertase
(*c*) Lipase (*d*) Maltose.

43. Which chemical substance acts as emulsifiers in lipid metabolism ?
(*a*) Hydrochloric acid (*b*) Bile acids
(*c*) Nucleoside (*d*) Fatty acids.

44. *Ergocalciferol* is name of vitamin
(*a*) A (*b*) B
(*c*) C (*d*) D.

45. Insulin belongs to the category of
(*a*) Peptide hormones (*b*) Vitamins
(*c*) Amine hormone (*d*) Steroid hormone.

46. Peptide linkage is present in
(*a*) Proteins (*b*) Polypeptides
(*c*) Both (*d*) None.

47. Classification of proteins into fibrous and globular forms is on the basis of

(a) Primary structure
(b) Secondary structure
(c) Tertiary structure
(d) Quarternary structure.

48. The enzyme invertase converts
(a) Starch into glucose
(b) Starch into maltose
(c) Maltose into glucose
(d) Sucrose into glucoese and fructose.

49. Deficiency of enzyme tyrosinase causes the disease
(a) Phenyl keto urea (b) Albinism
(c) Anaemia (d) Haemoglobin

50. Which of the following is not a derivative of pyrimidine ?
(a) Thymine (b) Cytosine
(c) Uracil (d) Guanine.

ANSWERS

1. (b)	**2.** (c)	**3.** (d)	**4.** (d)	**5.** (b)	**6.** (c)	**7.** (b)	**8.** (a)	**9.** (d)
10. (b)	**11.** (a)	**12.** (c)	**13.** (a)	**14.** (b)	**15.** (d)	**16.** (b)	**17.** (a)	**18.** (b)
	20. (b)	**21.** (b)	**22.** (a)	**23.** (d)	**24.** (a)	**25.** (c)	**26.** (b)	**27.** (d)
28. (b)	**29.** (d)	**30.** (c)	**31.** (c)	**32.** (c)	**33.** (b)	**34.** (a)	**35.** (d)	**36.** (a)
37. (b)	**38.** (c)	**39.** (b)	**40.** (c)	**41.** (d)	**42.** (a)	**43.** (b)	**44.** (d)	**45.** (a)
46. (c)	**47.** (c)	**48.** (d)	**49.** (b)	**50.** (d).				

HINTS/SOLUTIONS

Q. No.	Ans.	Hints & Solutions
1.	**(b)**	α-D-Glucose and β-D-Glucose are **anomers** because they differ in configuration about C-1.
2.	**(c)**	Both glucose and fructose give silver mirror test.
3.	**(d)**	Cellulose.
4.	**(d)**	Gum and pectins are examples of polysaccharides.
5.	**(b)**	Structural isomers.
6.	**(c)**	In nucleic acids both phosphate and base are attached to sugar molecule. Hence, the sequence Phosphate-Sugar-Base is correct.
7.	**(b)**	Uracil is not present in DNA but is present in RNA.
8.	**(a)**	A sugar-base combination is known as nucleoside.
9.	**(d)**	Adenine and guanine are purine bases.
10.	**(b)**	Peptide linkage, —CO—NH—.
11.	**(a)**	Maltose is formed by condensation of two α-D-glucose units.
12.	**(c)**	Glycine, being an α-amino acid, can exist as zwitter ion.
13.	**(a)**	GCUAUG. Corresponding to each A (adenine) in DNA there is U (uracil) in *m*-RNA.
14.	**(b)**	Fructose (II) is sweetest sugar while maltose (IV) is the least sweet sugar among the given compounds.
15.	**(d)**	Stearic acid.
16.	**(b)**	Ribose.
17.	**(a)**	Glucaric acid. HNO_3 oxidizes —CHO and primary —OH to carboxylic acid.
18.	**(b)**	During formation of osazone three molecules of phenylhdrazine are used.
19.	**(b)**	Three hydrogen bonds.
20.	**(b)**	Lipids are digested (after emulsification with bile juices) in small intestines.
21.	**(b)**	Enzyme pepsin is produced in stomach.
22.	**(a)**	Glycolysis.
23.	**(d)**	The pH of blood is about 7.4.
24.	**(a)**	31 kJ/mole. Factual question.
25.	**(c)**	Haemoglobin has four polypeptide chains.
26.	**(b)**	Vitamin B_6 is known as pyridoxine.
27.	**(d)**	During glycolysis two molecules of ADP change into ATP.
28.	**(b)**	One polypeptide chain.

Q. No.	Ans.	Hints & Solutions
29.	(*d*)	All monosaccharides and disaccharides are sugars.
30.	(*c*)	Two molecules of pyruvic acid are obtained from each molecule of glucose during glycolysis.
31.	(*c*)	Metabolism and anabolism are collectively known as catabolism.
32.	(*c*)	During photosynthesis solar energy is converted into chemical energy.
33.	(*b*)	Aerobic.
34.	(*a*)	Hyperglycemia means high blood sugar level.
35.	(*d*)	Vitamin A.
36.	(*a*)	Vitamin D is synthesized by our body in the presence of sun-rays.
37.	(*b*)	Coagulating property of blood is affected by vitamin K.
38.	(*c*)	ATP is adenosine triphosphate.
39.	(*b*)	Chloroplast.
40.	(*c*)	During cellular respiration 36 molecules of ADP change to ATP.
41.	(*d*)	Hormones are produced by endocrine glands.
42.	(*a*)	Peptidase.
43.	(*b*)	Bile juices emulsify fats before absorption.
44.	(*d*)	Ergocalciferol is vitamin D.
45.	(*a*)	Insulin is a peptide hormone.
46.	(*c*)	Both proteins and polypeptides contain peptide bond because both are polymers of α-amino acids.
47.	(*c*)	Tertiary structure.
48.	(*d*)	Invertase converts sucrose into glucose and fructose.
49.	(*b*)	Albinism is caused by deficiency of enzyme tyrosinase.
50.	(*d*)	Guanine is a pyrimidine base.

10 POLYMERS AND CHEMISTRY IN EVERYDAY LIFE

10.1. CONCEPTS IN BRIEF

Develop Logic

POLYMERS

POLYMERS

A polymer is a compound of high molecular mass formed by the combination of large number of small molecules.

- The small molecules which constitute the repeating units are called **monomer units.**
- The process by which monomers are transformed into polymers is called **polymerisation.**

CLASSIFICATION OF POLYMERS

Polymers can be classified in number of ways depending upon the criterion chosen for classification :

1. **Classification of Polymers on the basis of Nature of Repeating Structural Units.**

Homopolymers	*Co-polymers*
Polymers formed from one kind of monomers.	Polymer formed from more than one kind of monomer units.

2. Classification Based on Origin or Source

Natural Polymers	*Synthetic Polymers*
Polymers obtained from natural sources such as plant and animal kingdom. ***Examples :*** *Polysaccharides, proteins, nucleic acids, natural rubber* etc.	Polymers prepared in the laboratory by chemical reactions. ***Examples :*** *Polyethene, PVC, orlon bakelite, melmac* etc.

3. Classification Based on Structure

Linear Polymers	*Branched Chain Polymers*	*Cross-Linked Polymers*
• In these polymers, the monomer units are linked together to form *long straight chains*. • They have high tensile strength, high densities and high melting points. ***Examples :*** *Polythene, nylons, polyesters* etc.	• In these polymers, the monomer units link to form long chains (called main stem). At the same time, there are side chains of different lengths constituting branches. • They have low densities, low tensile strength and low melting points. ***Examples :*** *Low density polythene, amylopectin* etc.	• In these polymers, the monomer units constitute three dimensional network. The links involved are called cross links. • They are very hard and brittle. ***Examples*** : *Bakelite, melmac*, etc.

4. Classification Based on Mode of Polymerisation

Addition Polymers	*Condensation Polymers*
• These polymers are formed as a result of addition reaction between monomer units. • The monomer unit are unsaturated compounds containing C = C or C ≡ C bonds. • Molecular mass of polymer is whole number multiple of that of monomer. ***Examples :*** *Polythene, PVC, Orlon, Polypropylene,* etc.	• These polymers are formed as a result of condensation reaction between the starting compounds. • The monomer must contain two reactive groups and react by eliminating simple molecule such as H_2O, ROH, NH_3 etc. • Molecular mass of polymer is not the whole number multiple of molecular mass of monomer. ***Examples*** : *Nylon-66, Bakelite, Glyptal etc.*

A more systematic classification of polymers based on mode of synthesis is as follows :

Chain Growth Polymers	*Step Growth Polymers*
• Here, the monomers add on successively to the growing chain. • Polymerisation occurs through chain reaction in the presence of suitable initiators.	• Here, the monomers generally condense to form dimers, trimers and so on step wise. • Polymerisation generally occurs through condensation reaction.

5. Classification based on Molecular Forces

Elastomers	*Fibres*	*Thermoplastics*	*Thermosetting Polymers*
• Polymers in which the polymer chains are held by *weakest attractive forces.* • Amorphous polymers having high degree of elasticity. • They can be stretched to about 10 times their normal length but return to their original position once streching forces are removed. • *Natural rubber* is a common example.	• Polymers having *strong interparticle forces.* • They are thread like polymers and can be woven into fabric. • Common examples are *Nylon, Dacron Silk* etc.	• Polymers in which interparticle forces are in between *Elastomers* and *fibres.* • They can be moulded into desired shapes by heating and subsequent cooling. • There are no cross linking between polymeric chains. • They soften on heating and harden on cooling. • Examples are *polystyrene, pvc, polythene,* etc.	• Polymers having excessive cross linking in three dimensions. • They are made from semi-fluid substances by heating. Heating introduces cross-links between the chains producing three dimensional net-work of bonds. Thus, on cooling they become *hard* and donot melt on reheating. • *Bakelite* is a common example.

TYPES OF POLYMERISATION REACTIONS

There are two broad type of polymerisation reactions, *e.g.,* the addition or chain growth polymerization and condensation or step growth polymesisation.

• Addition polymerization or step growth polymerization

In this type of polymerization, the molecules of the same monomer or different monomers add together on a large scale to form a polymer. The monomers used are unsaturated compounds, *e.g.,* alkenes, alkadienes and their derivatives. This mode of polymerization leading to an increase in chain length or chain growth can take place through the formation of either free radicals or ionic species. However, the free radical governed addition or chain growth polymerization is the most common mode.

Free radical mechanism

A variety of alkenes or dienes and their derivatives are polymerized in the presence of a free radical generating initiator (catalyst) like benzoyl peroxide, acetyl peroxide, tert-butyl peroxide, etc. **For example,** the polymerization of ethene to polythene consists of heating or exposing to light a mixture of ethene with a small amount of benzoyl peroxide initiator. The process starts with the addition of phenyl free radical formed by the peroxide to the ethene double bond thus generating a new and larger free radical. This step is called chain initiating step. As this radical reacts with another molecule of ethene, another bigger sized radical is formed. The repetition of this sequence with new and bigger radicals carries the reaction forward and the step is termed as chain propagating step. Ultimately, at some stage the product radical thus formed reacts with another radical to form the polymerized product. This step is called the chain terminating step. The sequence of steps may be depicted as follows :

Chain initiation steps

$$C_6H_5-\overset{O}{\overset{\|}{C}}-O-O-\overset{O}{\overset{\|}{C}}-C_6H_5 \longrightarrow 2C_6H_5-\overset{O}{\overset{\|}{C}}-\dot{O} \longrightarrow 2\dot{C}_6H_5$$

Benzoyl peroxide

$$\dot{C}_6H_5 + CH_2 = CH_2 \longrightarrow C_6H_5-CH_2-\dot{C}H_2$$

Chain propagating step

$$C_6H_5-CH_2-\dot{C}H_2 + CH_2 = CH_2 \longrightarrow C_6H_5-CH_2-CH_2-CH_2-\dot{C}H_2$$

$$\downarrow$$

$$C_6H_5 \leftarrow\!\!\!\!\left(CH_2-CH_2 \right)_n CH_2-\dot{C}H_2$$

Chain terminating step

For termination of the long chain, these free radicals can combine in different ways to form polythene. One mode of termination of chain is shown as under :

$$C_6H_5\text{-}(CH_2\text{—}CH_2)_n\ CH_2\text{—}\dot{C}H_2 + C_6H_5\text{-}(CH_2\text{—}\overset{+}{C}H_2)_n\ CH_2\text{—}\dot{C}H_2 \longrightarrow \underset{\text{Polythene}}{C_6H_5\text{-}(CH_2\text{—}CH_2)_n\ CH_2\text{—}CH_2\text{—}CH_2\text{-}(CH_2\text{—}CH_2)_n\ C_6H_5}$$

- **Condensation polymerization or step growth polymerization**

This type of polymerization generally involves a repetitive condensation reaction between two bi-functional monomers. These polycondensation reactions may result in the loss of some simple molecules as water, alcohol, etc., and lead to the formation of high molecular mass condensation polymers.

In these reactions, the product of each step is again a bi-functional species and the sequence of condensation goes on. Since, each step produce a distinct functionalized species and is independent of each other, this process is also called as step growth polymerization. The formation of terylene or dacron by the interaction of ethylene glycol and terephthalic acid is an example of this type of polymerization.

MOLECULAR MASS OF POLYMERS

During the formation of polymers, the growth of the polymer chains depends upon the number of monomers available in its vicinity. As the number of available monomers differ from one place to another place in the reaction vessel, so, a polymer may contain chain of varying lengths and, therefore, its molecular mass is always taken as an average. This average can be taken either in terms of *number* or *weight*. In former case it is called *number average molecular mass* ($\overline{M}_n$) and in later case it is called *weight average molecular mass* ($\overline{M}_w$). Mathematically these are defined by the following equations :

$$\overline{M}_n = \frac{\Sigma N_x M_x}{\Sigma N_x}$$

$$\overline{M}_w = \frac{\Sigma N_x M_x^2}{\Sigma N_x M_x}$$

where N_x is the number of molecules of molecular mass M_x.

- The ratio of these types of molecular masses *i.e.,* $\frac{\overline{M}_w}{\overline{M}_n}$ is called **'Poly dispersity index'** (PDI).
- In natural polymers, the value of PDI is unity. But in synthetic polymers, the value of PDI is greater than unity.
- PDI is used to determine the homogeneity of a polymer.

BIODEGRADABLE POLYMERS

A large number of polymers are quite resistant to the environmental degradation processes and are thus responsible for the accumulation of polymeric solid waste materials. These solid wastes cause acute environmental problems and remain undegraded for quite a long time. In view of the general awareness and concern for the problems created by the polymeric solid wastes, certain new biodegradable synthetic polymers have been designed and developed. These polymers contain functional groups similar to the functional groups present in biopolymers.

Aliphatic polyesters are one of the important classes of biodegradable polymers. Some important examples are given below :

1. *Poly b-hydroxybutyrate-co-β-hydroxy valerate (PHBV)*

It is obtained by the copolymerisation of 3-hydroxybutanoic acid and 3-hydroxyentanoic acid. PHBV is used in speciality packaging, orthopaedic devices and in controlled release of drugs. PHBV undergoes bacterial degradation in the environment.

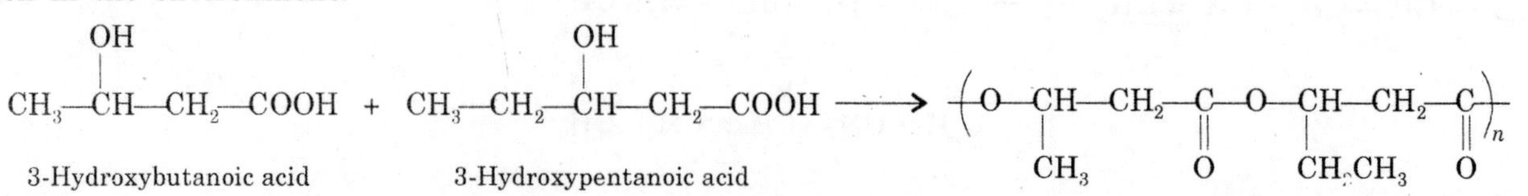

3-Hydroxybutanoic acid 3-Hydroxypentanoic acid PHBV

SOME COMMERCIAL POLYMERS

SOME ADDITION POLYMERS

Name of polymer	*Starting materials*	*Nature of polymer*	*Properties*	*Uses*
POLYOLEFINS				
• **Polyethylene** or **Polyethene**	$CH_2 = CH_2$ (Ethene)	• *Low density polythene* homopolymer (branched)	Transparent, moderate tensile strength, high toughness.	Packing material bags, insulation for electrical wires and cables.
		• *High density polythene* homopolymer (linear)	Transluscent, chemically inert, greater tensile strength, touchness.	Buckets, tubes, house ware pipes, bottles and toys.
• **Polypropylene** or **Polypropene**	$CH_3CH = CH_2$ (Propylene)	Homopolymer (linear)	Harder and stronger than polyethene.	Packing of textiles and foods, liners for bags, heat shrinkage wraps, carpet fibres, ropes, automobile mouldings, stronger pipes and bottles.
• **Polystyrene** or **Styron**	$C_6H_5CH = CH_2$ (Styrene)	Homopolymer, linear	Transparent.	Plastic toys, house hold wares, radio and television bodies, refrigerator linkings.
POLYDIENES				
• **Neoprene**	$CH_2—CH—C(Cl) = CH_2$ Chloroprene	Homopolymer	Rubber like, inferior to natural rubber and a superior resistant to aerial oxidation, and oils, gasoline etc.	Hoses, shoe heels, stoppers.
• **Buna-S-rubber** or **SBR**	$CH_2 = CH—CH = CH_2$ 1, 3-Butadiene and $C_6H_5CH = CH_2$ Styrene	Copolymer	Rubber like, inferior to natural rubber and a superior resistant to aerial oxidation, and oils, gasoline etc.	Manufacture of tyres, rubber sols, water proof shoes.

POLYACRYLATES				
Polymethylmethacrylate (*Plexiglas Lucite, Acrylite and Perspex*) **PMMA**	$CH_2=C(CH_3)—COOCH_3$	Copolymer	Hard, transparent, excellent light transmission. Optical clarity better than glass, takes up colours.	Lenses, light covers, light shades, signboards, transparent domes, skylights, air craft windows, dentures and plastic jewellery.
Polyethylacrylate	$CH_2=CH—COOC_2H_5$	Copolymer	Tough, rubber like product.	
Polyacrylonitrile (PAN)	$CH_2=CH—CN$	Copolymer	Hard, horny and high melting materials.	*Orlon, acrilon* used for making clothes, carpets blankets and preparation of other polymers.
POLYHALOOLEFINS				
• **Polyvinyl chloride** (PVC)	$CH_2=CH—Cl$ Vinyl chloride	Homopolymer	Pliable(easily moulded).	(*i*) Plasticised with polyester polymers used in rain coats, hand bags, shower curtains, fabrics, shoe soles, vinyl flooring (*ii*) Good electrical insulator, (*iii*) Hose pipes.
• **Polytetrafluoroethylene or Teflon (PCTFE)**	$F_2C=CF_2$ Tetrafluoroethylene	Homopolymer	Flexible and inert to solvents, boiling acids, even aqua regia. Stable upto 598 K.	(*i*) For nonstick utensils coating (*ii*) Making gaskets, pump packings, valves, seals, nonlubricated bearings.
• **Polymonochlorotrifluoro ethylene (PCTFE)**	$ClFC=CF_2$ Chlorotrifluoroethylene		Less resistant to heat and chemicals due to presence of chlorine atoms.	Similar to those of teflon.

SOME CONDENSATION POLYMERS

Name of polymer	*Starting materials*	*Nature of polymer*	*Properties*	*Uses*
POLYESTERS				
• **Terylene or Dacron**	$HO—CH_2—CH_2—OH$ Ethylene glycol or Ethane-1, 2-diol **and** 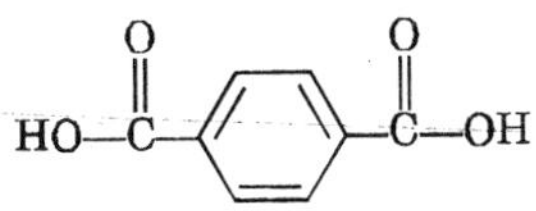 Terephthalic acid or Benzene-1, 4-dicarboxylicacid	Copolymer, (linear)	Fibre crease resistant, low moisture content, not damaged by pests like moths.	For wash and wear fabrics, type cords, sea belts and sails.
• **Glyptal or Alkyd resin**	$HO–CH_2–CH_2–OH$ Ethylene glycol **and** 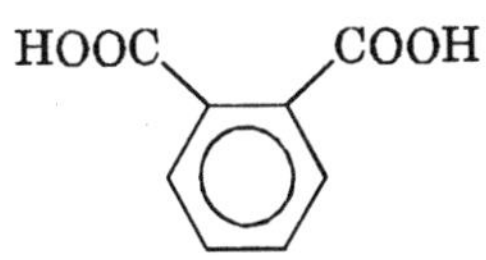 Phthalic acid or Benzene-1, 2-dicarboxylic acid	Copolymer, linear	Thermoplastic, dissolves in suitable solvents and solutions, on evaporation leaves a tough but not flexible film.	Paints and lacquers.
POLYAMIDES				
• **Nylon 66**	$HO—\overset{O}{\overset{\|\|}{C}}(CH_2)_4\overset{O}{\overset{\|\|}{C}}—OH$ Adipic acid and $H_2N—(CH_2)_6—NH_2$ Hexamethylenediamine	Copolymer, linear		
• **Nylon 6-10**	$H_2N(CH_2)_6—NH_2$ Hexamethylene diamine and $HOOC(CH_2)_8COOH$ Sebacic acid	Copolymer, linear	High tensile strength, abrasion resistant, some what elastic.	(*i*) Textile fabrics, carpets, bristles for brushes. (*ii*) Substitute of metals in bearings (*iii*) Gears, elastic hosiery.
• **Nylon 6 or Perlon**	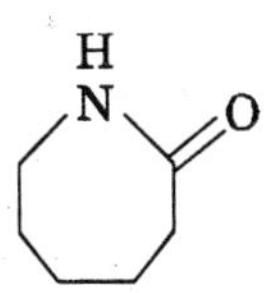 Caprolactum	Homopolymer, linear		Mountainering ropes, tyre cords, fabrics.

FORMALDEHYDE RESINS

• **Phenol fomaldehyde resin** or **Bakelite**	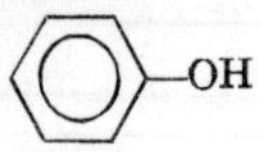 and HCHO	Copolymer		(*i*) With low degree polymerisation-for binding glue, wood, varnishes, lacquers. (*ii*) With high degree polymerisation-for combs, for mica table tops, fountain pen barrels, electrical goods (switches and plugs), gramophone records.
• **Melamine formaldehyde resin** or **Melmac**	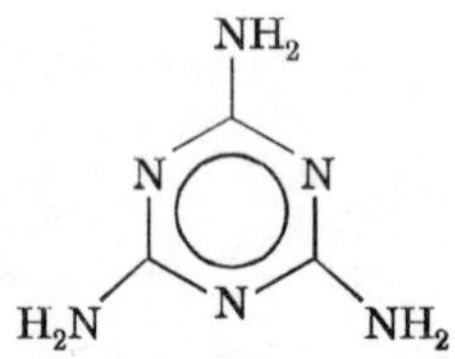 and HCHO	Copolymer		Non-breakable and non-plastic crockery.

SOME IMPORTANT POLYMERS

- **Saran.** It is a copolymer of *vinyl chloride* [$CH_2 = CHCl$] and *vinyledene chloride* ($CH_2 = CCl_2$).
- **Vinyon.** It is copolymer of *vinyl chloride* ($CH_2 = CHCl$) (90%) and vinyl acetate ($CH_2 = CHOCOCH_3$) (10%).
- **Dynol.** It is copolymer of *vinyl chloride* ($CH_2 = CHCl$) 60% and *acrylonitrile* ($CH_2 = CHCN$).
- **Buna-N rubber.** It is copolymer of *butadiene* ($CH_2 = CH—CH = CH_2$) and *acrylonitrile* ($CH_2 = CH—CH$).
- **Butyl rubber.** It is a copolymer of *isobutylene* [$(CH_3)_2\ C = CH_2$] and *isoprene.*
- **Gutta-percha.** It is a homopolymer and a hard horny material consisting of *trans polyisoprene.* Natural rubber ; on the other hand, consist of *cis polyisoprene.*
- **Super glue.** It is a polymer of *methyl, α-cyanoacrylate* ($CH_2 = \underset{\displaystyle CN}{\underset{|}{C}} - COOCH_3$). It has very high tensile strength. One drop of it can support a weight of 2000 lbs.
- **Lexan.** It is a polycarbonate and is condensation polymer of *diethylcarbonate* and *bisphenol.* Due to its unsually high impact strength, it is used in crash helmets.
- **Polyurethanes.** Condensation polymers of *toluene 2, 4-di isocyanate* and *ethylene glycol.*
- **Kevlar.** It is polyamide and condensation polymer of *terepthalic acid* and *p-phenylene diamine.* It is used in bullet proof jackets.
- **Nomex.** It is a polyamide and is a condensation polymer of *m-diaminobenzene* and *m-phthalic acid.* It is fire resistant substance and is used in *protective clothing for fire fighters.*
- **Plasticizers.** There are the substances such as *di-n-butylnepthalates* or *cresyl phosphates* which when added to plastics make then soft and readly workable on heating during moulding process.
- **Vulcanisation.** It is a process of heating of rubber with 3-5% sulphur. It introduces cross links of sulphur atoms between the polymeric chains thereby increasing the *elasticity, tensile strength* etc.
- **Resins.** These are amorphous organic solids or semi-solids which generally have typical lustre and are often transparent or transluscent.

CHEMISTRY IN EVERYDAY LIFE

THERAPEUTIC ACTION OF DIFFERENT CLASSES OF DRUGS

The chemical substances which are used for curing the diseases and for reducing the sufferings from pain are called **medicines.** The *branch of science that deals with the treatment of diseases* by using suitable chemicals is called **Chemotherapy.**

- **Antipyretics.** These are chemicals which are used to *bring down the body temperature in high fever,* but their use often leads to perspiration. Aspirin is a common example of antipyretic but should not be taken empty-stomach as it is hydrolysed to produce salicylic acid which may cause ulcer. Some other antipyretics are *Phenacetin* and *Paracetamol.*

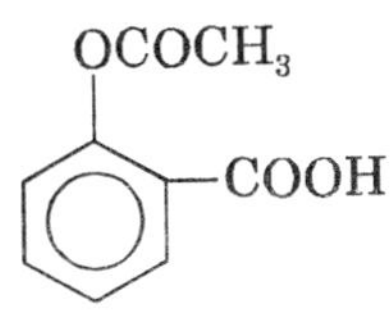

Aspirin

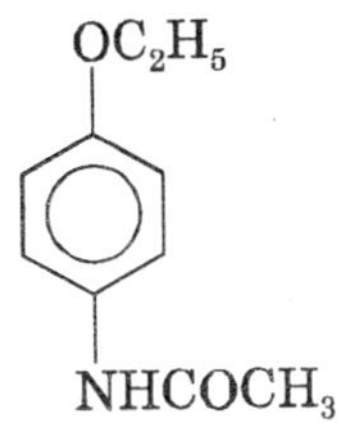

Phenacetin

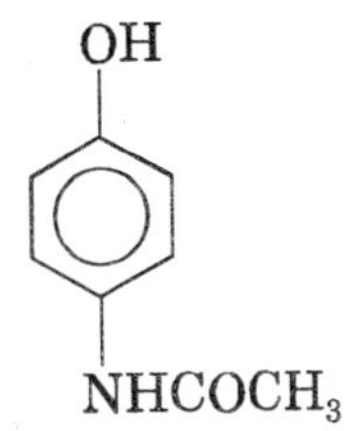

Paracetamol

- **Analgesics.** These are chemicals which are used for *relieving pain. Aspirin* and certain other antipyretics also works as analgesics. Some narcotics which produce sleep and unconsciousness) like *morphine, marijiuane, codeine* and *heroin* also act as analgesics. They are known to be habit-forming and hence their frequent use must be avoided as far as possible.
- **Antiseptics.** These are chemicals which are used to *kill or prevent the growth of micro-organisms.* These are not harmful to living tissues and can be applied to wounds, cuts or diseased skin surfaces. The antiseptics are incorporated into face powders, breath purifiers etc., to reduce *odours* resulting from bacterial decomposition on the body or in the mouth. **Dettol** (a mixture of *chloro-xylenol* and *terpeneol* in a suitable solent) and **Bithional** (added to soap) are common examples.
  (*i*) *Iodine* is a powerful antiseptic. It is used as *'tincture of iodine'* (2-3% I_2 solution in alcohol-water mixture) and iodoform, CHI_3 (it liberates iodine in coming in contact with organic matter).
  (*ii*) **Organic dyes** like *Gentian violet* and *Methylene blue* are also used as effective antiseptics.
- **Disinfectants.** These are chemicals which are used to *kill micro-organisms.* These are harmful to living tissues and hence should not be applied. These are applied only to inanimate objects such as floor, instruments, etc. The same chemical can be used as an antiseptic as well as a disinfectant by varying the concentration of the solution used. For example 2% solution of phenol is used as an antiseptic and its 1% solution as a disintfectant.
- **Transquilizers.** These are chemicals which are used for treatment of *mental diseases.* These are also called **psychotherapeutic** drugs. These are generally made the constituents of sleeping pills and act on higher centres of central nervous system. The administration of tranquilizers makes the patient to sleep and helps to control the emotional distress. **Barbituric acid** and its **derivatives** are used as tranquilizers since a long time. **Equanil** is another tranquilizer which is used in depression and hypertension.
- **Antibiotics.** These are chemical substances produced by micro-organisms (*bacteria, fungi and molds*) which prevent the growth or even destroy other micro-organisms.

  The first antibiotic, *penicillin,* was discovered by Alexander Fleming in 1929 from the mold *Pencillium notatun.* It is very much effective drug in case of pneumonia, bronchitis, sore throat and abscesses. Penicillin is used against diseases caused by gram + ve bacteria. The antibiotics like *streptomycin* (produced by streptomycetes) and tetracycline are used against diseases caused by gram – ve bacteria (*e.g.,* infection of the urinary tract). These may also be used against diseases caused by gram + ve bateria, which are penicillin-resistant.

  (*i*) **Streptomycin** is used for tuberculosis and **chloroamphenicol** for typhoid. Antibiotics which are effective against several different types of harmful micro-organisms are called **broad spectrum antibiotics,** *e.g.,* tetracycline, chloramphenicol, etc.
  (*ii*) Ampicillin and amoxicillin are semi-synthetic modifications of penicillin.

- **Sulpha drugs** also act against micro-organisms just like antibiotics. Some examples are : Sulphanilamide, sulphadiazine and sulphaguanidine, etc.

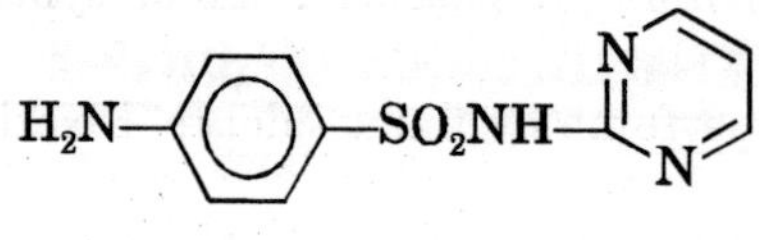

Sulphadiazine

In addition to above mentioned chemicals, there are a large number of other chemicals which are used as drugs.

- **Antihistamines** Antihistamines are amines which are used as drugs to control the allergy effects produced by histamines. The release of histamine induces allergic response in the body like tissue inflammation, asthma, itching, skin irritation, sneezing, nasal discharge, itching of eyes, etc. The drugs used to prevent the production of histamine and combat the allergy effects are known as *antihistamines.* Common drugs of this group are chlorpheniramine, promethazine, diphenylhydramine.

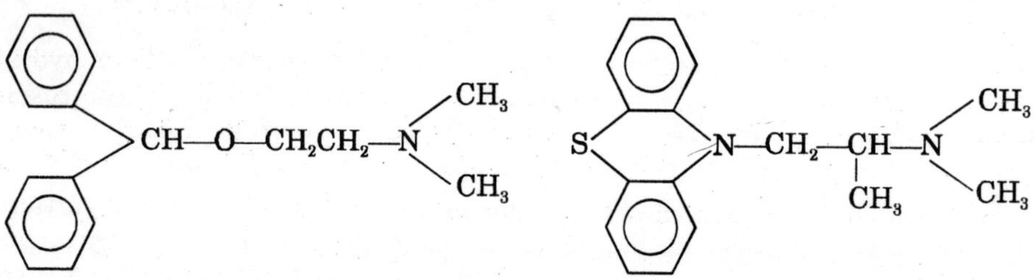

- **Antacids.** One of the most common ailments associated with digestion is ***acid gastritis.*** It is caused by excess of hydrochloric acid in the gastric juice. Substances which remove the excess of acid in the stomach are called **antacids.**

  Some common antacids are calcium carbonate, sodium bicarbonate, magnesium hydroxide and aluminium hydroxide. These are used in the form of tablets or aqueous suspensions. **Omeprazole** and **lansoprazole** are also used as antacids.

CHEMICALS IN COSMETICS

- **Creams.** Creams may be classified as cleansing creams, cold creams, vanishing creams, bleach creams and sunburn creams.
- **Cleansing creams and lotions.** A cleansing cream or lotion is used for the removal of facial make-up, surface grime and oil from the face. Cleansing creams are generally emulsions containing bees-wax, borax, mineral oil and water. Cleansing lotions are emulsions of some detergent in water along with some emulsifying agent. For example, an emulsion of triethanolamine lauryl sulphate in water is stabilized by triethanolamine stearate. Similarly, glyceryl monostearate, fatty alcohols, stearic acid and bees-wax are added as auxiliary emulsiying agents to thicken and stabilize the emulsion.
- **Cold creams.** Cold creams are applied to lubricate the skin and to prevent roughness and chapping. Cold cream is an emulsion in which oil component predominates. The oils present in the cold cream which may be almond oil, linolin oil, paraffin oil and bees wax. In addition to oils small amounts of perfume oil and borax is also added.
- **Vanishing creams.** Vanishing cream is applied to give a thin film on the skin, which is not greasy or conspicuous. The four common ingredients of a vanishing cream are : (*i*) stearic acid (*ii*) water (*iii*) alkali (*iv*) glycerine. Additional ingredients such as perfumes, preservatives and auxiliary emulsifying agents such as glyceryl monostearate are also added.
- **Bleach creams.** Blean creams are used to exert a lightening or bleaching effect on the dark skin. One or more of the following agents may be used in the preparation of bleach creams : Titanium oxide, zinc oxide, hydrogen peroxide sodium hypochloride, hydroquinone, ammoniated mercury etc.
- **Sunburn creams.** These creams save the skin from sunburn in summer.
- **Perfumes.** Perfumes are substances having pleasant odours. The essential characteristics of a good perfume are : harmonious odour, lasting fragrance, stability, volatility and ability to get fixed in the required cosmetic. Perfumes are generally obtained from the essential oils of flowers, leaves, fruits, roots or woods of a wide variety of plants. Perfumes are also obtained from animal sources (musk, ambergris, etc). Some perfumes are also made syn-

thetically. Chemically, a perfume is composed of three components : (*i*) **The odoriferous constituents** (*ii*) **Fixatives** and (*iii*) **Vehicle or Solvent.**

The **odoriferous constituents** of perfume are essential oils or synthetic substances or both. They impart the characteristic odour to the perfume.

The **fixatives** impart stability to the perfume by fixing odours. Fixatives commonly used as civet, musk, castor, vanillin etc.

A **solvent** is used for dilution of the odoriferous substance. Many terpenoids such as citronellon, nerol, geraniol, linalool etc. are the components of essential oils and are extensively used in perfumes and cosmetic preparations.

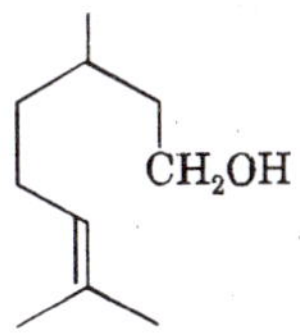

Citronellon

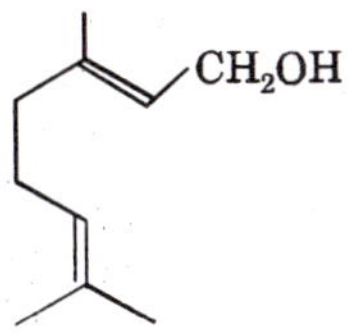

Geraniol

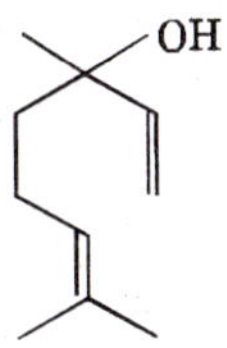

Linalool

- **Talcum powder.** Talcum powders, baby powders and face powders contains talc or hydrated magnesium silicate ($Mg_3(OH)_2Si_4O_{10}$ or $3MgO.4SiO_2.H_2O$) as the most important ingredient. These cosmetic powders absorb perspiration and superficial skin oils. They are suitably tinted and perfumed so that they make the skin look good, smell good and feel good. The common ingredients of talcum powder are :

  For absorbency — Calcium carbonate (chalk), magnesium carbonate.

  For opacity — Titanium dioxide, zinc oxide, magnesium oxide.

  For slipping — Talc, metallic soaps.

  For adherence — Magnesium and zinc stearate.

- **Deodorants.** Deodarants are preparations which are used to decrease or eliminate body odours due to perspiration. Body odour is generally the result of bacterial action on perspiration. Many substances which have astringent properties act as antiperspirants. For example, salts of aluminium, iron and zinc inhibit the flow or perspiration by their astringent action.

  Phenolic antibacterials such as parachlorometaxylenol and dichlorometaxylenol have been found to effective body deodorants.

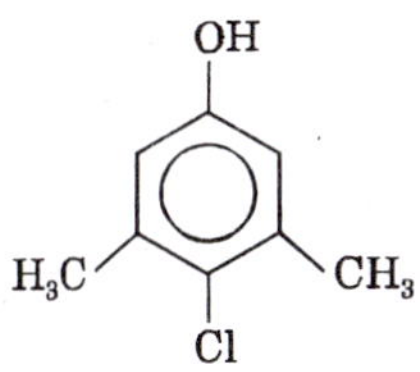

Parachlorometaxylenol

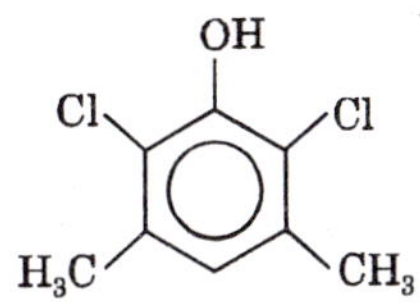

Dichlorometaxylenol

CHEMICALS IN FOOD

During processing of food, a number of chemicals are added to it to increase it shelf life and also to make it more attractive. Some of these chemicals which are present in food are discussed below :

1. Chemical Preservatives. Growth of micro-organisms in a food material can be inhibited by adding certain chemical substances. Such chemical substances which are added to food materials to prevent their spoilage are known as **chemical preservatives.** In our country, two chemical preservatives which are permitted for use are :

1. *Benzoic acid* (or sodium benzoate) 2. *Sulphur dioxide* (or potassium meta-bisulphite).

Benzoic acid or its sodium salt, ***sodium benzoate*** is commonly used for the preservation of fruits, fruit juices, squashes and jams.

Potassium meta bisulphite or ***Sodium meta-sulphite*** is used for the preservation of colourless food materials such a fruit juices, squashes, apples, lichies and raw mango *chutney.*

2. Edible Colours. There are three main classes of colours in foods. These are :

(*i*) Natural colours

(*ii*) Browning colours (produced during cooking)

(*iii*) Additives.

The principal natural colours are the carotenoids which give yellow to red colours, chlorophyll which gives green colour and the flavonoids (anthocyanins) which give red to blue colours. **Anthocyanins** are derived from sources such as red grapes, beet, sweet potato, cabbage etc.

3. Antioxidants. Antioxidants are added to the food to retard the action of oxygen on the food. These chemicals act as sacrificial materials because they are more easily attacked by oxygen than the food.

In order to prevent rancidity antioxidants are added to oils and fats. **Butylated hydroxy anisole (BHA)** is a widely used antioxidant used to preserve edible oils, fats, butter etc. Vitamin E is a natural antioxidant.

Another antioxidant which is commonly used is **butylated hydroxy toluene (BHT)**

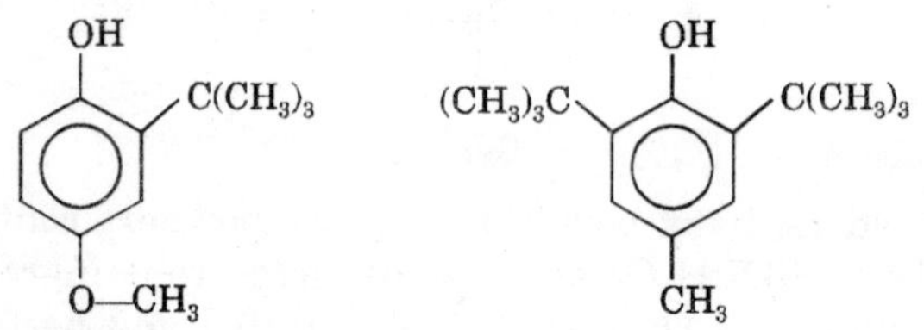

BHA BHT

4. Artificial Sweetening Agents. Sugar is the natural sweetening agent. However, excess of consumption of sugar leads to many diseases such as obesity, diabetes, coronary heart disease. Many artificial sweetening agents have been isolated which are much more sweeter than sugar. These artificial sweetening agents are non-nutritive in nature and are used as substitutes for sugar in foods and beverages especially soft drinks. Some examples of artificial sweetening agents are saccharin, cyclamates. Saccharin is about 300 times sweeter than canesugar or sucrose.

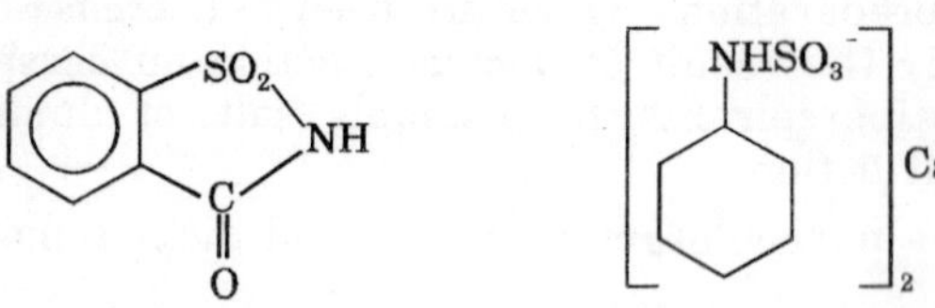

Saccharin Calcium cyclamate

The use of cyclamates as sweetening agent has been banned in many countries in view of suspected carcinogenic effects.

Aspartame is another artificial sweetener. It is methyl ester of the dipeptide aspartyl phenylalanine. It is about 180 times sweeter than sucrose.

PHEROMONES—INSECT SEX ATTRACTANTS

Animals, 'especially insects' communicate with other members of their species in a very interesting way. They release certain chemicals called pheromones which have characteristic odours. These odours are received by other members of their species and message is conveyed.

Pheromones are thus compounds produced by organisms for the purpose of communicating with the other members of the same species ; to attract members of the opposite sex ; to spread an alarm ; to mark the trail to food ; or to send the message to congregate.

For example, the female gypsy moth attracts the male by emitting the *pheromone disparlure.* On detecting the presence of disparlure, the male follows the scent to its origin and mates with the female.

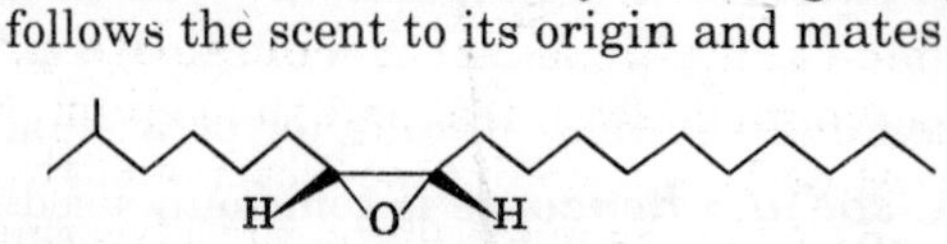

Disparlure

The pheromone *muscalure* is the **sex pheromone** of common housefly.

Bombykol is the sex pheromone of natural silk worm.

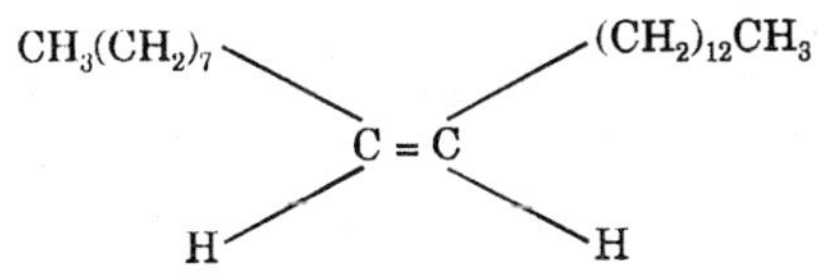

Muscalure
(Sex pheromone of housefly)

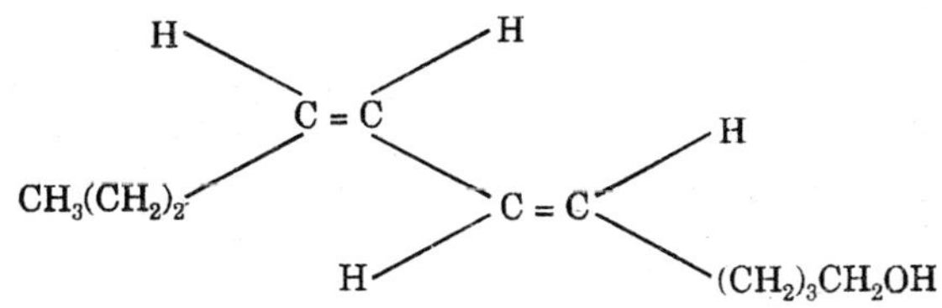

Bombykol
(Sex pheromone of natural silk worm)

2-Heptanone is a component of **alarm pheromone** of bees.

A species of cockroach uses undecane as an **aggregation pheromone**.

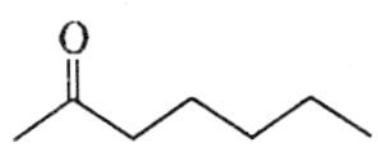

2-Heptanone
(Alarm pheromone of bees)

$CH_3(CH_2)_9CH_3$

Undecane
(Aggregation pheromone of cockroach)

Many sex attractants have been synthesized and are used to attract the insects into traps, as a means of insect control.

- **Insect repellents.** Contrary to insect attractants there are certain substances which act as insect repellents. Insect repellents keep insects away from the host which may be human being or some pet animal. There are some repellents available in the market. Dimethyl phthalate is a good mosquito repellent. N, N-Diethyl-meta-toluamide (*Deet*) is another insect repellent which is active against flies, mosquitoes and many other insects. N, N-Diethyl benzamide is the active component of many mosquito repellent creams.

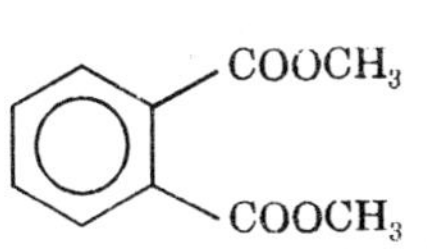

Dimethyl phthalate

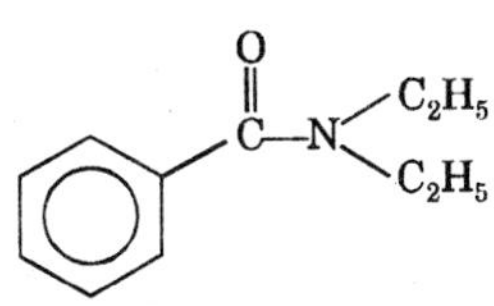

N, N-Diethyl benzamide

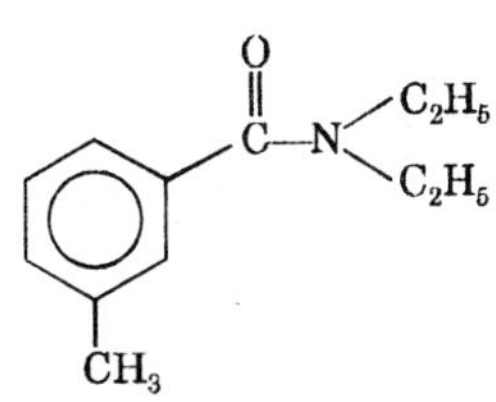

N, N-Diethyl-meta-toluamide
(Deet)

CHEMISTRY OF ROCKET PROPELLANTS

Rocket motors used in space vehicles has propulsion system, which consists of rocket engines powered by chemicals called **propellants. A chemical propellant** is a *combination of an* **oxidizing agent** and a **fuel** which when ignited undrgoes combustion to release large amount of hot gases. When these gases pass through the nozzle of rocket motor, it provides the necessary thrust for the rocket to move in forward direction (*Newton's 3rd law of motion*).

Classification of Propellants

The constituents of a propellant may be in a solid, a liquid or a gaseous state. Thus, depending upon the physical state, the propellants are classified into the following three types :

- **Solid propellants.** The most common solid propellant is **composite propellant** which is a mixture of polymeric binder like polyurethane or polybutadiene as fuel and ammonium perchlorate as an oxidizing agent along with an additive like finely divided Al or Mg to modify the working of the propellant. Another widely used solid propellant is **double-base propellant** which is made up to nitroglycerine and nitrocellulose. Solid propellants, once ignited, will burn with a pre-determined rate and their ignition cannot be stopped.
- **Liquid propellants.** These consists of a mixture of kerosene oil (alcohol hydrazine or liquid H_2) as fuel and liquid O_2 (liquid N_2O_4 or nitric acid) as an oxidizing agent. These **bi-liquid propellants** give higher thrusts

than solid propellants and their ignition and hence thrust can be controlled by switching on and off the flow of the liquid propellant.

In some cases, a single chemical compound, known as **monopropellants** act both the oxidizer and the fuel. Some examples are : methyl nitrate, nitromethane and hydrogen peroxide.

- **Hybrid propellant.** These propellants consists of a solid fuel like acrylic rubber and a liquid oxidizer like liquid N_2O_4.

Propellants used in different Space Vehicles

- **Saturn booster rocket of ASP**
  Initial stage ; Kerosene oil + Liquid O_2
  Upper stage : Liquid H_2 + Liquid O_2.
- **PROTON (Russian rocket).** Kerosene oil + Liquid O_2
- **SLV-3 and ASLV (Indian rockets).** Composite solid propellants
- **PSLV (Indian rocket)**
  1st stage : Solid propellants
  2nd stage : Unsymmetrical dimethyl hydrazine (UDMH) + Liquid N_2O_4
  3rd stage : Solid propellants
  4th stage : *Monomethyl hydrazine* (MMH) + Liquid N_2O_4
- **Space shuttles.** Liquid H_2 + Liquid O_2

Relation between the Specific Impulse and Critical Temperature

The relation bewteen the specific impulse of a propellant and the critical temperature attained in a rocket blast is as follows :

$$\textbf{Specific impulse, } I_s = \sqrt{\frac{T_c}{M}}$$

where T_c is the *critical temperature* inside the rocket motor and $\overline{M}$ is the *average molecular mass* of the product gases coming out of the rocket nozzle. The energy of a propellant is measured in terms of its **specific impulse.** Higher the value of I_s, the better is the propellant from energy point of view.

DYES

Dyes are coloured substances (generally organic) which are used for imparting permanent colour to fabrics, foods and other substances for their beautification and distinaction.

Characteristics of a dye

- It must have suitable colour.
- It must stick to the fabric by means of stable bonds.
- It should be resistant to the action of *light water, soap* and *alkalies*, etc.

CLASSIFICATION OF DYES BASED ON THEIR CONSTITUTION

The classification of dyes based on their constitution is given in the table given below. The characteristic structural unit present in dyes is also shown in the table.

Type	*Structural Unit*	*Examples with structures*
• **Nitro dyes**	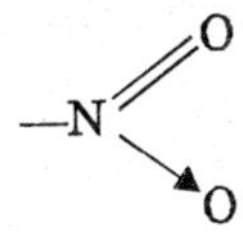 (*Nitro group*)	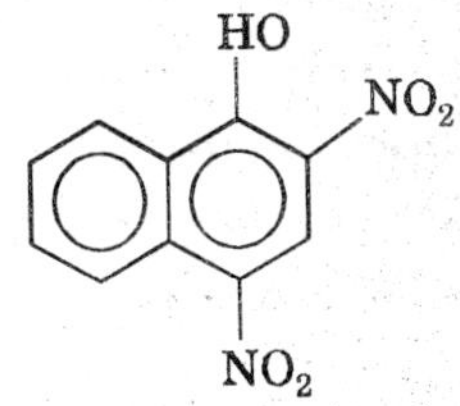 *Martius yellow*
• **Azo dyes**	—N = N— (*Azo group*)	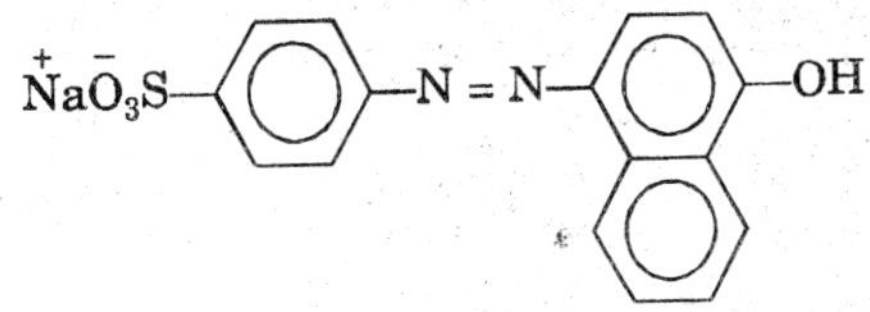 *Orange I*
• **Triphenyl-methane dyes**	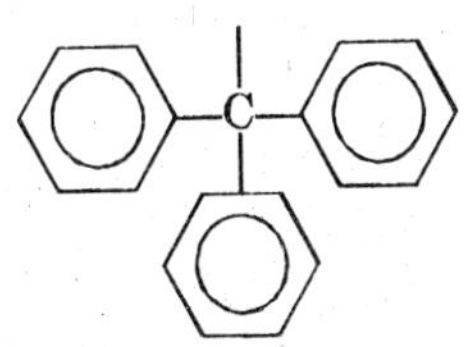 (*Triphenylmethane group*)	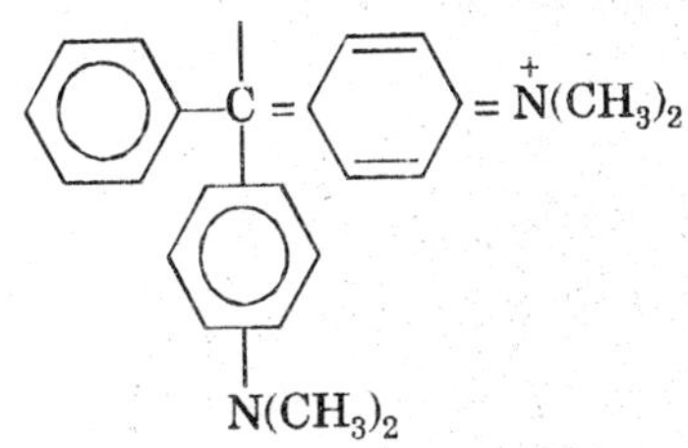 *Malachite green*
• **Phthalein dyes**	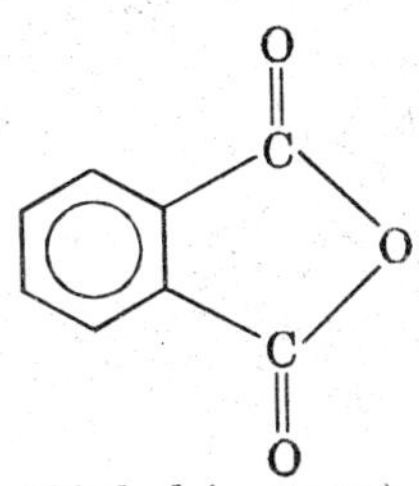 (*Phthalein group*)	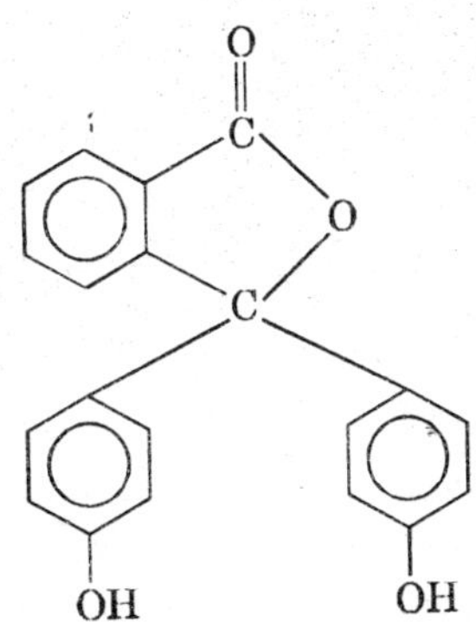 *Phenolphthalein*
• **Indigoid dyes**	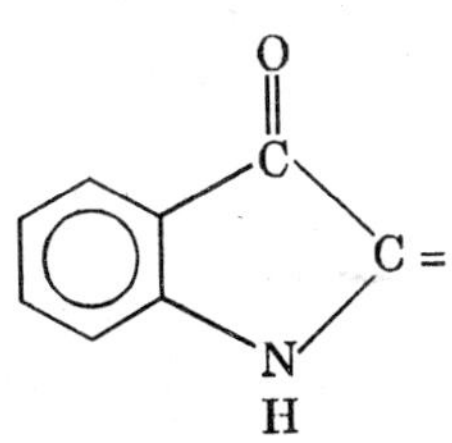 (*Indigoid group*)	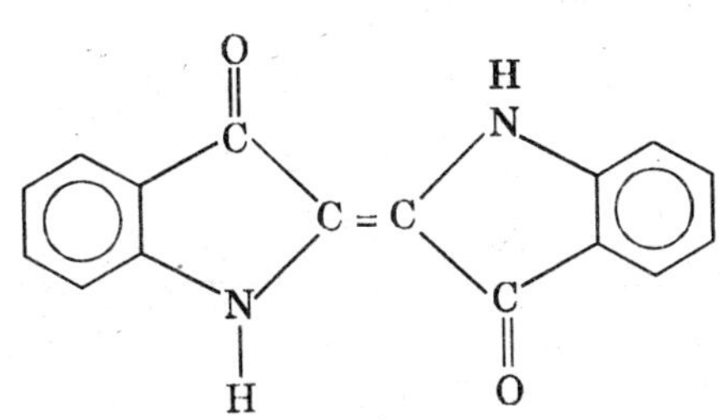 *Indigo*

• **Anthraquinone dyes**	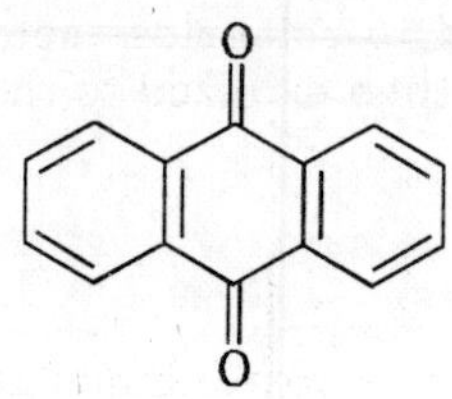 (*Anthraquinone group*)	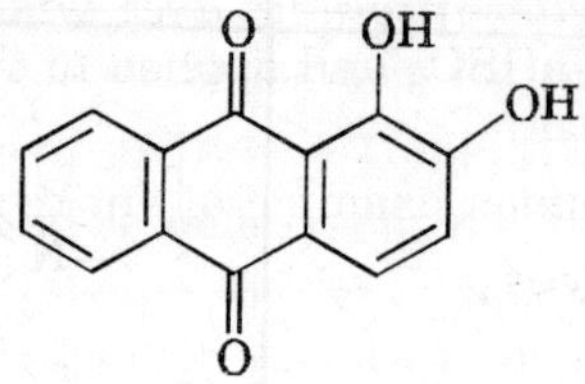 *Alizarin*

CLASSIFICATION OF DYES ON THE BASIS OF APPLICATIONS

The dyes are sold in market under the specification class, shaded and various other properties such as fastness, etc. Thus the dyes have also been classified on the basis of their characteristics (or application) as follows :

1. Acid Dyes

- These are generally *salts of sulphonic acids* and may also contain *carboxylic or/and phenolic group.*
- These don't have affinity for cotton but are applied to wool, silk and nylon.
- Some examples are : *orange-I* and *orange-II,* which can be obtained by coupling of diazotized sulphanilic acid with α– and β-naphthol.

2. Basic Dyes

- These contain *amino group* or dialkyl amino group as **chromophore** or **auxochrome.**
- These dyes in acid solution form water-soluble cations and they react with anionic sites present on the fabrics to fix themselves.
- Some examples are : *Aniline yellow, Butter yellow, Crysodine-G* and *Malachite* green.
- They are used to dye nylons, polyesters, wool, cotton, etc.

3. Direct Dyes

- These are directly applied to the fabrics from an aqueous solution.
- These are attached to the fabric through hydrogen bonding.
- These are used to dye cotton, rayon, wool, silk and nylon.
- Some examples of direct dyes are *Martius yellow* and *Congo red.*

4. Disperse Dyes

- These dyes as the name implies consists of minute particles which are *dispersed or spread* from a suspension (dispersion of minute dye particles in a soap solution) into the fabric.
- These are generally used for polyesters, nylon and polyacrylonitrile.
- Some examples of this class are : *Celliton fast blue B,* and *Celliton fast pink B.*

5. Fibre Reactive Dyes

- The reactive sites of these dyes combine with the hydroxy or amino group of the fibre by irreversible chemical reaction and hence are called fibre reactive dyes. As a result, dyeing is fast and practically permanent.
- These are generally used for cotton, wool and silk etc.

6. Insoluble Azo Dyes or Ingrain Dyes

- These are produced, *in situ*, by coupling a phenol or naphthol adsorbed on the surface of the fabric with a diazonium salt. As these act only through surface adsorption, so the colour is not very fast.
- These are useful for cotton, silk, polyester and nylon.
- Nitro aniline red is a common example.

7. Vat Dyes

- These are *water-insoluble dyes* which are reduced to colourless soluble form (**Leuco**) by alkaline reducing agents like $NaHSO_3$ and applied to the fabrics and then oxidized to the insoluble dye when exposed to air or an oxidizing agent.
- Indigo is common example of vat dyes.

8. Mordant Dyes

- These are the dyes which are applied to the fabric after treating it with a *metal ion*. The metal ion binds to the fabric, and the dye in turn is co-ordinated to the metal ion (called **mordant**). The colour of the same dye varies with the metal ion used. For example, **alizarin** with *aluminium ion* is *rose red* while with *barium ion* it is **blue** in colour.
- These dyes are usually applied for dyeing wool.

DRUGS AND THEIR CLASSIFICATION

Drugs are chemicals of low molecular masses (~ 100 – 500 μ). These interact with macromolecular targets and produce a biological response. When the biological response is therapeutic and useful, these chemicals are called **medicines** and are used in diagnosis, prevention and treatment of diseases. If taken in doses higher than those recommended, most of the drugs used as medicines are potential poisons. Use of chemicals for therapeutic effect is called **chemotherapy.**

Classification of Drugs

Drugs can be classified mainly on criteria outline as follows :

(a) On the basis of pharmacological effect

This classification is based on pharmacological effect of the drugs. It is useful for doctors because it provides them the whole range of drugs available for the treatment of a particular type of problem. For example, analagesics have pain killing effect, antiseptics kill or arrest the growth of micro-organisms.

(b) On the basis of drug action

It is based on the action of a drug on a particular biochemical process. For example, all antihistamines inhibit the action of the compound, histamine which causes inflammation in the body. There are various ways in which action of histamines can be blocked.

(c) On the basis of chemical structure

It is based on the chemical structure of the drug. Drugs classified in this way share common structural features and often have similar pharmacological activity. For example, sulphonamides have common structural feature, given below.

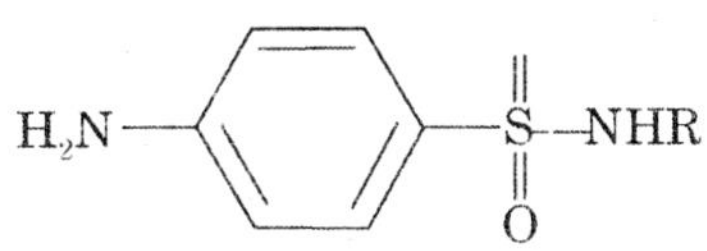

(d) On the basis of molecular targets

Drugs usually interact with biomolecules such as carbohydrates, lipids, proteins and nucleic acids. These are called target molecules or drug targets. Drugs possessing some common structural features may have the same mechanism of action on targets. The classification based on molecular targets is the most useful classification for medicinal chemists.

DRUG-TARGET INTERACTION

Macromolecules of biological origin perform various functions in the body. For example, proteins which perform the role of biological catalysts in the body are called **enzymes,** those which are crucial to communication system in the body are called **receptors.** Carrier proteins carry polar molecules across the cell membrane. Nucleic acids have coded genetic information for the cell. Lipids and carbohydrates are structural parts of the cell membrane. We shall explain the drug-target interaction with the examples of enzymes and receptors.

ADVANCED MATERIALS

- **Ceramics.** In general ceramics are inorganic, non-metallic, covalent network solids that can be made into a paste and shaped at normal temperatures which when fired at high temperatures gain strength. Clays, aluminium oxide, Al_2O_3 ; silicon nitride, Si_3N_4 ; silicon carbide, SiC ; crystalline and amorphous silicon dioxide, SiO_2 are some examples of this class of materials.

  The major groups of the ceramic industry are :

  1. Structural clay products such as bricks, tiles, terra cotta.
  2. White wares such as dinner ware, chemical and electrical porcelain sanitary ware, floor tile.
  3. Portland cement.
  4. Glass products.
  5. Abrasive materials such as fused alumina, silicon carbide.
  6. Aluminium silicate fibres and ceramic foams.
  7. Refractory materials *e.g.,* alumina, magnesia.

  Ceramics are lighter, stiffer and much more resistant to corrosion, as compared to metals.

  Most ceramics are electrical insulators. The manufacture of most ceramics involves four steps :

  1. **The preparation of the raw material**
  2. **The forming of the desired shape**
  3. **The drying and the firing of the piece.**

- **Carbon fibres.** Carbon fibres are made of long chains of carbon atoms. These fibres are *very strong,* very light and *resistant to corrosion.* Carbon fibres are obtained by heating regenerated fibres (such as rayon) or synthetic fibres (such as nylon or polyester), in the absence of oxygen. As a result of heating, the fibres decompose to form carbon fibres.

  Carbon fibres may be reinforced in light weight epoxy resin or polyester resin matrix. Such carbon fibres are known as **Carbon Fibre Reinforced Plastics (CFRP).** When carbon fibres are reinforced in a carbon matrix they are known as **Carbon Fibre Reinforced Carbon (CFRC).**

  Carbon fibres, CFRP and CFRC find applications in the following fields :

  1. High technology fields such as aerospace, military and nuclear.

  In the **aerospace field,** CFRP is used for aircraft wings, tail parts, and helicopter rotor blades. CFRC is used in brakes in heavy and fast jet air crafts. Carbon fibres are used as component of wall material of nuclear reactors.

  2. **General engineering** such as transportation, sports and chemical field.

  In the area of sports-goods, carbon fibres in the form of CFRP are used for making fishing rods, sky poles, tennis and badminton rackets and racing cycle frames.

  3. **Biomedical field.** In biomedical field carbon fibres are used as components of bone plates, ligaments and hydraulic motors for artificial heart implants.

  In India, carbon fibres are mainly used in defence sector and aerospace sector. In defence sector, it is used as nose tips and head shields of missiles and in aerospace sector, it is used for making components of rockets.

10.2. IMPORTANT FACTS AND TERMS

Add to Knowledge

- **Vulcanization** is the process of treating natural rubber under heat and sulphur. The sulphur cross links are introduced as a result of which the sticking nature of natural rubber is lost. Vulcanized rubber becomes more tensile and elastic. The process of vulcanization was introduced by *Charles Goodyear.*
- Certain additives like *diphenylamine* are added to the rubber to protect from deterioration or exposure to air.
- **Rayon** is known as *artificial silk.*
- **Natural silk** contains nitrogen as a constituent while artificial silk may not have nitrogen.
- **Copolymerisation** is possible in four different ways
  (*i*) *Alternating copolymerisation*
  (*ii*) *Block copolymerisation*
  (*iii*) *Random copolymerisation*
  (*iv*) *Graft copolymerisation.*
- **Some Commonly Used Polymers**
  (*i*) **Kelvar.** It is a ***polyamide***, which is obtained by condensation of *terephthalic acid* and *p-phenylenediamine.* The fibres of this polymer are very strong and are used in making light weight bullet proof vests.
  (*ii*) **Nomex.** It is also a ***polyamide*** made from *m-phthalic acid* and *m-phenylenediamine* (1, 3-benzenediamine). This polymer has fire resistance properties and is used in preparing fire proof coats for fire fighters. It is also used in clothings for astronaughts and race-car drivers.
  (*iii*) **Lexan.** It is a *polyester* and is obtained by condensation of *diethyl carbonate* and *bi-phenol-A.* It is used for making of crash helmets, bullet proof windows etc., because of its unusally high impact strength.
  (*iv*) **Polyurethanes.** These are the condensation polymers of *toluene-2, 4-diisocyanate* and *ethylene glycol.* During polymerisation a low boiling liquid such as Freon-11 are added to reaction mixture. The heat liberated during polymerisation causes vaporisation of low boiling liquid producing bubbles which converts viscose polymer to form *polyurethane foams.*
  (*v*) **Super glue.** It is polymer of *α-cyanoacrylate* and possess high tensile strength. One drop of super glue can support a mass of approximate 2000 lbs.
  (*vi*) **Ebonite.** It is highly vulcanized rubber having about 20—30% of sulphur.
  (*vii*) **Saran** is obtained by polymerisation of vinyl chloride and vinylidene chloride.
  (*viii*) **Vinyon** is obtained by polymerisation of vinyl chloride and vinyl acetate.
  (*ix*) **Dynel** is obtained by polymerisation of acrylonitrile and vinyl chloride.
  (*x*) **SAN** is obtained by polymerisation of styrene and acrylonitrile.
  (*xi*) **ABS** is obtained by polymerisation of acrylonitrile and 1, 3-butadiene.
  (*xii*) **Butyl rubber** is obtained by copolymerisation of isobutylene and isoprene.
- **Coordination Polymerisation**

It is the process in which the polymerisation occurs through formation of coordination complex. *For example* : In formation of high density polythene, ethylene first forms a coordination complex with titanium through π electrons. The π complex then reacts stepwise to form the polymer. During coordination polymerisation different stereochemical arrangements are likely to occur. For example, polymerisation of propylene gives three stereochemical arrangements.

(*i*) **Isotactic** (meaning—*same order*). In this type of arrangement all the methyl groups of propylene lie on the same side

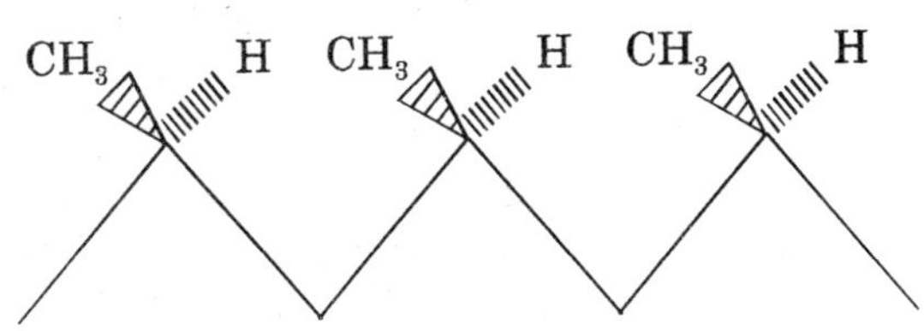

(*ii*) **Syndiotactic** (meaning *alternating order*). In this case the methyl group alternate regularly on the opposite side of chain.

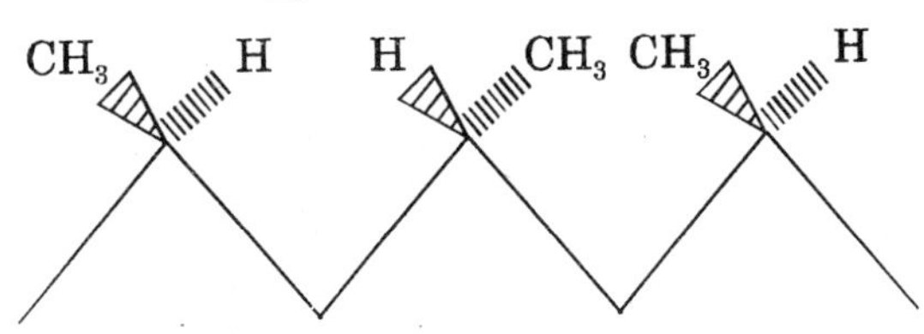

(*iii*) **Atactic** (meaning *no order*). In this type of arrangement no particular order is followed.

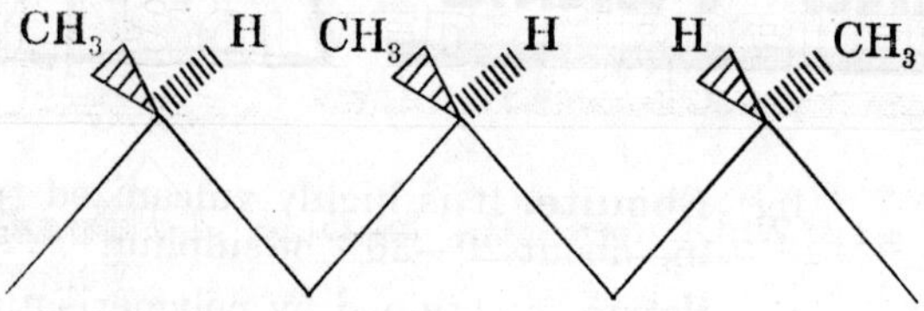

- **Resins and Plastics**

(*i*) **Resins.** Resins are amorphous organic solids or semisolids which usually have a typical metallic lustre and are often transparent or translucent. The examples are : shellac secreted by the lac insect.

(*ii*) **Plastics.** A plastic is a substance which is capable of being moulded whereas a resin lacks this property. Typical plastics on heating become soft but do not give mobile melts. In contrast, resins on heating give mobile melts. Plastics usually have much higher molecular mass than resins. Nevertheless, this differentiation is not very clear and the two terms are often used interchangeably.

- **Artificial Silk or Rayon**

The synthetic fibres which are manufactured from cellulose are known as rayon. It is manufactured by a number of different processes such as :

(*i*) *Cellulose nitrate process*
(*ii*) *Cellulose acetate process*
(*iii*) *Cupro ammonium process*
(*iv*) *Viscose process.*

Among these processes the most commonly used method for preparation is the viscose process and the rayon obtained by this method is called *viscose yarn or viscose rayon.*

- **Polymers and Environmental Pollution**

One of the common use of synthetic polymers is in the form of plastics. It is frequently used in abundance in the form of packing material and throw away bags. Now, the plastics thrown away after use are not *biodegradable, i.e.*, they do not disintegrate automatically in the atmosphere over the period of time. This stability and durability of plastics posess a serious waste disposal problem. With ever increasing use of plastics and its non biodegradable nature, the entire mankind may get buried under the pile of plastic debries.

Scientists these days are carrying on researches to develop such a plastic which is capable of biodegradability over the period of time.

Many of the country have put a ban on the use of polythene bags in daily routine.

10.3. QUESTION BANK

Develop Confidence

Polymers

1. Which among the following is a *crossed linked polymer ?*
(*a*) Amylopectin
(*b*) Melamine formaldehyde resin
(*c*) Glycogens
(*d*) Polyesters.

2. Which of the following is an *elastomer* ?
(*a*) Vulcanized rubber (*b*) Dacron
(*c*) Polystyrene (*d*) Malamine.

3. Bakelite is an example of
(*a*) Elastomer (*b*) Fibre
(*c*) Thermoplastic (*d*) Thermosetting polymer.

4. Dacron is an example of
(*a*) Elastomer (*b*) Fibre
(*c*) Thermoplastic (*d*) Thermosetting polymer.

5. Nylon-66 is obtained by condensation polymerization of
(*a*) Adipic acid and hexamethylene diamine
(*b*) Phenol and formaldehyde
(*c*) Terephthalic acid and ethylene glycol
(*d*) Sebacic acid and hexamethylene diamine.

6. The monomer of PMMA is
(*a*) Methyl methacrylate (*b*) Ethyl acrylate
(*c*) Acrylonitrile (*d*) Methyl acrylate.

7. The monomer of Acrilan is
(*a*) Methyl methcrylate (*b*) Ethyl acrylate
(*c*) Acrylonitrile (*d*) Methyl acrylate.

8. Which pair of polymers have similar properties ?
(*a*) PAN, PTFE (*b*) PCTFE
(*c*) Nylon, PVC (*d*) Bakelite, Alkyd resin.

9. Which of the following is a *natural polymer* ?
(*a*) Polythene (*b*) Polysaccharides
(*c*) Nylon (*d*) Terylene.

10. Which among the following is a synthetic polymer ?
(*a*) Proteins
(*b*) Polysaccharides
(*c*) Natural rubber
(*d*) Phenol-formaldehyde resin.

11. Which among the following is a *linear polymer* ?
(*a*) High density polythene
(*b*) Low density polythene
(*c*) Bakelite
(*d*) Vulcanized rubber.

12. Which among the following is a *branched chain polymer* ?
(*a*) Nylons (*b*) Polyesters
(*c*) Glycogens (*d*) Bakelite.

13. Which of the following type of forces are present in Nylon-66 ?
(*a*) Van der Waal's forces of attraction
(*b*) Hydrogen bonding
(*c*) Three-dimensional net work of bonds
(*d*) None of the above.

14. Which among the following is a *chain-growth polymer*?
(*a*) Nylon (*b*) Bakelite
(*c*) Terylene (*d*) Teflon.

15. Which among the following is a *step-growth polymer* ?
(*a*) PTFE (*b*) PVC
(*c*) Polyester (*d*) Polythene.

16. Polyacrylates contain the linkages of
(*a*) Amides (*b*) Esters
(*c*) Alcohols (*d*) Nitriles.

17. The synthetic polymer which resemble natural rubber is
(*a*) Neoprene (*b*) Buna-S
(*c*) Nylon (*d*) Rayon.

18. Terylene is produced by reaction of ethylene glycol and
(*a*) Phthalic acid (*b*) Terephthalic acid
(*c*) Metaphthalic acid (*d*) All of these.

19. Glyptal is a polymer of ethylene glycol and
(*a*) Phthalic acid (*b*) Terephthalic acid
(*c*) Metaphthalic acid (*d*) Adipic acid.

20. Which of the following is thermoplastic ?
(*a*) Dacron (*b*) Nylon
(*c*) Polythene (*d*) All of these.

21. The process of vulcanization of rubber makes it
(*a*) Hard (*b*) Soft
(*c*) Less elastic (*d*) None of these.

22. Polystyrene is an example of

(*a*) Elastomer (*b*) Fibre

(*c*) Thermoplastic (*d*) Thermosetting polymer.

23. Monomer of PTFE is

(*a*) Ethylene (*b*) Propylene

(*c*) Butadiene (*d*) Tetra fluoro thylene.

24. Which of the following type of forces are present in *vulcanized rubber* ?

(*a*) Weakest intermolecular forces

(*b*) Hydrogen-bonding

(*c*) Three-dimensional network of bonds

(*d*) None of these.

25. Which of the following is polycarbonate ?

(*a*) Acrilan (*b*) Lexan

(*c*) PAN (*d*) BuNa-S.

26. The starting materials for polyurethanes are *toluene 2, 4 dicyanate* and

(*a*) Ethylene glycol (*b*) Phenol

(*c*) Cresol (*d*) None of these.

27. The monomer of polyacrylonitrile is

(*a*) Vinyl cyanide (*b*) Vinyl chloride

(*c*) Vinyl alcohol (*d*) None of these.

28. Caprolactam is obtained from

(*a*) Cyclohexane

(*b*) Hexane

(*c*) Adipic acid

(*d*) Adipic acid and hexamethylene diamine.

29. Which of the following is not a polyamide polymer ?

(*a*) Nylon-66 (*b*) Nylon-6

(*c*) Nylon-610 (*d*) Glyptal.

30. Which of the following is not an addition polymer ?

(*a*) Polyethylene (*b*) Polypropylene

(*c*) Polystyrene (*d*) Dacron.

31. The monomer of *neoprene* is

(*a*) Butadiene (*b*) Chloroprene

(*c*) Isoprene (*d*) 2-methyl-1, 3-butadiene.

32. The monomer of *natural rubber* is

(*a*) Butadiene (*b*) Chloroprene

(*c*) Isoprene (*d*) Butadiene and Styrene.

33. For natural polymers PDi is generally

(*a*) 0 (*b*) 1

(*c*) 100 (*d*) 1000.

34. Buna-S is

(*a*) Natural rubber (*b*) Sulphur rubber

(*c*) Synthetic rubber (*d*) None of these.

35. Which is not a polyacrylate ?

(*a*) PMMA (*b*) Acrilan

(*c*) Poly acrylonitrile (*d*) PCTFE.

36. Which of the following is a polyester ?

(*a*) Nylon-66

(*b*) Nylon-610

(*c*) Alkyd resin

(*d*) Phenol-formaldehyde resin.

37. The alternative name of Glyptal is

(*a*) Alkyl resin

(*b*) Phenol-formaldehyde resin

(*c*) Melamine-formaldehyde resin

(*d*) Melmac.

38. The alternative name of PMMA is

(*a*) Alkyd resin

(*b*) Phenol-formaldehyde resin

(*c*) Plexiglass

(*d*) None of these.

39. Which of the following is intermediate in preparation of **Nylon-6** ?

(*a*) Cyclohexane

(*b*) Cyclohexanone oxime

(*c*) Caprolactum

(*d*) Adipic acid and hexamethylene diamine.

40. The monomer of PVC is

(*a*) Ethene (*b*) Chloroethene

(*c*) Dichloroethene (*d*) Tetra chloroethene.

41. The S in BuNa-S refers to

(*a*) Sulphur (*b*) Styrene

(*c*) Sodium (*d*) None of these.

42. The repeating units of Acrilan are

(*a*) $H_2C = \overset{\displaystyle H}{\overset{|}{C}} - COOCH_3$ (*b*) $H_2C = \overset{\displaystyle H}{\overset{|}{C}} - COOC_2H_5$

(*c*) $H_2C = \overset{\displaystyle H}{\overset{|}{C}} - CN$ (*d*) $H_2C = \overset{\displaystyle CH_3}{\overset{|}{C}} - COOCH_3$.

43. Which of the following is a polymer containing nitrogen ?

(*a*) Nylon (*b*) Polyester.

(*c*) Polydiene (*d*) Polyolefin.

44. Alkyd resins are polymers obtained by reaction of glycol or glycerol with

(*a*) Malonic acid (*b*) Maleic acid

(*c*) Phthalic acid (*d*) Acetic acid.

45. Teflon is an example of polymer which is

(*a*) A polyamide (*b*) An addition polymer
(*c*) A polyester (*d*) A formaldehyde resin.

46. *Perlon* is another name of

(*a*) Nylon-66 (*b*) Nylon-6
(*c*) Nylon-610 (*d*) Nylon-44.

47. Polymers can also be called

(*a*) Macromolecules (*b*) Micromolecules
(*c*) Sub-macromolecules (*d*) Sub-micromolecules.

48. Which of the following has an ester linkage ?

(*a*) Nylon-66 (*b*) Dacron
(*c*) Nylon-610 (*d*) Bakelite.

49. Isoprene is a monomer of

(*a*) Natural rubber (*b*) Synthetic rubber
(*c*) Salol (*d*) Saran.

50. The natural occurring polymer among the following is

(*a*) Rayon (*b*) Nylon
(*c*) Dacron (*d*) Proteins.

Chemistry in Action

51. Which of the following is an example of direct dye ?

(*a*) Alizarin (*b*) Congo red
(*c*) Martius green (*d*) Malachite green.

52. Aspirin is an

(*a*) Antipyretic (*b*) Antiseptic
(*c*) Antimalarial (*d*) Narcotic.

53. Antipyretics are medicinal compounds which

(*a*) Relieve pain
(*b*) Lower the body temperature
(*c*) Kill other organisms
(*d*) Control malaria.

54. The dyes which are originally present in reduced state and then oxidized in air are

(*a*) Azo dyes (*b*) Basic dye
(*c*) Disperse dyes (*d*) Vat dyes.

55. The compound extracted from the plant *Rauwolfia serpentina* is

(*a*) Aspirin (*b*) Bithional
(*c*) Quinine (*d*) Reserpine.

56. Which of the following is a natural dye ?

(*a*) Alizarin (*b*) Malachite green
(*c*) Martius yellow (*d*) Phenolphthalein.

57. Which of the following is an azo dye ?

(*a*) Alizarin (*b*) Indigo
(*c*) Malachite green (*d*) Orange-II.

58. To which class of dyes does Martius yellow belong ?

(*a*) Azo dyes (*b*) Nitro dyes
(*c*) Phthalein dyes (*d*) Indigoid dyes.

59. A chemical substance which can act both as antiseptic as well as disinfectant is

(*a*) Phenol (*b*) Pheromone
(*c*) Mestranol (*d*) Streptomycin.

60. Which of the following can bring down the body temperature ?

(*a*) Aspirin (*b*) Chloroquine
(*c*) Penicillin (*d*) Quimine.

61. Which one is a vat dye ?

(*a*) Alizarin (*b*) Congo red
(*c*) Indigo (*d*) Malachite.

62. Penicillin is an/a

(*a*) Antibiotic (*b*) Antipyretic
(*c*) Hormone (*d*) Vitamin.

63. Red ink is prepared from

(*a*) Aniline (*b*) Congo red
(*c*) Eosin (*d*) Phenol.

64. The blue print process involves the use of

(*a*) Indigo dyes (*b*) Iron compounds
(*c*) Vat dyes (*d*) Some other compounds.

65. Which one of the following is a **psychedilic agent** ?

(*a*) DDT (*b*) DNA
(*c*) LSD (*d*) TNT.

66. The compounds which possess general depressant action on the central nervous system are called

(*a*) Sedatives (*b*) Tranquilizeres
(*c*) Narcotics (*d*) Anaesthetics.

67. Compounds used to produce unconsciousness and muscular relaxation, sufficient to carry out surgical operations are called

(*a*) Sedatives (*b*) Tranquilizers
(*c*) Anaesthetics (*d*) Antibiotics.

68. A dye absorbs wavelength corresponding to the 480-490 mm colour. The observed colour is

(*a*) Blue (*b*) Red
(*c*) Green (*d*) Orange.

69. The drug used for treatment of typhoid is

(*a*) Chloromycetin (*b*) Novalgin
(*c*) Paracetamol (*d*) Quinine.

70. Chloroquine is a drug used to control

(*a*) Malaria (*b*) Oridinary fever
(*c*) Pneumonia (*d*) None of these.

71. What is the relation between the specific impulse and the critical temperature attained in a rocket blast ?

(*a*) $I_s \propto T_c$ (*b*) $I_s \propto T_c^2$

(*c*) $I_s \propto \frac{1}{T_c}$ (*d*) $I_s \propto \sqrt{T_c}$.

72. What type of propellant was used in SLV-3 ?

(*a*) UDMH + Liquid N_2O_4

(*b*) MMH + Liquid N_2O_4

(*c*) Composite solid propellant

(*d*) Kerosene + Liquid O_2.

73. Which of the following represents a biliquid propellant ?

(*a*) Nitroglycerine + Nitrocellulose

(*b*) Liquid N_2O_4 + Acrylic rubber

(*c*) Liquid N_2O_4 + Unsymmetrical dimethyl hydrazine (UDMH)

(*d*) Polyurethane + Ammonium perchlorate.

74. The chemicals used to destroy herbacious on the plant's body are called

(*a*) Herbicides (*b*) Fungicides

(*c*) Disinfectants (*d*) Antibiotics.

75. The chemicals used to destroy fungi on the plant's body are called

(*a*) Herbicides (*b*) Fungicides

(*c*) Both (*d*) None of these.

76. Which of the following is an alkaloid derived from the bark of cinchona tree ?

(*a*) Quinine (*b*) Atrabine

(*c*) Sulphadiazine (*d*) Penicillin.

77. Which of the following groups is an **auxo-chrome ?**

(*a*) —NO_2 (*b*) —NO

(*c*) —N = N— (*d*) —OH.

78. When a mixture of salicylic acid, acetic anhydride and acetic acid is refluxed, the product obtained is

(*a*) Aspirin (*b*) Acetophenone

(*c*) *p*-Acetamidophenol (*d*) Chloroxylenol.

79. With which cation alizarin will impart a red colour on the fabrics ?

(*a*) Fe^{+3} (*b*) Al^{+3}

(*c*) Cr^{+3} (*d*) Ba^{+2}.

80. Which one is a broad spectrum antibiotic ?

(*a*) Chloroamphenicol (*b*) Penicillin

(*c*) Paracetamol (*d*) Ampicillin.

81. Thrust imparted to the rocket is governed by the

(*a*) Newton's third law

(*b*) Third law of thermodynamics

(*c*) Gravitational law

(*d*) None of the above.

82. Which of the following term means pain killing ?

(*a*) Antibiotic (*b*) Analgesic

(*c*) Antipyretic (*d*) Penicillin.

83. Which of the following is a sedative ?

(*a*) Barbiturates (*b*) Bithional

(*c*) Penicillin (*d*) Benzyl penicillin.

84. Which of the following is an antidepressants ?

(*a*) Chloroxylenol (*b*) Bithional

(*c*) Cocain (*d*) Penicillin G.

85. Which of the following is an analgesic ?

(*a*) Paracetamol (*b*) Penicillin

(*c*) Chloromphenicol (*d*) Streptomycin.

86. Which of the following pairs of groups represent only chromophores ?

(*a*) —NO_2 and —NO (*b*) —OH and —COOH

(*c*) —COOH and —SO_3H (*d*) —OH and —NH_2.

87. **Veronal**, a barbituric drug is used as an/a

(*a*) Anaesthetic (*b*) Sedative

(*c*) Antiseptic (*d*) Antipyretic.

88. Which of the following is a hypnotic drug ?

(*a*) Catechol (*b*) Luminal

(*c*) Phenol (*d*) Salol.

89. What is the relation between the specific impulse and the average molecular mass ($\overline{M}$) of the product gases coming out of the rocket nozzle ?

(*a*) $I_s \propto \overline{M}$ (*b*) $I_s \propto \overline{M}^{-1}$

(*c*) $I_s \propto \overline{M}^{1/2}$ (*d*) $I_s \propto \sqrt{1/\overline{M}}$.

90. Nitroglycerine and nitrocellulose is

(*a*) Double base-propellant

(*b*) Composite propellant

(*c*) Biliquid propellant

(*d*) Hybrid propellant.

91. Which one of the following will impart a highest specific impulse ?

(*a*) H_2F_2 (*b*) H_2O_2

(*c*) Methyl nitrate (*d*) Nitromethane.

92. SLV-3 and ASLV rockets used

(*a*) Biliquid propellants

(*b*) Composite propellants

(*c*) Double base propellants

(*d*) Hybrid propellants.

93. Which of the following contains NO_2 group attached to aromatic nucleus in its structure ?

(*a*) Chloramphenicol (*b*) Luminal
(*c*) Seconal (*d*) Penicillin.

94. Streptomycin is an example of

(*a*) Antibiotic (*b*) Analgesic
(*c*) Antipyretic (*d*) Anaesthetic.

95. Which of the following medicines is required for malaria ?

(*a*) Aspirin (*b*) Penicillin
(*c*) Chloroquine (*d*) Paracetamol.

96. Which of the following is not a Narcotics ?

(*a*) Opium (*b*) Heroin
(*c*) Pethidine (*d*) Bithional.

97. Ampicillin is also called

(*a*) Penicillin G (*b*) Penicillin V
(*c*) Penicillin K (*d*) None of these.

98. Phenylbutazone is another name for

(*a*) Brufane (*b*) Phenacitin
(*c*) Butazolidine (*d*) None of these.

99. Chloroxylenol is an important component of

(*a*) Soap (*b*) Antibiotics
(*c*) Dettol (*d*) Pain killing Ointments.

100. Which of the following is not a narcotic ?

(*a*) Codeine (*b*) Pethidine hydrochloride
(*c*) Novalgin (*d*) Methadon.

ANSWERS

1. (*b*)	**2.** (*a*)	**3.** (*d*)	**4.** (*b*)	**5.** (*a*)	**6.** (*a*)	**7.** (*c*)	**8.** (*b*)	**9.** (*b*)
10. (*d*)	**11.** (*a*)	**12.** (*c*)	**13.** (*b*)	**14.** (*d*)	**15.** (*c*)	**16.** (*b*)	**17.** (*a*)	**18.** (*b*)
19. (*a*)	**20.** (*d*)	**21.** (*a*)	**22.** (*c*)	**23.** (*d*)	**24.** (*a*)	**25.** (*d*)	**26.** (*a*)	**27.** (*a*)
28. (*a*)	**29.** (*d*)	**30.** (*d*)	**31.** (*d*)	**32.** (*c*)	**33.** (*b*)	**34.** (*c*)	**35.** (*d*)	**36.** (*c*)
37. (*a*)	**38.** (*c*)	**39.** (*c*)	**40.** (*b*)	**41.** (*b*)	**42.** (*c*)	**43.** (*a*)	**44.** (*c*)	**45.** (*b*)
46. (*b*)	**47.** (*a*)	**48.** (*b*)	**49.** (*a*)	**50.** (*d*)	**51.** (*b*)	**52.** (*a*)	**53.** (*b*)	**54.** (*d*)
55. (*d*)	**56.** (*a*)	**57.** (*d*)	**58.** (*b*)	**59.** (*a*)	**60.** (*a*)	**61.** (*c*)	**62.** (*a*)	**63.** (*c*)
64. (*b*)	**65.** (*c*)	**66.** (*c*)	**67.** (*c*)	**68.** (*d*)	**69.** (*a*)	**70.** (*a*)	**71.** (*d*)	**72.** (*c*)
73. (*c*)	**74.** (*a*)	**75.** (*b*)	**76.** (*a*)	**77.** (*d*)	**78.** (*a*)	**79.** (*b*)	**80.** (*a*)	**81.** (*a*)
82. (*b*)	**83.** (*a*)	**84.** (*c*)	**85.** (*a*)	**86.** (*a*)	**87.** (*b*)	**88.** (*b*)	**89.** (*d*)	**90.** (*a*)
91. (*b*)	**92.** (*b*)	**93.** (*a*)	**94.** (*a*)	**95.** (*c*)	**96.** (*d*)	**97.** (*d*)	**98.** (*c*)	**99.** (*c*)
100. (*c*).								

HINTS/SOLUTIONS

Q. No.	*Ans.*	*Hints & Solutions*
1.	(*b*)	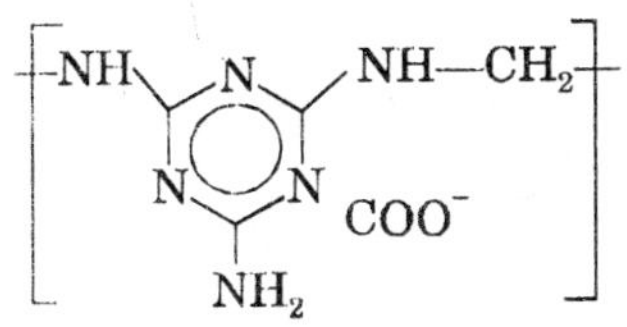 is melamine formaldehyde resin.
2.	(*a*)	Elastomers have an elastic strength *i.e.*, stretchable.
3.	(*d*)	A thermosetting polymer sets into a hard mass and cannot be remelted.
4.	(*a*)	Dacron is terylene and polyester textile fibres are made from *d*.
5.	(*a*)	$HOOC—(CH_2)_4—COOH + NH_2—(CH_2)_6—NH_2$ produces Nylon 66.
6.	(*a*)	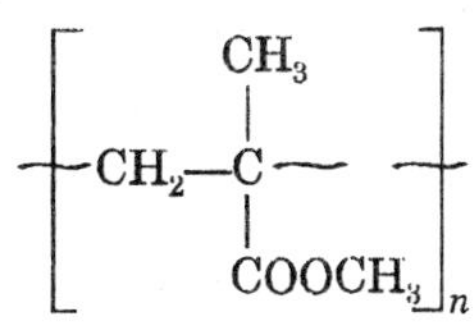 is PMMA.

Q. No.	Ans.	Hints & Solutions
7.	(c)	PAN is called Acrilan, $\left[-CH_2-\underset{\displaystyle CH}{\underset{\vert}{CH}}- \right]_n$
8.	(b)	Both these addition products of ethylene and have same halogen atom.
9.	(b)	Obtained from natural sources.
10.	(d)	Resins are synthesized in lab by carrying but reactions.
11.	(a)	Polythene has linear chains of ethylene and hence will have more density.
12.	(c)	Amylopectin and glycogen are examples.
13.	(b)	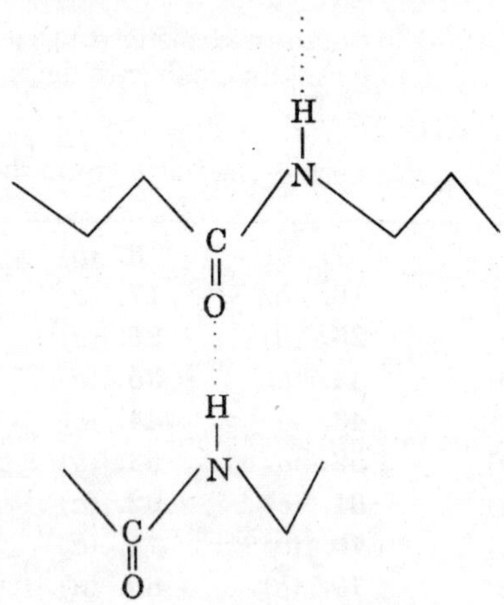
14.	(d)	In chain growth polymerisation monomers add without loss of molecules.
15.	(c)	Step growth involves lose of molecules at each step *i.e.,* condensation occurs.
16.	(b)	Polyacrylates are obtained from esters of acrylic acid ($CH_2 = CH-COOH$).
17.	(a)	Neoprene is obtained from $CH_2 = \overset{\displaystyle Cl}{\overset{\vert}{C}}-CH = CH_2$ polymerisation.
18.	(b)	Terylene is an ester and word ter is from terepthalic acid ($HOOC-C_6H_4-COOH$).
19.	(a)	Glyptal or alkyl resin are obtained from phthalic acid.
20.	(d)	Thermoplastics can be regenerated.
21.	(a)	Vulcanisation makes it hard and more elastic.
22.	(c)	$\left[-\underset{\displaystyle C_6H_5}{\underset{\vert}{CH}}-CH_2- \right]_n$ can be melted and recast again.
23.	(d)	$-[CF_2-CF_2]-$ is PTFE.
24.	(a)	Vulcanised rubber has 5-crosslinks.
25.	(d)	Lexan is a polyester obtained by condensation of diethyl carbonate and bi-phenol-A.
26.	(a)	Ethylene glycol is used.
27.	(a)	$CH_2 = CH-CN$ is used.
28.	(a)	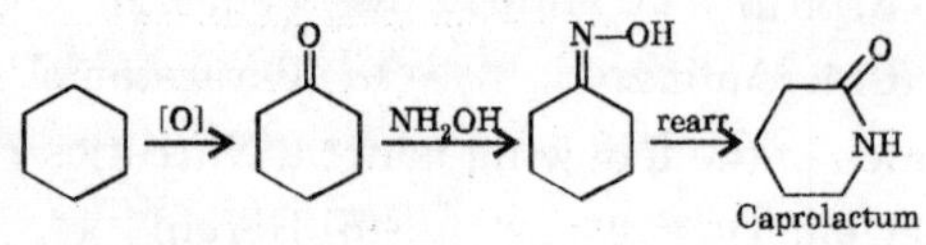
29.	(d)	Glyptal is a polyester.
30.	(d)	Dacron or terylene is a condensation polymer.
31.	(d)	It is obtained from $CH_2 = \overset{\displaystyle Cl}{\overset{\vert}{C}}-CH = CH_2$.
32.	(c)	Isoprene is $CH_2 = \overset{\displaystyle CH_3}{\overset{\vert}{C}}-CH = CH_2$.
33.	(b)	$PDI = \dfrac{\overline{M_w}}{M_n}$.
34.	(c)	It is a synthetic rubber obtained by addition of butadiene and styrene in the presence of Na.
35.	(d)	PCTFE is polymono chloro trifluoro ethylene.
36.	(c)	Alkyd resin is obtained by condensation of glycol with phthalic acid.
37.	(a)	See answer 36.
38.	(c)	It is called plexiglass.
39.	(c)	Caprolactum is formed.
40.	(b)	PVC is poly vinyl chloride.
41.	(b)	In BuNa-S the words Bu = butadiene, Na = sodium and S-styrene.
42.	(c)	Acrilan is also PAN.
43.	(a)	Nylon is a polyamide.
44.	(c)	Alkyd resin is a condensation polyme or a polyester.
45.	(b)	Teflon is PTFE or $-[CF_2-CF_2]_n-$. It is addition polymer.
46.	(b)	Nylon-6 is also called Perlon.
47.	(a)	Polymers (made up of many parts or monomeric units).
48.	(b)	Dacron is terylene with $-\overset{\displaystyle O}{\overset{\Vert}{C}}-O-R$ type of linkage.

Q. No.	Ans.	Hints & Solutions
49.	(*a*)	Isoprene is 2-Methyl-1, 3-butadiene.
50.	(*d*)	Proteins are natural occurring polymers with $-\overset{O}{\overset{\Vert}{C}}-NH-$ linkage.
51.	(*b*)	Direct dyes belong to class of azo dye.
52.	(*a*)	It is an antipyretic.
53.	(*b*)	Antipyretics lower body temperature.
54.	(*d*)	Vat dyes cannot be applied directly.
56.	(*a*)	These are obtained from plants.
57.	(*d*)	Azo dyes have —N = N— group.
58.	(*b*)	It contains $—NO_2$ structural unit.
59.	(a)	Phenol acts as both antiseptic and disinfectant.
60.	(*a*)	Antipyredics can lower body temperature.
61.	(*c*)	Such dyes cannot be applied directly.
62.	(*a*)	Antibiotics help in destroying micro-organisms.
63.	(*c*)	Eosin is used.
64.	(*b*)	Iron compounds are used.
65.	(*c*)	LSD is psychedilic agent.
66.	(*c*)	Narcotics reduce tension and anxiety.
67.	(*c*)	These are known as anaesthetics (induce deep sleep).
68.	(*d*)	It would reflect its complementary colour yellow or orange.
69.	(*a*)	Chloromycetin is used.
70.	(*a*)	Used in control of Malaria.
71.	(*d*)	$I_s \propto \sqrt{T_c}$.
72.	(*c*)	Composite solid propellants are used.
73.	(*c*)	Liquid N_2O_4 and UDMH.
74.	(*a*)	These are called herbicides.
75.	(*b*)	These are known as fungicides.
76.	(*a*)	Quinine.
77.	(*d*)	Auxochromtic groups are colour enhancers.
78.	(*a*)	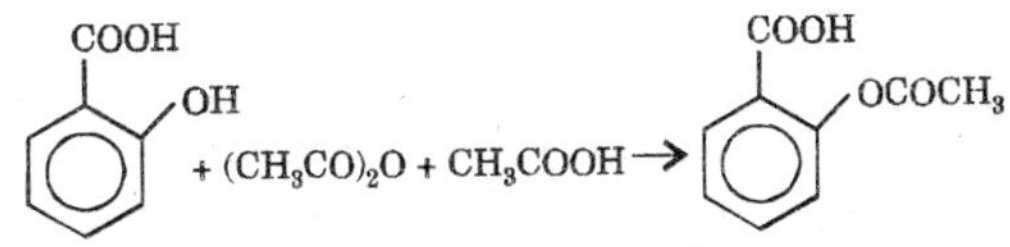
80.	(*a*)	These are antiseptics effective against different type of harmful micro-organisms.
81.	(*a*)	Newton's Law.
82.	(*b*)	Analgesics are pain killers.
83.	(*a*)	Barbiturate.
84.	(*c*)	Antidepresants are booster drugs which generate confidence in person with depressed mood.
85.	(*a*)	Paracetamol.
86.	(*a*)	Chromophore is the basic group in dyes.
87.	(*b*)	It is a sedative.
88.	(*b*)	Hypnotic help in inducing deep sleep.
89.	(*d*)	$I_s \propto \sqrt{\frac{1}{M}}$.
90.	(*a*)	Double base propellant.
91.	(*b*)	H_2O_2.
92.	(*b*)	Composite propellants.
93.	(*a*)	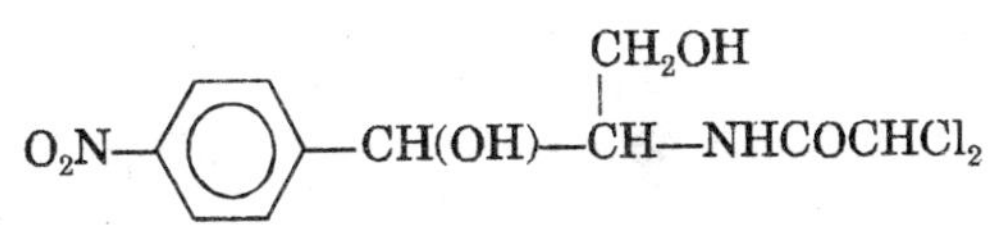
94.	(*a*)	It is antibiotic.
95.	(*c*)	Chloroquine is used.
96.	(*d*)	Bithional is an antiseptic.
97.	(*d*)	None of these.
98.	(*c*)	It is Butazolidine.
99.	(*c*)	It is component of Dettol.
100.	(*c*)	Novalgin is an analgesic.

10.4. EXAMINER'S VIEWPOINT

Test Your Confidence

1996

1. Which of the following contains *isoprene* unit ?
(*a*) Nylon-66 (*b*) Natural rubber
(*c*) Polyethylene (*d*) Dacron. ***(MP PMT)***

2. Which of the following is the correct representation of initial reaction for the preparation of bakelite ?
(*a*) Aromatic electrophilic substitution
(*b*) Aromatic nucleophilic substitution
(*c*) Free radical reaction
(*d*) Aldol reaction. ***(Delhi CEE)***

3. Among the following polymers, the strongest molecular forces are present in
(*a*) Elastomers (*b*) Fibres
(*c*) Thermoplastics (*d*) Thermosetting polymers. ***(Pb. CET)***

4. Glyptal polymer is obtained from glycerol on reacting with
(*a*) Malonic orange (*b*) Phthalic acid
(*c*) Maleic acid (*d*) Acetic acid. ***(MLNR)***

5. Which of the following is an azo dye ?
(*a*) Methyl orange (*b*) Phenolphthalein
(*c*) Malachite green (*d*) Methylene blue. ***(Delhi CEE)***

6. The following compound is used as

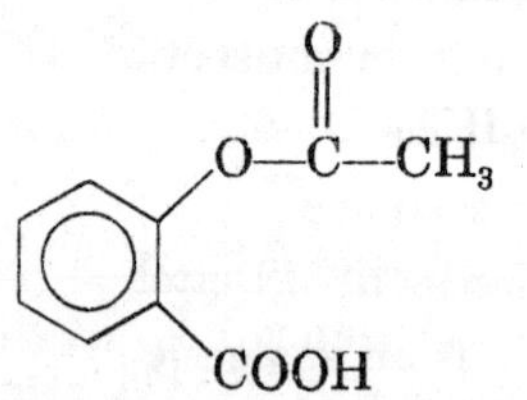

(*a*) An anti-inflammatory compound
(*b*) Analgesic
(*c*) Hypnotic
(*d*) Antiseptic. ***(CEET Kurukshetra)***

1997

7. Teflon is a polymer monomer of which is
(*a*) Difluoroethane (*b*) Monofluoroethane
(*c*) Tetrafluoroethene (*d*) Tetrafluoroethane. ***(Pb. CET)***

8. Which of the following is not an example of addition polymer ?
(*a*) Polyethene (*b*) Polystyrene
(*c*) Neoprene (*d*) Terylene. ***(Pb. CET)***

9. Which one of the following is used to make nonstick cookware ?
(*a*) PVC
(*b*) Polyestyrene
(*c*) Polythylene (terephthalate)
(*d*) Polytetrafluoroethylene. ***(CBSE PMT)***

10. Which one is a polymer compound ?
(*a*) SO_2 (*b*) CO_2
(*c*) CH_4 (*d*) PVC. ***(CPMT)***

11. Paracetamol is
(*a*) Both antipyritic and analgesic
(*b*) Analgesic
(*c*) Antipyritic
(*d*) Antimalaric. ***(Karnataka CET)***

12. Which of the following can possibly be used as analgesic without causing addiction and any modification ?
(*a*) Morphine
(*b*) N-acetylparaaminophenol
(*c*) Diazepam
(*d*) Tetra hydrocatenol. ***(CBSE PMT)***

1998

13. Example of addition polymer is
(*a*) BuNa-S (*b*) Bakelite
(*c*) Nylon-6 (*d*) Melamac. ***(Pb. CET)***

14. Copolymer is
(*a*) Nylon-6 (*b*) Nylon-66
(*c*) PMMA (*d*) Polyethene. ***(ICAR)***

15. Which of the following contains isoprene units ?
(*a*) Natural rubber (*b*) Nylon-66
(*c*) Polyethylene (*d*) Dacron. ***(MP PMT)***

16. Which of the following is not a synthetic polymer ?
(*a*) Polythylene (*b*) PVC
(*c*) Nylon (*d*) Cellophane. ***(MP PET)***

17. Acetoxy Benzoic acid is
(a) Antiseptic (b) Aspirin
(c) Antibiotic (d) Mordant dye.
(CEET Kurukshetra)

18. Indigo belongs to the class of
(a) Mordant dyes (b) Vat dye
(c) Direct dye (d) Disperse dye. **(Pb. CET)**

19. Which of the following is antipyretic ?
(a) Quinine (b) Paracetamol
(c) Luminol (d) Paprazine. **(Pb. CET)**

20. Which of the following is used as rocket fuel ?
(a) Cyanogen + O_3 (b) Cyanogen + O_2
(c) Water gas + O_3 (d) Nitrolim + O_3.
(EAM CET)

1999

21. Which is not true about polymers ?
(a) Polymers do not carry any charge
(b) Polymers have high viscosity
(c) Polymers scatter light
(d) Polymers have low molecular weight. **(MP PET)**

22. On the basis of the mode of their formation the polymers can be classified
(a) As addition polymers only
(b) As condensation polymers only
(c) As copolymers
(d) Both as addition and condensation polymers.
(MP PET)

23. Natural rubber is a polymer of
(a) Butadiene (b) Ethyne
(c) Styrene (d) Isoprene. **(CBSE PMT)**

24. Terylene is a condensation polymer of ethylene glycol and
(a) Benzoic acid (b) Phthallic acid
(c) Salicylic acid (d) Terephthalic acid.
(CBSE PMT)

25. Teflon, styron and neoprene are all
(a) Co-polymers (b) Condensation
(c) Homopolymers (d) Monomers.
(Kurukshetra CEET)

26. Which among the following is a board spectrum antibiotic ?
(a) Paracetamol (b) Penicillin
(c) Aspirin (d) Chloroamphenicol.
(Kurukshetra CEET)

27. Which of the following is an example of anthraquinone dye ?
(a) Alizarin (b) Methyl orange
(c) Methylene blue (d) Phenolphthalein.
(Kurukshetra CEET)

28. Which of the following is not a synthetic polymer ?
(a) Polyethylene (b) PVC
(c) Nylon (d) Cellophane. **(MP CET)**

29. Which one of the following is not an example of chain growth polymer ?
(a) Neoprene (b) BuNa-S
(c) PMMA (d) Glyptal. **(JIP MER)**

30. Ziegler-Natta catalyst is
(a) $K[PtCl_3(C_2H_4)]$ (b) $(Ph_3P)_3RhCl$
(c) $Al_2(C_2H_5)_6 + TiCl_6$ (d) $Fe(C_5H_5)_2$. **(JIP MER)**

31. Nylon-66 is made by using
(a) Phenol (b) Benzaldehyde
(c) Adipic acid (d) Succinic acid. **(AFMC)**

32. The process involving heating to rubber with sulphur is called
(a) Galvanisation (b) Vulcanization
(c) Bessemerisation (d) Sulphonation. **(ICAR)**

33. Terylene is made by polymerization of terephthalic acid with
(a) Ethylene glycol (b) Phenol
(c) Ethanol (d) Catechol. **(Pb. CET)**

34. Interparticle forces present in Nylon-66 are
(a) Van der Waals
(b) Hydrogen bonding
(c) Dipole-dipole interactions
(d) None of the above. **(Haryana CEET)**

2000

35. Soft drink bottles and baby feeders are generally made of
(a) Polyesters (b) Polyurethane
(c) Polyuria (d) Polyamides
(e) Polystyrene. **(Kerala MEE)**

36. $F_2C = CF_2$ is a monomer unit of
(a) Teflon (b) Glyptal
(c) BuNa-S (d) Nylon-6.

37. Which among the following is used in the preparation of *Nylon*-66 ?
(a) SF_6 (b) Adipic acid
(c) Sulphurous acid (d) Phthalic acid.
(MP CEE)

38. Polymer used in bullet proof glass is
(*a*) PMMA (*b*) Lexane
(*c*) Nomex (*d*) Kelvar. **(*DCE*)**

39. Which one among the following is not a correct match ?
(*a*) Silk (Polyamide) (*b*) Lipase (Ester)
(*c*) Indigo (Azo dye) (*d*) Karatin (Protein). **(*DCE*)**

40. SLV-3 uses propellants are
(*a*) Solid (*b*) Liquid
(*c*) Solid-Liquid (*d*) Biliquid. **(*MP CET*)**

41. $F_2C = CF_2$ is monomer of
(*a*) Teflon (*b*) Glyptal
(*c*) Nylon-6 (*d*) Buna-S. **(*CBSE PMT*)**

2001

42. Which of the following is a constituent of nylon-6 ?
(*a*) Adipic acid (*b*) Styrene
(*c*) Teflon (*d*) None of these.
(*MCIMS Wardha*)

43. Caprolactam polymerises to give
(*a*) Terylene (*b*) Teflon
(*c*) Glyptal (*d*) Nylon-6. **(*JIP MER*)**

44. Polyvinyl alcohol can be prepared by
(*a*) Polymerization of vinyl alcohol
(*b*) Alkaline hydrolysis of polyvinyl acetate
(*c*) Polymerization of acetylene
(*d*) Reaction of acetylene with H_2SO_4 in presence of $HgSO_4$. **(*UP SEAT*)**

2002

45. Which of the following is a polyamide molecule ?
(*a*) Terylene (*b*) Rayon
(*c*) Nylon-6 (*d*) Polystyrene. **(*UP SEAT*)**

46. Which of the following is not correctly matched ?
(*a*) Neoprene :

$$\left[-CH_2-\underset{\displaystyle Cl}{\underset{|}{C}}=CH-CH_2-\right]_n$$

(*b*) Nylon-66 :

$$\left[-NH-[CH_2]_6-NH-CO-[CH_2]_4-\overset{\displaystyle O}{\overset{\|}{C}}-O-\right]_n$$

(*c*) Terylene :

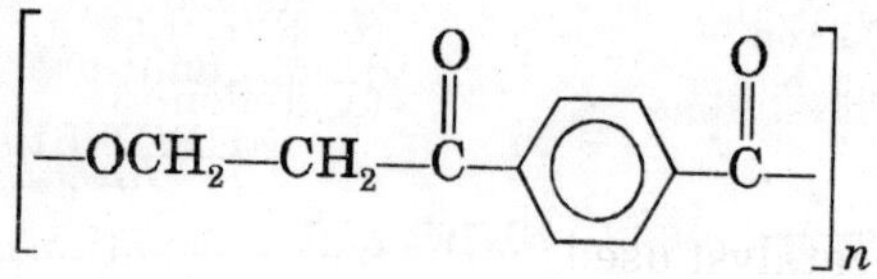

(*d*) PMMA :

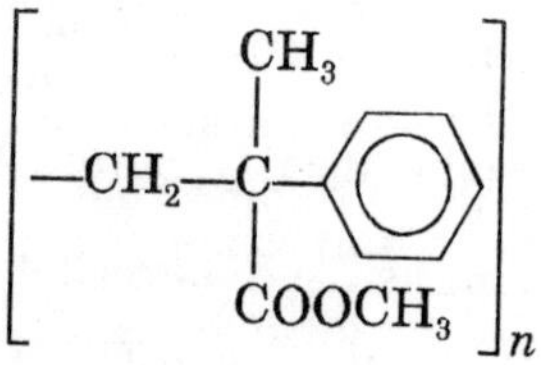

(*CBSE PMT*)

47. Monomer of 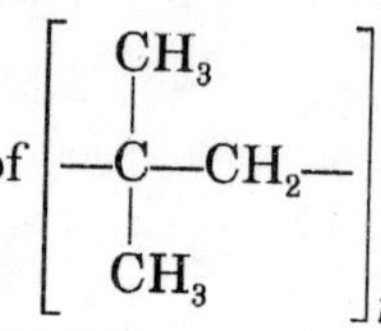 is
(*a*) 2-Methylpropene (*b*) Styrene
(*c*) Propylene (*d*) Ethene. **(*CBSE PMT*)**

48. Teflon is a polymer of
(*a*) Tetrafluorethylene (*b*) Tetraiodoethylene
(*c*) Tetrabromoethylene (*d*) Tetrachloroethylene.
(*AIIMS*)

49. Bakelite is a product formed from
(*a*) Reaction of formaldehyde with phenol
(*b*) Reaction of polyethylene with phenol
(*c*) Reaction of polypropylene with acid
(*d*) It is a natural product. **(*MP CET*)**

50. The components of nylon-66 are
(*a*) Hexamethylene diamine and adipic acid
(*b*) Hexamethylene diamine and sebaic acid
(*c*) Hexamethylene only
(*d*) None of the above. **(*DCE*)**

51. Natural rubber is which type of polymer ?
(*a*) Condensation of polymer
(*b*) Addition polymer
(*c*) Co-ordinate polymer
(*d*) None of the above. **(*DCE*)**

52. Which is a protein ?
(*a*) Nylon (*b*) Rayon
(*c*) Natural silk (*d*) Terylene. **(*Pb. CET*)**

53. Natural rubber is a polymer of
(*a*) Isoprene (*b*) Styrene
(*c*) Ethylene (*d*) Butadiene.
(*Pb. CET*)

54. A condensation polymer among the following is
(*a*) Dacron (*b*) PVC
(*c*) Polystyrene (*d*) Teflon.
(*Karnataka CET*)

55. The catalyst used for the polymerization of olefins is
(*a*) Ziegler-Natta catalyst (*b*) Wilkinson's catalyst
(*c*) Pd-catalyst (*d*) Zeise's complex
(*e*) Zeolite. **(*Kerala CET*)**

56. Which of the following is used in plants ?
(*a*) Terylene (*b*) Nylon
(*c*) Glyptal (*d*) Chloroprene. **(*DPMT*)**

57. Polymer formation from monomers starts by
(*a*) Condensation reaction between monomers
(*b*) Coordination reaction between monomers
(*c*) Conversion of monomer to monomer ions by protons
(*d*) Hydrolysis of monomers. **(*AIEEE*)**

58. Cellulose acetate is a
(*a*) Natural polymer (*b*) Semisynthetic polymer
(*c*) Synthetic polymer (*d*) Plasticiser. **(*JIP MER*)**

59. PVC is used for
(*a*) Manufacture of cosmetics
(*b*) Manufacture of tyres
(*c*) Manufacture of non stick pans
(*d*) Manufacture of plastic pipes. **(*Orissa JEE*)**

2003

60. Treatment of rubber with sulphur is
(*a*) Annealing (*b*) Vulcanisation
(*c*) Quenching (*d*) None of these. **(*AFMC*)**

61. Acrilan is a hard, horny and a high melting material. Which one of the following represents its structure ?

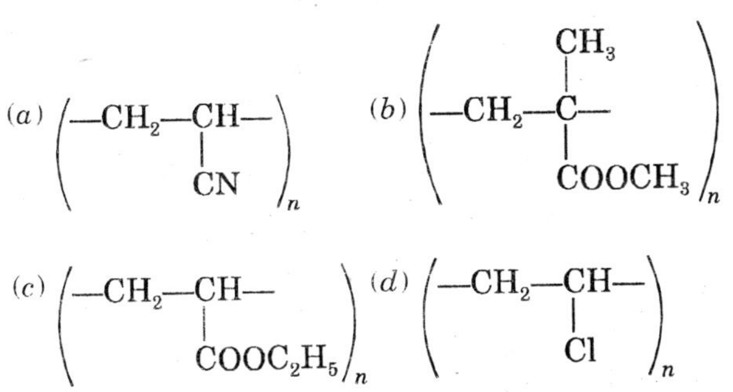

(*CBSE PMT*)

62. Which one of the following monomers gives the polymer neoprene on polymerization ?
(*a*) $CH_2 = CH—Cl$ (*b*) $CCl_2 = CCl_2$
(*c*) $CH_2 = C(Cl)—CH = CH_2$ (*d*) $CF_2 = CF_2$.
(*CBSE PMT*)

63. Nylon threads are made of
(*a*) Polyvinyl polymer (*b*) Polyester polymer
(*c*) Polyamide polymer (*d*) Polyethylene polymer.
(*AIEEE*)

2004

64. Insulin production and its action in human body are responsible for the level of diabetes. This compound belongs to which of the following categories ?
(*a*) A co-enzyme (*b*) An antibiotic
(*c*) An enzyme (*d*) A hormone. **(*AIEEE*)**

65. The nucleic and base having three possible binding sites is
(*a*) Thymine (*b*) Cytosine
(*c*) Guanine (*d*) Adenine. **(*AIIMS*)**

66. Which one of the following structures represents the peptide chain ?

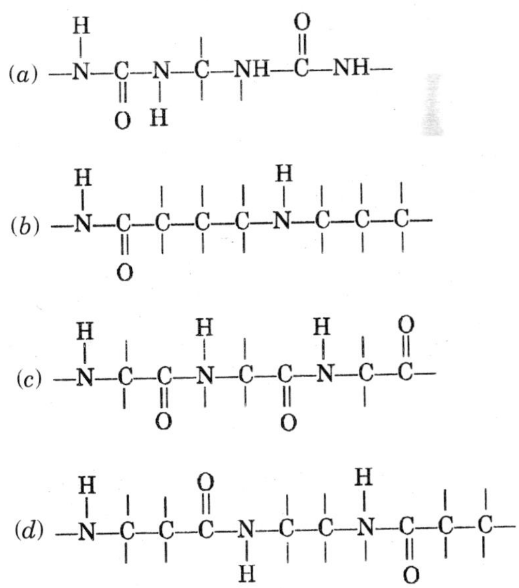

(*CBSE PMT*)

67. A sequence of how many nucleotides in messenger RNA makes a codon for an amino acid ?
(*a*) Three (*b*) Four
(*c*) One (*d*) Two. **(*CBSE PMT*)**

68. A hormone that helps in conversion of glucose to glycogen is
(a) Cortisone (b) Bile acids
(c) Adrenaliso (d) Insulin. **(CBSE PMT)**

2005

69. Which of the following is a polynude ?
(a) Bakelite (b) Terylene
(c) Nylon 66 (d) Teflon. **(AIEEE)**

70. Which of the following is fully fluorinated polymer ?
(a) PVC (b) Thiokol
(c) Teflon (d) Neoprene. **(AIIMS)**

71. On controlled hydrolysis and condensation, R_3SiCl yields
(a) $R_3Si—O—SiR_3$ (b) $-\!\!\left(R_3Si—O—SiR_3\right)\!\!-_n$
(c) R_3SiOH (d)

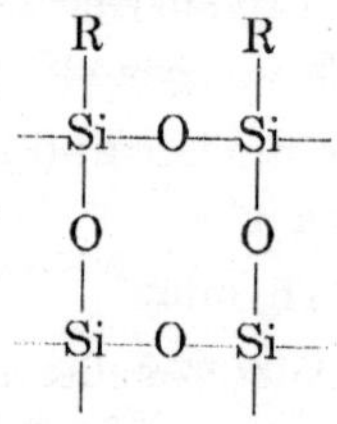

(Orissa JEE)

72. Which is not a polymer ?
(a) Sucrose (b) Enzyme
(c) Starch (d) Teflon. **(DPMT)**

73. Three dimensional molecules with cross links are formed in the case of a
(a) Thermoplastic (b) Thermosetting plastic
(c) Both (d) None. **(Karnataka CET)**

74. The number average molecular mass and mass average molecular mass of a polymer are respectively 30,000 and 40,000. The poly dispersity index of the polymer is
(a) < 1 (b) > 1
(c) 1 (d) 0
(e) – 1. **(Kerala PET)**

2006

75. $-\!\!\left(NH(CH_2)_6NHCO(CH_2)_4CO\right)\!\!-_n$ is a
(a) Homopolymer (b) Copolymer
(c) Addition polymer (d) Thermosetting polymer **(BSE PMT)**

76. A plot of log $\frac{x}{m}$ versus log p for the adsorption of a gas on a solid gives a straight line with slope equal to
(a) log K (b) – log K
(c) n (d) $\frac{1}{n}$. **(CBSE PMT)**

2007

77. Which of the following polymers is prepared by condensation polymerization ?
(a) Styrene (b) Nylon-66
(c) Teflon (d) Rubber. **(CBSE PMT)**

ANSWERS

1. (b)	2. (a)	3. (d)	4. (b)	5. (a)	6. (b)	7. (c)	8. (d)	9. (d)
10. (d)	11. (a)	12. (b)	13. (a)	14. (b)	15. (a)	16. (d)	17. (b)	18. (b)
19. (b)	20. (a)	21. (d)	22. (d)	23. (d)	24. (d)	25. (c)	26. (d)	27. (a)
28. (d)	29. (d)	30. (c)	31. (c)	32. (b)	33. (a)	34. (b)	35. (e)	36. (a)
37. (b)	38. (b)	39. (c)	40. (a)	41. (a)	42. (d)	43. (d)	44. (b)	45. (c)
46. (b)	47. (a)	48. (a)	49. (a)	50. (a)	51. (b)	52. (c)	53. (a)	54. (a)
55. (a)	56. (c)	57. (a)	58. (b)	59. (d)	60. (b)	61. (b)	62. (c)	63. (c)
64. (d)	65. (c)	66. (c)	67. (a)	68. (d)	69. (c)	70. (c)	71. (a)	72. (a)
73. (b)	74. (b)	75. (b)	76. (d)	77. (b).				

HINTS/SOLUTIONS

Q. No.	Ans.	Hints & Solutions
1.	(*b*)	Isoprene is $CH_2 = C(CH_3)—CH = CH_2$.
2.	(*a*)	Aromatic electrophilic substitution.
3.	(*b*)	Glyptal (gly = glycerol, ptal = phthalic acid).
6.	(*b*)	It is acetyl salicylic acid (aspirin).
8.	(*d*)	Terylene is a condensation polymer.
10.	(*d*)	$\left[CH_2—CH(Cl) \right]_n$.
13.	(*a*)	Buna-S is Butadiene-styrene rubber.
14.	(*b*)	It is formed by combination of different monomeric units.
16.	(*d*)	Cellophane.
17.	(*b*)	It is aspirin.
18.	(*b*)	These are insoluble in water and cannot be applied directly.
21.	(*d*)	Polymers have high molecular masses.
23.	(*d*)	It is made of isoprene unit.
24.	(*d*)	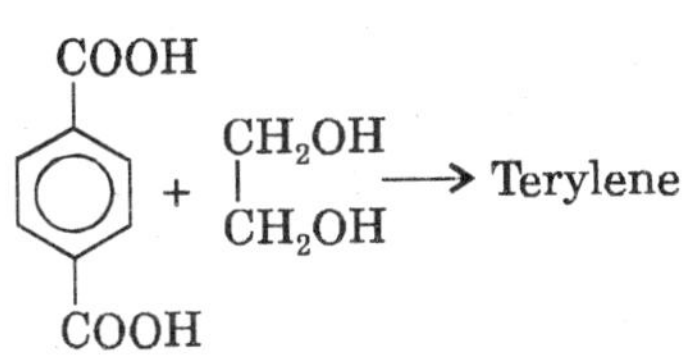
25.	(*c*)	Made up of same kind of monomeric units.
26.	(*d*)	Used for many micro-organism.
27.	(*a*)	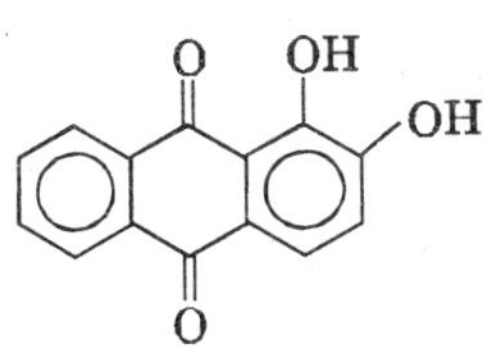
29.	(*d*)	It is a condensation polymer.
32.	(*b*)	Vulcanisation gives more tenale strength to the polymer.
34.	(*b*)	Hydrogen bonding.
35.	(*e*)	$\left[CH(C_6H_5)CH_2 \right]_n$.
36.	(*a*)	$\left[CF_2—CF_2 \right]_n$ is teflon.
37.	(*b*)	Nylon-66 is made of monomeric units of adipic acid and hexamethylene diammine.
39.	(*c*)	Indigo is indigoid dye.
42.	(*d*)	Nylon-6 is made from cyclohexane.
44.	(*b*)	$\left[CH_2—CH(OOCCH_3) \right]_n \xrightarrow[H_2O]{OH^-} \left[CH_2—CH(OH) \right]_n$
45.	(*c*)	Nylon-6 has -CO-NH linkage.
46.	(*b*)	Nylon-66 is $\left[NH—(CH_2)_6—NH—\overset{O}{\overset{\Vert}{C}}—(CH_2)_4—\overset{O}{\overset{\Vert}{C}} \right]_n$
49.	(*a*)	Bakelite is formed from condensation of phenol and formaldehyde.
50.	(*a*)	See answer Q. 56.
51.	(*b*)	Natural rubber has monomeric units of isoprene.
53.	(*a*)	$\left[CH_2—\overset{CH_3}{\overset{\vert}{C}} = C—CH_2 \right]_n$ is natural rubber.
57.	(*a*)	Condensation polymers are formed by loss of H_2O, C_2H_5OH or NH_3 molecules.
58.	(*b*)	It is prepared by treatment of cellulose.
60.	(*b*)	Vulcanisation renders elastic and stretchable properties to rubber.
62.	(*c*)	It is obtained from chloropene, $CH_2—\overset{Cl}{\overset{\vert}{C}}—CH = CH_2$
65.	(*c*)	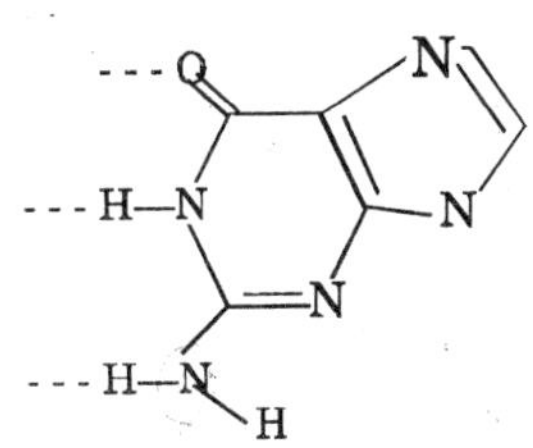

Q. No.	Ans.	Hints & Solutions
69.	(c)	$HOOC—(CH_2)—COOH + NH_2(CH_2)_6NH_2$ $\downarrow$ $\left[\overset{O}{\overset{\parallel}{C}}—(CH_2)_4—\overset{O}{\overset{\parallel}{C}}—NH—(CH_2)_6NH \right]_n$
70.	(c)	Teflon is, $\left[CF_2—CF_2 \right]_n$, a fully fluorinated polymer.
71.	(a)	R_3SiCl on hydrolysis forms only a dimer. $R_3SiOH + HOSiR_3 \rightarrow R_3Si—O—SiR_3$.
72.	(a)	Sucrose is a disaccharides which upon acid or enzymatic hydrolysis gives only two molecules of monosaccharides. Sucrose $\xrightarrow{H^+ \text{ or invertase}}$ D(+)-glucose + D(–)-fructose
73.	(b)	Thermosetting plastics have three dimensional cross-linked structure. Such polymers are prepared in two steps. The first step is the formation of long chain molecules which are capable of further reaction with each other. The second step is the application of heat which causes a reaction to occur between the chains, thus producing a complex cross-linked polymer.
74.	(b)	Number average molecular mass, $\overline{M_n} = 30{,}000$ Mass average molecular mass, $\overline{M_w} = 40{,}000$ Poly dispersity idex (PDI) $= \dfrac{\overline{M_w}}{\overline{M_n}} = \dfrac{40{,}000}{30{,}000} = 1.33$.
75.	(b)	It is nylon 6, 6. It is a copolymer of hexamethylene diammine and adeptic acid.
76.	(d)	The Freundlich adsorption isotherm is mathematically expressed as $\log \dfrac{x}{m} = \log k + \dfrac{1}{2} \log p$ Hence slope of st. line for plot between $\log \dfrac{x}{m}$ vs $\log p = \dfrac{1}{n}$.
77.	(b)	Nylon-66 is formed by condensation of adipic acid with hexamethylene diamine.

10.5. ASSERTION-REASON QUESTIONS

Quick Revision

In each of the following questions two statements are given as Assertion A and Reason R. Examine the statements carefully and answer the questions according to the instructions given below :

- *Mark **a** if both **A** and **R** are correct and **R** is the correct reason of **A**.*
- *Mark **b** if both **A** and **R** are correct and **R** is not the correct reason of **A**.*
- *Mark **c** if **A** is correct and **R** is wrong.*
- *Mark **d** if **A** is wrong and **R** is correct.*
- *Mark **e** if both **A** and **R** are wrong.*

1. ***Assertion A.*** Polypropylene is an addition polymer

Reason R. The monomer unit of polypropylene is $CH_3CH = CH_2$.

(a)	(b)	(c)	(d)	(e)

2. ***Assertion A.*** 1, 3-butadiene molecules add on to one another to form polybutadiene.

Reason R. Polybutadiene is a step growth polymer.

(a)	(b)	(c)	(d)	(e)

3. ***Assertion A.*** Glyptal is used in the manufacture of fibres.

Reason R. Glyptal is condensation polymer of ethane 1, 2-diol and terephthalic acid.

(a)	(b)	(c)	(d)	(e)

4. ***Assertion A.*** Teflon has high thermal stability and chemical inertness.

Reason R. Teflon is highly cross-linked polymer.

(a)	(b)	(c)	(d)	(e)

5. ***Assertion A.*** Nylon is a vinyl polymer.

Reason R. Nylon is a polyamide.

(a)	(b)	(c)	(d)	(e)

6. ***Assertion A.*** In the name *Buna-S,* S refers to sodium.

Reason R. Buna-S is also called SBR.

(a)	(b)	(c)	(d)	(e)

7. ***Assertion A.*** One drop of super glue is claimed to support a weight of 2000 lbs.

Reason R. Super glue is a polymer of α-methyl cyanoacrylate.

(a)	(b)	(c)	(d)	(e)

8. ***Assertion A.*** Aspirin can cause ulcers in the stomach.

Reason R. Ester group in aspirin gets hydrolysed to acid in the stomach where pH is about 12.

(a)	(b)	(c)	(d)	(e)

9. ***Assertion A.*** Penicilline is one of the tranquiliser.

Reason R. Penicilline is effective against gram +ve and gram –ve bacteria.

(a)	(b)	(c)	(d)	(e)

10. ***Assertion A.*** Alizrine dyes the fabric directly.

Reason R. Alizrine belongs to the family of mordant dyes.

(a)	(b)	(c)	(d)	(e)

11. ***Assertion A.*** The specific impulse of H_2F_2 is very high.

Reason R. NH_4ClO_4 is used as an oxidiser in composite propellants.

(a)	(b)	(c)	(d)	(e)

12. ***Assertion A.*** The active chemotherapeutic agent in *Disprin* is acetyl salicylic acid.

Reason R. Acetyl salicylic acid is also called 2-acetoxybenzoic acid.

(a)	(b)	(c)	(d)	(e)

13. ***Assertion A.*** Reserpine is a powerful tranquilizer.

Reason R. Reserpine in an alkaloid which is isolated from *Rauwolfia serpentina*.

(a)	(b)	(c)	(d)	(e)

14. ***Assertion A.*** Methylene blue is a basic dye and is also used in medicine as antiseptic.

Reason R. Methylene blue is used as indicator in acid-alkali titrations.

(a)	(b)	(c)	(d)	(e)

15. ***Assertion A.*** Crocin and Disprin are both antipyratics.

Reason R. Both crocin and disprin are ineffective against malaria.

(a)	(b)	(c)	(d)	(e)

ANSWERS

1. (b) **2.** (c) **3.** (c) **4.** (c) **5.** (d) **6.** (d) **7.** (b) **8.** (a) **9.** (d)
10. (b) **11.** (b) **12.** (b) **13.** (b) **14.** (c) **15.** (d).

10.6. SELF EVALUATION TEST

Choose the correct answer from the four alternatives given in each of the following questions :

1. Which of the following can absorb over 90% of its own mass of water and does not stick to wound ?
(*a*) Rayon (*b*) Gum cotton
(*c*) Cotton (*d*) Thiokol.

2. A co-polymer of isobutylene and isoprene is called
(*a*) Buna-S (*b*) Butyl rubber
(*c*) Thiokol (*d*) None of these.

3. A co-polymer of $CH_2 = CHCl$ and $CH_2 = CH(OCOCH_3)$ is called
(*a*) Saran (*b*) Orlon
(*c*) Dynal (*d*) Vinyon.

4. Which of following substance is employed in making explosive ?
(*a*) Oxalic acid (*b*) Glycerol
(*c*) Urea (*d*) All these.

5. Malamine plastic crockery is made of polymer material, the raw materials of which are
(*a*) CH_3CHO and CH_3NH_2
(*b*) HCHO and 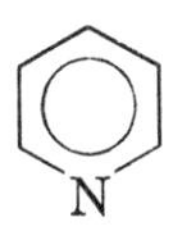
(*c*) HCHO and 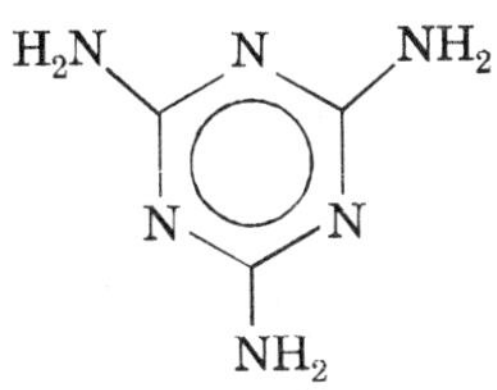
(*d*) HCHO and 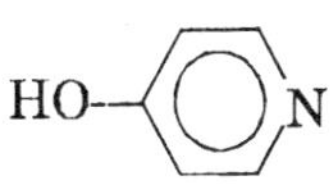

6. The polymer of 1, 3-butadiene and acrylonitrile is called
(*a*) Buna-S rubber (*b*) Buna-N rubber
(*c*) Acrilan (*d*) None of these.

7. Which one of the following is a protein fibre ?
(*a*) Polyester (*b*) Cotton
(*c*) Silk (*d*) Rayon.

8. Which of the following can be remelted time and again without producing any change ?
(*a*) Bakelite (*b*) Melmec
(*c*) Elastomers (*d*) Thermoplastics.

9. The polymerisation of ClC_2F_3 will produce
(*a*) PCTFE (*b*) PTFE
(*c*) PMMA (*d*) PAN.

10. Cellulose is a natural polymer of
(*a*) α-D glucose (*b*) β-D glucopyranose
(*c*) α-D fructofuranose (*d*) Maltose.

11. Chemical name of malamine is
(*a*) 2, 4-Diamino-1, 3, 5-triazine
(*b*) 2, 4, 6-Triamino-1, 3, 5-triazine
(*c*) 1, 3, 5-Triamino benzene
(*d*) 1, 3, 5-triamino 2, 5, 6-triazine.

12. Which of the following element is present in natural silk but not in artificial silk ?
(*a*) Sulphur (*b*) Chlorine
(*c*) Nitrogen (*d*) Hydrogen.

13. The additive used to make *PVC*, a flexible plastic is called
(*a*) Antioxidant (*b*) Plasticiser
(*c*) Stabilizer (*d*) Synthesizer.

14. Which of following rubber is not a polydiene ?
(*a*) Polyisoprene (*b*) Polychloroprene
(*c*) Thiokol rubber (*d*) Nitrile rubber.

15. Which of the following is not a cellulose product ?
(*a*) Gun cotton (*b*) Rayon
(*c*) Dacron (*d*) Celluloid.

16. Which of the following polymer has a monomer unit named *prop-2-ene-1-nitrile* ?
(*a*) Orlon (*b*) Saran
(*c*) Dacron (*d*) Tetron.

17. The process of vulcanisation of rubber was first introduced by
(*a*) Wohler (*b*) Charle, Goodyear
(*c*) A. Dunlop (*d*) J.K. Tyres.

18. Which of the following is not a biopolymer ?
(*a*) Starch (*b*) Proteins
(*c*) Nucleic acids (*d*) Sucrose.

19. Which of the following does not cause pollution ?
(*a*) Burning of coal (*b*) Burning of rubber
(*c*) Use of solar energy (*d*) Burning of petrol.

20. Silicone is a water repellant, heat resistant and acid resistant polymer. The monomer unit of silicone is

(a) Si atoms (b) SiO_2
(c) R_2SiO (d) $(SiO_4)^{4-}$.

21. The abbreviation PDi refers to
(a) Some application of polymers
(b) Poly dichloro isoprene
(c) Poly diphenyl icosane
(d) Poly dispersity index.

22. Which of the following is a branched chain polymer ?
(a) Polyester (b) Nylons
(c) PVC (d) LDP.

23. Which of the following is non-biodegradable ?
(a) Soap (b) Paper
(c) Polyethene (d) Wood.

24. Nomex is
(a) Polyester (b) Polyenes
(c) Polydiene (d) Polyamide.

25. DDT (dichlorodiphenyl trichloro ethane) is an example of
(a) Fungicide (b) Herbicide
(c) Insecticide (d) Rodenticide.

26. Which of the following is an auxochrome ?
(a) $-C=C-$ (b) $-NO_2$
(c) —OH (d) All the three.

27. Propellants may be
(a) Only solids
(b) Only liquids
(c) Both solids and liquids
(d) None of these.

28. Which of the following is used as anaesthesia ?
(a) CH_4 (b) CO_2
(c) N_2O (d) N_2.

29. Which of the following is used as an antiseptic ?
(a) Dettol (b) Iodoform
(c) Bithional (d) All the above.

30. 1% solution of phenol is used as a
(a) Antiseptic (b) Disinfectant
(c) Both (d) None.

31. Brufen is used to relieve
(a) Pain (b) Constipation
(c) Fungi infection (d) Bacterial infection.

32. Dettol consists of
(a) Xylenol + Terpeneol
(b) Chloroxylenol + terpeneol
(c) Cresol + ethanol
(d) None of the above.

33. With which cation alizarin will impact blue colour on the fabrics ?
(a) Al^{+3} (b) Ba^{+2}
(c) Cr^{+3} (d) Fe^{+3}.

34. A mordant binds the fabrics through
(a) Ionic bonds (b) Covalent bonds
(c) Co-ordinate bonds (d) Hydrogen bonds.

35. Which of the following is not a tranquilizer ?
(a) Barbituric acid (b) Seconal
(c) Luminal (d) Phenacetin.

36. Which one of the following is not used as fuel in rocket propellants ?
(a) Liquid hydrazine (b) Liquid hydrogen
(c) Liquid oxygen (d) Alcohol.

37. Which of the following can be used both as an antiseptic as well as a disinfectant ?
(a) Methanol (b) Phenol
(c) Chlorine (d) Dettol.

38. 'Ibuprofen' is a
(a) Propionic acid derivative
(b) Benzoic acid derivative
(c) Naphthoic acid derivative
(d) Sulphanilic acid derivative.

39. Which of following medicine was used to cause local anaesthesia ?
(a) Lignocaine (b) Nicotinic acid hydrazide
(c) Paracetamol (d) None of these.

40. Which of the following is a correct statement ?
(a) Every coloured compound can act as a dye
(b) Presence of a chromophore is necessary for a compound to act as dye
(c) Presence of chromophore as well as auxochrome group is necessary for a compound to act as dye
(d) All the above.

41. The colour bearing group in a compound is called
(a) Chromophore (b) Auxochrome
(c) Dye (d) None of these.

42. Antibiotic, penicillin-G is
(a) An alkaloid (b) An enzyme
(c) A heterocyclic (d) A steroid.

43. Bismarck brown is example of
(a) Phthalein dyes (b) Azo dyes
(c) Anthraquinone dyes (d) Nitro dyes.

44. Which dyes are synthesised and produced within the fibres and the colour so obtained is known as icecoloures
(a) Azoic dyes (b) Vat dyes
(c) Adjective dyes (d) Direct dyes.

45. The central nervous system depressants that produce sleep are called

40. What is the product 'C' in the following sequence of reaction

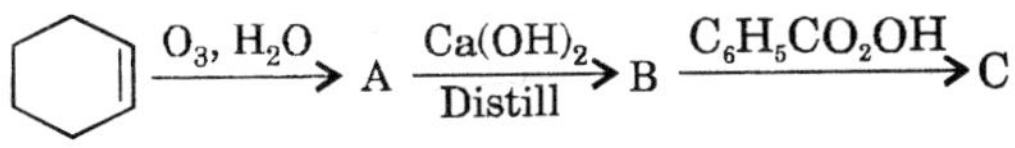

(*a*) 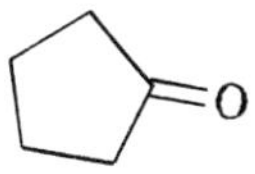 (*b*) 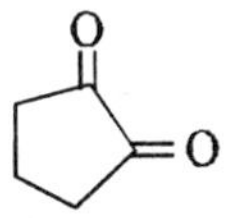

(*c*) 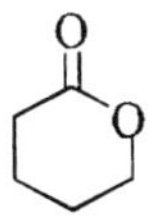 (*d*) 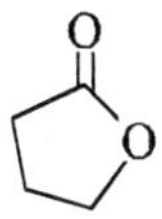

41. Which of the following will not give the same product on hydroboration-oxidation and acid-catalysed hydration ?

(*a*) 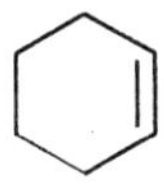 (*b*) 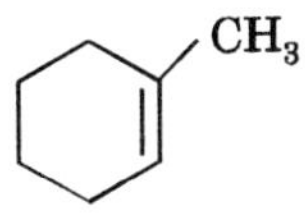

(*c*) $CH_2 = CH_2$ (*d*) $CH_3CH = CHCH_3$.

42. A hydrocarbon **A** of molecular formula (C_7H_{12}) on ozonolysis gives a compound **B** which undergoes aldol condensation to give 1-acetylcyclopentene. The hydrocarbon **A** is

(*a*) 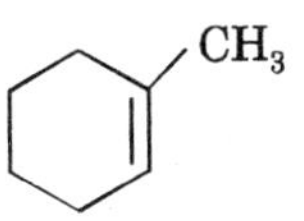 (*b*) 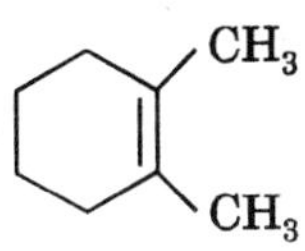

(*c*) 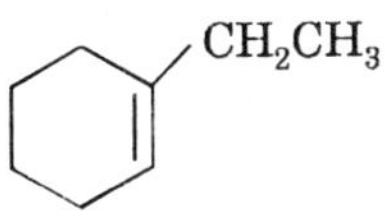 (*d*) 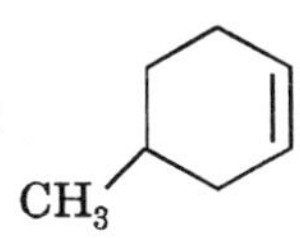

43. 0.175 g sample of an acid required 18.7 cm^3 of 0.09 M NaOH solution for neutralization. Which of the following acids, could it be ?

(*a*) 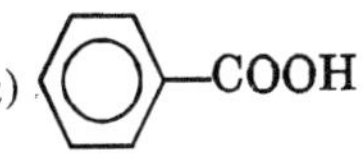 (*b*) 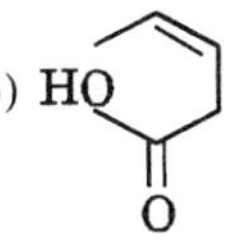

(*c*) 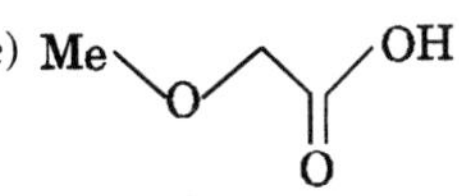 (*d*) 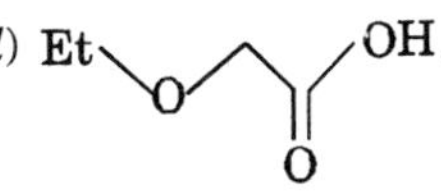

44. *trans*-2-butene is transformed into

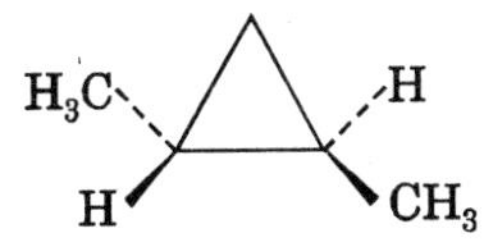

by reacting with

(*a*) CH_3^+ (*b*) $\dot{C}H_3$

(*c*) : CH_2 (singlet) (*d*) : CH_2 (triplet).

45. *Ethanenitrile* $\xrightarrow{\text{Na/EtOH}}$ A $\xrightarrow{\text{NaNO}_2\text{/HCl}}$ B $\xrightarrow{\text{NaOH/I}_2}$ C

In the above sequence **C** is

(*a*) CH_3COOH (*b*) C_2H_5NHOH

(*c*) CHI_3 (*d*) CH_3CONH_2.

ANSWERS

1. (*b*)	**2.** (*b*)	**3.** (*c*)	**4.** (*d*)	**5.** (*a*)	**6.** (*b*)	**7.** (*b*)	**8.** (*a*)	**9.** (*b*)
10. (*c*)	**11.** (*c*)	**12.** (*a*)	**13.** (*d*)	**14.** (*a*)	**15.** (*d*)	**16.** (*d*)	**17.** (*c*)	**18.** (*c*)
19. (*a*)	**20.** (*d*)	**21.** (*c*)	**22.** (*b*)	**23.** (*b*)	**24.** (*b*)	**25.** (*c*)	**26.** (*d*)	**27.** (*a*)
28. (*b*)	**29.** (*b*)	**30.** (*d*)	**31.** (*a*)	**32.** (*a*)	**33.** (*a*)	**34.** (*c*)	**35.** (*b*)	**36.** (*b*)
37. (*c*)	**38.** (*b*)	**39.** (*b*)	**40.** (*c*)	**41.** (*b*)	**42.** (*a*)	**43.** (*d*)	**44.** (*c*)	**45.** (*c*).

PRACTICE PAPER—1

$$RCH_2COOH \xrightarrow{1} RCH_2COCl \xrightarrow[NH_3]{2} RCH_2CONH_2$$
$$\downarrow 3$$
$$RCOOH \xleftarrow{5} RCH_2OH \xleftarrow{4} RCH_2NH_2$$

Consider the above sequence of steps and answer questions 1—7.

1. The appropriate reagent for step 3 is
   (*a*) Bromine/CCl_4 (*b*) Br_2/KOH
   (*c*) HBr (*d*) $LiAlH_4$.
2. Step 2 can be described as a/an
   (*a*) Neutralisation
   (*b*) Nucleophilic addition
   (*c*) Electrophilic substitution
   (*d*) Nucleophilic substitution.
3. Which reagent among the following is used for step 4 ?
   (*a*) HCl/H_2O (*b*) $NaNO_2$/HCl
   (*c*) NH_3 (*d*) N_2.
4. The other product(s) formed in step 4 is/are
   (*a*) NH_3, NO_2 (*b*) N_2, RCH_2NO_2
   (*c*) N_2
   (*d*) N_2 along with some alkene.
5. Which of the following is the intermediate involved in step 3 ?
   (*a*) RCH_2CO_2H (*b*) RCH_2CN
   (*c*) $RCH_2—N{=}C{=}O$ (*d*) $RCH_2—CH_2—N{=}O$
6. Reagent that can be used in step 1 is
   (*a*) KCl (*b*) Cl_2
   (*c*) NaOCl (*d*) PCl_5 or $SOCl_2$.
7. The process involved in step 5 is
   (*a*) Reduction (*b*) Oxidation
   (*c*) Substitution (*d*) Elimination.
8. A reacts with excess of methyl magnesium iodide and the adduct on hydrolysis produces tertiary butyl alcohol. **A** can be
   (*a*) Methylethanoate (*b*) Ethanol
   (*c*) Ethanal (*d*) Butanone
9. Reduction of *carbonyl group* to *methylene group* through alkaline decomposition of hydrazones is called
   (*a*) Birch reduction
   (*b*) Clemensen's reduction
   (*c*) Wolf Kishner reduction
   (*d*) M.P.V. reduction.
10. An acid **X** is neutralised with NaOH and the salt so produced is fused with sodalime when isobutane is produced. **X** can be
   (*a*) Butanoic acid (*b*) 3-methylbutanoic acid
   (*c*) Isobutyric acid (*d*) 2-methylbutanoic acid.
11. Electrophilic ring substitution is most difficult in :
   (*a*) 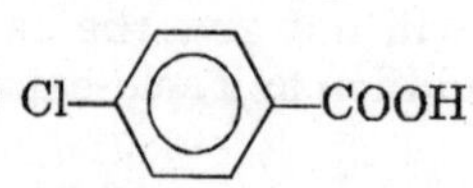 (*b*) 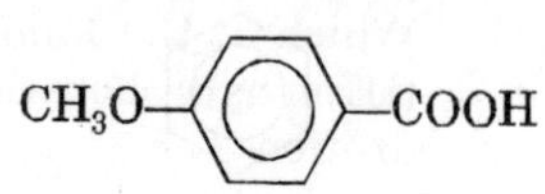
   (*c*) 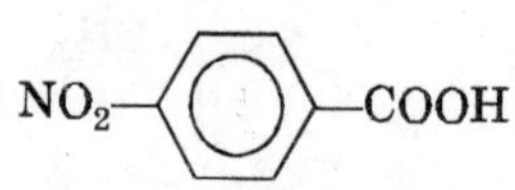 (*d*) 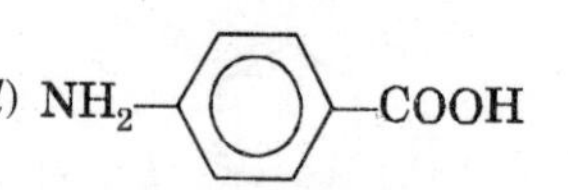
12. Benzene is refluxed with nitrating mixture at 413K for 1 hour and the product so formed is treated with sodium polysulphide. The end product will be
   (*a*) Aniline (*b*) *m*-phenylenediamine
   (*c*) *m*-nitroaniline (*d*) *m*-dinitrobenzene.
13. The correct IUPAC name of 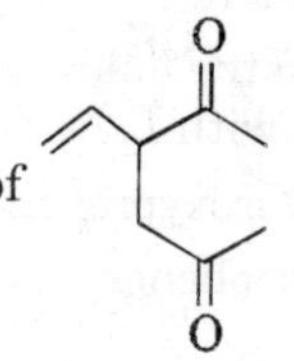 is
   (*a*) 3-Acetohex-5-ene-2-one
   (*b*) Acetylhex-1-ene-5-one
   (*c*) 3-(Prop-2-enyl)-2,5-pentanedione
   (*d*) None of these.
14. 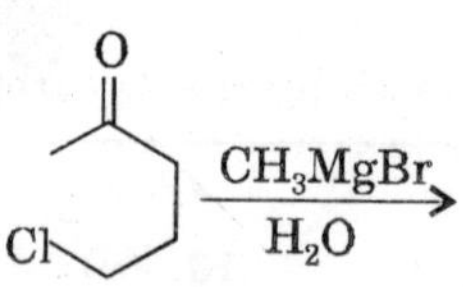
   The product of the reaction is
   (*a*) 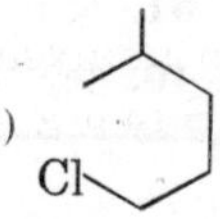 (*b*) 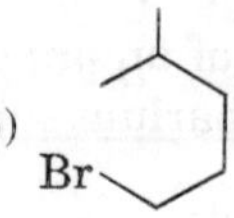
   (*c*) 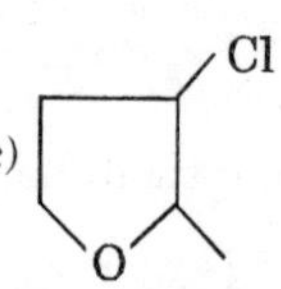 (*d*) 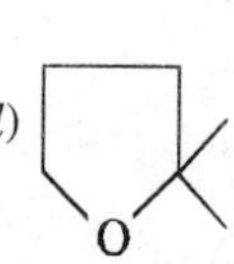
15. Which of the following statements is correct ?
   (*a*) Benzyl alcohol is more acidic than phenol
   (*b*) Ethanol is powerful oxidising agent

(c) Phenol is more acidic than 2-propanol

(d) Ethane has higher boiling point than ethanal.

16. The correct order of acid strength among the following compounds is

(I) 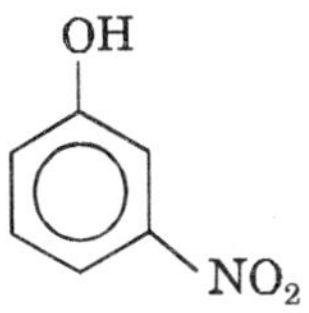 (II) 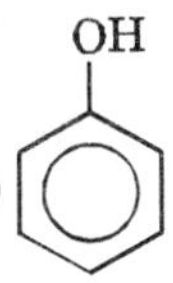

(III) 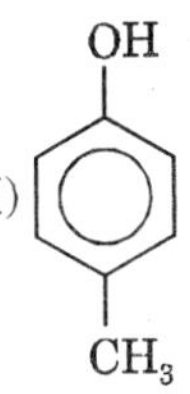 (IV) 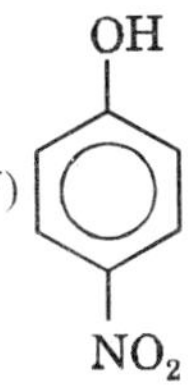

(a) II > IV > I > III (b) I > IV > II > III

(c) III > I > II > IV (d) IV > I > II > III.

17. Which of the following molecule can have different conformers ?

(a) MeOH (b) Azane

(c) Sulphane (d) Oxidane.

18. Which of the following compounds will give yellow precipitate with I_2/NaOH ?

(a) 2-Hydroxypropane (b) 3-Pentanone

(c) Benzophenone (d) Neopentyl alcohol.

19. $C_6H_6 \xrightarrow[(ii)\ CrO_2Cl_2]{(i)\ AlCl_3/CH_3Cl}$ **X. Here, X is**

(a) C_6H_5CHO (b) C_6H_5COOH

(c) Toluic acid (d) Chlorotoluene.

20. X $\xrightarrow[CCl_4]{Br_2}$ B $\xrightarrow{KOH(alc.)}$ **Propene**

Here, X can be

(a) Ethanoic acid (b) Silver isobutyrate

(c) Silver propanoate (d) Propene.

21. 0.2595 g of an organic compound on analysis yields, 0.35 g of barium sulphate. The % of S in organic compound is

(a) 81.52% (b) 28.42%

(c) 18.52% (d) 22.42%.

22. Which among the following is aromatic in nature ?

(a) 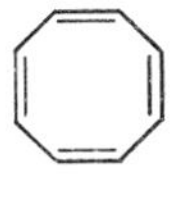 (b) 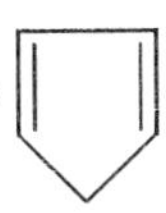

(c) 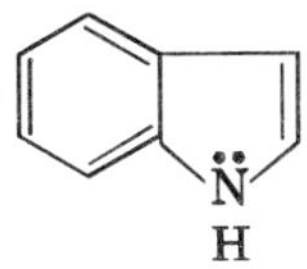 (d) 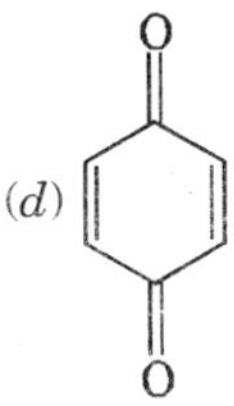

23. The number of optically active stereoisomers possible for pentane-2,3-diol is

(a) 1 (b) 2

(c) 3 (d) 4.

24. $(CH_3)_3CMgCl$ on reaction with D_2O produces

(a) $(CH_3)_3OD$ (b) $(CH_3)_3OD$

(c) $(CD_3)_3OD$ (d) $(CD_3)_3CD$.

25. The intermediate formed during the reaction of propene and HBr in the presence of benzoyl peroxide is

(a) $CH_3\dot{C}HCH_2Br$ (b) $CH_3CH_2\dot{C}H_2$

(c) $CH_3\overset{+}{C}HCH_2Br$ (d) $CH_3CH_2\overset{-}{C}H_2$.

26. Which of the following is the best leaving group ?

(a) – OAc (b) – OMe

(c) – OSO_2Me (d) – OSO_2CF_3.

27. Acrylic acid reacts with HBr to produce

(a) $CH_3(Br)CH(Br)CO_2H$ (b) $H_3CCH(Br)CO_2H$

(c) H_2C CH—COBr (d) $H_2C(Br)CH_2CO_2H$.

28. Which of the following is best starting material to synthesise 2-Methyl-2-butenoic acid ?

(a) 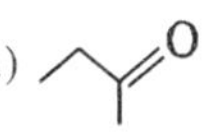 (b) 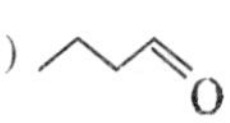

(c) 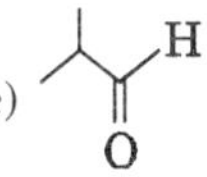 (d) 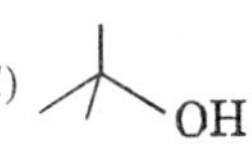

29. Which of the following pair of compounds undergo crossed aldol condensation followed by dehydration to give methyl vinyl ketone ?

(a) Methanal and ethanal

(b) Two moles of methanal

(c) Methanal and propanone

(d) Two moles of ethanal.

30. In general, the electron releasing effect among alkyl groups is found to be in order as CH_3— < CH_3CH_2— < $(CH_3)_2CH$— < $(CH_3)_3C$— but the order gets almost reversed when these groups are bonded to ethylenic bond. It is due to

(a) Electromeric effect (b) Mesomeric effect

(c) Tautomeric effect (d) Hyperconjugation effect.

31. In Duma's method, for estimation of nitrogen the gases produced in combustion tube are

(a) NH_3, $H_2O(g)$ (b) N_2, CO_2, $H_2O(g)$

(c) NO, NO_2, CO_2 (d) N_2, CO_2, CO.

32. Sec. butylchloride on reaction with sodium in ether yields

(*a*) 3,4-dimethylhexane (*b*) 2,5-Dimethylhexane
(*c*) *n*-octane (*d*) *iso*-octane.

33. Which of the following compounds has highest octane number ?

(*a*) 2,2,4-trimethylpentane
(*b*) *n*-nonane
(*c*) 2,2,3-trimethylpentane
(*d*) *n*-octane.

34. The number of allylic H atoms in cyclohexene is

(*a*) 4 (*b*) 2
(*c*) 6 (*d*) 8.

35. **Iodoform** $\xrightarrow{X}$ **Y** $\xrightarrow{Cl_2/H_2O}$ **dichloroethanal.**
Here X and Y are respectively

(*a*) Ag, ethanol (*b*) Ag, ethyne
(*c*) Zn, ethanol (*d*) Zn, ethene.

36. Most reactive cycloalkane among the following is

(*a*) Cyclopropane (*b*) Cyclohexane
(*c*) Cyclopentane (*d*) all are equally reactive.

37. 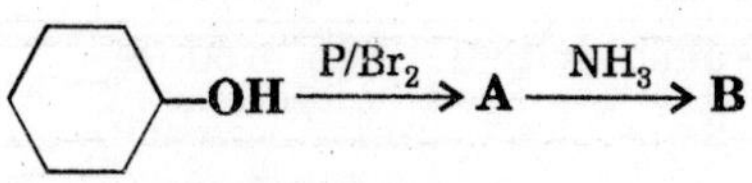

In the above sequence B is

(*a*) Cyclohexane
(*b*) Cyclohexylamine
(*c*) 2-Bromocyclohexylamine
(*d*) none of these.

38. *p*-nitroaniline is dissolved in glacial acetic acid and is treated with bromine. The main product is

(*a*) 2,6-Dibromo-4-nitroaniline
(*b*) 3,5-Dibromo-4-nitroaniline
(*c*) *p*-phenylenediamine
(*d*) 2,3,5,6-tetrabromo-4-nitroaniline.

39. The correct order of acidic strength among the following is

(I) 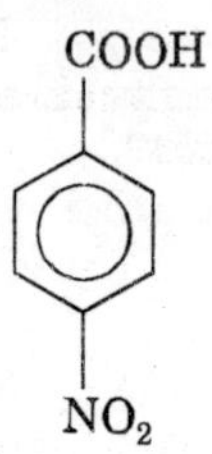

(II) 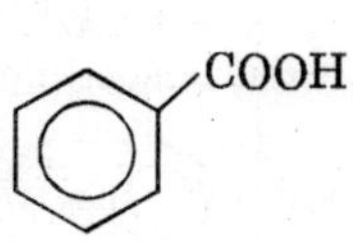

(III) 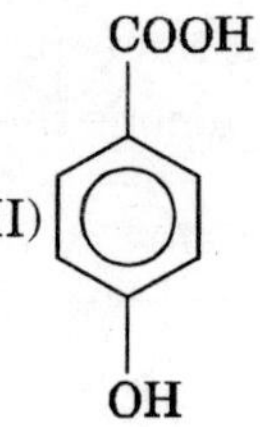

(IV) 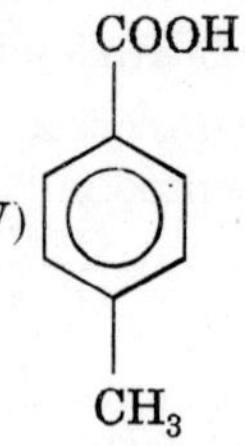

(*a*) I > II > III > IV (*b*) I > II > IV > III
(*c*) I > IV > III > II (*d*) IV > II > III > I.

40. 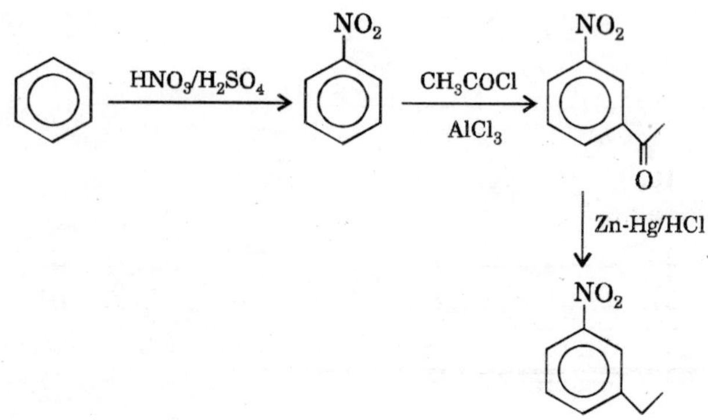

Which step is wrong in above synthesis ?

(*a*) (*i*) (*b*) (*ii*)
(*c*) (*iii*) (*d*) None.

41. The correct arrangement of increasing order of acid strength among the following is

(I) 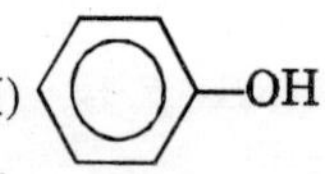

(II) 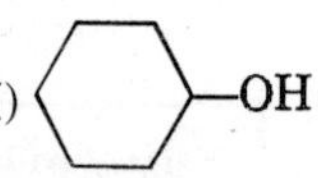

(III) 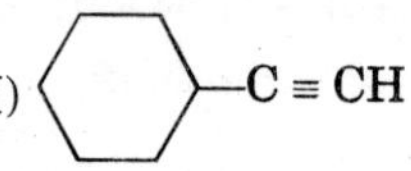

(IV) 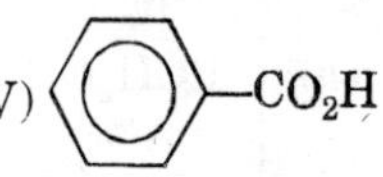

(*a*) III, II, I, IV (*b*) IV, III, II, III
(*c*) I, III, II, IV (*d*) I, II, III, IV.

42. The correct arrangement of increasing order of basic strength among the following **(I)** Aniline, **(II)** *p*-methoxyaniline **(III)** Benzylamine **(IV)** Acetanilide is

(*a*) IV, III, II, I (*b*) IV, I, II, III
(*c*) IV, II, III, I (*d*) III, IV, I, II.

43. In the reaction of HNO_2 with N,N-Dimethyl aniline the attacking reagent is

(*a*) NO_2^+ (*b*) NO_2^-
(*c*) NO^- (*d*) NO^+.

44. Which halide among the following undergoes nucleophilic substitution most readily ?

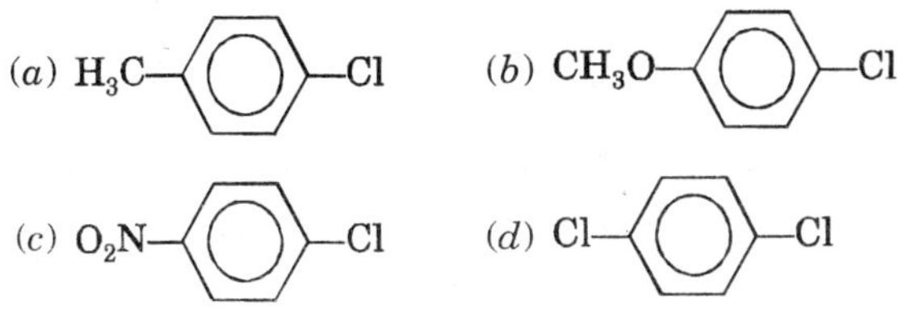

45. An optically active compound **X** ($C_4H_8O_3$) neutralises $NaHCO_3$ giving CO_2 gas. It also reacts with $LiAlH_4$ to give an achiral structure. X could be

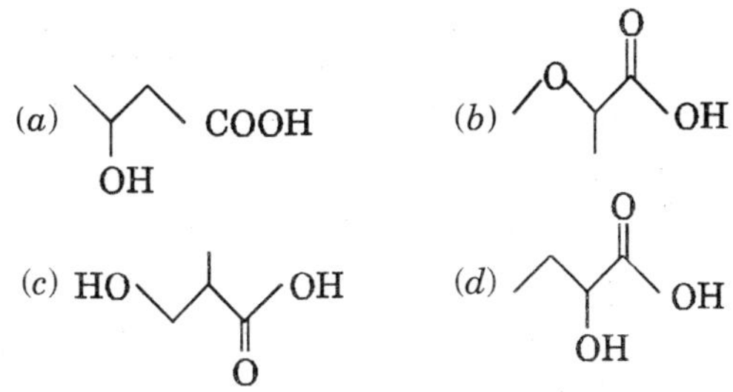

ANSWERS

1. (*b*)	**2.** (*d*)	**3.** (*b*)	**4.** (*d*)	**5.** (*c*)	**6.** (*d*)	**7.** (*b*)	**8.** (*a*)	**9.** (*c*)
10. (*b*)	**11.** (*c*)	**12.** (*c*)	**13.** (*c*)	**14.** (*d*)	**15.** (*c*)	**16.** (*d*)	**17.** (*a*)	**18.** (*a*)
19. (*a*)	**20.** (*b*)	**21.** (*c*)	**22.** (*c*)	**23.** (*d*)	**24.** (*a*)	**25.** (*a*)	**26.** (*d*)	**27.** (*d*)
28. (*c*)	**29.** (*c*)	**30.** (*d*)	**31.** (*b*)	**32.** (*a*)	**33.** (*c*)	**34.** (*a*)	**35.** (*b*)	**36.** (*a*)
37. (*b*)	**38.** (*a*)	**39.** (*b*)	**40.** (*b*)	**41.** (*a*)	**42.** (*b*)	**43.** (*d*)	**44.** (*d*)	**45.** (*c*).

HINTS/SOLUTIONS

Q. No.	Ans.	Hints & Solutions
1.	(*b*)	$RCH_2CONH_2 \xrightarrow{KOH/Br_2} RCH_2NH_2$
2.	(*d*)	Step 2 refers to nucleophilic acyl substitution.
3.	(*b*)	$RCH_2NH_2 \xrightarrow[(NaNO_2/HCl)]{HONO} RCH_2OH$
4.	(*d*)	Step IV occurs through the formation of carbocation ($RCCH_2^+$). It can give alkene or rearranged substitution products along with N_2.
5.	(*c*)	Step 4 is Hoffmann bromamide reaction. Hence, among the given species. R—CH_2—N = C = O is one of the intermediate of this reaction.
6.	(*d*)	Both PCl_5 or $SOCl_2$ can convert RCOOH to RCOCl.
7.	(*b*)	Step 5 involves oxidation of alcohol to acid.
8.	(*a*)	$CH_3—\overset{O}{\overset{\Vert}{C}}—OC_2H_5 + 2CH_3MgI + HOH \longrightarrow CH_3—\underset{CH_3}{\underset{\vert}{\overset{CH_3}{\overset{\vert}{C}}}}—OH + CH_3OMgI + Mg(CH)I$ Tert. butyl alcohol
9.	(*c*)	This occurs by use of reagent NH_2---NH_2/KOH at 450 K. The process is called Wolf Kishner reduction.
10.	(*b*)	$\underset{(X)}{CH_3\underset{CH_3}{\underset{\vert}{C}}HCH_2COOH} \xrightarrow{NaOH} CH_3\underset{CH_3}{\underset{\vert}{C}}HCH_2COONa \xrightarrow[Heat]{NaOH} \underset{\text{Isobutane}}{CH_3\underset{CH_3}{\underset{\vert}{C}}HCH_3}$
11.	(*c*)	Two electron withdrawing groups —NO_2 and —COOH deactivate the ring making it less susectible to electrophile attack.
12.	(*c*)	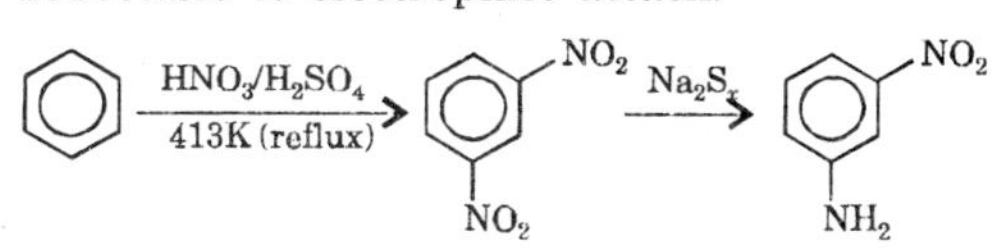
13.	(*c*)	The correct name is 3-(prop-2-enyl)2, 5-pentanedione.
14.	(*d*)	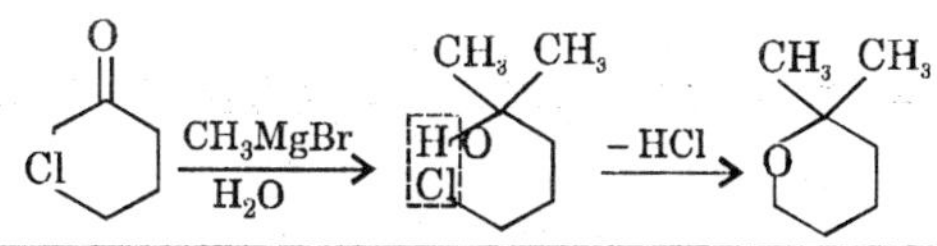

Q. No.	Ans.	Hints & Solutions
15.	(*c*)	Statements '*a*', '*b*', '*d*' are against the facts. But statement '*c*' is factual.
16.	(*d*)	The correct order of acid strength is : 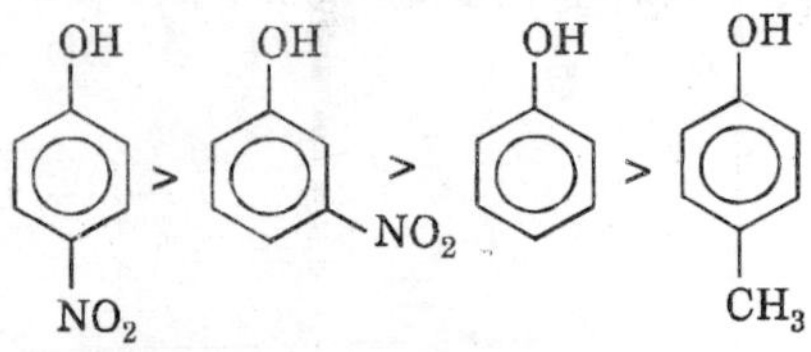 which is given in option (*d*).
17.	(*a*)	$CH_3—\ddot{O}—H$ can have different spatial arrangement of H atoms around the C—O bond. Azane (NH_3), sulphate (H_2S) and oxidane (H_2O) donot exhibit conformers.
18.	(*a*)	Among the given species only $(CH_3)_2CH—OH$ *i.e.*, 2-propanone will give idoform test positive. $(CH_3)_2CHOH + I_2 + NaOH \longrightarrow CHI_3 + CH_3COONa.$
19.	(*a*)	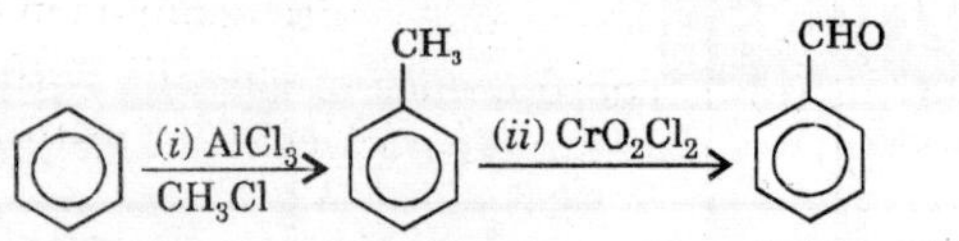 Benzene → Toluene → Benzaldehyde (X)
20.	(*b*)	$CH_3—CH(CH_3)—COOAg \xrightarrow{Br_2/CCl_4} CH_3—CH(CH_3)—Br + CO_2 + AgBr$ Silver isobutyrate (X); 2-Bromopropane (B) 2-Bromopropane (B) $\xrightarrow{KOH\ (alc.)}$ $CH_3CH = CH_2$ Propene
21.	(*c*)	$\% S = \frac{32 \times 0.35 \times 100}{0.2595 \times 233} = 18.52\%.$
22.	(*c*)	According to Huckel's rule, 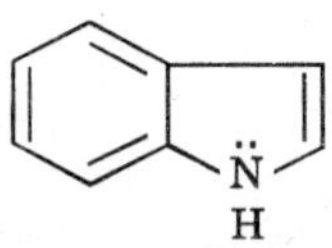 is aromatic in nature. It has 10π electrons delocalised over the entire stucture.
23.	(*d*)	It has two non-equivalent chirality centre. Thus, the number of optically active isomers is $2^2 = 4$
24.	(*a*)	$(CH_3)_3CMgCl + D_2O \longrightarrow (CH_3)_3C—OD + Mg(OD)Cl$
25.	(*a*)	The addition of HBr to propene in the presence of benzoyl peroxide follows free radical mechanism. It proceedes through the formation of 2° free radical, $(CH_3)(BrCH_2)\dot{C}H.$
26.	(*d*)	Among the given species $—OSO_2CF_3$ is the weakest base. Hence it is the best leaving group.
27.	(*d*)	$CH_2 = CH—COOH + HBr \longrightarrow CH_2(Br)—CH_2—COOH$ The addition occurs against the Markovnikoff's rule.
28.	(*a*)	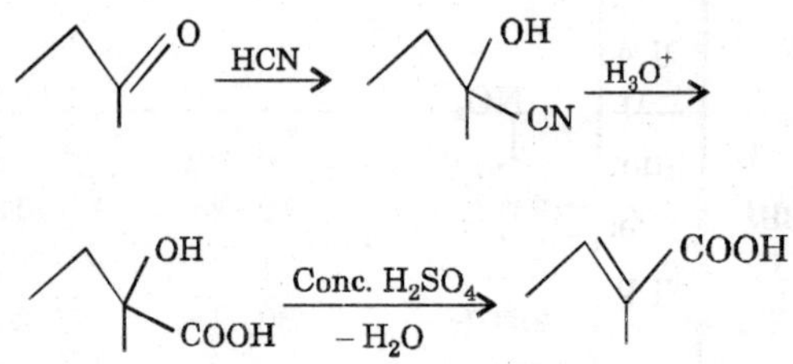 2 Methylbut-2-enoic acid
29.	(*c*)	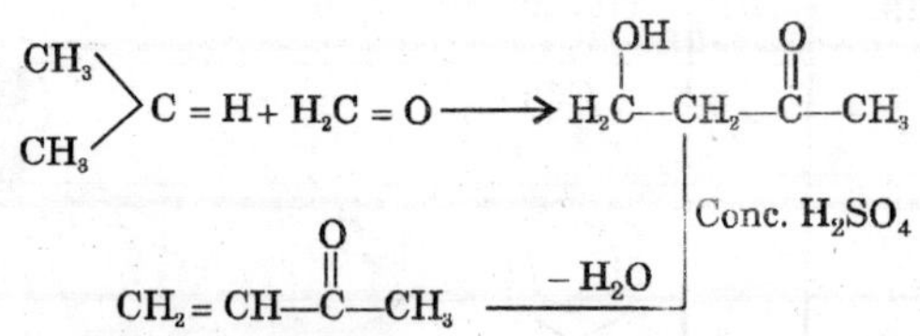 Methyl vinyl ketone
30.	(*d*)	This is due to hyperconjugation effect.
31.	(*b*)	The gases produced in the combustion tube are N_2, CO_2, $H_2O(g)$.
32.	(*a*)	$(CH_3)(C_2H_5)CH—Br + Br—CH(CH_3)(C_2H_5) \xrightarrow[Ether]{2Na} (CH_3)(C_2H_5)CH—CH(CH_3)(C_2H_5)$ 3, 4-Dimethylhexane
33.	(*c*)	2, 2, 3-Trimethylpentane has octane number 116.
34.	(*a*)	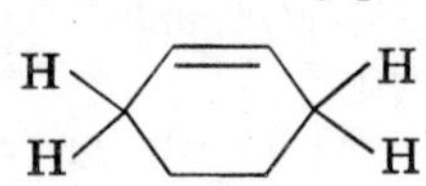 There are four allylic H-atoms as indicated.
35.	(*b*)	$2CHI_3 + Ag \longrightarrow CH \equiv CH$ (X) Ethyne (Y) $\xrightarrow{2Cl_2/H_2O} CHCl_2—CHO$ Dichloroethanal

Q. No.	Ans.	Hints & Solutions
36.	(*a*)	Cyclopropane is most reaction among the given cyclic compounds. Its unstable nature is related to angle strain.
37.	(*b*)	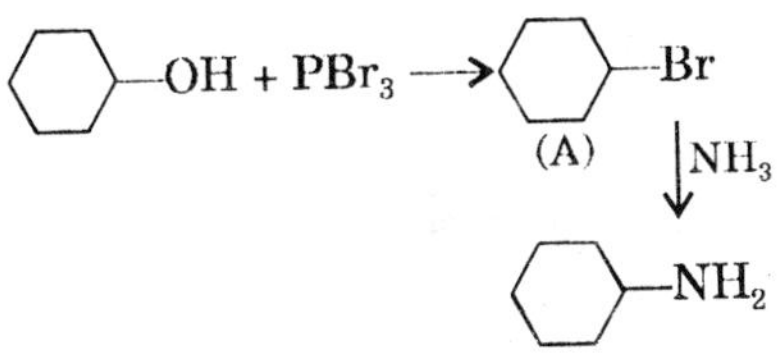
38.	(*a*)	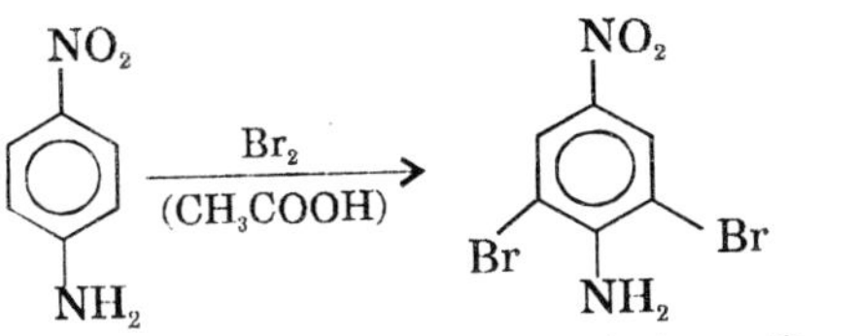 2, 6-Dibromo-4-nitroaniline
39.	(*b*)	The correct order of decreasing acid strength is 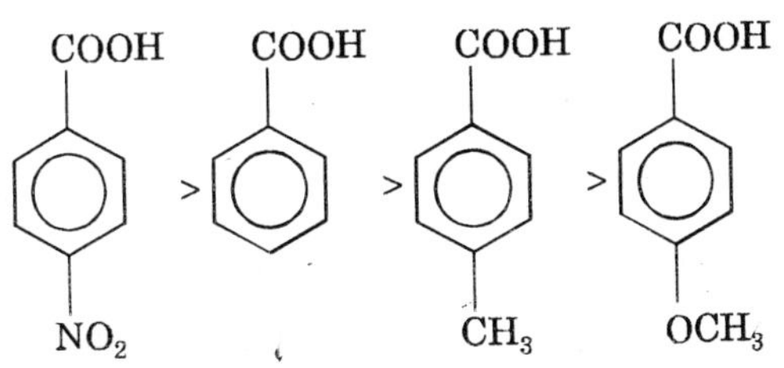 —NO_2 group is strong withdrawing group whereas — OCH_3 is strong electron releasing group.
40.	(*b*)	Step II in the above synthesis is incorrect because nitrobenzene does not undergo acylation.
41.	(*a*)	Correct arrangement of increasing acid strength is 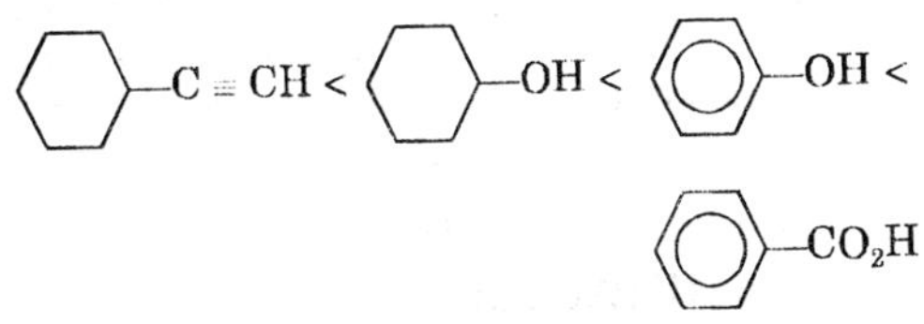
42.	(*b*)	The correct increasing order of basic strength is 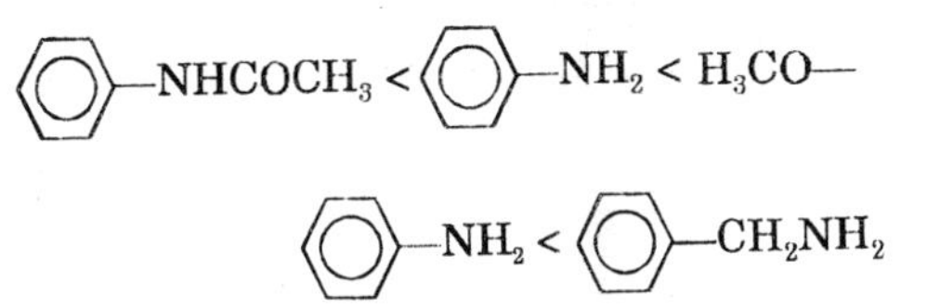
43.	(*d*)	During nitrosation of ring the attacking electrophile is nitrosonium ion NO^+.
44.	(*c*)	The presence of electron withdrawing group makes the nucleophilic substitution more easy.
45.	(*c*)	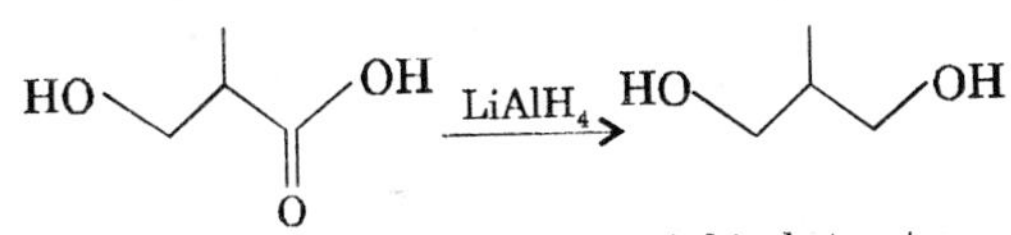 Optically active; Achiral structure (Optically inactive)

PRACTICE PAPER—2

1. Nitroethane is allowed to react with ethanal in basic medium, the product formed is

(*a*) 1-Nitrobut-1-ene (*b*) 2-Nitrobut-2-ene

(*c*) 3-Nitrobut-1-ene (*d*) 3-Nitrobut-3-ene.

2. A chloroderivative X on treatment with Zn/HCl gave a saturated hydrocarbon with molecular mass 72. X, on treatment with Na/Ether produces 2,2,5, 5-tetramethylhexane as one of the product. X is

(*a*) 1-chloropentane (*b*) *isoamylchloride*

(*c*) *neo* pentyl chloride (*d*) sec. pentyl chloride.

3. The IUPAC name of propargyl chloride is

(*a*) 1-chloropropene (*b*) 1-chloroprop-1-yne

(*c*) 3-chloroprop-1-yne (*d*) 3-chloroprop-1-ene.

4. Which of the following structures are chiral ?

I. 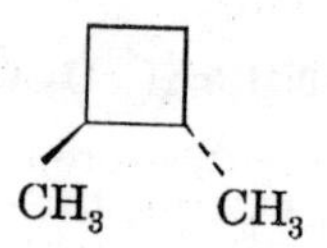

II. 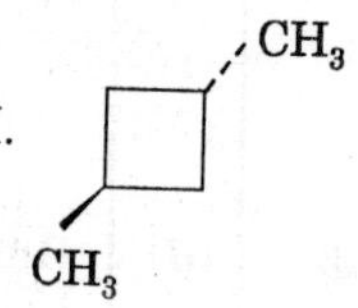

III. 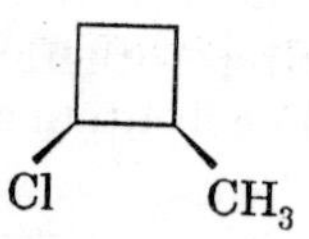

(*a*) I, II (*b*) I, II, III

(*c*) I, III (*d*) II, III.

5. 2,3-Dihydroxy-1,4-butanedioic acid is an IUPAC name of

(*a*) Caproic acid (*b*) Tartaric acid

(*c*) Melonic acid (*d*) Lactic acid.

6. The hybrid states of carbon 1 and 2 in acrylonitrile are respectively

(*a*) sp^2, sp^2 (*b*) sp^3, sp^2

(*c*) sp, sp^2 (*d*) sp^2, sp.

7. Which of the following reaction will give 2-Pentyne in good yield ?

(*a*) 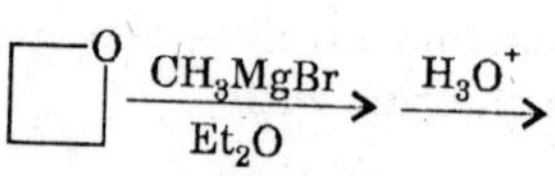

(*b*) $CH_3C \equiv CNa \xrightarrow{EtBr}$

(*c*) 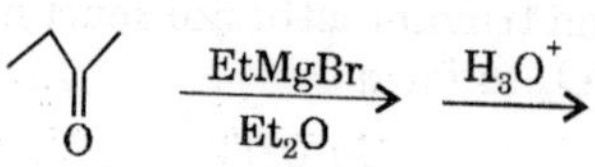

(*d*) 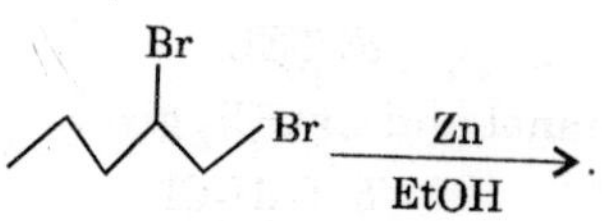

8. Which of the following will be able to affect separation of a mixture of benzene and aniline

(*a*) Water (*b*) dil. NaOH

(*c*) dil. HCl (*d*) dil. Na_2CO_3.

9. 300 mg of certain monohydric alcohol on treatment with ethyl magnesium iodide gave 112 ml of ethane at STP. The alcohol is

(*a*) Methanol (*b*) Ethanol

(*c*) *isopropyl* alcohol (*d*) Tertiary butyl alcohol.

10. Which of the following hydrocarbon will give double the volume of carbon dioxide on complete combustion

I. C_3H_6 II. C_2H_4

III. C_2H_6 IV. C_4H_{10}

(*a*) I (*b*) II, IV

(*c*) I, II (*d*) II, III.

11. 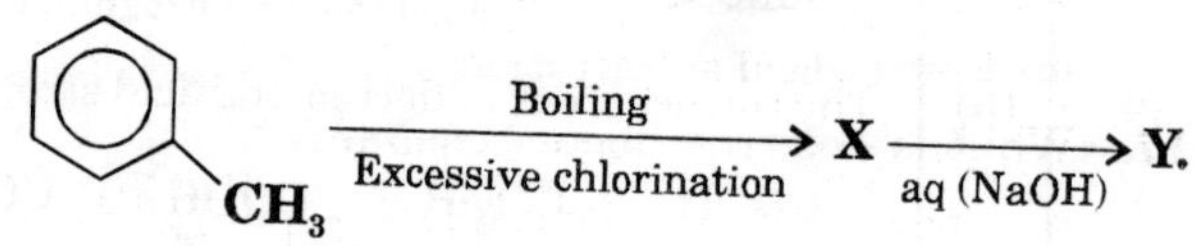

Here, Y is

(*a*) *o*-Cresol (*b*) 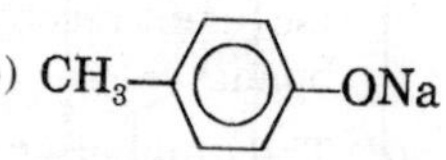

(*c*) C_6H_5COONa (*d*) 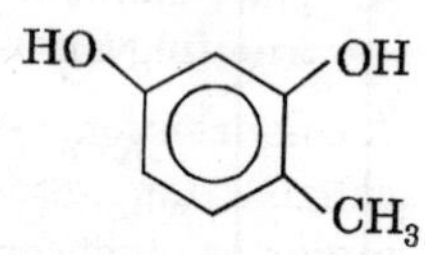

12. Which of the following is an enantiomer of

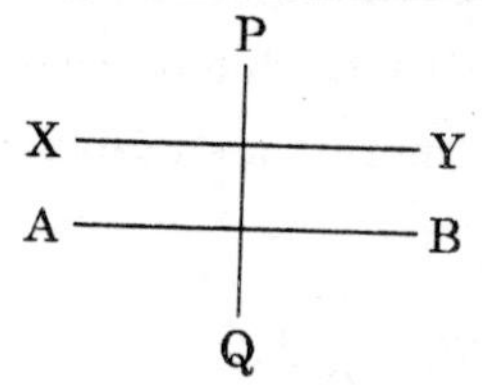 ?

(*a*) 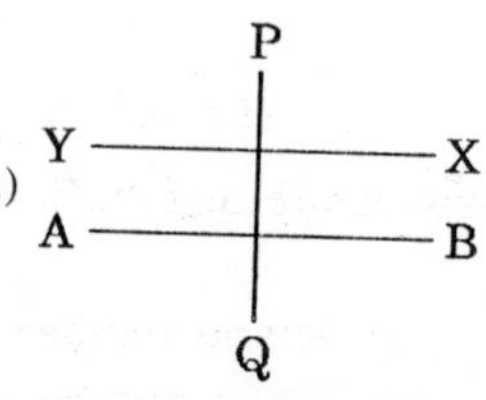 (*b*) 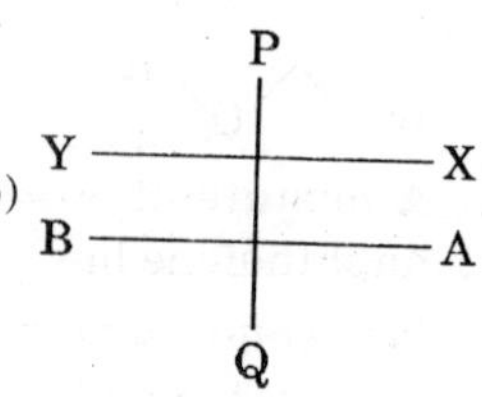

(*c*) 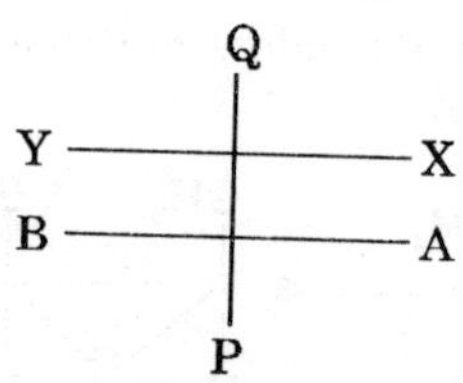 (*d*) 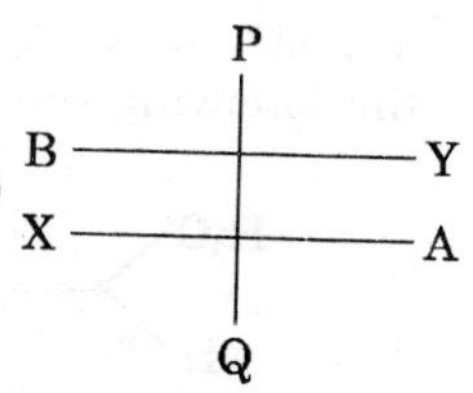

13. Maleic acid and Fumaric acid are examples of

(*a*) Functional isomers
(*b*) Geometrical as well as optical isomers
(*c*) Geometrical as well as diastereomers
(*d*) Geometrical as well as metamers.

14. In which of the following molecule, all carbon-carbon distances are alike ?

(*a*) 1,2-Butadiene (*b*) Benzene
(*c*) Butenyne (*d*) *m*-Xylene.

15. Which of the following molecule is aromatic as well as heterocyclic ?

(*a*) Tetrahydrofuran (*b*) Pyridine
(*c*) Naphthalene (*d*) Benzene.

16. The process of nitration of toluene is an example of

(*a*) Nucleophilic substitution
(*b*) Electrophilic addition
(*c*) Electrophilic substitution
(*d*) Free radical substitution.

17. What is not true about resonance ?

(*a*) Contributing cannonical forms are hypothetical
(*b*) Resonance hybrid is most energetic
(*c*) Isovalent resonating structures have same number of bonds
(*d*) The number of paired and unpaired electrons in contributing forms should be same.

18. A colourless gas obtained by the reaction of carbide of calcium and water was absorbed in chlorine water. The major product formed is

(*a*) Vinyl chloride (*b*) Dichloroethanal
(*c*) Chloral (*d*) Tetrachloroethylene.

19. Which of the following will not be a suitable solvent for Grignard's reagent ?

(*a*) 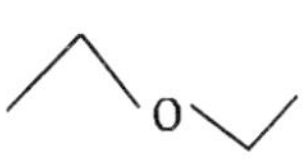 (*b*) 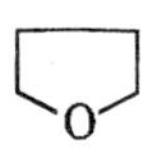

(*c*) 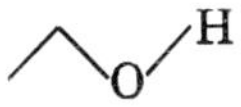 (*d*) All of these.

20. A mixture of 80% *n*-hexadecane and 20% α-methyl naphthalene has

(*a*) Cetane number = 80 (*b*) Octane number = 60
(*c*) Cetane number = 20 (*d*) Isomer number = 20.

21. Among the reagents given below which one cause the following conversion

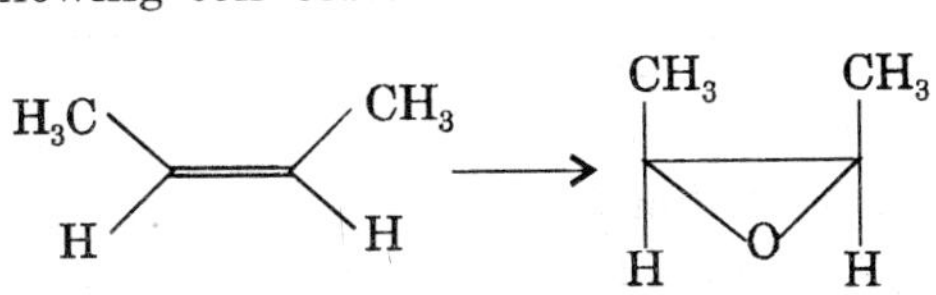

(*a*) H_3O^+ (*b*) $KMnO_4/OH^-$ cold
(*c*) (*i*) O_3 (*ii*) Zn/H^+ (*d*) 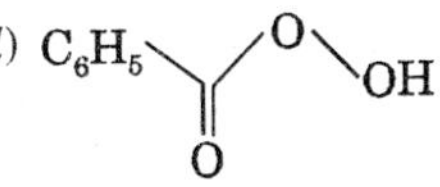

22. The given compound is an example of

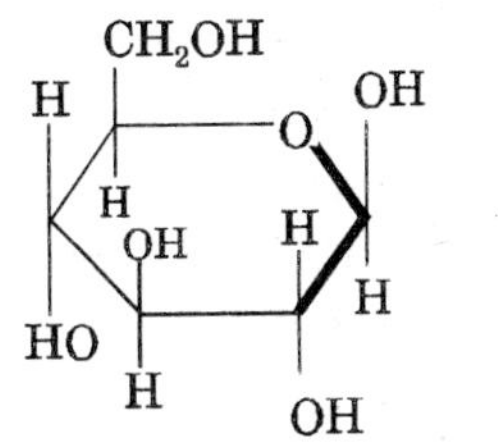

(*a*) Acetal (*b*) Hemiacetal
(*c*) Schiff' base (*d*) Hemiketal.

23. Which of the following reagent cannot be used to prepare an alkyl halide from monohydric alcohol

(*a*) $SOCl_2/Py$ (*b*) P/Br_2
(*c*) $NaBr(aq)$ (*d*) $NaBr/H_2SO_4$.

24. Which of the following will liberate propene gas ?

(*a*) 2-propanol and H_2SO_4 at 443 K
(*b*) Ethanolic KOH and mixture of *n*-propyl chloride and isopropyl chloride
(*c*) Propanone and $LiAlH_4$
(*d*) Both (*a*) and (*b*).

25. In order to prepare allyl bromide from propene, the reagent required is

(*a*) Br_2/CCl_4 (*b*) NBS
(*c*) $Br_2(g)/800$ K (*d*) Both (*b*) and (*c*).

26. Which of the following halide will give precipitate with (*i*) *aq.* NaOH and (*ii*) $AgNO_3$ at ordinary temperature

I. isopropyl chloride II. *p*-tolyl bromide
III. ethyl iodide IV. benzyl chloride
V. Chlorobenzene.

(*a*) I, III, IV (*b*) III, V
(*c*) I, III, V (*d*) II, IV.

27. Balz Schiemann reaction involves the treatment of benzene diazonium chloride with

(*a*) BF_3 (*b*) HBF_4
(*c*) CuCN (*d*) KI.

28. Chlorobenzene and trichloroethanal react in the presence of conc. H_2SO_4 to form

(*a*) Nylon-666 (*b*) DDT
(*c*) BHC (*d*) TEL.

29. Distillation of ethanol and $CaOCl_2$ gives

(*a*) CCl_4 (*b*) C_2H_5Cl
(*c*) $CHCl_3$ (*d*) $COCl_2$.

30. 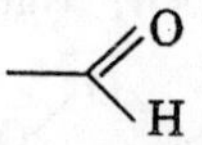 $\xrightarrow{H_2/Pd}$ **X** $\xrightarrow{Na}$ **Y** $\xrightarrow{(CH_3)_3CBr}$ **Z**. In this sequence Z is mainly

(a) Isobutylene (b) Isobutane
(c) Isobutyl acetate (d) Ethyl tert.butyl ether.

31. Hooch tragedies generally occur due to

(a) Lead poisoning in liquor
(b) Presence of C_2H_5OH in liquor
(c) Presence of MeOH in liquor
(d) Presence of dissolved CO_2 in liquor.

32. A compound with formula C_3H_8O is readily oxidised to a compound with formula $C_3H_6O_2$. The original compound is most likely to be

(a) An aldehyde (b) 1°-alcohol
(c) 2° alcohol (d) An ether.

33. Which statement is wrong ?

(a) For nucleophilic substitution, the reactivity order among alkyl halides is R-I > R-Br > R-Cl
(b) For nucleophilic substitution, the reactivity order among aryl halides is ϕ-I > ϕ-Br > ϕ-Cl
(c) For acylation, reactivity order among alochols is 1° > 2° > 3°
(d) The acid strength decreases as *trinitrophenol* > *phenol* > *ethanol* > C_2H_2.

34. The conversion of 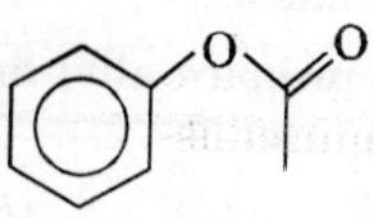 into a mixture of *o* and *p*-hydroxy acetophenone is called

(a) Backmann's rearrangement
(b) Schotten Bauman reaction
(c) Fries rearrangement
(d) Kolbe's reaction
(e) None of these.

35. Fenton's reagent converts glycerol to

(a) Glycerose (b) Tartonic acid
(c) Mesoxalic acid (d) Lactic acid.

36. 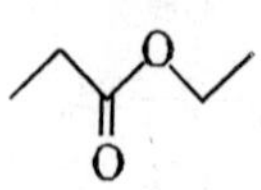 $\xrightarrow{NaOC_2H_5}$ X $\xrightarrow{H_3O^+}$ Y (β-*ketoacid*) $\xrightarrow{(i)\ Na,\ (ii)\ NaOH/CaO}$ Z

(a) 2-pentanone (b) 3-pentanone
(c) Ethyl acetoacetate (d) None of these.

37. An isomeric phenol of anisole gives tribromo product with aq. Br_2. The phenolic compound is

(a) *o*-cresol (b) *m*-cresol
(c) Benzyl alcohol (d) *p*-cresol.

38. $H_2C = CH_2 \xrightarrow{HI} X \xrightarrow[Alc.]{AgCN} Y \xrightarrow{S_8} Z.$

Here, Z is

(a) Ethyl isothiocyanate (b) Methyl isothiocyanate
(c) Methyl cyanate (d) Ethanethiol.

39. ***o*-toluene diazonium chloride** $\xrightarrow{CuCl}$ X $\xrightarrow{KMnO_4}$ Y.

Here Y is

(a) *o*-chlorobenzoic acid (b) *o*-chlorophenol
(c) Phenol (d) Chlorobenzene.

40. $C_2H_2 \xrightarrow[Hg^{2+}]{Dil.\ H_2SO_4}$ **P** $\xrightarrow{Cr_2O_7^{2-}/H^+}$ **Q** $\xrightarrow{Ca(OH)_2}$ **R** $\xrightarrow[Distill]{Dry}$ **S**

The end product **S** is

(a) 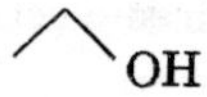 (b) 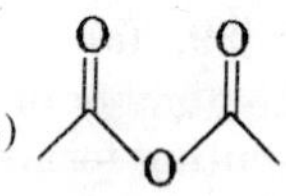
(c) 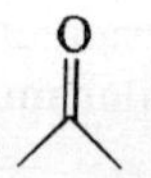 (d) 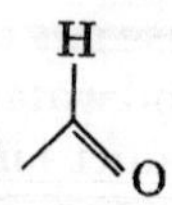

41. 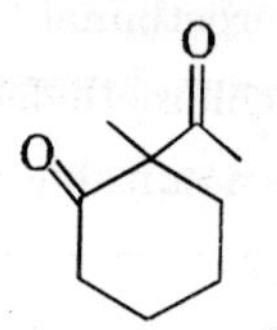 (A) $\xrightarrow{(i)\ \overset{+}{K}\overset{-}{O}Br,\ (ii)\ H_3O^+,\ (iii)\ Heat}$

The compound finally gets converted into

(a) 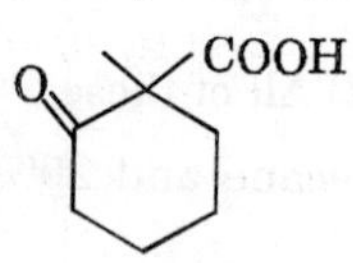 (b) 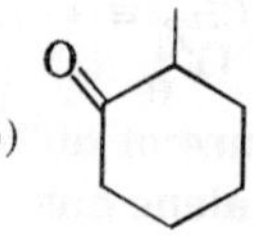
(c) 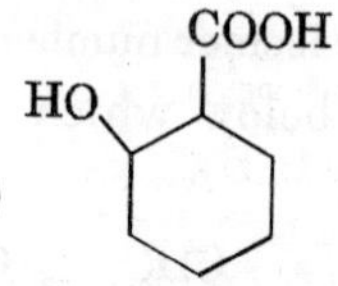 (d) 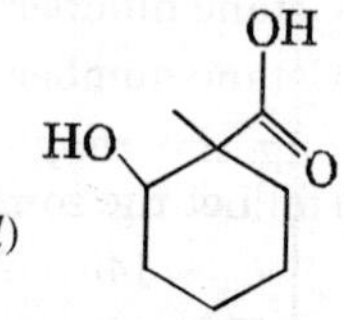

42. Which of the following is strongest acid ?

(a) Phthalimide (b) Acetamide
(c) Benzamide (d) Benzenamine.

43. 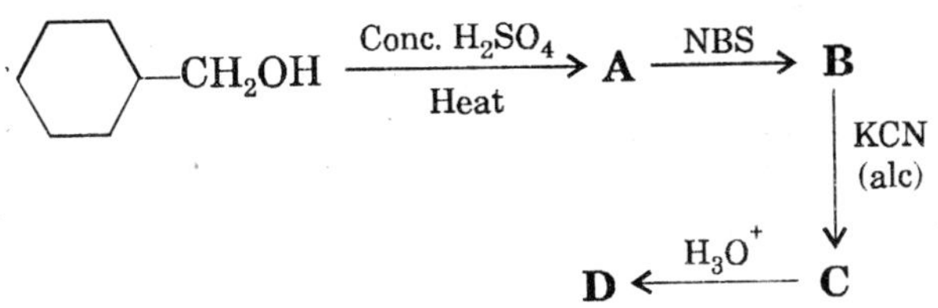

In the above sequence **D** is

(*a*) 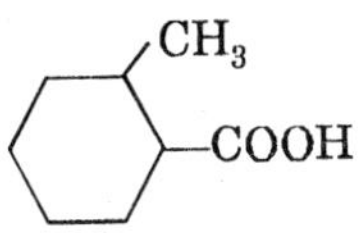

(*b*) 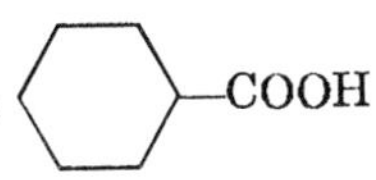

(*c*) 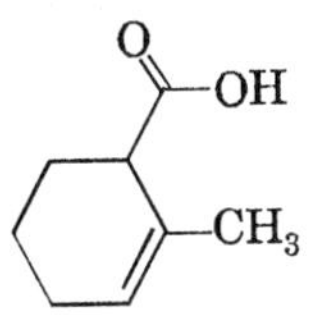

(*d*) 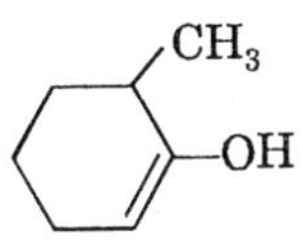

44. Which isomer of C_4H_8 on reaction with alkaline $KMnO_4$ gives a *d*, *l*-mixture of butane 2,3-diol ?

(*a*) 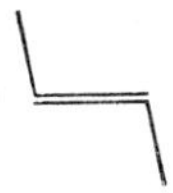

(*b*) 

(*c*) 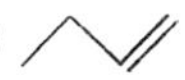

(*d*) 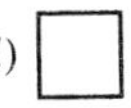

45. How many chirality centres are present in the structure ?

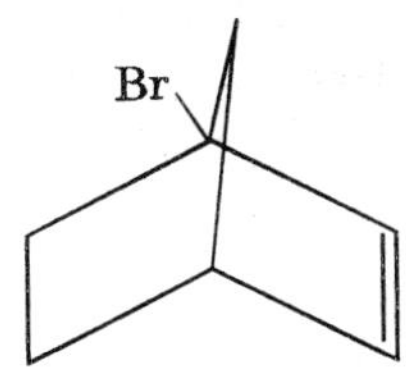

(*a*) 1 only (*b*) 3

(*c*) 0 (*d*) 2.

ANSWERS

1. (*b*)	**2.** (*c*)	**3.** (*c*)	**4.** (*c*)	**5.** (*b*)	**6.** (*c*)	**7.** (*b*)	**8.** (*c*)	**9.** (*c*)
10. (*d*)	**11.** (*c*)	**12.** (*b*)	**13.** (*c*)	**14.** (*b*)	**15.** (*b*)	**16.** (*c*)	**17.** (*b*)	**18.** (*b*)
19. (*c*)	**20.** (*a*)	**21.** (*d*)	**22.** (*b*)	**23.** (*c*)	**24.** (*d*)	**25.** (*d*)	**26.** (*a*)	**27.** (*b*)
28. (*b*)	**29.** (*c*)	**30.** (*a*)	**31.** (*c*)	**32.** (*b*)	**33.** (*b*)	**34.** (*c*)	**35.** (*a*)	**36.** (*b*)
37. (*b*)	**38.** (*a*)	**39.** (*a*)	**40.** (*c*)	**41.** (*b*)	**42.** (*a*)	**43.** (*c*)	**44.** (*a*)	**45.** (*d*).

HINTS/SOLUTIONS

Q. No.	Ans.	Hints & Solutions
1.	(*b*)	They undergo aldol type condensation reaction 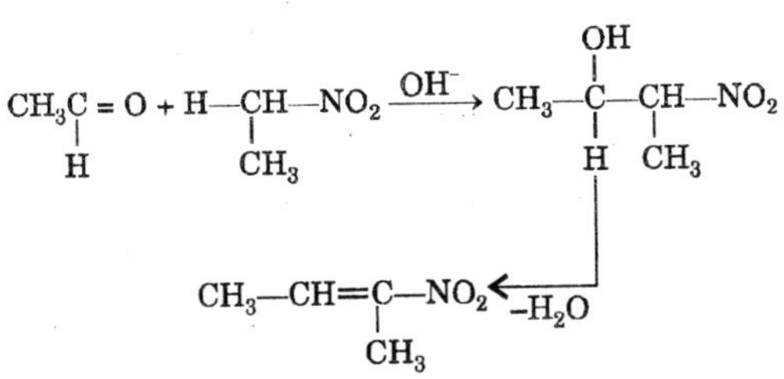 2-Nitrobut-2-ene
2.	(*c*)	Let the hydrocarbon be C_nH_{2n+2}. $14n + 2 = 72$ or $n = 5$. Since it gives 2, 2, 5, 5-tetramethylhexane as one of the product. Thus X must be neopentyl D chloride.

Q. No.	Ans.	Hints & Solutions
		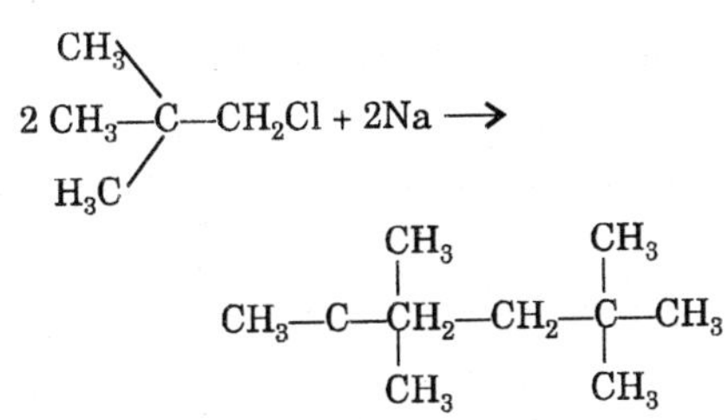 2, 2, 5, 5-Tetramethylhexane
3.	(*c*)	Propargyl chloride is CH ≡ C—CH_2Cl. Its IUPAC name is 3-chloroprop-1-yne.
4.	(*c*)	Structures I and III are chiral structures.
5.	(*b*)	The given name is on IUPAC name of tartaric acid

Q. No.	Ans.	Hints & Solutions
		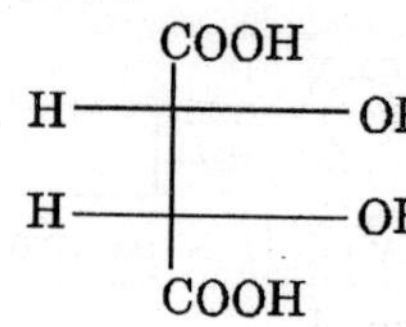
6.	(*c*)	Acrylonitrile is $\overset{3}{C}H_2 = \overset{2}{C}H—\overset{1}{C} \equiv N$. The hybride states of carbon 1 is *sp* whereas thta of carbon 2 is sp^2.
7.	(*b*)	$CH_3C \equiv Na - BrC_2H_5 \longrightarrow NaBr + CH_3C \equiv C—CH_2CH_3$ (2-Pentyne)
8.	(*c*)	Dilute HCl does not react with benzene, but it forms crystalline salt $C_6H_5N^+H_3Cl^-$ with aniline.
9.	(*c*)	Mass of alcohol which can produce 22400 mL of ethane $= \frac{300 \times 22400 \times 10^{-3}}{112}$ g = 60 g Let the alcohol be $C_nH_{2n+1}OH$ Molar mass = $14n + 18 = 60$ or $n = 3$ Thus, it is **isopropyl alcohol** which contains 3 carbon atoms.
10.	(*d*)	Both C_2H_4 and C_2H_6 will give double the volume of CO_2 on complete combustion $C_2H_4 + 2O_2 \longrightarrow 2CO_2 + 2H_2O$; $C_2H_4 + 7/2O_2 \longrightarrow 2CO_2 + 3H_2O$.
11.	(*c*)	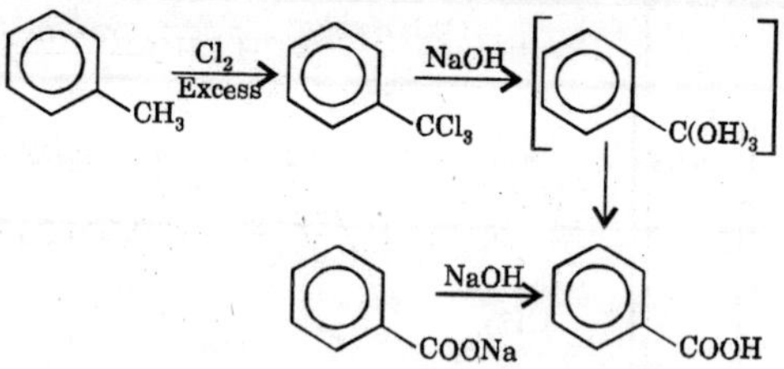
12.	(*b*)	The structure given in choice *b* is an enantiomer of the given stereoisomer.
13.	(*c*)	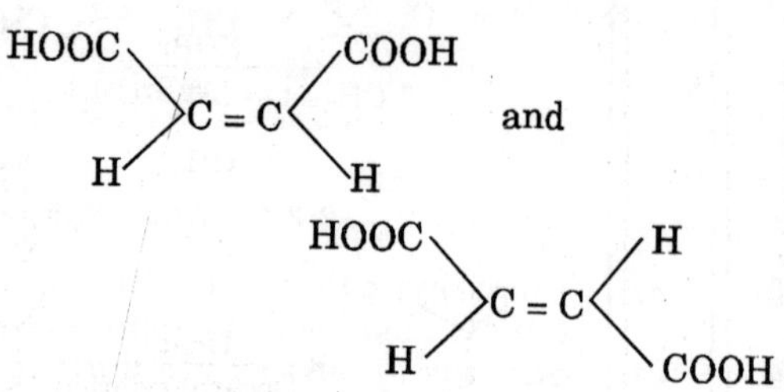 are geometrical isomers. they can also be called diastereomers.
14.	(*b*)	All carbon-carbon distances are same in benzene. It is because of delocalisation of π-electrons over the entire structure.
15.	(*b*)	Pyridine (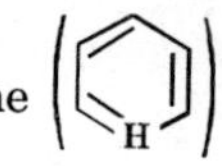) is aromatic as well as heterocyclic in nature.
16.	(*c*)	The nitration of toluene is an electrophilic substitution process.
17.	(*b*)	The statement '*b*' is wrong because resonance hybrid is most stable and thus least energetic structure.
18.	(*b*)	$CaC_2 + H_2O \longrightarrow Ca(OH)_2 + C_2H_2$ (Ethyne) $CH \equiv CH + 2HOCl \xrightarrow{Cl_2/H_2O} CHCl_2—CHO$ (Dichloroethanal)
19.	(*c*)	$CH_3CH_2OH + RMgX \longrightarrow RH + Mg(OC_2H_5)X$ (Ethoxy mag. halide)
20.	(*a*)	This is in accordance with the definition of cetane number.
21.	(*d*)	Peroxybenzoic acid will cause this change. It changes >C = C< to 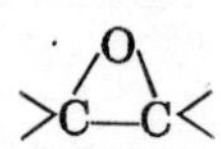
22.	(*b*)	The given compound is a cyclic hemicecetal.
23.	(*c*)	Aqueous NaBr cannot be used to prepare halide from R—OH.
24.	(*d*)	$CH_3CH(OH)CH_3$ (2-Propanol) $\xrightarrow[443K]{Conc.\ H_2SO_4} CH_3CH = CH_2$ (Propene) Both *n*-propyl chloride as well as isopropyl chloride on treatment with ethanolic solution of KOH will also yield propene. Hence correct choice is '*d*'.
25.	(*d*)	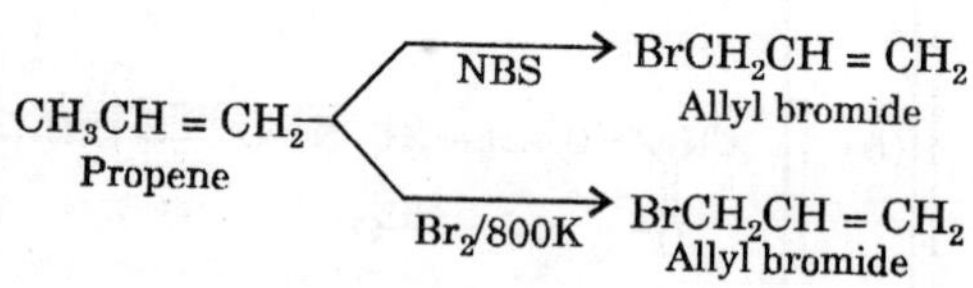
26.	(*a*)	$(CH_3)_2CHCl$, C_2H_5I and 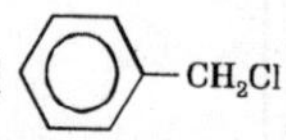 will give precipitate of siluer halide on treatment with (*i*) NaOH (*aq*) and (*ii*) $AgNO_3$. *p*-tolyl bromide and chlorobenzene will not form ppt. because of lower reactivity of aryl halides towards nucleophile substitution reaction.
27.	(*b*)	Balz Schiemann reaction involves the conversion of $C_6H_5N_2Cl$ to C_6H_5F by treatment with HBF_4 followed by heating.

Q. No.	Ans.	Hints & Solutions
28.	(*b*)	Chlorobenzene and trichloroethanal (chloral) react to form DDT.
29.	(*c*)	Distillation of ethanol and $CaOCl_2$ (bleaching powder) mixture results in the formation of chloroform ($CHCl_3$).
30.	(*a*)	$CH_3CHO \xrightarrow{H_2/Pd} \underset{(X)}{CH_3CH_2OH} \xrightarrow{Na}$ $\underset{Y}{CH_3CH_2O^-Na^+} \xrightarrow[\text{Elimination}]{(CH_3)_3CBr} \underset{\text{Isobutylene (Z)}}{CH_2 = C(CH_3)_2}$
31.	(*c*)	Presence of CH_3OH in denatured alcohol causes blindness and higher doses prove fatal.
32.	(*b*)	The compound C_3H_8O is likely to be 1° alcohol which is getting oxidised to $C_3H_6O_2$ $\underset{\text{1-Propanol}}{CH_3CH_2CH_2OH} \xrightarrow{K_2Cr_2O_7/H^+} \underset{\text{Propanoic acid}}{CH_3CH_2COOH} + H_2O$
33.	(*b*)	Statement '*b*' is wrong becuase the reactivity of aryl halide towards nucleophilic substitution is independent of the nature of halo group.
34.	(*c*)	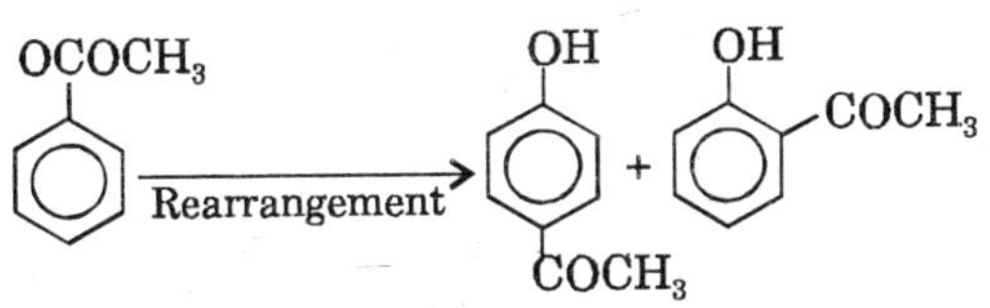 This rearrangement is known as Fries rearrangement.
35.	(*a*)	$\begin{matrix} CH_2OH \\ CHOH \\ CH_2OH \end{matrix} \xrightarrow[(FeSO_4 + H_2O_2)]{\text{Fenton's Reagent}} \begin{matrix} CHO \\ CHOH \\ CH_2OH \end{matrix} + \begin{matrix} CH_2OH \\ C=O \\ CH_2OH \end{matrix}$ Glycerose
36.	(*b*)	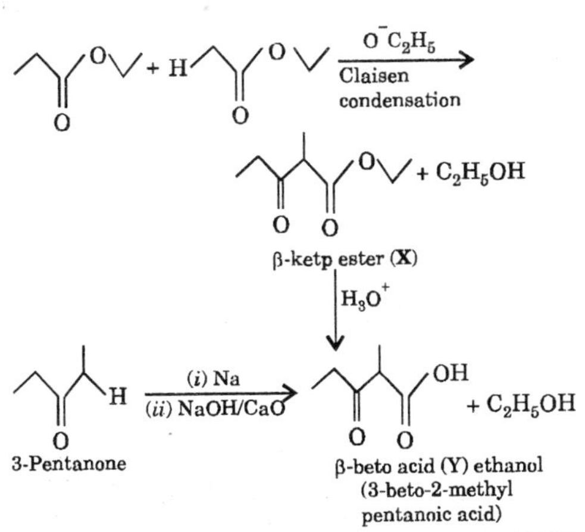
37.	(*b*)	Anisole is 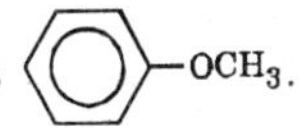. Its isomeric phenol could be 0, *m* or *p*-cresol. Since it gives tribromo product, therefore, it can only be *m*-cresol. 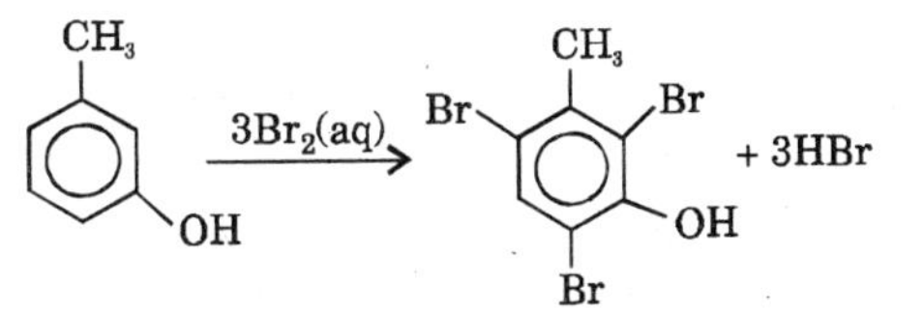
38.	(*a*)	$CH_2 = CH_2 \xrightarrow{HI} \underset{(X)}{CH_3—CH_2I} \xrightarrow{AgCN} \underset{(Y)}{CH_3CH_2NC} \xrightarrow{S_8} CH_3CH_2—N = C = S$ Ethyl isothiocyanate
39.	(*a*)	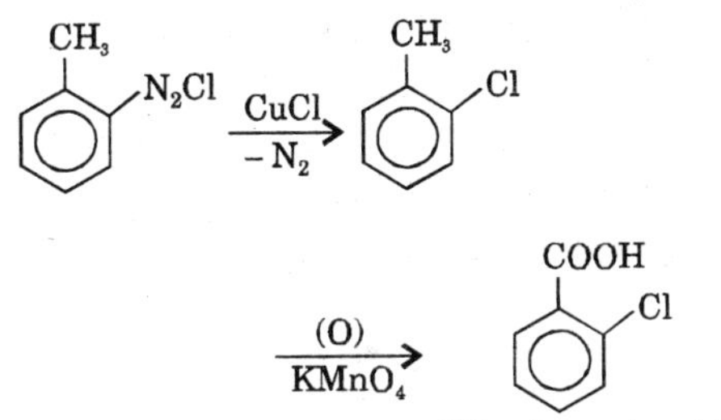 *o*-Chlorobenzoic acid
40.	(*a*)	$CH \equiv CH \xrightarrow[Hg^{2+}]{\text{Dil. } H_2SO_4} \underset{(P)}{CH_3CHO} \xrightarrow{Cr_2O_7^{2-}/H^+} \underset{(Q)}{CH_3COOH} \xrightarrow{Ca(OH)_2} \underset{(R)}{(CH_3COO)_2Ca} \xrightarrow[- CaCO_3]{\text{Dry Distill}} \underset{(S)}{(CH_3)_2C = O}$
41.	(*a*)	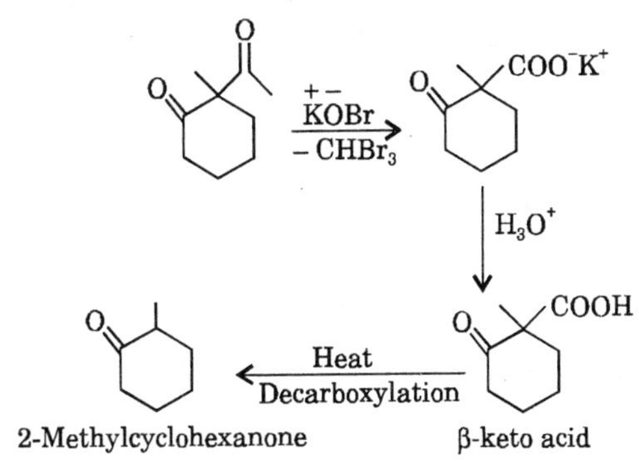

Q. No.	Ans.	Hints & Solutions
42.	(*a*)	Among the given species phthalimide 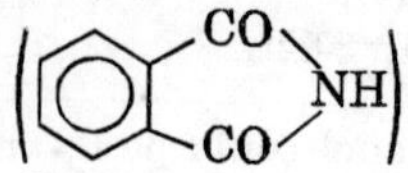 is strongest acid. If is due to electron withdrawing effect of two carbonyl group attached to NH part.
43.	(*a*)	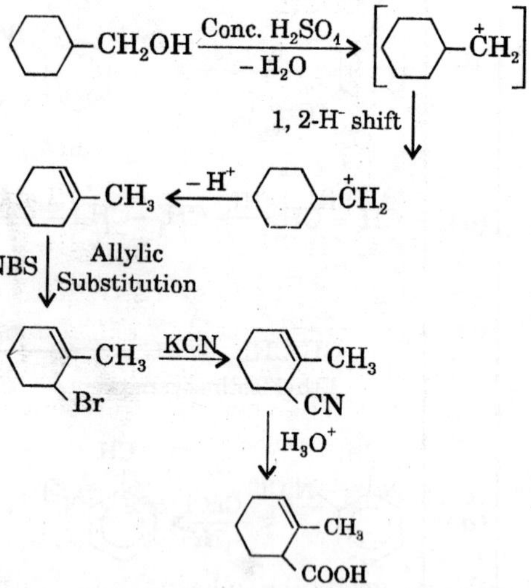
44.	(*a*)	Trans isomer of 2-butene on reaction with alkaline $KMnO_4$ gives *d*, *l* mixture of butane-2, 3-diol. Hydroxylation by $KMnO_4$ is *syn* in nature.
45.	(*d*)	The structure has two chirality centres marked as * 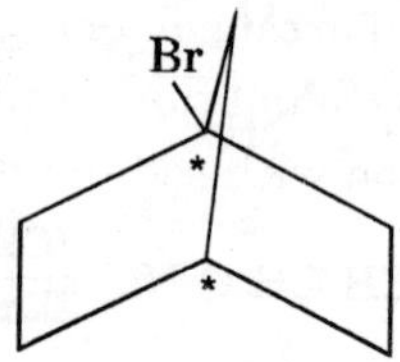

PRACTICE PAPER—3

1. Which of the following properties/characteristics are different for the compounds (+) 2-butanol and (–) 2-butanol ?

(I) Solubility in water

(II) Specific rotation

(III) Spatial arrangement around the chirality centre

(IV) Angle around the chirality centre

(V) Reaction with sodium metal

(*a*) I, II (*b*) II, III

(*c*) III, IV, V (*d*) I, III, IV.

2. An ester A($C_4H_8O_2$) on treatment with phenyl magnesium bromide in excess followed by hydrolysis gave B. B on dehydration gave alkene which on reductive ozonolysis produced equimolar mixture of benzophenone and methanal. The ester can be

(*a*) *n*-propyl formate (*b*) Ethyl acetate

(*c*) Isopropyl formate (*d*) Methyl propionate.

3. Which of the following reagents can provide distinction between 2-propanol and allyl alcohol ?

I : Br_2/CCl_4 II : $HBr/ZnCl_2$

III : $I_2/NaOH$ IV : Tollen's reagent

(*a*) All (*b*) I, II

(*c*) I, III, IV (*d*) III, IV only.

4. Which of the following is not optically active ?

(*a*) Sec. butyl amine (*b*)

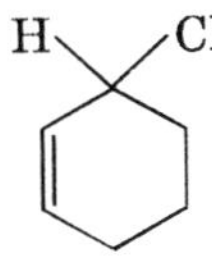

(*c*) $CH_3CH = C = CHCH_3$ (*d*)

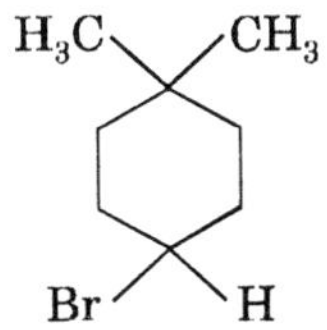

5. Which of the following structures have same configuration around the chirality centre ?

I.

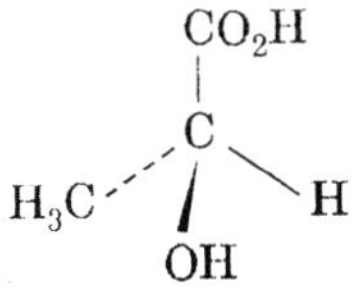

$\alpha = + 3.82°$

II.

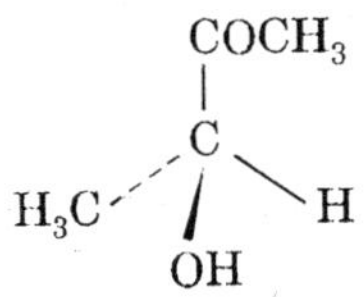

$\alpha = - 8.25°$

III.

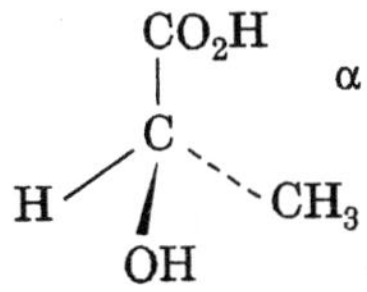

$\alpha = - 3.82°$

IV.

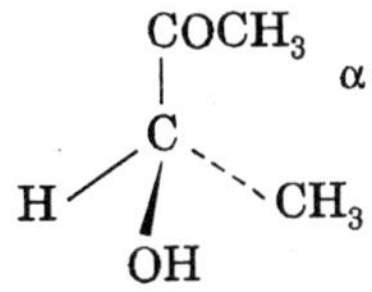

$\alpha = + 8.25°$

(*a*) I, II and III, IV (*b*) I, III and II, IV

(*c*) I, IV and II, III (*d*) I, III, II.

6. A mixture of RCHO and RCO_2H is shaken with dilute aqueous solution of $NaHCO_3$ in a separating funnel in the presence of ether and finally ether layer is separated. The compound present in ether layer is mainly

(*a*) RCOOH (*b*) RCHO

(*c*) $R_2C = O$ (*d*) R_2O.

7. Which among the following has lowest value of pKa_1 ?

(*a*) $CH_2(COOH)_2$ (*b*) $(COOH)_2$

(*c*) $(CH_2)_2(COOH)_2$ (*d*) $(CH_2)_3(COOH)_2$.

8.

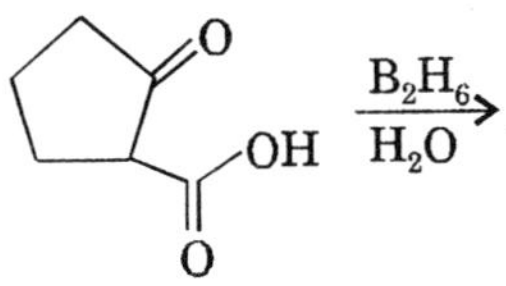

A. The product of the above reaction is

(*a*)

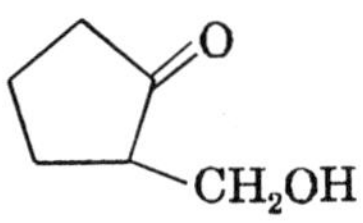

(*b*)

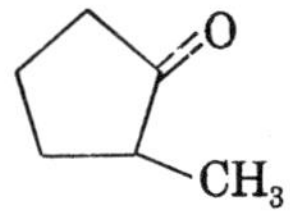

(*c*)

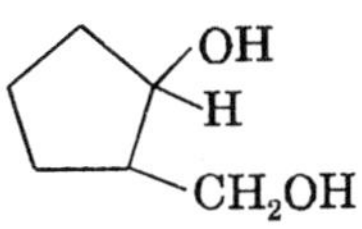

(*d*)

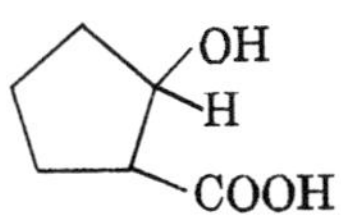

9. Which of the following pair of species are not enantiomers ?

I.

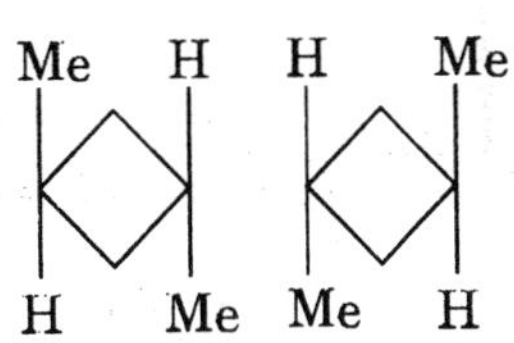

II. 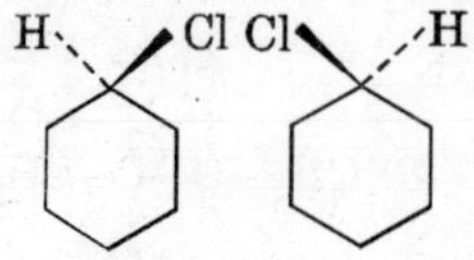

III. 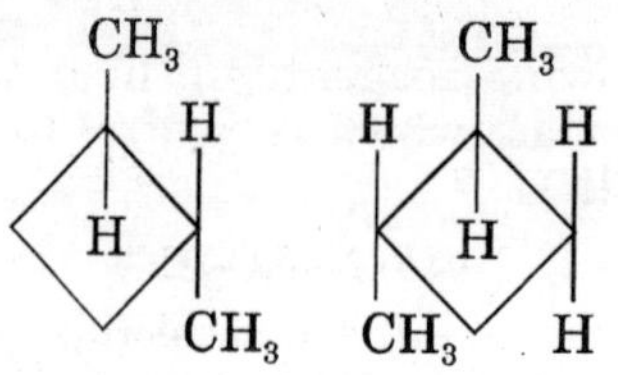

IV. 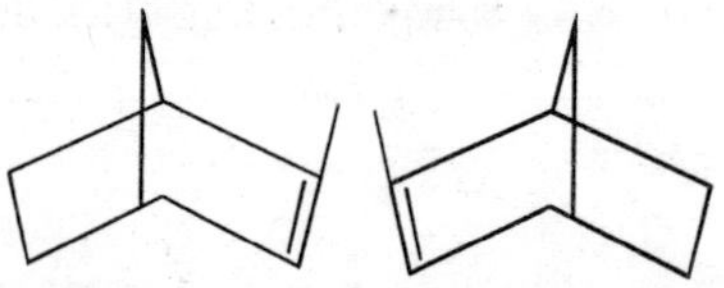

(*a*) II, IV (*b*) I, II
(*c*) II, III (*d*) I, II, III.

10. The number of possible functional isomers of CH_3CO_2H are

(*a*) 1 (*b*) 2
(*c*) 3 (*d*) 5.

11. The compound given below can exhibit

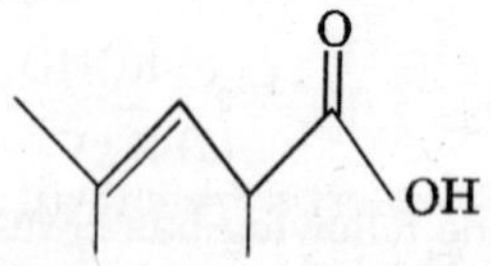

(*a*) Tautomerism
(*b*) Geometrical isomerism
(*c*) Enantiomerism
(*d*) Geometrical as well as optical isomerism

12. Which among the following is least basic ?

(*a*) 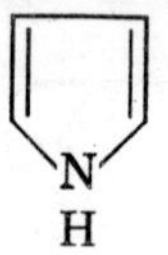 (*b*) 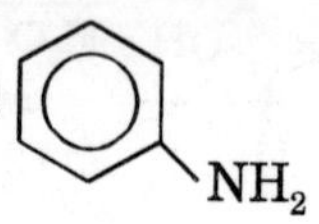

(*c*) $(CH_3)_2NH$ (*d*) 

13. Which of the following compounds on reductive ozonolysis will yield *5-keto-2-methylhexanal.*

(*a*) 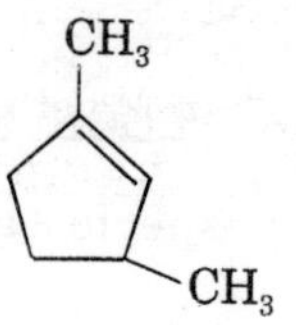 (*b*) 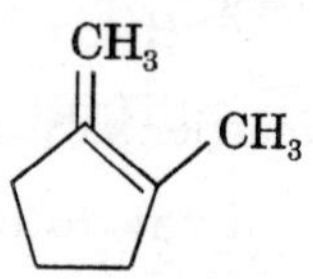

(*c*) 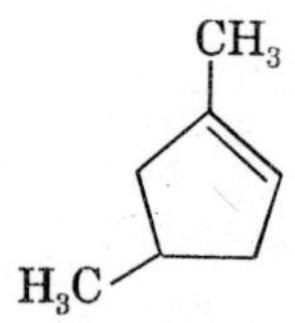 (*d*) 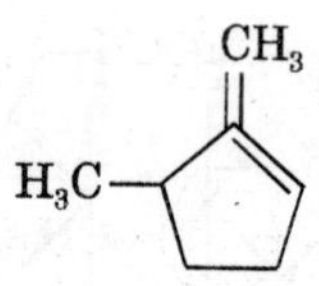

14. Which of the following pairs of species does not represent resonating forms ?

(*a*) 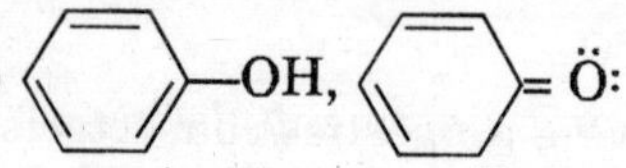

(*b*) $CH_3—\overset{O}{\overset{||}{C}}—NH_2,\ CH_3—\underset{O^-}{\underset{|}{C}}=\overset{+}{N}H_2,$

(*c*) 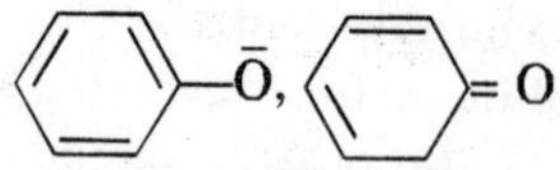

(*d*) $CH_2=CH—\underset{O}{\underset{||}{C}}—H,\ \overset{+}{C}H_2—CH=\overset{O^-}{\overset{|}{C}}—H.$

15. Which of the following matchings are correct ?

Type of reaction	**Intermediates**
I. Anti Markovnikoff's addition of HBr	A. Carbocations
II. Dehydration of alcohol	B. Carbanion
III. Reimer Tiemann Reaction	C. Carbene
IV. Clasien Condensation	D. Free radicals

(*a*) I-D, II-A, III-B, IV-C
(*b*) I-C, II-A, III-B, IV-D
(*c*) I-D, II-C, III-A, IV-B
(*d*) I-D, II-A, III-C, IV-B.

16. $(CH_3)_3C—OH \xrightarrow[(ii)\ KCN\ \ (iii)\ H_3O^+]{(i)\ H^+} A \xleftarrow[(ii)\ KCN\ \ (iii)\ H_3O^+]{(i)\ H^+} (CH_3)_2C=CH_2.$

The product A is

(*a*) $(CH_3)_3CNH_2$ (*b*) $(CH_3)_3C—CN$
(*c*) $(CH_2)_2C=NH$ (*d*) $(CH_3)_3C—COOH.$

17. 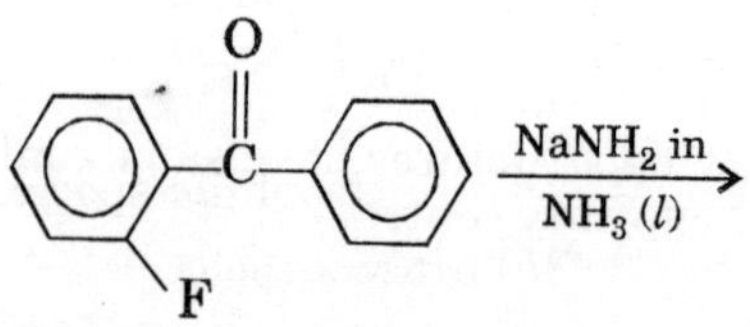

The products formed is/are

(*a*) 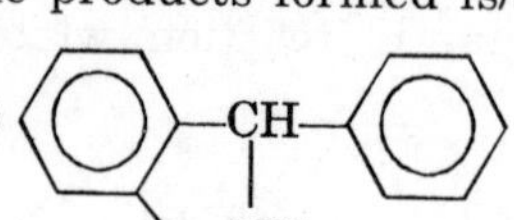

(*b*) 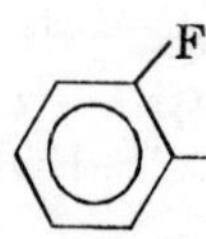 $—COOH + C_6H_6$

(c) 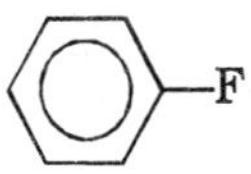 $-F + C_6H_5CONH_2$

(d) $C_6H_6 + C_6H_5F + CO_2$.

18. Which of the following statements is/are correct ?

(I) The relative ease of dehydration of alcohols follows the order 3° > 2° > 1°.

(II) 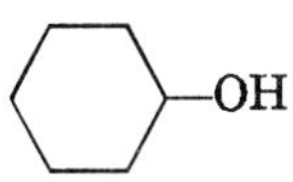 is relatively more easily dehydrated

than 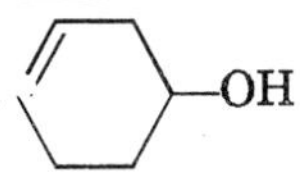

(III) 3° alcohols are generally dehydrated by E_2 mechanism.

(IV) β-hydroxy aldehyde undergoes dehydration very easily because it forms α, β unsaturated carbonyl compound

(a) I, II (b) II, IV

(c) I, IV (d) III, IV.

19. Among the three statements given below which are correct

I. Alkenes formed by Hoffman elimination and that formed by Saytzeff elimination must be more substituted.

II. Major product formed by elimination of *secondary butyl bromide* is *trans-2-butene*

III. All hydrogen atoms present in neopentyl amine are 1° hydrogens.

(a) II, III (b) only III

(c) I, III (d) I, II.

20. Which among the following, on boiling with aqueous solution of NaOH, gives a gas which can turn red litmus blue ?

(a) 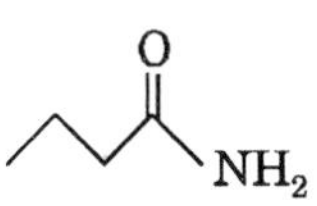 (b) 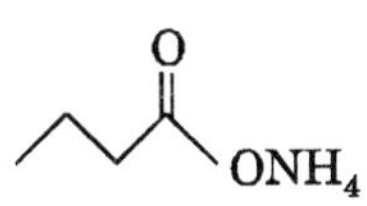

(c)  (d) All.

21. Which of the following is not a nucleophilic species ?

(a) Water molecule (b) benzene molecule

(c) Dichlorocarbene (d) acetate ion.

22. For a compound undergoing E_2 reaction, which step is correct

(a) 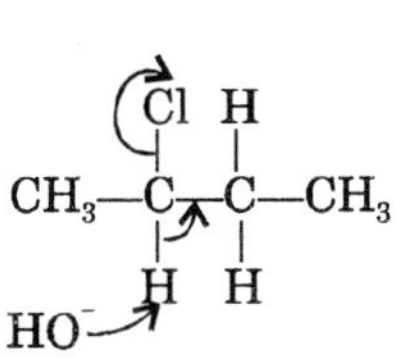 (b) 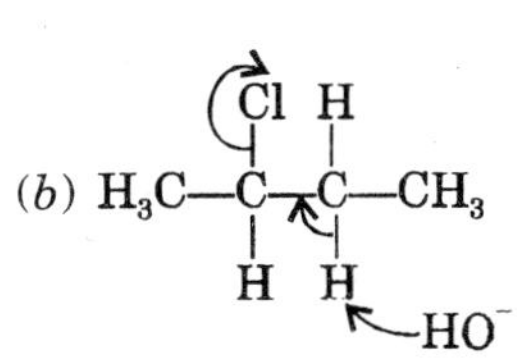

(c) 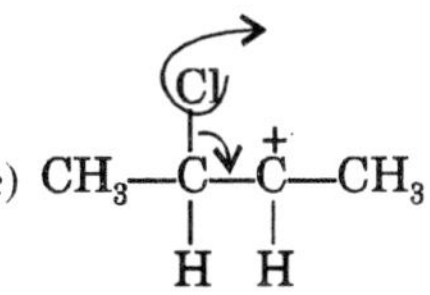 (d) 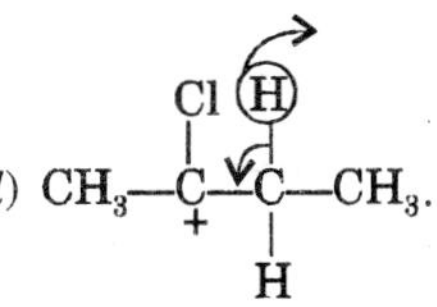

23. Which of the following reagent will liberate a colourless, odourless gas (which is neutral to litmus) by reaction with RNH_2 ?

(a) NaOH (b) $NaNO_2/HCl$

(c) Conc. H_2SO_4 (d) Acetyl chloride.

24. Which of the following compound has 1(E), 3(Z) configuration ?

(a) 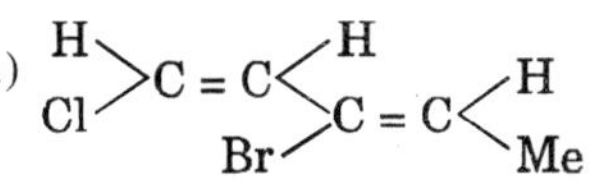

(b) 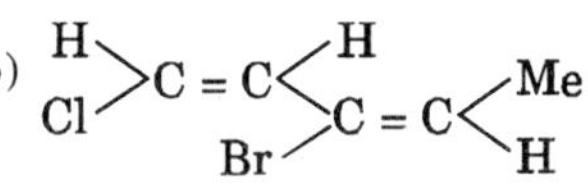

(c) 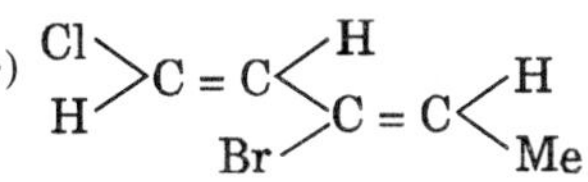

(d) 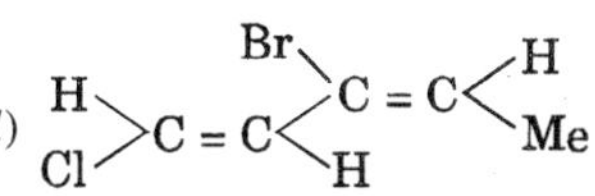

25. Out of *o*, *m* and *p*-xylenes which will react most readily will cold 80% H_2SO_4 ?

(a) *o*-xylene (b) *p*-xylene

(c) *m*-xylene (d) All.

26. Which one of the following compounds has 2S, 3S configuration ?

(a) 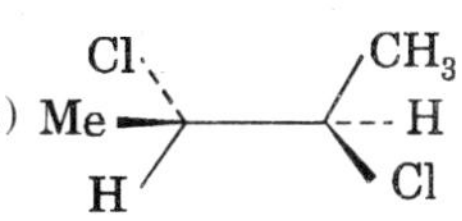 (b) 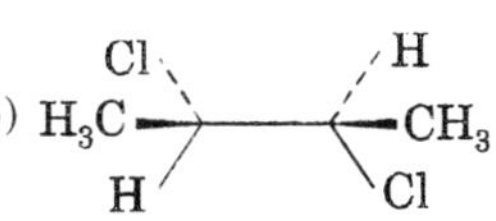

(c) 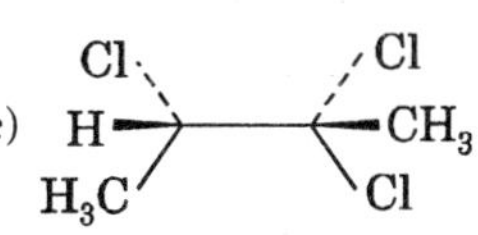 (d) 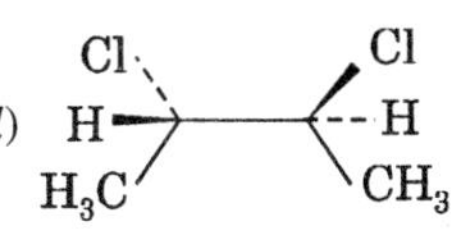

27. Which one of the following species is paramagnetic in nature ?

(a) Triplet carbene (b) Carbocation

(c) Carbanion (d) Singlet carbene.

28. In the reaction, 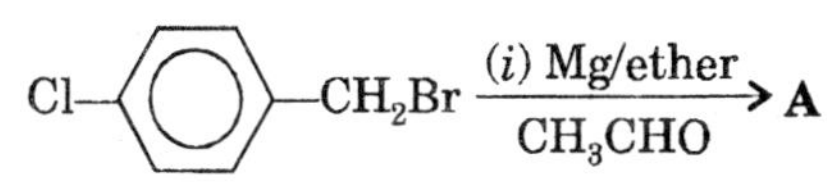

The product **A** is

(a) 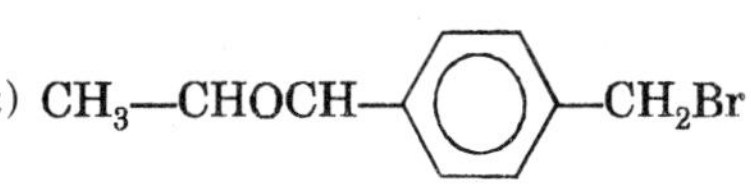

(b) 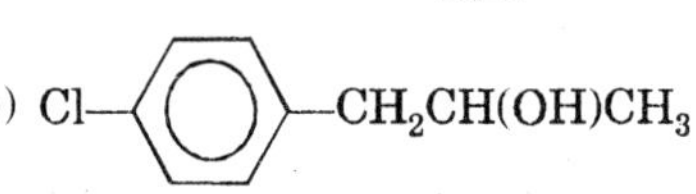

(c) 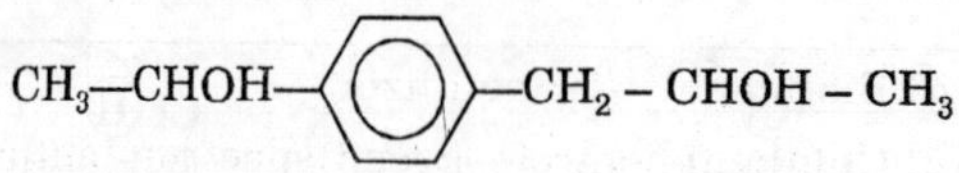

(d) 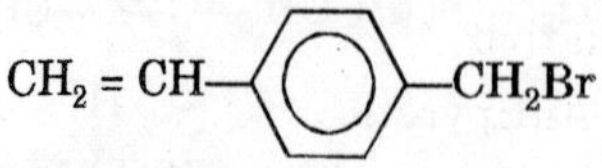

29. Which of the following process will convert

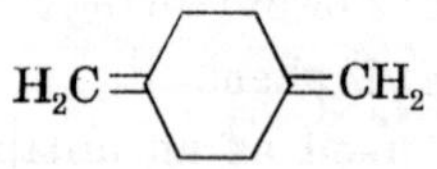 into terminal diol ?

(a) Oxymercuration-demercuration
(b) Hydroboration oxidation
(c) Acid catalysed hydration
(d) All the above pocesses.

30. Which of the following reagents can be used to reduce $CH_3 - NC$ to $(CH_3)_2$ NH ?

(a) H_2/Ni (b) $LiAlH_4$
(c) Na/alcohol (d) All the three.

31. The product X in the reaction

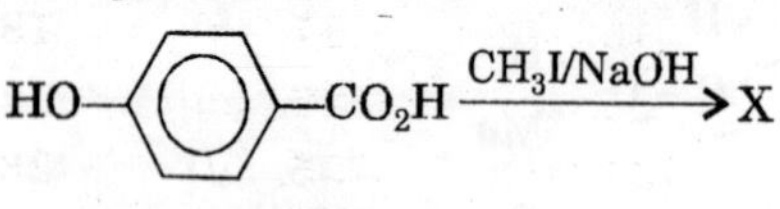 X is

(a) 

(b) 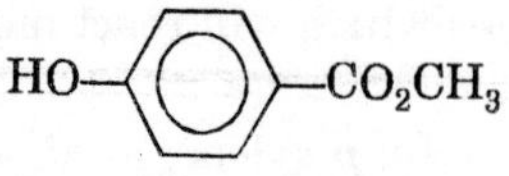

(c) 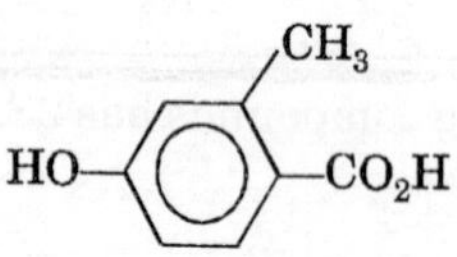

(d) 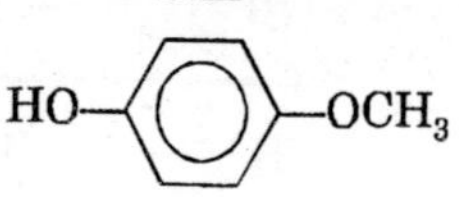

32. An aromatic compound 'A', $C_7H_6Cl_2$ gives AgCl on boiling with alcoholic $AgNO_3$ solution and yields C_7H_7OCI on treatment with NaOH. 'A' on oxidation gives a monochlorobenzoic acid which affords only one mononitro derivative. The compound 'A' is

(a) 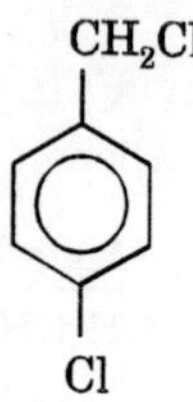 (b) 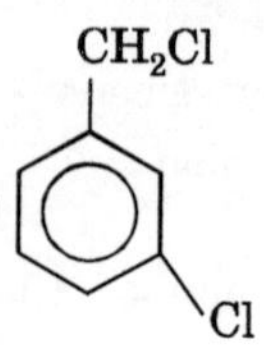

(c) 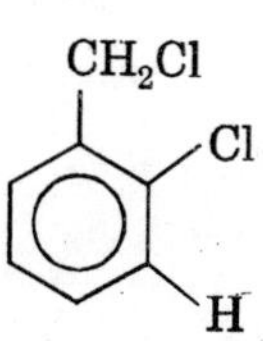 (d) 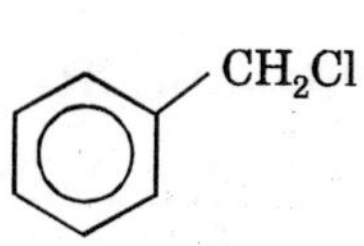

33. An alkyl hydroxylamines on warming with Tollen's reagent get oxidised to the corresponding nitroso compound and the Tollen's reagent is itself reduced to metallic silver. This reaction is called

(a) Baker-Mulliken's test
(b) Wacker synthesis
(c) Reimer-Tiemann reaction
(d) None of the above.

34. How many acyclic structural and stereoisomeric alkenes are possible for the molecular formula C_5H_{10} ?

(a) 4 (b) 5
(c) 6 (d) 7.

35. The compound 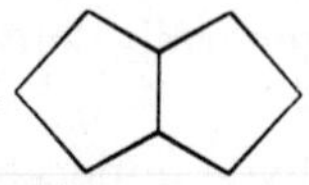 is known by which of the following names ?

(a) Bicyclo[3.3.1] nonane
(b) Bicyclo [3.3.0] octane
(c) Bicyclo [3.2.0] heptane
(d) Bicyclo [3.2.1] octane.

36. Which of the following reactions does not yield a primary amine with the same number of carbon atoms as the starting reactant ?

(a) $R—CONH_2 \xrightarrow{LiAlH_4}$

(b) $R—CONH_2 \xrightarrow{Br_2/KOH}$

(c) R—X (excess) $\xrightarrow{NH_3}$

(d) $R—CH = NOH \xrightarrow{Na\backslash C_2H_5OH}$

37. In the Cannizzaro reaction as given below :

$$2\ PhCHO \xrightarrow{OH^-} PhCH_2OH + PhCO_2^-,$$

the slowest step is

(a) Attack of OH^- at the carbonyl group
(b) Transfer of H^- ion to the carbony group
(c) Abstraction of proton from carboxyl group
(d) Deprotonation of Ph-CH_2OH.

38. In which of the following conditions benzaldehyde undergoes simultaneous oxidation and reduction in

(a) Aqueous $NaHCO_3$ solution
(b) Conc. KOH solution
(c) Dil. HCl solution
(d) Conc. H_2SO_4.

39. Which of the following reagents can be safely used to convert $CH_2 = CHCHO \rightarrow CH_2 = CHCOOH$

(a) Cold. dil. alk. $KMnO_4$ solution
(b) Diamminesilver (I) hydroxide
(c) Cold. dil. Cl_2/NaOH solution
(d) Br_2 in CCl_4.

(a) Sedatives (b) Hypnotics
(c) Anaesthetics (d) Antibiotics.

46. The central nervous system depresants that reduce nervous tension and promote relaxation without producing sleep are called
(a) Sedatives (b) Hypnotics
(c) Anaesthetics (d) Antibacterials.

47. 'Small pox vaccine' contains
(a) Chemical medicines (b) Immunoglobulins
(c) Killed micro-organism (d) Living micro-organism.

48. Which of the following drugs combination is not correct ?
(a) Biothional—Tranquilizer
(b) Chloramphenicol—Broad spectrum antibiotics
(c) Equanil—Sedative
(d) Phenacetin—Antipyretic.

49. The first antibiotic was discovered by
(a) Alexander Fleming (b) Howard Florey
(c) E.B. Chain (d) Graham.

50. Drug of the following is used as an antipyretic drug ?
(a) Aspirin (b) Paracetamol
(c) Phenacetin (d) All the above.

ANSWERS

1. (a)	**2.** (b)	**3.** (d)	**4.** (b)	**5.** (c)	**6.** (b)	**7.** (c)	**8.** (d)	**9.** (a)
10. (b)	**11.** (b)	**12.** (c)	**13.** (b)	**14.** (c)	**15.** (c)	**16.** (a)	**17.** (b)	**18.** (d)
19. (c)	**20.** (c)	**21.** (d)	**22.** (d)	**23.** (c)	**24.** (d)	**25.** (c)	**26.** (c)	**27.** (c)
28. (c)	**29.** (d)	**30.** (b)	**31.** (a)	**32.** (b)	**33.** (b)	**34.** (c)	**35.** (d)	**36.** (c)
37. (b)	**38.** (a)	**39.** (a)	**40.** (c)	**41.** (a)	**42.** (c)	**43.** (b)	**44.** (a)	**45.** (b)
46. (a)	**47.** (c)	**48.** (a)	**49.** (a)	**50.** (d).				

HINTS/SOLUTIONS

Note : The answers to all questions are self explanatory.

HINTS/SOLUTIONS

Q. No.	Ans.	Hints & Solutions
1.	(*b*)	(+) 2-Butanol and (–)2-Butanol are enantiomers of each other they will have different spatial arrangement of groups around the chirality centre and also their specific rotation will be equal in magnitude but opposite is sign.
2.	(*b*)	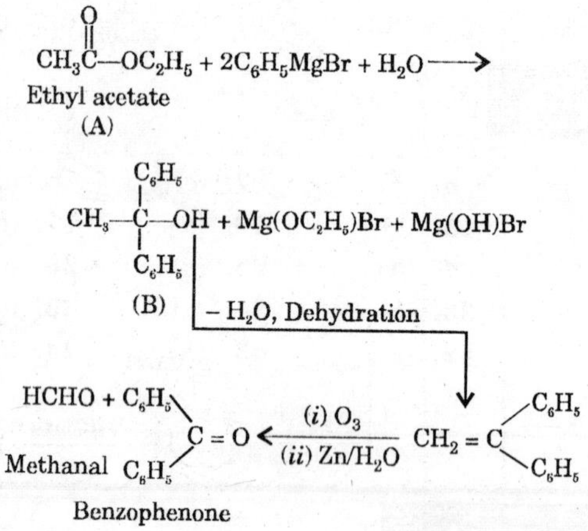
3.	(*c*)	Br_2/CCl_4 will react with $CH_2=CH—CH_2OH$ but not with 2-propanol. I_2/NaOH will give yellow ppt. of CHI_3 with 2-propanol but no reaction will occur with allyl alcohol. Hence correct choice is (*c*).
4.	(*d*)	Structures *a*, *b*, *c* are chiral while structure *d* is achiral.
5.	(*a*)	Structures I and II have S configuration while structures III and IV have R configuration.
6.	(*b*)	The compound dissolved in ether layer will be R—CHO whereas that which will go in aqueous layer is RCOONa.
7.	(*b*)	The strongest acid among the given species in COOH | COOH. Thus it will have lowest value of pK_a.
8.	(*a*)	B_2H_6 selectively reduces —COOH group to —CH_2OH in the presence of keto group.
9.	(*b*)	The pair of species given in I and II are not enomtioners. Then represent same structures.
10.	(*c*)	The various functional isomers are ; 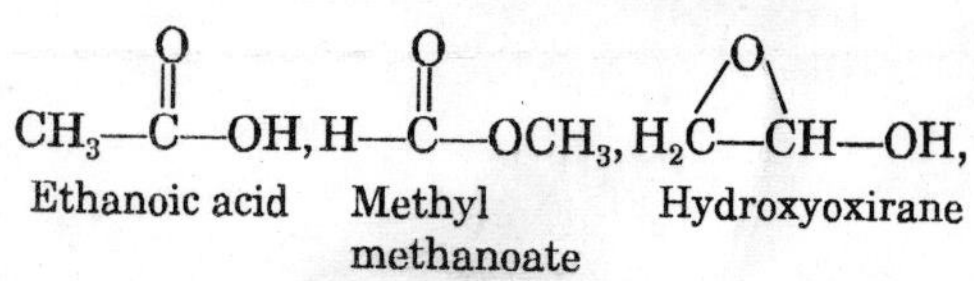
11.	(*c*)	The structure contains one chirality centre. Hence it can exhibit enantiomerism. One of the double bonded carbon carry two methyl groups. Hence it cannot show geometrical isomerism.
12.	(*a*)	Least basic among the given species in 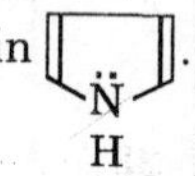. Its lone pair on N atom is involved in delocalised π cloud. Thus, its availability for protonation is very poor.
13.	(*a*)	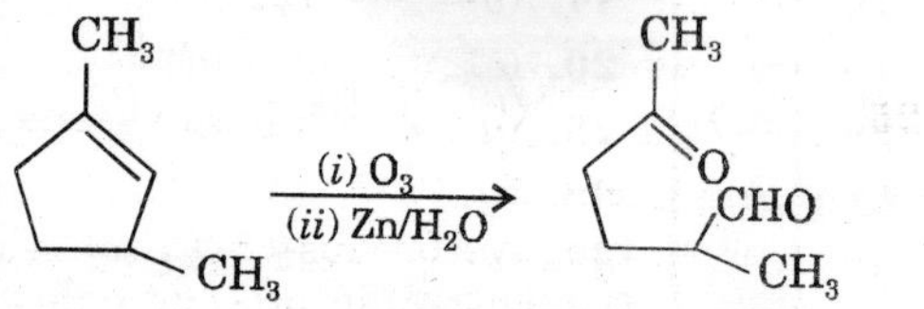 2-Keto-5-methylhexanal
14.	(*a*)	The pair of species given in choice '*a*' do not represent resonating forms. They donot have even same number of atoms.
15.	(*d*)	The correct matchings are I—D, II—A, III—C, IV—B. Thus, the correct choice is D.
16.	(*d*)	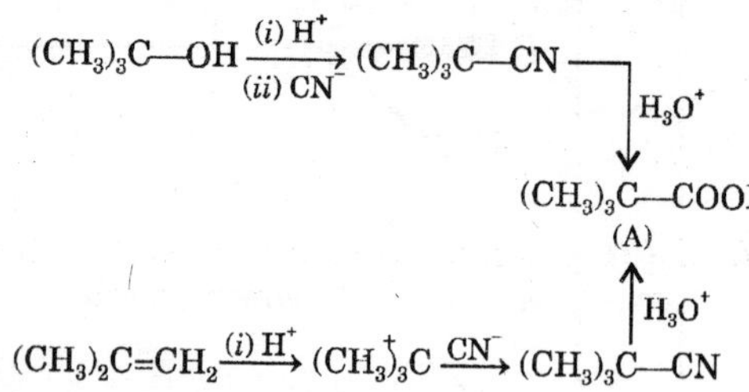
17.	(*c*)	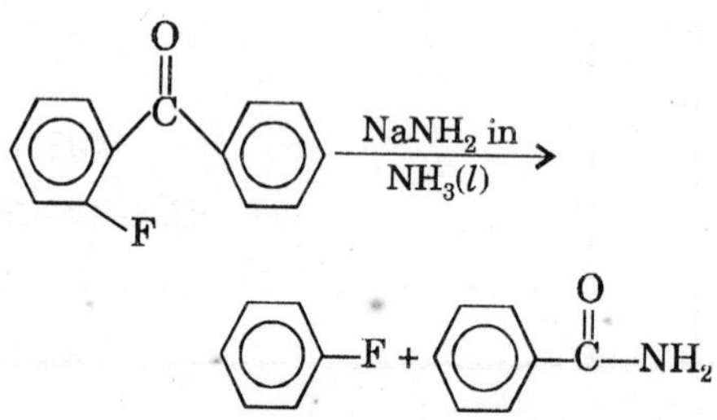
18.	(*c*)	Statement I and IV are based on facts whereas statements II and III are incorrect.

Q. No.	Ans.	Hints & Solutions
19.	(*a*)	Statements II and III are correct as they are factual.
20.	(*d*)	All the species namely amide (*a*), ammonium salt (*b*) as well as cyanides (*c*) on boiling with aqueous NaOH will liberate NH_3 gas which turns red litmus blue.
21.	(*c*)	Dichlorocarbene ($:CCl_2$) is an electrophilic species.
22.	(*b*)	For reaction undergoing bimolecular elimination (E_2) step shown in '*b*' is correct representation of β-Elimination process.
23.	(*b*)	$R—NH_2 + HNO_2 \xrightarrow{NaNO_2/HCl} R—OH + N_2 \uparrow$. Other reagents will not liberate any colourless, odorless and neutral gas.
24.	(*b*)	The structures given in choice '*c*' has 1(E), 3(Z) configuration.
25.	(*c*)	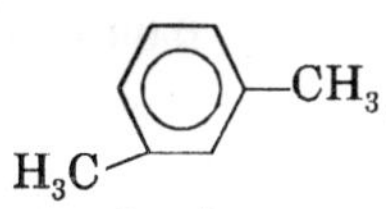 , *m*-xylene will be most reactive because both methyl groups cause concentration of – ve charge at the same positions making electrophilic attack relatively easier.
26.	(*d*)	2S, 3S configuration is shown by structure given in '*d*'. 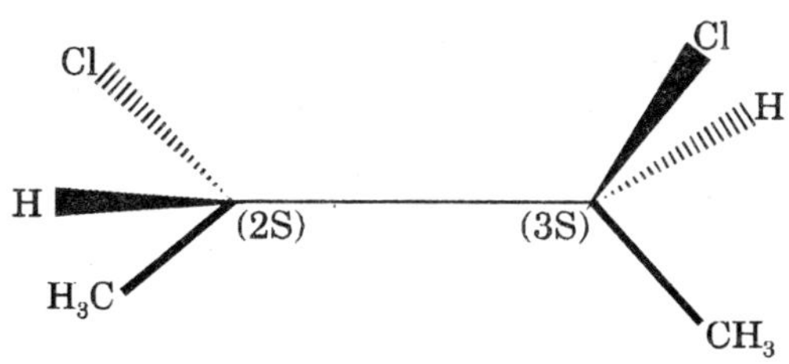
27.	(*a*)	Triplet carbene has two unpaired electrons. As a result it is paramagnetic.
28.	(*b*)	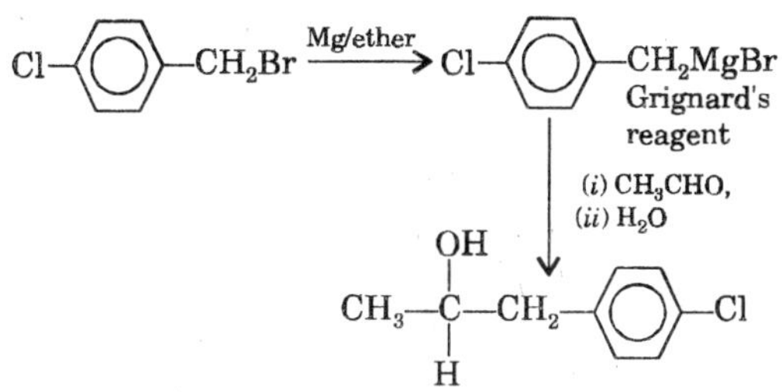
29.	(*b*)	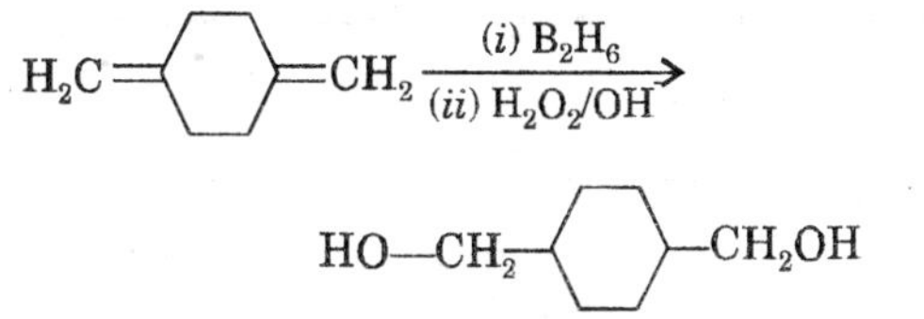 It represents anti Markovnikoff hydration.
30.	(*d*)	All the given reducing agents will reduce CH_3NC to CH_3NHCH_3.
31.	(*a*)	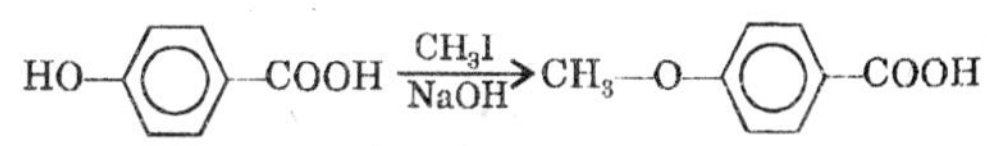
32.	(*a*)	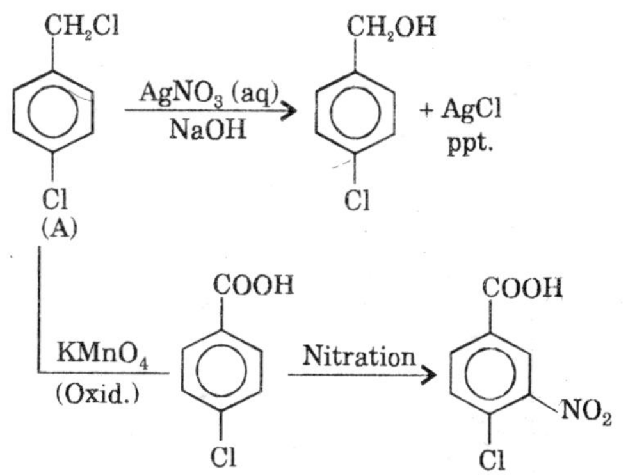 Only one mononitro product.
33.	(*a*)	The given reaction is called Baker–Mulliken's test.
34.	(*c*)	The acyclic isomeric structures are : 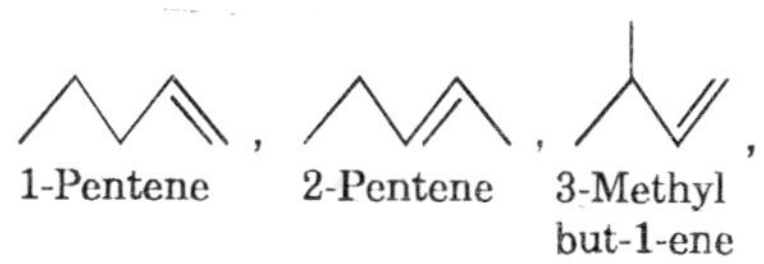 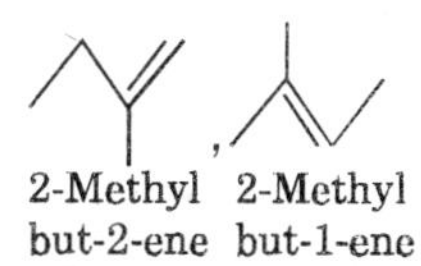 Among these 2-Pentene will exist in two geometrical isomers. Thus, total number of structural and steroisomers of alkenes will be **six**.
35.	(*b*)	The correct name of the compound is bicyclo [3, 3.0] octane.

Q. No.	Ans.	Hints & Solutions
36.	(*b*)	It refers to Hoffmann bromamide reaction and the amine formed in this case will have one C atom less than the parent amide.
37.	(*c*)	The slowest step in Cannizzaro reaction is transfer of hydride ion (H^-) to carbonyl carbon. (For details see mechenism of reaction given in chapter, Aldehyde and Ketones).
38.	(*b*)	Benzaldehyde undergoes disproportionation in 50% KOH to form benzyl alcohol and sodium benzoate, according to Cannizzaro's reaction.
39.	(*b*)	The selective oxidation of —CHO without affecting the C = C can be done by Tollen's reagent. Which is diammine silver (I) hydroxide.
40.	(*c*)	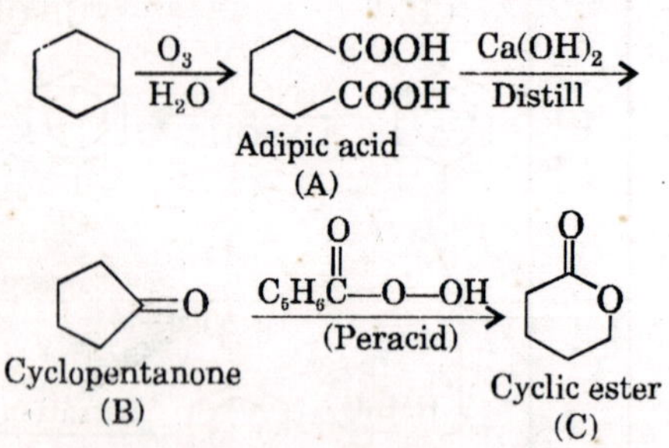
41.	(*b*)	Hydroboration oxidation will give product following anti-Markovnikoff's addition while acid-catalysed hydration will give product according to Markovnikoff's addition. 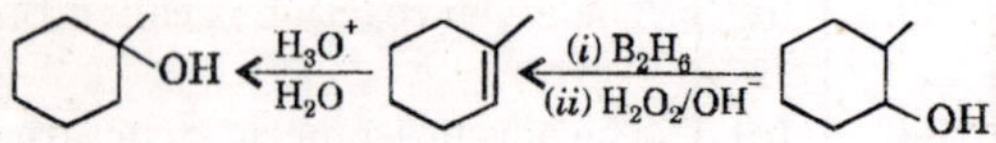
42.	(*a*)	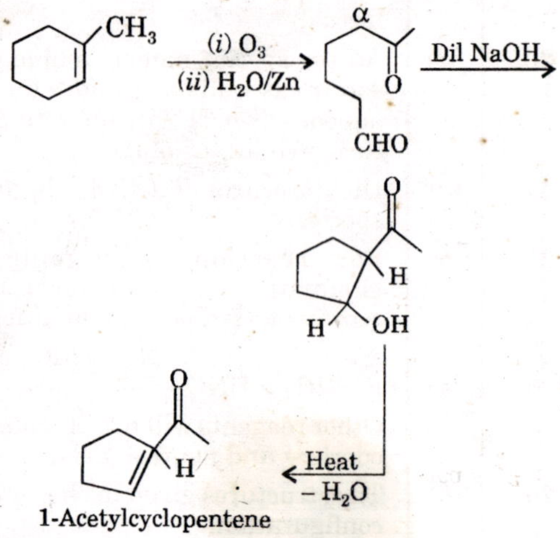
43.	(*d*)	Mol. of NaOH used = $0.09 \times 18.7 \times 10^{-3}$ Mass of acid that will consume 1 mol of NaOH $= \frac{0.175}{0.09 \times 18.7 \times 10^{-3}} = 103.98 \approx 104$ g Molar mass of acid = 104 g mol^{-1}. This molar mass matches only 1-ethoxyacetic acid given in choice '*d*'.
44.	(*c*)	Singlet carbene transforms trans-2-butene into desired product.
45.	(*c*)	$CH_3—CN \xrightarrow[\text{[H]}]{\text{Na/EtOH}} \underset{(A)}{CH_3CH_2NH_2}$ $\downarrow NaNO_2/HCl$ $\underset{\text{Yellow ppt.}}{CHI_3} + HCOONa \xleftarrow{NaOH/I_2} \underset{(B)}{CH_3CH_2OH}$